D1650933

BUILDINGS OF RHODE ISLAND

Buildings of
RHODE ISLAND

WILLIAM H. JORDY

Ronald J. Onorato and William McKenzie Woodward
CONTRIBUTING EDITORS

OXFORD
UNIVERSITY PRESS

2004

OXFORD
UNIVERSITY PRESS

Oxford   New York
Auckland   Bangkok   Buenos Aires   Cape Town   Chennai
Dar es Salaam   Delhi   Hong Kong   Istanbul   Karachi   Kolkata
Kuala Lumpur   Madrid   Melbourne   Mexico City   Mumbai   Nairobi
São Paulo   Shanghai   Taipei   Tokyo   Toronto

Copyright © 2004 by the Society of Architectural Historians

Published by Oxford University Press, Inc.
198 Madison Avenue, New York, New York 10016

www.oup.com

Oxford is a registered trademark of Oxford University Press

All rights reserved. No part of this publication may be reproduced,
stored in a retrieval system, or transmitted, in any form or by any means,
electronic, mechanical, photocopying, recording, or otherwise,
without the prior permission of Oxford University Press.

LIBRARY OF CONGRESS CATALOGING-IN-PUBLICATION DATA
Jordy, William H. Buildings of Rhode Island / William H. Jordy ;
contributing editors, Ronald J. Onorato and William McKenzie Woodward.
p.   cm. — (Buildings of the United States)
Includes bibliographical references and index.
ISBN 0-19-506147-0
1. Architecture–Rhode Island–Guidebooks.
2. Rhode Island–Guidebooks.   I. Onorato, Ronald J.
II. Woodward, William McKenzie. III. Title. IV. Series.
Arch. NA730.R5 J67 2004
720'.9745–dc22   2003022935

1 3 5 7 9 8 6 4 2

Printed in the United States of America
on acid-free paper

The Society of Architectural Historians gratefully acknowledges the support of the following, whose generosity helped bring *Buildings of Rhode Island* to publication:

National Endowment for the Humanities, an independent federal agency

Donald I. Perry

Alletta Morris McBean Charitable Trust

Champlin Foundations

Rhode Island Foundation

Felicia Fund

Prince Charitable Trusts

Rhode Island Committee for the Humanities

Mary Dexter Chafee Fund

Mr. and Mrs. A. L. Ballard

Hope Foundation

Southeast Chapter, Society of Architectural Historians

Many individual members of the SAH

Initial and ongoing support for the
Buildings of the United States series has come from:

National Endowment for the Humanities

Graham Foundation for Advanced Studies in the Fine Arts

Pew Charitable Trusts

University of Delaware

Ford Foundation

David Geffen Foundation

Furthermore, a program of the J. M. Kaplan Fund

University of Missouri

Samuel H. Kress Foundation

National Park Service, HABS/HAER/HALS

Richard H. Driehaus Foundation

Contents

List of Maps, xi

Guide for Users of This Volume, xiii

Foreword, xv

Editorial Note and Acknowledgments, xix

Preface, xxiii

Introduction: The Architectural Heritage of Rhode Island, 3

Providence (PR), 29
    Downtown Providence (Downcity) and Capital Center, 34;
    The Jewelry District and the Harborfront, 55;
    Main Street, South and North, 60; College Hill and East Side, 72;
    The State House and Northwest Providence, 122;
    Broadway and South Providence, 128

Providence Periphery, 135
    Pawtucket (PA), 135; Central Falls (CF), 153; North Providence (NP), 160;
    Johnston (JO), 168; Cranston (CR), 175

Northeast, 186
    Lincoln (LI), 186; Cumberland (CU), 206; Woonsocket (WO), 219;
    North Smithfield (NS), 238; Smithfield (SM), 249

Northwest, 263
    Scituate (SC), 263; Foster (FO), 274; Glocester (GL), 290;
    Burrillville (BU), 299

Mid-State, 316
    Warwick (WA), 316; West Warwick (WW), 324; Coventry (CO), 333;
    West Greenwich (WG), 341; East Greenwich (EG), 344

South, 351
    North Kingstown (NK), 351; Narragansett, (NA), 365;
    South Kingstown (SK), 374; Charlestown (CH), 397; Westerly (WE), 403;
    Hopkinton (HO), 418; Richmond (RI), 424; Exeter (EX), 429

East Bay, 433

    East Providence (EP), 433; Barrington (BA), 446; Warren (WA), 453; Bristol (BR), 464; Tiverton (TI), 480; Little Compton (LC), 489

The Islands, 498

    Portsmouth (PO), 498; Middletown (MI), 507; Newport (NE), 516; Jamestown (JA), 581; Block Island (New Shoreham) (BI), 604

Bibliography, 611

Glossary, 618

Illustration Credits, 643

Index, 645

List of Maps

Rhode Island Towns   5
Providence Overview (Including PR172–PR184, PR196–PR198)   30–31
Providence, Downtown and Capital Center (PR1–PR35)   35
Providence, Jewelry District and Harborfront (PR36–PR42)   56
Providence, Main Street (PR43–PR63)   62
Providence, Benefit Street, North (PR64–PR88)   74
Providence, Benefit Street, South (PR89–PR113)   86
Providence, Brown Center Campus and North (PR114–PR134)   98
Providence, Hope Street and Side Streets (PR135–PR171)   109
Providence, Broadway and the Armory (PR185–PR196)   129
Pawtucket   136
Central Falls   156
North Providence   163
Johnston   169
Cranston   176
Lincoln   188
Cumberland   207
Woonsocket   220
North Smithfield   240
Smithfield   250
Scituate   264
Foster   276
Glocester   291
Burrillville   300
Warwick   317
West Warwick   325
Coventry   336
West Greenwich   342
East Greenwich   344
North Kingstown   352–53
Narragansett   366
South Kingstown   376–77
Charlestown   398
Westerly   406
Hopkinton   420
Richmond   426
Exeter   430
East Providence   436
Barrington   448
Warren   456
Bristol   465
Tiverton   483
Little Compton   490
Portsmouth   501
Middletown   508
Newport Overview   517
Newport, Broadway and City Center (NE1–NE8, with Sites NE45–NE47)   518
Newport, Washington Street and the Point (NE9–NE44)   522
Newport, Harbor Area (NE48–NE88)   533
Newport, Kay–Catherine–Old Beach (NE89–NE128)   547
Newport, Upper Cliffs (NE129–NE139)   559
Newport, Bellevue Avenue (NE140–NE168)   563
Newport, Ocean Avenue and Spring Street (NE169–187)   579
Jamestown   584
Block Island (New Shoreham)   606

Guide for Users of This Volume

For touring purposes, *Buildings of Rhode Island* organizes the state's thirty-nine cities and towns into eight regional sections, beginning with Providence and its periphery, then moving north and west, south, and east. Series of entries for each city or town are denoted with a two-letter abbreviation for the name of the municipality. Entry heading information includes the current name (often followed in parentheses by an earlier name) of the building or site; the date of completion; the architect(s), if known; and the dates of major additions or alterations and their architects, if known. In location information, federal highways are labeled as interstates or as "U.S." followed by a number; the designation "Route" with a number applies to all state or town roads.

Within its small extent, Rhode Island presents an unusual range of navigational challenges, from very dense and irregular urban street networks to rural villages to isolated rural sites. For this reason, virtually every building or site included is keyed by number to an overview or street-level map. Sites in Providence and Newport are arranged by neighborhoods, each individually mapped. The maps included in this volume are, however, intended to be supplemented by commercially published highway and street maps.

Heading information for a number of entries includes the notation "open to the public." Other buildings not so described may be open at limited or irregular times or by arrangement. Many local historical societies and visitor information centers throughout the state offer additional information on historic buildings and sites in their towns, including times of opening.

Almost all the sites discussed in this book are visible from public roads or public property. Some properties that are not accessible to the public are included, however, because of their significance to understanding Rhode Island architecture. We know that readers will respect the property rights and privacy of others as they view the buildings.

Buildings of the United States volumes are intended to include only extant buildings. Some of the structures described here, however, will have been demolished or altered beyond recognition by the time this book is published.

Foreword

Buildings of Rhode Island is the ninth of a projected fifty-eight volumes in the series Buildings of the United States, which is sponsored by the Society of Architectural Historians (SAH). When the series is completed, it will provide a detailed survey and history of the architecture of the whole country, including both vernacular and high-style structures for a complete range of building types from skyscrapers to barns and everything in between.

The idea for such a series was in the minds of the founders of the SAH in the early 1940s, but it was not brought to fruition until Nikolaus Pevsner, the eminent British architectural historian who had conceived and carried out Buildings of England, originally published between 1951 and 1974, challenged the SAH to do for this country what he had done for his. That was in 1976, and it was another ten years before we were able to organize the effort, commission authors for the initial group of volumes, and secure the first funding, a grant from the National Endowment for the Humanities. Matched by grants from the Pew Charitable Trusts, the Graham Foundation, and the Michigan Bicentennial Commission, this enabled us to produce the first four volumes, *Buildings of Michigan, Buildings of Iowa, Buildings of Alaska,* and *Buildings of the District of Columbia,* all of which were published in 1993. *Buildings of Colorado* appeared in 1997, followed by *Buildings of Nevada* in 2000, *Buildings of Virginia: Tidewater and Piedmont* in 2002, and *Buildings of Louisiana* in 2003. *Buildings of West Virginia* will also be published in 2004, with many more volumes to follow.

Although Buildings of England provided the model, in both method and approach Buildings of the United States was to be as different as American architecture is from English. Pevsner was confronted by a coherent culture on a relatively small island, with an architectural history that spans more than two thousand years. Here we are dealing with a vast land of immense regional, geographic, climatic, and ethnic diversity, with most of its buildings—wide-ranging, exciting, and sometimes dramatic—essentially concentrated into the last four hundred years, although with significant Native American remains stretching back well beyond that. In contrast to the national integrity of English architecture, therefore, American architecture is marked by a dynamic heterogeneity, a heterogeneity woven of a thousand strands of originality, or, actually, a unity woven of a thousand strands of heterogeneity. It is this quality that Buildings of the United States reflects and records.

Unity born of heterogeneity was a condition of American architecture from the first European settlements of the sixteenth and seventeenth centuries. Not only did the buildings of the Russian, Spanish, French, Dutch, Swedish, and English colonies differ according to national origin (to say nothing of their differences from Native American structures), but in the translation to North America they also assumed a special scale and character, qualities that were largely determined by the aspirations and traditions of a people struggling to fashion a new world in an abundant but demanding land. Diversity marked even the English colonies of the Eastern Seaboard, though they shared a common architectural heritage. The brick mutations of English prototypes in the Virginia Colony were very different, for example, from the wooden architecture of the Massachusetts Bay Colony. They were different because Virginia was a farm and plantation society dominated by the Anglican church, whereas Massachusetts was a communal society nurtured entirely by Puritanism. But they were different also because of natural resources and the traditions of the parts of England from which the settlers had come. This is even more true of the buildings of French settlers in Louisiana or of the Spanish who came to Florida and the Southwest, which are different not only from each other but from those of the English in either Virginia or Massachusetts. As the colonies became a nation and developed westward, similar radical contrasts became the way of America's growth. The infinite variety of physical environment, together with the complex origins and motivations of the settlers, made it inevitable that each new state would have a character uniquely its own.

The primary objective of each volume, therefore, is to record, analyze, and evaluate the architecture of the state. The authors are trained architectural historians who are thoroughly informed in the local aspects of their subjects. In each volume, special conditions that shaped the state or part of the state, together with the building types necessary to meet those conditions, are identified and discussed; barns, silos, mining buildings, factories, warehouses, bridges, and transportation buildings take their places alongside the familiar building types conventional to the nation as a whole—churches, courthouses, city halls, commercial structures, and the infinite variety of domestic architecture. Although the great national and international masters of American architecture receive proper attention, especially in the volumes for the states in which they did their greatest work, outstanding local architects, as well as the buildings of skilled but often anonymous carpenter-builders, are also brought prominently into the picture. Each volume is thus a detailed and precise portrait of the architecture of the state that it represents. At the same time, however, all of these local issues are examined as they relate to architectural developments in the country at large. Volumes will continue to appear state by state until every state is represented. When the overview and inventory are completed, the series will form a comprehensive history of the architecture of the United States.

These volumes deal with more than the highlights and the high points of architecture in this country. They deal with the very fabric of American architecture, with the context in time and in place of each specific building, with the entirety of urban and rural America, with the whole architectural patrimony. This fabric includes modern architecture, as, on the other end of the scale, it includes pre-Columbian and Native American remains. But it must be said, regretfully, that the series cannot cover every building of merit; practical considerations have dictated some difficult choices in the buildings that are represented in this as in other volumes. There are, unavoidably, omissions from the abundance of structures built across the land, the thousands of modest but lovely edifices and the vernacular attempts that merit a second look but which by their very multitude cannot be included in even the thickest volume.

Thus it must be stated in the strongest possible terms that omission of a building from this or any volume of the series does not constitute an invitation to the bulldozers and the wrecking ball. In every community there will be structures not included in Buildings of the United States that are clearly deserving of being preserved. Indeed, it is hoped that the publication of this series will help to stop at least the worst destruction of architecture across the land by fostering a deeper appreciation of its beauty and richness and of its historic and associative importance.

The volumes of Buildings of the United States are meant to be tools of serious research in the study of American architecture. But they are also intended as guidebooks for everyone interested in the buildings of this country and are designed to facilitate such use; they can and should be used on the spot, indeed, should lead the user to the spot. It is our earnest hope that they will be not only on the shelves of every library from major research centers to neighborhood public libraries but that they will also be in a great many raincoat pockets, glove compartments, and backpacks.

During the long gestation process of the series, many have come forward with generous assistance. We are especially grateful, both for financial support for the series as a whole and for confidence in our efforts, to the National Endowment for the Humanities; the Graham Foundation for Advanced Studies in the Fine Arts; the University of Delaware; the Ford Foundation; the Samuel I. Newhouse Foundation; the David Geffen Foundation; the College of Fellows of the American Institute of Architects; Furthermore, a program of the J. M. Kaplan Fund; the Samuel H. Kress Foundation; the National Park Service, HABS/HAER/HALS; and the Richard H. Driehaus Foundation. For this volume, we are also enormously indebted to Donald I. Perry, the Alletta Morris McBean Charitable Trust, the Champlin Foundations, the Rhode Island Foundation, the Felicia Fund, the Prince Charitable Trusts, the Rhode Island Committee for the Humanities, the Mary Dexter Chafee Fund, Mr. and Mrs. A. L. Ballard, the Hope Foundation, the Southeast Chapter of the SAH, and many individual members of the SAH who have made unrestricted contributions to BUS and to a fund for this volume established in William Jordy's memory, including especially Brent and Elizabeth Edwards Harris, Lisa Koenigsberg, Carol and Robert Krinsky, Phyllis Lambert, William H. Pierson, Jr., and Barbara Wriston. We are also grateful to a number of individuals who helped us in our search for support; and in this regard we would like to thank Armin Allen, Mardges Bacon, Ralph Carpenter, Pauline Metcalf, Tom Michie, and Lynn Springer Roberts. We are thankful, too, to the members of our Leadership Development Committee: Madelyn Bell Ewing, Frances Fergusson, Elizabeth Harris, Ada Louise Huxtable, Philip Johnson, Keith Morgan, Victoria Newhouse, Robert Venturi, and the late J. Carter Brown.

We are very grateful to Dean Larry Clark of the University of Missouri and to

Provosts Mel Schiavelli and Dan Rich; Deans Mary Richards, Margaret Andersen, Thomas DiLorenzo, and Mark Huddleston of the College of Arts and Science; and Professor Ann Gibson, Chair of the Department of Art History, at the University of Delaware for providing institutional support for two successive editors in chief of the series.

Because Bill Jordy died when the book was almost, but not quite finished, we relied on the help of a number of people in completing his work. They are all mentioned in the editorial note and acknowledgments following the foreword.

Our gratitude extends to many other individuals. These include a large number of presidents, executive committees, and boards of directors of the SAH, going back at least to the adoption of the project by the Society in 1979, and a series of executive directors covering that same span. All of these individuals have supported the series in words and deeds, and without them it would not have seen the light of day.

We would also like to express our appreciation to the current members of our editorial board, listed earlier in this volume, and the following former members: Adolf K. Placzek, Richard Betts, Catherine Bishir, J. A. Chewning, S. Allen Chambers, Jr., Alex Cochran, Elizabeth Cromley, John Freeman, David Gebhard, Alan Gowans, Alison K. Hoagland, William H. Jordy, Robert Kapsch, Henry Magaziner, Tom Martinson, Sally Kress Tompkins, and Robert J. Winter. Of this group, we would like especially to single out Dolf Placzek, founding editor in chief, who served in that capacity from the early 1980s to 1989 and continued to play a major role in the project until his death in the spring of 2000.

We have tried to establish as far as possible a consistent terminology of architectural history, and we are especially appreciative of the efforts of J. A. Chewning in the creation of the series glossary included in every volume. The Art and Architecture Thesaurus, a comprehensive publication and database compiled by The Getty Art History Information Program and published by Oxford University Press, has also become an invaluable resource.

In our fundraising efforts we have benefited enormously from the dedicated services of our directors of development, first Anita Nowery Durel and then Barbara Reed, as well as associate director of development William Cosper, and of our administrative staff: first fiscal coordinator Hillary Stone and then Comptroller William Tyre; and the current executive director of the SAH, Pauline Saliga.

Finally, there are our former and present colleagues in this enterprise at Oxford University Press in New York and Cary, North Carolina, especially Laura Brown, Karen Day, Nancy Hoagland, Rebecca Seger, John Sollami, and Timothy DeWerff.

To all of these we are enormously grateful.

Damie Stillman
Osmund Overby
William H. Pierson, Jr.

Editorial Note and Acknowledgments

Although it is the ninth volume in the Buildings of the United States series, when the project was conceived in 1979 by William H. Pierson and William H. Jordy, *Buildings of Rhode Island* was to have been one of the first books published. Jordy (as friends and colleagues always called him), who had spent most of his professional career in Rhode Island, intended that he would write the volume for the state, and, indeed, he began it in the 1980s. But both because he knew so much about the architecture of his adopted state and because he wanted to make sure that this volume, above all, exemplified the aims of the series, he had not quite finished it when he died in the summer of 1997. At the time of his death, the manuscript, photographs, and research files were divided between his house and the office he had kept at Brown University. In addition, drafts for parts of the text were in the computers of various people who had assisted him. Thanks to these individuals and a number of others, we were able to put the book together according to the wishes he indicated to me in 1996. And that is what is presented here.

At the beginning of our efforts to ready the book for publication, we made the decision that we would not materially change Jordy's words or his concept nor add significantly to what he had written. As a result, the preface is based on the draft for a presentation that Jordy made at the BUS session at the Annual Meeting of the Society of Architectural Historians in Philadelphia in 1994, and the introduction, which combines a number of Jordy's drafts, is incomplete, as he did not finish the story beyond where we have left it. The selection of entries, except for the newly authored material noted below and a few deletions, is Jordy's, as is the overall organization of regions and towns. His sequencing of building entries in the interest of discussion sometimes resulted in oddities of touring order, but unless these proved impossible to follow when mapped, we have left them. Jordy had commissioned most of the photographs from John M. Miller, but we have supplemented these to fill in gaps, principally for buildings in Providence and Newport. Unfortunately, though we found almost everything else, there was no bibliography for the book in Jordy's files. We have put one together for the reader, using the bibliography for *Buildings on Paper: Rhode Island Architectural Drawings, 1825–1945*, which Jordy coauthored with Christopher Monkhouse, supplemented and edited for this volume.

Following Jordy's death, we asked Ronald J. Onorato of the University of Rhode Island and William McKenzie Woodward of the Rhode Island Historical Preservation and Heritage Commission, both of whom had worked with Jordy, to take the lead in helping complete the work. As Jordy's collaborators, Ron had coauthored the Newport section and Mack the sections on Providence and its periphery, East Greenwich, Tiverton, Little Compton, and Block Island and a portion of the Bristol entries. As contributing editors for BUS, they shared the enormous task of reviewing the manuscript, updated material to reflect changes and new information, and endured endless editorial queries. Mack filled in missing entries, principally for Providence and Warwick, and Ron completed Jamestown and some additional entries for Newport, Portsmouth, and Middletown, and authored the section on Charlestown, the one piece for which not even a fragment could be found in Jordy's files. In addition, both contributed entries to update the previously published bibliography. This book would never have reached publication without their efforts.

In preparing the volume, Jordy had asked a number of other scholars to contribute research or writing or to consult on specific subjects. I and my colleagues on the BUS Editorial Board want to acknowledge them, with their areas of responsibility, both on Jordy's behalf and on our own. They include Antoinette Downing (Newport), Jean Follett (North Kingstown), Richard Greenwood (industrial architecture), Elizabeth G. Grossman (Woonsocket), Robert O. Jones (Warwick and Barrington), and Walter A. Nebiker (Smithfield and Jamestown).

Jordy also had the invaluable assistance of Jane McIlmail, who entered his handwritten drafts and numerous revisions into the computer and did much to ensure order and consistency in the manuscript we received. She, as well as Mardges Bacon of Northeastern University, Irving Haynes of the Rhode Island School of Design, and Joachim A. Weissfeld of Hinckley, Allen and Snyder LLP, assisted with locating materials after Jordy's death.

A number of others have also helped us. BUS founding Coeditor in Chief Bill Pierson, a close friend and collaborator of Jordy's, reviewed portions of the manuscript and made suggestions. BUS Assistant Editor Michael J. Lewis of Williams College put Jordy's material on Bristol in order and pieced it together with Mack Woodward's Hope Street entries to complete that section as Jordy would, we think, have wanted. Robert O. Jones of the Rhode Island Historical Preservation and Heritage Commission helped review and update the section on Providence. Dietrich Neumann of Brown recruited research assistants there, including Douglas Klahr. At the University of Delaware, research assistants Jhennifer Amundson, Anna Andrzewjewski, Heather Campbell, Martha Hagood, Nancy Holst, Amy Johnson, Sarah Killinger, Ellen Menefee, Louis Nelson, and Molly Zillman inventoried research materials, helped ensure that the photographs commissioned by Jordy were correctly identified, and assisted in acquiring additional ones. Molly Zillman also contributed to the bibliography. Jean Pellam, research assistant at the University of Rhode Island, did field work to establish site locations for a number of maps and ground-proofed many more. Even so, the cartography presented unusual challenges for Andrew Dolan of the Geographic Resources Center at the University of Missouri. In production phases, Leslie Phillips designed the page layouts, Carol Wengler did the proofreading and Cynthia Crippen compiled the index. And, without BUS Managing Editor Cynthia Ware, this volume would probably never have been pulled together as a unified book. To all of them, we are deeply grateful.

Finally, on behalf of Ron Onorato and Mack Woodward, we would like to thank the following for their help: A. D. Goff, Bertram Lippincott III, Ronald Potvin,

Editorial Note and Acknowledgments xxi

Daniel Snydacker, and Joan Youngken of the Newport Historical Society; Pieter Roos of the Newport Restoration Foundation; Paul Miller and John Tschirch of the Preservation Society of Newport County; James C. Garman of the Program in Historic Preservation at Salve Regina University; David Maslin of the University of Rhode Island; and Gabrielle Bleeke-Byrne, Jane E. Carey, Frank McGirr, Elisabeth Marchi, Mary Miner, Giancarlo Onorato, and Henry Wood.

As part of Jordy's literary estate, the research materials for the book will be transmitted to Brown University for archiving. A substantial number of John Miller's large-format photographs will become part of the collections of the Historic American Buildings Survey in the Prints and Photographs Division, Library of Congress.

Damie Stillman
Editor in Chief
Buildings of the United States

Preface

When I thought of undertaking an architectural guide to Rhode Island, where I had lived for most of my career as a professor of architectural history, I hesitated. There are far more rewarding endeavors for an architectural historian. Admittedly, personal observations and reflections on places visited or, for that matter, works that fall between travelogues and essays of personal encounter, may be of substantive use to the traveler—can indeed occupy a distinguished, as well as a useful, place in literature. However, they are for armchair reading: to prepare for the trip or to savor its aftermath. One may be tempted to lug Ruskin to Italy, but *The Stones of Venice* was not written to serve in the field. A field guide, as this is intended to be, can be well written, perceptive, and thoughtful. Yet at its best, the genre is inherently fussy, factual, and fragmented. Its remorseless push from one attraction to the next and the next and the next exhausts the traveler before the start, anticipating the exhaustion that actually occurs after a long day of using it. So why do it?

For me it was, first, the desire to capture what Alexander Pope referred to as "the genius of the place." It is this care for a place—the wish to penetrate its core—that flags the serious writer of guidebooks on. I had always been interested in this theme. A guidebook to the place in which I had spent much of my life could provide the vehicle.

In the course of working, I became conscious that other, more personal reasons also drew me to the task. In addition to Nikolaus Pevsner, the patron saint of the BUS series, two Americans especially conditioned my approach: Henry James and Henry Adams. By chance, both were important influences on me at the very start of my professional career. It gradually dawned on me that perhaps, in taking this opportunity to investigate the meaning of the place in which I had lived, or at least an aspect of its meaning, I might also have been secretly returning to a piece of my own past, revisiting it.

Essentially a guidebook is a journey and a sequence of destinations: movement and confrontation. The author and his reader are continually off balance. It is paradoxically as though to place the building is to displace it. A nineteenth-century mill collides with an eighteenth-century house, and maybe the first view is disappointing (or, just as likely, overwhelming). Sometimes the guidebook needs to prepare the visitor—especially when looking at something unprepossessing. And here is where Henry Adams and Henry James returned from the past to haunt me. There was Henry James's *The American Scene*, among the most beautiful evocations of place—of American place. Virtually no people appear in his observations: just things, buildings, places, landscapes peopled (perhaps haunted) more by the resonance of occupancy than by people themselves—and, in truth, James's visit to America was more a return to the past than a survey of the present. For me, however, James as patron saint was balanced by his friend Henry Adams. For reasons too involved to discuss here, my first book turned out to be a study of Adams's up-to-then very much neglected *History of the United States during the Administrations of Jefferson and Madison*. I was interested in the phenomenon of "scientific history," which he embodied, in opposition to the romantic, or narrative, approach to history. It was not simply the notion of a guidebook that attracted me, but that of a format that might combine these two masters of nuanced observation: Adams for nuanced fact, James for nuanced evocation. If a guidebook has any "art" at all it mediates between these polarities. Evocation and facts—all, we hope, packaged in some seamless way.

Rhode Island is the smallest state in the Union, so small that one can be anywhere within it and reach the starting point for any day's architectural travels in a maximum of around two hours. It is also well covered: the Rhode Island Historical Preservation and Heritage Commission, under the leadership of Antoinette Downing, had published its own historical introductions and inventories for each of the thirty-nine towns in the state—no mean task because of the richness of Rhode Island's architecture, capped by the concentration of fine buildings in Newport and Providence. But both advantages proved illusory as far as time was concerned. The temptation to revisit and repeatedly recheck sites always within easy range, something which would have been impossible in larger states, as well as the abundance of possibilities offered by the commission's reports, made culling difficult. Sometimes my muses worked swiftly; at other times it was a day's work to get an entry right—only, on review, to start all over.

The first task is mechanical. How (out of a thousand possibilities) will the thing be organized? Archives at the RIHPHC are arranged by town. Why not make it easy for the serious user to pursue further research? So this organization started out as a matter of convenience, but, much to my surprise, as I worked into the project I observed that many of these little towns (after all these years) possessed discernible characteristics. So here was a "voice" that Rhode Island could contribute to the big picture, one that might be difficult to discern (except on a regional scale) in a state like Texas.

Take the town of Foster, in the most rural corner of northwestern Rhode Island. At the beginning of the twentieth century it was described as one of those places time had passed by. Poor lands, poor farms. But, astonishingly, it offers one of the biggest collections of Greek Revival churches in the state—six in all. In fact, the villages that once existed there have mostly dried up, and we know of them only by these churches. A theme suddenly occurs, very sophisticated in terms of architectural history: but can we engage the casual "architectural pilgrim" by this litany of precursors, styles, revivals, and revivals of revivals? Suddenly even the least of the churches has its significance, a Friends meeting house which looks like a school. How, I ask the architectural pilgrim, does one know when a school becomes a church? Not the most pressing issue to Quakers, perhaps, but we begin to inquire as to how monuments are born of plainness. One of the simplest of the churches has two long granite steps stacked as a pyramid running across the front of the wooden building. Oxen dragged them there as a start to monumentality. Through 1850, oxen were more plentiful than horses on Rhode Island farms. They gentled the way for horses, dragging huge granite pieces—the monoliths for the Providence Arcade, for example, twenty miles out of Johnston, just, it would seem, to make the monumental point.

Or another country example—and a problem for the writer of guidebooks. What does one do when there is too much of the same thing? Another five-bay, central-door, central-chimney house? There are a number of such houses in Smithfield—too many. Another theme: We begin to compare them—size, shape, distribution for their different characteristics. Then to notice that plenty are not five bays but four. How does the building handle an eight-spot elevation as compared to a ten? One other aspect makes the survey more interesting: Smithfield was the locale for one of the White Pine Series monographs, in 1931. So we are traveling where others have traveled, seeing the same things, looking in a new way.

Another characteristic of a small state is a certain inbred quality, a closeness of social organization. It has been said that seven families, led by the famous Brown family, controlled Providence in the eighteenth and nineteenth centuries. Supposedly, the Browns were the only family to become millionaires in the colonial period who were able to maintain their wealth into the twentieth century. When John Nicholas Brown was born at the beginning of the twentieth century, he was considered America's richest baby. Small states offer interesting networks, like evolving soap operas. Take one family: Metcalf.

We first meet them as mill owners in a corner of Providence known as Wanskuck, a company-built community of mill, church, social hall, and housing. Jesse Metcalf and his wife, Helen Adelia Rowe Metcalf, traveled and collected art. She became head of the Rhode Island Women's Commission for the Centennial Exposition of 1876, where she was impressed by a British exhibition on art and industrial design education. The Rhode Island women's exhibition ended with a profit of $1,675. Why not use it as a nest egg to begin a campaign, based on the example of the Victoria and Albert Museum, for an art school in Rhode Island to provide designers for local industry? The school started on the top floor of an office building downtown. But Jesse Metcalf soon raised money and contributed generously to build the first building, the Waterman Building, across from the First Baptist Church. The area around the church is a precinct which records the artistic culture of Providence in the 1880s: a studio building across the way and, uphill, colonial houses converted to the Providence Art Club. Now, up to Prospect Street. When the next generation of Metcalfs left Wanskuck to move into Providence, they built one of what I believe were the first three large-scale houses to make the Colonial Revival respectable. Of these, only the Stephen Metcalf House, now the RISD President's House, remains. Up the block, a street of Queen Anne houses, a Tudor Revival school building, and, diagonally opposite, a whimsical Froebel kindergarten. But then turn the corner, and there is a Federal house which used to hold one of the three great colonial antiques collections, the Pendleton collection. Edward Pendleton was something of a shady character who loved the beautiful life. He left the antiques to RISD on the condition that it would provide a building to house them (otherwise off to Boston). So, the Metcalfs to the rescue. They called on Edmund Willson to build Colonial House—the first replica house in any American museum designed to house a collection of furniture. Diagonally across the street is John Holden Greene's Truman Beckwith House. I compare the two as a kind of obituary to Willson—and a tribute to the Genius of the Place.

The trick, I think, is never to anticipate, but to let the story unfold. Only by taking the tour, only by standing in front of the monument, only by spending the time, does one come to the spirit of the place.

William H. Jordy
April 1994

BUILDINGS OF RHODE ISLAND

Introduction: The Architectural Heritage of Rhode Island

[The following essay was WHJ's draft, dated December 10, 1996, for the introduction to this volume. Given his personal style and his unique perception of what Alexander Pope called "the genius of the place" (a concept mentioned here and in WHJ's drafts for the preface), the editors felt that what is here speaks so perceptively and eloquently that we should not add to it or attempt to "complete" it. Instead, we present it as WHJ's thoughts on Rhode Island and its architecture eight months before his death.]

RHODE ISLAND IS THE SMALLEST STATE IN THE UNION. WITH AN area of slightly more than 1,200 square miles, 14 percent of which is taken up by the waters of Narragansett Bay, it would fit into the state of Alaska 483 times. Yet this tiny enclave contains one of the richest concentrations of important historical architecture to be found anywhere in the United States.

One thinks of it first as a coastal state, its shoreline generally facing south to the sea, but angled overall along a southwest-northeast diagonal between Watch Hill, at the Connecticut border, and Little Compton, against Massachusetts. At least the southwestern half of its coastal length rather nicely hews to a ragged diagonal, with a string of magnificent beaches fronting marshes and inlets, backed by a score of freshwater glacial ponds. As the schematic diagonal continues northeastward, however, the shape of this half of the coast is immensely complicated by the massive indentation of Narragansett Bay. In the western portion of the state, the bulk of its land mass is mostly contained within a tall rectangle, though even here its seeming straight edge against Connecticut splays out slightly as it approaches the coast, then breaks out of its bounds to follow the curl of the Pawcatuck as this river border defines the promontory at Watch Hill, as though the state were meant to escape its cramped quarters by expanding on its approach to the sea. Toward the east the state's boundaries swell from this box with a vengeance, crossing inlets, peninsulas, and open stretches of water and in the process expanding like a hungry sponge, or like the cast of a net, to take in virtually the whole of Narragansett Bay. The bay is Rhode Island's single most defining feature.

In the search for the genius of this place, we should begin here. Ask the stranger with only the vaguest knowledge of the state what comes to mind with its mention. The response, if any, is likely to be one or some combination of Narragansett Bay, Providence, and Newport. Few questioned in these circumstances will realize that the cities named are the important founding settlements of the eventual state, as well as marking respectively the head and the mouth of the bay. The distance between waterfronts of a little more than twenty miles measured along a straight line approximates the inland reach of the bay from the coast. So the bay and this pair of founding towns establishes the generative core for building and settlement in Rhode Island. The persistent significance in subsequent developments of these three elements and their relationship give them almost emblematic intensity for those seeking the special qualities of this place.

Driven successively from Boston to Plymouth, then to exile beyond the Puritan pale, the heretical Reverend Roger Williams eventually arrived at the head of Narragansett Bay in 1636. He found there a "sweet spring" (the presumed spot now aggrandized and sanctified by a walled garden that piously engulfs the well, even as a temple now smothers Plymouth Rock). There Williams also met with friendly natives, who welcomed him to this place at the mouth of a river opening to a broad bay. In gratitude he named the place Providence. Two years later, other dissenters, sharing convictions akin to Williams's point of view, but some with differences (including, most famously, Anne Hutchinson), followed his lead to settle in what is now the town of Portsmouth at the northern end of the bay's largest island, which the natives called Aquidneck. Here dissenters of one conviction quickly broke with dissenters of another, members of one faction removing themselves as far as possible from their objectionable cohorts to the southern end of the elongated island. They settled there, at the mouth of the bay, where there was also a splendid harbor sheltered by headlands, calling their place Newport. Providence and Newport: what hope implicit in this combination! (And "Hope," in fact, ribboned beneath an anchor, became the state seal and motto.) The settlers renamed the island. The full official name of the state from Williams's charter is Rhode Island and Providence Plantations, recognizing its independence as a colony from its neighbors around it. Hence the smallest state officially has the longest name. In its abbreviated form, the island tail wags the mainland dog.

So, from the first settlement in this sanctuary for freedom of religion and the liberty of conscience that went beyond religion, this colony-to-be staked out its two historic cities, the distance between them marking the inland reach of the bay.

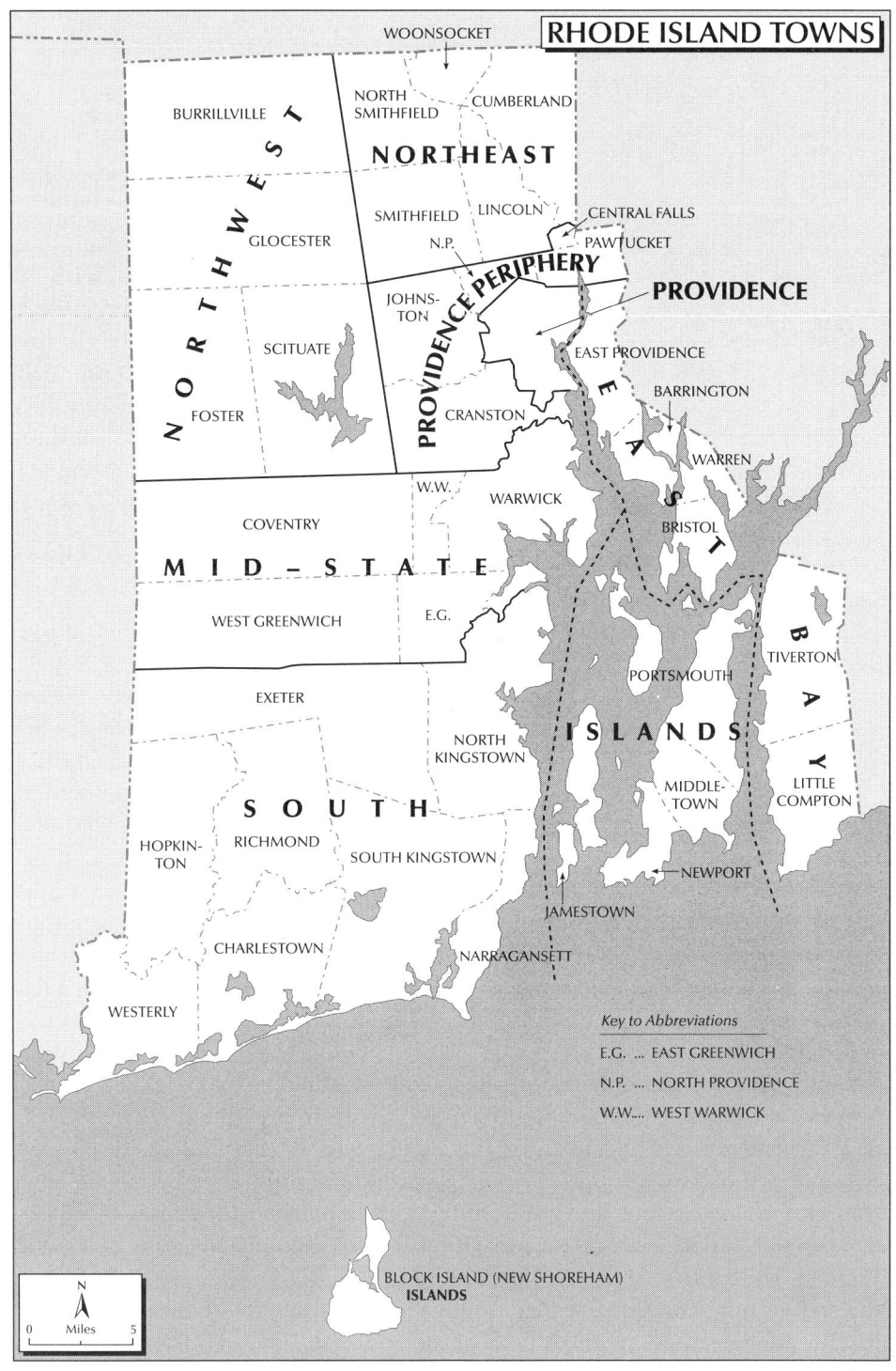

Some Consequences on Building of Williams's Spirit of Toleration

For the visitor, the long histories of these founding towns are doubly fascinating because they present continuous differences in both aspect and circumstance. Newport took the early lead, not only as the most important colonial city in the state, but also as among the major colonial ports of the time. Many of its inhabitants (and some from outside the city as well) believed that its maritime trade might eventually surpass that of all other colonial competition. (And in this connection it is interesting that the islands of Aquidneck and Manhattan are roughly equivalent in size, both also elongated in shape, with Aquidneck the more irregular of the two. With respect to its island, however, Newport's extent remained close to that of New Amsterdam.) Its early success as the port in Rhode Island resulted in Newport's unique legacy of eighteenth-century residential and public buildings and its position as a notable center of colonial culture and craftsmanship. Its location within Roger Williams's sanctuary meant that a number of its public buildings make manifest his ideals of religious toleration.

For example, the early affluence of Newport makes it the place to see residential architecture dating from the first three quarters of the eighteenth century, at least 200 [buildings] surviving from this period (while four predate 1700). Notable among them are gambrel-roofed houses, fashionable only into the 1760s. Their double-pitched roofs, folded to broaden the attic, give them top-heavy prominence and plasticity among their gabled counterparts. They enliven the visual effect of the crowd of houses pressed against the narrow streets of the slope down to the harbor, the heart of the old colonial town which became known as the Hill. A particularly impressive row of large gambrels lines the waterfront of the Point (an area extending from the Hill section at the northern end of the harbor, where the point itself has been obliterated by subsequent filling). Private docks and counting houses once competed for space in family gardens. Providence has few gambrel-roofed houses. The lesser economic status of the place until after the middle of the eighteenth century meant that fewer were built than in Newport, and most of those disappeared to later building.

Wealth also brought a splendor of colonial public architecture to the town of Newport and two fine designers representing successive generations: Richard Munday, carpenter-turned-designer working in the colonial tag end of the English Baroque style of Sir Christopher Wren; then Peter Harrison, a gentleman designer who used his fine architecture library to bring the scholarship of Renaissance Palladianism to Newport, even as Thomas Jefferson used the same means to bring it to Virginia. The contrast of the two generations in Newport design is vividly evident in the architectural contrast at either end of the central square of the town (named for Washington after his death): the Colony House, almost surely by Munday (1736–1739), and, at the opposite end, Harrison's Brick Market (1762–1763), beside the Long Wharf. The regularity of the ground-floor arcade of the Market House, rhythmically matched by the correctly pedimented windows and cornice entablature above, betokens the transit to America of the lyric purity of Renaissance classicism. Meanwhile, at the top of the Hill and surrounded by farmland when built, Harrison also designed the Redwood Library (1748–1750) in one of the earliest temple fronts in the American colonies reasonably true to classical precedent. It was only the second private subscription library in the country, and it is the oldest still inhabiting its original building (though several times enlarged). It also contains a treasure trove of books, manuscripts, and portraits from its long history of collecting, though its holdings before the Revolution were, unfortunately, a casualty of the British occupation.

But Newport's public architecture also displays the beneficent and stimulating effects of Roger Williams's ideal of religious tolerance, which attracted both Quakers and Sephardic Jews to Newport as early as the 1650s, the latter group driven from Spain and Portugal, along with Muslims, by the fervor of the Inquisition. Members of both groups became leaders in Newport's commercial prosperity and drew other dissenters and free-spirited individuals to the town's affluence and tolerance. The Great Meeting House of the Society of Friends went up as early as 1699. Revamped several times and enlarged into something of the restored compound one sees today, it long accommodated the yearly meeting for members throughout southern New England until the gatherings withered away in the nineteenth century. It is one of a number of Quaker meeting houses remaining throughout the state, especially in northern Rhode Island. Some continue as places with active meetings; others, like the meeting in Jamestown, where Philadelphians represent a strong seasonal presence, open during the summer only; still others, like the gabled and shingled meeting house in Little Compton, are retired to museum (or near-museum) status. The Little Compton meeting house, with its double-doored entrance dividing the genders, framed in two stories of simple sash, sits in a double field of grass, with a mowed path straight to the gate of its generous stone-walled enclosure, also mowed close, by which the lovely structure is framed and carpeted.

Near the Newport Friends Meeting House rose the first building of architectural significance in the American colonies intended as a synagogue. Again to Harrison's design, Touro Synagogue (1763) is austerely reticent on the exterior and most exquisitely wrought within. Mingling with these also stands the diminutive Seventh Day Baptist Meeting House (1729, now preserved as an attachment to the Newport Historical Society), probably by Munday and certainly in his manner.

Meanwhile, the strong Tory faction in town had already ensured the most conspicuous position in Newport for its religious convictions—Anglican, not Puritan Congregationalist—by calling on Munday (unquestionably his work in this instance) both to design and to build Trinity Church (1726) as the centerpiece of the town and one of the great early carpentered churches of New England. Later, when it came Providence's turn to mark the center of its town with a great wooden church (1774–1775), it was Baptist. (In fact, when Roger Williams arrived in Providence in

1636, his convictions were close enough to those of Baptists so that he was forced at one point to deny that he fully agreed with their position.) As the first architectural achievement of consequence for the denomination in this country, it has come to be known as the First Baptist Church of America—in effect, as the Mother Church of the denomination in the New World. No wonder that this Rhode Island potpourri of heresy affronted Massachusetts Puritans, together with the thought that tolerance could elevate any denomination (heretical or not) to the central position in the community, as, indeed, the spire that dominated Newport was Anglican. Outside Roger Williams's sanctuary, opponents commonly disparaged it as Rogue's Island.

Yet other observations and issues suggested by Providence's First Baptist Church deserve comment in grasping its position within the colonial culture of Rhode Island. First it should be noted that it is among the largest of New England's colonial churches; those who saw to its building also dared to give it the capacity to seat the population of the town at the time it went up. Like Harrison, its designer, Joseph Brown, was a gentleman architect with a fine private architecture library to inform his taste and a grand house capped by a distinctive ogee-curved pediment built to his design on Main Street as partial testimony to his skill. It is the earliest extant house of consequence in America designed by an architect for his own use. (When it was completed, in 1774, another gentleman smitten by architecture, Thomas Jefferson, was at work on his first version of Monticello.) In his design for the church, Brown ingeniously combined the old meeting house tradition of entrance doors centered in the side walls of the building (originally connected inside by a cross aisle) with the newer, longitudinal model of the spire on the end of the church as both the commanding marker and entrance to an aisle running the length of the building to the sanctuary. For one of the more ambitious colonial spires erected anywhere in the colonies, Brown turned to his copy of James Gibbs's *A Book of Architecture* (1728), selecting from it an alternate design for the spire Gibbs built for St. Martin-in-the-Fields, which graces Trafalgar Square as among London's most familiar churches.

The First Baptist Church, however, played a dual role. It reinforced the presence of a newly founded college (then called Rhode Island College), moved from a nearby town to the summit of the hill where the houses of the leading families of Providence clustered. So the Baptist Church acquired a dual purpose, looking to both town and gown. In the resonant statement of the time, it was built "for the publick Worship of Almighty God; and also for holding Commencement in." Topped by its college, the hill inevitably became College Hill. To honor contributions from the leading commercial family in town it was eventually, in 1804, renamed Brown College. So Baptists now joined Congregationalists, Anglicans, and Presbyterians in the sponsorship of one of the colonial colleges—seventh in line— that survive to the present, thereby in the process accounting for one of America's memorable urban images: the configuration of the spire thrusting from the commercial core of Providence, merging with the steep tiering of houses behind, up to the roomlike, rectangular Green at the heart of the campus.

To be sure, civic amenity paid a price for Williams's religious toleration. The dominating church overlooking the community domain of the common as the physical and spiritual core of the original New England town and as the essence of its eventual idealization was for Williams precisely the sign of hierarchical control and intolerance he meant to banish. So the early dissenters placed no emphasis on building churches, let alone setting aside a common. Meeting in private houses sufficed for those with such ardent conviction. As a matter of necessity, such is the case of all newly established settlements, but even sects preferring plain places of worship customarily anticipate early building for such purpose. For Williams, however, church building had low priority. Hence the architectural pilgrim who comes to Rhode Island to find here the expected "New England town" will be disappointed. Nevertheless, four of Rhode Island's thirty-nine towns do happen to have commons. All were former border towns in Massachusetts which jumped the state line as the result of a series of boundary adjustments (although one of the towns, Warren, did not set aside its common until long after it jumped to Rhode Island). The differing fates of these commons is another piece of the Rhode Island story.

There are also several instances of nostalgia for the missing common, as sentiment for this regional sign of the colonial past increased during the nineteenth century. Newport is the most evident. There, sentimental longing for the cachet of the common and the desire to "authenticate" Newport's New England character for the twentieth-century tourists streaming through it brought about the demolition, in the 1970s, of all the buildings between Trinity Church and the harbor and the moving of ancient houses to the perimeter of the space, to stimulate what Williams banned. The result, however, is not a space organic to the town, but a loss, in the void scraped out in the name of improvement, of part of what colonial Newport was.

Religious dissension combined with tolerance encouraged Williams's independence of thought in broader realms, which also had architectural consequences. His liberalism of spirit was also manifest in his organization of householders within Rhode Island towns in a manner which opposed Puritan reliance on hierarchical organization. It accounted for his success in obtaining a royal charter for his colony, one which in fact minimized British control. Rhode Island and Connecticut were the only two American colonies without royal governors. This meant more control of political and economic decisions by the colonists themselves, and it inevitably encouraged a degree of unruliness in reaching decisions that reinforced the sour, possibly partly envious conviction of certain outsiders that Rogue's Island was indeed the appropriate epithet for the place.

It was the first colony to strike an overt blow against the British, protesting what were considered unfair levies. In 1772, the year before British tea went overboard in Boston Harbor, a group of patriots rowed in darkness out to the British ship *Gaspee*, anchored off the coast of Warwick, and torched it to the waterline. Rhode Island was the first colony to declare its unilateral independence of Great Britain, on May 4, 1776. It was also the last to sign the Constitution, fearing its small size would make it vulnerable. Threats more than cajolery eventually made it the reluctant final star in the original flag. Indeed, the sculpture which crowns the present State House dome depicts the Independent Man, although the clunkish figure holding a spear in the manner of Roman emperors, but togaed in animal skins, hardly seems the messenger for the thought.

This independence of spirit characteristic of Rhode Islanders had its architectural consequence. The smallest state in the Union boasts no less than six state houses, all extant. This situation is sufficiently anomalous to warrant a short account. The proliferation of legislative chambers in such a tiny domain grew from the alternation of the original colonial legislative sessions between the colony houses at Newport and Providence, its legislature thereby literally embracing the colony's compound name of Rhode Island and Providence Plantations. The alternation proclaimed, in fact, that the colony had no existence apart from the act of its principal constituent towns and the sizable areas under their jurisdiction in voluntarily coming together to make it. As populations coalesced around other centers, and irritation at the time required to reach the centers, or at the neglect of local concerns from the center, increased, disgruntled settlers could appeal to the legislature as the source of ultimate power to establish full-fledged towns. Eventually the mainland state and the harbor islands became five counties, virtually nonexistent as governments, designed to serve merely as jurisdictions for the geographical rotation of the legislature, together with the governor and his staff, and also as districts for lower state courts. With occasional interruptions for short-term reasons, rotation continued through most of the nineteenth century. McKim, Mead and White provided the ultimate, or sixth, state house (1892–1904, and never officially called the capitol), the first in the line of those twentieth-century state capitol buildings to attempt a degree of Renaissance "correctness." Demoted, the five earlier state houses nonetheless remain as focal civic buildings in their communities. How astonishing that the smallest state in the Union should think it necessary to trundle the people's representatives about in such an excessive display of participatory politics, when even the largest states were content with a single capitol! But it was not primarily the functional reason of bringing government closer to the people that led to the traveling colonial government (although this was surely a benefit). The arrangement was in essence symbolic and again stemmed from Williams's repugnance for hierarchical authority. Each Rhode Island town was ruled, not by a hierarchy, but by its free householders regularly coming together to choose their leaders. So the pendulum move of the state government between Providence and Newport reinforced the notion that its authority was not overarching, but devolved from the towns (hence from the townspeople). The symbolic importance of Providence and Newport is clear from the power given to the state legislature to subdivide their two plantations into more towns as populations coalesced around other centers.

The eventual location of the State House in Providence also represented a capstone of sorts to Providence's rise during the nineteenth century as the state's principal city. After such splendor and promise during the eighteenth century, three blows in rapid succession struck Newport. Britain did not forget the *Gaspee*, and its occupation of Newport during the Revolution proved to be harsher than that of any other American colonial city. Meanwhile, Providence went virtually unscathed. Even before the Revolution its maritime economy had begun to gain against Newport's. Although the end of the war brought both ports a brief revival of maritime prosperity, Providence now had the greater economic momentum. Grand as were some of the houses built of Newport's maritime wealth, none quite attain the luxury of the four great extant Providence mansions which fortunes from trade erected on College Hill—the John Brown and Joseph Nightingale houses, from the late eighteenth century, and the Ives and Corliss-Carrington houses, from the first two decades of the nineteenth—and especially their cumulative impact as a cluster.

But the brief revival of maritime trade after the Revolution ceased with Jefferson's Embargo Acts and the War of 1812. Restricting both American exports and British imports, they decimated maritime trade in both directions. By the same token, the embargo, by temporarily shutting out British goods and British competition, favored the venture of American capital into manufacturing. Factories, however, required rivers for power and, ideally, direct connections with adjacent hinterlands. Newport, already economically shaken by its enemy occupation, had neither rivers nor land connections. Providence was advantaged in all these respects. From the end of the Revolution into the 1830s Newport mostly languished, its eighteenth-century heritage of buildings and decorative arts perpetuating the memory of past grandeur with increasing dilapidation. Buildings in the Federal and Greek Revival styles are barely evident in Newport. Providence abounds in examples of both. The attention of its entrepreneurs rapidly shifted from ships to factories.

Coastal Themes: The Grand Newport Scene

The pursuit of Rhode Island industrialization draws the visitor from the coast a little inland to its rivers and especially to the concentrations of textile mills that lined the two largest of these, the Blackstone and the Pawtuxet, north and south of Providence. We shall, however, stay a while longer with the coast before moving inland. For the rise of industry along Rhode Island rivers also paralleled the only slightly later beginnings of the discovery of the coast as a summer paradise. Of these two worlds, Rhode Island's nineteenth-century industrial activity is mostly gone, although built legacy recalls its glory days. Meanwhile, its carefree, escapist counterpoint has boomed.

Begin again at Newport. Even the colonial city had a considerable summer contingent, dominated by wealthy families from the major Mid-Atlantic cities and from the South, attracted by the ocean climate and scenery, but as well by the city's expansive affluence. When Newport faded, however, during its post-Revolutionary period of decline, most of the summer influx also drifted away. Only in the 1830s did Newport begin to make a noticeable return as a summer destination. Then the most conspicuous attraction was the fashionable Atlantic House Hotel (1844; burned and rebuilt 1845, designed by Russell Warren; demolished 1898), which looked into Touro Park, as did (and still does) Redwood Library, with an oblique view of one of its corners. A typical big, boxy, clapboard affair, the hotel derived its style (in the double sense of the word) from an Ionic-columned porch rising to a central three-story portico, the most conspicuous Greek Revival edifice in a town which otherwise boasted little more than a few houses in the idiom. The earlier Ocean House on Bellevue Avenue (1841; burned, quickly replaced and later demolished) had more extravagant porches at two levels, sparsely framed in a combination of posts and decorative bracketing that would probably be called Stick Style today but was then seen as a carpentered effort toward something "Gothic." Both hotels gained in visibility by their location at the top of the slope known as the Hill, where the core of the old colonial town crowded against the harbor. Hotel life, however, turned out not to be in Newport's future.

Houses, or, in the Newport euphemism, "cottages," became the accepted places in which both to live and to visit, as the evolving "summer season" for the steadily wealthier increasingly focused on a movable feast of privileged, extravagant, and competitive entertainment in opulent private settings, with admission only by well-guarded private invitations. In such affluent circumstances "cottages" were seldom modest. They more often approached the scale and comfort of small mansions, although the stiff propriety associated with their typical urban counterparts might here be somewhat relaxed by a carpentered look (if sufficiently suave) and, above all, by a generous veranda. In the end the use of the term "cottage," for many vacationers, really amounted to a form of hauteur, as it came to mean "for summer occupancy only"—also implying, however opulent the cottage might be, that its owners' winter establishment was grander still.

The area in Newport initially favored for the construction of such houses was close to the hotels, behind Redwood Library and stretching away from the side of Bellevue Avenue opposite to the colonial town sloping down to the harbor. Recently the neighborhood, known as Kay-Catherine for two of its streets, has also come to be known as the Top of the Hill. Though without historical sanction, it appropriately indicates its position in Newport. Summer places could of course go up anywhere and did in scattered locations throughout Newport. But development at the Top of the Hill, especially toward the end of the 1840s, occurred in such a concentrated manner as to make the trend away from hotel life toward cottages conspicuous.

As early as the 1820s ancient farmland, descending in families like the Eastons and Kays whose lineage extended back to the founding of the city or nearly so, began to come on the market for development. The earliest investors in the area tended to buy such large pieces of land for such sizable houses that the overall effect was of estates rather than cottages, more so because [the properties were] still interspersed with fields. The gravel Old Beach Road crossed the flattish crown of the hill, then dropped precipitously down to coved Easton's Beach, which is now the favorite town beach. Only gradually did further subdivision fill in the Top of the Hill, especially from the 1850s through the 1880s, until the enclave attained an urban density of medium-sized to large houses. From the beginning, moreover, the Top of the Hill became an area in which the leading local merchants and professionals mingled in year-round residences with the summer people whose influx steadily increased the local affluence.

The Top of the Hill is of interest for several reasons. It offers an extensive array and variety of nineteenth-century houses, mostly sizable and mostly ranging in style from Gothic Revival to Queen Anne shingle, designed by the best local practitioners as well as by distinguished outsiders. It introduces the summer development that continues southward along the crown of this rise, the steep hill descending to Easton's Beach becoming an eroded cliff drop which falls abruptly to the sea. Between Bellevue and the drop is the domain of the great mansions—and eventually the châteaux and palaces. Finally, the architectural pilgrim comes to the Top of the Hill because a number of works by two important New York firms, with a half generation between them, were built here. Richard Morris Hunt and McKim, Mead and White began their connections to Newport when they designed a number of their early houses for this area. These architects significantly contributed to the city and left here some of the defining buildings of their time.

Hunt was fairly well established by the time he bought an 1840 Newport house from his brother, the painter William Morris Hunt. A summer place, with a summer office in the yard, it was located near the town end of Bellevue Avenue, where it angles into Touro Street on part of the site now occupied by the Viking Hotel and almost directly across Bellevue from the Top of the Hill enclave. For a group of wealthy summer colonists from New York and Boston, most of whom were interested in some aspect of the arts, Hunt completed no less than five major houses for this enclave, only one of which (along with several lesser works) remains. This house for J. N. A. Griswold (1861–1864), one of his earliest and not as adventurous as some of the later, nevertheless fortunately still exists as the core of the Newport Art Museum, with some of its principal interiors intact. Like the Griswold House, all Hunt's Top of the Hill houses employ exposed framing in the manner of medieval half timbering and such vernacular building as Swiss chalets and certain Scandinavian wooden structures, all of which he knew firsthand from European travel, as well as from observing the nineteenth-century revival of such forms abroad and, finally, from its eventual reinterpretation in American carpentry. The conspicuous framework appears as the organizing entity. Set into the wall, slightly projecting, it contains within its paneling a variegated mix of clapboarding, shingles sawn in ornamented patterns, or brick and terra-cotta worked into polychromatic embellishments. Or, set free of the wall as a structural entity, the scaffoldlike framework can make a decorative wheel or extravagant bracket to support a projecting gable, or, standing even more free of the walls, skirt them with the post-and-bracket, trellised and latticed glory of the Victorian veranda.

The twentieth-century use of the term Stick Style has the merit of reawakening the modern eye to the structural and material expressiveness of such design, to which the rationalistic side of Hunt's sensibilities was certainly drawn. But he and his clients surely responded more (or at least as much) to the picturesque, ornamental, and evocative aspects of forms that ultimately alluded to history and nature. The disappearance of several of these lively, luxurious cottages—their plans so incisively organized by axes, their spaces so intricately interlocked—is surely the greatest of Newport's architectural losses. One wishes that a more substantial sampling of the work at the Top of the Hill were immediately at hand to present the full measure of Hunt's Newport achievement, this close to its beginning and a few blocks away from the grand finale of châteaux and palaces by which he virtually embraced the Gilded Age.

As for the work of Hunt's younger competitors at Top of the Hill, first as McKim, Mead and Bigelow, then as McKim, Mead and White, they produced seven houses for this enclave alone. All exist incredibly intact within two blocks of one another—another reason to wish that more of Hunt's were at hand for comparison. Bigelow's parents had long summered at Newport. In 1874 Charles McKim married Bigelow's sister. This double attachment to Newport brought the young firm to the city. The marriage lasted only four years, however, ending in divorce. Bigelow quit the firm, leaving his place to Stanford White, whose capabilities Charles McKim knew well because both had worked in Henry Hobson Richardson's New York office before its move to the Boston suburb of Brookline.

The radical shift in approach to design from Hunt's work at the Top of the Hill to that embraced by McKim, Mead and White was a shift from "Gothic" to the Queen Anne style (in the terms of the period), or from Stick Style to Shingle Style (using later terminology). Hunt's tall, angular, eruptive compositions, their linear structure patterned with various materials, gave way to the greater spread of McKim, Mead and White's compositions, composed of larger sculptural elements, their amplitude enhanced by their homogeneous enclosure in a single material, usually by the flicker of light over wood shingling. Most of their houses of the period (as here at the Top of the Hill), however, employed a brick lower story as the quiet horizontal base from which to lift the climactic manipulation of the shingled surfaces of the upper story. These were almost topological in character, wrapping the gentle angles or rounded swellings of bay windows and rising to envelop the gable with its dormers and the conical caps of towers. The lithe, linear enframement of large, multi-paned windows, recessed only shallowly, and the combination of these into abutted pairings, or in bands of three or more, permit the openings to stay with their wall surfaces, even to bend and curve with them.

Allusion to historic styles is, in these houses and at this stage in the firm's career, at least, so generalized that style is almost solely identified with elemental shape rather than with elaborated detail, an approach to architectural form derived from Richardson's example. The houses in their Top of the Hill cluster display the gamut of stylistic reference employed in their early shingle work: the looming shingled gables and gambrels of seventeenth- and early-eighteenth-century colonial houses, the latter especially prevalent in Newport; the mock conical towers and escutcheons of provincial French châteaux; hints of traditional Japanese houses and ornament, with screens and bracketing from India. It was, in sum, a mix of straightforward native vernacular with an elitist bow to European high culture on the one hand and, on the other, toward the romance of the exotic that gives such vitality and variety to the late-nineteenth-century Queen Anne Revival.

Meanwhile, somewhat earlier in the 1840s, Alfred Smith, a native Newporter who had already made a fortune in New York as a tailor, returned to his native city to make another in real estate by teaming with Joseph Bailey. They saw their golden opportunity in a mostly undeveloped site a little south of the area then being built upon in the immediate vicinity of the Top of the Hill. There, the steep pitch down to Easton's Beach below the Top of the Hill becomes a bluff with a long drop to the water due to wave action crossing the open mouth of the cove. The brink offers distant views across the water, up the Aquidneck shore, and out to sea. It long drew so many sightseers that they had worn a public pathway along the edge, captivated by the combined exhilaration of the coastal panorama and the sheer drop. Cliff Walk (what else could it be called?) gradually descends from its wind of the brink to end in a scramble over rock outcrop and surf, as the Aquidneck shore turns into its southern tip and directly confronts the full force of the open Atlantic. It was Smith and Bailey's idea to extend Bellevue Avenue southward, well back from the face of the bluff, while purchasing most of the land along its frontage and much between it and the precipice with its Cliff Walk. The walk now crosses the edges of the lawns of the great properties which bring them to public view—or loses sight of them where the path falls below the level of the brink. The developers terminated Bellevue at the ocean front.

Bellevue makes a sharp right angle and almost immediately feeds into Ocean Avenue, which winds along the outcrops and coves that face to the full the moods of the open Atlantic, from its crescendo splash on the outcrop and the swell and subsidence in its mini-coves on a calm day to its wild, thundering arcing of the roadway with water and cascades of mist. On the ledges rising behind are more houses scattered about (to which we shall immediately return). Upping the ante beyond the norm for the merely luxurious life at the Top of the Hill, Smith and Bailey envisioned a domain of greater extravagance, beyond the means of any but the wealthiest locals. They meant to cater to those who would command a grand summer "season." Even they did not imagine how grand it would become. So the domain of the mansions and, beyond their economic range, that of the châteaux and palaces was born. The site is sufficiently protected within the bay to escape the harshest blasts from the open Atlantic across the tip of the island. Ocean mists and moderated temperatures make this area an arcadia of trees. Here beeches of all varieties reign over all other species (however grand in their own right) by their sheer size and stateliness, adding their natural splendor to what the roster of architects brought. From New York there were Hunt and McKim, Mead and White, plus Ogden Codman and Carrère and Hastings; from Boston, Peabody and Stearns; from Philadelphia and Pittsburgh, Horace Trumbauer—to name merely the most renowned and among the most frequently commissioned architects who built here and those also with the good fortune to have done important work that remains more or less intact.

If the "season" for this gilded domain centered in entertainment within its great houses, it of course also involved the institutions of recreation. Three of them have long standing within the Bellevue–Ocean Avenue precincts, and another recent arrival has joined their company. At the city end of Bellevue is McKim, Mead and White's famed Casino, devoted not to gambling (as its name might suggest and its bylaws specifically prohibit), but to lawn tennis—and where, in fact, the International Tennis Hall of Fame has been located since 1954. From the very beginning of the final quarter of the nineteenth century, therefore, Newport reflected the growing athleticism of American life for both sexes. The Casino draws on the full array of design strategies and features apparent in the shingled houses at the Top of the Hill, but in a joyously expanded version. A garden court with fountain provided the original vestibule and pivot for the arc of an exceptionally generous screened and latticed veranda from which spectators viewed the competition on the grass courts beyond. The spread of the shingling absorbs into it the shapes and stylistic allusions of the nearby houses, but in this expanded context in a more encyclopedic manner. Writing at the time (1886), William Sheldon, in *Artistic Country Seats*, emphasized the Casino's importance in the culture of American recreation: "As a source of aesthetic pleasure, the country clubhouse in the United States is scarcely more than eight years old. Its beginning may be traced to the Newport Casino. . . ." And to its other virtues add the row of luxury shops that line its face to the street, giving it an urban aspect as well as one of exclusivity. These shops extend an earlier shopping block of Hunt's, making dramatic the shift from stick to shingle by the end of the 1870s.

At Bellevue's opposite end the developers saw to it that immediately after that street makes a right-angled turn and continues in effect as Ocean Avenue, a coved beach (called Bailey's) was set aside for the exclusive use of "members" of their exclusive community. Later, next to it came another, called Hazard's. Both were neatly rebuilt in shingle during the twentieth century. Ocean Avenue winds along Newport's Atlantic front for ten miles. The ecology abruptly changes from verdant to sparse. Not that its botanical aspect was always so spartan. Trees once covered much of this terrain, even if only of species that could take punishment or stand behind others that could. It was the colonial settlers who scalped the landscape, mostly for sheep pasture. Instead, the topological tumult of this coastal outcrop rises to ridges behind, with ground-hugging, crevice-seeking foliage, except in sheltered pockets, two containing large ponds with a long inlet from one of many covelets. This landscape has its own rugged, sculptural beauty, made all the more stimulating by the immediate recall of those lush beeches on broad lawn just left behind. There are big houses here, too, scattered about on irregular sites with overviews of the ocean, connected by winding roads around barriers and up and down slopes. In two instances, spacious subdivisions laid out by the Olmsted firm in the early twentieth century are beautifully adjusted to the rugged terrain. Lifted up to a middle height at the center of this eruptive topography, the fields of a onetime farm became the Newport Country Club, the third in the triangle of the main summertime institutions inhabiting this double domain of privilege, where the Ocean Avenue area extends that around Bellevue. Whitney Warren, himself from one of Newport's social families, designed the clubhouse. This originally accommodated a nine-hole golf course on one side and a polo field on the other. (Eventually polo moved elsewhere, and golf expanded to an eighteen-hole course.) The clubhouse is a V-shaped structure in Warren's own chunky, heavily framed, mock-monumental brand of clapboard and shingle Chateauesque—and very different from McKim, Mead and White's version of the style. A spur to the rear of the *V* once made the building Y-shaped, with a caryatid-supported "sun porch" facing south as its terminus (lost to Hurricane Carol in 1954 and sadly never rebuilt), while all three wings originally came to a hub in an oval mirrored and balconied ballroom.

Finally, to these well-established recreational facilities set down in the territory traversed by Bellevue and Ocean avenues add a much more recent addition, also situated off streets that continue Ocean Avenue back to the center of town. Like the seasonal colonists themselves, the New York Yacht Club is an outsider to Newport, but with a proprietary summer presence there. In a resort so dominated by New York society, its premier yacht club long sponsored the America's Cup Races in waters off Newport, which, after so many unbroken wins, came to be regarded by its members, and of course by Newporters, as the only rightful waters for the contest, until defeat at last in 1983 moved it elsewhere. But the club was determined to get it back. In 1987, it obtained its own Newport summer headquarters off Ocean Avenue, following the deaths of both John Nicholas Brown and his wife, Anne. He was a member of the club and onetime assistant secretary of the navy. The Yacht Club purchased their house, also château-inspired, this more conventionally in masonry by Ralph Adams Cram, the Boston architect much better known as the leading American designer of his time in ecclesiastical Gothic. Atop a lawned slope pitched at nearly 45 degrees (or so it seems in the vertigo of standing on it), this green swathe falls nearly a quarter of a mile from a terrace off the front of the house straight down to the harbor and its seasonal flotilla of pleasure boats.

And any accounting of the activities originally envisioned as part of Newport's social season should not forget the stream of horses and carriages with liveried attendants, all burnished, for which Bellevue and Ocean avenues were intended. The quintessential parades, of what was effectively a daily parade, were coaching competitions in which spit-and-polish display was part of the score, the rest being the consummate skill of managing the "four-in-hands," every horse matched and gaits coordinated, and the punctilious ceremonies of the liveried attendants in the various colors of their houses, showing their employers in and out of the vehicles. Stables from the past dot the Bellevue area (many converted to residences), although Cornelius Vanderbilt's barn and carriage house, with accompanying greenhouse, is open to visitors in a mews area a little removed from the Breakers. Except for revivals of the coaching tradition as occasional parades of antique carriages down Bellevue, the equestrian aspect of the gilded domain is no longer evident in Newport. It persists at a reduced scale in riding, horse shows, and polo on the remnants of former sizable model farms maintained in neighboring Middletown by a number of the wealthiest Newport residents to supply their tables. Two Vanderbilt families, for example, had such farms, both with large barns for indoor riding and competitions, one of which still exists.

So, to return to the phrase of the time, Newport before the end of the nineteenth century became widely acknowledged as the American "Queen of Resorts." For Newport in the 1890s, however, the regal domination of lavish entertainment, meticulously exclusive but widely publicized, conducted within an acknowledged shifting hierarchy, really was controlled not so much by *a* queen as by a bevy of competing near-queens, each struggling for acknowledgment as *the* queen. The mix of decorousness and ostentation on courts, links, decks, and horseback also made a realm for queenliness in palaces where they reigned over the rituals of receiving and cutting and knowing when and how to do each—and those of plotting the over-do and out-do of extravagant entertainment. Newport's queens often reigned alone, holding court through the whole of the Newport summer, while many of their consorts (even the millionaires) intermittently rushed off back to the city on business.

But today? Although this privileged precinct can be viewed (and too often is) as a collection of buildings out of the past, those concerned with the history of resorts should also consider its transformation as a precinct. Like other splendid resort areas with a fine heritage of architecture, Newport's "summer season" is today a duality. Tourists arrive to live vicariously the gilded summer seasons of the mythical past. Meanwhile, the privileged return to the same area to prolong the glitter and whirl inherited from it into the summer seasons of a somewhat diminished, or at least transformed, queen. Newport events are not necessarily diminished in opulence, but they occur in greater privacy, in a less hierarchical situation, with preference for luxurious possessions that enhance activity rather than monumentalize position.

Meanwhile, another major interloper has invaded the precinct. To tourists, add Salve Regina University, established in 1947, with its presidential headquarters in one of Hunt's châteaux (the Ogden Goelet House, Ochre Court, 1892) and its considerable spread to other historic buildings over a broad swathe of the area. At first adding new buildings in a quasi-modern style in yellow brick (seemingly the material of choice for Roman Catholic institutions through the mid-twentieth century), the university appeared unsympathetic to the context. But it has turned out to be otherwise. Salve (as locally abbreviated) has respected the exteriors of the historic buildings it owns and the best parts of their interiors, although most of the latter are bound to seem irrelevant for the kind of furnishings and uses that educational institutions bring to them. It has also generally respected the scale and spacing of the area it occupies, even to the overall maintenance of the outdoor appearance of things as they approximately were, thereby minimizing its presence as an obtrusive entity in the area. In its own more recent additions to the campus, thought has been given to the fitting of the new to what already exists. Moreover, the students who come to it seem in part to choose the school because of their awareness (which the university actively encourages) of its special quality. And it helps that the school operates at full pitch out of season, bringing life to an area much of which is then shut down, leaving for summer occupancy the quieter summer institutes and conferences that come to its vacated quarters. However intrusive Salve's fit to the area, nevertheless it provides something of a model for institutional intervention into an area of historic building. So far, though, the toughest decisions with respect to expansion lie ahead.

The alternative would have been the conversion of the larger buildings into apartment-sized units and the shift from space to higher density—of which the precinct shows ample evidence, though it has fortunately been checked by the spread of Salve and, more fortunately, by holdings that remain in private hands to perpetuate the glamorous season. And this is the more glamorous for the reverberation in it of echoes of its extravagant palatial past, as continually and brought to new luster through ongoing historical research and annual preservation campaigns by the Preservation Society of Newport County. Large houses of historic importance in the area are open to the public under its auspices and those of other organizations.

Of course, the balance of forces in the precinct is precarious. A challenge to the future of Newport, it has lessons to teach in all places where history, travel, changing uses, and new residents come together, each component potentially benefiting from respectful good will toward the others and ideally finding that just such reciprocity enhances rather than diminishes the experience that all take away from the "genius of the place." And for the architectural pilgrim who has, in whatever order, made the rounds of Newport's buildings, this precinct culminates the experience of the almost zoned run of fine residential architecture within a relatively small compass, which extends from the last quarter of the seventeenth century (with some interruptions) into the early twentieth. Add to this array of buildings its parallel in Providence, starting later, bridging the Newport interruptions of the architectural past untouched by Newport, with each city (eventually, at least) inflected toward quite opposite destinies, and one has reason enough to visit Rhode Island. Besides, there is much more to see.

Coastal Themes: Forts and Lighthouses

Another presence in Newport, the long tenancy of the United States Navy, introduces yet another coastal theme in Rhode Island. Although the size of the fleet stationed there varies depending upon circumstances, Newport has always been a favored naval base, with permanent installations on Aquidneck Island but also on Coaster's Harbor Island, just off shore of downtown Newport and reached by a short causeway at the north end of the harbor. There one of the components is the Naval War College, founded in 1880 as the navy's school for advanced officer training in tactics and strategy. The original row of buildings (including a rare Newport example of the Federal style (1817), now a museum, which predates the school's founding), looks across the harbor at Fort Adams (1824).

This, the second largest masonry coastal fortification in the country, occupies a promontory projecting off the top of the toe of the Newport boot, at the end of the narrower of the two main channels. Three of its faces command ranges across the channel, up the bay, and into the harbor, while the fourth looks across the toe of the boot out into the Atlantic and into Narragansett Bay between Aquidneck (Rhode) and Conanicut (Jamestown) islands, as a main protective barrier at the entrance to Newport Harbor. All ships leaving the channel pivot around this promontory to enter the harbor. Hence the fort. The massive masonry enclosure provides a double wall with many vaulted spaces, between enclosing storage areas and barracks, around a large drill field. Today it serves for popular music concerts and other mammoth festivals.

Fort Adams is the most conspicuous of a sequence of forts and other devices that have, from the Revolution onward, guarded the channels to the bay. The Revolutionary fort exists today only as a site on a tall outcrop at the mouth of the East Passage, across the channel from Newport on Jamestown, although the ovular ruins of the fort still existed at the end of the nineteenth century. Below it is a nondescript installation from World War II for setting out minefields around the channel mouth. Also at the mouth of the East Passage, on Jamestown, but facing out to the Atlantic, is Fort Wetherill, begun during the Spanish-American War and successively rebuilt during both world wars as reinforced concrete sunken gun emplacements and living quarters under bombproof slabs. Scooped into a rock ledge for heavy artillery, batteries are aimed out to sea. Now a park, the pitted shapes remain covered with graffiti, but offering elevated platforms from which to watch pleasure craft heading through the East Passage in and out of Newport.

And across to the opposite shore of Jamestown, where a long point swelling to the shape of a beaver tail (hence Beavertail Point) projects out from Jamestown Island as a kind of bowsprit to the island. It provides the leading edge for the eastern shore of the West Passage, which is the wide passage into the bay used by most big ships. There on the tip of the point beside Beavertail Light, a very advanced radar installation (now dismantled) huddled under a low dome during World War II. At the same time, out in the middle of West Channel, Dutch Island (where the Dutch once came from New Amsterdam to trade with the Wampanoag) was a veritable fixed battleship, loaded with artillery pieces, more mine-laying facilities, and an installation for closing the mouth of the West Passage with a submarine net.

Across the passage on the mainland are two adjacent naval bases at Quonset, the Naval Air Station and a Seabee Battalion for training naval aviators and engineers.

The architecture firm of Albert Kahn, famous for its Detroit factories, designed the principal administrative buildings and officers quarters for the air station, as well as its wide-span hangars, gawky but fascinating examples of adapting laminated wood arching partially suspended from cables off metal poles thrust through them, to save on scarce metals. Meanwhile, on their base the Seabees developed the ubiquitous Quonset hut, a few of which, semiderelict now, still exist. A museum to commemorate what occurred here now complements a quasi-museum at the air base, where antique planes are both repaired and displayed. And here and there along the coast a few beefy farmhouses or shingled summer cottages remain, which, on closer inspection, turn out to be more reinforced concrete bunkers in disguise.

Between wars the varied military installations that in Rhode Island cluster especially around the passages into the bay tend to molder, to be converted to parks, or to be sold off for commercial use. But threats of war have always reignited the beachhead syndrome here. And why not? If the opening engagement of the Revolution occurred with the *Gaspee* incident in Narragansett Bay and the subsequent occupation of Newport by the enemy, so roughly a century and a half later during World War II, the final official naval engagement in the Atlantic occurred in Block Island Sound, between the mainland and its offshore island. An American mine sweeper sank a German submarine and its twenty-one-man crew prowling the sound outside the entrances to the bay. Unbeknownst to the combatants, their confrontation occurred only hours after Germany had declared final surrender.

Lighthouses, too, abound, as peaceful protectors of the coast to set beside its warrior equivalents. Of the many that light the Rhode Island coast—or rather lit it, since all their positions have been automated—Beavertail Light warrants special regard. It is special for what and where it is. Character rather than comeliness attracts the visitor to this square-shaped tower in rough-faced granite, as bluntly shaped as a carton for a gift of liquor. Pairs of long, rectangular blocks run the full length of its square shape, built up as crib construction. First a pair laid down to establish front and back walls, then a superimposed pair at right angles to begin the rise of the side walls. Pair on pair, as the square crib rises, the corners of the tower interlock, as in those log houses built of timbers, where the logs have been squared so that both walls and interlocked corners are flush. Other long blocks shortened in length only by the width of the interlocks either end fill in the voids left by the cribbing to complete the walls. The beacon in its cast iron and glass caging sits as a perfunctory knob on its granite carton, with tower elements assertively self contained.

A row of gabled living quarters and storehouses extends from one side of the tower, all attached, all facing the sea. Its surfaces, concrete over stone, give a breadth to the shifts in light and shadow from shapes within the abrupt picturesqueness of its silhouette. It is the sort of motif to which the painter Edward Hopper would have been attracted. Completing and animating the composition, a rounded whale of a promontory flanks one side of it, the water on a calm day swelling up the curved rock toward the tower, then sliding away. But even when lulled by the breathing legato of this rhythm, one senses the surge and the undertow that a gale could bring.

Obviously, however, Beavertail owes its legendary status among the lights on this shore to its position as the marker between the two passages into the bay. To the mariner coming in from the sea, Beavertail proclaims as no other light, "You have arrived." Not only has one reached the protection of the grand bay, but the maritime entrance that, in this small state, opens directly to the very heart of its territory. More than this, it is also positioned close enough to the midpoint of the imagined diagonal of the Rhode Island coast as plotted between the boundaries of Connecticut and Massachusetts to seem centered.

For two other particularly favored lighthouses, different in character from Beavertail, go to those at opposite ends of Block Island, some fifteen miles out to sea from its nearest mainland ferry dock at Point Judith. Neither features the conventional tower lifting its light. Both employ what is usually the less expressive formula of emphasizing the keeper's quarters onto which or into which a squat tower hoists the light no more than it has to, leaving it a mere cupola to the house. Yet these are fascinating lighthouses, made more so by their locations on two sites of spectacular beauty, both differing dramatically in character.

Rising from what has become a wildlife sanctuary in the dunes, Sandy Point (now North) Light (1867) faces the sound between the island and the shore. It is a box of a building in rock-faced granite ashlar, gabled, with the light treated as a ridge-top cupola set on a sculptured pedestal, which is both bracketed off the seafront wall and fused with a heavy gable cornice, all cast iron and of such oppressive scale as even to overwhelm the overbearing stone box below.

Across the island, high on a sandstone bluff, South East Light (1873–1875) faces the Atlantic in all its furies. Of all Rhode Island lighthouses, it is farthest asea. Due east, across its longitude is open water clear to Portugal and Spain. A tall, steeply gabled, two-and-one- half-story house sets its flank toward the ocean. Smaller house-like ells, parallel to one another with equally steep gabling, butt its rear elevation. The smaller ells behind provide quarters for the first and second assistant keepers; the big house belongs to the keeper, except that he shares it with the light tower. The squat tower half protrudes from the ocean front of the main house, barely lifting its glass and iron lantern above the ridgepole; the bluff provides the altitude. A dowager composition this: bustle, body, corseted stance, with (shall we say it?) a glance flashing to meet the most formidable circumstances (with one exception: she did not anticipate the erosion). Her masonry body is clad in hard, cherry-colored brick, boldly trimmed with bluestone, flush with the walls as in the slightly arched and cornered lintels over the windows, or projecting in simple profiles to accent gables. The walls come down onto a rough granite base. Her style is functional Gothic of a bristling angularity, but with a Victorian feeling of what no-nonsense Gothic could be. Of all Rhode Island lighthouses based on house shapes, with an incorporated tower, this is both the most integrated and most compelling. Though it is rigidly symmetrical in composition, movement around reveals its picturesque possibilities. Terrifyingly dramatic when it sat on the brink of disaster, the lighthouse, geologists warned, could be saved from the sea only by moving it back from this edge. The warning became dire when the engineers estimated that were it not moved back within two years of their estimates, it might be too vulnerable to move at all. Thanks to heroic and frantic efforts of a committee of friends, in a campaign led by Gerald Abbott, a New York doctor so enamored of Block Island that he shifted his practice to Providence, South East Light was hauled on rails back from the abyss during the summer of 1990.

Grateful for such dedication to a landmark lighthouse, the Coast Guard returned to South East a Fresnel lens, like the one that was callously scrapped. The replacement, from a destroyed Connecticut lighthouse, came out of storage because the Coast Guard reasoned that no such lens should stay boxed. Among the great optical inventions of the nineteenth century, these structures of cast iron and glass are both handsome as objects and effective in intensifying the lighthouse beam. South East's Fresnel focused the beam to a visible range of forty miles in a clear sky. So the light again circles its traditional signal into the night. If to the mariner on his way to Rhode Island, Beavertail says, "You have arrived," South East says, "You are near." Committees of friends of the respective lighthouses have made museums of all three of these lights. Other lights have also found their friends. Lighthouses seem to attract them by the power of their human symbolism.

If the most important light to the mariner is the one immediately needed, to those living along the coast it is likely to be the local light as an abiding marker in their landscape. Forts, another protector of shores, must also be built to be abiding. But Rhode Island's have fared less well than the lighthouses. Even Fort Adams was ill used until recently. Yet what a fascinating story an open-air museum of them gathered around the passages could make on the sites they once commanded, in the parks they now occupy. But (except for Fort Adams) they are presently considered dispensable—until in the event of another dire emergency they may be recommandeered by the military and outfitted with the weaponry of some future war.

Coastal Themes: Other Coastal Towns

Newport as the coastal center and the nearby passage into the bay have thus far provided a core from which to examine certain coastal themes that have brought us elsewhere. Of course, there are other coastal towns remarkable for architecture; not on the coast, however, but gathered where most coastal towns and villages are found in Rhode Island, that is, within the protection of the bay. Again three, remarkably different in character, deserve special notice—Wickford, on the west shore of the bay located where the West Passage opens to the spread of the interior waters, and Warren and Bristol, adjacent towns well up toward Providence on its east shore. Wickford (originally Wickford Landing) presents itself at first encounter as a village street of mostly late-eighteenth-century and Federal houses extending into Greek Revival. Just enough modest Victorian commercial structures at one end to enliven, without disturbing the rows of earlier houses that characterize the place.

The street terminates in a cove full of small craft in summer, covelets and marshlands extending from it as part of the scene. Short spurs of more old houses line a few side lands, and a walkway leads back to one of the most elegant of Rhode Island's small eighteenth-century meeting houses.

Warren is a bigger place. It fronts on a narrow channel off the bay worn by two small tributaries as their cumulative waters spread to the bay. Views from the opposite shore show Warren as a dense picturesque cluster of colonial and early nineteenth-century buildings, pierced by spires and mixed with later additions, presenting perhaps the most compactly picturesque silhouette of any small Rhode Island town. The historic approach to the town over a pair of bridges in succession leads into its congestion, then divides immediately into Water Street, with Main Street behind it. No other coastal town more clearly typifies (despite the regrettable loss of fabric on both streets) the division within the small seaport town of waterfront and commercial activities. During the early nineteenth century Warren was a bustling port and center for boat building. It was also Rhode Island's principal whaling port, while Warren ships (many built there) were especially active in coastal trade. Wealth from its ships accounts for the extraordinary quality and interest of the houses packed into this town. Then, as maritime activity dwindled, the cotton industry appeared. It unmistakably announces its onetime presence as soon as one crosses the double bridging into town with the long, rhythmic stretch of one mill's handsome piered wall with three stories of very large windows inset between piers. Set directly on the water, it spreads over sites formerly given to marine enterprises. On the narrow neck of land into which Warren is squeezed (before being interrupted by yet a third small river), the houses of sailors mingled with those of sea captains and shipowners, giving Warren a democratic mix of elegance and grit, which the coming of mill workers with mill officers to the town continued. This, too, typifies the mingling in close proximity here of buildings indicating differing status of occupancy and use, which accounts for the lively aspect of the town, as though the picturesqueness in the variety and meld of its physical fabric repeatedly made and adapted were overlaid by a similar social picturesqueness.

Bristol, Warren's immediate neighbor, presents an altogether different aspect. Bristol is among the bevy of smaller queens of New England towns. It has almost always been wealthy. If Newport reigns at the mouth of the bay, Bristol's realm is its interior. At the very center of the bay, it sits on a double peninsula, shaped like an open lobster claw, even to one digit being shorter than the other. The harbor is between, and, like Newport, it commands a choice of passages. Sailing to starboard of the town brings one to Providence; sailing to port heads to Fall River, which was the home berth of the once famed Fall River Line with its deluxe overnight accommodations to New York. Having stopped at Newport, the boats steamed on to the bustling City of Spindles, once one of the important textile towns just over the Rhode Island border.

Bristol's stateliness depends in part on its regularized plan, a grid laid out by Massachusetts proprietors in 1680 on the larger of the two peninsulas against the harbor: four major streets running north and south intersected with nine cross streets. No other such regular planned community exists in the state. And, in the Puritan manner, an eight-acre public common appeared, roughly centered, between the third and fourth north-south streets back from the harborfront. The two in front of it (now called Thames and Hope) correspond to Warren's Water and Main. In fact the center of town life is focused toward the harbor, not the common, which seems bypassed, doubly so because the three buildings it contains fit on the harbor side of the common facing away from it. Moreover, if the Congregational Church is one of them, the place of honor goes to its charming Federal state house, with a Victorian public school beside it. So the common became with time more of a governmental than a theocratic precinct. It typifies the problems theocratic domination had in Rhode Island.

Bristol's Hope Street is everything that one imagines the ideal town center of the past to be—late-eighteenth-century and, even more, Federal houses, and some Italianate, with an Italianate customhouse (1857). These all blend with a fine Victorian Gothic church and a mix of Victorian and early-twentieth-century commercial buildings. Moreover, no other town is quite so centered in so large and lavish a Federal house as Linden Place, built for George DeWolf.

The smaller digit of the lobster claw is gated as a precinct for privilege. The road to the tip of the larger one rises to an escarpment on one side, where more large estate houses than now once looked down the bay and where the Roger Williams University campus occupies the other side of the road, to cross John Steinman's epoch engineering feat in the placement of the Mount Hope Suspension Bridge high above the fast-moving current of Mount Hope passage to Fall River below, the first major bridge (1929) to connect the mainland to Aquidneck Island.

All three of these bay towns were involved in shipbuilding. Warren and Bristol still are: Warren constructing such craft as ferries and excursion boats, Bristol sailboats and yachts. Bristol continues the grand tradition of pleasure boat building, both sailboats and yachts, associated with generations of the Herreschoff family.

*   *   *

[The rest of WHJ's introduction was not completed. What remains are tantalizing notes to himself, some virtually complete, others but notes and queries. Again, rather than elaborating on them or attempting to answer them, the editors have chosen to present them here as written—as suggestions of how WHJ might have continued the introduction.]

The Vacation Enclave

Whether catering to vacationists or retirees (often former vacationists grown old), resort areas are prone to enclaves. A vacation enclave is a semi-, often part-time, community within an established community, dependent upon the larger community for many of its needs but deliberately separating itself, the better to focus on recreational aims informally organized by social grouping. Insofar as enclaves have government, it is not more than necessary: an "association" which establishes community "rules" as needed and perhaps assesses "dues" for the use of community property or "assessments" for major improvements to improve the lot of the group. Except insofar as the enclave may approach the status of a mini-village with a convenience store, gas station, or such, it depends on the larger community for basic needs and on the surrounding area for certain recreational lacks (golf course), cultural institutions (summer theater, museum), and sightseeing. But its Rhode Island is essentially the piece of the coastal rim it inhabits. Rhode Island as Eden: a season and a euphoric state of mind.

The typical structure of coastal enclaves in the topological and road layouts that favor them: major interconnecting highway set back from the shore; the single road off it to the cluster of cottages; the cottage cluster on a U siding; the cottage cluster on a point, promontory, or other coastal feature which substantially bounds it.

The alternate vacation enclave clustered not necessarily within a structured enclave but gathering as an identifiable group within the larger community.

Geography and vacationers: Southerners to Newport view it as a summer colony within colonial Newport. Rhode Island captains active in slave trade until banned (1774 but continued by rogue traders until the 1820s). One-fifth of the population of colonial Newport were slavers. Large Narragansett colonial "plantations" in southern Rhode Island used slaves, as did Providence and Bristol wealthy (all three towns important in the slave trade). (Check on colonial Rhode Island and slaves vis-à-vis other New England states; believe Rhode Island had largest number of traders and largest slave population. But also very active abolition movement from the beginning.) Their return to Newport in the nineteenth century through contacts with the textile mills. The exceptionally close contact Rhode Island seemed to have had with Charleston and Savannah: easy coastal communication; contacts between manufacturers' representatives and southern brokers and factors; records of Rhode Island builders in both southern cities, especially Savannah (John Holden Greene provided design for one of Savannah's largest nineteenth-century churches). Similarities in scale, antiquity, and sophistication between these southern cities and Newport must have made Newport especially congenial (and for Charleston, the harbor setting (check on dates for forts in each). Southern settlement in Newport up to the Civil War.

Middle Atlantic metropole: New York most important, especially because wealthy New Yorkers brought most architects in the state and the most of the best ones. Newport and Watch Hill as enclaves (to what extent is nineteenth-century Newport a manifestation of New York?). Philadelphians attracted by the strong Quaker tradition in Rhode Island from Roger Williams's tradition of religious tolerance.

Paucity of Bostonians: On the East Coast it is north to Rhode Island more than south. Massachusetts has its own shore and probably moves "down east" to Maine when it vacations farther afield. But Boston architects were sometimes called to Rhode Island by the wealthy (in preference to local firms, probably for a "name" regional practitioner).

Scattered midwestern sources: The postcard syndrome: "Having a wonderful time. Wish you were here"—and the wish fulfilled. A Minneapolis example. St. Louis wealthy families replicating the private street enclaves in their city in the remarkable shingled enclave of Shoreby Hill on Jamestown. A New York residential enclave in Narragansett Pier.

Small house developments: (When does a development become an enclave?) The rows of enlarged "bathhouses" (about the size of the largest shipping containers) at Carpenter's Beach . . . East Providence.

Special-interest enclave: Shelter Harbor as a c. 1900 musical enclave with studios attached to all early houses. Down Wagner Road with a composer for every address (except those on Caruso road, which helps in dating). Portsmouth religious campground.

Naval officers' enclave on Jamestown: Large number of naval officer retirees on Jamestown. Fairly cheap; Shingle Style in its most vernacular manner, even for large houses, predominates—from bungalows to large houses wrapped in extravagant verandas. Close to navy activities, yet (until recent bridge), enclaved by the ferry trip. Lack of showiness and uniformity in Jamestown houses and the bounded quality of the place has comfortable quality of a military base.

Wealthy family enclave: Large acreages of cheap waterfront land (and other areas) available in late nineteenth century–early twentieth century. Many enclaves of wealthy local families. Examples of vacation enclaves are the large Beavertail and Highlands holdings of the intermarried Philadelphia families of Whartons (Bethlehem Steel) and Lippincotts (publishing). [On the other hand,] the Philadelphia Wideners chose to vacation with New Yorkers in Newport in a Trumbauer palace.

"Winterization" and the death of the vacation enclave?: Retiree "winterization." The growth of suburbanization in southern Rhode Island and the tendency for the "winterized" house to complicate the enclave by its necessary attachment to the larger community.

Rhode Island smallness and the enclaved nature of the state

Variations on the Shingled Cottage as a Paradigm: Rhode Island Vacation Habitation

Jamestown's rebuke to Newport: Piazzas, wicker, the roughstone fireplace, and the mostly ungardened landscape. The large luxury house in Jamestown versus its counterpoint (even when shingled) in Newport. The related vernacular Shingle Style. Weekapaug's row of overblown shingled colonials. Watch Hill's shingled "formalism." Little Compton: the vacationist as preservationist. Little Compton's conversion of small eighteenth-century and nineteenth-century farmhouses and outbuildings as the quest for the old agricultural coastal setting. Block Island: its preserved portion (one of Nature Conservancy's nominations as the "Last Great Places").

Providence and Industrial Rhode Island

The geography of Rhode Island's nineteenth-century industrial empire: The symbiotic relationship between the textile and machine tool industries. How an off-center city (close to its eastern boundary) may be seen as centered. The arc including Rhode Island's largest rivers, the Blackstone and the Pawtuxet, north and south of Providence, includes the bulk of its industry. The compact quality of the area; its proximity to the coast.

Colonial industry in Rhode Island: Windmills on Portsmouth and Jamestown. Gilbert Stuart's stepfather's snuff mill and the congeries of mills around it. Lime kilns at Lime Rock, etc.

Almy Brown, plus William Slater; the beginnings of the American textile industry and the founding of the American factory system. Old Slater Mill in Pawtucket, a prophetic colonial example of "research and development."

Early mill forms: Masonry mills on the Pawtuxet; brick mills on the Blackstone. Rhode Island's granite quarries. Zachariah Allen's case for rubble and plaster. James Bucklin, Thomas Tefft, and the brick mill: White Rock; Cannelton. Brick mill influence from Massachusetts via the Brown family mills on the Blackstone.

Zachariah Allen and fireproofing. Slow-burning wood construction. Founding of cooperative manufacturers' fire insurance—Allendale Mutual Insurance Company.

[December 1996]

Providence (PR)

AMONG AMERICA'S OLDEST CITIES, PROVIDENCE HAS A SPECIAL sense of place. Like its contemporaries, it has been built and rebuilt many times. Development, however, has not resulted in wholesale destruction and replacement, but in complex layering, which reveals the built legacies created by generations of earlier residents. Together, Providence, as Rhode Island's capital city, and Newport have since the American Revolution been the state's economic, political, social, and architectural pacesetters.

Providence is situated in a topographical bowl ringed by hills at the head of Narragansett Bay and at the confluence of the Seekonk, Moshassuck, and Woonasquatucket rivers. These bodies of water played important roles in the city's development, first as a shipping center, later as an industrial center. Initially, the encircling hills also limited development to the low-lying waterside areas; later they provided prominent sites for residential and institutional building. Today, the crowded area around the waterfront has been reclaimed. Its overlook from densely built neighborhoods on the surrounding hills reinforces the urban quality of the city.

Roger Williams and his followers settled Providence in 1636. Baptists (as most of them came to be called, although Williams eventually denied he was such), they were expelled from Plymouth Plantation because of religious beliefs heretical to the reigning Puritanism. Eventually, this band of dissenters settled on the east side of the Moshassuck River, along today's North and South Main Street (originally known as Town Street). They celebrated their gratitude for their haven by naming it Providence. In contrast to hierarchical New England towns, like Boston, Salem, and Plymouth, in which Roger Williams had successively resided, Providence did not center its new town on the typical common dominated by an authorized church. Nor, indeed, was there any church building until 1700. Because Roger Williams and his group favored religious toleration, initially they met in small groups to worship

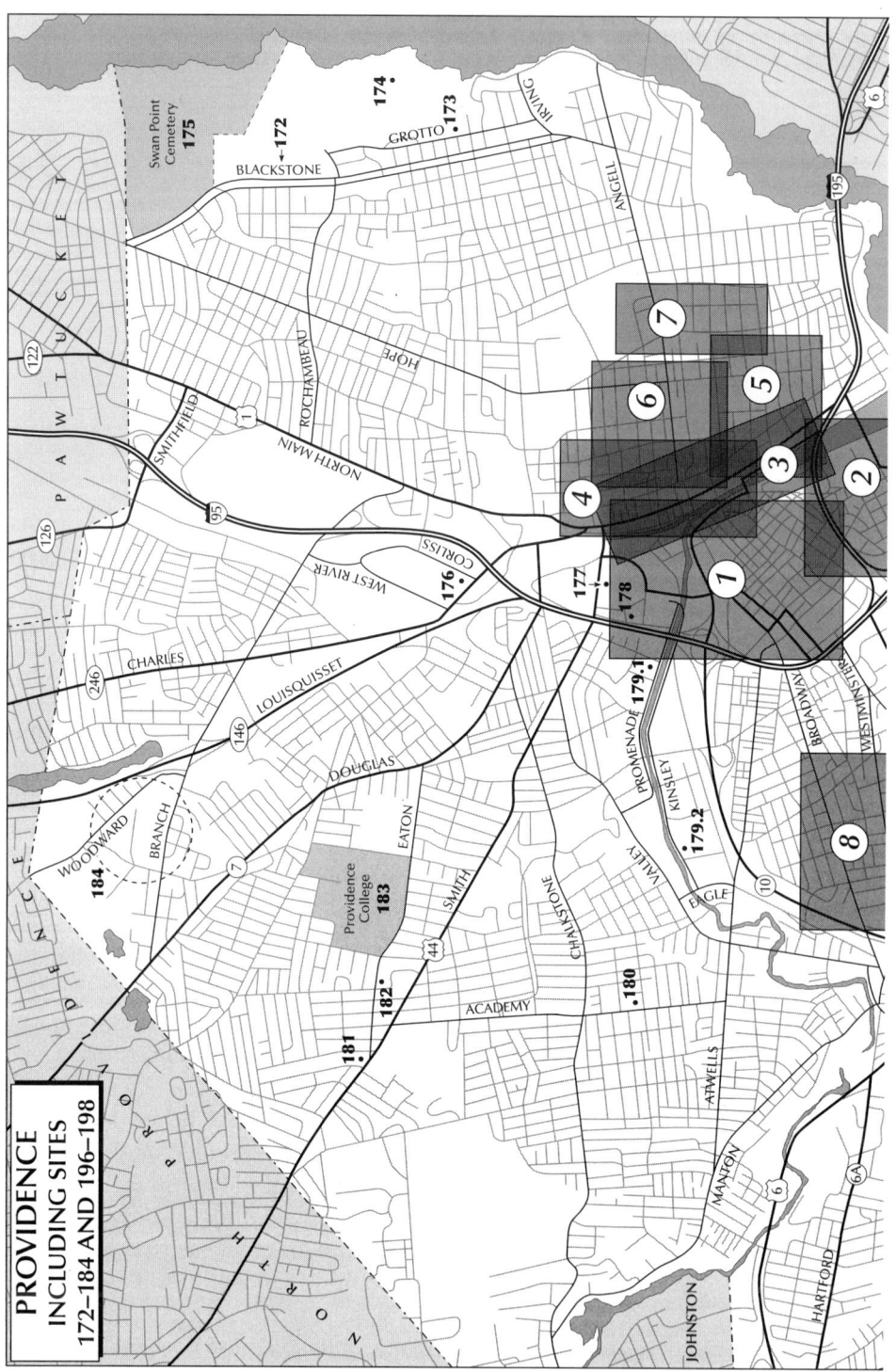

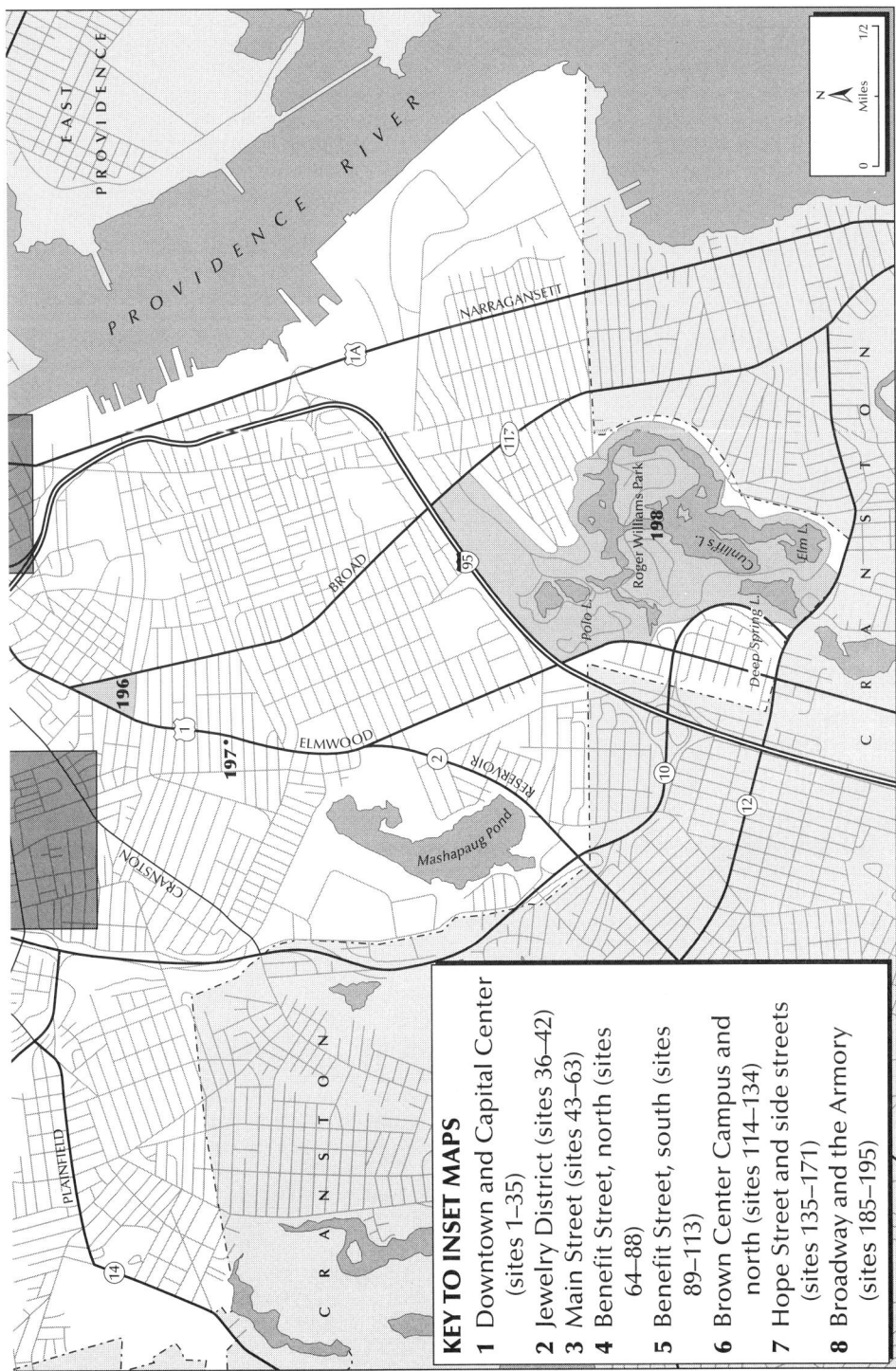

in private houses. They established a linear settlement along Main Street, from which very long, narrow house lots, in the Puritan manner of land division, stretched uphill east to Hope Street. None of the earliest buildings survives; all but one or two were burned by Indians in King Philip's War of 1675–1676. The original street and land development patterns, however, did guide future growth. Benefit Street, roughly parallel to North and South Main Street and perhaps the city's best-known street, was cut uphill of Main through the early house lots in 1756.

Until the eve of the American Revolution, Newport eclipsed Providence economically, politically, and culturally. By the mid-eighteenth century, however, local merchants had developed a strong maritime economy, and during the Revolution a protected position at the head of the bay gave Providence an advantage over exposed, British-occupied Newport. Newport never recovered its colonial political and economic power within Rhode Island vis-à-vis its rival town. Although shipping continued to flourish in Providence after the Revolution through the early 1800s, the town asserted its hegemony within the state as local merchants turned their attention to industry on the rivers that flowed into the city—an advantage that Newport did not possess. In 1790, Providence capital financed the country's first mechanized cotton factory, Slater Mill in nearby Pawtucket (see under Pawtucket). The costume jewelry industry was born here, too, in the early nineteenth century, when Nehemiah Dodge of Providence perfected the plating of precious metals to base metals. This and other manufacturing, in turn, encouraged the local development of the machine-tool industry. Initially focused on the production and improvement of machinery for textile and jewelry manufacturing, expertise in these areas quickly expanded to include a vast range of tools and machinery. For a century and a half, industry, led by textiles, costume jewelry, and tool making, propelled the city's prosperity.

Strong banking and insurance interests, which had developed in response to eighteenth-century shipping, were already in place to meet the new needs. To Providence's growing concentration of economic prowess within the state, add also politics. Paradoxically, especially for the smallest state in the union, its legislature long rotated through no less than five "state houses," one for each of the state's five counties. Ultimately Providence prevailed as the state's single capital, but only definitively at the beginning of the twentieth century with the completion of McKim, Mead and White's building.

The presence of industry attracted an increasingly diverse community of European immigrants from the 1840s through the 1920s, making Providence one of the country's most polyglot cities. The city grew and prospered almost unceasingly between 1780 and 1940, and many of the industrial, commercial, ecclesiastical, institutional, and residential buildings that document that growth remain. Then, economic decline, beginning in the textile industry after World War I, intensified after World War II. For the city's architecture, however, stagnation of growth proved to be a blessing in disguise. It preserved by default many buildings that might have fallen to "progress" in brisker economic circumstances. They remained long

Downtown Providence, Exchange Place as it looked in the 1880s, with Thomas A. Tefft's Union Depot (1847–1848; burned 1896) and, in the background, Providence City Hall (PR4)

enough for the development of a strong movement for preservation by the late 1950s, which reassessed the value of a neglected heritage.

The Providence tour begins at the heart of the modern city on the west side of the Providence River. This neighborhood offers an extraordinary survival of business and commercial buildings from the early nineteenth century onward. Here, too, is the site of Capital Center, the sizable public-private venture which is counted on to revitalize the city center into the twenty-first century. Immediately adjacent is a factory district where the city's important jewelry industry once concentrated.

From the center of the city our route crosses to the east bank of the Providence River, where the original colonial city once lined Main Street. From this we ascend College Hill, site of one of the outstanding residential enclaves in the country, most of its houses ranging from the colonial period through the Colonial Revival of the first decades of the twentieth century. College Hill is also home to both Brown University and the Rhode Island School of Design, nationally important schools with significant campuses. East of College Hill, the East Side demonstrates the later nineteenth- and twentieth-century expansion of this neighborhood of choice and, in addition to houses, offers the distinguished conjunction of a landscaped parkway, asylum, and cemetery, thereby providing an outstanding example of ways in which the Victorians and their early twentieth-century successors brought the country to the city.

Then, crossing to the west side of the city again, but north and west of the center, the tour continues with McKim, Mead and White's State House, in many ways a monument to the prosperity and the political and economic power of the city and state at the turn of the century. Architecturally it is among the more significant and influential state capitols in the country. Downhill from the State House is another major factory enclave, the Woonasquatucket Industrial District, dominated by the former plant of Brown and Sharpe, the largest American machine tool manufacturer until after World War II. Farther west and north are other residential neighborhoods, including what remains of Elmhurst, once a nineteenth-century sylvan retreat, in which the largest residences tended to be more spaciously conceived as mini–country villas with gardens and orchards, in contrast to the close-packed domiciles on College Hill. Two notable survivors remain on the campus of Providence College. Farther in the same direction at the very northwest corner of the city is the former textile mill village of Wanskuck.

The tour concludes southwest and then south of the city's center. Lined with Victorian mansions, Broadway was the address of choice from the 1870s onto the early twentieth century for many upper-income mercantile families. Next is a restored suburban enclave of late Victorian and Queen Anne Style houses around the Parade of Providence's formidable Cranston Street Armory. The tour concludes at the southern edge of the city with Roger Williams Park, an intact 230-acre landscape park, largely designed by the important nineteenth-century park and land planner H. W. S. Cleveland.

Existing architecture in Providence presents a microcosm of national developments, yet exhibits its own flavor with a number of distinctive characteristics. Here, in rare concentration, one can find colonial architecture and an especially lively, abundant, and autochthonous Federal style; much Greek Revival; a robust, long-lived development in the Italianate mode; and some fine High Victorian examples. Providence was an early and important center in the Colonial Revival, which left a number of distinctive buildings exhibiting a mix of Queen Anne and Colonial Revival modes. It boasts many handsome, if not radically innovative, industrial buildings, especially of the late nineteenth and early twentieth centuries; one of the country's most intact and evolved downtowns from the same period; and important institutional structures. Providence ranks with such medium-sized eastern cities as Charleston, Savannah, and Salem (to cite a few competitors at random) for the number and quality of extant historic buildings and for their urbanistic impact, but at the same time offers more variety of type and period than these other cities. Put more bluntly, no American cities of their size have more to offer the architectural pilgrim than Providence and Newport, each with its own distinctive ambience.

Downtown Providence (Downcity) and Capital Center

The old business district of Providence offers an exceptional array of fine historic buildings dating from the late Federal and Greek Revival periods, when stores and offices began in earnest to cross the Providence River from their original colonial center ranged along Main Street. As in most American cities, however, before the 1960s this core was in serious jeopardy from challenges by suburban malls and the move of a number of businesses out of the center city, leaving it vacant and increasingly dilapidated.

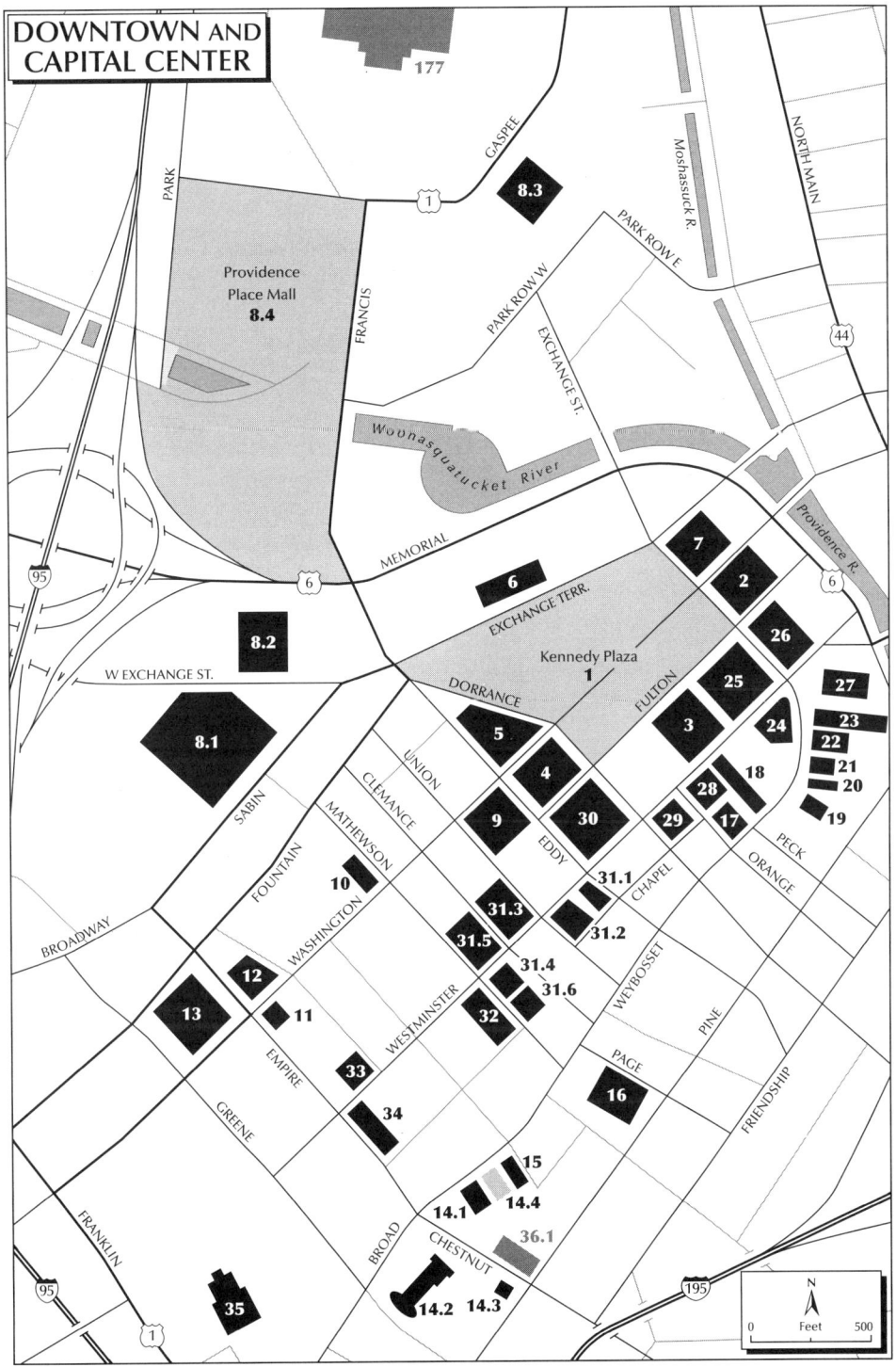

For the historic downtown this decade turned out to be something of a watershed. The tight sweep of the Interstate 95 and 195 interchange around two sides of the old downtown (fortunately narrowly outside it rather than through it), begun in 1958, was completed in 1964. Consonant with the interstate's streamlined zip as a sign of modernity, plans to demolish much of the architectural past for glassy replacements set on landscaped plazas appeared in 1961. Lack of funds at the time, together with the hesitations characteristic of this conservative city, permitted already strong counterforces for preservation to prevail—with ultimate victory still in the offing.

In preference to the "olde towne" label by which such "historic" districts are customarily banished from the workaday city into a gaslit touristland, this still active business district was christened in 1992, rather neutrally, Downcity. (That—and not "downtown"—seems to have been its local designation during its long heyday as the place for the city's shopping and entertainment.) Since the 1980s major redevelopment has augmented preservation efforts through a series of interrelated projects all around the Downcity core and the contiguous Jewelry District to the south.

The core business district is best traveled by foot, whereas most will prefer to drive around the surrounding orbit of redevelopment and the Jewelry District. Kennedy Plaza provides a civic vestibule to two major aspects of the civic center: from two sides into Downcity, of which the Plaza is historically a part; from the other two into the abutting Capital Center redevelopment. For an overview, stand first at the east end of Kennedy Plaza, in front of the Federal Building; from there this guide makes a clockwise examination of the most significant buildings individually.

PR1 **Kennedy Plaza** (Exchange Place)

1848, 1898, 1908, 1913, 1964, 1984, 2002. Bounded by (clockwise from northwest corner) Exchange Terrace, Exchange St., Fulton St., and Dorrance St.

A large combination park and transportation center, as well as the civic center, Exchange Place (renamed Kennedy Plaza after the president's assassination) was originally a smaller space. It has always been defined by a row of commercial buildings on its south side and, during most of the nineteenth century, by Thomas Tefft's famed Union Depot in the brick Lombard Style (1847–1848; burned 1896) on its north side. In 1878, the completion of Providence City Hall at the area's western end reinforced its importance as the city's heart. The plaza assumed its present configuration in 1898, with the completion of the second (PR6), which (as redeveloped for office and commercial use) is the area's present northern enclosure. The Federal Building (PR2), at the east end, is the symbolic counterweight to City Hall opposite. In the early twentieth century, an axial mall was completed through the center of the plaza across the street from City Hall, with a frilly cast iron and glass trolley shelter (1914, Martin and Hall), now attached to an enormous bus waiting room, whose obvious, heavy-handed inspiration by the earlier building mocks rather than complements. The Mall, as it was known originally, and the landscaped park to the north were designed in the aura of the City Beautiful movement, spawned by the World's Columbian Exposition in Chicago. Along the way, the square acquired two commemorations of the Civil War: a quaintly provincial monument to Civil War soldiers and sailors (1871) by Randolph Rogers, then a respectable equestrian monument (1887) by Launt Thompson, to General (and later Governor) Ambrose Burnside, Rhode Island's contribution to the top command of the Union Army. Sculpturally more interesting, however, is the Bajnotti Fountain, by Enid Yandell, a composition of allegorical figures inflected with Rodinesque Art Nouveau undulation. The plaza assumed its current use as a metropolitan bus interchange in 1984–1985, but the rerouting of buses and construction of waiting areas in 2002 significantly altered the scale of the open space.

PR2 Federal Building

1906–1908, Clarke and Howe, with Harvey Wiley Corbett as consultant designer. 25 Kennedy Plaza

A fitting pendant to City Hall at the opposite end of the mall (PR4), the gray granite Federal Building, built to hold the U.S. District Court and Customs Service, complements its predecessor in mass and scale. It is as telling an expression of attitudes toward civic architecture in 1908 as was City Hall in the mid-1870s. As a pair they illustrate the influence of French design on American architecture, from City Hall's Second Empire to the Federal Building's academic classicism, derived from Ecole des Beaux-Arts architectural training at the turn of the century. Despite interior renovations, the courtrooms remain intact, one overlooking the plaza with a stained glass window. Flanking the entrance are monumental seated female figures by John Massey Rhind: *Providence* on the left, *America* on the right.

PR3 Fleet Bank Building (Industrial National Bank Building)

1926–1928, Walker and Gillette. 1983–1985, Fleet Center addition, Hellmuth, Obata and Kassabaum. 55 and 28-44 Kennedy Plaza

To build what was then—and for many years remained—the tallest building in New England, the Industrial Trust Company looked to New York both for architects and for formal precedent and constructed what is one of the state's best-known landmarks and still its finest skyscraper. Walker and Gillette were established bank architects, and the stepped-back skyscraper form was just then emerging. The form, dictated in New York by the 1916 zoning laws to allow adequate light and air for adjacent buildings on crowded sites, was unnecessary here. What it provided instead was big-city bravado, a superbly sculptural Art Deco culmination of the Providence skyline in a major example of the setback. Above the ground story, sculptured reliefs celebrate Rhode Island's commercial and industrial history in the half-modernist, half-traditional mode of the period. The twenty-eight-story building is capped by a globe-topped lantern, once graced with masonry eagles (since removed) at each corner. At night, the building becomes the city's chandelier (or perhaps torchère), demonstrating the exceptional potential of setbacks for floodlighting. The original banking hall, surrounded by an Ionic colonnade and detailed in an Adamesque Art Deco mode, has been modified only by the removal of tellers' cages, installation of replacement chandeliers (which are nonetheless appropriate), and new lobby furniture.

Finally, a curiosity: a sharp eye and a distant vantage point reveal a windowed metal box sitting high on one of the eastern setbacks; its small interior space, which replicates that of the passenger cabin of a zeppelin, suggests its original use as an executive pied-à-terre and perhaps a private bar during Prohibition. (Documentation reveals that the Industrial tower was constructed partly with the intention of it being a mooring-mast for zeppelins, and that the penthouse was intended as the waiting room or terminal for passengers, hence the referential design.) The architects prepared it as their gift to the client. Later a radio station, then a private office, it is now vacant. The victim of requirements for "egress" of modern fire codes (and hence unvisitable), it remains as a dusty, leather-lined relic of a time when skyscrapers, zeppelins, and penthouse studios were glamorous signs of modernity. Kitchen and bar are fitted with meticulous cabinetry into closetlike spaces along the inside wall, and, as in the typical zeppelin lounge of the period, wicker provided its original furnishings. It anticipates the public relations images of the mast of New York's Empire State Building (1929–1931) as a mooring for dirigibles. Persistent rumor has it that a model and photographs of this skyscraper were used for King Kong's destructive ascent to the pinnacle of the 1920s metropolis with Fay Wray in his clutches. Not so: the honor belongs to a skyscraper in Minneapolis.

When Fleet Bank came to enlarge its original building in the early 1980s with the connected Fleet Center, "modern" skyscraper design was in part making a postmodern return to 1920s stepped massing with touches of Art Deco classicism. Architecturally, the new building suffers from mediocre detail. Urbanistically, it attempts, by a glossy arcade with "greenhouse" roofing, to extend the famous Arcade (PR18) across the street. Its interior is calculated to attract the noontime and late afternoon bustle within its famous predecessor, initially without its success, although the potential is there.

PR4 Providence City Hall (left) and Providence Biltmore Hotel (PR5) (right)

PR4 **Providence City Hall**

1874–1878, Samuel J. F. Thayer. 1914, attic reworked, Jackson, Robertson and Adams. Late 1970s, restoration, Irving B. Haynes and Associates; wall ornamentation in council and aldermen's chambers, Robert Dodge. 25 Dorrance St.

Samuel J. F. Thayer's scheme in the then modish Second Empire style was the outcome of a design competition that drew twenty-one entries—including a French Renaissance–inspired design from the young firm of McKim and Mead (before White joined the team). Thayer's competition entry featured a bombastic center tower which was eliminated in cost-cutting efforts during a crippling economic recession following the Panic of 1873. Nevertheless, as realized, his design introduced to Providence a degree of Victorian monumental grandeur hitherto unseen locally and established a new standard for the city's public and commercial buildings. Its severe, hierarchical granite exterior contains a remarkably intact original interior, including a silver metallic central stairwell and light court with gilt and polychrome trim and fine, elaborately stenciled council and aldermen's chambers on the third floor. The building was threatened with demolition and replacement by a glass box during urban renewal planning efforts in the 1950s and 1960s, and its restoration became a favorite project of Mayor Vincent A. Cianci, Jr., even before his election in 1975. It is one of the best-preserved buildings of its period and type in the country.

PR5 **Providence Biltmore Hotel**

1920–1922, Warren and Wetmore. 1978–1979 and later, remodeling, Philemon E. Sturges with Morris Nathanson for interiors. 11 Dorrance St.

Almost as symbolic a civic gesture as City Hall across the street, the Biltmore was constructed with funds raised by the Providence Chamber of Commerce to provide the city with modern, first-class hotel accommodations. Warren and Wetmore were well established as architects of stylish hotels, among them the Biltmore, Commodore, and Pennsylvania hotels in New York. The dryly detailed Renaissance ornamentation in brick and limestone, here on an L-plan tower rising from a trapezoidal base, is conventional for hotels of the period. The reworked lobby retains some fine, gilded Adamesque plasterwork at its second-floor level. The top-floor banquet hall and ballroom remains as an exceptional example of festive Neo-Colonial decor such as no eighteenth-century person ever laid eyes upon. The exterior now boasts an express elevator to the roof in a transparent shaft, a feature of a major renovation of the late 1970s that was then much in vogue for hotels, whether incorporated inside or out.

PR6 **Union Station Complex**

1896–1898, Stone, Carpenter and Willson. 30, 36, 40, 50, 56 Exchange Terrace

When Thomas Tefft's famous mid-century station in the Lombard Romanesque style burned in 1896, construction of this replacement was already underway. Like many other railroad stations designed following the World's Columbian Exposition, Union Station has a Roman triumphal arch as its centerpiece entrance to the passenger station, which also provides a symbolic gateway to the city. In this instance the grand Neo-Renaissance manner is leavened by construction in yellow brick and a utilitarian aspect overall (rather like that of its predecessor), but distinguished by fine craftsmanship and proportion. The spreading complex of five buildings, once linked by arcades, was reduced to four buildings when fire destroyed the easternmost pavilion in 1940. A simulacrum was completed in 1988 as part of the complex's rehabilitation after this station's abandonment in 1986 for another (PR8.3), required when trackage through Providence was shifted farther north. In 1987, fire also eliminated most of the original interior of the passenger terminal, which was never very sumptuous and was, in any event, scheduled for drastic alteration to accommodate new offices and shops, eventually offices only.

PR7 Federal Building

1938–1940, Jackson, Robertson and Adams. 2 Exchange Terrace

The Federal Building with post office in Depression-era Georgian Revival completes the circuit of civic buildings around Kennedy Plaza. More interesting than the building are some nice Art Deco sculptured panels displaying transportation by car, train, ocean liner, and airplane—a frequent theme for architectural adornment during the 1920s and 1930s when competition among these modes of transportation was at its height and the zip of movement was the very essence of modernity.

PR8 Convention Center and Capital Center Redevelopment Project

1978–, Skidmore, Owings and Merrill (New York office), Marilyn Taylor, project head, with preliminary planning by William Warner Associates (master plan for the area between Kennedy Plaza and the State House); William Warner Associates (river redevelopment plan and design)

Active planning for the Capital Center began in the late 1970s with discussion about the possibility of claiming a roughly 15-acre site stretching from the "Chinese Wall," made by the curve of the railroad tracks on an elevated embankment into the old Union Station northward to the lawn of the State House. Both the trackage through Providence and its station also required upgrading. What if the money for these improvements could be combined with additional funds to slide the curve of the tracks toward the northern and western edges of the site? The railroad could be given a shorter right of way on ground-level tracks, with a new station built over them at a new, higher artificial grade immediately below the State House. Then the "Chinese Wall" could be eliminated, thereby making available a sizable area for the expansion of the old center city. There were objections, two of particular importance. Some argued that the old station was ideally located at the heart of the city, rather than at some remove from it, as would be its proposed replacement. They also feared that development on the new site could occur only at the expense of the old commercial center. Although much of the old downtown was moribund and in part dilapidated from long and extensive vacancies, it had become the object of passionate concern; many saw that it contained an array of nineteenth- and twentieth-century buildings so exceptional as to be virtually (perhaps actually) unmatched by those in any other medium-sized American city. Not surprisingly, however, expansionist fervor prevailed.

To this initiative for Capital Center add two others, which developed in such an interlocked manner that in effect they became one. First, off the northwest corner of Kennedy Plaza and adjacent to Capital Center, the Convention Center, for trade shows and conferences (PR8.1; 1985 [planning]; 1991–1993, Howard, Needles, Tammen and Bergenhoff), on Sabin Street, with an adjacent Westin Hotel (PR8.2; 1992–1994, Nichols Partnership), went up beside an existing and lumpish Civic Center (1971–1972, Ellerbee Architects), for sports and entertainment spectaculars, and better inside than out. Meanwhile, parallel plans called for improving the banks of the two rivers that mark the southern and eastern boundaries of the Capital Center site—the Woonasquatucket and the Moshassuck—joining at the center of the city to make the Providence River.

Virtually the whole of the north end of the river was bridged by 1905 from a point opposite Old Stone Bank north to, and including, the junction of the Woonasquatucket and Moshassuck Rivers. No aspect of the redevelopment of the central city has been more significant. Liberating these rivers from their sewer-like bondage and bridging them into the light of day instantly revived, in a symbolic way at least, one of the great attractions of historic Providence. Precedent for the lower walkway quay close to the river level derived from San Antonio's comparable treatment of the winding course of its river through its downtown as the Riverwalk (although for the most part without its enlivening lower level of shops and restaurants). The prowlike point at the junction of rivers recalls Pittsburgh's Golden Triangle (although without the majestic volumes of water which mark the merging of the Monongahela and the Allegheny to make the Ohio). So Providence is San Antonio–cum–Pittsburgh as a delightful miniature. Given this fact, the heavy-handed, multi-piered postmodern neo-classicism of its new bridging is unfortunate—as though calling up a roll of drums and fanfare to herald a two-foot leap. More appropriate would be the light, elegant spanning of these rivers (in metal, perhaps) with arcs of lights tracing the leap and (if classical allusions are needed) possibly the delight of built-in illuminated obelisks instead of catalog street lamps stuck up on posts off the railings. At the junction of the two rivers, too, the tiny triangular plaza should have been extended by an open, festive lobby beneath the wedge-shaped Citizens Bank building (1989–1991; Jung/Brannen Associates). Yet this is closed and stodgy: the corner turrets at each of the glass building's angles underarticulated, as are the glass walls between; its liver color unprepossessing; its height both insufficiently low for a "palace" and insufficiently tall for a "tower." The miniature quality of the rivers called for a light touch, for common sense mingled with gaiety and finesse; but both the bridges and the principal building are pompous. The glory of small cities should be their smallness. The pressure of large-scale development to overinflate is not a measure of urbanity, but the opposite.

Moving from Kennedy Plaza into the Capital Center site requires crossing Memorial Boulevard and the parallel Woonasquatucket, which together curve around two sides of Kennedy Plaza. An integral part of Capital Center, the boulevard provides the principal access into it from I-95 and I-195. Ironically, its heavy traffic re-creates the Chinese Wall between the old downtown and its intended expansion, which the track relocation was supposed to have razed. As the umbilicus at the center of this redevelopment site, the Woonasquatucket has been puffed into a granite-lined circular pond to symbolize the ancient treatment of the so-called Cove, which once filled this hollow. Originally the Cove was a pond which, for smaller boats, marked the head of the Providence harbor. As the city grew toward the Cove, its containment within a circle of granite surrounded by a tree-shaded promenade converted its swampy stench and mucky squalor into a park (completed in 1856). Only briefly, however, for wastes from growing industrialization upstream soon returned the Cove to a nauseous, derelict condition. Now, this shrunken token of the larger circle of water that was once there resurrects the memory of a moment of elegance from the city's past and recalls the still larger body of water which was the Cove's original state.

Skidmore, Owings and Merrill's New York office supervised Capital Center's master plan. Its good intentions are evident in provisions for lower buildings close to the State House, with increasing height toward Kennedy Plaza, together with a few (but too few) axial vistas from the old downtown toward the State House dome. Such benefits are substantially undone, however, by the large size of most of the plots, reflecting the demands of developers for more extensive floor areas than those typical before the mechanical delivery of light and air made tolerable (if not always pleasant) deep spaces far from windows. Of the first buildings erected on the old Cove site, only Amtrak's Providence Station escapes the architectural deficiencies—especially chunky massing and banal detail—that beleaguer the Citizens Bank Tower, the American Express Building, and Center Place. Thinly detailed symmetrical schemes in a postmodernist classical mode seem stretched to the limit to wrap the bulk, like gift wrapping mammoth presents.

Of these buildings, Providence Station (PR8.3; 1981–1986, Marilyn Taylor of Skidmore, Owings and Merrill, chief designer), 1 Station Place, is architecturally the most significant. Like the Convention Center, it is postmodernist classical in style. Whereas the giant metal-framed glass front of the Convention Center alludes to a classical portico as a frontispiece to the immense box of space behind, the station is a full-round composition of severe geometrical shapes, possibly too unrelieved but nicely proportioned, which designate the essential functions within. A low, spreading box in reinforced concrete houses the waiting room, topped by a low, metal-clad dome. Inset porticoes, their columns simple cylinders, provide entrance from opposite directions. An up-ended box serves as a clock tower and civic marker. Inside, a monumental oak bench, a segmented circle, echoes the shape of the dome above. Seating on either side is built into its undulant profile, recalling the best of such monumental benches in late nineteenth- and early twentieth-century stations. The failure of the low dome to swell directly from the box of the interior space below is a fault. It exists instead in a compartment of space of its own, rather than expanding out of the space below. The rough, grottolike effect of its foamed undersurface, required for acoustics, but at odds with the smooth surfaces all around, is another flaw. Still, this is a worthy successor to the two fine Providence stations which preceded it, providing the city with an image that is distinctive and monumental yet scaled down to current railroad use.

Projected as the climax of Cove area building is Providence Place (PR8.4; 1997–2000, ADD, Inc., architects; Friedrich St. Florian, associate), a colossal shopping mall wedged into the intersection of I-95 and the River Boulevard, directly opposite the Convention Center and its hotel. It also contains four stories of warehouse-like space, compartmentalized into the usual department store "anchors" with shops, restaurants, and theaters in between, all cut through at one point by the mandatory climactic vertical atrium topped by a glass dome. The court and its bubble occur directly over the spot where the Woonasquatucket flows beneath the mall through an oversized, barrel-vaulted orifice. The outermost layer is fronted in classically derived embellishment, much at giant scale, and more seriously conceived than that stretched over earlier office and apartment buildings in Capital Center. The embellishment extends around the southern prow of the complex at the junction of I-95 and Memorial Boulevard. Toward I-95 the mall is walled by parking, like the Convention Center, both curving inward to make the boulevard gorge for access to Capital Center. Is this another Chinese Wall? Will lighting or some other adornment ameliorate the sullen grimness of these parking precipices and beguile the traveler between the Scylla and Charybdis of this new central city?

Even if the current plan is stunted or stymied by economic conditions or its late-1970s direction ultimately changed, Capital Center will interest the architectural pilgrim as a commentary on late twentieth-century urbanism. Has the central city taken its revenge on the suburban mall? Or is the suburban mall the real winner after all? The collection of shops and other enticements in this urban mall is decreed by the developer to be the city's "downtown." It is mostly sealed off from the city by basement, bridge, or arcade passages to the hotel, from the hotel to the Convention Center, and from there to the Civic Center. Against the overwhelming scale, self-containment, and floor space of this colossus and its satellites, what is the role of what *was* downtown? This is our next destination.

PR9 Slade Building and Adjoining Structures

Washington, Eddy, and Union sts.

These three commercial buildings dating from between 1880 and 1910, together with two others which fill the block, have been reworked as office and condominium space. Slade and Balcom long operated a large paint business in the earliest and most conspicuous of the group, the Slade Building (PR9.1; 1881), 38–52 Washington Street (at Eddy Street), a late Victorian brick building trimmed in stone, flush with the wall, making blunt allusions to "Gothic." Its design indicates a concern for well-lighted interiors: the attenuated cast iron, transomed store front (partially extant); the tall, close-packed windows upstairs; and especially the window-filled tower of wooden bays bracketed at the second-floor level off the corner of the building. At two points on each elevation outsized dormers from brick corbeling burst across the mini-mansard bearing stone plaques to celebrate the owner. The six-story Renaissance Revival building adjacent at 55 Eddy Street (PR9.2; 1908) boasts an even more open and attenuated cast iron storefront, beneath two stories of bay windows sheathed in sheet metal. The jutting bay sequence across the third and fourth floors is triangular-rectangular-triangular, all with rounded corners. The spiral stairs of an ornamental fire escape descend into the gullies between them, throwing out tendrils for support to the bays—a rare example of integrating what are customarily excrescences.

Around the block at 112 Union Street is the finest building of the three, although difficult to see in its narrow, shadowed location and too complexly conceived for easy description. The Providence Telephone Building (PR9.3; 1893, Stone, Carpenter and Willson; Norcross Bros., builders) was inspired by Stanford White's elaborately ornamented commercial and club buildings of the time for McKim, Mead and White in combinations of Roman brick and terra-cotta elaborately embellished with Neo-Renaissance ornament in low relief. Probably the first full-blown example of the turn-of-the-twentieth-century academic Renaissance Revival in downtown Providence, the Telephone Building initiated a series of designs in the same mode that Willson executed here between 1894 and 1900. Although documents suggest its construction in two stages, first to the top of the three-story base (masked as two), then, thirteen years later, three more, the facade seems so much of a piece that the whole elevation must substantially have been designed at the start. A bit overwrought, perhaps, in its ornamented and contrapuntal variety, this facade proclaims the youthful overstimulation of an architect naturally gifted as an ornamentalist, but its forceful organization is also evident. Its merits brought it national publication in *American Architect and Building News* (September 16, 1893).

Sobriety to either side characterizes the other properties in the block. At 56–70 Washington Street, the somber Earle Building (1895) displays a belated Victorian mansard. On Union against the alleylike Fulton Street, the Edwin A. Smith Building (1912, Martin and Hall), at 57–59 Eddy Street, modernizes a nominal Renaissance format with Chicago windows as rationalized for commercial buildings by a group of architects around the turn of the twentieth century in that midwestern metropolis. The successful conversion in 1999 of this building from commercial to residential use was the first of a precedent-setting trend transforming Downcity's traditional retail core.

Of several examples of terra-cotta in these blocks of Washington Street, first is the Hotel Dreyfuss, which has a colorful terra-cotta cornice with cartouches and other minor terra-cotta touches, part of a remodeling of a plain brick Neo-Renaissance building. The former Packard Motor Company showroom hardly suggests the aura of technological sheen and classy luxuriousness that it once conveyed, the latter still seen in the terra-cotta sheathing of the structure in a bone-white imitation of limestone bejeweled with bits of color. The nearby Johnson and Wales classroom building (PR15) employs the terra-cotta-clad frame to more daring effect; still, this claims attention as a remnant of a standard deluxe auto salon of its period. Its renovation as a restaurant was successfully completed in 1999 (Durkee and Brown, architects).

Best of all is the front of the Lederer Theatre, a fantasy version of the Roman triumphal arch motif. Pastel colors in bone white again, lemon yellow, and lime green make this an exceptionally subtle example of commercial terra-cotta. A delicately detailed two-story lobby topped with a stained glass oval dome has been partially restored. Providence's own George M. Cohan appeared in two productions here before it was converted from stage shows to movies in 1923, reverting to dramatic theater in 1971 as the home of the important regional Trinity Repertory Theatre. It was then that the cavernous, ornamented interior was gutted to provide for two replacement theaters, one above the other, in the stripped bare brick manner prevalent for theater reuse in the early 1970s.

PR10 Hotel Dreyfuss

1917, remodeling, William Walker and Son. 119 Washington St.

PR11 Packard Motor Company Showroom (Former)

1915, John Hutchins Cady. 202 Washington St.

PR12 Lederer Theatre (Emery's Majestic Theatre

1917, William Walker and Son. 1971–1973, partial restoration for reuse. 201 Washington St.

PR13 Providence Public Library

1896–1900, Stone, Carpenter and Willson. 225 Washington St. (at Greene St.) 1953, addition, Howe, Prout and Ekman. 150 Empire St. 1985–1988, interior restoration and remodeling of old section

It is not the bland and conspicuous 1953 addition at the corner of Washington and Empire streets that deserves attention here, but the original Renaissance Revival building behind it on Washington, restored in 1985–1988. Whereas the modernist monumentality of the addition makes the architectural image of the library easily accessible, the renovation of the approach into the original building gives precedence to the ceremony and aura of veneration which the early twentieth century considered appropriate for entry to a major cultural institution.

The example of the Boston Public Library was clearly in the architects' minds, although they saw it dressed not in its severe fifteenth-century Florentine garb, but in finery freely adapted from Jacopo Sansovino's sixteenth-century library in Venice. This is among the handsome translations of the much-imitated Boston building. Boston's granite walls have been translated into yellow brick with stone trim. The extravagant wrought-iron bracketing of McKim, Mead and White's famous clusters of lamps is drastically reduced, spikey and intricate, in keeping with scintillant delicacy and elaboration typical of Willson's ornament. These qualities are evident particularly in the treatment of the roof.

The main stair hall to what is now administrative offices is especially handsome. Columns in polychrome scagliola and brass sconces provide the approach to an upstairs entrance hall, columned and round-arched in the same quattrocento Florentine manner as that at the head of the grand staircase in Boston. Although the semirestored interiors of the original main reading rooms are spatially cramped and somewhat overwrought in detail, Willson, working closely with Librarian William E. Foster, effected a more convenient relationship between reading room and stacks than McKim managed in Boston. When the interior was completely revamped in the 1980s, however, the original relationship of interior arrangements was eradicated.

PR14 Beneficent Congregational Church Complex

PR14.1 Beneficent Congregational Church
(Round Top Church; formerly Second Congregational Church)

1809, Bernard Eddy and John Newman, builders. 1836, remodeling, James C. Bucklin. 1857, chancel. 1923–1924, chancel remodeling, William T. Aldrich. 300 Weybosset St. (at Chestnut St.)

Whether viewed theologically or stylistically, Beneficent Congregational Church has an extraordinarily interesting history. Theologically, it resulted from a split of College Hill's First Congregational Church (now First Unitarian Church), the first Congregational gathering established in Providence, when Joseph Snow, Jr., a fiery lay preacher who was also a carpenter by trade, began preaching a more evangelical

PR14.1 Beneficent Congregational Church

brand of Congregationalism than the staider members of the church could tolerate. Eventually, Father Snow, as his ardent adherents came to refer to their pastor of nearly fifty years, established a second congregation on the west side of the Providence River in what was then an area of scattered residences and farms. There, in 1744, Father Snow's background as a carpenter and the enthusiastic support of other builders and his congregation combined to erect a meeting house box of 36 feet by 40 feet, in clapboard with a spire, on the very site where the present church now stands. In 1791 the by then aging Father Snow was joined by the Reverend James Wilson as assistant pastor. Wilson was an Irish emigré imbued with the teachings and spirit of John Wesley. Now it was the turn of Father Snow and his adherents to be offended by evangelical intensity. History repeated itself, and the old rebel again exited with his flock and established yet another Congregational grouping, leaving Beneficent Church (as it came to be called in 1785) to almost another fifty years of preaching from the fiery "Paddy" Wilson.

Additions to the middle and rear of the original box, plus a new floor beneath for Sunday school and parish activities could not contain the audience for Wilson's ardor, augmented by the growth of a substantial village around the church. The old building was demolished. Finally, in 1809 a new church was erected (opening on New Year's Day of 1810), broadened and elongated from the meeting house formula to

75 by 91 feet, in the Adam style, in brick with stone trim (although the brick was originally whitewashed, then painted, until twentieth-century "restoration" returned the accumulated layers back to brick). The most radical change in its appearance occurred in the elimination of the old spire for a high, centered, once-gilded dome, which gave Beneficent its popular designation as Round Top. What accounted for the change? Legend has it that the Reverend Wilson wanted to recall the domed Custom House in Dublin, which had been dedicated shortly before his emigration to America. Round Top as originally designed, however, also had affinities to Charles Bulfinch's domes, such as that on the Massachusetts State House (1795–1798). So Irish memories and contemporary New England Adam style could have melded in its treatment.

The exterior as we now see it is the Adam style building Grecianized by James Bucklin in 1836, a modernization almost wholly confined to the exterior of the church, with the auditorium end inside left pretty much as the two-story balconied box of the 1809 remodeling. Bucklin transformed whatever was attenuated in the original Adamesque building into brusque, sculptural Greek Doric monumentality. "Greek" included here, as in other examples of the Greek Revival, allusions to Egyptian forms in the slight taper of the outside edges of the window frames to recall the similar shaping of openings in certain Egyptian temples. A comparable breadth of moldings and surface in the two styles wed the Nile to the Acropolis in the taste of the period, while the sense of both civilizations as remote founts of Western culture also appealed to the romantic sentiment of the times.

Bucklin's Greek modifications included the compactly proportioned Doric portico in wood, which dramatically responds to the curve and rise of Weybosset Street; the paneled effects around the base of the dome; the eaves-edge parapeting, also paneled but accented by bold Greek fret patterns; and, surmounting all, the cupola, penetrated by a Corinthian column based on the Choragic Monument of Lysicrates in Athens, which now crowns the dome.

Because music was a cardinal aspect of Beneficent's evangelical tradition, important interior renovations included a series of organ upgradings. The initial organ of 1825 went into the choir balcony at the entrance end of the church. It disappeared when, in 1857, the upper portion of the opposite wall was opened to a chancel to give a new Victorian organ and the choir greater prominence in the service. This was lost, in turn, when Mr. and Mrs. John D. Rockefeller, Jr., gave the present organ, a much grander instrument, in memory of her mother, Abby Pierce Aldrich. In 1923 the family called in architect William T. Aldrich, a brother of Mrs. Rockefeller, to design the organ case in a flamboyant manner, reminiscent of Sir Christopher Wren's work, as a backdrop to tiers of choir stalls stepped down and behind the opposed tiering and pyramiding of balustrades and stairs to a low, central pulpit at the front of the choir. Aldrich's embellishment is absolutely at odds with the pew area left over from the 1809 Adam style church, as though rebuking its plainness by way of advertising how much an architect with academic training and a sophisticated knowledge of "styles" could improve on the work of lesser-trained predecessors. Arrogance and disparity notwithstanding, however, no chancel design in the state is more expressively musical than this.

Three other items over the pew area are worth notice. Oldest is the 1826 clock on the gallery front, made by the famed Simon Willard, the "Babcock" on its face referring to a local mid-century repairer. The Victorian crystal chandelier of Austrian glass is one of several gifts from the textile magnate Henry J. Steere. He also donated another organ to a downstairs chapel. Finally, in 1959, Steere's gift was moved to the entrance gallery to serve as an echo for the Rockefeller instrument up front, but now wrapped in a new rosewood case which replicated that of the 1825 organ. So the present echo is not only sonic, but stylistic as well, suggesting (at least temporarily) a full circle back to the start of the to-and-fro shift of organ pipes which has contributed such an exhausting and lovely aspect to the history of this interior.

PR14.2 Beneficent House

1963–1969, Paul Rudolph. 1 Chestnut St.

As dean of the School of Architecture at Yale University, Paul Rudolph was influential in the design of large, multiple-unit housing projects during the 1960s. He and others rebelled against the monotony, inflexibility, and impersonality of the elemental building shapes preferred by earlier modernists for apartments and mass housing. Influenced especially by Le Corbusier's emphasis, in his late Unité apartment houses, on defining the unit within the whole, and by current theories on the design of megastructures from discrete increments, Rudolph had pioneered in this approach to housing for married students at Yale and for the elderly in New Haven. Hence the push and pull of form in this commission from the Beneficent Church for a housing project for senior citizens. Horizontal banding in concrete across the brick walls, in a manner sometimes used by Le Corbusier, marks off floor levels and serves, like a musical staff, as a means to organize the architectural elements. This banding also indicates a tentative return to favor of polychromatic walls in the manner of such Victorian examples as the Wilcox Building (PR20). Disagreements over the cost and convenience of Rudolph's original, more complex scheme led to its simplification. Even as compromised, however, the result is an innovative example of an important late modernist point of view toward housing design—as well as an attractive and comfortable apartment building, with a surprising reticence which encourages one to overlook what deserves notice.

PR14.3 Daniel Arnold House

c. 1826, John Holden Greene. 33 Chestnut St.

This cubical brick house in the Federal style with a monitor-on-hip-roof is typical for its architect. Now an adjunct to Beneficent Congregational Church, it is the only survivor of a row of four that stood nearby on Westminster Street. Moved and cursorily restored in 1967, it incorporates the modified Palladian window and chimneypieces from a now-demolished Greene house in Pawtucket.

PR14.4 Abbott Park

1746, 1873, 1927. 286 Weybosset St.

Unlike New England settlers in most communities outside Rhode Island, the founders of Providence set aside no centrally located town green. For the eighteenth-century predecessor to Round Top, one of the members of its congregation, Daniel Abbott, donated this as a "true New England common," albeit a tiny one, for many years no more than a grassy area extending from the church. A cast iron fountain surrounded by cast iron fencing (long since removed) was added in 1873 and restored toward its 1873 condition c. 1990. This was the first private donation of a public park to the city.

PR15 Johnson & Wales College Classroom Building (Summerfield Building)

1913, Albert Harkness. 274 Weybosset St.

Albert Harkness inaugurated his fifty-year Rhode Island architectural career with this extraordinary six-story building, which he designed as an investment property for his family. He seems never again to have done anything much like it. Its original name derived from a Boston furniture company which long operated a Providence branch. The state's first reinforced concrete-frame commercial and office building, it still astonishes for its expansive use of plate glass and the reduction of the "wall" to a skeletal minimum clad in tightly fitted, polychromatic terra-cotta impressed with delicate Neo-Renaissance ornament. A strongly projecting slab incisively terminates the elevations. Rising from the skimpiest spandrel clear to the ceiling, the huge expanses of glass work in concert with the minimal frame to flood the interior with light (a necessity before the advent of fluorescent fixtures). This skeletonized wall also reduces building weight for savings in foundation and materials. The glazed tile provides for colorful, inexpensive molded embellishments applied as prefabricated, modular units requiring, except in the sootiest conditions, no more than rainwater to restore their sparkle.

Such construction for commercial buildings was widespread from the 1880s onward but is associated with Chicago, where examples were ubiquitous—so much so that the three-part horizontal window unit consisting of a fixed center sheet, often the size of a shop window and flanked by narrower sections that open for ventilation, came to be known as the "Chicago window." Here, variants appear toward the rear of the side elevation. Up front, however, Harkness more than doubled the span of the Chicago window—so testing the limits of possibility for this feature that, here and there, thicker mullions have been added at halfway points to counteract threatened structural failure.

PR16 Providence Performing Arts Center (Loew's State Theatre)

1928, Rapp and Rapp. 1975–1978, restoration. 220 Weybosset St.

Concealed behind an embellished but unremarkable facade that shows a mild Spanish influence in what might be called 1920s Plateresque is one of the best-preserved monumental movie theater interiors of its period in New England. It was designed by a specialist firm notable across the country for movie houses in hybrid styles. The vestibule and main lobby are vintage movie-palace baroque, handled with more restraint (or a smaller budget) than in many other Rapp and Rapp theaters. Here the auditorium is the important feature. As the appropriate motif for a Providence theater, the decorous fantasy of New England's own Federal style envelops the space. Not the skimpy plaster ornaments of Bulfinch's prim Boston interiors, however: rather, the architects retreated to the source for Bulfinch's inspiration, engravings of Robert Adam's grandest salons for his most impressive town and country houses. Rapp and Rapp, however, had no qualms about consistency of style if an exotic interjection could heighten the final effect. So tent-roofed mini-buildings conjuring Eastern exoticism flank the stage as simulated boxes projecting from niches. They flatten out to become aediculated door frames around the fire exits on the side walls, topped by broken scroll pediments, between the stubs of which rococo foliage undulates to climactic wreaths around busts, much as Adam decorated many of his entrances into what he termed his "grand saloons." Still it is the spread of Adamesque ornament across the domical enclosure which prevails. Disks and half disks of various sizes, patterned in various ways and underpinned by an interwoven lattice (more Plateresque perhaps than Adamesque) skim the curved surfaces, which are edged by decorated banding, with the banding predominating as the space funnels toward the stage-screen focus. At the summit, an oval orifice breaks through to another smooth-surfaced saucer dome above it, which is circled by lights concealed in the coving. In this other realm, seemingly beyond the space it closes, a changing pool of iridescent light accompanies the moods of the mighty Wurlizter, risen from its tomb in the basement.

PR17 Atlantic Bank Building

c. 1866. 1977–1978, rehabilitation, James Estes. 75 Weybosset St.

This tiny "Elizabethan-style" stone building features round-arched windows and an ornate bracketed cornice surmounted by bulbous urns. That so diminutive a commercial build-

PR16 Providence Performing Arts Center (Loew's State Theatre)

PR18 The Arcade

ing of such antiquity should have survived in such pristine condition is miraculous.

PR18 The Arcade

1828, James C. Bucklin and Russell Warren. 1978–1980, restoration-renovation, Irving B. Haynes and Associates. 65 Weybosset St. through to 130 Westminster St.

The Arcade, financed and built by local merchant Cyrus Butler and the Arcade Corporation, remains one of the key Greek Revival monuments in the country. It is the only remaining American arcade in the style, although it followed the precedent of enclosed shopping arcades in the Greek manner in Europe and, closer to home, those by John Haviland in New York and Philadelphia. James Bucklin and Russell Warren studied Haviland's arcades, both of which are gone. The south (Weybosset Street) elevation retains the stepped parapeting of the portico originally intended for both elevations. Revisions to the building program after construction began resulted in the addition of a third story of shops and a triangular pediment to the north (Westminster Street) elevation facing toward the heart of the city. (Although the true reason for the change remains a mystery, the theory of a friendly "competition" between the two designers is now generally discounted.) Remarkably, the cylinders for the Ionic columns in antis are granite monoliths hauled by oxen from Johnston quarries some twenty miles away. Rubble party walls connect the granite porticoes. These walls are not contained within the narrow rectangle the Arcade appears to occupy, but project as stubby wings on either side near the center, making the plan of the Arcade not rectangular, but a stubby-armed cross.

The interior is little changed in essence, despite modifications in detail. Especially worth notice is the stepping back of the three floor levels so that each opens directly to the skylight above, with the space of the ground-floor corridor expanding outward as one looks up. The enlivening counterplay between Greek-inspired ornament and the broad wall surfaces typical of the Greek Revival style is brilliantly evident here in the filigree of cast iron and wrought iron hand railings set against the austere geometry of the masonry. The bridge across the center and upper-story shop fronts was added in the nineteenth century, modified in the twentieth, and retained in the 1978–1980 restoration partly because no images of the original interior have been found and partly to retain what is in itself a charming modification. In the restoration similar shop fronts on the ground floors were opened up by folding screens to create flow through a mix of eateries which merchandisers wanted; fortunately this arrangement could also be justified by accounts stating that a comparable marketlike openness did, in fact, characterize what was initially there.

The sensitivity of the restoration is evident not only in what was brought back and what was adapted, but also in what is evidently new, like the handsome redesign of the floor pattern. Especially challenging is the enclosure by glass walls of what, previous to the rehabilitation, was draftily open. Many feared the enclosures would become a coarse, disruptive barrier; but it is worth examining how elegantly and unobtrusively they are fitted to the building, even elegantly curved to the outside of the columns with which they are aligned in order to preserve the integrity of their cylindrical shapes. Making no apologies for their modernity and handsome in themselves, the glass enclosures deserve to be honored because of the honor they bring to their task.

Directly across the street, the reticent screening of a parking garage (1980s, Gilbane Building Company) by a local construction firm can be commended. Opposite this, at 45–53 Weybosset Street, is another restored Victorian commercial block, Halls Building (1876, refurbished 1981), severely brick and stone trimmed with another fine cast iron storefront.

PR19 Bank of North America (Former)

1856, Thomas A. Tefft. 48 Weybosset St.

This, the Wilcox and Equitable buildings, and the Federal Building (see the next three entries) represent a particularly handsome and varied sequence of mid-to-late Victorian commercial buildings in conjunction with an important public institution for the city's commerce. They also demonstrate the downtown scale of the medium-sized American city in the mid-nineteenth century. The Bank of North America is the only remaining commercial block by Tefft (save for a much-altered bank in Taunton, Massachusetts), although the substantial archive of his drawings owned by Brown University and the Rhode Island School of Design contains many similar designs for commercial buildings. Tefft was among the early proponents in America of the full-fledged Italian palace front. Growing European enthusiasm for the revival of the Italian Renaissance palace format in its various national manifestations began as early as the 1820s in Germany and the late 1820s in England, where it remained fashionable throughout the 1840s, when Tefft came to know of it; the A. T. Stewart Department Store in New York (1846) provided the first conspicuous American example of what became the ubiquitous nineteenth-century "commercial palace." From Tefft's elevation drawing for this bank we know that its ground story was originally heavily rusticated and penetrated by three large, arched openings for a central window flanked by entrances into the bank and up to offices above. This heavy treatment would seem to have been more appropriate to the ponderousness typical of much Victorian palatial detailing than to the delicacy of Tefft's treatment, still remaining in the upper stories. Screened by fire escapes, its brownstone face painted to cover grime, the elevation nevertheless reveals Tefft's refined and restrained attention to detail and his fine proportioning of wall openings with a full array of Renaissance window capping to provide mild variety across the surface, all incisively framed by the quoining and cornice. The identification of the long defunct bank appears beneath the cornice with the precision of an elegantly engraved Victorian letterhead.

PR20 Wilcox Building

1875, Edwin L. Howland. 42 Weybosset St.

This is the city's best High Victorian Gothic commercial building in what was sometimes derided at the time as the English "streaky bacon" mode of polychromatic contrast, employing the most commonplace combination of materials found in Victorian masonry buildings: red brick and near-white stone. The aggressive enlargement of the light, decorative trim vis-à-vis the dark walls intensifies the "streaky" aspect of the analogy, like bacon more white fat than red meat. The architectural result is ambiguity as to whether the "trim" is meant to predominate over the "wall." Actually, the use of the natural textures and colors of diverse materials for their intrinsic decorative effect—in this case stone, brick, and stubby column shafts in polished granite—follows John Ruskin's admonitions. Insofar as any style can be attached to this front, it would be the Italianate "Gothic" he recommended for urban buildings because of its pointed arches and cusped and crocketed forms. Here, however, these are stretched, butted, sliced, and, finally, pinched into prickly profiles to create what the mid-nineteenth century hoped might become a "modern Gothic." If the result seems eccentric today, the idea was conceived in a progressive spirit, with far-reaching implications for subsequent architectural design in emphasizing the use of materials for their texture and color effect. The Wilcox Building has an L plan which wraps around the slightly earlier and adjacent Equitable Building. Its subordinate elevation on Custom House Street is much more subdued. The interior was gutted by fire in early 1975, and the building's restoration helped to focus community concern on downtown preservation.

PR21 Equitable Building

1872, Walker and Gould and Builders Iron Foundry. 36 Weybosset St.

PR21 Equitable Building, center, with the Federal Building (U.S. Customhouse) (PR22), left, and the Wilcox Building (PR20) and former Bank of North America (PR19), right and far right

Few buildings with full facades in cast iron remain in Providence; this is by far the best of those that do and the best preserved. The "Venetian" facade was cast locally by the city's leading architectural foundry. The polychrome exterior was restored to what appears to have been something like its original paint scheme in the early 1980s. So this row of three buildings displays three stylistic variations popular for mid-nineteenth-century commercial blocks which also represent contrasting yet typical nineteenth-century approaches to the renovation of history to make modern buildings. The first identified the specific type from the past that most precisely met the new functional needs and made it a standard. The second adopted a preferred style from the past and manipulated it into a "modern" version of itself. The third chose a style from the past with characteristics, like the expansive windows and minimal walls of certain late Gothic and Renaissance Venetian palaces, which lent themselves to bold translation into new nineteenth-century materials and technology.

Equitable Insurance seems to have taken a progressive attitude toward its buildings at the time. It was in the vanguard in erecting a skyscraper for its New York company headquarters in the 1870s. For this local office it chose to build in a new material, which also offered easy assembly from bolting together prefabricated cast panels. The Equitable also preserves the Victorian custom of locating a shop a half flight down and the principal business floor a half flight up, thereby giving both levels visibility from the sidewalk and doubling the landlord's income from "street level" rentals.

PR22 Federal Building (U.S. Customhouse)
1855–1857, Ammi B. Young. 24 Weybosset St.

Built at the edge of the Providence River when port activity was immediately adjacent, the Federal Building originally housed customs, the federal district court, and other federal offices. As architect for the Treasury Department in the mid-nineteenth century, Ammi B. Young supervised the design of federal buildings from Maine to Texas, many of which employ some variant of a granite block in the format of an Italianate Renaissance palace. Like this one, they tend to be larger in scale than their neighbors and severely detailed with only such elements as quoined corners, belt courses as ledges on which to align the windows, and bold, simplified moldings. Custom houses were pioneer building types for architectural fireproofing. Hence construction is wholly in masonry, with interior iron beams carried on the masonry walls, arched brick floors between the beams, iron shutters against external conflagration, and handsome iron stairs to the upper stories lushly ornamented in motifs more Grecian than Italian. In the third-story federal courtroom, a finely carved, gilded wooden eagle, shrieking and poised for flight, remains in the round-arched niche above the erstwhile place for the judge's bench. From a distance, the contained quality of the block may appear to be somewhat compromised architecturally (if made more arresting) by the cast-iron dome perched atop its low hipped roof. The dome was conceived after construction began, as documents in the National Archives indicate, because of last-minute local Congressional pressure to dignify the upstairs courtroom and upgrade the building to exterior view above customhouses at less important ports, like that at Bristol (BR15).

In 1854 Thomas Tefft had prepared alternate schemes for the customhouse as either a Venetian or a Florentine palace, preserved in the collection of his drawings at Brown University. Although the building was delayed after Tefft's submission, he probably would have had little chance for the commission against the Office of the Supervising Architect.

PR23 Banigan Building
1896, Winslow and Wetherall. 10 Weybosset St.

Hailed as the first tall, steel-frame, "fireproof" building in Providence, the ten-and-one-half-story granite-sheathed Banigan Building was built on speculation by Joseph Banigan, founder of a large plant for rubber goods in Woonsocket and eventually also one of the founders of U.S. Rubber when his company went into the giant conglomerate. The layered vertical organization of its Neo-Renaissance exterior is typical of office buildings at the turn of the twentieth century. The rusticated base of the building shows the measurements of downtown flooding from Narragansett Bay in the "great" hurricanes of September 21, 1938 (the worst), and August 31, 1954 (merely horrendous).

PR24 Turk's Head Building
1913, Howells and Stokes. 17 Weybosset St.

The Turk's Head Building, erected by the Brown family as an investment property, is a squat version of Daniel Burnham's recently completed Flatiron Building (1902) in New York, although the angle here is less acute and the wall treatment less florid. Like the New York building, this uses its wedge-shaped corner site to dramatic advantage, interacting dynamically with the taller, boxier competition nearby. Originally, when it was the tallest building downtown, its seventeen stories served as a prow for the entire business district to a viewer descending into it from College Hill. A ship's figurehead located on a former building at the site was a landmark in this vicinity, as commemorated by the effigy at the prow of the Turk's Head Building's three-story base.

PR25 Exchange Bank Building
1888, Stone, Carpenter and Willson. 59 Westminster St.

A rare local commercial essay in the Queen Anne mode, the Exchange Bank has a fussy, bay-windowed exterior that fragments the mass and makes the building appear as a series of overblown late nineteenth-century china cabinets stacked one on another. Nonetheless, its whimsical charm ensured its rehabilitation as part of the Fleet Center complex (PR3), to which it is connected on its west side. It beautifully demonstrates the virtue of small buildings among tall ones in giving light, air, and (especially where the building is as delightfully detailed as this) human scale to the adjoining spaces of the street hub and the rectangular plaza of the skyscraper across the way.

PR26 Sovereign Bank Complex (Rhode Island Hospital Trust National Bank)
1917–1919, York and Sawyer. 15 Westminster St. 1971–1973, tower, John Carl Warnecke and Associates. 1 Financial Plaza

It is the late-twentieth-century tower that dominates, although its early-twentieth-century predecessor next door is architecturally more exceptional, especially inside. But pause momentarily to observe the quality of the architectural frieze that differentiates the base of the older building from the office floors above, and its even finer equivalent around Turk's Head. Both are doubtless machine carved and are unexceptional among the finest classical architectural ornament at the beginning of the twentieth century; but try to match such quality today. York and Sawyer provided this eleven-story palazzo in a vaguely Italianate Neo-Renaissance mode for the financial institution founded in the 1860s as the financial advisor and depository for the endowment of the state's largest hospital. Then also the state's largest bank, it went to New York for an architecture firm which probably designed more banks on a lavish scale than any other during the first decades of the twentieth century. A coffered, barrel-vaulted corridor, handsomely detailed, leads past a bank of elevators with exceptionally fine cast bronze doors in openwork depicting an arcadian setting in which partially draped female figures merge with lush vegetation. Reputedly, these are by Daniel Chester French. If so, this commission must owe something to the sculptor's earlier work for the nearby Union Trust Building (PR29). Halfway along the elevator bank, a right-angle turn gives entrance to the center of the rectangular banking hall at the midpoint of one of its long walls. Just inside each of these walls, ranges of luxurious Corinthian columns flank the banking space and support a coffered barrel vault between them. The original patterned pavement in marble was slightly lifted to provide space for subfloor wiring when the hall was converted to open-plan office space in 1974, and the original bronze and glass banking furniture has almost all been replaced.

Next door, the plaza setback from Westminster Street, as much a nod to New York zoning laws of the 1960s as the Industrial National Bank's stepped-back form was to the law of 1916, permits a sheer rise to the sleek, boxy, travertine-sheathed high-rise tower. Inside the banking room, outsized planes of polished marble in black and tan are encased in more travertine. The plaza, with benches and fountains by local sculptor Howard Ben Tré, was completed in 1998, the year in which Hospital Trust was acquired by Bank Boston.

PR27 Merchants Bank Building
1855–1857, Alpheus C. Morse and Clifton A. Hall. 20 Westminster St.

Diagonally opposite the old Hospital Trust Building on Westminster Street is another relic of the scale of Providence banking of an earlier day. This and the Bank of North America (PR19) introduced the Italianate palace as a popular mid-nineteenth-century image for commercial buildings, and the two remain the only pre–Civil War downtown survivors of a commercial type destined to have prolonged effect. At six stories, the Merchants Bank Building was long the tallest building in the city and the first to be retrofitted with an elevator. Like Thomas Tefft before them, Morse and Hall graduated the treatment of their features from ground floor to roof. Turk's Head nicely responds to its wedge shape as an opposed wedge. One creates a funnel into the plaza hub; the other, one out of it. And, like the Exchange Bank Building (PR25), this also demonstrates the urbanistic value of the pigmy building among giants.

PR28 Lauderdale Building and Francis Building

1894, Stone, Carpenter and Willson (both). 1977, rehabilitation of Lauderdale Building, Michael Ertel. 144 and 150 Westminster St.

Two unrelated clients engaged the same firm at the same time to design adjacent speculative office buildings. Edmund Willson did these buildings fresh from his design of the Providence Telephone Building in Stanford White's tawny, scintillant brick and terra-cotta Neo-Renaissance manner and just before he began to design the Providence Public Library. Of the two, the Lauderdale is the more completely clad in terra-cotta. Beneath its shimmering surface one also senses how lightly these materials cover the underlying steel frame, which appears at the ground floor in all its linear nakedness except that the supporting metal frame is also sheathed, here by ornamented metal plates. In a more conservative manner, the Lauderdale anticipates Albert Harkness's daring interpretation of the terra-cotta-sheathed skeleton in his Summerfield Building (PR15). The adjoining masonry elevation, predominantly stone with some brick, is still more conservative, both in the greater prominence of its masonry wall and in the greater restraint of its ornament, much of this in terra-cotta. Whereas Willson employed terra-cotta as overall sheathing for the Lauderdale Building, he used it more discreetly as trim for the Francis Building. The two are handsome complements, all the more for their different approaches to the ornamented facade.

PR29 Union Trust Company Building

1900–1901 and later, Stone, Carpenter and Willson, interior by Clarence Luce. 1981, restoration. 62 Dorrance St. (at Westminster St.)

For this twelve-story business block in limestone and red brick, the first skyscraper in Providence to use the classical base-shaft-capital organization for tall buildings, Stone, Carpenter and Willson ventured into Neo-Georgian with some French Rococo touches. Like the same firm's Union Station (PR6), this also features a triumphal arch as an entrance motif through the rusticated lower floors. *Indian* and *Puritan*, reclining figures by Daniel Chester French, cap the portal, giving a new iconography to Michelangelo's Medici Chapel figures *Night* and *Day*. But it is the tall screen of rococo-inspired windows, each with a stained glass medallion of one of the great international banking firms through history, which catches the eye both outside and in. The glitter of the emblems enhanced Clarence Luce's use of a variety of colored marbles in the banking hall, all nicely restored except for the elimination of tellers' cages and some changes of wall color. Luce probably garnered this commission from a number of designs for Newport houses. Marsden J. Perry, the president of Union Trust Company, who commissioned this building, gave other significant commissions to its principal architects.

PR30 Providence Journal Building (Former)

1906, Peabody and Stearns. 1983–1984, restoration, Estes-Burgin. 203 Westminster St. (northeast corner of Eddy St.)

When this building was "modernized" in 1955 by a flush encasement in enameled metal panels, the ornament and the original shop windows were buried, only to be resurrected in the mid-1980s. Fiberglass molded in casts made from existing details now patches whatever ornament the support structure for the modernist shrouding had defaced. The Flemish Baroque vocabulary, related to the firm's work in Boston for the Driscoll and Chandler stores, is wed to steel-frame construction. Originally, small electric lights outlined the elaborate frames for the shop windows at night, a not uncommon treatment at the time, reflecting early fascination with a new technology. In their choice of such a highly ornamented petite palace to celebrate Rhode Island's principal newspaper, its publishers and architects seem to have emulated McKim, Mead and White's famous landmark for the *New York Tribune*, which was also two-storied and also luxuriantly Neo-Renaissance (albeit Florentine rather than Flemish).

PR31 Westminster Street Shopping District

c. 1870–1940. Westminster St. from Dorrance St. to Mathewson St.; 139 Mathewson St.

It takes some peering and imagination to sense the role these few blocks, now faded and mostly vacant, once played in Providence. Above the renovated street level, the early Art Deco O'Gorman Building (PR31.1; 1925), 220–226 Westminster Street, in brick and terra-cotta, makes an effort at sophistication with a terracotta peacock perched on the ledge that demarcates the two-story base of shop windows from the upper stories. Too small for the message it means to convey, it is easily missed. Its "tail" extends as a thin blue line, like the mercury column of the thermometer, up the center support for the four upper stories to the tail feathers' "eyes," ranged along a curve of a halfovoid terra-cotta frieze, which the parallel curve of the cornice echoes immediately above. The fantastic ornamentation, ritualized eroticism and (to human eyes) dandyish decadence of the male peacock made him a favorite in Art Deco ornithology. Next door, at 228–232 Westminster Street, the Burgess Building (PR31.2; 1870, George Waterman Cady) displays a suavely Parisian cluster of bonnet dormers against its steep mansard. Triplet windows have Neo-Renaissance frames, their details variously stretched and pinched in the Victorian manner. Exceptional is the tubular, coppersheathed bay window with latticelike panes which pops from the center of the elevation, possibly a later Queen Anne addition to a Second Empire facade.

Next is the hulk of the former Shepard Department Store (PR31.3; 1870s; expansions and additions, 1880, 1885, 1896, 1903), at 259 Westminster, marked by two-story arched porches with entrance doors on an oblique angle inset into the corners of Union Street, at both Westminster and Washington streets, one block north. Like many other department stores of its period, Shepard's expanded higgledy-piggledy into additions and adjacent buildings until it virtually filled the whole of two city blocks by bridging over an alley. It was refurbished in 1994 for the University of Rhode Island College of Continuing Education, which lost its site to the Providence Mall at Capital Center.

In the early 1960s several blocks of Westminster Street were closed to vehicles to make a pedestrian mall with Shepard's as its focal point in a last-ditch effort to lure shoppers from suburban malls and return them downtown. But Shepard's was sold to a conglomerate, and soon slaughtered as a "cash cow" which was drying up. Westminster Mall, in two iterations, eventually gave way to a reopened street, gussied up with patterned brick sidewalks and stone pedestrian crossings. This crusty hulk stands as a reminder of its "anchor," the suavest of Providence's early twentieth-century department stores, with a street floor that featured fine commercial cabinetry to the very end.

The outstanding commercial building of this group, immaculately preserved by the most eminent jewelry retailer in what is sometimes called Jewelry City (see under Jewelry District, below), is the Tilden-Thurber Building (PR31.4; 1895, Shepley, Rutan and Coolidge), 292 Westminster Street (corner of Mathewson Street), which remained the final elitist holdout in the collapse of the street's former reputation. After more than a century in business, its original occupant surrendered to bankruptcy in 1990, and the successive owner sympathetically adapted it to another luxury business. The ornament of these terra-cotta surfaces is larger in scale and more deeply modeled than most of the examples described so far, with Venetian palaces as the source of inspiration. Its Neo-Renaissance sumptuousness appropriately invokes a jewel box. Appropriate too is its allusion to similar Fifth Avenue palazzi in New York, designed by McKim, Mead and White for Tiffany's and for Providence's own Gorham, where, as here, very open two-story showrooms contrast with the treatment for the floors devoted to offices and workshops upstairs. Across the street, at 291 Westminster (corner of Mathewson Street), the Burrill Building (PR31.5; 1891, Stone, Carpenter and Willson), also in the Neo-Renaissance style, borrows from the same McKim, Mead and White palazzo prototype, but with more modest means. The incrustation of Tilden-Thurber becomes a carefully proportioned interplay of brick and stone, culminating under the cornice in a simple but handsome inset checkerboard of brick as a decorative rectangular field. Here Edmund Willson shows his strong grounding in Charles McKim's more austere approach to design in his sure sense for the decisive, wellproportioned organization of the elevation. Intrinsic versus incrusted embellishment—or Florence and Venice on opposite sides of Westminster Street.

Half a block west into Mathewson Street, at number 139, is the Lederer Building (PR31.6; 1897, M. J. Houlihan, builder), an earlier and cruder interpretation of the Renaissance palace extended to mini-skyscraper height. The copper-clad bay windows of this Victorian holdover puff from the sheer, yellow brick walls, with more sheet metal work in the copper cornice and the tall, spindly cast iron front extravagantly infilled with plate glass at the ground. Commercial skyscrapers of this early vintage rarely exist so little changed, even to the terra-cotta corner cartouche which labels the building.

PR32 Grace Church

1845–1860, Richard Upjohn. 1912, remodeling and parish house addition, Cram, Goodhue and Ferguson. 1950, addition enlarged, Albert Harkness. 175 Mathewson St. (at Westminster St.)

Grace Church, which dominated a prestigious residential community when built, interrupts these architectural mementos of Providence's onetime shopping scene. With a tower off one corner, it is one of the country's earliest asymmetrical Gothic Revival churches (some have risked calling it *the* first). This was the first of several ecclesiastical commissions Richard Upjohn carried out in Rhode Island. As modified and expanded by the later chancel extension and parish house, the complex combines two predominant approaches to the Gothic in the nineteenth and early twentieth centuries: the doctrinal English parish Gothic of Upjohn and the more urbane Collegiate Gothic, a self-consciously aesthetic English Perpendicular, of Cram, Goodhue and Ferguson. Inside, Upjohn used an exposed timber roof construction to support a gable roof over the nave, with side aisles vaulted in plaster. Upjohn was an ardent Episcopalian himself and the most prominent architect member of the American offshoot of the British Ecclesiological Society, which called for a return to medieval forms as those of the "true" church. His Grace Church, with its long nave, side aisles, lack of balcony, and "dim religious light" from stained glass, opposes such luminous preaching boxes of the eighteenth and early nineteenth centuries as the First Baptist Church and the First Unitarian Church (PR56, PR92). A diverse and impressive collection of stained glass from the nineteenth and twentieth centuries fills the windows, including a Tiffany window (on the left, third from the entrance) and several by Reynolds, Francis and Rohnstock. The fifth window in the same wall, removed to this location when the chancel was deepened in 1911–1912, incorporates medallions from the original chancel window.

As often in Cram's revisions of earlier medieval revivalist efforts, his chancel alteration makes little attempt to blend with Upjohn's work. In its refinement of archaeological detail, it seems rather to rebuke what came before, as did William Aldrich with his later redesign of the choir and pulpit area of Beneficent Congregational Church. Medieval churches offer much precedent for this sort of one-upsmanship; but frequently, as here, it is the earlier work that holds most visitors' attention.

PR33 Conrad Building

1886, Stone, Carpenter and Willson. 1988, restoration. 375 Westminster St.

PR34 Caesar Misch Building

1903, Martin and Hall. 400 Westminster St.

Of all extant Queen Anne commercial facades in the city, the Conrad Building's beautifully restored front for the plain brick box behind is the most ambitious. The first impression it gives is of a multitude of windows, variously shaped and projected, above the handsome regularity of a classic Queen Anne cast iron storefront. Small-paned transoms increase the light into the high interiors and incidentally scale down the shop windows to accord with the horizontality of the street. As with many Queen Anne buildings, the basic compositional scheme is symmetrical, centered in the recessed entrance to the upper floors. An ornamented terra-cotta arch at the second-story level accents the entrance. Above it a three-story bay window, flanked by arched openings, all within a slightly projecting ornamental framework, makes a centerpiece for the elevation. But as is also typical of the Queen Anne Style, no sooner is a central axis established than asymmetrical "features" counter the balance. Another three-story bay projects south of the centerline. More sensationally, north of it a domed tower swells off one corner. Meaning to be "Saracenic," in accord with Victorian love of the exotic, it also suggests a stack of Victorian domestic conservatories looking toward the center of the city. Arched windows across the topmost floor also recall minarets.

These exotic touches probably represented the client's taste, since they are nearly unique in the architect's work. Jerothmul B. Barnaby was a self-made millionaire from his clothing store a few blocks away on Westminster Street. He presented this investment property as a self-renewing wedding present to his daughter and son-in-law. (Is the mysterious bust on the terra-cotta roundel toward the southern end of the building a portrait of him, of his son-in-law, or of some historical figure?) Barnaby's penchant for the Middle East seems confirmed by a similar tower which the same architects added two years later to his Broadway mansion (PR186). The Conrad Building was beautifully restored as luxury apartments, but the area was not ripe for gentrification, so the building became a dormitory for Johnson and Wales University students.

By comparison, Caesar Misch's commercial palace is tame indeed. Another example of Neo-Renaissance trim in terra-cotta which is boldly challenged by the logic of the steel-frame, plate-glass commercial building, the Misch Building's brick walls give it a more normative look than Harkness's reduction of "walls" to the bare bones skeleton of his Summerfield Building (PR15).

PR35 Cathedral of Saints Peter and Paul

1878–1889, Patrick C. Keely, with numerous interior renovations. Cathedral Sq.

Born in Ireland but arriving in the United States from England, Keely established a flourishing practice centered in work for the Roman Catholic Church. Among his more than 500 churches, one of his earliest is Providence's St. Joseph's (1853), at Hope and Arnold streets, with a well-preserved exterior but somewhat altered interiors. He provided designs for cathedrals for Buffalo, Chicago, Boston, Hartford,

PR35 Cathedral of Saints Peter and Paul

and Portland, Maine, as well as this for the Diocese of Providence. For the massive monumentality desired, Keely may have been peripherally influenced by the Romanesque of H. H. Richardson's newly completed Trinity Church in Boston, although he more likely depended directly on German medieval examples, crossed with detailing and overall forms from French cathedrals, as well as from his own earlier work in the Gothic Revival. The most remarkable aspect of the cathedral is less its forbidding exterior than the truly impressive size of its interior. It is interesting for the overall multicolored, small-scale pattern of gray marble moldings and stained glass. In contrast to the variety of styles and periods in the stained glass at Grace Church (PR32), here it is homogeneous. The tiers of windows, installed in 1886, were all made in the workshops of Tiroler Glasmalerei in a style which suggests that of the early nineteenth-century Nazarene painters in Germany. (This group archaized and abstracted in the early Renaissance manner of Raphael and his disciples with the aim of returning religious art to expression then deemed to have been exceptionally pure before its "corruption" by baroque flamboyance and histrionics.) Most extraordinary is the timber-and-board "vaulting" with elaborate polychrome paneling which hangs from roof supports, a tour de force of nineteenth-century carpentry. It is surface, multitudinously decorated, not structure, which dominates both walls and roof in this vast interior. The marble flooring and the altar area are obviously later additions; so is the present array of organ pipes in one of the transepts, giving to the original organ loft over the entrance the sense of void left by a large, uprooted tree.

The underused and rather forlorn plaza in front of the cathedral and the area of apartments around it result from a redevelopment effort (1967–1976; urban plan by I. M. Pei and Partners) to bring residents downtown. It was tied into the shopping mall created by the closing of Westminster Street to vehicular traffic, an attempt at reinvigoration which failed. Cathedral Square itself and its fine pyramidal fountain are the work of Pei with Zion and Breen. Consider here the outcome of a plan to "revitalize" a battered downtown: a failed shopping corridor; a downgraded apartment renovation; a vacant plaza; a mostly locked and empty cathedral. Where no community exists to inhabit them, even well-considered schemes for urban renewal offer no more than a slim hope for a turnaround.

The Jewelry District and the Harborfront

Beyond the interstate underpass, Chestnut Street becomes a street of industrial buildings which provides an introduction to the Jewelry District. If Samuel Slater represents the technological progenitor for Rhode Island's textile industry (see under Pawtucket), so Nehemiah Dodge stands in the same position for the Providence-centered costume jewelry industry. In his Providence jewelry shop in the late 1790s, Dodge discovered a process for plating silver and gold to base metals, thereby providing the industry's foundation. A tight area, the Jewelry District is mostly concentrated within the interchange loop of Interstates 95 and 195, although somewhat contracted from its onetime extent. The small size and high value of the product encouraged small-to-medium-sized factory units. Located here until well into the 1960s and 1970s in straightforward, handsomely proportioned brick construction, they are worth study as a group. As portions of this industry have disappeared or moved to more open sites, many of the buildings left behind have been converted to office, apartment, and studio use, although manufacturing continues in some of them. Notwithstanding the spread of the industry around the area (and increasingly outside it), Providence is still the national center for costume jewelry, to which buyers make biannual treks to inspect displays at two wholesale markets.

PR36 Jewelry Factories in the Chestnut Street Area

1888–1911. Late 1970s, mid-1980s, several converted to offices and apartments. Chestnut St. from Pine to Point sts.; 91 Friendship St.

The Waite-Thresher Building (PR36.1; 1911, Dwight Seabury; 1984–1985, reuse conversion), at 30–32 Chestnut Street (corner of Pine Street) introduces a line of brick factory buildings in the heart of what used to be the larger Jewelry District. These are straightforward loft buildings. Their functional beauty depends on the fine quality of their brickwork in planar or pier-and-spandrel walls, in conjunction with the regular rhythm of their generously proportioned sash windows, mostly in wood, all silled in granite and topped with shallow brick relieving arches. The window unit is usually (but not always) divided by a center mullion to form a pair, each half with two sash, the height often stretched by a fixed transom to increase interior light. Variations occur in dimensions and divisions into panes from four over four to ten over ten, with two to five for the transom.

In contrast to the spread of the three- and four-story brick textile mills (see especially examples in Pawtucket), these are compact blocks of five to seven stories. The projecting towers characteristic of the textile mills were unnecessary where rented loft space accommodated a patchwork of changing tenants and the products were tiny and precious. Nor is there either a reason or a place in these buildings for the assertive custom brick design and fancy capping found on many of the textile towers, through which their owners customarily proclaimed corporate identity and dominance. By contrast, the loft factory is essentially anonymous, its ornamentation mostly limited to the standard treatments of the corbeled eaves. But take a closer look at these beetling structures, which literally step out from the plane of the wall, terminating it while also integrally of it. Despite their apparent sameness, consider the variations among three buildings: the Waite-Thresher Building; the Horace Remington and Sons Company Building (PR36.2; 1888), at 91 Friendship Street; and the Irons and Russell Company Building (PR36.3; 1903–1904, Martin and Hall, 95 Chestnut (at Clifford Street). Corbeling in the eaves of the Waite-Thresher and Remington buildings occurs in two stages: immediately under the cornice edge in sheet copper or terra-cotta, below which is an entablature-like plane, then at or just below the topmost window arching. The effectiveness of the cornice is thereby doubled by providing projection and shadow both at the top of each stack of windows and again under the cornice. The Waite-Thresher Building also has a projecting topmost relieving arch, which further intensifies depth of shadow under the eaves. Alternatively, the corbeled cornice of the Irons and Russell building is lifted above the windows, as a separate element, which emphasizes the plane of the wall rather than its beetling termination.

In all three of these buildings, window elements within each stack are inset except for the bottommost sill. This projects a little beyond the plane of the piers and interlocks with them, so as to provide a baseline for each of the window stacks. In the unaltered storefronts set into the "front" walls of each of these buildings, the divisions between shop windows for the first two are framed in boxlike piers: those for the Waite-Thresher Building faced in wood and relatively narrow; those for Remington faced in cast iron and much wider, in accord with the exceptional width of the windows. For the Irons and Russell building, cast iron columns mark the divisions. These comparisons demonstrate the double nature of the most vital vernacular traditions: standardization based on "right" ways of building as these have evolved over time and variation in the possibilities of an infinity of particular expressions. Occasional details even break with anonymity. In the first building a wrought iron gate into which the original owners' names are worked closes a rear court off Pine Street. In the second, spare ornamentation occurs at the tops of the window frames where they fill the shallow curve of the relieving arch.

The final two examples display an alternate standard: simple windows punched into a planar wall, as opposed to paired windows inset within a piered wall. The wedge-shaped Champlin Manufacturing Company building (PR36.4; 1888, 1901; 1978, reuse conversion), 116 Chestnut (at Clifford, Ship, and Bassett streets) indicates how much is lost when, as was the custom during the early phase of "modernization" of these buildings, sheets of plate glass, allowing easier maintenance and improved climate control from double glazing, replaced the animation of grids made by panes, sashes, and transoms. The plane of the wall is lost across these voids. Scale is disrupted. The eyes of the building become empty sockets. However impressive its laconic acceptance of the functional imperatives of the brick wall regularly punched with identical openings, the potential for choice within the vernacular tradition is severely diminished in the Doran Building (PR36.5; 1907; late 1970s, reuse conversion), 150 Chestnut, addition at 70 Ship (at Elbow Street). The treatment is routinized, and the assertion of distinctive expression possible from formulaic anonymity diminished.

PR37 A. T. Wall Company Building

1908, Bowerman Brothers; Thomas F. Cullinan, builder. 162 Clifford St. (at Claverick St.)

This is not a pretty building, but it represents the dawning stage of a technology which would establish a new norm for twentieth-century factory design. It is the first known structure in Rhode Island to employ the "mushroom" column system of flat-slab construction in reinforced concrete developed by the American engineer C. A. P. Turner around 1905–1906. The interior retains its original octagonal columns, thirty inches in diameter, with hollow cores to accommodate vertical runs of utility wires and pipes. Toward their tops, the octagonal faces of the columns flare in all directions (shaped more like a day lily than a mushroom) to provide a broad pad (now more like the leaves of a water lily sustained by tapering membranes radiating from its center stem) in order to counteract the "shear" tendency of the weight of the machines above, which would bear down until their floor slabs were punched through by any small-diameter column below. Exterior walls are reduced to a grid of widely spaced piers bridged by spandrels and otherwise filled by glazing (here by rows of four window units butted together). The broad horizontals of glazing bring such luminosity to the interior that, with the coming of fluorescent lighting at the end of the 1920s, the top third of the windows was blocked out against the glare.

The primitive manifestations of a new technology inevitably suffer from a degree of awkwardness. Note the groping effort at using the new semifluid material, reinforced concrete, to create a dignified entrance on Clifford Street, but one based on a blunt version of shapes derived from masonry building. Or consider the indecisive expression of the principal vertical supporting columns of the frame in the slightness of their projection beyond the horizontals of the supported floors. Or, finally, a tiny detail: examine the angles from the vertical member to the horizontal at the upper corners of the window, braces inserted to counter shear surface cracks which would otherwise tend to develop from the right angle outward, precisely at the point of maximum stress where the horizontal weight of the floor has to be sustained by its vertical support. To the rear, over the parking lot, reinforcing rods stick out of the frame as a thoroughly pragmatic means of facilitating plant expansion by hooking onto the reinforcement within the frame of the existing building. As the type evolved toward a "standard," however, such awkward incidents would disappear (see the Coro Building [PR39]). Meanwhile, savor the special intensity and poignancy of the rawness at the time of discovery.

PR38 Imperial Place Factory Row

PR38.1 Imperial Cutlery Company
(Vesta Knitting Mills)

1893, 1903. 2 Imperial Pl. (corner of Bassett St.)

PR38.2 Phenix Iron Foundry, Elm Street Machine Shop

1848. 116 Elm St.

Vesta Knitting Mills, built for a manufacturer of cotton underwear and hosiery, was taken over by what was then the largest manufacturer of jackknives in the country, popularly known as

PR38. Imperial Cutlery Company (Vesta Knitting Mills), left and center, and Phenix Iron Foundry, Elm Street Machine Shop, far right

Imperial Knife, which enlarged the plant as it expanded to more diversified cutlery. This factory complex includes exceptionally handsome adjacent buildings on Imperial Place which enclose one side of an irregular court completed by lesser buildings behind. The initial Vesta building is the finest in the area and, in its detailing, probably the finest brick loft building in the state. It warrants careful looking. The rounded corners of the building, together with the breadth of wall between the corners and the outer limits of the window grid, contain the total mass as a shaped entity, not merely as the passive result of modular repetition, as in its next-door neighbor. The coved underside to the molding for the roof cornice echoes the corner curves. A band of brick dentils weds the cornice to the wall, and—incredible refinement—shifts from wide dentil–narrow interval over the piers to narrow dentil–wider interval over the windows. Below this, more corbeling occurs at the level of the springing of the segmental window arches of the topmost story. This provides a frieze for the cornice molding, thereby increasing the visual weight of the capping, while also preserving the wall's integrity in that the cap emerges from the wall itself. The windows are wide, their twelve-over-twelve sash filled with generously dimensioned panes. In the adjacent building, the same twelve-over-twelve sash occurs in more pinched dimensions, but the plane of the roof simply sits on the top-story windows. It is certainly the more economical of the pair, and its straightforward functionalism is admirable in other respects as well. But its predecessor is unique because of the sense it imparts of weight, breadth, shape, containment, the modulation of geometry by subtle curvature, and the impress of human intention. Thereby this functional building is elevated beyond the merely functional.

After exacting refinement, rude vigor. Of the Phenix Iron Foundry complex, founded in 1830, only this, one of its oldest buildings (together with a much lesser structure), remains. Phenix specialized in the manufacture of textile machinery for dyeing, bleaching, and print works, and by the late nineteenth century there was hardly a bleachery in the country that was not outfitted with its equipment. More decisively than any other structure in the city, the Elm Street Machine Shop recalls the thick masonry walls common in stone mills during the first half of the nineteenth century, here in beautifully textured cut stone randomly placed with some rubble patching. The gable roof and trapdoor monitor reflect established form for pre–Civil War mills. The wooden sash is original, with twenty panes over twenty for the ground floor and sixteen over sixteen above. As in many early industrial buildings, the openings for hoists at either end diminish in size as they climb. No other factory in the state makes a more forceful image of the feature than the rude, telescoped stack of brown granite arches here (not visible in the photograph). Long abandoned and neglected, the building was rehabilitated in 2000 and now serves as headquarters for Brown University's Development Office.

PR39 Coro Building

1929, Frank S. Perry. 1947, addition. 1988–1989, renovated for offices. 167 Point St. and the block bounded by Hoppin, South, and Hospital sts.

From rudeness, return to suavity. The Coro Buildiing is another reinforced concrete structure supported on flared, mushroom-capped columns, but it represents the "standard" factory type that evolved out of the dawning technology of the A. T. Wall Plant (PR37). Now the vertical piers clearly project as the primary structural entities. Spandrel spans are set back, the horizontal of the floor slab accentuated over the horizontal parapet capping above. The face of the parapet spandrel is sheathed in brick to differentiate it as an infill element subordinate to the primary structural frame. In the crisp, clean opposition of vertical against horizontal, the triangular elements designed to resist shear at the upper corners of the windows in the Wall factory are eliminated by concealed improvements in reinforcing. Whereas four traditional wooden windows once filled each of the voids, now a single metal unit, prefabricated for its purpose, suffices. (The present windows replace the original sash, and are both far more airtight and easier to clean than their predecessors; but the flatter profile of the metal grid that supports the glass diminishes the animation of the web of light and shadow across the sash. Seemingly inconsequential, this tendency to reduce and simplify profiles while enlarging the scale of parts in replacement components tends to lessen subtlety and animation, as here.)

The court of the initial U-plan building of 1929 opened south onto Point Street. It and the south front only were embellished with minimal Art Deco ornament, including a decorative stepped treatment of the roof parapet. The court was also planted with a lawn to provide the public entrance to the plant. Coro, unlike most of the enterprises in the area, meant its building to provide a corporate image in keeping with its position as one of the largest manufacturers of costume jewelry in the country. Eventually absorbed by another company, this plant was empty from 1979 until its conversion to office space in 1988–1989. It was then, when the original windows were removed prior to replacement, that one could revisit the time of its construction and see, through the empty rectangular voids, the orchards of mushroom columns inside. Then one could fully grasp the technological beauty of this factory type, outside and in.

PR40 Simmons Building

1880. Point St. at Eddy St.

Although far less prevalent among brick factory buildings than either the punch of windows in a wall or the projecting vertical pier with windows inset on spandrel panels, this type of wall construction, in which the windows sit on continuous granite coursings, does appear occasionally. Whereas the alternate organizations of the wall have justifications in structure, this banding of the wall in horizontals is essentially aesthetic, emphasized here by the restriction of the treatment to the street fronts. This building was absorbed into the Davol Rubber Company plant on the corner diagonally opposite.

PR41 Davol Square
(Davol Rubber Company)

1880, 1884–1889 and later, 1980–1982, conversion to shopping mall, Beckman, Blydenburgh and Associates. 1991, conversion to jewelry trade center. 69 Point St. (at Eddy St.)

Davol, a major manufacturer of medical rubber and plastic goods, abandoned its Providence operation in 1977. Its brick factory buildings are less interesting in themselves than for their 1980s conversion into a shopping mall, which unhappily closed in 1991. This was the city's first example of the many progeny across the country of San Francisco's Ghirardelli Square (1962), a pioneer conversion of a factory to boutique shopping in combination with office use. The most original aspect of Davol Square (although there are precedents) is the overhead glazing of an alley between three- and four-story factory buildings to create the core shopping arcade for the complex. The functional yet festive exposure of much of the structural and mechanical addenda was necessary to make the conversion. Although it was a well-conceived project, it did not sustain its initial commercial success. Hence its reconversion as a permanent display center for the wholesale jewelry trade (not open to the public, except for peripheral shops).

PR42 Narragansett Electric Company Power Plants

1903, George B. Francis, engineer. 1913. 1940–1950, United Engineers and Contractors, Inc. (Manchester Street plant). 1908, 1913, Narragansett Company staff engineers. 1924–1925, Narragansett Electric staff with Jenks and Ballou, engineers (South Street plant). 460 Eddy St. (at Manchester St.) and 342 Eddy St. (at South St.)

Bracketing Davol Square are two impressive early-twentieth-century powerhouses named for the side streets that access them off Eddy Street: the earlier opposite the entrance to Davol Square and set back from Point Street; the later stretching most of the length of South Street, marking the rear boundary of the Davol property. Both now belong to Narragansett Electric, although the earlier was originally built by the Rhode Island Electric Company to power its trolleys. The former trolley company is easily confused with the short-lived Rhode Island Electric Light Company, founded in 1882 as Providence's first power company. Three years later it disappeared into Narragansett Electric, originally organized as a consortium of businessmen under the leadership of Marsden J. Perry to build its own competing power plant, but with the long-range goal of consolidating the plethora of small electric companies that then existed throughout the state. In 1913 Narragansett swallowed the Rhode Island Company, in the process acquiring its trolleys as well as its powerhouse. Around the same time, in a burst of building activity, Narragansett simultaneously began the expansion of the 1903 powerhouse in a congruent style, while also demolishing its own initial generating plant to replace and enlarge it with the South Street Powerhouse in a different style. The fortune Perry acquired from these and other business and banking interests enabled him to purchase the grandest of Providence's colonial mansions and to embellish it with a sumptuous collection of eighteenth-century British and American furniture (see PR96.1).

Of the two, the earlier (Manchester Street) plant is more abstractly rationalized, and hence displays the more "industrial" aspect. Ranges of tall, narrow arches with metal sash along the side elevations light the interior, expanding to broader arched windows under stepped gables at either end. The chunky, close-fisted quality of this powerhouse is appropriate for the dynamos within and expressive of the concentration of energy at the point of its release. On the other hand, the 1913 building spreads as a classicized screen. Industry is self-consciously monumentalized toward the higher order of public grandeur associated with the classicism favored by the City Beautiful movement, here in an alternating rhythm of giant arches and pilasters ranged between a high base and an entablature. But this is stripped classicism, as though intended as well to celebrate the ideals of time-and-motion efficiencies, then also a conspicuous part of the progressive American industrial credo. Whatever Perry's role in the choice of this design (if any), one feels that he would have sympathized with its classical appeal.

Until 1991 an overhead conveyor belt linked the South Street powerhouse with coal piles adjacent to the Manchester Street complex two blocks away. It was removed in 1992 partly because of the closure of the South Street plant and partly to arcadianize as much as possible the linear park which now extends along the river frontage between the plants. As this book goes to press, the South Street building is poised for rehabilitation as Heritage Harbor, a consortium of historical museums, archives, and libraries.

As a complex, particularly as a spotlighted nighttime silhouette, the powerhouses are best viewed from the east bank of the Providence River. Cross on the Point Street Bridge (1928–1929, Boston Bridge Works), a trussed swing bridge (now fixed in place) with its onetime control shanty tucked into the trussing. Its sidewalks provide a good vantage point for views up the Providence River into the downtown.

Main Street, South and North

Main Street formed the spine of the original English colonial settlement of Providence. Along the street's west side lay the Providence River, and to its east, up the steep slope of College Hill, stretched the deep, narrow house lots of the early settlers. Main Street's designation in the 1770s as "the Towne Street" even more emphatically indicated its role as the principal street of the community. It is here, at the southern end of the street, that the remnants of the principal docks and warehouses along the Providence River are located, immediately below the Market House (where South Main becomes North Main).

This stretch of the Providence River, where the masts of the larger vessels thrust high above the buildings on the embankment, was the city's maritime core well into the nineteenth century. Smaller vessels could penetrate the center of the city all the way up to the Cove, as now marked by the circular pond at the heart of the Capital Center project (see PR8). Most of the river in the downtown was effectively buried in the late nineteenth century under continuous bridging, which acquired modest fame from a *Guinness Book of World Records* listing as the "world's widest bridge." The Providence River was returned to the city's downtown in the early 1990s, not for anything as grand as the boats which once docked here, but for small unmasted pleasure craft which at least bring a sense of the old harbor into the heart of the city.

The rehabilitation of the southernmost end of South Main Street beginning in the mid-1960s reclaimed what survived from the eighteenth and nineteenth centuries, mostly a row along the west (water) side of the street. Between this row and the river formerly existed the dockfront warehouses and other waterfront properties which accommodated shipping, matched by similar structures on the opposite bank.

PR43 Captain Joseph Tillinghast House

c. 1770. 403 South Main St. (at James St.)

PR44 Joseph Tillinghast House

c. 1800. 10 James St. (at South Main St.)

These two survivors opposite one another—an ample wood-frame house and a small brick house, built respectively for a father and his son—make a telling introduction to early domestic architecture in Providence. The size and center-hall plan of the earlier house reflect the town's growing prosperity and entry into the architectural mainstream on the eve of the Revolution; like most Providence houses, it is of wood. The later, Federal style house, of brick, severely plain both inside and out, is interesting for its siting. The facade is turned ninety degrees from the street and the residential entrance raised one story above South Main by use of a high basement with secondary entry into a basement office on Main Street at street level. Variations on this particular solution appear repeatedly along Main and Benefit streets on the steep western slope of College Hill, which rises from Water Street's eastern edge.

PR45 Earle's Block and Brick Rows

The Italianate brick Earle's Block (PR45.1; c. 1872), 369 South Main Street (at James Street) provides an astonishingly well-preserved example of the plain Victorian commercial palace type, with its commanding scale, high cast iron storefronts, and sandstone identification with bold, handsome lettering. Next in line, beginning with the Eddy Block (PR45.2; 1812), 283–297 Main Street, are three brick rows in the Federal style, one with a gabled roof, two with hips. They were originally designed as row houses, and interrupted by alley-like streets whose names—Guilder, Doubloon, Silver, Gold, Bullion—proclaimed this as the financial heart of the colonial city. In the nineteenth century brick rows ran to the river, breaking through the line of warehouses and dock facilities which then, and vestigially into the twentieth century, walled the riverfront. The middle of the three and most intact, Comstock Row (PR45.3; 1824), 263–273 South Main Street, which retains some of its original fanlighted doors, is known to be the work of John Holden Greene. The original owners of all three rows were residents of College Hill and had close business and social ties with each other. The owners of Clarke and Nightingale Row (PR45.4; between 1815 and 1823), at 247–257 South Main, for example, lived in substantial mansions a block above their investment property. Shop fronts, added during the nineteenth century, were "restored" beginning in the 1960s. Row houses are uncommon in Providence. Most were built between 1810 and 1845, their presence due to aspirations for urbanity more than population density.

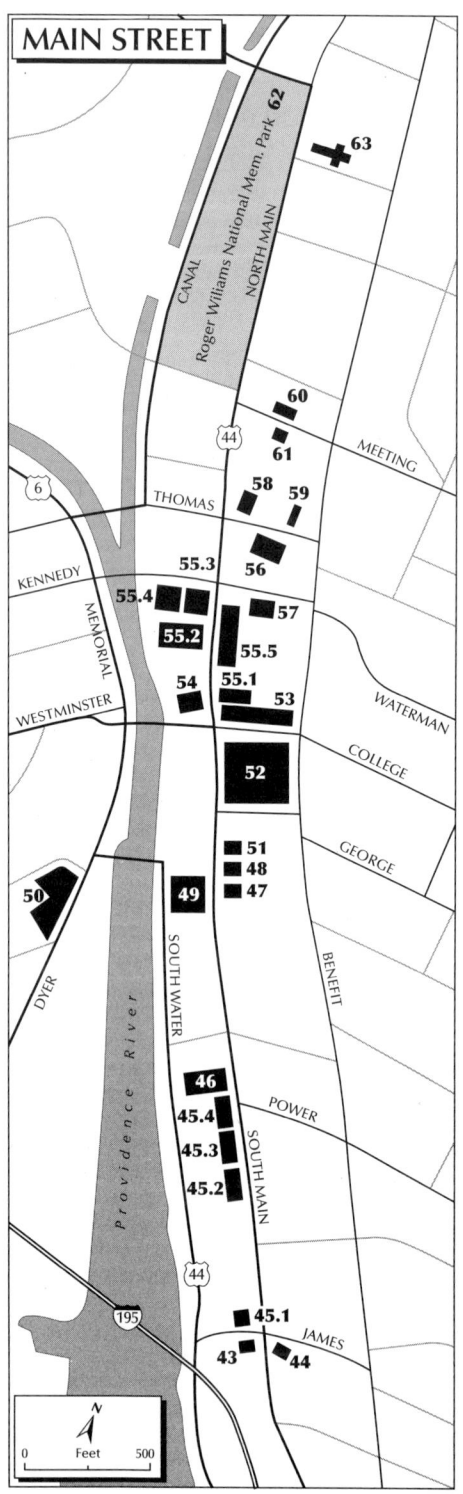

PR46 **Bayard Ewing Building, Division of Architecture, Rhode Island School of Design** (Fall River Iron Works Warehouse) 1848. 1978, restoration, Irving B. Haynes and Associates. 231 South Main St.

This monumental Greek Revival commercial building, with gable pediment facing the street and attic monitors, was built just as industrial construction shifted from stone to brick. Providence was once famous for Greek Revival storefronts deeply inset within just such simple granite framing. One of the few remaining and most impressive examples, it has been sympathetically rehabilitated for its present use. Others exist in North Smithfield (see NS16 and NS28).

PR47 **Benoni Cooke House**

1828, John Holden Greene; 20th-century basement door and arched windows. 110 South Main St.

When the Old Stone Bank closed its doors c. 1990, it was one of the state's largest and oldest financial institutions. The bank occupied the whole or part of three buildings. The oldest in the cluster is the Benoni Cooke House, a fine and typical example of John Holden Greene's 1820s work in the Federal style, which also begins a remarkable run of buildings and a lively streetscape north to the First Baptist Church (PR56). It is the remaining half of what was originally a pair of mirror-image houses, built for two business partners who were brothers-in-law, which once urbanely faced one another across a driveway leading to a large, well-preserved service court defined by connected outbuildings. Its twin for Isaac Brown was demolished c. 1896 for the enlargement of the bank. Almost cubic in mass, the Cooke House reflects Greene's typical approach to his larger houses. Windows in the hard-surfaced brick walls are generously proportioned and topped with flush stone lintels in a stepped shape which became something of a Greene hallmark. A square monitor tops the hipped roof. Greene's usual design would have been embellished with two stages of fretted and paneled parapeting; the removal of this embellishment here has left this house with a more severe aspect than one in which the roof is decked out as Greene intended (compare with the former Truman Beckwith House [PR84]). A delicately detailed Ionic porch, its door elaborately side- and fanlighted, with leaded ornament and an

PR48 Providence Institution for Savings Building, center, with the Joseph Brown House (PR51), left, and the Benoni Cooke House (PR47), right

ornamental upstairs stairhall window, complete the lacy overlay of the austere underpinnings.

PR48 **Providence Institution for Savings Building**

1854 (now north wing), C. G. and J. R. Hall. 1896–1898, remodeling and enlargements, Stone, Carpenter and Willson. 86 South Main St.

The bank, incorporated in 1819 as the Providence Institution for Savings, in 1854 built a smaller, three-bay gray granite building on the northerly portion of this building site. In the extensive mid-1890s rebuilding, the predecessor building was split in half, and its southern half was moved to the site of the demolished Isaac Brown House (mentioned above). Between these went the new centerpiece, with its Corinthian portico and low dome surfaced in gold leaf adornment against verdigris, set atop a hipped roof and an octagonal drum. Neither the scale nor the intricate ornamentation, outside or in, of this delightful interpretation of the Pantheon suggests the august character of its source. The bank sits back from the street behind a splayed perron. The sweeping curves of the severely balustered granite railings, their design derived from the original Italianate building, channels the rise up the stairs and through the portico. At their head, a delightful oval vestibule in tawny marble set within the larger space as a self-contained entity introduces the visitor to the circular banking hall centered in a circular banking counter with wainscoting in the same tawny material. It is climaxed by the coffering of the mini–Pantheon dome overhead, in an interlocked lattice of geometrical shapes, in green and gilt, which decrease in size as they rise. This design is more late baroque in the manner of Francesco Borromini's San Carlo alle Quattro Fontane in Rome than ancient Roman in spirit. In contrast to the rather archaeological aura of the sumptuous Renaissance interior of Hospital Trust (PR26), the Old Stone banking hall is sprightly, fanciful, and original. As an architectural design it is perhaps, except for the Arcade (PR18), the most important commercial interior in downtown Providence. As this book goes to press, the building stands vacant with its future completely undecided.

PR49 **Old Stone Square Building**

1982–1985, Edward Larrabee Barnes. 31 South Main St.

The sleek skyscraper across the street is carefully detailed and interesting as an exercise in squares and cubes which culminates in the big cube of space at the entrance. This frames the facade of the Benoni Cooke House opposite, while the squares organizing the elevation create taller than normal office spaces inside. But the building is also clumsily boxy overall and was vigorously opposed when built because it harshly abuts and obscures buildings on College Hill rising immediately behind it. The informal park beside it, however pleasant, might more forcefully have taken its cue from the geometry of this skyscraper and the axiality implicit in the domed centerpiece of the older buildings.

PR50 Hay–Owen Block

1866–1867, James C. Bucklin (Hay Block). 1866, Stone and Carpenter; cast iron storefronts, Builder's Industrial Foundry (Owen Block). 117–135 and 101 Dyer St.

Directly across the Crawford Street Bridge, on the opposite side of the Providence River from Old Stone's skyscraper, is the wedding of two handsome Second Empire commercial buildings to make one trapezoidal whole. Only a two-bay swatch of 1970s remodeling disrupts its pristine exterior. First James Bucklin designed the Hay Block on the southern portion of the site. The block, in red brick with granite trim, boasts cast iron storefronts. Then, a decade later, Stone and Carpenter continued Bucklin's format but varied it for the northern, rounded tip of the site to make a combined office and warehouse for the Owen brothers' business in worsted yarn. They topped their segment with an exuberant mansard and varied the repetitive quality of the earlier building by supplying their own treatment for the entrance and the stories above it at the prow of the site, changing cast iron columns to granite to either side of the (since altered) entrance door and giving it the dignity of an inset porch.

PR51 Joseph Brown House

1774, Joseph Brown. 50 South Main St.

This is the best—and best preserved—of several ample, stylish houses built in Providence on the eve of the Revolution. It is also one of the earliest extant self-designed architects' houses in America and, as such, probably goes out of its way to be a bit unusual, albeit, as betokens an amateur architect, in a conservative manner. Brown, a gentleman architect in the eighteenth-century tradition, was one of the four Brown brothers—Nicholas, Joseph, John, and Moses—who dominated the mercantile and maritime business of colonial Providence. For the design of his house, he withdrew to the early eighteenth century and the baroque of Sir Christopher Wren to find the exceptional feature of his house: its Dutch-inspired ogee-curved gable roof. Wren and his followers used this form in public buildings, but almost not at all for residences. Still, Brown could have located one example of the feature in a plate for a house in Colen Campbell's *Vitruvius Britannicus* (1717) or in Plate *P* of William Salmon's *Palladio Londinensis* (1748), for the design of a tiny garden structure. The novelty of the form makes it a local icon of the colonial. It particularly appealed to colonial revivalists, always on the lookout for unusual (or, as they would have put it, "picturesque") colonial detail, and was copied in the late nineteenth century (see PR55.4). Its exceptional quality also had earlier admirers, one of whom, Stephen Smith, in the early nineteenth century built a notable version adapted to country living (LI20).

The door was originally centered on the second story and reached from either direction by symmetrical flights of stairs. This entrance was moved to street level in the late eighteenth century, when, shortly after Joseph Brown's death, the house became home to the Providence Bank, founded in 1791 by Joseph's brothers John and Moses. The sort of trellised parapeting of which John Holden Greene was so fond adorns the summit of the ogee gable. The interior has been mostly rebuilt, save for its principal flight of stairs (now between floors two and three) with its twisted banisters, ram's-horn curl at the newel post, and another ram's horn curled vertically as the termination for the responding rail against the wall—a variant of what Joseph produced a few years later in a more splendid manner for his brother John. The bank moved to the financial district in the 1920s, but Brown family descendants still occupy the building for personal business use.

PR52 Providence County Courthouse

1924–1933, Jackson, Robertson and Adams. 250 Benefit St.

PR53 College Building, Rhode Island School of Design

1822. 1936, Jackson, Robertson and Adams. 2 College St. (corner of South Main St.)

A valuable lesson in urban design from the Colonial Revival, these two large complexes are deftly handled as a sequence of seemingly small buildings stepping up the slope to blend with the early nineteenth-century buildings on College Hill above. Whereas other competitors for the courthouse commission had suggested mini-skyscrapers, the winners, in effect, laid the stepped-back skyscraper on the slope. The evocation of smaller buildings emerging from the overall mass and the stretching of Adamesque-Federal ornament to cover new contingencies is handled with a finesse which a later generation of postmodernists, who have attempted similar effects, might envy, but rarely match. At the same time, the courthouse also reveals itself as a monumental entity rising from a courtyard in a grand axial composition which reflects the Beaux-Arts training of its principal designer, F. Ellis Jackson.

Shortly after the courthouse was completed, the same firm had the opportunity to orchestrate an appropriately plainer variation on the same fragmented, yet monumental, stepped theme in RISD's College Building, which even more closely approximates the old buildings it replaced. It actually incorporates a wall of the Franklin House, a hotel from the Federal era, at the South Main Street corner of the complex, and recreates a vehicular archway, using the original oak spanning beam, halfway up the hill, to preserve a right of way to the rear of the adjacent bank building on North Main Street. Above the Benefit Street entrance is Jackson's lovely paneled and alcoved reading room for the RISD library in a modernized mix of colonial and Federal styles, typical for 1930s Neo-Colonial.

The World War I Monument, fronting the courthouse (1927–1929, Paul P. Cret, architect; C. P. Jennewein and Janet de Coux, sculptors of the pinnacle figure and bas-relief, respectively) had to be moved to this location from a position near the present confluence of the three rivers. It resulted from a 1926 competition won by the leading Beaux-Arts designer and teacher. This 75-foot fluted granite shaft, derived partly from the Roman triumphal column, partly from the Roman cylindrical bundle of fasces, symbolic of authority, is topped by a too-small female figure described alternately as personifying Victory and Peace. In the relief at its base a soldier with lowered sword and furled flag confronts a woman and child. Embedded in the pavement are bronze reliefs of World War I weapons, each associated with one of the four branches of the armed services. These are realistic, whereas the neoclassicism of the shaft also shows influences from the contemporary Moderne. The monument anticipates Cret's attempt during the 1930s to modulate his Beaux-Arts-inspired design toward an emergent modernism. Adjacent is a monument, displaced to accommodate the column, to Giovanni da Verrazzano, who sailed into Narragansett Bay during his exploration of the North American coast in 1524. Completing this curious community of monuments is one of special interest to the architectural pilgrim—a rare memorial to a local leader in the profession during the first half of the twentieth century, Frederick Ellis Jackson, the architect of the courthouse.

PR54 Market Square, Rhode Island School of Design (Market House)

1773, Joseph Brown; 1797; 1865, James C. Bucklin; 1930s, 1950, John Hutchins Cady; subsequent interior remodeling. Market Sq.

Opposite the foot of the RISD complex and fronting the Providence River, where sailing ships could tie up close by, is this modest adaptation of Peter Harrison's slightly earlier Brick Market in Newport (NE8). The Market House is one of several architectural monuments that signaled the ascendancy of Providence as a commercial center. Originally a two-story building with an open arcade on the first story, it was used by vendors below and town officials above. A third story, added in 1797, housed the state's first Masonic hall. This later served as city hall until the completion of the present Providence City Hall (PR4) and then as the Board of Trade. The interior has been completely reworked several times since the 1930s for various RISD programs. Market Square demarcates South from North Main Street.

PR55 RISD Buildings

1869–1929, various architects. 27, 22–26, 28, and 55 North Main St.; 31 Canal St.

At 27 North Main is RISD's "Bank Building" (PR55.1), formerly People's Savings Bank (1913, Clarke and Howe), studio space converted from a typical early-twentieth-century banking hall fronted by a monumental two-columned Ionic portico. Just beyond and opposite are two Victorian commercial buildings, the Hope Block, at numbers 22–26 (PR55.2; 1869) and Cheapside, at number 28 (PR55.3; 1880, Stone and Carpenter), one mansarded Second Empire style, the other Victorian Gothic, both with fine cast iron storefronts. These two buildings and two earlier twentieth-century office buildings in the late 1920s Neo-Federal mode are now linked for RISD store, gallery, classroom, and office use in a sensitive and ingenious 1980s remodeling by Gordon Washburn, which merits study. Of the Canal Street Row, the Insurance Building, at 31 Canal Street (PR55.4; 1929, Clarke and Howe) is worth notice for nice detailing at the street and, crowning its eight stories, a pediment that is a replica of the undulant cap to the Joseph Brown House nearby (PR51). It is embellished by a relief, an allegory of business enterprise backed by a spectacular sky, presumably dawn, as a token of promise against the sunset reality toward which this building faces. In the Morris Plan Building, 25 Canal Street (PR55.5; 1926, Jackson, Robertson and Adams), metalwork balconies and a former board room enliven the interior of the RISD art supplies store, which extends through to Hope Block on North Main.

Across the street, at 29 North Main, the south half of the decorated reinforced concrete–framed building in the monumentalized factory style of the early twentieth century houses RISD's textile design studios (PR55.6; 1915, Charles Klauder). Named for Jesse Metcalf, the building fittingly memorializes the founder of the Wanskuck Company, one of Providence's principal textile concerns during the late nineteenth and early twentieth centuries. It was the gift of a family which, through three generations, contributed virtually all of RISD's buildings and gave much other support as well. The identical northern half, added five years later, provided for jewelry design and metalwork. When completed, the building therefore represented Rhode Island's three principal "art design industries," as they were called in the nineteenth century. Its completion also fulfilled, with proper quarters and equipment, a primary objective of the school's founding, which had been to provide designers in the decorative arts for regional industries. The Jesse Metcalf Building strikes a nice balance between severe functionalism and just enough concern for proportion, fine brick craftsmanship, and spare Renaissance ornament to suggest a monumentality appropriate for an academic building—somewhat marred, however, by replacements for the original standard metal industrial sash.

PR56 **First Baptist Church**

1774–1775, Joseph Brown; Jonathan Hammond, master carpenter; James Sumner, carpenter. 75 North Main St.

More than the Market House, the First Baptist Church architecturally marked Providence's growing importance. In 1638 Roger Williams founded the first Baptist church in America. For sixty years before the erection of a meeting

PR56 First Baptist Church, photo 1939

house it met in the parlors of members of the congregation. By the early 1770s a new building was needed, and it was conceived in a very large way. Built to accommodate 1,400 people (when the population of Providence stood at 4,300), it was intended both "for the publick Worship of Almighty God; and also for holding Commencement in." The commencement referred to was that of Rhode Island College (later Brown University), also founded under Baptist auspices. It had just been weaned away from Warren, Rhode Island, the place of its founding, largely by Brown family persuasion. The church that resulted from its avowed goal of combined benefit to town and gown is among the grandest built in the colonies, although not without a degree of provincialism in design.

Joseph Brown, the amateur architect, went to Boston, together with the master carpenter Jonathan Hammond, in search of ideas. In the end, however, most of these derived from several plates illustrating Sir Christopher Wren's London churches in James Gibbs's *Book of Architecture* (1728), a copy of which Brown had in his architecture library. Yet the result is especially interesting because Brown filtered the typical Anglican church schemes of Wren through what was then becoming an old-fashioned type. Not by accident is this church called a "meeting house." In contrast to the elongated space characteristic of Anglican churches, with an aisle running its length from the entrance to the altar as the place of sacred ritual, meeting houses tended to be square in plan, to combat this hierarchical aura with the immediacy of the congregation gathered around the pulpit to hear the Word. Doors centered on three sides of a typically square plan and the pulpit on the fourth, with cross aisles connecting these four features, further emphasized the nonhierarchical sense of gathering from all sides around the preacher. Here double-leaved doors do indeed appear, not only once, but centered in both of the side walls. A cross aisle originally connected them. The principal door, however, is moved to one end, the pulpit to the other opposite it. So a longitudinal axis is suggested. With time, additional pews obliterated the cross aisle, thereby intensifying the dominance of the longitudinal axis. The side doors became residual. On the other hand, the extraordinary breadth of this longitudinally oriented space inside derives from the dimensions of the meeting house square plan.

If any skyline feature gave special identity to the meeting house (and often none did), then the favored sign was a cupola. By substituting for the meeting house cupola a spired tower over the entrance end toward North Main Street, Brown further accentuated the churchly character of his design. So the First Baptist Church stands on the cusp of change: the old meeting house tradition fading with the emergence to dominance of the longitudinal church type even in situations where the older type had previously reigned.

As further indication of its transitional status, this church shows a sumptuous front grafted onto a plainer body, with the two aspects a bit unintegrated. From a plate in Gibbs which displayed three alternate designs for his own St. Martin-in-the-Fields on Trafalgar Square (1726), Brown (probably in consultation with the congregation's building committee) selected one of the rejected designs. They called on a carpenter from Boston, James Sumner, to handle this feature, which only barely interlocks with the body of the church. A square, quoined base with a clock is successively topped by a belfry with arched opening framed by paired Ionic pilasters and pedimented in four directions, then by two superimposed polygonal lanterns with paired Corinthian pilasters at the corners, all telescoped on the spire, with progressively smaller urns set on the ledging at each of the stages. Whereas the designs for St. Martin were made for a full temple front, the one-story porch flanked by paired Doric columns for the First Baptist Church came from another plate for a lesser church by Gibbs, Marylebone Chapel, with a much smaller tower, which Gibbs also illustrated.

A steep interior double stair folds within the base of the tower to bring the worshiper from North Main Street up to the level of the meeting hall. Simple hung plaster vaulting shapes the interior space, elliptical across the nave, groined over the aisles, in a visual and technical simplification of Marylebone's vaulting. Around three sides, balconies are supported midway by the giant Doric columns which separate nave from aisles. A double tier of round-arched windows lights each level. Although somewhat summarily wrought, the entablature cappings of the columns make a forceful visual presence, spreading as generous tables to receive the plaster vaulting.

Up front, the exquisite array of architectural elements against the plaster wall behind the raised pulpit recalls the layout of engraved models in architectural pattern books in Brown's collection. The crystal chandelier, imported from Britain (probably Waterford), was given to the church in 1791 by Hope Brown in memory of her father, Nicholas, and was first lighted on the occasion of her marriage to Thomas Poynton Ives. Changes occurred through time. In 1832 the original high box pews were removed for the present replacements (which, however, have doors to make them appear boxed). With this change, the aisle across the center of the church also disappeared; so, for a while, did the original high pulpit and its sounding board. In 1834 Nicholas Brown II donated the organ, which retains its original case and, despite two rebuildings, some of its original pipes. Finally, in 1884, the front went through a radical change. The baptismal pool was given more focus by placing it within a niche cut into the pulpit wall and backed by a stained glass image of St. John the Baptist. Then, in 1957, John D. Rockefeller, Jr., a graduate of Brown and a noted Baptist, provided for a near-complete restoration of the Mother Church of the American Baptist denomination. This restoration walled in the baptismal niche, recreated the high pulpit and its sounding board, and replicated the original Palladian window behind the pulpit as a shuttered screen. Colors were returned to the original white plaster with stony "sage green" for the architectural membering (the latter so prevalent for interior trim in the city through the Federal period that it came to be known locally as "Providence green"). This change of color magically intensified the light which streams through the clear glass, modulated by the subtle glints of pale amber and violet caused by the irregularities of the glass, especially on sunny days, when the interior becomes intensely luminous. Even the most hidebound preservationist, however, did not dare return the steeple to its colonial marbleized treatment in variegated colors, when the grand pretensions of the tower must have contrasted far more emphatically than is now the case with the plain clapboard siding of the box attached to it.

With two exceptions (1804 and 1832), Brown undergraduates have marched down the hill for commencement in the First Baptist Church every year since 1776—except that, as the university has grown, it is now an undergraduate baccalaureate ceremony that takes place in the church preceding commencement uphill on the Green. Of all eighteenth-century Rhode Island colonial churches, this and Trinity Episcopal Church in Newport (NE56) stand among the major ecclesiastical buildings of the period—Trinity toward the beginning of the eighteenth century, this at the threshold of independence.

PR57 Waterman Building, Rhode Island School of Design

1892–1893, Hoppin, Read and Hoppin. 11 Waterman St.

Providence was caught up in the enthusiasm for art that was widespread during the late nineteenth and early twentieth centuries, whether under the banner of art education, of the Aesthetic Movement, or of Arts and Crafts. The architectural enclave resulting from this pursuit of the muse clusters around the First Baptist Church. Its most important legacy is the first building erected for the Rhode Island School of Design. Another gift from Jesse and Helen Metcalf, it has always been known as the Waterman Building.

The idea for the school developed from a local coterie interested in the visual arts who founded the Rhode Island Art Association around the middle of the nineteenth century. Initially, its long-term goal was the establishment of an art museum for the city, with special emphasis on the "ornamental and useful arts." Helen Adelia Rowe Metcalf, wife of the textile industrialist Jesse Metcalf, was an active member. For the Philadelphia Centennial Exposition of 1876 she also headed the state Women's Centennial Commission, charged with producing the women's exhibition for the Rhode Island pavilion. At the exposition she was particularly stimulated by an extensive British exhibit on progressive approaches to art and design education which had developed in various ways in several European countries since the mid-nineteenth century. Among the movement's crowning achievements in Britain was the establishment of the technical schools for art education in South Kensington with the nearby Victoria and Albert Museum as a repository for the decorative arts. In addition to usual museum functions, the V&A was specifically charged with responsibility for improving current standards of British industrial design, the better to compete with foreign products. Concluding her leadership of the Rhode Island Centennial Commission with a surplus of $1,675, Helen Rowe Metcalf convinced her co-commissioners that this sum, augmented by a generous family donation, should go toward the establishment of a school for the training of designers for the "art industries" of the area, art teachers for the area's schools, and—because the Metcalfs were ardent collectors with a larger than merely pragmatic view as to what art education should be—for artists as well. From the moment the school got underway in 1877 on an upper floor of a downtown office building, Mrs. Metcalf, as the head of its Trustees' Executive Board, played a much more active role in the day-to-day management and educational policy of the school than is customary for board members. In this she established a pattern for two more generations of Metcalf women; next, her daughter Eliza Greene Metcalf Radeke; then her son's daughter Helen Metcalf Danforth—terminating with the death of the latter over a century later in 1984. Even then Metcalf dedication and generosity to the school did not cease, but anything like the earlier queenly reigns and the sense of the school as very nearly an exclusive preserve for family benefaction was over.

The Waterman Building, built in 1892–1893, inaugurated the school's physical presence on College Hill. It was the product of Hoppin, Read and Hoppin, the short-lived partnership of Providence-born brothers Howard and Fran-

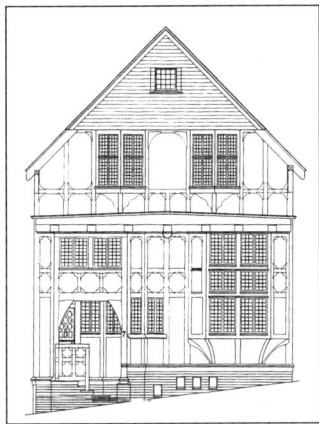

PR58 Fleur-de-Lys Studios, front elevation (HABS drawing)

PR59 Providence Art Club (Seril Dodge Houses), photo 1934

cis L. V. Hoppin and Spencer P. Read. In its initial layout, studios were located on the upper two stories (as they still are), lit by arched windows on the second floor (for the design arts, mechanical drawing, and architecture) and by roof skylights on the top floor (for drawing, painting, life, and "composition" classes. "Carving and modeling" were taught in the downhill, windowed portion of the basement. A museum and classrooms occupied the first floor. This handsomely crafted, Renaissance-inspired building is unique in Providence. Its hard-surfaced brick facade is tensely defined by shallow layering. Each stage of fenestration is progressively smaller: big arches, little arches, then, finally, mini-rectangle windows (peepholes for the uppermost floor of studios), set into progressively recessed planes. (Inspired by Renaissance palaces, the composition of openings is also a belated and loosened version of the elevational formula which H. H. Richardson made memorable in his Marshall Field Wholesale Store in Chicago [1885–1887], although he himself also had plenty of precedents for the formula in American commercial and institutional building from around 1880.) Relieving the big stretch of wall between the second- and third-floor openings are a row of shield-shaped terra-cotta reliefs, closed at either end by rectangular panels. Altered versions of plaques devoted to the guilds on the facade of Or San Michele in Florence, they show, left to right, keys (for the locksmiths' guild), an adze (stonecutters and carpenters), tongs (ironsmiths), and sword with breastplate (armorers). Most remarkable, however, is the inset band, with its Moorish lattice, at the level of the first-floor arches. The architects derived their inspiration from sketches and photographs supplied by Jesse Metcalf of buildings which he and his wife had admired in Seville (hence the Moorish lattice) and in Piacenza. It was intended that the lattice intervals be filled annually by student-designed terra-cotta plaques, and, in fact, prizes were awarded for at least two years of such competitions. Fortunately, it seems that none was ever wedged into the hexagonal grid. Interest in brick design was widespread at the time, and the careful and varied brick craftsmanship of this building, with its reference to Renaissance guilds, must have seemed especially appropriate for an art school.

PR58 Fleur-de-Lys Studios

1885, Edmund Willson of Stone, Carpenter and Willson in collaboration with artist-craftsman Sydney Burleigh. 7 Thomas St.

PR59 Providence Art Club
(Seril Dodge Houses)

Mid-1786–1789, 1791. 1886, second house converted to Art Club (expanded through 1983–1984 to include both houses), E. I. Nickerson, architect, with Sydney Burleigh, painter and craftsman, and Isaac Bates. 1906, first house raised over early 20th-century shop front. 10 and 11 Thomas St.

Thomas Street (the name for this single block of what is known as Angell Street for most of its length farther up the hill, and as Steeple Street for the next block downhill), has a fine row of eighteenth- and nineteenth-century buildings. Number 7, the Fleur-de-Lys Studios, now a studio building for the Art Club, is an example of Arts and Crafts collaborative design by the colorful artist-craftsman Sydney Burleigh, who occupied its ground-floor studio, and the architect Edmund Willson. The design derived from medieval half-timbered buildings with overhang and multiple bay windows which Burleigh had seen in Chester, England. He and his art workers decorated its walls in aqueous semi–Pre-Raphaelite, semi–Art Nouveau designs of muses, flora, and fauna, slightly raised and lightly tinted. Burleigh's own studio on the first floor remains essentially as he left it, as a whimsical tribute to the Aesthetic and Arts and Crafts movements in the decades before and after 1900. Inside (not open to the public) are more medieval-inspired woodwork, plaster ornamentation, and wrought iron fixtures, all centered on a capacious inglenook fireplace. Its plaster surround is charmingly modeled in a basket weave, rent at one point with water gushing through and a little Dutch boy, his finger in the dyke. This combined living room–studio, with balcony for canvas storage, kitchen, bedroom, and bath, provided the very ideal for the artist's life of the period. The Providence Art Club, which owns the building, carefully restored the decorative stucco in 1997–2000.

Up the hill, Seril Dodge built two adjacent houses in quick succession in the late eighteenth century, the two now joined and owned by the Providence Art Club, which was founded in 1880 and moved into the second of the houses in 1886. Dodge, a clockmaker and silversmith who came to Providence in 1784 from Norwich, Connecticut, built his first two-story clapboard house a short distance uphill from his shop. Success enabled him almost immediately to supplant it with a second next door. The special interest of these houses, however, lies in adaptations made of them at the height of the Aesthetic Movement—which were particularly appropriate for the reuse of quarters built for one of the city's postrevolutionary master craftsmen. The older (clapboard) house, was raised in 1906 so that a commercial floor with a pretty Neo-Colonial storefront could go under it, its old door becoming a balconied upstairs centerpiece. It retains, nevertheless, most of its interior woodwork on the original, upper stories. The second house was completely reworked as the initial quarters for the Art Club in 1886. The building's interior (gallery only open to visitors), has been minimally altered from the decoration of its first few decades. A fine period ensemble, the entrance hall, partially paneled with recycled shutters, offers a line of caned rocking chairs along the hearth on the way to a Nickerson-designed nook-and-crannied dining room, with a stair sweeping up from just behind the entrance to a spacious upstairs gallery. Burleigh presumably contrived the friezes of silhouettes of early members who keep company with their successors.

PR60 Brick Schoolhouse

1767. 24 Meeting St.

PR61 Shakespeare's Head

1772. 21 Meeting St.

The schoolhouse, remarkably persisting from the late eighteenth century, although remodeled several times, was adapted in the early 1960s for use by the Providence Preservation Society as its headquarters, shortly after a group of concerned "East Siders" banded together in 1956 to save College Hill from threats of demolition by institutional landlords and urban renewal. From this nucleus and its limited concern has emerged the core of citizen preservation efforts in the city today and its larger concern, to save Providence.

Across the street, "Shakespeare's Head," also occupied (in part) by the society, survives with few alterations. As a three-story wood-frame building, of its scale, at least, it is unusual for eighteenth-century Providence. John Carter published the *Providence Gazette*, sold books, and handled town mail here, "at the sign of Shakespeare's Head." Although it was founded in 1762 on Main Street as Providence's first newspaper by another editor, Carter took over in 1768, and the paper became strongly abolitionist. His presses occupied the basement; his office, the first story; and his living quarters, the second and third stories. Hence the unusual presence of more elaborate woodwork on the second story, where the family parlors were located, instead of the first. To the east side and rear of Shakespeare's Head is a "colonial" garden with plants typical for the period.

PR62 Roger Williams National Memorial Park

Early 1980s, National Park Service. North Main, Smith, and Canal sts. and Park Row

PR63 Cathedral of St. John (St. John's Church), exterior and interior detail

This park incorporates the site of the spring that supposedly attracted Providence's first European settlers. The conjectural spring itself is sited a little beyond the Old State House Parade, commemorated by a marble wellhead in a pretty garden, surrounded by marble parapeting, and reached from the higher grade at Main Street by an elaborate marble stair (1933, Norman M. Isham). This precinct speaks more of Colonial Revival pieties than of Roger Williams's original waterhole. The well no longer offers water, but the architectural pilgrim may find that this arcadian aggrandizement of the past slakes the thirst in a different way.

PR63 Cathedral of St. John
(St. John's Church)

1810, John Holden Greene. c. 1855, chapel, Richard Upjohn. 1866–1867, transept, Clifton A. Hall. 1904, chancel, Berkeley Updike. 1972, diocesan offices, Philemon E. Sturges of The Providence Partnership. 271 North Main St. (at Church St.)

This random masonry box with wooden tower is an early attempt by Greene to blend Federal form and proportion with Gothicizing detail. The stylistic conflation continued in several of his subsequent works (especially the Sullivan Dorr House [PR75] and the First Unitarian Church [PR92]). The traditional Wren-Gibbs formula of rectangular block intersecting a central, two-stage tower remains, but the thin, pointed-arched openings and flat decorative pattern purporting to be Gothic are in the spirit of Batty Langley's *Gothic Architecture Improved by Rules and Proportions* (1742). The design owes a debt to the comparable mix of styles in the second Trinity Church, New York (1788–1790; replaced by its famous extant successor). On the interior, the stairway which precipitously climbs from the street to the sanctuary within a narrow space is perhaps the most handsome, and certainly the most ingenious, example from the early Gothic Revival in the state. At the entrance end of the sanctuary, a "Gothic" organ loft on clustered colonnettes composed of metal pipes provides a delightful canopy as an entrance into the nave. Greene's saucer-dome plaster ceiling is more classical than Gothic in its Adamesque-Federal decoration (surprisingly so for one who sought a Gothic effect), but is also supported on "medieval" bundles of colonnettes, these of wood. This mix of Federal with nascent medieval recalls contemporary work by Charles Bulfinch in Boston. Up front, later interior remodelings, first medieval-inspired, then Neo-Georgian, finally quasi-modern (the latter somewhat apologetically mitigated by the spotlighting of a fine eighteenth-century Newport table) seem increasingly inappropriate to the charm of Greene's provincial medievalism. The late twentieth-century addition for the diocesan offices at the foot of the old burial ground does penance by at least attempting to recall, in precast reinforced concrete surfaced in granite aggregate, the kind of flat-patterned allusion typical of this phase of the Gothic Revival, but without Greene's sprightliness.

College Hill and East Side

College Hill is that part of the East Side residential district concentrated on the steep slope rising from Main Street out of the city's center and extending beyond the crest of the hill roughly to Hope Street—the cutoff point in the local unconscious for the baconlike slicing of properties in the original colonial plat of the city east from Main Street (see Providence introduction). Beyond Hope eastward (and in some places before), the local designation tends to shift to East Side, although some across the boundary align themselves and their property with the distinction of College Hill and are widely considered to enhance it.

College Hill is the locale of one of the country's outstanding collections of houses from the late eighteenth through the mid-twentieth century—all the more significant because of its immediate proximity to a compact downtown. As a historic urban residential district in a mid-size city, it ranks in interest with any such enclave of historic houses on the eastern seaboard. At its heart are two educational institutions: at the top of the hill, Brown University; halfway up, the Rhode Island School of Design. Benefit Street provides the spine of the area. It runs roughly parallel to North and South Main. To the uninitiated, "Benefit" may carry charitable or moral overtones, especially in conjunction with "Providence." The facts are more mundane. It was opened between 1756 and 1758 "for the common benefit of all," to relieve Main Street congestion. A few houses were built along it before the Revolution, but only in the 1780s did the street begin to fill with the simple two-and-one-half-story Federal houses that popularly characterize it. On the west side of the street, numbers ranging south from 40 through 118 form a superb, subtle row of variations on the traditional late colonial or Federal format of the five-bay, center-door front, with scattered Greek Revival examples. The architect Alpheus C. Morse lived in the south side of the Earle Pearce House, at numbers 42–44 (1827) after coming to Providence and marrying Pearce's daughter Caroline. When photographers want to capsulate "Benefit Street" in a single image, this is where they usually come.

The perception that Benefit Street is almost wholly a late-eighteenth–early-nineteenth-century street is, however, mistaken. The street remained fashionable through the Civil War, during which time a number of important mid-nineteenth-century houses were built. At the opposite end of Benefit Street and on Jenckes Street, triple-deckers represent the incursion of early twentieth-century working-class housing into the area, as its formerly elite aspect turned shabby.

The area's gradual deterioration between the turn of the twentieth century and the late 1950s encouraged its preservation because few changes were made to building exteriors during that time, and many remarkable interiors also retained all or most of their original appearance. The threat of extensive demolition during the 1950s from urban renewal programs and some destruction of old houses for the expansion of Brown University galvanized a dedicated nucleus to form the Providence Preservation Society. Its first major project was the publication of *College Hill: A Demonstration Study of Historic Area Renewal* (1959). The report, unprecedented as a systematic listing and eval-

College Hill and East Side: Benefit Street, North

Federal houses on Benefit Street

uation of such a large number of buildings in a historic area, provided the basis for its restoration. As we see College Hill today, it has the obvious patina of age and generally avoids the preciosity of over-restoration—save perhaps for the fakery of electrified "gas" lights along Benefit Street. On parallel Pratt Street (which was itself lined with triple-deckers during the doldrum interlude on College Hill), new houses (or condominiums), mostly of the 1980s, stilted off the steep slope, really take their cue from the humble triple-decker, crossed with neo–Queen Anne and neo–Neo-Colonial detailing. They pile up in a mixture of clapboard and shingle, tall, narrow and angular, with porches and decks, and a miscellany of openings of all shapes to take advantage of the view. So in the very heart of this "historic district," heralded as a privileged place, consider the revenge of the once despised triple-decker, by which turn-of-the-twentieth-century immigrants managed to squeeze into this classy enclave while its luck was down. Now the least wanted of its array of "historic" types and styles proclaims its potential for regeneration.

Benefit Street, North

The sequence follows Benefit Street north to south with diversions along side streets. Guided and more abbreviated self-guided tours are also available at the Providence Preservation Society (21 Meeting Street). What follows is only a sampling of what may be seen.

PR64–PR69 Italianate Houses on Benefit Street

This cluster of Italianate houses illustrates some of the style's range locally and immediately demonstrates that the Benefit Street area offers more than late-eighteenth-century and Federal houses such as the modest Staples House (1825) at number 24, which boasts a fine vernacular version of a Federal entrance characteristic of Providence.

PR64 Allen Greene House

1854. 27–29 Benefit St.

PR65 Duty Evans House

31–33 Benefit St.

These adjacent cubic double houses show typical mid-nineteenth-century Italianate qualities

74 Providence (PR)

in an engaging vernacular manner, especially in the prominently projecting entablature over windows and the same quality in the modillioned cornices. The porches of both of these are holdovers from the Greek Revival. So are the paneled corner piers in number 27–29, whereas the quoining of 31–33 is more characteristic of developed Italianate.

PR66 William G. Angell House
1864–1867. 30 Benefit St.

PR67 Mary M. Gorham House
1863–1865. 34 Benefit St.

The houses across the street show imposing, full-blown Italianate palace forms from the mid-1860s, one of hard-surfaced brick trimmed in brownstone, the other clapboarded. Both were designed by Alpheus Morse, who became the leading designer of large Italian-Renaissance-inspired houses for the city's mercantile and manufacturing elite, much as he also used the same style commercially downtown. Projecting heads again emphasize windows and eaves. The five-bay format for such houses typical of the colonial period is reduced to three (as is also true of the duplexes just observed across the street). The block as a whole and its detailing are blown up to a new level of monumentality, even to the handsome arrowhead belligerence of the cast iron fencing in front of the Gorham House. The widow of the head of the famed Gorham company, manufacturers of precious art objects in metal from silver tableware to public monuments in bronze, built this house about ten years after her husband's death. It attracted her nephew to build a grander palace in hard, smooth-surfaced brick next door.

PR68 Reverend Francis Smith House
1850. 35 Benefit St.

The set-back Francis Smith House, more modest in mien, invokes a country villa rather than a palace. Its bracketed door hood, playful and pretty, provides a sort of bonnet for the entrance. The flat, carpentered window frames have paneled overlays across their tops and drops (more than brackets) partway down the sides, which are sawn at the ends ornamentally. The cornice, too, is so abundantly bracketed

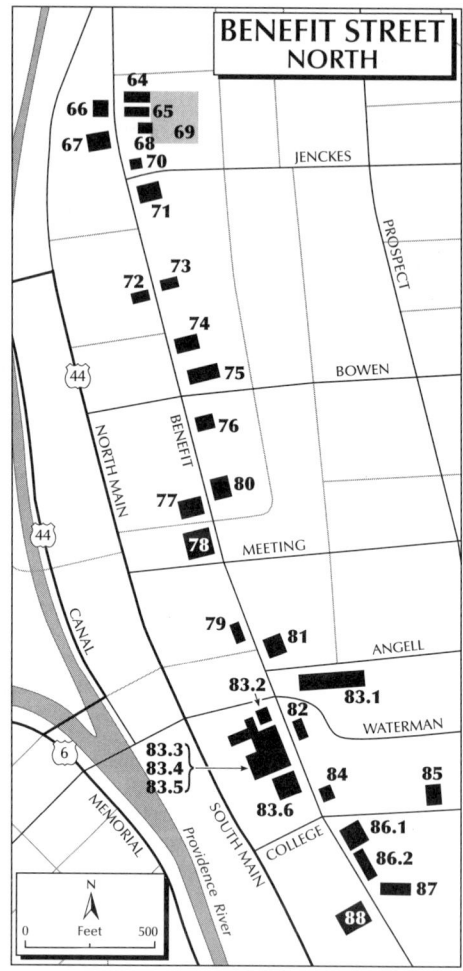

and the brackets are so flatly handled as almost to become a tasseled edge. Thomas Tefft, who pioneered in Italianate styles with round-arched windows and frequently handled villas of this type with just such reductive inventiveness, may be responsible for this design. The last three examples all show the low, flaring, flat-topped mansard which was popular in early Victorian examples of the feature.

PR69 Garden
1960s, Christopher Tunnard, landscape architect

At the center of the block is a neighborhood garden, hidden from the street behind surrounding restored houses on Benefit, Jenckes, and Halsey streets and, at the top on Platt

Street, by new row housing. With restricted access through the private yards of the bordering houses, it provides a handsome sequence of landscaped terraces that accommodate a variety of recreational activities for area residents, including play space for young children. It also exemplifies a communitarian alternative to the usual development of such blocks. Each house has its own limited yard and garden, beyond which each has its gate and key to the walled space shared in common.

Both the "secret garden" and the restoration of much of northern Benefit and its side streets depended on the vision and courage of an enterprising woman, Beatrice O. Chace, who dared to risk the purchase en masse of more than seventy decrepit old houses in the area, at a time when all were slated for demolition in the advance of urban renewal. She gradually sold the restored shells (some having the remnants of original interiors, others not, with owners given latitude inside as to how to adapt them). In this way she virtually salvaged this entire section of the Hill singlehandedly.

PR70 Joseph Jenckes House

1774. 43 Benefit St. (corner of Jenckes St.)

PR71 Joanna Jenckes Barnes House

c. 1795. 49 Benefit St. (corner of Jenckes St.)

The gambrel roof of the Jenckes House is rare in Providence, because few were built in the second half of the eighteenth century, and earlier ones have disappeared (as they have not in Newport). This is probably the largest and most handsomely preserved in the center city. Across Jenckes Street, the rusticated siding on the Barnes House is also rare among extant Providence houses and apparently never was popular here, whereas again the opposite seems to have been true for mid-to-late eighteenth-century Newport.

PR72 Hallworth House (Enoch W. Clarke–John Slater House)

1828, James C. Bucklin. 1904, 1967–1968, additions. 66 Benefit St.

PR73 Dustin Lacey House

1847; dormers added, 20th century. 65 Benefit St.

These houses represent two aspects of the Greek Revival style. James Bucklin designed the Doric Greek Revival Slater House (so called for its most famous owner; now known as Hallworth House) in the same year as the Arcade. He monumentalized the standard two-story, hip-roofed, brick Federal house in the usual Greek Revival manner, by enlarging and simplifying detail, but here also by emphasizing the blocky mass through tightly grouping the windows toward the center of the elevation around the door, then leaving a stretch of wall separating these from the outermost windows. Finally, and most decisively, the scale and simple geometry of the portico announce this as an unmistakable example of the Greek Revival. Providence appears to have had few such full-blown temple-fronted Greek Revival porticoes. This one is turned sideways to the street, so that the front faces south, as we shall see elsewhere to an unusual degree on Benefit. Although south-facing, rather than road-facing, houses are common among eighteenth-century houses, they are rare in cities. Their orientation recalls the most familiar American example of sideways urban siting, that of houses in Charleston, South Carolina. Because of its textile industry, Rhode Island's connections with southern plantation owners and especially with brokers in Charleston and Savannah were close. Does the sideways siting of a number of houses on College Hill indicate Charleston influence?

Enoch Clarke, the original owner, almost immediately sold the house to John Slater, brother of the famous Samuel Slater, associated with the Pawtucket mill that marked the start of New England's textile industry. A descendent, Horatio N. Slater, who had given Slater Hall to Brown University as a men's dormitory, in 1901 made a gender-neutral gesture worthy of a later day. He gave his own house to Pembroke College (then the women's college of the university) as its first dormitory. Twice enlarged (remodeling by Brown, then extensive further remodeling and downhill additions by the Episcopal church), it serves today as the frontispiece of a home for the elderly. The small, vernacular interpretation opposite, its full "portico" very nearly a "porch," came from the west side of town in 1976. Charming, it is really too small for the scale and spacing of what is around it.

PR74 Henry Rhodes House

1860–1862, Alpheus C. Morse. 101 Benefit St.

Another brick and brownstone palazzo by Alpheus Morse and proof (by comparison with the others) that within a relatively proscribed formula Morse achieved considerable variety and distinction in his designs. Located just at the point on the east side of Benefit Street where the land begins to rise more steeply, the Rhodes House fits comfortably into the slope, in part because of the street-level brownstone fence, one of few such to have survived.

PR75 Sullivan Dorr House

1809–1810. John Holden Greene. 109 Benefit St.

PR75 Sullivan Dorr House

The Dorr House, in the Federal style, is probably John Holden Greene's most innovative and is among his finest large houses. At the client's unusual request, Greene followed the format of the river facade for Alexander Pope's villa at Twickenham (England) for the design of this facade. Set at a ninety-degree angle to the street and presenting an almost solid brick wall on the northern edge of its narrow lot, it has a south-facing windowed front, like that of the Hallworth House. The three-story center section flanked by two-story elements inaugurated no trend in either Greene's work or in that of other Providence architects (although Russell Warren seems to have copied the formula for his famous Bosworth House, c. 1815, now demolished, in Bristol). The main body of the house connects through the service wing to an ell for stables—a solution Greene repeated for similar conditions elsewhere on College Hill (the former Truman Beckwith House [PR84]). Barely preceding his use of Batty Langley's version of Gothic at St. John's, here, too, Greene blends Federal classical ornament with Gothic touches: in the clustered colonnettes to support the porch and in the cusped, pointed arches of its cornice frieze, reappearing in the second-story Palladian window, and again in the cornice molding of the main roof. The interior (not open to the public) follows a center-hall, four-room plan. Its somewhat severe detail, as in the simple, almost Greek Revival mantels, serves as a foil for the lush detail of the painted murals, by Michel Félice Corné, in the parlor and the main hall (and originally in the dining room as well). Commissioned when the house was built, they represent the Bay of Naples, Corné's home, and are similar to murals this emigré Newport artist also executed in Salem, Boston, and Bristol houses.

PR76 George R. Drowne House

1862, Gridley J. F. Bryant. 119 Benefit St.

High-style Second Empire houses are uncommon in Providence, and especially on College Hill. This is the first in the city, among the most commanding, and one of the few remaining domestic commissions anywhere by Bryant, who is most widely known for his Boston City Hall in the same style. Here the imperious elevation is all for show, however, as the $35,000 price tag for the house suggests and the original plain interiors proved. The house is now altered for apartments.

PR77 Old State House (Providence Colony House, later Rhode Island State House)

1762. Alterations, 1850–1851, Thomas A. Tefft. 1867, James C. Bucklin. 1906, Banning and Thornton. 150 Benefit St.

Rhode Island's five early state houses, one for each of the state's five counties, were not successors to one another, but all designed as equivalents, with the governor and legislature meant to rotate among them. They were built, in the order of their construction, in Newport, Providence, South Kingstown, East Greenwich, and Bristol. All are preserved. Providence became the first among equals as it assumed economic and political dominance in the state. It was not always so. During the colonial period, when there were only two colony houses, the one at Newport took precedence and is decidedly the grandest of the lot. Over time, during the nineteenth century, rotation ceased to include the minor state houses and oscillated between Providence and Newport, with the initial preeminence reversed. Eventually, Newport became more the locus of ceremonial events than legislation. But the policy of rotation did not officially end until the legislature finally moved into McKim, Mead and White's State House (PR176) after its opening in 1901.

The main block of the old Providence State House went up at two times. The western half of it, toward the long, narrow green terminating at North Main Street, was completed in 1762 as a replacement for an earlier building of 1732 which burned in 1758. It is a much simplified version of Richard Munday's 1739 Colony House in Newport. To this in 1850–1851 Thomas Tefft added an entrance tower facing the narrow green down to Main Street, which necessitated the removal of a central gable. In this addition, he conscientiously followed the style of the original building—an especially early example of respect for then mostly despised colonial buildings on the part of an architect who, as early as 1853, read a pioneering paper in Providence in praise of colonial architecture. In this talk, "Architecture Ancient and Modern" (where "ancient" meant "colonial"), Tefft opposed the meretriciousness that he found in most "modern" building while praising the straightforwardness of colonial design and the integrity of its workmanship. His architectural emphasis differed significantly from that of occasional contemporary pleas from antiquarians to respect the "houses of our forebears" merely for genealogical, historical, or sentimental reasons. A century after the erection of the original building, James Bucklin roughly doubled its size, also in the original style, eastward to Benefit Street.

After completion of McKim, Mead and White's building and following a thorough interior revamping in the Colonial Revival style by Banning and Thornton, what was now the Old State House served as a courthouse until 1975. (The downstairs courtroom best reveals Banning and Thornton's Neo-Colonial design.) Thereafter it became state offices, among them quarters for the Rhode Island Historical Preservation and Heritage Commission, which is the repository for much of the information in this volume. In the commission's present meeting room, once the secretary of state's room, in the southwest corner of the upper floor, is the only substantial remnant of the original interiors: bolection paneling, the most important surviving example of this mid-eighteenth century type of paneling remaining in the city. It was in this building that Rhode Island declared its independence from Great Britain, the first colony to do so, two months before the Declaration of Independence. Here, too, Washington, Adams, Jefferson, and Lafayette were received as notable guests of the state.

PR78 What Cheer Garage

1910. Conversion, 1990, Gordon Washburn. 156 Benefit St. (at South Court St.)

Adjacent to the Old State House is the reinforced concrete What Cheer Garage, now a studio building for RISD. It proclaimed the arrival of the automobile on College Hill and the need for garage space in a locale too close-packed with buildings to provide adequately for private parking. The reduction of the motif of the classical triumphal arch to a minimal functional statement recognizes the industrial origins of reinforced concrete technology while also according the garage a degree of "class" appropriate for the limousine set to whom it catered (much as Auguste Perret's famous reinforced concrete garages off the Champs-Elysées in Paris used the same formula to the same end at about the same time). The handsome proportions and blunt vigor of its elevations, resurfaced when adapted to studio use, have only been enhanced by the reinstallation of factory-style metal windows, happily with a deeper sash than was originally used, reversing an earlier renovation in which the window openings were insensitively infilled. ("What Cheer" is not the rueful response to a mechanic's bill, but was, according to tradition, the Narragansett Indians' greeting to Roger Williams on his landing at Providence [a contraction of "What cheer with you?," the seventeenth-century equivalent of "How are you?"]. Many Rhode Island businesses perpetuate the historic greeting.)

PR79 Benefit Street Arsenal, Providence Marine Corps of Artillery

1839–1840, James C. Bucklin. 176 Benefit St.

The Benefit Street Arsenal is another version of the elevation as aggrandized orifice. If St.

John's, the Sullivan Dorr House, and the Unitarian Church show John Holden Greene as a designer venturing into "Gothick" from the Federal style, the Arsenal shows a Greek Revival designer making the same stylistic move. A rubblestone, stucco-covered box, it presents to Benefit Street a giant, hobnailed portal as a pointed arch between crenellated towers with slotted. Lancet windows fortified by iron shutters line the side elevations. Not nicety of form, but the resonance with which "signs" convert the box to an "armory" accounts for its expressive charm. No sooner was it built than the battery had to rally to "defend" its castle against the Dorr Rebellion in 1842, although what little military confrontation there was occurred in Chepachet, close to the western boundary of the state. Bronze tablets placard its front with more serious claims to glory. As a type it compares with the Newport Artillery Company, built four years earlier (NE79). Fortunately it was not so impregnable that it could not be moved one lot to the north when, in 1906, a tunnel for a now defunct Providence, Warren and Bristol Railroad was bored through College Hill beneath its old site.

PR80 Knowles Row

c. 1862. 149–155 Benefit St.

The four-story clapboarded Italianate Knowles Row belongs to a type of housing that is rare and scattered in Providence, here particularly impressive for the forceful projection of repetitive bracketed entrance heads. In the interior planning of these houses, the parlor is lifted to the piano nobile, above the entrance floor.

PR81 Thomas Jenckes House

1856. 2 Angell St. (corner of Benefit St.)

A fine Italianate block in brick and brownstone, the Thomas Jenckes House is notable for the handsome double curve of its stair up to a raised arched entrance porch. More arching on Benefit Street makes an entrance porch for Jenckes's law office.

PR82 University Club (Rufus and Emily Waterman House)

1830, John Holden Greene. 1866, renovation and enlargement, Alpheus C. Morse. 1900 and later, south addition, Stone, Carpenter and Willson. 1901, dining room addition, P. O. Clarke. 219 Benefit St. (southeast corner of Waterman St.)

Across Waterman Street, the University Club began as a two-and-one-half-story Federal house designed by John Holden Greene for Rufus Waterman. The tall stair and porch are his. Rufus's daughter Emily called on Alpheus Morse to blow the house up into an Italian palazzo. Hence the full third story, the modillion cornice, and the low hipped roof. After the University Club acquired the house, in 1900, Stone, Carpenter and Willson enlarged it with a wing to the south, which is most evident in Neo-Federal touches inside. Finally, P. O. Clarke added the Neo-Federal one-story dining wing with its large bow window toward Waterman Street. Although architecturally something of a comfortably elegant mishmash after so much renovation (which is not necessarily a deficit for a club), it holds its triangular site and its front on Benefit Street with distinction.

PR83 Rhode Island School of Design

The block on Benefit Street between Angell and College Streets marks the center of the Rhode Island School of Design, with dormitories up the hill between Angell and Waterman streets and an interesting sequence of buildings along the west side of Benefit. (The RISD College Building is discussed in the urban design context of the Providence County Courthouse complex; see PR53.)

PR83.1 Homer and Nickerson Halls, Metcalf Refectory

1955–1957, Robinson Green Beretta with Warren A. Peterson and Pietro Belluschi, consulting designers. 1985–1987, East and South Halls addition, William D. Warner. 60 Waterman St. and 55 Angell St.

This dormitory complex was originally conceived to blend with the historic neighborhood as a stepped, high-gabled grouping which opened to the slope by platforms and stairs. Although somewhat bland, it sensitively responded to a prevalent desire in the 1950s to reconcile modernism with tradition in a manner associated with the work of Pietro Belluschi. It won an award from *Progressive Architecture* in 1959. Pity, therefore, that its postmodernist expansion jams what used to be the openness of the complex on its downhill, Waterman Street edge.

PR83.2 Dr. George W. Carr House

1885, Edward I. Nickerson. 29 Waterman St.

On the opposite side of Benefit Street, at the southwest corner of Waterman and Benefit, the Carr House typifies the idiosyncratic eclecticism of Edward Nickerson's essays in the Queen Anne mode, in which masonry, wood shingles, slate, and sheet copper are skillfully crafted and combined in a picturesque composition. Its rooftop bristle of gables, dormers, and tall, sculptured chimneys in brick culminates in a corner turret. This nicely turns the composition around the corner; its curvature and the complexity of its rooftop forms are echoed on both elevations by bowed bay windows of varying size, roundness, and projection. It was designed for a doctor, whose basement office was entered at the downhill Waterman Street corner of the house. It was acquired by RISD c. 1908. Although it is battered by generations of student and office use, much of its interior woodwork remains intact.

PR83.3–PR83.5 Museum of Art, Rhode Island School of Design

224 Benefit St.

Next in line come the three major buildings which make up the Museum of the Rhode Island School of Design. The first (set back from Benefit Street) is the Farago Wing; then the Pendleton House and, attached to it, the entrance facade to the Radeke Building (both set back only slightly from the sidewalk). We shall consider them chronologically. (See also the WatermanBuilding [PR57].)

PR83.3 Pendleton House (The Colonial House) and Waterman Galleries

1904–1906, Stone, Carpenter and Willson (Pendleton House). 1905, Charles Platt (Waterman Galleries). 222 Benefit St.; stair and gallery connector on north side of Pendleton House to Waterman Galleries

Stephen O. Metcalf, son of Jesse and Helen and successor to the helm at the family's Wanskuck Mill, financed the building once generally called the Colonial House. It was Edmund Willson's last major commission. Appropriately, it was also his final salute to a style and an aura from the American past which had meant much to him in his career. Derived partly from the great Federal buildings of his own native Salem and partly from those of John Holden Greene, the local genius of the style, it is the first American example of a "period house" specifically designed as a museum setting for an installation of furniture and decorative arts.

The initial collection came to the museum from one of the major collections of American and English eighteenth- and early nineteenth-century furniture amassed during the initial late nineteenth-century craze for colonial and Federal-period "antiques," together with their English antecedents. Charles Pendleton demanded a special wing for his collection, which otherwise would have gone to the Museum of Fine Arts in Boston. Willson's choice of the Federal Revival style for Pendleton House was all the more appropriate because Pendleton had displayed his collection in his home, one of Providence's more elaborate Federal houses, a few blocks away (PR134); in fact, Willson used the interiors of the real Pendleton House and the provision they made for the arrangement of Pendleton's collection as partial inspiration for the interiors of the museum version. (Willson's decision poses interesting questions for architects and curators. Should the museum's Pendleton House simply use the settings to display the collection and subsequent acquisitions with as much historical accuracy as possible? Or, now that the Colonial Revival is itself of interest, should a bifocal view of the past prevail, one which also acknowledges the way in which the late nineteenth century saw the American past by exhibiting its taste in the arrangement of these furnishings, even including some handsome examples of Colonial Revival fakes? Some of the latter have themselves become candidates for inclusion in the Museum as superb, or at least revealing, revival pieces.)

For the exterior of Pendleton House, Willson totally abandoned the rather overelaborated ornamentation of Pendleton's own wooden house to create a more monumental version of the Federal style in brick, over a fireproof, reinforced concrete frame. He also transformed the carpentered provincialism of John Holden Greene's Federal ornament toward something more monumental and worldly. But comparison between these two should wait for a major house by Greene, the Truman Beckwith House (PR84).

Attached to the north side of Pendleton House is a tiny arched, columned hexagonal adjunct no larger than a gazebo, with an interior dome. Exquisitely designed in the Neo-Renaissance manner, it provides the vestibule (out of

PR83.3 Pendleton House (The Colonial House)

PR83.6 Memorial Hall, RISD (Central Congregational Church)

the Pendleton House dining room) for Charles Platt's stairway connector down the slope, through a (subsequently revamped) corridor gallery, also designed by Platt, to link the house with the so-called Waterman Galleries below. These had been added to the back of the Waterman Building (PR57) in the initial enlargement of the museum's gallery space beyond its original ground-floor quarters in RISD's first permanent building in 1896–1897 as a memorial gift by Jesse Metcalf following his wife's death. (At the time of his addition to Pendleton House, Platt was working on an important house in nearby Bristol, now demolished.)

PR83.4 Eliza G. Radeke Building, Museum of Art

1925–1926, William T. Aldrich

The Eliza Radeke Building, the major body of the museum, permitted the conversion of museum space on the ground floor of the Waterman Building to classroom space while incorporating the later galleries which had been added to it. This time Stephen O. Metcalf joined with his brothers to contribute the building in honor of their sister, who represented the second generation of women in the Metcalf family in the role of quasi-manager, contributor, and intermediary between the family and the School of Design. Continuing in the Neo-Federal style, Aldrich solved a difficult design problem so adroitly that no problem seems ever to have existed. How to manage his addition so that it would not downplay the Pendleton House while nevertheless making clear that the new building was henceforth the principal entrance to the museum? Aldrich set his addition slightly back from the Willson block as a one-story addition, but (measured against Pendleton House) actually one and one-half stories in height. Within this height he proceeded to blow up the domestic aspects of Pendleton House to more institutional scale. He articulated his wall with brick pilasters the full height of the wing, enlarged on its fanlighted door and projected it slightly from the wall (versus the inset Pendleton entrance). He simplified and enlarged all ornament and, finally, provided a wide, courtlike approach. To balance Pendleton House he projected a plain, two-windowed block at the opposite end of his addition. Preserving the residential feel of Benefit Street, the screenlike wall conceals what eventually becomes a five-story building as it falls down the hill, plus another which is set back above. The lobby behind this screen goes to the sources of the Federal style in suggesting an Adamesque salon (or "saloon," as Robert Adam would have called it), but in a constricted and restrained manner. It even contains classical sculpture, like Adam's Kedelston Hall, although without Kedelston's overwhelming opulence. Windows along the opposite wall look down to a garden court three levels below, completed a few years later, after Mrs. Radeke's death, and dedicated to her memory. The galleries around it look into it as well. Again, institutional and residential qualities blend to increase the pleasure of viewing one of the outstanding small-city museum collections in the country.

PR83.5 Peter and Daphne Farago Wing, Museum of Art

1990–1991, Tony Atkin and Associates

Peter Farago, a RISD graduate, and his wife, Daphne, shared their good fortune with his alma mater, finally interrupting significantly the cumulative contributions of the Metcalf family to the museum building complex. Building in brick out from the rear of Pendleton House, and behind Platt's tempietto so as to leave a garden court in front, the Philadelphia architect Tony Atkin shifted to an elegant but somewhat industrial mien. Square red glazed tile blocks combine with heavily framed, square metal windows, each squared again in a lighter grid. These tile and glass units make a glazed, screenlike elevation, out in front of the boxy mass with a copper-clad, barrel-vaulted roof, containing the principal galleries on two levels.

The vault diffuses light from a monitor down the walls of the two-story upper gallery for large modern works. Inside, as on the exterior, faintly industrial and overtly structural elements deliberately contrast with the insistent geometrical abstraction of other components and the spaces made by them. Intended as sophisticated counterpoint, it misses a little, to produce an effect which is not so much disharmonious as uncertain. A stair, leading to a balcony corridor, folds around the core box of galleries and projects as a high, overhanging element to the outside. It provides elevated views down into the big gallery in one direction and vignetted views of Providence in the other, with a bridge into the topmost floor of Aldrich's building.

PR83.6 Memorial Hall, RISD (Central Congregational Church)

1853–1856, Thomas A. Tefft. 1903 and later, renovations, Hoppin and Ely. 1950, towers removed. 226 Benefit St.

Next in line is Thomas Tefft's brownstone-fronted, but otherwise brick-walled, Central Congregational Church, vacated by the congregation for a replacement on the East Side in 1893 (see PR151). RISD purchased it in 1902, thanks again to the second generation of Metcalf benefactors, Eliza Radeke and her brothers, who dedicated it to a recently deceased member of the family and renamed it Memorial Hall. Again they called on a Hoppin—this time of Hoppin and Ely—to alter an interior which has itself been much realtered.

What at first sight may be dismissed as a rather forlornly truncated fragment, the one-time smoothness of its brownstone walls somewhat scarred by spalling, is far more important than first appearances indicate. It lost a pair of identical tall, skinny bell towers in 1950, after structural damage from the 1938 hurricane. A drawing by Tefft at Brown University shows their attenuation magnified by chamfered corners and arched openings even taller than any of the elongated openings which remain, while another shows them topped by spires in polychromed slate, which would have more than doubled the present height of the elevation had the budget permitted. Even if perfectly preserved and fully completed, however, the church can never have been beguiling. It means instead to be compelling and it still is, even in its diminished condition. Attenuated arched openings and the three-part division of the elevation provide the sole relief for the once assertive planarity of its wall, its only ornamentation being the overlarge band of arcuated corbeling which stretches across the elevation at the base of its towers and under the eaves of the gable between them. The full importance of this church, however, is its position within the vanguard of those buildings which introduced the Neo-Romanesque style, especially from German precedents (but also with some British influence), to the United States just before mid-century. Then called variously Romanesque or Lombard, it was also known more generically as the "round-arched style" (from the German Rundbogenstil), which Tefft himself preferred to call it. The nexus of circumstances explaining its importance requires more than usual comment.

From German books in his quite extensive library, Tefft was aware of the early nineteenth-century origins of the revived round-arched style in works by Friedrich von Gaertner and Karl Friedrich Schinkel and their followers, who, in turn, patriotically revived German medieval Romanesque. This predominantly brick-built style had, in its turn, largely derived from precedents around the northern Italian region of Lombardy. Beginning around 1848 at the latest (or possibly earlier) and while he was still employed as a designer for Tallman and Bucklin, five years previous to his commission for the Central Congregational Church, Tefft had experimented with the round-arched style, which rapidly became a favorite with him. (A "broad and bright" style, he called it.) It was in this round-arched brick style, ornamented by corbeling, that he realized his best-known commission, the original Providence Union Station (1847–1848), which was also his finale as a designer in the Tallman and Bucklin office before setting up on his own. Until it burned in 1896, the long, narrow station block with its attenuated towers, flanking the entrance like tall tapers, responded to the pair of towers which fronted the church in the new style on College Hill and provided the monumental secular counterpart to its monumental ecclesiastical equivalent.

But the choice of Romanesque for Central Congregational was also reinforced by current debates about church architecture. As the most notable church designer of his day, Richard Upjohn built what is generally regarded as the first important American Neo-Romanesque church. The style of his Congregationalist Church of the Pilgrims in Brooklyn, New York (1844–1846; demolished), seems to have been influenced by German emigrés in Upjohn's employ. Almost simultaneously, for Bowdoin College in New Brunswick, Maine (1844–1855), and in an unbuilt project for Harvard College (1846), both also under Congregational auspices, Upjohn designed two more Romanesque-inspired chapels, the first of these still extant among the American masterpieces in the style, the second rejected for another design. Upjohn was himself a High Church Episcopalian and, as such, an ardent supporter of the Ecclesiological Movement within the Church of England, which peaked from the late 1830s through the early 1850s, during a period marked by widespread religious fervor. The journal of the movement, *The Ecclesiologist* (first published in 1841, with a New York offshoot begun in 1848) assailed as impious the flimsy boxes typical of recent construction for Anglican worship, and their classical ornamentation as pagan. To Ecclesiologists, both evils resulted from the baleful effect of a long period of skepticism, secularism, and an overweening faith in science and rationalism reaching back through most of the eighteenth century. Reform required a return to the true church and the "Christian Style." This demanded a liturgy devoted to the mysteries of the high mass celebrated on an altar set deep in a chancel, partially veiled from the lay congregation by a rood screen across its front and by the dim religious light of stained glass, together with the panoply of incense, clerical processions, and sumptuous trappings by which the High Church now meant to overturn years of excessive Low Church tolerance. Caught up in the piety of the mid-nineteenth century, even Low Church members sought more "churchliness" in buildings and ritual. So did Protestant sects, but especially the Congregationalists. Clearly, however, the degree of churchliness acceptable for their buildings and ritual had to be a cut below that of Low Episcopalians; hence the choice of Romanesque. Its forms were ampler and more stolid; its ornamentation was blunter; its round-arched vaulting was more elemental; its kinship with secular building so close that Romanesque boasted great religious architecture in common brick as well as stone—witness German Romanesque, and behind it Lombard Romanesque. Or so the arguments at the time simplified the true nature of Romanesque. Congregationalist discussion on this issue came to something of a culmination in the unpretentiously titled *A Book of Plans* (1853, the very year which saw the start of Central Congregational), the result of a "Central Committee" of clergy and businessman who recorded their selections for a portfolio of ideal churches as a guide for local building committees. Other styles were included, but Romanesque much predominated.

So both secular and ecclesiastic developments seem to have coalesced in making Central Congregational among the earliest monumental statements of mid-nineteenth-century Neo-Romanesque. If Central Congregational responded on the secular side to the Romanesque Revival railroad station at the foot of the hill, on the ecclesiastical side it countered Upjohn's slightly earlier Gothic Revival Grace Church downtown (PR32) and, later, farther uphill, St. Stephen's (PR110). Minutes of the building committee for Central Congregational inform us that, as in both of these Upjohn churches, and almost certainly influenced by what was then the radical example of Grace Church, Tefft also had originally intended a single, asymmetrically placed steeple like that on the corner of Grace. The committee objected, however, allegedly on the basis that two towers and double stairs to the sanctuary would better suit the site—although another reason may have been their discomfort with the lopsidedness they saw in an unfamiliar asymmetrical composition. As the bellwether for monumental Neo-Romanesque churches in Rhode Island, Central Congregational anticipates two other important mid-century examples in the state, Pawtucket Congregational Church (PA22) and Newport Congregational Church (NE61) and a number of country examples. Hence the significance of this battered remnant.

PR84 Handicraft Club (Truman Beckwith House)

1826, John Holden Greene. 42 College St.

The Beckwith House ranks with the Sullivan Dorr and Benoni Cooke Houses (PR75 and PR47) as one of the best preserved among Greene's more ambitious Providence residences. This house is in fact based on the Cooke House, which had been completed during the previous year. Here Greene elongated the more cubic quality of the brick box for Cooke into a rectangular volume. He organized service and stables as he had done for Sullivan Dorr: the service wing set back from the main block of the house, to which the stable area is joined as an ell. As in the Dorr House, too, Greene angled this front south to the sun. The attenuated Ionic porch and door with fanlight and side lights overlaid with lead ornament, the embellished window centered in the second story, and the distinctive stepped lintels over windows all reappear as variants of the same features in the Cooke House. However, the Cooke House has lost the three stages of decorative fretted and paneled parapeting which once capped the roof of the entrance porch, eaves, and monitor. Here this parapeting, of which Greene was so fond (perhaps overly so), has been restored after disappearing in the 1920s.

The comparison of Greene's carpentered stick and panel ornamental treatment can be measured against the more sculptural and monumental balustered parapets in Edmund Willson's nearby Pendleton House (PR83.3), with their exquisite profiling subtly varied between the two levels. Willson's entrance, its porch, and the stair window above are both broader and more lavish in treatment; the stepped marble lintels over his windows are more complicated in their Greek key patterning. Indeed, the breadth and suavity of the Pendleton House, informed by Willson's knowledge of Salem's Federal style and his training in Beaux-Arts academicism, call attention to the plainer, more pinched and provincial qualities of Greene's work. His ornamentation seems to be more an attempt to dress up plainness than to provide Willson's finishing grace. As though to prove the fact, Greene left the interiors of this house, where one expects the committed ornamentalist to be even more indulgent, typically spare, if elegant. (The Dufour landscape wallpapers in some rooms are twentieth-century installations of paper removed from the Corliss-Carrington House [PR96.4].) Yet there is an intensity about the Beckwith House as an image, due partly to the sure proportioning of all elements, partly to very directness of the ornamentation, that makes it the more indelible of the two. The price of Willson's greater sophistication may be that the Pendleton House "fits in" so comfortably that one too easily passes it by and overlooks its quality. The Beckwith House captures our attention with the force of Greene's discovery, as he reached out to regional versions of the Federal style, making his own synthesis and building so much in his personal manner that his work came to characterize the Federal moment in Providence. At the end of his career, Willson reached back in Pendleton House to epitomize in a worldly manner a composite memory of Greene's house as the essence of Providence, the city where both architects realized the bulk of their accomplishment.

PR84 Handicraft Club (Truman Beckwith House)

PR85 Albert and Vera List Art Building, Brown University

1969–1971, Philip Johnson. 64 College St.

The List Art Building represents an early stage in the reincorporation of classical allusion into modern architecture at a time when Philip Johnson was in the vanguard of the movement. The colossal portico recalls classical precedent, but as unadorned piers blunted toward the elemental rectangularities of modernism. What is initially perceived as a "portico" through which one normally passes into a centered doorway in a classical building becomes an "arcade" for a walkway from the street to the entrance at a right angle to the approach. Such conflation of classical syntax, together with its modernist reduction of form, came to characterize much postmodernist design. Johnson also envisioned the front of the List Art Building as something of a portico to the principal approach of Brown University in its orientation toward the center of the city. The architect wanted marble cladding or, at least, an aggregate of marble chips. Economic restraints compelled plain concrete surfaces, but so well detailed and crafted that they dignify the monumental aspirations of the building. The incised geometry molded into the building measures its surfaces in the manner of "masonry" although the circular holes centered in each of the rectangles indicate the location of the bolts for the concrete formwork and thereby record "reinforced concrete" as the actual material of the wall. A return to masonry, or to the simulated effects of masonry, would become another aspect of postmodern classicism. The angled sun screens which front each of the instructors' studios up in the "entablature" have more sculptural than functional effect.

PR86 Athenaeum Block

PR86.1 Providence Athenaeum

1836–1838, William Strickland. 1868, remodeling (interior and main staircase), James C. Bucklin. 1871, exterior drinking fountain, Ware and Van Brunt. 1917, southeast rear wing, Norman M. Isham. 1977–1979, rare book room, Warren Platner. 251 Benefit St. (corner of College St.)

William Strickland's only New England commission brought this rugged gray granite temple to Providence for a private subscription library, which was originally also to have included quarters for the local Franklin Society, a scientific group, and the Rhode Island Historical Society. The latter two organizations abandoned the venture, although the Franklin Society's involvement may have accounted for the award of the commission to the major architect in the Greek Revival style, who was himself a member of the Franklin Society in Philadelphia. The slope above Benefit Street provides its mini-Acropolis, reached by an elaborate stair with split approach, then a steep finale between two Doric columns (distyle) in a porch inset between wings (in antis). All this invokes the Goddess of Wisdom for the books behind, her precinct prettily bounded by cast iron fencing with fine Greek Revival detailing based on anthemion and palmette themes. The incongruous outdoor drinking fountain of 1871, by Ware and Van Brunt, is a fine example of Victorian outdoor embellishment.

Inside, James Bucklin's plain but comfortable renovation of 1868, with open stacks ranged beside reading tables, predominates, creating the kind of invitation to browsing and reading which eludes most post-Victorian libraries completely. A fantastic library table modeled on an Egyptian temple and painted with hieroglyphs is tucked under the stairs at the lower level. In front of a small earlier (1917) addition by Norman Isham toward the rear of the south wall, Warren Platner's late 1970s addition for rare books tries to be sympathetic outside, but nevertheless impinges on the original independence of Strickland's box. Its plate glass window proclaims a modern interior, well detailed and pleasant in itself, but at odds with the Pickwickian environment elsewhere.

PR86.2 Athenaeum Row

c. 1845. 257–267 Benefit St.

The Thomas Poynton Ives family, which contributed the site for the Athenaeum, appropriately built Athenaeum Row as an investment property beside it. This Ionic-fronted row, originally five attached houses, allegedly introduced the "English plan" for row houses to Providence. The principal rooms were not at entrance level, as was then common here, but at the next level, corresponding to the English first story or the Continental piano nobile, which the steep slope made sensible. Innovative and sophisticated for Providence, En-

College Hill and East Side: Benefit Street, North

PR85 Albert and Vera List Art Building, Brown University (above, left)

PR86.1 Providence Athenaeum (left)

PR88 John Field–Stephen Hopkins House (above right)

glish planning little affected the city's building patterns at the time, although there are Victorian successors—for example, the Knowles Row, nearby, which provides an interesting comparison (PR80).

PR87 Eliza Ward–Marsden J. Perry House

1814. 1892, remodeling, Stone, Carpenter and Willson 2 George St. (corner of Benefit St.)

Immediately south of Athenaeum Row is the Eliza Ward House. The owner was Joseph Brown's daughter and the early widow of Richard Ward, from a prominent Newport political family. This Federal house in brick, raised on a high granite base, originally had an entrance one story above street level. After purchasing the house from Mrs. Ward's collateral descendants, the banking, electric company, and traction line tycoon Marsden J. Perry commissioned extensive interior changes from the firm that would shortly design his bank and office building. These involved, among others, the creation at street level of a library-study to the left of a new entrance and a stair up to the elevated principal floor with a mid-story oriel window where the original entrance had been. All reveal Stone, Carpenter and Willson's sure, delicate touch, tending to be more piquant than its sources. The interior (not open to the public) contains a double parlor papered in Dufour landscape scenes. Here, for a while, Perry installed his splendid and growing collection of American and English eighteenth-century furniture, before moving on to much grander quarters (PR96.1).

PR88 John Field–Stephen Hopkins House

1707, 1743, 1804; 1927, restoration, Norman M. Isham; landscaping, Alden Hopkins. 15 Hopkins St. (corner of Benefit St.) (open to the public at particular times or by appointment)

John Field's one-and-one-story gabled cottage comprises the rear ell for the two-and-one-half-story, four-bay enlargement (also quite small) which Stephen Hopkins placed in front of it. The latter is a narrow, two-room house with center hall on the ground floor. A tiny setting for large accomplishment: Hopkins was a merchant who was active in some of John Brown's enterprises, championed independence, signed the Declaration of Independence for Rhode Island, and served ten terms as governor. His house is twice removed from its original location on South Main Street at the foot of Hopkins, most recently for the construction of the Providence County Courthouse across Hopkins Street. Now owned by the state of Rhode Island and managed by the National Society of the Colonial Dames of America, it stands as Norman Isham interpreted it in the 1920s for a house museum. Although it is remarkably intact inside, the front entrance is Isham's, based on equivalent examples from the 1740s. Alden Hopkins, a descendent, who designed gardens for the early phases of restoration at Colonial Williamsburg, produced the formal, terraced garden, characteristic of the eighteenth century, but not of Providence.

A diminutive thing in a diminutive garden, the house seems somehow unreal in its idealization and slightly aloft on its terrace, as though wafted on a magic carpet out of a Colonial Revival dream.

Benefit Street, South

On and off Benefit from George Street south to Williams Street stands a remarkable ensemble of nineteenth-century buildings, especially of the Federal period, which deserve attention, although few can be individually discussed. Farther south, Benefit and its side streets—Williams, John, James, Arnold, Transit, and Sheldon—are an equally outstanding architectural district in which two-and-one-half-story Federal dwellings predominate. Formally, their chief variations lie in entrances and roof forms. Sociologically, they tend to become smaller and plainer toward the bottom of the hill, indicating a shift in original ownership from business and professional men to skilled artisans. A few later buildings filled vacant lots, but the

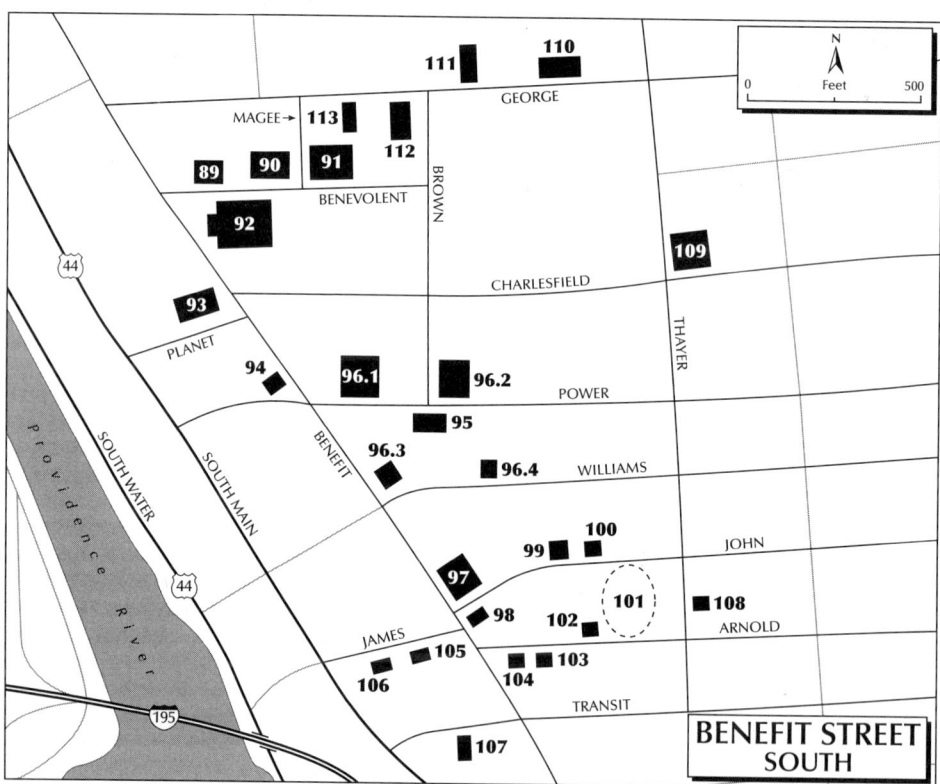

PR92 First Unitarian Church (First Congregational Church), interior and exterior (photo 1937)

area has remarkably consistent and little altered street views. The choice of John and Arnold Streets as the route to leave Benefit Street is almost arbitrary, and so are the few houses featured here. The entire district is worth exploring.

PR89 **Hope Club**

1885, Gould and Angell. 1911, 1912. 6 Benevolent St.

The Hope Club's mix of Queen Anne, Richardsonian Romanesque, and Scottish Baronial styles illustrates the preference for domestic forms in club architecture. The shift in size without a shift in scale both fragments the whole and enhances its appeal.

PR90 **Candace Allen House**

1819, John Holden Greene. c. 1860. 1996, interior restoration. 12 Benevolent St.

PR91 **Brown University Faculty Club** (Zachariah Allen House)

1864, Alfred Stone. 1980, dining room wing, Ira Rakatansky. 1 Megee St. (corner of Benevolent St.)

The Candace Allen House, in the Federal style and similar in form to Greene's Beckwith and Cooke houses, but with detached stable, has unusual triple-hung sash on the first story and recently restored roof balustrades. Inside (not open to the public) the first-floor double parlors, each with its fireplace, are little altered, and the stair hall retains a Venetian glass chandelier and encaustic tile floor from the 1860 remodeling executed for Candace's brother Crawford, who inherited the house upon her death. Across Megee Street is the Brown University Faculty Club, once the Italianate palace of Miss Allen's brother Zachariah, the prominent textile industrialist of the early nineteenth century (see Allendale under North Providence and Georgiaville under Smithfield).

PR92 **First Unitarian Church** (First Congregational Church)

1815-1816, John Holden Greene. 1877–1878, Parish House, Stone and Carpenter. Restoration, 1966, Irving B. Haynes and Associates. 2001–2002, Parish House addition, Centerbrook with James Barnes Architects. 1 Benevolent St. (corner of Benefit St.)

Caleb Ormsbee's two-towered church of 1795 on this site burned in 1814, necessitating this grand replacement. A member of the congregation, Greene considered this to be his finest work. Certainly it is his most ambitious. Sited at an angle well above the street, which shows it to advantage, this monumental, almost square, granite building is dominated by a 200-foot tower and spire above a colossal front of engaged Doric columns and broken pediment. A large, round-arched window inset with lancet tracery arches cuts into the pediment somewhat awkwardly, while similar, narrower versions continue around the other sides. In these touches, the church represents a stylistic retreat from the more marked version of early Gothic Revival that characterizes Greene's slightly earlier St. John's. It follows the Wren-Gibbs form of the First Baptist Church, although, at the congregation's request, its design looks specifically to that of Bulfinch's then recently completed New South Church (demolished 1868) in Boston. However, Greene's handling of the facade is both original and more robust than Bulfinch's (notwithstanding the small scale and fussy detail of much of its ornamentation), while its steeple is more emphatically vertical. This church and the First Baptist Church boast the finest early church interiors in Providence. First Unitarian, like St. John's, features a plaster saucer dome, here adorned with cofferlike panels. Ornamental fans fill the corners of the ceiling. English Adamesque ornament of the preceding generation provided Greene's inspiration, via Bulfinch's work in Boston and probably through plates from Asher Benjamin, who publicized such ornament in a series of pattern books published in the United States beginning in 1797. As in St. John's, too, the dome is again supported on four freestanding columns, here Corinthian. The saucer concavity relieves the majestically scaled cubic quality of the luminous space, while the curvature of the balcony also counters its boxiness. The balcony catches two of the columns and skirts the other pair, leaving them freestanding to either side of the curved mahogany pulpit, also designed by Greene.

Irving Haynes's meticulous restoration followed a disastrous fire in 1966 after lightning struck the steeple. The curved prow at the center of the balcony is his. In his belief, it was part of Greene's design but was frustrated by an inadequate cantilever construction to support it. Haynes designed to his predecessor's presumed intention and (intended or not), improved on the flat bulge that was there.

Of all his works, Greene was proudest of this church. He repeated it in an enlarged version for the First Presbyterian Church in Savannah (1819)—one of many interchanges between centers for cotton growing in the Carolinas and Georgia and cotton manufacture in Rhode Island. Its fate—destruction by fire and rebuilding in the nineteenth century—prophesied what was to happen at Providence.

PR93 General Ambrose Burnside House

1866–1867, Alfred Stone. 314 Benefit St. (corner of Planet St.)

PR94 Mrs. Edward Brooks Hall House

1866, Alfred Stone. 336 Benefit St.

Here are variant contemporaneous solutions in different styles by the same architect for nearly identical corner sites a block apart. Ambrose Burnside's eclectic brick mansard-roofed house in the Second Empire manner is the showier and more ambitious of the two architecturally and programmatically. Stone dared to locate the entrance of the house at its apex, in a curved bay, sweeping an entrance porch around it and out over the steeply sloping corner, while trailing the service wing down the hill along Planet Street. It was a conspicuous house for Rhode Island's contribution to the general staff of the Union Army, who continues to sit (in bronze, on his horse) in Kennedy Plaza. A rifle manufacturer, he returned to civilian life as an active politician, first as governor, then as United States senator.

The restrained Hall House would probably have been designated as Modern Gothic at the time. It was built for the widow of a beloved pastor of the Unitarian Church with funds contributed by grateful parishioners. Clipped on a diagonal at the corner of its ground story, the upper story extends on bracketing to a full right angle, while downhill the mass nicely adjusts to the street diagonal. The inset door with its second-story bay poking into a corner of the front gable is delightfully quirky.

PR95 President's House, Brown University (Rush Sturges House)

1922, William T. Aldrich; garden, Beatrix Farrand. 55 Power St. (facing Brown St.)

The Sturges House is early-twentieth-century Georgian Revival, recalling its large neighbors in scale and material but, compared to their simple blocks, more complexly organized than they and revealing a more worldly knowledge of British Renaissance sources, albeit with some loss of their compelling effect. The horizontal cornice cutting across the elevation above the first two floors gives a decided "attic" aspect to the top story. As in much large-scale, early twentieth-century Colonial Revival design, it also gives the house a spreading effect, more typical for Middle Atlantic or southern colonial examples than for the high compactness characteristic of New England, but responding even more perhaps to the genial expansiveness of twentieth-century suburbanism. The horizontal aspect of the lowered cornice also seems sympathetic to the interior plan of the Sturges House. This reflects the challenge at the beginning of the twentieth century to the traditional colonial entrance hall as a central axis through the house front to rear from the typical French Renaissance variant of a lateral hall (as here) running the width of the house immediately behind the entrance, its French origin reflecting the popularity of American professional study at the Ecole des Beaux-Arts at the time. The lateral hall fronts a suite of rooms ranged side by side immediately behind it, *en filade* (in file, or in a line), each room opening to the hall and, laterally, into the next, and all opening (ideally through French doors) onto a terrace and garden stretching across the rear of the house. For Sturgis, Beatrix Farrand, a leading garden designer of the period, established the original scheme, some of its basic structure still remaining.

PR96 China Trade Mansions

Up the hill on the east side of Benefit Street is a remarkable and distinguished group of four large houses built in a cluster by rich China Trade merchants during the heyday of Providence's sailing ships, before manufacturing became the source of huge fortunes. It is one of only a few similar clusters of houses of this age, scale, quality, and integrity that remain in such close proximity in this country. All stand on ample, landscaped lots, their main blocks three-storied, cubic in massing, covered with hipped roofs, with later attached ells and outbuildings.

PR96.1 Rhode Island Historical Society (John Brown House)

1786–1788 and later, attributed to Joseph Brown. 1901, renovation, Stone, Carpenter and Willson. 52 Power St. (open to the public)

The John Brown House, earliest of the group, is first in line. It is splendidly sited well up the

PR96.1 Rhode Island Historical Society (John Brown House)

hill on Power Street with an almost full block of lawn sweeping up to it from Benefit Street. (The sweep is the more impressive because of the early-twentieth-century demolition of another mid-nineteenth-century family house, allegedly by Richard Upjohn, which once occupied the northwest corner of the site.) Brown's brother Joseph, who died before the beginning of construction, probably had a hand in the building's design (although, unlike other buildings associated with him, this one has no close association with the pattern books which he is known to have owned). The scale of the house inside and out, together with its interior elaboration, set new standards for Providence when it was built and today mark it as among the grand eighteenth-century American houses. John Quincy Adams called it "the most magnificent and elegant private house I have ever seen on this continent." Stylistically it is somewhat conservative, although with innovative features, at least for Rhode Island. Its dominating three-story brick block has projecting brick courses marking off tiers of six-over-six sash windows, the topmost diminished in size hard against the modillioned cornice, over which its hipped roof is balustered and decked. Exceptional for Providence at the time, although previously familiar elsewhere, was its projecting and pedimented center block. Never widely used in Rhode Island, it is timidly employed here. Indeed, the projecting center of University Hall at Brown University (1770), with which Joseph Brown was also associated, is the first known example of this feature in

Rhode Island. Like the upstairs Palladian window under an inset brick relieving arch, the one-story Doric porch, especially impressive in stone, became popular only in the first three decades of the nineteenth century. The secondary entrance, on the west side and now closed permanently, was also both original and unusual for the time, although its reverberation in Victorian mansions on College Hill would be considerable.

The interior is based on a conventional central hall plan flanked by four rooms reached by double doors from the hall and more double doors connecting front and rear chambers on the east side. Four chimneys against the exterior side walls provide each of the principal rooms with its fireplace, their designs derived mostly from William Salmon's *Palladio Londinensis*. On entering, one immediately senses that the scale is grand, the embellishment lavish. The use of pilasters and of entablatures with full pediments over the hall doors makes a parade of academic correctness which was more fashionable for mid-eighteenth-century grandeur than for the 1780s. Unusual Ionic pilasters with shelves for decorative busts flank the entrance hall and are repeated to either side on the arched door which connects the principal front parlor to its rear counterpart. The stair, rising in a long flight from the back of the hall to a landing and a shorter return flight above, is railed in natural mahogany with elaborately turned balusters and a ram's-horn curl as a newel, which is echoed in another grand ram's-horn spiral against the wall as the terminus of the inside railing. Low, paneled chair rails run around all rooms. Otherwise the walls are plastered. Their brocaded wall coverings were discarded in the late 1960s when remnants of the original papers were discovered and reproduced. Those expecting Williamsburg discreetness in wall treatment as the hallmark of "Colonial propriety and good taste" will be shocked by the intensity of color and pattern which is the vibrant result. The rear ell reached its present extent after successive additions during the mid- and late nineteenth century.

Upon purchasing the house from collateral Brown descendants, Marsden J. Perry, self-made multimillionaire, an executive of the local electric company and of its principal trolley system, and president of a bank as well, moved his by then truly impressive collection of eighteenth-century furniture from the Ward House at the corner of Benefit and George, where he had previously lived, to these much grander premises. Grandeur was called for, since Perry's taste in furnishings inclined toward the most monumental and lavishly ornamented Chippendale. As he had earlier commissioned Edmund Willson's firm for the renovation of the Ward House, and even as the firm worked on Perry's bank, Willson came to make Colonial Revival alterations to the John Brown House and to design a lavish brick carriage house with clock cupola, which still stands, converted to residences, below the John Brown House diagonally across the corner of Benefit and Power streets.

Willson's most telling changes to the John Brown House are a basement library, new plaster ceiling ornamentation throughout, woodwork for the northwest room on the main floor (the only one which had lost its original woodwork), improvements in the service area, and bathrooms—including one in white tile with Bougereau nudes in grisaille. John Nicholas Brown (1900–1979), John Brown's great-great-grandnephew, bought the house from Perry's estate and donated it to the Rhode Island Historical Society in 1941. It now contains a display of extraordinary furniture, including many Rhode Island pieces—not Perry's collection, which was sold after his death, but possessing a splendor of which Perry would have approved.

PR96.2 Thomas Poynton Ives House

1806, Caleb Ormsbee. 1882–1883, exterior renovation, Stone, Carpenter and Willson. 1885–1890, garden, Frederick Law Olmsted. 66 Power St.

Up the hill at the corner of Brown Street is the Thomas Poynton Ives House, built for a niece of John Brown and her husband, who early became a partner in Brown family enterprises. It uses the same exterior format, without the projecting center; but walls and detailing are in the flatter and more delicate Federal style. Although the barn at the rear is contemporary with the main block of the house, the rear ell was added in the early 1880s and decorated by a porch and screening from Edmund Willson's design. At the same time he added the very fine semicircular porch at the front of the house (as he had redone the porch for Marsden Perry downhill), which immediately became *the* Providence model for a plethora of Colonial Revival variations, none as handsome as this. Willson also raised and extended a one-story semicircular bay on the garden side the full height of the elevation. Its walled Olmsted garden remains substantially intact. Inside (not open to the public), its Federal-period rooms, with some Neo-Federal remodeling, are interrupted by a library installed in the 1870s by the New York firm of Marcotte, a leading Victorian furniture and interior designer.

PR96.3 John Nicholas Brown Center for the Study of American Culture (Joseph Nightingale–Nicholas Brown House)

1791, Caleb Ormsbee. 1853, stable, Thomas A. Tefft. 1862–1864, library, Richard Upjohn. 1890, garden, Frederick Law Olmsted. c. 1923, renovation, Jackson, Robertson and Adams. 1987–1991, restoration, Irving

PR96.3 John Nicholas Brown Center for the Study of American Culture (Joseph Nightingale–Nicholas Brown House)

B. Haynes and Associates. 357 Benefit St. (open to the public by appointment)

Joseph Nightingale and his brother-in-law and partner, John Innis Clarke, built nearly identical houses beside one another on Benefit Street; Clarke's, one block south, burned in 1849. The Nightingale-Brown House is the largest extant wood-frame eighteenth-century house in the country. It is elaborately decked out in bold quoins and lintels, heavy turned balusters, and gable-on-hip roof, all retardataire for 1790. Such could be expected of Caleb Ormsbee, who was more a builder than a designer, but seems to have mustered all his resources to give Nightingale and his partner the images they wanted.

Nicholas Brown acquired the house in 1814, and his descendants remodeled and enlarged it in several campaigns. John Carter Brown (1797–1874) went to Tefft in 1853 for a brick barn and bowling alley in his round-arched style. He also commissioned an addition on the south side in 1858 and, finally, called on Upjohn in 1862–1864 for a library on the northeast to house his collection of Americana. Olmsted redid the grounds in 1890. John Nicholas Brown (1900–1979) oversaw the extensive remodeling of the interior by Jackson, Robertson and Adams in the 1920s. They left the center hall intact with its scenic wallpaper but changed the configuration of the northeast parlor and the southwest dining room. They also reworked much of the detail: the pedimented doorways and stair balusters in the hall are theirs (with inspiration from the interiors of the John Brown House); so is the refacing of the walls of the dining room with imported 1720s English paneling—stripped of paint, as was then stylish. Finally, in the late 1980s, it was adapted to its new use as a study center honoring the memory of John Nicholas Brown. The restoration, by Irving B. Haynes and Associates, involved extensive repairs to stabilize what was by then a seriously endangered fabric—in part because of natural causes, but as much because of the cumulative effect on the structure of piecemeal construction.

PR96.4 John Corliss–Edward Carrington House

1810. 1812, renovation and porch. 66 Williams St.

Finally, left into Williams Street, for the remaining China trade mansion: the Corliss-Carrington House, which stands immediately behind the Nightingale-Brown House. Built in 1810 for John Corliss and originally two stories high, the house was purchased as early as 1812 by Edward Carrington, who added the third story, kitchen ell, and barn. Most conspicuously, he provided the distinctive four-bay, two-tier porch, with its unusual balustrading of circles and compasslike stars, as the centerpiece for the front elevation. Southern in flavor, it is another reminder of connections between Rhode Island and the plantation South. A small wing on the building's northeast corner served as Carrington's office. It communicated with the house but segregated business callers from family and guests. The interior (not open to the public) is little altered and retains most of its original Dufour and Chinese wallpapers installed by Carrington. Like all the so-called China Trade houses, this still has its stables and outbuildings intact, itself a remarkable phenomenon.

PR97 **Brown University Offices** (Thomas F. Hoppin House)

1852–1855, Alpheus C. Morse. Early 1980s, conversion to offices, Robinson Green Beretta. 383 Benefit St. (corner of John St.)

PR98 **Tully D. Bowen House**

1850–1851, Thomas A. Tefft. 389 Benefit St. (corner of John St.)

On Benefit Street at the corner of John are two important Italianate houses: the Hoppin House, on the northeast corner, and the Bowen House (see next entry) on the southeast. Alpheus Morse's showier Hoppin House, fronted by a generous lawn which was once enclosed in a brownstone fence, looks to English examples, specifically to the garden facade of London's Travellers Club (1829, Sir Charles Barry), an early example of the Italian palazzo revival and one that popularized the form in the Anglo-American world. Morse adapted this elevation for his own east elevation, seemingly the rear, although it contains the carriage entrance to the house through a triple-arched center porch inserted between two projections too slight to be properly termed "wings." Beyond the carriage turnaround, the attached stable block projects as an ell from the main house, its court reached through an arched opening. A second major entrance for those arriving by foot, centered in the side elevation (the John Street front), gives entry to a hallway which runs the length of the house, making a *T* with the corridor from the carriage approach. The entrance to the restored drawing room, with its arched triplet window overlooking Benefit Street, occurs at the junction of the *T*. Although such elaborate planning for city houses is rare, it indicates that carriage entrances for the grandest houses were, by the early Victorian period, complicating what had, from the colonial period onward, been the conventional plan of a central hall with flanking parlors. In keeping with its approach, the Hoppin House, like the John Brown House before it, established a new standard of sophistication for the large houses of its day, and henceforth we shall observe variations on this plan in a number of ambitious Victorian mansions on College Hill. Despite conversion to offices, much interior woodwork remains, while semi-restoration of the principal parlor preserves a sense of the mansion's onetime splendor.

Prominent both socially and in the art culture of the day, Thomas Hoppin was himself a respectable painter and sculptor. While the Hoppins owned the house, a life-size bronze of a Labrador adorned the front lawn; it recalled the family dog whose barking saved Mrs. Hoppin's life when her family's home on this site was destroyed by fire in 1849. The statue eventually went to Roger Williams Park, where the hero of the disaster today greets visitors near the zoo's menagerie. The social éclat of the Hoppin palace attained a climax after he left it, with a reception and dinner for President Rutherford B. Hayes, after which it came to be known locally as the House of a Thousand Candles.

Nephews of the original owner became architects. Howard Hoppin remained in Providence as a principal in the firm of Hoppin and Ely (later Hoppin, Read and Hoppin); his more famous brother, Francis L. V. Hoppin, left for New York, where Hoppin and Koen became one of the leading academic firms in the city around the turn of the twentieth century.

Thomas Tefft's three-bay brownstone house for Tully Bowen, a cotton manufacturer (now converted to apartments), is less ambitiously composed, but is architecturally an even finer example of the Italian palazzo type. Henry-Russell Hitchcock, in his *Rhode Island Architecture* (1939), suggested that this might be the finest Italianate design in the state. It displays the full vocabulary in a compact, concise com-

PR96.4 John Corliss–Edward Carrington House

PR98 Tully D. Bowen House

position which is quite grandly yet personally proportioned. The recessed arched entrance, surrounded by a flat-headed, Doric-pilastered frame, is flanked by pedimented first-floor windows resting on brackets with the alternation of flat and pedimented heads at the second story, the windows at both levels resting on projections from the wall. Squarish, segmental-arched windows under a modillion cornice close the composition at the top, while quoining frames the corners. Taut restraint with unobtrusive variety is the essence here, where the house maintains its near cubic intensity by relegating service activities to a brick ell off the rear. The iron gate, too, is Tefft's design, one of many such included among his drawings at Brown University. Both Morse and Tefft, especially the latter, magnify the impact of their palazzi by elevating their sites as platforms raised above the sidewalk.

PR99 Moses Lippitt–Cornelia Burges Green House

1805; 1857; 1865, Clifton A. Hall. 14 John St.

PR100 John D. Jones House

1844. 16 John St.

Turn into John Street for yet another Italianate house which, because of its elevated siting, sweeping double stair to the entrance, and large scale, commands the lower end of the street. The Lippitt-Green House began as a two-story Federal house. It was enlarged to three stories and remodeled at mid-century. Both interior (not open to the public) and exterior details show how well the more robust versions of Providence Federal forms marry with those of the Italianate.

PR101 Double Houses

1820. 20–22 and 25–27 John St.; 26–28 Arnold St. (all corner of Roome Ln.)

Facing Federal duplex houses at 20–22 and 25–27 John Street back up on yet a third identical duplex on the next street south, 26–28 Arnold Street. This is an early development linked by an alley, Neighbors Lane, for access to the yards. The main house blocks are hip-roofed with a monitor (except that the Arnold Street example is gabled with a trap door dormer); small service wings with their own hips project at right angles off the rear corners, giving each duplex a U-shaped configuration with a utility court between.

PR102 Josiah Baker House

c. 1800. 23 Arnold St.

The two-story, five-bay Baker House is unique in Providence for an extant Federal house of its small scale in that its petite ground-floor windows, barely projecting from the plane of their brick wall, are pedimented over crossetted architraves in very elegant detailing, which the pedimented and fanlighted entrance completes. Its construction is also unusual for its day. The wall is one brick thick ("brickplated," in the terminology of its day, or "brick veneered" in ours) against boards supported on a post-and-beam frame. The delicacy of the woodwork outside is matched by that inside (not open to the public). Evidence of a fire inside suggests that it may be a little later than the exterior (possibly c. 1825).

PR103 Menzies Sweet House

1850, Thomas A. Tefft. 12 Arnold St.

Diagonally opposite the Josiah Baker House is the three-story, three-bay Sweet House, the home of a builder responsible for other houses in the area. This demure version of the Italianate palazzo in wood, flushboarded and quoined in front and clapboarded on all other sides, which conveniently appears grown up, in stone, around the corner in the Tully Bowen House (PR98). Its windows crowd the tiny front. Again handsomely detailed, they are more original than their Bowen counterparts. The window itself is framed in an Italianate molding, with sills and pediments projecting on brackets below and above, the pediments lifted as independent entities off their frames.

PR104 William and George Bucklin Houses

c. 1824, John Holden Greene. 8–10 Arnold St.

PR105 William Smith House

1824–1828, attributed to John Holden Greene. 18 James St.

PR106 William Woodward, Jr., House

1828, attributed to John Holden Greene. 22 James St.

Incongruously abutting the Menzies Sweet House is a duplex built for the Bucklin brothers, one a barber, the other a grocer. It demonstrates how elegant Greene can be even in his most modest work. The projecting entablature and door hoods on brackets shelter transoms with another bit of Greene "Gothick" in the pointed arches of its leaded ornamentation.

In the immediate vicinity, on James Street, are two quite similar Federal brick houses on granite bases with fanlighted doors and monitors. They are either by Greene or excellent examples in his manner. The delicacy and sophistication of the woodwork in number 22 especially suggest that Greene must have done it for William Woodward, the grocer who commissioned it; whereas William Smith, the carpenter next door, may have been able to proceed more on his own—and possibly built both houses.

PR107 Daniel Pierce House

1781, c. 1850. 53 Transit St.

Here it is not the little eighteenth-century house itself that counts so much as its mid-nineteenth-century enlargement—vertically, into a tall gable. Reminiscent of a twentieth-century A-frame, the type was then known as a "lightning splitter." Several other examples of this rare mid-nineteenth-century vernacular type exist in the state (see the Bicknell-Armington House in East Providence [EP24]).

PR108 John Holden Greene House I

1806, John Holden Greene. 33 Thayer St.

The first of two houses which Greene built for himself, this Federal house is exceptional as a mini-palace amid simple, much smaller gabled examples. The mass of the house is not as Greene built it; in the late nineteenth century the original five-bay front was extended by one and the third story was added in order to fit a door and stair to a later attic apartment. The house is unusual for its all-over carpentered rustication and for its delicately detailed entrance with an elliptical fanlight and exceptional paneling on either side where narrow windows are customary. (The same entrance reappears next door at number 29.)

PR109 Watson Institute for International Studies, Brown University

2000–2002, Rafael Viñoly. 111 Thayer St.

Built to house an interdisciplinary center for foreign studies and international relations established in the 1980s, the Watson Institute's center brings together offices, classrooms, and resources previously spread across the Brown campus. The selection of a prominent architect of considerable talent and international repute (for the first time since the selection of Philip Johnson to design the Albert and Vera List Art Building thirty years earlier) was deemed necessary because of the international scope of the institute's work and, no doubt, a growing awareness of (if not actual embarrassment about) the mediocrity of the university's recent buildings. Stretched across the full width of a block behind Wriston Quadrangle, the U-plan complex comprises three components around an interior court: a linear three-story brick-clad building along Thayer Street and two glassy cubic pavilions that extend behind it on the north and south sides of the block. The brick section, which houses offices, classrooms, and meeting areas used for research, publications, and administration, is modulated on its exterior with recesses in the mass, notably at the entrance around a venerable old elm (carefully retained as part of this project), and regularly spaced projecting prismatic windows. The glazed pavilions, used for library and assembly spaces, are dazzling, especially at night. The light, open interior of the office block is especially compelling, with balconies around a three-story atrium the full width of the building yet connected midway for ease of access across it. Detail work and finishes are especially well handled. But how does it succeed? Overall, it is a welcome addition to a community that has very little of even moderate-quality modern design. The use of brick, a seeming attempt at contextualism, makes little sense when the context is no longer the College Hill community but the Brown University campus. The windows, which admittedly increased the openness of small offices, nevertheless come across as awkward from the outside. Ultimately, the building is more successful on the inside than out, a curious self-absorption for an outward-looking organization.

PR110 St. Stephen's Episcopal Church

1860–1862, Richard Upjohn. 1882–1883, chancel remodeling, Henry Vaughan. 1899–1900, upper-stage tower and steeple, Hoppin and Ely. 1900, Guild House, Martin and Hall. 1933, vestry alterations, Cram and Ferguson. 114 George St.

In a mix of coursed graystone with brownstone trim and a polychromatic slate roof, St. Stephen's sits sideways on its mid-block site, entrance at one end, its tower at the other, with a row of four gabled pointed windows connecting the two. Its interior format is one that Richard Upjohn used on other occasions for tight sites, where he placed a space for Sunday school and guild hall (here a chapel) parallel to the nave with a glazed screen separating the two. As in his earlier Grace Church, Upjohn employed a vigorous version of English Gothic.

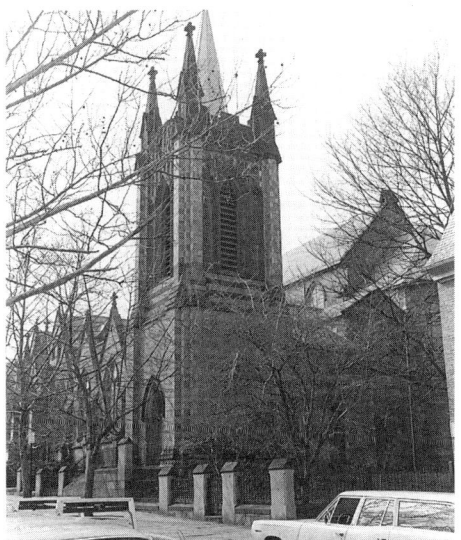

PR110 St. Stephen's Episcopal Church

The Reverend Henry Waterman, parish rector from 1850 to 1874, was the leading Tractarian of his time in the state. Adherents of this movement within the church believed that the true church was rooted in a high Catholic liturgy devolving from medieval precedent, minus the authority of the pope—a belief in accord with Upjohn's own inclinations. If Ralph Adams Cram had to be called in to deepen the original shallow chancel of Grace, there was no question but that the mass at St. Stephen's would be celebrated in a space of sufficient depth and remove from the congregation in order to intensify its ritualistic sanctity (although it too had eventually to be enlarged by eliminating a row of pews and by moving the rood screen out toward the congregation).

As at Grace, Henry Vaughan, a medieval specialist from a later generation and another English emigré, was called in to redo the more decorative aspects of Upjohn's work, or, as a church history puts it, "Vaughan represented a later and purer school of Gothic design than the one often called 'Victorian.'" In the chancel, the reredos, choir stalls, pulpit, and the lovely rood screen to further veil the sanctity of the ritual are all essentially Vaughan's, although his full program was not completed until a decade after the work began, with the placement of the Calvary atop the screen. Meanwhile, Vaughan moved the old altar into Upjohn's guild house and Sunday school, redesigning it toward the conversion of the area to a Lady Chapel, although this work seems mostly to have disappeared in a radical redesign, most recently in 1965–1966. At the same time, Vaughan apparently also designed stenciling for a polychromatic treatment of the nave, carried out in 1883 (an effect to which many hope the interior can return).

Although an engraving of the church records Upjohn's original intention for a 180-foot stone tower, this had to be much reduced. Later Hoppin and Ely lifted the upper stage of the tower and capped it with a copper-clad conical spire, partly because of the exhaustion of the quarry used for the stone, but doubtless mostly because of limited resources. A new Guild Hall, built almost simultaneously by another Providence firm, substantially increased the bulk of the building, compared to which Cram and Ferguson's still later add-on to the tower of a new sacristy and vestment room is nothing more than a closet addition. It must have required the intervention of John Nicholas Brown, a member of the congregation, with Cram, a close friend, to involve the firm in such a tiny renovation. Still, the architectural pilgrim will find satisfaction in the fact that all three of these eminent medievalists, representing three phases of the American Gothic Revival, came together in this enterprise.

Other accretions are worth notice, particularly the gift in 1933 of an ancient Swabian altarpiece, to the left of the main one, as well as the transformation of many of Upjohn's original windows in grisaille patterns to stained glass, as also happened at Grace Church. Notable among the later windows are a Tiffany window in the Lady Chapel and, best of all, a Feast of Cana in the narthex. Here, however, fully half of Upjohn's old windows remain, giving a far better notion of their effect than the minimal remnants remaining at Grace.

PR111 Martin Page–Horace Buffum–George W. Gardner House

1806, Daniel Hale. 1920s. 106 George St.

Adjacent to the church is a two-and-one-half-story brick house in the Greene manner. The five-bay house has end chimneys and a monitor which may originally have been screened by an ornamental parapet at the eaves. A mason who worked for John Holden Greene built the house on speculation and sold it to a mariner, who passed it on to his son-in-law partner. It ul-

PR112 Maddock Alumni Center, Brown University (William Giles Goddard–Hope Goddard Iselin House)

timately became the repository for a collection of American antiques owned by the George W. Gardners, who left the house and its contents to Brown University. It now serves as a guest house for distinguished visitors to the campus.

PR112 Maddock Alumni Center, Brown University (William Giles Goddard–Hope Goddard Iselin House)

c. 1830. 1882, enlargement, Stone, Carpenter and Willson. 30 Brown St. (corner of George St.) (open to the public only by permission of the university)

PR113 Nicholson House (Francis W. Goddard–Samuel C. Nicholson House)

1878, Stone and Carpenter. 1991, fire stair tower, Clifford M. Renshaw. 71 George St.

Two brothers from a long-established and wealthy Providence family once occupied these adjacent houses. From their mother, Charlotte Rhoda Ives Goddard, William (son of William Giles Goddard) inherited the three-story, late Federal house in red brick with graystone trim. Francis was given a frame house on the property next door, which the family also owned. About the same time each turned to the same architecture firm. William chose to enlarge his heirloom; Francis to demolish and start over.

The Goddard-Iselin House, now Brown University's Maddock Alumni Center, originally faced George Street with a central entrance. When Stone, Carpenter and Willson made a substantial addition to the south, they moved the original side- and fanlighted door with colonnettes so that it faced Brown Street at the juncture of the two pieces of the house, and fronted it with a shallow Roman Doric porch. The older, L-shaped section contains Federal fireplaces and early Victorian marble replacements. It is the later interiors, however, that are especially interesting. Filling in the L shape of the original house, the paneled hallway leads to an ornamented Neo-Colonial stair, backed by a fine opalescent stained-glass window. In the opposite direction are two well-restored Queen Anne interiors: the library, backed by a dining room, both with handsome fireplaces and fixtures, the latter with walls of pressed papier mâché to resemble embossed leather above paneled wainscoting and the usual built-in serving sideboard.

Next door, the mansarded Goddard-Nicholson House employs its polychromatic mix of stone, hard brick, slate, and wood to make a new image for the late-nineteenth-century house based on structural expression and the use of materials for their inherent textures and color, with medieval-inspired ornament. Among the finest examples in the city of the kind of Victorian Gothic associated with Ruskinian principles, the style is very rare on College Hill, where most preferred at the time to play it safe with the colonial and early national past by extending its format and ornamental forms to the Italianate palazzo. The Goddard-Nicholson House also contains interesting original fireplaces, woodwork, and appurtenances, but as scattered remnants rather than restoration. During most of the first half of the twentieth century Samuel C. Nicholson, president of the Nicholson File Company, lived here; hence the name Brown University has given to the house in its current use as administrative offices.

At 47–49 George Street (corner of Megee Street) is a fairly plain brick Italianate duplex, the Seth Adams, Jr., Houses, now a department building owned by Brown University (1852–1854, Richard Upjohn). At the time Upjohn was building a house for Adams (now demolished), and this was an investment property. A luxury duplex, it initially attracted an elite bachelor clientele.

Brown Center Campus and North

PR114 Brown University Center Campus

1770 and later. Front entrance to the campus on Prospect St. at the termination of College St.

Brown University's central campus is enclosed by a brick-piered, wrought iron fence (1900–1901; Hoppin and Ely and Hoppin and Koen) inspired by a similar fence enclosing Harvard Yard. Augustus Stout Van Wickle Gate (PR114.1), on axis with College Street, is popularly known as simply the "front" or "main" gate. The ornamental centerpiece opens only twice a year: inward, on the opening day of instruction, to receive the new freshman class: outward, at commencement, for the procession down College Street to baccalaureate ceremonies in the First Baptist Church. Through 1930, the fence enclosed the university's main campus and most of its buildings (except the then separate Pembroke campus for women). Established as Rhode Island College in 1764, the school was lured to Providence from Warren by Providence financial capital. The location of the campus here, on what was then an isolated hillcrest, symbolically set the school aloof from everyday activity but overlooking it. To this day the axial relationship of the elevated university to downtown Providence two blocks below is essential to the city's symbolic image.

Unlike most other institutions of its age, Brown University has added, and seldom replaced, buildings. Except for the demolition of the original President's House, its central campus retains all its original buildings. In line with College Street and Van Wickle Gate is University Hall (PR114.2; 1770 and later), which derives from Nassau Hall at Princeton University (1754–1756). It centers the row of buildings which delimits the Front Campus. Initially known as the College Edifice, it then housed the entire college: classrooms, library, office, refectory, and dormitory. James Manning, the college's first president, had been trained at Princeton, and it could have been his recollection of the founding building for his alma mater that inspired this more provincial version. Indeed, Robert Smith, the designer of Nassau Hall, may have visited Providence in conjunction with the planning of the College Edifice, or may have made sketches for it. In any event, the traditional attribution of its design to Joseph Brown, the local architectural luminary, is now generally rejected, although he was active on the building committee and could have contributed to the result. University Hall simplifies the long, four-story mass of Nassau Hall. Its projecting and pedimented center is considered to be the first such centerpiece on a Rhode Island building, in a state where

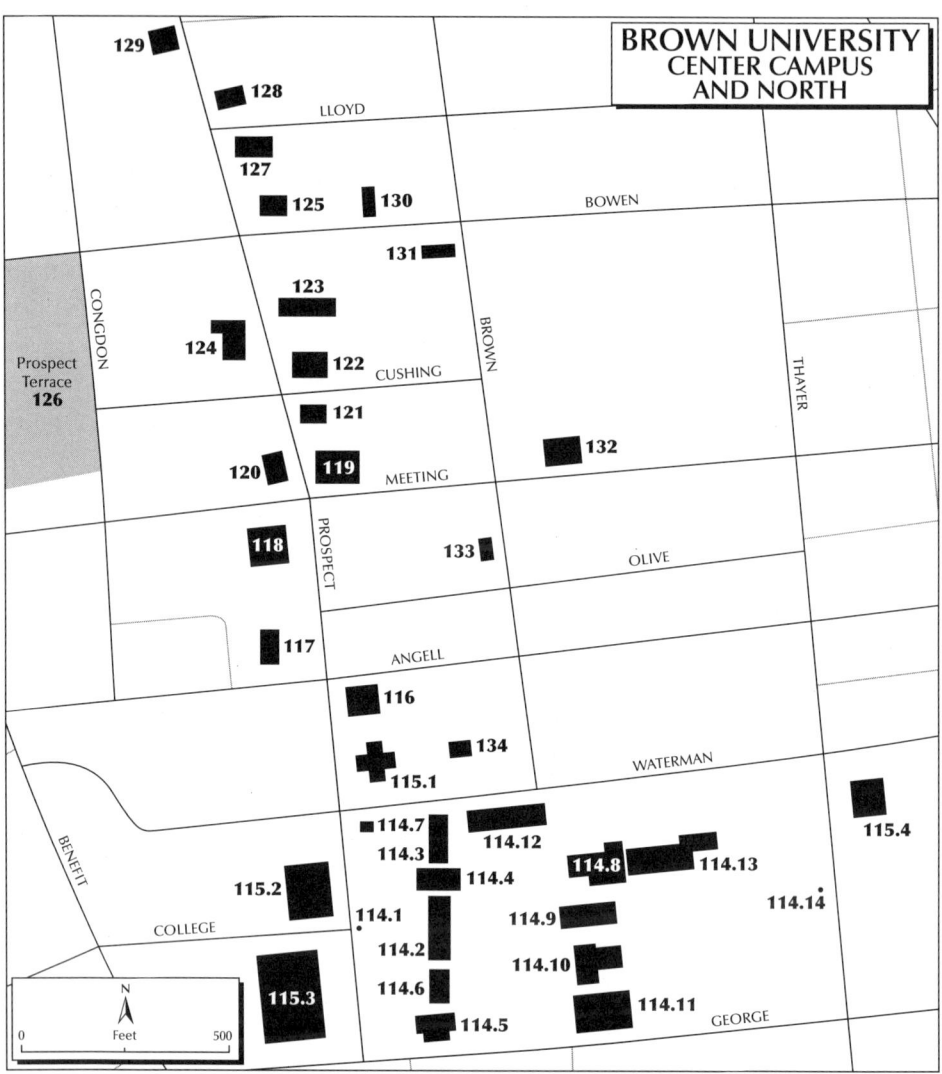

such treatment is rare through the Federal period, however unremarkable it may have been in some other states. Here the open bell cupola climax to its hipped roof, surrounded by a fretwork parapet, is far reduced from Princeton's, too much so for the size of the building, let alone for its position at the end of the College Street axis. In fact, the building is "monumental" more because of the extent of its wall and the repetitive accumulation of its openings, domestic in scale and functional in treatment, than because of any overt effort at monumental effect. That perhaps is the reason for its charm, proclaiming as it does the dignity meant for the future institution, while also acknowledging the college's simple, even penurious, beginnings—that and the lovely, soft, generously mortared brick, its textural quality sustained over the openings by the veiling of the small-paned sash over the narrow, brick-arched windows—six-over-six reduced to four-over-six for the topmost story. Look at the corner of the building and observe how the masons used the molding courses between floors to disguise the incremental thickening of the wall as it was called upon to sustain increasing loads toward the ground. The building has been reworked several times, most recently in 1939–1940

(Perry, Shaw and Hepburn), when the exterior was restored to its original brick surface following a period (from 1834) of Greek Revival and Victorian stucco and paint. At the same time, the repeatedly altered interior was completely rebuilt to its present state.

In the line of buildings flanking University Hall, next in date, to the extreme north, is Hope College (PR114.3; 1823), the first building serving wholly as a dormitory. This is a late Federal version of University Hall and an interesting comparison: soft brick to hard; smaller-scale elements to larger; more crafted appearance to a more rationalized aspect. The second Nicholas Brown financed the building, and it is named for his sister. The college had already (in 1804) changed its name in gratitude for Brown's benefactions. The campus pump still stands in front of the opposite elevation of Hope College.

Between these two is the thunderclap of Manning Hall (PR114.4; 1834, Warren, Tallman and Bucklin; 1959, upper floor restored to original chapel use, Perry, Shaw and Hepburn), where Greek Revival aspirations to monumentality explode with contempt for the overblown domesticity and vernacular quality of its predecessors. Stuffed between the brick blocks as the new institutional centerpiece, Manning originally housed the library downstairs and the chapel upstairs. By contrast with its neighbors, note especially the grandeur achieved in the bulk of the columns, the large scale of the coffering of the porch ceiling overhead, and the aggrandizement of the actual doors into a giant "portal" by a false paneled extension above. Beneath a stucco surface (the scoring to imitate granite blocks now barely visible), the body of the temple is rubblestone; the columns are brick. The upstairs chapel is a simple but forceful barrel-vaulted space, the plaster vaulting hung from wooden roof trusses overhead. Paint colors approximate the original treatment, although the invented fixtures and some other furnishings of the restoration seem out of key with the style, and certainly more pretentious than anything that was once there.

Next, to the extreme south, in a more vernacular Greek Revival, is Rhode Island Hall (PR114.5; 1840, James C. Bucklin), a cruciform Greek Revival classroom building employing the same stuccoed rubblestone construction. Finally, between University Hall and Rhode Island Hall is Slater Hall (PR114.6; 1879, Stone and Carpenter), named for its donor, Horatio N. Slater, a Victorian dormitory with some nice decorative brickwork and touches of naturalistic ornament, especially in capitals flanking the doorways, which are mindful of Ruskin's exhortations on the use of local flora and fauna as sources of inspiration. (The architects repeated this design for a building at the University of Maine at about the same time.) At the north end of the front campus is Carrie Tower (PR114.7; 1904, Guy Lowell), a clock tower in academic brick and limestone English Baroque, tapered to its cupola cap from rolled base molding with a flamboyance which epitomizes the care lavished by Beaux-Arts-trained architects on "profile."

East beyond University Hall is the College Green, the physical and symbolic heart of the campus. Part of its charm is the surprise with which one comes upon it. There is no special or formal preparation for the entry. One simply emerges between the buildings either side, close to, but not quite on the center of, the long east-west dimensions of the rectangular space. A row of mostly Victorian buildings closes the opposite (east) side of the Green, laid out in the straightforward manner of the front row of buildings, as the houses along a suburban street—building, gap, building, gap—but around a preordained space. Their accumulation finally defined the roomlike Green, in the piecemeal way in which American campuses have traditionally evolved. Much is achieved by the most unprepossessing means; in fact, the very simplicity of the resources that create the experience substantially accounts for our delight—this, and several aspects of the space itself. First, there is its central location within the university, resulting in the constant need to pass through it; then its position at the crest of College Hill and its sensed relation to the downtown core of the city. "Town" is precipitously downhill two blocks away; "gown" is up on the local acropolis, and the tension between dependency and separation is vividly evident.

For any "room," inside or out, physical dimensions and dispositions are important. Geometrically, this space roughly occupies two squares. For this reason it is inherently harmonious. Yet subtle asymmetries animate its rectangularity, apart from the off-center position of all points of entry into it. The principal cross axis of the space, both formally and symbolically, is that between University Hall in the front range of buildings and Sayles Hall (described below) in the back range. The bulkiest building of the lot, as the University's historic audito-

PR114.4 Manning Hall with Hope College (PR114.3), left, and University Hall (PR114.2), right

rium, Sayles projects forward from its row to emphasize the axis. But from the front range it is the other historic focus of congregation, Manning Hall, that thrusts beyond the alignment of its row into the space of the Green. Thus the principal cross axis is asserted, then challenged by the suggested alternative just off it (in a juxtaposition that would have been even more effective had Manning been double-porticoed, and its projection onto the Green not so plainly its behind). One final ingredient in the asymmetries which animate the space: it spreads slightly to the south, like double doors pushed beyond the width of their jambs, toward the open end of the Green into George Street (with the further complication of the alignment of the front walk with the rear building range, while the front range and the rear walk are splayed). So the space of the campus gently expands to that of the street in another confrontation of town and gown (even though the houses off the Green now mask university operations). In how many reinforcing ways has this space stumbled into its modest greatness, all the more satisfying for the prosaic and mostly happenstantial means by which it occurred!

Across the space, with its devastated rows of elms attempting a comeback, is the back range of Victorian buildings, beginning with the first in line from the north, and originally the smallest, Rogers Hall (PR114.8; 1862, Alpheus C. Morse). A fairly plain example of early Victorian polychromy, it limits the use of variegated materials to Gothic arching in the "Venetian" manner. It was the first building to house a single discipline (chemistry). Now gutted, it serves as a two-story vestibule to an attached auditorium and classroom building, the whole rechristened the Solomon Center for Teaching (1988, Goody, Clancy and Associates).

Next, Sayles Hall, the centerpiece of its row (PR114.9; 1878–1881, Alpheus C. Morse). Morse had designed Sayles in polychromatic brick, but switched to the tan and brown granite that came to characterize Richardsonian Romanesque. He did so, however, even as the fountainhead of this style, Richardson's Trinity Church (1873–1877) approached completion. So Sayles represents a surprisingly prompt and mature response to the innovative Boston church, despite the high, angular awkwardness of Sayles's silhouette and less finesse in detail. The interior, including a major central auditorium for the campus (and long the site of "daily chapel"), retains most of its original Victorian character. Given as a memorial by the textile magnate William Francis Sayles, whose son, William Clark Sayles, died while a student at Brown, the entrance vestibule contains flanking portraits by the prominent Victorian portrait painter George Peter Alexander Healy. The father, hearty and stalwart, sits before what appears to be a vague representation of the family's Pawtucket estate); the son, pale and brooding, is surrounded by books, implying that the business acumen of one generation permits the cultural and professional achievements of the next. Town and gown once more.

Completing the Victorian legacy on the Green is Wilson Hall (PR114.10; 1891, Gould and Angell), which shows more influence from Richardson, the absolute symmetry of Sayles here becoming somewhat asymmetrical, with more concern for the individualistic design of "features" and the "artistic" handling of materials consonant with contemporary art move-

PR114.11 John Carter Brown Library

ments). It served as the second departmental building, for physics, until renovated as a classroom building.

Concluding the row is the John Carter Brown Library (PR114.11; 1898–1904, Shepley, Rutan and Coolidge; 1988–1990, addition to rear, Hartman-Cox Architects). It houses Mr. Brown's Bibliotheca Americana, comprising publications and manuscripts about the exploration, settlement, and colonization of the Western Hemisphere dating from 1493 through approximately 1825, which were moved from the Upjohn-designed library at the Nightingale-Brown House to this "Germanic Ionic" temple (as it was then called). This label apparently derived from the extensive use of Greek precedent in German Renaissance Revival buildings during the nineteenth century, with a comparably clear and luxuriant, yet restrained, linear ornamentation derived from Greek precedent and set against the austerely shaped mass. The cornice cresting mixes classical palmettes and scallop shells with headdresses of Brazilian Indians in a hybrid Americanization of classical Greek ornament. The same exquisite elaboration and sense of profile appears in the beautiful bronze lamp standards flanking the entrance. The reading room, with its oriental carpets and Tiffany reading lamps, is rather more colonial–English Renaissance in effect than Grecian but is also restrained in overall effect. The use of Greek rather than Roman and Italian Renaissance elements is rarer in classical revival buildings of the early twentieth century than might be expected, and especially in Rhode Island. (Curious that the tremendous popularity of the Greek Revival in American architecture of the early nineteenth century finds minimal response in the early twentieth-century return of classicism, when Roman and Renaissance overwhelmingly prevail.) The addition to the rear extends stack and office space by continuing the exterior style of the original too obsequiously so some may think.

Finally, at the north end of the green, Faunce House (PR114.12; originally Rockefeller Hall; 1903, McKim, Mead and White; 1929–1930, addition to east, Clarke and Howe) repeats the compositional format of University Hall, but in the expansive—or, looking at the big central window, even explosive—manner of colonial and Federal detailing which typifies the Colonial Revival. McKim, Mead and White planned Rockefeller Hall, only the third student union building to be erected in the United States, as something between their gentlemen's clubs in New York and a YMCA meeting place. John D. Rockefeller, Jr., funded the 1930 addition and requested that the entire complex be renamed as a memorial to Brown president William H. P. Faunce. This antiquated idea of what a student union should be accounts for adjustments attempted in a series of interior renovations to meet changed student demands (the latest, c. 1988, by Goody, Clancy, and Associates). Then terrace steps were expanded to make a kind of bleacher or a stage from which to survey the central green. Only here, by exiting the door to the fast food, can one enter the green in a centered position.

The part of the campus behind the Victorian buildings on the Green is called Lincoln Field, about which the Olmsted Brothers consulted. There, the most interesting building, to the rear and north of Rogers Hall, is Lyman Gymnasium (PR114.13; 1890–1892, Stone, Carpenter and Willson; 1980s, renovated as Leeds Theater for drama and Ashamu Theater for dance). The first building on campus specifically designed as a gymnasium, it reflected the beginnings of intercollegiate athletics in American colleges, and, together with Faunce House, the initial stages of the institutionalization of student extracurricular life. It is also the finale to Richardsonian Romanesque on campus, delightfully reinterpreting the design of Richardson's Crane Library and Converse Library in Massachusetts, completed almost a decade earlier. The firm also submitted an alternate design maintaining the low, turreted form but substituting frilly Georgian trim; indeed, the Richardsonian format barely disguises the sharper, more prismatic handling of brick and stone and delicacy of detail which the

Colonial and Renaissance revivals (and this firm in particular) were bringing to Providence at the time. Its former swimming pool is now buried beneath the flooring of a dance studio. At the eastern end of this part of the campus, known as Lincoln Field, is Soldier's Memorial Gateway (PR114.14; 1921, Shepley, Rutan and Coolidge), a somewhat thin Roman triumphal arch dedicated to the Brown dead of World War I, with a fine replica of the famous Roman bronze of Marcus Aurelius at the opposite end of Lincoln Field to keep it company. (Reputedly the finest replica anywhere of the version on the Roman Campigdoglio, it was consulted when the Campigdoglio version, acid-eaten by Roman pollution, was retired to a museum and itself replaced by a replica.)

PR115 Brown University Library Buildings

In addition to a good portion of the stylistic history of collegiate building, the Brown central campus also offers a typological study of the development of library buildings. In the beginning there was a book room in University Hall, succeeded by the room below the chapel in Manning Hall (now a classroom). Next, in succession, are the three buildings in an L-plan cluster just outside the campus fence and a fourth of a block east on Waterman Street.

PR115.1 Robinson Hall (University Library)

1875–1878, Walker and Gould. 1989–1991, partial restoration, Frederick Love of Robinson Green Beretta. 64 Waterman St.

Of the four buildings, Walker and Gould's octagon in Ruskinian Gothic deserves the most attention as an exceptional library building—called simply the New Library while it functioned as such, then the Old Library (after the building of John Hay), until it eventually became Robinson Hall, for departmental use, in honor of Brown's seventh president, under whose administration it was built. In fact, it represents an unusual and especially progressive approach to library design for its day. The conception of the tall octagonal reading room at the center, with three radiating wings for three-story stacks, originated in the late 1840s, with the university's distinguished librarian Charles C. Jewett, who later became librarian for the Smithsonian, then for the Boston Public Library. He was captivated by the so-called panoptic concept for libraries, which envisioned the ideal shape for a library as one which provided for simultaneous supervision from the core of both the entrance from the street and the radiating stacks. The idea had developed in the late eighteenth century for such building types as prisons and hospitals. During European book-buying trips Jewett became enamored of the idea from a project (not built) for a library in Paris. In the late 1840s Jewett's advice led to the first panoptic college library in America, at Williams College, a modest octagonal affair (1846–1847) which is now incorporated into the college art gallery.

Although Jewett wanted the same thing for Brown, funding for its library had to wait for roughly a quarter of a century, by which time Jewett's successor and protégé called for the same sort of facility. By then, too, the type was already enshrined in several prominent examples, most notably in Sydney Smirke's circular library attached to the British Museum (1854–1857) and a derivative version for Ottawa's Houses of Parliament (1859–1867). Moreover, just before the start of construction of Brown's building, Princeton had completed its polygonal Chancellor Green Library (1873). Both Brown librarians, however, viewed the panoptic principle as more than a tool for supervision. At a time when college librarians generally conceived of their function as protecting the books *from* the students in locked closets, these librarians meant to stimulate the student with the spectacle of the books: the balconied stacks visibly at hand and easily accessible to the reading tables at the core. Moreover, the polygonal shape permitted an upper circle of windows for light above the reading space and ventilation from all sides, with air exhaust through the small trefoil windows (operated by pulleys) up in the conical cap of the reading chamber.

The building provides the most elaborate demonstration of Ruskinian Gothic on campus and one of the most interesting in the state. The style derived from Ruskin's enthrallment with the use of patterned brick and stone in much late Italian Gothic building. Brick and stone were to be used in their natural colors and textures as integral ornamentation which literally celebrated the diversity of nature's abundance and symbolically abstracted the gorgeous striations of rock, plumage of birds, and splotchings of leaves. Sculpted images of botanical, animal, and sometimes (but not here) human form elaborated on God's cornucopia. Here the local sculptor fitted brownstone plaques (some now badly weathered) into the

College Hill and East Side: Brown Center Campus and North 103

PR115.1 Robinson Hall (University Library), interior of dome (above) and exterior (below)

walls with images of local flora and fauna, extending the latter to the imaginary griffins on either side of the entrance porch, with its inevitable owl pressing a book under one wing. The polychromatics outside extend within, in stained glass and multicolored ornamented walls and metal balconies. Within the tall center space, these have recently been restored, although the wings for books are now partitioned for offices.

Next to Robinson Hall, at 68 Waterman Street, is a reduced Greek Revival temple which originally served as the "Cabinet" of the Rhode Island Historical Society and is now owned by Brown University (1844, James C. Bucklin; 1891, rear addition, Stone, Carpenter and Willson).

PR115.2 John Hay Library

1910, Shepley, Rutan and Coolidge. 20 Prospect St.

The white marble John Hay Library, now converted for rare books by the firm that originally designed it, recalls the baroque style current in the mother country at the time of the university's founding. In the history of library design, the Hay Library's metal stacks with floors of structural glass as a self-sustaining unit within a multistory space typify progressive book storage design at the beginning of the twentieth century. What was once the third-floor reading room now offers a substantial array of Ann S. K. Brown's collection of lead soldiers, from her larger collection devoted to the history of military costume. The lead soldiers extend back to Renaissance examples and represent armies worldwide.

PR115.3 John D. Rockefeller, Jr., Humanities Library

1962–1964, Warner, Burns, Toan and Lund. 10 Prospect St.

By the time of the seven-story Rockefeller Library (in campus lingo, "The Rock"), with three floors sunk into the slope of College Hill and four out of it, stacks were placed on warehouse floors for flexibility, which is denied by the metal and glass structure built into the Hay Library. The thin formality of the Rock's symmetry, with an approach on broad stairs to the equivalent of a classical stylobate, all treated as "floating slabs" in a manner which Ludwig Mies van der Rohe developed, proclaims it as typical of the 1960s revival of traditional neoclassical monumentality. There is here the same compromise between modernism and neoclassicism which also appears in Philip Johnson's List Art Building (PR85) across the street.

PR115.4 Sciences Library

1967–1971, Warner, Burns, Toan and Lund. 197 Thayer St.

For committed library enthusiasts, a short diversion a block east on Waterman takes in the

Sciences Library, where the architects of Rockefeller Library shifted from horizontal to vertical. They seem to have been more attracted to the glory that might accrue to a tower at the true summit of College Hill than concerned with the practicality of a stack of relatively small floors for an ever-expanding collection. The design derives from a series of contemporaneous buildings in which floors were bridged between corner towers containing stairs, elevators, and utility stacks, possibly the first and the most influential being a classroom and laboratory tower by I. M. Pei at Massachusetts Institute of Technology.

PR116 Admissions Office, Brown University (George Corliss House)

1878, George Corliss. 2000-2002, exterior and interior restoration, Durkee, Brown, Viveiros and Werenfels. 45 Prospect St.

George Corliss was internationally famous as an inventor and manufacturer of steam engines. The huge Corliss engine which powered the Centennial Exposition in Philadelphia was among the event's most popular and awesome attractions. In the mid-1870s, doctors informed Corliss that his young hypochondriacal second wife must winter in Bermuda. A busy man who thought he could ill afford the indolence of this remedy, Corliss replied, "I will *build* a Bermuda for Mrs. Corliss." This house was that Bermuda, as functionally advanced for its time as it was stylistically retardataire. A subterranean porte-cochere, thermostatically controlled radiant heating and cooling system, hydraulic elevator, direct tap in the dining room to an artesian well, and concealed sliding insect screens were all incorporated into this austere, towering Italianate villa, a rigid interpretation of a picturesque form introduced nearly forty years earlier. When it was built, another, similar Italianate villa existed opposite, creating a twin-towered approach to an important residential street, where now most of the survivors have been converted to professional offices. Some interior woodwork and fireplaces remain after conversion to offices.

PR117 Captain George Benson House

1794. 65 Angell St. (corner of Prospect St.)

Diagonally opposite the George Corliss House, this ample, two-story Federal house, built for the partner of John Brown and Thomas Poynton Ives in China Trade ventures, typifies the substantial wooden houses built in Providence in the 1790s: correct one-story entrance porch, balustraded hipped roof, and paired interior chimneys. Like his partners, George Benson chose to locate his new house on what was then an isolated, commanding site, overlooking the town, with distant views of its waterfront.

PR118 Woods–Gerry House

1860–1863, Richard Upjohn. 1931, entrance and interior reconfiguration; driveway redesign, Fletcher Steele. 62 Prospect St. (first-floor galleries open to the public)

Marshall Woods, who married into the Brown family and was active in the affairs of Brown University (eventually as treasurer), was also involved on the building committee for St. Stephen's Church—thus this Upjohn commission. The house, designed in a Neo-Renaissance style, is probably better known as the residence of Rhode Island's longtime United States senator Peter Goelet Gerry (1917–1929 and 1935–1947). The exterior of the three-story brick building is almost abstract in its reductivist geometry, but with a bowed centerpiece on its east elevation for the handsome new front entrance installed in 1931. Its appearance today somewhat belies the original configuration, with combined pedestrian and vehicular entrance centered on the south elevation. The central stair hall terminates on the north with a transverse stair rising to the east. A row of three rooms (now used for gallery space for student exhibitions), arranged en suite, lined the west side of the building, each communicating directly with the terrace to the west, perhaps the earliest example in Providence of siting a large house on an elevated site to exploit the view to the west. On the east side of the hall are a parlor on the south and the new pedestrian entrance, through a circular vestibule with saucer-dome ceiling; the effect here is of the T- or L-plan hallway system introduced during the mid-years of the nineteenth century, but here a retrofit rather than an original treatment. (Such complex circulation systems were more typically simplified in the mid-twentieth century, as in the James Kimball House [PR129].) The formal terraced garden, now informally planted, is far more wooded than originally intended.

PR118 Woods-Gerry House

portant early nineteenth-century cotton manufacturer and his sister, bears a strong resemblance in form and plan to an 1852 design by Thomas Tefft for Smith Owen. Hence the attribution, although the house was built three years after Tefft left Providence for a lengthy European sojourn, where he died in Florence in 1859 of typhoid. (Was the design adapted by a builder? If so, how did Almy belatedly come to build a design originally intended for Owen?) The mansard roof may be a later addition, but the combination of mansard roof and flushboard siding typifies early examples of the Second Empire. When Owen, a prominent jeweler, eventually did build his house, after Tefft's death, he turned to Morse. Like the Binney House, the Owen House has a rigidly symmetrical Italianate facade, whereas the massing becomes irregular toward the garden side.

PR119 Christian Science Church

1906–1913. Hoppin and Field. 71 Prospect St.

Based on sixteenth-century Italian Renaissance prototypes, this building follows the stylistic lead of the Mother Church at Huntington and Massachusetts avenues in Boston, but its giant interior space is too economically detailed to live up to the majestic promise outside. Strangers, viewing it from afar, often assume it to be the state capitol.

PR120 William Binney House

1859, Alpheus C. Morse. 72 Prospect St.

PR121 Sampson and Eliza Almy House

1852(?) (design), 1859, Thomas A. Tefft(?). 75 Prospect St.

PR122 Smith Owen House

1861, Alpheus C. Morse. 79 Prospect St.

Alpheus Morse excelled in combining the palazzo and villa forms into ample detached Italianate houses. In the William Binney House the symmetrical three-bay facade, restrained and elegant, holds the front; but the massing appropriately erodes picturesquely on the garden side. The formula proved especially useful for College Hill.

More than proximity relates the other two Italianate houses. The Almy House, for an im-

PR123 Machado House, Brown University (Ellen Dexter Sharpe House)

1912, Parker, Thomas and Rice. 1974–1975, dormitory wing, Steven Lerner. 84 Prospect St.

PR124 Rochambeau House, Brown University (Henry Dexter Sharpe House)

1928, Parker, Thomas and Rice. 87 Prospect St.

Built across the street from one another, for a sister and brother associated with the important Brown and Sharpe machine tool industry, these two houses—the earlier Neo-Tudor, the later of eighteenth-century French inspiration—illustrate the capabilities of Parker, Thomas and Rice in accommodating the eclectic desires of early twentieth-century patrons. Both houses have fine interiors, which are substantially retained on the ground floors (especially in the Henry Sharpe residence), albeit without the furniture that completed the ensemble. Ellen Dexter Sharpe's house, now a dormitory, has a large yet well-designed and well-sited wing gently added at the rear. Both properties included extensive gardens, Ellen Sharpe's now fragmentary, but, appropriately, Mary Elizabeth (Mrs. Henry) Sharpe's substantially intact. She was an avid gardener who made many garden and landscape benefactions to Providence and the state, including an active role in their design; among them were most of the landscaping for Brown University and most of the initial financing for India Point

Park on the Providence waterfront). A fund in her name endows planting of city street trees. She collected French art and furniture and appropriately left her house to the French Studies and Hispanic Studies departments at Brown. A meeting room retains some of her furniture, and the ground floor especially is sympathetically adapted, with fireplaces and stuccowork intact in major offices.

PR125 **RISD President's House** (Stephen O. Metcalf House)

1889–1891. Andrews, Jacques and Rantoul. 132 Bowen St. (corner of Prospect St.).

Of the three large houses in Rhode Island which most conspicuously announced the social "arrival" of Neo-Colonial as a style, only the Stephen Metcalf House survives; the others were McKim, Mead and White's H. A. C. Taylor House (1882–1886) in Newport and Stone, Carpenter and Willson's Henry J. Steere House (1886) in Barrington. It was also the first grand house in Providence thoroughly committed to the style, notwithstanding its un-colonial horizontality, which characterizes other houses at this phase of the Colonial Revival. In contrast to the pretentious quality of the other two houses, the relative plainness and low spread of this gambrel-roofed mass may have encouraged a relaxed, expansive version of the style even for luxurious houses. However, the bay windows on the ground floor, the aggressive spread of the double-columned porch, the width of the windows upstairs, and the assertiveness of the dormers are all far from "colonial." The magnificent terracing of the house, lifting it above its site, belies the first impression of its relative modesty. So does a circuit around the property, which discloses the extensive service areas extending from the conventional gambrel-roofed core. Whereas the first generation of Metcalfs to head the Wanskuck Mill had lived close to the factory, the second- generation head of the corporation removed his family to College Hill. As the family gave much to RISD, so they also made a gift of this house, which has served as the presidential mansion for RISD.

PR126 **Prospect Terrace**

1867, 1877, 1925–1929. 1936–1939, Ralph Walker of Voorhees, Gmelin and Walker with Leo Friedlander, sculptor. 70 Congdon St.

Given to the city in 1867, the park was improved with a fence and retaining wall a decade later. Through a series of transactions between 1925 and 1929 the park was enlarged to its present size. This spot provides a fine overview of the central city and especially of the State House. From the portal-cum-triumphal-arch the granite figure of Roger Williams, arrayed in New Deal cubism, steps out to commemorate the tricentennial of his founding of the city. Walker, reared in Providence, was one of five architects—all of whom were Rhode Islanders "by birth and heritage"—invited to submit proposals. His scheme also projected an unrealized terrace level below the park connected by an elaborate stair continuing down to the street below.

Congdon Street is worth the walk for a number of medium-sized Federal and Greek Revival houses, all enjoying in private the public view from Prospect Park.

PR127 **Henry A. Dike House**

1850–1852, Russell Warren(?); subsequent enlargement and renovations; 1889–1891, landscape renovation, office of Frederick Law Olmsted. 101 Prospect St. (corner of Lloyd Ave.)

Henry A. Dike, a shoe manufacturer, occupied this house only briefly. For Providence, it is an early example of the Italianate villa, as indicated by the way in which it incorporates the simple, abstract forms and large scale of Greek Revival elements into a new vocabulary of arches, heavy balconies and brackets—all shapes more attuned to primitivistic stonecutting than to wood. The shapes make sharp-edged layers down to the rusticated, flushboarded wall. Individual parts are also aggrandized into larger shapes: the square bays and deep-set door of the first story joined to the window triplets above by intervening balconies; the arching over the second-story windows merged into a strange scalloped shape; the cornice lifted into a center pediment with no distinction between the two. Whoever designed this elevation did so with sophistication, but also with a residual awkwardness not due to lack of capacity, but perhaps to a venture into new territory. Such occurred in Warren's late work, when he moved from his long familiarity with Greek Revival to embrace the new vogue for Italianate forms. Portions of the Olmsted landscaping remain, done for Albert Harkness, Brown classics professor and father of the identically named Providence architect. It subsequently became the residence of

Murray S. Danforth, whose wife, Helen Metcalf Danforth, continued benefactions to the Rhode Island School of Design as the third in the triumvirate which included her grandmother and her aunt.

PR128 William F. Sayles House

1878. Alpheus C. Morse. 103 Prospect St. (at Lloyd Ave.)

PR129 James M. Kimball House

1873, Alpheus C. Morse?. 108 Prospect St.

Numbers 103 and 108 are Second Empire mansarded houses, one certainly and one probably by Alpheus Morse, for two textile manufacturers. The relative clapboarded modesty and rather plain treatment of the William Sayles house in the company of other Prospect Street residences belies his powerful business position as head of one of the largest textile companies in the country, with a sizable family estate in Pawtucket, shared as an enclave with other members of the family. However, the grand scale of the house and its marble-paved terrace in front, as a platform for the entrance porch, magnify its impact. It was he who gave Sayles Hall to Brown to memorialize the death of his son while a student there.

Across the street, James Kimball, a banker as well as a textile industrialist, lived in a brick block with brownstone trim which exploded the popular East Side three-bay mansard type with tall openings into what may be the largest-scale house on the East Side. Scroll keystones, rosetted entablature, and bracketed eaves sparingly but tellingly embellish the trim. The side carriage entrance with porte-cochere (and a two-story brick carriage house behind) originally signaled another of those Victorian L-shaped corridor plans. The front entrance, however, has been replaced by a frankly modern bow window, and the onetime corridor space behind it merged with its flanking parlors to create one large room across the front. Fences in front of both mansions remind us of the importance of these mostly vanished Victorian features, and so does the effect of the meticulous maintenance of the Kimball landscaping.

PR130 Stephen A. Cook, Jr., House

1889, Edward I. Nickerson. 158 Bowen St.

PR131 John D. Lewis House

1891. 134 Brown St. (at Bowen St.)

Two large houses, 158 Bowen Street, built for a partner in a leading law firm, and 134 Brown Street, are first in a cluster of five Queen Anne buildings. Both are horizontally divided into a red brick ground floor with red slate above. Although English precedents for such houses typically have slated second stories, American derivations are overwhelmingly wood shingled. To find two slate-covered examples of such quality (and, nearby, a third) is exceptional; to find both slate walls and roofs so splendidly preserved is doubly so. In Edward Nickerson's design, cross gables and dormers emerge from a tall, central, hip-roofed core, their verticality countered by a series of horizontals, whether bands of wooden panels or groups of windows. Ornament appears as paneled accent here and there. The porch is asymmetrically tucked under the big slate-sheathed volume above, while a variety of differently shaped bay windows protrude from the side wall. In contrast to the relative frontality of the Cook House, the Lewis House, on a corner, makes an in-facing *L* on its lot line, so that the Bowen Street wall is really its back—with a tantalizing glimpse of a large stained-glass stair window. In contrast, too, with the surface quality of Nickerson's articulation of his large massing with panels and ornament, in this house treatment of structure and mass is more severe.

PR132 Pembroke Hall, Brown University

1896–1897, Stone, Carpenter and Willson. 172 Meeting St.

Pembroke Hall was the initial building for Pembroke College, Brown University's adjunct campus for women, which fully joined the university in 1971. The college, founded in 1891, took its name from Pembroke College at Cambridge University, Roger Williams's alma mater. Pembroke Hall's version of Elizabethan-Jacobean red brick and terra-cotta became popular as a Queen Anne mode for progressive school design after the firm of E. R. Robertson and J. J. Stevenson turned to the style for a series of London school buildings in the early 1880s. The ranges of tall, transomed windows, natural to the style, were ideal for lighting classrooms. In addition to these, Pembroke Hall displays a gamut of other effects which characterize these schools, including a cavernous shelter up to multipaned doors which anticipate the amplitude of the entrance corridor inside and a tall, projecting stair bay fitted inside with window seats on the landings. Outsized gable windows illuminate the high slated and gabled attic, while (here especially well handled), the sculptured chimneys are each embossed with sunflowers, a favorite Queen Anne transliteration of the chrysanthemum disks on Japanese prints. Such features were meant to ameliorate institutional grimness. They also fit neighborhoods of bayed, gabled, and chimneyed houses, while asserting the school as the centered monument of its community. This Elizabethan-Jacobean mix was flexible as well, lending itself to as much symmetry or asymmetry, compactness or spread, plainness or adornment, as any imaginable combination of site, function, budget, and expression might dictate, while its approach to form mirrored in the nineteenth century an earlier shift in preference from medieval to Renaissance.

Like Brown's University Hall, Pembroke Hall, except for its lack of living arrangements, was designed to be multipurpose. It contained administrative offices, classrooms, reception rooms, and, in the attic, a still extant library which is rather cozier than other reading rooms on campus. In a second-floor meeting room (off the original library, and now a classroom) is a bronze plaster frieze depicting the multiple roles of educated women in modern society. All the allegorical figures are clad in flowing, liberating gowns, which also descend, in keeping with the Queen Anne style, from the somewhat overwrought medieval–early Renaissance inspiration of Pre-Raphaelitism into these ampler costumes of an idealized classical Renaissance.

PR133 Hillel House, Brown University (Froebel Hall)

1879, Stone and Carpenter. 80 Brown St. (at Olive St.)

Nearby is another architectural memento of progressive educational currents of the late nineteenth century. This cottagey ramble, with latticed gables front and rear butting a long wing with what was once a narrow, open-balconied porch, was a pioneer venture in Friedrich Froebel's kindergarten education. Only the second American school (after one in Boston) for training kindergarten teachers, it was specifically inspired by an exhibit at the 1876 Centennial that gave wide publicity to Froebel's emphasis on early education through systematic play. Only two years after the exposition, this whimsical school for kindergarten pedagogy, with facilities for an experimental class, opened here. The demure, intimate, and "artistic" aspects of much Queen Anne architecture made the style ideal for the purpose. (Either kindergarden rumpus or, more likely, subsequent stresses from the use of the building as a dancing school for the East Side elite required the timber bridging with tension rods which now pops through the roof to counteract the sagging floors.) The building now serves as the university's Hillel House. A substantial expansion is planned.

PR134 Edward Dexter–Charles Pendleton House

1795–1797. 1860, moved. c. 1925, ballroom. 72 Waterman St.

Commissioned by Edward Dexter, a prominent merchant of the period, the house was moved to its present site from Prospect and George streets in 1860. Now converted to apartments, it is best known as Charles Pendleton's residence during the late nineteenth and early twentieth centuries. It was here that he displayed his pioneer collection of colonial furniture, which ultimately found its way to Pendleton House at the Museum of the Rhode Island School of Design.

Of all extant Federal houses in the city, the Dexter House is perhaps the most elaborately embellished. It is clapboarded front and back, with brick end walls that accommodate double fireplaces and provide some protection against fire from neighbors. Its Doric-columned entrance porch with balustraded roof shows the customary fanlighted entrance of the period, but with paneling instead of side lights. Giant pilasters (with uncanonically decorated capitals) flank the porch to support a pediment at the roofline which breaks the cornice. The upper stair hall Palladian window unusually thrusts into this pediment. This, in turn, breaks into the balustraded and paneled parapeting with exaggerated finials around the perimeter of the hipped roof. Joined dormers break from the roof slant with more parapeting around the roof deck above and behind. Exceptional, too, its crossetted ground-floor windows, which are nearly unique among Federal houses in Providence for their pediments. Corner quoining and intricate moldings complete this array of Federal decorative effects.

Hope Street and Side Streets

PR135 John A. Mitchell House
1865–1867, William R. Walker. 190 Hope St.

PR136 Horatio N. Campbell House
1877, William R. Walker. 141 Waterman St.

These two Second Empire houses, built by the same architect ten years apart, offer plain and fancy versions of the mansard-roofed house. In its severe boxiness, the Mitchell House, for a prominent merchant, is typical of William Walker's work of the mid-1860s, a sober version of the newly stylish mansard roof placed atop a traditional center-hall, four-room-plan mass. A semicircular three-window bay with loggia provides a flourish at the top of the south elevation and looks down to Narragansett Bay. Across the street, his later and larger essay in the French mode for Campbell, a wool wholesaler, is more assured, far more ornamental, and doubtless influenced by the new city hall, then building. Although it is now aluminum sided, this fortunately occurred without injury to its trim, which provides perhaps the most elaborate example of Victorian French Renaissance architectural ornament in the city. Both houses are now in institutional use: the first as a Brown University departmental building; the second as a center for transcendental meditation. Both contain interesting interior fragments.

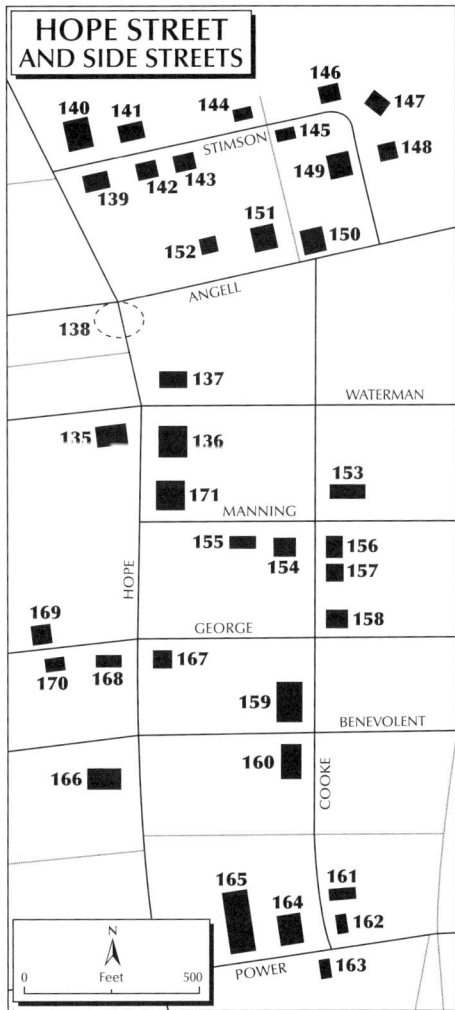

PR137 Robert Lincoln Lippitt House
1852, Thomas A. Tefft. 193 Hope St. (corner of Waterman St.)

PR138 Henry Lippitt Houses
1855–1857, Russell Warren. 198 Hope St.
1862–1865, Henry Childs (builder). 1981–1984, refurbished as museum and apartments, Kite, Palmer Associates. 199 Hope St. (opposite corners of Angell St.). Owned by Preserve Rhode Island; open to the public by appointment.

These three Italianate houses were built for two brothers of a family early involved in cotton manufacturing and later prominent in state politics. Thomas Tefft's brick cube for Robert Lippitt is extremely restrained, even for this restrained architect, who is also responsible for the comparably severe, contemporary Tully Bowen House, in brownstone (PR98). As with all Tefft's works, however, his acute sensitivity to proportion deserves notice. The stable (now converted to residential use) is also his. Russell Warren's double house for Henry Lippitt is among his last commissions and an early example of the Italian Villa Style, restrained in manner, with projecting window hoods over some windows and a deep, bracketed cornice. It would hardly be a villa, however, without some picturesque asymmetry. A columned and porticoed porch set back from the front gives entrance to one side; a long, arched loggia coming up to the front to the other. Service ells and mews occur to the rear.

When he again built across the street, Henry Lippitt set aside whatever allusion to informality his early villa residence might have made and opted for the formality and grandeur of an Italianate palace set high on its corner site. Like the Thomas Poynton Ives House (PR96.2), whose Federal style this may have attempted to "update," this has an elevated semicircular porch toward Hope Street and a semicircular, two-story bay to the side garden elevation, plus a small hooded cast-iron porch into the garden. But what a difference between the attenuated restraint of the Ives House and the insistent plasticity and overblown scale of this house, where a typical three-bay Victorian elevation attempts to do work earlier allotted to five! In the arrangement seen in other Victorian examples, a carriage entrance faces Angell Street with an L-shaped corridor connecting the two approaches.

The interiors of the second Lippitt house are now minimally furnished, but remarkable for their extravagant Victorian decorative painting. The hall and billiard room present an astonishing display of exotic woods, all *faux bois*. More wood graining occurs on the "paneled" ceiling of the dining room, which is complete with robust sideboards, paintings of gustatory abundance in the Dutch and Flemish manner, and an antlered deer head. The library and one of the front parlors are sumptuously stenciled with the kind of motifs found in Owen Jones's encyclopedic *Grammar of Ornament*—the library in Jones's favorite style, "Saracenic." All rooms are enlivened with fine brass gas chandeliers. Generations of the family resisted the beguilements of future styles, except for redecorating one parlor as a ballroom in a French eighteenth-century style around 1903–1904 after a family visit to France. A textile magnate, Lippitt served as governor of Rhode Island from 1875 to 1877, when his house served as the de facto governor's mansion in a state that has none.

PR138 Second Henry Lippitt House, interior

story with gray-green slate on the second, the overall composition relying more on juxtaposition of form than on application of ornament. The local Art Workers Guild designed and executed some elaborate interiors (not open to the public). The entrance hall, staircased, fireplaced, and paneled in the typical Queen Anne way, is especially delightful, with a mural in the Pre-Raphaelite manner by Sydney Burleigh, decorative painter also for the Fleur-de-Lys Studios and the Providence Art Club (PR58, PR59), on the theme of "May and December" romance, tucked into the fireplace inglenook. Stained glass windows which could portray a young wife and older husband, also in medieval costume, are angled into the stair landing. Whether or not the images make a biographical allusion, Lyman Klapp, owner of a local oil firm, died before living a year in his new house. The barn, elaborately decked out as a sculptural, shingled, and overtly charming pendant to the house, stands at the rear. In fact, it seems less a barn than an overblown playhouse masquerading in a grown-up role.

PR140 Amos Beckwith House

1861–1862, Alpheus C. Morse. 1867, enlargement, Alfred Stone. 2 Stimson Ave. (corner of Hope St.)

On Stimson Avenue are about two dozen houses, most of them stylish, architect-designed dwellings from the late nineteenth century for the well-to-do to affluent. There is no better place to observe the transformation of Queen Anne into early Colonial Revival, with a few baronial ventures along the way. The clientele in

PR139 Lyman Klapp House

1886, Stone, Carpenter and Willson. 217 Hope St. (corner of Stimson Ave.)

A fine example of this firm's work in the Queen Anne mode, the Klapp House has uncoursed Seekonk stone set in pink mortar on the first this neighborhood tended to be upper-middle-class local businessman rather than the manufacturing barons and scions of old families who inhabited the largest houses farther west on College Hill.

First, the grandfather of the street. When Amos Beckwith (son of Truman Beckwith and, like his father, a cotton broker) built his house, this sparsely settled part of town was an ideal setting for a picturesque, towered, Italianate villa. Though impressive, the Beckwith House appears somewhat clumsy and underembellished, in part because of the rather plain enlargement and the infill of its porch. The irregular profile notwithstanding, the house follows a cross-axial center-hall plan much like those of Alpheus Morse's palazzos. In 1882, Beckwith subdivided his family's holdings in the area to make half the building lots on Stimson Avenue. John J. Stimson owned the rest of the land, from which the other half of the building lots on the street were subdivided.

PR141 International House (Byron Thomas Potter House)

1894–1899, Edward I. Nickerson. 8 Stimson Ave.

Despite its high hipped roof, the Potter House, built for a real estate and insurance broker, has a strong horizontal quality, emphasized by the thin ochre Roman brick that McKim, Mead and White popularized beginning in the 1880s. Detail is effectively limited to sheet-copper trim for bay windows, balustrades, and dormers, which, in its weathered verdigris, contrasts handsomely with the brick. The interior shows Nickerson's interest in interlocking spaces in a rich variety of woods.

PR142 Henry A. Waldron House

1893–1894, Hoppin, Read and Hoppin. 9 Stimson Ave.

Henry Waldron, like Byron Potter, was a real estate and insurance broker. His house typifies a number of Providence houses of the 1890s in its use of a gambrel roof pulled down from the attic over the second story, set flank side to the street, and intersected by an asymmetrically placed polygonal tower with steep roof. The tower is typically Queen Anne, the gambrel typically colonial. Most astonishingly, against the tower the entrance slice alludes to a more developed colonial classicism, but with a coyness, asymmetry, and strangeness of cartouche against paneled ornamentation unknown to the eighteenth century. This lively and original eclecticism delightfully epitomizes the dawning phase of the full-blown Colonial Revival to come. The cozy one-and-one-half-story cottagey look of the front becomes undisguisedly a full two and one-half stories behind.

PR143 Joseph Fletcher House

1890, Stone, Carpenter and Willson. 1911, wing off southwest corner. 19 Stimson Ave.

Like the same firm's Klapp House on the corner (PR139) and the Waldron House behind it, this is another brick-based, slate-topped Queen Anne design, although the colonial mix in the others hardly exists here. But more like the Potters opposite, Charles Fletcher, president of the Providence Worsted Mills, wanted a touch of mansion grandeur for this house, which he built for his son. Hence his architects mixed what can best be called Scottish Baronial touches into the prevalent Queen Anne vocabulary. As in many works by this firm, however, the principal interest of the exterior is less its style than its forceful shaping. Semicircular bay windows at the ground floor subtly shift to polygonal bays upstairs. Arched dormer hoods boldly project off stepped bracketing. Most daring is the giant gabled overhang on the side elevation toward the street embracing both second story and attic. Although largely supported on a downstairs bay window projection, the overhang appears to thrust from sinuous lion's-head brackets. Skewed off the top of this gable, a piquant jerkinheaded bay tucked under the ridgepole enlivens the larger form in the pretty and intimate manner which substantially accounts for Queen Anne charm. An extra-large lot and the sideways siting of the house, so that its porch overlooks an expanse of lawn, also contribute to the grandeur.

The lavish interiors (not open to the public) in an eclectic range of period styles are among the finest to have survived in the city from their period: Louis XV parlor, Italian Renaissance library with tooled leather walls, and Tudor dining room, all elaborately and exquisitely carved, and set off by marble fireplace facings in varied colors. They open from a central, cross-axial entrance hall with its own fireplace, which is Queen Anne in character, with some allusion to colonial. A splendidly expansive stair with a giant stained-glass landing window of an autumn landscape in the browns, tans, and ambers of the interiors, lightened by clear glass, completes the entrance hall.

Finally, a tall, hip-roofed carriage house vernacularizes the forms of the dwelling in the sculptural manner of the same firm's stable behind the Klapp House, here less coy in aspect.

Opposite, at 30 Stimson Avenue, the Frederick Condit–William Benedict House (1884) exemplifies the older Queen Anne manner of the high, angular, exposed-frame variety, beautifully maintained but a bit spoiled by the infill of much of its original porch. Built as an investment by Condit, it was purchased by Benedict, an oil dealer.

PR144 Newton D. Arnold House

1888, Edward I. Nickerson. 24 Stimson Ave.

PR145 William P. Goodwin House

1885, J. B. Goodwin. 31 Stimson Ave.

Diagonally opposite one another are two versions of Queen Anne in the shingled manner, built for the corporate secretaries respectively of the Rumford Chemical Company in East Providence and of a local insurance company. Here, the exteriors are hardly touched by the emerging Colonial Revival. In both, lower stories in brick provide the base for the plastic manipulation of wood-shingled forms above. Edward Nickerson's treatment of his base makes an overt statement about brick craftsmanship through its textural hobnailing, while J. B. Goodwin lets the precision of his plain brick surfaces speak more reticently for its workmanship.

The Arnold House spreads in a deep-gabled mass, its broad slope to the street. This slope, the eruptions from it, and even more the shapes butting into it to modify its front edge most vividly characterize the house. Over the entrance porch a semicircular bay with a conical roof jostles a polygonal bay, which counters a two-story cylindrical corner turret capped by another conical roof. They complicate the smooth flow of the slope with the topological turbulence of the moraine edge of a glacier. Up on the slope itself, outcroppings poke through the flow—chimneys and a perky dormer which echoes the cap of the turret under a Japanese parasol hat. The trellised porch supports are vaguely Japanese in inspirations. Though tightly sited for its size, the Arnold House spreads in the typical suburban manner.

The Goodwin House adopts a more urban approach. Tall and compact, its front rises just behind the site line, its door reached by stairs directly from the sidewalk in the eighteenth-century manner. Especially urban is the extension of the brick base of the house to enclose the property from the sidewalk view within a high garden wall. This house derives from contemporary English examples in the Chelsea district of London, the late nineteenth-century "aesthetic" enclave of the city where artists and the artistic had similar houses, but in masonry rather than in wood shingle. Whereas the eye-catching features of the Arnold House are scattered, here they gravitate to a vertical axis marked by the projecting entrance box and the Juliet balcony which overhangs the eaves. Off to one side, another of those half-submerged and tapering, conical- capped turrets links the two. The boldness of Queen Anne window organization is evident in the nudge of the second-story triplet against the spin of the landing windows of the adjacent stair tower and in the dissolution of the corner, where two upstairs windows butt at right angles while their downstairs counterpart follows the bias of a wall which diagonally clips the corner. Different levels among the upper-story windows indicate changes of floor level inside to invoke the "picturesque" effect, say, of some old, rambling English inn with floor levels repeatedly interrupted by a few steps up or down. Latticed openings relieve the closed quality of the garden walls, but so placed as to protect the secrets within. In a city full of Queen Anne houses this is among the most sophisticated in the style, as well as being perhaps the most urbane.

PR146 Charles H. Sprague House

1894, Edward I. Nickerson. 44 Stimson Ave.

PR147 Albert L. Calder–R. Austin Robertson House

1892, William Chamberlain. 50 Stimson Ave.

PR148 Louis E. Robertson House

1892, Gould and Angell. 60 Stimson Ave.

In the Sprague House, as in his Cook House (PR130]), Edward Nickerson walled his upper floors and gable in rosy-red slate, with brick and brownstone trim below, a palette both unusual and handsome. Whereas the earlier house spread in the Shingle Style manner, however, the mass here contracts toward a rectangular compactness meant to be sympathetic to the Federal style. But it is only Nickerson's elegant detailing that makes it such, especially the semicircular entrance porch alluding to the frontispiece of the Thomas Poynton Ives House (PR96.2), Providence's premier Federal mansion. (But recall that Edmund Willson redesigned its porch as an 1880s restoration!) As for the near-mansard roof, the L-shaped mass, and the handsome echo of the front porch in the Neo-Federal gazebolike piazza off the side elevation, these aspects of the house are remote indeed from the style Nickerson so lightheartedly embraced.

The Calder-Robertson House takes a particularly quirky, but quite sophisticated, approach to the colonial past. As in some preceding houses and in the next, the traditional gambrel roof form is manipulated to produce unusual sculptural effects. A gambrel roof, flank side toward the street, folds over the capacious but plain body of the house. Instead of stopping at the eaves as a respectable gambrel should, however, this continues its front slope down the elevation to become a shed-roofed porch. Not quite so simply, however, because the porch extends beyond the side of the house and appears to interlock with it as a two-story gable, its end open to a large arch. The melding elisions give contradictory cues. The porch seems at once an appendage to the house and drawn back under the roof as part of its interior. From the slope above the porch pop oversized dormers (originally two only), their shape, scale, and entablature embellishment suggesting European Renaissance more than colonial American sources. Then one notices the homey clapboarding up in the steep pediments capping the arched end of the porch and those topping the dormers, and the European allusion slips. Ionic columns embellish the porch, again in the intricate Federal Revival manner, combining with the arch at the end of the porch to make a Palladian motif. The infill of the double-columned intervals of space with latticing and their oval-shaped "view frames" to the landscape, however, also recall the pervasive garden influences on the Queen Anne Style. The recombinations of architectural features as well as the shifts in scale and stylistic allusion result in architectural ambiguities suggestive of postmodernist strategies of the 1980s—albeit without postmodernist self-consciousness and irony. Albert Calder, a druggist, built this house for his daughter Mary Calder Robertson, whose husband was treasurer of the Builders Iron Foundry, then (and long) the leading fabricator for decorative architectural ironwork in the area.

The Louis Robertson House is more representative of the transition from Queen Anne to Colonial Revival than either of the previous two, and an exceptionally fine example. Here the gambrel-roofed mass, set gable end to the street, folds over, leaving only the ground story and side elevations in clapboard. Through this hillock of green-gray slate poke gables and bays of varied shape and size. Another of those porches detailed in the delicate Neo-Federal manner projects tentatively out from under the weight of the roof's protection. On the other hand, the polygonal half-turret attached to the side elevation and its paneled grouping of windows retains the older medieval allusion of the early Queen Anne style, but with a spidery, broken-scroll finial topping the stair hall window to remind the stylistically befuddled that this creative mix of periods really wants to be "colonial."

PR149 Charles H. Baker House

1898, Gould and Angell. 67 Stimson Ave.

Across Stimson Avenue is one more baronial gesture. Charles Baker was a superintendent for the famed Gorham silver works. His house shows a Flemish gable in brick stepping its way through a fissure in Queen Anne shingling. The result is delightfully brazen—long before anyone had imagined the postmodernist ironies possible from such disruptive stylistic composition.

PR150 Ebenezer Knight Dexter–John J. Stimson House (Rose Farm)

c. 1799; mid-19th century, addition to east. 300 Angell St. (corner of Stimson Ave.)

When Ebenezer Dexter built this country retreat it stood at the eastern edge of settlement in Providence. Several of the city's wealthy residents maintained country seats on the then-rural outskirts of the city, but this is the only remaining gentleman's farmhouse from the period in the vicinity. This brick-ended Federal house, with exceptionally tall, paired end chimneys to accompany its high roof hipped to a balustraded deck behind paired dormers, recalls other substantial local houses from the 1790s. Its original plan, with a transverse hall crossing the center hall to doors on either side of the building, may have been more commonly used in the country than in town houses. Dexter, a businessman and philanthropist, left the bulk of his farm to establish a home for the poor on land immediately north of present-day Stimson Avenue. He gave more land on the west side of the city to establish a parade ground for the local militia, at one end of which now rises Cranston Street Armory. John Stimson bought the farmhouse and the residue of land around it. His grandson sold off the house lots for the eastern half of Stimson Avenue. So the Stimson farmhouse at one entrance to Stimson Avenue and the Beckwith villa (PR140) at the other—rural versus suburban country types—indicate something of the scale and density of buildings in this part of Providence into the 1870s, as well as the fate of farmlands and villa estates with later city expansion.

PR151 Central Congregational Church

1893, Carrère and Hastings. 296 Angell St.

If Central Congregational had earlier created an exceptional edifice on Benefit Street in calling on Thomas Tefft in 1850 to design one of the first examples of monumental American Romanesque Revival (PR83.6), this replacement is equally so—and not only as a surprising bit of Spanish baroque for New England Congregationalism. Carrère and Hastings came to Providence at the request of Francis W. Carpenter soon after the architects completed commissions in St. Augustine, Florida, that drew on local historical traditions, including the Ponce de León Hotel (1885–1887) and, more to the point, Memorial Presbyterian Church (1889–1890), where their patron, the famous railroad tycoon and real estate developer Charles Flagler, had a domical memorial chapel for his family built as a semidetached structure. It may have been momentum from the immediately previous work, or simply Carpenter's enthusiasm for what he had seen during a Florida sojourn, that accounts for this unusual conjunction of New England Congregationalism with a southern European and Counter-Reformation style, albeit here treated in a restrained manner. The Spanish Floridian origins of the church were more evident before neglect and a 1954 hurricane eliminated the elaborately ornamented, cupola-ed mini-towers, mostly in terra-cotta, from the corners of the elevations, where painfully rudimentary replacements now exist. Yet, even missing these culminating ornaments, the front is exquisitely wrought and proportioned in the tawny combination of narrow Roman brick and terra-cotta trim which McKim, Mead and White had popularized.

Of primary interest, however, is the interior, which unites concurrent trends in American architecture of the time: Renaissance monumentality and the earthy and naturalistic aspects of the Arts and Crafts Movement. In this cross-shaped space, the broad, stubby arms for the nave and their more extended (perhaps too extended) transepts are barrel-vaulted in Guastavino tiles which swell upward into the central dome at the crossing, and outward into the broad, semicircular apse. This indigenous Catalan technique of vaulting by means of thin, overlapped tiles set in concrete, which Rafael Guastavino brought to the United States from his native Barcelona, had three structural advantages. Because the vaults were built up of progressively projecting layers of tile embedded in concrete, they required minimal scaffolding and centering to erect. Because the materials did not have to be custom shaped and fitted, like cut stones, fabrication was cheaper and quicker. Finally, because the finished vaults were much lighter than those built of cut stone, they could be more economically supported. (The company Guastavino organized played a cardinal role in spanning the grand public spaces which the American Renaissance called forth from the 1890s through roughly the 1930s—and indeed made most of them economically possible.)

The apse is painted with illusionistic panels decked with jewellike patterning and entwined grape leaves in muted mauves, pale greens, and ivory, set against the clay tan of the tiles, all of which were colors popular in Arts and Crafts interiors of the time. Stained glass windows designed by J. A. Holzer (an artist-craftsman who had previously been one of Tiffany's top designers) and fabricated by the Duffner Kimberly Company fill either end of the arched transepts. In their proscenium scale, theatrical brilliance, and technical virtuosity, these are unsurpassed in the state. The theme is divine life unfolding in the universe: the Creation toward the east and the vision of the heavenly city of the New Jerusalem toward the west, with lesser windows

PR151 Central Congregational Church

in minor spaces representing various virtues. In these windows it is not the small figures (a bit sketchily realized) that count, but the light of the sacred and eschatological landscape in layerings of opalescent glass.

The architects were probably responsible for the design of the handsome pulpit lifted to one side of the apse and reached by a semicircular stair, along with most of the other furniture. They may also have designed the carefully shaped illuminated cross suspended in the space from the center of the dome, boxed in metal, faced with translucent glass, and outlined with tiny exposed bulbs, as a testament to the wonder of the time at the emerging technology of electricity.

The building committee specifically wanted a unified, "artistic" interior which would not depend on generations of gifts for its completion. It all comes together with a certain bluntness and spareness which *is* Congregational, as though this spare faith drew the threatened opulence of the building back in the nick of time from anything too Episcopalian or Catholic. Among all church interiors in Providence this probably stands next in architectural interest after the colonial and early national interiors of the First Baptist and First Unitarian churches, and is more unusual than either of its predecessors.

PR152 Francis W. Carpenter House

1896. Carrère and Hastings. 276 Angell St.

Adjacent to and now owned and operated by the church for community programs is the house of its principal benefactor, a partner in an iron and steel supply firm. He called on the same architects for his quite overwhelming brick and limestone château in the Louis XIII style, which marked Carrère and Hasting's signature style for the 1890s, as opposed to their exotic venture next door. Incongruous as a pair, together the church and house display the suave shift from style to style of which Carrère and Hastings were capable. Much of the interior of the house remains intact.

PR153 Donald E. Jackson House

1929–1930, William T. Aldrich. 66 Cooke St.

PR154 Jeannette B. Huntoon House

1925, Jackson, Robertson and Adams. 63 Manning St.

PR155 Frank B. Lisle House

1923, Clarke and Howe. 59 Manning St.

Cooke Street and the streets immediately off it provide a fine sampling of the Colonial Revival in Providence, with more developed examples than those on Stimson Avenue. (The entire East Side bears study in this regard, which is impossible here.) This cluster of Neo-Colonial residences dating from 1890 through 1930 exhibits a range of possibilities in the style, from the catch-as-catch-can use of miscellaneous colonial features to quite archaeological approaches.

First, three prepossessing examples of the Colonial Revival, beginning with 66 Cooke Street. The president of a real estate firm commissioned the house from William T. Aldrich, a master of eighteenth-century styles—this example, on the exterior at least, more specifically beholden to English than to French example. High garden walls, plantings, and its sideways siting on the lot combine to obscure the house. At quick glance, it appears to be organized as a typical late colonial New England brick mansion, an elongated rectangular box with central hall front to back and major rooms, two deep, on either side. Yet this is not the case. The massing of the block steps back from its projection as a pedimented entrance. Then it spreads to embrace a corridor at right angles to the entrance axis, with ancillary rooms ranged along it on the side toward the porch, especially for reception. Finally, the block of the house spreads again to embrace the suite of principal entertainment rooms arranged in a line between their successive entrances from the cross hall and openings to a continuous garden terrace beyond. What appears from the front as typical Anglo-American eighteenth-century planning becomes typically French for the climatic rooms and their relationship to the garden—a blend in planning schemes which reflects the popularity in the United States, from the 1880s through the 1930s, of architectural training at the Ecole des Beaux-Arts as the finishing touch to professional education. The porch is Federal style of exquisite delicacy (with a glance perhaps at contemporary English Regency, which was especially popular in the 1920s and 1930s, as American architects sought to expand the repertoire of the Federal Revival style, while increasing both its grandeur and its sophistication). The fanlight is a veritable peacock's tail of leaded ornament, which characterizes the revivalist attitude of either outdoing the sources or choosing the most extravagant precedents possible.

Just off Cooke Street on Manning Street are two more sophisticated examples of the Colonial Revival by major early-twentieth-century Providence practitioners: one for the widow of a textile manufacturer, the other for a partner in a brokerage establishment. For number 63, presumably F. Ellis Jackson, as the principal designer in his firm, recalled John Holden Greene's early-nineteenth-century balustraded eaves and fretted monitors in designing the hipped roof. However, the entrance detail derives from the late eighteenth century, while the L-plan interior hall connecting central entrances on two streets is (if not unknown earlier) far more common to post–Civil War houses, as some described earlier make clear. Next door, for number 59, two decades earlier, Wallis Howe (a Providence designer who frequently changed partners) preferred a narrow box to Jackson's fatter massing. This stretches the traditional five-bay colonial front to eleven bays by placing the additional windows in slight projections at the ends which are meant to be seen as wings. At the center, a huge stair window enlivens the regular window rhythms of the conventional colonial front with a touch of Neo-Colonial picturesqueness. Neither double entrance doors nor their deep boxing are at all common in American eighteenth-century residences, although some occur in large English Georgian examples. The inset door does appear in America with the Federal style, but not much before 1810, then infrequently, and rarely doubled—another indication of Colonial Revival tendencies to enrich, enliven, and aggrandize colonial and Federal precedent.

PR156 William H. Wood–Walter Ward House

1896. 56 Cooke St.

PR157 Caroline Bliss House

1895–1896, Franklin J. Sawtelle. 46 Cooke St.

PR158 Edward A. Greene House

c. 1850. 1889–1890, remodeling, Stone, Carpenter and Willson. 38 Cooke St.

These three houses of more modest mien, all of the 1890s, represent earlier manifestations of the Colonial Revival than the big houses just seen. All three have flank gambrel roofs. A popular roof shape for early examples of the revival, as we have already seen on Stimson Avenue, the gambrel continued the high, angular silhouettes of such preceding late Victorian styles as Second Empire and the steep, sculptural roof silhouettes of Queen Anne (as just seen on Stimson Avenue). All three also stretch the dimensions typical for colonial and Federal door and window openings horizontally—so much so that they convince the casual observer to accept three-bay fronts (which are more typical of Victorian Neo-Renaissance blocks) as the equivalents in magnitude of their five-bay colonial precedents.

Number 56 (which the initial owner, a real estate agent, almost immediately sold to a plant superintendent) shows a Queen Anne porch fronting a broad door flanked by "side lights" as wide as full windows (with later minor alterations). Semicircular bays, especially popular in the Colonial Revival around 1900, balloon from the side elevations. Next, in number 46, the colonial allusion is more explicit. Its source of inspiration is the eighteenth-century John Vassall House in Cambridge, Massachusetts (1759), known to early Neo-Colonialists as "Longfellow's House" because the poet long lived there and was himself an abettor of the colonial revivalists' cause in poems which called up spinning wheels and grandfather clocks as momentoes of "olden tymes." The dainty semicircular entrance porch is Federal Revival (far removed from the mid-eighteenth-century Vassall House). This doubtless, again, derived from prototypes such as that fronting the Thomas Poynton Ives House (PR96.2). Here it shelters another double door, with side lights and a stretched fanlight, which is utterly foreign to the stern, transomed and entablatured Vassall portal.

In the third in the trio, number 38, Willson himself used a variant of his own Ives semicircular porch, from the year previous, to colonialize an Italianate house of the 1860s. He also back-dated the double Victorian doors to Federal style, perhaps raising a mansard to a tall slate gambrel up top, and poked colonialized dormers through it. Two of the three dormers have balconies with attenuated screens of balusters, supported on large, ornamented, S-curved brackets. Presumably, he also framed the result in the pretty fence with roundel picketing of the sort at which colonial revivalists were adept.

In this trio of houses, then, the first modulates Queen Anne toward Neo-Colonial. The second uses the bare bones of a well-known colonial prototype as its organizing image and cue for sentiment. The third lyrically recreates pieces from the colonial past, to transform Victorian ugliness (as it would then have been viewed) into colonial loveliness, with all the enchantment of a children's book illustration. Kate Greenaway would have adored it.

PR159 Museum of Rhode Island History, Rhode Island Historical Society
(Robert S. Burrough, Jr.–Senator Nelson W. Aldrich House)

1822, John Holden Greene; 1905, Stone, Carpenter and Willson; 1975, museum conversion. 110 Benevolent St. (corner of Cooke St.) (open to the public)

Built as a two-story house for a customs officer, this must then have resembled Greene's other houses of the period. It was probably expanded somewhat later in the century for Samuel Noyes. Its present appearance, however, records its occupancy for seventy-five years by Senator Nelson W. Aldrich and his family as their Providence residence. For the senator, Edmund Willson added the third floor with its modified repeat of the centered Palladian window below and the Tuscan-columned side porch to echo the front porch, and did considerable interior work as well. (Much of this in the principal downstairs rooms was revamped for museum use, but Willson's ballroom addition and remodeled upstairs bedrooms remain mostly intact.) Around the same time Willson also worked on Senator Aldrich's vast bayfront estate in Warwick, Indian Oaks, which eventually provided the senator with an even grander summer escape.

PR160 Mrs. Herbert A. Rice House
1931–1932, Albert Harkness. 25 Cooke St.

In the late 1920s and 1930s, Albert Harkness, like other architects at the time, turned frequently to the English Regency and especially to the informal, playful aspects of resort and country villa architecture. Hence the piquant quality to the bow windows and the lacy metal porch on the garden side of this house, whereas the interior (not open to the public) retains Regency-inspired ornament inflected toward Moderne. The integration of the house with its walled garden and the street merits special notice.

PR161 Robert S. Burrough House
c. 1816, John Holden Greene. 6 Cooke St.

A "real" Federal house, this has the era's distinctive monitor-on-hip roof (with balustrades at the eaves of both), wooden quoining, and a modified Palladian-motif entrance with broad elliptical fanlight and side lights and attenuated ornamentation.

PR162 Patrick Monroney House
c. 1892. 2 Cooke St.

PR163 Samuel Gerald House
c. 1844; cornice trim added. 169 Power St.

At the corner of Cooke and Power streets, number 2 Cooke, built for an importer of wines and liquors, merits attention, first as a loosely organized Queen Anne house with unconventional colonial details scattered about, such as a pediment folded over a corner window. Then, too, the porch capitals and the paneling of the front gable show curvilinear ornament of exceptional energy (exhibiting influence from Art Nouveau, which is unusual for Providence). In fact, the "Corinthian" capitals for the sweeping veranda seem to have been derived, not from architectural plates, but from direct observation of fern fronds.

Facing the intersection is 169 Power Street, a

fine three-bay, end-gable Greek Revival house with an Ionic portico, built by Samuel Gerald, a carpenter, as his own house. He had earlier lived next door, at number 171 (1828), presumably having built this late Federal house as well. If so, then this carpenter literally moved with the style changes of his time.

PR164 James Burrough House

1818, John Holden Greene. 160 Power St.

PR165 Dyer Hall, Brown University
(John Holden Greene House II)

1821–1822, John Holden Greene. 154 Power St.

The block of Power Street between Cooke and Hope presents a nice group of smaller, mostly early-nineteenth-century houses. They include two houses by John Holden Greene: minimal and modestly grand, standard and unique. James Burrough, like others nearby, managed a livelihood from the customhouse and commissioned a house from Greene. The house shows Greene's design at its most basic: small—and small-scaled—with a paneled monitor-on-hip roof, a simple console-supported architrave over its leaded, Gothic-paned transom light, and a lateral ell with porch.

As for Greene's second, and larger, house for himself (see PR108 for the first), it originally stood next to the James Burrough House, one lot to the east. (The business-oriented Bryant College moved it in 1950 to make way for the dormitory which now bludgeons its way between the two Greene houses, and which Brown University subsequently acquired, along with all Bryant's Providence holdings, when the latter moved to a country site in Smithfield.) Squeezed between two lumpish institutional buildings, raised on a high, concrete-faced foundation, and abused as a dormitory, it has suffered a triply cruel fate. Yet it deserves notice and has been substantially restored on the exterior, but baldly so, without much of Greene's original ornament, which presumably once completed it.

To his customary monitor-on-hip roof, Greene here added a cupola. A veranda extends across the full length of the house, supported on postlike Doric columns, their irregular spacing echoing the placement of the openings behind them. A wide central stair (eliminated) once led to the side- and fanlighted entrance. Greene echoed the side lighting of the entrance door by extending the same three-part rhythm to his windows in the front elevation—wide center opening flanked by narrow side lights, united under a light entablature. Downstairs the three parts are separated by colonnettes standing on attenuated bases; upstairs, by paneled pilasters. Typical for Greene, balustrading once topped the delicately bracketed cornice, and the monitor as well. The stepped quality of the massing, along with the balustrading and the animated detailing, suggests so-called southern riverboat house types found along the Mississippi more than in Providence, where this delightful design is exceptional. Finally, the main block is complicated by stubby setback wings on either side. If Greene strove for originality in his first house he outdid his earlier efforts here.

The owner designation of Greene's house is often hyphenated as the Greene-Fenner-Dyer House. No less than three state governors lived here: James Fenner during the last of three widely separated terms of office (1843–1845), Elisha Dyer (1857–1859), and Elisha Dyer, Jr. (1897–1900).

PR166 King Hall, Brown University
(Robert W. Taft House)

1895, Stone, Carpenter and Willson. 154 Hope St.

Here Edmund Willson applied the curved gable of the Joseph Brown House not only to the gable ends, but to the front porch, the three dormers, and the conservatory on the south side as well. The house is a cardinal example of the colonial revivalist's focusing on a particularly distinctive motif from the past and exaggerating it to new effect. Significantly, in 1877 Alfred Stone, Willson's partner, acquired the very copy of William Salmon's *Palladio Londinensis* which had originally been owned by Joseph Brown's carpenter and had been passed down through other Providence carpenters since the eighteenth century. Salmon illustrated the curved gable for use in a summer house, so Willson's use of the design for the conservatory at least is doubly "correct."

PR167 George M. Smith House

1888, Stone, Carpenter and Willson. 165 Hope St.

Slightly earlier than the Taft House, this typifies the same firm's work in the late 1880s as it moved from Queen Anne asymmetry and idiosyncratic form to Colonial Revival symmetry and classical detail. Apparent already is the recent arrival into the firm of Stone and Carpenter of young Edmund Willson, then fresh from the Ecole des Beaux-Arts and McKim, Mead and White's office, and caught up in the fervor for rediscovering the colonial past. The abrupt variety and contrast in shapes of the gambrel-roofed Queen Anne house remain, even to the skewed relationship of the entrance and its second-story Palladian window, but in a far more balanced and contained way than in houses built just a few years earlier. Consider how the polygonal turret at one end of the elevation is balanced at the other by the two-story bay topped by an enlarged dormer capped with a broken scroll. But, typical of the Queen Anne Style, what is tentatively balanced in shape becomes unbalanced in detail. So the "colonial" dormer opposes the flaring "medieval" peaked turret with its miniature balcony. Still, Neo-Colonial symmetry is making its way, as indicated by the much greater asymmetries and virtual absence of classical allusion in a slightly earlier house by the firm (PR171).

PR168 Eugene Graves House

1924, Albert Harkness. 195 George St.

In the French Norman farmhouse style, especially popular in the 1920s, number 195 George Street is probably the best exemplar of the style in the city and among Albert Harkness's finest houses. He skillfully combined picturesque massing, textural manipulation of surface (mostly in brick and slate), and scattered openings of different shapes and sizes with plenty of wall between to assert the primacy and physicality of the overall shape. Then, axial balance occurs at the entrance to discipline the asymmetries and celebrate the principal rooms. Such imagery and form recalled the rambles of the architecture student about the French countryside as respite from the ateliers of the Ecole des Beaux-Arts, and the "charm" discovered in the formality of minor country châteaux giving way through time to functional adjustments and additions. Ceremony and manners married to relaxation and nature: precisely the milieu sought by suburbanites of the 1920s (or by city dwellers who wished to embrace country life). As in Harkness's nearby Rice House (PR160), the integration of the house to its walled site intensifies the allusion. So does the wrought iron balcony embellishment over the entrance, as the chef d'oeuvre of the village blacksmith. It was a time for fine work in wrought iron in the United States, as many in this defunct trade hung onto their livelihoods by making the hinges, andirons, ornamental house numbers, and so on for which the diverse architectural "revivals" of the period hungered.

PR169 Division of Applied Mathematics, Brown University (Henry Pearce House)

1898, Angell and Swift. 182 George St.

Number 182 George, built for a banker, is a late full-blown, rather diffuse and, for Providence, surprisingly rare example of residential Richardsonian Romanesque in customary tan and red-brown quarry-faced granite, with a giant carriage entrance with porte-cochère hidden off a rear corner. Stylistically, however, the house is curiously dichotomous. What is a belated style on the exterior becomes quite up-to-date Colonial Revival inside, predominantly of an intricate Neo-Federal inspiration. It is as though *he* sought dominating monumentality, while *she* craved glittering refinement. Much of the interior remains despite its academic use, including a copper-clad conservatory embellished as a grape arbor in stained glass, like a live-in Tiffany lamp. (The more impressively compact Neo-Romanesque carriage house for this house stands around the corner, right, on Brook Street.)

PR170 Frederick M. Sackett House

1894, Stone, Carpenter and Willson. 177 George St.

Number 177 George Street is an exceptionally ebullient example of early Willsonian Colonial Revival among other near-contemporaneous works by his firm. The picturesque, asymmetrical massing left over from the Queen Anne Style has disappeared; the front is symmetrical and the motifs are exclusively colonial in derivation. But with what un-colonial exuberance of scale, invention, and combination! The ogee curve of the porch anticipates the somewhat more (but not overly) archaeological flavor of the multiplicity of ogee-curved gables and dormers of Willson's Taft House around the corner (PR166), of one year later. The delight of the early Colonial Revival in such curved forms also appears in the pregnant bulge of another of those semicircular bowed spaces derived from the Federal period, here projecting from a side elevation.

To follow Willson's evolution from Queen Anne into Colonial through a neighborhood sequence, consider this chronology: PR139 (1886); PR167 (1888); this (1894); PR166 (1895). And finally (with the next entry), back to his Queen Anne, pre–Colonial Revival starting point.

PR171 Esther Baker House

1882–1883, Stone, Carpenter and Willson. 179 Hope St. (corner of Manning St.)

Like the Burnside and Hall houses on Benefit Street by Stone and Carpenter, the Baker House exploits its corner site to good effect, here with a dominating diagonal bay that thrusts its way toward the corner and separates the very disparate street elevations to either side. Despite this disparity the playful mix of materials and delightfully conceived features in this Queen Anne design still manages a tenuous whole, while leaving the parts to captivate the wandering eye by their diversity and threatened discord. The two entrances—one for carriages, the other for arrivals on foot—again require an interior organization around an L-plan hall. No part of Rhode Island, other than College Hill—not even Newport, it would seem—boasted so many mansions with this double approach for visitors. Inside (not open to the public), woodwork and fireplaces are substantially intact.

Note: For locations of sites PR172–PR184, see the Providence.overview map on pp. 30–31.

PR172 Blackstone Boulevard

1892–1904, beginning of construction to completion of original landscaping. Horace W. S. Cleveland and Olmsted Brothers for landscaping

Here, ranged along a bluff above a stretch of the Seekonk River, the arcadian legacy of Victorian culture and its metropolitan aftermath reigns supreme in the linear conjunction of a park, an asylum, and a cemetery, each beautifully landscaped and all connected by a landscaped boulevard 200 feet wide and 2.2 miles in length, with opposed traffic lanes separated by a linear park. The Proprietors of Swan Point Cemetery (PR175) commissioned this grand approach to its entrance gate. The Swan Point Proprietors offered the landscape improvement as an inducement for the city to shift the route of the boulevard from its previous course, which had curved down to the shore of the Seekonk, thereby separating the cemetery from the river. Construction of Horace Cleveland's design began in 1892, and the Olmsted firm's planting scheme was completed in 1904. A trolley line ran through the middle until 1948. Blackstone Boulevard emulates the connective strand first projected for the "emerald necklace" of Boston parks and boulevards by Olmsted and eventually given regional dimensions by Charles Eliot, Jr., in his famous Metropolitan Plan of 1893. These examples from Boston, in turn, inspired a similar plan for greater Providence, developed by the Metropolitan Park Commission in 1903, in which Blackstone Boulevard played a stellar role. Few of the boulevards in the plan were realized, however, and no other approached the beauty of this. Its park strip, minus the trolley tracks, is today alive with joggers and walkers. Of all the highly desirable and mostly luxurious residences which either line the parkway or are close by, one of unusual interest requires a momentary detour.

PR173 Frank Mauran, Jr., House

1929, Edmund Gilchrist. 137 Grotto Ave.

A family from Philadelphia called in a Philadelphia architect for this house. Set within a gently sloping bowl, the Mauran House recalls a vignette from old brick-built streets left behind. It appears as a large Neo-Federal house with powerfully massed paired chimneys at either end, flanked by a stable in the same style and a smaller, "colonial" building. It is exquisitely placed within its landscape and beautifully crafted. Once more the hoary five-bay, two-story, center-door arrangement; but it appears allusively in a split six-bay variant. Thus four bays go to the main block (with two windows, instead of the usual one, over the entrance portal), with the balancing two to complete the conventional arrangement out of alignment in the wing. This wing, in turn, masquerades as a smaller, more vernacular house of earlier date butted to the formal centerpiece, just as ancient houses often jostled newcomers in Philadelphia's old streets. The result is picturesqueness as the Colonial Revival liked it, with variety of massing incorporating seeming differences of period. There are also such typical Colonial Revival exaggerations of shape and scale as the outsized fanlight with its splendidly elaborated lead ornamentation, which may take its cue from frequent Philadelphia lavishness in this respect, but goes precedent one better. Such sophisticated and mannered playfulness anticipates the subtlety and wit in compositions of the most creative designers in the subsequent postmodern generation of colonial revivalists, but with less irony and more quality. The principal entertainment rooms are ranged along the back of the house *en filade*, opening onto a garden terrace.

PR174 Butler Hospital

1847, Thomas A. Tefft, original building. 1859, Horace W. S. Cleveland, original landscape plan. 1903–1912, Olmsted Brothers, additional landscaping. 345 Blackstone Blvd.

Among the oldest psychiatric institutions in the United States, Butler Hospital was conceived and designed in the spirit of humanitarian reform for care of mental disorders. It was isolated from the city in the 1840s in a location on the Seekonk River providing a calming "asylum" removed from the pressures of urban life, then understood as a major cause of mental disturbance and derangement. The site has been somewhat reduced by the sale of portions of it for development, and expansion of facilities has taken its toll. Still, the grounds and the altered building core remain as a beautiful legacy of nineteenth-century planning for social and medical reform. Near the present entrance is the Richard Brown House (c. 1731). This gambrel-roofed farmhouse, one of the city's oldest—and, rare for Rhode Island, of brick—conveniently met the hospital's need for groundskeeper's housing. But such relics also played a modest role, as something ancient and rural and comforting, in supporting the healing process. Farther along, on the right, is the original entrance to the hospital, planted out but still visible with its gate intact.

The Neo-Tudor Gothic of the original hospital building, known as Center House, complemented the romantically wooded landscape and established the tone of subsequent architectural enlargements. Its program, which emphasized the need for abundant light and fresh air, resulted in a symmetrical E-plan composition of three-story hip-roofed pavilions (the middle wing extended a little beyond the others), with all three wings connected toward the rear of the complex by two stories of passageways. The original character of Center House can still be discerned, although it has been enlarged, both by raising its height to four stories and by the addition of more wings, while another recent addition has reversed the entrance front of the building from its original location facing the river. Later buildings follow its medievalizing lead: first Stone and Carpenter for David Duncan Ward (1873–1875); then Stone, Carpenter and Willson for Kane Gymnasium (1882), Sawyer House (1886–1888), and Duncan Lodge (1889); Hoppin and Ely for T. P. I. Goddard House (1897–1898) and Weld House (1900); and J. Robert Hillier for the enlargement of the initial Center House (1977).

PR175 Swan Point Cemetery

1847 and later. 1847–1863, original southeast quarter, Atwater and Schubarth. 1894–1913, landscaping, Frederick Law Olmsted and Olmsted Brothers. 585 Blackstone Blvd.

Swan Point Cemetery adjoins Butler Hospital. Quiet contemplation in a picturesque, rural landscape was considered appropriate for cemeteries as well as mental hospitals in the mid-nineteenth century. Before Swan Point, leading families in Providence tended to have their burial plots in the North Burial Ground, which, when established in 1700, was in open country, well beyond the effective end of building on North Main Street. Typical for its time, it was essentially a field with minimal landscaping and with geometric regularity as the basic scheme for its drives. Swan Point evolved as a reaction, led by members of an active intellectual circle sympathetic to progressive early-nineteenth-century romantic ideas as to the proper setting for bereavement and commemoration. It is an early and distinguished rural landscaped cemetery—the handsomest of its type and scale in the state and among the finest in the country. At its principal gate, through a tumbled wall of cyclopean boulders completed in 1899–1900, Stone, Carpenter and Willson erected a simple trolley shelter (1904). This, too, is a pile of boulders pierced with ragged openings, under a low, shingled roof which is at once hipped and cross-gabled with deep, flaring eaves.

The 210-acre cemetery is composed of two distinct parts: the eastern section, acquired and developed between 1847 and 1870 (along the river), and the western section (toward the entrance gate), developed during the mid-twentieth century. The undulating eastern section, along the Seekonk, is heavily planted and laced with curving roads, which, with vegetation and monuments, reinforce the picturesque landscape. The western section is flatter, more simply arranged, and designed for lower maintenance. The northwest quarter of the cemetery is undeveloped woods. Several important structures by Providence architects serve the cemetery itself. Thomas Tefft's round-arched receiving tomb (1847) remains on Forest Avenue. The Neo-Gothic office and chapel was built in three stages, by Stone, Carpenter and Willson (1905) and John Hutchins Cady (1932 and 1945). Monuments in the older section of the cemetery are of fine quality. They include many by Thomas Tefft, a number by Alpheus Morse, and a wealth by the prominent local stonecutting firm, Tingley Brothers. Although some funerary sculpture exists—a fine example is the Sprague *gisants*, carved by Charles Hemenway, close to the original entrance road toward the middle of the cemetery—Swan Point's monuments are generally geometric in quality. Of special interest to the architectural pilgrim will be the mausoleum for Marsden Perry, onetime resident of the John Brown House. This is located at the cemetery's northeasternmost corner, and nearly as close to the river as it can be, its precinct surrounded by evergreens. His hillside mausoleum is fronted by an elevation of a two-story Georgian brick house which replicates as its entrance, at full scale, the portico of his Power Street mansion—as he commissioned Edmund Willson to remodel it. Scattered through the cemetery are the burial sites of many of the Providence architects whose work we have seen, mingled with grander monuments to their clients.

Cemetery, asylum, park, and parkway all are characterized by effulgent landscape in the naturalistic mode. Few places offer in such proximity the experience of four such handsome exemplars of the Victorian conviction of the beneficent effect for the city dweller of escape to nature. Together they induce "repose"—that nirvana for Victorian aesthetics, psychology, and redemption.

PR176 U.S. Post Office, Providence Main Post Office

1960 with extensive later additions. 24 Corliss St.

The post office, a large, triple-cross-vaulted structure in reinforced concrete, is enclosed by a purely functional wall which diminishes any sense of its true scale and makes it look unimpressive. While under construction as three vast, billowing enclosures of space coming to the ground without any interior supports, however, it was impressive indeed. It was built at a time when such vaulting in reinforced concrete, developed by European engineers such as Pier Luigi Nervi in Italy, was just being discovered in the United States and similarly employed in the near-contemporaneous three-vaulted concourse for the St. Louis Airport (1951–1958, Hellmuth, Obata and Kassabaum). This was the first fully automated post office in the United States, employing then radical German technology, since modified, as an experiment toward speeding the mail. It was also in the vanguard of the warehouse type of postal operation in an industrial park outside the city center, which funnels all mail coming to or originating in a region through a large-scale facility. As such, it marks the beginning of the end of the downtown city post office for pride of place in the postal system.

The State House and Northwest Providence

PR177 Rhode Island State House

1892–1904, McKim, Mead and White. 90 Smith St. (open to the public)

To local residents it is always the "State House," never the "Capitol," in deference to the old multi-state house system which prevailed, in theory if not quite in practice, in Rhode Island until this building opened in 1901. A key monument of the American Renaissance, the Rhode Island State House is the product of an elaborate—and probably "arranged"—competition. Its design came from McKim, Mead and White's office contemporaneously with the firm's Beaux-Arts classical work for the World's Columbian Exposition, and the rather unelaborated plans for the integration of the new State House with the central city were among the first applications of the ideals essayed in the "White City." As the principal designer for the State House, McKim scorned the gaudy and idiosyncratic ornamentation of capitols designed during the Victorian period for a "correct," if reduced, variation of Sir Christopher Wren's dome for St. Paul's. This he visually enlarged by the addition of corner turret domes in the manner of Michelangelo's St. Peter's.

As both a type and an ideal, the Rhode Island State House informed the designs of subsequent new state capitols for a generation. Programmatically, the building follows the established domed capitol scheme with legislative chambers to either side. However, the clarity of McKim's massing and plan, together with the "chasteness" of its Renaissance detail at a time when classical propriety had begun to assert its hegemony over Victorian picturesqueness and originality, gave the format new life. If fault is to be found with the exterior massing it would be, above all, in the wings. The projections to mark the legislative chambers are too unemphatic in wings a bit too long, which equally unemphatic detail emphasizes. (In the original design the wings were more compact, but with uniform windows which muted exterior articulation of the legislative chambers.) As a result of the extended wings, the centerpiece with its dome does not so much build from the mass below as interrupt a bland expanse of wall in a sudden eruptive ascent.

On the inside, the most dramatic aspects of the building are the trapezoidal-plan entrance halls and the central stair. Wide toward the outer walls, narrowing toward the stairs, the halls funnel the visitor from the north and south entrance fronts of the building to flights which make a pyramid of stairs within the rotunda. These rise to a landing, elevated in the center of the space between the ground and legislative floors, then break at right angles into another pair ascending east and west toward the two chambers. Actually, the space possesses some of the quality of a tight vertical shaft characteristic of Victorian staircases, a verticality which the raised stair landing halfway between the principal floors partially mitigates by raising the floor below the dome well above the ground level. The severe ornamentation of white marble in the lower portions of the space gives way to the lively color of coffers and murals in the upper reaches of the space and dome, the latter of which received the planned mural decoration only in 1947 with *The Settling of Providence*, by James Allen King. (The virtues painted in the squinches below the dome are apparently of earlier date.) What surely accords with McKim, Mead and White's intentions are the coffers in green and gilt, as well as the corridors in Pompeian red, gray, and ochre. The architects must have expected that sculpture would go into or onto places in the lower zone of the staircase which seem designed for it and appear a bit bare compared to the ornamentation in the dome above. Working drawings (now in the New-York Historical Society), however, give no indication of any intended embellishment for this part of the building.

The centrality of the elevated landing between the first and second stories makes one physically conscious of a plan organized around a central point: legislative chambers, House of Representatives and Senate to the west and east respectively, library and the Governor's Reception Room to the north and

PR177 Rhode Island State House

south, offices along the peripheral corridors. In contrast to the elaborate Neo-Renaissance embellishment of the library and the Governor's Reception Room, the legislative chambers are especially severe. That for the Senate, a half-hemisphere, is influenced by Benjamin Henry Latrobe's Greek Revival design for the Supreme Court Chamber in the national Capitol; that for the House of Representatives is rectangular, softened a bit by illusionistic tapestries by William Baumgarten and Company. Their severity is doubtless due to Henry Bacon's role as one of McKim's principal design assistants on the State House. Bacon was committed to Greek rather than the Roman and Renaissance forms which characterize McKim, Mead and White's design. (Later Bacon designed the principal Neo-Grecian American monument of the early twentieth century, the Lincoln Memorial in Washington, D.C., 1912–1922.) The austerity of portions of the staircase may be attributable to him as well. In any event, the contrasting decoration of the library and the reception room from Renaissance sources is a more elaborate and important indication of the firm's attitudes to interior design at this time. With the Rhode Island State House and the preceding Boston Public Library, McKim, Mead and White began its ascent to become the leading architecture firm of the then new American Renaissance.

The State House library is "leathery" Neo-Renaissance in tan and dull gold, relieved by touches of color in the printers' marks used decoratively in the coffered ceiling. Shelving around the room provides the equivalent of three to four stories of books as its principal embellishment. This space handsomely anticipates the libraries the firm later designed for a series of New York clubs (notably that for the University Club). The Governor's Reception Room, featuring a fine full-length portrait of George Washington by Rhode Island native Gilbert Stuart, shows more glitter, more color, more sumptuousness. It is a major example of Neo-Renaissance taste at a time when the transition from Victorian styles to the full-blown Renaissance Revival had just begun. It is more chaste than Hunt's contemporary Newport interiors, but McKim himself later criticized it (correctly, but too dismissively for such a splendid chamber) for being "pink, white, and gold, too liney, and too ballroomish." (The decorative schemes for the library and reception room, together with the colors originally specified for the corridors, were restored in the mid- 1990s.)

The south terrace of the State House offers an overview of Capital Center in the foreground, with the old downtown (or Downcity) behind. It also provides a panorama of the dramatic immediacy with which the tiers of College Hill houses and institutions rise out of the city's center, in summer poking through foliage, in winter starkly exposed.

PR178 Gloria Dei Evangelical Lutheran Church

1925–1928, Martin Hedmark in collaboration with Jackson, Robertson and Adams. 15 Hayes St. (near corner of Park St.)

Gloria Dei is unique in Providence. Built for a primarily Scandinavian congregation, it was designed by Martin Hedmark, a Swedish emigré, who arrived in the United States in 1923 with Gloria Dei as his first commission. It is one of the best of a number of churches he designed, mostly for Swedish Lutheran congregations. Its style is most simply described as Swedish Moderne, a national variant of international Art Deco or Art Moderne, for which the native designation is the Swedish National Romantic style. It is most prominently exemplified by two of Stockholm's landmarks, the City Hall (1911–1923) and Engelbrecht Church (1906–1914), on the latter of which Hedmark assisted. Gloria Dei's shaped forms in brick with limestone trim are meant to recall a famous Swedish castle, Gripsholm, in a free, decorative adaptation of its style and some of its motifs. The exterior is dominated by three towers: one with an onion dome, one with a spire, and the largest square-shaped to a stepped gable (somewhat in the manner of Austrian architect Joseph Maria Olbrich's tower for the Darmstadt Exhibition Hall).

Inside, the plaster walls of a boxy space are relieved by the semi-oval-plan apse, subtly echoed by curved walls to the arched inset windows, curved recesses to suggest transepts, and a cove molding from walls to flat ceiling. The craftsman's touch is conspicuous. The painted wooden chandeliers are aggressively sculptural as pendant, inverted obelisks from which stacks of gilded wings extend out of the four faces to symbolize the spread of the gospel by the four evangelists, each with a winged and flaming disk as the climactic eye of God. The wooden altar furniture is also exceptional. Chunky stained glass mingles with clear glass in the windows. All culminates in a mural by Newport artist Benny Collins, which conflates the symbolism of Good Friday, Easter, and the Ascension. But the symbolism is (except for the chandeliers, windows, and apse mural) covert, allowing the chasteness of the white plaster and simple woodwork to prevail.

Hedmark, who seems to have moved about, lived in Newport during the early 1950s, where he designed another Lutheran church. Record of the impressive proposed interior exists in a pamphlet on Gloria Dei, but the Newport church was never built. Churches erected by Swedish congregations are more numerous than one might expect in Rhode Island. This, the largest and most impressive Swedish-inspired church in the state, contrasts with the small, plainly carpentered buildings usually found in the neighborhoods of machine tool factories, to which highly skilled Swedish tool makers gravitated. Downhill of Gloria Dei and the State House were the former sites of the American Screw Company, to the east, and the Brown and Sharpe Machine Tool Company, to the west. Both ranked among the largest producers in their respective industries. Both have vanished from Providence: American Screw absorbed into another corporation, Brown and Sharpe to a smaller suburban plant in southern Rhode Island, the scope of its production reduced to a specialty within the broad spectrum of machine tool design that it once commanded. Its cast-off complex remains (see next entry). The vacated quarters of American Screw were, except for a peripheral building, lost to arson in the 1970s, even as the complex was slated for conversion to commercial and apartment use.

PR179 **Woonasquatucket River Industrial Corridor**

The Woonasquatucket River has served as the city's principal industrial corridor since the mid-nineteenth century. Industries at the east end of the river, nearer its mouth, are base-metal industries, such as Brown and Sharpe and Nicholson File, both major companies in the tool-making industry. Those farther out—enjoying cleaner water and usually making it dirtier—tended to be textile related. After 1848, a railroad parallel to the river reinforced the advantages of the location for industry.

The most extensive industrial complex in the area is Brown and Sharpe (PR179.1; 1872–c. 1940), in the vicinity of Holden and Promenade streets). These are typical of factories built in the late nineteenth and early twentieth centuries: brick buildings with large, multipane sash set between barely projecting piers. The finest and most recent of these brick pier-and-spandrel buildings are at the top of the hill and halfway down it along Holden Street. They are six-storied and are handsomely, even nobly, proportioned in the relation of the square-headed windows to the skeletal pier-and-spandrel walls. Pier corners are rounded against damage. These buildings are unusually corniced with stepped courses of quarry-faced light tan granite topped by a narrow banding in brown. The same light tan serves for windowsills and lintels.

The entire complex is visible from Kinsley Avenue on the opposite bank of the river. From there the tiered effect of the buildings up the slope makes visible the extent of the plant. The oldest unit (completed in 1872) is at the front, and east from Holden along Promenade Street, with wings extending back to the higher building behind (which fronts on Beech Street). Here segmental-arched windows punch the continuous surface of flat brick walls. The original treatment of three stories on a half-story basement (as the stepped cornice corbeling indicates, with a fourth story added later) is retained across the street. In the earliest buildings, cast iron columns support wrought iron beams, which, with brick arching, sustain concrete and wood floors. In the later buildings immediately behind (1895–1916 and later[?]), cast iron columns support concrete beams. The complex is largely intact except for the loss of its foundry (1902), which once existed front and center in what is now a void at the heart of the complex, presenting a massive, extended pyramidal effect created by a hipped monitor on a tall and steep hipped roof. This plant was converted to mixed industrial use, then to office and studio space, with light industrial use, and ironically renamed "The Foundry," the one component missing from the complex.

At Kinsley Avenue and Acorn Street is Nicholson File, which is architecturally of less interest than the factories just described, but was long the world's largest manufacturer of files. At 530 Kinsley, close to its termination at Eagle Street, is the Monohasset Mill (PR179.2; 1866, James C. Bucklin). It was built as a textile mill and later served the Armington and Sims Engine Company. By repute, Bucklin designed more than a hundred industrial buildings in the state during the early nineteenth century and, in Rhode Island, seems to have pioneered in their construction in brick. Specific factories documented as being definitely by him, however, are rare. This is one, and of special interest for this reason. However, it has curious aspects which partly seem to look back to earlier factory buildings and are partly puzzling. It is unusually wide for its length, giving it a somewhat chunky quality, which the very broad projection of one of two stair towers intensifies. The peak of its gable roof is clipped, with narrow trapdoor monitors into the attic.

PR180 Church of the Blessed Sacrament

1899–1905, Heins and La Farge. 169 Academy Ave. (at Regent Ave.)

From the 1890s through the first decades of the twentieth century, late medieval and early Renaissance Italian churches in brick such as those in Siena and Verona became very popular models for American Roman Catholic parishes. Interest in artistic brickwork and terra-cotta was substantial at the time—so much so that *The Brickbuilder* became a popular professional journal. Moreover, this type of church, with its exposed timber construction to support a gabled roof, avoided the problems and expense of vaulting in stone, and its square-planned campanile also eliminated the complexities of spires. In any event, the western part of Providence and other industrial cities around it are dotted with parish campaniles in brick, all of the period of the Church of the Blessed Sacrament. This is the finest example of the type in the city for both exterior and interior. Apart from the fine quality of the brickwork, the triple-arched Neo-Romanesque portal and wheel window above it are handsomely trimmed in red granite, and the staging of the arched windows at the top of the almost free-standing campanile is nicely proportioned. But the special attraction of this church is its richly finished basilica-plan interior, with fittings and fixtures by Heins and La Farge together with several windows by the architect's father, John La Farge. Changes at the altar occasioned by Vatican II have little harmed it.

PR181 Zachariah Allen House

c. 1789, Amos Allen, builder-designer. 1093 Smith Street (at Modena Avenue and Eaton St.)

Sea captain Zachariah Allen commissioned this elaborate if somewhat charmingly provincial house of his carpenter brother for a conspicuous site. The entry is especially ambitious, with its Ionic pilasters and pediment, set against a rusticated panel (its projection as a vestibule probably later). Somewhat crude lintel rustication and keystones over the ground-floor windows and quoining enlarge the display. The erratic placement of the central upstairs window may be a later alteration; but the off-center chimney and (most interesting) the belated "medievalism" of a rare projecting eave over a side elevation with more windows out of alignment also dilute the sophistication elsewhere. The famous early-nineteenth-century industrialist of the same name was raised in this house.

PR182 Winsor–Swan–Whitman Farm

c. 1750, center section. c. 1800–1810, street front and kitchen wings. 416 Eaton St. (at Sharon St.)

This Federal house is one of the few remaining within the city limits of the small farmhouses built during the eighteenth and early nineteenth centuries in the hinterlands beyond the compact part of Providence. Its two-stage construction is best understood by viewing the three parts which step down the hill from the corner of Sharon Street. Initially the house consisted of the one-story, center-chimneyed middle section facing south. At the beginning of the nineteenth century an in-line kitchen ell extended it toward the rear of the property, and a narrow, two-story addition, right angled to the mid-eighteenth-century nucleus, enlarged it toward Eaton Street. The new front is plainly treated, with a transom-lit, entablature-capped door, but with moldings of some delicacy. Its complement of auxiliary structures is unusually complete, including a well, a corncrib, a shop, a wagon and woodshed, and even a schoolhouse at the rear of the lot.

PR183 Providence College Campus

Corner of Eaton St. and River Ave.

Providence College was established in 1917 by the Dominican order of friars at the invitation of the Roman Catholic Bishop of Providence. At the end of an axis diagonal to the intersection of the two streets is the original building in Collegiate Gothic, Bishop Harkins Hall (1919, Matthew Sullivan). Of most interest on the campus are two houses, survivors from

Mount Pleasant's nineteenth-century country estates, both now used as academic buildings.

PR183.1 Charles Bradley House

c. 1853. Thomas A. Tefft. 235 Eaton St. (.2 mile from college entrance gate)

PR183.2 William Bailey House (Hillwood)

c. 1855. (.25 mile from college entrance gate)

These two houses in the Italian Villa Style are the only large country retreats remaining of half a dozen built here on the rolling slopes of Mount Pleasant in the mid-nineteenth century, when this was a fashionable area for large estates. They tended to supersede farms such as the Winsor-Swan-Whitman homestead, described in the preceding entry. Both are timber-framed, coursed-ashlar-faced buildings with picturesque profiles, cross-gable roofs, and prominent corner towers. Tefft's square-towered Bradley House is much the better designed of the two—in fact, the finest extant example of the style in Providence, thanks in part to the example of Richard Upjohn's well-published Edward King House in Newport, which Tefft certainly knew first hand. A projecting square bay window with a balustraded roof makes an outdoor balcony for the triplet window pair above. Another balconied pair is canopied with replicated awning fringe. Tripled loggia-like windows open the tower in four directions. It is as though Tefft sought every possible means to express the scenic intention of his elevated villa in what was then an open rolling panorama with a view to Providence. Despite this varied treatment of rectangular and arched openings, however, they are superbly placed and fused with one another in units of two and three openings both horizontally and vertically from floor to floor. Hence the picturesque diversity also has the gravity of order and compactness. The Bradleys, who lost their only child, left the house and an endowment to found a children's hospital named in her memory. The Emma Pendleton Bradley Hospital had its beginnings here as a pioneer institution in child psychiatry, before moving to larger and better-equipped quarters in East Providence.

The nearby Bailey House is a broad-eaved, L-shaped farmhouse aggrandized by its masonry walls. The mostly narrow, slotted windows are more spottily placed, although charmingly "shaded" by louvered caps. Whereas the Bradley House is all planes, the front of this L-shaped mass breaks out into angles. With a bay window, a polygonal tower loosely attached to the mass behind, and a polygonal porch interlocked between, it seeks picturesqueness more hectically—and more venturesomely.

PR184 Wanskuck Mill and Village

Founded for the manufacture of woolen worsteds by three major textile industrialists, among whom Jesse H. Metcalf became dominant, the Wanskuck Company eventually took over other Providence plants, as well as country mills in Oakland and Mohegan (see Burrillville). The focal point of the brick mill (PR184.1; 1862–1864, 1874–1875; major building, c. 1885), 725 Branch Avenue, which now houses various operations, is its bottle-shaped tower, in which the square plan below tapers to an octagonal belfry, with an ogee-curved copper roof and arched openings in every face, each of which contains a (fast deteriorating) double-arched wooden frame. Such bottle-shaped towers derive from brick steeples and industrial chimneys. In some of these towers the tapered transition is "shouldered" with curved shapes; in others, as here, it is wholly angular and more easily built. Several extant Rhode Island mills have towers of this type. Of all of them, Wanskuck and the Pawtucket Hair Cloth Mill in Central Falls (CF2) are probably the finest: this one partly for the superb proportioning of the segmental-arched loading doors within their recessed arched field; partly for the handsome circular inset for the oculus at the point of transition, with its bronze letters proclaiming the mill and its date; partly for the commanding scale and shape of the belfry. For the rest, the five-story design is handsomely straightforward. Its plain brick wall is punched with windows silled in granite and capped with segmental brick arches, which miraculously retain their original multipaned sash. Across the way at 754 Branch Avenue is Wanskuck Hall (PR184.2; 1881), in minimal brick polychrome with spotty granite trim and widely bracketed eaves, which served for community meetings and as a social center.

Up the hill behind Wanskuck Hall, at 21–28 Winchester and 29–36 Vicksburg streets, is the company's housing (PR184.3; 1864). Rows of four brick duplexes in granite trim with end chimneys face each of these two streets. Although a few other examples of piered wall construction may be found in what little brick industrial housing exists in the state, no other so consistently employs this kind of wall to frame the openings. Unusual, too, and perhaps unique is the lift of the flank gable into a cross gable over the paired entrances. (The precedent is probably English.) Other nearby streets show less well-preserved wooden housing of various vintages, which is more characteristic of Rhode Island mill housing (but see especially Lonsdale, Berkeley, and Ashton, in Cumberland, for other brick examples). The mill superintendent's house (PR184.4; 1880, Stone and Carpenter), 158 Woodward Road, is a fine example of high, angular, multigabled Stick Style, with structural framing (or its surrogate) revealed on the exterior, which is covered in a mix of clapboard and shingle. Awning-like shingled hoods emphasize windows toward the street. The peaks of the principal cross gables step out slightly, with medievalizing ornament in their pinnacles under the projected eaves.

As a blunt evocation of English rural precedents, the Gothic Revival Roger Williams Baptist Church, the company church for the village of Wanskuck (PR184.5; 1866; 1892 addition, Stone, Carpenter and Willson), 201 Woodward Road, could well have been built by mill masons. Before the diffuse addition to the rear, it was quite compact—truly a "chapel" consisting only of the chunky bell tower with its flaring, pyramidal cap and cross-gabled meeting hall behind. Curled over the simplest of pointed windows on both side gables is the church's name in a variant of the bronze letters on the mill tower. The cornered entrance into the western gable seems to have been part of the original building.

Across Woodward Road from the church is Wanskuck Park (PR184.6; opened 1949), originally the site of the Jesse Metcalf estate. Metcalf's wife, Helen Adelia Rowe Metcalf, was the force behind the founding of the Rhode Island School of Design. Their son, with the same name as his father, and a United States senator, succeeded to the house. It was his widow, Louisa Sharpe Metcalf, who gave the property to the city as a park with the proviso that the family mansion be demolished. Wanskuck Park had barely opened to the public before the closing of the mill as a Wanskuck operation in the 1950s.

Broadway and South Providence

Broadway really became such in 1854, when it reached its present length and breadth of 80 feet for the full distance, making it the broadest way in Providence at the time. Lined with trees, and from the 1860s served by trolleys, it became a street of choice for Victorians with new fortunes. In contrast to the retention of colonial types in the conservative Italianate palazzi on College Hill, here clients ventured into the more flamboyant and individualistic display of true High Victorian styles and into showy versions of their Queen Anne aftermath, from roughly the 1860s through the beginning of the twentieth century. As happened to most such radial arteries in other cities, decline occurred with the coming of the automobile, as Broadway became a major traffic corridor. Although most of the mansions became degraded as places of business, Broadway fortunately avoided the commercial excesses which tended to occur on most other streets of its kind. Enough Victorian grandeur remained to make possible its redemption as a local historic zoning district in 1982. What follows samples a few highlights, minus the canopy of trees which once dignified the street.

PR185 Colin C. Baker Duplex
1872. 243–245 Broadway

Alterations in the mansard spoil this otherwise impressive investment Second Empire duplex—although a side view reveals the nature of what was changed. Stacked two-storied, semipentagonal (almost semicircular) bays with tall, narrow windows between bracketed pilasters flank the double porch and its cascade of granite steps. Arches above the second-story windows make a transition to the deeply bracketed cornice and the (original) roll of the culminating dormers (with mini-gables between on the side elevations).

PR186 Jerothmul B. Barnaby House

1875. 1888, conservatory, rear tower and addition, carriage house, Stone, Carpenter and Willson. 299 Broadway and 159 Sutton St.

Rambunctiousness more than form stops us at number 299 Broadway. Jerothmul Barnaby was a self-made millionaire in ready-to-wear. His gawkily tall and flaring mansarded house began as a somewhat pinched mass, which breaks out at one corner up front into a square pyramidal tower, and off one side into a clumsy bonnet gable. It is the bracketed "Modern Gothic," or Eastlake, carpentry overall (somewhat at odds with the more monumental, masonry-derived Gothicism of its porch) that makes this interesting. To this start, Stone, Carpenter and Willson added a copper-clad circular conservatory to the west with a frieze of lancet openings filled with stained glass floral ornament and, behind this, at the rear of the house, a swelling turret with a glass and metal umbrella roof over the garden terrace and a loggia on top. The tower echoes that of the remarkable Queen Anne style Conrad Building (PR33) downtown, which Barnaby had commissioned from the same firm a little earlier and named for his son-in-law. There the feature is embellished with "Saracenic" detail. Does this indicate that Stone, Carpenter and Willson also designed the unusual frontispiece in the same style for Barnaby's carriage house on Sutton Street to the rear of his house? It is a mixed affair with keyhole window triplets on either side of the entrance and, above it, a wheel window (derived from Islamic ornament rather than from Gothic design) set within another keyhole inset, which is surrounded by a field of colored tile and terracotta. In both instances, the fervor for Saracenic motifs seems to have been more Barnaby's than his architects', since the style is exceptional in the firm's work. So the house is a strange but fascinating conglomeration.

Shortly after her husband's death Mrs. Barnaby died from a poisoned bottle of whiskey sent to her while she was on vacation. The gift was traced to her physician, who stood to benefit from her will. In the coloratura of press coverage at the time, the turreted grandiosity of the house became "Barnaby's Castle." As for the convicted physician, he cheated execution by committing suicide in his cell at the state prison—by Stone and Carpenter—in Cranston.

PR187 A. A. Spitz Duplex

1902. 409–411 Broadway

Another duplex presents a late example of Queen Anne style in which the push-pull of cross gables, dormers, and bays out from a cubic, hip-roofed core seems to be restained by incipient classical enframement. The cage, which is composed of paneled pilasters at key corners butting exaggerated entablatures at each story, and the balustrading of important bay windows below, can also be viewed as a last hurrah for the kind of revealed framing characteristic of the Stick Style. As if to underscore the incipient classicism, the complicated shapes of the entrance porches at either end display vernacularized versions of grand Beaux-Arts planning in which columns and piers work in concert to stake out complicated shapes. Both the dainty garlands capping the second-story bay fronts and the Palladian window in the attic "pediment" also proclaim the arrival of the classical dispensation at the turn of the century.

PR188 Colin C. Baker Row Houses

1868. 412–428 Broadway

The same Colin Baker who commissioned the duplex at 243–245 also invested in this row, which is perhaps the best-preserved example

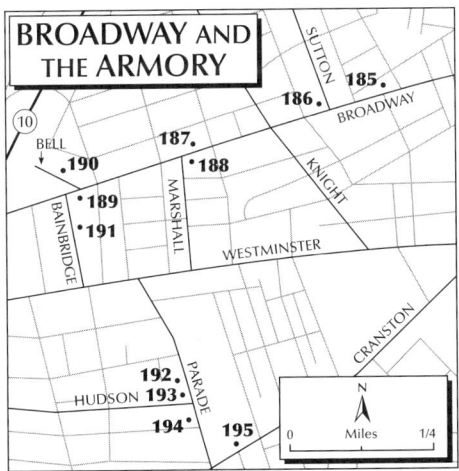

in the city of monumental Second Empire row housing in the manner of what was then the latest fashion in Back Bay Boston. The *ABAABAABA* rhythm of the mansard dormers indicates that the row was conceived as three palatial houses abutting one another, each duplexed up the middle. The impact of the mansard is increased by its double slope and the breaking of the cornice by tiny round-headed attic windows fronted by decorative iron balustrading over each of the principal bays.

PR189 John K. Kendrick–George W. Prentice–Anna Tirocchi House

1867, Perez Mason. c. 1881, c. 1931. 514 Broadway

More rambunctiousness appears at number 514, built for a loom harness manufacturer and designed by Perez Mason, an architect from Fall River, Massachusetts, who eventually lived on Broadway himself. A loose, clapboarded hybrid, part mansarded Second Empire palace, part towered villa, the house is uniquely set back from Broadway on a large, landscaped property. Details more than composition attract the eye here: a flaring mansard with cupola topping a spindly tower; bonnet gables of two types; a curious sagging, tentlike hood molding over accent windows; and a plethora of varied brackets and pendant ornamentation. Later owners George W. Prentice, a buttonhook manufacturer, and Anna Tirocchi, a very prestigious dressmaker to the Providence carriage trade, made additions and alterations to the house, which may account for its particular character.

PR190 Bell Street Chapel

1875, William R. Walker. 5 Bell St.

At the end of one-block Bell Street is the anomaly of the classical brick and brownstone Bell Street Chapel. Surprisingly both for its time and its architect, who more typically designed in High Victorian styles, it features the portico of the Roman Temple of Nîmes (A.D. 212) elevated on a rusticated ground-floor base. Originally the chapel sat at the entrance to James Eddy's estate, Pine Grove (the entrance turnaround for the demolished house and the pines are still extant behind the chapel). Eddy, an art dealer, erected the chapel for a nondogmatic, liberal congregation (and doubtless requested that it be modeled on the Temple of Nîmes). The interior (most unclassical) still retains its plain, round-arched Victorian flavor, with landscape paintings intended as revelations of divine emanation as much as works of art. Eddy's collection of rocks and minerals, in the front hall to the chapel, was meant to work to the same end. Now used by a Unitarian Universalist congregation, the temple is still a venue for exhibitions by local artists, as Eddy would have wished.

PR191 George E. Boyden House

1882, Gould and Angell. 20 Bainbridge Ave.

Dry goods and a penny arcade financed the George E. Boyden House. From the center of the flank roof projects a cross gable, to which small Queen Anne porches cling. Into one L-shaped corner is fitted a double sled-roofed entrance porch; into the other, a conical, semioctagonal sitting porch, with a bedroom balcony porch above and behind. Yet another shed-roofed variant off one side provides entrance to what is apparently a small adjunct apartment.

PR192–PR195 The Armory and Parade Street

The following group of Queen Anne houses on Parade Street comprises a few of those rehabilitated in the area known as the Armory District (for the Cranston Street Armory), a focus of the Providence Preservation Society, which established its Revolving Fund to buy up abandoned houses of quality in order to facilitate their restoration. In the late 1960s and early 1970s the society, with backing from other preservation groups, expanded its activities to embrace more than an elitist and parochial concern with College Hill, thereby justifying its name as a society concerned with preservation citywide. For other restored nineteenth-century houses in the area explore the side streets back from Parade Street for a couple of blocks.

PR192 Frederick W. Hartwell and Joseph C. Hartshorn Houses

1883–1884, E. I. Nickerson. 77 and 81 Parade St.

In 1824, heirless Ebenezer Knight Dexter, owner of large parcels of land across the city, left Providence ten acres here for a drill field and park. In the late nineteenth century, the streets on either side of the Dexter Parade filled with stylish houses, many in the Queen Anne manner. Notable among them are numbers 77 and 81 Parade St. They were developed as a pair, with the same clapboarded ground floor and shingled upper story, and sharing the same metal fence. Joseph Hartshorn, secretary for a steam and gas pipe company, commissioned number 81 for his wife and himself, and shortly thereafter number 77 for his daughter and her husband, who served as the company's treasurer. Mrs. Hartshorn died before their house was completed, and her husband never moved in.

At the core of both houses is a high, gabled box, against which peripheral gabled and shed-roofed shapes butt, or past which they slide—all complicated by the variety of features, texture, and ornament customary in the style. In number 81, one angle of the front gable extends across the front elevation in a long diagonal to become the slightly inset shed roof of the cornered porch, while the L-gable of its principal mass remains high. Next door, the slope is back to front. A pyramidal hipped roof spills down (originally in slate) over an L-shaped porch. The gabled entrance to the porch projects forward, just off center of the projection of the principal front gable, thus adding to the multitude of diagonal relationships, as also does the second-story window placement above. Note, too, the bracketed shelter over a canted window set just behind the lower left angle of the principal elevational gable. Its echo occurs over another canted window on the ground floor at the opposite corner of the house, thereby ghosting the literal diagonal which slashes across the front of its neighbor. What satisfaction both architect and carpenters must have experienced in working out these disciplined but lively counterpoints in compositional diagonals, revealed structure, and varied surface treatments in a style that, more than any other in America, glorifies virtuoso carpentry!

PR193 Benjamin F. Arnold House

1883. 89–91 Parade St.

PR194 Charles A. Hopkins House

1875. 103 Parade St.

In numbers 89–91 Parade Street, owned by a partner in a grocery concern, the two units appear as different halves of an asymmetrical composition. The simulated structure enframes openings and decorative panels. As we look at such elevations the pseudostructure alternately combines the features it frames in various combinations, or frame and feature may fall apart into their individual components. Number 103, built for an insurance agent, takes us farther back in time, from the early 1880s to 1875, the structural framework now tending to exert more restraint. Overall, this house is the simplest of L-shaped mansarded boxes (although originally with iron cresting) into which an entrance porch is fitted. Yet the splaying of its bracketed corners above the ground story transforms the shape, as does the concentration of the windows to the center of the front elevation as a kind of compositional updraft to the dormer. Daringly inverted scroll brackets serve as open dampers to draw the ascending columns of form through the bracketed cornice.

PR192 Frederick W. Hartwell and Joseph C. Hartshorn Houses

PR195 Cranston Street Armory

1907, William R. Walker and Son. 375 Cranston St.

The looming centerpiece of the district is the Cranston Street Armory. General William Walker used his high commission during the Civil War and active continuing interest in military affairs to good professional effect in garnering four large armory commissions in the state during the later nineteenth century (see PA18 and WO17), plus one in Westerly not included in this volume). This is by far the grandest of the four and is impressive measured against any American armory. The drill hall, lighted by a monitor, sits between identical castellated endpieces. These culminate in square, turreted towers centered on the (north-south) front elevations, and semiturrets swell from the (east-west) side elevations. The building is superbly crafted of yellow brick, penetrated by triplets of tall slit windows, arched and square-headed, with plenty of smooth-faced wall between, on a two-story base of rough gray granite, which also serves as spare trim above. Cylindrical turrets, bracketed from granite bases carved with leaves and projecting at all the principal corners of the endpieces, echo the half-buried cylindrical towers that bulge from the side walls. Along the side elevations, a cross gable midway and piers to reinforce the drill hall walls provide further excuse for mini-towers, square instead of round. The walls rise to corbeled moldings and crenellations, those in the principal cornice slotted for arrows, with a final capping in sheet copper, easily dented by the enemy response, but handsome in its verdigris weathering against the yellow brick. Where the slopes of the drill hall roof spread beyond the endpieces, stepped granite caps the brick walls. Cavernous, round-arched entrances at either extremity (each with loggias flanked by another pair of turrets) provide views, immediately inside, up a circular shaft through four floors of work and meeting spaces. The impressive free-span space of the drill hall—235 feet by 165 feet and 90 feet from the floor to the crown of its arching—depends on a mixed, light-metal structure of archlike girders tapered to the floor, stabilized by a plethora of cross bracing. Abandoned by the National Guard in 1996, this spectacular behemoth awaits a decision by the state on several proposals for future use.

(Note: For locations of sites PR196–PR198, see the Providence overview map on pp. 30–31.)

PR196 Grace Church Cemetery

c. 1848, platted by Cushing and Wallace. Elmwood Ave. and Broad St.

At the corner of Elmwood Avenue and Broad Street is the Gothic Revival gatekeeper's house (c. 1860) of Grace Church Cemetery, which now serves as headquarters for the Elmwood Foundation. Continue south on Elmwood Avenue. This, another area of restoration, is known as the Elmwood District. Those interested might make a detour, from the end of the cemetery three blocks left on Parkis Avenue to see a Victorian street of grand houses, including the Davol House (c. 1872), number 48. This entire street is largely owned by a single individual who restored the exteriors of these houses while improving their previous predominant use mostly as rooming houses.

PR197 Knight Memorial Library

1923–1924, Edward S. Tilton. 271 Elmwood Ave.

This must be one of the last library buildings to pay such pious obeisance to McKim, Mead and White's Neo-Renaissance Boston Public Library, a good thirty years after its building and twenty-five years after Willson used the same inspiration for the main Providence Public Library building (PR13). Both employed yellow brick with stone trim, but whereas Willson's sparkled with ornament, Tilton's is severe and chaste in aspect.

PR198 Roger Williams Park

1871–1872. 1878 and later, H. W. S. Cleveland, landscape architect. 950 Elmwood Ave.

This 430-acre public park evolved from a smaller, 100-acre parcel bequeathed to the city in 1871 by Betsy Williams, a descendant of Roger Williams. H. W. S. Cleveland began his design in 1878. Cleveland was the chief competitor of the principal nineteenth-century park designer, Frederick Law Olmsted, and his work deserves to be better known than it is. As perhaps one of Cleveland's best park designs (the other being a plan for Minneapolis), Roger Williams Park has national, as well as local, import. Cleveland envisioned this park, in Olmsted's concept, as a bead in an "emerald necklace" of green spaces linked by landscaped parkways comparable to his own Blackstone Boulevard. As a start toward such a circuit of

PR195 Cranston Street Armory

green spaces to make a park "system," he specifically recommended that Roger Williams be linked by a short parkway to open space at Field's Point, which would have anchored the system to Narragansett Bay. But it was not to be. Ironically, Field's Point became the site of Providence's sewerage plant, and one must today visit Minneapolis to experience the result of Cleveland's larger dream.

His original plan for Roger Williams, which provided for creating lakes from a swampy site, remains in the area around the main entrance from Elmwood Avenue, although later additions to the park largely follow his scheme or its precepts. Its most notable feature is the meander of connected artificial ponds which so extensively snakes through the park that one perceives much of it as islands, shoreline, and ridges between water.

Following a period of decline and dilapidation, especially after World War II, the park was systematically refurbished and restored during the 1980s and early 1990s as the result of energetic and enlightened park management in combination with public and governmental support. Its condition today is worthy of its status among the notable nineteenth-century park designs.

Navigation of the park's winding and branching circuits can follow many well-marked routes. From the Elmwood gate, Betsy Williams's own tiny homestead (c. 1773) is conveniently at hand, as the beginning in a double sense. A one-and-one-half-story gambrel-roofed cottage with some minor additions, it has a simple transom door extending from ground to eaves, flanked by only two nine-by-six sash windows. None of the openings is quite symmetrically placed; nor is the "center" chimney quite centered. Incongruously, but perhaps appropriately in this context, it is enshrined in a diminutive picketed and gardened precinct.

Close at hand is the Casino (1896–1897, Edwin T. Banning), for receptions and banquets. An imposingly long and high Neo-Colonial brick box, hip-roofed and cross-gabled, it would be very plain except for the broad wraparound porch of its first floor, with an open and parapeted deck at the second, which becomes at the center, front and rear, two-story porches swelling as semicircular volumes ringed with giant columns. In sheer size, this is probably the city's most spectacular paean to the American porch. Its restored interiors contain a mix of varnished and painted woodwork with brick fireplaces appropriate for a park casino of its period. Upstairs, however, is a creamy-white banqueting room, embellished in a dainty porcelain manner in gilt and painted flowers. This is extended all around by its splendid open deck, except for the roofed semicircles front and back. Adjacent is a wooden bridge (on metal construction) over a ravine, with a tall, cross-gabled gazebo at the center, decked out in scroll-sawn stenciled ornament. The pond below contains the bandstand (1915, John Hutchins Cady), a classical revival structure with an open, colonnaded Ionic rotunda as the climax of a metal-fenced, masonry platform for folding chairs projecting out into the water, but with plenty of grassy slope for audience overflow.

Farther south, in its own natural dell, and also backed by water, is the classical Temple of Music (PR198.1; 1924, William T. Aldrich), for larger performances. Formally, it ranks first among the park's monuments, done by Aldrich while he was also at work on his contribution to the Museum of the Rhode Island School of Design. All in marble, it is a tall, narrow, hip-roofed block. It also frames the rectangular opening cut through its core to provide a landscape backdrop. Across both faces of the block, ranges of four handsome Ionic columns screen the void, with more columns paired in inset panels on each of the narrow ends. Stairs from the bases of the columns spill downward and spread onto the marble performance platform below, from the ends of which a low, semicircular, marble-faced retaining wall defines a slightly depressed half-circle of lawn (with another short flight of stairs from the platform down to it) to mark the privileged segment of the audience space. It defines as well the base of the half bowl of the fan of slopes around it, while also coupling the realms of nature and of the larger audience gathered on its slopes, to that of abstract geometry.

The Dalrymple Boathouse (1895–1896, Martin and Hall) is a thin version of English half timbering in Queen Anne style, but a pleasant control center for the paddle boats' wanderings. Nearby are a nice metal footbridge, probably from the 1890s, and a modern merry-go-round with plastic horses. Looping north from this is the Museum of Natural History (1894–1895, Martin and Hall), in the Chateauesque, or François Premier, manner in buff brick with a mixture of tan stone and delicate terra-cotta detailing. Among its exhibitions of local flora, fauna, and geography, one section is devoted to the history of Roger Williams Park.

Near the sizable zoo's menagerie (1891–1892 and later) is the statue of the canine hero of the Hoppin House fire (PR97). The zoo's accretions over time have been too ad hoc to

PR198 Roger Williams Park, Temple of Music

make much of an impression as an overall plan. But many of its habitat displays are handsome, most notable perhaps that for polar bears. The most recent (opened 1991) evokes an African savanna for elephants, giraffes, and zebras with wooden and rush architecture as an accompaniment. A Victorian brick barn converted to administrative use for the zoo is close to the Elmwood gate.

Providence Periphery
PAWTUCKET, CENTRAL FALLS, NORTH PROVIDENCE, JOHNSTON, CRANSTON

Pawtucket (PA)

PAWTUCKET TAKES ITS NAME FROM AN ALGONQUIAN WORD MEANing "at the falls." The area below the falls of the Blackstone River—a major ford long before the arrival of white colonists—was a place of good fishing in the aerated pools and a place for informal meetings, sometimes tense, of tribes. Here the river created a boundary for the Narraganset, Wampanoag, and Nipmuc tribes, where the vaguely determined corners of their territories came together and where paths from one territory to another converged. Today one must hunt for the falls, hidden as they are beneath a double width of highway bridge, which conceals both a primeval site of importance and the primal cause for the city's being. It was not just the falls, but the incline of the river bottom and the fast-moving water above it that initially drew industry to Pawtucket. Symbolically, this final drop at the mouth of the Blackstone, while only middling in height, roars with the impressive volume of water that slides over it in a twisting sheet and drops to the tidal calm of the few final miles of its course, inexplicably designated in quick succession as the Pawtucket and Seekonk rivers. It is as though the roar would celebrate, in as grand a finale as Rhode Island topography permits, the importance of the Blackstone and of Pawtucket in the history of nineteenth-century industry. Insofar as the beginnings of the American Industrial Revolution can be pinpointed to a particular place, the honor traditionally goes to Slater Mill (as it has come to be known), beside its own spill of water over a modest dam a few hundred feet above the "fall of the water."

Here, American textile technology began to evolve as a mechanized factory operation, largely from the knowledge of the secret processes of Arkwright's invention which an ambitious overseer for a British textile manufacturer, Samuel Slater, brought, mostly in his head, as an immigrant to the United States. He teamed with William Almy and

Moses Brown, Providence merchants and traders, who were looking for new investment ventures. Brown had gradually gathered at an old fulling mill in Pawtucket examples of all the pioneer machines for the mechanized spinning of cotton that he could locate in and around Rhode Island in order to test and combine their capabilities. It was a pioneering example of systematic industrial research and development. He chose Pawtucket because of its waterpower and because the waterpower had attracted skilled mechanics. In the early eighteenth century the leading mechanic-industrialists in the area were Joseph Jenks and his son, a family destined for continued prominence in the machine tool industry during the nineteenth century. By the late eighteenth century, however, it was Oziel Wilkinson and, in time, his five sons who took the lead. Wilkinson's furnaces and forges produced a variety of products, whatever he was asked to come up with or saw a market for: nails, screws, barrel hoops, oil presses, and, during the Revolution, cannon. But Wilkinson's most important products—as also was the case for other mechanic-industrialists like him at this time—were the machines he devised or adapted for what he made. Wilkinson, together with his mechanic assistant Sylvanus Brown (no relation to Moses) tackled the collection of primitive spinning machines. Recent research gives this native group credit for progressing further toward mechanized textile production than earlier accounts had been given them, and thereby somewhat diminishes the impact of William Slater's arrival on the outcome. Still, his knowledge certainly sped the process and facilitated the coordination of the technology and its organization for production. After his arrival in January 1790, Slater provided the information that Wilkinson and Sylvanus Brown needed. As part of the bargain for sharing it with them, the Providence investors made Slater a partner, and the firm became Almy, Brown and Slater. On December 20, 1790, cotton spinning operations got underway. Three years later the firm moved into its own newly built mill, which, as later enlarged and much rebuilt, exists today at the center of the city as a museum to the beginnings of the American textile industry. In 1799 Slater, who had married one of Wilkinson's daughters, left Almy and Brown to team up with the Wilkinson clan in establishing a second cotton manufactory across the river, Samuel Slater and Company, which started operations in 1801. By 1813, thirteen mills operated in Pawtucket, the only other still standing being another Wilkinson mill, of 1810–1811, which stands beside the original clapboard mill as part of the museum complex.

In 1829 a depression was initiated in Rhode Island by the collapse of what was then known as A. and I. Wilkinson and, in its wake, the ruin of other Pawtucket manufacturers and banks. It took a decade, until the early 1840s, for Pawtucket manufacturing to begin its revival, with textile and machine tool plants predominating. Most of the pre–Civil War factory buildings have disappeared, but no Rhode Island city offers a collection of brick factory buildings of greater range and quality from the end of the 1860s to around 1920. These include the industrial complex which, better than any other in the state, still preserves a sense of the impressive size of the grandest Rhode Island textile operations of the nineteenth and twentieth centuries. It had its beginnings in 1869 when the local Conant Thread Company combined with the British firm of J. and P. Coats in what eventually resulted in the largest thread factory in the world. The plants of other very sizable industrial complexes remain in Pawtucket—that for the Royal Weaving Mill, for example, once among the largest silk manufacturers in the world. None extant, however, makes its size as visibly apparent as the premises of the former Conant-Coats operation, although it is also true that the lesser volumes of water in Rhode Island's rivers prohibited any of its mills from attaining the scale of the brick mills which lined the Merrimac at Lowell and Lawrence in Massachusetts, or those at Manchester in New Hampshire.

Scattered about the city are an interesting collection of houses of mill executives and business leaders. More often they are substantial rather than truly palatial, although several would have been considered mansions in their day. Company-built housing for workers is rare in Rhode Island cities. Wherever private builders provided rental housing, mill operators tended to relieve themselves of this responsibility. The most distinctive worker housing in Pawtucket, as in Providence and Woonsocket, was provided by the triple-decker, eventually despised but more recently viewed with renewed respect. Pawtucket and adjacent Central Falls once boasted many of these dwellings, but they have been so extensively disfigured and destroyed in both cities that they are best seen in Woonsocket.

Aside from its brick mills, Pawtucket also provides the preeminent Rhode Island example of 1960s downtown planning. No other downtown in the state was as badly shattered by an interstate expressway as Pawtucket. No other city more wholeheartedly embraced the opportunity to "improve" itself from the resulting clearance. The best and the worst of urban planning in this period are more evident in Pawtucket than in any other Rhode Island city. The 1960s vision of what tomorrow's city might be has itself become "historic."

PA1 **Louis Kotzow House**

c. 1875. 641 East Ave. (at Progress St.)

This beautifully maintained house, built for a jewelry manufacturer, illustrates the delightful manner in which the Queen Anne designer frequently gave vernacular carpentry exotic forms. Across an L-shaped mass, roofed with steep pitches, an oversized dormer, and deep overhangs, he fitted a Japanese-inspired porch. Its Japanese aura depends on its ensemble of heavy posts, the exaggerated horizontality of its scroll bracketing, the frieze with alternating panels of revealed structure and scroll-cut ornament, the simulated strapping of the ends of the bracketing to the structure above (like the metal straps which bind adjacent timbers in Japanese building), the flare of the roof and the upward slant of its underside plane. In the frieze panels, too, the delicate leaves and berries sawn and drilled from stencil patterns may also have been meant to recall Japanese flower painting on paper wall panels (although the same can hardly be said of the stencil-sawn ornamentation of the board balustrading). The simple elegance of the ornamented frames of the upstairs windows also suggests Japanese precedent. This and the house described in the next entry were built on sites purchased from the German Cooperative Land Association of Providence, with right of first refusal for the association should an owner sell.

PA2 **Frederick Scholze House**

c. 1874. 625 East Ave.

Larger than the Louis Kotzow House, this was built for a cabinetmaker in what was then called the "modern Gothic" manner, with the same L-shaped massing around a porch, but with more numerous dormers and bay windows. Similarities in the paneled style suggest that the same designer did both. The porch and front bay window show the same basic design, if somewhat simplified, except that the curved flare to the projecting eaves and the rounding of window corners are additional Oriental incidents. This and the Katzow House also retain identical pairs of terra-cotta chimney pots in the "Gothic" or "Tudor" manner.

PA3 Charles W. Shea Senior High School (Pawtucket West High School)
1938–1939, John F. O'Malley. 485 East Ave.

This ambitious school design in the Moderne manner (as it was then called, or Art Deco, as it is more usually known today) is one of two ambitious civic designs in the same style which cap O'Malley's career (see PA16). In limestone and yellow brick, trimmed with green glazed brick, the school sits high above the street. Its three symmetrically ordered entrances, at the recessed center and the projecting ends of the building, are entered by flights of stairs connected by terracing and tied to their embankment by an elaboration of retaining walls in a composition of cubes which extend the cuboid embellishment of the building. This cast stone parapeting provides surfaces for what may well be a record number of inspirational and exhortatory quotations—so numerous that inscription approaches graffiti. More messages are scattered on the building, together with charming reliefs of student life around the entrance and under the cornice—while across the frieze the five virtues of Determination, Ambition, Effort, Activity, and Endurance seem unfairly pitted against the six vices of Disaster, Insolence, Indifference, Lawlessness, Carelessness, and Fear. A belated but fine example of Moderne built under the Public Works Administration program, this dates from the time when the style had itself become an establishment means of endowing public buildings with the kind of forward-looking optimism which the inscriptions make explicit. The entrance hall and auditorium also retain their Moderne treatment.

PA4 Modern Diner
1940. 364 East Ave. (opposite Patt St.)

By chance, downtown redevelopment forced the relocation of this diner from Dexter Street appropriately close to the Moderne high school. Here is the vintage "streamlined" diner of its period. Its sheet metal siding with rounded roof and end invoke train transportation. So does the two-toned effect of windows contained in a band of red-brown with a field of cream and the stripe of red-brown tan below. The image is conflated: big windows, simulating those of the passenger car, give way at one rounded end to horizontal slots to simulate the engineer's cab. Chrome around the windows and chrome stripes bent around the engineer's cockpit give sparkle and zip. The brown stripe across the lower field dips to a point at the cab end of the car. Its apex rests on a glass block cowcatcher (which lights basement space designed for customer overflow and once served by dumbwaiter, but now relegated to storage). From the beginning an appendage contained most of the kitchen functions, rest rooms, and a stair to the lower level (now replaced by an above-ground dining room extension).

Remarkably, this diner retains most of its original interior fittings, revealing a charming mix of wooden carpentry and metal components which are more at ease with modernity. Curved sheet metal marks the ceiling and the hood over the cooking area, the latter painted dark brown and striped with aluminum. Booths along the windows are unupholstered, their plywood backs scroll cut into curves at the top, to which tall wooden posts are attached along the aisle with peg hangers for coats. The effect is more of bedsteads and bedposts than of Super Chief seating. All is tan and brown, perhaps to recall the woody and leathery look of the classic railroad "club car" of the period.

The Modern Diner is only partway to the luxurious "banquette" upholstery and metallic glitter of more convincing streamlined successors. Rhode Island is fortunate in having this early example, however—all the more because the diner as an American type supposedly originated with horsedrawn windowed wagons in Providence in the 1880s. There, their automo-

PA4 Modern Diner

bile successors are still to be found near the city hall, appearing in the evening and serving until dawn.

PA5 American Textile Mill

1900–1901, Martin, Hall and Howe/Clarke, Spaulding and Howe. 250 Esten Ave.

It is the square, well-proportioned, campanile-type corner tower of this onetime lace mill which is important here. Above a rusticated brick base, an inset brick panel contains the loading doors, scalloped at the top with corbeled arching to frame three small, round-headed windows. The water tower is screened with paneling containing brick diamond patterns, each centered by a bull's-eye light, and all topped by a flaring, tentlike roof.

PA6 Hope Webbing Company

1889 et seq. 1005 Maine St. (between Warren Ave. and Dudley St.)

This factory has an unusual layout. Three-story elements run around three sides of the perimeter of the site, in a U-shaped configuration, revealing the depth of the *U* on the front elevation, before terminating in chunky four-story towers. A long two-story connector, containing a centered vehicular entrance signaled by a false-front elevated cornice, stretches between the two towers. This opens to a large interior loading court (now cluttered by later additions). Note especially the articulation of the handsome brick cornice of the terminal three-story blocks. A corbeled plane roughly bisects the topmost row of windows; then, at the springing of the segmental arching of the window tops, more corbeling projects to receive the eaves. The low loggia arching which tops the towers is also nicely handled, within its own inset panel, which is corbeled at the top, the whole of the tower accented with spare elegance by granite trim. The bracketing of the entrance hoods is done with gusto. As the American Textile Mill was allegedly the largest lace manufacturer in the country early in the twentieth century, so this was then supposedly the largest manufacturer of narrow braided goods for such uses as upholstery webbing. Both indicate that, as mills elsewhere tended to take over the staples of textile manufacture, the smaller Rhode Island mills, with a skilled work force, frequently found niches for themselves in specialty textile goods.

PA7 St. Jean-Baptiste Church

1925–1927, Ernest Cormier. 68 Slater St.

The yellow brick exterior of this church, with an attached campanile to one side and well to the rear, derives from late medieval brick Italian churches. Although the outside is skillfully handled, as might be expected of Quebec's leading designer during the early twentieth century, it is the interior—among the grand Catholic churches in the state—that especially impresses. Giant engaged arcades contain the space and provide frames for the stained glass windows. Clusters of pilasters mark the corners and frame the altar, which is set in a deep apse ringed by four more full-round columns, these supporting an entablature instead of arching. The tawny hues of the nave, complemented by gray scagliola column shafts, shift to white at the entrance to the apse, then to color again—scarlet, blue, and gold—at the altar (which reform theology has moved forward, closer to the congregation). Binding this wall of giant columns and pilasters is a frieze beneath a paneled ceiling with an encircling inscription in letters scaled to proclaim their message over the faithful, like that in the frieze of Il Jesù in Rome. Such a space calls for a huge congregation; but, as in other such grand center-city churches, attendance is small. Former parishioners have died or moved to the suburbs; Interstate 95 smashes through the parish and runs in a cut immediately beside St. Jean.

PA8 Sayles New Village

1882. 3–18 Lockridge St.

Sayles New Village is the only extant example of company-built housing in Pawtucket, where private investors provided rental quarters for workers. When the Sayles Corporation purchased the nearby Lorraine Mill (1868, 1881, 1919) in 1881 from a worsted manufacturer and converted it to cotton, this housing was also provided. Exterior changes through time notwithstanding, the basic configuration of these one-and-one-half-story duplexes is clear, and most retain, at either end, their sheltering door hoods supported on ornamental brackets.

PA9 Coats and Clark (Conant Thread Mill Complex, later J. and P Coats Ltd.)

1868 and later (most buildings erected in two campaigns, 1870–1881 and 1917–1921). 366 Pine St.

Implacably austere and domineering, with scant ornamentation, these buildings derive their impressiveness from the cumulative impact of segmental-arched windows in plain walls and unadorned towers, mostly mansarded. In 1868 Hezekiah Conant began what ultimately became the largest thread company in the world in a small wooden mill (Mill 1), which no longer exists. The following year he made an alliance with J. and P. Coats of Paisley, Scotland, under the terms of which he would manufacture their six-cord thread in the United States. Eventually Conant Thread (known as J. and P. Coats Ltd. after 1913) occupied a factory complex which extended over 55 acres. During World War II it employed over 4,000 workers. Thereafter, in 1951, southern textile interests bought the company and (in a familiar story) gradually phased out the complex. By 1964 the Pawtucket operation had closed. But, exceptionally, this huge mill complex survives, pretty much intact, although now with multiple tenants. As nowhere else in the state—except possibly the Valley Queen Mill in West Warwick (WW9)—this gives some idea of the onetime scale of its largest textile operations and, by extension, the immensity of the destruction which has obliterated a number of plants of comparable size as well as hundreds of smaller operations.

The earliest mills were set as three major blocks at right angles to Pine Street; the later ones, behind the first group, broadside to them and angled in a looser arrangement, but also parallel to one another. As a result, although planned purely for function, the corridorlike spaces between the buildings possess a dynamic directional quality, which the diagonal shift in building placement toward the rear of the site intensifies.

The sequence of buildings begins with the three big blocks at right angles to Pine Street from south (closest to Conant Street) to north. Three-storied Mill 2 (1870) has one of the longest extant monitors in Rhode Island in a fine state of preservation. The main block of the building stretches between a chunky tower at either end, which are more fused with the main building mass than given individual identity. Bloated, flattish mansards, topped by their original finials, set them off. (The brick blocks now attached to the towers are later additions; so is the gabled building toward Pine Street.) In Mill 3 (1872), the monitor disappears. The tower (minus mansard) is now centered and projects from the building mass in the canonical formula for nineteenth-century factory stair towers. With this mill the company began to insert in the brick field at the top tower a brownstone plaque announcing the year that construction began.

Mill 4 (1875) is the building that first catches the eye and establishes the image of the place. The scale is upped to four stories, with half windows at the basement. Like that for the 1872 plant, the two towers set on either end of the building project as nearly independent entities from the front elevation. But being doubled and taller and retaining their tall mansards, like high hats with sharp brims, they are vastly more monumental. In the plainness of the tower the simple sunburst pattern of the arched transoms over the loading doors stands out as decorative relief. The original sash is immaculately preserved (as it is generally throughout the complex).

After a pause, major construction shifted to the rear of the site, with the even larger Mill 5 of 1881. At a time when brick mill construction was tending toward pier and spandrel, its walls continue the planarity and punched windows of the earlier plants, as though the owners considered a unified image for the complex as a whole (and doubtless the convenience and economy of duplicating what was already drawn) as more important than keeping up with the latest developments in factory design. Even in Mills 6 and 7 (1919), on parallel angles to either side of Mill 5, the wall formula remained the same. Now the towers are low, their flat roofs reaching only up to the eaves, and topped by a minimal sort of crenellation. (One of the two towers at either end of Mill 5, which lost its tall mansard, now has the same capping as the 1919 towers.) Smaller buildings scattered about complete this magnificently maintained monument to one of Rhode Island's textile behemoths.

PA10 Conant Street Mill
1919. 179–225 Conant St.

Named for the street, this factory had no business connection with Conant-Coats; nor did it continue that company's conservative approach to mill construction. By contrast, the Conant Street Mill is pier-and-spandrel brick construction pared to its near structural limit. Minimal projecting round-cornered piers (even dissolved at their tops by beveling into the wall below the cornice) oppose skimpy inset spandrel horizontals designed to maximize interior light through the most extensive openings possible. Now tatterdemalion screening in brick, insulation board, and aluminum, as determined by various tenants, counters what had once been prized. Imagine, however, the original impact of this stretch of four floors of paired wooden windows, each containing together twelve-over-twenty-four sash (as a few still do). The skeleton may indeed be pared to spindliness: the Conant-Coats walls maintain a better equilibrium between solid and void. Viewing these mills together makes vivid the trajectory of brick mill construction out of the mid-nineteenth century into the opening decades of the twentieth.

PA11 Slater Cotton Company New Mill

1881–1882, Frank P. Sheldon(?). 40–50 Church St. (at Pine St.)

PA11 Slater Cotton Company New Mill

This mill was substantially spoiled when most of its windows were remorselessly blocked. Still, it is among the most powerfully expressive brick mills extant in Rhode Island. A Sayles family operation (see the introduction to Lincoln), the plant had since 1869 produced twills, sateens, and other fine cotton goods across the street in a converted factory before the firm erected this "new mill." One immediately senses the magnificent scale and rhythm of the segmental-arched windows even without their original wooden sash bisected by a single central mullion. (A few openings still have the old sash.) The windows rest on continous coursings of rough-faced granite, uninterrupted by projections in the wall. For Rhode Island brick factory elevations this treatment is exceptional, if not quite unique. Hence, though the very size of the windows tends to skeletonize the wall consonant with trends in early twentieth-century mill construction, the planarity of the surface works against this perception by recalling an older tradition of the wall as dominant element, merely punched with windows. This conservative treatment is reinforced by the possibility of viewing the window ranges as a stack of arcades set on granite shelves which count the floors within. Seen this way, the wall appears as a superposition of additive weights, in contrast to the then pervasive skeleton mill construction, in which projecting vertical piers oppose inset spandrel bridging elements over and under the window openings.

The beefiness of the identical towers set at either end of the long Church Street elevation is even more at odds with skeletal expression. Severely rectangular, the towers rise to deep, beetling cornices, displaying as they do so the virtuosity possible in standard brick ornamentation, as well as the grandeur of effect which exacting workmanship and noble proportions can create. To the front are stacks of loading doors framed by slightly inset panels, with windows up the side elevations, all resting on the unbroken granite floor markers. The segmental arching which caps these tower openings is excessively overbuilt as hoods from corbeled brackets, indicating a conscious effort at monumentality, as though brick tended toward design in stone. The treatment of the boxy tops is truer to brick. At the lowest stage, rectangular panels of brick are set out in a checkerboard pattern. Above these, the deep cornice begins with corbeling, a coursing of bricks inset at an angle, then vertical slots of brick, and a finale of still more corbeling. A tour de force in brick craftsmanship! Now pity its blindness.

Who is responsible for its design? The Sheldon firm seems a likely candidate. At least one other example known to be theirs, the Lonsdale Mill in Cumberland, shows the same continuous granite coursings as visual shelving for the windows (CU5), while other Sheldon mills at this time emphasize planar walls climaxed by massive towers articulated with brick ornamentation of comparable precision and virtuosity. This firm seems just then to have sought to combine the sharp-edged openness of aspect expected for emerging twentieth-century factory design with monumental assertiveness as a last hurrah to the image of overwhelming power that nineteenth-century industry came to cherish. Whoever designed this brick mill, no other in Rhode Island deserves more notice.

PA12 Jonathan Baker House

1823. 20th century, dormers and side door. 67 Park Pl.

PA13 Park Place Congregational Church

1935, Arland Dirlam. 71 Park Pl.

Park Place is the starting point for a look at Pawtucket's downtown. The Jonathan Baker House is a fine brick house in the Federal style, with its entrance porch on attenuated columns and a fan- and side-lighted door. Remodeled inside, it has lost its paired northern chimneys, but retains the two to the south and some fireplaces. Adjacent is the pinched and verticalized Park Place Congregational Church, a good example of the colonial revivalist penchant for composing in terms of "features" (in this instance from the Federal style) rather than the subordination of these within a larger geometrical scheme.

PA14 Apex

1969, Raymond Loewy–William Snaith, Inc. 100 Main St.

The parking lot for the Old Slater Mill museum complex (next entry) is the place to consider the center of Pawtucket. Across the river stands the most conspicuous modern structure at the center of Pawtucket. Apex, as it is known, is a popular discount department store. Its designer celebrated its name by burying the center third of the store beneath a stepped, white-stucco-surfaced pyramid with flanking entrances. This conceals the visual clutter of mechanical equipment as though it were the booty of a pharaoh's tomb. The shrouding of the mechanical aspects of the building within a shaped container also epitomizes the approach of the industrial designer, which marks the beginnings of this design firm. Then it was known simply by the name of its founder, Raymond Loewy, one of the key participants in the revolution in American industrial design that occurred in the late 1920s and 1930s. Whatever its considerable merit as a mall building (without the mall), however, Apex is, in scale, shape, isolation, and stridency, at variance with every-

thing around it. For all of these reasons this pyramid in its desert of macadam virtually *is* downtown Pawtucket. The heart of the city is ceded to the reigning pharaoh as his domain. So at least it appears to the visitor, who must poke around the perimeter of this monument to discover what remains of the rest of "downtown"—if, in fact, in its withered state, it really exists at all anymore. It is too easy in hindsight to be critical—and a bit unfair too, because here one senses a genuine effort to renew the city through thoughtful redevelopment rather than merely to exploit opportunity in the crassest way possible. The caring effort shines through and accounts for the value of examining the result. It also adds to the regret at the loss of a number of Pawtucket's oldest buildings and some of its nineteenth-century mills, most then dilapidated and despised, to be sure, but waiting to shine again.

It could be argued that since it is here, even the pyramid deserves consideration as perhaps the most conspicuous example in the state of what the French theorists of the eighteenth century referred to as *l'architecture parlante*, or architecture which speaks directly to the spectator by illustrating what it is, something like a picture in a dictionary. "Apex" may not be the subtlest theme for architectural illustration, even when rendered as decisively as here by a stepped pyramid and bold graphics. Yet the problem of design for roadside communication was very much a concern in conjunction with the spread of the automobile for several decades after World War II, until, by the 1980s, cute fairy-tale villages, clumsily assembled of crude classical design, began to appear beside every cloverleaf. Better to have spared what is partly credible than to have demolished with the thought that a porticoed replacement or faked village nostalgia might be the key to downtown urbanity for Pawtucket.

PA15 Slater Mill Complex

1758 and later. 67–71 Roosevelt Ave.

The demolition carried out in order to permit I-95 (the main Boston–New York expressway) to cross the center of Pawtucket was, when it occurred during the 1960s, widely viewed as a blessing. Clearance was seen as an opportunity not only for new construction but also for the celebration of Slater Mill and its companion Wilkinson Mill by their isolation in a park. The return of the river to public consciousness is admirable. The park is pretty, yet with a functional plainness about it that accords with its centerpiece. But the mills never knew such a bucolic setting. Some retention of older buildings would have given a modicum of context, at very least providing appropriate places for such too familiar tourist expectations as restaurants and boutiques and, with luck, have preserved some bits and pieces in which to have housed the "real" downtown that was so thoroughly flattened. As it stands, the entire area is too transparent, lacking in mysteries to explore. The mill complex appears as a precious object spotlighted in its urban vitrine.

Now for Old Slater Mill, the focus of this urban clearing (1793 and later, Benjamin Talcott, builder; Sylvanus Brown, wheel. 1924–1945, restoration to conjectural 1835 condition incorporating original building). Here the firm of Almy and Brown, consisting of two Providence businessmen, William Almy and Moses Brown, set up the operation which is generally regarded as the fountainhead of the New England textile industry and of the development of manufacturing in the United States. They teamed up with the foundryman-mechanic Oziel Wilkinson and with another mechanic, Sylvanus Brown, before adding as a third partner the English emigré mechanic William Slater. The original two-and-one-half-story clapboard mill with a narrow trapdoor monitor to light the attic was merely 43 feet long and 27 feet wide. It is now buried toward the center of a further elongated structure, to which a right-angled wing was added in the nineteenth century. So the total configuration is now a *T*, albeit with a stubby stem and a lopsided crosspiece. The cupola at the west end of the initial building was later replaced and relocated to the front of the projecting stub, which now provides the main entrance to the museum. Even as aggrandized, the initial factory building derives (except for its attic monitor) from typical colonial public buildings (though for these, the principal door is typically centered in the long elevation, while the cupola is usually centered on the roof, astride the ridgepole).

When it was threatened with demolition in the early twentieth century, the mill's rescue depended on a private group which organized in 1921 as the Old Slater Mill Association, among the first preservation groups to be concerned with industrial archaeology. The association faced a formidable job. Much remodeled, badly battered, with a flat roof at the time of its rescue, the mill required such drastic overhaul that little of the original fabric remains. As a museum it now presents exhibits that cover the entire history of the industry in Rhode Island and the lives of factory workers and owners.

Adjacent is the Wilkinson Mill (1810, with later brick and tower; 1970s, restoration, Irving Haynes), built roughly a decade and a half later than its predecessor. Like the Slater Mill, it has trapdoor monitors sliced into the roof slope, but it is of masonry instead of clapboard and an additional story in height: already the burly scale of the factory begins to assert itself against the almost domestic mien of the Slater Mill. The original mill must have continued the tradition of loading doors on one of its gable-ended walls with a cupola set a little back on the roof. The later brick tower addition defines the industrial type further.

Oziel Wilkinson, who built this mill, together with several of his five sons, also variously gifted with the talents of their father, operated a cotton mill on the upper floors. David Wilkinson, the most mechanical among the sons, ran a machine shop on the ground floor. Today, powered partly by a rebuilt waterwheel and partly by electricity, this portion of the mill is a museum of early, belt-run machine tools associated with the textile industry.

The third building in the cluster, now known as the Sylvanus Brown House (1758. 1970s, restoration, Irving Haynes), was moved to its present site. It was originally built for Nathan Jenks, who was Pawtucket's first master mechanic and industrialist, and a leader of the same sort as Oziel Wilkinson. Jenks's descendants came to the fore in the machine tool industry in the next generation. It was appropriate that another skilled mechanic, Sylvanus Brown, should later live in the house. His will contains a complete listing of his possessions. The house has been furnished in accord with this inventory, so its interiors provide an accounting of the spare possessions accumulated in the late eighteenth and early nineteenth century from a lifetime of skilled mechanic toil.

Slater Mill is a focal point along the Blackstone River Valley National Heritage Corridor, a combined pedestrian path and bikeway along much of the length of the Blackstone River valley and parallel to remnants of the Blackstone Canal between Providence and Worcester, Massachusetts. Sites along the roughly 46-mile route also include the Museum of Work and Culture in Woonsocket.

PA15 Old Slater Mill, right, and Wilkinson Mill, left, background

PA16 Pawtucket City Hall, Police Department, and Fire Department

1935, O'Malley and Richards. 137 Roosevelt Ave.

In addition to his high school (PA3), John F. O'Malley made this other impressive, and earlier, Art Deco contribution to Pawtucket, a municipal complex in a blown-up and Modernized Neo-Federal style. It contains typical Moderne elements, all well executed: stylized reliefs beneath the ground-story windows; plain limestone blocks with eagles carved into corner panels on either side of the entrance; a stepped-back frame to the entrance recess; over these, a fine ornamental iron balcony in a Neo-Federal diamond pattern stretched and rhythmically counterpointed into modernity; then, farther up, fluted pilasters without base or capital and, finally, a stone grille in more diamonds immediately under the tower. Original ornamentation of the cuboid capping of the tower included more eagles (these in metal) at the four corners, but deterioration forced their removal. Inside, the ornamentation is sparser. Astonishingly, the tower interior is one big, unused hollow (except for a small storeroom at the top), revealing inside the steel frame and cinder block over which the brick and limestone walls were laid. It may have been justified as grossly inefficient space for record storage, but ended up as what it really was—an extravagant folly to civic grandeur.

PA17 Lebanon Mill Company

c. 1900. 10 Front St.

Acting as a screen behind City Hall, the brick factory rising directly from the river's edge, built originally for the Lebanon Mill Company, reveals the everyday straightforwardness by which standard segmental-arched windows minimize the enclosure of a planar brick wall—the effect all the finer because the windows contain their original wooden sash. Note, too, how the slight additional height of the first-floor windows in each stack gives a visual boost to the vertical effect—which, however, terminates too abruptly in the collision of topmost windows against the cornice. An adjacent factory, which extended this rhythmic walling of the river, was lost to a 1978 fire.

PA18 Pawtucket Armory

1894–1895, William R. Walker and Son. 172 Exchange St. (at Fountain St.)

This is one of four armories in Rhode Island designed by the same firm at the late nineteenth and early twentieth centuries; the others are Westerly's, the Cranston Street Armory in Providence (PR195) and (the smallest) Woonsocket's (WO17). The symmetrical Providence building looms much larger, but a sloping site on a skewed city block presented subtler massing challenges in Pawtucket. Hence the slight differences in seemingly symmetrical towers at either end of the front elevation. One has a slightly larger diameter than the other, and comes to the ground at the low corner of the block toward the downtown; uphill, the one of smaller diameter is an oriel that emerges from the first story at the corner of the building. One is corpulent and relatively static, the other lean and more dynamic. A rough-faced, high granite base serves as an eminence from which a brick castle rises, as in Providence screening a free-span drill hall supported on metal arches.

PA19 Deborah Cook Sayles Memorial Library

1899–1902, Cram, Goodhue and Ferguson. 1962–1967, remodeling, Millman and Sturgis. 13 Summer St.

As the first mayor of Pawtucket, the industrialist Frederic C. Sayles, whose Oak Hill estate was on East Avenue, donated the land and building for this library as a memorial to his wife. An architecture firm that would become famous for its Neo-Gothic design in this early work employed Greek classicism. It is one of very few Greek-inspired buildings among the plethora of early-twentieth-century classical designs in Rhode Island. An extended, severely plain block of gray granite raised on a high base above the street is interrupted at the center by an Ionic portico and a portal, both based on the Erechtheion and reached by an impressive pyramid of steps. To either side are similarly concentrated groupings of three large windows, framed by pilasters and topped by handsome marble reliefs in a decorative adaptation of archaized Greek precedent. They depict the history of world civilization, oddly from right to left: Egypt, Greece, Rome, then Dante, King Arthur, and Shakespeare squeezed into a single panel, and, finally, the *Nibelungenlied*. They demonstrate why their creator, Lee Lawrie, would become the most popular architectural sculptor in America during the 1920s and 1930s, his most familiar work appearing on the skyscraper Nebraska state capitol and at Rockefeller Center in New York. A scintillant cresting of alternating palmettes and lion's heads, again elegantly detailed, unifies the top of the building ornamentally, as the austere base unifies it below. The way in which plain surfaces and carefully distributed, meticulously linear Greek ornament combine to enhance one another is worth attention.

Behind the portico, a square center space with more Ionic columns ranged around it and a coffered, clear-lighted ceiling once provided the focal point for reading tables and metal stacks, all exhibiting the same "Greek" restraint. Remodeling added a recessed wing to the south and placed the entrance there at ground level, providing at once accessibility for those who cannot climb, better control for the circulating collection, and much additional space for all library functions. To reach the "entrance hall," however, one follows a tortuous route of unintegrated corridors and stairs before finally coming upon it all but empty, with an unhappy metal mezzanine inserted behind the columns. The mezzanine seems to offer little additional space for the damage it inflicts on the original interior. However, the modern bridging of classical interiors was something of a passion in the 1960s, when the collision of now with then, each true to itself, was very much at the center of architectural theory. Too bad that Pawtucket cannot take a cue from the Providence Public Library (PR13), where precisely the same sort of 1950s–1960s "modern" renovation was reversed in the 1980s in order to restore the importance of the original entrance.

PA20 Municipal Welfare Building
(U.S. Post Office)

1896–1897, William Martin Aiken, James Knox Taylor. 1 Summer St. (corner of High St.)

Next door to the restrained use of ornament for the library, this former post office seems lushly ebullient. Like the armory (PA18), it rises on a skewed, sloping site, its most prominent corner celebrated by a dome at its lowest point. The picturesque balancing of two towers which differ slightly in size and detail in William Walker's armory contrasts with the absolute symmetry of this building—symmetrical at least if one stands on axis with the dome, so that the side elevations pivot from it like the covers of a book from its spine. This design coincided with a change of guard in the Office of the Supervising Architect in Washington and a transition from Aiken's essentially Victorian design to Taylor's Beaux-Arts classicism. The skillful employment of classical form here appears to reflect Taylor's intervention, although the momentum of Victorian design seems to have continued in the ebullient combination of shape and polychromatic use of materials. As Supervising Architect, Taylor was responsible for a number of fine turn-of-the-twentieth-century public buildings in the classical manner. The finest building in the state from his office, built a decade and a half later, is the Westerly Post Office (WE17). Here the style is a lush "Venetian" Neo-Renaissance; in the Westerly building it is far more restrained, another rare Rhode Island example of twentieth-century Greek-inspired classicism. The contrast between the two reveals a broad trend toward more restraint in ornament and more simplicity in composition in the design of federal buildings under Taylor's aegis.

PA21 St. George Maronite Catholic Church
(St. Paul's Episcopal Church)

1852–1853, Samuel J. Ladd(?). 1865, west aisle, Clifton A. Hall; redecoration, Cattanach and Cliff. 50 Main St. (at Lennon Ln.)

So rural in aspect is this rough-hewn, broad-gabled Gothic Revival church that it seems misplaced in the city. A local mason, more familiar with mills than churches, is its most likely de-

PA19 Deborah Cook Sayles Memorial Library

PA22 Pawtucket Congregational Church

signer. Indeed, the helmet capping of its stocky spire, drawn down tightly around the bell, relates most closely to the treatment of certain mill towers, like that for the Valley Falls Mill in nearby Central Falls (see CF10). However, the forceful shaping of the tower, from stepped corner buttressing at its base to corners sliced into a semioctagon above to contain the bell, as well as the combination of rough texture and meticulous craftsmanship in the handling of the stone, indicate more than ordinary competence in its design. Clifton Hall's 1865 addition of the aisle (on the entrance side of the church only) with its dormered "clerestory" provides the other notable exterior feature. At the same time, the firm of Cattanach and Cliff redecorated the interior. Although repainting has lightened the walls and disturbed the intended tonality of the space, some decorative patterning remains. More important, symbols of the four Evangelists and of the Apocalypse pierce the shadow with heraldic intensity, as a rare reminder of High Church ceiling decoration from the mid-nineteenth century. Stained glass and Victorian marble plaques complete the setting.

PA22 **Pawtucket Congregational Church**

1867–1868, John Stevens (Boston). 1915, interior remodeled. 1938, steeple replaced. 1980s, pulpit area altered. 40 Walcott St. (at the fork of Broadway)

The most commanding building in downtown Pawtucket is also among the most important Victorian churches in the state, not only for its design, but for the reasons it looks as it does. This, the earlier Newport Congregational Church (NE61), and the still earlier first Central Congregational Church in Providence (PR83.6) are the most ambitious extant examples in the state of the Congregational round-arched style (see the discussion under Central Congregational Church).

Of the three this is the most assertive. While the Newport and Providence examples are masonry, this is wood. In the heft of its buttressing and molding, however, it means to evoke stone. The viewer will be struck by the wall-like width of the tower, which tapers slightly with each of the four steps of its simulated corner buttresses. They measure off the tower's four levels, up to a cornice with centered round arches facing in each of four directions. The arching of the cornice at this point echoes the arching of doors and upper-story windows and prepares for more curves in the belfry above. Tall, curved corner brackets meld into more arches to shape the belfry and its four clock faces. Viewed head-on, each face appears as a giant Victorian mantel clock, but architectural in character and exceptionally well integrated to the tower. In this clock-faced cupola and the molding below from which it emerges, as well as in the ogee swell of the second-story roofing which flanks the tower, one senses a flamboyant, bloated quality suggesting influence from the current mansarded Second Empire style. That style affects the ornamental treatment throughout, and gives to this church an exuberance of aspect, setting this church apart from other, more severe Romanesque contemporaries. The loss of a steeple in the 1938 hurricane diminished what was surely the most exciting Victorian tower on any extant church in the state. The repair of the tower as a cupola-on-cupola might have been a fair compromise were it not in the namby-pamby Neo-Federal style toward which New England Congregationalism all too readily gravitates.

All corner buttressing, window frames, and cornice moldings mean to be weighty in their evocation of masonry. Pendant molding elements hang like heavy Victorian tassels and fringes from the arched capping of the windows, from moldings beneath the eaves and, most extravagantly, at the fourth stage of the tower. Their source is the sort of corbeling, natural to brick construction, found in Pawtucket's brick mills, but here specifically derived from round-arched brick medieval churches in Italy and Germany. For this reason the painting of the church (and of many other mid-nineteenth-century churches like it) in "Congregational white" during most of the twentieth century was historically, and even aesthetically, wrong. Its bulky, masonry-inspired detailing needs the weight of its original stony colors, recommended in the Congregationalist *Book of Plans*, to which it was returned in the late 1970s, albeit possibly in too timid a shade.

Inside, the generous vestibule offers a double flight of stairs on either side to a platform which opens to the church interior. Although mildly "colonialized" to the Federal period in 1915, when the Victorian pews were replaced, the interior remains basically intact, with balconies on three sides. What is remarkable, however, is access to the balconies on either side of the pulpit by curved flights of stairs from the preaching platform. Choir processions came down the aisles, flanking the pulpit in their ascent to the balcony. If the stairs sweep in concave curves away from the congregation, the pulpit curves toward the congregation as a prow. So the round-arched monumentality of the exterior reverberates in the curves inside. The interior is now cream and white, but its original colors seem to have been the ochre and deep browns recommended in *A Book of Plans*. The windows are stained glass of a later period than the church, although the original geometric treatment in pale yellow and clear glass exists in windows in the vestibule. Among the charming Colonial Revival additions are handsome stained glass windows in the opalescent pictorial treatment popular around 1915, showing a Puritan couple (John Alden and Priscilla perhaps) standing on the shore of the New World, perturbed and thoughtful, watching the vessel that brought them disappear over the horizon; then, in the pendant window, the same couple resolutely on their way to church.

PA23 Ellis B. Pitcher–Lyman Goff House

c. 1845. 1881, remodeled. 58 Walcott St.

When built for a cotton-mill owner, this Italianate house was tightly limited by the porch across the front, the block behind it, and the flanking one-story wings in line with its front elevation. Its roof was then hipped and topped by an octagonal cupola. Bracketed eaves at two levels and forceful corner pilasters for the wings underscored the contained nature of the house. In the capitals of the porch columns, the mix of Egyptian lotus and Greek acanthus motifs indicates a response to the romantic interest of the Greek Revival in whatever was exotic in ancient Greek precedent. Passed on, it became a component of the Victorians' deeper and broader appreciation of the exotic. It was Lyman Goff, the second owner of the house and a partner in the cotton braiding business, who added the two-story additions to either side of the rear. One side features a porte-cochere, making a new entrance at right angles to the entrance from the front porch, as already seen in so many Victorian houses on Providence's College Hill. Opposite, an added two-story bay steps down to a one-story semicircular conservatory, thereby nicely adapting the mass to the slope, while providing another sign of Victorian status. Goff also redid the interior extensively to make a display of Queen Anne woodwork. Much of this still remains, especially in the entrance hall, with its fireplace, despite recent repeated renovation of the house, first as the local headquarters for the Red Cross and then for the Children's Museum, which moved to Providence in 1998.

PA24 Apartments (First Free Will Baptist Church, later Independent Eastern Orthodox Church of the Resurrection)

1884, William R. Walker and Son. 130 Broadway

The wide-gabled body of this nicely preserved Queen Anne church exterior accepts a cluster of appendages to its front. An entrance porch, domestic in mien but overly wide (in sympathy with the big gable behind), butts against a shed-roofed entrance vestibule, which, in turn, butts against a spire off the corner opposite from the porch. The collision of shapes, which can be piquant in the best of Queen Anne, is here somewhat clumsy. So perhaps is the exceptional shape of the spire, in which a swelling, mansardlike base is abruptly pinched and capped by an octagon steeple (doubtless an awkward echo of the original spire cap to the round-topped belfry of the nearby Pawtucket Congregational Church [PA22]). Typical Queen Anne details are the most delightful aspect of this design.

PA25 Summit Street Houses

Summit Street runs across the brow of what was known locally as Quality Hill. From roughly 1880 through 1920, the eminence came to be filled with residences of the wealthy. Here they overlooked, symbolically if not actually, both the city center and those less fortunate. Quality Hill's status had seriously eroded before the nearby arrival of I-95 nearly administered the coup de grâce. Since then houses have been renovated, but not before many had been converted to apartments.

PA25.1 Oliver Starkweather House

c. 1800. 60 Summit St.

Normal attrition and redevelopment have left few Federal houses in Pawtucket. The twice-moved Starkweather House, considered to be the largest and finest in the city in its day, is one of them. Its front boasts an exceptional amount of quoining—not only at the corners of the elevations, but around the windows as well. The capping of first-story windows by deep lintels of rustication surmounted by projecting moldings suggests a possible influence from the Joseph Nightingale House in Providence (see PR96.3). If such excessive and robust quoining and rustication were a bit old-fashioned for the Providence house, built in 1791, they were more so for this later one. The same can be said of the heavy character of the entablatured entrance porch, made heavier by its parapeted balcony, onto which opens a floor-to-ceiling Palladian window at the center of the second story. Only the arched fanlight over the door and the urn finials on the parapet posts suggest the lightness of handling and certain specific motifs which would characterize the Federal style. Whatever the sources for this facade, clearly its owner intended high visibility for his success as carriage builder, storekeeper, and local politician.

PA25.2 Albert A. Jenks House

1903–1904. 90 Summit St. (corner of Potter St.)

Toward the top of the hill is the largest of the Summit Street houses, commissioned by a descendant of the Jenks family, pioneer mechanic-industrialists in eighteenth-century Pawtucket. Albert Jenks was president of Fales and Jenks, the textile machinery plant in adjacent Central Falls. By the turn of the century the revival of colonial and Federal styles was well underway, and the designer of this house attempted to catch up with the new fashion. He remained, however, hopelessly mired in the Victorian past. The "colonial" hipped roof is here really a mansard. The uniformity and reticence of colonial dormers here become boisterous with strangely scaled broken scroll pediments and other variations. The two-story semicircular porch toward Summit and the bay toward Potter are overblown, while the spindly "classical" columns and balusters of the sprawl of porches and porte-cochere recall Victorian posts. And so on. But such misunderstandings make for a lively hybrid effect.

PA25.3 Ellis Pearce House

1871–1872. 98 Summit St.

PA25.2 Albert A. Jenks House

PA25.4 Thomas P. Barnefield House
1872. 99 Summit St.

The Ellis Pearce House, built for the owner of a grocery, flour, and grain store, once stood alone as the first house in the street. It is hip-roofed, with a heavy porch and with pilaster elements at every corner topped by handsome ornamental brackets. Across the street (now resided, but unobtrusively so), the Thomas P. Barnefield House, built for a prominent lawyer, sets the varied pretensions of Summit Street to one side. A straightforward, cross-gabled, T-shaped mass is infilled with a three-sided porch: open form against closed form in a simple interlock. The arched colonnading of the porch seems almost too formal for the vernacular quality of the rest, and for the sawcut, stenciled ornamentation between each of its arches—as if a farmhouse had wandered onto this classy summit and put on costume jewelry for the occasion.

PA26 Edward J. McCaughley House
c. 1917. 51 Arlington St.

Built for an executive of a dye and bleach works, this is a fine example of English Arts and Crafts in the manner of C. F. A. Voysey, as popularized in English periodicals like *The International Studio*, which had a considerable American following among the artistically inclined. Spreading roofs and deep porches characterize this long, thin house, in stucco with the first floor in brick. What could be an abortive hipped roof over the first story barely gets beyond the stage of eaves stretched across the front as a protective overhang for the entrance and ground-floor windows, before a pair of wide cross gables, side by side, cuts through it to make the second floor, above which the hip concludes. Porches supported by chunky Doric columns cut into the mass at either end, over which broad dormers poke through the roof slopes. A cozy vernacular image is invigorated by the formal symmetry and forceful geometrical organization of the blown-up mass.

PA27 William A. Ingraham House
1850. 112 Walcott St.

This Italianate house for a partner in a cotton firm has a low hipped roof with spreading brackets and cupola (here octagonal) characteristic of the style. Distinctive in its unelaborated vernacular treatment is the pairing of elements in the front elevation: paired posts for the porch and paired windows behind, all nicely spaced (except for a too massive door, which may have been somewhat altered).

PA28 Walter Stearns House
1892–1893, Albert H. Humes. 22 Walnut St.

A local architect with a peculiarly personal style designed this high, angular gambrel-roofed shingle and clapboard Queen Anne house for Walter Stearns, an executive in a cardboard box company. Elegant details, each sharply defined as a self-contained entity, are scattered about somewhat disjointedly: in the front gable, a pseudo-Palladian motif consisting of an oculus window resting on a window triplet; below, at the second floor, an off-center loggia; along the side elevation a brick chimney framed at the corners and paneled with gray stone and a cluster of triangles to mark the dormers and the bay that contains the fireplace. Interspersed with this precise and disjunctive geometry are panels and cartouches of lush ornament, and what is left of cushion-capped porch columns, all embellished with Romanesque leafage, but given a mild Art Nouveau or Sullivanian undulation. For good measure, a bracketed Neo-Colonial shell hood surmounts a side door.

PA29 Joseph T. Cullen House
1929–1930. 12 Walnut St.

Formally more resolved than the Walter Stearns House is a fine, if modest, example in what might be called the pergola mode of the early twentieth century, in which Arts and Crafts interests in revealed structure and in garden design are intermixed. In California this development culminated in such extraordinary examples as the glorified bungalows in shingle and redwood by Greene and Greene. In New England the mode is often at its best in a tighter, more reticent Neo-Colonial format of Doric columns and projecting scroll-sawn bracketing, which is more ornamental than structural, as here in its decorative attachment to the walls. Compositionally, this house is admirably abstract, however unaware its builder probably was of the result. The rectangular, hip-roofed mass of the house is enlivened by three features: an inset entrance porch at the center and flanking projections on either end for a sunroom and open porch. Each is compactly framed by chunky columns with the scroll-cut ends of simulated pergolas floating above them.

PA30 Asa Carpenter House
1890s, Stone, Carpenter and Willson. 97 Cottage St.

This expansive Colonial Revival house is interesting because its original owner was the father of a principal in the firm which designed it. The family relationship suggests that Charles Carpenter, trained in Victorian ways, worked in the shadow of the youngest partner, Edmund Willson, Rhode Island's most creative Colonial Revival designer. The result is bloated "colonial," with the ballooning curves of the porch and second-story window bays as culminating features. The contrast between the over-assertive center dormer and its tiny accompaniments with pinched scrolls on either side, the medieval casements upstairs, the spectacular spread of the entrance and its glazed enframement—all further suggest a venture into unfamiliar territory. But the design fascinates as an ebullient and even touching statement of stylistic betwixt-and-between.

PA31 Omar Currier House
1889. 12 Howard St.

This house, built for a grocer, is a charming carpenter's version of Queen Anne. All upper window frames are cornered with disks. To the modern eye it appears just the place for Mickey and Minnie Mouse. More disks occur in paneling on the bays and in a delightful variation on the colonial scroll pediment in the attic window.

PA32 Slater Memorial Park
1903–c. 1930. Entrance on Armistice Blvd. opposite Parkside Ave.

Like Providence at the same time, Pawtucket was influenced by late-nineteenth-century visions of the City Beautiful which included public landscape parks as an arcadian escape from Victorian industrial squalor. Slater Park provides a setting for a loose cluster of buildings—pleasant and appropriate rather than architecturally outstanding—around a lake. The John Daggett, Jr., House (PA32.1; 1685[?]) is the only extant seventeenth-century house in Pawtucket, a fact of more sentimental than architectural interest, because it has been much altered through time, although its "restoration" in 1907 is part of Rhode Island's colonial revival. As with the Betsy Williams House in Providence's Roger Williams Park, its existence here gives a historic dimension to the park and reminds its visitors that through the nineteenth century these urban arcadias existed as relatively large farms extending back to the beginnings of their cities. Their survival at the city's boundaries in the 1890s made the parks possible. A sequence of eighteenth-century additions to the Daggett house brought the original one-room house to two stories and five bays, after which a nineteenth-century wing was added to the rear. Of special interest is the carousel (PA32.2; c. 1880; set up in Slater Park, 1910), one of two in the state by the important Rhode Island manufacturer Charles I. D. Looff. The other example, in East Providence (see EP22), is more elaborate, and (as this is written) the animals here are crudely painted. The boathouse (PA32.3), in brick with trellising, is gracefully fitted to its site; but its transformation to a community art gallery deprives the lake of animation. At the far end of the lake is a nicely proportioned reinforced concrete bandstand (PA32.4; 1917) in the guise of a circular classical garden temple.

PA33 James Potter–Ralph A. Lumb House
c. 1917. 1008 Newport Ave. (corner of Carter Ave.)

PA34 Potter and Johnstone Machine Company
1899 and later. 1001 Newport Ave. (at Saratoga Ave.)

The Potter-Lumb House is another version of the Edward McCaughley House (PA26), of the same date, but much grander (too grand for the format), with three stories rather than two fitted under its spreading hipped roof. James Potter, a partner in a machine tool company, built the house diagonally across the street from his factory, for his daughter Mae on her marriage to Ralph Lumb, who was treasurer of the Lumb Knitting Company. Earlier, c. 1915, Potter had made a similar gift of the stucco Neo-Colonial house next door, number 1012 and also handsome, to his daughter Elizabeth at the time of her marriage. He himself lived at number 1042, in a Queen Anne house built for Herbert S. Jenks, a superintendent for a cotton company (1897–1898, purchased by Potter in 1902). All three houses are now nicely maintained as funeral homes. Johnson was also among the most prominent backers of Slater Park.

Potter's factory contains an alternating pier-and-window wall with corbeled inset at the top of each tier of windows. Sheets of plate glass now replace the multipaned wooden sash of the original windows. Now headquarters for Hasbro, a major toy manufacturer, it has been playfully modernized on the interior by a group of artists and interior designers. In a more architectural vein, a walled garden court off the cafeteria employs lawn, flowering trees, and terracing patterned in black and gray granite to make a restrained but beautifully proportioned space (court by Richard Fleischner, environmental sculptor, 1980s; visitable only with permission).

PA35 George Hadfield, Jr., House

c. 1950, General Housing Company. 190 Hunts Ave.

A printer for the *Pawtucket Times* ordered this prefabricated house of polished enamel steel panels in two tones of tan, of a type then being developed for commercial structures. The house takes its form from the elongated box with low-pitched gable typical of the developers' "Cape Cod" (or simply "Cape") ubiquitous after World War II. Rippled panels in dull gray are intended to give the roof a shakelike appearance. A corrugated metal shelter over the entrance and a slight boxy projection to make a living room bay window provide the only breaks in a simple shape which was designed to be assembled in various configurations. Lustrously preserved, it commemorates a postwar dream of converting aircraft manufacturing capacity from defense to housing production. It was hoped (but in vain) that the reassuring image would sell the homebuyer on the unconventional material and technology of the exterior shell.

PA36 Royal Weaving Mill

1900, 1905, 1909, 1914. 300 Cottage St. (intersection of Central Ave. and Cottage St.)

A German immigrant, Joseph Ott, introduced silk weaving to Pawtucket in 1888. By 1900 he began building this plant, which, by World War I, boasted the largest weaving shed in the world. Prolonged labor troubles from the 1930s onward eventually closed the plant in 1949, as an aspect of the widening collapse of Rhode Island textile operations. Cotton manufacturing moved to the South first, followed by manufacturers of the other basic fibers. As is frequently and disappointingly the case, most windows have been blinded. The brick walls are pier and spandrel, and, as in the Potter and Johnstone Machine Company (PA32) are corbeled over the topmost windows between the piers. Here, a cornice zone above contains more corbeling with a rhythmic accent over each of the piers. The tower, at a major intersection, displays yet another battlemented treatment in brick to frame handsome clock faces in four directions. The extent of the side wall along Sabin Street reveals the plant's considerable size, where a few windows from different phases of the plant's enlargement have been left exposed.

PA37 John F. Adams House

1867, Nathan Crowell, carpenter. 11 Allen Ave. (corner of Broadway)

Provincially exuberant ornament adorns the front of this otherwise severe Italianate house, a mid-nineteenth-century industrialist's mansion (now apartments) which originally faced Broadway. The projection of the undulant hoods over the ground-floor windows and the drops off the end are carried to such extremes as to seem more appropriate for portière overhangs in a Victorian parlor than for architectural capping. The Corinthian porch columns which support the shelter to the arched entrance are treated at their bases as cast iron lamp standards and have cushiony capitals. Above, embellished triangular and segmental pediments on brackets appear over the windows, with more brackets under the eaves and a chunky, bracketed cupola on top, lighted by triplets of arched and bull's-eye windows. The house was built for a manufacturer of print cloth who also served as city councilman and, in 1898, as mayor.

PA38 Apartments

1896–1897. 339–341 and 343–345 Broadway (at Kossuth Ave.)

This pair of identical two-and-one-half-story gabled Queen Anne flats were built as investment properties for Susan E. and Abby N.

Fuller. Each has a gazebolike circular porch off one corner and a polygonal bay worked into a conical turret off the other. Too-rarely-preserved spindlework decorates both the balustrade and frieze of the principal porch, augmented by the same lacey treatment for entrance and side porches.

PA39 **Offices-Condominiums** (Greene and Daniels Mill)

1860–1866. 10 Front St. (at Central Ave.)

Fronting the Blackstone and the backs of other mills in Central Falls on the opposite bank, this was once a thread mill. One of the partners in the business lived a few blocks away in a grand Victorian house in Central Falls (see CF5). Windows topped with segmental arches and silled in granite are conventionally set into unbroken brick walls in both buildings, Mill 1 (five stories) and Mill 2 (the three-story wing). The detail of special interest is the elaborate corbeled cornice of Mill One, with corbeled bracketing extended down between the arching of the uppermost windows. The two towers are differently designed. As variants on inset arched panels with arched windows, however, both provide examples of the round-arched industrial mode. Some cast iron balconies salvaged from old fire escapes serve as ornaments for the mix of offices and apartments which now occupy the mill. As such conversions go, this is well done. In fact, the developers rebuilt the rooftop mansarded floor and capped and restored the clock to the principal tower of Mill One, gaining another floor of revenue by returning it to its original form. In replacing old, drafty windows with tight modern ones, they also sought out paned sash to approximate what was there. All appropriate and admirable, yet the bland textures of synthetic modern building materials and the lack of decisive reveal in the window frames simply cannot match the vigor and resonance of their Victorian equivalents. Grateful as we are for the effort, this is a pale, well-manor-ed reflection of the mill that was.

Central Falls (CF)

Central Falls is a mere knob of a community and the most densely populated of all Rhode Island towns. The Blackstone River curls around the northern and eastern sides of the town, marking these boundaries with three falls along this portion of its course. As in precolonial Pawtucket, good fishing at the falls, fords, and the junction of territories made this a meeting place for Native American tribes. Once the colonists arrived, like Pawtucket, this area was part of Providence's "North Woods." Like Pawtucket, too, it shared in the final Native American battles of the area and in a particularly bloody way. In one of the first encounters of King Philip's War of 1674–1675, a band of Wampanoag, led by their chief, Canonchet, spied the approach of a group of armed colonists from the abrupt outcrop which is now Jenks Park and ambushed them in a major massacre. Thereafter the area remained sparsely populated until the usual cluster of tiny mills gradually appeared after 1751, especially around the upper and middle falls.

Farther upstream, where the river flows in a narrow channel through high banks, the falls came to be known as Valley Falls. By the late eighteenth century this had been exploited for manufacturing by Oziel Wilkinson, who had begun his extraordinary career as mechanic-industrialist with a forge in Smithfield, then moved to Valley Falls, and moved again around 1780 to the climax of his achievement in Pawtucket. But a family presence remained at Valley Falls, eventually incorporating as I. and A. Wilkinson, which prospered only after one of the partners, Isaac, built a turnpike in 1813 and a bridge over the Blackstone into what is now Lonsdale, roughly along the route of present Broad Street. This permitted the movement of goods north, but especially eased the route south to Pawtucket and on to Providence. Although the opening of the Blackstone Canal from Providence all the way to Woonsocket by 1828 promised another easy connection between Providence and the mills to the north of it, the promise was early frustrated. Not until the Providence and Worcester railroad in 1848, with a spur to Boston the following year, did transport radically improve over what turnpikes and roads provided. The railroad was built mostly by the first wave of Irish immigrants to the area, many of whom stayed on to work in the mills, providing the nucleus for the Irish immigration which followed.

Meanwhile, a more significant early industrial development occurred at the middle falls. As early as 1780, Sylvanus Brown, the same mechanic who assisted Oziel Wilkinson and Samuel Slater in setting up the pioneer mechanized cotton mill in Pawtucket, had built a dam there. But it was not until after 1823, with the improvement of the power canal and the division of the water rights into six privileges controlled by mill owners who incorporated as the Central Falls Mill Owners Association, that the industrial history of the city truly began. By 1825 eight textile factories for cloth weaving and thread manufacture had clustered in the area, which was then still known as Chocolateville or Chocolate Mills because of the popular appeal of a local chocolate plant. Two years later, however, the village became Central Falls. But both this village and Valley Falls upstream remained part of the town of Smithfield, to which they had been assigned when Smithfield was severed from Providence in 1731. As elsewhere in Rhode Island, growing antagonism between urban and rural interests led to prolonged and often bitter demands for separation, which were temporarily assuaged by the creation of the Central Falls Fire District in 1847. This district, which also encompassed the village of Valley Falls, gradually took on more municipal functions, in effect becoming a town government in embryo.

Whereas a cluster of entrepreneurs formed the Central Falls industrial area, at the northern end of town in Valley Falls the Wilkinson Mill, which failed around 1830, and others in its vicinity were eventually all acquired by a single company dominated by the brothers Samuel B. and Harvey Chace. From 1849 onward, as the Valley Falls Company and, later, S. B. and H. Chace, from their headquarters mill in Central Falls they built one of the great Rhode Island textile dynasties of the second half of the nineteenth century. Ultimately they also occupied the opposite bank of the river in Cumberland. Between the two clusters of mills on the river at Valley Falls and Central Falls, the textile machinery firm of Fales and Jenks, which had originally occupied a mill at the middle falls, commissioned a new mill on Foundry Street in 1863. Almost immediately, in 1865, the giant Sprague textile organization in Cranston took it over as the United States Flax Company. Only briefly, however: the collapse of the Sprague empire in 1871 (see Cranston introduction) with its holdings in mills, railroads, and banks, was sufficiently momentous to contribute in a major way to the depression of 1873. Mill building pretty much stopped in the city for nearly two decades thereafter. When prosperity returned in the early 1890s, new entrepreneurs moved into existing factories; and, together with a few twentieth-century buildings (which will not concern us), the same buildings continue to serve manufacturing in Central Falls today.

From the 1870s onward, immigrant nationalities other than Irish began to augment the local labor force, especially French Canadians. In fact, the first French-speaking parish church in Rhode Island was built in what is now Central Falls in 1875. And the town status of the Central Falls Fire District changed yet again when, in 1871, a division of the oversized town of Smithfield occurred and the fire district landed in Lincoln. Indeed, Central Falls was so much a center of population in its new town that the Lincoln Town Hall initially located in what is now Central Falls! But the debate as to whether Central Falls should be independent, remain part of Lincoln, or join Pawtucket continued. Only after a ballot in 1895 did the fire district finally become an independent town—on the brink of its era of greatest prosperity from 1890 to 1920.

Although new types of industry supplemented textiles and textile-related products to a degree, there was no countering the effects of the collapse of the Rhode Island textile industry in the 1920s. The effects remain to this day. Indeed, the state of Central Falls's economy, supported only by a small tax base and a mostly antiquated industrial plant, has revived discussions as to whether the town should be joined to Pawtucket or Lincoln, with little enthusiasm for any alternative on the part of any of the three towns involved.

CF1 **Central Falls** (Roosevelt Avenue) **Mill Complex**

Roosevelt Ave. north of Cross St.; visible from Pawtucket side of the Blackstone River or from the Roosevelt Avenue Bridge

This group of mills around the middle of the three falls accounts for the town's name. The rear range of older factories, screened by later building in front of them, are visible from Pawtucket, on the shore opposite the rear of the Central Falls mill complex, which faces on Roosevelt Avenue, or—the only alternative when trees along the Pawtucket bank are in leaf—from the bridge, on the town border, looking upstream, where Pawtucket's Central Avenue becomes Central Falls's Cross Street. (Central Falls has its own Central Street nearby.)

The view upriver from the bridge includes three old industrial buildings worth special attention. Closest, or southmost, of these is the Central Falls Woolen Mill (CF1.1; 1870, Phetteplace and Seagrave), 523 Roosevelt Avenue, a long, gabled factory building in brick which extends its narrow end toward the river; above (north of) this, the Pawtucket Thread Manufacturing Company (CF1.2; 1825), 527 Roosevelt Avenue, a much smaller 1825 masonry block; finally, farther upstream (north), the Stafford Mill extension (CF1.3), 581 Roosevelt Avenue, another long, gabled brick factory, this from the 1860s (excluding for now its front monitored section, to which we shall come shortly; see CF3). The brick buildings exhibit the trend in brick factory construction typical of (but not universal for) progressive change from pre– to post–Civil War examples. Whereas the rear extension of the 1860s Stafford Mill shows granite lintel slabs sustaining the loads of the brick wall over window openings, the 1870s Central Falls Woolen Mill uses segmental arches to deflect these loads to either side of the window openings. The degree to which planar brick walls under factory eaves characterize prewar work is uncertain; but corbeled eaves surely flourished after the war, as Pawtucket's brick factories have already revealed. This 1870 mill also boasts a well-preserved and, by this date, rather old-fashioned trapdoor monitor with narrow transom lights. It lost its tower toward Roosevelt Avenue when connected to a later building. The two brick mills flank the little gabled stone Greek Revival thread mill of 1865, which originally also had a trapdoor monitor, as well as a belfry on the roof. (This early building was later slightly enlarged toward Roosevelt Avenue with a purely functional tower in front which did not project above the roof.) Although it is much battered, and most of its windows are filled with successors makes immediately evident the jump in the average scale of operations between the first quarter of the nineteenth century and the post–Civil War period.

Behind this granite block is the brick extension of 1860 to the original Stafford Mill. As a factory unit it is intermediate in size between the 1825 and 1870 mills, and less developed than the later mill as an industrial type. Its windows, still domestic in scale and handling, in the 1870 building become larger, nonresidential in character, and more abstractly organized to a geometric grid.

Finally, immediately upstream of the Stafford Mill is this company's replacement (CF1.4; 1863) for Sylvester Brown's original 1780 dam. It is a curved dam, but virtually right-angled with a rounded corner in midstream. The cooperative power canal which once ran from it beneath these three mills and others was filled in 1965.

CF2 Pawtucket Hair Cloth Mill

1864, William R. Walker. 501 Roosevelt Ave. (corner of Cross St.)

Begun in a small factory across the street in 1856, this business became successful after the acquisition of patents for weaving haircloth (most of the raw material for which originally came from Russian horse markets) for upholstery, crinolines, and inner linings. Growing success called for this new, three-and-one-half-story plant (four and one-half stories downhill). The original entrance has been shifted from the base of the tower to Roosevelt Avenue, and the tower has lost its low pyramidal cap. Otherwise the mill stands pretty much as built, and as an important early work by William Walker. Its tower is worth noting for its bottlelike adjustment of shape from the lower stages to the generous belfry. The corbeling at the eaves is rhythmically interrupted at the piers by corbeled planes, each ornamented by an inset cross. Like deep brackets, they descend almost to the level of the springing of the segmental arches of the upper-story windows. Once the largest producer of haircloth in the world, the plant continues this production today (with plastic thread instead of animal hair), as another example of the way in which a number of small Rhode Island mills have survived within a specialty niche in the textile industry.

CF3 Stafford Manufacturing Company Mill (John Kennedy Mill)

1825. 581 Roosevelt Ave.

To the rear of this three-story mill with monitor roof, the Stafford cotton firm built its extension and other additions after acquiring the John Kennedy Mill in the middle of the century. The

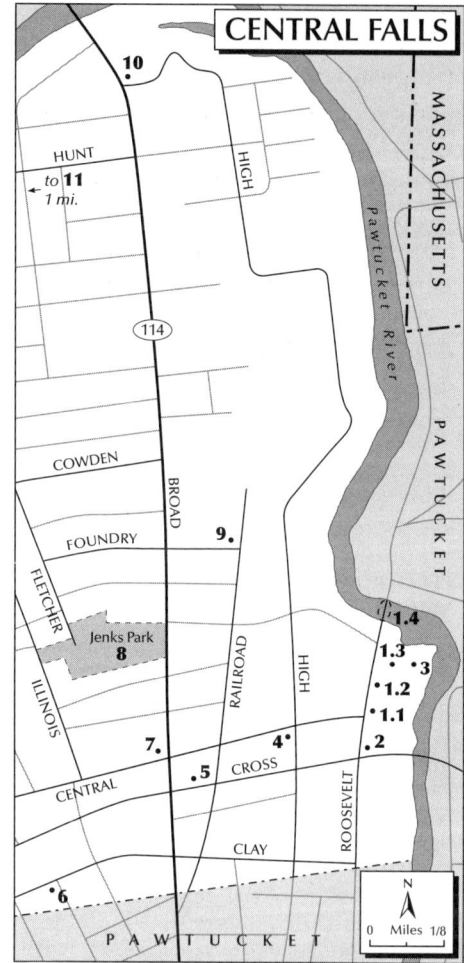

original Kennedy Mill, built at a time when factory construction in the state was generally moving from wood to masonry, is important as among the earliest extant brick mills in Rhode Island. In contrast to the early use of brick in Massachusetts factories, it is rare in Rhode Island before the Civil War, despite a few sizable outriders from mid-century and possibly James Bucklin's still unexplored early use of brick for some of his mills. Hence the extraordinary significance of this very early example, when the masonry thread mill behind it (CF1), of exactly the same date, represented the progressive norm, while clapboarding was also widely employed. Masonry does occur in the granite trim of the broad, wall-like tower, as it continues in many other brick industrial walls after 1860 for functional and decorative purposes. Granite

Central Falls (CF) 157

CF2 Pawtucket Hair Cloth Mill

CF5 Benjamin F. Greene House

sills protect against weather; granite lintels span openings; granite frames resist hard use. But old drawings for fire insurance also show that the material was used here in a more purely decorative manner: a demolished setback bell cupola with trapezoidal openings was also in granite, capped by stepped block parapeting. Still ruggedly handsome, how much more so it must have been with its original capping, which also proclaimed its Greek Revival allegiance. The preservation of its full-windowed attic monitor marks the extent of the original Kennedy Mill because the Stafford Company extension did not continue it.

CF4 D. G. Fales House

1858. 1867, remodeling, Clifton A. Hall. 476 High St. (corner of Central St.)

The now dilapidated mansarded Fales House once belonged to one of the partners of the nearby Fales and Jenks machine works. Clifton Hall's principal work on the exterior consisted of the addition of a mansard and its cupola to a very plain house. It displays Hall's exceptional sensitivity to shape, in enlivening Victorian features which are typically ponderous. The mansard has a pagodalike flare, the cupola a piquant aspect in its wide-eyed, triple-arched window and visored overhang which epitomizes the act of peering out. Few mansarded buildings by Hall remain, but some fine wash drawings indicate that he shaped the usually cumbrous mansard toward lively effect unmatched by his local contemporaries. Hence this example deserves special notice.

CF5 Benjamin F. Greene House

1868, Clifton A. Hall. 85 Cross St.

Another mill owner's house, again by Hall, this originally belonged to a partner in the Greene and Daniels Thread Company. It had operated on the site which Greene sold, a few years earlier, to the Pawtucket Hair Cloth Company (CF2), after relocating his own much enlarged plant in Pawtucket directly across the river from its old site. The house is among the finest examples of preserved Victorian Second Empire in the state, and it is something of a miracle that such an elaborate panoply of ornamentation has survived in an area where most architecture of its vintage has been badly treated. The curvature of a bonnet gable in the flared mansard is reinforced, first by its arched and balconied window, then below in a tripleted variation (like the Fales House cupola [CF4]) by way of aggrandizing the arching entrance and its ornamental porch. Tripled postlike Corinthian columns support the corners of the porch, which is further embellished with a mix of bracketing, paneling, leafage, and crosses of Lorraine. Heavily framed windows on either side have segmental-arched hoods on the first story, and flat hoods above. Two first-story window bays project toward what used to be a lawn to the corner of Dexter Avenue and extending most of a block along it. From the east, a recessed service wing has its own porch in the manner of the other.

CF6 Samuel M. Conant House

1895. 104 Clay St.

This elegantly detailed Neo-Federal house was built for the head of a printing firm, who was also a son of the owner of the Conant Thread Company. A tower of the Conant plant can be seen as the terminal vista of Clay Street a few blocks away in Pawtucket. (A block away, and within equally easy walking distance of the High Victorian splendor of his residence, Benjamin Greene located his Greene and Daniels thread works, which the Conant-Coates behemoth eventually did in.)

The house is brick-faced below, clapboarded above, its front symmetrically organized around a pair of two-story semicircular bays which swell out to embrace an entrance porch that is ample enough to serve as a small veranda. This shelters a spreading, double-doored portal framed by leaded glass side lights and a stretched fanlight. But apart from the swollen grandeur of the mass, the meticulous and copious detailing attracts attention: attenuated Corinthian columns as porch supports, an elaborate Palladian window-door centered in the upstairs story, a broad and elegant modillion cornice, all topped by a row of four dormers, two capped in pinched and austere pediments, two in pinched and elongated broken scrolls. At every level closely packed, elegantly turned balustrading animates the elevation: at the second level as a porch parapet, at the cornice as bay parapets, at the roof pinnacle as a "widow's walk" between chimneys which flank the central halls cutting through the house downstairs and up. Much excellent woodwork remains inside.

In the time separating this and the Greene House (CF5), styles changed, along with the attitude toward the proper display of wealth—assertive effects giving way to refined. But the joy in exuberant display continued. Up to the time of writing, this house has survived the usual fate of such houses converted to nursing homes. It would be unfortunate to lose such a fine example of the early Colonial Revival attempting to outdo its Federal precedent.

CF7 Adams Public Library

1910, McLean and Wright from a Carnegie Corporation prototype design. 205 Central St.

The small-town box-of-books library is here blown up into a monumental scale in this yellow brick building with forceful stone trim. The basic design is one of a number of prototypes for small-town libraries supplied to set standards by the Carnegie Corporation in conjunction with its endowment of such facilities in cities and towns throughout the United States during the early decades of the twentieth century—although Carnegie shared none of its largesse in Rhode Island. After choosing this design from the portfolio of possibilities, for example, local architects could work out their own variations in scale, materials, individual choices within the recommended Greek detailing, and so on. It is a rare Rhode Island example of Greek detailing within an early-twentieth-century Neo-Renaissance format, at a time when the fervor for classical detail overwhelmingly favored Roman and Renaissance sources. Only three other early-twentieth-century buildings in the state show such significant Greek influence, three of the four being libraries, and this the smallest in the group (see PR114.11 and PA19). Breadth of planes and moldings; emphatic allusion to structure; linear clarity; severity of ornament enlivened by sumptuous, even exotic, climax; and, of course, specific historic references to Greek monuments: these qualities make it "Greek." Here the portal and flanking windows sit with such grand assertion on their base and are so broadly enframed by pilasters that they reduce the wall to mere panels for their display. Such monumental emphasis on framing as partly actual, partly simulated structure also typifies the Greek Revival even in its most vernacular reduction in the early nineteenth century. The framing is so intense and iterated around the entrance—door frame, column, and pilaster on either side; projecting lintel, entablature, and pediment overhead—that it can be imagined as a mini-building within the strong enframement of the semi-hip-roofed block.

Inside, the monumentality continues, but with a modesty of mien appropriate to a small library. A top-lighted and columned rotunda provides for both entrance lobby and circulation desk. This circular theme at the center is echoed by shallow bow windows at the ends of flanking reading rooms. Each has its fireplace, and all woodwork is varnished, including the columns. Such homey touches reduce the monumental aura of the rotunda as though the building would reassure its borrowers, "Monumental, yes; but homelike too. Be impressed, but not intimidated."

CF8 Jenks Park

1890. 1904, Caroline Cogswell Tower, Albert Humes. Broad St. between Summit and Fales sts.

A modest memento of the civic consciousness spurred by the City Beautiful movement of the American Renaissance, these four acres are Central Falls's only park. The gift of Alvin Jenks, a descendant of the family which had been among the town's earliest settlers and one of its leading industrial families, this park contains the abrupt rock outcrop from which the Wampanoag chief Canonchet and his band spied upon the approach of their unsuspecting colonist victims in 1674. The visual interest of this much loved but underkept park is primarily folkish. Cement-paved paths snake around the ledge past two gazebos sheltered by huge iron umbrellas fabricated at the nearby Fales and Jenks machinery works. The climax is the Caroline Cogswell Tower, a clock and observation tower given by a former resident of this city and designed by the Pawtucket architect Albert Humes, who, as mayor of the city at the time, doubtless had an inside track to the commission. It is a simple, tapered form in rough stone with a pyramidal roof once surmounted by an eagle. Its most remarkable aspect, however, is the circular structure of thin metal pipe with a glass roof which surrounds the base of the tower. Its spindly fragility is so much at odds with the boulder tower as to make the confrontation interesting—appearing from below as a sort of Hula-Hoop to the tower. Lights strung from the top of the tower to the hoop make the community's Christmas tree. For the architectural pilgrim in search of a panoramic view, the tower, if open, would improve on Canonchet's advantage.

CF9 **United States Flax Company** (Fales and Jenks Machine Works)

1863 and later. 27 Foundry St. (at Railroad St.)

Consider the location: roughly halfway between the clusters of factories at Central Falls and Valley Falls, this was the first major factory in the town to ignore the river for the railroad. Because it fronts the railroad cut rather than the street, from Foundry and Railroad streets one clearly sees an end of the plant only; enough, however, to ascertain its high quality. (The main front of the plant is visible from High Street, at the truncated stub of Foundry Street.) The long brick building is beautifully preserved, with granite trim, a narrow trapdoor monitor (now mostly blinded) running its full length, a crenellated tower, and virtually all sash intact. It belongs to the same generation as the post–Civil War brick plants to the rear of the Roosevelt Avenue cluster (CF1)—and is the finest of the lot. The stack of loading doors is divided only by heavy granite spanning elements which simultaneously provide a sill for the door above while capping the door below. As the stack reaches the gable, the width of the topmost door is slightly reduced, and one can discern the filled-in trace of its vertical extension by a window-sized opening. So originally the stack of openings tapered to accord with the gable pitch. Inside, diagonal bracing in four directions off many of the timber supports augments customary post-and-beam construction. It is designed to counter the bending moments of heavy machinery on the floor beams by providing the columns with intermediate props. The complex barely functioned as a machinery plant, however, before its conversion to textiles (see Central Falls introduction).

CF10 **Apartments** (Valley Falls Mill, also known as Blackstone Mill)

1849. 1853, dam. c. 1853, raceway gatehouse. 1363 Broad St. (at High St.)

Here the conversion from near ruin to apartments has so denatured the mill that one might not stop, except that this was once headquarters for a major textile dynasty. Controlled by the brothers Samuel B. and Harvey Chace under the corporate names of the Valley Falls Company and S. B. and H. Chace, in the twentieth century it was taken over piecemeal by other companies (this factory was absorbed by the nearby and, by then, even larger Sayles Corporation; see under Saylesville in Lincoln).

What remains from a much larger complex is a medium-sized brick mill, with windowsills and lintels trimmed in granite. It is important as another major early brick factory building. Its central tower is unusually capped with a cross-gabled roof in which each of the ridgelines slopes from its central point, the front gable less so than the others—a so-called helm roof, from an archaic term for "helmet." If one can mentally strip the apartment "beautification" and imagine the original treatment of the tower, the proportions of the arched openings to the tower and their placement are especially handsome: a stack of wide, segmental-arched loading doors flanked by arched window slots, then a nice arrangement of company name, oculus, hoist, and, hard above it, the triplet of arches which make the tower's "face"—all openings underscored by flush granite sills. The top floor of the mill was specially constructed as clear-span space to accommodate the first American version of the British Sharp and Roberts self-actor mule-spinning frames, made in Pawtucket by James Brown. With the addition of a new dam, visible behind the mill from the Broad Street Bridge over the Blackstone, the Chace brothers were able to expand their plant on both sides of the river. Eventually most of it occupied the opposite (Cumberland) bank of the river, where now nothing exists of it but ruins (CU1).

CF11 St. Matthew's Roman Catholic Church

1929, Walter Fontaine. 1018 Dexter St. (at West Hunt St.)

However meritorious the masonry craftsmanship of the somewhat Frenchified exterior of this English Perpendicular Gothic church for a French Canadian parish, it is like many others of its period. Its interior is more interesting. As in his much larger and earlier (1914) St. Ann's in Woonsocket (WO1), Fontaine saturated a wide, spacious nave with color. Lightweight Guastavino tiles in the beiges and tans of the stonework, some with decorative stampings, span the width of the aisle, crossed with ribbing which is quite obviously more decorative than structural. Decorative in emphasis, too, are the piers of the aisle walls, their colonnettes treated more as frames to panel the wall than as cues to structure. Color in cartouches to either side of small clerestory windows and in the zone of the arching for the aisles (painted by Guido Nincheri of Montreal) is climaxed in the elaborate and unified program of stained glass (by the Paris studio of Maumejeane). Compare this with a similarly ambitious program of stained glass in the Cathedral of Saints Peter and Paul in Providence (PR35), by Munich designers. There, influence from the Nazarene painters accounts for the picturelike quality and milky glow of the windows. Here, the glassy nature of the windows is more evident. The design is more abstract; color is more intense and transparent, with much clear glass introduced around the periphery of the windows. Influenced by modern ideas of truth to materials and by a modern preference for a less illustrative, more symbolic expression, the craft returns full circle toward the conditions of its origin.

North Providence (NP)

Not until 1874 did North Providence acquire the boundaries it has today. Previously, what is now its eastern third had been part of Pawtucket and its middle third part of Providence. Agriculture long dominated its economic life. Fairly good soils and relatively easy access to Providence made farming profitable even before the coming of turnpikes. Only along its western boundary, made by the Woonasquatucket River, did major mill towns develop: from south to north, Lymansville, Allendale, Centredale, and Greystone, established between 1807 and 1822. Of these, Allendale and Greystone remain the most interesting architecturally, so only they are included here.

Lymansville is important, however, in the history of textile technology. So-called Scotch looms were first employed there, in a long-disappeared stone mill, in 1817, when Daniel Lyman brought together patterns supplied by a Scottish mechanic, William Gilmour, and the practical skills of David Wilkinson of Pawtucket. The "Scotch loom," with variations, became the standard power loom in the Rhode Island textile industry. Moreover, the Lymansville plant was the first in Rhode Island to process cotton from raw material to finished cloth under one roof. So, if the contribution of Pawtucket to the early textile industry is well known, then the importance of Lymansville in its slightly later development deserves more recognition. Unfortunately, the oldest mill which now exists in the village is a brick building erected in 1885 by a German emigré, Albert Sack. It is of some interest in itself: built to his specifications, it is apparently exceptional among Rhode Island factories for its cross-shaped design. In the socioeconomic history of Rhode Island, moreover, Sack's venture also pioneered as a factory store. Allegedly for the first time in the state, at least as an avowed premise for the business, clothing produced in the factory was primarily sold on the site—making Sack's mill a forerunner of the numerous clothing "mill stores" which now occupy pieces of old factories throughout New England.

As for Centredale—the other important North Providence mill village omitted from consideration here—it centered the row of Woonasquatucket factory villages. It was also a traffic "centre" for its area (the spelling still deferring to the initial predominance of English workers). Modern road widenings and turnabouts have neatly done in the village, and its remaining factory buildings are fairly recent and of no special distinction.

Surprisingly, in view of its closeness to Providence, railroads did not reach North Providence until 1877. Privately financed turnpikes were built in the nineteenth century—in all, five of them, three of which involved Centredale. From the perspective of Providence, one terminated in the village; one passed through it; one originated there. Mineral Spring Turnpike (now Route 15 and an "avenue"), chartered in 1826, runs east–west virtually across the center of the town from Pawtucket to Centredale, and does indeed pass a mineral spring near Orchard Avenue in Pawtucket. The others run generally northwest to north with reference to Providence. Powder Mill Pike (Route 44, now Smith Street, built 1810–1815), which passes through Centredale, was named for the first powder mill in Rhode Island, established in the village during the Revolution. It was the last privately owned turnpike in Rhode Island, becoming public only in 1874. Its closing ended private toll roads in the state. Finally, Farnum Pike (Route 106, known as Waterman Avenue in North Providence, built 1808–1828) took off in a northwesterly direction from Centredale. North-south movement among the four mill villages along Woonasquatucket Avenue also passed through Centredale. Only two of North Providence's old turnpikes avoid Centredale: Old Louisquisset (now roughly Route 246, chartered 1805), which hugs the eastern border in a north-south direction, and Douglas Pike (Route 7, chartered 1806), which cuts across the center of the town on a northeast-southwest diagonal.

Because the bulk of manufacturing in North Providence was at its western edge, paradoxically the area to the east of the Woonasquatucket mill villages and closer to the heart of Providence remained lightly developed longer. Looking at the map of North Providence, one might expect that the easternmost section of the town, closest to the peripheral expansion of Providence, would have become a premier location for suburban houses. Following the Civil War, however, fashionable development, including some sizable "country seats" moved out onto the northwestern hills of Providence, along the axis of Smith Street (which focuses on the capitol dome) in the direction of Fruit Hill at the center of North Providence. Fruit Hill Avenue winds in a long diagonal across the crests of a series of hills with extensive views of the countryside to the west and over Providence to the east. The old farms and their orchards, which once lined it, many owned by the Olney family, were substantially transformed to building lots by Samuel Hedley in the 1880s according to a scheme grandly known as the International Park Plan Plat. By the late nineteenth century the reasonably affluent, in search of country living, together with a sizable contingent of artists and aesthetes, were commuting on the Smith Street trolley (which began operation in 1893) to Fruit Hill Avenue. To the present day these hills retain an attractive enclave of mostly nineteenth- and early-twentieth-century houses.

NP1 Joseph Smith–John Jenckes–Cushing House

Pre-1675–1676, 1713, c. 1750, mid-19th century. 109 Smithfield Rd.

This house evolved by a complex but interesting development. It began as a one-and-one-half-story structure with one room at ground level and another under the roof. An exposed masonry chimney on its north end made it a "stone-ender." Barely completed, it was destroyed during King Philip's War (1675–1676). Joseph Smith built a new house on the old foundation and the stub of its ruined chimney (hence the change in the exposed chimney at the north end of the house from exposed masonry up to the top of the ground floor to brick above). The new house, raised to two and one-half stories, with a long saltbox slope to the rear, extended in four irregularly and widely spaced twelve-over-twelve windows at the second-floor level through what was then an entrance at the end of the elevation.

Around 1750, John Jenckes, the next occupant, lengthened the house to the south by another two bays (making six in all), thereby bringing a rough symmetry to the front. The stretched-out saltbox profile of the new side elevation shows a fascinating play between symmetrical and asymmetrical configurations within the pattern of openings spread over it. Finally, around 1865, idiosyncratic porches were added to the front door and at the north end of the house. Their pierlike supports suggest pilasters with pierced fluting, a charming example of provincial playfulness with current Italianate form. At this time, too, the first-story windows (and some in the back) were enlarged with six-over-six sash (those in the front elevations later reconverted to twelve-over-twelve in a twentieth-century effort to relate their scale to the old windows above). As a result the house now has three different window treatments: the oldest twelve-over-twelve sash projecting in boxed frames beyond the plane of the clapboarding; then the six-over-six interlopers of 1865 inset into wall; finally, their partial twentieth-century reinstallation to the twelve-over-twelve "correctness" of the original. So the house reveals successive efforts to give more formality and dignity to an older fabric deemed lacking in these qualities. The result is a "picturesqueness" which enlivens the formulas from the past and records the modification of generations all the way back to the period of earliest settlement.

Diagonally across the way, at 128 Smithfield Road, the Captain Stephen Olney House (1802), with a five-bay, one-and-one-half-story format, offers the traveler a less complicated bonus.

Fruit Hill

The combination of rural and aesthetic in this country suburb and the success of the mix of late Victorian–Queen Anne design in expressing these sentiments is evident in two nearby

Fruit Hill, St. James Episcopal Church

buildings, both unfortunately altered. The first, at 474 Fruit Hill Road, is St. James Episcopal Church (1879, Howard Hoppin; twentieth-century side addition). Reported in the Providence *Journal* of the time as being in the "Dutch style," it is a picturesque composition which is both monumental and self-consciously pretty—a prettiness rather exaggerated in an otherwise commendable restoration. At 354 Fruit Hill Avenue is the Fruit Hill School, now an American Legion Post (1879, William E. Colwell[?]), in the Queen Anne style, with an intimate scale, snug gables, deep overhangs, and fanciful ornament, but insensitively altered for its present use.

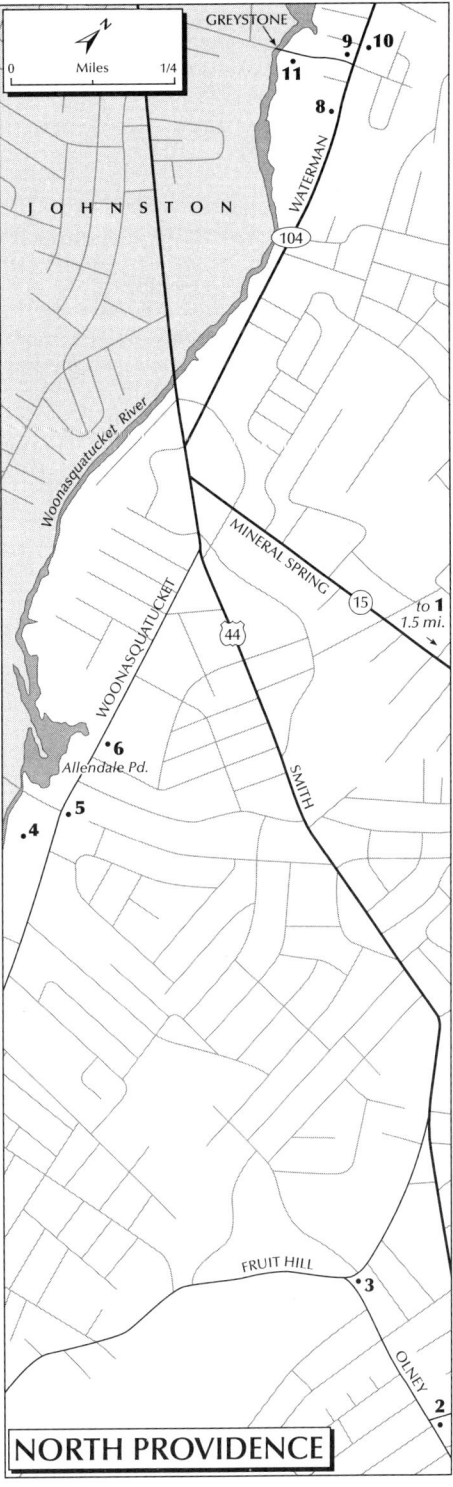

NP2 Bungalow

c. 1905. 95 Olney St.

The best of a group of bungalows that gave the area its onetime artistic reputation originally belonged to the locally prominent artist H. Cyrus Farnum. The long pitch of the roof to the front, with its trapdoor dormer, shelters a deep porch, columned to acknowledge the Colonial Revival. But note especially the subtlety of the side profile, where the expected gable is gentled by the double pitch of its rear slope. Such double pitches nicely abstract the qualities of an "olden" sag and (when to the rear) invoke the ancient custom of a saltbox addition. The sagging roof profile accords as well with the tapered chimney. The chimney slices into the roof plane rather than penetrating through it. Paint now intensifies the Neo-Colonial aura, although originally the shingled walls and chimney brick may have shown their natural colors and textures, with trim and columns only in white.

NP3 Whipple–Angell–Bennett House

c. 1767, c. 1820, c. 1850, c. 1940. 157 Olney St. (at Fruit Hill Ave.)

Down the street from Farnum's bungalow is the oldest extant house in the area. This small, gambrel-roofed, one-and-one-half-story farmhouse with a transom-lit door and a commanding central chimney exhibits the casual approach to symmetry typical for its period, especially in vernacular carpentry. It was built by a member of the family who settled and originally owned much of the land in the area. Other onetime farmhouses from the eighteenth and early nineteenth centuries are scattered through the area, all but overwhelmed by later suburban developments. Of these ancient survivors, this is the best preserved.

Allendale

As late as 1978 a report of the Rhode Island Historical Preservation Commission asserted that Allendale was the best preserved of North Providence mill villages. Though dilapidated by the 1980s, its mill was also picturesque—until its conversion to condominiums resulted in gross changes of sash, entrance doors, and hardware which are completely out of keeping—plus, of course, the necessary evisceration of the interior. Similar insensitive alteration and outright demolition of much of Allendale's mill housing occurred at the same time. So dispiriting are these changes that one is tempted to hurry by without looking; but the mill calls for more than dismissal. Allendale is a significant mill village which was dominated by one of the most important early Rhode Island industrialists, Zachariah Allen. So what remains deserves attention.

NP4 **Condominiums and Offices**
(Allendale Mill and Company Store)

1822, Zachariah Allen (designer), John Holden Greene (builder). 1844, storage buildings near tail race. 1864, engine room, south side. 1880, masonry addition. 1910, brick addition. 1988, conversion to condominiums. 494 Woonasquatucket Ave.

The original (1822) four-story masonry mill is among the historically significant mills in the state. In the technology of the textile industry, it is important as the site of the first use of power looms for the manufacture of broadcloth, and the first mill to use a rolling process to impart a glossy finish to cloth. Traditionally considered to be the first use of slow-burning wood construction (see WO15), it is now viewed as merely a pioneer exemplar of a mode of wood interior construction emphatically championed by Zachariah Allen as a means of reducing constant mill losses by fire. When his substantial efforts to create fireproof construction here and in other of his factories led to no reduction in Allen's fire insurance rates, however, in the 1830s he organized a consortium of industrialists to self-insure against fire. The Allendale Mutual Insurance Company took its name from this place, although its headquarters building is in Johnston (JO10).

Allen's contribution to building and manufacturing technology aside, however, the 1822 portion of this mill is an example of the early masonry type, with massive walls of stuccoed random rubble masonry, domestic-sized windows, and (originally) six-over-six sash. Projecting in front is a compact tower (long missing its original cylindrical wooden belfry) and, at the ridge of the roof, a unique dormered monitor (which seems to have been early altered from the narrow slit of a trapdoor monitor to increase light in the top story). It is not as pretty a mill as its contemporaries in cut stone; but Allen specifically ridiculed owners who wasted money on cut stone for factories when rubble walls surfaced in stucco would do.

The upstream (north) end of the mill, mentioned in Henry-Russell Hitchcock's *Rhode Island Architecture* (1939) as a memorable image, is unique, although somewhat spoiled by careless window changes. It is best viewed from the bridge between the mill and its dam, to the north of the factory. The stepped and angled profile of the dormered roof combines with bulky, stepped buttressing at the corners of the building to make an intensely shaped image in both silhouette and three dimensions, its masonry mass relieved only by two widely spaced tiers of paired windows. These corner buttresses were not, as is customarily assumed, installed to dampen the vibration of the machinery inside, but to dampen a freakish sympathetic vibration induced in the wall structure from the frequency of the waterfall—a mystery which Allen himself eventually solved. A third buttress midway along the river wall completes the only example in Rhode Island of a building correction for vibration. Close by is a fine example of the almost vanished craft of large-scale drywalling for the tail race.

Together with its two additions, the three sections of this mill clearly reveal the typical progression of the factory type toward bigger units, larger windows, thinner walls, wider spans in column placement, and greater uniformity in membering as machines increasingly took over from hand craftsmanship, and the eventual shift from masonry to brick as the dominant walling material for the nineteenth-century mill. Steel framing, let alone reinforced concrete structures for factory buildings, came slowly to the United States. The 1910 addition is still timber supported. Although steel reinforcing hardware became more widely used over time, "slow-burning" wood construction was so highly regarded by American insurers that it persisted in much factory design well into the twentieth century, long after it was anachronistic in most of Europe. It is these tall, open timber interiors, with their industrial grime sanded away, which attract a generation enthralled by "loft living," although studio spaciousness, let alone loft spaciousness, tends to disappear when spaces are, as here, conventionally divided.

Broad, gambrel-roofed workers' houses (c. 1824) ranged across the mill front, until demolished for apartment parking. Allen's company store (1822), in stucco-covered rubble masonry like the mill, but piered, still stands immediately to the north of the factory. It has been noted as among the earliest extant Greek Revival buildings in the state, although with such a rudimentary portico of piers and pediment that, were it in brick, its "Greek" aspect might disappear.

North of the mill along the river, in various states of preservation, are several gambrel-roofed workers' houses of the type which also once existed in front of the mill, as well as other types of housing (for example, 500, 512, 518, 522–524, and 535 Woonasquatucket Avenue, all c. 1824). Allen laid the houses out with the provision of space for subsistence gardens. At 530 and 542 (c. 1850), two-story Greek Revival houses, originally identical, with a rhythm of simple two-story pilasters running around the clapboard walls, served as boarding-houses for single workers. They repeat the pilasters of the store, perhaps indicating that this treatment, in Allen's mind, gave greater dignity to community buildings.

NP5 **E. B. Olney House**

c. 1848. 515 Woonasquatucket Ave.

Of several Gothic Revival cottages built in Allendale, this is the most interesting. High-pitched roofs and Tudor moldings over windows give a vaguely "Gothic" aura to this plainly carpentered house, with specific reference mostly concentrated in the charming entry porch. Note also, however, the paired diamond-shaped terra-cotta chimney flues molded to "medieval" taste. The design seems to have been nominally inspired by one of Andrew Jackson Downing's many books, which, in the 1840s and 1850s, popularized medieval styles as ideal for suburban and country houses. Extant Gothic Revival houses are rare in Rhode Island and seem never to have been very popular. Allendale is the only extant mill village in the state with several of them, all seemingly allocated to those with superior status in the community hierarchy.

NP6 **Baptist Church** (Allendale Schoolhouse and Community Hall)

1847, Thomas Tefft. Belfry later. 545 Woonasquatucket Ave.

Zachariah Allen originally commissioned this building as a combined Sunday school and community meeting hall, with a library projected for the basement. Such multipurpose meeting halls in mill towns were also used for church meetings on Sunday; here the church took over as early as 1850. The exceptional nature of a Rhode Island schoolhouse–meeting hall in the Tudor period style then designated as "Elizabethan" resulted from an English trip during which Allen's fancy was taken by such a building. The trip doubtless also inspired Allendale's Gothic Revival houses as well. Tefft used some such source as the "Elizabethan" design number 4 in H. E. Kendall's *Designs for Schools and School Houses* (1847) as his model, possibly by way of his friend and mentor, Henry Barnard, with whom he collaborated as designer. Barnard's *Reports and Documents Relating to Public Schools in Rhode Island for 1848* (published in 1849) includes Tefft's Allendale schoolhouse with the comment that its novel style afforded relief from "the dull monotony of wretched perversions which characterize the village and country schoolhouses of New England." Influential as Barnard's practical ideas on schoolhouses may have been, however, his encouragement of more stylistic diversity in their exteriors had little effect in his day. The architectural success of this unique schoolhouse depends upon the skillful shaping of its stuccoed masonry masses (probably originally rough cast like the mill) and placing of window openings so as to maintain the force of the massing (altered, however, by added stained glass windows). Such details as the corbeled, earlike projections of the end walls at the eaves of the steeply pitched roof, the stepped roofline of the entrance gable, and the ledged projection to establish a base enliven the outlines of the simple massing. The judicious placement of smallish windows in symmetrical and asymmetrical arrangements gives precedence to the wall surfaces, and, through them, to the shaped mass. The original design (now in Tefft's papers at Brown University) shows the gable topped by a masonry bellcote in the English Decorated Gothic manner, which may not have been built, and the first floor set up as a classroom. The present wooden belfry and the stained glass windows are later Baptist additions.

Greystone

Although textile manufacture began in Greystone as early as 1813 and continued through a series of operations punctuated by fires and many changes of ownership, most of what one now sees is the result of the twentieth-century location there of the English firm of Joseph Benn and Company to avoid heavy import duties on wool. From 1904, the firm built a large mohair and alpaca plant, workers' housing for the 1,500 Yorkshire and Lancashire workers who came with it, and, of course, a cricket pitch. Most exceptionally, Benn created what might today be called (with some exaggeration) a housing megastructure which is unparalleled in the state. Although the Benn company sold out in the 1930s, textile production persists in the big mill, with some other industries in the rest of the complex.

Immediately north of the mill, a view down Oakleigh Avenue shows ranges of clapboard mill housing (1904–1910), the small, carpentered Greystone Primitive Methodist Church (1904), 1 Oakleigh Avenue, and, behind these buildings, the looming presence of one of the two mammoth brick towers of the principal mill building. Only the re-siding of the foreground buildings mars the picture. Hardly the tourist's image; but for those who take the time to look, it is among the memorable images of Rhode Island mill villages—and made more so by the war memorial in the block between Oakleigh and Greystone avenues.

NP7 Duplexes

1910–1913. 112–134 Waterman Ave.

This mostly well-preserved row of clapboard mill duplexes with porches the length of their fronts was built by the Benn company. They contrast with most of the company's housing and, because of their special character and relative spaciousness, may have been overseers' houses, or possibly intended for very skilled workers.

NP8 Whitehall Building

1911, Frank P. Sheldon. 158–178 Waterman Ave.

A commanding three-story block (with four floors on the downhill side toward the plant behind it), the Whitehall Building effectively marks the heart of the village. The buff-tan brick is mere veneer for underlying reinforced concrete construction; walls and floors are extravagantly seven inches thick. Whitehall was built as an apartment and store block for company workers, with quarters on the second and third floors for factory overseers. The apartments are reached by an exterior balcony running around the building at the second story, providing access to a series of entryways, each serving two apartments—a door directly inside to the second floor apartment, with a stair to the apartment above. The cantilevered concrete slab which makes the balcony also permits an unsupported canopy for stores below (the shop windows now bricked up to provide additional housing). An external stair at the north end (complemented today by a handicapped access ramp) leads to a second-story auditorium. Although Rhode Island mill towns offer some tentative examples of multifunctional community buildings, none is more comprehensive than this, nor does any other as boldly and uniquely proclaim its combined uses.

It is the building's unusual program and its overexpression that impress, more than architectural acumen. The principal visible elements—balcony, outside stairs, and third-story apartments popping out at the top—are more juxtaposed than integrated. Details are gauche: curlicues stretched across the metal balcony railing, and against these the utterly disparate classical allusion of the balcony supports and the entablature facing of the roof, both overly attenuated and crudely realized. Above, equally crude pilasters (grossly overscaled compared to the skinny columns) make inset panels for alternate groupings of double and triple windows to signify the difference in apartment size. Sixteen bays long and three wide, the narrow box is modestly "centered" by three bays which are not inset—one bay actually off center, its window rhythm 3-2-3 instead of the expected 2-3-2, which would have more emphatically centered this feature. Finally, the factory-inspired corbeling of the cornice is decidedly at odds with the classical effort below.

The Sheldon firm allegedly designed this in conjunction with the mill behind (see NP11). But how could the design of one be so conscientiously uncertain, whereas the other is so incisively right? Could this sophisticated firm of functional designers have been so thoroughly cowed by the demand for an expression of civic monumentality? Or did the firm pass on this responsibility to a local builder, perhaps with a rough sketch of what was wanted, so that it could concentrate on the magnificent mill? Given the proportions of the box, moreover, its elevated aspect, and the folkish pretensions of curlicues and classical allusion, could "Whitehall" have been intended to invoke Inigo Jones's great banqueting room for his contemplated royal palace in London? Was it meant to conjure up a bit of nostalgic majesty for this remote English place, even as George Berkeley, another displaced Englishman, had earlier called on Whitehall when christening his conversion of an eighteenth-century Rhode Island farmhouse into an intellectual retreat (see MI3)? More important, what accounted for the exceptional program for mill worker housing in the first place: an English precedent, company philosophy, or another motivation?

NP9 Veterans Monument

1923, F. F. Ziegler, sculptor. Waterman Ave. between Oakleigh and Greystone aves.

A semicircular plot is partially screened from the ranges of worker housing behind it by a wall in reinforced concrete, primitively paneled and capped with scrolls which support the flagpole at its center, all painted white. In front, on center, the roster of honor for World War I veterans on a bronze plaque is attached to a natural boulder topped by a bronze eagle with outstretched wings (with rosters for subsequent wars set beside it). A semicircular sidewalk frames this centerpiece; then radiating sidewalks out from it to an arc of flanking posts in reinforced concrete topped with spheres, and again white paint serves for poor man's marble. Piers topped with more spheres at adjacent street corners attempt (not successfully, but touchingly) to aggrandize the aura of the place. Is it the folk contribution to the shrine and the sense of authenticity which automatically inheres in the vernacular, together with the obvious care given to the place, that make it seem that here the remembered may be better memorialized than at most monuments professionally designed?

NP10 Richard Anthony House

c. 1822. 20th century, veranda. 201 Waterman Ave.

This two-and-one-half-story Federal house, set back from the street on a rise, with quoining at

NP11 Joseph Benn and Company (Greystone Mill)

the corners and an unusual quoined arch around its side-lighted entrance, is the glorified farmhouse residence of the owner of the Greystone mills from 1816 to 1835 (when they were, in fact, gray stone). Richard Anthony's son James, a partner in the business, lived in a house nearby (c. 1822), at 154–156 Waterman Avenue. Richard Anthony also helped found the Coventry Company in Anthony in 1805 (see CO6).

NP11 Joseph Benn and Company
(Greystone Mill)

1904 and later, Frank P. Sheldon. Waterman Ave. at Greystone Ave.

Although Greystone Avenue brings one to the mill entrance, the rear porch of the Whitehall Building provides the best overall view of the front of it. This is among the most complete plants remaining in Rhode Island designed by the leading early twentieth-century firm of industrial engineers in the state. The long spinning mill is a brick-piered block, four stories in front, five to the rear. Tall, inset segmental-arched windows on simple rough-faced granite sills fill the full width between the piers, those for the top story reduced a little in height with roof brackets between to terminate the elevation visually with some sense of the horizontality which a cornice would provide. The immense stair towers divide the elevation into thirds. They are articulated with a double tier of arched windows climaxed by now blocked circular windows or clock faces, all openings framed in inset panels, with "medieval" crenellation and capping, handsomely simplified. In the higher rear elevation, visible on Greystone Avenue, the pier-spandrel / pier-spandrel rhythm is unbroken and inexorable. All is crisply planar, incisively edged, grandly proportioned, in the elegantly modular, simple, and modern style which the Sheldon organization had developed by 1900. Both towers and walls represent a further simplification and aggrandizement of the Sheldon-designed type used for the earlier Ann and Hope Mill at Lonsdale (CU5).

The great loss is the blinded windows, felt particularly in the blinded sawtooth roof of the low weaving plant in front of the wall of spinning machines—a type of building specifically *made* for light. The foreground buildings are not easily seen because of the topography, but an overall view of the complex shows the basic array of nineteenth-century architectural inventions for bringing light to the factory: tall windows and minimal walls in the spinning mill; square monitors popping from the flat roof of an old weaving shed (casually oriented east and west); finally, the crisply executed sawtooth monitor scientifically oriented to the steady coolness of northern light. One-story sawtooth-roofed buildings for weaving sheds and other industrial processes with exacting requirements for illumination (before the invention of fluorescent lighting) were a Sheldon specialty. The firm's engineers published several scientific studies on sawtooth design to maximize the most intense and unvarying light by natural means. Although this roof type, with its multiple north-facing skylights, was widely used in weaving sheds in England, where clear steady light was especially prized beginning as early as the 1830s, New England mill architects were slow to adopt it, fearing that the sawtooth could not withstand snow loads. In the late 1890s, however, American architects and engineers, the Sheldon firm in the vanguard, successfully naturalized this serrated roof form by building it in the sturdy plank and timber style of American mill construction, and only later in metal framing.

Johnston (JO)

Like New Hampshire's Old Man of the Mountain, the eastern border of Johnston, where it butts against Providence, might, with some imagination, be seen as a profile emergent from a skewed block. The Woonasquatucket River defines the profile. As in North Providence, Johnston's first intensive settlement occurred along the river or its tributaries. The town is still most heavily populated there, on its eastern edge.

Originally part of the "inner woods" west of Providence, Johnston was set off as a separate town in 1759, and named for the current state attorney general, Augustus John-

Johnston (JO) 169

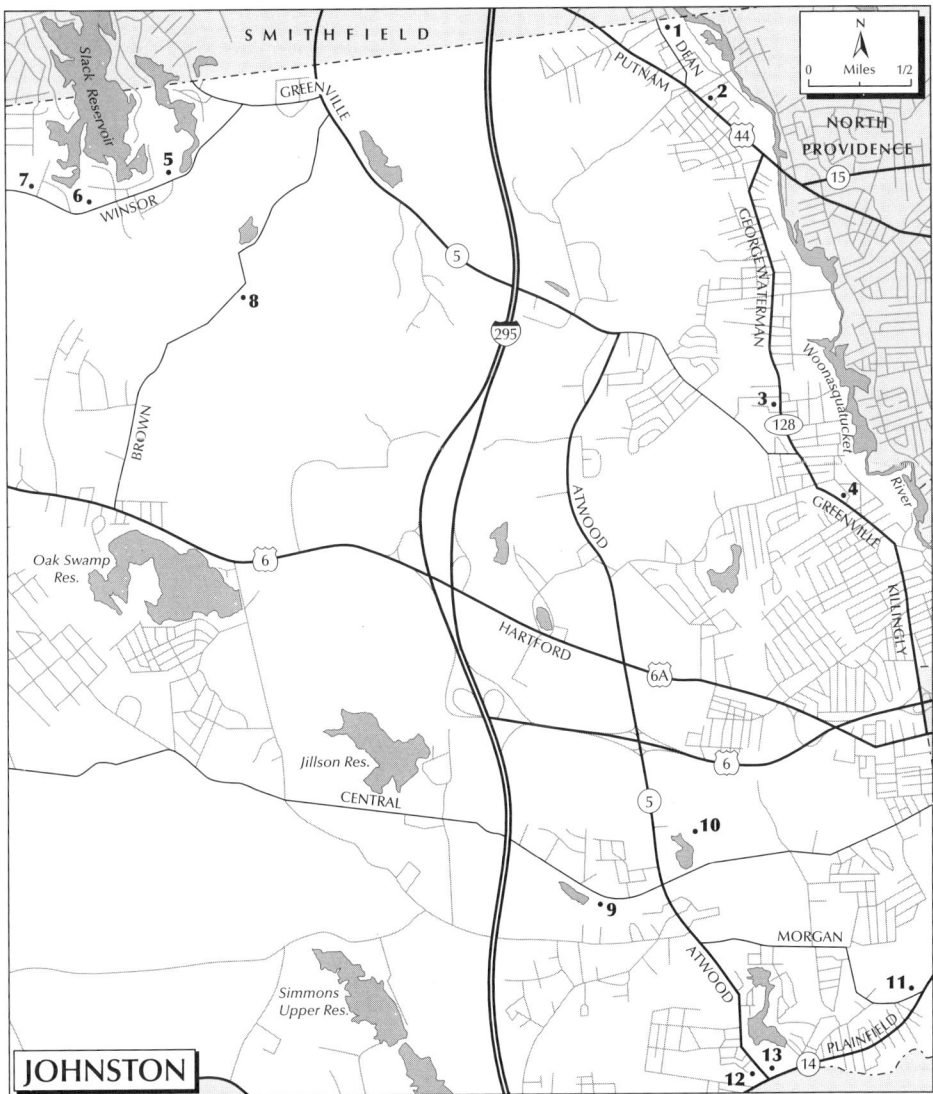

ston, who came from the area. Mills began along the river in the eighteenth century; during the course of the nineteenth century they concentrated at three points, some so dimly defined today that they no longer show on the official state tourist map: in Centerville, in the northeast corner of the town (not to be confused with the similarly named village nearby in North Providence); farther south at Manton and Olneyville (which were mostly absorbed by Providence in a boundary adjustment at the turn of the twentieth century); finally, in the southeast corner in the now merged cluster of Hughesdale, Simmonsville, Morgans Mills, and Thornton. Quarrying, particularly at Graniteville (close to Centerville) and at Snake Den Quarry in the northwest corner of the town, was another major industry of the town into the twentieth century. From

Graniteville, oxen dragged the monolithic granite columns for the Providence Arcade (PR18); from Snake Den, they hauled blocks for the First Unitarian Church in Providence (PR92). But this industry, too, has perished.

Johnston and Scituate, the next town west, are crossed by more nineteenth-century turnpikes than any Rhode Island towns. Angling from Providence northwest to due west are Putnam Pike (1810), Hartford Pike (originally the Rhode Island and Connecticut Pike, 1803), Central Pike (originally Foster and Scituate, 1822) and, oldest of all, Plainfield Pike (originally Providence and Norwich, 1714), which marks most of the southern boundary of Johnston, and the short Shun Pike in its southwest corner. Agriculture spread with the turnpikes. Although the relative distribution of population in Johnston has remained the same since settlement in the eighteenth century along the river, the visitor will experience its more densely populated eastern part as the westward sprawl of Providence. This suburbanization of old farms intensified following the north–south bisection of the town by Interstate 295, the circumferential of Providence, and the T-junction at its near center for east–west Interstate 195.

Unexpectedly for a town which is part of the metropolitan perimeter of Providence, virtually all nineteenth-century mill buildings of any consequence have disappeared, except for two late-nineteenth-century brick mills still operating in Thornton. Ironically, the destruction of most of Johnston's historic factories and the loss or modification of most of the old fabric around them make scattered remnants from the town's remotest past our principal concern, including one of Rhode Island's most important colonial icons, the Clemence-Irons House. They are pieces of a past spared by some miracle, but mostly wholly out of context or fast losing it. Johnston also contains, in the headquarters built for Allendale Mutual Insurance, Rhode Island's architecturally most important corporate headquarters building.

JO1 Daniel Angell House

c. 1725, c. 1775. 15 Dean Ave.

The incongruous setting which presses in on this house tends to obscure its quality. Built in two sections, this one-and-one-half-story house with a steep gambrel roof originally consisted of its five western bays (closest to Dean Avenue) with typically narrow windows and a central stone chimney. This was apparently extended for a tavern before the Revolution by an in-line addition with a second door and chimney. Further minor changes around 1800 left the house pretty much as we now see it. Descended from Thomas Angell, who accompanied Roger Williams to Rhode Island, the Angell family were among the first settlers of Johnston.

JO2 Johnston Historical Society (Elijah Angell House)

1765, c. 1815. 101 Putnam Pk. (open to the public)

In this two-and-one-half-story gabled house, the side-lighted door with a blind fan, suggesting the early-nineteenth-century Federal style, first catches the eye. But the central chimney and the small six-over-six sash boxed out beyond the plane of the clapboarding, as well as the window placement, which just misses symmetry, all indicate its mid-eighteenth-century origin with another member of the pioneering Angell family. Like the preceding house, this also served briefly as a tavern for travelers along the turnpike between Providence and Putnam, Connecticut. It was early (c. 1815) converted into two-family housing with apartments downstairs and up for workers at a forerunner of the present Greystone Mill, immediately behind and across the river in North Providence (NP11). The downstairs, restored for use as a museum and meeting place, displays fireplaces and woodwork in plastered rooms typical of country work in the late eighteenth and early nineteenth centuries.

JO3 Thomas Clemence–Irons House   JO5 S. Winsor House

JO3 Thomas Clemence–Irons House

1650s? 1691, rebuilt. Early 18th century, saltbox extension. 38 George Waterman Rd. (opp. Irons Ave.) (owned by the Society for the Preservation of New England Antiquities; open by appointment)

This and (even more) the grander Eleazer Arnold House (LI14) are the classic northern Rhode Island "stone-enders" which most frequently appear as representatives of the type in surveys of seventeenth-century architecture. Disconcertingly surrounded as the Clemence-Irons House is by a suburban housing tract, on a busy highway, with its most distinguishing aspect—the exposed flank of its pilastered stone chimney—oriented away from the road, it takes a degree of imagination to recapture its aura. The front slope of the gable projects roughly half its length beyond the edge of the chimney block (unlike that of the Eleazer Arnold House, which originally filled the entire end elevation). The chimney may date from just after 1654, when Thomas Clemence, another close friend of Roger Williams, settled here, considerably earlier than the Angells (JO1). If so, it provided the nucleus for the rebuilding of the house immediately after it was destroyed, along with virtually all other isolated farmhouses, during the course of King Philip's War of 1675–1676. If not, then chimney and all date from 1691. This was originally a one-room house with a garret chamber above, under what was then a roof with identical slopes front and rear. The clapboarded section of the end wall forward of the chimney block contained the boxed stairs which connected the two communal spaces. The lean-to, or saltbox, slope to the rear provided a kitchen behind the original keeping room, or fire room. Although the broad exterior surfaces of clapboard and shakes are almost wholly restored, and the placement of the small battened door and the tiny, diamond-leaded windows is conjectural, they nevertheless make vivid the introverted, protective nature of this shelter against an environment so substantially experienced by the early settlers as harsh and threatening.

JO4 Killey House

c. 1860. 193 Greenville Ave.

Here the late colonial five-bay, two-and-one-half-story format becomes early Victorian. A heavily posted, shallow-arched porch shelters an arched entrance. Square bay windows, topped by scalloping, project on either side. Upstairs, bracketed table hoods project over upstairs windows, with a segmental-arch-on-lintel motif at the center reminiscent of a Palladian window, but squashed. Brackets under deep eaves and bristling corner quoining complete the assertive and idiosyncratic features which characterize Victorian sensibility. Scroll-cut curlicued bracketing connects the square posts of a side porch. The original stable exists behind the house.

JO5 S. Winsor House

Early 18th century. 29 Winsor Ave. (opposite Meadowbrook Dr.)

This began as a so-called end-chimney half house with a customarily simple transomed door. The brick chimney end is, however, clapboarded over (in contrast to its exposure in the Clemence-Irons House; see JO3). "Half house" means that its three-bay front roughly halves that of the standard two-and-one-half-story, five-bay, central-chimney type. Often such a house was intended as a partial building, to be completed when required by family size and permitted by financial conditions. More often, probably, it was meant to stand as is, and if an addition was made it might well be a one-and-one-half-story ell rather than completion as a full-height, five-bay block. (Two other early eighteenth-century half houses in Johnston, neither as handsomely restored as this, make the point: the East House, 325 Cherry Hill Road, has a one-and-one-half-story gabled addition to a modified half house; the Clemence House, 475 Greenville Avenue, has a one-and-one-half-story gambrel addition.) Inherently ad hoc as compositions, most such additions are awkward. This, however, with its complications of a gabled door shed at one end and the still later addition of a second floor at a new pitch to the rear, is quite satisfying, offering up a picturesque record of successive enlargement.

JO6 House

c. 1845; later wing. 77 Winsor Ave.

This typical one-and-one-half-story Greek Revival house is a fine country version of the style. In this instance the entablature treatment under the eaves and over the recessed, sidelighted door is particularly prominent. The plain pilasters which flank the door and corner the elevation may seem a trifle thin for the visual weight of the eaves entablature and the entrance lintel. What they lack in girth, however, they almost make up for in the gusto with which the capitals flare to take the assertive cornice at the roof and the projection of the lintel over the door. The provincial builder often grasps a style by exaggerating certain of its salient features.

JO7 Choice Acres

Mid-to-late 18th century. 115 Winsor Ave.

This plain five-bay house, later expanded to seven bays, acquired its handsome projecting Greek Revival–revival entrance vestibule sometime in the early twentieth century. The house has been re-sided, unfortunately, but at least this change is not obtrusive, except for the clip-on thinness at the corners. It is part of what is now a horse farm, which also includes a masonry barn, rare for Rhode Island.

JO8 Dame Farm

Late 18th century. 1910, barn. 1925, silos. 91 Brown Ave. (operated by the Rhode Island Department of Environmental Management; open to the public)

As farms go, no one could make a spectacular claim for this one. It has special interest, however, as a family farm continuously worked by the Brown family from the eighteenth century through the 1970s, when it passed to the state Department of Environmental Management, which continues its operation as a typical family farm for educational purposes on the edge of Snake Den State Park. It is worth noting that something once so common has acquired "museum" status.

The house is a perfectly plain, eighteenth-century type, with windows regularly placed in asymmetric arrangements; but it is commanding in size, and made more so by an in-line one-and-one-half-story ell off one end. A large gambrel dairy barn, shingled overall, with a pair of cylindrical wooden silos (now rare in the state), dominates the farm complex. But what is it that makes the complex seem so integral as a group? The house is set in the middle of a large field in an angled relationship to the roads around it, while all the outbuildings roughly respond to the orientation of the house. Hence the dynamic quality of these basic but varied shapes: overtly shifting in diagonal relationships but covertly ordered by their underlying alignment.

JO9 Duplex Workers' House, Thomas Hughes Chemical Works

c. 1849. 468–470 Central Pk.

This one-and-one-half-story duplex of stuccoed rubblestone is an important survivor from a no longer extant textile chemical works built in 1849 behind it on the precipitous slope down to Dry Creek. Few workers' houses built of stone exist in the state, very few as nicely maintained as this. Some have been shrouded, like the next-door companion of this pair. (At least one other nearby in Upper Simmonsville has also been buried under renovation.)

JO10 FM Global (Allendale Mutual Insurance Company Headquarters Building)
1970–1973, Maguire Associates, engineer-architects; Patrick Gushue, landscape architect. Entrance to grounds at 1301 Atwood Ave. (northeast corner of Atwood Ave. [Route 5] and Central Pk.) (open only by permission; visitors not normally encouraged except on business)

Among major corporate headquarters buildings set in large landscaped grounds, this may be the largest and is architecturally perhaps the most interesting in the state. How astonished Zachariah Allen would likely be to come across this evidence of the success of his scheme for industrial coinsurance as a way to reduce fire insurance premiums at his mill (NP4) in nearby Allendale! Allendale Mutual was the largest American insurer of industrial properties when it became part of the FM Global conglomerate in the late 1990s.

The principal elevation of the headquarters building, facing south toward Atwood Avenue, is visible on its 240-acre site through a screen of trees across a pond and lawn. Horizontal and extended, from a distance this elevation seems to be composed of huge, boxlike increments lined up lengthwise, ranged in two- and three-story stacks and open to deep shadow. These are punctuated on each of the long elevations (south and north) by four boxes on end (eight in all), which are completely sealed as windowless stair towers clad in dark brown glazed block. On closer view the open-ended "boxes" that accumulate to make the south elevation turn out to be projecting frames composed of a widespread pair of cantilevered beams, roughly 148 feet apart and extended 16 feet beyond the plane of the office window wall. Their ends, connected with narrower stiffening face beams, at a distance contribute to the effect of an elevation made of an accumulation of boxes. Behind the face beams, reinforced concrete louvers cross the frame, angled so as to shield the floor-to-ceiling amber-colored thermal sheet glass windows against the direct heat and glare of the sun, while also permitting the diffuse light of the sky to reach the offices through the slots in the louvers. (The north wall has no need for these built-in awnings, while the narrow east and west ends of the building are virtually closed against the rising and setting sun.)

The folded profile observed outside at the ends of the sun screens continues, but to different structural effect, inside. Here the cantilevered beams on the outside extend inside as "edge beams" folded down from floor and roof slabs. They provide beams integral with their slabs, while also stiffening the slabs along their edges. A structure folded in this manner makes possible the extreme longitudinal span of 148 feet supported only by rows of four columns (two in the interior, plus one each in the outer walls) across the building at either end. The columns, in turn, are themselves beefed up as outsized boxlike entities. They accept the spanning folded planes for floor and roof on either side of their hollow cores. The cores can then receive the vertical runs for the wires, pipes, and channels which service the building's machinery, as well as for service closets where needed for circuit breakers, valves, and control panels. From these hollow trunks, horizontal channels within the hollow-cored construction of combined floor and ceiling slabs bring mechanical services to the offices: wiring raceways for lights, telephones, and computers; ducts for heating and cooling; pipes for plumbing and sprinkler systems.

Hence the building as a whole is a crisscrossed, hierarchical system of hollow linear containers, each theoretically infinitely extendable, for which the varied raceways of the mechanical systems provide the conceptual starting point. Exploded from their status as minichannels for bringing utilities and services and stood on end, they become hollow supports and stair towers. Further exploded, and hung at right angles from their box supports, they become inhabitable boxes of space, which are celebrated along the major (south) front of the building as jutting sunscreens. These tubular boxes, each with its own integrity, come together to make the whole. So the machinery of the building becomes the building as machine; yet, because of its encasement, the machinery is as invisible as the innards of the computer. This ideal of the integration of the machine with the architecture (instead of forcing the machine into the building after the fact as a necessary but often disruptive adjunct) was very much at the center of architectural and systems thought in the 1970s. This headquarters building epitomizes the ideal, its handsomeness born of the integrity and elegance with which the design adheres to its premise. The machine purrs away unobtrusively at the center of its garden paradise. Machine and garden are the polar ingredients of the corporate castle. Combined, they protect against outside disorder and, like the moats and walls of their feudal predecessors, provide a grand, elitist, and dominating image for those who pass by.

JO11 **Frank Bernardis House** (Casarino)
1930s and later. 9–19 Morgan Ave.

This house and the two adjacent exemplify ethnic discontent with the perceived ephemerality of American wooden construction, particularly among the many Rhode Islanders of Portuguese and Italian descent, who often modify wooden houses in ways which accord with their Mediterranean stone building heritage. This preference for stone is even more commonly displayed in landscaping, where masonry terracing, gravel, painted stone borders, and statuary replace lawns, while arbors and spare, geometrically positioned plants, often clipped to balls and pyramids, rebuke the Anglo-American tradition of the clumping of foliage left more or less to run wild.

Frank Bernardis, himself a mason, modified his own wooden house and the two adjacent for his children. (Number 19 does not seem to have been quite completed, and, anyway, is the most conventional. The middle house was radically re-renovated, in part, back to wood. So number 9, Bernardis's own "little castle," provides the focus.) Bernardis favored stone veneers in varied colors, set up in what are sometimes referred to as "crazy" patterns, reminiscent of broken crockery, with the linear ramblings of the joints (here tinted in bright blue) as important as the stone. This typically Mediterranean love of flat pattern and intense color is very much at variance with the dour naturalism of traditional New England approaches to masonry. The crazy-patterned stone surface spreads upward and outward to conceal the shingled house raised close to the road against a precipitous slope. Eventually it spreads beyond the bounds of the house as a freestanding masonry screen. On one side, it becomes a garden wall punched with pointed and round-arched openings.

Bernardis countered the drooping amorphousness of the outer boundaries of his facade by a fierce assertion of vertical symmetry up its center. At the ground, he gave the main entrance more importance with a projecting masonry frame between existing windows. An axial element rises from its stone lintel, topped by a ceramic plaque emblazoned "Casarino," like a sign on a post. This projects into a field of plain stucco at the second-story level, an island in the surrounding sea of crazy masonry. Its mostly rectangular shape takes its dimensions from outside cues: horizontally, from the width of the original house; vertically, from the desire to "sit" the second-story window pair and the attic balcony above on its lower and upper boundaries respectively. But then the lower corners of this stuccoed field are also charmingly drawn down to give a pedimented aura to the principal entrance, while also echoing the cross gable over the attic balcony. Toward the periphery, other openings are fixed by more occult symmetries (if a little askew) within their own territories. So the flat patterning of the crazy masonry is echoed in the flat patterning of fields drawn across the original elevation, like the tailor's cut of a suit from a bolt of cloth on a table. In the process Bernardis's work reorders the prosaic ordering of the original elevation in accord with a grander vision for his Casarino.

Meanwhile, oblivious of this upgrading, doors and windows maintain their commonplaceness, and oblique views reveal the ordinariness of the original shingled house behind the stage-set frontispiece. One senses here the endlessly expanding dream of the weekend craftsman, whose ambition grows with every completed phase of his project. Bernardis's ambitions extended to masonry conversions inside as well as out. Thus the Old World critiques the New.

Thornton

JO12 **James F. Simmons House**
c. 1845. 928 Atwood Ave.

James F. Simmons, the principal early industrialist in this area, founded the dual mill villages of Upper and Lower Simmonsville in 1822 and 1831 respectively. (When a British immigrant entrepreneur later took over Simmons's holdings, he changed the name of Lower Simmonsville to Thornton in honor of his birthplace.) Simmons was also a leading Whig in the state and eventually, after holding state offices, was elected to the U.S. Senate. (He became the first ever to be removed by that body for conduct deemed unsuitable for his office.) His house, among the more elaborate extant Greek Revival houses in the state, is sited with

JO12 James F. Simmons House

its side elevation to the street, so that its two-story Ionic-porticoed porch once commanded the slope down to the stream that drained the twin reservoirs that powered his upper and lower factories. One-story entrance porches with simple Corinthian columns must have originally flanked the main portico. (Although only the porch toward the street now exists, the alteration to the rear of the house appears to have been made to a similar arrangement.) Toward the street the Corinthian porch shelters the principal entrance, while a similar porch at the rear provides for service, with a square bay window tucked between. The surroundings are so chaotic that it is difficult to imagine the original effect; but then, Simmons's factories initiated the process that one day overran his idyll.

JO13 St. Rocco's Church

1951, Oreste Di Saia. 927 Atwood Ave.

St. Rocco's Church is the most effective monument in the area to the first appearance of Italian immigrants in the Rhode Island work force. Reputedly Italians first appeared in substantial numbers during the 1870s at the Upper Simmonsville Mill, brought there by B. F. Almy, a successor to James Simmons. The church "modernizes" traditional early Italian Renaissance brick churches with campanile, but seems also to have been affected by forms characteristic of Art Deco—as though resuming the style after the interruption of World War II. Inside, the space is quite grand, its arcaded aisles and flat, paneled ceiling articulated by simple geometry in a warm color palette. Its major attraction, however, is its unified program of stained glass in the "modern" manner by Guido Nincheri, in which the principal episodes are depicted in brilliant color surrounded by areas of clear glass comparable to those in St. Matthew's Roman Catholic Church in Central Falls (CF11).

Cranston (CR)

Cranston, located on the west side of Narragansett Bay, has only a short coastline. Topographically the town has two principal sections, a flat outwash plain to the east and rolling uplands to the west; the two are divided in the middle of town by a granite scarp.

Cranston's development followed a slow, steady pace from the seventeenth through the nineteenth century. Occupying land purchased by English colonists in 1638 and 1662, Cranston was set off from Providence as a separate town in 1754. As was the case elsewhere in the state, its earliest buildings were destroyed during King Philip's War of 1675–1676, but the Thomas Fenner House (1677), built immediately after the war, remains as one of the state's earliest dwellings. Pawtuxet, on the Post Road, grew into a trading center, where the most notable early remaining house is the George L. Tucker House (c. 1790). Population has traditionally centered in the eastern part of the town. Farms spread across the rest of the area. Although these have virtually disappeared in suburban development, scattered farmhouses remain, like the Christopher and William Lippitt houses (c. 1735 and 1805).

Small textile mills began in scattered villages throughout Cranston in the early nineteenth century. Earlier, during the colonial period, a considerable iron founding industry was established at the center of the town along Furnace Brook. But most of this industry died away during the nineteenth century. Although the lower portion of the Pawtuxet River comprises a substantial portion of the southern boundary of Cranston, its slow-moving waters failed to attract the factories which made the Pawtuxet farther upstream (especially in West Warwick) Rhode Island's second industrial river, after the Blackstone. The one exception to the lack of sizable industry in Cranston during the nineteenth century was the Cranston Print Works on the Pocasset River in the northwest corner of the town. The Spragues, during the first three-quarters of the nineteenth century this town's wealthiest and most influential family, began textile production here in 1807. The large mill village around the Cranston Print Works, developed between 1807 and 1873, became the hub of the vast A. and W. Sprague Company manufacturing empire, which eventually stretched from Maine to North Carolina. No leading industrial family in Rhode Island experienced a decline which was both so dramatic and so

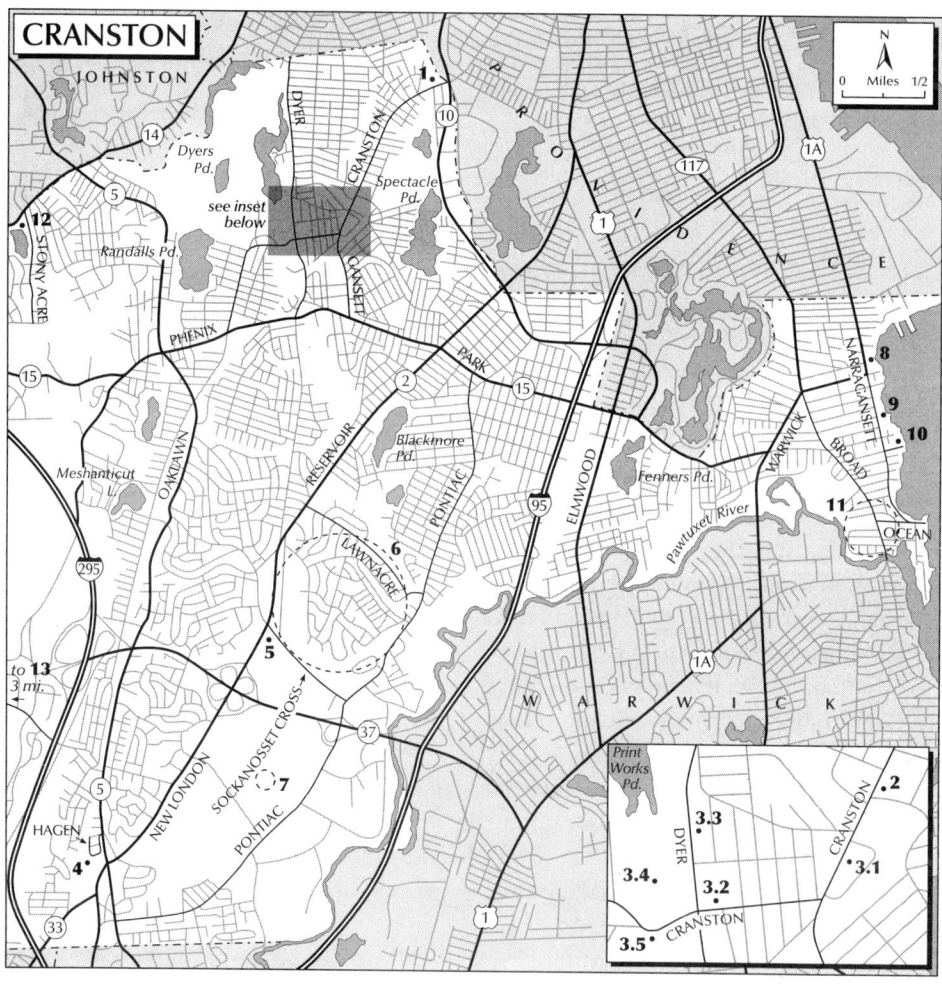

precipitous. Overextended by holdings in mills, banks, transportation, and speculative ventures of various sorts, the Sprague fortunes collapsed in 1873; so spectacularly, in fact, that the family's misfortune became a major precipitant of the Panic of 1873. Two other brothers picked up most of the textile pieces, and their firm, B. B. and R. Knight, became, in its turn, one of the textile giants in the state. Around the turn of the twentieth century, several new industries spilled over from Providence along Cranston Street, most conspicuously, the Narragansett Brewery (now defunct), which located near the clear waters of Tongue Pond. The completion of the Providence & Stonington Railroad in 1837 and of the Hartford, Providence & Fishkill Railroad in 1852 linked Cranston with regional centers to the south and west.

Toward the heart of town, on the Howard Reservation, in what was then open country, a state institutional complex developed beginning in 1870 with the construction of the first buildings of the State Asylum for the Insane (now the Institute for Mental Health). Howard grew to include the state prison (the Adult Correctional Institution) from 1873, the Oak Lawn Girls School from 1880, and the Sockanosset School for Boys from 1881. The development of this complex documents changing attitudes toward facilities for the treatment and incarceration of the mentally ill and the criminal, although we shall be concerned only with early buildings for the prison and the boys' school.

The growth after 1865 of streetcar routes made the town more accessible to Providence as a suburban residential community. Edgewood, located between Roger Williams Park and Narragansett Bay, throve because of the street railway from Providence to Pawtuxet, just south of Edgewood. This accessibility and its location close to a park, as well as to the bay and three yacht clubs, made it attractive for middle- and upper-middle-income suburbanites who built ample, stylish dwellings there in the late nineteenth and early twentieth centuries.

CR1 United Traction Company Trolley Barn

1900. 833 Cranston St. (opposite Depot St.)

Dilapidation does not obscure the impressiveness of this massive brick barn erected for one of the Rhode Island Company's electric trolley lines. Around 1900 industrial buildings associated with emerging electrical technology tended to receive special architectural attention. Such was also the case with the Narragansett Electric Company's slightly later Manchester Street generating plant (PR042). There, Neo-Renaissance detail notwithstanding, the treatment anticipated a more modern approach to the industrial structure. Here the image is more archaic— powerfully so, and especially in the spreading central gable with its tiered arching and its rugged cap in rough granite. This feature is slightly stepped from a blunt pinnacle down to corner blocks which the brick wall below receives in the subtle swelling of its upper corners.

The Rhode Island Company's ownership of both the trolley line and the electric company indicates the convenient monopoly it enjoyed. Marsden J. Perry headed both United Traction *and* the Union Trust Bank in Providence, accounting in part for the fortune that could buy the prestige of the John Brown House and fill it with a magnificent collection of colonial American and eighteenth-century British furniture.

When trolley service declined in the late 1930s, the barn became a warehouse for the Narragansett Brewery, located across the street until its demolition in 1999. One only hopes that the fate of this vast, abandoned barn could be more secure.

CR1 United Traction Company Trolley Barn

CR2 **Rhode Island Company Trolley Barn**

1912, H. H. Bronsdon (engineer), William H. Hamlyn and Son (builder). 1160 Cranston St.

Whereas United Traction's primary area was Providence, the Rhode Island Company focused on statewide service. This 400-by-200-foot building served as its barn and repair shop. Cars entered the building along twelve tracks through large openings ranging the length of the building's south side (the original doors have been lost to modern replacements). More utilitarian than the United Traction Company barn, it is nevertheless impressive for its size, its handsome brickwork (especially the layering of its corbel work), and its large tower, a landmark in the area. It now houses buses.

CR3 **Cranston Print Works Village**

Bounded by Cranston St., Dyer Ave., Queen St., and the Pocasset River; offices in the mill, 1381 Cranston St.

Despite the later development which now engulfs it, the Greek Revival core of this important textile village, situated around the intersection of Cranston Street and Dyer Avenue, is still evident. The close juxtaposition of mill, mill owner's house, workers' housing, and church invokes the tightly knit, paternalistic organization of early industry, and suggests the orderliness of the best-planned mill villages. Although the complex appears somewhat quaint today, it was once the heart of one of the world's largest nineteenth-century textile empires. The Sprague family, creators of both company and village, were pioneers in cotton manufacturing and printing from 1807, until their company's financial collapse in 1873.

The village was known locally as the White City: factories, houses, and public buildings were uniformly white. Although so much whiteness was hardly exceptional during the Greek Revival, it became a mark of distinction in Victorian America. Yet one does not know how many New England factories and villages might have persisted in painting their buildings white throughout the Victorian period—until the architecture of the World's Columbian Exposition of 1893 made it clear that whiteness and the brightness of floral pastels had to be reckoned with by even the most hidebound adherents of the shadow and glitter of the "brown decades."

From the east on Cranston Street, first is an end-gable Greek Revival supervisor's house, at 1230 Cranston (CR3.1; 1840), one of what was a row of five, now aluminum sided and merely a token of what was there. Next is the Governor Sprague Mansion, formerly the Sprague House (CR3.2; c. 1790; additions early nineteenth century, early twentieth century, stable probably 1864), 1351 Cranston Street (near Dyer Avenue). The Sprague family seat began in 1790 as a five-bay Federal dwelling, commissioned by the company founder, William Sprague. The next generation extended it three bays to the east; then, in 1864, the third generation much enlarged the house to the west, in the Italianate manner, giving it an addition higher than the original, with a glazed octagonal cupola and a second, larger porch—paired clusters of three columns rather than two, and a Victorian aggrandizement of the initial side- and fanlighted door. The 1864 addition is more public than private in character—more innlike than houselike—and suggests that the house had by then assumed a ceremonial and symbolic role for the family, several of whom by then owned substantial houses in Providence, Warwick, and Narragansett. Set behind a picket fence amid ample gardens but close to the road, the Sprague mansion provided an appropriate recall of first- and second-generation entrepreneurship, while expanding to accommodate important visitors to the plant and others on business and civic missions (William Sprague served as state governor from 1860 to 1863).

All the other principal Sprague mansions, some of the grandest of the mid-nineteenth century, have disappeared, leaving only the ancestral, somewhat rustic manse to celebrate the family's restrained prosperity toward the start

CR3.3 Cranston Print Works Village, mill housing

of undertakings, as opposed to the spectacular display in dispersed settings which surrounded the Spragues' desperation and humiliation at the end. It, too, might have been demolished had it not become the headquarters of the Cranston Historical Society. It is now open to the public.

Ranged along Dyer Avenue, which bounds one end of the factory property, stands a long row of some of the best-preserved Greek Revival mill housing in the state, all duplexes with small trapdoor dormers and end chimneys (CR3.3; 1844, 1864). At the south end of Dyer Street (crossing Cranston Street) approximately one hundred more workers' houses, identical to their predecessors, went up in 1864 along an enclave of four streets named for trees. Much more altered than the Dyer Street row, the later workers' houses nevertheless preserve the sense of what was there.

Beyond Dyer Street, at 1381 Cranston, is the mill (CR3.4; 1844, 1852, 1864, 1921). The earliest buildings were replaced by the stuccoed rubblestone mill complex, erected along the Pocasset River, in a plain style with segmental-arched windows, all painted over in white except for red granite corner quoining. Its most striking component is the U-plan printing and engraving building, with a prominent bell tower, at its eastern outside corner. The tower is simple, but gracefully shaped by stepped, chamfered corners. These steadily widen, whittling the tower down to its culminating stage with bull's-eye windows, at least one of which may once have contained the company clock. The roof picks up the chamfering, absorbing it into a low pyramidal cap, its tentlike concave planes extended to flaring eaves. Following the collapse of the Sprague fortunes, the plant was idled for a decade and a half—the only such time in its history—before B. B. and R. Knight reopened it. Under different ownership, the plant continues to print and dye textiles.

Opposite, moved from a site nearby to 1390 Cranston, is the simple Greek Revival Sprague Meeting House (formerly St. Bartholomew's Episcopal Church) (CR3.5; 1825, burned in 1924, rebuilt in 1927, when the present cupola cap was added). Erected for workers on company-donated land, it was later used by the historical society and is now devoted to commercial use. (Incongruously adjacent, for aficionados of roadside pop modernism, is a vintage Mobil "Flying Horse" gas station in enameled metal [c. 1955]).

CR4 Temple Sinai

1961, Isidor Richmond and Carry Goldberg. 30 Hagen Ave.

Architecturally this is interesting as a period piece of 1960s modernism in poured concrete and white brick. Among the most conspicuous hallmarks of its time is its boomerang-shaped massing, which partakes of a 1950s and 1960s quest for a more "organic" approach designed to mitigate the flat planes and sharp angles preferred in earlier modernism. Its overall shape doubtless owes something to the similarly shaped auditorium at the base of the United Nations Building (1947–1950) in New York. The boomerang shape "floats"—in the then current terminology—off a recessed basement. The sloping site is excavated so as to separate the building mass from its approach drive by a dry moat, necessitating a vestibule-bridge and enhancing the floating quality of the building.

CR4 Temple Sinai

So do the three curved planes projecting bannerlike off the wall on pipe supports to shade the tall window triplet which lights the hall of worship. Another warped plane floats off the climactic end wall of the temple (appearing concave without, convex within). A surround of glass between it and the wall from which it projects provides a curved backdrop of radiance behind the sacred ark inside. At the opposite end of the boomerang, devoted to offices and adjunct activities, a long window slot disengages the ceiling plane from the wall. Below it, radically off-centered and slightly canted, raised aluminum letters in Futura type announce the synagogue's identity.

The interior is, with a significant exception, a well-preserved space of its period and style. It is a box, tall but very broad, expanding toward the front with a culminating platform stretched full width behind it, the sanctuary, with its Holy Ark, jutting as another, niched box, which its revelation outside leads us to expect. Changes of surface for walls, floor, and ceiling give the illusion that the box originates in large, primal planes lifted like stage flats in making a set. In front, flush-fitted, polished wood panels in a basketweave opposition of grain combine to make the climactic wall a super-plane, pierced by tiny rectangular windows in a punchcard scatter of pastel-colored light. Side walls are ruddy brick, randomly "hobnailed" with projecting units. Sliced by the slots of paired floor-to-ceiling windows, these cuts intensify the effect of planes making the box. Pews in polished wood and tan carpeting complete the autumnal hues of the setting, much favored in the 1960s as a "naturalistic" foil to the alleged starkness of earlier modernism. At the east end of the worship space the luminous, white-plastered sanctuary shatters this earthly palette, its light, from a mix of concealed natural and artificial sources, seeming otherworldly. The window slots ranged along the side walls were not always filled with pale yellow opaque glass. Once they enlivened the now somewhat bland interior with the changing color and allusive meanings of symbols and images in stained glass. Shortly after the temple's completion, however, vandals so thoroughly smashed the windows that only a single compartment from one of them could be salvaged. Now it adorns the entrance vestibule as a disquieting memento.

CR5 Sockanosset School for Boys Administration Building, Cottages, and Chapel

1881–1895, Stone, Carpenter and Willson (Stone probable designer for cottages, Willson for chapel); Allen and Browne, masons; French and MacKenzie, carpenters. Sockanosset Cross Rd. (at New London Ave.; intersection of Routes 37 and 2)

In this area in which the state has concentrated its penal and mental institutions, the setting is now substantially spoiled by major traffic arteries bordering it, and the buildings are abandoned and so much deteriorated (as this is written) that their future is doubtful. In the nineteenth century the site was a removed, quite pretty plateau commanding a long slope toward Narragansett Bay. In accord with nineteenth-century views of the power of nature to heal both illness and what was then usually termed "wayward" conduct, the Providence Reform School changed both its name and something of its barrackslike image on moving here. Ranged around an elliptical drive is a "village" of three double-cross-gabled and dormered buildings (remaining from five) in

CR5 Sockanosset School for Boys, cottages

stucco-covered rubble with corners and window edges quoined in red granite, which gives a bristling effect. The administration building stands at one end of the ellipse, with a combined chapel-hospital downhill and a bit removed from the rest. Although still formidable, the "cottages" vainly attempt to disguise the institutional scale of the operation by their medium size, with the barest domestic gestures of gabled and dormered roofs and gabled entrance porches embellished with a lattice of cross bracing. Even in village dress, however, Victorian agencies for reform and charity usually manage an assertive image of authority.

The most exceptional building in the complex is the combined chapel and hospital (1891), the chapel in front, hospital wing to the rear. It is a low-slung, cross-gabled granite structure, nominally Neo-Romanesque in style and cruciform in plan, with a squat crossing tower capped by a polygonal cone with flaring eaves. Like St. Colomba's in Middletown (MI5), it evokes the English parish church as strained through 1880s sensibilities. Edmund Willson here seems to have attempted to blend a somewhat severe aspect appropriate for a Victorian institution of incarceration with the redemptive promise implicit in allusion to solace and charm. How to justify this mishmash of contradictory reaction?

The chapel front is no more than a granite gable, brought low to the ground, punched only by a pair of small, arched windows with a tiny wheel between them and the porch. All three openings pull tensely apart from one another. A cozy porch, firmly framed by timbers at the corners and sparely braced, projects from the wall, prettily roofed as a shingled, gabled shelter curved out to flaring eaves. More shingling pulled down the front of the gable in-

creases the sense of shelter. It is pierced by an arch with spoked timbering. The masonry is powerfully but awkwardly handled: in part squared and neat, in part random and rough. Inmates with varying degrees of building skill may have been responsible for the execution of the building from the architect's drawings. Although also deteriorated, the plain wooden interior retains Arts and Crafts touches, including exposed roof bracing, some faded painted stenciling, and elaborate wrought-iron chandeliers (all perhaps products of the institution's shop). One senses an effort to balance severe Romanesque with healing Arts and Crafts, as though the combined styles were meant to convey a subliminal admonition to the inmates: "Choose your path." Unique, with decided character and even touching aspects were it shown some care, of all the buildings it at least merits conservation, which may be hard to come by.

CR6 Garden City

1946 (housing), Nazzareno Meloccaro. c. 1970, N. Robert Meloccaro; land planning, Federal Housing Authority (Peter Cipolla directing). 1947 (Garden City Shopping Center); after 1956, expanded; c. 1988, redesigned, Flatley Corporation. Bounded roughly by Route 2, Reservoir Ave., Route 37, and Sockanosset Cross Rd.; to the east and north, by the Pocasset and Pawtucket rivers

Across Sockanosset Cross Road from the Sockanosset School for Boys is Garden City, incorporating Rhode Island's first shopping mall. Small by later mall standards (originally 18 acres, now 40), around 1988 it received a new look when it was tweaked into a fashionable postmodernist village image with gables, dormers, and a mix of bull's-eye and Palladian windows. There are unintended ironies in this juxtaposition of "villages" designed, on one side of the highway, to reform Victorian waywardness and, on the other, to captivate the post–World War II consumer's heart. Behind the mall to the east as far as the Pocassett River are the serpentine streets of its residential component, the Levittown of Rhode Island, intended to provide the first line of customers for the shops.

Nazzareno Meloccaro emigrated from Pontecorvo, Italy, in 1920, taught himself to build houses and, three years later, set up his own successful contracting firm. In 1941 he bought a 233-acre farm and, with help from the Federal Housing Authority's land planning division, laid out a "city within a city." In addition to the shopping center, the town originally contained a Roman Catholic church, parochial school, post office, some ninety-four apartments in seven brick buildings, and, when completed, more than 700 houses. Nazzareno built these in blocks of fifty in an assembly-line operation, mixing Cape Cod, colonial, and ranch exteriors, which were popular for development houses at the time. During the post–World War II boom they sold for $8,000–$30,000 as fast as he could pour foundations. Following his death in 1955, his son, N. Robert Meloccaro, completed the community, before eventually selling the shopping center. Although many houses have been altered and enlarged, the nature of the original community is still evident. As with Levitt's towns, the current generation of owners (1990 population, 3,189) is remarkably stable and satisfied. At the beginning of the 1990s, houses went for $98,000–$178,000, depending on the accretion of owner add-ons.

CR7 Adult Correctional Institution

1873, State Workhouse, Christopher Dexter. 1878, Providence County Jail and State Prison, Stone and Carpenter. Pontiac Ave. (near intersection with Sockanosset Cross Rd.)

In contrast to the openness of the boys' correctional facility, the slightly earlier facilities for adult prisoners reflect the nineteenth century's approach to the prison as a fortress. To the south, Christopher Dexter's T-plan gabled barracks building recalls the format of such institutional buildings from earlier in the century, including its predecessor, the demolished 1845 state prison in Providence. To the north, Stone and Carpenter's massive granite building also originates in a T-shaped core of cell blocks, but centered in an octagon as the heart of a sophisticated massing which means to suggest the basic functions of a complex institution. The octagonal core was designed to house a reception room at ground level and centralized guard stations at the levels above, all topped by a chapel, its interior rising within the conical roof into the domed cupola. It is laterally flanked by two cell blocks on either side, stepped wings composed of the outermost block stepped back from its companion. Another cell block projects from the hub building rearward. Together, they give the prisoners' portion of the complex an overall T-configuration with extended arms and a stubby stem. Walls cornered by guard towers slide a

CR7 Adult Correctional Institution

little beyond the outermost cell blocks, to make a rectangular enclosure for exercise courts divided by the rear cell block. Finally, the attached building for the warden's house and prison office fronts everything. A near cube, it is gabled across its width and doubly cross-gabled toward its porte-cochere-covered entrance. The tall, multistoried arching of the cell blocks gives way to a livelier pattern of domestically scaled windows. Each gable in the warden's house has the fillip of a chimney, and the motif is echoed at the centered gables of each of the prison blocks. These accents, in concert with the guard towers and the culminating octagonal cupola at the core of the composition, animate the skyline against the sullen spread of cells and enclosing walls. Axial, hierarchical, visibly ordered to its function, sparely detailed, expressively shaped, and very, very solid: such buildings proclaim the nineteenth-century faith in institutions and belief in the virtue of strong institutions in society.

Edgewood

Edgewood extends south from the Cranston-Providence line to somewhere around Massasoit Avenue, from the bay roughly west to Broad Street, and to Roger Williams Park above Westwood and Edgewood streets. The area has always maintained a degree of residential stability because of the advantages of the bay, the park, three nearby yacht clubs, and "The Boulevard" (as natives refer to it), which serves as its spine. Narragansett Boulevard was at one time more distinguishable as such because of its rows of magnificent trees, many of which have been lost to successive hurricanes and only spottily replaced. It and the more impressive stretch of Veterans Memorial Parkway in East Providence (EP18) were the only substantially built pieces of the projected boulevard system in and around Providence as recommended by the *Report of the Metropolitan Park Commission* (1906)—unless one also includes a short stretch of Mount Pleasant Avenue and (most important) Blackstone Boulevard (PR172), even though the latter had been previously built by private initiative.

Edgewood developed first as a summer resort, then, especially after a trolley line came to the boulevard, as a year-round residential community. It is among the most intact upper-middle-class Providence suburbs dating from the trolley-car era, restoration having reversed some recent dilapidation. Its houses, predominantly Queen Anne and Colonial Revival, may be appreciated as much for a sense of the whole as for outstanding examples.

CR8 Rosedale Apartments

1939. 1180 Narragansett Blvd.

Among the few Art Deco–inspired buildings in the state, Rosedale Apartments is also among the last in the mode. In what was then a suburban area with relatively few apartment houses, and these mostly small, the Rosedale is remarkable for the extent of its U-shaped mass. Its Art Deco flourishes are sparsely if strategically distributed against the spartan expanse of its speckled chalk-white brick walls with unadorned windows. Presumably, its developer meant to recall the sophistication of movie-Miami modernity as economically as possible. The principal entrance is surmounted by a stainless steel canopy and a glass brick oriel. Minimally patterned brick on two window bays in front, with additional stainless steel canopies over some exits onto the rear lawn. Commandingly sited overlooking the upper reach of Narragansett Bay, it replaced a large house similar to others that once lined this artery. It typifies the deluxe quasi-hotel/apartment house of its period, which tended (originally at least) to be an isolated exception to its elitist suburban surroundings. Designed to attract older people, bachelors, cosmopolites, and any others who preferred not to mow the lawn, the Rosedale contributes to the period quality of the suburban image of Edgewood and to its viability as the kind of community it is. As a type, the building anticipates the present plethora of condominiums and "lifetime care" alternatives to the suburban house.

CR9 Edgewood Yacht Club

1908, Murphy, Hindle and Wright. 3 Shaw Ave. (open by private invitation only)

As the oldest extant yacht clubhouse in this boat-oriented state, the Edgewood Yacht Club is also the only remaining example of a type common around 1900, but decimated by fires, hurricanes, and updatings—all of which it fortuitously survived (although its nearby rivals did not). It epitomizes the classic type of its vintage. Lifted on pilings from the water, it is clad in shingles, with two levels of circumferential porches (random portions of which seem to have been enclosed from the beginning). A gable-on-hip roof crowns the club. From the hipped portion of the roof, gabled dormers project port and starboard, flagpoled mini-decks fore and aft, all topped at the center by yet another parapeted mini-platform on which the culminating cupola sits, capped by a high, broad-brimmed dome which slicing at the four corners modulates to a semipolygon. Alternately wide and narrow intervals in the porch supports take their measure from events on the roof. The imagery is of wharf shacks, launch decks, and pilothouses and, for good measure (with the cupola in mind), lighthouses, all bluntly and crudely assembled, but with connotations of elegance. An ancient anchor at the entrance anticipates the clutter of sailing and yachting memorabilia spread over the matchboarded partitioning of the thoroughly functional interior.

CR10 Arthur J. Levy House

1935. 30 Fairview Ave.

The *Providence Sunday Journal* for December 8, 1935, announced: "The inevitable has happened. Modernist architecture has at last invaded the Rhode Island home building field.... One of those fantastic dwellings championed by such radicals as Wright, Lescaze and Le Corbusier has sprung up in a neighborhood of highly conservative homes." The lawyer client had, while on a Florida vacation, seen something like what he got. And what he got was a long, one-story block elegantly crafted in unpainted cinder block, abruptly punctuated at the center by another, slightly projecting and two stories. This contains the door, framed in a recessed stepped molding, with a semicircular plane projecting above it. A one-car garage (with later door), slightly set back, is attached. The metal-framed windows (so important to early modern design, but then so vulnerable to rust and the desire for improved insulation) seem to be original. Inside (not open to the public) are original wooden kitchen cupboards, structural glass bathroom accoutrements, decorative metal stair railing, and more relics of the Moderne. The stair leads to a small rooftop sunroom. A lot of stair for a mini-destination, perhaps, but the sunroom also serves as the vestibule for the modernist pièce de résistance: the rooftop terrace, from which parts of the bay are visible. Tenative "modern" here urbanely works itself out of a sort of Regency formality.

CR11 Pawtuxet Village

18th century and later. Bounded by the Pawtuxet River, Pawtuxet Cove, and Ocean Ave.

Settled in 1638, Pawtuxet Village, at the mouth of the Pawtuxet River, is the oldest node of development in Cranston. Only a handful of houses, now mostly altered, survive from the eighteenth through the mid-nineteenth centuries, including the Remington Arnold House (c. 1730; 12 Bridge Street); the George L. Tucker House (c. 1790; 27 Tucker Avenue); and the William C. Rhodes House (1857; 143 Sheldon Street). They are intermixed with later houses built as vacation cottages and now mostly occupied year round. A point of land which shelters the river mouth and creates a small harbor, while not really part of the colonial village, is also tightly covered with small to medium-sized houses. It contributes to the village and seems today to extend it. The only section of the immediate greater Providence metropolitan area that retains the feel of a small New England waterfront village, it is distinguished more by its density, topography, and ambience than by its architecture. The village continues on the Warwick side of the Pawtuxet, where more early houses are extant.

CR12 Thomas Fenner House

1677, c. 1835 and later. 43 Stony Acre Dr. (not visible)

One of the few surviving examples of postmedieval "stone-ender" construction peculiar to Rhode Island, the earliest, northern section of the Fenner House, dominated by a large stone chimney on the north elevation, originally contained two rooms, one on each story. As passed down through the Fenner and collateral families, the house was extended south of the present entrance and received a large shed dormer to the rear (west) side for its entire length. These later changes are, however, relatively minor compared to alterations of other surviving seventeenth-century Rhode Island houses and have never obscured the obvious antiquity of the house, recognized as a landmark since the 1880s. (Other stone-enders, including the familiar Clemence-Irons House [JO3] and the Eleazer Arnold House [LI14], have undergone much more extensive restoration in the twentieth century to return them to something like their original appearance.) The remarkable aspect of this house is its truly grand keeping room, perhaps the most dramatic seventeenth-century interior in the state. Both larger and higher than the norm, it is climaxed at one end by one of the largest of seventeenth-century fireplaces, 10 feet wide, 5 feet 8 inches high, and 3 feet 9 inches deep.

CR13 Lippitt Hill Farm
c. 1735; 1805 and later. 1231 and 1281 Hope Rd.

Uphill from the Lippitt Mills (WW7), the Lippitt Hill Farm complex includes two farmhouses, numerous outbuildings, extensive stonewall-lined fields, and an unusual family cemetery. Both the Christopher Lippitt House (c. 1735) and the William Lippitt House (1805) are south-facing center-chimney dwellings typical of their respective periods of construction, the earlier with a pedimented Ionic door frame, the latter with a semicircular fanlight beneath a pedimented frame. The numerous barns, stables, well houses, and sheds bear testimony to the long agricultural history of the complex, while twentieth-century additions illustrate the transformation of the gentleman's farm into a country retreat. The family cemetery is now enclosed by a unique heart-shaped ring of pines planted by Julia Lippitt Mauran in the early twentieth century. Densely surrounded by a tangled thicket and now unintelligible from the ground, it can be discerned only from the air. But it was partly meant, after all, for angels' eyes.

Northeast

LINCOLN, CUMBERLAND, WOONSOCKET, NORTH SMITHFIELD, SMITHFIELD

Lincoln (LI)

LINCOLN AND CUMBERLAND FIT TOGETHER IN THE NORTHEAST corner of the state roughly as two right-angled triangles, except that they are askew with a few nontriangular jogs and curves and with the Blackstone River making an undulant hypotenuse between them. Lincoln is broad at its southern base and tapers toward the north; Cumberland (a part of which was originally joined with Attleboro and Rehoboth as part of Massachusetts) essentially inverts this shape, being wide at the north and tapering southward. The long evolution of what is now Lincoln from a remote piece of the wilderness fringe of the extensive "Providence" which Roger Williams negotiated from the Wampanoag chief Massasoit in 1636 was part of the evolution of the political identity of Central Falls (see Central Falls introduction). It belonged to the area labeled on the earliest maps of the colony as the "North Woods," or, more specifically for the locale in which most of Lincoln is located, as "Louisquisset" (a corruption of the phonetic rendering of the Native American name "Loquasquisuck").

Complaints of neglect and of the travel required to attend town meetings and courts received temporary redress when, in 1731, the sizable new town of Smithfield was split off from Providence—with a territory which included the present Rhode Island towns of Smithfield, North Smithfield, Woonsocket, Lincoln, and Central Falls. Then, as the same complaints gradually mounted, Smithfield was partitioned in 1871, and Lincoln (including Central Falls) was separated from it. Finally, in a referendum of 1900, Central Falls gained its independence. This status owed less to the zeal for separation in Central Falls, where the vote was almost even, than to lopsided support for severance from the rest of Lincoln, whose Yankee population was either rural or benefited from the largesse of paternalistic owners in the mill villages and regarded demands for public services in immigrant-dominated Central Falls as an unwarranted drain on town taxes. At last, then, in 1900, Lincoln became officially what it now is.

The earliest substantial colonial settlement of Lincoln around 1680 is, above all, associated with Thomas Arnold, a Quaker, who came to Providence with Roger Williams. His house early disappeared; but his son Eleazer's looming stone-ender house still stands close to the southern boundary of Lincoln near the Mohassuck River as the most familiar image of seventeenth-century Rhode Island. The Arnold family, with Eleazer as its most prominent early member, so dominated the area in the beginning that it came to be called Arnoldia. For his sizable holdings, Eleazer Arnold gave the site for the Quaker meeting house. He was a leading sponsor of Great Road, which passes his house and still approximates in its windings the path of least resistance through the wilderness, with occasional deflections for special interests like Arnold's. Ultimately it provided the principal connection between Providence and Worcester, Massachusetts, during the colonial period. Because the Rhode Island principle of freedom of religion encouraged a reaction against the Puritan form of communitarian settlement, early houses were not grouped into villages as in Massachusetts, but were strung along Great Road, each within its individual clearing, as a number of survivors testify to this day. Other early houses owned by prosperous farmers, even though a little removed from Great Road, also tended to be located within easy access to it.

Close by Great Road, too, is the exceptional community of Lime Rock. Its importance in the colonial economy is evident from a report of 1665 by the governor of New York to the King's Commissioners of Rhode Island: "Here only [in the colonies] yet is Limestone found." Lime for mortar, plaster, and cement was shipped from here the length of the eastern seaboard. It played a major role in the American building industry through the Civil War, when hydraulic cement began to usurp its dominance. Lime was also a key chemical for tanning and bleaching. The survival of the lime industry here—in continuous operation from colonial times to the present—is extraordinary, though it now primarily produces lime for agriculture, and at last there are signs that the quarries are beginning to give out. Equally extraordinary, however, is the preservation of old farmhouses and their walled fields in the immediate proximity of the works, and of a lovely village as well. This symbiosis of industry and countryside through more than three centuries may be Lime Rock's most significant lesson. Not until around 1990 did the invasion of development housing begin to impinge on this community in a significant way.

Yet it is also true that countryside and industry comfortably coexisted in most of the rural Rhode Island towns into the twentieth century. Where mill villages were relatively small and either scattered or clustered, this balance is not so surprising. But by the second half of the nineteenth century, Lincoln and Cumberland contained some of the largest textile operations in the country along the hypotenuse of the Blackstone, which they shared. The coming of the large mills to Lincoln and Cumberland, in fact, at first rather strengthened than spoiled the predominant rural character of these towns. Now the local farmers depended less on transporting their surpluses south to the industrial and commercial cities of Central Falls, Pawtucket, and Providence or (later) north to

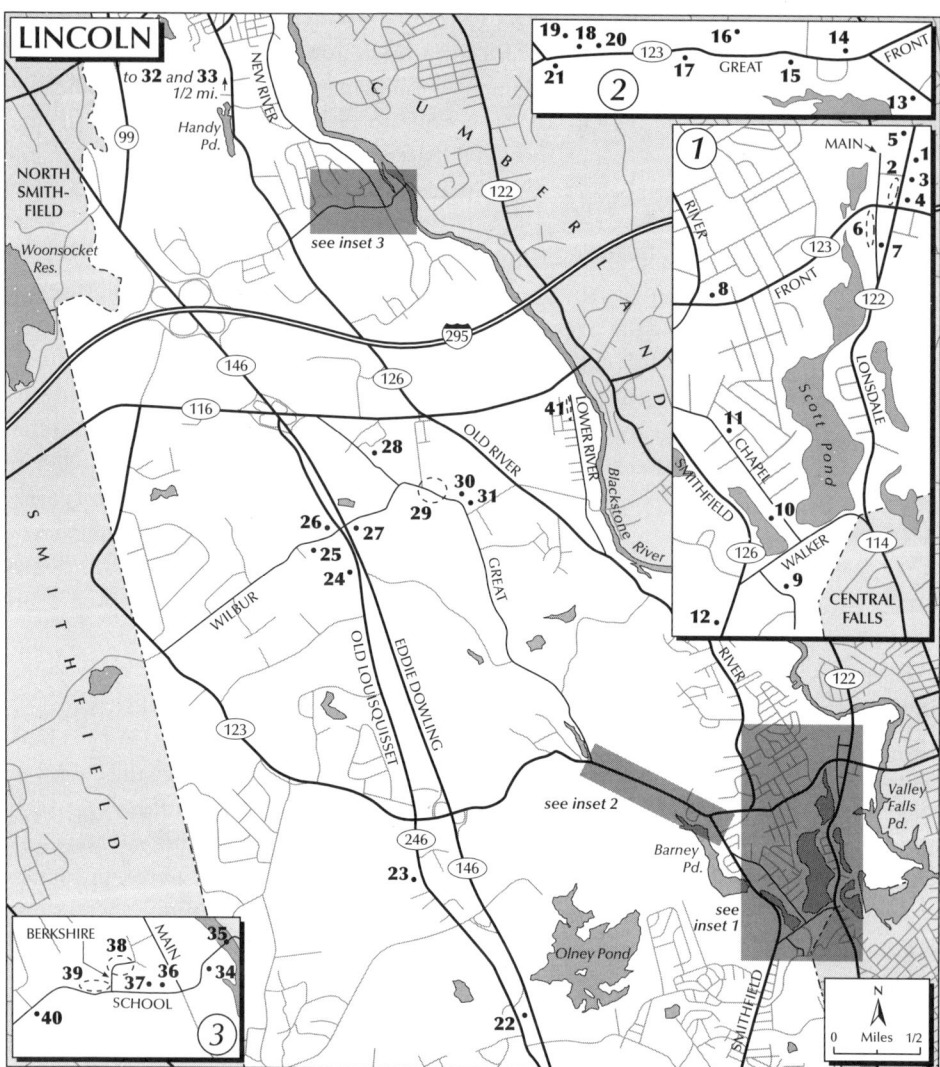

Woonsocket. Alternate markets existed close at hand, in the mill villages of Manville, Albion, Ashton, Berkeley, Lonsdale, and Saylesville, from north to south along the Blackstone.

By the mid-nineteenth century, three groups of entrepreneurs—two families and one consortium—came to dominate these mill villages in the rural stretch of the Blackstone between Woonsocket and Central Falls. Headquarters for two of them and the flagship plant of the third existed in a tight triangle of villages near the southeast corner of Lincoln—Valley Falls, Lonsdale, and Saylesville. This cluster included the Valley Falls headquarters factory of the brothers Samuel and Harvey Chace at the northern border of Central Falls (see CF10), which spilled across the river into Cumberland. Two years before their incorporation as the Valley Falls Company in 1856 they had acquired the mills at Albion and at Manville, also major operations for their day. Before the Chace incorporation, in the early 1830s, a consortium of Providence families, united by business and marriage ties, had founded at nearby Lonsdale the flagship plant for a corporation which also used the village name as the corporate label. Moses Brown had led his family from merchant trading into textile manufacturing in the 1790s in Pawtucket; Nicholas Brown, in the next generation took the lead in extending the family's commitment in this direction. First he participated in a successful mill on the Blackstone just over the Massachusetts border. Then he combined with his brother-in-law, Thomas P. Ives, and later Thomas's daughter Charlotte R. Goddard, to establish the Lonsdale Company. This corporation subsequently built mills at Berkeley and Ashton. In the early 1840s, two brothers, William and Frederic Sayles, emerged from small mill operations in northwestern Rhode Island to establish their headquarters south of Lonsdale. They named their village Saylesville. By the early twentieth century the Sayles Corporation was a textile colossus, with mills in Rhode Island and in the South, together with several locations in between.

So the Providence group that included—in order of their incorporation—the Brown family, the Sayles brothers, and the Chace brothers together came to control all the large mills on the Blackstone between Woonsocket and Central Falls. The corporate headquarters for the Sayles and Chace organizations and the flagship operation for the Brown consortium (which had its corporate headquarters in Providence) were all located within a short carriage drive of one another.

One other aspect of these Blackstone mills is of special interest to the architectural pilgrim. Brick construction for both industrial buildings and workers' housing was far more uniform in these villages than elsewhere in Rhode Island. Especially was this the case with the Lonsdale organization, where brick became the company standard—most of all in the towns it built on the Cumberland side of the river (see under Cumberland). One immediately expects that this preference for brick was due to influence from Massachusetts, where brick housing was much more common than in Rhode Island—and this may be the case, the more so as Nicholas Brown had initiated his industrial enterprise on the Blackstone across the border in Massachusetts. But the labor force in the Lonsdale Company remained exceptionally English and Scottish through the 1870s. There is evidence to suggest that housing typical of the British Midlands industrial area provided a familiar image the mill owners cultivated. Although initial costs for brick were higher than for clapboard construction, greater permanence and low maintenance may also have been factored into the choice. In any event, brick housing resists the blandishments of the re-sider, as it also makes such alterations as picture windows and prefabricated bow windows more complicated. Hence, much of the mill housing in Lincoln, and even more in Cumberland, is among the best preserved in the state.

As the glory days of Rhode Island's textile industry faded in the 1920s and 1930s, some of the most violent encounters between dispossessed workers and scrimping owners occurred in the Blackstone River mill towns, but especially in Saylesville, where the National Guard was called out in 1934 to quell the rioting. The idled mills preserved for several decades the shell of the longtime symbiosis between industry and a declining agriculture in both Lincoln and Cumberland. But as mills shut down, so the farms withered in the face of the centralization of agriculture in larger units elsewhere in the country. Yet adversity prolonged the image of what had been in Lincoln and Cumberland until well after World War II. Since then, suburbanization has transformed them.

Lonsdale

The three affluent closely related Providence families who controlled the Lonsdale Company—the Browns, the Iveses, and the Goddards—brought a relatively new mode of ownership to the state's industry. Operating initially as the investment firm of Brown and Ives, this was the first major venture in Rhode Island to abandon joint-stock partnerships in favor of the corporate form of management which Boston capitalists had introduced at Waltham around 1812 and at Lowell in 1822. No longer did the owners live close to the mill and have their offices at the gate of one of their key factories. They lived in their Providence mansions and there maintained their counting house in what had once been Joseph Brown's house (PR51). Plant superintendents handled operations at the factory. The corporation moved quickly to establish additional mills at Berkeley and Ashton farther up the Blackstone and on the Cumberland side of the river (see under Cumberland). They also bought other mills south of Providence at Hope and Phenix on the Pawtuxet (see under Scituate and West Warwick), which they managed as the Hope Corporation. Brown and Ives also innovated by operating their mills as a unit, with centralized purchasing and marketing, as well as specialization among all the plants. Thus the Lonsdale plant originally specialized in fine cotton cloth for such items as umbrellas and eventually handled bleaching and dyeing for all Blackstone operations.

The village of Lonsdale was also conceived in a larger, more integrated way than previous mill villages, although the measure of this integration is better seen today at Berkeley and Ashton, and across the river in the Cumberland half of Lonsdale. Mills No. 1 and 2 at Lonsdale (1831 and 1832) were both demolished. Mill No. 3 (1833), on Cook Street, still exists, although radically altered, not only in details, but reduced from four stories to two. The buildings for bleaching and dyeing are dilapidated. Important later mills in Lonsdale, which do remain, are concentrated across the river on the Cumberland side of this bifurcated community.

So it is essentially aspects of the life of the village apart from the factory that remain to the Lincoln side of Lonsdale.

LI1 H. S. Magoun House

c. 1885. 1693 Lonsdale Ave.

In this shingled Queen Anne house, the slope of an elongated gable (longer in front than behind) slides over a pretty corner porch with turned supports and interlocks with a low, truncated, squarish mass with hipped roof which suggests a stunted tower. Another small entrance porch off a side elevation has a gable roof which curves like an awning sag. The sophisticated counterpoint of petite windows in front is lighthearted. So, too, is the treatment of the most prominent chimney on the side elevation, revealed at the first floor, while the second puffs out as a shingled bay with a sloping roof to hide it. Wood shingles above (originally covering the roof as well), with burnt brick below, provide textures which further animate the irregularity and collision of the varied shapes. Such overt celebration of cheerful domesticity could easily have become cloying, except that here a skillful sense of composition and proportion saves it from sentimentality. As with Christ Church, one would like to know who designed it.

LI2 Four-Unit and Duplex Mill Housing

c. 1830 (1662–1664 Lonsdale Ave.), c. 1865 (1672–1674 Lonsdale Ave.), c. 1910 (1700–1702 Lonsdale Ave. and nearby)

LI3 Lonsdale Hall

1869. 1661 Lonsdale Ave.

North on Lonsdale Avenue is one of the more extensive collections in the state of a wide range of mill housing types, from Greek Revival through the early twentieth century, much of it quite well preserved. The three Greek Revival units diagonally across the street from Christ Church and closest to the corner, at 1662–1664 Lonsdale Avenue (LI2.1), have been resurfaced but are recognizable, with transomed doors and narrow trapdoor monitors. Although they appear from the front to be duplexes, like many such houses each contains four apartments, including an attic bedroom for each of the upstairs units, designed to accommodate larger families than those below. Then, at 1672–1674 Lonsdale Avenue, is a Victorian duplex, with bracketed doors (LI2.2).

Opposite is the forceful brick block of Lonsdale Hall, built by the corporation as a community center to house a meeting space, library, reading room, and shops, the last seriously compromised by remodelings. Opposite this and farther along (see especially 1700–1702 Lonsdale Avenue) are Queen Anne duplexes and four-unit brick houses (LI2.3), well preserved even to their slate roofs, with long roof dormers and pitched roof porches, which recall attempts at cozy picturesqueness in early twentieth-century housing derived from model industrial towns in England.

LI4 Christ Church

1883–1884. 1659 Lonsdale Ave. (northeast corner of John St.)

The Lonsdale Company financed the rebuilding of this church after a fire. Typically, a book-length history of the church commemorates its pastors and deacons, but not its architect. Whoever the designer, the composition of the sweeping curve of the pyramidal roof, which rises between flanking gables for arched portals to the culminating bell cupola, comes as a happy juxtaposition to St. Jude. (Was the shape of the later church conditioned by this nearby Victorian precedent?) Designed at the height of the popularity of Richardsonian Romanesque, while the master himself still lived, this manages, as a composition especially, more originality than most of its type. Its massing seems to be unique in the state. Like Richardson's village Romanesque, too, it maintains a nice balance between monumental effect and village allusion, although its impact was more impressive before asphalt shingles replaced slate after a 1950 hurricane. A connecting row of windows unites the portal gables; above this, a three-light stained glass window, penetrating the roof under a simple trapdoor opening, completes a composition in which every element speaks to its purpose without affectation.

At the opposite end of the building, the same bold shaping of mass appears in the large, semi-circular apse, which swells directly from the upper part of the gable. It appears, too, in the buttressing of the side walls, and in the directness with which bands of windows (similar to those between the front portals) are fitted between. Although competently handled, neither the detailing of the red brick trim against gray granite nor the proportion of the openings to the whole quite lives up to the massing of the building. But it teaches the big lessons. The interior, with timber trussing suggestive of millwork in its heft, is quite well preserved. The mill relinquished control of the church only in 1947; until then, the company took care of all repairs, and even the minister was on the company payroll.

LI5 Four-Unit Mill Housing

1851–1862. 1742–1748 Lonsdale Ave. and 9–11 Cook St.

Downhill from the joyous image of the Magoun House, a grimmer cluster: four-unit brick housing, built in the severe brick and granite mode of the factories. It typifies quarters built by the Lonsdale Company for its predominantly Irish, English, and Scottish work force here as well as in Berkeley and Ashton. But to those who came from housing of this kind in the English Midlands, these may have seemed at the time to be model quarters in having more space and more modern equipment than their British equivalents, as well as in the provision for a modicum of greenery in company-maintained lawns with trees bordering the street (the typical Midlands village had none). Here each family has its own narrow two-and-one-half-story slice through the housing block. The middle units are entered from two doors toward the center of the front elevation, with dormers to light the attic, while the outer units have entrances at the ends of the buildings and attic windows in the gable ends.

LI6 Mill Housing

1920s (148–168 Main St.); c. 1860 (85–99 Main St.)

Main Street presents another instructive cross section of mill housing. The most interesting examples are, first, brick houses with porches and dormers in Arts and Crafts manner of English model factory towns, at 148–168, all duplexes except that 156–162 contains four units. Then, across Front Street, at 85–99 Main, is four-unit housing in blunt clapboard with simple transom windows and projecting moldings as window hoods. One of these, returned to something close to what it was, demonstrates the impact of their original plainness.

LI7 Dance Studio (Lonsdale Baptist Church)

1843–1847. 1572 Lonsdale Ave.

The Lonsdale Baptist Church, later the town hall, now a dance studio, was once located elsewhere. Like Christ Church, the Episcopal church, it was funded by the Lonsdale Company, the two representing the majority denominations in a village which, for longer than most mill towns, had few Catholics. When voters separated from Central Falls in 1900, Lincoln lost its town hall, and the old Baptist church was pressed into service for a number of years to meet the need. Although it is now somewhat dilapidated, the scale of its Greek Revival transomed door, windows, and pediment is still impressive.

LI8 St. Jude's Roman Catholic Church

1967, Robinson Green Beretta. 301 Front St. (corner of Old River Rd.)

The 1960s and 1970s saw many experiments in the use of engineered structures to create dramatic, monumental interior spaces. In the United States, buildings by such architects as Eero Saarinen, Matthew Nowicki, and Eduardo Catalano generated interest in such structures. They in turn were inspired by such leading European engineer-architects as the Swiss Robert Maillart, the Italians Pier Luigi Nervi and Giuseppe Terragni, and by the Mexican Felix Candela, all brought to public attention when publication of their work proliferated around 1960. Here huge roof beams of laminated wood shaped as segments of a parabolic curve are inverted off concrete anchoring piers arranged in a near-circle to make the tentlike enclosure, its aspirant climax a funnel of light down to the altar. The principal beams project beyond their column supports and are cut at their ends so as to terminate in a perimeter wall which is square in plan. The result is arresting outside, although the building, which, like most such structures, makes a long ascent from the ground, is too low in profile to be truly commanding. Moreover, if the exceptional shape of the roof were to make a compelling architectural and symbolic expression, it should have been of a material with more physical presence than asphalt shingling. Finally, the openwork of the bell tower over the entry and the similar climax to the upward swoop of the roof seem too spindly and frilly for the shape they are meant to enhance, and more attached to than integrated with the larger structural elements.

The interior is more impressive, especially in the intimate relationship between the altar and the semicircular sweep of the pews. This seating plan reflects theological efforts of the period to supplant the remoteness of high altars in deep choirs with an environment permitting a rapport between communicant and priest comparable to that which theater in the round provides between audience and actor. The suggested simple circle of supports on the perimeter is subtly complicated by alternating roof bends of shorter lengths and steeper tilts to provide a circumferential aisle inside the perimeter wall. Similar variations in other bends permit the shift of the seemingly centered altar to a position which is actually off center, providing more space for worshippers in front of the altar and less for the sanctuary behind. Like the exterior, however, the interior shows problems endemic to such free-form spaces. How, for example, can walls, windows, doors, and ornament be integrated with the roof when the curvature of the larger forms calls for kindred shapes? Criticism notwithstanding, however, St. Jude is perhaps the most serious attempt in the state at modern free-form structural expressionism in ecclesiastical architecture.

Saylesville

What became the textile manufacturing town of Saylesville began with a small cotton print works in the 1830s. William F. Sayles initiated its important history in the textile industry, however, when he purchased the bankrupt print works at auction in 1847. He converted it to a bleachery. A fire in 1854 forced the building of a new plant which, after 1863, when Frederic C. Sayles joined his brother in the firm, operated as the W. F. and F. C. Sayles Company. The Sayles brothers steadily expanded operations and from their home base acquired control of the Slater Cotton Company and the Lorraine Mill in Pawtucket, as well as the Glenlyon Dye Works in Phillipsdale, a village in East Providence, together with plants in Connecticut, New York, North Carolina, and elsewhere. In 1894, Frank Sayles, son of William, bought out his uncle's share of the business. Under his leadership, the business (finally known simply as the Sayles Finishing Company) became one of the largest cotton finishers in the country, the first to mercerize cotton thread, the first to print cotton and silk in fast color, and the first to finish organdies without starch and sizing which washed out. When Frank Sayles died in 1920, his company had 3,000 employees in widely dispersed plants throughout the eastern United States. His Saylesville operation was reputedly the largest bleachery in the world.

LI9 Sayles Mill Complex and Administration Building

1855, earliest mill building. 1870s and later, dye house. c. 1875, mill office. c. 1900–1909, great warehouse. Industrial Circle, Pond Ave., Walker St.

Although the core of this industrial complex appears at first sight to be arranged in a deep, ramified *U* of mostly brick buildings around one end of its mill pond, these in fact tend to have their principal entrances away from the pond, facing the workaday bustle in the irregular courts that linked the complex. As in other industrial buildings of the late nineteenth and early twentieth centuries, the quality and variety of the brick here reveal its potential for handsome, functional building. Worth special attention are, first, the Italianate administration building at the southeast end of the pond (c. 1875), with arched openings and, toward the front, a long, low spreading gable crossed by two others, which a onetime bracketed cornice converts to pediments, each with an oculus (now blinded). The pediment has been aluminum sheathed, and the windows have been spoiled by aluminum replacements. Still, the laconic virtues of the building are evident: the quality of the brickwork, the well-proportioned openings, the animating effect of the arcs and circles, and the austere ornamentation of the entrance gable with its insets of brick in checkerboard patterns for roundels and as a frame around the granite slab with the date of the company's establishment.

Immediately behind is a hybrid building. At its front is a rather battered tower, all of architectural consequence that remains from the original (1855) mill. It is interesting for the pointed windows set into the big, arched openings of its side elevations—a rare fragment of early Gothic Revival for industry in Rhode Island, looking more churchly than industrial. Attached to this is perhaps the most handsome building in the complex, the 1870s dye house, with a range of exquisitely proportioned tall arches over semicircular basement openings, all framed by the double layering of the wall, with all edges elegantly incisive. From the end of this building, looking south across the mill yard, are the looming warehouses, their tiers of low, segmental-arched windows topped by corbeled cornices. Clark Sayles, the father of William and Frederic and a builder, was responsible for some (possibly most) of the buildings dating from before his death in 1885.

LI10 Workers' Housing

c. 1915. 90–96 Chapel St.

This pair of duplex workers' houses, wood shingle and siding above brick with multiple gables and porches, is a well-preserved example of what was regarded at the time as an enlightened approach to housing. Influenced by the picturesque recall of images of traditional houses which British designers were then using in "model" industrial towns, such dwellings were meant to counteract the barrackslike, institutional quality typical of most earlier industrial housing. If the upper-story cladding were stucco and the framed porches were timbered, they would begin to approach the Tudor vernacular which inspired much contemporary new town industrial housing in Britain. There are other houses like these in Saylesville and, in fact, this area contains much housing for the Saylesville plant, built at various times in various types, which bears examination.

LI11 Sayles Memorial Chapel

1873, Clark Sayles. Original spire, 1876–1877. 185 Chapel St.

This rugged chapel in Westerly granite gives the street its name. It may seem unprepossessing on the exterior, a verdict which the cumbersome twentieth-century addition intensifies. Originally, it made a much more forceful appearance. Then, what is now a one-story gabled entrance off one corner of the church was topped by the equivalent of three more stories with pointed arches, from which rose a tall steeple in polychromed slate. Despite some crudity in the rigorous detailing, this was once among the grand Victorian steeples in the state, partly because of the height of the nearly freestanding tower, mostly because of the shaping of the tower in accord with its diagonal placement. Cornered buttress slopes rose to chamfered corners which were, finally, crimped into the polygonal spire. Damage from the 1938 hurricane, then the discovery of defective foundations forced the progressive diminution of this fiercely conspicuous tower to its present stunted mildness. Inside, however, much of the original fabric remains: some woodwork (but not that at the altar end, unfortunately), stone memorial tablets, stained glass, and polished marble colonnettes rising to bracketed roof supports with metal tie rods. Returning its walls to a more somber tonality with the stenciled ornamentation evident in old photographs could restore the solemn splendor which the brothers William F. and Frederic C. Sayles intended for this memorial to their children—a son and a daughter lost to illness in William's family, a son in Frederic's. Shortly thereafter, in 1878, William again memorialized his son, who died while a student at Brown University, with Sayles Hall (PR114.9). Whereas he turned to a Providence professional for the Brown commission, the brothers here employed their builder father, Clark, who, among many other buildings in an itinerant career, built country churches in the Federal and Greek Revival styles for Protestant congregations in Rhode Island. Beginning his career with the Federal style, he took on High Victorian before its end.

LI12 Bungalow

c. 1910. 1012 Smithfield Ave. (opposite Sutcliff Ave.)

This bungalow in one of the classic modes of the type, with an elongated trapdoor dormer, stained glass in transoms over the first-floor windows, and slat railings, is worth notice, although it has been re-sided.

Great Road

It was on this road near what is now Saylesville that Thomas Arnold, a Quaker, settled in the 1680s to establish one of the founding families of Lincoln. They called their substantial holding "World's End." Not surprisingly, Great Road is notable for a number of historic structures along its course.

LI13 Saylesville Meeting House

1704–1705, c. 1740. 374 Great Rd. (Route 126) (opposite River Rd.)

Quakerism was the dominant sect in the Blackstone Valley during the seventeenth and eighteenth centuries. Hence this meeting house, built on land contributed by Eleazer Arnold in 1708, was a center of community as well as of Quaker life in colonial Lincoln. The initial building was what is now the one-story ell. Although it has been much altered, its original timber frame is still evident. Thirty-five years later the present two-story meeting house, also clapboarded, eclipsed the original building. Its entrance and stair to the top balcony were originally on the east elevation toward Great Road. These were early moved to the south (side) elevation. The interior, with its U-shaped balcony, its exposed timber frame against the plaster wall, and the plain furniture arranged in a *U* configuration around the elder's dais, is relatively unchanged. Adjacent is the cemetery where many members of Lincoln's oldest families in the Great Road area are buried.

LI14 Eleazer Arnold House

c. 1693. 1920, restoration, research by Norman M. Isham. 487 Great Rd. (Route 123) (owned by the Society for the Preservation of New England Antiquities; open by appointment)

Much restored, the Eleazer Arnold House is nevertheless among the classics in any history of colonial architecture. The scale of its exterior makes it the most impressive of seventeenth-century Rhode Island "stone-enders," although the less familiar Thomas Fenner House (CR12) competes for the more striking interior. The chimney of the Arnold House once filled the entire end wall at the west end of the building, providing for two in-line fireplaces, one for the keeping room or great room in front, the other for the kitchen under what was originally a long saltbox slope to the rear. It seems that such in-line fireplaces are usually incorporated in two building phases: first, the front fireplace at the end of a one-room house; later, the rear fireplace for the kitchen in a lean-to extension. Here the whole may have been built at once, although some changes in the manner of handling may mean that this was not the case. Moreover, in addition to these two rooms, there were two chambers to the east, adjacent to the principal rooms. (Hence, if the lean-to dates from the time of the keeping room, then the original house had four rooms on the ground floor.) Originally, too (though omitted from the reconstruction) the front slope of the roof was cross gabled as an immense dormer which extended perhaps two-thirds of the length of the eaves, giving more space and light to the attic chamber. So this was for its time a grand house indeed.

The fireplace flues ascended within the stone mass through the bulky pilastered chimney, which is among the most elaborate of its medieval type in New England. Later the roof was raised to the rear for still more space in the second story. Hence the wedge of clapboarding which now appears above the saltbox slope to the rear of the masonry chimney block and equalizes the roof gable pitches either side. This extravagant embodiment of the domestic hearth anchors the big, but compressed, gabled box clad in clapboards and shingles, which has been extensively restored. It seems all the bigger for the tiny diamond-leaded fixed windows, a few in triplets (all mostly restoration), distributed (mostly conjecturally) over the walls in a manner dictated by conventionalized function rather than by formal order. The door, now paneled, was originally batten.

The keeping room fireplace may be the widest in any New England house—10 feet 8 inches wide, 5 feet 3 inches high, and 3 feet 10 inches deep—all spanned by a heavy oak timber more than 12 feet long, flush with the wall. (Equivalent dimensions for the Fenner House hearth are 10 feet, 5 feet 8 inches, and possibly an inch less in depth.) Although timber frames were usually exposed in such houses at this time, in the Arnold House, ceiling timbers only are exposed. Wide vertical boarding, which is plastered over, apparently concealed the rest of the frame from the beginning.

Behind the Arnold House is the gambrel-roofed Croade Tavern (c. 1700), moved here from downtown Pawtucket and now used to house the caretaker. Its most notable feature is the medieval overhang of its upper story to the front and, more extravagantly, to the side (it was originally located on a corner site). There, curved ship's-knee bracing on the extension supports the projecting attic.

LI15 Richard Scott House

c. 1808. 528 Great Rd.

The Richard Scott House is a Federal "half house" (three bays of an ideal five) with the door at one end and (rare in such a small house) double chimneys at the other. Next door is an interesting Greek Revival house at number 530, and immediately adjacent is one of the entrances to Lincoln Woods Park, among the major recreation areas in the state and part of a scheme for regional parks around Providence, it was developed from 1909 onward, at a time when the creation of regional systems of parks around major metropolitan areas was both popular and possible.

LI16 George Olney–Arnold Moffitt Mill

c. 1812. 595 Great Rd.

Squeezed between the road and the Mohassuck River, the Moffitt Mill is among the early machine shops in a state which eventually became as famous for machine tools as for textiles—although the subsequent history of the mill's varied production also included shoelaces, wagons, wagon wheels and, finally, blacksmithing. Built by George Olney, it is especially associated with Arnold Moffitt, who purchased it in 1850. Dilapidated though it is, it characterizes the mix in early mills of barn characteristics (on the front) and house characteristics (on the rear). Its sagging clapboards nailed to vertical undersiding are sustained by a heavy pegged frame, the second floor being additionally supported by metal tie rods attached (probably later) to the roof beams. The adjacent mill pond and its dam complete this old-time vignette. Moffitt replaced the original log dam with masonry and the original waterwheel with a turbine. He lived in the mildly Italianate house (1862; visible only in winter), with handsome barn, on the hill behind his factory.

LI17 Olney–Israel Arnold House

c. 1740. 600 Great Rd.

Although the Olney family erected this house, Israel Arnold married into this family in the early nineteenth century, and it remained with his descendants well into the twentieth century. This important mid-eighteenth-century house is exceptionally well preserved. It is a typical two-and-one-half-story, five-bay, central-chimney house with an in-line one-and-one-half-story, one-room gambrel ell with its own equally substantial end chimney in brick like that of the main house. It shows on the exterior as a brick panel to the height of the eaves which is roughly the size of the fireplace. The rest of the end elevation is clapboard. It was long believed that the ell was built first, c. 1720, the main house following at mid-century. Recent evidence, however, suggests that they were built at the same time.

LI16 George Olney–Arnold Moffitt Mill

LI20 Stephen H. Smith House (Hearthside)

The disposition of the narrow nine-over-nine windows in the front elevation of the main house is especially effective. A stretch of clapboarding separates the centered, transomed door with its own window above from the remaining windows. Clustered close to the outside corners of the elevation, they vaguely recall a four-spot domino, but form compact clusters that exert peripheral tugs against the central axis. The most interesting colonial and Federal elevations of this type are those, like this, in which the exceptional placement or proportioning of the openings lifts a symmetrical format out of the expected. The simple hooding of the first-floor windows and door by a projecting board also relieves the otherwise planar quality of the elevation and subtly emphasizes this floor as the principal one.

Characteristically, the main block contains five rooms with fireplaces all working off the central chimney. Much of the original character of the interiors remains, especially in the huge stone fireplace in the keeping room (kitchen). On a corner of the property, a so-called Dutch Colonial playhouse, a miniature version of the house fashionable during the 1920s and 1930s, is an incongruous delight.

LI18 Chase Farm

Probably 1910–1925. 667 Great Rd.

Across from the mill is a fine masonry and shingle dairy farm complex, the Chase Farm (also known as the Smith Farm; most of it probably 1910–1925), its most conspicuous elements the gambrel-roofed cow barn and twin wooden cylindrical silos, both agricultural icons of the period. The exposed metal hoops around the silos become closer together toward the ground, their increase beautifully recording the ever-increasing, bursting force of the harvest inside. The farm continued in operation until the early 1980s as one of the last working farms in Lincoln.

LI19 Hanaway Blacksmith Shop

Between 1870 and 1895. 671 Great Rd.

Built when the owner moved from Moffitt Mill, this shop remained in business well into the twentieth century; thereafter, abandoned and dilapidated, it nevertheless survived, until it was moved across the road from its original site and piously restored as a blacksmithing establishment with its original sign—but so overrestored as to appear new, at least until it weathers. Survivals of small workshops like this little board-and-batten building are rare, and rarer still when they continue to house the ancient craft (itself here revived) which they long served.

LI20 Stephen H. Smith House (Hearthside)

1810–1811. Great Rd.

Hearthside is among the most interesting Federal houses in Rhode Island, vernacular rather than high style, and very idiosyncratic. Stephen Smith owned the Butterfly Mill, across the road (see next entry). He was also a Blackstone Canal commissioner and a prominent member of the Friends Meeting; he is buried in the cemetery of the Saylesville Meeting House (LI13). He reportedly built his house from the winnings of a Louisiana lottery. Its two-story porch suggests southern connections. By legend—but legend of such antiquity that it may approach the truth, and, indeed, some unusual explanation seems necessary to account for this unusual house—he built it to win a reluctant Providence lady. Its magnificence notwithstanding, she supposedly decided against settling in such a remote location, and Smith lived out his active life in Hearthside as a bachelor, together with his brother's family.

The romantic circumstances apparently surrounding its inception may have contributed to the theatricality of this house, surpassed in this respect among extant Federal houses in Rhode Island only by the unique Linden Place in Bristol (BR13). Instead of the customary gable, the side elevations of this random stone house are topped in a heaving curve derived from scroll-shaped pediments. Between these end walls the roof rolls down over a two-story porch supported on tapered square piers (another rarity in New England Federal houses). The entrance is conspicuous by its generous enframement in glass from side lights and fanlight, but made more so by the echoing curvature of the commanding scroll-topped dormer on the roof above. A wheel window for a door to what was originally a parapeted upstairs porch combines with two rectangular openings on either side in a configuration which, with wit (or at least with country ingenuity) invokes a Palladian window. The unusual multiple scroll pediment curves which crown Hearthside doubtless derive from the exceptional use of this element by Joseph Brown in his 1774 house in Providence (PR51), where the wheel window is also to be found. Aside from the attraction a conspicuous building in the metropolis may have had for a provincial, Smith may have settled on this Providence landmark as a lure for the lady who was so strongly attached to the city. The interior preserves its original wainscoting and fireplaces (two of them in marble). Both its fine state of preservation and its picturesque eccentricity made it a favored house in Colonial Revival publications, whose authors and editors were always on the hunt for whatever was piquant in colonial and early national architecture.

LI21 Butterfly Mill

1811–1813, c. 1950. 700 Great Rd.

The stone Neo-Colonial house diagonally opposite Hearthside is what remains of Stephen Smith's Butterfly Mill. Once two stories, long abandoned and very dilapidated by 1950, it was important as among the earliest stone cotton mills in the state. It is revealing of the taste of the times that the mill should then have been whittled down into a "colonial" house, whereas today enthusiasm for vernacular conversions and for loft living spaces might have preserved its mill image (although possibly subdivided into condominiums!). The mill's name derived from a configuration made by dark stones in the random masonry, once embedded in its front elevation and now moved to a new chimney.

LI22 **State Police Barracks**

1931, Jackson, Robertson and Adams. 1575 Old Louisquisset Pk.

The State Police Barracks were the first of four such facilities which this firm designed in variations on a Colonial Revival format. In every instance, a two-story central office and dormitory block is flanked by garage wings masked as office, or perhaps stable, extensions. As with much 1930s Neo-Colonial design, the prototypes were from the Mid-Atlantic rather than the New England colonies. Examples from the middle colonies seem to have offered the picturesque and functional possibilities of spreading central blocks and wings, which are rarer in New England precedent, as well as more genial detail.

LI23 **Jenks House**

c. 1790. 1730 Louisquisset Pk. (barely visible in winter)

This and the following three entries describe houses, built over seventy years, which employ the familiar two-and-one-half-story, five-bay, central-chimney formula in different ways. The first is visible only when the trees are bare and even then with difficulty. This has narrow windows with nine-over-nine sash and a plainly framed, transomed door. The windows in this and the other three houses are boxed out from the plane of the clapboarding in the manner which prevailed throughout most of the eighteenth century. Typically, too, the lower-story windows are simply capped; those on the upper story cut across the cornice board under the eaves in a series of juts. Symmetry, the quality which first strikes us about these houses, is often compromised, as here, to accommodate parlors with slightly different dimensions. The Jenks House was commissioned by a family prominent in the area; Daniel Jenks was especially important as a large landowner, a charter member of the Mt. Moriah Masonic Lodge, and a major shareholder and officer in the company that financed the turnpike which passes his house. But the house, as now seen, is also interesting as a Colonial Revival dream, vintage 1920–1930. The severity of the old farmhouse homestead with slightly quirky "picturesqueness" has acquired the additional charm of shutters, an arched arbor of roses around the door, foundation planting, and a comfortable sprawl of additions behind. Such details speak more to the early twentieth-century romance with things colonial than to the past itself—the comfortable expansion of authentic colonial, made "livable" by a fairy-tale sprawl of wings and gardens.

LI24 **Simon Aldrich Farm**

c. 1760. 1882 Old Louisquisset Pk. (pole 146, opposite the ramp to Route 146)

The second in this string of houses sits back from the road and boasts a complement of barns and outbuildings, all of heavy frame construction. Some, at least, probably date to the eighteenth century. It is the most completely preserved early farm complex in Lincoln, including an array of stone walls which bound and cross the property. Again the symmetry of its front is proximate. The six-over-six windows are larger than those of the Jenks House. The door exhibits a greater awareness of the classical tradition, especially in its simple, pedimented cap. This could have been a later improvement, in view of the plain block capitals of the flanking pilasters; the obtrusive, board-like quality of the splayed lintels over the first-story windows; and the crowding of the apex of the pediment against the window above. These are evidences of a country sensibility, which, together with the vigorous handling throughout, also accounts for the character and charm of the house.

LI25 **Jeremiah Smith House**

c. 1790; ell, 20th century. 42 Wilbur Rd.

The third house in this sequence (known for a gunsmith who lived here in the 1850s) is without the confusing exterior shutters which are generally additions from the late nineteenth century onward. The effect is more sophisticated than that of the earlier examples just described. Symmetry is maintained. The spacing of all elements, including the pediment, is more adroitly managed. More correctness and virtuosity are apparent in the treatment of the pilasters and the broken pediment. The fanlight in many variations (this one handsomely ornamented in lead, and doubtless compiled from catalog parts) was immensely popular during the last decades of the eighteenth and first of the nineteenth century. Hence this builder was alert to fashion. Even so, certain of past practices linger on: the clapboards narrowing toward the base of the elevation; the boxing of window frames out from the plane of the clapboards; the simple splayed boards for downstairs window lintels; and the repetitive jut of the cornice board under the eaves as the frames of the upper-story windows cut across it. Whereas the earlier elevations show vernacular tradition grappling with the impact of cosmopolitan influence, this displays an easier familiarity with modern tendencies mingled with continuity with the past.

LI26 Jonathan Harris House

1742. c. 1810, remodeling. Additions, 20th century. 1896 Louisquisset Pk. (northwest corner of Louisquisset Pk. and Wilbur Rd.)

The fourth house in the sequence goes back to the earliest date but was modernized seventy years after it was built. Then the simple window splays of the ground floor received ornamental keystones; the cornice, a continuous dentiled molding (without the breaks of the preceding example); the old transomed door, a new frame under a porch with pediment and freestanding Doric columns. Jonathan Harris was the brother of David Harris, one of the principal entrepreneurs in the local lime business. A prosperous farmer, like others in the area, he owned stock in the company and sold the lime on his farm to the company.

Lime Rock

A short distance north of the Jonathan Harris House on Louisquisset Pike are remnants of the early layout of Lime Rock: the remains of a stone furnace for burning limestone, which is dug into a hillside, then a rusting metal boiler, which is a later furnace, and just beyond it a pond that inundates an old quarry site. (Neither furnace is easily visible in summer.) The name of the community advertises its principal business from the 1660s down to the present. As indicated in the introduction to Lincoln, Lime Rock was a major source for building lime, as well as for lime in tanning and agriculture for the entire eastern seaboard from the colonial period through the mid-nineteenth century. The core of the industry still clusters along Wilbur Road, where the quarry, crushers, and furnaces which are used today mingle with the antiquated remains of past operation. Gregory Dexter and Thomas Harris began two separate operations in the vicinity. Their combined descendants, through business deals and intermarriage, dominated all aspects of lime production through the nineteenth century—including lumbering operations to provide charcoal for the furnaces and barrels for shipping, as well as a substantial fleet of wagons by which their lime was hauled to Providence and Boston for distribution. Lime Rock entrepreneurs were, in fact, among the heaviest investors in the Louisquisset Turnpike as a means of shortening the roundabout route by Great Road into Providence. The revenue provided by the parade of lime wagons from Lime Rock to Providence was a major reason for the longtime success of the Louisquisset Turnpike as a toll road, until 1870, well after all but one other in the state had ceased as private operations.

Although of major importance as a source of lime for the building trades, tanning, bleaching, and agriculture, through the mid-nineteenth century especially, the operation remained compact, while its owners continued to participate as an extended family in the farming and village life of the community. In fact, although the Harris family controlled the lime industry, much of their work force was contracted labor from local farmers who worked part-time lumbering and teaming for the company and in some instances even operated their own furnaces under marketing agreements with the Harris organization. On the site, one is struck by the juxtaposition of the factory with the farms and the village, a balance which was nearly perfectly maintained here up to the late 1980s, when suburban building increasingly impinged on the ancient village fabric.

LI27 Lime Rock Crushing Plant

20th century. Wilbur Rd. at Eddie Dowling Hwy. (Route 146)

Into the 1990s, this was an especially impressive pyramid of sheds, containers, and elongated ducts made of wood or corrugated metal for conveyor belts. It fascinated as a record of the successive functional adaptations and improvements essential in any such structure, where machine and factory meet most directly. The ghosting of its thinly sheathed forms by white dust fused the parts and intensified their photogenic possibilities, as perhaps so did its rise out of the hollow of prior operations. As depletion of resources begins to alter operations, however, dismantling has reduced some of the complexity that characterized this industrial heap. Yet this old ghost, recalling continuous operation from the colonial period, still merits attention.

LI127 Lime Rock Crushing Plant

LI28 Valentine Whitman, Jr., House

c. 1694. 1129 Great Rd.

Like the Eleazer Arnold House (LI14), this is another important "stone-ender." The original pilastered stone chimney has become brick above the roof ridge and alterations (both old and new) have occurred on the elevations. But a sense of the original house remains. The short shingle extension beyond the stone chimney end contained the box which housed the entrance vestibule (its door now topped by a pretty nineteenth-century bracketed hood) at one corner of the house, together with a tight boxed stair to the upper story. Both fitted into the thickness of the fireplace wall. Like the Arnold house, this has a four-room plan, but the frame, all of a piece, indicates that the four-room enclosure was built all at once. Whereas the Arnold House has a lean-to to the rear, here the rear, like the front, is fully two stories—making what was described at the time as an "upright house." The parlor, with fireplace in the end wall, originally faced a realigned road, with the kitchen, its fireplace in the same end wall, immediately behind. The three major rooms claim three of the four (later) windows ranged along the elevations front and rear, while two tight bedrooms claim the others. The original owner was married in 1694, and this house may have been a wedding gift from his father. The first town meeting of what was then part of Smithfield occurred here.

LI29 Lime Rock Village Center

Intersection of Great and Anna Sayles rds.

Extending uphill from the intersection of Great Road and Anna Sayles roads is a line of country buildings which mark the heart of the village, interesting less for their individual merit (although this is considerable) than as a village group with varied functions. First, at the junction, is a plain, well-crafted brick building of two stories with a hooded, transomed door in one corner and Masonic symbols in the gable roundels at either end, which was enlarged in 1804 from a one-room schoolhouse by the local Masons as Mt. Moriah Lodge Number Eight (LI29.1; 1804; 1093 Great Road). Diagonally across Great Road, at 1092, what was formerly Mowry Tavern (LI29.2; c. 1800 and later; gabled unit to east, late 1980s) began as a small, plain house, which expanded linearly as the inn prospered. Sometime during the mid-nineteenth century, a pretty porch on open supports with scroll bracketing was added across the front of the elongated building. Opposite, at 1091 Great Road, is another fine standard two-and-one-half-story, five-bay, end-chimney Federal house (LI29.3; c. 1820) with a transomed door which is actually off center, framed in plainly paneled pilasters and a plain pediment. Plainness is the rule here, but forcefully so, as is especially evident in the impact of the largish twelve-over-twelve sashed windows. Modern eyes, accustomed to the postmodernist's penchant for adapting regular colonial schemes, then disrupting them for taut, witty effects, will delight in the congestion of four-fifths of the expected pattern of openings at one end of the elevation, then a wider interval before its termination. What originated from functional considerations (and possibly as a later addition) eventually inspires conscious aesthetic response. Next to this, and up the hill at 1089 Great Road, is a one-and-one-half-story Greek Revival building with Doric portico that originally housed the Smithfield Lime Rock

Lincoln (LI) 201

LI30 Lime Rock Baptist Church

LI31 Jesse Whipple House

Bank (LI29.4; c. 1835). Chartered in 1823 by leaders of the lime processing industry, the bank went to Providence in 1847, and its old headquarters became a house. Its handsomely straightforward Doric porch reappears farther along (see LI31).

LI30 Lime Rock Baptist Church
1975, Irving Haynes Associates. 1075 Great Rd.

Bold shaping of mass is the essence of this church. Two giant dormers, serving as light sources for the interior, jut from the long slope of the shed-roofed, shingled box. The smaller, well down on the slope, lights the vestibule; the larger, at the pinnacle, the pulpit area. This building, designed at a time of intense concern for melding the values of both, combines the modernist commitment to large, abstract shape, volume, and luminosity with the ancient country vernacular.

LI31 Jesse Whipple House
1840s; dormer later. 1073 Great Rd. (at Simon Sayles Rd.)

It appears that the builder of the Doric distyle porch for the Smithfield Lime Rock Bank, with its simple side-lighted and transomed door, duplicated his effort here in a slightly grander manner. Whereas the bank is three bays, the Whipple House is five. Whereas the principal entablature under the bank's eaves appears only on front and rear elevations, here it occurs on all four sides, converting the simple side elevation gables into "pediments." Whereas the hipped roof of the bank's entrance porch butts into its roof entablature, here more "correct" treatment of that element as a flat roof permits porch and eaves entablatures to fuse, the porch becoming more spacious thereby. The hipped dormer of the Whipple House (shingled, whereas the rest of the house is clapboarded) is a later enlargement. Hence the unusual total effect of the massing, more Virginia than New England, is accidental.

Manville

Cotton manufacturing started in a small way in Manville in 1812; eventually, after several corporate reorganizations, the Mann family came to play the leading role in the mill and town, by the 1830s under the leadership of Samuel Mann. The Manns were bought out in the 1850s by Harvey and Samuel B. Chace. The mills were located on the Cumberland side of the Blackstone immediately below the Main Street bridge. With the construction of a new granite dam (1868, the third on the site) and a huge brick mill (1872), 350 feet by 76 feet, with a 76-foot-by-26-foot ell, on the Cumberland side of the river, the Manville Company became a very sizable textile plant indeed, to which Frank P. Sheldon in the early twentieth century made substantial additions, including one of his most extensive one-story sawtooth-roofed weaving sheds, measuring 560 feet by 240 feet. Meanwhile, the Manville Company expanded outside the village by mergers with the Social Manufacturing Company in Woonsocket and the Bernon Mills in Georgiaville). Today this giant plant is totally gone, destroyed after a period of abandonment following a combination of flood and fire in 1955, leaving only two tailrace arches and the dam as relics of what was there. Surviving mill housing is of special interest; much of it is altered by renovation, but one project is unusual.

LI32 Manville Mill Tenements

c. 1890. Crossed by Winter, Summer, and Spring sts., between Fall and Park sts. and River Rd.

Although long, continuous blocks of row housing for mill workers are not unknown to Rhode Island, they are much less common than independent blocks containing duplexes or at most four units (which are often disguised as duplexes). Not only is the housing here in long brick rows; some of it employs a format which is unique in the state. Many of these rows show low, narrow, two-story units containing paired doors alternating with full two-and-one-half-story, end-chimney, four-bay blocks—thereby presenting something of a compromise between the continuous row and the discontinuous block. The blocks are split across the center, side to side, and longitudinally, end to end, to divide them into two-story quadrants. Each door serves as the entry to its flanking quadrant. Exterior alternation of door-unit/house-unit gives the long blocks a more human scale. Seven in a total of eight rows were purchased in the 1960s by a developer, who transformed them from mill housing to apartments, prettifying them with varied facings and nondescript Neo-Colonial detail. The one remaining row between Summer and Spring streets in the northeast corner of the site maintains its original brick-with-stone-trim character, although it does not show the alternation of lower door element with taller window block of the most interesting of these rows. Trees, grass, and drying yards between the rows have been substantially lost to parking. Even so, the intent of the project is still evident. It was the Manville Company's final effort at large-scale mill housing.

LI33 Triple-Deckers

c. 1900. Winter St.

On Winter Street, facing into the project, are remarkable triple-decker flats, much knocked about, with a fourth story at street level which once housed stores. One, at number 19, with a heavy cornice over three tiers of porch which front apartments, above a store with a cast iron front, is exceptionally monumental. In Manville, of all the Lincoln/Cumberland mill villages, provision of worker housing depended

LI32 Manville Mill Tenements

on private enterprise rather than on the paternalism of mill owners. These triple-deckers are conspicuous examples of privately built housing face to face with the company housing across the street.

Other housing in the area includes a beautifully preserved one-and-one-half-story Greek Revival duplex workers' house with transomed doors at either end and four pedimented dormers (c. 1840), at 2–4 Cottage Street.

Albion

Albion is another mill community which came to be controlled by the Chace brothers. The designation of this village by the ancient name for Britain indicates the predominant nationality of its original work force, as do the names of the neighboring villages of Ashton, Berkeley, and Lonsdale. Mill operation by a group of local business leaders began in 1823 in buildings which the Chace brothers eventually demolished. They acquired the mill in 1854, adding it to the acquisitions in Manville four years earlier, and incorporated it as part of the Valley Falls Company (see Central Falls) in 1856. They replaced the old stone and wooden mills with a brick mill, which, with the sum of its accretions, is among the larger brick mills surviving in the state.

LI34 Apartments (Albion Mill, Valley Falls Mill)

c. 1856, 1874, 1909, 1921. 1987 and later, converted to apartments. School St. (on the Blackstone River)

LI34 Apartments (Albion Mill, Valley Falls Mill)

LI35 **Pony Truss Bridges**

1885, Boston Bridge Works. School St. over the Albion Mill Canal and the Blackstone River

The sizable brick Albion Mill went up by increments on a tight site in a scenic stretch of the Blackstone, where the river is channeled through a narrow valley with steeply rising and wooded banks. Built in four major campaigns over seventy years, it finally reached a length of nearly 400 feet. The scenic potential of the site is enhanced by the broad spill of water over the curved slope of the fine granite dam upstream, as well as by what is probably the finest display of metal pony-truss bridging in the state, its original lacy quality only slightly altered by low safety rails around 1990. One riveted Platt pony truss spans the granite-lined power canal; then, two pin-connected pony trusses, resting on a midstream granite pier, carry School Street over the Blackstone, with the spill over the dam visible through the latticing. They enhance the character of the mill and dramatize the double waterways typical for all old mill sites—the river and the canal off it into the underbelly of the factory.

The size of the mill is hard to grasp because of the perspective views of it forced by the fit of the plant into the length of its narrow site, with its end elevation on School Street. What exists today began with the construction c. 1850 of a five-story section 120 feet long (later augmented by a sixth story), with double scroll bracketing under the eaves and segmental-arched windows, at the center of the long, narrow block. The Chace brothers gained control of the plant in 1854. They nearly doubled its length in 1874 with a second five-story (later six-story) addition, attached to the south end (in the direction away from School Street) of the original structure. The addition uses the same bracketing under the eaves and the same window rhythms as the older section but, in place of the segmental brick arches that cap the earlier window openings, has rectangular windows with cast iron lintels and sills. One of the most forcefully proportioned brick industrial stair towers in the state was also built at this time, on the side away from the river, close to the precipitous rise of the hill. In a wide-faced, looming tower, the segmental-arched loading doors climb between flanking windows to a brick-paneled parapet at the top story which serves as a base for a hierarchical ordering of narrow arches in a 2-3-1-3-2 arrangement by height and size, all closed by wooden louvers. They screen the water storage tank. But also observe how the fifth story (immediately below the level for the water tank) prepares for the finale by also being inset as a panel with a projecting molding course to set it apart from the floors below. It is all topped by a projecting cornice. Unfortunately, manicured alterations to fit its new role as a foyer to condominiums vitiates some of its original force. Successive additions to the north of the original block in 1904 and 1921 (this time toward School Street) show larger, segmental-arched windows with granite trim, and another, thinner tower, this one facing the river. Although well proportioned, it lacks the impact of the other. (The 1904 extension swept away a stone mill from the early nineteenth century.) Finally, to the south are more twentieth-century additions, with walls in the more modern pier-and-spandrel formula and floors spread horizontally, in contrast to the

LI35 Pony truss bridge, Albion

narrow, elongated quality of water-driven mills during the nineteenth century.

The Albion Mill continued in textile manufacturing until 1962, eventually as part of the giant Berkshire-Hathaway operation. After a period as a luggage firm, beginning in 1987 it was converted into apartments, the old wooden sash replaced by plastic similacra. Although more of the rugged integrity of the old mill has been retained than in most such conversions, it is still sad to see this great industrial whale diced into condominium sushi.

LI36 **Apartments** (Green Mill)
1830. 27–33 School St.

LI37 **Duplexes**
1870. 55–75 School St.

At numbers 27–33 School Street is what remains of the so-called Green Mill, a piece of one of the earliest clapboard textile mill buildings on the site. It was reduced to half of its 120-foot length when it was moved up the hill to accommodate successor buildings and converted to workers' apartments. One can still sense its two-story scale, the rhythm of its windows, and its gable-top monitor—now immured in aluminum siding. Farther along on the same side, at numbers 55–75, are four mansarded duplexes, also re-sided and much more altered, yet still providing a good sense of what they were.

LI38 **Berkshire–Hathaway Mill Housing**
1949, Bernard J. Harrison. 15-80 Berkshire Dr.

This is among the last company-built housing to be built by the textile industry in Rhode Island, while Berkshire-Hathaway operated the Albion Mill. The institutional quality of nineteenth-century housing, and even of prettier early twentieth-century versions, here typically gives way to the small individual house, which assumes a suburban mien. These are plain versions of the prevalent "Cape Cod" merging into what would become "ranch," both characteristic subdivision modes at the time. The first two houses on Berkshire Drive, though both are now vinyl sided, nonetheless provide some sense of the original design.

Entrance to these houses is from the side, not through what at first appears to be the curious anomaly of a very narrow door centered in the front elevation and covered by the sheathing material of the wall. This turns out to be a service door to a compact central heating unit, which services rooms around it through a run of hot air ducts above dropped ceilings. Such features represent widespread thinking of the time about compact central utility units to reduce building and maintenance costs. The erection of these basementless houses on poured concrete slabs is further evidence of the rationalization of the small house at the time. So is the orientation of all service doors toward the north insofar as possible, to give living rooms opposite a southern exposure. (This also means that almost half the houses derive the perhaps unintended additional innovation bonus of living rooms facing away from the street toward the yard.)

Eventually sold to the occupants by Berkshire-Hathaway, the houses continued to be popular, and their value steadily increased. Complaints of tight storage and cold floors have been met by alterations. Changes through time to increase both the amenities and individuality of the houses also merit attention.

LI39 **Duplexes**
1908 and later. 91, 98, 104, 103, and 41 School St.

LI40 **Triple-Decker**
c. 1900. 178 School St.

Opposite and near the intersection of School Street and Berkshire Drive are brick duplexes fronted by porches and topped by various gable treatments which represent American adaptations of comparable housing in British model factory towns of the early twentieth century. They, and other similar housing along Main Street, are characteristic of efforts by the Albion plant superintendent William Erskine to modernize the village by replacing deteriorated stock. Number 178 offers a nicely preserved four-story triple-decker, with three stories of flats fronted by spindled porching, all supported on metal brackets above the same sort of storefront we have seen in Manville—again an example of private speculation amidst mill-built housing.

Old Ashton–Quinnville

On Lower River Road, along what still remains as a beautiful stretch of the Blackstone Canal, are the remnants of the mill village of Old Ashton, which steadily lost visibility after the brief heyday of the canal. This mere speck of a village has had many names. But it was once important, not only as a mill community—the first textile village in Lincoln, in fact—but also as a major crossing point of the Blackstone. Both the mill and the bridge have gone; the mill as an economic operation around 1870, the bridge replaced by the towering Ashton Viaduct of 1935, which now sails over it. And to complete its bypassed status, Old Ashton is located at the very end of a dead-end road. It got its start in 1810–1815 when the Smithfield Cotton and Woolen Company, established by a local consortium of leading commercial and landholding families, financed the building of a masonry mill on a site which eventually had the river on one side of it and the canal on the other. After a fitful commercial history (which is partially recorded in the earlier names Factory, Olney Factory, and Sinking Fund Factory Village), the Brown family acquired the mill and its village, which it renamed Ashton, as part of the expansion of its textile enterprise at Lonsdale, and assigned that mill to cotton sheeting. A new, enlarged mill, however, was built on the other (Cumberland) side of the river, and the village name went with it. What was Ashton then became Old Ashton, and its mill eventually a mere warehouse to the new factory. But even this reduced economic role had disappeared before the remnants of the vacated Old Ashton mill were demolished to facilitate the construction of the Ashton Viaduct.

Old Ashton may receive a new lease on life, however. Old River Road terminates in a forested glade deep in a gorgelike constriction and depression of the Blackstone, through which the canal squeezes beside the river, and out of which leaps the reinforced concrete arch of the Ashton Viaduct. As a scenic stretch of the national park known as the Blackstone River Valley National Heritage Corridor, it provides the kind of confrontation of nature and technology in which Victorians would have delighted. From the verdant and watery shadow springs the rainbow of progress. This is the place to sense the scale and structure of the viaduct. Immediately downstream from the viaduct stands a simple Greek Revival house, which was restored with mostly new clapboarding as a feature of the national park. It was built by Captain Wilbur Kelley between the canal and the river and immediately adjacent to the mill, which he owned during part of the 1820s and 1830s. Kelley had skippered the Brown family's *Ann and Hope* on European and Chinese voyages before he followed the example of his employers and ventured into the textile industry. His difficulties in business provided an opening for the Brown family's Lonsdale Company to take over, and Captain Kelley went to work for a time as a manager for the company.

Quinnville, another name for this village, is actually a community of mostly early twentieth-century mill housing farther south with which Old Ashton has come to be lumped. But this final assault on the old village's identity is unwarranted, because Quinnville is oriented to the opposite side of the river. Most of its residents crossed another bridge to work in another Brown-owned mill at Berkeley (see CU8).

LI41 Old Ashton Housing
c. 1810–1815. 1014, 1016, 1018, and 1027 Lower River Rd.

Only these four mill workers' houses are left to testify to the original village of Old Ashton. Despite their early nineteenth-century date, they are traditional eighteenth-century vernacular buildings of their type, all one and one-half stories and five bays with central entrances, two with gable roofs and two with gambrels. Three are in good condition, one re-sided. All have central chimneys except number 1027. The two chimneys through its gambrel roof indicate its construction as a duplex, its front resting on the shoulder of the road, its back dug into a steep slope as a full-story rubble masonry basement which opens, across a yard, to the canal. Now single family, it is handsomely restored.

Cumberland (CU)

Cumberland marks the northeast corner of Rhode Island. It was here that William Blackstone arrived in 1635 (before his friend Roger Williams) as the first white settler in what is now Rhode Island. A well-educated minister and among the first settlers of Boston, he left because of religious intolerance and settled in present-day Lonsdale on the east bank of the river that eventually bore his name, bringing with him a library of some two hundred books. He lived there for forty years, in a house he called Study Hill, and, when not immersed in his books, experimented with modern farming practices, all in relative solitude. He died in 1675 on the very brink of King Philip's War, during which Wampanoag Indians totally destroyed Study Hill and its library, together with whatever other houses had been built up to that time by white settlers in present-day Cumberland. An unprepossessing monument commemorates the site of his house on Mendon Road, which parallels the Blackstone as colonial Cumberland's principal highway and still perhaps its most heavily traveled. One of the principal factories of the Lonsdale Mill occupies the site of Blackstone's sylvan retreat.

When Blackstone settled in Cumberland it was part of Massachusetts—the western edge of the town of Plymouth, but so far west that, even as a dissenter, he was permitted to stay. In 1660 the Wampanoag tribe sold a great portion of present-day Cumberland to settlers in the border Massachusetts town of Rehoboth, which, in a redrawing of town boundaries, became a piece of another Massachusetts border town, North Attleboro, in 1694. Not until 1746 was this piece of Massachusetts annexed to Rhode Island. At the time, it took its name from the Duke of Cumberland. It then included a chunk of the present city of Woonsocket as its northwest corner, which it retained until 1867, when Cumberland finally attained the boundaries it has today.

Cumberland's mixture of hilly woodland and pasture with valley farming was much like Lincoln's. It also had a variety of mineral deposits. It was early famous, in fact, among mineralogists as "the mineral pocket of New England." Among its more usable deposits were iron, coal, copper, and gold, all of which, together with quarrying (limestone, soapstone, and granite), were exploited in its colonial economy, up to and in some cases a little beyond the Revolution. Surprisingly for a town with neither bay nor ocean waterfront, it was also early known for small boat building, there being no less than nineteen such shops in 1815. Partly floated and mostly carted to the sea, they served many purposes, among them boats lowered from whaling ships to make the final kill. In many cases such work, together with that in gristmills and sawmills, provided farmers with winter occupation. Small mills were particularly prevalent along the Abbott Run, a small tributary of the Blackstone, and the streams that fed it.

But in Cumberland, as in Lincoln, it was the Blackstone which provided for big industry. Generally industry began on the west side of the river, in Central Falls and Lincoln. Transportation to and from Providence was easier from this side; the Blackstone Canal also followed this bank. As these early plants expanded, however, their later and larger additions located in Cumberland beginning around 1860. By then more vacant land existed there than on the west side of the river and so, after 1850, did a substantial section

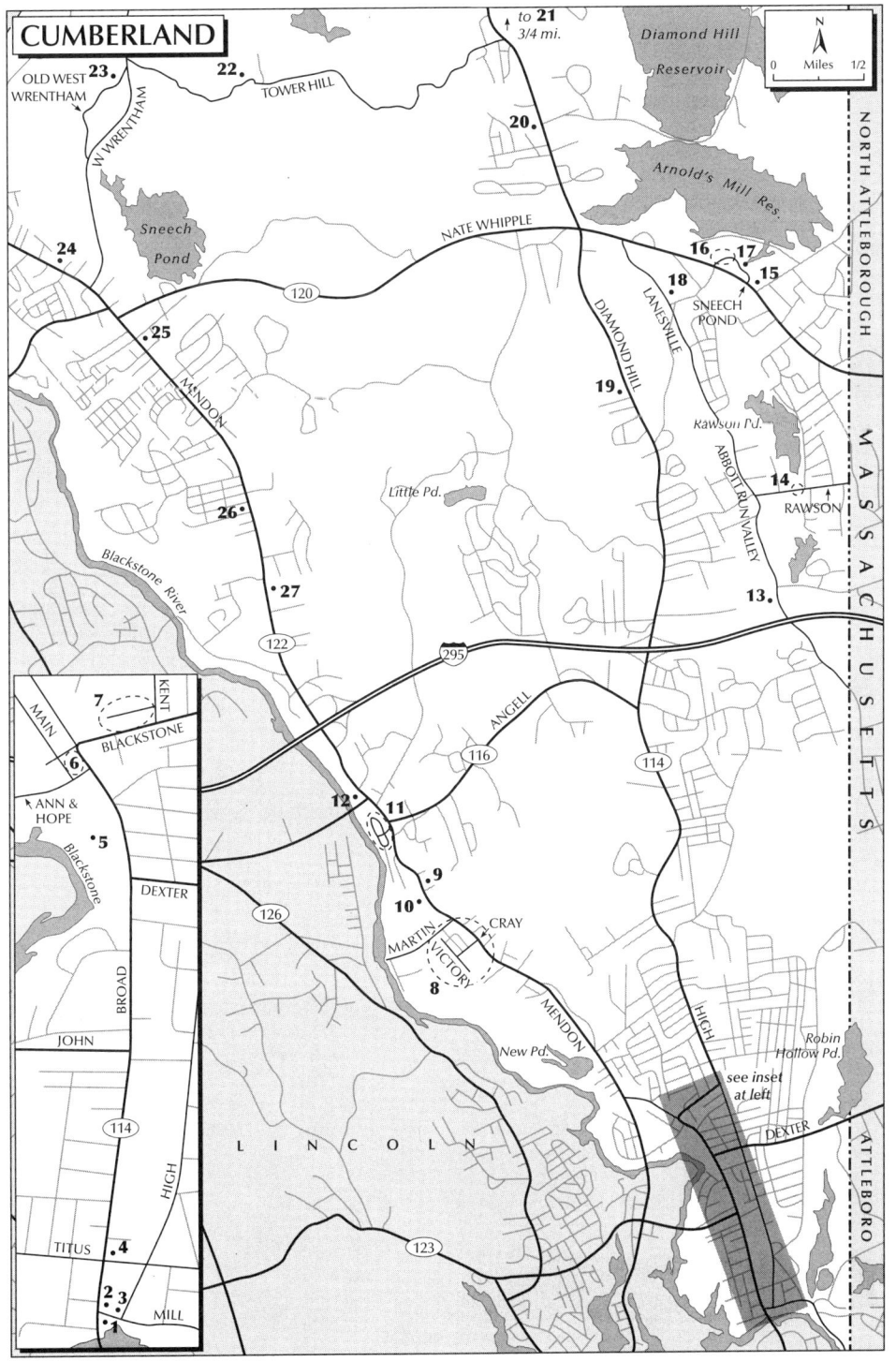

of the main line to Worcester off the New York–Boston track, crossing to the Cumberland side of the river at Central Falls through Lonsdale, Berkeley, and Ashton before recrossing to Lincoln south of Albion. So, by the 1860s, the two leading textile firms on the stretch of the Blackstone below Woonsocket—the Chace brothers' Valley Falls Company and Brown and Ives's Lonsdale Company—looked for expansion toward the Cumberland side of the river. The Chaces jumped the river at Central Falls to the opposite bank, as did Brown and Ives at Lonsdale and Ashton, leaving "Old" Ashton to wither while also establishing the new village of Berkeley between their other Cumberland properties.

The fate of the Cumberland mills is familiar. The Chace mill at Valley Falls is a picturesque and informative ruin. Of the Brown and Ives enterprises, both mills and villages fared better. The mill at Lonsdale on the Cumberland side now accommodates a discount store. Those at Berkeley and Ashton continue in manufacturing, leased, after their original operations failed following World War II, to various and changing firms. Together they provide the best preserved and most coherent of all the brick mill villages in Rhode Island, where mill housing is overwhelmingly of wood. As for Cumberland's farms and landscape, they share in Lincoln's suburbanization.

Valley Falls

The Cumberland side of Valley Falls, like the part that is in Central Falls (see the Central Falls introduction and especially CF10), was, during the latter half of the nineteenth century, the fiefdom of the Valley Falls Company and the brothers Samuel B. and Harvey Chace, who controlled it.

CU1 **Ruins of the Valley Falls Mill, Cumberland Operation**

Before 1850 and later. 1930s, burned. 1993–1995, conversion of foundation ruins to a park, Gates, Leighton and Associates, landscape architects. Corner of Broad and Mill sts.

What little remains of the Cumberland piece of the Chace brothers' operation can be seen at the former mill site immediately upon crossing Broad Street, which links the two towns. Long before the 1980s the ruins of its foundations, its gate machinery, and the canals and raceways lay under a tangle of foliage, with the roar of the falls over its granite dam immediately at hand. At the end of the 1980s the mill ruins began to be excavated for their combined value as an industrial archaeological site and a park within the Blackstone River Valley National Heritage Corridor designed to interpret the massive underpinnings of a great nineteenth-century textile mill. Although the Chace brothers continued to maintain their corporate headquarters on the Central Falls side of the river, the Cumberland mill became the larger plant. The brothers also lived in Cumberland only a few blocks from the mill.

CU2 **Cumberland Town Hall**

1894, William Walker and Sons. 45 Broad St.

The economic importance of Valley Falls and the weighting of the town's population toward its southernmost point doubtless determined the location of the town hall a block north of the mill, even though a site farther removed from the geographical center of Cumberland could hardly be found. It is one of several town halls by William Walker and Sons, which managed to control a substantial portion of late-nineteenth-century civic commissions. This and Warwick City Hall most grandly epitomize the mix of Victorian flamboyance and Neo-Colonial detail in the firm's work of the 1890s—the result being a Victorian dream of what "Colonial Revival" might be at a time when the revival was coming into vogue. The clock tower with cupola vaguely derives from that of Independence Hall in Philadelphia, a feature much adapted for civic buildings of the time because its ebullience and bulk appealed to Victorians groping their way toward the

Colonial Revival. The assertive, even aggressive, Victorian quality of its Cumberland translation and the overblown scale of its windows within a very vertical mass are positive aspects of a building which fairly shouts its civic pride. Typical of the Walker's civic work, too, is exuberant naturalistic terra-cotta relief, here seen in a stretched-out, wreath-centered lunette in a side gable. Although substantial refurbishing of the exterior at the end of the 1980s indicates a renewed sense of commitment to the building after a period of neglect, callous window alterations remain, and so does the thoroughly insensitive interior renovation.

CU3 **Store** (Post Office)

c. 1870. 12–14 Mill St.

This nicely preserved brick store building, once the village post office, has a corbeled cornice and what appears to be a pair of cast iron shop fronts. But they are of wood. If cast iron often imitated stone, the imitation was sometimes reversed, with stone and (more rarely) wood taking their attenuation and detailing from cast iron.

CU4 **John F. Clark House**

c. 1882, William Walker and Sons. 95 Broad St. (at Titus St.)

This nicely preserved Queen Anne house displays a typical mix of clapboard and cut shingle sheathing. Typically, too, its entrance porch, with turned columns and rondels in paneled friezes, nestles against a projecting gable with an exposed chimney. The chimney serves as a spine for an inventive stack of enclosing shapes, varied from floor to floor—polygonal for a downstairs bay, rectangular for an upstairs bay, triangular for the attic. This stack also inverts an expected order of size. Thus the gable at the top is the largest element, the polygonal bay at the bottom the smallest. Just so with the expansion of forms in front, from smaller to larger as the building rises, with an inset arc in the attic gable to increase its apparent swell. Unexpected in the normal order of architectural composition, such interaction among forms is typical for the Queen Anne style. Throughout, Victorian showiness is tempered by combinations of form which invoke domestic intimacy and informality.

Lonsdale

With the support of its Providence backers, the Lonsdale Company expanded from Lincoln across the Blackstone River to Cumberland in 1860, building its first mill on the new site in 1871. This is gone, like most of the company's mills on the Lincoln side of the village. What remains of later mills on the Cumberland side is far more substantial. Moreover, the brick housing in the "new village" on the Cumberland side is exceptionally intact, most of it built in two stages during the 1870s and 1880s (presumably in concert with two stages of mill building). It also provides an introduction to similar housing by the company in Berkeley and Ashton, the next two mill villages north, which were also under Lonsdale Company control.

CU5 **Ann and Hope Store Complex** (Ann and Hope Mill, Lonsdale Mill)

1886, 1901, Frank P. Sheldon. 1 Ann and Hope Way

These are the second and third mills of the Lonsdale Company in Cumberland, diminished but not destroyed by foreground encumbrances. The four-story 1886 mill (now a warehouse) was named for the wives of the founders of the Lonsdale Company, Ann Brown and Hope Ives. Previously, while still merchants in shipping, and before their conversion to manufacturing, Nicholas Brown and Thomas P. Ives had christened two sailing vessels in succession with the same compound name. The 448-foot length of the 1886 mill allegedly made it among the largest of its time. The projecting segmental arches over narrow windows which are punched directly into the plane of the wall depend on Victorian precedent and mark this factory as an early Frank P. Sheldon design. So do the brick moldings and corbelings of its squat, deeply projecting tower. Yet already there is an abstraction and precision about the thin layering of these residual castellated motifs which anticipates qualities developed in subsequent Sheldon industrial buildings.

The same tendencies pushed a bit further are evident in the tower embellishment of the later two-story addition (now a discount store), whose similarity to the earlier building may reflect the desire to harmonize the two units. Both have lost their low pyramidal roofs (a similar one still exists at Ashton [CU11]). The 1901 building, however, shows the transformation away from the narrow-windowed, thick walling of the 1886 mill to the skeletal pier-and-spandrel walling typical of early twentieth-century factories, but consummately proportioned and articulated in accord with evolving Sheldon standards for this new design approach. Note, too, in the older mill, the refinement of the projecting corbeled band under the cornice at the level of the springing of the topmost row of windows to create a frieze that contains the bracketing, while also marking a transition to the eaves above—a detail which is not unique to Sheldon buildings, but is handled here with admirable straightforwardness and elegance. Inside the discount store, much of the "slow-burning" heavy timber construction of the original mill is still visible.

CU6 Ann and Hope Mill Housing, Enclave North of the Mill

c. 1861–1870s. Adjacent to the mill on Broad, Main, and Cross sts.

CU7 Ann and Hope Mill Housing, Enclave Uphill from Blackstone St.

1880, final row 1920–1921. Blackstone St. and Blackstone Ct., across Broad St. from the mill

A major portion of the "new village" on the Cumberland side of the Lonsdale operation is situated off the slope of Mill Street immediately north of the mill in a close-packed, gridded enclave which contrasts with the more scattered arrangement across the river in Lincoln. Granite-trimmed at sill and lintel, with corbeled eaves and gables parallel to the street, these brick houses are predominantly of two types: one-and-one-half-story duplexes with paired entrances centered in the front elevation and two-and-one-half-story, four-family units with doors to stair halls at either end. Construction occurred in two campaigns: during the 1870s, duplexes at 2–12 Main, 550–560 and 566–572 Broad, and quadripartite units at 13–55 Main; during the 1880s, duplexes at 574–600 Main, quadripartites at 562–564 Broad. Exceptional are two long nineteen-bay units at the foot of the enclave on Cross Street, with entrances at every three bays, except that a four-bay unit closes one end toward the mill and a two-bay unit the opposite end, with another door which doubtless gave entrance to maintenance facilities. These long units seem to have provided semidormitory quarters for single workers; now all partitions are gone and the buildings have been converted to warehouses. At 602–606 Broad are examples of another exceptional type with Greek Revival characteristics which seem to be the only survivors of 1860s housing. Like the brick buildings across the river, to a modern-day eye they have a severe mien that masks their status as model housing at the time they were built. Their substantiality, their orderliness, the exceptional attention to such amenities as yard space, the high standard of sanitation and an effort at community beautification, especially through tree planting, were all praised when they were built.

The 1880 construction campaign crossed Broad Street to produce one of the more attractive enclaves of workers' housing in the state. Nor does any other provide a more striking image of the hierarchy of mill work as is here revealed in the adjacency of the residence for the superintendent with those for overseers and lesser employees. On the corner of Broad and Blackstone, now much altered, is the large, spaciously sited, brick house for the superintendent, liege lord of the mill. Behind it, at 3–17 Blackstone Street, is a beautifully crafted row of cross-gabled, T-shaped brick duplexes for overseers lifted a little above the street on a terraced green. Their steep gables face the street with a commanding gravity appropriate to their inhabitants' position. The granite-silled and granite-linteled window tiers for each of the duplex units pull away from the center of the elevation toward flanking entrance porches set into the right angles of the T. A blind inset arch caps each attic window, with a corbeled frame of exceptional refinement under the eaves, its stepped coursing curved beneath the gable's apex. Cross dormers with clipped peaks also light the attic floor.

CU7 Ann and Hope Mill housing, overseers' duplexes on Blackstone Street (foreground) and four-family house on Blackstone Court (to rear)

Behind this row, on Blackstone Court, with a stretch of green and yard between, is a row of four-family houses for workers of lesser status, akin to those already seen. Also two and one-half stories, they set gable broadside against the road without terracing. Windows organized in regular rhythms across the facade terminate in doors at either end, each of which opens to two apartments, the smaller downstairs, the larger up, including dormered attic space. They appear more as a uniform wall than as the series of independent entities assigned to overseers. The still surviving parklike environment was one of the amenities which led the progressive-minded to applaud Lonsdale Company housing when it was built.

Across Blackstone Street from the overseers' duplexes is a later row of equivalent houses erected in 1920–1921. By comparison with the Victorian assertiveness of the earlier buildings, these are vaguely Neo-Colonial, with arch-roofed and (originally) latticed entrance porches. They are more domestic in mien and, measured by their predecessors, aspire to be demurely pretty.

Berkeley

With the success of the Lonsdale mills in Lincoln (and in conjunction with their expansion across the river into Cumberland), Brown and Ives purchased sites farther upriver for two more mill villages, in 1867 for Ashton and in 1872 for Berkeley, where they continued the company policy of brick for mills and housing. This village and the separate incorporation of the plat as the Berkeley Company commemorate the British philosopher George Berkeley and his famous eighteenth-century sojourn in Middletown, Rhode Island.

CU8 Berkeley Mill and Mill Village

1872, mill, Frank P. Sheldon(?); additions to south in the original style. 1872 and later, village. Cray, Lawrence, Woodward, Victory, Drummond, Silis, and Martin sts.

Berkeley is another well-preserved brick mill village in a gridded enclave four streets wide, all crossed by Cray Street. The housing here fronts the mill on a slope above it, rather than off one end of it as at Lonsdale. The first impression may be of identical houses, but a closer look shows considerable variety among them. In Lonsdale, as in most mill villages, variety in workers' accommodations resulted principally from building campaigns carried out over a long period of time. Here, housing appears to have gone up within a more restricted span of time. If this is the case, then how to account for the variation in treatment seen in Berkeley's housing?

On Woodward, the middle street of the first three off of and paralleling Mendon, are two-and-one-half-story four-family units comparable to those seen of Blackstone Court in Lonsdale, with entrances to either end and paired dormers and chimneys toward the center on either side of a party wall. The flanking streets (Lawrence and Victory) contain one-and-one-half-story two-family units with pairs of small eaves windows to light the attic stories from the front. At the north end of the cross streets are two larger units that seem to have been dormitories for single workers. But this distribution of differing accommodations is only roughly maintained. Moreover, while the walling of most houses is planar, some have piers, with a puzzling randomness about their location, making it uncertain whether successive stages of building or comparative experiment within the same stage account for variations among them.

The four-story brick mill stretches much of the length of the community at the foot of the slope. Window and wall treatment are identical to that for the later Ann and Hope Mill in Lonsdale, suggesting that this may be an earlier Sheldon-designed building. The towers of both show handsome proportions and comparable corbeling, but this is more conventionally Romanesque-inspired, whereas Lonsdale is more uniquely and abstractly ornamented. This mill, like that at Lonsdale, has lost its original low pyramidal roof. Miraculously, Berkeley preserves most of its twenty-over-twenty window sash, more than Lonsdale, as well as two tiers of cast iron–decorated fire escapes.

At 9 Martin Street, near the junction of Victory and Martin streets, is the village school, eight classrooms fitted into a tall brick cube, with boys' and girls' entrance porches on either side (now much battered from commercial use). Next to this is the brick house for the superintendent. Then, at the junction of Martin Street and Mendon Road, on the northwest corner, is a pair of overseers' houses in brick, altered in varying degrees. Diagonally across Mendon, the once charming Queen Anne Berkeley Methodist Church (probably 1870s) has been skewered into a photographer's studio. (Norman Vincent Peale, famous for *The Power of Positive Thinking*, honed his preaching prowess for Marble Collegiate Church in New York in this pulpit while still a theological student.)

CU9 St. Joseph's Church

1888, F. E. Page. 1303–1317 Mendon Rd.

This is the second Roman Catholic parish church in Cumberland, replacing a previous church of 1872 on the site which was established to minister to increasing numbers of Irish immigrants in the area. St. Joseph's exhibits Carpenter Gothic at its most assertive: in fact, it is perhaps the most monumental example of the type extant in Rhode Island. Pointed and wheel windows appear as simple shapes cut out of the clapboard walls. Worth special notice is the balancing of the visual impact of the tall, broach-spired bell tower at one corner of the front elevation by a shorter tower and an intermediate turret. So the seeming commitment to symmetry manages to become asymmetrical. The variety in the adornment of the towers is spirited, if a bit crude.

Inside, too, some unfortunate (if well-intended) modernization has occurred, especially in the redesign of the altar and the introduction of modern stained glass into the ground-story windows. But much remains. Instead of creating the structural wedding of parts that might be expected of a Neo-Gothic building, the Gothic revivalist often scattered traditional architectural elements as decorative pieces, in this case emphasizing the plastered, rather than masonry, quality of the interior. Skinny clustered columns in *faux marbre* break out in luxuriantly ornamented capitals, but the projecting moldings that frame the arcade are lifted off the capitals. The colonettes which support the vaulting cling as brackets to the plastered walls of the nave, floating the plaster vaults above the arcade rather than building from it. Above each of the nave arches are half-length figures of saints in quatrefoils against gold-leaf backgrounds; over these, tiny trefoil windows make a clerestory of sorts. A little stubby in length for its height, this is nevertheless an impressive space, and possibly more interesting for the tension between the two dimensions. Within the space, each of the elements possesses its own assertive shape, thereby continuing the play of resonant shape found among the windows on the exterior. Doubtless a darker palette and some stenciled patterning once helped to meld the pieces. Now the creamy white repainting sets them loose in the big space, and one feels the architect pasting the pieces into this belligerent yet charming expression of faith.

CU10 Fiberglas Furnace

1950s. 1970s, closed. Mendon Rd.

From the parking lot of the church, assertive shape appears in another context. This tall cubic industrial structure in sheet metal is remarkable for the giant, wrenchlike pincer shape at its top, rising from the Blackstone River declivity. This rooftop industrial sculpture for functional effect recalls the now antiquated sawtooth treatments for top-lighting expansive floors in single-story factories during the early twentieth century, especially prevalent in Rhode Island weave sheds and machine shops built at the time. Here, housing for venti-

CU8 Berkeley Mill

CU9 St. Joseph's Church

lating fans accounts for the startling shape. Plastic compounds fed from the rectangular silos beside the furnace shed were converted to glass pellets as the raw material for Fiberglas, until the plant was closed after vain attempts to meet environmental emission standards. It now serves as a cavernous shelter for construction equipment, the impressive ventilating shafts (for this use, at least) now as obsolete as sawtooth skylights after the advent of fluorescent lighting. The thin cardboard quality of corrugated sheet metal at this scale is also impressive in the unreal (perhaps surreal) manner in which it appears weightlessly to envelope space, more as a shroud than a wall.

Ashton

CU11 Ashton Mill and Mill Village

1867 (mill), 1872 (village). Store Hill Rd. and Middle, Front, Wiggin, and Weber sts.

Ashton is the third in the row of Lonsdale Company brick mills and villages on the Blackstone River (in date, slightly earlier than Berkeley). Unlike the Berkeley mill, this suffers by the infill of most of its windows, although its belfry retains its slate pyramidal roof. As in Berkeley, the tower is abstracted from historical precedent; in this instance, not a battlemented tower, but the campanile of an Italianate Romanesque brick church. The bell to call its congregation can still be seen behind the slotted, round-arched openings. Instead of the refinement of a slightly projecting frieze under the eaves, this has a series of inset panels forming symbolic parapets under each of the top-story windows, to prepare for the mansard which once topped this mill (providing a fifth story, where now there are four). Only the old company office building still retains its mansard. The horizontality of the inset panels also accords with granite moldings which run uninterrupted around the building, even crossing projecting piers. They provide a series of visual shelves on which the segmental-arched windows sit. This countering of the verticality of the window stacks by such extensive stone banding is infrequent in Rhode Island, and is probably exceptional everywhere. Uniquely, entrance to the mill is through a boxlike entrance building into a tunnel and under the railroad tracks which traverse the front of the factory.

The village at Ashton is as well preserved as that at Berkeley and, like it, is stretched across the mill front—but here closer to the factory, thereby making the most dramatic confrontation of mill and housing in the state. Again, apparent identity among the houses at first glance subsequently reveals considerable variety. Ashton housing, however, shows more uniformity than that at Berkeley: one-and-one-half-story duplexes to the south side of the town; two-and-one-half-story four-unit houses toward the northern end; two long row houses, one and one-half stories and two and one-half stories, along one side of Middle Street. Most interesting are four-unit houses fronting the factory with entrances at either end of the fronts for downstairs units and entrances at either side for two units above, like those on the Lincoln side of Lonsdale (see LI5). The roofs of these units were later raised to provide for more attic space; hence the tiny square lights under the eaves.

Ashton village provides a fine elevational view of the highway viaduct (see the following entry), which is unfortunately obscured by a metal-sheathed shed for a conveyor belt that runs through one of the arches—a relic of the time when this mill served as the fabrication operation for the Fiberglas furnace nearby.

CU12 Ashton Viaduct (New Bridge)

1934, design, Samuel A. Engdahl. 1934, substructure. 1942–1945, superstructure. George Washington Hwy. crossing the Blackstone River and Canal

CU11 Ashton Mill and mill village

Paired reinforced concrete arches, in parallel, are fused with vertical supports and the roadway deck to make this span between the precipitous banks of the Blackstone, finished more than a decade after it was designed because of delays in funding. Although it shows vestiges of neoclassical detailing in the railing and in the suggestion of capitals in the spandrel uprights of the arching, as well as some heaviness in the bracketing under the roadbed and the piers, it is nonetheless an impressive example of reinforced concrete engineering for its period. Its designer, Samuel Engdahl, was subsequently chief bridge engineer for the Rhode Island Department of Public Works.

The Cumberland end of the bridge provides a splendid overview downstream of the Ashton Mill and its village. The view upstream shows the dam and remnants of water control machinery which once powered the Ashton Mill, the canal from dam to mill still evident even if filled in. On the opposite (Lincoln) side of the river, traces can still be discerned upriver of a similar canal which once ran from an earlier dam to the long demolished mill (c. 1810–1815) at Old Ashton. Downstream from this end of the bridge is the overpassed original site of Ashton (see under Lincoln), mostly obscured, even in winter, by woods.

CU13 Elisha Waterman House and Replica Addition, Brieswood Farm Riding Ring

1757. 1970s, addition, Larry Thibideau. Riding ring, c. 1983. 100 Abbott Run Valley (Lanesville) Rd. (pole 79)

The original one-and-one-half-story house with central chimney has been restored. Sliding behind it and connected to it is a 1970s replica addition: a more commanding version of the original, twice as long and twice as large, with a much grander, sculptured chimney and, at close view, making a virtuoso display of nailheads and wood craftsmanship. The sequel at once gives obeisance to the original and means to surpass it in meticulous "reproduction." Its designer apprenticed with Armand La Montagne, the North Scituate craftsman-architect who is famous for his replication of seventeenth- and early-eighteenth-century wooden houses (see SC6). The nineteenth-century barn complex, with its mix of clapboard and shingle, is typical. Not so the riding ring behind it. Its column-free interior (visible only to riders and their invited audiences) is spanned by triangulated trussing interconnected with a semi–space frame of light boards and timbers. Like a number of farms in an evolving suburban situation, this one shares in the suburbanization. Dairying and meat production are out; horseback riding is in—even as subdivisions eliminate the riders' trails.

CU14 Bridges and Dam

c. 1886, Boston Bridge Works. Rawson Rd. a short way in from Abbott Run Valley (Lanesville) Rd.

Of three similar late nineteenth-century pony truss bridges over Abbott Run—one to the south on Howard Road, another farther north at Arnold Mills—this has the additional interest of a small lattice-girder metal bridge over a mill raceway as a prelude. The granite-lined raceway, which once served a small mill for cotton cloth, now runs under the corner of a large shingled mid-twentieth-century house where the Rawson Mill once stood. The site is now landscaped to the mill pond. Its granite dam and the sylvan setting enhance the setting for the principal bridge over the run itself, which nearby suburbanization thus far threatens but does not spoil. This bridge and others like it over Abbott Run were completely unaltered until 1989 when safety upgrading compelled the addition of low aluminum guard rails inside the trusswork, thereby somewhat disrupting a little the lacy character of the original engineering.

Arnold Mills

Until a fire in 1987 destroyed it, a dilapidated but picturesque wooden factory of 1825 provided the climax for this early-nineteenth-century mill village—the best preserved of its date in the state—and just as plans were underway for the mill's restoration. So the centerpiece of the village is lost, although the rest remains.

This was long an open farm community, its very openness making it vulnerable to development. We are concerned only with a tight cluster of houses and community buildings around a small mill site on Abbott Run, off a corner of the porous spread of the village; most specifically, along a curve of old Sneech Pond Road bypassed by Route 120. This bypassed section of the village was further protected from development by the incorporation of its millpond into the Pawtucket reservoir system, providing a wooded background for what is there.

CU12 Ashton Viaduct (New Bridge)

CU14 Pony truss bridge, Ashton

CU15 Arnold Mills Methodist Church (United Baptist Church)

1825–1827. 1846–1847, remodeled. 1956, chancel wing. 1961–1962, chancel. 1–3 Nate Whipple Hwy.

Seven years after it was built as a Baptist church, this became the second Methodist church in the state. It has been much altered—from Federal to Greek Revival, then twice enlarged and altered, not very happily, in the twentieth century, as well as re-sided. What results is a gabled two-story box topped by a chunky octagonal cupola, appearing the more boxy for the sparseness and small scale of both its openings and its rebuilt porch. Inside, it originally had balconies along either side and toward the doors, with the pulpit placed between the doors but out in front of the balcony, which housed the choir. This unusual, but not unique, positioning of the pulpit at the entrance end of the church was reversed in 1846, when the floor level of the audience room was also raised to the level of the balconies, with a vestry (now offices) added beneath. The extant Greek Revival woodwork and furniture date from this time. Finally, in the course of the twentieth-century renovations, the chancel was added.

CU16 Houses

Late 18th to mid-19th century. Sneech Pond Rd.

The first of the arc of adjacent houses, at number 300, is an early nineteenth-century, one-and-one-half-story dwelling with doors at either end under narrow transoms. It was originally a duplex and probably workers' housing. Then, at number 302, is a standard Federal two-and-one-half-story house with center door, center hall, and off-center chimney, initially shingled, but clapboarded in 1930 (when the porch was probably added). Sometime after 1825 it belonged to Davis Metcalf, one of two Metcalf brothers who built the mill. The third house (number 304), which long housed both Dr. Halsey Walcott's office and his brother William's general store in addition to their shared residence, was built for Louis Arnold (1819–1924). A basic one-and-one-half-story Greek Revival house with a flank gable comes next (number 306). It provides a rare example of the attenuated Federal style edging into the motifs and format of the Greek Revival. Then, the largest house in the arc (number 308), a Federal house gussied up with a 1913 Neo-Colonial veranda and projecting centerpiece with Palladian window. The remodeling was commissioned by Neil MacKenzie, a twentieth-century owner of the Metcalf plant.

Another basic one-and-one-half-story, flank-gabled Greek Revival house, this for Dr. Addison Knight (1843; rear ell, twentieth century), at number 312, is the best of all. Instead of a portico, a porch supported on wide-spaced Doric columns stretches across the front, with a fine example of the basic Greek Revival side-lighted door. Angled toward Abbott Run is number 314, another one-and-one-half-story house with shed dormer (c. 1800), which received a tentative Greek Revival facelift (1837), especially to the entrance, with an attenuated quality which again suggests Federal sensibilities groping toward something new. Opposite this row, and close to the stream, is the oldest house in the village, a much-altered eighteenth-century, one-and-one-half-story gambrel (1745, 1773) at number 315, probably built by Amos Arnold. He operated a sawmill on the west side of the stream and a gristmill on the east bank, and gave the place its name. The adjacent barn is mid-nineteenth century.

CU17 **Bridge and Dam**

1886, bridge, Boston Bridge Works. 1875, dam. Sneech Pond Rd. at Abbott Run

This metal bridge with Pratt pony truss construction crosses Abbott Run, completing the Boston Bridge triad. Above it is a handsomely crafted, cut stone dam (1875, later raised), with bursts of water pushing through the crevasses here and there. Only the foundations of the three-story clapboard mill (1825, addition to east) now exist, on which a modern restaurant with touristy "general store" has risen. Across the road are the foundations and race of Amos Arnold's 1747 gristmill. Joseph Metcalf purchased the mill rights in 1818–1819. With his brother Ebenezer (who eventually bought out the partnership) he built the mill and operated it as a textile machine shop. It remained in the Metcalf family until 1896, then housed a series of lesser enterprises—wagon repair and grain mill, blacksmithing, even straw hat manufacture, among others, before its purchase in 1975 by the City of Pawtucket, which continued to rent it until the late 1980s. By the late 1980s it had become a picturesque ensemble of patches and adjustments skirting the edge of outright decay. Then (when it was momentarily vacant) flames did it in—and the village enclave which perhaps best represented the rural phase of the small-scale early Rhode Island industrial entrepreneur substantially lost its reason for being.

CU18 **House**

Late 18th century. 331 Abbott Run Valley (Lanesville) Rd.

This Federal house in the full two-and-one-half-story format, genially enlarged at various times, boasts a fine doorway with fluted pilasters and heavy entablature. It once belonged to Ebenezer Metcalf, Sr., the first Metcalf to settle in Arnold Mills. Although he came as a farmer, he was mechanically inclined, a predisposition which he passed on to his sons.

CU19 **House**

1809; later additions. 2944 Diamond Hill Rd.

This five-bay, two-story, center-chimney Federal house is interesting for an unusually commanding entrance. A broad door with the familiar Federal treatment of fanlights and side lights (its leading probably early twentieth century) is surrounded by a wide frame with provincial carpentered ornament applied both to pilaster panels and to the spandrel intervals between frame and fan. The eyecatcher is the scroll-cut, silhouetted embellishment of the spandrel areas. Plants? Plumes? Shells? Fragmented sunbursts? Hard to tell. It's the exuberance that counts.

Diamond Hill

CU20 G. Whipple Commercial Block
1895. 3782 Diamond Hill Rd.

This late Victorian store is perfectly preserved and, so far, retains an appropriate setting. A severe, gabled, two-and-one-half-story frame block with paired bracketing in the eaves and tiny dormers, it sits flank wall to the highway. Four six-pane shop windows alternate with doors to provide for two shops with office space upstairs. The diffuse village still contains some interesting nineteenth-century houses intermixed with recent building. Farms and the once important Diamond Hill quarry nearby once provided the hamlet's livelihood. Suburbanization and passing patronage to and from Diamond Hill State Park now make it viable.

CU21 Grants Mills
c. 1818. Near intersection of Diamond Hill and Wrentham rds. (visible in winter)

On private property and virtually impossible to see through the trees in summer, this one-and-one-half-story wooden mill in a mix of shingles and vertical boarding, with its dam at the end of Lake Miscoe, is a well-preserved (and partly restored) example—among the best—of the earliest rural industry in Rhode Island. So is the restoration of the handsome masonry craftsmanship of its raceway. It was built by Joseph Grant as a combined sawmill and gristmill. The sawmill was housed in a long, narrow gabled shed beside the raceway, the gristmill in a chunky gabled ell set at right angles to the dominant element at one end and a bit lower with the raceway underneath.

CU22 Miller House
1797. 161 Tower Hill Rd.

This important house is best seen when the leaves from nearby trees do not press in on it. It follows the two-and-one-half-story, five-bay, central-chimney formula; but sometimes the formula results in a house of particular character and charm. Why here? The obvious answer is the unexpected bird with outstretched wings flying out of the pediment, like the cuckoo from its clock, flanked by pretty chestnut blossoms on the supporting pediment blocks. Less obvious are the effects of the distribution of

CU22 Miller House

openings over the elevation, together with the high quality of both the design and execution of the entrance. Closely paired on either side, the first-floor windows seem low and pulled well away from the door. The small-paned texture of their twelve-over-twelve sash is a bit at odds with the blunt breadth and abruptness of the window caps. Those above hug the eaves with exceptional intensity, not just because they butt it (as is standard for most such elevations through the early nineteenth century), but because the *fact* of their butting is transformed into the sensation of the very *act* of butting by the stretch of the interval between downstairs and upstairs windows, the arrowlike thrust of the door in its height and narrowness, and even the explosive flaring of the window caps downstairs. The windows leave a generous field to the entrance centerpiece. Its slotlike proportions—those of a grandfather clock or of a coffin—accentuate its axiality with the surprise of an exclamation mark.

The designer of the handsome door frame (presumably other than the master builder of the house) was sufficiently immersed in the classical canons to handle its elements with an assured fluency which extends to the subtle modeling of all elements. These qualities contribute to the lyric and intimate overtones of this portal. They encourage examination up close. There we can better experience the exceptional precision in the rendering of the formulaic chestnut blossom motif, or the resiliency of the "Gothic" arching in the semicircular overdoor light, or the meticulous beveling of the inset door paneling and the animated play of verticals against horizontals between the four panels above and the four below. All these clues suggest the refining sensibilities of the skilled cabinetmaker. So does the bucolic, illustrative sentiment of the bird and the flowers, although the bird may have been added later. Legend has it that a carver-craftsman related by marriage to the locally prominent Tower family made the bird. In any event, it is in the spirit of carved figures which break in finial-like climax through the broken-pedimented cappings of highboys and (yes) grandfather clocks. Perhaps the entire entrance was his.

The present siting of the house back from the road in swath cut on axis from the woods accentuates the axial verticality of the door. Some sort of axial withdrawal from the road seems always to have been intended, although its original handling doubtless exposed the house more completely. So the image projected by this house, though rural in character, possesses much sophistication. One senses it (rightly or not) as the idyll of a country gentleman who may have had a farmer on his property, rather than the homestead of a working farmer, however wealthy.

CU23 **Cyrus Cook House** (Orchard House)

1810. 12 Old West Wrentham Rd.

Old West Wrentham Road is another bypassed segment of a winding road with an enclave of houses extending back to the colonial period. Most notable is the Cyrus Cook House, known as Orchard House, a large, relatively plain Federal house, handsomely walled and set among trees. It has been considerably altered by twentieth-century Colonial Revival additions. Its primary interest is its entrance, which is unique in the state. This broad, strongly framed side- and transom-lighted door is topped by a very wide entablature surmounted by a vigorously projecting cornice. The entablature is embellished over its entire extent by small incised squares, making a dense lattice of the surface which may be related to basketweave patterns fashionable at the time for interior fireplaces. But this carpenter-designer delighted in repetitive pattern: bands of diamonds, dentils, and dog teeth appear above and below the latticing, with rope moldings on either side of the door. As provincial, no doubt, as the door of the Miller House is sophisticated, yet to forceful and joyful effect. A little farther, on the left, are two more Federal houses, all three ancient neighbors.

CU24 **Ornando Remington Vose House**

c. 1890. 3533 Mendon Rd.

Vose began as a farmer, then established a hardware store and then a florist shop before building this Queen Anne house, the largest in the area when it went up. The roof sweeps down in a long slope around a balconied dormer, then continues in a curve out over a deep porch with turned posts and balusters, which embraces two sides of the house. It is all anchored by a polygonal corner tower, with the usual Queen Anne mix of clapboard and shaped shingle siding.

CU25 **Burlingame–Noon House**

c. 1800–1815. 1835–1840, remodeled. 3261 Mendon Rd. (south of Nate Whipple Hwy.)

The Burlingame-Noon House was originally a five-bay Federal house of one and one-half stories only (the trace of its original roofline can be seen against the chimney). When enlarged to two stories, it was restyled in the Greek Revival mode by the first member of the Burlingame family to occupy it, as testified by the change from country Federal trim in downstairs rooms to Greek Revival upstairs (not open to the public). The rear ell dates to the same time. The Noon family's long tenancy in the house, from 1866 to 1971, accounts for the hyphenated label.

CU26 **Luke Jillson House**

c. 1775–1792. 19th century, additions. 2510 Mendon Rd. (at Boardman Ave.)

This imposing, somewhat austere house is possibly the most sophisticated residential design in Cumberland up to its time. The five window bays, with generous six-over-six sash, are evenly distributed over an expansive front. The entrance is broad and quite severe in its restricted ornamentation: plain pilasters supporting a cushioned entablature, which is surmounted by a pediment. The Jillson House proclaims a concern for propriety and impressiveness, so different from the beguiling idiosyncrasies of the nearby Miller House. No cabinetmaker participated here. The master builder prevailed. The same handsome proportions and severity of treatment characterize the Jillson interior (not open to the public), which has a broad central hall with rooms front and back on either side and chimneys centered in the wall between them. Fragments of stenciled patterns discovered in upstairs bedrooms have been reconstructed by Shirley Houde Armstrong. A rear ell attached to an Italianate tower is obviously from the mid-nineteenth century.

CU27 Philip Tomas(?)–Lewis Tower House

c. 1825. 2199 Mendon Rd.

Masonry below, clapboard above, this L-shaped, low-gabled house is built into the side of a hill at two levels. Exceptional for Rhode Island is the complete screening of the two outside walls of the *L* by porches, giving an effect more expected in the mid-Atlantic or southern states. Of the two tiers to the front, the upper porch only runs the full length of the *L*: across the front, then folded along the ell addition on the upper terrace, for which the extension of the ground-story masonry provides the base. The crudeness of the simple, widely spaced, square-posted porch supports takes on a degree of elegance from the lightness of membering characteristic of Federal design. Elegant, too, is the right-angled junction of the gables. They make the L-shaped mass not a simple collision, but an interlock, in which the front gable (as a gable-on-hip-roof) intersects a simple cross gable coming in over the rear ell, but so seamlessly as to maintain a continuous pitch for the L-plan roof of the upper porch. As a final grace note, the carpenter capped the upstairs porch posts with strips of wood to make a primitive Tuscan or Doric allusion under the eaves.

The principal living floor has always been upstairs—another rarity for the period. Each leg of the gabled *L* has its central chimney core, with back-to-back hearths for parlor and dining room on the upper level in front, for kitchen and (originally) a bedroom behind. Aligned with each chimney is an entrance, one at the ground into the basement level at the front of the house, the other off the side terrace. This house fits both site and function with the nicety and abstraction of a morticed joint. In virtually every respect its qualities oppose those of the Jillson House (preceding entry). Although it is customarily known as the Tower House because of this family's long ownership, the Towers purchased the property only in 1833, when the house already existed on it.

Woonsocket (WO)

Legend has it that Woonsocket means "thundering mist," in celebration of the largest waterfall on the Blackstone River. But mist hovers as much over the legend as over the falls. The name seems to have been given to nearby Woonsocket Hill (in the adjacent town of North Smithfield) before it was extended to the area around the fall. The architectural pilgrim who nevertheless chooses to embrace the myth and seeks out the Thundering Mist as a spot sacred to the Industrial Revolution will again be frustrated. If Pawtucket's fall is buried beneath a highway bridge, Woonsocket's grander fall has, since an early 1950s flood, been dammed to flow over metal gates which can be lifted when there is danger of flooding.

Woonsocket began as a cluster of six independent mill villages along the southern border of Massachusetts: Bernon, Globe, Hamlet, Jencksville, Social, and, at the falls itself, Woonsocket Falls. The villages depended not only on the water of the Blackstone, but, to a lesser extent, on that from the confluence in the vicinity of three tributaries: the Mill River, Peter's River, and Cherry Brook. Except for Jencksville, located on one of the tributaries, the villages clustered in a rough arc on the Blackstone at and below the fall. They had more riverfront because the Blackstone here interrupts its generally southerly course by making an abrupt eastward shift in a loose, winding *W*—a sign, perhaps, to those searching for the magic of legend in "Woonsocket," that, whatever the name for this place, it was fated to begin with this letter.

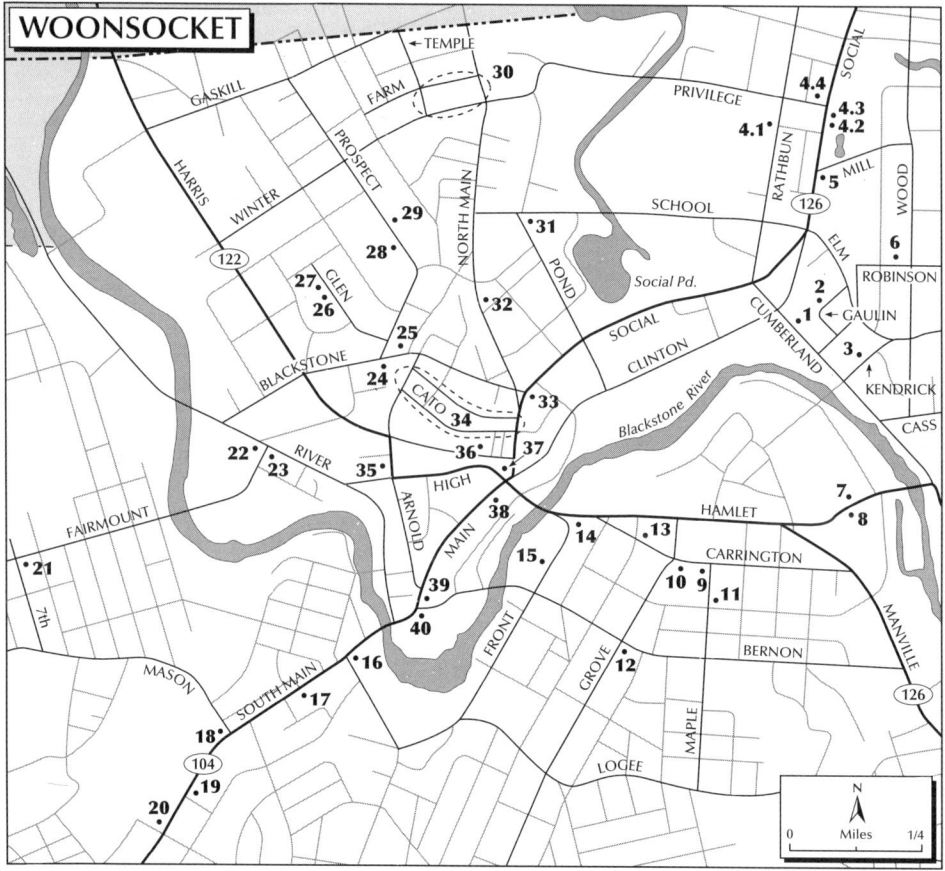

At least until the latter quarter of the nineteenth century, the six villages were distinguishable as entities. Five took their names from the companies which dominated them, except that the owners of the Woonsocket Manufacturing Company named their village Bernon (formerly Danville, a derivative of the name of a former owner) and the mills there were commonly known as the Bernon Mills. Woonsocket Falls, however, contained a cluster of mills with various owners. Here, the commercial and professional center for the future city developed.

During much of the nineteenth century, what is now Woonsocket belonged to two other towns. East of the Blackstone it was part of Cumberland; west of the river, part of a very much larger Smithfield than the Smithfield of today. Public clamor for independence in what was becoming an urban enclave amidst agricultural surroundings severed Woonsocket first from Cumberland, in 1867, then from Smithfield, in 1871. In 1888 it was incorporated as a city.

Still, the six villages long maintained their identity as sections of the city centered in the grandly grim prominence of their leading mills, which also elicited competitive pride from those who lived around them. At the beginning of the twentieth century, no

Rhode Island industrial center better displayed the emergence of the mill city from a cluster of mill villages. Unfortunately, the greatest mills are mostly gone. The old village labels linger on among the city's oldest inhabitants, but disembodied, as though searching for something to identify—and for how much longer?

Mention of a few of the principals in the changing cast of owners among the mills can wait until we reach the buildings they commissioned. But two towering figures among Woonsocket's industrial elite deserve special notice in connection with this tour. By the mid-nineteenth century Edward Harris, a member of the enterprising quarrying family in Lime Rock (see under Lincoln), rose to power as both an industrial and a civic leader. Starting as a clerk in a relative's firm, he soon owned mills in Woonsocket Falls which introduced woolen manufacture into what had been up to then a cotton manufacturing town and, by his example, ultimately made Woonsocket a major center for woolens. After buying up a number of nearby mills, he eventually built a new complex on a nearby site on a tributary of the Blackstone, which he called the Privilege Mill. On the slope above it, he also provided an extensive village of brick housing in the manner pioneered by the Blackstone industrialists downstream (see under Lincoln and Cumberland). From an Italianate mansion, eventually mansarded (and long gone) at the center of an extensive hillside estate in the northwest corner of the city between his two plants, he overlooked his factories, where the bulk of production gradually shifted to the Privilege plant and away from the congestion of the older operations. The site of his estate—roughly bounded by Harris Avenue and Winter, Prospect, Spring, and Blackstone streets—as subdivided after his death became, around the turn of the century, the city's choicest residential district. It maintains this character today. As a philanthropist, Harris sponsored a number of projects in Woonsocket. The most conspicuous, the Harris Institute, served as a combined educational and social institution for mechanics.

During the latter nineteenth century and into the twentieth, Joseph Banigan was Woonsocket's leading industrialist. Beginning as an Irish immigrant, he enjoyed a Horatio Alger rise to wealth and prominence even more substantial than Harris's. From a small rubber operation which he established with other partners to provide rollers for a clothes wringer manufacturer in the city, he reportedly became the leading manufacturer of rubber footwear in the world. Eventually, he was among the founding members of the consortium which organized U.S. Rubber, of which his Woonsocket Rubber Company was a major component. He also celebrated his success by building the first "skyscraper" in Providence (see PR023).

As elsewhere in industrial Rhode Island, Woonsocket shared in the successive waves of immigration beginning with the Irish, who came to work on the Blackstone Canal and stayed to work in the factories. But French Canadians, who began to arrive in substantial numbers beginning in the late 1860s, have played a particularly conspicuous role in Woonsocket's history. By 1930 those of French Canadian descent made up 70 percent of Woonsocket's population. As the city with the largest French-speaking population in the state by 1880, it long supported the only French-language newspaper in Rhode Island. By 1900 French predominated as the language of the city. The Union Saint-Jean-Baptiste, founded in 1900 and eventually located in a well-designed Neo-

Renaissance commercial and club block (1926), was organized primarily to provide vigorous support for French language and culture and for Catholicism. Designed by an important local architect of French Canadian descent, Walter Fontaine, the building still stands (with a 1975 entrance and front on what had been its rear elevation in order to orient it to the new downtown). In 1924–1929 the Sentinelle movement (abetted by the Crusaders, a secret society organized in 1920) came close to violence in upholding French Canadian interests. It was not surprising, then, that earlier, Aram J. Pothier, banker, manufacturer, staunch Republican, the first French Canadian mayor of Woonsocket in the 1890s, and later governor of the state, also became the United States commissioner to the Paris expositions of 1889 and 1915. In this capacity, he convinced French and Belgian manufacturers to build worsted mills in his "French" city, which resulted in several French-influenced mill designs, as well as two very French-inspired mill office buildings that are unique to the state. The French hegemony in Woonsocket did not abate until the 1950s, although French language and culture remain substantial but dwindling forces in the city.

Because it began as a cluster of self-contained and self-aware villages in a fairly isolated location, Woonsocket probably had more company-built housing than either Providence or Pawtucket during the early decades of textile manufacturing. What remains of consequence are fragments of housing erected by two companies only, the Globe Mills and Harris's Privilege Mill, the latter especially well preserved. As in other areas which became urbanized, Woonsocket manufacturers increasingly left their workers to fend for themselves in the free rental market. Two-family houses were the rule until the appearance of the triple-decker around 1885. This Massachusetts import appeared in all northern Rhode Island industrial towns, but seems to have been most prevalent, in terms of total residential building, in Woonsocket and Central Falls, followed by Pawtucket and Providence. Most have been painfully modified, especially their difficult-to-maintain stacks of spindly porches, which, of course, are their preeminent mark of distinction. Even so, Woonsocket, where they were built through 1930, is still the best place in the state to see the type in all its variations, and especially its extra-large versions designed for more than the typical three-family capacity.

Although many of the old commercial buildings along Main Street dating from around 1870 through the 1930s survive, with some gaps, the street had already become moribund before the displacement of the city's downtown. This occurred with the demolition of a run-down area of what was once the village of Social (including two blocks of fine mill housing shown in Henry-Russell Hitchcock's *Rhode Island Architecture*) to set up a listless mall type of commercial district which is neither downtown nor mall. Too bad, because Woonsocket's Main Street was enlivened by a dogleg angle with public squares at either end and another at the angled joint. Each celebrated a community function: Monument Square with its memorial of 1870 to the Civil War dead (the first such monument erected in Rhode Island) at one end; Market Square, with its backdrop of mills, at the other; Depot Square, with its Victorian station and the nearby Harris Institute, at the hinge. Just a shade too late, there is revived interest in the old Main Street as a possible tourist attraction and as a visible center to the city.

With the demolition of the majority of Woonsocket's most interesting nineteenth-century mills and the substantial loss of its commercial buildings, the largest Roman Catholic churches assume exceptional visibility as monuments in the city. They are threatened, too, by their grand size and declining congregations, and their adjunct complexes of rectories, convents, schools, and community centers are, in many cases, already substantially devoted to other uses. New waves of immigrants may replenish them.

Our tour circles first through what remains of the village clusters, then some of the big houses up on "the hill," and concludes downtown near the tamed fall.

WO1 St. Ann's Church

1914, Walter F. Fontaine. 98 Cumberland St. (at Locust St.)

This is the parish church for the neighborhood known in the early twentieth century as "the Coin" because its French Canadian population was centered on the *corner* of Cumberland, Social, and Rathbun streets. The area, which now resembles a suburban mall of parking lots dotted by office and commercial buildings, is also the obliterated site of Social Village, one of the constituent villages of Woonsocket. By the late nineteenth century, the Social Flatlands, as it was called, had become the commercial center of the French Canadian neighborhood. The widening of Cumberland Street from two moving lanes to four, with demolition along the route, has seriously undermined the sense of the old neighborhood, into which St. Ann's was tightly fitted, its twin towers with cupolas marking the spot within the city. The conspicuous towers also bear witness to the size and wealth of the French Canadian community in Woonsocket. The architect of St. Ann's was responsible for many of the important buildings in the city during the early twentieth century.

Unlike earlier prominent Roman Catholic churches in Woonsocket which display variants of the Gothic Revival, St. Ann's exemplifies the turn for precedent to the towered brick churches of late medieval and early Renaissance Italy, which became popular for Rhode Island Roman Catholic churches during the first decades of the twentieth century. Over medieval massing appears classical detail. The cream-colored stone trim of the Ionic-columned entrance porch and the Tuscan-columned embellishment of the bell towers only slightly relieve the looming expanse of yellow-beige brick walls. As with other Roman Catholic churches in Woonsocket, however, it is the interior that most claims attention. The first impression, of ballooning space, is derived from the deeply compartmented, barrel-vaulted nave, which culminates in a domed crossing surrounded by semidomes for apse and transepts—all extended in their spatial impact by tall, broadly arcaded aisles. Then, color. The tawny colors of the marbleized columns and pilasters capped with gilded Corinthian capitals were enhanced over time by French stained glass windows, designed by Charles Lorin of Chartres (installed in 1925). Later still, between 1941 and 1953, Guido Nincheri of Montreal filled the compartments of the barrel-vaulted ceiling with murals. The sacristy also contains a very large arched stained glass window (in shape, size, and, no doubt, intent reminiscent of Raphael's frescoes in the Stanze) which contains life-size, full-length portraits of a priest, curates, and altar boys made in the 1970s to honor a beloved pastor on his retirement from the church. This subject in this medium and at this scale seems to be unique in Rhode Island religious architecture.

WO2 Triple-Decker Tenements

c. 1900. 57, 67, 77, 10, and 24 Gaulin Ave.

Woonsocket is famous for the triple-decker wooden tenements which constituted, in the late nineteenth and early twentieth centuries, its most plentiful housing. Conventional three-family units were frequently expanded horizontally to contain six or even nine or more rental units; often the owner of the building occupied one of the tenement units, renting the others out, sometimes to relatives. Really small apartment houses, these larger versions generally maintained a three-story height, although a few occasionally rose another story, as seen in number 24. As here, they often assume a monumental character, with heavy cornices and flat roofs replacing the usual gable.

The tiers of porches ("piazzas" to their inhabitants) are particularly vulnerable to alteration. Despite some missing porches the tenements behind St. Ann's preserve one of the best surviving images in the city of a triple-decker neighborhood. The porch variously wraps a corner, repeating the contour of the bay windows behind, creates a centered, three-story pseudo-portico, or is set into or projects from corner stair towers. Number 10 has an especially long porch—six supports in width across its front, with a centered attic gable unusually popping from its otherwise flat roof. Not only do the piazzas relieve the overall boxiness of these large buildings, which the flat roofs emphasize, but in expanding the dwelling beyond its perimeter walls, they also provide a place outdoors that is public as well as private, thereby augmenting the sociability of the street. In the fragile openness of their form they contrast with the massiveness of the church, the parish house, and the stern red brick building that was formerly the parish school; but they also open to these institutions that once focused the life of this neighborhood. The little one-story building on the corner at the very top of the street used to be the neighborhood store.

A block over on Locust Street is a view of the subsidized housing that, in the early 1980s, replaced blocks of tenements of the sort that still remain on Gaulin Street, and thereby contributed to physical and social change in this area. Whatever the amenities of the new housing, a "project" image replaces that of the old neighborhood.

WO3 Kendrick Avenue School

1895, Forbush and Hathaway. 64–72 Kendrick Ave.

The schoolhouse as central block, with ranges of windows for classrooms, flanked by set-back entrance wings—originally one for each sex—is among the ubiquitous institutional building types of the late nineteenth century. Here, the expected symmetry of composition is undercut by the asymmetry of the silhouette of the bell tower over the left entrance and by variations in the wood paneling and door placement within the round-arched entrances. Both are typical devices of the Queen Anne style for creating playful surprise. The dark red brick of the walls is set off by the yellow-orange capping of doors and windows: arched over the entrance doors and second-floor windows, splayed lintels over the first-floor windows. Both red and orange brick maintain the continuous plane of the wall, except that a projecting rim relieves the planarity a little around the outer arc of the

WO2 Triple-decker tenements

arching, while the slightly stepped and pleated keystone at the center of the lintel works to the same end. So does the rough granite coursing which tops the set-out basement story, on which the entrance arches and first floor sit.

For the modern visitor, however, the special surprise of the school is the interior (open to the public only with permission from the principal's office). This is an almost perfectly intact six-room schoolhouse of its period. The central block is actually T-shaped to accommodate one high-ceilinged classroom on each of the two floors to the front, two per floor to the rear, with wide corridors the length of the building between. All rooms are wainscoted in varnished tongue-and-groove board with plaster walls, except that corridor ceilings are also of wood. Sunlight through windows at either end of the corridors turns the varnished wood of the wainscoting, the row of wooden pegs for coats, and the wood ceiling golden brown. The width and shortness of the halls make them not just passageways, but roomlike meeting places at the heart of the building. It is a far cry from the narrow, locker-lined, fluorescent-lit halls that are the bane of too many modern schools. After a century of teaching and learning, the building still seems to be appreciated by those who use it, its spacious classrooms apparently having adapted comfortably over time to changes in lesson techniques and equipment. Meanwhile, the basement contains the original boiler—a Fuller and Warner of Boston and Troy, New York, as announced with a flourish on its cast iron door.

Across the street is subsidized housing, designed as a linked "village" of gabled dwellings in stained wood. Here, however, the commendable 1970s effort to provide living conditions more congenial than the "projects" seems to have achieved only the look of informality rather than actual intimacy and privacy, and buildings which are neither urban nor rural in character.

WO4 Triple-Deckers

The neighborhood offers more triple-decker variety, beginning with a pair of prim, double-posted, hip-roofed triple-deckers at 330 and 336 Rathbun Street (WO4.1; c. 1923). On Social Avenue and on Privilege Street are three other treatments within the doubled triple-decker formula (although missing most of the original turned balustering). At 913 Social Street (WO4.2; c. 1915–1920), fully balustered porches run around an octagonal corner, the upper two at least once also topped by a spindled band. Next door is a standard triple-decker with its stack of (rebuilt) porches in front. Since both also have stacks of porches to the rear, the combined building contains twelve apartments. At 939 Social (WO4.4; 1915–1920) a six-unit triple decker raised over a rebuilt base for stores shows porches wrapping three walls—front, side and rear—interrupted only by a stair tower centered in a side elevation. Diagonally across the intersection, at 570 Privilege Street (WO4.4; c. 1895), porches run the width of the broad front and almost half of one of the side elevations, this time retaining a transomlike band of spindling for the top porch. Doors for three units are in front, with three more off the end of the leg of the *L* for units to the rear.

WO5 Jenckes Mills

1822, addition 1901 (No. 1 Mill). 1828 (No. 2 Mill and Mill St. Bridge). 96 Mill St., 767 Social St.

Though much obscured by remodeling and/or subsequent enlargement, these two stone industrial buildings, with their clerestory monitor lighting in the roofs, are the oldest extant remaining Woonsocket mill buildings. Built by three Jenckes brothers, who had earlier invested in the Social Mills, these buildings formed the nucleus of the village of Jenckesville on Peter's River.

Located at what was the upper privilege of Peter's River and built originally to produce cotton cloth, No. 1 Mill (WO5.1), with its random masonry walls, must have seemed, even before its near eclipse by the larger brick addition of 1901, much less imposing than the only slightly later No. 2 Mill (WO5.2), built of near-rectangular granite blocks. (About 1901 Joseph Guerin, a Belgian who was among the French-speaking European entrepreneurs attracted to Woonsocket in the early twentieth century, established his Guerin Spinning Company in No. 2 Mill.) Although the Social Street side of the second building and three sides of its tower have been covered with vinyl siding, it is still possible to appreciate the quality of the rugged construction of this mill from the rear, where three-quarters of the old building pops from this sheathing. All windows and the narrow monitor on the gable ridge are blinded, but the walls and overall shape remain. The size and regularity of the granite blocks is emphasized by the length of the corner quoins and of the headers over the windows. The two mills together with the dam (subsequently redone in concrete) that powered the No. 2 Mill and the bridge, whose stone-arched substructure still supports the revamped roadway, give some idea of the disposition of this early mill complex.

The three-story brick double house at 837–839 Social Street was a Jenckes mansion. Beneath the tier of porches which marks its conversion to a triple-decker is a grand Federal house. Housing at 752 and 842 Social Street also remains from the original village.

WO6 Triple-Deckers

c. 1905. 310–314 Wood Ave. (at Robinson St.)

WO6 Triple-deckers

These re-sided, roughly L-shaped triple-deckers are conceived as a pair, making the asymmetrical massing of each symmetrical. With their weighty wooden cornices scaled, it seems, to the pair of buildings, and their attenuated Tuscan porch columns, which become mini-Ionic at the topmost story, these buildings incline toward somewhat greater architectural pretension than most already described. This middle-class elegance seems evident, too, in the paired entrances with full-length oval windows and the shaped newel post, which appears to be a truncated version of the elongated porch columns. Visually, the two tiers of bay windows at the outer corners create a folded transition between the two stacks of porches, one for three apartments in front, the other for three behind. More important, they maximize light and air in the apartments. A third, similar but solitary tenement in this format exists on Robinson Street. Paired triple-deckers of this type may once have filled the lot to the corner, or at least that may have been their owners' intention.

WO7 Lafayette Worsted Company Mills

c. 1889–1930, Charles Loridan, engineer. c. 1900, company office. 134–160 Hamlet Ave. (mills), 150 Hamilton Ave. (at Hamlet Ave.) (office)

WO8 French Worsted Company Mills

c. 1907. 153 Hamlet Ave.

The remaining Woonsocket mills are late nineteenth- and mostly early-twentieth-century brick mills on either side of Hamlet Avenue, stretched along the Blackstone River, three to five stories, almost flat roofed, constructed in the pier-and-spandrel mode of the period. The extant twentieth-century mills built on the north side of Hamlet Avenue for the Lafayette Worsted Company and those on the south side for the French Worsted Company both had French owners who imported the latest in French spinning technology. They are the most substantial evidence of Aram Pothier's success in wooing French and Belgian industry to Woonsocket. In the case of the Lafayette mill, the complex itself is known to be of French design. In contrast to Charles Loridan's straightforward treatment of the mill, his design for the red brick company office with lavish granite trim offers French-inspired chateauesque flourishes which, by local mill office standards, are decidedly lavish: a basement striped in the two materials, pilastered walls with the suggestion of end pavilions, panels, and parapet, all topped by a mansard with decorated dormers, bull's-eye windows, and chimneys. It flaunts French expertise in matters of taste. Across the street, the French Worsted Company mill office, now a restaurant, was another nationalistic effort to show how things *should* be done—this more sculptural, with rusticated brick piers, giant keystones and brackets, and a floating pediment. Beside the very rational design of the mills, the offices seem intended to remind all who enter that here is a bit of French culture, here is style.

WO9 Charles Welles House

c. 1901, 215 Carrington Ave. (at Maple St.)

WO10 George Welles House

c. 1903. 199 Carrington Ave.

WO11 House

c. 1885. 74 Maple St. (at Willow St.)

Among the substantial late nineteenth- and early twentieth-century single-family residences on Carrington Avenue and the blocks around it are these three stylistically interrelated medium- sized houses representing early "stick" and late "shingle" Queen Anne and the Colonial Revival. The Charles Welles House displays typical shingle elements in particularly plastic combinations swelling from a first-story base in brick. A pyramidal-roofed near-cube at the heart of the house spins off a three-story conical turret to the front, which is wrapped by a deep, semicircular porch with angled entrance stairs, while a two-story bay with rounded corners supported from a bracketed base thrusts from the side elevation. Next door, at the George Welles House, home of a local lumber dealer, Queen Anne bays, wide doors, and a dormer with a bloated semicircular cap evocative of the Colonial Revival invade the bungalow form with its cobblestone chimney. At the corner of Maple and Willow streets is a restored example of Queen Anne style, in an earlier phase of this sequential development. It is angular, with exposed framing, spindles, a mix of shingles and siding, and radiating fan or ray motifs in the peaks of its gables. Taken together, these houses display how the Queen Anne style and its aftermath generated playful features in fanciful combinations to make homey images: from the angular exploitation of the structural frame, through the asymmetrical sculptural manipulation of shingled volumes, to incipient Colonial Revival in the introduction of colonial-derived features into an asymmetrical, essentially noncolonial context.

WO12 Grove Street School

1874–1876, E. L. Angell. 312 Grove St. (at Bernon St.)

In a characteristically Victorian manner, the simple shed-roofed porch of this school makes rhetoric of simple structure and function. The shed roof folds into a gable hood over the stairs; pairs of chamfered posts and bracketing extend the hood toward the visitor; diagonal struts on either side provide triangular fields for cut-stencil ornament and tongue-and-groove boarding. It makes a little ceremony of the act of entering this otherwise plain brick building, relieved, except at its door, only by a bracketed cornice and a minuscule belfry. Whereas the later Kendrick Avenue School (WO3) responded to a specific program of school requirements—differentiating the central classroom block from the entrances, articulating current preoccupations with sexual difference, and providing a conspicuous tower for the school bell—the older Grove Street School proclaims no more than its general institutional character. Still in use, it has been thoroughly remodeled inside.

WO13 Précieux Sang (Church of the Precious Blood)

1881. Copper belfry, early 20th century, Walter F. Fontaine. 92 Carrington Ave. (at Park Ave.)

Until this church was built, the French Canadians of Woonsocket had no parish of their own. Instead they worshiped at the Irish Catholic church of St. Charles Borromeo (WO32). Although Précieux Sang has a similar asymmetrical elevation with corner tower and gabled nave, unlike the earlier church, which is built in gray stone, this is of more economical brick with desultory white masonry trim. As in much polychromatic Victorian design, color changes emphasize the plethora of small-scale features—now a gable, now a pointed-arched opening, now a diminutive rose window—that seem to have been assembled piece by piece, as a collection of motifs, rather than to have been composed as a whole. The result, at this scale and with this degree of contrast in colors and materials, is quaint, but also compelling in the manner in which the components assert themselves as primal signs of "architecture" and "Gothic church," like pieces in a box of Victorian polychromatic building blocks.

Allegedly the design of Précieux Sang derived from that of St. Charles Borromeo, the composition of this facade generally recalling that of the earlier church. Inside are far greater similarities in the comparable handling of the pointed ribbed plaster vaults and their focus on a delightful Gothic Revival altarpiece. Here the saints stand in niches outside a miniature replication of a Gothic church which seems to have been lifted from an illuminated medieval manuscript. The balconies (once also in St. Charles Borromeo, but there removed) are here squashed under the aisle arcades with their railings nudging the capitals. They betray an unsophisticated designer, as do the tiny arched windows where a generous clerestory should be. So also the voussoirs of the vaults taking off from bracketed capitals tucked into the forkings of the arcade arches, the capitals recalling bird nests in the branching of a tree. Clearly the vaults do not consistently rest on supports that come to the ground, but arc as a plaster canopy, its lathing hung from wood or metal ties off the roof structure above. Such, indeed, is the case with most Gothic Revival interiors of the period—St. Charles Borromeo and St. Joseph's in Cumberland (CU9) providing two nearby examples. But one also senses in this interior the commitment of the parish and its designer to creating an impressive ecclesiastical setting which, like that at St. Charles, is today among the best-preserved mid-to-late-nineteenth-century Gothic Revival interiors in the state. Guido Nincheri, who much later painted the murals at St. Ann's (WO1), here executed the stained glass windows. Across the street, at 61 Park Avenue, stand St. Clair High School and Convent, the latter especially of considerable architectural quality, and now converted to housing. Taken together, they suggest the extent and visibility of the parish and measure the rise of French Canadian dominance in the culture and politics of Woonsocket.

WO14 **Woonsocket Courthouse** (Former)

1896, William R. Walker and Son. 24 Front St. (corner of Court St.)

This former courthouse is another public building by the Walker dynasty, whose members must have spent half their professional lives wooing politicians. Or it may be fairer to say that they understood how to serve the architectural ambitions of cities like Woonsocket by designing small-scale versions of big-city civic buildings, albeit in somewhat retardataire styles. By this date, classicism had become the sign of progressive institutions, and in civic buildings elsewhere, the Walker firm had tried its hand at the new style. The shift in taste toward classicism is here limited to some details of the courthouse porch and the columns in its tower loggia. Yet even they accord with the Richardsonian character of the design as a whole. In the Walkers' accomplished version of the style, a well-proportioned tower acts as a vertical anchor at the fold of the L-shaped, hip-roofed mass, between the curved termination of the courtrooms toward the rear and the dormered front. Banding in lighter stone links these quarry-faced volumes together. Inside, the original, plainly handled wooden paneling in courtrooms and stairs is unexpectedly modest given the pomp and massiveness promised outside.

WO15 **Woonsocket Manufacturing Company Mill Buildings** (Bernon Mills)

1827–1828 (No. 1 Mill), 1833 (No. 2 Mill). 110–115 Front St.

These two mills are the best extant examples in Woonsocket of early nineteenth-century stone mills. Owners who very shortly went bankrupt built the first mill, begun just a little before the Jenckes No. 2 Mill (WO5). It is the less imposing of the Bernon pair in its masonry construction, with uncoursed granite walls, rough stone headers, and corner quoins. But the profile of its slit monitor roof is handsome. Moreover, it is the first known mill in the United States (if present scholarship holds) to have been built to "slow-burning" structural specifications: that is, with floors and beams thick enough to char without burning through, thus sparing the machinery from a destructive crash into the basement in the event of fire. It became the standard for American mill construction into the first decades of the twentieth century, thereby hampering the development of iron and steel framing for American factories, because conservative fire insurance companies preferred to stay with what had been tested.

If the Bernon No. 1 Mill was straightforwardly committed to function, No. 2 had overt architectural pretensions. Sullivan Dorr and Crawford Allen, Providence entrepreneurs who were brothers-in-law, acquired the bankrupt mill and surrounding property as a result of the depression of 1829. The new owners rejected the functionally useful projecting tower, just then emerging as a typological feature for factory design (and visible at the Jenckes Mills, where the first mill was also rudely wrought, the second built with more refinement). They retreated to the traditional warehouse and barn placement of loading doors directly into the working floors. They also sacrificed the more useful top story of the monitored earlier mill in order to dignify their factory with the low pitch of a Greek pediment, even to attic blocks over the pediments at either end. Regularized quoining frames the handsome random-cut masonry walls. Most revealing, however, of the classical pretensions of this mill, because most subtle, may be the granite uprights for the loading doors. They appear with shadow capitals supporting entablatures, which are expressed as such by the decisive extension of the lintels beyond their supports, whereas most such supports are plain and part of a flush frame.

The image of this mill must be related to the vision that Dorr and Allen had for their mill village as a model community. It was they who named the village Bernon, after Huguenot Gabriel Bernon, a much admired ancestor of the Allen family. Not only was their mill yard (now a parking lot) once planted with trees and grass all the way to Front Street, but the cottages they provided up on the hillside, only one of which remains (208 Park Street), were reputed to be of above-average quality. Most remarkable, Dorr and Allen's vision of their model mill village was somewhat less paternalistic than that of most mill owners. They provided the opportunity for certifiably sober workers to purchase house lots in the village. How many of their employees took this temperance carrot for the apparent security of home ownership is unknown. As early as 1887 these mills ceased to be used for textile production. Even earlier, the ideal village had passed into oblivion, leaving scant evidence today of its appearance.

WO16 **Globe Mills Housing**

1874, boardinghouse. c. 1830 and c. 1865, tenements. 805–807 and 810–816 Front St., lower Lincoln St.

The mansarding of both towers and top floors of the tall Globe Mill complex at the horseshoe bend of the river directly opposite Market Square made it a dominating element in nineteenth-century panoramas of Woonsocket. It culminated the achievement of the brothers Dexter and George C. Ballou, who became the first big textile barons of the city. It is all gone now, except for scraps of its housing. The remnants of the Globe Mill village offer one of the best surviving remains of company-built mill housing dating from the early through the mid-nineteenth century.

At 805–807 Front Street is the mansarded brick Globe boardinghouse of 1874 (now run down), for single workers. Across the street, at 810–816 Front Street, is a much earlier example of the company's housing. Built in 1830 as part of the first development of the Globe Mills (begun in 1827), this tenement, though remodeled, clearly was designed in the Federal vernacular typical in this period. Around the corner on Lincoln Street, the two-and-one-half-story tenements (1865) that line both sides of the street reveal the longevity of this basic gable-roofed dwelling type.

WO17 **Woonsocket Armory**

1912, William R. Walker and Son. 316 South Main St. (at Providence St.)

In the Walker firm's clean sweep of the largest armory commissions in Rhode Island, this was the last—after those for Providence (PR195), Pawtucket (PA18), and one identical to this in Westerly (not included in this volume). Whereas the site encouraged an asymmetrical treatment for Pawtucket, this building, like Providence's on a relatively level rectangular plot, could be symmetrical. In the much larger Providence armory, the tall, castellated office and meeting blocks act as bookends for the drill hall between. Here the compact, cubic block of the armory's frontispiece rises above the low drill hall behind, which extends beyond it. Its hipped roof encloses a conventional metal-trussed spanning system. In addition to the blocky entrance tower in front, a turret guards every corner of the mass. Those for the entrance building are polygonal but near circular, and bracketed off the second story; those for the drill hall are square echoes of the main tower, but with chamfered corners and, like it, brought to the ground. Here again Walker demonstrates a nice sensibility for the way in which changes of tower shape and base enhance the expressive quality of these architectural chess pieces. In different ways, however, both the Providence and Pawtucket examples are sculpturally more impressive than the literalistic cornering of all towers here. The usual crenellation in sheet copper completes the masquerade of fortification. Sparing use of rough granite trim against the red brick walls provides accent and definition to the mass. It gathers around the entrance to invigorate the arched opening with a martial presence, although more bulk would have been more impressive.

WO18 Stone House

c. 1835. 383 South Main St. (corner of Mason St.)

One of the prestigious (and still handsome) residential districts of the turn of the twentieth century extends south of the armory along South Main Street. At the corner of South Main and Mason is what remains of Stone House and of its property, on which some of the later houses were built. It deserves notice only because it was once among the grandest residences in the city and among the most impressive Greek Revival houses in the state. Although it has been cruelly altered for apartments and re-sided in vinyl, some sense of the portico and of its monitor remains. Built by a Massachusetts mill owner who was also proprietor of the Woonsocket Hotel, by 1850 it belonged to George C. Ballou, half owner of the Globe Mills in the valley below.

WO19 Church of the Holy Family

1909. 424 South Main St.

Earlier than St. Ann's (WO1), Walter Fontaine's Church of the Holy Family is mildly Romanesque Revival, with a soaring brick tower piled on top of its triple-arched portico. The cavernous, barrel-vaulted space within again shows Fontaine's interest in color; but, in comparison with St. Ann's, is underelaborated.

WO20 George H. Baker House

c. 1890. 473 South Main St. (corner of North Ballou St.)

Most interesting among the houses in this area may be this Queen Anne house, with its patterned shingling of the second story over clapboarding below, its porch wrapped around a conical tower, and its projecting stair bay, which features a stained glass window at the half-story landing. It is more formal, more classical, and compositionally more subdued than its counterparts on Carrington Avenue (WO9, WO10). George Baker, a manufacturer of ladies' and children's cotton "ribbed" underwear, by 1900 celebrated his mounting prosperity by moving his family to a bigger house on Prospect Street.

WO21 Swedish-Finnish Lutheran Church (Former)

c. 1917. 531 Fairmount St. (corner of 7th Ave.)

Fairmount Street is the focal street of Fairmount, laid out in the late 1870s and early 1880s as a working-class subdivision. Like Gloria Dei Church in Providence, this is another of the Scandinavian (usually Lutheran or Methodist) churches located near machine tool plants, to which these skilled immigrants gravitated. (Several such factories existed nearby.) And just as Gloria Dei demonstrates in a large way, and at a slightly later time, the Scandinavian craft sensibility, so this tiny church does so in a small way. Both are exceptions to the plainness of most Scandinavian mill community churches; but whereas the craftsmanship of Gloria Dei derives exclusively from sophisticated Swedish inspiration, this seems to mix more folkish aspects of native handiwork with borrowings from American Arts and Crafts.

Evidences of the craft approach in this curious but charming church appear in the sawn stencil ornamentation at the eaves, the patterning of the roof slate, the turned porch posts, and the coy domestication of medieval allusion. But this prettiness is countered by strange proportions and unexpected combinations, for example, in the swollen dumbbell contours of the stubby porch supports, which are spread farther apart than the allusion to a small domestic "stoop" leads us to expect. The porch is also overly deep. So the little image acquires grand dimensions. Or consider the various uses to which the miniature buttressing in brick against stuccoed walls is put. It steps up the walls to the front, dividing a long band of windows, and supports the roof as one expects it to do. But buttresses also rise on either side of the porch through the roof to become the supports of a bellcote. Again a diminutive image is puffed in scale. On the side elevation, the but-

WO20 George H. Baker House, left, and WO21 Swedish-Finnish Lutheran Church (former), right

tresses step up the wall to either side of a wheel window, taking the measure of its circumference as they merge with it, to become its frame (or perhaps, like calipers, to measure it). This extraordinary building served several other Protestant congregations before its conversion to apartments.

WO22 **Woonsocket Rubber Company, Alice Mill**

1889. 85–87 Fairmount St.

WO23 **Desurmont Worsted Yarn Mills**

1907. 50 Water St. (corner of Fairmount St.)

Here are two handsome brick pier-and-spandrel factory buildings, both in the manner which became standard for mill construction at the end of the nineteenth and the beginning of the twentieth century. Joseph Banigan's rubber plant, named for his mother, is fronted by two severely handsome, hip-roofed Italianate towers, one with an open loggia for the company bell, the other closed around its water tower. Wings to the rear more than double the space of the front block. Unhappily, distant views are best because closer inspection reveals that all the windows have been altered.

The nearby Desurmont plant was another of Aram Pothier's French or Belgian imports. About it there is a faintly alien air. This may be wishful vision, as there is no present evidence that Gallic clients or engineers conditioned the result. But in contrast to the typical American preference for verticality seen in the elevation of the Alice Mill, in which piers rise to the eaves, here corbeled cornicing stops the piers beneath the eaves. As the segmental arching of the windows bites into the underside of the corbeling, they take on a marked horizontal rhythm, like stones skipped on a pond. Within the windows, too, the horizontal divisions of the sash are more emphasized. So it would seem that whereas the typical American treatment for brick factory construction at the time skeletalizes and verticalizes the wall, the treatment of the Desurmont holds more to the block of the building and the extent of the wall. In the Desurmont chimney also note the elegant simplicity of transition from its boxy base to its tapered cylinder.

WO24 **First Baptist Church**

1891, (?) Butterfield, Darling Brothers, builders. 298 Blackstone St. (opposite Spring St.)

The soaring, sheer-surfaced brick Italianate tower captures the view down Blackstone Street, making this Protestant church a landmark in the city (and a dramatic contrast to the onion-domed tower flanked by similarly domed turrets, of St. Michael's Ukranian Catholic Church (1923), nearby at 394 Blackstone Street. An open-arched belfry and four clock faces climax the Baptist tower, both capped by the broad flaring eaves of the tower's pyramidal roof. Its dramatic vertical rise is emphasized by the exaggerated contrast with the low, roughly L-shaped cluster of gables and other roof shapes which make up the body of the church, making the tower seem more appropriate for a mill than for a church. Sheer surfaces throughout are sparingly but effectively interrupted by simply shaped openings, slight changes in plane, and occasional granite trim. The rough finish of the granite provides startling accent to the hard smoothness of brick overall. One of the two entrances is located at the base of the tower as the terminus for a run of three round-arched windows; the other is set back into an enclosed porch, its gable outlined in a ragged silhouette of the rough granite trim. Between them is a single, enormous stained glass window.

Surprisingly, the radical asymmetry in the positioning of the entrances is resolved on the interior, where they give access to a pair of diagonally placed doors which symmetrically fit into the entrance corners of a near-cubic auditorium space. This focuses on a two-tiered niche for the pulpit, backed by an elevated organ. At the pulpit level the space curves inward in plan as though to make a semicircular apse; above, at the level of the organist's bridge, a circle cuts into the inward curve of the apse to make a frame for the pipes, which suggests a Japanese moon gate. It may, indeed, reflect the Japanese vogue in Queen Anne design. In keeping with such taste, the auditorium of the church is spanned by a lightweight structure of finely turned and polished wood, with huge openwork brackets at the four corners, each with its own large circle. They support the ceiling (or so it appears—actually girders and tension rods are probably concealed overhead). Taken together, apse and bracketed structure seem to make this interior unique in the state.

Knowing nothing about the architect, one is tempted to relate the church to some of the more idiosyncratic examples of the English Queen Anne Revival; it may have been designed by a British emigré or inspired by a plate from an English publication such as *Builder* or *Studio*. Since the architect's office was in Manchester, New Hampshire, he may have been inspired by brick mills as well, although probably more by osmosis than intent.

Next to the church, at 312 Blackstone Street, is yet another unusual building, a one-and-one-half-story Greek Revival house (c. 1840) in which what should be a pediment becomes a full half story, pulled out over its row of Doric columns, for an effect at once ingenious, charming, and provincial.

WO25 Duplexes

c. 1915. 325–365 Blackstone St.

This row of three shingled duplexes with a double-gabled front elevation subsumed within a spreading hipped roof and set well back from the street up the slope were probably overseers' houses for the Social Mill down the hill. In setting, design, and spaciousness there may be no more livable mill housing in the state, especially in an urban situation.

WO26 Charles A. Proulx House

c. 1915. 183 Glen Rd.

Glen Road, in the area that was Woonsocket's most exclusive residential district from the mid-nineteenth through the early twentieth century, contains a number of fine residences. Four overlapping influences, plus elegance of execution, account for the exceptional interest of this house. To dispose of those of least significance first: the tiled roof and stuccoed upper floor suggest at least a glance toward the widespread influence of Spanish Colonial Revival (or Mission Revival) architecture, which was popular in other regions at this time, and occasionally seen as an exotic import in Rhode Island. But in this house Mediterranean influences are naturalized by such Neo-Colonial allusion as the prim symmetry, the generalized Palladian motif for downstairs windows, and the Neo-Federal fanlight over the door.

WO26 Charles A. Proulx House

WO29 Thomas Thurber–Rachel Harris Rathbun House

Of more consequence, however, are two other influences. This is among the rare examples of midwestern Prairie School design in the state. Prairie School hallmarks appear in the hip-roofed, markedly horizontal block, its horizontality accentuated by the layered treatment of the wall. A high brick wall, defining the base of the block, rises to the bottom of the second-story windows. There the wall surface changes to stucco, making a friezelike band which contains the upstairs windows between the top of the base below and deeply projecting eaves above—the treatment akin to that of Frank Lloyd Wright's Winslow House (1893). Here, the brick was originally tawny yellow, the stucco probably left in a sand finish. This "truth to natural materials" would have increased its midwestern tang, before white paint colonized the effect in favor of New England propriety. Typically Prairie, too, are the brick pier supports for the flanking sunroom and screened porch. Finally, a filigree of pergola elements over the three porches relieves the blocky qualities of the house. Other houses of the period display what might be characterized as the pergola mode, but none more elegantly and conspicuously. This, too, could have come from the mid- (or far) west, but is broadly associated with the garden interests of the Arts and Crafts Movement. With the return of pergolas and trellises to popularity in the 1970s and 1980s, this handsome example of the pergola mode takes on renewed interest.

WO27 **Walter F. Fontaine House**

1925, Walter F. Fontaine. 211 Glen Rd.

Built in the latter part of his career as Woonsocket's premier architect, Walter Fontaine's own residence confirms his interest in understated stylistic elements (here nominally Neo-Tudor), his understanding of brick, and his skill in designing picturesque compositions. A two-story bay window locked into the entrance shelter is the fulcrum for a symmetrical vertical composition of openings under a tightly contained gable to the right and horizontal asymmetry under a skewed gable to the left, with one slope making a long slide over a glazed porch or garden room.

WO28 **Congregation B'nai Israel Synagogue**

c. 1967, Harry Ramsay. 224 Prospect St.

This synagogue exemplifies 1960s use of shaped structure and modern stained glass for what might be seen as a revival of design in a modern expressionist vein for religious architecture. Prevalent during the early 1920s, this tendency was mostly eclipsed by the rationalism at the core of influential modern architecture through World War II, then reemerged, perhaps especially during the 1960s. The synagogue sits high above the street, with stairs up to a terrace sheltered by an ovoid reinforced concrete slab lifted on seven branched concrete supports which symbolize the seven tribes of Israel. Behind, on either side of an axial entrance foyer, are temple and social rooms contained within a low box with angled ends. Composed of tilted reinforced concrete slabs, it is reminiscent in its overall shape of an overturned boat (or ark). On either side of the entrance (complicating the boat metaphor), the walls are slashed by rows of narrow, floor-to-roof gables filled with chunky stained glass in a modern, semiabstract design of superb color. Color and sheltering shape are the expressive ingredients of the plainly handled but sumptuous wood and glass interior.

WO29 **Thomas Thurber–Rachel Harris Rathbun House**

1867, Michael Volk. c. 1915, porch addition. 289 Prospect St.

Of brick with brownstone trim and a low, flaring roof, this may initially appear as a typical Italianate house. But none in the state—not Richard Upjohn's Edward King House (NE145) or Thomas Tefft's Charles Bradley House (PR183.2), however outstanding—presents as subtle an example of Italianate design as this tighter composition. With reason perhaps; it was a German emigré who designed it, fresh from the fountainhead of the Italianate style. For it was especially German fascination with the dream of a Mediterranean arcadia that led architects from other countries to appreciate the picturesqueness of medieval villas in the Italian countryside as they had been remodeled piecemeal toward Renaissance ideals. The wandering eye associated with the picturesque sensibility delighted in the mix of towers and porches, of round-arched and rectangular openings of diverse sizes, and of suavity of detail set against severe essentiality, as the total experience of the building emerged from cumulative comprehension of its fragmented ensemble. Here the projection of the freestanding arched mini-porch which fronts the square bay window for the parlor strikes the leitmotif for the design. Its combined prominence and relative uselessness give it symbolic intensity, as though its sole function were that of proclaiming the conceptual kernel from which the design emerged.

Consider the two porches: the symbolic porch against the real entrance porch. The sober elegance of the treatment conceals adroit variations within the theme. Both have wide arches toward the front with narrow arches either side. The gabled roof of the symbolic porch with its entablature of enlarged dentils is, in turn, set off against the flat cap over the entrance with its entablature of small modillions. Doric columns for the entrance porch play visually across space with a Corinthian column centered in the symbolic porch, but the latter is framed by severe right-angled piers. The severe forms frame the highly ornamented one, as a cushioned box intensifies our focus on a jewel. The contrast also alludes to evolution: from primitive stages civilization elaborates.

Upstairs, the triple arching, slightly projecting in its own field, is framed in primitive but elegant pilasters, the outside edges of which barely skim the adjacent quoining. Against the horizontality of this triple arching, the cartouche-like panel punctuated by the oval window over the entrance porch is vertical and exclamatory. So the melodious shift of parts occurs across the elevation and back and forth in depth.

Little is known of the client, who, in any event, owned the house for barely two years. It then became the property of the daughter of Edward Harris, Woonsocket's leading industrialist during the middle of the nineteenth century. Her husband, Oscar J. Rathbun, a banker, succeeded his father-in-law as president of the Harris Woolen Company and, in the 1880s, served as lieutenant governor of Rhode Island. Also significant is the semicircular appendage to the northwest added by a later generation. Although obscured by foliage, this is another fine example of the pergola mode, more monumentally conceived than the Proulx House (WO26).

WO30 Privilege Mill Housing

1864–1865. 26 and 74–112 Farm St. (between Temple and North Main sts.), 714 and 724–730 North Main St., 625 Winter St.

The Harris Woolen Company built eighty of these units to provide housing for its workers in the Privilege Mill complex, which then existed at the foot of the hill. What remains is the finest extant mill housing for nonsupervisory workers in Woonsocket. The houses continue the tradition of the Federal vernacular type found at the earlier housing of the Globe Mills (WO16), but in brick rather than wood. The granite foundations, the stone lintels over the windows, and the deep recesses of the doorways, lighted by very narrow transoms and side lights, reinforce the sense of solidity. They seem to depend on the Massachusetts tradition of brick mill housing that extended along the Blackstone River in Rhode Island and is today most vividly evident in Lonsdale, Berkeley, and Ashton (see under Cumberland). The six-bay houses that line both sides of Farm Street originally con-

WO30 Privilege Mill housing

tained three rental units each, one on the first floor and two units above cutting through the attic. Number 26 Farm Street (now separated from the others) differs from them. It is smaller and set gable end to the street. Below on North Main Street is a row of small, one and one-half-story double house variants. At 625 Winter Street is a single-family house with cross gables, probably for an overseer. The latter two were built a little later than the others. (Beyond the Winter Street intersection, at 373 North Main, is a brick building which served as the company store and eventually as a warehouse [1865]. Housing aside, it is architecturally the most important extant building from the original building campaign.)

To build his mill and housing, Edward Harris established his own brick kilns and sawmill on the site. Although buildings of the Privilege Mill complex still exist down the Farm Street hill, they are of later date or so altered that the crown of Harris's career can no longer be said to exist. The Privilege Mill is important, however, not only for its housing, but also because Harris specifically recruited his workers from French Canada, thereby initiating the influx of French Canadians into Woonsocket, although Irish apparently continued to predominate in his work force.

WO31 Gasometer

1865. 313 Pond St. (between East School St. and Mechanic Ave.)

This twelve-sided structure with hipped roof was once a new type of fuel storage building. Originally a windowless, paneled-brick polygon set on a stone rubble base, it was designed as a shelter against ice formation for an iron drum that held coal gas. Woonsocket once had three of these; gas was used for lighting the mills as early as the 1850s and for most houses and streetlights in the 1860s. Many of the larger or more isolated mills had their own gasometers during the late nineteenth century; one exists at the Wanskuck Mill in Providence). The Woonsocket structure is the only extant example in Rhode Island built by a public utility. With the tank removed, its brick shell now serves as offices.

WO32 St. Charles Borromeo

1862–1871. Patrick C. Keely. 189 North Main St. (corner of Daniels St.)

The exterior of this church is not ingratiating; nor is it meant to be. Its rugged rough-surfaced granite walls are unadorned. Its offset corner tower is belligerently buttressed and crocketed. Its smallish lancet windows and even smaller rose window poke into the walls without transitional moldings and no more than rudimentary tracery. Its deeply inset portals are similarly bare. Its stoniness is unrelieved by planting. As though by its formidable presence, St. Charles Borromeo announces the Irish as a major economic as well as an ethnic presence in the city's population (the cornerstone for this church was laid less than two decades after the erection in 1844 of a minimal clapboard building on the same site. In choosing Keely as its architect, the parish turned to the leading nineteenth-century designer of American Roman Catholic churches and cathedrals, who was, like most of the parishioners a first-generation Irish immigrant. Keely brought to America a solid knowledge of medieval building—specifically, of twelfth-century form. As was the norm in nineteenth-century Ireland (and during the Middle Ages for that matter), provincialism, relative poverty, and religious faith all combined in favor of unequivocal assertion over beguiling refinement.

Keely's architectural conditioning as well as his nationality proved advantageous for the commissions he garnered from working-class parishes in his new country. But so did his ingenuity. Consider how he met the demands of the sloping, triangular site by splitting the front elevation into three vertical folds: the tower at the apex of the plot, then a stepped fold back to the gabled nave, and another to the shed-roofed east aisle. Each component in the pleating received its portal, and each portal its individual calvary of stairs (the flight for the tower entrance being inside). The varied sizes and levels of the entrances are equalized elevationally by the common heights of their pointed arches. The interior is more congenial—another of those delightful mid-nineteenth-century plaster vaulted naves, and surer than its imitation at Précieux Sang (WO13). As there, the vaulting (actually suspended from the concealed roof structure) seems to float off decorated brackets above the arcade without even a pretense that the ribs of the vaulting evolve as a structural extension of the columns below them. The interiors of the two churches are similar, except that Précieux Sang has retained its balconies over the aisles. Both churches possess their Gothic Revival altars virtually intact. Again, the saint images are housed in an exquisitely detailed, mini–Gothic church, this one being interior (or cross-sectional) in quality, whereas at Précieux Sang the saints seem to stand outside their church. Both St. Charles Borromeo and Précieux Sang boast unified programs of stained glass. Both have suffered no more than minor alterations, except that, to its benefit, St. Charles lost its balconies in 1928. Much of the lumber went to enlarge the organ loft over the entrance, and some into wainscoting for the nave. Spatially, St. Charles is ampler than Précieux Sang, which the elimination of the balconies enhances. Possibly both churches are now lighter in color than they once were because of repainting, which doubtless also erased some stenciled ornament. Even so, Woonsocket is fortunate in having two such nineteenth-century ecclesiastical interiors so beautifully preserved, one evolved from the other.

WO33 **Monument Square**

Junction of Main, Blackstone, and Social sts.

Monument Square is the northernmost of the three street nodes—Monument, Depot, and Market squares—which mark the ends and middle of an angled Main Street, in what was a nice urban sequence of open space / corridor / open space /corridor / open space. The corridors are now sadly gutted by demolition and neglect, although the revival of Main Street is presently a city goal. The 1870 Civil War memorial (by the Hartford sculptor J. G. Batterson) at the center of Monument Square was the first erected in the state. At 14–22 Monument Square is the Stadium-Costa Building, a combined theater-store-office complex commissioned by Arthur Darman (1926, Perry and Whipple). In the lobby, Dutch tiles and a mural, *The Progress of Woonsocket*, by Maurice Compris, provide the introduction to the well-preserved Neo-Renaissance fantasy of the theater.

WO34 **Cato Hill**

1834 and later. Church St., Cato St., and connecting cross streets

Greek Revival houses probably designed for the upper levels of the working class, skilled artisans, or small entrepreneurs densely line these narrow streets. This enclave was laid out by an African American, Cato Willard, and his wife, Lydia Brayton Willard, beginning in 1834. Despite the development of the central city around it, Cato Hill still remains a residential enclave, no doubt in part because of the topography, more recently because of an active preservation program. No house deserves special mention; the significance is the ensemble and the resurrection of such a charming community enclave at the very heart of the city.

WO35 **Gas Station**

c. 1933. 339 Arnold St.

This diminutive stone slate-roofed building with wings extended to either side of a steeply capped octagonal tower, began as a gas station, its fancy architectural styling recalling a bygone era of American car culture. At one end of the building, where there are now offices and shops, was originally a garage door leading into a tiny service bay.

WO36 **Harris Warehouse**

1855. 61 Railroad St.

Built by Edward Harris, who subsequently commissioned the Privilege Mill complex and the city's first major civic building (now Woonsocket City Hall), this gabled structure of rubble masonry with quoined corners and a brick cornice is exceptional for its unusual shape. Not the rectangular building one would expect, it is curved to accommodate a railroad spur that entered the building through a large opening made by the brick relieving arch to unload bales of raw wool, which were stored in the upper floors of this warehouse. Dentils under the cornice are a minimal architectural elegance in this otherwise functional building. It served a cluster of factory buildings in the immediate vicinity, most purchased from previous owners, which housed Harris's business when he began his career, before he built his Privilege Mill.

WO37 **Providence & Worcester Railroad Station**

1882, John W. Ellis. 242–246 Main St.

This building, now converted to offices, is the third train station on this site, the first having been built in 1844. Although it has been unfortunately remodeled at its west end, walking up High Street provides a less spoiled view of its Queen Anne design, as well as an elevated perspective on the curved Harris Warehouse below.

WO37 Providence & Worcester Railroad Station

The station's most notable feature is the lightweight curved and chamfered wooden brackets that arch well beyond the walls in a series of spoked struts and extend the roof into a parasol shelter. The low brick gable with spired clock cupola over the interior waiting room exemplifies late Victorian manipulation of commonplace materials for decorative purposes. Brick in various patterns combines with carved stone, ornamental terra-cotta and, in the windows, panes in geometric patterns of greenish opalescent glass. In its simultaneous appeal to practicality and fantasy, this mix of materials and treatment complements the roof bracketing below. A locomotive once graced the weathervane.

WO38 Woonsocket City Hall
(Harris Institute)

1856, 1891. Mid-20th century, streetfront remodeling. 169 Main St.

Above and behind its partial encasement by the Neo-Romanesque addition (built when this, the city's first large-scale civic building, officially became the city hall), can be seen the tall, paired, round-arched windows of the original facade of the Harris Institute. The obscured building was commissioned by the industrialist Edward Harris. In both design and function the institute combined the pragmatic economic attitudes of a good businessman with the philanthropy of industrial paternalism. Built to look as if it were of fashionable and expensive brownstone with stone trim, the high-style Venetian Renaissance–inspired facade is in fact of stuccolike mastic, while the columns and surrounds of the windows are cast iron. On its top floor the building housed an assembly hall with 1,100 seats that placed Woonsocket on the national lecture circuit. (Abraham Lincoln spoke here in 1860 during his presidential campaign.) The floor below contained a "Sunday school"—not a religious institution, but a free school for mill workers, where they could learn to read and write and otherwise improve themselves on their days off. Street-level stores generated income to help finance the social and civic programs of the building. Modern streetfront excrescences to accommodate the city's bureaucracy seriously deface a building which deserves a better fate, for both architectural and historical reasons.

WO39 Market Square

1820s and later. Junction of Main, Bernon, Arnold, and River sts.

This square fits into one of the W-bends of the Blackstone, where the towered and mansarded Globe Mills directly across the river once provided a looming backdrop. Main Street enters at one corner, echoing the curve of the river's bend around two sides (and becoming South Main Street as it does so). Following this bend, and on alleylike spurs from it (notably on nearby Allen Street), is a huddle of old mill buildings (many altered) which were once substantially occupied by Edward Harris's woolen enterprise. Where there is now parking, an open-air market once existed at the center of mill activity, making it an area of intense but colorful congestion. At 1 Main Street (intersection of Main and Bernon streets) is the Lippitt (later Hanora) Mill office and warehouse (WO39.1; 1836), a fine, mansarded factory building in which the wall is virtually reduced to sheer brick piers crossed by continuous banding of rough granite into which expansive windows are fitted. Attached to it is the older Ballou-Harris-Lippitt rubble masonry mill (1827), centered by a low, clapboarded loading tower with a Greek Revival pediment. Built by George Ballou, it passed to successive owners who provide a roster of Woonsocket's most prominent woolen manufacturers. Fronting the complex is a piece of the granite-lined Lyman-Arnold Power Trench (c. 1827), which once served a number of mills in the vicinity. The image of the mansarded corner building, which Henry-Russell Hitchcock illustrated in his *Rhode Island Architecture* (1939), has inevitably lost some of its force through its conversion to senior citizen apartments, with unhappy window replacements and their elimination from the mansard. Farther around the semicircular Market Square is the Falls Yarn Mill (1850), another early stone-rubble mill, now restored.

WO40 Museum of Work and Culture
(Barnai Worsted Company Dye Works, later Lincoln Textile Corporation)

1919. 1993–1997, conversion to museum, Christopher Chadbourne Associates. 42 South Main St.

Discussion about such a museum arose during the city's centennial in 1988 and a designer was selected the following year; the economic recession of the early 1990s delayed funding and completion. The museum was originally slated to occupy the picturesque Falls Yarn Mill (see above), but as plans evolved the project was moved to this plain brick, flat-roofed, pier-and-spandrel building. A site along the Blackstone River Valley National Heritage Corridor, the museum organizes exhibitions and programs that focus on Blackstone Valley mill workers and their ethnic and cultural backgrounds.

North Smithfield (NS)

North Smithfield emerged as a separate town in 1871. Like Central Falls, Lincoln, and a portion of Woonsocket, it too had until then been submerged within a much larger Smithfield. The first settlement of consequence within present-day North Smithfield was established toward the end of the seventeenth century, when the area was politically a piece of Smithfield, but in reality a remote clearing known as the "North Woods" of Providence. Quakers settled what is now Union Village. A meeting of major highways in and near the village made it a professional and cultural center for the area, which it remained throughout the first half of the nineteenth century. A street of fine late eighteenth- and early nineteenth-century houses is its architectural legacy. Farther west and north, the Branch River provided for the manufacturing villages of Waterford (at the Massachusetts border where the Branch joins the Blackstone), then (moving upriver) Branch Village, Forestdale, and Slatersville. Of these, Waterford (now virtually merged with the village of Blackstone, Massachusetts) has little left of architectural interest, while Branch Village has all but disappeared. Forestdale contains some interesting architectural survivors (though not its mill). Slatersville, however, merits special attention. Of all attempts to create model mill villages in Rhode Island, it remains architecturally the most attractive, not only for its buildings, but, as important, for their arrangement as a community on a steeply sloping site. This band of villages across the northern edge of North Smithfield, just south of the Massachusetts border, contrasts with the rest of the town, which was formerly agricultural and, since the 1950s, has been largely suburbanized. Some sense of the agricultural past remains in an occasional early farmhouse, or, more rarely, along stretches of lesser-traveled roads—most evidently in a line of modest 1780s–1840s farmhouses along Grange Road.

NS1 Daniel Smith–David Andrew–Nelson Taft–Albert Todd Farm

c. 1740; c. 1800 and later. 241 Farnum Pk.

This is more impressive as a remarkably long house with a saltbox extension than for the nicety of its ad hoc additions. The center section is oldest, its in-line continuation on either side added about sixty years later. Each segment has its chimney, the first centered, the others at either end of the elongated structure. Windows are rather casually ordered, and some are odd-sized. Still, the ensemble and its reduced complement of outbuildings are both exceptional and picturesque (see FO31 for a comparable example of early in-line additions over time).

The successive occupants of this house present a generational saga of country trades through the entire nineteenth century (except for one city venture). Smith, a blacksmith, sold to Andrews, a millwright; his widow married Taft, a carpenter; their daughter chose Todd, a printer (the exception, who worked for the Woonsocket paper); his son became a lumber dealer and operated a sawmill on the property.

NS2 Tyler Mowry Farmhouse

c. 1825. 112 Sayles Hill Rd. (.5 mile north of junction with Iron Mine Hill Rd.)

Neglected through most of the twentieth century but consequently one of the least altered, most intact late Federal farmhouses in the state, this handsome house has an exquisite entrance: a high entablature stretched the breadth of a side-lighted door, each side light with flanking pilasters (paired at each end), each pilaster supporting a block which boldly divides the entablature into three parts, with an oval medallion centered in the wide section over the door. The facade of the ell has a similarly bold but much simplified entrance. The framing of the windows, all with original twelve-over-twelve sash, is as elegantly attenuated as the entrance and well proportioned to the facade. Extant examples of this Federal type are rare in Rhode Island, and the painstaking restoration of this house in 1996–1998 has recaptured a great treasure.

NS3 Milk Can

c. 1931. Eddie Dowling Hwy., .5 mile from intersection with Manville Rd.

This giant icon fashioned of flushboarding and sheet metal sits atop a circular dispensary for cones and shakes—or "cabinets" to native Rhode Islanders. More *architecture parlante*. Whereas Apex (PA14) illustrates a concept, here depiction is literal. The Milk Can was moved from a site in Lincoln several miles south which went to new highway construction.

Union Village

Woonsocket before Woonsocket! This little village seems originally to have gone by that name from its location on Great Road at the intersection of Woonsocket Hill Road. The first houses in the future village were built around 1690 by Richard Arnold, Jr., and his brother-in-law Samuel Comstock. They and other early settlers were mostly Quakers, and, in 1721, the Society of Friends built a meeting house (the predecessor to the present building). It was the first, and for a century, the only one in the Woonsocket area. Though on the Woonsocket side of Smithfield Road, which marks the boundary between Woonsocket and North Smithfield, it is associated historically with Union Village and so is described here (see NS4).

As the main street of the village, Smithfield Road (Route 146A) appropriately retains its colonial name of Great Road (in the eighteenth century, the road between Providence and Worcester). Its intersection in the village with Woonsocket Hill Road and Pound Hill Road also brought traffic from eastern Connecticut and from central Rhode Island. Moreover, a short jog at the village from either of these routes connected to highways linking Providence and Boston. Well before the end of the eighteenth century, the village at this junction of major crossroads became known as Smithfield.

Innkeeping began with the Arnold family during the eighteenth century. During the first half of the nineteenth century, which saw the pinnacle of the village's prosperity, at least three inns, which survive in residential use today (see NS10–NS12), occupied corners at the principal intersections. In addition to stores and blacksmith shops, the village was home to the Smithfield Union Bank, the first bank chartered in northern Rhode Island. Its establishment in 1805 was an event of such magnitude that village recognition gravitated toward the name of the bank, and Smithfield became Bank Village, then Union Village. (The bank building, much modified as a house, is located at 21 Pound Hill Road.) By the mid-nineteenth century two academies also existed in the village: Smithfield Academy for boys and Linden Hall Seminary for girls. Prosperity from all these sources, together with the professional and cultural life of the village during the first two decades of the nineteenth century, encouraged the construction of large houses. Most are set well back from the highway—so far that many retain generous lawns, even after the widening of the highway during the twentieth century. During its peak building period, Union Village was fortunate to have an outstanding resident master builder in

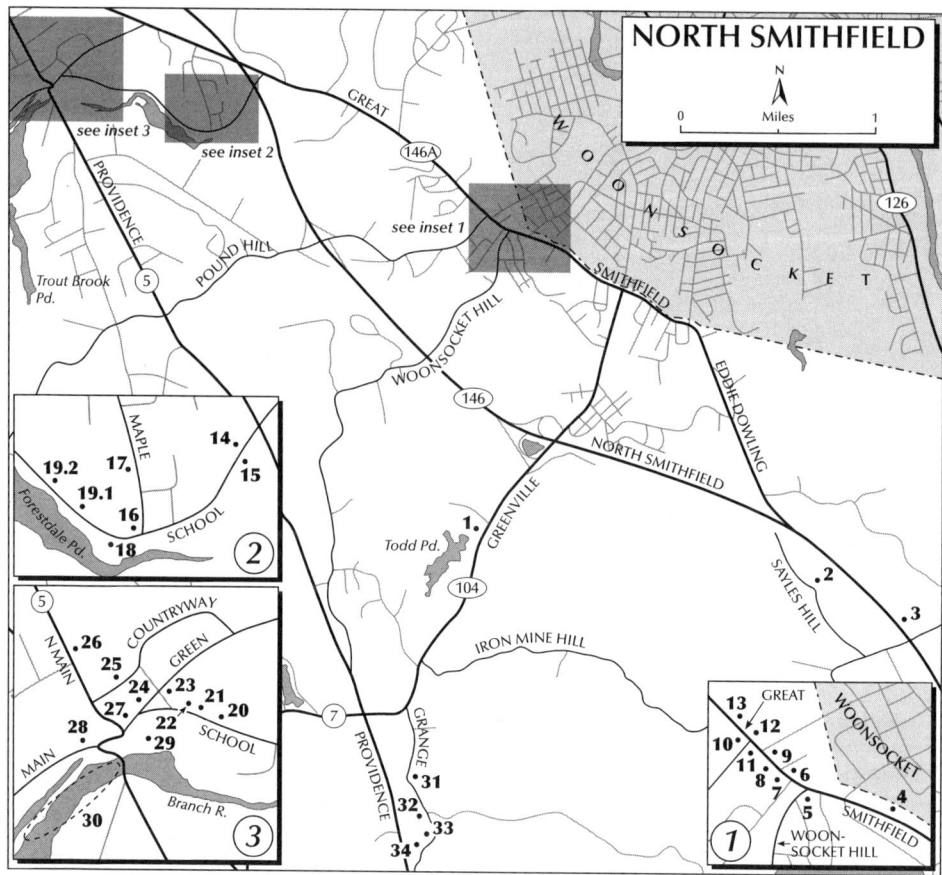

Walter Allen. His Federal houses and those of his disciples characterize the village and account for its appeal.

Bypassed by the railroad, which also snatched the bulk of Providence–Worcester traffic from Great Road, and without waterways, Union Village stagnated after the mid-nineteenth century. The bank left for Woonsocket in 1851; Linden Hall Seminary went to the same city (where it eventually folded). Smithfield Academy withered away. This economic decline, during which "old families" resolutely held onto ancestral homesteads, preserved Union Village for twentieth-century rediscovery as a suburban town, when colonial and Federal houses were once again prized. Thus, in addition to work by Walter Allen and other houses described in the following entries, houses in Union Village are worth examination because, with few exceptions, its building history virtually vaults the latter half of the nineteenth century. At the west end of the village are the Lapham House (1790, 212 Great Road) and the Anson Arnold House (1806), 188 Great Road, one-and-one-half-story Federal structures which provide convenient small house comparisons to the predominating two-and-one-half-story variety. The Southwick House (c. 1825), 171 Great Road, shows Greek Revival monumentality in mass, but still in combination with Federal refinement of detail. The full-blown Greek Revival appears in a barn (now a house; c. 1830), at 6 Pound Hill Road. After the Jacob Morse House (NS9) and another Italianate house (c. 1860), at 202 Great Road, building virtually ceased for half a century. Then, twentieth-century suburbanization brought the rediscovery of the old-fashioned charm of colonial and Federal houses: first shingle houses with white trim and a "colonial" feel to details, as in the Charles H. Stebbins House (1909), 61 Great Road, the George W. Lathrop House (1916), 91 Great Road, and a third at number 115); then two fine colonialized bungalows (c. 1920), 108 and 115 Great Road; and finally full-fledged Georgian Revival (c. 1935), 150 Great Road.

NS4 Friends Meeting House

1881. 108 Smithfield Rd.

John Arnold and other Quakers in the vicinity of Union Village erected the first religious building in northern Rhode Island on this site (then on Great Road) in 1721. It burned in the early 1880s, and this is its replacement. Plain for its period, it appears to be at least twenty or thirty years older than it is. But compared to the even more modest mien of earlier meeting houses, this one is unusual in Rhode Island for the scale of its windows and especially for the bold projection of its almost flat-roofed, sparely braced and bracketed shelter over the entrance. This assertiveness is countered by a certain gentleness in the sheltering quality of the hood over the pyramid of granite stairs and in the restrained handling of even the most Victorian aspects of the design. The delicacy of molding for cornice, paneling of corner boards, and the gentled labels is positively un-Victorian, as is the taut linearity of the window sash. Quaker decorum holds Victorian ostentation in check. An unprepossessing cemetery is gathered around the building. Opposite, Union Cemetery serves the village. Between them they contain graves extending back to the early settlement of this place and through the generations who subsequently inhabited it.

NS5 Apartments (Richard Arnold–Pelog Arnold Tavern)

1690; late 18th century. 4 Woonsocket Hill Rd. (southeast corner of Great Rd.)

The south end of this structure was the 20-by-20-foot Richard Arnold, Jr., House, which was the beginning of the village and of English settlement in the Woonsocket area. It was extended to its present size at the end of the eighteenth century by the patriot Pelog Arnold, who operated the house as an inn. He used its tavern as a recruitment center and depository for arms during the Revolution. Pelog Arnold rose to the rank of lieutenant colonel, served as a delegate to the Continental Congress, and eventually became chief justice successively of the Rhode Island Superior Court and Supreme Court. The building remained a tavern into the twentieth century, until its conversion into apartments.

NS6 Daniel Arnold House

1714, Walter Allen, master builder (or in his manner); c. 1800; later side wing and bay. 71 Great Rd. (northeast corner of Great Rd. and Warren Ave.)

This central-chimney house, among the earliest houses in the village, was radically enlarged and modernized in the Federal style, acquiring a two-and-one-half-story, five-bay facade. The exceptional entrance treatment is duplicated in two houses across the street, as well as in others in the village, giving it a distinctive quality. The door is topped by a semicircular light, usually set against a field of rustication bounded by fluted pilasters, all sheltered by a wide, pedimented porch supported on attenuated Tuscan columns. This entrance motif is associated with Walter Allen, the local master builder at the time; but the detailing seems somewhat less refined than that in houses known to be by him. So the house may merely reflect his influence. (In Rhode Island, this porch type seems to be limited to a few northern towns; wherever it appears, it can be assumed that there is some connection with Union Village.) The wing off one side and a bay window on the other are of indeterminate date.

NS7 David Aldrich House

c. 1808, Walter Allen. 20th century, porch and wing. 76 Great Rd. (diagonally opposite Daniel Arnold House)

NS8 Stephen Brownell House

1806 (rear portion earlier), documented as by Walter Allen. 86 Great Rd. (diagonally opposite Daniel Arnold House)

These are two more examples of the Walter Allen porch. In the exceptionally refined Aldrich House, the typical Allen porch appears at its best. The wide spacing of its tapered Tuscan columns and their responding pilasters tautly stake out the stretched space of the entrance shelter, with the rusticated wall further separating it from the clapboard mass behind. Within (not open to the public) is an outstanding stairhall that must have seemed a perfect setting for the young ladies of Linden Hall Seminary when this house served as their dormitory. For the Brownell House, next door, Allen chose

NS7, NS8 David Aldrich House, left, and Stephen Brownell House, right

a hipped roof instead of a gable set lengthwise to the street, which prevailed in Union Village, and was generally more conventional for large Federal houses. (Gables were cheaper to build, and, unless a monitor was added, provided more usable attic space. Hipped roofs, however, gave more elegance and monumentality.) The Brownell chimneys are centered to either side of the entrance hall, so the fireplaces are on the interior rather than the exterior walls of the principal rooms, as in the Aldrich House (where, however, chimney alterations have been made on the east wall). Here the treatment of the back wall of the porch is clapboard instead of rustication. Apparently exceptional for Allen's porches, it may represent the client's wish for economy, or even a later modification. Its roof forms a half-hip echo of the main roof, rather than the pediment echo of the main gable found in porches of related houses. Typical Federal twelve-over-twelve sash, like that of the Aldrich House, here were replaced by single sheets of glass in the lower sash, doubtless reflecting a preference popularized in Queen Anne houses at the end of the nineteenth century for an unobstructed "view" window with transomlike panes above. The small, arched window centered in the second story of the Aldrich House must also be a later alteration.

NS9 Jacob Morse House

c. 1851. 101 Great Rd. (northeast corner of Morse Ave., diagonally opposite Stephen Brownell House)

This Italianate house, the last major nineteenth-century house built in the village, displays scroll bracketing of various sorts under the eaves, over the heads of paired windows, and around the cornice of the unusual entrance porch. Paneled polygonal columns in a tapered curve support an undulant entablature in keeping with the curves of the bracketing. In this ambience of Federal reticence and attenuation, Victorian preferences for rambunctious showiness and plasticity are magnified. Even though this house is different in style from others in the village, its similarity of format and scale saves it from disrupting the community harmony. But observe how the Victorian sensibility typically concentrates the characteristic five-bay organization of colonial and Federal elevations into three grander openings, here by coupling windows to either side of a residual "Palladian" centerpiece.

NS10 Walter Allen House

1802, Walter Allen. Early 20th century, wing, altered rear elevation, and stable. 138 Great Rd. (northwest corner of Pound Hill Rd.)

NS11 Apartments (Seth Allen House)

c. 1804, Walter Allen. Early 20th century, basement story and front. 120 Great Rd. (southeast corner of Pound Hill Rd.)

NS12 Apartments (George Aldrich Inn)

1809. 1830 and 1902, cross gable, dormers, and bay window added. 127 Great Rd. (facing intersection with Pound Hill Rd.)

Walter Allen's house, the first he built in Union Village, established his basic design idiosyncracies. It was the site of the first Roman Catholic services in the area, held in 1829, and was later a tavern. Edgar M. Slocomb, who purchased the house early in the twentieth century, shifted the principal entrance to the rear, adding a Colonial Revival veranda and an upper-story Palladian window to that elevation. So Federal and Colonial Revival are front to back here. The simply detailed shingled barn was probably added at the same time.

In the nineteenth century the Walter Allen House was one of three that functioned as inns and taverns at the intersection of Pound Hill Road and Great Road. The two others, now much renovated, were the Seth Allen House and the George Aldrich Inn. The latter converted to the Smithfield Academy when, in the 1830s, headmaster James Busbee moved in and began to teach classes there. The original house was probably expanded for this purpose. Its porches are reduced variants of the Allen type.

NS13 **John Osborne House**

1812. 20th century, ell addition. 137 Great Rd.

Opposite the Walter Allen House is the John Osborne House, built by Walter Allen's son-in-law. It reflects the older tradition of center chimney and boxed-out six-over-six windows. Its projecting weather porch is certainly later, its door either much remodeled or possibly Neo-Federal.

Forestdale

Nothing but the footprint of Forestdale Mill remains, set deep in a hollow against the Branch River and obscured by growth. Built in 1860, it was one of the grandest stone mills in the state, boasting a monumental tower and an open cupola of superb proportion; a brick extension was added in the late nineteenth century. Its predecessor was a small scythe manufactory founded in 1824 by Newton Darling. Mansfield and Lamb built the large stone cotton mill, incorporating the adjacent scythe works. It was later taken over by the J. and W. Slater firm of nearby Slatersville, when French Canadians were brought to the town as its principal labor force, and still later, in the twentieth century, by a succession of owners. Textile production continued, sporadically toward the end, until the early 1970s. After the plant closed, it survived vacant for several years, until 1978. Then, with plans for rehabilitation for new uses underway, vandals torched it.

NS14 **North Smithfield Heritage Association** (Forestdale School, officially Branch School, District 3)

1877. 190 School St.

The Forestdale Manufacturing Company (headquartered in neighboring Slatersville) contributed the land for what is now the most handsome extant example in the state of the one-room clapboarded High Victorian schoolhouse. A pair of replacement stairs climb to flanking doors capped by heavy, projecting hoods on ornamental brackets. Lesser bracketed heads enhance and decorate the windows, while sparse double brackets embellish the cornice. By such accent the Victorian monumentalized even the simplest building forms. The building's present tenant, the North Smithfield Heritage Association, beautifully maintains it.

NS15 **Duplexes**

c. 1850. 175–183 School St.

This row of five Greek Revival duplex mill workers' houses dates from the period of the scythe works.

NS16 **Commercial Block**

1858. School St. at Maple St.

The two-and-one-half-story Forestdale commercial block compares with two similar and larger blocks in nearby Slatersville (see NS28). In these examples plain granite piers and lintels

NS14 North Smithfield Heritage Association (Forestdale School)

frame a row of shops, with red brick walls above (now painted white at Forestdale). If smaller than its Slatersville equivalent, the Forestdale version is more complex. It is actually two buildings fused together. One, with its gable end fronting on Main Street, provides an excuse for a stepped screening wall to mark the corner as the partial false-fronted climax of the building. The other, also gabled and slightly recessed, turns its flank to Main Street, with dormers punched into the roof slope. Whereas the front elevations are brick above the granite-piered storefronts, the Maple Street elevation is wholly of random stone. Nice, too, is the contrast between the flush granite capping of the storefronts and the patterned play of light and shadow in the brick corbeling which marks the cornice horizontals overhead. For a village building this one has surprisingly sophisticated lessons to teach about making a center, marking a corner, and giving the prosaic a touch of monumentality.

NS17 **Forestdale Mill Superintendent's House**

c. 1860; later alterations. 47 Maple St.

Behind the store block, a mildly Italianate, quite plain superintendent's house proclaims its ascendancy over the lesser workers' houses along School Street by its greater size and relative isolation. Paired brackets under the eaves and paired windows with peaked labels (except for a round-arched couplet window in the gable) are the sole concessions of this clapboarded box to historic style, although a replaced front porch and portal may once have further embellished it.

NS18 **Forestdale Mill Office**

1860. 125 School St.

The early Victorian office of the mill is heavily ornamented with double brackets and modillions at the eaves and corner quoins. A pedestrian bridge beside it once led to the main entrance of the mill. The sole survivor of the fire, it now overlooks the ruins, itself falling into ruin.

NS19 **Forestdale Mill Housing**

Opposite the mill office, at 126 School Street, is a one-and-one-half-story duplex (NS19.1; 1865), unusually in gray stone block striped with narrow blocks of darker color. It once served as a duplex residence for mill officers. At 102–110 School Street are more mill tenements (NS19.2; c. 1865), two-and-one-half-story this time and more altered than the earlier Greek Revival group on School Street (NS15). Unusually, paired central entrances for these face away from the road toward the Forestdale Mill Pond behind them.

Slatersville

Most would agree that of all Rhode Island mill towns, Slatersville is the prettiest, as its final owner intended it to be. The village, which is today the town seat for North Smithfield, began around 1800 as Buffum's Mills, named for the family who owned most of the land in the area, including a grist- and sawmill, as well as a halter's shop. Richard Buffum's house (1786) still exists at 9 Main Street, but much altered. After reconnaisance by Pawtucket's John Slater, William Almy and Obadiah Brown of Providence began purchasing land here in 1805 under the firm name of Almy and Brown, with the Slater brothers (John and Samuel) joining the company partnership in 1806. They thereby pretty much duplicated the situation which had brought this group into partnership in Pawtucket, where Samuel Slater, then Almy and Brown's employee, had set up America's pioneer mechanized cotton mill in 1790 (see Pawtucket introduction and PA15). By 1807 their cotton mill was operating, at its start one of only fifteen cotton-spinning mills in the country. Their mill village was the first in Rhode Island and among the earliest mill villages in the nation. Although fire in 1826 destroyed the original mill, it was immediately replaced with the present masonry mill. On the deaths of Almy and Brown in the 1830s, the Slaters bought the mill. John Slater became resident manager of the business until his death in 1843 when his sons John F. and William acquired full title.

Through the middle of the nineteenth century, Slatersville was the purely Yankee town it appears to be today. After the Civil War a small influx of workers from overseas, then a much larger immigration of French Canadians to work in the mills made Slatersville 82 percent Catholic by 1872.

The Slater family owned Slatersville until 1900, when a Boston banker, James R. Hooper, bought the village and used the pure water at the upper reaches of the Branch for dyeing, bleaching, and mercerizing cotton cloth. (Like other such plants at river headwaters, this assured that the water downstream was considerably less clean than the water it took in). It was the purchase of the village by Henry P. Kendall in 1915 and his interest in preservation and improvement which made it what it is today. What Austin Levy was to nearby Harrisville in his effort to create a model mill village (see under Burrillville), Kendall was to Slatersville. Although both sought to resurrect the old-time village community, Levy became the slightly more progressive of the two architecturally and Kendall remained the more nostalgic. Improving on the high quality of existing building and a handsome site on a ridge, Kendall and his longtime superintendent, Arthur Blane, built, restored, modified, and moved extant mill houses to make the picture they sought, while also attempting to disguise the institutional character of the housing so as to give more the appearance of individual houses. Tree planting and landscaping, including the area around the mill, were other Kendall contributions. His idyllic vision for Slatersville did not, however, prevent him from moving his operations to the South in 1956. Subsequent owners gradually let the splendid stone mill deteriorate. Recent building in the valley around the mill and on surrounding hills looking into the town is also mostly unfortunate. Even so, Slatersville retains so much of Kendall's vision of the ideal New England village that it is now a sought-after residential community, and preservationist sentiment, never wholly dimmed, intensifies.

NS20 George Johnson House, Caretaker's Cottage for the William Slater Estate

c. 1855, attributed to Thomas A. Tefft. 30 School St.

It is known that around 1850 the Providence architect Thomas Tefft built for William Slater a large, mansarded house with a four-story tower, which has disappeared. It therefore seems likely that this caretaker's cottage was Tefft's, as the high quality of the design also suggests. An early Victorian house with the principal bracketed roof gable paralleling the road, it has an abrupt cross gable at the center that provides a mildly "Gothic" quality, even though the windows in it are not pointed, but round-arched, which Tefft much favored. More round-arched triplet windows appear in the gables of the side elevations. The bracketed porch running around the corner of the main block to the rear ell is supported on especially handsome trellising, which is probably a later addition.

NS21 House

c. 1810. 20 School St.

NS22 John Slater House

c. 1810. 16 School St.

These two Federal houses with the typical two-and-one-half-story, five-bay, central-chimney format were both moved from Green Street near the Congregational Church. The move of number 20 was part of Kendall's reshuffling of village residences to his liking. Its door has been much photographed. Although ill fitted to its elevation, it is a charmingly provincial version of the side- and fanlighted type, where the side lights are set in flanking pilasters—thereby making windowed pilasters! (The Thomas Cutler Farm in Glocester [GL20] presents another example.) These are further ornamented in an amateurish manner: three meaningless lines below and panels with funerary urns above. The latter interrupt the simulated structural connection which should exist between pilasters and the entablature they are supposed to support. On this, too, ornament is inconsistent. The folk designer rummages through images and recollection, indulging in whatever pleases, and his pleasure at the result is infectious. Number 16, its near neighbor, is similar, but more prosaic. Its principal interest is, in fact, its plainness for the house of one of the Slater mill founders and eventual owner; John Slater's mode of living was far simpler than that of his brother, Samuel. The reasons for its removal from Green Street are explained in connection with the Dr. Elisha Bartlett House (NS27).

NS23 Slatersville Green and Congregational Church

1838. Later enlargement and several alterations, most recent 1966. Corner of School and Green sts.

NS24 Mill Overseers' Housing

1810–1820. After 1915, restored and altered. 4–40 Green St.

What one expects of the idyllic New England village! A white carpentered church dominates

NS23 Congregational Church

NS28 Commercial block

its green, overlooked by white clapboard houses. The sight, however, is rare in Rhode Island, because the tradition of freedom of worship eliminated both the communitarian basis for a village green and its dominion by any single church. So the green is actually no more than a small village park, and the Congregational church was built by the Episcopalian Slater family for the work force at the factory. But the image remains. Moreover, no other Rhode Island village has more appropriately placed in a compact gathering, and in such essentialist architectural form, the symbols of its being. The green is a level place close to the edge of the ridge with roads angling steeply up and more gently down to it. At its base the green triangle opens to the Greek Revival church with double-doored Doric portico and simple three-stage steeple, now too stubby, but taller and more commanding before its rebuilding after the 1938 hurricane. As is often the case, the medium-steep pitch of the roof lifts the pediment toward a gable. As is also characteristic of the Greek Revival, tongue-and-groove boarding (and putty) do their best to give a stonelike planarity to the front, whereas the rest of the walls are clapboarded. The light brought into the simple, preserved, but partly altered interior by rows of four windows in each of the side elevations is blocked from one of them, where a window has been converted to a door in order to link the church to a well intentioned, but clumsily obtrusive, parish house of the early 1980s.

A row of six identical five-bay, center-chimney Federal houses lines one leg of the triangular park on Green Street. They accommodated mill overseers. The entrance porches with half-hipped roofs are Kendall additions to add charm to barebones mill housing.

NS25 Duplex Mill Housing

c. 1915. 21–43 Ridge Rd.

This row of duplex housing dates from Henry Kendall's remodeling. These elaborate on the Colonial Revival detail of the remodeled houses on Green Street. The porches are now set toward the rear of the side walls and made large enough for summer living. They reflect a prevailing Neo-Colonial modification of precedent. Unfortunately, the whole row has been re-sided. What is today a fine allée of maples is also Kendall's doing.

NS26 J. H. Parkis House

c. 1885. 177–179 North Main St.

The Parkis House provides a stiff example of the Queen Anne style with fine decorative shingling in its gables, ornamental banding at bay windows and eaves, and a front porch of turned columns and spindlework—this for the postmaster and proprietor of the village general store.

NS27 Dr. Elisha Bartlett House

c. 1850. Later additions to the rear. 2 Green St.

The "big house" at the village center appropriately occupies a dominant site and is conspicuously adorned with Greek Revival regalia: a distyle (paired-columned) Doric entrance porch; a row of six more Doric columns supporting the entablatured roof of the sitting porch on the North Main Street side of the house under an attic pediment; two-story, paneled Doric pilasters as decorative corner posts for another entablature under the eaves; and pedimented gable ends.

Elisha Bartlett deserves a parenthesis. A Brown University graduate, he was a famous physician who held professorships in no less than nine medical schools during his career, the first mayor of Lowell, Massachusetts, and a member of the Massachusetts state legislature. His *History of the Diagnosis and Treatment of Typhoid and Typhus Fever* (1842) was the first comprehensive account of the disease in the English language. In *An Essay on the Philosophy of Medicine* (1844) he emphasized the impact of environmental factors on human illness. He had married Elizabeth Slater, daughter of John Slater, and ill health forced his retirement around 1850. They came to the family village, moving her father's house (NS22) to build their own more impressive replacement. Shortly after its completion, Dr. Bartlett died in 1855 at the age of fifty-one. He is buried in the Slatersville Cemetery.

Across the street from the Bartlett House, the brick North Smithfield Town Hall (1921) is rightly placed to enhance the village symbolism, but is architecturally too inconsequential to hold the eye.

NS28 Commercial Blocks

1850, 1870. 7–9 and 11–13 Main St.

Beside the Town Hall, two commanding, flank-gabled blocks of three stories mark the commercial center of the town. As in the smaller, two-story Forestdale block, granite piers and lintels frame the storefronts. (At Forestdale, however, shop windows and doors are more deeply recessed, thereby intensifying the structural and sculptural properties of their enframement.) The upper stories are red brick (painted on one building and corbeled at the cornice), their windows trimmed in granite at sill and lintel. The brick wall up front is neatly morticed into side walls of random masonry which also enclose the rear.

Extending in a long curve away from the commercial blocks are six two-and-one-half-story, gable-roofed, four- and five-bay duplexes, now re-sided. Some or all of them probably predate Slatersville as a textile village, but all were used as workers' housing.

NS29 Slatersville Mill

1806. 1826, replaced. 1843, second mill building. 1894, weaving shed. Off Railroad St.

In front of the lineup of church, green, elite houses, town hall, stores, and mill housing, the land drops precipitously to the mill. During the

NS29 Slatersville Mill

nineteenth century, a mill complex of the same scale also existed on the other (west) side of Railroad Street. What now remains was then locally referred to as "Center Mill" to differentiate it from the "Western Mill." (When cotton manufacturing in Slatersville gave way to bleaching and dyeing after 1900, the Western Mill became redundant. It was eventually destroyed, except for two minor stone buildings, one of which (c. 1850) now houses the North Smithfield Public Library, on Main Street.)

From the original 1806 Center Mill complex, which burned in 1826, all that remains is a long, one-and-one-half-story clapboard adjunct building in the millyard with the trapdoor type of roof monitor typical of early mills. The owners immediately replaced the destroyed mill, on the same site apparently, with a four-story block (five at the rear), which now dominates the factory cluster. It has a slightly pitched gable roof and attached entrance and stair tower, all handsomely crafted in dressed granite blocks, with a well-proportioned relationship between wall and window. The rugged granite contrasts with the very refined, if straightforward, carpentry of the window sash. During the 1980s especially, when this mill was left to storage and eventually vacated, much of the sash seriously deteriorated. The wooden cupola capping the mill's projecting stair tower once contained the other bell in town: the workday bell in the hollow, the sabbath bell above.

In 1842 a stone building behind the 1826 mill also burned. It was replaced in the following year by a sizable, if somewhat less refined, three-story granite building on the same site, which went through the same process of deterioration as its companion in front. Finally, the large two-story brick weave shed of 1894 completed the Center Mill complex. The beautifully crafted raceway with substantial remnants of Kendall's landscaping, some of it now overgrown, is also evident. At this writing, new owners promise restoration of the granite buildings, and none too soon to rescue the very heart of this handsome village.

NS30 **Mill Dams**

1849 and later. Railroad St.

The Western Mill site on the west side of Railroad Street provides the best view of a sequence of dams on the Branch River that successively hold back the water for the raceway. Behind a relatively small dam is the principal barrier, Middle Dam (1849), with yet another built behind it late in the nineteenth century. Middle Dam is an especially splendid masonry structure of very large blocks, in all 300 feet long with a rollway 160 feet wide for a twenty-foot fall of water.

NS31–NS34 **Grange Road Farmhouses**

18th century. Grange Rd. between Rocky Hill Rd. and Providence Pk.

Grange Road is an eighteenth-century road, its antiquity evident in its winding course and in the stone walls along it, but more remarkably in the preservation of a string of four small eighteenth-century farmhouses and an accompanying accumulation of fields, family cemeteries, mostly nineteenth-century outbuildings, and a late-nineteenth-century grange. All the houses are one and one-half stories and of clapboard; all but one have transom lights over their doors. The progression begins with the oldest of the group (just north of the intersection of Grange Road and Rocky Hill Road), the gambrel-roofed John Durran's Farm (NS31; 152 Grange Road) with its door at one end of the front elevation, asymmetrically placed windows, and later dormers. A large barn, smaller storage building, former sawmill (all nineteenth century), and the family cemetery complete the farm. Opposite the intersection is Primrose Grange (NS32; 1887), with another small family cemetery on the property. Nondescriptly plain, it became more awkward looking when it was raised in the early twentieth century to make another story. (Why are all Rhode Island granges almost without architectural appeal, when starkness in country schoolhouses and churches has left its share of memorable images?)

Next of interest, on the east side is a farmhouse (350 Grange Road) set well back. Nearby to its north is the former Andrews Schoolhouse, moved from across the road near the Rocky Hill corner and now used as an outbuilding. On a rise at a sharp bend and set very close to the road, is the Grayson Phillips Farm, number 391 (NS33; c. 1790) with another family cemetery, and its barn hard by the road opposite. Unusually for the original date, the eaves are lifted away from the windows to provide a quarter story, with a pair of transomlike windows under the eaves. It could have resulted from a roof raising done at a later date when the motif would have been common. The front is five bays, although asymmetrical. But look again: its peripheral elements—end windows and centered openings under the eaves—are in fact symmetrical. It is as though an atavistic will for formal order framed the casual functionalism of the placement of the door and windows at the center.

Last is number 445, the N. Baker house (NS34; c. 1780), where prim symmetry prevails at last, with a central chimney and eaves drawn snugly to the window tops. A small shingled barn (c. 1890) with an attached shingled shed (c. 1780, moved from another site nearby), and a three-hole privy seem to stake out the flat rectangular platform on which this farm sits. Two of its boundaries are further delimited by a picket fence. Again the road rises suddenly, turns and dips at the house. The beautiful modesty of these buildings, as well as their sensitive and varied relationship to fields and road, rebuke the strident heedlessness of the present. May this cluster of farmhouses, their fields, and the road itself survive to remind us of what much of rural Rhode Island once was.

Smithfield (SM)

Of Smithfield's seventeenth-century settlement, almost nothing remains. Much was destroyed in King Philip's War of 1675–1676, the rest by attrition. Hence the oldest extant buildings in the town date from the eighteenth century—the Mowry House, which appears to be the oldest, from 1701, and all others probably after 1725. As settlement progressed in the "Woods" north and west of Providence, demands for independent government grew because of the difficulties of reaching the central city. In 1731, Providence was forced to set up three large towns: to the north, Smithfield; to the west, Scituate and Glocester. Smithfield then included all of North Smithfield, Lincoln, and Central Falls, as well as the western part of Woonsocket—roughly all the territory between its present western boundary, the Massachusetts border to the north, and the Blackstone River on the east. What is now Cumberland on its east bank was then part of Massachusetts. With time, the clamor by some of this population for independence from Smithfield repeated earlier protestations to Providence and led to further division in 1871, leaving town boundaries as they are today.

Farming and logging on the choppy topography of glacial soils were inevitably the principal occupations, with plenty of boulders at hand for the walling of fields. As in other hinterland towns, transportation to markets was an early concern and led to the building of three turnpikes which ran across the town west and northwest from Providence. About 1733, the Providence-Woodstock Road, also known as Great County Road, was laid out along what is now the route of Putnam Pike (Route 44). Other turnpikes were not built until the early nineteenth century. The Douglas Turnpike Company, incorporated in 1805, was responsible (eventually under another corporate name) for Douglas Pike (Route 7). Finally, just as the Harris family in Lincoln initiated the turnpike from Lime Rock to Providence to transport their product to market, so the Farnum family, which had a forge at Georgiaville, began the effort, finally realized in 1819 by the Farnum and Providence Turnpike Company, to build what is now Farnum Pike (Route 104). This connects with Putnam Pike (Route 44) at Centredale (in North Providence), and thence goes on to Providence.

A surprising number of fine eighteenth-century farmhouses remain in Smithfield, their size—they are typically two-and-one-half-story, five-bay, central chimney houses—

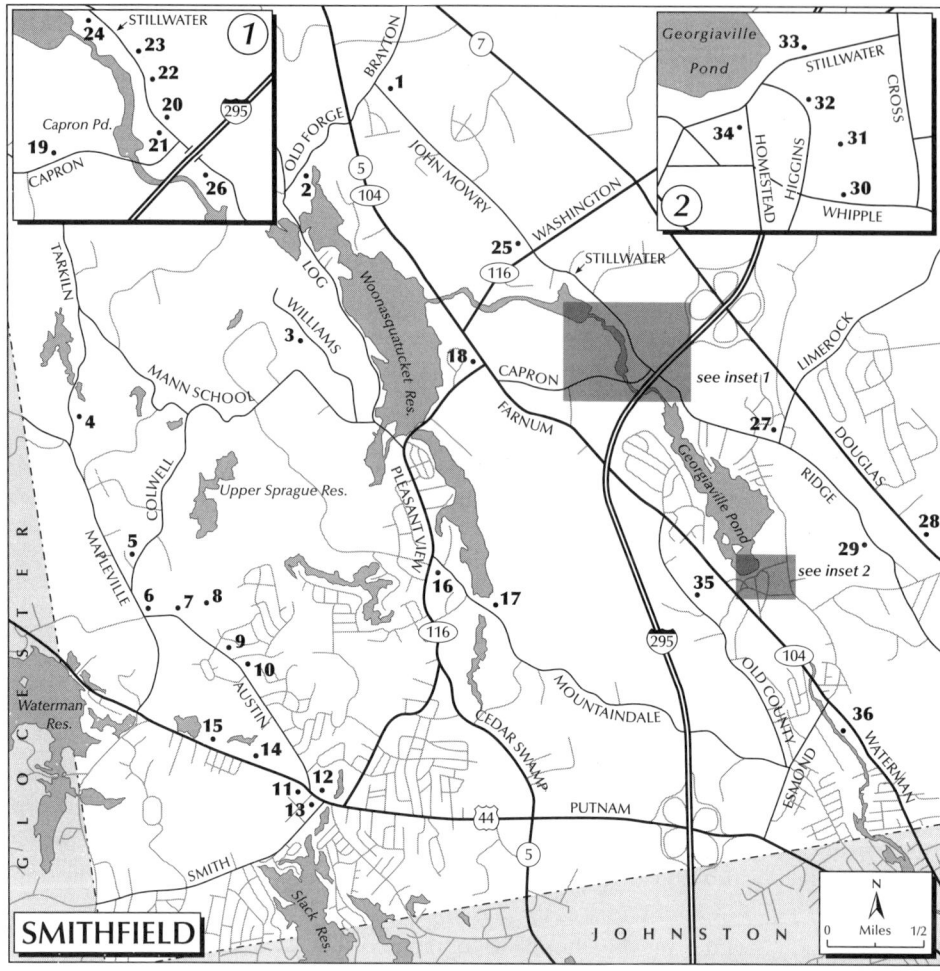

and their embellished doorways testimony to the relative prosperity of the town's early agriculture. When scouts for the White Pine Series combed New England in the late 1920s and early 1930s in search of colonial houses for an important corpus of publications designed to serve architects as a source for Neo-Colonial design (and also to sell lumber), they found no less than five examples in Smithfield. The examples described here, roughly twice that number, may seem to be too much of the same thing. But Smithfield provides a case study of ways in which a persistent architectural formula can be subtly manipulated to provide differences within the sameness.

Like neighboring towns where the water was sufficient, Smithfield had its share of textile mill villages along its streams and rivers—principally the Woonasquatucket River and its tributary, the Stillwater River. The largest villages ranged across the town just above its southern boundary: from west to east, in the order we will be visiting them, Greenville (with West Greenville), Georgiaville, and Esmond. Above these, where the stream of water was smaller, the villages, too, were smaller: Spragueville and Mountaindale withered away before the end of the nineteenth century, as larger industrial operations became necessary to turn a profit; Stillwater was finally done in as late as 1980 when a fire destroyed its mill, long after textile production had left it. Although small textile plants appear in Smithfield around 1810, nothing of much size appeared until the 1840s, which was fairly late. Doubtless Smithfield had to wait for the exhaustion of better sites elsewhere and the delayed arrival of the railroad. Two villages are particularly interesting for their early factory buildings, West Greenville and, especially, Georgiaville—although it is a sign of the suburban times that the most interesting factory buildings in Georgiaville have been converted to condominiums.

SM1 Mowry House

1/01 and later. 239 John Mowry Rd.

Our tour of Smithfield begins with its oldest extant house. The Mowry family lived in this elongated, two-part house for two centuries. As is typical of early houses, it is sited diagonally to the road so that its front elevation faces south. Three sides now covered with asbestos shingles may originally have been wood shingled, with clapboards reserved for the front. The section closest to the road is the earliest portion of the house: originally probably one room below and one above, each with two windows on the front elevation, the door with its stair in the corner, and an end chimney. To this was early added a four-bay extension with a second door and another chimney at the opposite end of the house. Both doors are simple and elegant: tall and narrow in opening and frame, rising like grandfather clocks to five-light transoms capped with moldings. The earlier door has eight panels, the later four. The White Pine Series recorded them and the ancient well sweep (which no longer exists) in measured drawings.

SM2 Cobblestone House

1903. 40 Old Forge Rd. (.35 mile from Farnum Pk.)

Cobblestone construction is unusual for Rhode Island, but this one-story cobblestone house with a hipped roof was designed and built by an emigré from Rome, Joseph Octaviana (1903). It seems to be another example of Mediterranean scorn for the apparent impermanence of a wooden house where stone is possible (see also JO11).

SM3 Windy Brow Farm

c. 1800–1810 with later wing. Early 20th century, barns. 82 Williams Rd. (pole 29)

Raised above the road on a grass terrace retained by a handsome granite wall, this house is a typical two-and-one-half-story, central-chimney farmhouse. It is plain and boxy, with plenty of clapboarding on its front and perhaps a little too much space between the center elements and the peripheral windows on either side to contain the spread of its windows. It is graced by a typical Federal entrance with side lights and a blind elliptical fan. Beside it are a group of late-nineteenth- or early-twentieth-century farm buildings, including a handsomely shaped shingled barn with a sharply pitched mini-gable at the ridgepole, then long, gentle inclines in either direction. From their elevated site, house and outbuildings overlook a field with a steep decline across the road.

SM4 Evans House

1805. 11 Tarkiln Rd. (pole 2)

The Evans House takes its name from a family who occupied it beginning in the late nineteenth century. Whereas Windy Brow (preceding entry) is aligned with the road on a terrace, this only slightly earlier house in the same format is angled to the road in the old-fashioned manner, south facing and raised on a mound, probably augmented with the dirt of its cellar excavation. The front, which at Windy Brow seems generously stretched in keeping with a trend of the Federal style, is here compact, drawn in toward its central brick chimney. Both have twelve-over-twelve sash on the first story and, unusually, twelve-over-six above; but these windows are smaller. The framing is also heavier, with decisively projecting sills and heavy boards cut in exaggerated diagonals on either end as lintels. No elegant fan over the door, nor side lights to illuminate the hall; this one is narrow, set deep within a broad, simple frame and capped by a small transom with a strongly projecting door head. "Elegance," says Windy Brow, elegance more attempted than fully achieved, to be sure, as though the country house went to town in quest of the latest fashionable graces. "Shelter," says the Evans House, in its traditional carpentered vigor, content to stay with country ways.

SM5 Tucker–Stone–Colwell House

1815; modified early 20th century. 38 Colwell Rd. (pole 70)

This plain Federal house, set well back on a lawn, was modified in the early twentieth century by a one-story columned porch across the front and the double-tiered screening of a later ell, so that now it almost appears to date from 1915 rather than 1815.

SM6–SM9 Austin Avenue Farms

Among Smithfield's oldest roads, Austin Avenue contains a row of farmhouses, all but one going back to the eighteenth century. Their owners, mostly members of the Winsor family, came to dominate the apple business around Greenville. Like the others we have seen, they are plain houses; but each is large and conspicuously sited, with some mark of distinction to elevate it above the simplest farmhouses as an indication of the prosperity and social status of its owner.

SM6 Jesse Foster House

Pre-1750; side porch probably 20th century. 147 Austin Ave. (corner of Mapleville Rd.)

Both this and the Daniel Winsor House (next entry) show the familiar five-bay, two-and- one-half-story format. The narrow, six-over-six boxed windows of the Foster House, with their fairly crude splayed lintels, as well as differences of interval between the window openings left and right of the door (and on the left side of the elevation even a lack of alignment between first- and second-story windows) all seem at variance with the expansive and sophisticated door. This displays a modillioned pediment with entablature raised on fluted pilasters. It is probably later. It takes over the elevation, and overwhelms the flaws around it.

SM7 Daniel Winsor House

c. 1739–1749. 129 Austin Ave.

The Daniel Winsor House is one of three associated with the Winsor family. Again, an apparently later door takes visual precedence over an elevation which includes a discrepancy between window intervals either side of the door. The door typifies what seems to have been most common in Smithfield from the mid-eighteenth through the early nineteenth century. A semicircular transom (here blind) breaks through the base of a pedimented cap (here with dentils) which is lifted on blocks above fluted pilasters. Its narrowness and elaboration better accord with the twelve-over-twelve windows, which are more compactly organized than in the previous example. Again, upper windows are twelve over six. Contrast between the shingled walls and the openings, however, gives the elevation a somewhat spotty, unintegrated effect, which clapboarding would help, and perhaps originally did, on the front elevation at least.

SM8 Stephen Winsor House

Mid-19th century. 113 Austin Ave. (not visible)

One more Winsor property in the area, the Stephen Winsor House. This is a tall house based on an asymmetrical composition of cross gables, giving it a compact L-shape in a carpentered version of the Italianate manner. Its detailing, which is blunt and forceful, consists of a broad, bracketed shelter over the entrance, more bracketing under the projected eaves, and paneled corner boards. Unusual for its style, however, is the use of seemingly Neo-Tudor molding heads, which typically fold around the tops of windows and doors, but are here lifted above the windows around their lintels. (Or perhaps the intention was not stylistic allusion, but emphatic framing for the lintels derived from stone building.) Stephen Winsor was a banker who operated his country place as a gentleman's farm. It is still surrounded by the remnants of orchards and stone walls.

SM9 Reverend Resolved Waterman, Sr.–James Winsor House

Pre-1735. c. 1780, gambrel-roofed ell to rear. Late 19th century–early 20th century, barns. Late 20th century, restoration. 85 Austin Ave.

The oldest and most commanding of the Austin Avenue houses, the so-called Waterman–Winsor House, has a four-bay front with a large, slightly off-center chimney which has been reduced in size and a gambrel-roofed ell to the rear, curiously offset from one corner of the house. It again displays the tendency for early houses to face south on a slight rise regardless of orientation to the road. The jutting attic gables over both side elevations are a surprising anachronism going back to the seventeenth century. Then, and into the early eighteenth century, it continued a medieval tradition of attic projections seemingly designed to squeeze extra space from tight sites in crowded towns, before also becoming something of a style feature. But what does it mean for a house of 1774? Is it a gratuitous later elaboration to make "restoration" more colorful? The handsome, pedimented door may also have been part of an extensive late twentieth-century restoration, when an added Colonial Revival porch was ripped from the front. A group of much later outbuildings, probably from the nineteenth century, angled to one another for reasons of site and function, make a picturesque grouping.

The house came into James Winsor's ownership in 1861. His son, Thomas, expanded his father's property and orchards after 1903, making it the largest apple business in Rhode Island until the mid-1950s. His death at this time, followed by severe hurricane damage to the orchard in 1954, led to the gradual sale of most of the farm for the suburban development which now impinges on the ancient manse.

SM10 **House**

c. 1860(?). 73 Austin Ave.

This modest and well-preserved early Victorian residence, set gable end to the street, is distinguished by an entrance with a flat hood consisting of a full modillioned and dentiled cornice, brackets with drop pendants, and jigsawn spandrel ornamentation, all sheltering a door surround of side lights and transom. When built, this was a village type for the middle-class businessman or professional and marked the bounds of Greenville, before one reached the orchards immediately beyond.

Greenville

Although it had a short-lived academy, begun in 1814, and a bank, established by 1822, Greenville was long little more than a farm and tavern village with some importance as a local transportation hub, the usual rural industry, and a small thread and twine mill. Then, in the 1840s, three manufacturers descended upon it and built mills that used the Stillwater River for power. One of these, a woolen mill on Austin Avenue close to the village center, eventually, in the twentieth century became a piece of Austin T. Levy's Stillwater Worsted empire (see Harrisville under Burrillville), before its later use by a series of firms for various purposes. Two cotton factories, one long demolished, located a little apart from the village on Putnam Pike, together created the satellite village of West Greenville. The one remaining includes buildings going back to the beginnings of Greenville's life as a textile village.

Prosperity over a long period left an impressive and varied architectural heritage of houses ranging from the colonial period through the early twentieth century. But Greenville's importance as a regional hub has wreaked havoc. Highway widening and heedless demolition or alteration for drive-in business have substantially diminished the coherence and charm that Greenville once possessed.

SM11 **Freewill Baptist Church**

1820, Clark Sayles. 1866, basement story and forward stair tower. 582 Putnam Pk.

Essentially this is the typical Wren-Gibbs church as filtered through the Federal designs in Asher Benjamin's handbooks. Its approximate original appearance is apparent from Clark Sayles's nearby building for the Free Will Baptists at Chepachet (see GL05). As there, the tower once rose from a pedimented frontispiece which projected from the main body of the meeting house. The tower styles are identical: a tall, windowed box for a base, stepped back to a belfry (here with clipped corners), and a small octagonal transitional element to the spire. The later raising of this meeting house on a brick basement story for parish house functions and the consequent stair enclosure in front of the frontispiece encumbered the building's facade but elevated the spire to a better command of the enlarged village. The interior has been altered several times.

SM12 **Greenville Fire Company**

1939, Public Works Administration. Later addition. 611 Putnam Pk.

The Greenville Fire Company is a good example of Public Works Administration Neo-

Colonial design. On the upper story the brick wall provides a textured background against which to display a crisply detailed Palladian motif fronted by a balcony, the latter (as was often the case in PWA work) a decorative feature in metal. Above is an arched "belfry" for the siren. The dialogue between civic embellishment and efficiency is nicely maintained in the best PWA work by such spare and sprightly use of Neo-Colonial detailing. Simulated end chimneys front and back provide an ingenious means aggrandizing the facade with a false-front capping.

SM13 St. Thomas Episcopal Church

1851, Thomas A. Tefft. 1891, tower. 1950, partial interior renovation. 578 Putnam Pk. (at Smith Ave.)

This is the only church in the Gothic Revival mode which Thomas Tefft certainly executed, although another in nearby Georgiaville (now demolished, but illustrated in Henry-Russell Hitchcock's *Rhode Island Architecture*) is attributed to him by some writers. It is comparable to Richard Upjohn's little masonry Calvary Church in Stonington, Connecticut, erected 1847–1849, just before St. Thomas. Like Calvary, this originally had a bellcote instead of its present steeple. We know that Tefft admired Upjohn; in an essay, "The Cultivation of True Taste" (1851), he praised Upjohn's Grace Church in Providence. Furthermore, St. Thomas had close ties with the other Upjohn church in Providence, St. Stephen's. The Reverend Henry Waterman, pastor of St. Stephen's, was among the early and ardent converts to the Oxford Movement, which called for a revitalization of the Anglican Church through a return to the high ritual and architecture of the medieval past. The pastor of St. Thomas at the time of its building, the Reverend James Eames, was also close to this movement. So had Tefft been unaware of Calvary in Stonington, these clerics could have pointed it out. Though Tefft himself was a Baptist and pioneered in the Romanesque as the medieval revival style for church building, his library contained *The Ecclesiologist* and books by A. W. N. Pugin, both of which were key sources for the design of buildings and ritualistic objects proper to the High Church convictions of the Oxford Movement.

The church Tefft built—with its diminutive and austerely simple buttresses, its tiny pointed windows, and its handsome coursed-rubble masonry, partly smooth and partly rough surfaced—gives the sense of a civilized style developing from rude beginnings. It was an appropriate image for a country church. Whoever later designed the tower respected Tefft's metaphor of architecture emergent from primitive sources. It rises from a brute block at the base, through minimal transitional shaping, to the specific Neo-Gothic detail at the top, although Tefft might have been a little bolder and less literal had he designed it. Inside, the little box of space (as partially renovated in 1950) is simply plastered, including the gabled ceiling, with its black-brown linear roof structure of rafters, wall brackets, and cross spans its only embellishment aside from the pews and the slots of stained glass (although the walls were probably originally darker, with perhaps stenciled ornament). Off the rear wall is a small chancel, that preeminent symbol of ecclesiology, which banished the old-fashioned reading desk in the clear light of the meeting house for shadowy choirs in which, at some remove from the congregation, the mysteries of the mass were properly respected. So the two churches at the center of the village embrace opposed views of the religious experience (although most village worshippers would have been oblivious to the disputation which brought such diametrically opposed buildings to this place). One embodies the Word delivered from the pulpit, luminous in the transcendence of clear light and geometry (even in an altered interior); the other, the mass received through sacred ritual as reinforced by appeals to nature, the primitive, and the romance of remote historical styles and institutions.

SM14 Tucker-Quinn Funeral Home (Richard Waterhouse House)

c. 1900. 649 Putnam Pk.

This hip-roofed near-cube of a house is complicated by three features, all characteristic of the Queen Anne style. An elaborate dormer centered on the front elevation thrusts a simulated balcony on brackets hooped with a "moon" circle toward the visitor. A polygonal tower at one corner rises to a bulbous gourd dome. A veranda, supported on pairs of classical columns with an off-center pediment to mark the entrance steps, extends across the front and jogs around the corner in a polygon before butting the tower. Richard Waterhouse, who probably commissioned the house, served as superintendent in the nearby woolen mill on Austin Avenue. In 1903, the local paper called it the "finest modern dwelling" in Greenville.

SM15 West Greenville Mill (Pooke and Steere Mill)

1844 and later. Putnam Pk.

Two cotton factories supplanted earlier, smaller mills in this area: Elisha Steere built this one in 1844; Stephen and Albert Winsor, with William F. Brown, the second around 1845. The latter, the "lower mill," separated from Steere's operations a little way downstream, was destroyed in the early twentieth century along with a community of mill housing. Steere's mill also went through a series of name changes and shifts from cotton to wool and back to cotton. It takes its most familiar name from William Pooke and Anthony Steere, who controlled it from 1855 to 1873. The mill is visible today behind a hodgepodge of late-nineteenth- and twentieth-century accretions. A three-story mill in random masonry, it was originally sheathed in stucco. The roof is still surmounted by its original trapdoor monitor, now completely encased by artificial siding, but the four-story tower has lost its belfry. A granite dam backs up the Stillwater River into a millpond from which a raceway once powered the mill. After textile operations ceased in the early twentieth century, a variety of commercial and industrial users leased pieces of the complex.

SM16 Mill Duplexes

c. 1825. 316–322 Mountaindale Rd.

In 1825 near what is now the intersection of Mountaindale Road and Pleasant View Avenue, Thomas Sprague built a cotton mill on family property, naming the place Spragueville. Neither his stone mill (destroyed in the early twentieth century) nor much of the village still exists, although the vicinity, renamed by subsequent owners, has returned to its original designation. Of interest only are two clapboard workers' duplexes near the corner, the one closest to the corner in the best condition. Mill housing of this antiquity and as well preserved is extremely rare. Both duplexes maintain their tiny trapdoor monitors and paired interior chimneys, and one of them has its original paired doors at the center, with narrow transom lights topped by a molding in common. The delicacy typical of Federal carpentry is apparent even in so plain and utilitarian a structure as this.

SM17 Mill Duplex

1860s. 261 Mountaindale Rd.

Mountaindale Road once led from Spragueville to a small village centered in a machine shop for the textile industry, before its conversion to various textile operations, which ceased around 1880. From these enterprises another duplex remains, one of such commanding scale that it may have housed mill overseers or served as a boardinghouse. The gentleness of the Federal scale and detailing in the preceding example intensifies awareness of the bristling magnification of these aspects here. All windows are doubled and decisively capped, with doubled doors at the center coupled by a minimal but broad and strongly projecting Greek Revival cornice. Transoms and side lights glazed down to the floor surround the butted doors, so that they share the middle side light. The entrance's broad frame and entablature appear to be exploded away from the doors by the intervening transparency in a manner which incongruously calls to mind modernist effects more than Greek Revival. The tall, narrow proportions of the windows and their doubling, however, are typical for the early Victorian Italianate style just then emergent.

SM18 Steere-Harris House

c. 1760; wing probably later. c. 1840, door. 310 Pleasant View Ave. (near the junction of Farnum Pk.)

The Steere-Harris House is yet another instance of the five-bay, two-and-one-half-story format with six-over-six sash, with an off-center chimney and asymmetrical window distribution. The windows are stretched out about as far as the formula can be managed, with the intervals of clapboarding separating the center door and its window from the four windows to either side a little overextended. Yet the relatively wide spacing between the outer window pairs suggests some effort to regularize the rhythm of openings across the front, but here left unresolved. The door is a Greek Revival modernization.

Arranged in an arc around the former millpond behind the farmhouse, a colony of clapboard apartment units with multipitched roofs comprises the farm's final crop. Countering what would otherwise be barrackslike plainness, up-and-down gable pitches, in-and-out massing, and picturesque grouping, together with a perfunctory "green," do their best to transform suburban apartments into a "village."

SM19 Stephen Steere Farm

1825–1830. 56 Capron Rd.

Plain rusticity persists in this house, although it is among the latest examples in Smithfield of the two-and-one-half-story, five-bay formula. Only the simple order of its elevation and the splayed-board lintels for the first-floor windows hint at higher aspirations to "architecture." Except, of course, for the door. The formula is familiar—pilastered, with a broken and modillioned pediment over a fanlight. The ornamentation of the fan in lead provides the most elaborate embellishment of this feature we have yet seen in Smithfield, and characterizes the delicate elegance favored during the Federal period. It is too elaborate and elegant, perhaps, for what surrounds it, yet it elevated a farmer to a landowner in public estimation and captivated the White Pine investigator who recorded this entrance in a measured drawing. As he indicated, the delicate leaded work in the fan is not likely to have been the original treatment, or to have been done locally. It probably came from the shop of a specialist who combined elements from a catalog of cast lead ornaments and brass ruling in various ways to customize the design. The same was doubtless the case for the sophisticated door frame.

The Steere farmhouse belongs to a group of Smithfield country houses displaying identical door designs (see SM29 and SM35), which were allegedly installed (sometimes into earlier houses) between roughly 1780 and 1825. The nearly identical frames are the most standardized and stable element in the ensemble. They confer class. The ever-changing fanlight treatments are the most volatile element. They give individuality and define the period (or, for the clients who bought them, the "modernity") of the installation. Did the changes in the fanlights catch the White Pine draftsmen off guard? In their twentieth-century compendium they delineated three of these entrances as though all differed in design, when essentially only the decorative division of the fanlights changed.

Stillwater

This linear mill village, the northernmost on the Woonasquatucket River (which eventually flows through downtown Providence), twice saw its livelihood consumed in factory fires. A cotton mill built here in 1824 was demolished in 1866 for a new woolen mill. It burned in 1872 and was immediately rebuilt. Fire took this building in 1980, after textiles had given way to furniture manufacturing. What remains of special interest are a row of workers' houses, mostly lined along the west side of Capron Road, which are recognizable despite abuse and renovation; the combined company store and post office; and houses, presumably for the owner and head management, with the burned-over mill site between them.

SM20 Four-Family Mill Housing

c. 1867. 283–295 Stillwater Rd.

SM21 Mill Owner's House

c. 1830. 294 Stillwater Rd.

SM22 Duplex Mill Housing

c. 1836. 297 and 299 Stillwater Rd.

The row of six two-and-one-half-story, five-bay, hip-roofed units for four families were originally quite plain. The center door at the front rises between the two interior chimneys to the upstairs units. Side doors open into the downstairs units. The house once occupied by the mill owner dates back to the original settlement of the place. Its most conspicuous feature is its early nineteenth-century entrance, a fairly rare type in Rhode Island, although a standard Federal design. Spread like a three-part altarpiece, entrance and flanking side lights are united by a wide lintel curved as a cove molding at the top. Slender colonnettes flank the sidelights, supporting skinny blocks. They cross the lintel, then fan out as they merge with the cove. Diagonally across the road are a pair of two-and-one-half-story frame gabled workers duplexes, again with two interior chimneys, erected in conjunction with the first mill. Transomed doors at either end reverse the central positions of those at Spragueville (SM16), while the slit openings of the trapdoor monitors there are here expanded to shed dormer proportions.

SM23 **Day Care Center** (Company Store and Post Office)

c. 1867. c. 1908?, renovated. 311 Stillwater Rd.

Records date a company store here from at least the late 1860s, so this pretty Queen Anne–Colonial Revival front may represent a remodeling around 1908, when the village was relandscaped, of what was probably an Italianate design. (Even before this refurbishing the village was praised as being especially attractive for its neatness, trees and amenities.) One of Stillwater's attractions was a small linear park of uncertain date which once split the road between the entrance to the mill and this commercial building opposite. A double stair turned sideways to the street rises to a platform before the store, which has a shop window partially framed by a multipaned transom and sidelights. A single stair fronting on the street passes between paired Doric porch columns into the post office through double doors, which are also side lighted and transomed with small panes. A stepped "false front" punched with an oculus window unites the two elements and masks the gable roof. Few Queen Anne commercial fronts have survived as well as this unique example, which now serves as a day care center.

SM24 **Mill Superintendent's House**

c. 1867. 320 Stillwater Rd.

Somewhat dilapidated this compact mansard "cottage" with a porch fitted into its stubby T-shape nevertheless retains most of its ornamentation, especially the big dormers in the mansard, which are decorated with a frame of scroll brackets, capped by scroll pediments, and culminate in pinnacles topped with spheres.

SM25 **Elisha Mowry House**

1759. 10 John Mowry Rd.

Any four-bay ("three-quarters") house poses the problem of the placement of the odd bay in the format. The elevation of this house extends as needed, with the odd bay centered in the resulting space. The small panes of the twelve-over-twelve sash magnify the impact of the emphatic framing. The simply molded caps of the first-story windows mitigate the bluntness of their boxed-out board enframement while subtly marking the importance of the principal rooms. The typical transomed door of the period is given an austere, well-proportioned pediment, its base aligned with the tops of the adjacent windows. Like the Evans House (SM4), the Mowry House gives the sense of a compact, near-cubic mass. All is vigorous, ordered, and restrained. Hence the intensity of this elevation.

SM26 **Smithfield Historical Society** (Elisha Smith House, known as the Smith–Appleby House)

1713. 1750, ell. Late 18th and early 19th century, additions. Early 19th century, principal entry. 220 Stillwater Rd. (open to the public)

Elisha Smith built his house as a saltbox, two stories in front, one behind. The roof was raised around 1750 to make a two-and-one-half-story block, and an ell was added. The "house" we see on arrival is really the house-sized ell, a full two and one-half stories in height and the equivalent of five bays in length. Instead of discreetly projecting from the side or rear of the original house, this insubordinate addition butts headlong into one end of the south-facing front, making an *L* of the total mass. So we must circle the addition and enter the space made by the arms of the *L* in order to reach what is still the principal entrance. Although this entrance front was probably a full five bays, we cannot be certain because the collision of the addition occurs immediately to the right of the fourth window bay, squeezing these windows against the angle of the *L*. Assuming a full five-bay elevation from what is visible, here is yet another (and rare) placement for the peripheral windows as a "four-spot" configuration which is very nearly centered between the entrance axis of the elevation and its corner, instead of the usual placement of the peripheral windows closer to the edge. As a result, the mass of the house seems swollen. The rhythm of openings does not so much control the elevation as sustain itself against the swell. (Still, the precise visual effect of the original elevation must remain uncertain because the intervals between the central axis and the peripheral windows on either side are unequal.) The principal door is enframed in the same sort of cove-capped, side-lighted, triptych arrangement just seen in the mill owner's colonial house in Stillwater (SM21). Like that, it is an early-nineteenth-century remodeling, and probably by the same carpenter.

Other minor accretions further complicate the amplitude of the house. The end result (despite its good repair) is the kind of old-manse picturesque which nineteenth-century writers like Nathaniel Hawthorne savored: gently sagging clapboarded masses with varied roof slants, tilted chimneys, and smallish, multi-paned windows scattered about, on the verge of fixing on some simple, ordered arrangement, which then falls off toward functional randomness. In short, the setting has possibilities for the ambiguities in which legend flourishes: overall a surface of feckless simplicity, with a touch of grandeur in the would-be formality of the front, frustrated by the functional vernacularism which extended the house. Above all, there is a sense of generation, as much in conflict as in harmony.

The same fascinating mix of periods and intentions occurs within. The keeping room shows the heavy framing of the 1713 house, its original beaded vertical planking of wide boards never painted. Upstairs, the bedrooms show a collection of fireplace moldings of the 1750s, and the southwest chamber has a marbleized painted floor of the same time. The northeast chamber is stenciled with an alternating pattern of weeping willows and flowers climbing the plaster wall between rope moldings—Federal motifs, perhaps of the same date as the front door. More of this rare stenciling appears faintly in the tight, boxed stair of 1713.

A forge operated here from around 1750 until the 1840s, and a later sawmill into the early 1870s. Early on, an Appleby married a Smith daughter, and, as their descendents continued to occupy the house for generations (until 1959), it gradually acquired its present name.

The railroad station on the grounds is a former Providence & Springfield depot built after 1867 and moved here from Capron Road. It is a minimal example of its type: a gabled box sided in vertical boarding, plainly bracketed with diagonal bracing on either side and radiating strutwork in the gabled ends. Hard by, and elevated on a high embankment, Interstate 295 shatters the seclusion—a memorial to a time when highwaymen brooked little interference from preservationists, and getting there was *all* the fun.

SM27 Asahel Angell House
1780. 4 Limerock Rd.

Mounds of clipped yews and a flush in-line addition added in the early nineteenth century interfere with a proper view of this five-bay elevation. Unfortunately, too, the large center chimney is gone. But the elevation is among the finest in the state for its period, and another included in the White Pine Series. As already noted, Smithfield houses based on the same elevational type show differences in character. But for this and the next, let us look more closely, with a backward glance at the Elisha Mowry House.

As in the Mowry House, and typical of eighteenth-century design, the box-framed windows have twelve-over-twelve sash projecting from the plane of the clapboards; also as in the Mowry House, the window tops are aligned with the base of the pediment over the door, and the design has the same vigor and gravity over all and in detail. However, the format of this elevation is grander. Observe the spacing of the openings, which permit a fluctuating view of the arrangement: as an entity of ten openings, or alternatively as a sequence of four-two-four. The molding of the eaves is simple but elegant. It is decisively differentiated from the projecting boxed window frames aligned beneath it. The somewhat attenuated proportions of the windows draw the eyes up and down the elevation, even as they move across the row. The first-story window lintels are subtly shaped, while retaining the sense of the blunt board whence they came. This restrained but decisive shaping and placing of elements extends to the door frame, with its plain broken pediment (no dentils, no modillions) over a spoked fanlight, flanked by plain pilasters. The entrance is slightly pinched in accord with the window proportions. Reticent quoining frames the elevation at both ends. No detail obtrudes as an eye-catching embellishment. The elevation has been deliberated as a whole—not self-consciously, perhaps, but with the intuitive sensibility of the carpenter-designer who maximizes the possibilities inherent in an agreed-upon formula.

SM28 **Thomas Burbank House**
Early to mid-19th century. 495 Douglas Pk.

Deep into the mid-nineteenth century now, yet the elevational formula from the eighteenth century through the Federal period persists, virtually untouched by the intervening Greek Revival and Italianate styles. Here it is bare-bones, the six-over-six windows, sash again inset in the modern manner behind the clapboarding. But how tensely rectangular is the effect in this elevation, with its insistence on the framing elements! The upper windows are set against the plain, decisive curve of the cornice molding; the tops of the lower ones, capped with the simplest of board projections, all aligned with the bottom of a basic rectangular transom. If the format survives from the Federal period, the beguiling graciousness of an elitist past seems to have been altered by the more aggressive, democratic sensibilities implicit in Greek Revival and Italianate taste. The cluster of barns and outbuildings across the way is worth notice, especially the corncrib (now rare in Rhode Island).

SM29 **Angell–Ballou House**
Early 19th century. 49 Ridge Rd. (poles 21–22)

This Federal house was another of Smithfield buildings that received notice from the White Pine editors. Two chimneys, a bit awkwardly down on the front slope of the roof, are set in slightly from the ends of the building. The top edges of the splayed board lintels over the ground-story windows are unaligned with the major features of the door frame, but the spoked fanlight, in lead, and the door frame's parts are decisively articulated. The wall clarifies the near-regular rhythm of window openings across the elevation, except for the barest increase of interval on either side of the entrance and its upstairs window above in order to emphasize the center. Given the symmetrical arrangement of ten openings on a rectangular plane, the carpenter-designer for this house opted for their near equilibrated spread, whereas his counterpart for the Asahel Angell House (SM27) set up a tension between the center and the tight clusters of four windows at either end by widening the interval of wall which separates them.

Georgiaville

Architecturally, this is the most interesting mill village in Smithfield, and among the most interesting in the state. Manufacturing at this location began with an iron forge shortly before the Revolution. John Farnum, together with his sons Joseph and Noah, set up the forge at a site on what is now Old Forge Road, using ore from Cranston. Not until 1813, however, when the Georgia Manufacturing Company erected a textile factory (expanded in 1828 and 1846) did Georgiaville's history as an industrial town properly begin. And not until the arrival of Zachariah Allen, one of the leading and most progressive industrialists of the early and mid-nineteenth century, did Georgiaville attain high visibility as an important textile village. He expanded his Allendale operations (see under Allendale in North Providence) by purchasing the original Georgiaville textile plant and using it as an adjunct to the new mill he erected. In front of this mill he built two four-story tenement buildings as well as other housing, and rehabilitated what was already there. He also commissioned a charming stone Gothic Revival chapel for his workers, perhaps from Thomas Tefft, who had designed a Tudor Revival schoolhouse

SM30, SM31 Georgiaville Mill boardinghouses, left, and mill complex

for him at Allendale (NP4). The chapel was demolished sometime after World War II, but Henry-Russell Hitchcock's *Rhode Island Architecture* preserves a view of it. Allen envisioned a verdant village: trees along the streets and a two-acre meadow to be converted into a "very ornamental central village square" with houses along one side. At or just below the dam, he even transformed a wooded ravine by quarrying the bedrock to create a series of cascades.

But the idyll soon perished. Allen, overextended financially, was bankrupted by the Panic of 1857. A series of other manufacturers took over his plant and eventually built another of brick along with more houses for a succession of Irish, French Canadian, and, finally, Italian and Portuguese immigrants. This expansion fortunately left the original nucleus of the village a little apart. Later structures made minimal intrusion—until the location nearby of a sizable housing complex for the elderly in the mid-1980s. The factory complex itself became condominiums after 1987.

SM30 Mill Workers' Boardinghouses

1854, 1855, possibly Thomas A. Tefft with James Bucklin. 34, 38 Whipple Rd.

This pair of nearly identical Greek Revival buildings, four and one-half stories in stuccoed rubblestone, set gable end to the road, with the mill as backdrop in the same material, provide the distinctive image of the village. Zachariah Allen had used the same material in Allendale. Then he had chided manufacturers who wasted money on exposed cut stone except where structural need or hard usage made it expedient, as for windowsills or lintels and door frames. Those who invested in exposed masonry did so, he believed, merely to aggrandize the public image of their factories and themselves. Of the two lodging houses, number 34 (to the left, viewed from the street), which housed male workers, is a perfectly plain box, except for a slight projection of the wall at the gable ends of the attic. A pair of arched windows framed by a larger arch opened into these gable ends (now partially blocked in this building), with a raised frame around it within which flanking piers, primitively capped, simulate support for the arch.

Curiously, however, its companion building for women, number 38, is somewhat more grandly articulated. A projecting base for the ground floor provides a ledge from which piers rise between window stacks, those at the corners being further articulated by rudimentary capitals. Moreover, unlike the uniformly rectangular windows throughout the men's quarters, those around the top story next door are arched. Did Allen intend this modicum of grace as an appeal to his female workers? Or was the change occasioned by functional or (more likely) economic considerations—where plain walls might have been cheaper than piered; or the reverse, where the reinforcement by piers may have permitted thinner walls overall? Or does the more elaborate version represent architectural intervention? Whatever the reason, as early as 1822, Allen had erected a piered store and boardinghouse amidst the plainer walls of family housing in Allendale (see NP).

SM31 Georgiaville Mill Complex

1813, dam and first mill (mill demolished). 1828, 1846, additions. 1853, Zachariah Allen's mill, possibly Thomas A. Tefft, architect, and James Bucklin, builder. 1865, tower. 1871, enlargement. 1880, brick warehouse; 1970, reinforced concrete addition. 1987 and later, conversion to apartments. 15 Higgins Ln.

The principal building of the mill complex incorporates as adjuncts two of the earlier mills, all in stuccoed rubblestone. Although its formal front faces north, toward the mill court, where a later owner added a tower (originally mansarded, now flat-roofed), the south front, toward one of the principal approaches to the village and toward Allen's projected green, is the more visible of the two. There Allen's designer (possibly Tefft) concentrated his aesthetic effects. Notwithstanding the serious damage to this elevation by the incongruous addition in 1871 of a top story in brick, the original intent is evident. The center of the building projected as a piered wall, topped by a broad pediment with a stepped triplet of arched windows (sliced off at either end into a "gabled dormer" when the fourth floor was added). (The piered centerpiece of the mill matches the treatment of the women's boardinghouse. Hence another hypothesis: was the men's boardinghouse erected first in a simple vernacular for use initially as a barracks for construction workers, and the women's dormitory then built to the higher architectural standard of the factory centerpiece?) From this piered and pedimented seven-bay centerpiece, plain walls extended ten bays in either direction, with a simple cornice molding under the eaves and cornered by crude pilasters. Pediments similar to that on the south elevation also capped the end walls. As built, it was the most ambitious of the Greek Revival mills, with the barest suggestion of Italianate influences which were just then coming into vogue, and more self-consciously designed for aesthetic effect than the Allendale mill, as, it seems, was the village at Georgiaville. In his *Rhode Island Architecture*, Henry-Russell Hitchcock described the original building as possessing "something of the grandeur of a baroque palace and the solemnity of Greek Revival public buildings."

The architectural ambitions evident in the mill, the use of piers for visual effect, and the nascent Italianate influence in the arched windows at points of compositional climax all suggest the influence of Thomas Tefft as designer and James Bucklin as builder. As yet no documentation proves it, but Tefft's Cannelton Mill (1848–1849), still standing in Cannelton, Indiana, and shown in a drawing by Tefft at Brown University has similar features. A drawing in the Zachariah Allen Papers at the Rhode Island Historical Society proves that Allen laid out the machinery floors. In the Georgiaville factory Allen continued his inventive ways by introducing long, hollow wooden shafts (instead of many individual pulleys), which turned three revolutions per minute faster than current speeds. Along these he spaced the belting to run his machines. Widely noticed at the time, the system was little followed. After Allen's 1857 bankruptcy, a series of textile and machine tool entrepreneurs operated the plant until its conversion to apartments in 1989. A village once dedicated to production is now given over to bedrooms.

SM32 Georgiaville Mill Office

c. 1871. 27 Higgins Ln.

This delightful, diminutive building is distinguished by the boldly ornamented enframement of all openings. Flat table hoods on brackets cap the ground-story windows; gable pediments with scroll-cut stenciling decorate those in the steep mansard. Especially inventive are the bracketed door hoods with unusual diamond-shaped panels, also displaying sawn decoration, which are fitted under the peaks of the hoods on the front, and were echoed in glazed diamond-shaped windows (now infilled) behind.

SM33 Granite Mill Housing

c. 1813. 23–29 and 6–18 Stillwater Rd.

On Stillwater Road near its intersection with Higgins Lane are two groups of granite mill housing predating Zachariah Allen's arrival, which are among the earliest and best preserved of their vintage still extant in Rhode Island. First, to the north of the Higgins Lane intersection, is a U-shaped court of three one-and-one-half-story houses in rubble masonry set into a slope. Probably rough-stuccoed originally, they show hard use. Each of those with gable ends to the road has two entrances: one facing the road for a semibasement apartment permitted by the fall of the site, the other at the center of the flank walls facing onto the court. The unit which closes the back of the court has two entrances centered side by side and a later dormer. Brick end chimneys serve each unit. At least the rear one (and probably all of them) had been converted to warehouse

SM33 Granite Mill Housing

use when Allen took them over and returned them to housing.

To the south of Higgins Lane are slightly later rubblestone houses of various sizes, also associated with the 1813 mill. Number 18 (with later additions), sited on a larger lot than the others, is given the special distinction of a jerkinhead clip to the front peak of the gable, which is repeated in the door hood. It seems to mark the special status of an overseer.

SM34 Mill Supervisors' Housing

c. 1871. 24–34 Homestead St.

One of these clapboarded one-and-one-half-story duplexes, number 24–26, with its porch extended the width of the house, was converted from the company store; the others, built originally as houses, have individual porticoes over their entrances. The corner house at number 20 served the mill superintendent.

SM35 Joseph Farnum–Brown House

c. 1720. c. 1780(?), door. 243 Old County Rd.

Here the two-and-one-half-story house with slightly off-center chimney has four bays, but in a house of nearly five-bay length, this time the window in the fourth odd bay is slid out nearly (but not quite) to the position of the outermost bay at the opposite end of the house. Window capping at the ground floor is quite elegant. Once more the familiar pilastered entrance, with broken-pedimented door and fanlight, reappears. This one is ornamented with a simple interlace of pointed arches in wood—an outrider of the impending Gothic Revival by way of such "Gothic" furniture motifs such as those popularized by Chippendale.

SM36 Esmond Mill Housing

320–338 Waterman Ave. (Farnum Pk.)

Esmond was also known at various times as Allenville and Enfield for a series of mill owners (the Allen being Philip, brother of Zachariah). Esmond, the final textile operator in this village, famous for blankets, pulled down the old mills and most of the old housing shortly after buying the village in 1905. So what exists is a sizable pier-and-spandrel plant in brick completed in 1907, standard for its period. It would nevertheless be impressive for its extended ranges of tall, spandrel-arched windows, except that the openings have been walled in. Workers' housing from the period is plentiful and generally well preserved. Along the west side of Waterman Avenue (or Farnum Pike) are duplexes designed in variations on Tudor themes lifted from the English model industrial towns of the first two decades of the twentieth century.

Northwest
SCITUATE, FOSTER,
GLOCESTER, BURRILLVILLE

Scituate (SC)

ON THE MAP OF THE TOWN IT IS THE SCITUATE RESERVOIR which dominates. Construction of the reservoir, which supplies water for the Providence metropolitan area, got underway in 1915 with the appointment by the Rhode Island General Assembly of the Water Supply Board. Up to its time, it was the largest public works project ever undertaken in the state. Construction extended from 1917 to 1926. In the process, most of Scituate's villages disappeared. Glen Rock, Elmdale, and Harrisdale in the northeast corner; Saundersville in the east; Ashland, South Scituate, and Richmond in the center of the township; Kent in the southeast corner; Ponagansett and Richmond (the former seat of the town clerk's office) in the west: all were submerged or eliminated to create a forest surround for the reservoir. Of Scituate's manufacturing past of small mills in scattered towns, only Hope remains, in the southeast corner of the town, and it really relates historically to the line of mills south of it along the Pawtuxet in the towns of Coventry and West Warwick. A folk painter of the early twentieth century has left an idyllic image of Clayville as a typical Scituate mill village—a pair of stone mills and their cluster of houses invading the fields of farmers. This village, too, was spared the flood, but its mills are gone, along with most of its farms, although many of the houses remain as a semisuburban enclave.

In Scituate's nineteenth-century balance between manufacturing and agriculture, turnpikes played a major role in getting product to market. As a tall, rectangular area due west of Providence, Scituate intercepted its share of turnpikes radiating from the city. Moreover, it was sufficiently removed from the state metropolis so that their angling as radii from a point had space enough to settle into more determinate east-west courses, and at such regular intervals that they seem almost to mark off the town into

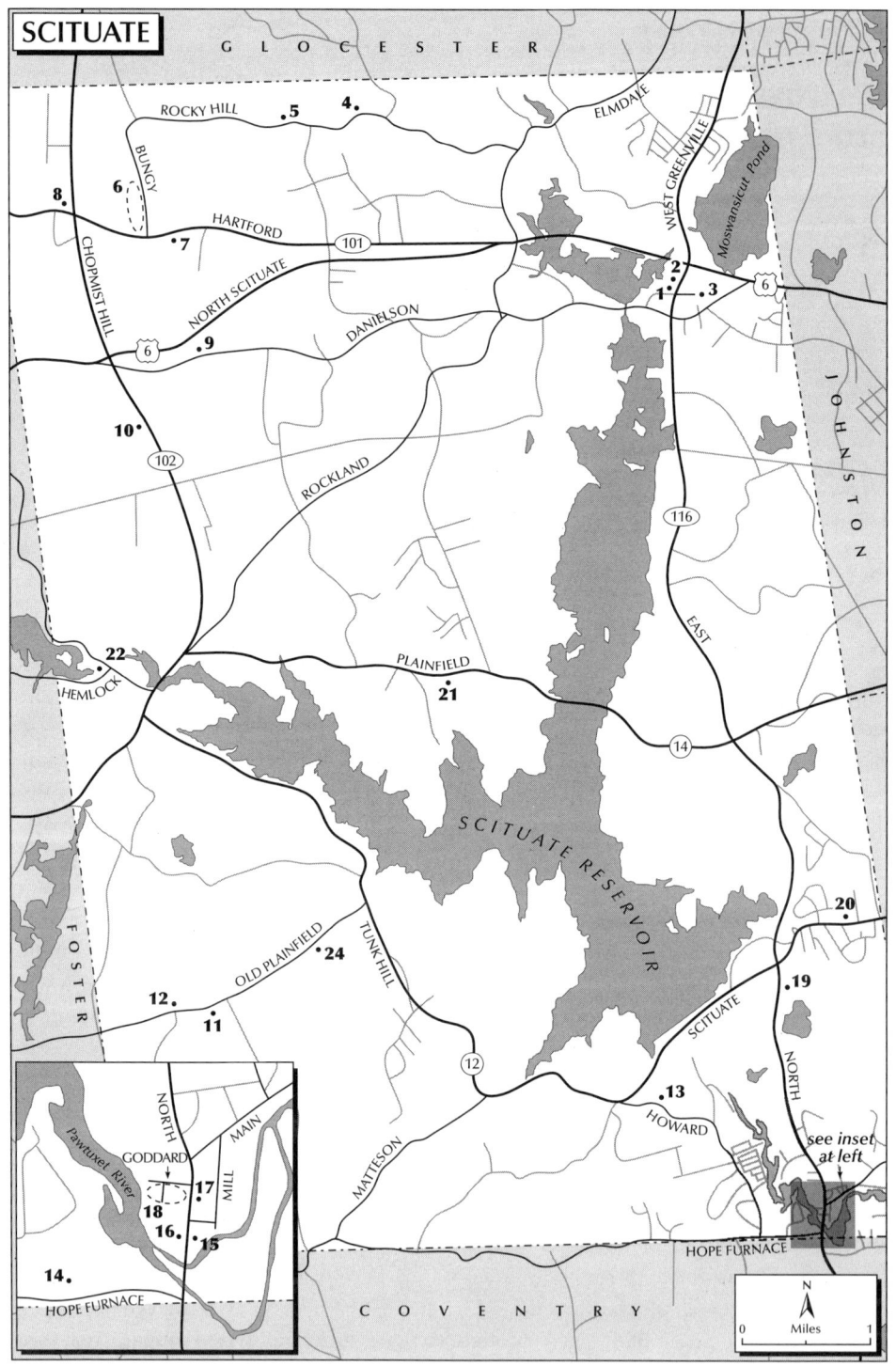

broad stripes in crossing it, before each fixes on a different nearby Connecticut destination. North to south, the pikes are Hartford, Danielson, Central (a local affair which runs into Danielson just over the Rhode Island border), and Plainfield. The reservoir has severed the western part of Central Pike from the rest. Similarly the rerouting of the western portion of Plainfield Pike relegated a disconnected stretch of this to the status of *Old* Plainfield. The relatively untraveled quiet of these severed western sections of Central and Old Plainfield pikes through sparsely populated areas (at least as this is written) evoke, to those receptive to it, some sense of the aura of the ancient pikes and their importance to the nineteenth-century town.

North Scituate

After the construction of the Scituate Reservoir, the village of North Scituate, located toward its northern end, became the principal surviving town and new town seat. Although there is nothing outstanding about Main Street, the white clapboarded ensemble, with some nice country examples of the Federal and Greek Revival styles, is sprucely attractive. Conversions of many of these buildings to antique and gift shops modestly capitalize on the image of a typical nineteenth-century New England rural town.

SC1 North Scituate Congregational Church

1831, Clark Sayles. 619 West Greenville Rd. (near intersection with Main St.)

This fine white clapboard, gable-roofed country church in the Federal style, designed by a carpenter-builder, is notable for the wooden fans which run around all four elevations. Topping the windows along three sides, they continue across the front, floating well above the flanking entrance doors, but reconnecting with the central window. Its three-story tower consists of a tall, unwindowed, quoined box as base, supporting an open polygonal belfry, then a short octagonal transition to a polygonal needle spire. The tower sits a little back from the front elevation, astride the ridgepole rather than projecting in an interlocking relationship with the gable, and hence comes all the way to the ground, as in similar churches by Sayles in nearby Chepachet and (originally) Greenville (GL5 and SM11). Of the trio, this is the latest and the most sophisticated of these designs. Well proportioned in itself, it is also nicely scaled to the box below and maintains its presence as the climax to the ample space before it.

Inside, unusuallly, the pulpit was originally against the entrance wall, in a niche made by flanking boxed vestibules behind the entrance doors, with a shallow choir loft above supported on two columns capped in the same manner as those in the First Baptist Church in Providence (PR56). Hence, in the local phraseology, the congregation sat on "backward [facing] pews." Legend gives two explanations for the exceptional arrangement, one surely fictitious, the other dubious. Allegedly, it was meant to keep the congregation watchful against the break-in of marauding Indians (whose marauding had ceased in this area more than a century before the church was built). Alternatively, it was meant to censure latecomers. More likely, the arrangement could have resulted from some perceived benefit for church function. The symbolic significance of morning light from its eastern orientation for the Sabbath service? The heightened import of the preacher's message delivered from the stage made by the niche? Its convenience as a pocket for the containment of an immersion bath (Baptists having joined Congregationalists in sponsoring the church, although they separated a year after its completion to build one of their own)? The desire to keep preacher and choir at the same end of the church? The positioning of the preacher at the doors in a strategic position to mingle with the departing congregation? Two churches in Foster (FO20 and FO34) have the same orientation, but as a result of renovations; and, of course, so do some late nineteenth-century "auditorium" churches; but there the arrangement reflects a later, different ecclesiastical and architectural approach.

Unexpectedly, too, at least to those who expect whiteness within and without in all early nineteenth-century country Congregational churches, paint analysis carried out in the 1980s indicated that the interior of this church originally contrasted with the whiteness outside. Light green ("stone") trim against white walls set off pews painted in dark maroon trimmed in medium gray to match the painted floor.

Overall, this is among the most captivating and commanding of Rhode Island's early nineteenth-century rural churches, although it no longer serves as such. The church continued in the Congregational denomination until 1897 or 1898, thereafter leading a more indeterminate and dwindling existence until, by 1912, it was without congregation. It then continued as a town landmark and meeting place until, in 1940, it was legally deeded to the town. Today its support comes partly from town funds, partly from a share of such fund-raising activities as the popular art market held every Columbus Day weekend on the common since 1968.

SC2 **North Scituate Community House**
(North Scituate Academy)

1825, c. 1942. 546 West Greenville Rd.

The former North Scituate Academy, adjacent to and north of the church, is a hip-roofed, white-clapboarded structure with chunky belfry. Built for an organization known as the Central Society for the Establishment of a School as a tuition academy, it served as a school until the twentieth century. As built it was a two-room structure with entrance porches, each supported on two square piers, simply paneled and capped, notched into either end. The cupola was then centered between the porches. Its utilitarian style seems somewhere between Federal and Greek Revival, although repairs may have increased the Grecian aspect. The twentieth-century addition to the rear added a third in-line room (with folding partition to combine the two rear spaces as needed). This renovation also placed the rear porch toward the center of the now extended elevation. The ceiling of the square room in front contains a circular field marked off with an illustration of the solar system, dating from the period of the building's design, in which the planets and their moons appear as colorful half-spheres against blue infinity, their orbits in red.

SC3 **Smithville Seminary** (Former)

1839–1840, Russell Warren. End of Institute Ln.

Whereas the adjacency of church and academy represents a utilitarian conjunction of the two institutions, the spatial relationship which once existed between the church and Russell Warren's Smithfield Seminary was more dramatic. The institutional confrontation was the more resonant because Warren's composition, across the green from the church and up on a rise beyond it, when built probably represented the most scenographic array of Greek temple forms in the state.

Converted to "luxury apartments" today, the central block, and its flanking wings represent a very diminished version of what was once here. The flat-roofed, recessed hyphens which flank the centerpiece with its three-story portico have lost their original square-piered colonnades. The terminal pavilions are gone, along with all three crowning cupolas which once topped the principal blocks. The present cross gabling of the central temple block is a 1920s alteration, after a fire which also destroyed both of the outermost wings. But it is astonishing that anything at all remains, given the succession of occupants (including educational institutions and a hotel) which led hand-to-mouth existences behind its noble, and increasingly dilapidated, front.

The Rhode Island Association of Freewill Baptists opened the seminary in September 1840 to serve as a middle academy between common school and college, with a curriculum based on literature and science. Warren's five-part massing of the building may owe something to Thomas U. Walter's analogous linear array of Greek temple forms for Girard College in Philadelphia (1833). In any event, it elegantly recorded the functional program of the school. At the center, marked by an Ionic portico derived from the Temple of Artemis at Ephesus (a model which Warren frequently used), were community and ceremonial spaces above a first-floor reception room, office, and living quarters for faculty: on the second floor, classrooms, library, "apparatus" and "specimen" rooms (the latter for minerals, shells, and other natural history collections); on the third floor, a combined auditorium and chapel, with folding doors to halve it for less grand occasions. The hyphens housed recitation class rooms. Strangely, for such a pure Greek Revival front, the classrooms to the rear projected in half octagons with twin entrances into each of the angled walls. At either end were dormitories, one for boys, the other for girls. This was an exceptional example of private school coeducation for its time, although the two sexes met only for meals, worship, and recitation. (As an old engraving shows, even the front lawn was informally divided into gender territories.)

The school attracted both day and boarding students, the latter from all over New England and, thanks to contacts between textile manufacturers and plantation owners, from the South as well. The first Rhode Island "teachers' institute" was held here in 1845 by Henry Barnard, the nationally famous educator whose reform efforts included school buildings as well as curricula. Its glory years were brief, however. In 1850 financial difficulties forced the sale of the seminary. Thereafter, from 1850 until 1974, a series of religious and vocational schools (one of the latter for "Indians"), a hotel (very briefly), and a summer camp occupied the premises. All were poorly funded; all eventually failed; each (along with fires and a hurricane) took its toll on the building. Eventually, abandoned and derelict, its remains received a last-minute reprieve for conversion to housing. Although the building is sadly reduced, at least it still boasts its portico by Rhode Island's greatest Greek Revival designer. (A footnote, lest one believe that the Ionic scroll which prettily tops the window in the central pediment is an exceptional treatment within the Greek Revival: this was a fragment from a fallen portico column which was converted to its present use in the twentieth century.)

SC4 Brown Homestead

c. 1745; early 19th-century entrance porch and later wing. 253 Rocky Hill Rd.

SC5 J. Aldrich House

Late 18th century; altered sash and later wing. 439 Rocky Hill Rd.

Both of these five-bay, two-and-one-half-story, central-chimney houses have the type of attached entrance porch with a pediment lightly supported on flanking columns which is most particularly associated with Union Village in North Smithfield. In both the supports are something between posts and columns: more post at the Brown Homestead, where the porch was added over an earlier door capped with a simple rectangular transom; more column at the Aldrich House, where side lights in addition to the transom and a lighter treatment throughout indicate more sophisticated awareness of the aesthetics of the Federal style. (The Rhode Island Historical Preservation and Heritage Commission has located two other examples in Scituate, the first of late date and the second early: the Tanner House on Trimtown Road and the Westcott Farm on Westcott Road.)

SC6 Seventeenth-Century Replica Houses

1962–1967, Armand La Montagne. 381, 391, 399, and 405 Bungy Rd.

This astonishing row of four replicas of seventeenth-century houses by the builder-craftsman Armand La Montagne recalls, in order of building, a gambrel-roofed house demolished around 1960 in nearby Glocester, Rhode Island; the Parson-Capon House in Topsfield, Massachusetts; the Ashley House in Deerfield, Massachusetts; and, best of all, La Montagne's final house in this group, a personal adaptation of a Rhode Island "stone-ender" derived from the Eleazer Arnold House in Lincoln (LI6).

La Montagne's building career began with his desire to restore a seventeenth-century house; but the right house was either unavailable or in the wrong place. So, using as a model what had become a drearily sited, dilapidated house in Glocester which was about to be destroyed, he built his own "seventeenth-century house" on a site in Scituate. Trees from his property provided most of the timbers and boarding. He used old tools and techniques (as well as power tools and modern construction equipment) to make *de novo* what he was unable to attain by restoration. After living in his Glocester gambrel for a while (number 381), he replicated the Parson-Capon House next door (number 399), moved into the second house, and sold the first. (A subsequent owner augmented the Parson-Capon replica with an attached two-car garage which can loosely be described as in the seventeenth-century manner.) While living in this house, La Montagne successively built the Ashley replica (number 391) on speculation and the Arnold stone-ender (number 405) as his third and final house for himself. The latter is a more vertical, compact version of the original, with an added workshop wing.

What elevates these four houses above routine replication and technical achievement is the intensity of La Montagne's feeling for the originals, the sensitivity of his craftsmanship, his eye for detail, and a subtle theatricality. This last results, it would seem, from the heightened sense given to even the smallest details when they are so meticulously and affectionately valued as signs of a past way of life which he meant to regain and relive against all odds. Inside as well, he employed the ancient ways, but inflected original plan types to more open and picturesque treatment derived from modern planning. His own house is furnished throughout by his replicas and variations of seventeenth-century furniture. The handsome fieldstone walls, a full three feet thick and frequently photographed as splendid examples of colonial walling in a state famous for them, are, like everything else, La Montagne's doing. "I like the masculinity of seventeenth-century building," he says. (Jack Sobon and Roger Schroeder's *Timber Frame Construction* [1984] uses La Montagne's houses and methods in discussing a modern revival of this ancient mode of construction, and credits him with an important role in initiating it.)

Eventually, having built twelve houses in his favorite style (eight at scattered sites, mostly on commission), La Montagne abruptly stopped house building altogether, although he has some disciples. He turned to sculpture in painted wood, contemporary celebrities meticulously depicted life size from multiple photographs, down to the crinkled surface of the skin: General George Patton for the Patton Museum; President Gerald Ford for the presidential library, Ted Williams for the Baseball Hall of Fame, along with sporting greats for other

SC6 Seventeenth-century replica house (adaptation of the Eleazer Arnold House in Lincoln; see LI14)

halls of fame. Meticulous replication and admiration for masculine qualities seem to link La Montagne's disparate worlds. Still, it is disconcerting to step into a seventeenth-century Rhode Island stone-ender and confront Larry Bird crouched, life size, a basketball on his fingertips moments before delivering another two points to the Celtics, being readied for immortality in the Basketball Hall of Fame.

sc7 **Black Estate**

Mid-20th century. 960 Hartford Pk. (barely visible through evergreens)

Through the evergreens that nearly obscure this house, it is the extravagance of the wavy cut of wooden shingles, piled up to simulate thick thatching, which first catches the eye. The matted texture of the roof responds with picturesque effect to the projections and recessions of the supporting white stucco walls and the heave of dormers and gables. This house is saved from coyness by the severity of the opposition of roof to wall, by the skillful composition of its plastic massing, and by the press of all first-floor casement bands against the eaves of a roof extravagantly pulled down to them. Windows for the second floor and attic peek through the roof under trapdoor dormers at two levels, one of which is crossed by a projecting chimney. A peppering of carefully placed small windows and plaques sparingly relieves such plain surfaces as remain, like the cylindrical swell of the turret beside the entrance. The roughcast and thatched cottage that inspired English Arts and Crafts designers is blown up to

SC7 Black Estate

country manor house size to accommodate a twentieth-century suburban fantasy that combines the cozy charm of a Cotswold cottage with the rustic pomp of a turreted Norman manor house.

SC8 **Christopher Smith–Fenner House**

c. 1790. Northwest corner of Hartford Pk. and Chopmist Rd.

A two-and-one-half-story, central-chimney house typical of the plain country Federal treatment is here restored to natural boarding with a roof of cedar (rather than the usual asphalt) shingles. As early as 1807 it was sold to the Fenner family and has served as an inn and an antique shop, and is now again a residence.

SC9 **James Aldrich–Florence Price Grant House**

c. 1830–1835; wing and outbuildings mostly 19th century. 853 Danielson Pk.

This two-and-one-half-story Federal house has unusually impressive features for its rural location: a hipped roof with paneled monitor; a recessed fanlighted and side-lighted door set into recess, framed by a rusticated arch; corner quoining; and the commanding spread of its five windows, almost equally spaced across its front. The unusual depth of the house permits four similarly spaced windows along the sides. Interior chimneys flank the monitor. James Aldrich, for whom it apparently was built, served for a period in the state senate. Another house, the S. P. Taylor House (SC25), virtually identical with this, is believed to have been erected by a Providence builder for himself as a country retreat. The houses are surely related. The Aldrich-Grant House is now owned by the Episcopal Diocese of Rhode Island.

SC10 **Dexter Arnold House**

1813; contemporary and later 19th-century farm buildings. 1335 Chopmist Rd.

This two-and-one-half-story, five-bay house with central chimney has a plain clapboard elevation with windows a little small for its expanse. It is given pretension, however, by a door made tall by a transom topped with a bold hood carried on brackets, and by pilasters (unexpected in a farmhouse) at the corners. A fine nineteenth-century carriage house with cupola, a combined woodshed and carpenter shop, and a corn crib and privy also remain, making this among the finest surviving farm complexes in Scituate—and absolutely the finest in conjunction with a house of this quality. It remained in the Arnold family until 1975.

Potterville

SC11 **Community Center** (Potterville District 7 Schoolhouse)

c. 1852. 320 Old Plainfield Pk. (at Carpenter Rd.)

SC12 **Tavern** (Former)

c. 1711, c. 1783 (western end), c. 1800 (eastern end). 373 Old Plainfield Pk.

What remains of Potterville is a small cluster of buildings on the original western segment of Plainfield Pike as it passed through Scituate before its truncation by the reservoir and relocation. Two structures, somewhat separated, stand out: the former Potterville District 7 Schoolhouse, now a community center, and the tavern. Both are plain carpentered structures. Probably at the time the schoolhouse was converted to a community center the entrance was shifted from its expected place in front to the rear of the building, perhaps to be within easier reach of its still extant privy. The tavern began as a house for an E. Fish or Fiske; by 1730–1731 it was functioning as a tavern. The eastern end contained a kitchen in the basement open to the downhill side, dining room on the first floor, and ballroom above. It ceased being a tavern when Abner Potter bought the house and built two dams and a mill for the manufacture of wooden textile bobbins on the stream behind his property. It remained in the Potter family until 1949. The door frames appear to be Greek Revival alterations. This withered village is additionally significant for building because of the nearby Nipmuc quarries, once a source for granite, now also inoperative.

SC13 House

c. 1800(?) 339 Howard Ave.

This late-eighteenth-century house with the same format as the Arnold House (SC10) also reveals a provincial effort to lift a plain house to an unexpected degree of sophistication and monumentality. Whereas the typical house in Rhode Island by 1800 shows door and window openings as separate entities, this elevation features a wide, projecting board molding (really too wide) across the front serving as a unifying band for downstairs openings. To it the base of the entrance pediment over the transom light is aligned; against it the exaggerated splayed lintels of the first-floor windows abut. Such unifying moldings are more typical of ambitious late colonial design than of the fully developed Federal style, and indicate a builder who holds with the past. The blunt vigor of his interpretation, however, has a charm of its own, although the suspiciously small pediment may in fact be a twentieth-century alteration.

SC14 Breezy Hill Farm

1793. 71 Hope Furnace Rd.

Again the familiar five-bay central-chimney format, with the traditional four-spot grouping of the peripheral windows pulled just sufficiently far from the entrance axis to maintain a tense equilibrium among the openings. The pilastered and pedimented door with leaded fanlight and, more unusually, a molded and dentiled cornice are handled with such discrimination that nothing appears meretricious. Downstairs window tops align with the height of the entrance pilaster caps, their splayed lintels projecting above. Commanding but not overwhelming, the door seems perfectly proportioned to the clapboard field that immediately surrounds it and, like the windows, beautifully equilibrated within the pattern of openings. In short, here are no surprises; just conservative building design so exquisitely and sensibly considered that it could hardly be bettered.

Hope

As a manufacturing village, Hope was first important for its ironworks. A site for the furnace was purchased in 1766 by the four Brown brothers of Providence—Nicholas, Joseph, John, and Moses—together with other investors, including Stephen Hopkins, a former governor, and Israel Wilkinson, former owner of Unity Furnace. The venture was named Hope Furnace for the mother of the Brown brothers. Initially producing such items as potash, kitchenware, nails, and cask hinges, the furnace turned to cannon during the Revolution. Between 1778 and 1781 some 270 cannon were cast and bored at Hope, most to arm American privateers. Thereafter the profitability of such a small operation declined rapidly and the furnace property was sold in 1806. The very same year saw the first cotton manufacturing at Hope by other entrepreneurs. But the Brown absence from Hope was brief.

Brown investment (now as Brown and Ives) returned to Hope in 1821 when these Providence entrepreneurs bought out the original owners of the cotton mill and built another. Most of this early complex burned in 1844. (From this period, however, a small, one-story clapboard mill building, a mere three windows in length, with a gabled monitor, existed in the mill yard until 1989, when it too, succumbed to fire.) With minor reorganization, in which John Carter Brown and two Ives brothers (Moses Brown and Robert H.) were joined by yet another Providence investor, Charlotte R. Goddard, the ruined mill was replaced by the centerpiece of the existing complex.

SC15 Hope Mills

1844, Mill No. 1, David Whitman, mill engineer, and Thomas Sharpe, mason. 1871, Mill No. 2. 1916, weaving shed. 1960, 1972, additions. 1 Main St. (Route 116)

The original four-and-one-half-story masonry structure of random mixed with cut blocks, 183 by 55 feet with a projecting stair tower, is well preserved even to its sixteen-over-sixteen sash, except for the flattening of what was once a

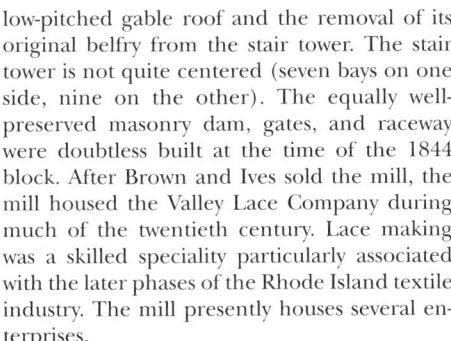

SC15 Hope Mills (above, left)

SC16 Samuel Allen House (above, right)

SC17 Mill workers' housing (left)

low-pitched gable roof and the removal of its original belfry from the stair tower. The stair tower is not quite centered (seven bays on one side, nine on the other). The equally well-preserved masonry dam, gates, and raceway were doubtless built at the time of the 1844 block. After Brown and Ives sold the mill, the mill housed the Valley Lace Company during much of the twentieth century. Lace making was a skilled speciality particularly associated with the later phases of the Rhode Island textile industry. The mill presently houses several enterprises.

SC16 Samuel Allen House

1878. 15 Main St. (opposite the Hope Mill)

Stubby two-story wings project from the near-cubic core of this late Victorian house with Queen Anne detail. Porches fill in the L-shaped void made by the rear ell extension; in front, a lattice-skirted, unroofed deck, surrounded by an ornamental railing divided by a band of rectangles beneath a row of stubby balusters, ties the house to its sloping site. Against the side elevation, a roofed section shelters the entrance and shades the area for sitting. Topping this, a small outlook porch off an upstairs bedroom repeats a reduced version of the railing below. Turned supports and spindles, together with arched and diagonal bracing, complete this cascade of porching. In fact, they provide the principal adornment for what would otherwise be a quite plain clapboard house, thereby emphasizing these useful and decorative appendages as characteristically Victorian. The Allen House served a series of mill superintendents. Its pivotal location with respect to the mill and to workers' housing is almost precisely duplicated at the Brown and Ives mill at Berkeley in Cumberland (CU8). It seems likely that the Hope Furnace was sited somewhere toward the rear of this property.

SC17 Mill Housing

c. 1872; some buildings possibly earlier. Mill St.

Toward the foot of this sloping street of well-preserved mill housing are gabled, one-and-one-half-story duplexes with doors at the third points of the elevations, which are doubtless of earlier date than the double row of mansarded duplexes, with doors at either end, farther up the hill. The latter, shown in Henry-Russell Hitchcock's *Rhode Island Architecture* (1939), were probably built at the time of the mill's 1871 enlargement, and remain perhaps the best-preserved street of mansard housing in the state.

SC18 Mill Housing

Probably 1920s. 1, 2, 3, 4 Brown St.; 7, 9, 11, 13 Goddard St.

On a hill above Main Street in a cul-de-sac enclave with streets named for and doubtless by the mill owners (Ives Street is nearby) are four beautifully maintained early twentieth-century duplexes, probably for overseers. Brown and Ives built roughly similar housing at the same time at Lonsdale in Cumberland (see CU7). In a mildly Neo-Colonial treatment, upper-story windows are arranged under a double-sloped roof which also seemingly invokes the trapdoor dormers associated with some of the earliest workers' housing. Entrance porches contain a typical cozy touch of the period: high-backed flanking benches topped with decorative latticing, some roofed with arches, others with arching under pediment-inspired gables. Numbers 66 and 68 Main Street offer variant versions of these duplexes. So Hope provides well-preserved examples of mill housing from all periods of its long history of manufacturing.

(Aficionados of mill housing may remember a charming view in Henry-Russell Hitchcock's *Rhode Island Architecture*, taken in nearby Fiskeville, of a curving row of workers' cottages shaded by a row of elms. These are of the earliest corner-door and two- or three-window type but, exceptionally, in rubble masonry rather than wood. The elms, of course, have perished, but some of the houses still stand, behind other buildings on an alleylike lane opposite the Fiskeville firehouse, all resurfaced except one, and all enlarged in various ways.)

SC19 City of Providence Water Supply Board, Philip J. Holton Water Purification Plant, Scituate Reservoir

c. 1930. North Rd.

The Moderne monumentality of this water filtration plant recalls such public works of the time as those built in conjunction with dams by the Bureau of Reclamation and the Tennessee Valley Authority, as well as many New Deal projects erected under the Public Works Administration. Slightly varying arrangements of cubes in sheer planes of yellow limestone, broken only by the skimpiest projections of vestigial cornices, make slightly asymmetrical flanking wings behind an entrance centerpiece. This consists of an unadorned projecting gray marble frame, infilled with plate glass and thin metal mullions, which rises nearly the full height and full length of the center block. An austere frame for the transparent wall to the public reception lobby, it also appears as the most elemental of classical porticos, meant to recall the traditional neoclassical public building even as it nudges the past toward streamlined modernity. Like the fountains and the manicured, parklike landscaping of its surroundings, this temple to progress was intended to attract the public. Visitors were invited into the tall, luminous reception hall, adorned with such typical features of the period as a metal-railed stair making a slashing diagonal in the space, up to a balcony (both stodgier and less open than in the most dramatic Moderne interiors) and set off against a curved wall, which displays a map and a cross section of the plant. By now enthusiasm for this intended destination of Sunday drives seems to have waned. The reception desk is empty. The slab-capped entrance door through the metal and glass wall is locked. A sign points the infrequent (even unwanted) sightseer to a side entrance behind the sans serif aluminum identification of the facility on one of the wall planes. The arcadian ambience is magnified by appearing so pristine, so benign in its promise of uncomplicated technological benefit, and so deserted.

SC20 Knight Farms Subdivision and Knight Farm

1970s–1980s, c. 1900–1930s. Route 12 (about 1 mile east of Route 116)

One-half of the once very sizable Knight Farm was sold off in the 1970s for a subdivision. A sign at the brick-piered entrance announces it as Knight Farms. Here is a handsome array of variations on the good life as these are manifest in expensive, custom-designed suburban housing of the 1970s and 1980s, edging into the 1990s. Houses with a relaxed formality, tending toward symmetry, with "Mediterranean" derivations from French and Italian precedents, predominate. (Conversely, American colonial, French Provençal, English medieval, and suburban ranch are minimally represented.) Most are masonry, as a veneer if not as a wall, especially brick, indicating the effort here to suggest the luxury of permanence and substance. (This is in marked contrast to the more informal but equally posh use of wood in a picturesque angularity of gables and dormers interspersed with decks and allusion to American vernacular or American colonial, either traditionalized or modernized in varying degrees, which characterizes another type of expensive development during the same period.) The predominance of masonry and "Mediterranean" at Knight Farms indicates its special appeal to the substantial population of Mediterranean descent in Rhode Island.

Although most of these houses derive from older suburban housing, all are subtly altered from earlier suburban precedents by the intervention of modern architecture. The same could be said of such accoutrements as mailboxes, light standards, and flanking piers at some driveway entrances, which indicate a wide range of possibilities for the external finishing touches of affluent houses toward the end of the twentieth century. Handsome landscaping, as posh as the houses, returns the cleared fields toward a wooded environment.

At the next entrance east on Route 12, a rutted gravel road leads back to the plain frame house and barns of Knight Farm (which can be viewed only from the highway). The barns present a pretty complex from afar: the cow barn with a big gambrel roof of terra-cotta block typical of the 1920s and 1930s, smaller gambrel calving barn behind, and, behind both, a tall wooden hay barn of earlier, possibly late-nineteenth-century vintage. Somewhat shabby on closer view, it is, nevertheless, in lively operation. It, too, has been suburbanized: riding horses have replaced cattle. There are complaints at the farm that places to ride are fast disappearing. An abrupt dip in the intervening field reveals only the turrets and dormers of the adjacent development through the tops of the neighboring suburban reforestation. How long can Knight Farm hold out against Knight Farms?

SC21 Horace Battey–Joe Barden House

Between 1815 and 1830. 710 Plainfield Pk. (just beyond Trimtown Rd.)

This fine country Federal house, two and one-half stories with a large center chimney, is mostly screened from the road by a high hedge. It is intensely boxy, tautly and sparingly linear. A wooden fan over a side-lighted door mitigates the plainness a little, but is itself tensely linear. Inside (not open to the public), original painted stenciling from 1821 remains in the front hall, augmented by the former owner who is principally responsible for its restoration. An artist who specializes in the medium, she has restored and recreated stenciling in a number of old Rhode Island houses. Horace Battey, proprietor of the local general store, was also a justice of the peace. Sold to Joe W. Barden in 1867, the house remained in the Barden family until 1962.

SC22 Site of Ponaganset Village (Barden's Mills, Bettyville)

1883–1884, dam. Mid-19th century, granite bridge and burial ground. Intersection of Ponaganset Rd. and Hemlock Rd.

This intersection, surrounded by the forest of the Reservoir Management Area, was the site of one of Scituate's lost villages—and is a place to remember them all. A little up Hemlock Road is a handsome granite dam in an arc, rare in Rhode Island, but appropriate where rock on either side of a narrow channel gives solid abutment to the curve. Now part of the Scituate Reservoir, it is a beautiful spectacle when an overflow brings an arc of water against the tumble of native rock in this remote and sylvan setting. John Barden erected an ironworks at the fall in 1760, and later a gristmill. (Betty was his wife.) Eventually Providence entrepreneurs moved in to establish a small cotton mill in 1826. The mill burned and was rebuilt in 1854. Both company and town then became Ponaganset. The dam, built to hold a reservoir on the Ponaganset River, followed in 1883–1884. Back to Ponaganset Road, where another view of the dam and its random masonry abutments is available from a reinforced-concrete bridge below the dam. Around the road intersection and bridge the mill and the village once clustered.

At .4 mile from the bridge is the Ponaganset burial ground, possibly the loveliest small cemetery in Rhode Island and certainly the one with the most architectural approach, but of a type that occurs elsewhere in Rhode Island. A wide granite wall encloses three sides of a rectangle, the front left open as a high granite threshold, notched in the front by two stairs. In this roomlike enclosure, open to the sky and pressed by the reservoir woods, the generations who lived here are remembered. The granite dam, the old bridge abutments, and this granite enclosure bear witness.

Foster (FO)

A town of poor soils and small farms, rocky and, by Rhode Island standards, hilly, Foster contains the state's highest elevation. Jerimoth Hill in northwestern Foster is 812 feet above sea level. The area of the future town was isolated in the colonial period and retained a sense of remoteness, along with other western towns in the state, into the mid-twentieth century. Well into the "Outer Woods" beyond the Seven Mile Boundary measured from Fox Point in Providence, its sparse population early complained of lack of representation and difficulties in conducting town business. Split off as part of Scituate in 1731, it separated from Scituate in 1781.

The four principal turnpikes which cross Scituate west from Providence also cross Foster. Central Pike, officially the Foster and Scituate Central Turnpike, built 1814–c. 1824, was once an important east-west road across central Foster. Located between Danielson and Plainfield pikes, it was a favorite route for Connecticut drivers to get their cattle to Providence markets but had fallen on hard times as early as the 1840s. It was severed at Balcom Road toward the west and some time thereafter by the Scituate Reservoir toward the east. Although it is macadamized now, its bypassed quality preserves some aura of an early turnpike, matched only by sections of New London Pike (see under West Greenwich).

But no village transportation hub, like Greenville in Scituate or Union Village in North Smithfield, developed. No major north-south routes cut across the major turnpikes. Nor did Foster seem to be placed to catch many overnight travelers. Its hamlets were of the smallest sort, except Clayville, a one-time mill village which straddles the Scituate border, and perhaps Hopkins Mills for a few decades before its small mill folded around 1850. Foster's headwater brooks and streams fed rivers large enough for serious manufacturing only outside its boundaries. No railroad touched it except for the very tip of a spur to Clayville. The trolley line from Providence to Danielson, Connecticut, completed in 1902, affected more important change in Foster than any other means of transport before the popular use of the automobile. But condemnation for the Scituate Reservoir severed the "P. & D." and its brief existence ceased in 1919. The bulk of Foster's population increase and the height of its prosperity before the late twentieth century occurred between 1780 and 1830, after which a long period of decline set in. Those who stayed in Foster turned mostly to livestock as the best living from the poor land, or to sawmill operations. Most farmers got by only with a second occupation.

Building in Foster was minimal after 1830. One result has been the often dilapidated preservation (until well after the mid-twentieth century at least) of ancient farms and outbuildings. No other town in the state has a larger number of rural Greek Revival church buildings. The type survived here through the Victorian period. Although the peak for Greek Revival building is the decade of the 1840s into the 1850s, four of the six examples in Foster date from between 1864 and 1882. Virtually every Foster hamlet has one. Without them, indeed, these tiny hamlets might be invisible to outsiders. Already by the 1890s local writers commented on the delights of a town that "time had passed by." Not until 1970 did the population of Foster exceed the peak it had previously reached in 1830 as a new breed of country dwellers, followed by out-and-out suburbanites, moved into the lean arcadian void.

FO1 P. Rounds Farm
c. 1820, additions c. 1870. 3 Ponaganset Rd. (pole 4; 1 mile north from Scituate line)

This secluded farm is sited at the top of a sloping field with a fine complement of three small, shingled late-nineteenth-century barns, an outhouse, and icehouse, all as well fitted to the landscape as the house itself, and all handsomely maintained with the prettiness of a Currier and Ives vision of rural America. The Greek Revival house is a plain, carpentered one-and-one-half-story, five-bay, gabled building with central chimney, to which was added a set-back ell with a Victorian porch fitted into the setback. Across the road from the house is a rustic twentieth-century sugar house. Maple sugar and Christmas trees are now this farm's suburban-oriented crops.

FO2 Reproduction Covered Bridge
1992, Charles Borders and Jed Dixon, carpenters, with a volunteer crew. 1993, destroyed by arson. 1994, rebuilt. Central Pk. over Hemlock Brook, west of Foster Center Rd.

From 1920, when Rhode Island lost its only covered wooden highway bridge, until 1992 it was the only New England state without one. A shame, thought Robert Salisbury, a Foster resident, after his young son, enthralled by pictures of covered bridges on a glossy calendar, asked to see one. Over a period of better than six years, Salisbury spearheaded a drive to gather funds for such a bridge in Foster, to obtain permission from the state Department of Transportation for its erection on a public highway, and to find skilled carpenters capable of directing a volunteer weekend crew for its erection. Triumph at last; then arsonists destroyed the result close to the first anniversary of its opening. A stunned community rallied to rebuild it (with the convicted culprits compelled by court order to join the work force). Because DOT insisted on the safety of a steel-reinforced road deck, purists might deny the validity of the result. But the shed derived from Ithiel Town's design, patented in 1820, for lattice-trussed bridging composed of X-configurated membering. This simplest of standard bridge trusses assembled from the simplest elements was overwhelmingly popular in the nineteenth century for bridges of modest span, whether in wood or metal. The same advantages account for the choice of this construction by the novice builders of this 38-foot crossing of Hemlock Brook. (Rhode Island boasts two other smaller reproduction covered bridges: one on private property, the other in a state park, neither part of the public highway system.)

FO3 Deacon Daniel Hopkins House
c. 1790, c. 1810. 1980, ell. 123 Central Pk. (at Balcom Rd.)

Given the handsome elaboration of the pedimented and fanlighted doorway, the rest of this typical two-and-one-half-story, five-bay, central-chimney Federal house seems rather too plainly carpentered, although impressively ample because of the pull of the flanking windows away from the center. It is also remarkable for the quality of its well-preserved front parlors to either side of the squarish entrance hall, and most especially for the southwest parlor. (The interiors are not open to the public.) The latter has teak woodwork allegedly carved at sea by the original owner's ship captain brother, which, to this day, has never been painted. The parlor contains what are, by country standards, a complex cornice and chair molding, crossetted paneled doors, and a mantel flanked by columns, ornamented with an incised pattern of alternating triglyphs and stars. Moreover, both parlors contain original wall stenciling of entwined garlands and flowers by one "J. Gleason"—a rare signature in what is customarily an anonymous art. Daniel Hopkins owned a local sawmill and was a deacon of the meeting house in Foster Center. Reportedly, it was this family which sealed off the failing pike by building a chicken coop across it. The Balcom (also spelled Bolkcom) family purchased this property in 1856; hence the name of the road.

FO4 International Lead and Zinc Research (ILZRO) House
1973–1976, Marc Harrison, Kent Keegan, and Barry Fitzpatrick, designers. 22 Balcom Rd. (not visible)

This prefabricated house, hidden in the trees, uses steel framing and modular metal insulated panels to demonstrate the virtues of lead and

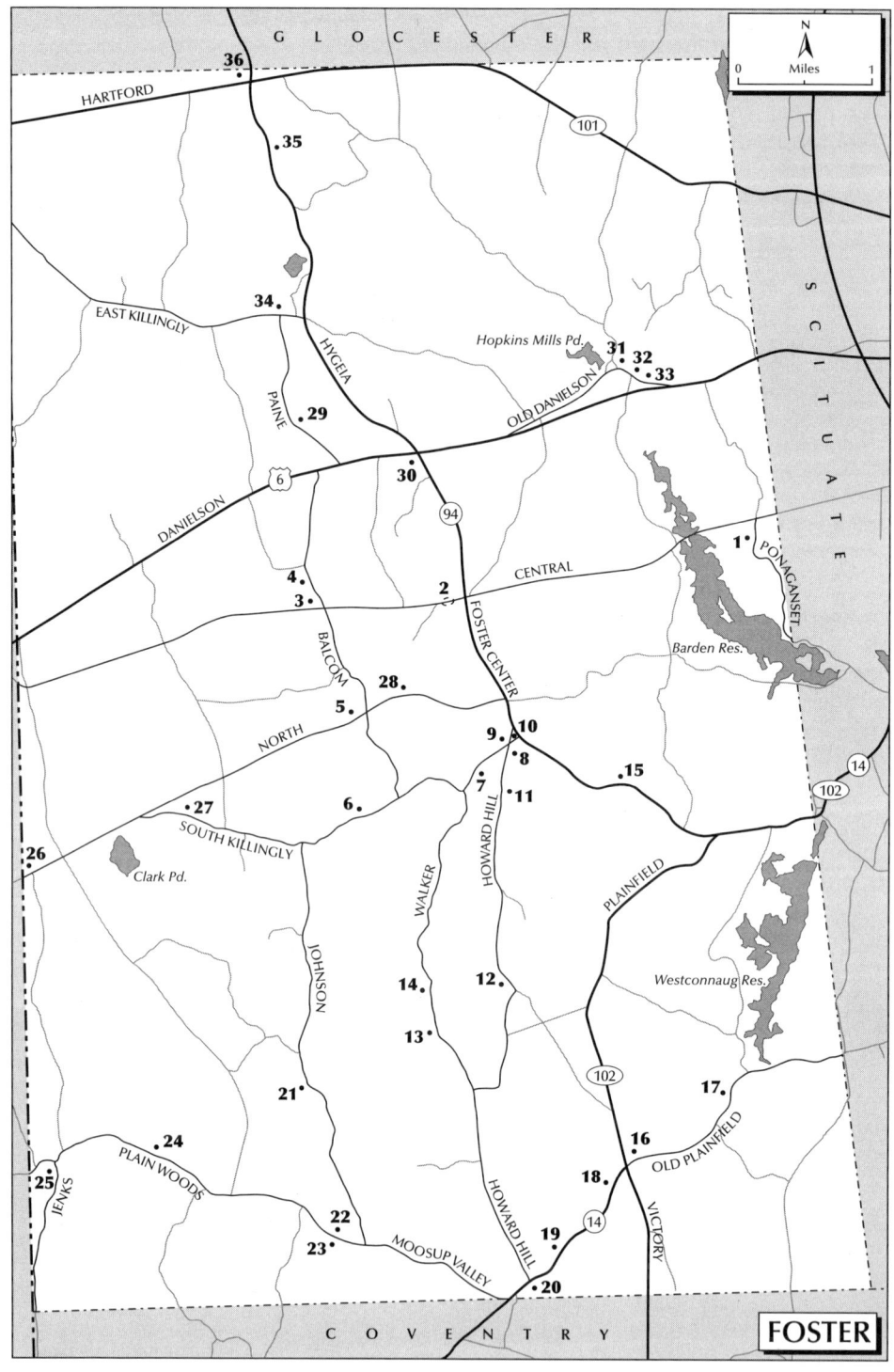

zinc for home building. A team at the Rhode Island School of Design in Providence captained by a professor of industrial design, Mark Harrison, developed it as a prototype that could be built from a kit of parts allowing innumerable variations (and even a traditional appearance, if desired). Using the pewter gray natural to lead as a surface for the interlocked panels, and openings banded in silver natural for zinc (with corners rounded as in the windows of cars and trains), this industrial house fits nicely in its woodsy setting. The slight elevation of the house on a wooden deck, three steps above the forest floor, gives a sense of being slightly suspended in the woods. Modular packages for heating plant, closets, and a "greenhouse bar" are attached to the basic rectangular volume of the house like saddlebags to a horse. Other boxlike service packages form pent-roof monitors for top lighting. The house contains all of its original interior fittings, including furniture custom designed by the RISD team. This modest retreat, which makes uncompromising use of high-tech components for rustic effect, mediating between the realms of architecture and industrial design, is among the modern buildings of consequence in the state. It thus far remains in the realms of experiment and promotion, although variants had been built at the time on sites as far apart as Pittsburgh and Peru.

FO5 George Phillips Farm

1840; original ell later lengthened. 31 North Rd. (pole 31; .1 mile from intersection with Balcom Rd.)

This very late one-and-one-half-story, central chimney Federal house is interesting for its porch, an elliptical vaulted enclosure beneath a pedimented roof supported on attenuated, turned, postlike Tuscan columns. Two other houses in Foster, both twenty-five years earlier than this house, have comparable porches (see FO16 and FO19). Records indicate that an earlier house on this site burned. Could the porch be a sentimental reconstruction or a salvaged relic? In any event, the country carpenter who put it up used an augur to drill a series of holes in diagonal bands across a quarter-round molding to edge the support under the pediment, hoping thereby to give an illusion of rope molding. It was a familiar country dodge, using the tools at hand to suggest a grand effect which was economical in both time and skill compared to what its carved equivalent would have required. Country, too, in that the two bands of sparkle from the drill holes running the depth of the porch under its eaves are discordant with the severity of the rest of the treatment. The same molding occurs under the eaves of the main mass. (Its continuation partway along the eaves of the ell indicates that a portion of this was original to the house, which was subsequently further extended without the trim.) The placement of fussy ornament along the edges of forms to mitigate their starkness—like a ruffle on a sleeve, the lace of a doily, or a row of petunias outlining flower beds—is particularly attractive to the provincial. The ornament does not participate in *articulating* the larger composition, but rather plays an *alleviating* role. Yet there is charm and poignancy in such ornament too: charm in the directness of its handling; poignancy in the carpenter's intent to elevate his craft toward "art," and in his client's willingness to pay for the "extra" for the pleasure it brings, while perhaps also elevating his image within his world.

Inside (not open to the public), the main body of the house has a five-room arrangement quite common to Foster houses. Whereas the normative arrangement for such a plan is a double parlor in front separated by the entrance hall, with bedroom, central kitchen, and pantry in back, here one parlor and the kitchen are in front, with two bedrooms and a pantry behind the kitchen in back. It is a practical use of a conventional plan and was common in late-eighteenth- and early-nineteenth-century farmhouses where a second parlor would have been superfluous. Like most farmers in areas where soils are poor, Phillips had an adjunct job; he also operated a sawmill making furniture and shingles.

FO6 Jeremiah Bennett Farm

c. 1790; c. 1840; barns and outbuildings probably mid- to late 19th century. 50 South Killingly Rd. (pole 50; .45 mile from junction with Balcom Rd.)

This farmhouse is a typical two-and-one-half-story, five-bay, central-chimney Federal house inflected toward Greek Revival by a remodeling which accounts for the heavy, corniced door surround and broad corner pilasters. The ell with pier supports and temple pediment on the side elevation was added during the change. The site and surroundings of this well-maintained property are also important. The house occupies a commanding location on a knoll above the road, with a nineteenth-century board-and-batten barn behind, a slatted well house in front, and a combined cattle barn and equipment shed hard by the opposite side of the road, where a field slopes steeply down to woods.

FO7 **Foster Town Pound**

1845. South Killingly Rd. (on the west edge of Foster Center Village)

This tumble-down enclosure, nearly 48 feet square, stone walled and with a stone lintel over a narrow entrance, held stray animals. A crude modern wood gate replaces the iron original. (On town pounds, see GL27.)

Foster Center

The grove of trees from which this place took its older name of Hemlock has mostly disappeared, leaving to local geography and government the renaming of this well-preserved country village of eighteenth- and nineteenth-century buildings. Although it is the administrative heart of Foster, its existence as a village is now disappointingly truncated. Its commercial buildings—which once consisted of three taverns, a general store, a grain store, and a couple of blacksmithing shops, to name only its most prominent enterprises—are now all gone to houses, leaving it for the most part a place of residence and town business. The center of the town is marked by the triangular intersection of three roads where South Killingly Road crosses a little south of the fork of Howard Hill and Foster Center roads.

FO8 **Public Library** (Foster Center School, Hemlock School)

c. 1822 with later ells. 1957. 1964, 1970, additions, Richard Colwell. 184 Howard Hill Rd.

FO9 **Foster Center Baptist Church** (Foster Center Christian Church)

1882. 1959, 1972, additions. 185 Howard Hill Rd.

Architecturally, the buildings at the heart of Foster Center are more rewarding as a pretty ensemble adapted to a varied topography at a major rural intersection than they are individually. But both separately and taken together, they suggest interesting architectural reflections, particularly the 1822 school and the 1882 church.

Although the school (or so it functioned until 1952) might be considered Greek Revival because of the simplified entablatures over the transomed doors and the breadth of their frames, which are minimally capped by a molding, its date is very early for the style in such a remote location, and, in fact, little about it suggests the Greek Revival. No breadth of corner treatment or window frames, nor sculptural emphasis of the gable, marks the building as anything but carpenter vernacular of the period, while the fussy, rather flimsy cupola seems to be mostly a twentieth-century injection. Could it be that the most prominent features of an older vernacular structure were later redone in the Greek Revival manner (and perhaps even renewed when the library conversion occurred in the 1950s)? In any event, let the doors at least stand for Greek Revival, so that we can set them against those on the church across the way.

The church is not exactly an architectural gem, but is fascinating as Greek Survival. It means to be a typical Greek Revival country church—one of seven we shall look at in Foster, which alone among Rhode Island towns has preserved so many of them. This, the first in the itinerary, is the latest in the series. It boasts the basic hallmarks of the Greek Revival country church: doors broad-framed and heavily capped; corner boards with the barest suggestion of a capital; gable, cornice, molding, and side elevation entablatures classically derived; above all, the platform on which it stolidly sits, the most telling sign of a country church in this style. But look again. The memory persists of the typical chunky cubical cupola, yet no such survival escapes its time. In contrast to the firm containment of the simple rectangular shape of the entrances into the school under their "Greek" entablatures, here Renaissance hoods do the capping. They appear to be lifted off their frames, an effect which the stretched transoms reinforce. Hence it is not geometrical containment that prevails here, but rather the further exaggeration of an already elongated feature. The equally basic rectangle of the window between the school doors, and the stretch of the blank clapboarding which makes all three shapes resonant, is here converted to a contrast of square-headed and arched shapes rather closely packed. To variety of shape therefore add variety of style, with allusions not only to Greek and Renaissance, but also to round-arched Romanesque, a style which began to be popularized for Protestant churches during the mid-nineteenth century. As for the cupola, it, too, hardly possesses the bulk and command of its typical Greek Revival counterpart. Relatively small and thinly detailed, with four-directional pediments more elegant than forceful, it apparently derives from the roof ventilator of a Victorian carriage house. Corner posts and roof moldings are likewise thinned from the bulk that authentic Greek Revival demands.

As in some other village churches (see SC1, FO20, and FO34), the pulpit stands against the entrance wall, in an arched niche made by double-doored vestibules on either side. Though much changed, with little original furniture, its rear wall now completely opened to the parish hall behind, the interior exudes a comfortable community quality. The church front was shifted from South Killingly to Howard Hill, presumably the better to accommodate the overblown parish house to its site.

Finally, look across South Killingly Road to the tiny cottage which once housed the town clerk (1904). If this, too, can be construed as Greek Revival, then this corner boasts Greek Revival, Greek Survival, and Greek Revival Revival (or perhaps Greek Re-Revival).

FO10 Welcome Rood Tavern

c. 1780, 1824. 3A South Killingly Rd. (at Howard Hill Rd.)

The oldest building at the intersection is fitted to the triangular site made by the meeting of the three roads. It is a rambling combination of gabled elements which includes a one-and-one-half-story house with a one-and-one-half-story ell (c. 1780, facing on South Killingly Road, the ell appearing to be older than the house), plus a two-and-one-half-story house (added in 1824, and facing Howard Hill Road), plus another one-story ell (probably for grain storage). Welcome Rood built the 1824 section as a tavern, with a barrel-vaulted Masonic meeting hall upstairs (not open to the public). It was originally decorated with stenciling, the scanty remnants of which show only standard willow and flower designs. Masons met here for ten years, before they were evicted in 1834 at the height of anti-Masonic sentiment. The building was later variously used as the town clerk's office and residence, a post office, and a store. So it was clearly a center, with changing functions, for village life during the nineteenth century. It now serves as a pottery studio, shop, and residence. Other outbuildings, including an arcaded carriage shed built for the inn, which once further complicated this complex, are gone.

FO11 Foster Town House
(Second Baptist Church, later Elder Hammond's Meeting House)

1796–1797. 181 Howard Hill Rd.

If Foster residents were to designate a single building to symbolize their town, this important meeting house would almost certainly be it. Until 1990 it stood in relative isolation, with field and woods as backdrop. Then a new town hall and adjunct building went up next to it. Stylistically, the new town hall blows up the clapboarded spareness of the meeting hall, meaning to complement what it subtly undermines by its outsized mimicry.

The meeting house again indicates the architectural impact of an elemental building shape, careful proportions, straightforward carpentry, and a commanding site. As is typical of the eighteenth-century meeting house, the west-facing entrance is placed at the center of one of the longer elevations. Here it is a double door, its dignity magnified by a broad frame with paneled pilasters and an entablature, while the equivalent side entrances are thoroughly vernacular in character. The two-story, five-bay format of the front elevation invokes that of a typical substantial house of the period, but made more monumental both by increased height and by stretches of wall at either corner which screen two interior stairs flanking the entrance, as well as by the entrance itself. A subtlety is the

FO11 Foster Town House (Second Baptist Church, later Elder Hammond's Meeting House)

slightly smaller size of windows upstairs: twelve-over-eight panes, in contrast to twelve-over-twelve downstairs. Consider the difference in visual effect on the front and the south side elevations because of the proportional relationships between the clapboarded wall surface and their openings: in front, the five-bay format of ten openings seems subtly stretched by the eaves immediately overhead and the expanse of wall at either end; on the south side, the reduced three-bay format of six openings seems further compressed around the entrance by its engulfment beneath the immense blankness of the gable wall above. The typical window of the period is not inset, but framed to project the sash forward of the clapboarding. This, too, intensifies awareness of the volume of the whole.

The meeting house formula decrees that the principal entrance on the long side face the platform for preaching, originally with a high pulpit for the elder and a lower one for the deacon. The stairs to either side of the entrance rise against the front wall, across the outer, downstairs windows of the elevation, to a landing fitted into the corner; then up another flight against the side wall to a balcony around three sides of the space. This is fronted by solid plank facing. It and the roof are supported on vigorously turned supports which, at their base, proclaim their origin as sizable square timbers. The simple, rugged structure is intensified against the white plaster walls and ceiling. Plank benches with open rail backs focus on the speaker. Light from all directions makes this an inverted lantern. Concentration on such essentiality intensifies the architectural experience.

The proceeds of a lottery originally financed this meeting house, after the Foster Center Baptist Church, under Elder John Hammond, split on doctrinal principles from the Hopkins Mills Baptist Church. Receipts from the lottery, however, proved insufficient, so the interior remained incomplete during the period of Hammond's ministry. With Hammond's departure from Foster in 1815, the rival Christian Church in nearby Rice City usurped his Calvinist fervor and enticed most of the congregation away. The decimated congregation of the Second Baptist Church approached the town in 1822 about the possibility of a transfer of ownership, especially since the building had been used for town meetings since 1801. (Town residents boast that in no other case have town meetings been held continually in the same place over such a stretch of time.) The town seems to have struck a hard bargain, since the church and town halved the cost of repairs and the completion of the interior—$86 apiece—before the transfer occurred. It has served as the Foster Town House ever since. By 1900, however, neglect had brought it again to serious disrepair. In 1904 the Ladies Home Mission Society of the Foster Center Christian Church held an Old Home Day to raise funds for restoration—the kind of action which was popular in the revival of enthusiasm for colonial architecture at the turn of the century. These festivities inaugurated a series of annual Old Home Days and fairs on the grounds of the Town House. They persist to the present, as another contribution to the life of the community centered in this building.

FO12 Daniel Howard–Judge Daniel Howard House

c. 1805, 1856, and later. 102 Howard Hill Rd. (gravel at this point) (pole 102; 1.8 miles south of Foster Center)

Another example of a colonial or early national farmhouse angled to the road to make the best use of its site, this typical two-and-one-half-story, five-bay, center-door Federal house presents its gable side to the road, which (from this direction) curves toward it and away. The house takes advantage of a southern exposure for the front windows and of a slight rise in the ground, with possible adjustment to trees existing when it was built. The windows are stretched quite regularly across the elevation, though are not precisely symmetrical. The rebuilt "central" chimney is also a bit off center. Recent modifications have occurred in the door frame and ell. The elliptical wooden fan over the door is original, however. It contains an incised shortcut, as a kind of signboard to the fully developed wooden fan of slatted boards which appears elsewhere in Foster.

Daniel Howard was a farmer and the longtime town clerk, a position inherited by his son, who parlayed it into a notable political career, including the local offices of tax assessor and justice of the peace, thirteen terms in the state House of Representatives, and, eventually, judgeship, first on the Court of Common Appeals, finally as an associate justice of the state Supreme Court. He died at ninety-three and is buried in the small family cemetery lifted above the road opposite his house.

FO13 Hopkins–Bennett House

c. 1800 or 1828(?); ell and porch late 19th and 20th century. 13 Walker Rd. (pole 13; .65 mile north of Briggs Rd.)

Mostly screened by trees in summer, this one-and-one-half-story, central-chimney Federal house with a gable roof appears to be a typical late eighteenth-century type—except for a puzzle which is no longer visible. Until the 1960s it had a Greek Revival transomed doorway which was considered to be a later alteration. When the Greek Revival door frame was removed during a restoration of the house to return it to its "original" condition, it became clear from the way in which the frame tied into the rest of the structure that it was, in fact, original. Was this, then, built as an old-fashioned house type by a traditionalist carpenter who was encouraged by the whim of the client or some other cause to add a newfangled door? The interior (not open to the public) exemplifies the five-room arrangement frequent in Foster at the time: parlor balancing the kitchen, with its large cooking fireplace, in front; a sizable pantry and two bedrooms behind. The white plaster walls and ceilings of this beautifully restored interior emphasize the combination of spareness and elegance in the country-carpentered woodwork. Stenciling in the small entrance hall and parlor, customary for the better houses of the period, is the work of Susan Hibbert, a local specialist in the art, who here combined traditional Rhode Island motifs for her design. From the road one expects this house to be tightly enclosed by woods; but, high on a ridge, it commands a surprising panorama down the slope of a field and off to Connecticut hills.

FO14 Daniel Wood–Jerah Hill Farm

1771, c. 1825–1830(?). 19 Walker Rd. (1 mile north of Briggs Rd.)

This one-and-one-half-story, center-door house turns its gable end to the road that curves between it and the barn (c. 1870). Jerah Hill bought the house and remodeled it, doubtless introducing the Greek Revival treatment of the door. (Could it have influenced the preceding example?)

FO15 Phillips–Wright House

c. 1765 with 20th-century additions. 2 Foster Center Rd. (at Plainfield Pk. [Route 14/102])

Unprepossessing on the exterior, this is nevertheless one of Foster's best-preserved one-and-one-half-story half houses with end chimney. (A deed of 1801 from its probable first owner, Ezekiel Phillips, to his son Augustus, specifically refers to the "1/2 house where I now live.") The essential preservation of the interior (not open to the public) and its original plan is notable in making clear the nature of this small, mid-eighteenth-century house type. On the first floor, the entry, with a straight flight of stairs to upper bedrooms, opens into the side of the principal keeping or great room opposite that of the fireplace. This arrangement varies the alternative squeeze of the stair into the box of space between the side of the fireplace and the outside wall as a tight twist of short stair runs and landings, familiar from the Eleazer Arnold and Thomas Clemence–Irons houses (LI14 and JO3). In place of the stair beside the dominating fireplace, here the nook is fitted with shelves and a cupboard. As in some other half houses already visited, clapboarding conceals the masonry back of the fireplace outside. Two bedrooms in back off the keeping room, plus upstairs sleeping quarters, complete the compact living arrangements in the original house.

FO16 Paine–Bennett Farm

c. 1815. 174 Old Plainfield Pk. (near the junction of Plainfield Pk. and Victory Hwy.)

This farm is a fine example of a country-carpentered Federal house with a well-preserved complex of nineteenth-century barns and outbuildings. A plain, typical two-and-one-half-story, five-bay house is here relieved by a very elegant entrance porch and boasts the refinement, unusual for a farmhouse, of a fanlighted transom, its spoked leaded ornamentation ordered from a catalog. The hooded porch, with a delicate modillion and dentil cornice typical of the Federal style, is barrel vaulted beneath in response to the semicircular fanlight. Reductive Doric pilasters to either side of the door have column responses which are a charming cross between a classical column and a turned porch post, rather more the latter. This entrance seems to be related to those of two other houses in Foster (FO5, FO19) and ultimately to the cluster of such entrances in Union Village (North Smithfield). A shingled, gambrel-roofed barn with long projecting ell terminating in a stone milk room (where "1888" is carved into the masonry) is augmented by a carriage barn, shed, and privy.

FO17 Twin Pond Farm (Beriah Collins House)

c. 1760, c. 1790, 1989. 142A Old Plainfield Pk. (pole 142; .8 mile east of Route 102)

This farmhouse undoubtedly started out as a two-and-one-half-story half house with an end chimney (now rebuilt), to which a one-and-one-half-story ell was added. The graduated siding of the front elevation is exceptional in the narrowness of the clapboards. All, however, is seriously compromised by an inappropriate modern addition.

FO18 Brown–Whidden–Fuller Farm

c. 1770, c. 1800. 189 Plainfield Pk. (pole 189; .4 mile west of Route 102)

The massing of this house recalls that of the house described in the preceding entry, but its building history is inverted. This farm began with the one-and-one-half-story ell, probably built by Major James Brown (the owner's sign says Thomas Brown). When Charles Whidden, a blacksmith, purchased the property c. 1800, he added the two-and-one-half-story component with graduated siding, bringing the old chimney to the center of his extended residence and making the old house the kitchen ell. It was sold again in 1857 to a new owner, Thomas Fuller, who grew corn for brooms and made them here until the business moved to Providence after 1870. His broom-making facility, joined to a carriage shed, is now a garage and guest quarters. Adjacent to the house is a large, handsome horse barn of the 1940s built on the site of a nineteenth-century predecessor.

FO19 Mount Vernon Tavern

c. 1760, c. 1814. 199 Plainfield Pk.

When Pardon Holden purchased this two-and-one-half-story, five-bay, central-chimney house for a tavern in 1814, he added the ell, a bar on the first floor of the original house, and the pretty entrance porch. The hood roof of the porch with ceiling arched to echo the curve of the fanlight compares to that of the Paine-Bennett Farm (FO16) and ultimately probably

FO19 Mount Vernon Tavern

also derived from the Union Village type. Were the two porches by the same master? The Mount Vernon version is classically more correct, its ornament more severe in comparison with the rather ebullient, sketchy quality of that of its neighbor. The indication of capitals on these postlike columns is handled with greater sophistication, while the addition of rosettes, a keystone over the door, and Ionic (rather than rudimentary Doric) pilasters also lead to the same conclusion. Hence it seems likely that this may have been the inspiration, which the other two Foster porches in this manner (FO16 and FO5) progressively countrified. Whereas spoked arrows were ordered from the catalog of lead designs for the nearby fanlight, here it was garlands.

The ell later functioned as a general store and after 1828 as a post office. The tavern also housed (in the upstairs front west chamber) the Mount Vernon Bank during its first year of operation. Holden was one of the original six petitioners for the bank and was its second president from 1828 until his death in 1831. The enterprising Holden also operated a plow manufactory on his property. Both the foundry and the bank eventually moved to Providence, in 1850 and 1853 respectively. There the foundry metamorphosed from the High Street Foundry to the Builders Iron Foundry (B.I.F.), which played an important role in the casting and fabrication of custom metal design for buildings throughout southern New England until it was swallowed up in a conglomerate takeover during the 1960s. Finally, another owner, George Fry, ran a printing press in the southeast parlor, before this center of village commercial activity settled down to becoming a residence pure and simple in 1888.

FO20 Mount Vernon Christian Church
(Mount Vernon Baptist Church, Friends Meeting House)
1795, 1887. 210 Plainfield Pk. (at Howard Hill Rd.)

Merely by looking at this simple carpentered structure, how might one know that it began as a Quaker meeting house? Its exterior format suggests a school. If blown up, it clearly becomes a church or some secular meeting place like a meeting hall or a grange (although both of these building types are more likely to have contained a single door). To reinforce the churchly image, in the mind's eye add an inflated cupola, similar to those of Foster's most ambitious Greek Revival churches. But presented as here with the typical identifying features, diminutive scale, and plainness of a typical nineteenth-century one-room schoolhouse of the most modest sort (lacking even a bell cupola), how might one possibly construe this as a church, even if its original Quaker congregation might have been content with the ambiguity?

First, a negative observation favoring church over school: no privies. Then two positive hints, too slight to shake the impression of "school" in this instance, but revealing as to how this image might more certainly be viewed as that of a church. The building confronts a T-junction. Two other Foster churches described here do so as well (FO26 and FO34) in situations which magnify their authority. Schools rarely do, although there are exceptions (most memorably in Rhode Island, the Woody Hill School in Exeter [EX2]). Schools usually occupy wayside sites (as perforce do most churches) because sites at intersections are limited. Or they are also frequently found at the corner of a secondary intersection with a major highway, which, in effect, is also a wayside location. If at a major junction, they occur more often in a crossroads cluster than at a T, but usually in a deferential position. They are not so much "at" the crossroads as "near" it, leaving the more conspicuous places for uses which benefit most from traffic: to churches first and foremost, then to a community hall, the country store, the tavern or inn, with perhaps room from a few houses of the village elite at a time when the elite enjoyed showing themselves off to the traffic more than is customary today. Not that these observations on church and school location are surely right, but mulling the probable outcome of some reasonable sampling at least provides a key to one way in which even the simplest sort of building may begin to accrue to itself a modicum of meaning sufficient to elevate its position above mere utility.

A second hint of churchliness here is more intrinsic. The window head is barely lighted above the heights of the flanking doors. In schoolhouses, the heights of all three elements are typically aligned. Occasionally, but rarely in the authoritarian aura of the nineteenth-century schoolhouse, the window may drop a little below the height of the doors in acknowledgment of the lesser stature of its pupils. When the window is elevated even as slightly as this one (let alone as extravagantly as those in the North Foster Free Baptist Church [FO34] or the Chepachet Union Church in Glocester [GL14]), this most dematerialized of architectural elements encourages an aspirant frame of mind against the earthly function of the doors. Here one witnesses no more than the stirrings of such symbolic possibility, as though the loosening of geometric bonds between doors and window were more an intuitive act than a conscious decision.

As if these were not puzzles enough to lay on such an unpretentious building, here is another. What style is it? Having a rough date for its erection, late eighteenth-century vernacular is the easy response, and one hard to fault. Yet the broad, simple door frames suggest a Greek Revival resonance, as in the schoolhouse at Foster Center (FO8). This could reflect later replacement or remodeling, perhaps sometime close to 1843, just before the original congregation abandoned Foster for want of members and merged with the Coventry meeting. Not likely, however, when they were in such dire financial straits; and perhaps not likely anyway for Friends, who were notably immune to changes of fashion in their meeting houses (if not always in their homes). The building then stood empty for a period before being sold as a barn, which use it served from 1850 to 1887. The religious revivalism of the 1880s encouraged the Christian Church to purchase this by then run-down building and extensively remodel it (before lack of funds to pay the minister encouraged Baptists to take over in 1895). Is what one now sees, therefore, a mix of vernacular building from the time of the c. 1887 renovation that combines (or eclectically draws upon) a compendium of vernacular practice derived partly from the late colonial period, partly from the Greek Revival, and partly from practice at the end of the nineteenth century (or later)? Blank simplicity can be perplexing.

By now the architectural pilgrim has most likely overstayed his or her patience with such a minor building, and willingly leaves to others the probing of this structure and its comparison to nearby buildings from the 1890s in carpentered styles. Still another puzzle, however, exists inside. As in a few other churches in Foster and Scituate (FO9, FO34, and SC1), the pulpit is exceptionally located against the entrance wall in a niche made by separated entrance vestibules. This placement must result from the final remodeling or later because it is difficult to imagine it as the original Quaker solution.

FO21 Barn, Obadiah Harrington–William Blanchard Farm

1892. 46 Johnson Rd. (pole 38; 1.1 miles north of Moosup Valley Rd.)

Although the Harrington Farm dates back to c. 1762, its present architectural distinction is its capacious, now somewhat dilapidated shingled barn dominating a hill on this farm. The barn was probably built by the Blanchard family, who purchased the farm in the mid-nineteenth century. The siting of the barn and its attached auxiliary sheds make the view from Johnson Road no more than a starting point for a structure which invites a full circuit. Foundations only remain of three silos (probably later additions) which were clustered around the entrance and once made the play of building shapes even richer.

FO22 Moosup Valley Congregational-Christian Church and Cemetery

1864–1865 (church). 81 Moosup Valley Rd.

The hamlet of Moosup Valley contains, in a clapboarded cluster, church, cemetery, library, and grange as the civic center for an area which still presents much of its traditional farming image, although actual farming is minimal. This church is almost as vernacular in quality as the Mount Vernon Christian Church (FO20), except that its scale, the puffed-up belfry, and perhaps the aggressive quality of the sheltering door hoods all clearly proclaim its function. Nor is there any doubt of its being Greek Revival, however belated. Here, in fact, the country Greek Revival church displays all its principal elements, unadorned except for the simplest moldings. Broad, flat door frames and building corners and the same qualities in the paneling of the inset doors speak directly of the boards which make them. Most "Greek" are the projecting door entablatures, but only allusively. If the positioning of the domestically scaled window slightly above the level of the projecting entablature gives it a degree of levitated independence, the weighty quality of the doors on either side restrains its animation as by a gravitational force. The box-on-platform belfry crowns the ridgepole with its louvered openings, each stage marked by strongly projecting eaves. And at the base, a low, stepped platform of rough-quarried granite, running the length of the elevation, serves as the country stylobate. Pieced together with long granite blocks eight to ten feet in length, this was the work of oxen, who also hauled the comparably seated granite pieces for the cemetery wall. Prime movers for most of the nineteenth century, oxen outnumbered horses on Rhode Island farms until well after the middle of the century. They worked the rocky fields, plodded through mired roads, and made possible the heaviest chores of building. Only when oxen had gentled the land could horses take over.

This church, however, was not always so plain. Acroteria of signboard scale and flatness once embellished its belfry moldings with forceful undulation. It would cost little to return this feature so that the church could again flaunt its Grecian message to the countryside with an audacity unmatched by any other extant in the state. In the process, why not strip the belfry of aluminum siding?

FO23 Banquet Shed, Moosup Valley Grange

1928. 81B Moosup Valley Rd.

Opposite the church, the grange; beside the grange, shelter at its simplest. This is no more than a large, open, gabled pavilion supported on rows of square posts over a concrete floor. Each post is lettered, with a board cross piece at table height. Three planks the full width of the pavilion resting on the cross pieces make the tables. They have their own bit of folkish ingenuity. There is a deep slot the width of the post supports for the roof at one end of the middle plank of the tabletop and a shallower slot at the other. The deep slot is rammed home on one

FO23 Banquet shed, Moosup Valley Grange

FO24 Iri Brown Farm

side, allowing the opposite end to fit inside its post. Now the shorter slot is rammed home, leaving a void at the inside of the deep slot. A block set into the void locks the tabletop in place. The blunt simplicity of everything appeals here: the elemental shelter, the almost crude trussing on either end, the integral fit of building and furniture, the line of letters on the posts assigning each diner a place. The culminating summer feast was the famed annual Moosup Valley clambake, which the grange sponsored in early autumn from 1928 into the early 1990s, when rising costs forced at least a temporary halt. Other, less ambitious banquets continued here; but consider this a temple to the official "state animal"—the quahog, the humble, retiring dweller of the bay and ocean sands around Rhode Island and mainstay of its clambakes.

FO24 **Iri Brown Farm**

1815, c. 1850, c. 1875, house. 1885, barn. 111 Plain Woods Rd. (.5 mile from Moosup Valley Rd.) (barn partially visible in winter)

The Iri Brown farmhouse is a prettily maintained one-and-one-half-story Federal house with a tiny trap-door monitor, and comfortable Greek Revival and plain Victorian additions to the rear. It is the spectacular barn, however, among the grandest extant in Rhode Island (except for estate barns) and beautifully restored, which is of exceptional interest. (Unfortunately, its position behind the house and trees make viewing difficult except in winter, and then only partially.) A four-story barn under one immense spreading gable set broadside against a rock ledge slope, it contains entrances on three levels. One entrance cuts to the cow stalls through the masonry base at the front of the barn facing the road. Up the slope, entrances at the second level into either side of the barn provide for carriages and horses. Up again, an entrance to the rear at the third level serves the hay wagon and the loft above. Three square shafts from the third level to the ground (two of masonry, one of wood, and all plaster-lined) provide for interior silage. The fitting of the barn to the rock ledge of the hill with a combination of masonry walls and pegged timber construction is especially impressive. An altogether unusual barn type for Rhode Island, it recalls the Shakers' multilevel "bank" barns set into hill slopes; and, in fact, the Brown family had Shaker connections. (Still in the barn is an exceptionally large oxen-drawn wagon for hauling the granite blocks used to build it. The blocks were suspended beneath an oak timber construction resembling a giant sawhorse on wheels.)

Iri Brown, a schoolteacher as well as a farmer, was a director of the Mount Vernon Bank and founded, with his brother-in-law, the Moosup Valley Church. He very likely provided the means of moving the granite blocks which underpin the church and form its cemetery walls. Family legend has it that the magnificent barn was built as an inducement to convince Iri's grandson Curtis Foster to remain on the farm, as, indeed, he did. It is now primarily used as a barn for riding horses.

FO25 **Dorrance House**

1720, c. 1750. 1971, restoration, Peter Flint. 2 Jencks Rd. (not visible)

Close to the road, this house has been palisaded by planking to thwart even minimal intrusion. Samuel Dorrance, the first minister in the border Connecticut town of Voluntown,

FO29 Zuriel Paine–Andrew Paine Farm

once owned a very large tract, part of which he gave to his brothers George and John, who built this house near their saw- and gristmill in the 1720s. In 1728 a boundary adjudication placed this portion of the Dorrance holdings in Rhode Island. Originally a narrow two-story, five-bay shingle house one room wide with a large stone central chimney, the house had only two rooms on the ground floor—parlor to the west, kitchen (keeping room) to the east with a stair between (here quite wide) to two bedrooms upstairs. Although this five-bay, one-room-wide type was fairly common in the seventeenth century, Rhode Island has few survivors. Like many of its type, it was expanded around 1750 by three rooms to the rear of the original two by extending the rear half of the gable into a typical saltbox massing. Parlor woodwork which survives from this remodeling and much other evidence provided the basis for a thorough and knowledgeable restoration by the architect Peter Flint, who lived here and was active in Rhode Island preservation work until he moved to Keene, New Hampshire. The Dorrance House is among the important seventeenth-century houses in Rhode Island.

FO26 Union Free Will Baptist Church
(Line Baptist Church)

1867. 175 South Killingly Rd. (at Kennedy Rd.)

This church, long known as Line Baptist because the Connecticut boundary runs immediately behind it, might be too unassuming as a destination in itself to warrant a visit, being less prepossessing architecturally (although considerably larger) even than the Mount Vernon Christian Church (FO20). It is the most modest example in the remarkably preserved series of rural churches in Foster inspired by the basic Greek Revival formula for the type. Its size within the village, location at the principle T-junction, and very simple cupola (once open, now screened against birds, unfortunately with aluminum shuttering) are its sole attributes to monumentality. No projecting moldings mark the doors, only slotlike transoms. In such simple, anonymous buildings, the slightest efforts to rise to the occasion make visible the ways in which grander architecture accumulates more complex meanings. Here each entrance gets its mini-pyramid of steps from a single granite slab (now half buried) which has been dragged across the entire front and left uncemented to the low foundation behind, topped by two shorter and narrower, though still considerable slabs beneath the thresholds. A diagonal pipe affixed to one corner of the building with plumbing joints provides a handrail.

The tiny, as yet undisturbed village is as prosaically appealing in its setting of lawns and trees as the church. To the west, the ell of the H. Smith House (c. 1820) connects with what was once C. Cory's (or the Line) General Store (c. 1885). A stone marker diagonally across the street from the store in front of the parsonage (c. 1890), with "RI" on one face and "CT" opposite, indicates that the state line passes through the parson's house. (At the north end of Kennedy Road, 600 feet east, another quartzite marker near the church indicates the earlier state boundary, before adjudication.)

FO27 Sweet Farm (Abjah Weaver Farm)

1809 with later ell. 124 South Killingly Rd. (.25 mile from Kennedy Rd.)

FO28 A. Bennett Farm

c. 1849. 50 North Rd. (west of intersection with Boswell Rd.)

Angled to the road, the Sweet farmhouse is a typical one-and-one-half-story, central-chimney house of the Federal period which retains a later barn and outbuildings. (A "1790" sign on the house probably dates it too early.) The A. Bennett farmhouse is a typical one-and-one-half-story, five-bay, central-chimney Greek Revival house with an unusual flush ell which seems to be original. The framing of the front elevation by a heavy entablature cornice (interrupted by a mini-transom over the principal door) and paneled corner boards is vigorously handsome. The main section of the house exemplifies the common Foster five-room plan with a front kitchen to the right of the entrance.

FO29 Zuriel Paine–Andrew Paine Farm

c. 1785 (original house), c. 1835. 22 Paine Rd. (pole 22) (.7 mile north of Danielson Pk.)

Although small and tightly constricted on either side of the road, this is a remarkably well-preserved operating family farm with a full complement of farm buildings, most of which date from the Greek Revival period. The oldest portion of the house is its three-bay, set-back ell, fronted by a later porch. To this, Andrew Paine added a one-and-one-half-story, five-bay, center-chimney Greek Revival house with a narrow trap-door monitor dormer, which fronts on a sloping apple orchard.

The extensive outbuildings are laid out as two opposed arrangements of buildings on opposite sides of the road—those relating to traditionally female-dominated tasks gathered around the house, with the male domain around the barn across the road. On the house side, all outbuildings occur in right-angled relationships, taking their alignment from the house. On the barn side, two clusters converge toward the house in angular relationship. The angles, in turn, respond to a curve in the road. The three buildings that make up the barn cluster touch one another at right angles, with the carriage shed as intermediary. It comes corner-to-corner with the barn, while the garage butts one of its long sides.

Around the bend is the shop, now converted to an apple salesroom, which once housed Andrew Paine's alternative livelihood of coffin maker and undertaker. His long, narrow workbench remains. The hearse had its own coffin-like shed, a little removed from the shop but so placed that its walls parallel those of the shop in front. How functional this siting of the farm complex! And how the instinctual spacing, grouping, and ordering of the functions of this farm in relation to its site enhance the meaning of this place! In his role as carpenter, Andrew Paine assisted in the building of nearby North Foster Free Baptist Church (see next entry), as he was also a leader in founding it.

FO30 Church of the Messiah

1966, William D. Warner, project designer, William Kite. Danielson Pk. (corner of Foster Center Rd.) (Route 94)

This board-and-batten church echoes the dramatic shaping of vernacular shed and gable elements which was conspicuously associated with the Sea Ranch development north of San Francisco, designed by Moore, Lyndon, Turnbull, Whitaker (1968) and widely popular in the late 1960s. By comparison with Irving Haynes's more vernacular shaping of Lime Rock Baptist Church (LI30), this is abstract. The opposed tilts of the shed roofs of its separated towers for light and the fall of the shed roof away from them also suggest reverberations from Le Corbusier's then very popular chapel of Notre-Dame-du-Haut, Ronchamp (1950–1954). The long shed roof molds an interior defined by its exposed structure and enclosed in the natural finishes of its flush wood siding. The lighting of this space is its outstanding feature. A row of square clerestory windows under the eaves along the high side of the shed space culminates in a diffuse shower of reflected light to the right of the altar area from a concealed opening in the towered light box, with direct light from an off-center arched window behind the altar itself as its climactic focus. A responding row of slightly smaller square windows close to the floor lights the low side. The result is a shadowy, but also luminous, space which is particularly receptive to changes in outdoor light. Multidirectional sheathing of the various walls further animates the space: for the high side, vertical boarding; for the low side and ceiling, horizontal; for the altar, boarding in opposed diagonals with their center line at the focal arched window.

South Foster

The mills which once gave this place the name Hopkins Mills are gone. What remains is a stretched-out straggle of houses and other structures, including a number from the late eighteenth century through the early Victorian period, along a hilly, winding road with the original millpond in one of the hollows.

FO31 Ezekiel Hopkins–William Potter House

1720 and later. 45 Old Danielson Pike (at Windsor Rd.)

Opposite the millpond is the house of the first mill owner, who established a grist- and sawmill in the eighteenth century. The house seems to have begun at the center as a one-and-one-half-story one-room stone-ender (although the initial section may have been larger). Additions were made to either end, toward the west (left), extending the original house linearly, and toward the east (right), terminating in a two-and-one-half-story gable set at right angles to the rest of the elevation, as well as later additions. The house remained in the Hopkins family until William Potter of Warwick acquired it and the mill and converted the factories to coordinate with his Foster Woolen Mill. The next Potter generation converted part of the house to a general store.

FO32 Henry Davis House (Seamans-Davis House)

c. 1865. 39 Old Danielson Pk. (pole 39)

This mixed Greek Revival–Italianate early Victorian house with its gable end to the road behind a picket fence has a long, porch-fronted ell added to one side. The pediment treatment of the gable, the corner pilasters, and the piers of the ell porch are Greek Revival. But these are overlaid with such Italianate details as the round-headed window in the gable peak and flat hood roof installed atop a standard Greek Revival entrance with pilaster and broad entablature with dentil cornice and scroll-cut ornamental brackets. Henry Davis and a succession of other owners ran a store (now a residence) in the clapboard Greek Revival building immediately adjacent to his house (1843) and hard by the road. It also served as the post office and was once fronted by the first gasoline pumps in the village.

FO33 Hopkins Mills Union Church (South Foster Union Chapel)

1869–1871. 36 Old Danielson Pk. (pole 35)

The South Foster Union Chapel Society was founded in 1868 to provide a combined nonsectarian church and town meeting hall. What resulted is yet another Victorian design which bows to Greek Revival precedent. Smaller than its North Foster predecessor (FO34), this contains the same essential ingredients, with some lessening of specific Greek allusion in detail, except that the gable is here treated as a full triangular "pediment" by extending a horizontal molding across the front elevation as its base. The gable, however, maintains its everyday pitch instead of the low pitch of a true Greek pediment, while the doubled pilasters of the cupola (borrowed from North Foster, no doubt) are here simply left as slats. Compared to North Foster, Greek allusion and vernacular exist in an easier relationship here. The vigorous framing of the doors and their sensible paneling are examples: the mundane lifted just sufficiently toward the monumental so that the entrances exist in both realms, although again the uncapped quality of these entries seems at odds with the projecting pediment overhead. The Moosup Valley church (FO22) excepted, of all the other interiors of the Foster churches included in this guide, this best retains the qualities of the average country church of the late nineteenth and early twentieth centuries. In this instance, it is not a matter of restoration, but of arrested accretion. Original pews, foot-pedaled organ, potbellied stove, and ornamental pressed metal ceiling are all part of its un-Grecian ambience.

FO34 North Foster Free Baptist Church

1848, Nelson B. Bowen, builder. 1868, remodeling of the interior and addition to the north. 1955, addition and further remodeling. 158 East Killingly Rd. (at Paine Rd.)

Here on this elevated site, at a T-intersection, Greek Revival makes a self-conscious effort to be dramatically monumental. All features are exploded, especially the center window with its stack of eight-over-eight-over-sixteen-paned sash, which become twelve-over-twelve along the sides. (The too-narrow shutters, later addenda, are ribboned distractions.) In keeping with this grandeur, detailing is more sophisticated. Door frames are paneled with flush cor-

FO33 Hopkins Mills Union Church (South Foster Union Chapel)

ner squares. The paneling of the doors themselves derives from monumental rather than domestic precedent, unlike the churches just visited. A specific allusion to Greek entablatures caps the giant paneled pilasters which bound the front, and the entablature extends the length of the sides. There is some incongruity, perhaps, in the uncapped entrances vis-à-vis the emphatic capping of the corner piers and in the full-fledged entablature at the sides vis-à-vis the simple vernacular gable in front (where a more sophisticated practitioner in the style might have extended the entablature across the front, topped it by a pediment, and halved the verticality of the window). The intensity of the "Greek" inflection doubtless also owes much to the date of this building at the height of the movement, before the explicit stylistic character became diluted and reduced to an expected image of monumentality. But there is a poignant charm about the rather explosive and inconsistent attempt at monumentality in the Grecian manner that makes this, the earliest in the series, the grandest by far of Foster's collection of Greek Revival churches.

The vernacular loses its own sure instincts in the belfry. Although it is well scaled, its paired pilasters are underarticulated, and the horizontal clapboarded band under its cap is so awkward that it appears to be a careless later repair. During the 1955 renovation which connected the parish house (1868) to the church itself, the major entrance was shifted from the front of the church to a covered entrance at the junction of the church and its wing to serve the entire complex. The church entrance is now on the former rear wall, and the pulpit was moved to the wall behind the front elevation, thus effectively sealing it. Although evidence of the original interior exists, successive renovations have not been kind to its preservation.

FO35 **Dr. Solomon Drown House**
(Mount Hygeia)

1807–1808, c. 1845. 83 Mt. Hygeia Rd. (1.3 miles north of East Killingly Rd.) (distantly visible in winter)

This fine Federal house, shingled on three sides and clapboarded in front, shows a conventional two-and-one-half-story, five-bay format, with paired chimneys centered in the lateral interior walls of the principal rooms. It boasts a door with broken pediment and fanlight and is grand for its locale at the time. Its special interest is the man who commissioned it and the use he made of its grounds (though his landscaped garden no longer exists). After an active practice in several states and extensive travel to study medicine in Europe, Solomon Drown settled in Foster in 1801 to a life of scholarship and reflection in literature, botany, and medicine, calling his house Mt. Hygeia for the Greek goddess of health. He surrounded it with an elaborate botanical garden which featured plants with medicinal properties. He served Brown University as its first professor of botany and materia medica from 1811 until 1827, and, with his son William, published a treatise on progressive agriculture, *Compendium of Agriculture, or the Farmer's Guide*, in 1824.

After Drown's death in 1834, his heirs opened the house and gardens part time as a museum, making this perhaps the first museum house in the state. The museum led an increasingly languishing existence until around 1950. Thereafter the house fell vacant and rapidly deteriorated until new owners began its restoration. Little remains of the garden, except for the continued growth of some of its specimens and a circular stone mound, the base for a projected study which Drown referred to as a "Rotunda of Worthies." Like English garden prototypes, it was doubtless intended to contain busts or portraits of admired predecessors in the various professions which Dr. Drown pursued. Joseph B. Gay, who married into the family, designed a combined gazebo and waiting shelter near the road in the late nineteenth century. The shingled barn was built in the twentieth century.

FO36 Mt. Hygeia Schoolhouse

c. 1840. 194 Hartford Pk. (near Mt. Hygeia Rd.)

The building type which seems to be most entwined with the churches of this rural town is the one-room schoolhouse of the simplest sort with paired doors flanking a centered window in its gabled entrance elevation. The Mt. Hygeia Shoolhouse continued in use until the completion of school consolidation in Foster in 1952. It was the last one-room schoolhouse to close in the town and, as this is written, the only one to have escaped renovation, although it is neglected.

Glocester (GL)

"Foster-Glocester": this euphonious territory serves local weather forecasters on TV and radio as the "inland" comparison for conditions along Narragansett Bay. In this sense the towns are popularly linked, becoming a shorthand for the northwestern portion of the state. More substantial similarities also link the two towns, although there are differences between them, too—the first being that, when in 1731, the area of the Providence "outlands" was divided into separate towns, what is now Foster was part of Scituate, while what is now Glocester also included Burrillville, to the north.

Although Foster and Glocester have similar geological and soil conditions, Glocester has more extensive areas of relatively good soils. This advantage made the proportion of planting to livestock raising considerably higher in Glocester than in Foster during the eighteenth and early nineteenth centuries when both were primarily agricultural. On the other hand, Foster had an advantage over Glocester in the number of turnpikes which passed through it. Because Scituate and Foster stand due west of Providence, four major east-west turnpikes crossed them by 1820. Being farther north, Glocester really had only one (except that the Hartford Pike barely nicks its southern boundary in crossing northern Foster). But Glocester's single turnpike is a key east-west artery, now Putnam Pike (Route 44), which runs diagonally across the middle of the town. Although it existed as a wagon trail as early as 1722, it was an effort to improve a stretch of this road that called forth, in 1764, the "Society for establishing and supporting a Turnpike Road from Chepachet Bridge in Glocester to the Connecticut Line." The seven-mile West Glocester Turnpike (as it was then called) which resulted became the first private toll road in New England—and was later extended eastward to Providence.

Unlike Foster, Glocester did have a significant regional transportation hub: Chepachet, just north of the center of town. Moreover, though most of Glocester's waterways, like Foster's, originated within the town and were generally too small for mills of any size, the flow of the Chepachet River was sufficient to make Chepachet village a mill village of consequence during the nineteenth century. In addition to the usual rural types of mills which initially clustered in the village, textile factories began operation in 1810. By 1848 they had come under the control of Henry White (later "and Son"), who expanded on what he acquired. The addition of mills to its existing colonial economy as a center catering to local farmers and turnpike travelers brought a prosperity to Chepachet which resulted in a number of buildings of considerable architectural interest, many of which survive.

As in Foster, doldrums followed the steady decline of agriculture during the nine-

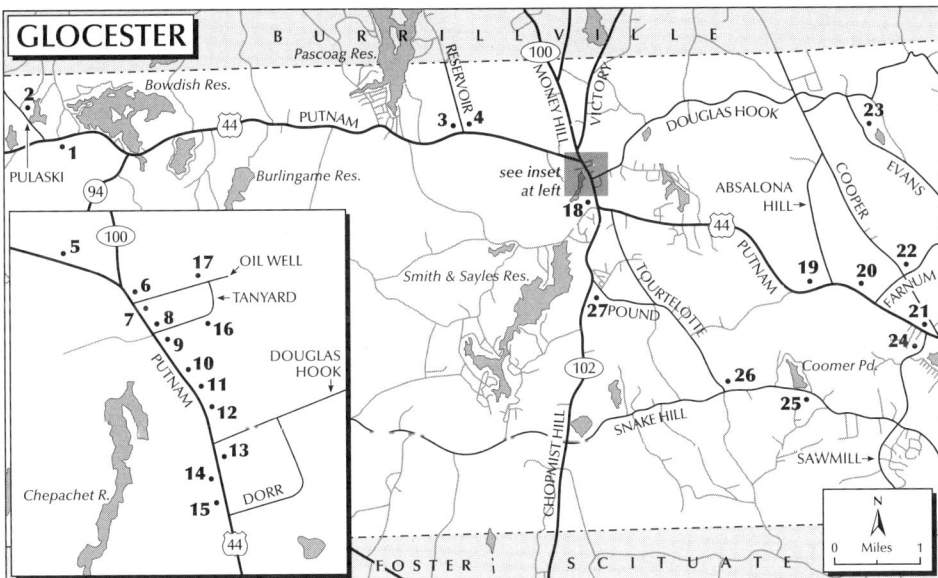

teenth century, the fields returning to woods. A fire destroyed most of the White holdings in Chepachet precisely fifty years after the founding of the company. They were never rebuilt. Eventually into the moribund town came the Sunday tourist, encouraged by the trolley line from Foster to Burrillville, but only briefly, from 1902 until 1924, when the line was severed by the Scituate Reservoir. Then the automobile began to take over. Increasingly (much more than in Foster), a summer population began to cluster in the woods and around a number of ponds, many made by dams from vanished industry. And especially after 1960, as everywhere, suburbanization occurred in patches throughout the town.

GL1 **Clarkville School**

1913. Putnam Pk. (.8 mile west of the intersection of Routes 44 and 94)

Glocester has two extant one-room schoolhouses, both standing in open countryside, both now vacant, and this one rapidly deteriorating. Clarkville School, a very late version of the type, is located at the northwestern corner of the town, and the Greek Revival Evans School (GL23) at its northeastern corner. Of the two this is the larger and more impersonal in character. One feels it as the replication of a prototype of the ideal one-room schoolhouse already rationalized and institutionalized. In fact, it pretty much copied a predecessor on the site, only enlarging its model a little. Guided by an approved standard, the carpenter seems secure in his laconic distribution of the openings in the front elevation: the pair of transomed doors with a window between and a louvered oculus in the gable above. The rows of classroom windows placed toward the rear of the side elevations indicate the space left for entrance vestibule and coatrooms in front. This was among the last one-room schoolhouses to close in Glocester after school consolidation in 1936; it reopened briefly during World War II because of gasoline shortages for busses, before closing for good in 1944.

Clarkville

GL2 **Apartments** (Clarkville Mill)

1864 with later additions. 57 Pulaski Rd.

This early clapboard mill, with a narrow trapdoor monitor which has been boarded over centered in the roof probably originated as the center section, to which additions were made at either end. Much rebuilt, it replaced a predecessor dating back to 1818 which burned. If the factory as it now exists looks older than 1864, this appearance could well result from the partial retention or imitation of an earlier phase of the building. After a succession of owners, it became part of the White organization, based in Chepachet. As now remodeled for apartments it has lost much of its architectural interest, but it does indicate the nature and scale of small industrial operations during the early nineteenth century in a setting which is still isolated. Moreover, the millpond dam and raceway are close by and substantially intact, though overgrown. Two clapboard workers' duplexes exist up the hill farther north on Pulaski Road, but the village church has disappeared. Still other textile manufacturers took over from White, then a furniture factory, before the mill was revamped, and this tiny mill village lost its initial reason for being.

GL3 **Joseph Eddy Farm**

c. 1820; wing added. 1503 Putnam Pk. (west of Reservoir Rd.)

This is a five-bay, central-chimney Federal house, close to the road, though screened by trees. It is a fine example of the country version of its type, with a carpenter-designed triptych type of doorway under a continuous deep cornice over door and side lights, all divided by attenuated paneled pilasters.

GL4 **Smith Farm**

Mid-18th century. Putnam Pk. (at Reservoir Rd.)

The massive gambrel folds over a one-and-one-half-story central-chimney house, displaying this roof type in a particularly impressive manner. The frame and side lights at the entrance are certainly later (perhaps a Federal alteration, which apparently has itself been altered). The additions to the east, which date partly from the nineteenth century, seem to have been redone in a twentieth-century picturesque version of cozy storybook "colonial." Meticulous maintenance and conscious prettification alter the character of the house into an idealized picture of itself, so that it appears to be almost a replica of what it really is. But the eighteenth-century massing of the house and the bulk of its central chimney prevail, and there is visual and intellectual stimulation in the play between the forceful assertion of the past and the charm that the present means to give it.

Chepachet

Chepachet, the principal town in Glocester, is located where the major east-west highway (variantly known by its original name of Killingly Road or, more popularly, as Great County Road and, west of Chepachet, as West Chepachet Pike before becoming Putnam Pike and Route 44) crosses both the Chepachet River and what was then a lesser north-south road (now Route 102). It grew as a combined agricultural marketing town, mill center, and transportation hub. Especially after the pike was improved toward the west as a private toll road in 1794, Chepachet became a major coaching stop for travel between Providence and Hartford. A number of stores, hotels, and taverns dating back to the early nineteenth century remain.

The mills have fared less well. Although gristmills, sawmills, and, most important, a tannery and a linseed oil mill existed in Chepachet from the eighteenth century, the town's real development into the bustling rural mill center that it became during the nineteenth century and remained up to the twentieth began with the establishment of textile mills. They started in 1809 with a cotton carding mill at the bridge in the center of the town, followed by another mill in 1810, both probably small clapboard affairs. The third of these mills, located in a building of random masonry construction, partially stuccoed, was the Lawton Owen Mill, established sometime between 1814 and 1820. It still stands beside the bridge, suggesting the scale of early operations. Eventually, this grew, through several changes of ownership and name, into the substantial H. C. White woolen mills. It is an indication of the tremendous economic devastation brought to such Rhode Island communities as Chepachet in the twentieth century that, of this substantial complex as recorded in 1870 fire insurance records, only the original Owens block, a clapboard boiler building immediately behind it, and the mill office building on the opposite bank remain. The rest, along with other textile mills downstream and the oil mill and tannery have vanished. Chepachet suffered, too, as agriculture declined in the Glocester area after 1900, and fields returned to woodlands.

Through the first four decades of the twentieth century, the village persisted marginally, becoming a regional center and a hub for tourist travel. Starting in the 1960s suburbanization threatened to alter radically the contained quality of the town and the fabric of its center—already invaded by gas stations, a supermarket, and a fire station in elephantine versions of modern design along Route 44. But local historical and preservation groups have also begun to revive the substantial remnants of the mostly early-nineteenth-century town which remains.

GL5 Chepachet Freewill Baptist Church and Parsonage

1821, church, Clark Sayles, builder. c. 1821, parsonage. 1213 Putnam Pk.

Chepachet is fortunate in having the two finest nineteenth-century churches in Glocester, this one on the west end of the town and the Chepachet Union Church (GL14) on the east end. They handsomely epitomize country churches in the respective styles in which they were built.

Typical of the Federal continuation of the Wren-Gibbs ecclesiastical format, Freewill Baptist is a tall, gabled box above a stepped granite platform, fronted by a projecting pedimented centerpiece which frames three recessed doors, the center one dominant, with three twelve-over-twelve-paned windows above. Simple fanlights decorated with wooden spoking over the doors, each capped with moldings and keystones and echoed by another fan in the pediment, are the only adornment. The spire intersects the planes of the projecting front elevation and the body of the meeting hall in a straightforward statement of the triple function of the church spire: as base (with another twelve-over-twelve window), belfry (with open arches), and stubby, polygonal spire. The principal pieces of the composition combine with a certain abrupt awkwardness, but with compelling vigor, too.

Although the interior retains some of its original features, they are diminished by an accumulation of unsympathetic changes. To the rear, carriage barns remain. The itinerant builder, Clark Sayles, also designed two nearby churches in this style: one begun a year earlier for Greenville, and one for North Scituate (where Baptists briefly joined with Congregationalists), begun a decade later (SM11 and SC1).

The adjacent parsonage has a transomed and side-lighted door, recessed like those of the church. This entrance displays a delightfully simple reeded treatment in its door frame, broken at the top corners, and again in between, for square and rectangular blocks inscribed with rosettes. Precedent for this entrance design appears to derive more from mantel motifs than from door frames. Enough variants occur among Federal door frames, however, to make it a minor type in the repertoire of Federal design, more especially for interior doorways. Reeded corner pilasters and a lunette window (the latter like that in the church pediment) complete the elevation.

Enlarge the framing elements, substitute plain paneling for reeding and Greek ornamental allusion for rosettes, with the gable ideally transformed to a pediment, and the Federal parsonage would become Greek Revival; some of the commonplace stylistic devices of the later style are already anticipated here. Reeded corner pilasters and a lunette window in the pediment (again like that in the church) complete the elevation.

GL6–GL15 Main Street, Chepachet

Putnam Pike is Chepachet's main street. The east side of the street especially preserves much of the Federal legacy of the town in an array of boxy buildings, which is all the more remarkable in that a number of them maintain their original uses. The fairly unprepossessing, clapboard-fronted, otherwise shingled house with its gable end to the street at 1189 Putnam Pike (corner of Oil Mill Lane) was originally the Franklin Bank (GL6; c. 1818, front door obviously modern) until closed in 1865. Two plain side-lighted doors open to the lane. The building with the long veranda next to the bank (GL7; original building, 1814; number 1187) received this addition when a store and harness shop became the Central Hotel. Farther along, at 1181 Putnam Pike, is the Job Armstrong Store, now the Art Center and Glocester Heritage Society (GL8; 1814). It retains the general quality of an early nineteenth-century store, although its shop windows appear to be a restoration. Adjacent, at 1179 Putnam Pike, is the Brown and Hopkins Store (GL9; c. 1809), which advertises itself as the oldest continuously operated store in the United States. It is now a country store and antique shop with competitors in other nineteenth-century buildings along the street.

Just across the bridge of the Chepachet River (where the West Glocester Turnpike originated), at 1169 Putnam Pike, is the random stone and stucco building of the Lawton Owens Mill (GL10; 1814 according to the inscription carved over the door, but this building seems to be 1820). Ironically, the original building of what became the sizable White complex and the clapboard boiler building behind were pretty much the survivors of the operation. The latter was saved by the present owner of Owens Mill even as the wrecking ball was poised in 1986 to smash this too. He hopes to approximate the clapboard mill buildings which were once attached in two stages to the original stone block. Interior flooring and framework in the stone building were rebuilt from a gutted interior and do not duplicate the original construction. The Masonic Friendship Lodge No. 7 (GL11; 1814), hard beside the stone mill, at 1167 Putnam Pike, boasts a fine Federal-period door with a broken pediment over a semicircular transom light. Its upstairs lodge room is considered to be the oldest such hall in continuous use in the state. Next, a gas station craters the row followed by what is now called the Stagecoach Tavern, originally Cyrus Cook's Tavern (GL12; c. 1800; 1157 Putnam Pike), in fact built as a coaching stop shortly after the completion of the toll road. The long piazza is another later nineteenth-century addition.

GL13 Chepachet School

1936, PWA. 1145 Putnam Pk.

This typical brick Neo-Colonial village school built by the New Deal Public Works Administration has a variant in nearby Harmony (see GL21). Both display a nice balance between sense of craft and straightforward function, as well as an appealingly optimistic pride in the American past and confidence about the future which characterize PWA work. In fact, these schools initiated modern school building in Glocester.

GL14 Chepachet Union Church

1846. 1138 Putnam Pk.

Now for the Greek Revival counterpart to the Federal style of Chepachet's Freewill Baptist Church (GL5). A low, spreading block is fronted by a classical Doric portico, but with its four columns pulled apart to make more of a deep porch than a portico, lifted on a platform reached by a broad flight of stairs. In contrast to Freewill Baptist, the entrance is now within the hollow of the porch, enclosed in the enveloping mass of the building in a way which exploits the dramatic contrast between solid and void. The paneled door rising the full height of the porch (its upper part false) is an astonishing gesture toward monumentality for a country church. Originally the windows on either side were comparably scaled, extending from the top of the porch the full width of the vertical frames still evident on the facade, down to the level of the sills of the tiny replacements necessitated by vestibule renovations inside. Originally, too, the tower was grander: not one cupola, but another slightly smaller on top of it in a two-tiered, telescoped composition. The top tier was lost to the 1938 hurricane. Still, the well-proportioned lower cupola which remains nicely complements the squat density of the block below, all of which is piquantly relieved by the arched window tensely crowded into the apex of the pediment, with its shutters sprung open as if releasing the cuckoo from its clockcase. Like the Baptist church, the interior retains much that goes back to the original design, but also with incongruous additions. Union Church suffers, too, from a recent parish house addition which, although well considered in itself and set back, is incompatible in form and scale with the church and might be better screened by planting.

GL15 **H. A. Sayles House**

c. 1850. 1132 Putnam Pk.

A fine early Victorian Italianate villa, nicely set back on a deep lawn on the south side of the pike, the Sayles House was commissioned by a member of the mill-owning family. Fronted by a porch with trellised supports, it is especially remarkable for the way in which the paired windows of the tall ground story seem to merge with those above to create assertive vertical bands of glazing within the compact block. This composition is decisively enframed by a deep cornice with paired bracketing and broad corner treatment.

GL16 **Mill Workers' Housing or Mill Offices and H. C. White Company Mill Office**

c. 1860. 7–9 and 15 Tanyard Ln.

One of these mementos of Chepachet's mill heritage presents a puzzle. It appears to be a four-family workers' house with a pair of centered doors serving two apartments at the ground and stairs to two more which extend into the attic. (The building presently contains six family units, and may always have done so.) If, however, it is housing, it is unusually monumental in two respects. First, its double doors are each individually capped by a lintel which extends beyond the door frame, then these are doubly capped by a projecting table molding which sails over both lintels and unifies the doors. Moreover, its elevation is extended at either end by broad, unbroken expanses of clapboarding. Was this building therefore intended as adjunct offices to the H. C. White Company Mill office which once occupied the bracketed Victorian hip-roofed structure nearby that is now a house?

Between the tenements and the company office a view down the precipitous wooded slopes to the mill dam makes one marvel that a sizable industrial complex, plus smaller factories, once choked this narrow depression.

GL17 **Fiske House**

c. 1850. 12 Oil Mill Ln.

The Fiske House is a large, bracketed early Victorian mansion with an unrailed porch extending across the front, one side and half of a third supported on paired turned posts with more bracketing. A wing and barn are joined in a stepped-back relationship on one side. Another well-preserved, one-and-one-half-story mid-nineteenth-century workers' duplex is opposite.

GL18 **Captain Israel Inman's Inn** (Simon Sweet House)

c. 1790; c. 1840; c. 1870, barn. 1096 Putnam Pk.

This Federal house, set side elevation to the road, with some Greek Revival alterations and an ample Victorian barn, is the most substantial house in town. It appears on a 1790 map as Captain Inman's Inn. Although Simon Sweet was its best-known owner locally, the house is famous in Rhode Island history as a headquarters for state militia men during the so-called Dorr Rebellion, or "Dorr War," which came to a climax in the Battle of Acote's Hill, now a cemetery across the pike. There the episode is marked by a bronze-tableted boulder within sight of the highway. Thomas W. Dorr, a wealthy Providence lawyer who identified with populist causes, beginning in 1841 led a paramilitary movement for more liberal suffrage. In 1842 the Dorrites took possession of Chepachet, where several converging forces were to have rendezvoused. State militia dispatched from Providence easily won a "war" in which the defenders had effectively disbanded before the engagement. The movement faded away, and its leader was briefly imprisoned and financially ruined; but many Dorrite demands were eventually realized.

GL19 Olney Manton–Pardon Hunt–Cyrus Farnum Farm

c. 1795. 479 Putnam Pk.

The classical door frame of this two-and-one-half-story, central-chimney house is exceptionally sophisticated for a farm. The fluted Doric pilasters, cushion entablature, and denticulated pediment are all correct and handled with great breadth. (It is so elegant that the missing capitals are easily overlooked! Were they never installed, or did they rot away without replacement?) Elegant, too, are the simple framing of the five-bay arrangement of windows in the compact elevation, the careful alignment of the first-floor window caps with the base of the swelling cushion molding for the entrance entablature, and the molded roof cornice. Such sophistication implies a cultivated client who doubtless considered his "farm" a "country house." Of the three owners with whom this house is associated, Cyrus Farnum is the best known, as a large landowner in the area who served in both houses of the state legislature. To the rear of the typical five-room plan are ad hoc additions—a storeroom, well shelter, and second flight of stairs. Hence the functional, workaday aspect behind contrasts with the formal, ceremonial approach in front, as is usually the case, but particularly so here. The same dialogue between functional arrangement and formal order appears in the distribution of the mostly later nineteenth-century outbuildings in weathered shingle and white trim—two barns (one containing facilities for slaughtering, the other for blacksmithing), two henhouses, a corn crib, and a privy. They are conveniently clustered, but all roughly ordered in parallel alignment to the basic rectangularity of the house. So are the stone walls, which beautifully partition the site into working and living spaces. The interiors (not open to the public) have seen only minor modifications and contain much original woodwork and hardware. Of all the early upper-middle-class farms in Glocester this is probably the most handsome, and self-consciously so.

GL20 Thomas Cutler Farm

c. 1860. 367 Putnam Pk.

Now a rougher, and later, approach to the comfortable middle-class farm, Greek Revival rather than Federal in style. Thomas Cutler purchased property on both sides of the pike. The southern section of his property already boasted a house with a well-established tavern adjacent which served as a stage stop; house, tavern, and an adjacent barn have all disappeared. Wishing to insulate his family a little from the tavern, Cutler built a house on the opposite side of the pike. This second house exhibits the familiar characteristics of the one-and-one-half-story country-built Greek Revival type, except for the delightful gaucherie of the entrance side lights capped as pilasters in the same manner as those flanking the door—to make side lights into something like a windowed pier.

Among the extant outbuildings to the east of the house is a charming dovecote with two narrow doors at either end tightly squeezing a row of three windows between. The doors boast the unusual treatment of transoms which float mini-fanlights. This is a twentieth-century design by architectural students at the Rhode Island School of Design in Providence. In contrast to the clustered organization of the outbuildings around the just visited Manton-Hunt-Farnum farmhouse, here they stretch out along the highway, set back from it a little farther than the house and roughly parallel to both.

GL21–GL23 Cooper Road–Farnum Road

Cooper Road and Farnum Road offer a scenic loop of intact farms (many now of the gentleman sort) with a sampling of Glocester's barns and stone walls, mixed with woodlands and views from high ridges. For as long as it survives, it provides as nice a sense of nineteenth-century rural Glocester as can be found, and it deserves preservation.

GL21 Northwest Community Nursing and Health Center (Ada Hawkins School)

1938. Intersection of Putnam Pk. and Cooper Rd.

GL19 Olney Manton–Pardon Hunt–Cyrus Farnum Farm

GL20 Thomas Cutler Farm, dovecote

At the start of the loop is the former Ada Hawkins School, now converted into the Northwest Community Nursing and Health Center, a finer variant of the Public Works Administration Neo-Colonial elementary school at Chepachet (GL13), apparently by the same architects, showing the same virtues, but better composed, with less disruptive variation in window size. Altered only superficially for its present use as a medical facility, the interiors give a pleasant sense of traditional building crafts coming to terms with modernism. Red tile (not vinyl) floors, yellow brick dados (not sheet paneling), stained oak (not metal) doors, wooden sash (not aluminum) in the transoms to light the halls from borrowed classroom light. Is it this quality which substantially accounts for the charm of such "Depression" buildings? They mediate between tradition and modernist functionalism, their point of view being less "modern" than progressive. Everything seems so solid, so luminously matter-of-fact, so cheerfully derived from everyday experience—as though an unproblematic modernity would automatically emerge from the progressive application of the familiar.

GL22 Farnum Farm

1820s or 1830s with later porch and outbuildings. 96 Farnum Rd. (at Cooper Rd.)

This Currier and Ives image of the idealized early American farmstead is completed by a handsome complement of mostly later nineteenth-century outbuildings set out in clustered alignment to the house, a picket fence, and a splendid row of mature maples in front (which obscures the house in summer). Although built roughly a quarter of a century later, it is, like the Manton-Hunt-Farnum farmhouse (GL19), a simple gabled box, fronted by a similar elevation in the Federal style. This house, however, has a more distinctly vernacular quality. Instead of placement precisely in accord with an abstract symmetrical scheme, here the paired windows on either side of the center door are slightly adjusted to more functional considerations. Instead of symmetrical window placement in the side elevations, here asymmetries of use determine the arrangement of a pair of openings to light the parlor, to the front, while a single opening suffices for the smaller, less ceremonial room to the rear. Instead of the classical embellishment of the pedimented door frame, here it is simply handled with a basic transom, and a bit askew of the "center" chimney. The cornice is also plainly handled rather than molded. The restrained pier supports of the unrailed veranda stretched across the front appear to be Victorian—although early twentieth century is also possible.

GL23 Evans Schoolhouse

c. 1855. Jim Evans Rd.

This is the very ideal of the one-room schoolhouse. It is moving to observe how much civic dignity a simple wooden structure on a low rubble foundation can attain from a compact, elemental mass, straightforward carpentry, and an intuitive sense of the just placement of openings within the elevation and of the graduated accent of openings for their relative importance by simple moldings. Whereas the Clarkville School (GL1), in the opposite corner of Glocester, possesses the impersonality of a carpenter expertly working his way through a

GL23 Evans Schoolhouse (top, left)

GL24 Chase House (top, right)

GL27 Glocester Town Pound (above)

specified and standardized program, here the builder seems to be finding his own way in accord with an agreed-upon model. The school, an "institution" in Clarkville, is more a "house" here. The original precinct of the abandoned school is also preserved, spreading tree in front and privy behind.

Modified mid-nineteenth-century and Greek Revival farmhouses (the Place and D. Evans farms) are opposite. The latter, delightfully known as Seldom Seen Farm, which operates as a sheep farm, contains a string of outbuildings and barns connected to one another and to the house, which is rare in Rhode Island.

GL24 **Chase House**

c. 1915. 21 Saw Mill Rd.

A wide veranda on tapering cobblestone supports crosses the front and one side of this rather grand shingled, gambrel-roofed bungalow with a cobblestone chimney. Beyond the Chase House the road dips quickly to the stream where the sawmill once was.

GL25–GL26 **Snake Hill Road**

Aside from Putnam Pike, the other important east-west road in Glocester is Snake Hill Road. It was laid out as South Killingly Road in 1733 across Glocester into Connecticut, apparently following the route of an old Indian trail, but in the early twentieth century was severed by a tributary reservoir of the Scituate Reservoir (see Scituate introduction). Unlike the pike, Snake Hill Road has remained a well-traveled local thoroughfare as far west as Anan Wade Road, because of the suburbanization of its wooded sites. Increasing traffic threatens its narrow, winding, undulating quality, overhung with trees, bounded in many stretches by stone walls and by occasional farms, which conjure up the colonial condition of the road.

GL25 **Irons Homestead**

Probably 1840s; ell probably 20th century. 688 Snake Hill (west of Sandy Brook Rd.)

Among all the common Greek Revival house types, the one which is exemplified in this fine example may be the most distinctive to the style. Others, with gable ends to the road to invoke a classical format and a two-columned porch shelter over the entrance, are more pervasive, but hardly as original. Still others with various allusions to full-fledged Greek temple elevations may be more sensational, but are also more derivative and less pervasive. In this type, Greek Revival allusion is adapted to the one-and-one-half-story clapboard house. Broad corner piers seem to support the equally broad entablature under the eaves, making maximum use of wide boards to recall heavy stone Doric precedents. Projecting from the plane of the wall, they also forcefully enframe it. Echoing piers and entablature frame the entrance, with the door deeply inset into its box of space. Taken together, the projecting elements recall in an abstract way a stocky colonnade set in front of the plane of the wall, which thereby appears to be reduced to clapboarded panels for pairs of windows. This facade clearly indicates the tendency of developed Greek Revival design toward a more sculptural and (even when merely simulated) structural approach to the architectural components of the elevation, as compared to the earlier colonial and Federal emphasis on openings and ornament as pattern positioned on a plane. This house is enhanced by barns and outbuildings which give a sense of the longtime farm, even though they now serve conferees.

GL26 Barn, Steere–Angell–Colwell Farm

1930s. 1987, addition to west. Snake Hill Rd.

The house on this property replicates a modified eighteenth-century predecessor destroyed by fire in 1938, but the barn is interesting in its own right. Elizabeth Colwell commissioned professional engineers to build this gambrel-roofed barn, said to represent a model animal barn of the period. As a barn image, it is a type that became one of the most popular of the twentieth century. In the 1920s and 1930s it was the barn for model trains; in the 1970s and 1980s it was the barn for the storage of garden tools and the lawn tractor. Its replication is a favorite in roadside restaurants, antique outlets, and tourist general stores where a barn image is sought.

When a particular image has such currency as *the* stand-in for its type, a good example is worth notice and thought. The shape swells to the load of hay it contains much more efficiently than a gable permits, while its bloated appearance seems a metaphor of fecundity. At this scale, moreover, the gambrel also invokes hills.

Beyond its volumetric shape, however, surely the image of its end elevations also makes it memorable. Rectangular openings of various sizes stretch around the perimeter as a monster mask, the center line emphasized by the largest openings and their aggressive diagonals. In the most handsome examples of this type, as here, the abrupt angularity of the gambrel silhouette is modified by a flare at the eaves and, over the hay loading door, by the projection of the roof into a visor for the lifting pulley.

GL27 Glocester Town Pound

1749. Victory Highway at Pound Hill Rd.

In addition to this survivor, an unusual number of colonial pounds are extant in Rhode Island—fairly well-preserved examples here and in Exeter (see EX5); tumbled relics in Foster (see FO7), Hopkinton, and Richmond. This is not only the best preserved of the group, but the most carefully wrought. Unfortunately, its effect is compromised by the crowding of recent residences around it. Stray animals were corraled in such pounds and auctioned unless redeemed for a fee within a certain time limit. Pounds were simple walled structures, roughly rectangular, usually of a height at least twice that of ordinary field walls. This, approximately six feet high, is an impressive example of colonial drywall stonework, displaying a rugged mix of rounded boulders and roughly cut lintel-like stones, all capped by long, flat stones with a narrow iron-gated, lintel-topped entrance in one wall. It is approximately fifty feet square, with a trapezoidal deviation to adjust to the fork at the intersection.

Burrillville (BU)

The biggest of the towns in the northwestern corner of Rhode Island—making the corner, in fact—is Burrillville. Its qualities and its broad historical development compare with those in Foster and Glocester. Its eastern border, too, follows the Seven Mile Line which marked the boundary between the Providence "Inlands" and the "Outlands" or

"Providence Woods." It, too, gained its separation in two stages, first, as part of Glocester in 1731, then its present identity in 1806. Indeed, Burrillville is named for James Burrill, a resident of Providence and holder of a succession of state offices before becoming a United States senator, who was foremost among those in power who supported the town in its move for independence. As in Foster and Glocester, poor soils made farming precarious, though it was dominant economically at least through 1825. Wood products and the usual sorts of small mills and forges, tavern (or inn) keepers, merchants, and a sprinkling of professionals more or less filled the roster of Burrillville's early occupations.

Transportation was more difficult than in towns to the south. A single turnpike, Douglas Pike, merely clipped the northeast corner of the town on its way to Douglas, Massachusetts. Rail service arrived only in 1873, after William Tinkham, who then owned the mills in Harrisville, led a drive to connect Burrillville's principal mill towns with the Providence-Springfield line. A spur to Woonsocket in 1893 also connected with the main New York–Boston line after 1893. Trolleys came in 1902, through Pascoag to Wallum Lake in the northwestern corner of the town. They brought the first substantial body of summer excursionists and vacationists to Burrillville's lakes, ponds, and woods. Today, sizable state forest preserves (Buck Hill and Casimir Pulaski) occupy the western edge of Burrillville (the latter extending south into Glocester), while another preserve (Black Hut) occupies an off-center position to the northeast. As in Foster and Glocester, the automobile increased the number of vacationers. And, as a result of the automobile, rail service vanished from the town: the trolleys in 1922, the rail lines in the 1930s—the latter as much because of the decline of the mills as the coming of trucks.

In one important respect, however, Burrillville did differ from Foster and Glocester. Whereas these latter towns had mere brooks and streams, save for a short stretch of the Chepachet in Glocester which made the mill village of Chepachet possible, by the time the rivulets reached Burrillville they had accumulated a considerable volume of water. The Pascoag and Nipmuc rivers join the Clear. The Clear and Chepachet rivers join at Oakland to form the Branch, which then flows eastward as a major tributary of the Blackstone. So Burrillville had many more mills than any of the other northwestern towns (even more, and many of these larger, than those which dotted Scituate before the coming of the reservoir took them out). They were located toward the center of the town in a sprawling and curving *Y*. The most important, and those with which this guide is principally concerned, are Mapleville and Oakland, at the stem of the *Y*; Glendale, Nasonville, and Mohegan, on its eastern arm; and, to the west, Harrisville and Pascoag. Although cotton mills predominated during the early decades of the century, Daniel Sayles's woolen mill, started in Pascoag in 1815, proved to be the bellwether to Burrillville's future in textiles. Woolen manufacture, which had been concentrated in the southern part of the state before 1840, thereafter shifted to the north, and especially to Burrillville and Woonsocket. As elsewhere, owners of single mills during the first half of the nineteenth century tended to be bought out by larger operators during the second half. In this town there were three. The Wanskuck Company, owned by the Metcalf family in Providence, came to control the mills in Oakland and Mohegan. Although various members of the Sayles family owned the mills at Pascoag until 1850, after this date Albert L. Sayles became dominant. Finally, in the early twentieth century, Austin T. Levy augmented his Stillwater Worsted Company from its base in Harrisville on the Stillwater River by purchasing the mills in Glendale and Nasonville.

Although this tour of Burrillville includes a number of farmhouses from various periods, its primary focus is the mill villages. Not the mills, however. Unfortunately, like most of those in Scituate and all but a tiny relic in Chepachet in Glocester, the mills of architectural interest in Burrillville are virtually gone. Two of its notable masonry mills—those in Graniteville, a satellite village to Harrisville, and the "Granite Mill" at Pascoag survived in dilapidated condition until the 1980s. Then, even as plans for reuse were underway, arson claimed both of them. So it is literally the villages that we shall see: the housing for what became a predominantly Irish, then French Canadian, immigrant work force; the mansions of some of the mill owners; a few institutional structures; and efforts at village planning. The most conspicuous evidence of such planning remains in Oakland, Harrisville, and Glendale New Town. Indeed, Henry Kendall's plans for Slatersville in nearby North Smithfield and Austin Levy's in Harrisville represent the most extensive efforts in Rhode Island to elevate the mill village to an early New England image, at least as the early twentieth century conceived of this possibility. Meanwhile, Levy's housing at Glendale New Village represents a tentative venture toward modernism.

BU1 Smith House

c. 1850. 264 Victory Hwy.

Another one-and-one-half-story Greek Revival house (plus ell) in the most developed vernacular phase of the style, with strongly accentuated paneled corner piers, deep entablature, and the repetition of the pier-entablature motif to frame a deeply inset door with side lights. As a type, it has already been discussed in connection with the Irons Homestead (GL25), which has the end chimneys typical for a plan with central hall and fireplaces pushed to the exterior walls. Here, however, the chimney is central, indicating the older type of hearth core with its cluster of fireplaces. In this house the structural image of the elevation is exceptionally compelling: in its intense commitment to Greek precedent; in the harmonious scale of the double measure of its front elevation by the structural surround of its recessed door inset within the slightly larger sealed embrace of the broad cornice and corner pilasters; finally, in the elegance of its laconic ornamentation by moldings and paneling. No finer example of this Greek Revival vernacular type exists in the state. Siting on an elevation and immaculate maintenance enhance its presence.

Mapleville

Mapleville and the adjacent mill village of Oakland are at the mouths of the Chepachet and Clear rivers respectively, where they join to make the Branch River (a major tributary of the Blackstone). From the south, Victory Highway enters Mapleville at a triangle crammed with buildings, among them (now hidden behind a garage) a virtually unrecognizable Friends Meeting House (1791 and later; much altered as a residence). An abrupt dip brings one past workers' housing of various vintages, provided by various mill owners, which, though mostly altered, is interesting in part for this reason; then on down to the mills and river. The "upper mill" is a hodgepodge of buildings extending back to the 1860s and possibly earlier, but dominated by early twentieth-century building, while the "lower mill" is wholly c. 1900; neither is of great architectural interest. The history of mill ownership in Mapleville is varied and vexed. Suffice it to say, in 1841 Darius Lawton built the original textile mill at Mapleville, a small affair for cotton yarn and warps, which turned to wool in 1853 and burned in 1871. Another owner built the lower mill. Both came under the control of James Legg in 1867, then of Joseph E. Fletcher in 1900. Enlightened owners, they enlarged the mills and were so much concerned about village amenities that Mapleville came to be regarded as something of a model town, although it is not easily decipherable as such today. Indeed, it was never envisioned as an entity, but grew as a sequence of piecemeal "improvements." After Fletcher's death in 1924, Austin Levy (see under Harrisville) brought the mills into his Stillwater Worsted complex.

BU2 Apartments (Mapleville School)

c. 1890. End of Sand Hill Rd. (off Victory Highway at intersection with Cooper Hill Rd.)

Despite its conversion, this school retains its exterior image as a handsome Queen Anne schoolhouse, the best preserved of three in the style we shall look at in Burrillville. Because of the exceptional focus on schools among Queen Anne buildings and their importance in the origins of progressive attitudes toward school building in the twentieth century, it is a pity that so few in this once popular style still exist. So it is worthwhile to examine what little remains from a campaign for new schoolhouses around 1890 which brought Queen Anne to Burrillville.

The compelling triangularity of the front elevation, reinforced by the equally forceful triangles of the flanking porches, is enhanced by the scale and placement of the openings. The tall, functional windows on the first floor have a coupled syncopation that, on first glance, may seem regular. They are set against the more decorative Palladian window above, with an elongated keystone making an exclamation mark toward the climactic bell cupola. This decorative quality of the upstairs window becomes more explicit in the ribanded and foliated reliefs in the porch gables. Such a playful approach to style, combined with the conspicuous use of ornamentation in the conviction of the importance of art in early education, made Queen Anne especially congenial for schools—and for the overt prettiness of buildings in the remarkable illustration of children's books of the period. Here the ornamentation of the porch gables means to suggest the low triangularity of classical pediments, but in fact their steepness of pitch makes them gables, sharing the slant of the larger roof. The vertical picturesqueness of the massing is characteristically Victorian, while the classical and Renaissance detail anticipates the American Renaissance to come—a combination typifying the Queen Anne moment of the 1880s and early 1890s, when, in American culture generally, the flamboyance associated with Victorianism gave way to the decorousness associated with the American Renaissance. A view of the side of the building reveals how the shallower pitch of the abutting cross roof to the rear permits a full two stories of classrooms behind. The Queen Anne style schools in Bridgeton (BU33) and Pascoag (BU34) are doubtless the work of the same, as yet unidentified, designer.

BU3 Darius P. Lawton House

1840s; porch trellising 1880s(?). Main St. (Victory Hwy.) at Gazza Rd.

This one-and-one-half-story Gothic Revival clapboard cottage in a picturesque composition of gables, dormers, and bays is ornamented with scrollwork bargeboards, a crested bay window, and a "Tudor"-capped window in the gable. (The Japanesque trellising of the side porch appears to have been a later renovation influenced by the exotic strain within the Queen Anne movement.) An unusual feature is the lack of vertical corner boards where the clapboarding meets at right angles. Surrounding the house is a rare and handsome, if somewhat derelict, example of wire and cast iron fencing, which dates from a time close to its completion.

Darius Lawton, who introduced textile manufacturing to Mapleville, almost surely, if not quite certainly, commissioned the house. He occupied it in any event, and, after him, a succession of Mapleville mill owners—Joseph Smith, James Legg, and Joseph Fletcher—all able to survey their mills and village from its hilltop site. The original one-room Mapleville schoolhouse, built in 1847, was moved to this site around 1890, after the new school (BU2) was built, to make a servant wing for the house. Gothic Revival houses are few in Rhode Island, especially in such fine condition. During Mapleville's apogee as a textile village this was its "great house." Immediately opposite is a fine brick duplex (c. 1860[?]), probably for managers.

BU4 Seventh-Day Adventist Church
(Mapleville Methodist Episcopal Church)
1908. 854 Victory Hwy.

Concrete blocks cast to resemble rough-hewn granite were particularly popular for foundations, small commercial buildings, and the occasional house during the early decades of the twentieth century. Their use for more monumental work, as here, reflects aspiration well beyond budget. Implacably textured, implacably gray, the assertive blocks build a series of ponderous shapes out in front of the gabled body of the building. These are graduated (left to right) from big to little, with flat and angled planes bluntly opposed. At one corner, a chunky tower with prominent angled buttresses; then a swollen bay window; finally, a miniature octagonal turret projecting three quarters out of the opposite corner. The shapes are clumsily impressive; the technique is folkishly quaint. Dull glints of color from opalescent stained glass windows strain to make an impression against the grayness.

BU5 Our Lady of Good Help Church, Rectory, and Parish House–School
(Notre Dame de Bon Secours)
1905–1907, Walter Fontaine. 1063 Victory Hwy.

Again the texture and color of materials are important. Whereas the Methodist Episcopal church emphasizes folk design attempting a sophisticated result, this Roman Catholic church for a French Canadian parish shows a sophisticated architect borrowing folkish effects for a country church under the influence of contemporary Arts and Crafts inspiration. If the massing is a bit disjointed and the detailing somewhat heavyhanded, the church overall exhibits a straightforwardness in the use of a random fieldstone base and shingling, with wooden post-and-bracket construction which gives it exceptional character. The complex of church, rectory, and school is beautifully preserved, except that around 1989 the complex was resurfaced in larger shingles—reducing the cost of repair, no doubt, but blunting scale

BU5 Our Lady of Good Help (Notre Dame de Bon Secours) Church and Rectory

and texture. The exposed stained wood and revealed timber structure inside the church mostly retains its original character.

The rectory is a gambrel-roofed, shingled house with classical allusions edging it out of Queen Anne or Arts and Crafts toward incipient Neo-Colonial. By the date of the combined parish house and school, the Neo-Colonial and Neo-Renaissance styles were at their height of popularity. It is difficult to know which label to apply to the rustic minimalism of this shingled neoclassical front. It is so abstractly composed, in fact, that only by looking at the very functional side and rear elevations does one realize that it *is* a school.

Oakland

Mapleville is connected to Oakland by the linear arrangement of buildings along Victory Highway. Whereas Mapleville grew in higgledy-piggledy clusters, Oakland exhibits a gridded street pattern with well-spaced houses off the main road (Victory Highway). It remains among the more ordered mill villages, partly because of its slightly elevated site looking into wooded hills rising from the opposite side of the Branch River (which up to this writing have miraculously remained wooded), partly because of the unusually high architectural ambitions of owners who successively contributed to the town, and partly because of better than average maintenance, which has persisted down to the present. The town began when John L. Ross built a stone mill here in 1850. He continued in ownership (sometimes leasing to oth-

BU10 Oakland Assembly, Recreation, and Market Hall

ers) until 1892, when the Wanskuck Company, owned by the Metcalf family in Providence, took over. They ran the mill as the Oakland Worsted Company until 1957. Textile manufacture continued through most of the 1960s, until in 1973 the mill was sold to a used machine dealer. Like Mapleville, Oakland was long known as a model mill village; it had gravity-fed drinking water, sewers, and lights installed by the Wanskuck Company in 1894—much like the improvements Joseph Fletcher brought to Mapleville after 1900. Under Wanskuck auspices, too, there was considerable new building and upgrading of old.

BU6 Daniel Cooper House

1820, later ells. 1168 Victory Hwy.

This two-and-one-half-story Federal house with central chimney features another variant of the Union Village porch type (see North Smithfield). Whereas the examples in Foster tend to reduce the wide-spaced columns to posts (FO5, FO16, and FO19), these are definitely columns with somewhat swollen entases. All other examples thus far seen use flat pilasters as responds; here they are half-round engaged columns. A gouged carpenter's version of rope molding occurs at the eaves of the half-hipped roof, with a delightfully primitive frieze of alternating projected squares and diamonds fashioned from latticing below it.

BU7 Mill Housing

c. 1850. 1287 and 1289 Victory Hwy.

A row of well-maintained workers' tenements indicates how handsome early duplex workers' housing can be when left more or less as built (except in this case for such detracting addenda as shutters and ornamented aluminum storm doors). Plain clapboard walls are animated by a syncopated rhythm of well-proportioned openings on the ground story common to this type—window/window/tall door (emphasized by a projecting molding cap)/paired window (one half shared with the adjoining tenant)—which then repeats in an inverted sequence. Above, the wall is absolutely plain, except for the punctuation of paired low bathroom windows at the center of the entablature which echo the doubled window below. All other light for upstairs rooms comes from windows in side elevations. The austerity of treatment and even basic molding shapes suggest the Greek Revival, but so nominally that the ground is prepared for early Victorian forms to take over. This housing seems to have been built when John Ross built Oakland's first mill.

BU8 Dairy Barn

c. 1885. 1288 Victory Hwy.

Across the way is a fine late-nineteenth-century dairy barn with bracketed eaves. It housed the Wanskuck Company herd (and probably that of the preceding mill owner) to supply the village with milk—a common patriarchal benefit in the better-managed mill villages, although the dairy barns are seldom preserved.

BU9 Duplex

c. 1900. 1290 Victory Hwy.

Adjacent to the dairy barn is a Neo-Colonial cross-gambrel-roofed duplex built by the Wanskuck Company for its supervisory personnel. Joined like Siamese twins to make the front elevation, the gambrel profiles at either side become a pair of shallow gables on top. William White, the superintendent who was directly responsible for village improvement around 1900 for the Wanskuck Company, lived here.

BU10 Oakland Assembly, Recreation, and Market Hall

1898. 1422 Victory Hwy.

This brick recreation hall for community functions, with store space below, was the Wanskuck

Company's most conspicuous benefaction to Oakland. Its precedent appears to have been the colonial brick box for meeting hall and town offices atop an arcaded, open-air market such as the Brick Market in Newport and the Market House in Providence (NE8 and PR54), but here the ground-level arching of ancient market buildings is raised and enlarged as windows for a two-story hall. To the market house precedent, moreover, a stair and gabled entrance vestibule have been appended toward the road with something of a baroque flourish, as though the owners meant to go beyond providing mere access to the hall, so that this overstated approach would also advertise the cohesiveness of the village and make visible the patriarchal largesse from which all blessings flowed. At the top of the climb, the entrance arch with its extravagance of radiating voussoirs seems to allude to such bristling displays of stone against brick as the provincial William and Mary detailing of Newport's Colony House. This entrance dramatizes the arching of the tall auditorium windows behind, these more modestly trimmed in stone and recessed between piers. Off each pier scroll brackets support the flared eaves of the hipped roof. The mix of modesty and pretension which this building somewhat awkwardly displays in its engagement of colonial precedent makes it among the most expressive of community buildings in Rhode Island's mill villages.

BU11 Workers' Housing

Mostly c. 1852. 7–9, 8–10, 11–13, 12–14 Mill St. (off Victory Hwy.)

Mill Street, which leads to the Oakland Mill, is lined with clapboarded Greek Revival duplexes similar to those just seen (BU9), but with slight variations in door placement and moldings. Despite many alterations, the overall effect remains.

Glendale

As a manufacturing village, Glendale began, like most others, with a cluster of sawmills and gristmills in the 1790s. They were converted to cotton mills in 1841, when the place was known as Newells Mills. After a fire in 1850, a new mill was built in 1853, this time for woolens, following a pattern of conversion typical for Burrillville. In 1889, William Orrell bought the mills and made a number of town and mill improvements. He sold out in 1934 to Austin Levy's Stillwater Company, which was headquartered in nearby Harrisville. Weaving continued at Glendale until the 1970s. Today plastics, manufactured in twentieth-century buildings of little architectural interest, are the principal industry. The more important nineteenth-century mills are gone.

BU12 Burrillville United Methodist Church

1893. 185 Joslin Rd.

This shingled Queen Anne church would be more interesting had it not lost its steeple to the 1938 hurricane. Although it is certainly a provincial production, the use of the Palladian format for the stained glass windows in the side elevations and the halving of a medieval wheel window over the entrance into something like a classical arched window indicate how "medieval" forms could be inflected toward Renaissance and Neo-Colonial forms in accord with the trend of informed taste at the end of the nineteenth century. Such classicizing strains within high, angular massing, coupled with the coziness of sheltering hoods over openings, makes this a vivid and charming, if awkward and belated, example of Queen Anne. Its interior retains little of its original character.

BU13 Glendale New Village

c. 1936, Jackson, Robertson and Adams. Elm, Maple Leaf, Woodside, and Stockwell rds.

The most interesting architecture in Glendale is outside the village proper. This gridded enclave of some thirty one-story houses was self-consciously set a little apart as a new village. Having tested three variant examples of modern vernacular mill housing in Harrisville (see BU21 and BU22), Austin Levy built this village shortly after acquiring the Glendale plant. (On Levy and his more extensive benefactions for his headquarters mill village, see Harrisville, immediately below.) When Henry-Russell Hitchcock wrote his *Rhode Island Architecture* (1939), he judged it to be the most significant example of contemporary group housing in the state, although he frowned on its conception as a collection of individual houses, which hampered its comprehension as an entity. (At the time, progressive modernist architectural theory emphasized collective planning even as enlightened mill owners in Rhode Island

BU13 Glendale New Village

sought to eliminate the stigma of earlier collective housing.) Smooth stucco wall surfaces, unbroken and unaccented by moldings; the elegant attenuation of windows and door elements; the absence of specific allusion to past styles, except for a touch of nostalgia in hipped or low-pitched gable roofs and trellising at front doors: these qualities indicate Levy's attempt to effect a compromise between modern architecture and traditional house types, of which he was fond. Although it is tightly and unimaginatively sited, the prim maintenance of New Village preserves something of the aura of hopeful enlightenment which brought it into being. As with all such communities, which were models for their time, typical alterations through the years will also interest those concerned with architectural and social history: the closing of porches to make winter vestibules; programs for enlargement and garaging; efforts to individualize houses (which Hitchcock had criticized as being conceptually *too* individualistic). But happily the ideal of a unified community of related houses persists.

BU14 Smith–Darling House

c. 1810–1820. 370 Barnes Rd.

This two-and-one-half-story, five-bay farmhouse with a central chimney, plainly carpentered in the Federal style, is nicely related to the winding road, at once close to it but also a little aloof from it on a slightly raised stone-edged terrace which gives a degree of privacy. Its pilastered door (which may be of a later date) looks to the barn, across the road and hard by it, which reinforces the wedding of road to farm. Adjacent to the house is a stone retaining wall, an outstanding example of drywall construction. It once closed the rear of what appears to have been a blacksmith shop, and continues as the high basement of a no longer extant barn. The exposed smithing fireplace still stands as a relic. In the fields behind the barn is the family burial ground, which contains eighteenth-century stones.

BU15 Henry Stafford Nichols House

c. 1860. 1450 Tarkiln Rd.

This one-and-one-half-story house indicates how a typical Greek Revival formula could be modulated toward Victorian preferences for showier ornamentation and more plastic handling of space and mass. The typically plain Greek Revival entablature acquires bracketed ornament. The severity of a heavily framed door is lessened by a projecting door hood on decorated brackets. The planar Greek Revival elevation swells with symmetrically positioned bay windows. A handsome carriage barn accompanies the house. It displays a pair of double-doored portals and alternating windows, all with low arching, above which float strongly projecting and bracketed flat table hoods. So conventional round-arched Italianate treatment gets a lively rhythmic counterpoint. This was originally the residence of the son of Joseph P. Nichols, who owned the nearby Oak Valley mills (now disappeared). Henry served as superintendent for the mills from 1857 to 1888.

Nasonville–Mohegan

These adjacent mill villages began with the manufacture of such edge tools as scythes, axes, and hoes. Leonard Nason's mill, founded in 1821 and in operation until destroyed by fire in 1881, was renowned for the high quality of its products. Textile manufacturing appeared around 1820 in Mohegan and 1838 in Nasonville. In both places, the first product was a coarse cloth known as "Negro cloth" because its principal market was southern plantations, where it was used for slave clothing. Both mills graduated to satinettes, cashmeres, and worsteds, eventually under the control of two of the largest twentieth-century woolen operators in the state: Levy's Harrisville-based Stillwater Company came to Nasonville; the Metcalfs' Providence-based Wanskuck Company to Mohegan. Both operated until around 1960. A series of fires and demolitions have taken all the interesting nineteenth-century mills, leaving twentieth-century replacements of little architectural interest, now predominantly serving plastics manufacturers.

BU16 Western Hotel (Walling Hotel)

c. 1810, c. 1850–1860. Douglas Pk. at Victory Hwy.

Nasonville boasts a prominent regional landmark in the Western Hotel, originally named for the Walling family, which established it. It is the double porch of this much revamped establishment which catches the eye—a feature which seems to go back to its Victorian remodeling. But behind the porch its origins in a plain, five-bay Federal building with center chimney are discernible. Probably opened as an inn when the turnpike was built, it served as a coach stop and tavern for the Providence-Douglas stage. Across the pike from the hotel is a fine stone-arched bridge (1907; reinforced concrete guard rails added in 1924), which carries Victory Highway over the Branch River.

BU17 Sweet Family Houses

1870s. 620, 650, 668, 675 East Ave. (Route 107)

On East Avenue, formerly Sweet's Hill Road is a group of houses built on the hill by the Sweet family in the 1870s and 1880s in variant versions of rural Italianate (numbers 625, 650, and 675, as well as a Greek Revival house at number 668, also occupied by members of the family). David Mathewson, a grandson of the original settler, lived in a Greek Revival house at the foot of the hill. He became a major builder in the area in the mid-nineteenth-century and was reportedly responsible for some hundred buildings. Almost nothing precise is known about them, except that they include a number of mills, of which the A. L. Sayles Mill in Pascoag and the Plainville Mill in Massachusetts, both demolished, are known to be his. His sister married Henry Sweet, an expert in machine installation whose work took him to many states and to Cuba. It was their three sons who came to dominate the summit of the hill, as dairy farmers, one also becoming a meat packer, another a lumber dealer.

The Thomas Sweet House, at number 675, displays another variant of the way in which the one-and-one-half-story Greek Revival format can be Victorianized by bracketing (here widespaced and doubled); a small, centered cross gable (which might equally be termed a large dormer); and nominal "Renaissance" projecting lintels over the windows, together with a bracketed entrance and side porches, supported on columns with bases which have become rather like posts. A curious anomaly is the continuation of the bracketed cornice at the base of the gable, which is interrupted by a round-arched Italianate window. It seems to be the relic of a Greek Revival pediment holding fast to old ways in a new stylistic situation, although it could be a later renovation cut through the line of the old eave.

Harrisville-Graniteville

Harrisville, the administrative center of Burrillville, is among the most interesting rural mill towns in the state. It is not its extant mill complex, however, that makes it so. Nineteenth-century mill buildings have been mostly superseded by nondescript twentieth-century replacements, although the dominating building in the complex is a fascinating example of early reinforced concrete factory construction. Harrisville today is more exceptional as the center of the community activities of Austin T. Levy, the enlightened head of the sizable Stillwater Worsted Company, which had mills in several towns in northwestern Rhode Island. Levy leased several Harrisville mills in 1912 (purchasing all of them in 1921), after their operation for two generations by William Tinkham and his son Ernest. Stillwater Worsted continued operations until 1972, but with increasing difficulty in the face of the general collapse of the textile industry in the state. Levy was among the leaders of the last generation of Rhode Island textile barons.

BU18 Ernest Tinkham House

1880–1882, 1902. 169 East Ave.

This imposing Queen Anne mansion invokes, as immediately as any in the state, the image of Victorian mill ownership at its most lordly. Lifted a little above and away from the mill on a piece of Sweet's Hill, it proudly displays itself at the far edge of a deep oval lawn framed by magnificent trees. It is at once remote from the highway and conspicuous when viewed from it. Up to the roof, it is handled fairly plainly, despite the puff of its amplitude by a generous porch and porte-cochere. The rhetoric of power erupts above: an open turret, a dormer, and a gable, side by side. As treated here, these three features epitomize four stylistic influences which coexist in architectural details during the Queen Anne interlude of the 1880s; Victorian, exotic, Neo-Colonial classicism, and vernacular. The turret is Victorian in its aggressively plastic quality, mingled with the exoticism of "Saracenic" openwork around the elevated circular porch as a kind of aerial gazebo and in the inflated quality of its beehive or bell-shaped crown. The dormer, as a broken scroll pediment, heralds the taste for Neo-Colonial classicism during the American Renaissance to come, although the pediment motif is also bloated in the Victorian manner. Finally, the gable, like most of the house, is shingled vernacular, pulling the grandeur back to a level consistent with houses in the rest of the village.

BU19 William Tinkham–Austin T. Levy House (Southmeadow)

c. 1856, altered 1918. 169 East Ave.

Diagonally across the road from the Ernest Tinkham house is what was apparently his father's house, built by William Tinkham on his arrival in Harrisville, and later occupied by Ernest's brother Henry, before Levy bought it in 1915. Originally, the house was a rather plain dwelling with a mixture of late Greek Revival and Italian Villa influences. When Austin and June Levy remodeled it and approximately doubled its size in 1918, they sought a comfortable "colonial" effect, insofar as this was possible. The favored style for Levy's generation (loosely incorporating early national work as well), it also characterizes most of his many architectural and restoration benefactions in Harrisville. His most conspicuous additions to the exterior of this house were the side ell and the tall staircase window on its east side. Levy christened his house Southmeadow for the climactic feature in its spacious grounds: a terraced landscape, with an Italian garden and tennis courts on a south-facing slope down to the Clear River. As an ensemble, house and landscape possess a more reticent and secluded aspect than the conspicuous display of Levy's predecessor across the street. It proclaims the "refinement" in which early-twentieth-century wealth often chose to shroud itself under the aegis of Renaissance and colonial inspiration.

BU20 Mill Housing

Possibly 1890–1900. 113–151 East Ave.

West of Levy's house, East Avenue dips past a row of minimally altered Colonial Revival duplexes erected by William and Ernest Tinkham. They are starkly looming, with broad stretches of clapboard wall between wide-spaced, sparsely detailed windows. Attenuated framing accentuates their severe aspect. So does the plainness of their centered porches, with roofs supported on spindly turned supports sheltering paired entrances. Unusually, the upper windows cut through a broad, frieze-like band under the eaves. These houses are the more interesting because of their proximity to housing built under Levy's aegis in the New Village immediately opposite.

BU21 New Village Experimental Prefabricated Mill Housing

1935, Jackson, Robertson and Adams. 2 and 11 Stewart Ct., 124 East Ave.

BU22 New Village Mill Housing

1918, Jackson, Robertson and Adams. Off East Ave., including 1–19 Steere St. (odd numbers) and 1–4 and 6–12 Park Ave. (even numbers)

For a site off East Avenue, Austin Levy commissioned twenty-two Neo-Colonial houses from a leading Providence firm. It was the first of Levy's projects designed to give his village a colonial aura. Already in this tract, in 1911, along one side of Park Avenue, the preceding Tinkham dynasty had erected a row of plain, clapboarded workers' houses (1902) similar to those across East Avenue (see preceding entry). When Levy and his architects added to the stolid Tinkham quarters, they envisioned a cozier image of housing, one less institutional and more suburban in aspect. The enclave came to be known as the New Village. The newer houses are single family (instead of the typical nineteenth-century two- or four-family houses), and were doubtless designed for managerial and supervisory personnel). Basically all are simple two-and-one-half-story five-bay boxes, clapboarded, with living rooms opening onto side porches, but given a modicum of individuality by mixing gable and hipped roofs and by varying door treatments derived from various late colonial and Federal sources. Subsequent alterations have not obscured Levy's intent.

Then, built thirteen years later, the prefiguration of the newer village to come: flanking the entrance to the development are three variant versions of prefabricated, one-story, hip-roofed houses in stucco-covered, steel-plate, panel construction, as "modernized" versions of what had gone before. They were experimental models for the larger community of houses completed in 1936 (also known as New Village) which Levy commissioned for a site close to nearby Glendale (see BU13, where the qualities of these houses are discussed).

Also in this New Village enclave are a scattering of houses unrelated to either the Tinkham or Levy building initiatives: one of them going as far back as 1806, with extensive Greek Revival and later additions, which originally housed a member of the Tinkham family; others built independently in the early twentieth century. Of these, a shingled bungalow (c. 1915), at 9 Stewart Court, is perhaps the most interesting.

BU23 Workers' Housing

c. 1920, John Hutchins Cady. 93–95 East Ave.

Architectural Record (December 1924) praised this four-family row housing in nominal Colonial Revival style as an enlightened example of low-rental housing. The entrances to the downstairs units are on the side elevations. The front entrances open to stairs to the upstairs apartments only, although the elevation is so organized as to give the appearance of duplex units side by side. The pairs of entrances on the front elevations are each unusually coupled with a window under a bracketed hood. It is just such playful, picturesque novelty used as a means of generating "charm" that characterizes much of the design of the period—and offers further evidence of the effort among enlightened mill owners at the beginning of the twentieth century to replace the institutional appearance of earlier housing with more endearing effects. Especially creative is the placement of living rooms to the rear of these units, where they are extended by tiers of screened porches. They look toward the Clear River, providing the workers with their own "south meadow."

BU24 Harrisville Dam and Mill Pond

1857. Clear River at East Ave.

East Avenue crosses the Clear River on Stone Arch Bridge (1902; steel-deck sidewalks added 1952), a single segmental arch in wet-laid rubble with dry-laid abutments. It was a replacement for a metal-truss bridge (the trusses from which are now incorporated in the Pratt Pony trusswork of the Shippee Bridge across the Clear River north of Harrisville). Why? Did Pinkham, too, long to alleviate the "ugliness" of technology from his village by this appeal to the "beauty" of the old-fashioned craftsmanship of the adjacent dam?

Also spanning the river at East Avenue is the Harrisville Dam, built by William Tinkham and his then partner Job S. Steere immediately after Tinkham and Steere enlarged their Mapleville operations by buying the Harrisville mills. It is a stepped construction of roughly dressed granite blocks which epitomizes the very nature of a gravity dam. Spillage and seepage over and through the granite blocks now convert it to a beautiful water wall. Levy created a village park around the pond. In it, just across the bridge, near the junction of East Avenue and Main Street, he located The Assembly, which represents one component in the cluster of buildings which he provided as the civic core of Harrisville.

BU25 Harrisville Civic Buildings

1933–1937, 1950, Jackson, Robertson and Adams. 1990–1991, addition to library, Providence Partnership. East Ave. at Main St. (The Assembly); all others on Main St. from East Ave. to the corner of Chapel St.

Austin Levy's civic complex serves as the administrative core for the whole of Burrillville. For his own project, Levy had the site cleared of several commercial buildings. The buildings that replaced them are all in brick, all in the Colonial Revival style (the post office of 1950 only nominally so). As was often the case at the time, the colonial sources were more typical of the mid-Atlantic states than of New England. The underlying appeal of mid-Atlantic colonial seems to have been its greater emphasis on picturesque compositional detail and mellow surface texture, as compared to the more austere emphasis on geometry, plainness, and harder surfaces typical of colonial building in New England. In the attempt to achieve a genial village colonial, the exteriors of The Assembly and the library are the most successful, dignified yet with a demureness appropriate to the idyll which Harrisville was meant to become. The library's enlargement has altered its impact on the village; so The Assembly best characterizes the intent of Levy and his architects, and with the more original design.

The Assembly (BU25.1; c. 1933), at 116 Main Street, not only served the usual civic function of a meeting hall, but also housed regional drama and chamber music groups. (Levy was himself a fine violinist who frequently participated in chamber music activities.) Its simple brick mass is an architectural blank against which doors, and windows, and a portico occur in rudimentary but pretty versions, which are also exploded or reduced beyond the scale expected for the extent of the wall they occupy and in relation to one another. One thinks of the endearing images of village architecture in children's books of the period, where just such an archaized, overblown portico as this becomes the immediately grasped "sign" for the building's function. "Here the village gathers," it proclaims—its rude quality identifying with the folkish life of the village, while its puffed-up allusion to historic grandeur asserts that even villagers need civic dignity. Six tall, bulky, square pillars capped with only the narrowest of moldings support a simple beam as the "entablature" for a shed roof in slate. The portico is not so much integrated into the brick wall behind as set against it. Its disconnected quality reinforces the emblematic quality inherent in its archaic and bloated aspects. Centered in three of the portico's five openings are a pair of overblown doors, between them an overblown window, excessively paned, all topped with overblown transom lights fretted in diamond-shaped trelliswork. Then, between the outermost pillars, the scale plummets to domestic-sized windows, with a row of itsy-bitsy windows up in the gable sitting on the upper edge of the portico roof like three birds on a wire, and a tiny, round-capped cupola over all. So the mock pomp of the portico yields to the sweetness of the village. Postmodernist designers at the end of the twentieth century, addicted to the use of historic allusion to create just such ironic shifts in architectural meaning, might well pause before this unpretentious anticipation of effects they seek.

The Jesse M. Smith Memorial Library (BU25.2; 1937, enlarged 1990–1991), 144 Main Street, is another of those tiny, village-sized gabled boxes for books of the period, with shuttered windows subtly enlarged beyond their domestic aspect against the grander Neo-Colonial door they flank while its scale is subtly diminished. The stage-set quality of the facade also appears in the paneling of its entrance, which is not inset, but made of moldings tacked onto what would otherwise be a flush door. From the side elevation a bow window marks the original children's room. Extensively paned, the bow derives not from colonial shop windows (and certainly not at all from colonial houses), but from the late nineteenth-century quest for colonial charm. The early 1990s enlargement swells behind and down-slope of the original building, which fortunately retains its frontispiece. The addition, however, is overassertive on the exterior and makes an awkward angled connection with the axial symmetry of the old plan inside. In moving from village to suburban scale, moreover, the interiors have lost much of their intimacy, even though original elements: plain plastered walls and barrel vaulting in the reading rooms flanking the circulation desk, with fireplaces and the spare adornment of wrought iron fixtures from the village smithy. The Assembly now shows the effect of the original interiors to better advantage.

The Universalist Church (BU25.3; c. 1886, remodeled 1933), 134 Main Street, began as a tall, shingled Queen Anne building, regarded by Levy (and most of his generation would have agreed) as a serpent in his colonial Eden. Following a fire, which only partially destroyed the church, Levy offered to donate a colonial renovation. Too much of what was then generally regarded as the "ugly" Victorian past disappeared in such reform. Yet sometimes such renovation did produce a building of character. Here, Neo-Colonial detail stretches and multiplies itself in swollen elements to cover the Victorian mass. Panes accumulate in large openings until they appear almost as latticing. Such colonial elements as pilasters, panels and fans over the

BU25.3 Universalist Church

windows are mobilized and enlarged to "fill" the space left over. (The fans which top the windows of the side elevations were probably inspired by the similar exterior treatment of the Congregational church in nearby North Scituate; see SC1.) The pediment strives to assure us that it is really what it pretends to be; but the humiliated Queen Anne gable, with its tall proportions, takes its revenge on Levy's masquerade, and certainly on the too stubby, but quite elaborate, spire as well. It is the joust of the styles—one meant to triumph, the other resisting defeat—and the quality of workmanship that accounts for the interest of this building, and behind it of course the poignant sense of what a "colonial" Harrisville meant to Levy.

The interior is reached from an entrance porch added to a corner of one of the side elevations, which cuts through the ungainly height of the raised basement. Stairs mount to a simply plastered box of space spanned by plaster barrel vault and straightforwardly fitted in light-stained furnishings. This spareness provides the ideal setting for relics of the old church. At the head of the stairs is a pretty Queen Anne stained glass window; on the platform, the ornate black furniture which always adorned it. Best of all is what may be the finest Queen Anne chandelier in the state suspended at the center of the space. A polished brass structure of hoops with punched ornament shapes the undulations and curlicues of bent brass tubing into a cage to catch and concentrate the light. Unelectrified, it is lighted once a year for an evening candlelight service at Christmas.

The government buildings, including the Harrisville Town Hall (BU25.4; 1934 and the North District Courthouse (BU25.5; 1934), both at 105 Main, are less interesting individually, and, in any event, depend on variants of the sort of design already seen, except that the post office (BU25.6; 1950), 131 Main Street, built roughly a decade and a half after the other buildings, modestly modernizes the Neo-Colonial idiom. (Exceptionally, Levy paid for this building and donated it to the federal government.) The Council Chamber in the town hall provides a sense of the cheerful, understated Neo-Colonial functionalism of the interiors of all the public buildings, with their repertoire of wavy plaster, wrought iron fittings, fireplaces, and simple vaulting.

Notwithstanding the considerable accomplishment of this civic center for a mill village, it has a scattered quality. The individual buildings do not cohere as a group. To this deficit, add the shattering intrusion of the plate glass, metal, and brick modern high school immediately behind the post office—shattering not for its modernity (Levy favored modest modernity, as witness his late model housing and the post office), but for its scale and lack of sympathy for what surrounds it. One serpent slain, another appears, ironically bearing Levy's name.

BU26 Central Hotel (Former)

1837. 173 Main St.

The block of Main Street immediately south of East Avenue completes Austin Levy's Harrisville. It contains a number of nineteenth-century houses of various vintages, some remodeled from eighteenth-century structures, most of which Levy purchased and renovated. Most became mill housing. The Greek Revival Smith-Wood House, however, became Stillwater House (176 Main), which was used for company offices. The Daniel Mowry–Joktam Putney House includes at least two eighteenth-century houses, and may include pieces of other old structures. At some time the Putney House was moved to this site, apparently from the library site when the library was built, and joined to the Mowry House. The front of the Mowry House early became Keach's Tavern, and Levy continued the tradition, using the complex as a restaurant in conjunction with the adjacent

Central Hotel, a large Greek Revival structure built for this use. Under Levy, in 1922 this became the Loom and Shuttle Inn, a combined village inn and guest facility for visitors to the mill across the street. It has been converted to elderly housing.

BU27 **Harrisville Mill Complex**
1895–1926. 1911, Mill No. 4. Adolph Suck. Off Main St. south of East Ave.

This became the headquarters mill for Austin Levy's extensive Stillwater Worsted operations. Despite Levy's activity in town building at Harrisville, he contributed nothing of architectural consequence to the nondescript cluster of fourteen buildings which make up the present mill complex. The dominant building in the complex, Mill No. 4, built in 1911, at the very end of Ernest Tinkham's regime, is (or was) exceptional, however. It has been so altered that its original state can be seen only in photographs. It was one of the earliest reinforced-concrete-framed mills in Rhode Island and, at the time of its construction, was alleged to be the largest concrete-framed mill in New England. Windows, now mostly blind with cinder block, once boldly filled the frames with wooden sash (as the view of the rear wall best reveals). Under the circumstances, the incongruity of the spindly, medievalized tower with its crudely cast climax of balconies and corner crenellations seems more appropriate to the battered remains which now exist than a herald of a new approach to the twentieth-century factory.

BU28 **Sumner Sherman Farm** (Othonie Young Farm)
c. 1865. 325 Sherman Farm Rd.

This residence, now surrounded by development, is connected with a third principal mill-owning family in the area. The Sumner Sherman Farm was purchased by the son of Syra Sherman, who, with his brother Stephen, had founded what was then called the Shermanville (later the Graniteville) Woolen Mill. Sumner eventually left the mill for farming, which his son Ernest continued very grandly as a breeder of Hambletonian trotting horses. For this he built an impressive barn which was lost to a 1970s fire. Mildly interesting as another example of the transition from Greek Revival to early Victorian Italianate, the house is perhaps more so as a third hillside setting for residences of Harrisville-Graniteville mill-owning families. All are large and comfortable, if architecturally somewhat provincial houses: the grand farm of the Shermans; the lordly mansion of the Tinkhams; the expansive suburban country house of the Levys.

Pascoag–Bridgeton

In Pascoag, once the largest of Burrillville's woolen manufacturing villages, generations of the Sayles family, going back to the end of the eighteenth century, were the mill owners. Wool manufacturing in this area began with Daniel Sayles's mill in 1814, augmented by other mills under various members of the family, until Albert L. Sayles came to head the dynasty in 1853, and dominated family operations during the latter half of the nineteenth century. Under his leadership, a large mill known as the Sayles Mill, or more popularly as the Granite Mill for the forceful presence of its masonry walls, was built in 1865 (and enlarged, along with a mill office, in 1880) at the principal intersection in town, Main and Grove streets. The Granite Mill provided the economic as well as the visual focus for a line of mills along the Pascoag River, into which the Clear River flows at the neighboring village of Bridgeton. During the last decades of the nineteenth and first decades of the twentieth century, however, an extraordinary number of fires destroyed six mills within a radius of a mile and a half. Most were never rebuilt. Others foundered for economic reasons, until, finally, only the big Granite Mill remained, among the handsome cut-stone masonry mills in the state. It operated into the middle of the twentieth century. Eventually, it, too, closed. Abandoned for a number of years, it was also slated for conversion to elderly housing, when—in a too-familiar finale—another of Rhode Island's stone mills was lost to arson in 1981. A small, derelict stone warehouse building is all that remains from the complex. The centers of Pascoag and Bridgeton are topographically much more picturesque than that of Harrisville because of the precipitous banks worn by the flows of the Pascoag and Clear rivers through these merged villages. But compared to Harrisville, the center of Pascoag especially presents a disorderly appearance; the nineteenth-century wooden and brick commercial buildings along Main Street, winding beside the river, are interesting, if dilapidated.

Likewise dilapidated and much altered mill housing from various epochs of the village's history on the adjacent slopes south of the river offers some rewards for the architectural tourist and, one hopes, someday for at least modest refurbishment, because the village possesses considerable character. The most interesting architecture in town today, however, tends to be located north of Main Street on the slopes and plateau made by the wear of the rivers. There mill owners, supervisors, town businessmen, and professionals concentrated their houses, although mill housing also exists for those who worked the mills in both villages.

BU29 Albert L. Sayles House

BU29 **Albert L. Sayles House**

c. 1880. 43 Church St.

This Queen Anne residence, built for one of the mill-owning Sayles brothers, the dominant millowner in the village from 1853 until his death in 1898, exemplifies the usual characteristics of the style in its asymmetrical massing and sheathing in clapboard below with an exceptional mix of cut shingle ornamentation above to animate upper-story surfaces. A plain Palladian attic window in the front gable tops this layered elevation. Its grandest gesture, however, is a wraparound porch with a latticed railing and spindled arching under the eaves, with a balconied porch upstairs off a bedroom, fronted by more latticed railing in a contrasting pattern. This porch structure reveals influences from the traditional Japanese house as filtered through American carpentry and provides subtler evidence of the exotic strain in Queen Anne than the "Saracenic" porch of the Ernest Tinkham House in Harrisville (BU18). Like the Tinkham House, this also closes a corridor edged with planting, which is here more roughly treated in accord with its rugged terrain. It once gave an overview of the mill below (as, in a less direct manner, did the elevated placement of Harrisville mill owners' houses over their factories).

BU30 **Pascoag Public Library**

1924, Jackson, Robertson and Adams. 57 Church St.

The Pascoag Public Library is another village Neo-Colonial brick, hip-roofed box by the same Providence architects who later designed Levy's more elaborate building in Harrisville. Its recessed entrance porch is flanked by a pair of sizable windows designed to maximize light inside, as though designed for a schoolhouse in contrast to the diminutive shuttered quaintness Levy wanted for Harrisville. Here the elevation is more prosaic, but also more inventive.

BU31 **Pascoag Community Baptist Church** (First Baptist Church)

1839, c. 1865. 113 Church St.

The church that names the street at first sight appears to be wholly Greek Revival, its portico supported by four wide-spaced Doric columns, but with the delightful provincial variation of a center pulled forward to combine the dual functions of entrance porch (with a too-steep pediment) and attenuated stilting for the steeple. But look again. The entrance, with its Romanesque dogtooth ornamentation, probably a later modification, calls upon the round-arched style. Provincial in detail, strange in scale, its front elevation disconcertingly windowless, still this church, with its giant Doric order, is decidedly commanding. Its interior has been considerably altered. The Victorian parsonage (c. 1880) deserves notice too.

BU32 **Calvary Episcopal Church**

1894. 165 Church St.

This is a well-preserved Queen Anne country church, although its raising for another story and alterations to the front steps spoil the effect. It was built as an Adventist Church, bought by the Episcopal Church in 1896, and moved from another site on Church Street. The interior has been somewhat altered.

BU33 Joseph C. Sweeney School

c. 1896. Centennial and Laurel Hill Grove sts.

Here the Mapleville School (see BU2) is duplicated for Bridgeton. More altered than its Mapleville equivalent, it was cruelly stripped of ornament and other subtleties in the process, and now stands abandoned. The cast iron fountain and water trough on the school lawn used to be at the intersection.

BU34 Apartments (Pascoag No. 11 School)

1893–1894. 83 Sayles Rd.

The projecting gabled centerpiece of this Queen Anne schoolhouse features the delicately scaled Palladian window already seen in counterparts (BU2, BU53), and another of those exclamatory keystones. Here the hip-roofed cube topped by a louvered octagonal belfry goes back to older schoolhouse prototypes.

Mid-State

WARWICK, WEST WARWICK, COVENTRY, WEST GREENWICH, EAST GREENWICH

Warwick (WA)

[Editorial note: WHJ left no introduction and relatively few entries for Warwick. The following is offered as a brief overview of a community more thoroughly shaped by the automobile than any other in the state.]

DEVELOPED IRREGULARLY FROM THE TIME ITS FIRST EUROPEAN settlers arrived in the 1640s, Warwick never had a strong town center. Mostly found here were scattered farms, a few summer retreats established by Providence's wealthy in the eighteenth and nineteenth centuries, and some modest residential neighborhoods developed in precincts closest to Providence in the early twentieth century. Now by population the second largest city in the state, Warwick has also shown the most recent and prolific growth. The 35-square-mile city grew from almost 29,000 in 1940 to more than 85,000 in 2000. Much of its growth, which occurred largely between 1950 and 1980, took the form of tract residential development in its eastern half, along the irregular and once picturesque coastline of upper Narragansett Bay, which forms the city's eastern boundary. Near the heart of the community is the vast open space surrounding the state's principal airport, T. F. Green. Significant amounts of open space or low-density neighborhoods exist only on remote Warwick Neck, in the rising elevations to the southwest, and along Potowomut Neck, physically separated from the rest of the community by the town center of East Greenwich.

Warwick's development was reinforced by the construction of interstate highways 95 in the 1960s and 295 in the 1970s. Route 2, as it stretches some ten miles from the Cranston border to its intersection with I-95, is now Rhode Island's main street, lined with large shopping plazas, the late-twentieth-century byproducts of two en-

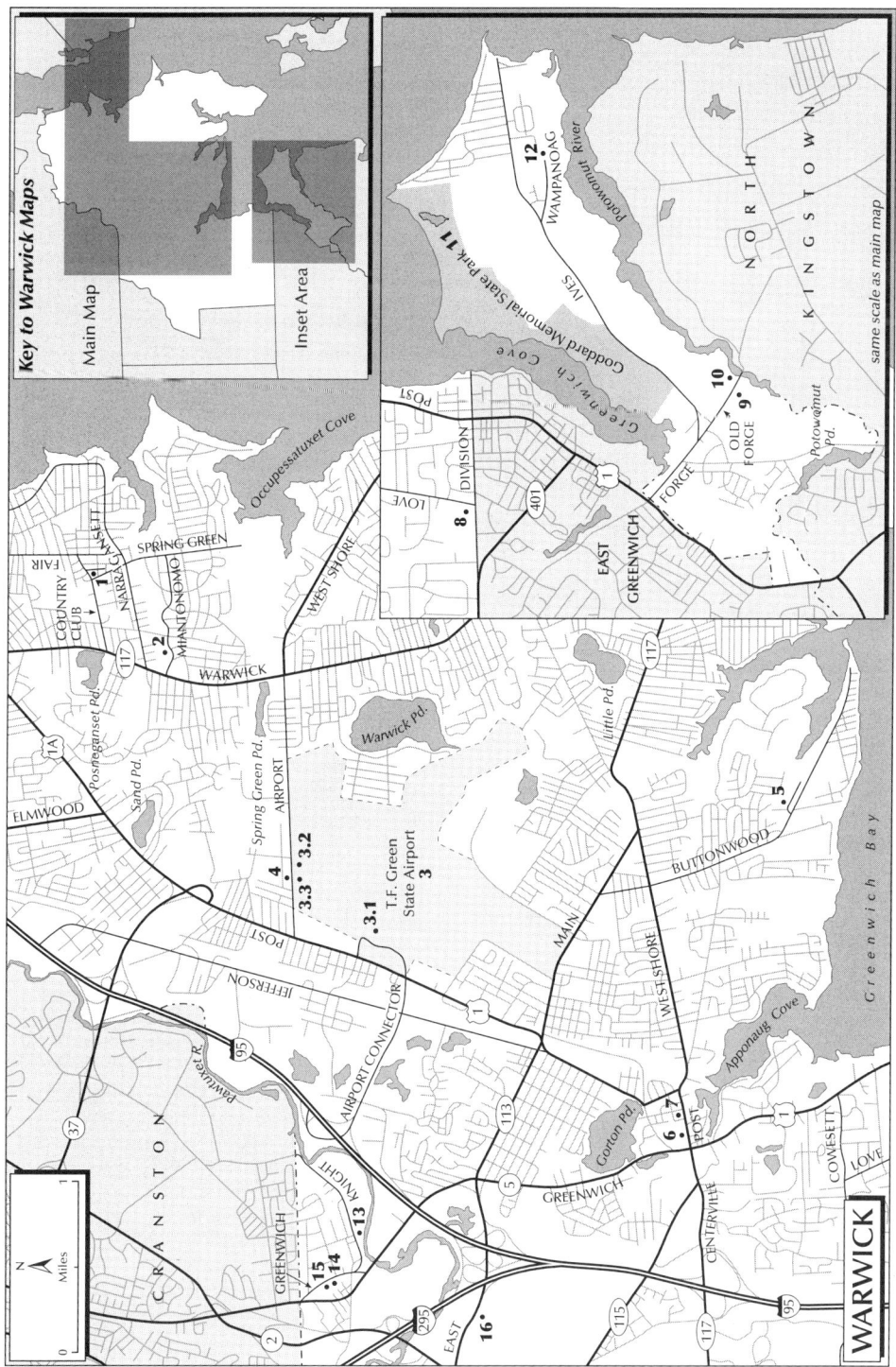

317

closed shopping malls created around 1970 at the intersection of Route 2 and I-295. Near these is one of the state's most compelling-late twentieth-century buildings, constructed, appropriately enough, as the heart of a college campus planned for a commuter population.

WA1 Mary Rose Ross House

1938. 27 Druid Rd. (northwest corner of Clearwater Ave.)

This essay in modernism never fails to astonish visitors to Gaspee Plateau, a pleasant but thoroughly conventional suburban subdivision of dwellings in revival styles, colonial or medieval in flavor. Local tradition holds that the owner had this house constructed after returning from a trip to Florida, where she had seen similar ones. To date, the identity of the architect or builder remains a mystery, though the character of the design suggests that it was produced by a contractor (as was Rhode Island's first modernist residence, the Arthur J. Levy House in Cranston (see entry). The house has all the hallmarks of the early International Style: the juxtaposition of simple, bold, geometric masses; plain white stucco walls punctuated with unadorned punched openings; corner windows; steel-frame sash; and steel pipe railing crowning the roofline. Built almost to the lot line on the west, the house is designed almost as a stage set, with the interplay between its tall, rectilinear block and lower, asymmetrically placed bow-front unit dominating the east facade. At the northwest corner, a window sports a tiny, nearly vestigial cantilevered balcony, bringing a little bit of "architecture" to what is definitely a secondary elevation. The exterior masks, rather than fully expresses, the fact that the interior is organized on a split-level floor plan with five distinct levels. Modernist dwellings were not popular in Rhode Island and are rare; this is a particularly curious and engaging example of the type.

WA2 John Brown Francis School

1954, MacConnell and Walker. 325 Miantonomo Dr.

One of the innumerable post–World War II suburban elementary schools cast in the mold of Eliel and Eero Saarinen's Crow Island School in Winnetka, Illinois, this structure is a particularly nice example of the type. The local firm of MacConnell and Walker designed nearly all of the schools built in Warwick after the war. They eventually settled on a formula for turning out one-story brick-clad structures with far-flung, asymmetrically arranged wings counterbalanced by tall, broad slab chimneys. In the Francis School, the most distinctive example, the firm embellished the formula with an angled assembly room/auditorium wing and a sundial on the chimney. Classrooms, scaled to the size of children five to eleven years old, are strung along one or both sides of long corridors. They have high, canted roofs that leave space for glass-block clerestories rising above the flat-roofed corridors. Together with the glass-block and awning window strips on the exterior walls, this arrangement was intended to provide all classrooms with bilateral natural light. Sadly, maintenance problems caused by cracking glass blocks and leaky roof joints have caused the clerestories as well as the glass strips on the outer walls to be blocked up.

Though the building was economically constructed using concrete block, glazed tile, linoleum, steel, and glass, its planning shows a generosity of spirit toward those who spend their daytime hours inside, an effort to make the surroundings as comfortable and pleasant as possible. MacConnell and Walker's schools were the product of an ideal for accommodating and educating schoolchildren and a testament to the community's care for and commitment to its youngest citizens. The real quality of these buildings, which won recognition in the architectural press at the time of their construction (*Progressive Architecture*, August 1951 and March 1955), stands out in marked contrast to the humdrum additions that were stuck onto many of them in the 1960s and 1970s. Perhaps their true historical and architectural value will be recognized and appreciated before additional alterations detract further from this marvelous body of work.

WA3 T. F. Green State Airport
(Hillsgrove Airport)

1932, first terminal. 1938, second terminal and Hangar No. 1. 1953, second terminal altered. Jackson, Robertson and Adams. 1996, Sundlun Terminal, HNTB. 562 and 660 Airport Rd.; Post Rd.

Airport terminals must be among the most mutable of building types. They are regularly and routinely subject to demolition or disfiguring alterations. Rhode Island's current facility (WA3.1), named the Bruce G. Sundlun Terminal for the governor who implemented its construction, on Post Road around the corner from the original terminal, was completed in 1996 to designs by HNTB, a New Jersey firm specializing in such projects. It handsomely replaces a somewhat ackward and inadequate 1950s terminal. All the more miraculous that two previous terminals stand virtually untouched, on the other side of the airfield. The original terminal (WA3.2) is a rare survivor from the 1930s. This small gem is simply composed of flat-roofed, cream-colored stucco boxes: a central block flanked by one-story wings, fronted by a two-story entrance tower with a monumental geometric Art Deco portal and topped with a polygonal glass control room. Except for the deep, stepped-back reveal of the main portal, door and window openings are unornamented cutouts, and the roofs are ringed with steel tubular industrial railing. It is somewhat remarkable that this was executed by a Providence architecture firm that made red brick, white-trimmed Neo-Georgian structures its stock in trade, though it seems clear that to these architects modernism was yet another style to be used eclectically, as here, where the program called for imagery that was contemporary, even futuristic. The somewhat more conservative second terminal (WA3.2), known as Hanger No. 1, has a central entrance block articulated with pilasters. At the parapet, a relief depicts the state seal—an anchor on a shield—flanked by stylized bird wings and parts of airplane propellers.

The original Hillsgrove Airport, as the core of the much larger T. F. Green Airport (renamed for a governor, then a very long-termed U.S. senator) was the first state-owned airport in the country. It was opened in 1931 at the behest of the Providence business community, which thought that a publicly owned facility would serve the metropolitan area better than the collection of small, privately operated airfields that then ringed the city. The airport and its original terminal represent the state's commitment to providing up-to-date facilities for future economic development. They reflect the culmination of the optimistic outlook of the Roaring Twenties carried into the early years of the Great Depression.

WA4 Maintenance Center

1961, Castellucci and Galli; Occupasstuxet Rd.

Diagonally opposite the former airport terminals, this geodesic dome in aluminum tubing and sheets demonstrates the ability of Buck-

WA3.2 T. F. Green Airport, Old Terminal

WA4 Maintenance Center

minster Fuller's tetrahedral structures in light materials to provide vast free-span spaces. As is the case with many of his domes, the more usual arrangement of buried structure and exposed protective cover is here inverted so that the structure is exterior to the covering although both are fully visible. The dome comes to the ground on sixteen elongated tilted piers, providing openings all around for access by the airport's larger vehicles.

WA5 Greene–Bowen House

Between 1687 and 1715; altered 1758. 100 Mill Wheel Rd.

This is one of the eight identifiable surviving structures of the stone-ender type peculiar to Rhode Island (not counting a few houses which started out as stone-enders but which now have later additions that cover their original stone ends). Here, however, the end chimney is actually constructed of brick. Among stone-enders the Greene-Bowen House has long been especially interesting to scholars because it contains features derived from both Providence and Newport building practices, perhaps not totally surprising, since Warwick is located between the state's two major colonial centers. The main block of the house conforms to the standard two-room plan with side-by-side fireplaces on the end chimney, found also in the Eleazer Arnold and Valentine Whitman houses in Lincoln (see entries), although the plan is slightly modified here by the one-story addition across the chimney's end. It was once thought that the house was built for Fones Greene about 1715, but research done in the 1970s suggests that the "original" house may have been built in two sections, one of which may be the "old house" built for Fones's father, James Greene, perhaps as early as 1687. Except for the installation of some interior partitions and sash windows in the mid-eighteenth century and the addition of a west lean-to and a narrow extension along the back (north) side of the house, the structure remained virtually untouched until the 1970s, never having had plumbing, central heating, or electricity. Since then the house, once in danger of collapsing, has been stabilized and restored in a manner that has had very little impact on the historic fabric.

WA6 Warwick City Hall

1893–1894, William R. Walker and Son. 3275 Post Rd.

This landmark is one of the most elaborate—and curious—of the town and city halls erected in Rhode Island during the late nineteenth century. Its somewhat peculiar design characteristics are typical of the Walker firm, neither of whose principals received academic training. It is a classic example of a building assembled from the elements of a fashionable style without regard for, or perhaps without knowledge of, the compositional rules for using such elements. The result is original, if idiosyncratic. Though the overall massing has nothing to do with Georgian or Federal precedent, the building is nominally colonial because of its detailing. An Ionic porch shelters an overscaled main entrance featuring enormous side lights and fanlight. Above this, a tower rises, topped by a belfry with Palladian motifs and a domical octagonal-plan roof. The two other surviving Walker town halls in Rhode Island (in Warren and Cumberland) also have central towers; however, Warwick City Hall, with its strict symmetrical composition and columned porch, has a greater air of formal grandeur and dignity. Inside, an entrance hall leads to a double-loaded central corridor on the ground floor and contains paired staircases leading to a two-story council chamber occupying the second and third stories.

At the time City Hall was constructed, Warwick was a prosperous mill town, the sixth most populous municipality in the state. This grand structure stands as an important symbol of Warwick's status at that time. Twenty years later much of the population and industrial base were lost when West Warwick was set off as a separate town. By the 1950s and 1960s City Hall was considered hopelessly outmoded. The council chamber was cut up into a smaller chamber and offices under a dropped ceiling. Plans to construct a modern government center were shelved as the city became preoccupied with providing schools, libraries, sewers, and other services for a burgeoning post–World War II suburban population. By the late 1970s city officials and residents alike had come to appreciate this seat of government as a historical focal point for a community built up largely after 1945. Since then City Hall has undergone a rehabilitation which has included reopening the original council chamber. This monument to Warwick's nineteenth-century propserity and pride stands today as a fitting emblem of the civic identity of Rhode Island's second largest city.

WA7 U.S. Post Office

1940. Louis A. Simon, supervising architect. 3205 Post Rd.

This post office is one of hundreds erected during the 1930s and early 1940s, symbols of the federal presence at the local level. It reflects the federal government's efforts to standardize the design and construction of small public buildings and provide facilities architecturally compatible with their locales. Here, a spare vocabulary of traditional Georgian colonial forms has been used to create an image which, while not particularly specific to Apponaug, is generally evocative of New England. What does give this building a distinct identity is the mural in the lobby entitled *Scenes of Apponaug Cove and Village*, painted in 1942 by Paul Sample, artist in residence at Dartmouth College, under the auspices of the Department of the Treasury's Section of Fine Arts. The mural, in a Geometric Realist style, depicts shell fishermen under the Apponaug railroad viaduct, with a panorama of the village, dominated by City Hall, in the background. This is one of seven surviving examples in Rhode Island of the public building decorative painting known popularly, if not entirely accurately, as "WPA art."

WA8 Gorton–Greene House

c. 1685, c. 1758, and later. 777 Love Ln.

Prominently located at the intersection of Division Street and Love Lane, the Gorton-Greene House has at its core a prominent stone chimney, testimony, even more visible on its interior (not open to the public), of its origins as a stone-ender, built just following King Philip's War (1676–1677). The massive chimney is, appropriately, the most readily visible element of this much-expanded house set in a well-manicured precinct. In the course of occupancy by eight generations of related families, the original small, single-cell house evolved into an amiably rambling structure surrounded by handsomely landscaped gardens. At the rear and mostly hidden by vegetation is a cider house, recently restored with great care, that probably dates from the late seventeenth or early eighteenth century.

WA9 Greene House (Forge Farm)

c. 1684, mid-18th century, 1862–1863. 40 Forge Rd.

One of the state's most remarkable survivors, this evolved seventeenth-century house occupies a 165-acre parcel in Warwick's noncontiguous southern section. The original (1684), or south, portion of this house, like that of the Gorton-Greene House (both named for relatives of James Greene, who first settled here), is no longer immediately visible: what we see appears as an eighteenth-century center-chimney house with charming mid-Victorian accretions. The birthplace of Revolutionary War General Nathanael Greene (1742–1789), this farm remains in the hands of the original family of owners. Most remarkable, though, is the survival of such a large parcel with house and attendant outbuildings.

WA10 Greene House (The Grange)

c. 1776, c. 1860. 57 Forge Rd.

Set far back from Forge Road on a large, amply treed lot, The Grange, like Forge Farm, is essentially an eighteenth-century farmhouse remodeled in the mid-nineteenth century, in this instance by the granddaughter of the builder, Elihu Greene (brother of Nathanael Greene, above), Emily Greene Waterman, who used this as a summer retreat from her house in Providence.

WA11 Goddard Memorial Park (Russell Estate)

1875, 1928. Ives Rd.

Providence residents Henry G. and Hope Brown Ives Russell developed this marvelous bayside property as their country retreat in the mid-1870s. After their deaths in the early twentieth century, the heirs of this childless couple gave the property to the state for use as a public park, a much-appreciated amenity at the southern extent of the densely developed metropolitan Providence area. Although the High Victorian Gothic main house succumbed to fire exactly a century after its construction, the outbuildings and circulation system from the country-retreat era remain, enhanced by 1920s designs created by Olmsted Brothers' Percival Gallagher and carried out by the Metropolitan District Commission, predecessor of the Department of Environmental Management. Today this is a pleasant rural park, with all the attendant facilities expected for recreation, including golf, tennis, picnic sites, and swimming.

WA12 Rocky Hill School (Hopelands)

1686, 1793, c. 1885. 1 Wampanoag Rd.

Like Forge Farm and The Grange, this began its life as a Greene family house, a small stone-ender. The original section is now part of the ell that extends north from the main block of the house, added in 1793, after Hope Brown married Thomas Poynton Ives. The Ives family and their descendants, the Goddards, used this as a country retreat from Providence from the late eighteenth through the early twentieth century. Hope Brown Ives added the more-or-less-square-plan main block with wraparound colossal porch. In the mid-1880s her grandson Moses Brown Ives Goddard extended the main block to the east and also enlarged the service wing to the north. The interior largely retains Goddard's Queen Anne–Colonial Revival revisions. Rocky Hill, a private middle and upper school that acquired the property in 1948, does an admirable job in maintaining this fascinatingly evolved property.

WA13 Pontiac Mill

1863 and later. 333 Knight St.

This expansive complex, the chief relic of Warwick's important role in nineteenth-century textile manufacturing, is one of the most visually striking factories in a state noted for its many architecturally significant industrial buildings. The mill is squeezed onto a wedge-shaped site in a serried range of rectangular masses strung on a long east-west axis. The centerpiece is Mill No. 1 (1863), a brick block identified by a fine Neo-Romanesque stair tower with blind arcading and chamfered corners. Mills No. 2 and No. 3 (1870), to the west, and Storehouse No. 1 (1874), to the east, are constructed of rubblestone covered in stucco scored to imitate ashlar masonry, with rusticated door and window surrounds of brick, a treatment very baroque in character. The river frontage contrasts dramatically with the wide street side. Here the triangular plot bordering the Pawtuxet is given over to a picturesque jumble of structures in assorted states of disrepair, diverse in height, roof form, and material, variously oriented to the river or the main block of the mill. The rich display of angles, shapes, and textures, coupled with the aura of decay, creates an atmospheric panorama.

The present mill replaced a series of factories on this site dating back as far as 1810. It formed part of the empire of B. B. and R. Knight, one of the largest and most prosperous cotton textile firms in Rhode Island. Here the Knight brothers manufactured an exceptionally fine fabric marketed under the trademark Fruit of the Loom. The name survives, though the operation originally responsible for it has long since passed away.

WA14 All Saints Episcopal Church

1887, Howard Hoppin. 111 Greenwich Ave.

All Saints is one of a series of small village chapels, most Neo-Gothic in style and Episcopalian in affiliation, executed by Providence architect Howard Hoppin. His signature is the asymmetrically placed side tower—here polygonal in form—engulfed by the gable-roofed main mass of the building. A number of these Hoppin churches have been destroyed, which makes the rather forlorn condition of this building all the sadder. Despite the effects of aluminum siding and deferred maintenance, the basic strong geometry of the structure shines through. The serial imagery of the steep gables along the east facade, extending from the end of the nave along the length of the parish house, is particularly splendid.

WA15 Pontiac Free Library

1956, Jackson, Robertson and Adams. 101 Greenwich Ave.

This small village library is an adept, straightforward example of American modernism as it developed out of the International Style. Following Walter Gropius's emigration to the United States in the late 1930s, the Bauhaus aesthetic came to be reinterpreted in this country as an architecture of abstract, clearly defined forms rendered in wood, brick, or stone, materials more common here than the uncompromising white stucco favored in Europe (and used here when the real International Style effect was desired). The Pontiac Library follows in this vein. It consists of a plain, sharp-edged rectangular main block clad in common-bond Roman brick that emphasizes the building's horizontality. The steel-frame vertical strip windows with horizontal mullions both counter and reinforce this emphasis. A plain, flat-top entrance portico, generous rectangular bay window on the south, and a subsidiary block to the rear complete the composition. Inside, the bay window is balanced by an unornamented fireplace whose domestic air somewhat mitigates the stark qualities of the building. Jackson, Robertson, and Adams themselves had all died by 1950; by the time this structure was commissioned, the firm was conducted by George Fraser and Raymond J. Henthorne, who shortly thereafter changed the name of the practice to Fraser and Henthorne.

WA16 Community College of Rhode Island Knight Campus (Rhode Island Junior College)

1968–1972, Perkins & Will Partnership with Harkness & Geddes and Robinson Green Beretta, associated architects. 400 East Ave. (Route 113)

CCRI Knight Campus is the state's preeminent megastructure—and an important exemplar in the American development of this building type. The concept attained the height of its popularity during the 1950s and 1960s, when supersized buildings (formerly associated mostly with

WA16 Community College of Rhode Island Knight Campus (Rhode Island Junior College)

the largest industrial plants) began to proliferate for such uses as shopping centers and suburban corporate headquarters. Properly speaking, however, megastructures require more than super size and single purpose. They are giant buildings designed to contain the diverse functions of a community. Their immediate progenitor was, above all, Le Corbusier's Unité d'Habitation in Marseilles (1946–1952). This megastructural type seemed to many especially advantageous for colleges and universities, a number of which, like this, were then being established. For this community college, where commuting eliminated the important residential aspect of the full-blown megastructure, a reduced version seemed especially appropriate. A huge parking lot would counterbalance a huge building in the overall site plan, with the remainder of the landscape of its hilltop location considered as a panorama—more for overlooking than for living in and walking through. Like Le Corbusier's Unité, it is a long, narrow, slab-like or liner-like shape in reinforced concrete—here 475 feet long, 225 feet wide, and five stories high. Whereas the windowed and balconied walls of the Unité are outwardly oriented, the community college walls in its interior spaces. Here a sculptural accent is the two-story expansive cylindrical element that both makes a sort of pilot house at the prow of this "liner" and marks the library's reading room; then, more dramatically, the wall peels away to permit a wide ramp on stilts (all Corbusian motifs) to rise into the very heart of the megastructure.

West Warwick (ww)

Until the twentieth century, West Warwick was part of Warwick. Its intensive industrialization and the increasingly polyglot composition of its immigrant work force, however, set the western portion of the town apart from the rest of it. Well into the twentieth century, the eastern part of the town was predominantly farms, suburbs, and seaside cottages, mostly inhabited by those who considered themselves "native stock." What was then a Republican- dominated state legislature feared that an independent industrialized West Warwick would elect Democrats; so not until 1913 did West Warwick split from Warwick.

The north and south branches of the Pawtuxet River wind in a tortuous *Y* through the center of West Warwick—substantially determining its curious shape, in fact—until the river crosses into Warwick. This principal fact of West Warwick's geography has also been the principal determinant of its economic and cultural life. Together with the mills spread all along the Blackstone and its tributaries in the northern part of the state, those lining the Pawtuxet and its tributaries accounted for the bulk of Rhode Island's once vaunted textile industry. Nowhere do they cluster as tightly as they do in West Warwick and on a short stretch of the river immediately upstream of its northwest corner, where West Warwick meets Scituate and Coventry. From Hope, Jackson, and Fiskeville in Scituate and Arkwright and Harris in Coventry, the mill towns arc across upper West Warwick along the North Branch (Phenix, Lippitt, Clyde, Riverpoint, and Natick) and south along the South Branch (Arctic, Centerville, and Crompton).

Phenix

Once known as Wales for a property owner in the vicinity, Phenix acquired its present name from a fire which in 1821 destroyed the Roger Williams cotton mill, a wooden structure of 1810–1811, along with much of the center of the town. Parts of the existing mill "rose from the flames" as the Phenix Manufacturing Company in two stone buildings in 1823 and 1825, along with a second raceway—and survived five other substantial downtown conflagrations dur-

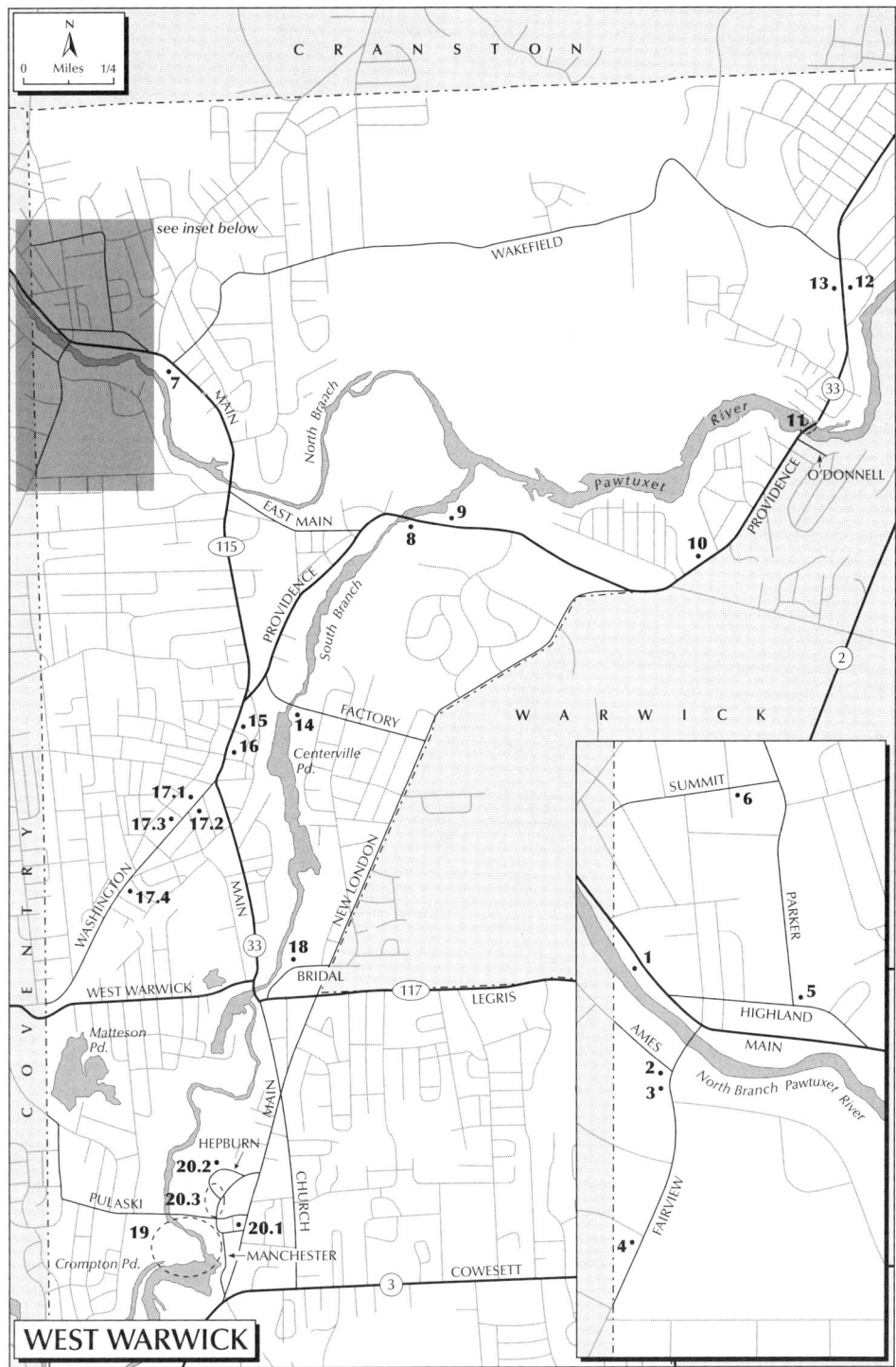

ing the late nineteenth century. Shortly after this initial rebuilding, around 1830, the mill came under the control of Crawford Allen, eventually joined by his brother Zachariah Allen, especially prominent as an early textile entrepreneur and inventor who operated mills in the villages of Allendale (North Providence) and Georgiaville (Smithfield). They joined with the engineer David Whitman, a technological innovator who planned and altered some of the great Massachusetts mills in Lowell, Lawrence, and Fall River as well as others in Maine. At Phenix this group allegedly manufactured the first sheeting in three-yard widths—that is, wide enough for a double bed without seaming. In a succession of moves, the Lonsdale Company (see under Lincoln), which also owned the mill at nearby Hope (Scituate), gradually took control of this mill, obtaining full ownership in 1861 when they incorporated it, together with their nearby operation, as the Hope Manufacturing Company. As such it continued into the middle of the twentieth century.

WW1 Phenix Mill

1823, 1825, original buildings. 1860, addition to east building. 1882, link between original buildings. 1882, remodeling. weaving shed, 1902. 771 Main St.

Over time, from two stucco-surfaced rubble masonry buildings of four stories, this mill was expanded and linked by harmonious additions to its present length. Granite trim relieves the textureless stucco surface as windowsills and quoining for the projecting tower. The survival of the Greek Revival belfry, with classical balustrades, pilasters, and pediments in four directions, is rare among industrial structures. The plant now produces pharmaceuticals.

WW2–WW4 Fairview Avenue Houses

At the end of the plant site on Fairview Avenue is a fine enclave of houses ranged along a steep hill. Most are worth examination, but three especially so. The scale of these hilltop houses reflects the importance of Phenix as the nineteenth-century commercial and shopping center for the cluster of mill villages in this corner of West Warwick, spilling over into Coventry and Scituate. On the avenue are examples from the mid-nineteenth century through the Colonial Revival, as well as the farmhouse of the original owners of the land, the Carr Levalley House (c. 1722), at 42 Fairview Avenue.

WW2 William B. Spencer House I

WW3 William B. Spencer House II

WW2 William B. Spencer House I

1847. 2 Ames St. (at Fairview Ave.)

Made more commanding by its siting on a steep rise over the Pawtuxet, the first of two houses built by William B. Spencer is among the finest and most elaborate Greek Revival houses in the state. It boasts a two-story, four-columned Corinthian temple front, with flanking hip-roofed wings. Downhill, toward Fairview Avenue, at a lower level, yet another hip-roofed appendage with its own porch supported at its outside corner by a single Doric column animates the composition in an exceptionally picturesque manner for the style. The client was a merchant and store owner in Phenix and Lippitt, as well as a bank president and undertaker. The house is the more attractively sited because a pathway from this point to Harris (the next village upstream, in Coventry) winds through a linear park, which suggests the great potential for wooded walkways along much of the Pawtuxet.

WW3 William B. Spencer House II

1869–1870, Augustus Truesdale(?). 11 Fairview Ave.

Having built one handsome house, William Spencer stylistically updated his living arrangements by building another uphill and next door, in the then more fashionable Italian Villa manner, with a particularly fine two-storied, arched-windowed frontispiece united by a sequence of balconied porch, bracketed roof, and undulant canopy. (Did he remarry?) Although the architect of the house is unknown, Spencer contributed the land across the street and possibly some funds for the Phenix Baptist Church, also in the Italianate style (1860, following plans drawn by a Connecticut architect, Augustus Truesdale—and demolished for a modern structure in 1978). The similarity of style suggests that Truesdale may have been the architect for the house. It later became the residence of Robert Reich, first manager, and eventually president of the mill. The scale of the granite blocks of the stepped retaining wall in front of the house and the rolling, S-shaped step of its capping are also remarkable, although other instances of this kind of walling make it appear to have been a standard rather than a custom design.

WW4 House

c. 1910. 37 Fairview Ave.

Of all the other fine houses on Fairview Avenue worth examining, The most exceptional is number 37. The way in which the volumes of this high-roofed shingle house with bungalow intimations interlock with a crenellated tower in masonry incorporating an arched porch is unique. What could be an absurd collision of form with implications of intimacy on the one hand, and of pomposity on the other, is here reconciled by the forceful abstraction of the treatment.

WW5 Bungalow

c. 1915. 44 Highland Ave. (corner of Parker St.)

WW6 Morel House

c. 1915. 92 Summit Ave. (corner of Chamberland Ct.)

Whereas, beginning in the mid-nineteenth century, the very well off of Phenix lined up on the summit of Fairview Avenue, in the late nineteenth and early twentieth centuries, those of more modest means tended to live on the lesser mound of Highland Avenue in a subdivision cluster. The houses here also include two worth notice, one a bungalow in natural shingles on Highland Avenue, exceptional for its California look: elaboration of brackets beneath widely projecting eaves; long window bands for the "sun parlor," brick chimney, and, most especially, the scroll cutting of purlins projecting well beyond the plane of the roof. Whoever the Master of the Phenix Bungalow may have been, he apparently built another nearby, the Morel House, on Summit Avenue.

Lippitt

Small-scale industrial development at this site under various owners extended back to a sawmill and gristmill established in 1739 by Joseph Edmonds, for whom the village was originally named. A consortium of businessmen under the leadership of Christopher and Charles Lippitt finally erected a shingled mill, which stands today substantially as built.

WW7 Lippitt Mills

1809–1810, Mill No. 1. 1830, Mill No. 2 (bleaching). 1865–1871, other buildings. Dam rebuilt, 1889. Main St. at Wakefield St.

To understand the nature of the most ambitious textile mills in Rhode Island at the threshold of the Industrial Revolution, Slater Mill in Pawtucket (PA15) and this one are the obvious starting points. Whereas Slater Mill (1793) is a museum, Lippitt Mills may be the oldest American textile mill still in industrial operation (and, what is more, until around 1980, in continuous textile production). By comparison with the near-domestic scale and mien of Slater Mill, this represents a jump in size and a sharpened typological distinction between early industrial buildings, which tended to use domestic and institutional prototypes, and the more specifically "industrial" aspect of a specialized mill type which was just then beginning to be defined. Originally planned for two stories plus a trapdoor monitor story in the roof—then a new feature for lighting the attic spaces of industrial buildings—the Lippitt mill was lifted another story during the course of construction. Its right-angled siting to Main Street places its belfry on the gable roof at the entrance end of the mill (as at Slater), more in the manner of eighteenth-century institutional buildings than the projecting stair-towered belfries in later mills, which are typically centered between flanking walls. Here the stairs were originally inside the building (the brick tower on the north side being an obvious addition). The open, octagonal belfry itself, miraculously preserved, shows the attenuated elegance of the Federal style, which is also more appropriate to institutional than industrial buildings. A stack of loading doors, topped by a hoist down the front of the building, unfortunately removed in a 1984 reconstruction, once animated this elevation, and provided vertical reinforcement to the belfry on the roof. Although the wooden interior construction with post supports and plank flooring was apparently altered around 1830 to improve fireproofing, this too remains essentially true to the original. It was built for spinning only, as was usually the case for textile mills of this period; the Lippitt organization initially farmed out weaving to home industry, some of it, innovatively, to inmates of the Vermont state prison.

Riverpoint

Riverpoint is located at the confluence of the North Branch and the South Branch of the Pawtuxet. Here are two mills, the Royal Mills, at the bridge crossing the South Branch, and, a bit farther east, Valley Queen, each with a dam. Originally under separate ownership, both came under the ownership of B. B. & R. Knight in the 1880s.

ww8 Royal Mills (Greene Mills)

1919–1921, D. M. Thompson, engineer. 125 Providence St.

The most impressive view of this mill, which straddles the Pawtuxet, is upstream from the highway bridge, where mill walls in random stone and brick, together with an overhead walkway, tightly frame the dam spillway. The masonry buildings are what is left of an old plant which dated back to pre–Civil War operations—and ultimately to a vanished wooden cotton mill which began operations in 1812. Before the war the principal product of this plant was "Kentucky jeans" or "negro cloth," a rough twill which was mostly shipped to the South for slave clothing and was something of a

WW7 Lippitt Mills, Mill No. 1

Rhode Island specialty. The gradual drop in demand for this cloth after the war placed the plant in a precarious economic situation and enabled the omnivorous B. B. & R. Knight organization to take it over in 1884–1885. After enlargement and modernization, the product shifted to cambrics and sheeting. Fire swept away most of the existing plant in 1919 and forced substantial rebuilding in brick under the engineer responsible for most of the redesign of Knight-acquired mills. Architecturally, the best single building in the complex may be a slot-windowed warehouse across Providence Street from the mill, in random masonry with brick trim for loading doors and corbeled cornice. For the rest, the ensemble is more impressive than the individual parts. (The area to the north of the warehouse contains a scattering of much knocked-about, small-scale, mid-nineteenth-century industrial buildings of superb masonry construction.) Knight operation of Royal Mills continued until 1935. Since then, Royal Mills has housed multiple tenants in leased spaces.

ww9 Valley Queen Mill

1834–1835, original block at center (3 stories, 22 bays), Stephen Norton and Thomas Beck, masons. 1888–1889, 2 stories, 9 bays to south, and wing to north, D. M. Thompson, engineer. 200 Providence St.

Although it was built in two stages more than fifty years apart, Valley Queen, in contrast to the clustering of many buildings at Royal Mills, appears as a single elongated entity, right-angled to the river rather than parallel to it. This positioning gives it high visibility, showing

WW9 Valley Queen Mill

off its ashlar block and granite walls, with a nice equivalence in the relationship between wall and window openings. The extensive enlargement and modernization of Valley Queen, principally the north and south extension of the facade and ell on the east, occurred shortly after B. B. & R. Knight took control of the mill in the 1880s and used it to produce their top-of-the-line cambrics and sheetings marketed under the Fruit of the Loom label. During that building campaign the monitor roof was replaced and the tower Victorianized. As part of the collapse of the Knight textile empire in Rhode Island, the Original Bradford Soap Works acquired the mill in 1931. A firm established by two English immigrants in the 1870s to produce soaps for the textile industry, it today markets many kinds of industrial soaps, and is also a leading producer of specialty and private-brand luxury soaps.

A cast iron footbridge from the street at the top of the sloping site unusually gives entrance to the mill office at the center of the third floor. The banner which caps the tower is a more outstanding example of Victorian metalwork. The raceway and dam are beautifully maintained. The relation of the mill to the water is best observed from the precipitous wooded hill on the opposite bank, which, together with the well-kept grounds, isolates this mill from the town. The same vantage point also provides a reciprocal view of the Royal Mills upstream, its towers giving it a castlelike appearance (one of them is even crenellated). What an opportunity for a linear park, making visible the beauty of the ignored river and these monuments to an industrial past—the more compelling here because these mills continue to operate.

WW10 New Village Mill Housing

1921, Osgood Construction Company. 476–528 Providence St.

This row of ten houses in the manner of model "new town" cottages in English industrial towns was to have been part of a planned community of fifty such houses erected by B. B. & R. Knight in connection with the 1919–1921 rebuilding of the Royal Mills. The Boston engineering firm of Lockwood, Greene laid out the site plan for the project, which was acclaimed at the time for its size and its progressive nature. Hard times in the textile industry, beginning at this time, forced early abandonment.

Natick

WW11 Natick Mill Site and Mill Dam

Visible from O'Donnell Ave. overlooking the Pawtuxet

Only scanty ruins remain of the range of four mills which once lined the Pawtuxet at Natick and made up a sizable component of the mid-nineteenth-century textile empire owned by the Sprague family. When the Knight brothers took over in 1882–1883, some years following the Sprague bankruptcy, their building engineer, D. M. Thompson, joined the separate buildings into a single entity and further extended it until it stretched 1,350 feet with a uniform height of six stories. It was Rhode Island's largest single mill structure, comparable in the magnitude of its elongation to the Lowell and Lawrence mills in Massachusetts or those in Manchester, New Hampshire. The Knights declared bankruptcy in 1935. The vacant mill was eventually torched on Fourth of July eve in 1941. What remains is best viewed from a wooded bluff at the Warwick border (reached by following O'Donnell Avenue to the top of the hill). This overlooks the largest mill dam in West Warwick, a granite barrier 20 feet high and 166 feet across (final rebuilding 1886), with an impressive waterfall. A wall in random rubble masonry, also impressive in height and extent, lines the opposite shore. It once retained a 50-foot-wide raceway to the mill. Crumbling walls, dimly discernible behind trees downstream, are what is left of the mill itself, much of the site now serving an auto junking business. Again one is struck, here on the bluff, by the civic design potential of an industrial site.

WW12 Providence Street School

1914. 835 Providence St.

This clapboard Neo-Colonial schoolhouse accommodates sixteen rooms with entrance vestibules attached to either end. The clustering on the facade of expansive windows for each classroom creates double pavilions, each capped by a large, decorated gable dormer projecting from a complex roof composed of two hipped sections and culminating in a delightful open belfry. The splendid state of preservation of an early-twentieth-century wooden school building of such quality in this style is extraordinary and should be maintained.

WW13 Sacred Heart Church

1928–1929. 1951, convent (now rectory). 1958, school (now offices). 840 Providence St.

Across the street, the Sacred Heart complex of stuccoed church flanked by complementary buildings is an exceptional example of small-town grandeur evolved through time by simple means. Plain, unadorned buildings derived from houselike shapes, with elemental gables and hoods over doors, are sited in a symmetrical scheme, well set back from the street on a carefully considered composition of stairs and terracing. The homogeneity of the complex, remarkable considering the time elapsed from start to completion, is substantially due to the essential nature of the principles which conditioned its design.

Arctic

Originally named Rice Hollow and Wakefield, Arctic, one of the three principal West Warwick mill villages lined along the South Branch of the Pawtuxet, received the curious name it has today from the Spragues in the mid-nineteenth century. The location of their mill at the bottom of a hollow that reportedly collected the coldest air in West Warwick suggested "Arctic," which also rhymed with the names of other Sprague-dominated factory towns at Natick (above), Quidnick in Coventry, and Baltic (in Connecticut). The history of mill operations at Arctic follows a familiar West Warwick pattern: small plants bought up by Sprague in 1852 and aggrandized, passed on to the Knights in 1884 for further aggrandizement, then splintered into smaller and changing operations after 1935, when the Knight firm closed down. But Arctic is important architecturally in another sense. Its location close to the hub of the double arc of mill town, east-west across the top of West Warwick, and north-south through its center, made the commercial district of its merged Main Street–Washington Street the main street for the area from the final decades of the nineteenth century until the advent of major shopping centers in nearby Warwick during the 1960s and 1970s. Although its commercial glory has faded, Arctic remains one of the finest examples of an early-twentieth-century "Main Street" in the state, with the typical holdover of some late-nineteenth-century commercial blocks. Restoration and marketing could give this still lively area a fresh competitive edge. As the commercial center for West Warwick, Arctic also became its town center.

WW14 Arctic Mill Complex

1834, 1865, 1875. 21 and 33 Factory St.

Two mills in this complex, on the east and west banks of the Pawtuxet, typify local textile mill architecture. The 1834 stuccoed rubblestone mill is a low four-story plant that hunkers down close to the river (all the less assertive after a fire in 1875 destroyed the roof and occasioned the installation of the present low gable roof). Across the Pawtuxet, retained here by a handsome stepped granite-block structure, and on land higher than that on the west bank, the magnificent five-story L-plan 1865 mill rises forcefully along Factory Street as well as perpendicular to it: the main block, more than 300 feet wide, is fronted by an emphatic eight-story stair and bell tower with two-story belfry. The stonework here is particularly fine: random-coursed granite ashlar with stones of varying

WW15 Centreville National Bank

size and beautifully struck mortar joints. Stonemason Rufus Wakefield built the first mill and leased it to a number of woolen manufacturers before the Sprague Manufacturing Company bought it to use as a warehouse. After the breakup of the Sprague empire, B. B. & R. Knight bought the complex and continued textile manufacture until 1935.

WW15 Centreville National Bank

1938, Hutchins and French. 1218 Main St. (at Curson St.)

Steps up to a planted terrace provide the ideal Beaux-Arts foreground to this handsome Neo-Federal elevation—handsome in its division between Ionic portico and wings, in the relative sizes of the openings, and in the play between circular and rectangular shapes. Inside, a shallow central dome combines with a deeply coffered ceiling to make the principal structural members, severely veiled as classical elements, the basis for the articulation of the space. As the principal ornamentation, column capitals mix gold-leafed motifs abstracted from the Ionic and Corinthian orders, in an original manner that makes what is mostly flat pattern suggest three-dimensional form. This compact interior is among the fine banking spaces in the state, testimony to the glory days of the West Warwick mills, even as these were rapidly waning.

WW16 U.S. Post Office

1932. 1190 Main St.

Here is another example of Beaux-Arts classicism, also arcadianized by its setback from the sidewalk with steps to landscaped terracing. In contrast to the projecting portico of the bank as the centerpiece to an elaborated wall, here the severe Doric order is locked into a more severely treated block, with the arched entrances recessed. One feels the austerities here of an awareness of modern architecture—explicitly in the abstract flat patterns of the colorful tile plaques.

WW17 Washington Street Commercial Buildings

J. J. Newberry (WW17.1; 1921), 37–43 Washington Street, is a classic 1920s dime store. Flamboyant brackets support an elegant decorated cornice of urns, leafy panels, and dentils as the culmination of the terra-cotta frame for the one-story stretch of shop windows with its familiar red and gold sign. The raised Roman letters, stretched out to squarish proportions, are among the classics of commercial lettering. Rarely does a store building of this type and quality exist in such mint condition. Two turn-of-the-century commercial buildings are worth looking at: the Alice Building (WW17.2; c. 1898) and especially the pristinely preserved Donant-Archambault Building (WW17.3; c. 1910), at 63–65, 64–74, and 115–119 Washington Street, respectively. Finally, at 268 Washington Street is the cadaver of a Texaco station which is derived from a pioneer prototype in enameled metal panels designed for "streamlined" modern gas stations by the prominent industrial designer Walter Dorwin Teague (WW17.4; 1930s).

Centerville

From the complicated history of early mill building on both sides of the Pawtuxet at Greenville, no more than a single important structure remains, and that in precarious condition. Although the Providence entrepreneurs Almy and Brown united all the small mills into a single operation before 1851, after this date the mills on either side of the river split into separate firms. The brick plant on the west bank (c. 1896, c. 1907), designed by the Providence mill specialist Frank P. Sheldon and erected by J. W. Bishop, replaced an earlier one destroyed by fire. The stone mill, originally the Lapham Mill, on the east bank, is somewhat older and architecturally more interesting.

WW18 Lapham's Cotton Mill

1873–1874 and later, Benedict Lapham with Horace Foster, builder. 1807, Green Mill. Bridal Ave.

The brothers Benedict and Enos Lapham, manufacturers from Scituate, bought the mill buildings on the east side of the Pawtuxet in 1851, enlarged them in 1861, and finally virtually replaced what existed before by the four-story existing building in masonry with a corbeled brick cornice. Horace Foster, a well-known mill builder, served as supervisor of construction for the owner-designed mill. Except for the way in which it is elbowed into a curve of the river, it is not among Rhode Island's most compelling mills visually—and what quality it once had has been all but destroyed by recent, utterly incongruous one-story additions. But the mill is also interesting for its 14.5-foot-high arched dam. These are rare in Rhode Island, where dams are mostly gravity barriers that cut straight across the flow. High banks (here artificially heightened) and rock that is suitable for abutments, however, allow an arched form, which can be thinner because its curved ends thrust into natural rock in line with the flow. The visual result is a pretty horseshoe falls. Following the deaths of the Lapham brothers and an interval under another owner, the Lapham Mill came under the control of the Knight brothers in 1903—their final acquisition. They enlarged the facility and eventually converted it, in 1920, into a major producer of book cloths, until the company's collapse in the mid-1930s.

When Enos Lapham built his mill he pushed to one side one of the oldest mill buildings in the area, converting it as a warehouse. This, known as the Green Mill because of its original color, exists in dilapidated condition east of the main plant on the edge of what is now a parking lot. For historians of technology it may be more significant than the big mill. It is a long, two-and-one-half-story clapboarded structure with a high, windowed, rubble basement on the side away from the parking lot. It has lost its sheathing on one side, revealing its structure as an an irregular grid of timbers with brick infill, each square irregularly punched with its window. How long it can survive without care is questionable, but it is an important relic of Rhode Island's earliest industrial past.

Crompton

WW19 Crompton Mill

1897, Mill No. 1. 1828, Mill No. 2. 1832, Mill No. 3. 1876, mansard roofs added to Mills No. 2 and No. 3, L. & C. Walker. 1882–1885, Mill No. 4 (Velvet Mill), Stone and Carpenter. Main St. (Mill No. 1), others off Pulaski St.

Architecturally, the Crompton Mill is the finest of West Warwick's mills, and one of the splendid mill spectacles in the state. Crompton is another famous Rhode Island brand name: What Fruit of the Loom was to sheetings and percales, Crompton was to corduroys, velvets (marketed under the Century label), and velveteens. Crompton and Century are still leading brand names, but the fabrics are no longer manufactured in Rhode Island. The place was first known as Stone Factory for the original stone mill (which came to be known as Mill No. 1 in the eventual Crompton complex), built on the

WW19 Crompton Mill

east bank of the river in 1807. This and other buildings were purchased in 1823 by a business consortium, reorganized from a previous Crompton company. The corporation was named for Samuel Crompton, the inventor of the spinning mule. Its specialty was bleaching, calico printing, and dyeing. It built Mills No. 2 and No. 3. The Providence calico printer George M. Richmond gained the controlling interest in the Crompton Company in 1866 and shifted production to velvets and corduroys. His was the first American venture into domestic manufacture of what had hitherto been imported from Europe. Richmond capped his mill complex with the dominating Velvet Mill in 1882–1885.

The 1807 mill, on the east bank of the river, a three-story stone mill with a squat projecting tower, has been all but swallowed up by surrounding additions. It has been alleged to be the first stone mill in Rhode Island; it is probably not, but it is among the earliest. On Pulaski Street a rise gives an overall view of the grandest part of the complex, on the west side of the river with the dam between. It is one of the most picturesque mill views in the state. Two low stone buildings (four and two stories) in an angled relationship with one another, each with its later mansarded roof and tower, are topped by the five-story Velvet Mill, 70 feet wide and 260 feet long, looming on the hill behind. The block is sufficiently large and its site sufficiently conspicuous so that it overwhelms the potential of the smooth, textureless stucco surface and the small windows to diminish its impact. It is, however, the force of its projecting tower, cornered in giant solitary crenellations, more like animal horns than anything architectural, and quoined like the block itself in red granite, that makes it memorable. The ensemble appears as an industrial castle. Openings in the front face and one side of the tower are formally and compellingly composed; on the other side they are distributed more functionally. (Only a few years earlier, its architects, Stone & Carpenter, had completed another great Victorian masonry mill in nearby Anthony (see entry for Coventry Mill).

WW20 Crompton Mill Housing

Mid-19th century, 1876–1877, 1921. Remington St., Hepburn St., Manchester St.

Of the mill housing of varied vintage, mostly much altered, in Crompton, three closely neighboring samples are perhaps most interesting. The long, one-and-one-half-story Greek Revival multifamily tenements with low monitor lights in the roofs (WW21.1; mid-nineteenth century) at 6–8 and 10–12 Remington Street typify the earliest workers' housing. In 1876–1877, L & C Walker provided a "new village" (as it was then called) of workers' duplexes in an unusual design around the semicircular loop of Hepburn Street (19–25 Hepburn Street) (WW21.2). Shingled hoods over doors and their adjacent windows combine with a projecting central gable to give a touch of Queen Anne variety to the basic box. As Hepburn curves back into Manchester, two stucco duplexes at 24–26 and 28–30 Hepburn Street and a six-unit tenement at 19–29 Manchester Street (WW21.3; 1921) display a variety of slate covered gables and picturesque manipulation of mass derived from English "Cotswold" industrial housing—more purely English in character than the Americanized version in nearby Westcott (see the entry for New Village Mill Housing, above). Immediately after these last examples were built, Crompton Velvet and Corduroy began its move to the South.

WW21.3 Crompton Mill Housing, 28–30 Hepburn Street

Coventry (co)

Squeezed into the northeast corner of Coventry are two mill towns, Arkwright and Harris, links in the chain of Pawtuxet River mill towns beginning at Hope and Fiskville immediately across the Scituate border and extending downstream in West Warwick to

Phenix, Lippitt, Clyde, and Riverpoint. Together these villages occupy what was once, at least, a beautiful stretch of the Pawtuxet, which is evident when one pokes along its shore behind the mills and mill houses.

Arkwright

The village, earlier known as Remington's Run and Burlingame's Mills, received its present name when the Arkwright Manufacturing Company (in honor of Richard Arkwright, the British inventor and mill owner) got underway in 1809 and opened its cotton mill during the following year. Nothing remains of this initial phase of textile operations except for altered mill housing across the river (over the Arkwright Bridge). The present mills are standard brick buildings of the late nineteenth and early twentieth centuries. The glories of the village are two late-nineteenth-century metal bridges exhibiting different technologies.

CO1 Arkwright Bridge

1888, Dean and Westbrook, engineers. Hill St. off Main St.

This 125-foot Pratt through-truss bridge is a structure so lightly constructed that it appears as the geometrical diagram of itself. It is both visually and technologically fascinating for its mixture of pipes, rods, and straps—each performing different structural functions. Phoenix columns make the basic trapezoidal frame. Invented in the 1880s and only briefly used as a basic structural element, Phoenix columns are composed of quarter-circle elements with projecting flanges for bolting into tubular members. This is the only extant bridge using Phoenix columns in the state. Vertical rods take up the tensile stresses for the roadway suspended within the trapezoidal structure. More rods crossed overhead in X-configurations at the top (upper chord) of the trusses prevent their lateral deformation, while diagonal straps within the trusses prevent comparable deformation vertically. Plaques and fanlike corner embellishments over either entrance to the bridge provide just the right amount of relief to the geometry to satisfy the late Victorian taste for ornamentation of even the severest functional structures—and they captivate us, too.

CO2 Interlocken Mill Bridge

c. 1885, Berlin Iron Bridge Co., engineers

This disused lenticular truss bridge, a little downstream from its near contemporary, is concealed from easy view behind the mill buildings of what is now the Interlocken Arkwright Company. Originally built to connect a new dye house and bleachery (which no longer exist) on the opposite side of the river from the weaving plant, this is the only example in Rhode Island of a bridge form invented by William O. Douglas of New York in 1878 and for which the Berlin Bridge Company virtually controlled the patents. The roadway crosses between lozenge-shaped trusses which swell to their maximum width at the center, where the most tensile support is needed to counteract the sag and collapse of the roadway. The trusses taper to points at either embankment. More than technologically fascinating, both bridges are a visual delight.

Harris

Cotton manufacture, which began here after the War of 1812, was more impressively established when Elisha Harris, later a governor of Rhode Island, bought the mill and a large farm and erected a new stone mill. It is the third mill, that of 1851, which stands today in unsympathetic surroundings (Main Street at Harris Street, facing on Harris). This mill, of stucco-surfaced rubble masonry, is an important late Greek Revival survivor, even though it has lost the belfry from its projecting tower. Harris Street leads to an enclave of mid-nineteenth-century mill housing, well preserved but all aluminum sheathed. The approach into the village contains several mill owners' mansions, two of special architectural merit.

CO3 Christopher Greene House

1882. 2 Potter Ct. (corner of Main St.) (visible with difficulty)

The owner of the Clyde Print Works, which once operated in nearby Clyde, West Warwick, commissioned this splendid country house, well screened on a rise at the end of a winding drive, with a matching carriage house. It exemplifies the Queen Anne style at its best, in a

CO1 Arkwright Bridge

CO2 Interlocken Mill Bridge

CO3 Christopher Greene House

panoply of angular massing, extravagant porch, and half-timbered walls filled with a mix of clapboard, cut shingle, and stucco, all climaxed by tall, shaped brick chimneys and remarkable shingle patterns. The display here, however, celebrates domestic function in ornamental forms derived from the qualities inherent in the materials used to build them.

CO4 Riverview Nursing Home (Elisha Harris–Henry Howard House)

c. 1840, c. 1870, c. 1890. 546 Main St. (corner of Potter Ct.)

Christopher Greene located his Queen Anne house adjacent to the grand Greek Revival residence commissioned by the industrialist who founded the town as a major mill center. Elisha Harris's house is as conspicuously set on axis at the top of its sloping lawn as the other is skewed on its site and concealed. Here the three-story central block is flanked by two-story wings. Each of these is topped by a windowed cupola, forcefully cornered by paired piers and capped by a projecting roof. This lavish Greek Revival house underwent some Victorian renovation before Harris's son-in-law, Henry Howard, made major Neo-Colonial alterations, most notably the swelling semicircular porch, which once had a square porch on top of it that formed a belvedere shelter in front of the central window. Rooftop balustrades, like those over the front porch and side bays, once also crowned the upper porch and wings above the second story. The earlier alterations and the barn may be of the same date. Clearly the house was intended to provide multilevel views of a pretty stretch of the Pawtuxet, so much of which these mill owners regarded simply as a commodity. Perpetuating the "Riverview" theme for its own commercial end, the nursing home preserves a decayed ghost of the old image in front, with extensive modern additions behind which incorporate the barn. Even in its present knocked-about state, this house can be appreciated as among the grand Greek Revival efforts in the state, with an equally grand Neo-Colonial finale.

Quidnick

Stephen Taft converted a paper mill to cotton in 1811, then sold out to A. and W. Sprague in 1846. They in effect built a new town, centered in two stone mills set in line with one another, parallel to the river, and at right angles to Washington Street. Of all the Spragues' vast

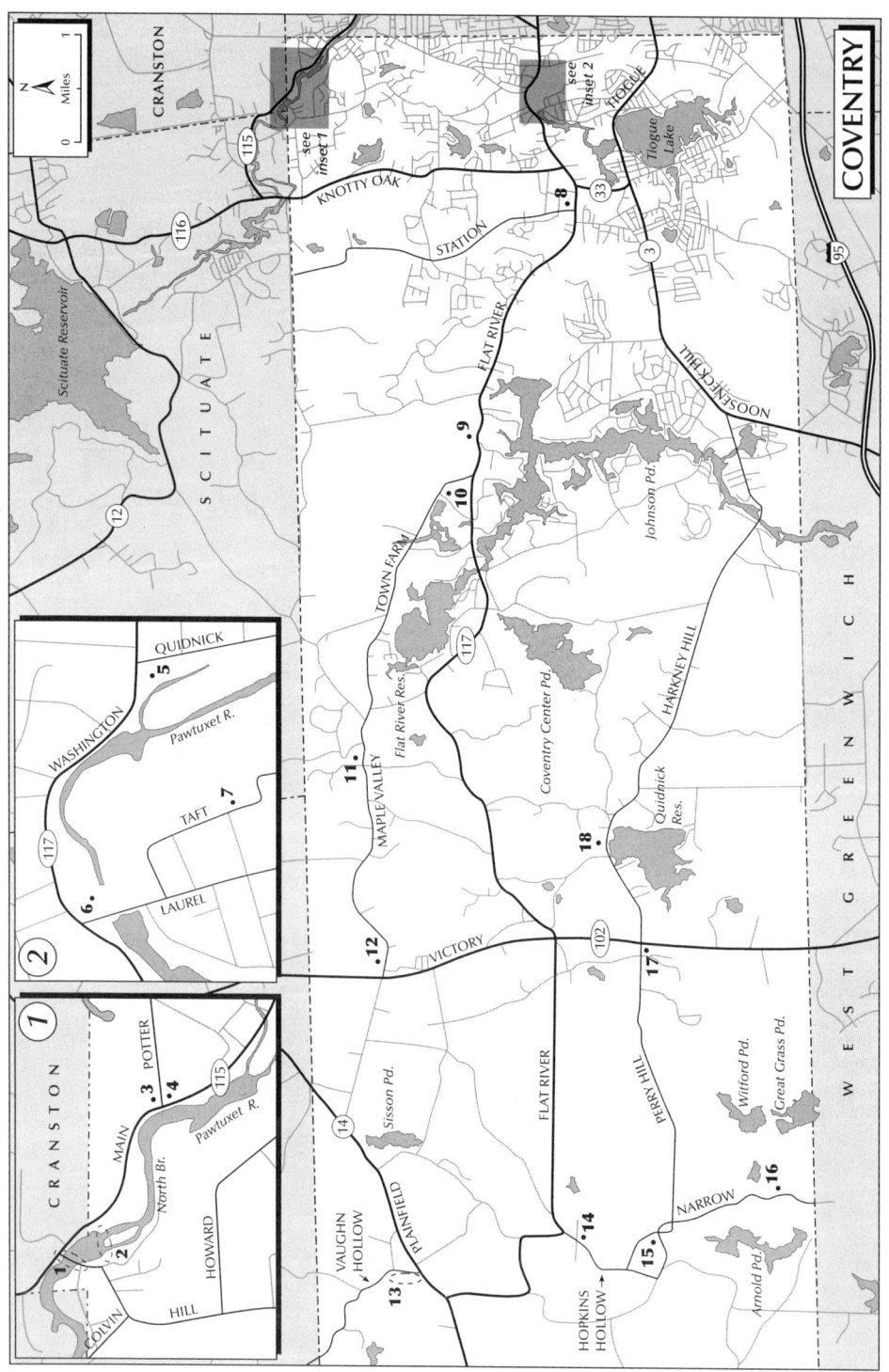

mill holdings, Quidnick, as built, was the corporate jewel. The public entrance to the mills was (and still is) on Washington Street, directly opposite the mansarded house of the mill superintendent (495 Washington Street, now much altered as a restaurant). Mill workers' houses, many destroyed and most of what remains much altered, lined Washington Street, together with the brick company store (487 Washington Street, now a garage). From Washington Street, Quidnick Street runs across a rise parallel to the mill fronts and offers the best view of them. On North and South streets respectively, mill owners built single-family and duplex houses, a few recognizable, though all have been altered and most demolished by the present owner of the mill. The single-family houses go back to the old Taft mill (c. 1815); the double on South Street, in better condition, were built by the Spragues (c. 1848). A company farm, long disappeared, was adjacent.

CO5 **American Hoechst Chemical Company** (Quidnick Mills)

1848–1849, No. 1 Mill. 1868, No. 2 Mill, Horace Foster, master mason. Quidnick and Washington sts.

Although the tower of Mill No. 1 has lost its Greek Revival belfry and the windows have been badly altered, this sizable four-story granite mill, with a continuous monitor to make a fifth story under the roof, is impressive in scale and workmanship and is beautifully maintained. At the time of its building, it was the largest mill in the state—and the largest Greek Revival structure by far. The same master mason built the smaller Mill No. 2—three stories with a simple, low-pitched gable roof but no tower—which now serves for storage. Stone for both mills came from a company-owned quarry in Oneco, Connecticut. A portion of the mill race behind the building, edged in woods (not readily accessible) remains in excellent condition.

Anthony

Originally this area and neighboring Quidnick were generally known as Greeneville, for the farming family who owned most of the land in the vicinity. A member of the Greene family opened a cotton mill here in 1806, among the first in the state. The six-story mill which replaced this original wooden structure in 1810 was said to be the largest in Rhode Island at the time. Small mills persisted here until Richard Anthony bought out previous owners and established the large mill which now dominates the town.

CO6 **Coventry Mill** (Anthony Mill)

1874, Stone and Carpenter. Washington St. at Laurel Ave.

A five-story mill built of stucco-covered rubble masonry, with red granite quoining, brick window frames, and granite window sills, this is an elaborated version of the nearby Crompton Mill, built by the same firm at roughly the same time. It was later elaborated by a more sculptural tower and ornamental frames for more grandly scaled windows. Unfortunately, most of the openings have been walled over. As in the Crompton Mill, the glory of this factory is in its tower. Here rounded shoulders meld into the forcefully scaled and sculptural belfry with giant arched openings. The "back wall" along the river is impressive in its own right, with an-

CO6 Coventry Mill (Anthony Mill)

other, more modest but well-proportioned tower. The machinery for the mill gates to the dam exit across Laurel Avenue is exceptionally well preserved.

CO7 General Nathanael Greene Homestead

1774. 50 Taft St. (open to the public)

This dignified, if rather bland, two-and-one-half-story, five-bay house with central hall flanked by four rooms and two interior chimneys is the home of the general who is usually credited with being General Washington's most effective commander during the Revolution. It was from here, after the war, that the general became the leader in shifting the economic focus of the family from its extensive farming holdings, first into a forge, then, in 1810, into the progenitor for cotton milling in Anthony.

Washington

Washington, which had six mills by the end of the nineteenth century, was once the biggest industrial town in Coventry. By 1935, all of its mills had been demolished. Today the town seems (except to its residents) a continuation of Anthony. But the oldest house in town is of interest.

CO8 Paine House (Francis Brayton House)

1748. 7 Station St., approximately 100 yards from Main St. (owned by the Western Rhode Island Historical Society; open to the public)

Francis Brayton, who originally acquired much farmland in the vicinity, eventually also owned saw-, grist-, and fulling mills in what was then (inevitably) called Braytontown. Above the masonry basement of this now resurfaced house, which is set into a hillside, is a narrow four-bay, two-story shingled front, to which, early on, a saltbox extension was added. Although the basement level has an entrance on the road, it is the stretched-out "side" elevations, where the principal entrances are, that attract attention. These differ radically from one another. The most public side, toward Main Street, aspires to regularity with a fulsome array of windows in two sizes planted across the elevation like rows of corn in an irregularly bounded field. Whereas the tail of the saltbox slope is cleaved at an angle on this side, on the opposite side the saltbox heaves out in an angular billow of shingles as it slopes to the ground. Here windows of three sizes are irregularly scattered around two doors. The warping of conventional imagery by this combination of the odd shaping of the house, the visual tension of openings of various sizes, and the counterpoint between regularity and asymmetry should delight postmodernist sensibilities at the end of the twentieth century—however removed such compositional niceties from the concerns of early occupants.

CO9 Thomas Arnold House, Ruins of the Arnold Acid Factory, and Arnold Cemetery (Historic Cemetery 42)

c. 1826. 1949 Flat River Rd.

The Arnold family cemetery enclosure is raised against a sloping field on a terrace fronted by a beautifully fitted, smooth-faced random stone retaining wall topped with long capping stones. Above this, walls of the same sort enclose the markers on three sides with the terminations sloped down to the wall below. The stone ruins beside it are the remains of the factory, where wood was processed into acid for the calico industry. Thomas Arnold returned to the family farm from Warwick to establish his factory, building his house—Greek Revival with additions—immediately adjacent. House, workplace, and cemetery exist in a spread cluster and recall a time when a venturesome farmer might make the transition to small industrialist.

Coventry Center

Although cotton manufacturing came to this village in 1809, it was Pardon Peckham, who arrived in 1848, and his Peckham Manufacturing Company that eventually dominated the village. To the now disappeared Upper Mill (originally on the corner), he added the three-storied Lower Mill, in stuccoed random masonry, which still stands on Flat River Road at Hill Farm Road, more a pleasant functional building than an architectural wonder. The most interesting surviving buildings in the village are two rows of three well-preserved Greek Revival duplex mill workers' houses, all built by Peckham.

CO10 Joseph Briggs Farmhouse (Coventry Poor Farm–Town Asylum)

c. 1790. 195 Town Farm Rd.

This two-and-one-half-story dwelling has a typical five-bay format and, for a farmhouse, an unusually fine fanlighted and pedimented doorway flanked by fluted pilasters. The addition to the rear undoubtedly provided more room for the poor and mentally ill who worked the farm while the property served as the town poor farm and asylum, from sometime after 1853 until the late 1930s. It is now a residence, and its crop is a subdivision of sizable Neo-Colonial houses of the late 1970s that aspire to its dignity.

CO11 Waterman Tavern

1747 or earlier. 468 Maple Valley Rd. (pole 36)

Historically important because the field opposite was a campsite for French forces commanded by General Rochambeau on the way to and from the battle of Yorktown, this tavern is architecturally interesting as an earlier, country-cousin version of the five-bay formula just seen in a later, more sophisticated version. The elements of the facade are more loosely arranged here, with a simple, transomed door slightly askew both of the midpoint of the elevation and of the center chimney.

CO12 Isaac Bowen House

c. 1755, ell. 1795, main house. 1670 Maple Valley Rd. (pole 96)

The third in a string of examples of the pervasive eighteenth-century two-and-one-half-story, five-bay formula, this is the best of the trio—the finest, in fact, of its type in Coventry. The main house here dates from the same time as the Briggs Farmhouse–Poor Farm (see

CO12 Isaac Bowen House

entry, above) and has exactly the same kind of entrance. Comparison, however, reveals that the arrangement of the openings of the Bowen house is a bit more compact than that for the Briggs Farmhouse (even though the contrast between the shutters here and the more authentic lack of shutters there obscures the comparative relationship of openings to wall). Here, too, the door is more elegantly elaborated: Ionic half columns instead of Doric pilasters and more intricate leading in the fanlight.

CO13 Rice City

The handful of buildings at the head of Vaughn Hollow Road hardly makes a "city." Rice City, however, does take in much countryside, for which a booster like the tavernkeeper Samuel Rice in the 1790s imagined a grand future. Halfway between Providence and Norwich, Connecticut, it was the location of a turnpike tollbooth (which still stands on Plainfield Pike as a rare example of the type), several inns and taverns, and a number of small mills. Today the ensemble counts for more than the individual buildings. First in the village cluster, which barely denotes a center today, comes number 25 Vaughn Hollow Road, a pretty but much restored house of 1804. Adjacent on a corner of the site is a very early school building. The Democrat School (before 1812), moved from another site to 25 Vaughn Hollow Road, is the most important structure in the village simply because one-room schoolhouses of such an early date are extremely rare. Originating as a private school and built by subscription, it was sold to the public in 1817 and served as the village schoolhouse until 1846. The petiteness of its functional detailing and overall scale give it a charming vulnerability. By happy coincidence, its intimate quality as a child's house can be immediately measured against the institutionalized quality of its successor, the Rice City School (1846; 63 Vaughn Hollow Road), both in the latter's rationalization of the double-doored entrance to separate the sexes and in its slightly larger, more public scale. The contrast between the school as child's house and a full-fledged one-room schoolhouse would have been more evident had the latter retained its original bell cupola. Restoration would be especially appropriate because the school and the First Christian Church (1846), also at number 63, rose beside one another simultaneously; both employed two doors, three windows along the side elevations, and a cubic belfry as characteristic minimal Greek Revival expressions of their respective institutional types. Bluntly assembled and ponderous on its gable, the church belfry explodes its modest secular underpinning into the rural majesty appropriate for its calling.

CO14 **Hopkins Hollow Church**

1862. 186 Hopkins Hollow Rd.

Originally a Christian Union church founded from the Rice City congregation, Hopkins Hollow Church became Baptist in 1894. In essence, it is the Rice City First Christian Church reduced and vernacularized: the same pair of transomed and hooded doors, but here without flanking pilasters and more modest in scale; the same gable treatment, but without the horizontal molding which invokes a "pediment"; the same three windows on either side elevation, but domestic in scale and more compressed. The biggest difference here, however, is the omission of the belfry. As a result, the church could be mistaken for a school, except for its somewhat larger size (but again much smaller than First Christian in Rice City) and its churchly accoutrements of a well-preserved carriage shed and cemetery. On the positive side, this little successor to the Rice City church seems tauter in the justness of the proportions of openings to clapboard walls and the greater compactness of their placement. Surrounded by a handsome wall of stone which is partially hand hewn to create squared surfaces, the beautifully maintained cemetery (established earlier than the church, probably c.

CO14 Hopkins Hollow Church

CO15 Arnold Home Farm, Barn

1840) is notable for the prevalence of the favorite Greek Revival funeral emblem, urn and weeping willow, carved repetitively on a number of tombstones as a standardized ornament. The spanking maintenance of everything here gives this church a crisp, manicured beauty which is its own country testimony to the glory of God.

CO15 **Arnold Home Farm**

c. 1838 and later. 1905, principal barn. 375 Hopkins Hollow Rd. and Narrow Ln.

For a house with what seem to be a number of additions accreted over a century (after the original house burned in 1838), the ultimate, vaguely Neo-Colonial result, with its bays, gables, and spacious porch, is surprisingly impressive. But the principal barn is exceptional—especially for Rhode Island. A large,

CO17 Coventry Town Highway Sand Storage Facility

clapboarded barn with cupola, also vaguely Neo-Colonial, this possesses an unusual cobblestone front, in the manner of some of the fencing on the property. Impressive too is the view down to the cluster of farm buildings from Narrow Lane, which runs beside the farm on a ridge. Set in a lovely vista of stone-walled fields and distant woods, the farm buildings stretch out behind the big barn in a linear cluster, dominated by another sizable barn in random masonry, which houses equipment and upstairs living quarters. Perhaps because of the dominance of the masonry barn, or more subtly because of the relatively large scale of all the buildings here, the view suggests Pennsylvania more than Rhode Island. In any event, this is among the idyllic farm views in the state. Will it remain?

CO16 **The Greene Company** (Coventry Cranberry Company)

Mid-19th c. 1065 Narrow Ln. (south end of lane)

A clearing in this woodsy lane opens to the only extant commercial cranberry operation of consequence in Rhode Island. The complex is dominated by a rectangular diked land shape which contains the bog. The stream from the bog is bridged by a fairly sizable random masonry arch topped by a cobblestone railing, (perhaps the work of the same mason who flaunted his craft at nearby Arnold Farm). The reworked wooden barns for processing appear to date from the end of the nineteenth century. (Recent "improvements" have spoiled a cluster of houses on the property going back to the Greek Revival.) Come to this tucked-away operation in early October, when the bogs are flooded and paddlewheel threshers beat the cranberries off the bushes, after which they are skimmed from the surface. Then patches of shimmering cranberry crimson on the pond are in tune with the New England autumn.

CO17 **Coventry Town Highway Sand Storage Facility**

Southwest corner of Victory Hwy. and Flat River Rd.

Buildings which make a process as visible as this are always fascinating. Here a well-designed conveyor belt lifts highway sand into the conical, asphalt-shingled building shaped to the master pile, with a cavernous entrance for trucks.

CO18 **Windy Parks Farm** (Israel Wilson House)

1814. Northeast corner of Harkney Hill and Camp Westwood rds.

Despite additions, this is yet another typical two-and-one-half-story, five-bay, central-chimney Federal house, which is very grand for such a remote country location. A commanding pilastered and fanlighted door centers an elevation in which the four openings to either side are set off from the central axis about as far as this compositional format permits, thus increasing its grand presence. Compare the spread quality of this elevational format to the compactness of the distribution of the same elements in the Isaac Bowen House (CO12).

West Greenwich (WG)

Although West Greenwich shows on the map as a considerable rectangular swatch of west central Rhode Island, it offers less of architectural interest than any town in the state. Historically, it has been the most isolated of all Rhode Island towns. Poor soils discouraged any but the barest subsistence farming, woodcutting, and gravel operations.

Railroads completely bypassed it, thereby thwarting industrial development along the Big River and its tributaries in the eastern half of the town or along the tributaries of the Flat River in the west.

Industrial development was further limited by lack of connection with the prosperous eastern shore of the state. Even today, West Greenwich is the only western town from Glocester to Exeter without a major east-west highway; its principal highways slash diagonally across it north and south, as though the town were less a destination than a stretch to get through. Its greatest transportation improvement during the nineteenth century was the New London Turnpike (1821), a toll road to the New London harbor. The turnpike still exists as a gravel road paralleling Interstate 95 and continues south across Exeter and Richmond. The section through West Greenwich, however, which has been only minimally regraded, together with parts of the Central Turnpike in Foster, affords the best sense of the turnpikes when they were much-heralded improvements. Within a decade of its completion, the West Greenwich section of the turnpike, lined with taverns and inns and notorious for prostitution, robberies, and occasional murders, even anticipated the seediest of modern strip development.

The undeveloped quality of the town encouraged the substantial acquisition of acreage in the western portion for state forest preserves. Then, in the late 1960s, an extensive area in the eastern portion of the town was condemned for development of a giant supplement to the existing reservoir in Scituate by damming the Big River. Most of what was of any architectural significance in West Greenwich has been lost to the reservoir project, including Nooseneck, its only village of historic consequence. Population levels confirm its historic isolation. From a high point of 2,054 in 1790, its population declined to a mere 367 in 1920. By then, its unexpensive wooded land had already begun to attract summer colonists and those who preferred to "rough it." By the 1970s

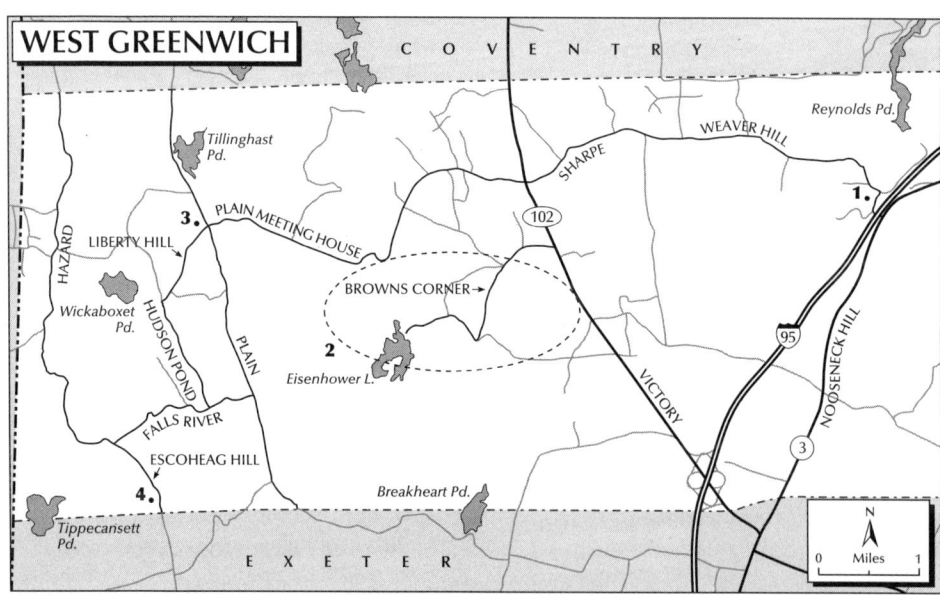

suburban development was well underway, and in the 1990s the population of West Greenwich finally surpassed the level reached two hundred years earlier.

WG1 Kitt Matteson Tavern

1741; raised porch a later addition. 58 Weaver Hill Rd. (.2 mile beyond I-95 underpass)

A sloping site gives this five-bay, central-chimney house two and one-half stories uphill, three and one-half down. Sturdy and plain, it typifies the most prosperous West Greenwich architecture, here magnified by a generous site with rolling fields and fine nineteenth-century shingled barns and outbuildings. It continues to command a fork known locally as Kitt's Corner.

WG2 Alton Jones Campus, University of Rhode Island

c. 1830, Parker Farm. 1930s with c. 1965 additions, W. Alton Jones hunting and fishing estate (now University of Rhode Island Conference Center). End of Matteson Plain Rd. (center open only to conferees)

This campus covers 2,300 acres of woods, fields, and ponds and includes two complexes of architectural interest. First, the road winds past Parker Farm, including a two-and-one-half-story Greek Revival house and an impressive complement of early twentieth-century farm buildings. Once the Parker District School (destroyed by fire) and facilities for the town poor farm were also on the preserve. Finally, the road terminates at the luxurious hideaway where W. Alton Jones relaxed from his duties as president of Cities Service Petroleum. The estate, now a University of Rhode Island conference center, focuses on two lodges: Whispering Pines, built for the Jones family, and Nettles, for their visitors, both overlooking a scenic pond. Whispering Pines, a blown-up version of a log bungalow, is among the best examples of rustic work in the state. A plaque in the upstairs master bedroom celebrates a visit by President Eisenhower. The university has made a few sympathetic additions since acquiring the estate in 1965 as a gift from the foundation Jones established in his name. A spur road leads off to an environmental studies complex, which wins no architectural prizes for environmental concern. Finally, the estate also contains the barely visible remains of the Bela Clapp Acid Factory, which once derived acetic acid from wood harvested in local forests. Trundled by wagon to the nearest railroad station in Greene, the hogsheads of acid were shipped

WG2 Alton Jones Campus, University of Rhode Island, Conference Center

principally to the B. B. and R. Knight textile mills in Riverpoint, Natick, and Arctic. (This history explains how the university's environmental studies program eventually found a location on Acid Factory Creek.) Jones's fishing lake was once the factory's millpond.

WG3 West Greenwich Center Baptist Church

c. 1830. Plain Meeting House and Liberty Hill rds.

The two doors of this meeting house are dignified above by cornices and below by two granite slab steps in the simplest of pyramids—the lower step spreading to the ground, the upper narrowing to the width of the door. Between the horizontal stretch of the doors, a stack of two commonplace windows creates a vertical into the gable between the two doors. Everything is duplicated, very plain, but this image of bare essentials burns into the memory.

WG4 Escoheag Advent Church

1870. Escoheag Hill Rd.

Jason P. Hazard, a deacon of West Greenwich Center Baptist Church and a farmer of importance in the area, broke with the congregation and built this church. It is a slightly more pinched version of the building he left. Plainer than plain, inside it contains what appear to be its original Victorian fittings: chairs, gas chandeliers (electrified), and organ.

East Greenwich (EG)

Division Street marks the town line between Warwick to the north and East Greenwich to the south. South of Division Street is the compact part of East Greenwich, divided by the north-south axis of Post Road, which is Main Street to the local population. Most Main Street commercial buildings either never were of much architectural consequence or have been reduced to little consequence by insensitive alteration. Some public buildings are the exception. The area to the east of Main Street toward the waterfront contains the town's densest concentration of early houses (from the eighteenth through the mid-nineteenth century), a few early industrial structures, and a bustling fringe of marinas and restaurants catering mostly to leisure boating and those who seek its ambiance. The area west of Main Street rises steeply into what is known locally simply as the Hill. It is here that most of what is architecturally significant is to be found.

EG1 Varnum Memorial Armory
1914. 6 Main St.

Just south of Division Street two public buildings almost opposite one another make an auspicious entry into the northern end of Main Street. The Varnum Memorial Armory exploits the typical battlemented type to achieve a degree of monumentality that is remarkable here, in what is for armories a tiny building. The wide, shallow buttressing on the facade and battlemented rooflines are particularly effec-

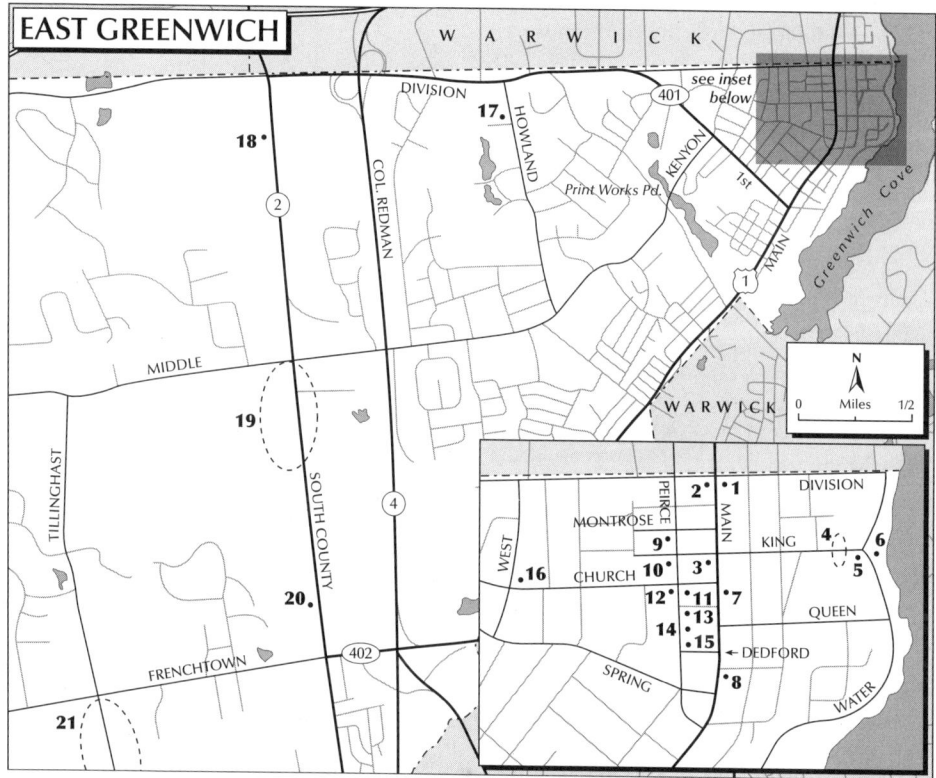

tive. The asymmetry created by the height and the staggered windows of the stair tower effectively anchors the corner of the building at Division Street. Built by the Varnum Continentals, a filiopietistic military organization founded in 1907, the armory houses the organization's headquarters, a drill hall, and a military museum.

EG2 Offices (East Greenwich Post Office)
1934, Frank J. Anthony. 11 Main St.

This Neo-Federal post office in red brick and cast stone, with a high basement and landscaped terrace several feet above Main Street, is a particularly well-proportioned and handsomely detailed example of New Deal enthusiasm for colonial and early national imagery for historic towns. At the center of the elevation, the five tall glazed arches framed by a pilastered portico recall eighteenth-century orangeries. Insufficient parking forced the removal of the post office to a vastly inferior building and conversion of this one to a restaurant. But the resonance of its original use persists, compelling the present owners to warn off potential letter droppers. The lobby interior retains its spatial configuration and finishes.

EG3 East Greenwich Town Hall (Kent County Courthouse; East Greenwich State House)
1804, Oliver Wickes, builder. 1908–1909, remodeled, William R. Walker and Son. 1993–1995, addition, William Kite Architects. 127 Main St.

Commandingly sited on a hill above Main Street, the former State House is located near the geographic center of the village of East Greenwich at the intersection of what were the village's principal thoroughfares, King and Main streets. One of five state houses in different towns which were originally used in rotation, this is comparable to the Colony House in Newport (NE6) in marking the end of an axis from the middle of the town to the waterfront. Its form draws heavily on the example of the Newport building as well as the Old State House in Providence: a high sandstone basement, a pedimented center pavilion framing the principal entrance, and a centrally located cupola. But here the building is rendered in clapboard, and the Wren Baroque detail of the mid-eighteenth century has been supplanted by the more delicate Federal forms of the early nineteenth century. The building was used as a state house for Rhode Island's peripatetic government only through the 1850s. Its most remarkable interior is the original two-story Assembly Room. It is extraordinarily short for its width, with the axis from the entrance to speaker's platform at center across the narrow dimension. In an ingeniously provincial manner, Oliver Wickes crossed the width of the room at every window with a steep plaster vault, giving the ceiling a deeply scalloped effect. A nice paneled backdrop with scroll capping contributes to the speaker's (judge's) dignity. Very little used after the county court moved to new quarters, it was turned over to the Town of East Greenwich and converted to the Town Hall. A large addition to the rear, which provides a second principal entrance on Peirce Street, hovers somewhere between contextual and postmodern.

EG4 King Street Railroad Bridge
1835–1836, Major William Gibbs McNeill, chief engineer. King St.

At the Greenwich Cove end of King Street is the best sample of what remains of eighteenth- and early-nineteenth-century houses and commercial buildings in East Greenwich. More interesting are two industrial structures and the old town jail.

The powerfully massed, rock-faced ashlar King Street Railroad Bridge was built for the Providence & Stonington Railroad as part of the original rail system in the state. Unlike most of its contemporaries, it retains its original function and continues to carry the main Amtrak line on the Northeast Corridor. Its designer, the uncle of painter James Whistler, had just completed, with the painter's father, construction of the Canton viaduct for the Boston & Providence Railroad. His subsequent work included construction of the Moscow–St. Petersburg Railroad in the late 1830s. This is probably the state's earliest extant bridge for a major transportation use which still serves its original purpose.

EG5 Bay Mill (later Shore Mill)
c. 1840 and later. 1986–1987, converted to condominiums. 83 King St.

This three-story, stuccoed stone mill with a trapdoor monitor is typical of early-nineteenth-century Rhode Island industrial structures; additions on the south and the mansard-roof office are late nineteenth-century additions. The only documented textile mill in East Greenwich, it stands on the site of one of the first fully mechanized textile mills in the state.

EG6 East Greenwich Preservation Society (Kent County Jail)
1795–1804. 110 King St.

Located at the eastern terminus of King Street, the jail has two parts: a two-and-one-half-story clapboard front section for an office and housing for the jailer and his family, and a two-story brick rear section for the inmates. The two-part program and the domesticity of the front section typify late eighteenth- and early nineteenth-century town jails in Rhode Island. The jail was built in conjunction with the East Greenwich State House, a relationship suggested by the situation of the two buildings at either end of King Street. It was used as a jail as late as 1957.

EG7 Greenwich Hotel (Updike Hotel)
1896. 162–168 Main St.

A series of hotels occupied this site as far back as the 1740s. The Greenwich Hotel, with its two-story Ionic-columned porch, was a late nineteenth-century classical revival effort to provide a grand focus for the commercial district, today frustrated by nondescript shop fronts.

EG8 Offices (The Brick House; Colonel Micah Whitmarsh House)
1767, John Reynolds, builder. 294 Main St. (corner of Dedford and Main sts.)

Around the time of the Revolution, John Reynolds built a group of houses that introduced more sophisticated planning and design to East Greenwich. Local tradition maintains that this is the first brick building in East Greenwich. Although the gambrel roof and low ceilings were retardataire by the late 1760s, the center-hall plan was a stylish change from the traditional center-chimney plan. The plan is particularly interesting for its paired chimneys on the outer walls with angled corner fireplaces in each of the four rooms on both floors.

EG5 Bay Mill (later Shore Mill)

EG9 First Baptist Church

EG9 First Baptist Church
1884–1887, J. B. Goodwin; Lodowick C. Shippee, builder. Peirce St. at Montrose St.

Queen Anne design typically alters the expected scale of colonial ornamentation. Here the alteration is excessive. On the two principal elevations for the corner location, ranges of tall windows are topped at their centers by semicircular openings to suggest giant Palladian elements. Bloated broken scroll motifs topped by diminutive urns surround the semicircles. The entrance occurs at the corner, which was originally capped by a steeple. The loss leaves a mini-cupola over the center of the church as its "tower," plus another octagonal turret on Melrose Street as the compositional anchor for a picturesque cluster of windows at the entrance to the attached parish house.

The space inside seems influenced by both the seventeenth-century square-plan preaching box and the nineteenth-century preaching au-

EG10 General James M. Varnum House

EG12 St. Luke's Episcopal Church

ditorium. But the corner entrance vestibule establishes a spatial bite from the corner of the square which, when duplicated around the square, converts it to a stubby Greek cross. Except for the substitution of stained glass for clear across the lower range of windows and for repainting, most of the interior remains intact, including a pretty Queen Anne structural X across the center of the cross, sprung from wall brackets and more mock than real. Somewhat heavy-handed in its whimsy, this church nevertheless fascinates for its invention.

EG10 General James M. Varnum House

c. 1774, John Reynolds, builder. 57 Peirce St. (open seasonally to the public)

Built by Reynolds for a noted lawyer and Revolutionary War hero, this is one of several large, elaborate houses in East Greenwich dating from the early 1770s that were the product of a talented builder and the taste of cosmopolitan patrons. The proportions of the front elevation are exceptional, as is the correctness of the pedimented entrance porch with Ionic columns. The same qualities appear in detailing inside as well, where particular note should be taken of the dining room fireplace with its broken pediment overmantel. The armory on Main Street (EG1) commemorates James Varnum's military exploits.

EG11 East Greenwich Free Library

1914–1915, Angell & Swift. 82 Peirce St.

The churchly look of this rock-faced granite library building in nominal Neo-Gothic probably shows the influences of nearby St. Luke's (see next entry), to which it nicely relates. Its conservatism also appears in its general recall of the tradition of rock-faced granite libraries H. H. Richardson had established forty years earlier, even though their stylistic precedent was Romanesque. Originally this library hearkened back further, to mid-nineteenth-century panoptic plans, which the architects could have known from the organization of the Brown University Library. Such plans placed the librarian's desk at the center of the space with book stacks radiating out from it, in this case housed within a stubby semioctagonal wing to the rear. Recent renovations have eliminated that organization by moving the desk to one side near the front door.

EG12 St. Luke's Episcopal Church

1875–1876. 1920–1923, spire, Clarke and Howe. 1960–1962, parish house, McConnell and Walker. 2001, parish house rebuilding. 101 Peirce St.

Few churches were built in Rhode Island during the economically stagnant mid-1870s. Rather than exhibiting markedly High Victorian characteristics, as might be expected of an up-to-the-minute design of the 1870s, this looks back to the revival of the Gothic English parish church which had developed in the 1850s and 1860s as a deliberate effort on the part of ecclesiologists to return to the ancient medieval pieties of Anglican ritual. On the exterior, shapes, simple and austere but also picturesque, count for more than detail. The high gabled mass in rock-faced granite is minimally interrupted by a tiny trefoil clerestory before the roof slides out at a lesser pitch over wide aisles. The charm of the textured mass is enhanced by an attached burial ground at the church's east end with a lych-gate—rare in Rhode Island, although St. Colomba's Chapel in Middletown provides another example. And of course the much later addition of the squat tower topped by a stone steeple at one corner of the nave works to the same picturesque end.

But the principal interest is the well-preserved interior, the too-white repainting and some later fittings notwithstanding. Worth special notice are the handsome straightforwardness of the bracketed structure which fans from the capitals of the polished granite columns to support clerestory windows the length of the space and roofing for aisles and nave (the latter by scissor framing) across it. Remarkable too is the stained glass, dating from the 1870s through the 1980s. Most immediately noticeable are two sets of Tiffany glass, opposite each other in the second bay west of the chancel: an opalescent, scenographic set (1910) on the north, and an organic, geometric set (c. 1890) on the south. Also especially deserving of attention is the Wright Goodhue *Tobias and the Angel* (1924), the northernmost window on the west wall. In the entrance wall, a pair of windows celebrates the two Greenwiches: Sir Christopher Wren's palace at Greenwich appears beside the State House at East Greenwich, with appropriate nautical embellishments around both. St. Luke's is among the finest examples of country Gothic in the state—countrified, but in the sophisticated manner of a worldly town.

EG13 **Kentish Guards Armory**

c. 1842. 90 Peirce St.

Following the Dorr Rebellion in 1842, several independent military organizations around the state built new armories. They had two notable exemplars, both from the 1830s: the Greek Revival Newport Artillery Company (NE79) and the Gothic Revival Benefit Street Arsenal (PR79). Warren chose to follow the Gothic lead, while Warwick, Bristol, and East Greenwich followed the Greek. Of the four, this is the most elaborate, with two Doric columns in an inset porch—technically a distyle in antis portico. From this hobnailed portal one expects the march of the toy soldiers rather than a serious military force.

EG14 **Samuel Knowles House**

1850. 100 Peirce St.

EG13 Kentish Guards Armory with the Samuel Knowles House (EG14) to the right

EG16 Greene House (Hilltop Cottage)

EG15 **Rose Cottage**

1859. 112 Peirce St.

For a commanding sight with a view across East Greenwich Cove toward Potowomut Neck, two contemporaneous houses illustrate diverging architectural aesthetics and orientation of building to site at mid-century. The foursquare Knowles House—its architectonic quality now somewhat diminished by the removal of a one-story, three-bay front porch—uses the monitor-on-hip roof form seen across the state in buildings from the first half of the nineteenth century; the monitor here is remarkably tall, doubtless to exploit the view. Rose Cottage typifies the picturesque, asymmetrical villas just coming into vogue—especially for suburban, picturesque locations—and its glazed octagonal cupola is another response to the view.

EG16 Greene House (Hilltop Cottage)

c. 1872. 143 Church St.

This is one of several suburban villas built in the late 1860s and early 1870s on ample lots away from the Post Road, both here and just north, across the Warwick city line. The form hearkens back to the Downingesque cottages of the 1840s and 1850s, but the profusion of detail—elaborate bargeboards, exposed pseudo-framing—link it to the cottage orné, then at the height of its popularity for suburban and resort dwellings.

EG17 Clement Weaver House

1679 and later. 1937–1940, restored by Norman M. Isham, Edwin E. Cull, associate. 125 Howland Rd.

The Weaver House is the oldest extant house in East Greenwich and one of the few remaining seventeenth-century buildings in the state. In the 1930s, analysis of its complex history by Rhode Island's preeminent restoration architect, Norman M. Isham, identified four separate phases of construction, all probably completed by approximately 1750. A drawing by Isham shows how the house, which originally comprised a great room with upstairs bedroom on the south side of the chimney and a lean-to kitchen to the north, was expanded successively above the kitchen, then by a row of rooms in a saltbox extension to the rear, and finally by a new kitchen in an ell to the south. The simple interior includes large stone and brick fireplaces and an unusual pair of staircases in front of the main chimney, straight runs of seven steps on the south and eight steps on the north, which meet at a wedge-shaped landing and rise separately to the north and south garret chambers.

EG18 Mary Ellis House

1937–1939, William Wilde. 1629 South County Trail. Not visible.

Set back from the road in the woods is one of the few, and one of the best, International Style houses in the state. *Architectural Forum* published the house in 1939, and Henry-Russell Hitchcock praised it in his survey of Rhode Island architecture, published the same year. Its vertical cypress siding, typical for New England modernist houses after World War II, would appear to date it to c. 1950; but it is a replacement for the original siding of smooth composition board, which weathered badly. Originally, too, the house was wholly one story, except for a second-story artist's studio at its garden end. It has been substantially enlarged for additional bedrooms, probably at the time of the residing. Mary Ellis's fiancé, the painter Albert Gold, and the architect had met when they both lived in the U.S.S.R. Their reunion in America doubtless accounted for Wilde's commission. The substantial alterations have upscaled (and somewhat upset) what must, at the time of completion, have appeared as a more modest and more avant-garde getaway from the New York art scene.

EG19 Fry's Hamlet Historic District

Late 17th century through mid-19th century. 2068, 2153, 2196, and 2233 South County Trail (south of Middle Rd.)

Extending well back from either side of Middle Road between Fry's Corner and Greene's Corner, Fry's Hamlet comprises three adjoining farmsteads spread across approximately 272 acres. Remarkable as a small node of early houses still owned by the family that built them, this group comprises three eighteenth-century farmhouses, each with several outbuildings, including barns, privies, sheds, corncribs, milk houses, silos, icehouses, washhouses, and a cider mill. Few complexes in the state remain as primary documentation of the agricultural activity that predominated in Rhode Island during its first two centuries. This is among the best.

EG20 Bostitch

1955–1957, Charles T. Main, Inc., architects and engineers; E. Turgeon Construction Co., builders. South County Trail, north of Frenchtown Rd.

Founded by Thomas A. Briggs in Arlington, Massachusetts, in 1896 as the Boston Wire Stitcher Company, and eventually coming to dominate the stapling industry as Bostitch, this company located at various places in New England before settling on East Greenwich in the mid-1950s. It is one of the state's earliest sprawling, one-story modern industrial plants of impressive size located on a large acreage. The Bostitch "park" minimizes planting to feature mowed lawn and pond, using the same ingredients as the proudest of its manicured predecessors in the nineteenth century to display the plant rather than screen it—in this instance, made particularly conspicuous by a sweep of field up to the hilltop site for the building. As a pioneer venture for large-scale manufacturing operations in the southern part of the state, Bostitch heralded an end to the agricultural economy of the area as tenuously epitomized in Fry's Hamlet.

EG21 Tillinghast Road Historic District, Steam and Wireless Museum

c. 1702–present. 679 Tillinghast Rd. (south of Frenchtown Rd.)

A linear settlement of seven eighteenth- and early-nineteenth-century farm complexes and a church along both sides of a winding country road, this district, also known as Place's Corner, shows a remarkably coherent relationship among historic structures in an agrarian setting, one which could be threatened by suburban development. Each of the farms retains several outbuildings, including barns, sheds, chicken coops, and well houses. The houses illustrate the persistence of the five-bay, one-and-one-half-story format for rural vernacular buildings. Only two, those of Daniel Briggs (c. 1702–1717, 1725) and Thomas Tillinghast (1760), are two and one-half stories. The Tillinghast House is unusual for the asymmetry of its four-bay front elevation, and even more so for its paired center chimneys, a form also seen in the old section of the Whitridge House in Tiverton (TI11).

The vernacular Greek Revival Frenchtown Baptist Church (1822), at the corner of Frenchtown Road and Tillinghast Road, was moved to the Tillinghast Farm in 1972 to prevent its demolition. Its projecting vestibule is rare in Rhode Island. It is now used as a lecture hall by the Steam and Wireless Museum. The pioneer Massie Wireless Station "PJ" (1920s, formerly located at Point Judith and moved here also to prevent demolition) is a tiny gabled box mounted as a mini–viewing "tower" out to sea, at one end of a larger gabled box, all shingled. The Steam and Wireless Museum, established in 1963, features several steam engines, including a 150-horsepower Harris-Corliss engine with a twelve-foot flywheel, together with steam rollers and other antique steam-operated construction and farming machines.

South

NORTH KINGSTOWN, NARRAGANSETT,
SOUTH KINGSTOWN, CHARLESTOWN, WESTERLY,
HOPKINTON, RICHMOND, EXETER

North Kingstown (NK)

THE HISTORY OF NORTH KINGSTOWN IS CLOSELY ALLIED WITH two aspects of its topography: its thirty-mile coastline and its three river systems. The town's irregular northern border follows the Hunts River as it flows northeasterly into the Potowomut River estuary. In the central portion of the town a series of small mill villages follow the line of the Annaquatucket River through a chain of glacial kettle holes. The smaller Mattatuxet and Pettaquamscutt River system winds through the irregular, hilly terrain in the southern part of town. In the west, around the village of Slocum, is a level glacial outwash plain with a rich topsoil ideal for agriculture.

By the seventeenth century the sheltered coastal inlet north of present-day Wickford had become a tribal center for the 25,000 Narragansett Indians who lived in the area from Narragansett Bay west to the present-day Connecticut border and south to what is now Westerly. Early in the seventeenth century the Dutch came to this cove to trade with the Narragansett. They were shortly followed by the English. Roger Williams spent the summer of 1636 at the cove making peace with the friendly Narragansett chief Miantonomi. Subsequently Richard Smith built a trading post, Cocumscussoc (Smith's Castle) on the cove near the home of Chief Canonicus. Between them, Smith and Williams made Cocumscussoc the social, political, and religious capital of the region. However, the growth of the white population created increasing tensions. A series of small skirmishes led up to The Great Swamp Fight of 1675, in which the Narragansett were all but annihilated. The survivors are said to have burned every house south of Warwick early in 1676. Cocumscussoc was quickly rebuilt, and the English population reestablished its farming, fishing, and trading operations.

Beginning in the seventeenth century the area's prosperity fostered the growth of a

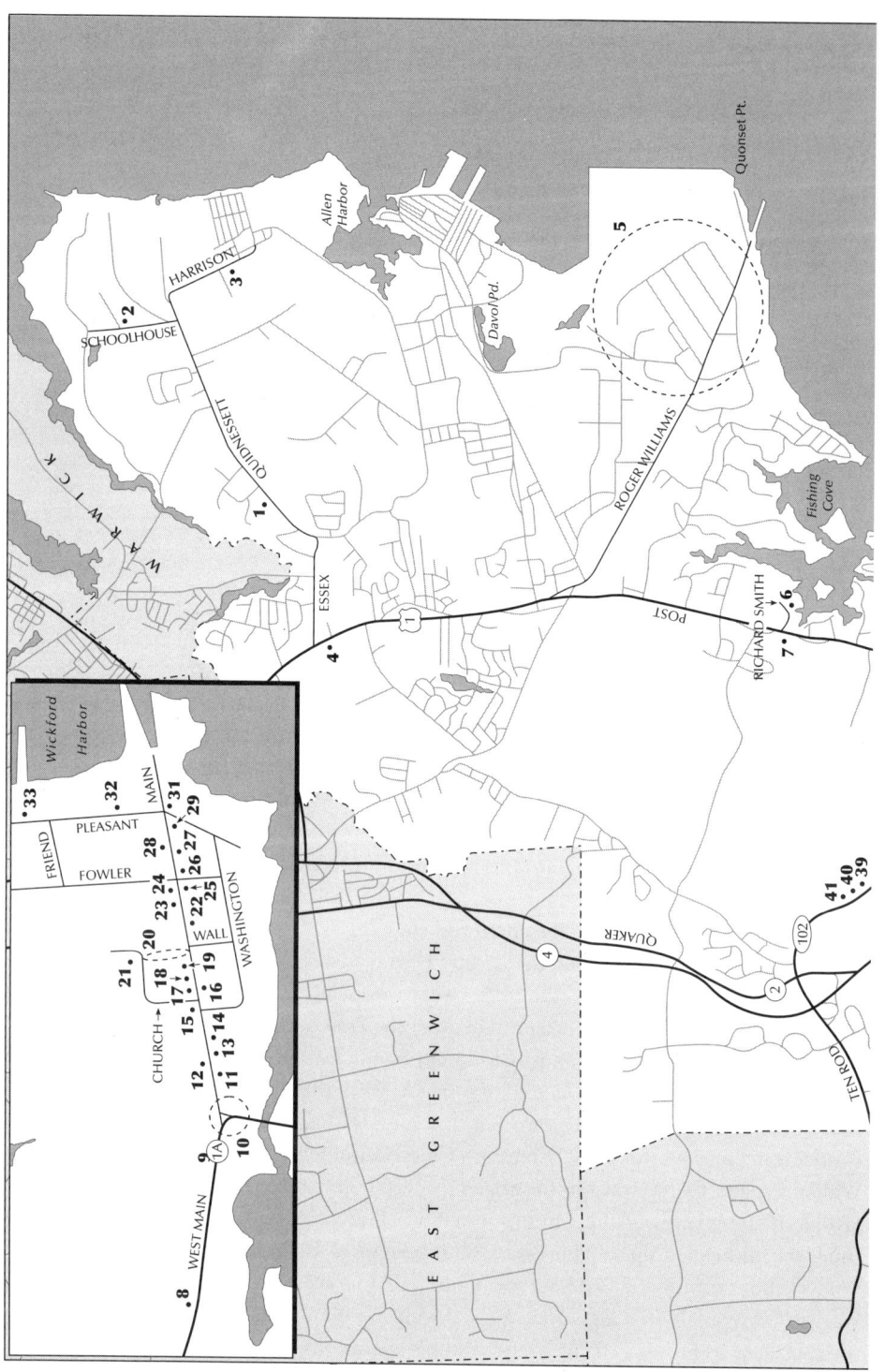

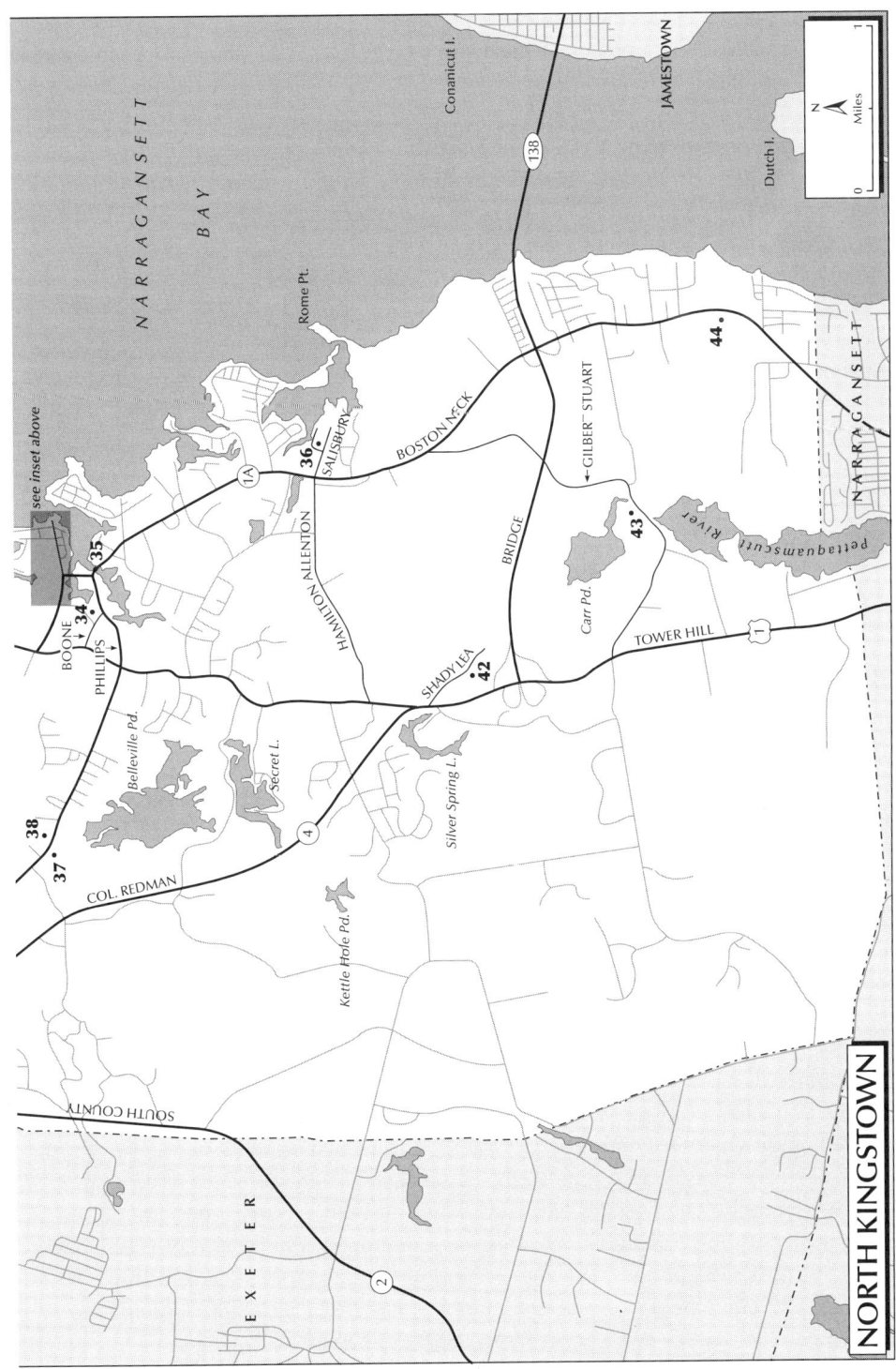

society unique in New England, the Narragansett Planters. Unlike New England's subsistence farmers, the Planters made extensive use of slaves in raising livestock and dairy cattle on large farms in the temperate climate of southwestern Rhode Island. The stock, dairy products, and wool were then sent on Planter-owned ships to be sold along the eastern seaboard and in the West Indies. It was this plantation society which gave Rhode Island the highest proportion of slaves of any New England colony in the eighteenth century. Today, the outlines of the Planters' large, rectangular, east-west landholdings can be seen in the long stone walls abutting Route 1A (Boston Neck Road) south of Wickford.

Immediately following the Revolution most of Rhode Island's slaves were freed, and the loss of this labor force, coupled with the war's long interruption of trade, caused the plantation economy to wither. Economic activity shifted increasingly to the villages and towns, where, after 1790, there was a resurgence in the shipping trade. From 1790 until it was again halted by the War of 1812, shipping was a tremendous source of wealth for all of Rhode Island's ports. The many houses from this period still standing along Wickford's tree-lined streets are eloquent testimony to this rapid and short-lived prosperity. Following the War of 1812, however, Wickford's fortunes began to fade and the town was given a serious financial blow when it was bypassed by the new railroad line in the 1830s.

While Wickford was declining, the rest of North Kingstown was experiencing a boom in the construction of textile mills. Although the town's three rivers had helped to power a number of sawmills, gristmills, and fulling mills in the eighteenth century, the first cotton mill appeared on a tributary of the Annaquatucket around 1800. Cotton and woolen mills were quickly established in the early 1800s: nine mills existed by 1832 and twelve by 1870. Manufacturing soon became the town's primary source of wealth. The mills inevitably generated the growth of nearby villages to house the workers, foremen, and owners.

The latter part of the nineteenth century brought a wave of summer residents, most of them settling either to the far north on the broad, rolling fields of Quidnesset or south in the tightly knit communities of shingled cottages at Plum Beach and Saunderstown. This influx of summer visitors was greatly increased in 1900 with the construction of the Seaview Trolley line from Providence to Narragansett. The trolley brought day trippers from Providence's large working and middle class communities. They eventually established their own sheltered summer colonies at Shore Acres, Bay View, and elsewhere north of Wickford.

After 1939 North Kingstown's fortunes became increasingly tied to those of military installations at Quonset Point. The Quonset Naval Air Station, designed to protect the entire northeast coast, covered more than 1,000 acres (some of which were cleared of houses just rebuilt following the 1938 hurricane). Its rapid construction was followed in 1942 by that of the Davisville Naval Construction Training Center and the accompanying Advance Base Depot and Proving Ground for the prefabrication and shipment of buildings and equipment to overseas military bases. It was here that the Quonset hut was developed. The navy ceased its operations at Quonset in 1974, leaving much of central and northern North Kingstown an abandoned hulk.

Increasingly the town has been subsumed into the larger metropolitan area of Providence, and its entire eastern half is now heavily suburbanized. Fertile farmlands of the seventeenth, eighteenth, and nineteenth centuries are now cultivated as sod for suburban lawns, providing an unusual New England landscape of level fields of lush, manicured grass. North Kingstown has nevertheless retained the deep woods in its interior portion, as well as much of its character as an assortment of villages built up at different periods and for different purposes. Although strip commercial development along Route 1 has diverted attention from many of the villages that lie along it, recent interest in old houses and cohesive pedestrian communities has given new life to long-dormant Wickford. A continued interest in historic district zoning and open space preservation will help North Kingstown retain the complex fabric of its history.

NK1 **House**

Late 18th–early 19th century. 120 North Quidnesset Rd.

This rather grand five-bay, two-and-one-half-story farmhouse is made grander by its elevation on an earth platform. Owners' houses of this size are characteristic of the very large farms established in southern coastal Rhode Island during the eighteenth century. That they were known at the time as "plantations" in itself suggests their scale. Today subdivision colonial crowds the real thing.

NK2 **Scalabrini Villa** (John Carter Brown House)

1872. 860 North Quidnesset Rd. (west of Harrison St.)

This ambitious Second Empire villa was built by John Carter Brown on land adjoining his father-in-law's summer home as a retreat from his late-eighteenth-century mansion in Providence. It is an unusually high-style house for nineteenth-century North Kingstown, with pressed brick walls, fine dressed granite trim, and a convex mansard. The symmetry of its central pavilion is offset by an aggressively corbeled and patterned chimney set in a swelling easterly bay and a delicate stickwork porte-cochere to the west. A bracketed porch running the length of the house once provided views southward over sloping lawns, now filled with sprawling one-story residential development. The first floor interiors retain some of their original detail, including tiled fireplace surrounds and an unusually fine brass newel post lamp. To the rear is a handsome Colonial Revival barn. In 1907 the Brown family gave the house and 100 acres to the Rhode Island Hospital for use as a summer home for crippled children. In 1957 it passed to the Society of St. Charles, which operates it as a nursing home.

NK3 **Wightman Farm** (Swanholme)

Late 17th century and later. 215 Harrison St.

Now a housing subdivision for which construction began in 1988, this was until then the site of a grand "plantation" founded by George Wightman. Wightman settled in North Kingstown in the 1660s, married well, and eventually amassed 2,000 acres comprising parts of what is now North Kingstown, Exeter, and Westerly. The family residence began as a three-bay, two-story "chimney-ender": chimney to the west, with the original second-story overhang still visible under the gable of the eastern end. The addition of two more bays to the east, probably in the late eighteenth century, converted the three-bay house to the standard five-bay house, with the chimney now approximately centered. Another two bays added to the west in the nineteenth century brought the house to its present form. The original seventeenth-century interior contains an unusual treatment of the overmantel of the great room as a series of ornamental shelves, with much paneling of the period.

NK4 Quidnesset Baptist Church

1906, Murphy and Hindle. c. 1960, addition. 6345 Post Rd. (.1 mile south of Essex Rd.)

Following a fire in 1906, Quidnesset Baptist Church was rebuilt by a firm particularly well known for Catholic churches. Although the building's profile supposedly duplicates that of the original (1842) Gothic Revival structure, the extraordinary use of fireproof rock-faced concrete blocks and the blunt layering of the cuboid massing gives this building a hard-edged, looming silhouette which, while certainly crude, is nevertheless surprisingly commanding. The effect is now compromised by a large parish house addition to the south, also of concrete block, but smooth faced in contrast to the church.

NK5 Quonset Point Industrial Park
(Quonset Point Naval Air Station)

1939–1943 and later, Albert Kahn and others. Quonset Access Rd., Belver Ave.

In 1939 the cottage colonies and farms on the southern portion of Quidnesset Peninsula were leveled and the land was extended into the bay to form the 925-acre Quonset Point Naval Air Station. During World War II, together with the adjoining Davisville Construction Battalion Center, the base became a major center for Seabee training, equipment testing and supply, and the design and construction of prefabricated shelter for military purposes—including, most conspicuously, the Quonset hut. The navy removed its operations in 1974. Since then, the state and the town have converted the area to an industrial park and airport, but pieces of the past (steadily decreasing) linger on.

Beyond the original entrance can be seen the first of what were once fields of Quonset huts. Design of the famous prefabs began at Quonset in 1941, by designers at the base challenged to improve on the World War I British Nissen hut. During the war over 32,000 Quonset huts, designed in forty-eight variations for different military programs, were shipped from the base.

The Naval Air Station buildings (1939–1941) are by Albert Kahn, who built his design office on commissions from the Detroit automobile industry, to become the leading designer of American factories before World War II. The station's central complex consists of a grouping of enlisted men's barracks fronted by administration buildings and officers' quarters. Although the complex is unified by an underlying treatment of steel frames, screened by brick walls with bands of stock metal-framed windows, Kahn and his design team individualized the buildings by varying window groupings and by chunky entrance towers as occasional accent. He distinguished the administration building from the rest by a two-story concrete porch with decorative trellis beside an extra-large window with conspicuous framing. The whole complex shows typical late 1930s factory or institutional modernism, here well proportioned and detailed despite wartime austerity standards for construction.

At the end of Belver Avenue lies the airfield, with a row of three Kahn-designed hangars (1942–1943) for land planes. Another, for seaplanes, has been demolished. More obviously the product of rationed building materials than the administrative and living quarters, the hangars have low, spreading steel frames hung with asbestos, metal sheeting, and steel sash that steps up the shallow roof pitch for monitor lighting. Size gives monumentality; so, more specifically here, does the extended pediment-like capping of the portal front which contains the hangar doors. As with the Quidnesset Baptist Church, these hangars are architecturally interesting for the degree of monumentality attained through the severe use of banal materials. Just beyond the hangars, four windowless, polygonal, silolike towers in composition wallboard project ominously from a minimal shed. Here pilots learned celestial navigation and how to cope with cockpit emergencies.

NK6 Cocumscussoc (Smith's Castle)

1678 and later. 55 Richard Smith Dr. (open to the public)

Built on the foundation of an earlier house, Cocumscussoc was originally a fortified trading post on Mill Cove and later became the center of the eighteenth-century Updike plantation. Picturesque, but too much altered to be of first-rate architectural significance, it is nevertheless an important seventeenth-century house, interesting both as a rare seventeenth-century survivor and as the only extant trading post of the period in the state, which played a considerable role at the time in the South County economy. In 1948 the house was threatened with demolition, prompting a group of local citizens to buy and restore it for use as a museum.

NK7 State Police Barracks
1935, Jackson, Robertson and Adams. 7875 Post Rd.

The Colonial Revival barracks is given a Mid-Atlantic formality by tall end chimneys and low flanking wings with shallow recessed arches and connecting hyphens to pedimented end pavilions. The typical compression of New England colonial architecture often encouraged escape to the expansive massing of this alien colonial precedent, with its hyphens and wings, especially during the interwar period, when such massing was popular both for functional expression and for picturesque silhouette. The architects designed a series of state police barracks in the style (see those in Lincoln [LI22] and Portsmouth [PO10]).

Wickford

Wickford was platted c. 1700 by Lodowick Updike, the owner of Cocumscussoc, who proudly proclaimed his achievement in giving the town its original name of Updike's Newton. Like many Narragansett Bay ports, Wickford enjoyed its greatest period of growth between the end of the Revolution and the War of 1812. However, it remained an active secondary port and became a center for a variety of manufacturing ventures during the nineteenth century. At the end of the nineteenth century the town became a stop on the Sea-View trolley line, which extended along the west shore of the bay, and a popular tourist spot. Despite its continued prosperity, Wickford's location east of the main north-south railroad line and early local appreciation of its "antiquities" served to protect much of the town's small seaport character, and enough of its building heritage to give it a place, with Bristol, among Rhode Island's smaller towns with architectural parallels to Providence and Newport. By comparison with the grander quality of Bristol, Wickford's architecture displays a more modest mien—the charm and elegance which late eighteenth- and early nineteenth-century country carpenters could produce where means and taste encouraged the creation of a lovely town. In Bristol, architectural style is varied. Here it is homogeneous, with the homogeneity of the provincial place which knew a moment of minor glory in history, then stagnated until rediscovered by those who had enough discretion to want to keep it as it was.

NK8 Rufus and Mary Congdon House
1843. 115 West Main St.

NK9 Stephen Cooper House
1728. 17 West Main St.

West Main Street (Route 1A) winds downhill into the center of town, passing a fine Greek Revival house, the Rufus and Mary Congdon House, one of the few placarded houses in the state with the wife's name beside her husband's. The elaborate fence with its tall urns and the elegantly spare side porch appear to be the work of the famous Rhode Island restoration architect Norman Isham. At numbers 79–103 is a row of eighteenth-century houses. Farther down is the oldest in a town of old houses, the Stephen Cooper House, a shingled, gambrel-roofed house, gable end to the street, which may originally have served as a store or tavern, as it certainly did later.

NK10 West Main Street Commercial Area

The traditional heart of commercial Wickford is tightly gathered around the junction of West Main with its continuation as Main Street and a fork south (Brown Street, which continues Route 1A). This commercial area consists of the Wickford Standard-Times (originally the Wickford National Bank Building; 1871), the Avis Block, a brick commercial block built in 1850, and, tucked behind this, a rare surviving example of the minimal store building from the Greek Revival, now an art gallery, unfeelingly altered but still compelling (1840s). Best of the commercial center is the Gregory Building (1891), at 1 Main Street (corner of Brown and Main streets), a commercial block of brick, brownstone, and iron whose ambitious scale was never again matched in Wickford. The original plate glass storefront wraps around a monumental corner entrance. The round-arched windows, monumental brick pilasters, and deep cut-brick entablature with a diaper-work frieze beneath a molded wood cornice are typical of more urban work of this period.

MAIN STREET HISTORIC DISTRICT

The Gregory Building imposingly marks the entrance to one of the most endearing town views in the state. The prospect east on Main Street is a nearly uninterrupted tree-lined vista to the town dock at the opposite end, approximately half a mile away. Most of the houses, dating from 1780 to 1820, press close to the sidewalk, opening directly onto it with a single granite step. They are based mostly on a five-bay, central-chimney model, varying in door surrounds, cornices, trim, quoining, window treatments, and the presence or absence of shutters (most of which are later additions). Preservation began in the early years of the twentieth century, and the Main Street Association emerged in 1932 to regulate signage, plant trees, lay sidewalks, start a historic marker program, and draw up an early historic district zoning law. The work of the association led to the 1959 designation of Wickford's Main Street as Rhode Island's second historic district.

NK11 15–17 Main St.

1802 and later

Number 17 is a typical five-bay, central-chimney Federal house (the chimney now reduced in size) with a cushioned-frieze door frame and original paneled door. Number 15, attached to the west end of number 17, is a late Greek Revival five-bay house with central entrance; its twentieth-century owners, amused by the abundance of historic markers on Main Street, have erected a small oval plaque commemorating the residence of "Increase Heartburn, 1971."

NK12 Noel Freeborn House

1817. 24 Main St.

Although typical in its plan, this is the only brick house on Main Street. Its window frames and chimneys, unlike those of its wooden companions, are flush with the wall, giving it a greater austerity. In contrast also to the gabled roofs with central chimneys that characterize most other Main Street houses, this has a low hipped roof with two end chimneys, elements which contribute to its greater formality. So do the handsome proportioning of windows to wall and their regularity of rhythm across the elevation, which contrast with the prevalence on Main Street of the more countrified double, four-spot domino effect (where the four windows on either end of the elevation are separated from the entrance and the center window overhead by a wider interval of wall). The entrance fanlight, too, with its sophisticated sawtooth leading and star-shaped joint fasteners, is exceptional on the street. Noel Freeborn was a silversmith who owned a shop in Newport, and it is doubtless the influence of the grander town that one sees here.

NK13 Nicholas Hart House

1762 and later. 31 Main St.

The Hart House is a good example of one that has been added on to repeatedly, including the Colonial Revival pergola and pedimented side porch. The barn to the rear is a late example of flushboarding (perhaps 1830s or 1840s).

NK14 Immanuel Case House

1786. 41 Main St.

Built for a retired tavernkeeper, this was one of five houses chosen to represent Rhode Island in the 1933 Historic American Buildings Survey, which initiated this important national program. The doorway, with full pilasters supporting a cushioned frieze and a finely molded pediment, is typical of the 1780s and 1790s. The house has extremely fine paneled interiors.

NK15 First Baptist Church Parish House (Masonic Hall)

1828. 44 Main St.

The Masons who built this doubled the form of a Federal three-bay house for their meeting hall. The simple top-lighted doors with flat bracketed caps are expressive of its institutional use.

NK16 St. Paul's Church

1847, Thomas A. Tefft. 1851, chapel. 1872, steeple. 55 Main St.

NK17 Allen Mason Thomas House, left foreground, with the Samuel Thomas House (NK18, next door) to the right

The campanile spire, round-arched windows, and arcaded trim of Wickford's little church are typical of Thomas Tefft's use of the round-arched forms of northern Italian architecture, probably by way of German sources, whence the style was self-consciously reintroduced in the early nineteenth century as the Rundbogenstil. Tefft was among the first American architects to employ the style, which became widespread in this country from the 1850s into the 1870s. The thinness of the framed detail and the flushboard siding (to simulate the flat planes of masonry or stucco) invoke American carpentry more than masonry. Tan paint for the siding would better suggest the original intended effect rather than the white main body of the church. The sanctuary contains fine examples of Victorian ecclesiastical brass fittings.

NK17 Allen Mason Thomas House

1825, Jabez Bullock, builder. 1927, restoration, Norman M. Isham. 56 Main St.

Having been heavily Victorianized, this Greek Revival house was restored by Isham for R. G. Clark, heir to the Singer sewing machine fortune. The Clarks had previously been involved in restoration work at Cooperstown, New York. Isham probably added or heightened the monitor and enhanced the parapet and front portico. (No other house on Main Street boasts any of these features.) The east porch, with its extraordinary shaped-plank chairs, is also his work, as, probably, is the connecting fence.

NK18 Samuel Thomas House

1795. c. 1870s and 1880s, additions. 64 Main St.

This house seems to have begun as a standard five-bay house with central chimney. Raised at some point with an addition off one end, it was Victorianized by a bay window and a new bracketed entrance into its side elevation. Finally, a Queen Anne stoop was added where the original entrance had been, probably to split the house into two units, as it is today. This is the single extant record on the street of the nadir of interest in colonial architecture, when these houses seemed shabby antiques to be updated as well as one could. But even by the time of the stoop alteration, the Colonial Revival was getting underway.

NK19 Wickford House (Alexander Huling House)

1769. 68 Main St.

Wickford House, originally a tavern, has an integral lean-to which is exceptionally extended. The door frame appears to be derived from that of the Benjamin Reynolds House (see entry below) and may be another Isham replication. The carved ornamentation in the entablature on the theme of the flowering chestnut is typical for early nineteenth-century embellishment; but the motif occurs with such frequency on Main Street that it is almost an emblem of the place. A crumbling barn with cupola provides an added dash of the picturesque.

NK20 The Greeneway

Early 1930s. Adjacent to St. Paul's Parish House

This landscaped path was given to St. Paul's Church in 1944 by the Greene family. Stones carved by Newport sculptor John Howard Benson name the rectors of St. Paul's. The path provides an arch of greenery as a transition from densely packed Main Street to the church with its open surround of fields and marshlands.

NK21 Old Narragansett Church
(Old St. Paul's Church)
1707 and later. 60 Church Ln.

Originally built in 1707 some five miles away on Shermantown Road, this famous church was moved in 1800 when Narragansett plantation society began to disperse and Wickford became the center of trade in the region. In its utter simplicity the building speaks eloquently of the influence of English classicism on the New England meeting house form at a very early date. The round-arched windows lie directly in the plane of the thin walls. Simple pilasters and a resounding segmental-arched pediment mark the double entry door. By placing the windows high on the wall and enlarging both the structure and the details, the builder has given monumentality to the basic five-bay house form. The interior reflects this same combination of classicism and New England tradition. The plan is that of a meeting house, with the pulpit against the long wall, but the ceiling is a plastered barrel vault. The gallery dates from 1723 and the pews were installed c. 1800. St. Paul's celebrates the summer season by moving its services from Thomas Tefft's "new" St. Paul's to its unheated ancient quarters.

NK22 Daniel Wall and George Bailey Houses
1802, 1809. 79 Main St.

This is an unusually fine Federal double house. The eastern portion of the house, although slightly later than the rest, retains its original door, set within a band of cross reeding. Fluted pilasters support a pediment which is broken through to make room for a relieving arch above the fanlight with a scrolled keystone.

NK23 Cyrus Northrup House
1803. 90 Main St.

NK24 Benjamin Reynolds House
1804. 94 Main St.

Built a year apart, these two Federal houses show the kind of variation that was possible within the standard framework of a pilastered and fanlighted entrance. That of the Cyrus Northrup House has unusual rustication inside the pilasters; the Reynolds House has a more conventional treatment enhanced by quite delicate reeding.

NK25 Ye Old Narragansett Bank
1805. 99 Main St.

The Benjamin Fowler House (1768) was incorporated as an ell into the rear of this building when the bank was established. At some point the original entrance appears to have received a Greek Revival replacement. It later became the home of Colonel Hunter C. White, the local historian and a good friend of Norman Isham. The fine turned fence is probably Isham's work.

NK26 Matthew Cooper House
1750. 109 Main St.

The reeded diamonds in this door entablature are so unusual that one suspects they may be Norman Isham's work.

NK27 Richard Barney House
1804. 115 Main St.

This is one of the only two houses on Main Street with quoining, another indication that Wickford tends to put limits on grandeur. Number 166, at the end of the street, is the other.

NK28 120 Main Street

1840s

This pristine example of mid-nineteenth-century Greek Revival infill on this street of eighteenth-century houses has a pedimented gable to the front rather than a characteristic flank gable front.

NK29 Stephen Heffernan and Israel Williams House

1795. 121 Main St.

The scroll-cut tracery of the side porches and Greek Revival door, added in the 1860s, are, for Wickford's Main Street houses, rare instances of Victorian incursion.

NK30 Daniel Fones House

1770. 126 Main St.

This unusually broad building, set with its gable end to the street, was built as a tavern and public hall. It is merely an enlarged version of the standard Wickford house of the period.

NK31 Aaron Peck House

1785. 143 Main St.

This small gambrel-roofed house with its narrow clapboards and heavily molded dormer abuts the former site of Wickford's fish shacks and shipyards. It has been enlarged by a narrow, sharply pitched lean-to off the rear—an awkward addition for a gambrel, although there is a certain charm in the directness with which the need for expansion was met.

NK32 John Updike House

1745. 19 Pleasant St.

Wickford's grandest eighteenth-century house was built by the grandson of the town's founder. It typifies the American translation into wood of the blocky five-bay house of England's middle class. When its elaborate detail is compared to the plainness of Old St. Paul's, one can see what great strides classicism had taken on this side of the Atlantic during the intervening forty years. In the 1920s the house was moved back on its site and remodeled by Norman Isham for Alonzo T. Cross, the inventor of the Cross pen and a prominent Rhode Is-

NK92 John Updike House

land industrialist. The effect of the paired trees and the tall turned fence with a central bow framing the elegant pedimented entrance clearly represent the romantic viewpoint of Isham's colonial revivalism. So does the trellising toward the front of the side elevation, meant perhaps for a frame of vines. The door surround has a copybook precision and elegance that suggest Isham's hand. He is definitely responsible for the ell and the sunroom/porch to the rear overlooking the harbor. Farther along, number 95, a simpler variant of the Updike House, contains a door with much the same qualities as that of its neighbor, together with a scale and sophistication against the straightforward builder's elevation that again points to Isham. And there is more of Isham's fencing.

NK33 E. E. Young House

1895. 71 Pleasant St.

After so much sober symmetry it is something of a relief to come upon the abandon of this superb Queen Anne double cottage, which is built right on the water. Its steep gambrels cross each other, swinging their sides out over the entrance porches on opposite sides of the house. Within the crisply folded and peaked front elevation, upper-story windows are organized in a tense asymmetry. At ground level a diverse array of architectural elements play with the horizontal molding which differentiates the base of the building from its exaggerated roof gables: between its dissimilar porches at either end, a Palladian motif jams into and through the molding; a pair of rectangular windows

NK33 E. E. Young House

hugs it; an oval window within a confined field sends out thrusts in four directions. In the 1980s, such freewheeling use of traditional motifs in playful counterpoint strongly influenced postmodernist design.

SOUTH OF MAIN STREET

NK34 North Kingstown Free Library

1974–1975, The Architects Collaborative. Boone St.

The library is thus far the finest of North Kingstown's twentieth-century public buildings. It is sited both to minimize its scale and to take advantage of the view over Academy Cove to downtown Wickford. By car, one approaches it on the uphill side, where large-scale brick, angled wings, and low-reaching shed roofs mask the building's size. A sunken outdoor reading area with stone tables and benches is located just outside the entrance. Inside, the space responds to the slope of the hill—an upstairs reading area opening as a balcony to the stack level, with more reading space below, open through generous plate glass walls to grass and trees. An auditorium in one of the wings can be sealed off from the rest of the building for after-hours events.

NK35 Hussey Bridge

1925, Clarence L. Hussey, engineer for the Rhode Island State Board of Public Roads. Boston Neck Rd. (Route 1A) over Wickford Cove.

The Hussey Bridge celebrates concrete as a plastic building material by fusing the spanning arches with the hexagonal concrete tie overhead and even integrating the corner lampposts as part of the sculptural structure. Cables from the arches support the roadway slab.

South of the bridge, Route 1A passes the North Kingstown Town Hall (Abel Peck, 1888) and, at 125 Boston Neck Road (corner of Beach Road), a magnificent Colonial Revival house and barn (c. 1910) set behind stone walls. The nondescript shingle house with colonialized porch partially infilled by recent renovation, on the opposite corner of Beach Road (143 Boston Neck Road), was Norman Isham's summer cottage.

NK36 Hamilton Harbour Condominiums
(Hamilton Web Company)

Mid-19th century and later. 40 Web Ave.

For nearly three hundred years this site on Bissell's Cove was used for manufacturing purposes. The present clapboard mill, hunkered down by the cove at the bottom of a bank, and village complex typify the small scale of southern Rhode Island's textile communities. The dramatic double clerestory topping the main building may be unique for Rhode Island; it is certainly the only extant example. A mill with single clerestory and an office were later added to the southern end. Across the road lies the mansard-towered brick weaving shed with brick pilasters extending right up into the gable. Following the closing of the mill in 1979, the entire complex was converted to housing, with greater sensitivity to the exterior of the build-

NK36 Hamilton Harbour Condominiums (Hamilton Web Company)

ing and its beautiful setting than is the case with many such conversions.

Lafayette

This community, built along the old market road to Wickford (which was famous for being ten rods wide), is typical of southern Rhode Island's highly paternalistic mill towns. Robert Rodman, a relative of the Hazard family of mill owners (see Peace Dale under South Kingstown), purchased the site in 1847 for woolen manufacture. His descendants owned and operated the company until 1962.

NK37 **Lafayette Mill**

1877–1878. 650 Ten Rod Rd.

NK38 **Mill Boardinghouse**

1878. 611 Ten Rod Rd.

In addition to windows that are unusually large for the period, the handsome brick mill has the unusual feature of a tower at each end of its short central hall. The towers have mansards so high as to be almost peaked and are topped with wood shingles and delicate wrought iron cresting. It was converted to commercial space in 1987.

Various types of mill housing flank Ten Rod Road, the most interesting of which is the now dilapidated boardinghouse directly across from the mill. Farther up the road lie the grand houses which Robert Rodman built for himself and his three children. Number 681, the Franklin Allen House (1882), has been severely altered for use as a nursing home.

NK39 **Condominiums** (Walter Rodman House)

c. 1879. 715 Ten Rod Rd.

Renovated in 1986 for condominiums, this is the most intact of the four Rodman houses. Walter, the artistic Rodman son, is said to have designed each interior doorway differently and kept a kiln in the basement for local pottery classes. Set behind a granite and iron fence on broad lawns and replete with porches, the house carries its striped mansard on deep sawn brackets, like a broad-brimmed hat.

NK40 **McKay's Front Porch** (Robert Rodman House)

1864. 731 Ten Rod Rd.

Despite alterations which include the removal of its dormers and chimneys and the intrusion of a fire stair, the fine porches on this house help it to cling to its dignity.

NK41 **George and Hortense Rodman Allen House**

c. 1865, 1882. 781 Ten Rod Rd.

This house received an extensive Queen Anne addition with Palladian window, ogee-capped square tower, and porte-cochere in 1882. Still

NK37 Lafayette Mill

NK44 Silas Casey Farm

owned by Rodman descendants, it is said to contain fine Queen Anne interiors.

NK42 Shady Lea Mill

Most structures c. 1870. 215-225 Shady Lea Rd.

Shady Lea was a small arm of the Rodman family's paternalistic North Kingstown textile operations. Charles Rodman's mansarded mansion (c. 1870) still stands at the entrance to this tiny textile community, and mill housing lines the lane. Tucked into a hollow beside the stream at the end of the road is the mill itself, some of which dates from the eighteenth century—a low and spare building, modestly towered, part brick, part clapboard. The diminutive scale of Shady Lea and the survival of many of the auxiliary mill buildings set it apart. Although small operations such as this were typical of rural Rhode Island from the late eighteenth century on, few survive. Though in dilapidated condition, this continues in manufacturing.

NK43 Gilbert Stuart Birthplace

1753, 1757. 1930, restoration, Norman M. Isham. 815 Gilbert Stuart Rd.

Gilbert Stuart, one of America's finest eighteenth-century portrait painters, was born in this gambrel-roofed mill in 1755. Although milling had taken place on this site since the seventeenth century, Stuart's father had emigrated from Scotland to establish New England's first snuff mill. No doubt Norman Isham was attracted to its restoration by the beauty of its decaying antiquity, made more picturesque by its siting in a deep hollow. Surely prettier than its reality, this conjunction of the water-powered mill and the miller's quarters nevertheless makes vivid the earliest stage of rural industry.

NK44 Silas Casey Farm

c. 1750. 2325 Boston Neck Rd. (now owned by the Society for the Preservation of New England Antiquities)

Built as a summer farm by Daniel Coggeshall of Newport, this property is especially significant as one of the last survivors of Rhode Island's eighteenth-century plantation economy. Set on a hilltop overlooking stone-walled fields, the house, with gable-on-hip roof, is typical of contemporary Newport buildings. During the nineteenth century the farm became the summer home of the Caseys, Coggeshall's descendants. Silas Casey was a general in the Civil War, and his son Thomas Lincoln Casey was the engineer for the completion of the Washington Monument; the Library of Congress; the State, War, and Navy Building; and numerous other commissions in Washington and in New York City. Memorabilia from Thomas Lincoln Casey's building career make the house doubly interesting to the architecturally inclined traveler. A range of mostly nineteenth-century barns and the use of the property as an operating farm invoke, as nowhere else, the plantation life of so much of coastal South Kingstown. As early as the 1880s the Caseys, who spent much of their time in other parts of the country, began to worry about the farm's preservation. In 1955 Thomas's son Edward gave the house and 330 acres to the Society for the Preservation of New England Antiquities.

Narragansett (NA)

Narragansett is long and narrow, with water borders. Along its east flank the shore extends beyond the entrance to the West Channel of Narragansett Bay to face the open ocean. Off its narrow southern end is more open ocean. Up its west flank, ponds and the ponding of the Pettaquamscutt River almost disengage this strip of Rhode Island from the rest. Europeans originally settled it as part of the Pettaquamscutt Purchase, which included the shoreline and an area extending well inland in what is now much of North Kingstown, South Kingstown, all of Narragansett, and land to the south into Charlestown. It included some of the richest agricultural land in New England, the economic basis for the social, hunting, and horse-breeding culture associated with the so-called Narragansett Planters. The Narragansett Pacer, bred in these plantations, was a prized export.

For over two centuries a ferry that sailed from a landing point near the end of South Ferry Road to Jamestown and thence, in conjunction with another ferry, to Newport, was a prime connection between the rural life of southern Rhode Island and the world of sophistication in Newport, especially significant during the colonial and early national period. Not only was it a conduit for agricultural goods and high-style finery between the mainland and the island, but also a shuttle for Newporters who owned country property on the mainland and for mainlanders who operated outlets in Newport, conducted business there, or sought the glitter of its social life.

Around 1780, John Robinson, son of William, who owned the largest of the plantations in Narragansett, built a stone pier for shipping a little south of the mouth of the channel into the bay. It was the start of Narragansett Pier, which eventually became one of the premier seaside resorts in New England. A chain of beaches extending from Narragansett Pier, together with those south and west to Watch Hill on the Connecticut border, are among the best on the East Coast. To the south, flanking the entrance to Point Judith Pond, are the twin villages of Jerusalem and Galilee, fishing communities where the local fleet parades for its annual blessing. North of the Pier are clustered marine research facilities, among them the oceanographic campus of the University of Rhode Island.

NA1 Rowland Robinson House

1710. 1755, remodeling and enlargement, Rowland Robinson II. c. 1928, restoration and addition of rear servants' ell, Norman M. Isham. 450 Old Boston Neck Rd.

Few Narragansett Planter houses exist. This is the finest and very much the largest. It must be the largest eighteenth-century central-chimney gambrel house of its size in the state, testing the limits to which this inherently cozy format can be inflated. Indeed, its impressive scale combines with austerity of treatment to give it a rather barnlike quality, which is enhanced by its uncompromising situation in a skewed position at the center of a handsomely stone-walled field. Large as it is, however, at its grandest it was approximately 110 feet long, or nearly twice the 60-foot length that now exists; the extension, added by Rowland Robinson II, grandson of the original owner, was given over to service facilities and slave quarters. The grand door with its broken-segmental-arch pediment, an echo of that on the second-story balcony of the Colony House in Newport, is the sole adornment of this severity. It was no doubt added during the late 1920s restoration by Norman Isham, who knew the Colony House intimately after his work there in 1917.

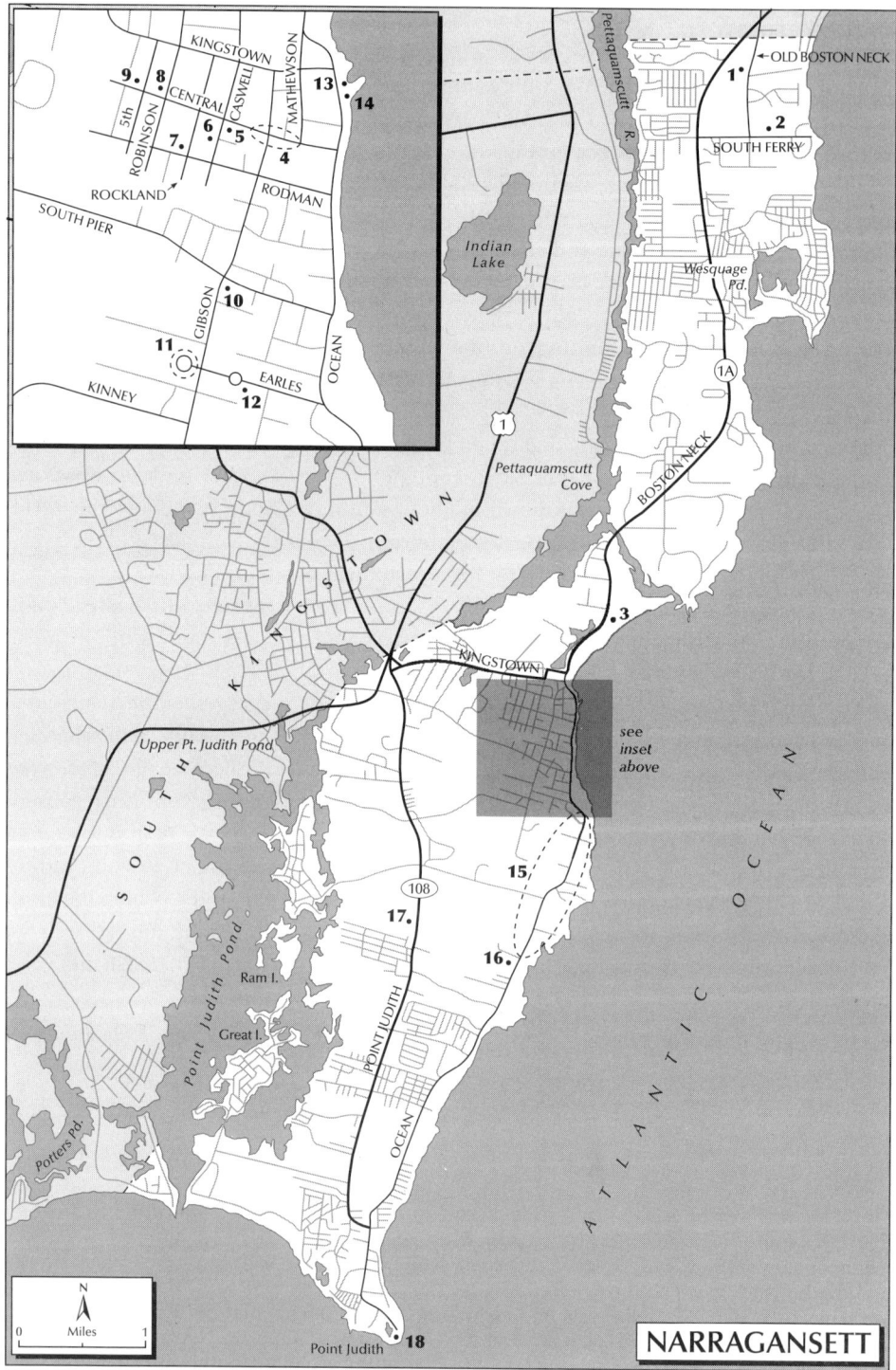

NARRAGANSETT

NA1 Roland Robinson House

NA2 **South Ferry Congregational Church**
(Narragansett Congregational Church)
1851, Thomas A. Tefft. 170 South Ferry Rd.

Of all the wooden Gothic country churches from the mid-nineteenth century in Rhode Island, this and Richard Upjohn's Church of the Holy Cross in Middletown (MI1) are probably the finest. As in all Thomas Tefft's work, this is characterized by compelling shape, austere detailing, and rightness of placement and scale. Consider merely the shaping of the three-story tower, where the door level is a cube, but the corners of the second story above are inset so that the cube below becomes four planes, from which the third story slides like an object past the open flaps of a cardboard box. Note, too, the straightforward but graceful manner in which the flared hipped roof at the eaves accommodates the octagon steeple, with tall, arched, louvered windows mediating the transition. Here Tefft employs his favorite round-arched "Lombard" style, which he was one of the first American architects to adapt from progressive German theory of the time. In fact, the steeple shape derives from German precedent. Moldings around the arched windows are utterly simple and broad. They project from the plane of the wall, contrasting with the texture of hexagonal shingling, which decorates without "decoration." This is a spectacle for an early morning or a late evening visit, when raking light enhances the hexagonal patterning. A charming watercolor rendering from Tefft's office at Brown University shows penciled experimentation in the margin as Tefft weighed pointed shingles as an alternate to the pattern which he finally selected. It also shows a dovecote on the ridge of the gable at the pulpit end of the church and indicates that his color preference was probably soft gray rather than white, a tint more in keeping with mid-nineteenth-century preferences.

The original congregation left the church for nearby Saunderstown in 1908, taking the Victorian pulpit furniture with it. The building was finally sold to the Roman Catholic Diocese of Providence, which planned to move it; a group of former members of the congregation purchased it with plans to preserve it, but it remained vacant until the late 1920s, when a successor organization realized restoration of the then deteriorated building. Its simple plaster interior with board wainscoting retains the original pews, with re-created fittings in front. Now owned by the University of Rhode Island, which has located its campus for oceanography a little farther along the road, the church now houses occasional meetings and weddings.

Narragansett Pier

Bathing came to Narragansett Pier in the 1840s, when local farmers took in boarders. The first hotel appeared in 1856. Between 1880 and 1890 the village boasted almost fifteen of them—the most impressive concentration in Rhode Island. Large wooden structures, most of them three and one-half stories, they were rather plain for the most part, but typically enlivened by extensive porches (ideally two stories), with sufficient mansards, turrets, dormers, and flags to present a festive image. After 1872, a railroad spur from the main New York–Boston line, built by the Hazard family, woolen manufacturers in nearby Peace Dale (see under South Kingstown), swelled the crowds. Substantial cottages were built on the streets behind the hotels, especially after 1870. The most substantial ranged along the cliffs overlooking the sea south of the village center on property which was also originally owned by Hazards. They opened a portion of their vast domain to real estate development, with Ocean Road as a carriage drive from the pier to connect with the road south to Point Judith. Grandest of all the houses in the Victorian heyday of the pier's popularity, the estates of William Sprague and John Peace Hazard presented a study in contrasts. Sprague's Canonchet Farm (c. 1870–

1880, William Walker), destroyed by fire in 1909, was as extroverted in appearance as its owner's lifestyle, both dependent on a grandson's inheritance from one of the huge nineteenth-century Rhode Island textile fortunes. It was a rambling, mansarded Victorian mansion, cornered by four heavily ornamented mansarded towers and featuring a telescoped tier of swelling conservatories topped by a belvedere at the end of a grand porch. Its owner eventually faced the sheriff with a hunting gun when the bubble burst. And then, embodying a very different lifestyle, there was (still is) Hazard's Castle, with its ascetic tower for mystical meditation (see entry, below).

Although Narragansett Pier boasted a few other very large houses by the end of the nineteenth century (now all destroyed), it never managed, though it tried, to attract residents who would build palaces in the quantity and magnitude of Newport's. Rather, social life here centered in the hotels and the sort of substantial cottages which still command the scene, as well as McKim, Mead and White's Casino, at least until 1905, when most of this burned to the ground. Casino life continued in makeshift quarters until the opening in the 1920s of the Dunes Club, the largest of all the exclusive beach clubs in the state. The period between the wars saw the scuttling of the last of the once proud armada of grand wooden hotels. Not one exists today. By the end of World War II, the center of Narragansett Pier had declined into semi-dilapidation. "Rescue" occurred in the early 1970s with the wholesale demolition of eleven blocks of the old town center. A bland board-sided and gabled style of commercial building set in parking lots characterizes the new mall-like center—which curiously seems like no "center" at all.

NA3 Dunes Club Gatehouse and Cottages

1928–1929, Kenneth Murchison. 1939–1940, rebuilding of club after 1938 hurricane, Purves, Cope and Stewart. 137 Boston Neck Rd.

Where Boston Neck Road curves in toward the beach at Narragansett Pier, a stuccoed gatehouse, featuring a stubby turret at one end of a rectangular block, is all that remains of Kenneth Murchison's original Dunes Club. Murchison, an architect best known for the Moderne Beaux-Arts Apartments in New York, which he designed with Raymond Hood, was part of the Narragansett summer scene. For the Dunes Club he concocted a sprawling stage set in creamy stucco with a variegated roof which hybridized Provençal château turrets with Spanish haciendas, splashed with patches of hand-wrought brick and colored decorative tile. This pretty dream was lost to the 1938 hurricane, whereupon, much to Murchison's vexation, another firm received the commission for its present replacement. This is a more severe design in a plain wooden vernacular made monumental and semimodernist. Even more extensive than the original, it boasts a grand array of leisure facilities: glass-fronted dining and ball room, decks, boardwalks, horseshoe enclaves of luxury cabanas—and reassures by appearing to be much more hurricane proof than Murchison's sandcastle. Fortunately, what may be the best of Murchison's designs for the Dunes Club complex still remains. At the northern end of the public section of the beach (and visible from it) are two houses, privately owned but built in conjunction with the club, which show his Mediterranean forms abstracted. Here Mission Revival massing of turrets and winding stairs creates cubes, cylinders, and spirals organized in sculptural compositions, with dramatic roof silhouettes pierced by two tall chimneys.

NA3 Cottage, Dunes Club

NA4 Central Street Houses

1880s, 1890s. 40, 45, 49, 50, and 60 Central St.

Central Street was platted in 1867, but most building occurred between 1880 and 1900, so the street provides an overview of the kinds of cottages built on good-sized lots which characterized upper middle-class housing during the height of Narragansett Pier's popularity as a summer resort. Numbers 40, 45, and 49, which represent the first phase of cottage building at the Pier, were all built for Charles E. Boon, a Providence dealer in drugs, dyestuffs, paints, and other chemicals, who sold this business to go into real estate in Providence and Narragansett Pier. Number 40, Idlewild (1869), a one-and-one-half-story mansarded villa with curved mansard surmounted on three sides by a balustrade, lost its original porch with jigsaw ornament to a Tuscan-columned replacement. It was among the first summer villas specifically designed as such. Number 50, the Brander Matthews House, known as Shingle-nook (c. 1887, George A. Freeman, Jr.), is a plain, shingled, hip-roofed house with a generous veranda on two sides supported on turned, cigar-shaped posts. Matthews, a professor of literature at Columbia University who was also a playwright and freelance writer, wrote an article on summer life at the Pier for *Harper's Weekly* in 1906. The best of the cottages on the street is number 60, the Emma B. Carver House, formerly named Kabyun, now Sonnenschein (1884–1885), an eclectic mix of Stick Style framing, chalet jigsaw ornament, and typical Queen Anne wall cladding.

NA5–NA7 Summer Chapels

The "summer chapel," an informal type of church building designed for a vacation community, is frequently found in Rhode Island resorts, especially in upper-class communities, where summer weddings were (and are) a special part of the season's festivities. Uniquely, the Pier has three fine examples within two blocks of one another. Whereas the Portsmouth-Middletown chapels are mid- to late nineteenth century, those at the Pier continue the story from the late nineteenth into the early twentieth century.

NA5 St. Peter's by the Sea

1870, Edwin L. Howland. 1879, porch. 1889, tower. 72 Central Ave.

The first St. Peter's Episcopal Church, a wood-frame structure (1869), was almost immediately lost to a gale. Its more substantial replacement on the same site suggests a plain English country church in random masonry, an L-shaped mass with high gabled nave butted by low roofs over the aisles. The rugged, vernacular plainness and expressive roof silhouette, complemented by the somewhat later simple wooden porch with a massive and stubby tower set into the L, are appropriate for a summer church. Legend has it that Stanford White designed the tower, but this is probably not the case. On the interior are nice structural woodwork and some Tiffany windows. Most remarkable, however, is the chancel, with all its original furniture and brass fittings, and plaster walls stenciled in dull green, tan, and gold.

NA6 Residence (Baptist Church)

1889–1890. 101 Caswell St.

Just around the corner from St. Peter's on Caswell Street is a former Baptist church (now a residence), which retains an extravagantly bracketed entrance porch, topped by a sculptural tower. Bulbous shingled elements call up the shapes and textures of basketry. A bell shape pops from the slope of the porch roof, making a delightful transition to a cupola for the bell, all topped by a steeple.

NA7 St. Thomas More Roman Catholic Church (St. Philomene's Roman Catholic Church)

1908, Murphy, Hindle and Wright. Rockland and Rodman sts.

Recent painting has cost St. Thomas More some sense of its original rustic shingled texture, and the interior has been redone. But it is exceptional for the way in which picturesque cosiness is embodied in large abstract forms which dignify what might otherwise seem overly sentimental. A big shingled gable is crossed by another with triple-arched window. This is flanked by tiny dormers. Below, the repetitive slants of simple buttressing rhythmically frame paired windows and enliven the wall. At the corner, steep stairs climb to a gabled porch entrance into an austere bell tower with Lombard arching under the capping hip, giving monumental punctuation to the sheltering shapes stretched out below it.

NA11 Sherry Cottages

NA8 Kate Lane Richardson House
(Yellow Patch)

c. 1916, George F. Hall. 115 Central St.

A real "cottage" at last, straight out of the illustrations in Kate Greenaway's books! For all its overt fairy-tale prettiness, this is as disciplined an example as can be found in Rhode Island of the sort of artistic "Cotswold" cottage featured in the English publication *International Studio*. Plain board shutters with abstract flower cutouts frame multipaned sash, each of the upstairs windows with a boat-shaped window box waiting for petunias. Asphalt thatch bonnets the windows and eaves, and gives way to little eye-shaped windows lighting the attic on either side. (Some window alterations have occurred downstairs.)

NA9 Narragansett Town Hall
(Fifth Avenue School)

1924, J. Winfield Church. Remodeling as town hall by Raymond W. Schwab. 25 5th Ave. (at Central St.)

Another surprise! This giant brick gable broadside to the street, topped by a chimneyed endpiece, contains three stories of windows grouped around a three-arched opening. The date is astonishing. One would guess it to be a postmodernist adaptation of a motif made famous by Robert Venturi. No other Rhode Island school resembles it. How did its architect come to this solution?

NA10 Gibson Court Condominiums
(Charles H. Pope House, Gardencourt)

c. 1888, William Gibbons Preston c. 1928, garage and caretaker's cottage. 10 Gibson Ave.

What was planned as a cottage compound ended up as a single-family house which has been considerably altered and surrounded by other shingled condominium units—thereby finally becoming something like what it was originally intended to be. Immediately to the left of the entrance, the later shingled garage, with attached caretaker's house, particularly well composed, retains more of its original character than the main house.

NA11 Sherry Cottages

c. 1888. Gibson Ave. (end of Earles Court Rd.)

Earles Court Road terminates in a grassy green, the heart of a symmetrical grouping of Shingle Style cottages, all differently composed, built for Louis Sherry. Sherry, the famous New York restaurateur and owner of Delmonico's, was also the caterer for the Narragansett Casino. He developed this compound as a rental property. Originally there were six houses. Two front on Gibson Avenue and flank the open axis into the court (on the corners of Westmoreland and

NA15 Seaside Colony, Stonelea

Woodward), each with a turret to mark the front corners of the property. Two more behind these also remain. They are set more deeply into the property at right angles to the other two (facing Westmoreland and Woodward respectively). Behind these, two more at the back of the property (no longer extant), again frontally oriented and still more deeply set into the property, filled in the steplike arc of the houses around the green. And a little behind the center of the green a symmetrical structure with tall pyramidal roof flanked by two more turrets completed the composition. This housed a community restaurant, with some single guest rooms. The effect was relaxed and picturesque, yet sophisticated and ordered. Today it is difficult to discern the arrangement in the summer because of the luxuriant planting.

NA12 Earles Court Houses and Water Tower

1886–1887, J. J. Jarding and Constable Brothers. Earles Court Rd.

It was the New York lawyer Edward Earle who encouraged Charles Pope and Louis Sherry to undertake similar enterprises. He envisioned a road with piered entrances at either end, along which he planned a double row of Queen Anne houses with a fantastic water tower on an oval island at the midpoint of the road. Only two houses were built: number 36, Edward Earle's own house, much altered, and number 46, immediately beyond the tower to the right, a gabled frame house with conical-roofed polygonal tower and veranda. (Houses at numbers 37 and 50 are visually related and contemporary but not part of the Earles development.) Most sadly, only the circular stone base of the water tower remains. It was once topped by a shingled "castle" to shroud the water tank, which rose from a bracketed balcony. The castle consisted of a castellated polygonal enclosure fronted by a conical turret. Attached to the turret, like a bowsprit, was a copper griffin with wings spread around the turret.

NA13 The Towers, Narragansett Pier Casino

1883–1886, McKim, Mead and White. 36 Ocean Rd.

McKim, Mead and White's original casino in Rhode Island (NE141) was a center for Newport life; Narragansett Pier had to follow suit. Of the elaborate complex, only The Towers, which have become the Chateauesque emblem of the community, remain. Stubby flanking turrets with conical caps, they support a stone arch over the road which contains a loggia, topped by a steep gable. The single row of dormers now existing was once doubled, and a tower cupola rose from the center of the ridge post, so that the effect was far more picturesque. What remains is merely the most incombustible piece of what was once a mostly long, thin, shingled building. A square tower with a mansard cap was attached to the ocean end of The Towers. A long tail, paralleling Ocean Road, was attached to the opposite end. This was folded once at right angles along the street, and folded again at an oblique to adjust to the jog. At the first fold were private dining rooms, then a long billiard room fronted by a deep porch overlooking the carriage traffic on Ocean Road, with a garden café below along the sidewalk. Next in line, the circular Palm Room and a hexagonal room made the hinge for the second fold, accommodating lounging, receptions, or balls according to the occasion. Then, finally, along the oblique of the jog, was a playhouse. After the 1905 fire the casino moved to improvised quarters. But it was never the same. Eventually the Dunes Club filled the void—but with a generational change in the kind of recreation which such a facility was supposed to provide. Ironically, the destruction of the original Dunes Club left the same emblematic relic of past glamour.

NA14 Coast Guard House (U.S. Lifesaving Station)

1888, McKim, Mead and White. 40 Ocean Rd.

Through the arch, and adjacent to it, is McKim, Mead and White's Romanesque Revival lifesaving station, a severe, beautifully crafted, and compact granite box with the Rhode Island anchor emblazoned in low relief over the entrance as its only embellishment. Its grand austerity is all but swamped by unsympathetic appendages for its present use as a restaurant.

NA15 Seaside Colony

Most houses 1885–1910. Bluffs along Ocean Rd.

Off season is the only time to take in this group of substantial summer houses, many of which are now year-round residences—although the narrowness of Ocean Road and the press of traffic make even off-season viewing difficult. What is visible, tight against the road, are a row of gatehouses, gate lodges, and stables which provide a period overview of these architectural types, despite their modification to a great extent to residential use. The most interesting of these are probably the gatehouses for Dunsmere and Suwanee Villa, at 560 and 380 Ocean Road, respectively, both very large houses which, like most others of their size, have been sacrificed for a greater number of somewhat more modest replacements. Dunsmere was originally the summer estate of Rupert Dun, one of the founding partners of the New York financial firm of Dun and Bradstreet. This random masonry gate and gate lodge (1895) are as pretentious as they get in Rhode Island. The lodge makes an arc of turrets and gables (much larger than it appears from the road), from which a thin arch of rough boulders leaps over the approach road to another tower in a structure which appears ominously ready for collapse. The ironwork of the gates is also interesting. The equivalent gate lodge for Suwanee Villa, the property of David Stevenson, another New Yorker (c. 1889, James H. Taft), a handsome mix of granite and shingle rising to a stepped gable for the carriage barn, butts a fat turret, its shingled upper story extended by gables and bays as living quarters.

One of the public streets, Newton Avenue, approaches the cliff edge, a vantage point for an oblique view of typical cottages, especially to the south. In the foreground is McKim, Mead and White's Stonelea, designed for George V. Cresson (1883–1884, altered 1940s), at 55 Newton Avenue, again granite below, shingles above. A stone-piered porch projecting across the ocean front and the second story with an exceptionally tall double-pitched roof rising above it, are broadside to the view. Changes to the roof, originally quite exotic, were the most pronounced of exterior alterations made in the 1940s. Three tall dormers in a double row once looked to the sea, with bell-shaped caps rising to mini-pinnacles, and two mini-dormers above these. The side elevation at the third-floor level under the eaves was scooped out for a balcony with elaborately turned supports and a sunflower frieze. Another bell-shaped dormer projected from the roof of the rear ell, which was repeated in the capping for a clock tower on the barn (now another residence). All of these eccentric and exotic shapes, which constitute a secondary theme in much of McKim, Mead and White's early work, have been tamed to prosaic "colonial." But the extraordinary altitude of the roof, rising above the forceful containment of the granite base, remains.

In the distance, north to south (near to far), are a gambrel Colonial Revival house all but concealed by a modern addition, and more sizable Shingle Style houses: Turnberry (1910–1911), for Emma R. Sinnickson of Philadelphia; Stone Croft (1890–1891, William Gibbons Preston), for Francis H. Dewey, a lawyer from Worcester, Massachusetts; Over Cliff (1884–1885), for Charles H. Pope, a cotton merchant from Providence who was active in trading Pier real estate and also built Gardencourt (see entry, above); finally, barely visible around the bend, Fair Lawn, the Jeffrey Davis–Charles H. Pope house, and Indian Rock (c. 1880–1890), for the Reverend William Babcock.

NA16 Joseph Peace Hazard House
(Hazard's Castle, Seaside Farm)

1846–1849, 1882–1884. 6343 Ocean Rd. (now a retreat house maintained by the Roman Catholic diocese of Providence)

An eccentric of a type more British than American, Joseph Peace Hazard was the odd man out of his generation in the powerful Hazard family (see Peace Dale under South Kingstown). Large as it was, Seaside Farm was a small piece of his massive holdings along the cliff. He was a principal force in developing Ocean Road and the grand oceanfront properties between it and the sea, with the proviso that the area never go into commercial use. But aside from farming, his personal passion was mystical

NA17 Kinney's Casino

metaphysics, in contrast to the interests of his intellectual brother in Peace Dale, Rowland Gibson Hazard, whose more worldly speculation inclined him toward economics and social theory. The house he built of random masonry is a sprawling, nominally Neo-Gothic mansion of gables, bays, and tower organized with an eye for picturesque silhouette, sober and weighty rather than ostentatious and restless. The culmination of his building pursuits occurred only in 1887, however, when he added a massive square, 105-foot-high stone tower with battered walls rising to a crenellated cap—this to bring him closer to the spirit world. (An isolated cottage at 100 Gibson Avenue known as Druidsdream is the other architectural result of Hazard's cosmic speculations. Druids appeared to him in a dream and commanded him to build a stone house where it still stands on what was once part of his farm.)

Magnificent specimen planting completes the mood, making this among the outstanding examples of mid-nineteenth-century romantic vision in the state. Not even the addition of brick dormitory blocks (in the tawny yellow which Catholic institutions seem to favor) dispels the effect. (Visitors should, however, be respectful of the purpose of the place.)

NA17 Kinney's Casino

1899, Clarke and Spaulding. 505 Point Judith Rd. (near intersection with Windemere and Farm rds.)

The casino appears to be a super-sized hip-roofed bungalow, raised on stilts, with a fireplace chimney at one end. In fact, it is also known as Kinney's Bungalow and is truer to the Indian source for the bungalow than what customarily passes as such in America, especially when its porch-sized band of windows running around all four sides of the building are covered by green louvered shutters. Just as the Newport Casino was founded in a pique, when a member of the exclusive Reading Club, ousted for ungentlemanly behavior, retaliated by establishing his own club, Francis Kinney took the same action after a falling out with the nearby Windemere Country Club. A field across the highway was a heart of South County polo, and his casino became the center for after-polo parties. Architecturally, it is unusual as a design and remarkable for the severe manner in which the structural supports mark off the daring treatment of the well-proportioned windows as generous squares. The polo field is now condo-ized as—what else?—Polo Estates.

NA18 Point Judith Lighthouse and Coast Guard Station

1857. 1470 Ocean Rd. (end of Ocean Rd.)

Point Judith is the place where the mostly east-facing coast of the biggest chunk of the state abruptly shifts to face south. Its lighthouse was the third established in the state, after Beavertail and the Watch Hill, and the existing light in this is the third on the point, preceded by towers completed in 1810 and 1815. These exemplify the development of typical lighthouse structures. The first, of wood, lasted less than five years before a gale took it down. The second, of rubble coated with cement, lasted forty-two years. The present tower of cut granite blocks fitted to its polygonal and tapering shape represented standard building technology for lighthouses by the mid-nineteenth century, although the first such tower in the state was built as early as 1823, at the north end of Goat Island in Newport Harbor. Point Judith is 51 feet high, with a 24-foot spread at the base tapering to 13 feet and a typical cast iron lantern cupola surrounded by a metal-railed balcony at the top. The keeper's dwelling, originally attached by a passageway to the tower, was torn down after automation in 1954. A scattering of utilitarian Coast Guard structures completes the photographer's composition.

South Kingstown (SK)

South Kingstown (as for North Kingstown, the final syllable should not be slurred) is shaped like a skewed axehead with beaches and open ocean at its cutting edge, but with water marking its eastern boundary as well, along Point Judith Pond to the south and the Pettaquamscutt River to the north. (Until 1901, in fact, when Narragansett became an independent town, South Kingstown had by far the largest stretch of shoreline in Rhode Island.) Fronting on its watery eastern and southern edge, and extending back two to four miles, in an arc from the northern boundary of South Kingstown into Charlestown, is some of the richest agricultural land in New England. It encouraged early colonization, and, in 1657–1658, the Pettaquamscutt Purchase from the native Narragansett tribe. The land along the water became meadows, and behind this sweeping arc (in the triangle roughly bounded by present-day Ministerial and Tuckertown roads with Route 1) are the Matunuck Hills, a choppy wooded area dotted with small ponds marking the melting edge of the glacier. Inland from these colonial land holdings, Worden Pond, which is the largest natural lake in Rhode Island, serves as the catch-basin to the Great Swamp, by far the most extensive swamp in the state and now state-protected. There, in 1675, after colonist victories over the Pequot and Wampanoag tribes to the north and east, the Narragansett were driven. Although neutral in skirmishes with the other tribes, the Narragansett refused to relinquish Pequot and Wampanoag refugees, which was excuse enough for the colonists to turn against their former allies. In the Battle of the Great Swamp, the Narragansett leader, King Philip, was killed, and further native resistance to colonial settlement collapsed. A monument in the heart of the Great Swamp off South County Trail (Route 2) commemorates the colonial usurpation.

With the Pettaquamscutt Purchase, the area became known as "the King's province," later Kings County, and, eventually, Kingstown (North and South). Each of the Pettaquamscutt investors received a huge tract of land extending back from the water. Several began to aggrandize on their already large territories, above all Robert Hazard, who owned a farm at Little Rest (now the village of Kingston). Before his death in 1718, Hazard was the largest landowner in the area. Building on their inheritance, subsequent generations made the Hazard family the largest landholders in colonial New England. They were, however, only one of a group of large landowners who, from around 1660, developed the remarkable plantation culture of the so-called Narragansett Planters. By the mid-eighteenth century, African and American Indian slaves amounted to almost half the white population of what is now South Kingstown. Fertile meadowlands supported corn and cattle as the principal crops. The ocean moderated harsh New England winters and provided transport for agricultural products down the East Coast and to the West Indies. Sheep grazed in higher pastures, especially in the Matunuck Hills, making the area the most important area for sheep in New England through most of the nineteenth century.

Around the time of the Revolution, the glory days of the system collapsed in a combination of setbacks. Trade changed because of agricultural competition from other areas as new frontiers opened up closer to what had been markets for the Narragansett Planters. The Revolution also altered trade patterns with the Indies. Land holdings were divided by successive generations, and the supply of labor was disrupted by antislavery legislation in the state (supported especially by Quakers who had been attracted to Rhode Island by Roger Williams's doctrine of religious tolerance, prominent among them members of the Hazard family, which was traditionally Quaker). By the time of the Revolution, the wanton export of breeding stock for quick profit had even decimated the trade in the renowned Narragansett Pacer. Prosperous, large-sized farms continued where the Narrangansett Planters had reigned; but of a more modest sort, and without their global reach.

Inevitably, some of the leading planter families turned to industry: the Robinsons in Wakefield; the Rodmans in Rocky Brook; most spectacularly, the Hazards in Peace Dale (see under these villages). These three villages on the Saugatucket River, closely clustered but clearly defined into the twentieth century, have more recently tended to merge as the principal urban area in South Kingstown. North of this cluster is the village of Kingston, rescued from a period of decline from its early preeminence in South Kingstown's professional, political, and cultural life, when it became the home of the state university in 1883. The rest of South Kingstown to the north and west long remained mostly in small farming, with occasional village industry which generally collapsed in the late nineteenth century. Granite quarrying, centered in Westerly and Charlestown, extended into South Kingstown and, in fact, for reasons which are obscure, South Kingstown probably boasts more stone buildings from the eighteenth and nineteenth centuries than any other Rhode Island town. As for the former realm of the Narragansett Planters, a flurry of demand for gentlemen's estates during the late nineteenth century revived the legends of old plantation life. Only temporarily, however: Vacationers and, increasingly, year-round residents have swarmed over their holdings, taking advantage of industries which have moved into the area, of relatively easy connections to Boston and New York and, of course, some of the finest beaches and sailing on the East Coast. They have planted their lawns and ornamental shrubs where the great agricultural domains once existed. Where large farms still exist, the principal crop, in this rich, relatively stoneless soil, tends to be turf.

SK1 Shadow Farm (Samuel A. Strang–Job J. Welsh House)

1884, Douglas Smyth. By 1885, two outbuildings and greenhouse. 1904, enlargement and renovation. 1986–1987, condominiums with additional buildings, Newport Collaborative. 80 Shadow Farm Way

This site was originally a small portion of a 3,000-acre property accumulated by Rowland Robinson before his death in 1716. His son William built a farmhouse on what was then known as Kit's Pond as the center of one of the major South County colonial plantations. It remained intact until descendants sold 60 acres

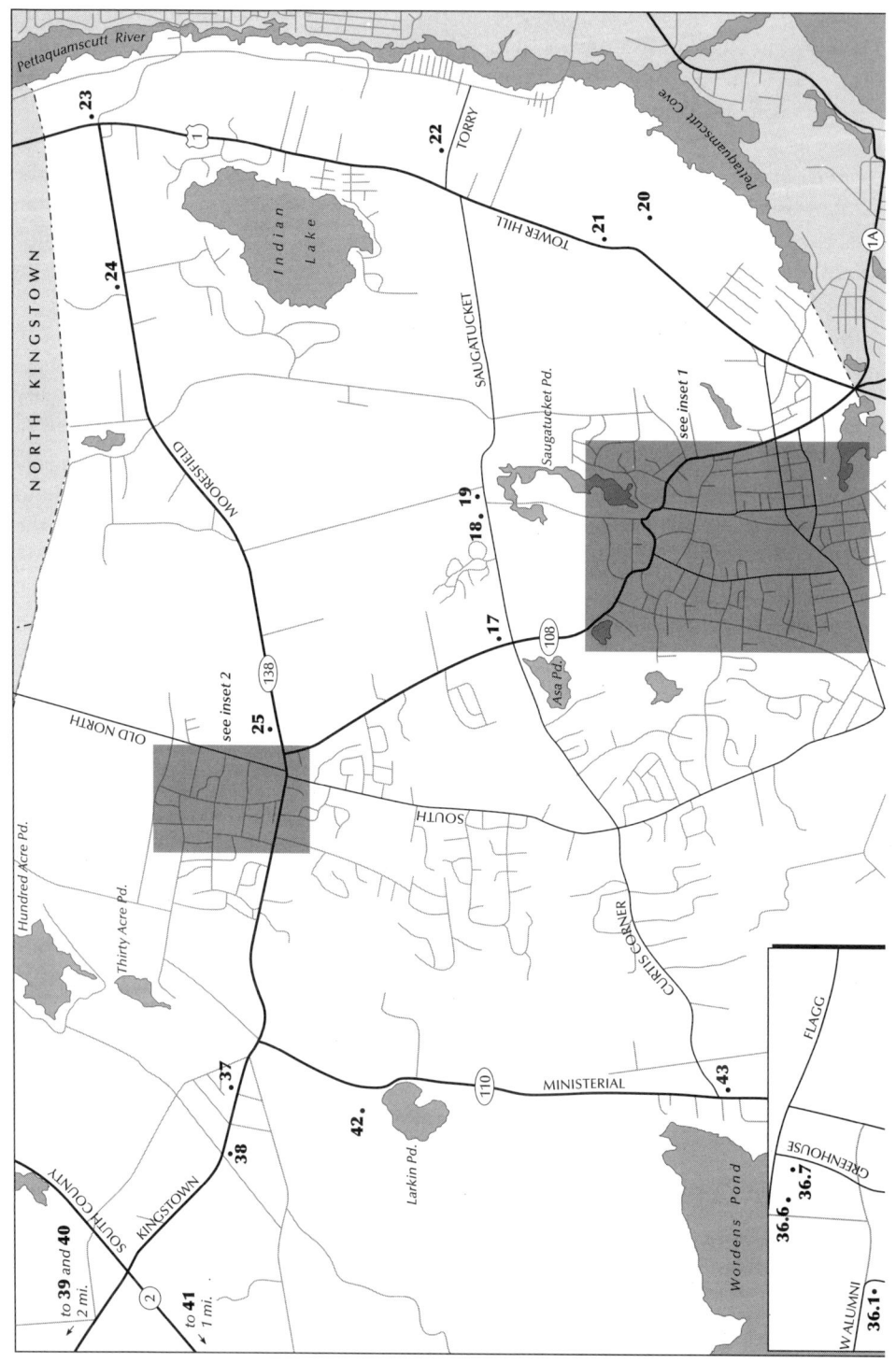

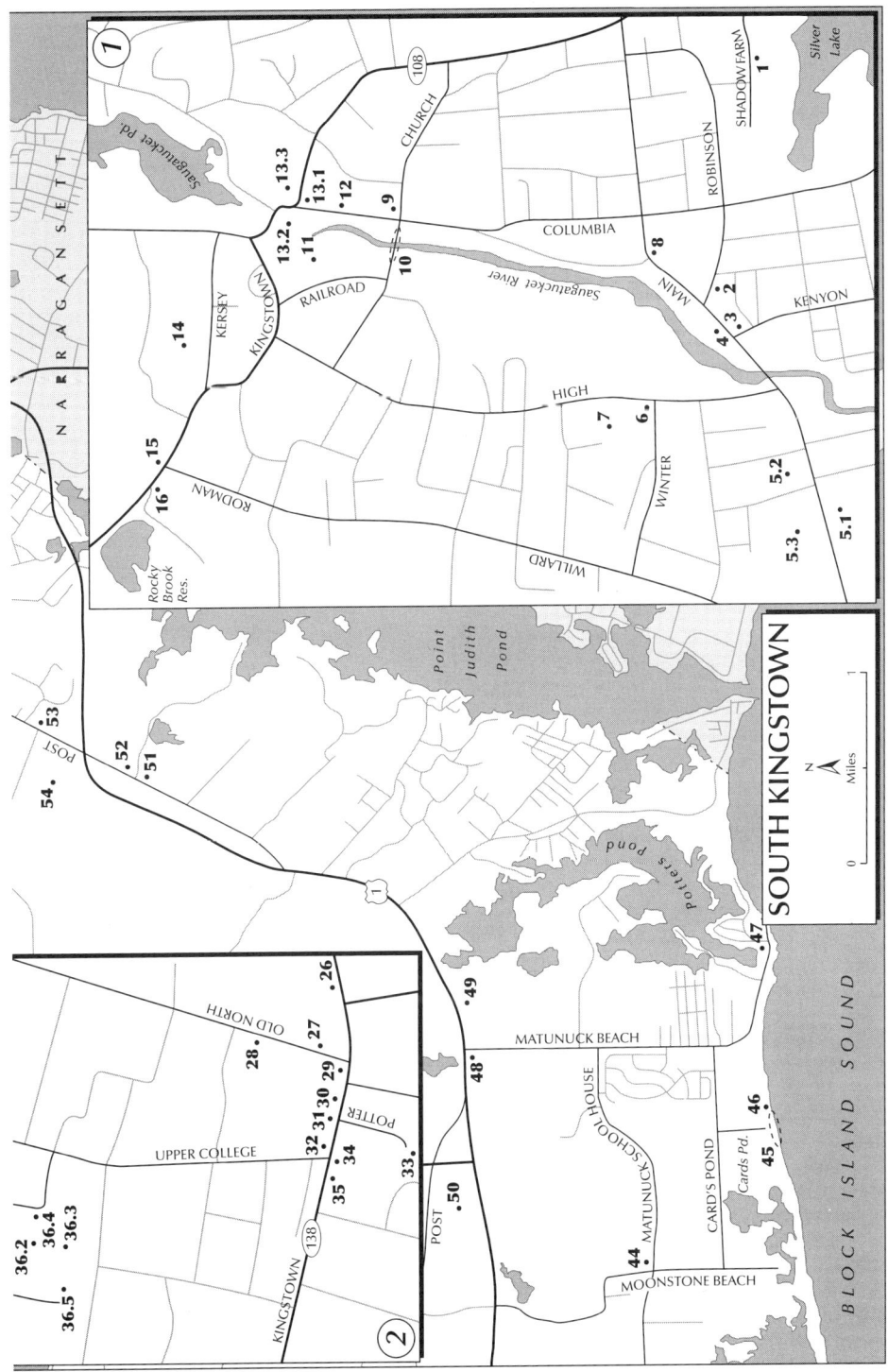

on the north bank to the New Yorker Samuel Strang, who made it a summer residence and gentleman's farm. Three years after Strang's death the property went to John J. Welsh from Philadelphia, who somewhat "colonialized" the original Queen Anne house (especially inside) in the process of enlarging it. Because *American Architect and Building News* (April 14, 1884) published the original design, the extent of the sensitive remodeling is known, although its architect is not. He or she extended the front of the house in both directions, stretching it out; stripped fake half timbering and other Victorian details from the second story; replaced a turned-post entrance porte-cochere with a simpler version in stone; restyled the porch toward the lake from Victorian turned supports and latticing to classical Tuscan; and removed a sculptural chimney toward the front of the house and built new chimneys at either end of new extensions in a symmetrical arrangement. In short, the alterations followed the prescriptions of the Colonial Revival in chastening and broadening what was then regarded as Victorian excess and frippery. The architect also partitioned the Victorian plan, especially its big stair hall and central fireplace, extravagantly open in three directions to the main living areas, and replaced Victorian stained woodwork with Neo-Colonial detail painted white, which is especially fine in the new entrance hall. Still the sense of the original house remains. There is no better example of the transition from Queen Anne to Neo-Colonial in Rhode Island. The exterior, with a shingled upper story above a tan granite ground story with brick trim, reveals a long, gable-roofed core bisected by a gabled axis with gabled wings to either end. It possesses the panoply of forceful window grouping, varied dormers, and sculptured chimneys associated with the best of late-nineteenth-century Shingle Style houses. The shingled barn with silo, its cobblestoned base set into a slope, combines picturesqueness with forceful organization in a particularly creative manner, marking it as among the finest barns of its style in the state.

All has been condominiumized, with the addition of other shingled buildings. Granting the inevitable squeeze on such properties by the developer to maximize number of "units" and parking, the additions are well adjusted to the original buildings and to their handsome setting. The sensitive site plan owes something to consultation by William Shopson, author of *Saving Large Estates: Conservation, Historic Preservation, Adaptive Re-Use*. So much cannot be said

SK1 Shadow Farm, barn

for the evisceration and crowding of a handsome Neo-Colonial house on the lake, with more condos by another developer. In the larger picture, Shadow Farm provides a paradigmatic introduction to the successive fates of South Kingstown's large-scale colonial land holdings: from plantation scale, to still very ample gentlemen's farms, to subdivision and condos.

Wakefield

The village of Wakefield is at a turning point. The sense of its traditional existence as a village is still there, and could be reinforced if there were the will to regulate the hyped-up commercialization which suburban growth and a hub of major regional highways have brought to it. Its Main Street still possesses the mix of some exceptionally fine nineteenth-century houses and commercial buildings with early twentieth-century commercial additions. But the promise these might offer for a handsome business district which preserves the "village" while catering to the "mall" are fast disappearing through thoughtless ad hoc remodeling and the continuing growth of a nearby regional shopping strip which siphons off business and institutional activity.

Just off Main Street, on High, the present Wakefield Mill began with a stone mill constructed around 1867 following the destruction by fire of an earlier wooden mill (one of several farther upstream, which have also disappeared). The names of its successive owners—James and William Robinson from 1821, Gideon Reynolds from around 1862, Robert Rodman from around 1875 until 1903—are a roster of South County first families. The architectural character of its early buildings has been veiled by twentieth-century remodeling and accretions, although a portion of the compartmented plant still produces textiles. In the vicinity of Main Street, from among a number of buildings or their remnants dating from the Greek Revival to the 1930s, a number merit special notice.

SK2 Wakefield Post Office (Former)

1934–1936, Albert Harkness. 220 Robinson St. (just off Main St.)

The former post office is a good example of New Deal Federal Revival design in brick with stone trim. Located, inconveniently for the postal service, on a side street with no on-site parking, it closed in the mid-1990s, and its services were removed to a nearby shopping center with ample free parking. More familiar than the building itself is an often reproduced mural by Ernest Hamlin Baker, *Economic Activities in the Days of the Narragansett Planters,* which was originally installed here in the lobby. It features heroic images of the contributions of African slaves to the economy of the Narragansett Planters—although these appear, by now, intolerably complacent. In the center, a black man controls a rearing Narragansett Pacer; others stack bales, drive an ox team and hold a lantern for crewmen rowing in from a sailing ship, while the whites assume supervisory roles. (Baker made his mark as the longtime artist for the realistic portraits which once adorned the covers of *Time* magazine.) Happily, the mural is now reinstalled at the Pettaquamscutt Historical Society (SK35), where it can be interpreted more fully in a setting that reflects its Colonial Revival tone.

SK3 Kenyon's Department Store

1888–1891, Charles Chase. Facade, c. 1915?. 505 Main St.

A Wakefield commercial landmark since 1857, this interesting hybrid is a Queen Anne building in wood partially sheathed and partially expanded by a fine Neo-Renaissance sheet metal front which has miraculously escaped alteration.

SK4 Bell Commercial Block

1899. 345 Main St.

Diagonally across the street from Kenyon's Department Store, Louis F. Bell built his Queen Anne store and office block, infilling its yellow brick piers with rambunctious bay windows stamped from copper sheet metal. This late nineteenth-century fervor for picturesque (as well as functional) bays (and for boastful lettering) is countered by elegant swags, panels, and a scroll pediment as signs of the nascent Neoclassical Revival. The unusual preservation and restoration of the storefronts to an approximation of their original character completes the picture of first-class commercial accommodations for Main Street in 1900. The post office once occupied one of the stores. Exceptionally, but more frequently then than now for such buildings in small towns, the top floor was originally given over to a hall—this one for dancing, roller skating, and even school graduations. The adjacent mansarded Sheldon Block (c. 1875), originally at another site on Main Street, was moved next to the Bell Block and raised one story by the insertion of storefronts. The same family has maintained it as a furniture store down to the present.

SK5 Robinson Family Houses

The Dr. R. R. Robinson House (SK5.1; 1904, Hilton and Jackson), at 600 Main Street, is not readily visible behind its screen of planting. Restraint is the key to the quality of this house, in which the Shingle Style edges into Neo-Colonial. Openings are beautifully proportioned to the elevation, which centers in a handsome, if austere, Neo-Colonial door with side lights, set into a recessed arch. The house appears to be no more than a very prosperous village property on no more than an ample lot. But behind (not visible) is a Doric-columned porch with pergola stretched across the back and overlooking a formal garden and an expanse of fields, making the house the frontispiece to a fairly sizable estate. (By comparison with the Watson house next door, note how the Neo-Colonial style, and desire for greater privacy, tended to push porches to the side of the house or, more rarely, to the back—forerunners of the eventual twentieth-century banishment of the front porch as a hallmark of nineteenth-century domesticity.)

A descendant of one of the brothers James and William Robinson, who originally owned Wakefield Mill, Dr. Robinson built his house on a small portion of his family's holdings. James Robinson, then William, originally lived across the street, at number 521, in a Greek Revival house known as The Larches (SK5.2; 1831 and later). The third owner of the mill, Stephen Wright, acquired it around 1850. Wright was a local boy who left for San Francisco to make a fortune as a merchant and banker. Returning home in grand style, he celebrated his good fortune by enlarging The Larches and making it the showplace of the vicinity before he sold everything and returned to San Francisco to retire. The old grandeur of house and grounds, which has long operated as the Larchwood Inn, is still feebly visible. Next door, at 571 Main Street, was another Robinson house, which was sold to William David Miller. He tore it down for a large brick house (SK5.3; 1934–1935, Albert Harkness; barely visible from the street) in a hybrid style concocted of two of his architect's favorite styles, a French-Provincialized American Neo-Colonial. A former navy commander, Miller was a bibliophile and local historian. His principal books, on the Narragansett Planters and on Kingston silversmiths, indicate the interest of the new generation of owners of gentleman's estates in the life and culture of their precedessors. This cluster of estates, together with Shadow Farm (see entry, above) is part of a line of country estates and large houses which existed along Post Road with Wakefield at its center. They occupied the lands of the colonial Narragansett Planters and, with the Colonial Revival, basked in the ancient glory.

SK6 St. Francis of Assisi Roman Catholic Church

1932. High and Winter sts.

The Arts and Crafts look of these buildings suggests a date around 1915. What accounts for the late date of this church? And who designed it? Rustic and diminutive qualities are exaggerated in the church, conveying an appealing sense of humility and protection consonant with the life of the saint it celebrates. Random masonry walls are laid up to emphasize their textural qualities. All openings are tiny: the rose window in front a mere eyeball; the tower openings slits; the nave triplets squeezed under the eaves and marked off by miniature buttresses; the "clerestory" a row of minuscule gables. The three entrances are protected by a wooden porch for the center door, flanked by bracketed hoods for those to either side, all rustic in character and domestic in scale. The rustic effect is increased by the change from the

SK6 St. Francis of Assisi Roman Catholic Church, auditorium building

masonry to wood to mark the gable end, with the lower truss bent in an arc over the little rose window. The bell cupola is a domical mini-building clasped, like a gem in a ring, by prong-like crenellations at each corner of the tower. Although this expressive craftsmanship is very appealing, some streak of conventionality restrains the architect from forcing his shapes and proportions to their full effect. Likewise in the too barren interior, where, however, the elements of the exposed wooden structure, which comprises the arcade columns and rafters for nave and side aisles, meld into one another as continuous curves rather than in the usual angular manner.

But what astonishes is the community hall behind the church, a barnlike building of textured brick with random splashes of masonry and tiny metal casement windows here and there. The brick wall, edged in stone, seems to part like a stage curtain around a nondescript entrance. Vertical stained boarding fills the resulting orifice (as in the gable front over the entrances to the church). At the pinnacle, the stone border continues as a cross inset in the wall up to the gable ridge. It seems to be a composite emblem of the stage within and of a cave or mountain with a cross on top. A sculptural stone chimney sits low on the gable. There is nothing like it in Rhode Island. So the final questions: how did it come to Wakefield? After this frontispiece one expects more than a quite functional interior, with exposed wood, brick, and steel spanning girders and a few craftsy touches. It is as though its designer counted on the nondescript factualism of the building to magnify the magic of the emblem.

SK7 South Kingstown Town Hall

1877 and later, Rowland Hazard II. 180 High St.

This forceful towered and mansarded masonry building with vigorous arched moldings topping its upper-story windows is factory design transformed to civic use. If it can be said to have a style, it would be a vernacular version of the Lombard, which in some ways anticipated the Romanesque Revival. Its designer was, in fact, a mill owner in nearby Peace Dale who loved building and contributed this hall to the town. (More on Rowland Hazard II shortly when we reach his village.) Fortunately, the iron roof cresting remains; for the rest, "and later" refers mostly to the destruction of the second-floor porch over the boldly bracketed entrance and of the interior.

SK8 Wakefield Baptist Church

1852, Thomas A. Tefft. 1891, remodeling. 236 Main St.

The Wakefield Baptist Church was radically altered when the pulpit was shifted to the south side and the entrance to the north. But any building by the important nineteenth-century Rhode Island architect Thomas Tefft is worth recording.

Peace Dale

Although Wakefield, Peace Dale, and Rocky Brook are today merged into a regional urban area, Peace Dale still retains the clearest sense of its village character. It has its shambly commercial district, and incessant traffic winds through it. Recent development has spoiled much, partly by remodeling, partly by prettifying; but at least efforts have been made to preserve the village identity, which remains substantial. In fact, Peace Dale, Slatersville, and Harrisville (see under Burrillville) provide the most visible surviving remnants of enlightened efforts to create ideal village environments for industry in Rhode Island. Of these, Slatersville and Harrisville present predominantly Colonial, Greek Revival, and Neo-Colonial images. Peace Dale is Victorian. It is largely the work and vision of one of the third generation of the mill-owning Hazard family, Rowland Hazard II. Together with the mill mason Kneeland Portelow and his firm, and occasionally working with the Providence professional Frank Angell, this amateur architect was responsible for several masonry mill buildings; the village center complex of office and store buildings, library, and meeting hall; five stone-arched bridges over the unusually well-preserved water system which webs the village; a church; the railroad station; and much worker housing. Rowland Hazard favored the plain style of his South Kingstown Town Hall (SK7) but employed it with a fine sense of building and an intuitive awareness of the expressive possibilities inherent in a sensible, solid style. His work, which dates from roughly the mid-1850s to the late 1870s, was augmented by benefactions of other members of the family. The family lived in a parklike enclave of houses, screened from the mill, but using the millpond as a focus.

The Hazard dynasty at Peace Dale began around 1805 when Rowland Hazard I acquired a gambrel-roofed farmhouse of c. 1790 on a site between a millpond and a small wool fulling mill. It was he who gave the village its pretty name—not, however, solely to signal a benign social order, although the Hazards were originally Quakers; Peace was his wife's maiden name. The psychic heart of the enclave, the farmhouse, came to be know as "The Cottage" or "The Homestead." With enlargements in the nineteenth and early twentieth centuries, it remains a charming, rambling mélange reflecting changes in style and function over the generations which have inhabited it. Eventually, the enclave contained six major houses, only three now extant. Among those lost were one known as Oakwood (1853-1857), designed in the Italian Villa Style by Thomas Tefft, and Holly House (1892), a mid-century Gothic Revival house remodeled by McKim, Mead and White in a severe Tudor Revival manner—a style rare for this firm, with its classical predispositions. The family would have known the firm for their design of the Casino at Narragansett Pier (see under Narragansett), where Hazards owned much of the real estate, and for their selection as the architects of the Rhode Island State House, for which a Hazard was on the building committee. Of the extant buildings, Lily Pad (partly 1860, partly later) is another of the survivors—but just barely, having been ruthlessly altered, first as a Catholic convent, then as an office center. Aside from The Homestead, only The Acorns (an offshoot of demolished Oakwood) can properly be said to survive. A subdivision now occupies the former domain of the Hazards.

Immediately adjacent to the domestic enclave, Rowland Hazard enlarged the fulling mill and added processes to realize, in 1813, the first integrated woolen mill in the state, handling all phases of manufacture from raw material to finished product. (Similar integration in Rhode Island cotton manufacturing occurred two years later.) The destruction of the original plant by fire in 1844, and its subsequent rebuilding encouraged the second generation of Hazard owners, Isaac Peace and Rowland Gibson, to shift from coarse linsey-woolsey and jeans to fine woolens. The family bought into the chemical business toward the end of the nineteenth century, becoming the American patentees for the Belgium Solvay ammonia process. Their chemical plants were built ouside Rhode Island, close to supplies. The Hazard interests increasingly moved toward chemicals and away from textiles. A major strike in 1906 shook a family which had always seen itself as having a high-minded view of workers and their welfare. Rowland Gibson, of the second generation, who preferred his considerable intellectual activities to business, wrote essays on the ideal relationship between owners and workers. Rowland II, of the third generation, the amateur architect, introduced plans for workers' sick care, a pension fund, and, finally, in the early 1890s, an unsuccessful plan for profit sharing based on the Rochdale principle. Embitterment, new business interests, and the difficulty of competing with new, large-scale operations in the textile industry led the Hazards to sell their mills in 1918 after the death of the fourth generation of its management. Benefactions to the village from family members who continued to live in the area did not end until around 1930. The family enclave was sold for development only in the late 1960s. The legacy of Hazard endeavors, however, continues to be pervasive in Peace Dale.

SK9 **Peace Dale Congregational Church**

1870–1872, Rowland Hazard II. 1895, transept for organ and choir. 1958, 1983, restorations (the latter after a fire seriously damaged the interior). Columbia and Church sts.

Although raised a Quaker, Rowland Hazard II became a Congregationalist. In fact, this congregation held its first services in the 1850s in Hazard's own house. One of the most charming country churches in the state, of all his building designs this is the most self-consciously architectural. The pitched roof, slated in polychrome stripings of red, green, and gray,

SK9 Peace Dale Congregational Church

is echoed below in four more gables forming bracketed shelters over doors. They accentuate the massing of the body of the church and its nearly freestanding tower as an assemblage of forthright shapes, somewhat abruptly linked, with minimal modulating elements from one to the other. Instead simple geometry brings the parts together, the brusque angles of the gables countered by the arcs and circles of windows, doors, clock faces, and the curved scoop of the transition from tower to steeple. Ornament is integral with the building processes of sturdy carpentry and basic masonry. The original bargeboarding of all gables, drilled with quatrefoils, has unfortunately been lost. The design derives in part from the most rationalistic Victorian Gothic practice, but also substantially from Rowland Hazard's Quaker upbringing and engineering proclivities, here challenged by a special occasion. "Had I known how difficult it would be," Hazard confessed after the church was finished, "I wouldn't have started."

The inside affords the pleasure of seeing a Victorian interior which retains something approaching its original autumnal coloring of varnished woodwork, mustardy tans, and muddy greens, here sparingly relieved by stenciled borders. The roof's dark-stained truss structure, punched with more quatrefoils, is supported on brackets off the walls. In 1895, after his wife's death, Rowland Hazard designed the organ and choir transept with much more elaborate trusswork and commissioned a new treatment of the apse by C. Grant La Farge of the New York firm of Heins and La Farge, called from his vacation house in nearby Saundersville for the job. His father, John La Farge, designed the apse window in concert with the geometric windows along the nave, which he had already installed. Its opalescent subtlety contrasts with the strident color and assertive geometry of the Victorian rose window over the entrance, marking the change in taste and values in the quarter century of Rowland Hazard's participation in the building. He himself is memorialized in the tablet inside the entrance: "He loved this people with a father's love and did with his might what his hands found to do for the good of his fellow men and the glory of God."

SK10 Church Street Bridge

1883, Rowland Hazard II. Over the Saugatucket River

When built, the Church Street Bridge, generally considered to be the most handsome of the five stone-arched bridges that Rowland Hazard designed for Peace Dale, had the longest span of any bridge of this type in Rhode Island. It was built in documented collaboration with the mason Kneeland Portelow, as most of Hazard's masonry structures must have been.

SK11 Peace Dale Railroad Station

1876, Frank W. Angell (Providence), in collaboration with Rowland Hazard II. 1985, conversion to office and residential use. Railroad St.

This and the Kingston Railroad Station (SK38) are the popular images of what country Victorian railroad stations should be. Minus its tracks and abandoned, the little station retains it parasol roof supported on all four sides by elaborate bracketing. In the manner categorized as Stick Style, its display of wooden structure, partly functional, partly fanciful, is really a vernacularization of mid-century Gothic, crossed perhaps with some inspiration from Swiss chalets. The station was locally referred to as the Narragansett Pier station, named for the railroad rather than the village. The Hazards were primary backers of this spur from Kingston Station on the main New York–Boston line. It stopped at Peace Dale, Wakefield, and Narragansett Pier, providing connections for the Hazards' mill to rail lines at one end and shipping at the other, where the family owned one of the principal piers. It was this line, too, that vacationers took for over a century to Narragansett Pier, where Hazards owned much of the land. The first passenger run occurred on July 18, 1876, the last in 1951; the tracks were abandoned in 1955. Old photographs show the station mobbed by villagers in their Sunday best, headed for the beach two stations away.

SK12 Peace Dale Neighborhood Guild

1908–1909, R. C. Sturges. 131 Columbia St.

This masonry building resembling a Neo-Colonial schoolhouse was the benefaction of a member of a later generation of Hazards, Augusta Hazard in memory of her husband, John Newbold. She had founded the guild in 1903, at the height of the Arts and Crafts Movement in America, as a center to teach industrial arts to women and children. It met initially in the upstairs hall of the store block (see immediately below). It continues as a very active community center for the region. Russell Sturgis, a Boston architect who married into the Hazard family, is responsible for the design both of the guild and of the pretty park fronting it at the center of the village. Stemming from some of the same impulses as the guild movement is the kindergarten movement, embodied in the Stepping Stone Kindergarten (1916–1917), nearby at 30 Spring Street. Margaret Hazard became fascinated with the kindergarten movement during a trip to Germany in 1891. She brought a German teacher to Peace Dale and set up a kindergarten at The Homestead. After her death, her daughter, Caroline, continued to support it, eventually providing for the kindergarten on Spring Street, a square, stuccoed building with a hipped roof and entrance projection which is typical of functional school buildings of its period.

SK13 Peace Dale Center

1840s–1902. Junction of Kingstown Rd. (Route 108) and North Rd.

Another of Rowland Hazard's bridges of the 1880s, this one double-arched, carries Columbia Street into the center of Peace Dale. The buildings at the village center are, with one exception, of granite, matching the granite of the mills, and all built by the mill owners. Of all such village centers in the state, none presents a more condensed image of the civic dimensions of nineteenth-century industrial paternalism. Only an owner-built church and school are missing, and these are not far away. The mill buildings take the preeminent position along the top of a T-intersection, with an office and store block on one side of the stem, a library and hall on the other, and the stem aimed straight at the mill office, or Counting House.

We arrive at the center off axis, at the principal corner of the business block (SK13.1; 1856), 1058 Kingstown Road. It still contains the village post office. Upstairs, in a former meeting hall, a museum of Indian and Alaskan relics collected by one of the Hazards (now the Museum of Primitive History) continues the tradition of the partial use of the upper floors for civic purpose. In addition to supporting all sorts of organization and village meetings in the hall, the company also sponsored programs of educational and cultural interest, perhaps under the influence of the first Rowland Gibson Hazard (in the second generation of management). Avocationally, he was a political economist and philosopher of sufficient stature to have attracted the friendship of John Stuart Mill. His venture into worker education, with the opening in 1856 of this hall hard by the mill entrance, is among the pioneer projects of its sort in America.

Although the mill employed 600 workers by 1900 (and 1,100 during World War I), the factory buildings (SK13.2; 1844–1902), Columbia Street and Kingstown Road, are congenially scaled and clustered to the village. Except for the original mill, lost to fire, all the principal buildings remain, with some alterations, together with an exceptionally complete system of waterways. They have been nicely maintained, save for the unfortunate replacement of paned windows with plate glass, as rental property for small industry and business. Following the fire, the second generation of Hazards built a replacement on the original site. This was a low two-story building which they fronted with a towered elevation which remained as a relic of the burned building, letting this project above the new building as something of a false front. (The recent openwork at the top of the tower is coyly inappropriate; but the shuttle weathervane was always the mill emblem.) This second generation of managers used the new plant as an opportunity to shift production from rough to fine woolens. Shawls were the new specialty. Peace Dale shawls wrapped Victorian shoulders for thirty years, until the fashion waned in the 1880s and the mills turned to the production of high-quality woolen suitings.

Except for the replacement building, the front tier of granite blocks are the work of Rowland Hazard II (of the third generation of owners), together with his masons. From the replacement building west (left to right facing them), these are the 1856 mill, the 1872 worsted mill, the 1880 weaving shed behind (all with altered roofs and some top stories added by the early twentieth century, and a mansard added to top the southeast tower of the 1856 block c. 1880). These were all built under Rowland Hazard's supervision, as well as the mill office building, where he had his office. He distinguished the Counting House, as it was known, from the row behind by its hipped roof and the ornamentation of the granite by the polychromatic contrast of the somewhat disruptive brick corbeled cornice. The bracketed roof shelter over the entrance is a variant of those on the Congregational Church. Most of the brick buildings behind went up around 1902, after his death.

On a slight elevation opposite, at 1057 Kingstown Road, is the Peace Dale Library, originally Narragansett Reading Room and Library (SK13.3; 1891, Angell and Swift; landscaping, Charles Eliot), for which Rowland Hazard and his brother John Newbold Hazard provided to honor their scholarly father, the first Rowland Gibson Hazard. For this they called on Frank W. Angell, who had already worked with Rowland Hazard II on the railroad station and presumably on workers' housing. The result was a late example of Richardsonian Romanesque immediately following Angell's design of Wilson Hall at Brown University in the same style. Looser in organization and not as finely wrought as Richardson's own work, this is nevertheless an impressive example of the style, especially in its forceful employment of Richardson's familiar centerpiece features of gable bay (here shingled at the upper story), hipped dormer, and swelling turret bay all grouped around an entrance arch, with granite steps spilling and spreading out of the cavern. The substantially intact interiors show a comfortable mixture of architectural and mill design, which suggests that the local building team was as much responsible as the architect. Attached to the rear of the library is Hazard Memorial Hall, a plain but well-proportioned shingled building on a cobblestone basement with an interior which is straightforward and sturdy timber and tie-rod mill construction. Its long-underused formal, ceremonial space, now nicely refurbished and incorporated programmatically into the library, adapts well to the

SK13.3 Peace Dale Library

wider variety of activities characteristic of libraries at the turn of the twenty-first century.

The three-tiered granite watering trough (1890) on the library grounds was originally in front of the store and office block. Water from the horse basin drains into a second-tier basin at mid-height for oxen and again to the smallest basin close to the ground for small animals. An awkward but expressive design by Rowland Hazard, it can be seen as an unintended allegory to paternalistic benevolence, whereby benefits flow from larger animals to smaller— and it is now, significantly, dry.

Finally, to this cluster of Hazard benefactions commemorating Hazard forebears, Caroline Hazard added a memorial to her father, Rowland Hazard II (the amateur architect) and her two brothers (the final, fourth-generation Hazard management team). *The Weaver* (1902, Daniel Chester French) is an allegorical bronze relief of life-sized figures which is among the sculptor's finest. Herself distinguished, Caroline Hazard had been president of Wellesley College and is generally credited with lifting what had been a floundering seminary into leadership in women's education. (Her presidency, incidentally, saw the addition of six new buildings to the campus, some covertly financed by her, a legacy which doubtless reflects her father's example.) Derived from Roman funerary reliefs, the memorial depicts time as a young maiden with distaff who hands the thread to a weaver, clad in sheepskin, seated before an ancient loom. The inscription reads: "Life spins the thread. Time weaves the pattern God designed. The fabric of the stuff he left to men of noble mind." In setting, distinction, and sentiment, this is the most poignant memorial that exists to the once vaunted Rhode Island textile industry. Completed two years after the Hazards sold their mills, in effect it commemorates a century of involvement by four generations of Hazards in the industry and the community they fostered at a time when the New England industry as a whole was on the brink of disaster.

SK14 Peace Dale Elementary School

1923. 109 Kersey Rd.

The elementary school is a nicely rationalized neoclassical building in brick, which, as the last of the series of memorial donations by different Hazard families, represents the effective end of their architectural contributions to Peace Dale.

Rocky Brook

From a Hazard village to a Rodman village, established by another of the major nineteenth-century mill-owning families in southern Rhode Island. The Rodman tract of some 1,000 acres, including Rocky Creek, a tributary of the Sauguatucket, goes back to the late seventeenth century. Samuel Rodman, who began by managing the Hazard mills while operating a shipping business at Narragansett Pier, early in the nineteenth century purchased a piece of the ancestral property (from a Hazard who then owned it) and built a masonry mill complex which once included three mills for yarn production, two gutted by fire in the 1870s. Prosperous before the Civil War, the Rodman mills fell on hard times afterward, and finally ceased operation completely in 1895. Thereafter, the millpond became a local reservoir with a cobblestone pumping station. In its heyday during the 1850s, Rocky Brook won praise in J. R. Cole's *History of Washington County* (1889): Samuel Rodman "built pretty cottages for his operatives, made roads, set out trees and beautified the place until it became ... one of the thriftiest as well as one of the most picturesque villages in New England."

SK15 Rodman Mill

1847, c. 1853 and later. 1877, gutted by fire. c. 1917, rebuilt. 1425 Kingstown Rd. (at Rodman St.)

Although rebuilt only after a long period of abandonment, this well-maintained mill with gable roof, monitor, and tower on a gable end could be among the state's most handsome examples of small mid-nineteenth-century granite mills, but extensive, insensitive additions destroy its effect. Along Rodman Street especially but also along Kingstown Road are the one- and two-family mill workers' cottages which Cole praised in his history.

SK16 Pump House Restaurant (Reservoir Pumping Station)

1889, Willard Kent; built by Partelow and Bullock with Louis F. Bell. 1464 Kingstown Rd.

The pumping station is a picturesque attempt at industrial rusticity which contrasts with the functional approach of the mill opposite. Porte-cochere and rustic stack dominate this melding composition of hipped and gable roofs, which slope to a flaring curve at the eaves. The predominance of curved openings, arches, and oculi enhances the naturalistic contours of the building and suggests Queen Anne influence. So do small-paned windows, the interior woodwork, and even a fireplace in the pump room. Such deliberate efforts at charm in an industrial structure, more typical of country estates than of town reservoirs, suit its present use. The country builders responsible for the Peace Dale Congregational Church and probably for the Peace Dale Library also collaborated on this.

SK17 Isaac Peace Rodman House (The Stone House)

1855. 1789 Kingstown Rd.

Mostly screened from the road, this beautifully crafted house of quarry-faced, square-cut gray granite has the severity of mill construction, relieved by a broad porch across the front, broad eaves, a row of tiny dormers, and, capping the hipped roof, a cupola monitor—all features of the Italian Villa Style. A service ell behind abuts what is otherwise a compact, squarish block.

SK18 William C. Watson House

1838. Mid-19th century, additions. 957 Saugatuck Rd. (pole 837)

The forceful presence of this Greek Revival house depends on the reinforcement of its contained cubic quality by a hipped roof rising to a full monitor from which jut flanking chimneys.

Plain corner piers, cornices, and a Doric entrance porch emphasize the severity. By way of playful contrast, the ornamentation of the Gothic Revival porch added to one side of the house subtly rounds the customary angularity of the roof bracing with complementary handlelike, semicircular elements halfway up the post supports. A pretty Gothic Revival wellhead near the rear kitchen ell must date from the same time.

SK19 Elisha Watson House

1812–1820. 845 Saugatucket Rd. (at Rose Hill Rd.)

This standard five-bay, central-door Federal house with a sculptural brick central chimney has a door flanked by fluted pilasters under an exceptionally large three-light transom. This raises its capping cornice and aggrandizes the effect of the door. Corner quoining and a modillion-and-dentil course at the cornice complete the ornamentation.

Northeast

SK20 Convent of the Sisters of the Cross and Passion (Rush Sturges Mansion, Shepherd's Run)

1933. 4640 Tower Hill Rd.

Wrapped as a spreading, L-shaped mass around a landscaped entrance turnaround, this is a fine example of the large "Norman" house popular during the period. Ornament is minimal. Its effect depends on intricately textured masonry walls and a complex composition of gables and turret caps, with openings informally disposed but so small and scattered as to suggest a slightly fortified effect. Although the house is substantially spoiled by additions made since 1960, the original integration of the house with its gardens, which were designed by Beatrix Jones Farrand, is still apparent. Sturges was a Providence lawyer; his wife was the daughter of Rowland Gibson Hazard.

SK21 Shadblow Farm

1810 with later ell addition. 4638 Tower Hill Rd. (U.S. 1)

The five-bay, center-door format used half a century earlier in the Marchant House (SK41) is perpetuated in this comparably grand farmhouse. One central chimney becomes two to accommodate a central hall. Windows are larger in proportion to walls. As is characteristic of Federal houses, details are lighter. The heavy, spreading frame of the Marchant pedimented door, for example, becomes taller and thinner with the addition of a lighted transom and Ionic pilasters instead of the Marchant's Doric. Two pilastered dormers are inserted into the roof. The house sits high, still overlooking a panoramic landscape of fields down to the ocean.

SK22 Governor Charles Dean Kimball House (Kymbolde)

1901–1903, Stanford White for McKim, Mead and White. 173 Torry Rd.

For a former Rhode Island governor, Stanford White designed this combined country and summer estate in a crisply linear Neo-Federal style in which formal and informal qualities are nicely balanced. Sited on the crest of a meadow sloping down to a distant view of the ocean, on the sea side the house presents a screen of porches; on the land side, an entrance porte-cochere. This is centered in a five-bay, two-story format typical for late colonial and Federal houses, giving a formal axis to the house. This formality is countered by certain cozy, relaxed aspects of the design, pervasive in late nineteenth-century houses, such as the sheltering quality of the porte-cochere and the horizontal spread of the low entrance porch and windows. Spareness of ornament gives a chaste aspect to the house, but also suggests that a plain farmhouse lurks within this "country seat." At the opposite end of the central hall, the ocean elevation provides the climax. An attenuated two-story Doric portico is flanked by one-story Doric porches, with parapeted decks on top. The motif of high and low porches with decks recalls Thomas Jefferson's University of Virginia, for which McKim, Mead and White had designed alterations and additions. Only a swoosh of trellis ornamentation in an arc over the fanlighted entrance accents the otherwise restrained embellishment. Kymbolde lives in its porches: a porch for ceremony, with broad steps down to a sunken formal garden, flanked by porches for living, and topped by decks for viewing and sunning. (The stable, also by White, has been converted into another house.)

SK23 Observation Tower, Hannah Robinson Park

1936, Rhode Island Department of Public Works. 1986–1987, rebuilt. Junction of Tower Hill Rd. (U.S. 1) and Route 138

A stone lookout tower existed at this elevated site for some time from around 1759, and there are reports of later short-lived towers as well. The state erected its tower with the completion of Route 138, shortly after the donation of the area as a small state recreational area. It was named for the legend of a woman who regularly met her lover at the big rock at the base of the tower, with ultimately tragic consequences. A burly, tapering frame of tree trunks (or telephone poles) envelopes several tiers of viewing platforms, with the metal connecting devices as much a part of visible action as the logs. This is a completely rebuilt and re-engineered version of the original, which deteriorated over time. It was closed to the public during World War II and manned as a lookout. On a clear day, the principal view across Narragansett Bay over Conanicut Island (Jamestown) to Aquidneck Island (Newport) is impressive. It also provides a fine sense of the land- and seascape which the Narragansett Planters and then estates like Kymbolde along the Post Road enjoyed—minus the subdivisions which now patch the view. Added attraction: its only cost is the climb.

The separation of the double traffic lanes of Bridgetown Road by a median landscape strip in 1930 was the first instance of this kind of highway treatment in Rhode Island. Designed both to separate opposing lines of traffic and as a dignified approach to the Jamestown Bridge, it was referred to at the time as "pairway pavement construction.")

SK22 Gov. Charles Dean Kimball House (Kymbolde)

SK24 Palmer Gardner House

18th century. 527 Mooresfield Rd.

An almost picture-perfect complex with a large and seemingly typical Rhode Island farmhouse and handsome barns, the farm is compelling for both its remarkably intact rural setting and its unusually framed center entrance. Although front entrances in eighteenth-century houses characteristically receive comparably elaborate embellishment, the frame we see here far outdoes the usual: set atop fluted pilasters and otherwise simple entablature, a broken segmental arch with carved rosettes at each end frames a plinth with a foliated vase centered over a three-dimensional scallop shell above the five-light transom. The treatment echoes, slightly more simply, that of the doorway to the balcony on the second floor of the Colony House in Newport and even more closely resembles the frame on the Rowland Robinson House in Narragansett (see entries). The close ties between South County and Newport in the eighteenth century might easily account for the adaptation of the Colony House frame in both houses, but far more likely is the twentieth-century involvement of Norman Isham, who had restored the Colony House in 1917 and the Robinson House in 1928.

Kingston

Kingston (note that the village is Kings*ton*, as opposed to South Kings*town*, both pronounced as spelled) was Little Rest until around 1825. Apparently no farmer's grumble accounted for its wry designation, but soldiers who camped here before the Battle of the Great Swamp (1675), which ended the hegemony of the Narragansett tribe in the area. By the early decades of the nineteenth century, in appearance and history, Kingston was reminiscent of two other inland Rhode Island villages—Union Village in North Smithfield and Hopkinton City in Hopkinton (see under those towns). Like them, Kingston came to flourish as a regional transportation hub with the attendant bustle of shops, blacksmithing, inns, taverns, and the sort of well-heeled rural professional elite of lawyers, bankers, and doctors typically drawn to such centers. Kingston's function as a transportation hub and the legal and banking center for the Narragansett plantation culture was augmented by the move in 1752 of the Washington County Courthouse from Tower Hill on the Post Road to Little Rest (where it originally stood directly across the road from its extant 1775 replacement). During the late eighteenth century and early decades of the nineteenth, all this activity supported relatively lavish building. An academy followed the courthouse in migrating from Tower Hill to Little Rest; originally the Pettaquamscutt Academy, it later changed its name with the village. The *Rhode Island Advocate* appeared in the 1830s as the first newspaper in the southern part of the state. Then, without water power to attract industry and bypassed by railroads, Kingston, like Union Village and Hopkinton City, stagnated economically, and for the same reasons. In Kingston's case, railroad building inflicted a double insult. First, the main New York–Boston line went west of the village, necessitating the creation of West Kingston. Then the Hazards ran their spur from the Kingston Railroad Station, well south of Kingston, directly to Peace Dale, Wakefield, and Narragansett Pier. The removal of the county court to West Kingston in 1876 was a further blow to the village. Then, the already foundering Kingston Academy finally collapsed after a major fire in 1882. But throughout these adversities a sense of gentility hovered over Kingston, as well as its two sister towns, all three precariously sustained by old money, the sense of old culture, and the threadbare economy of certain transactions which continued by custom. Such conditions also protected the old buildings, since there was little call for replacements. So, there they sat for decades, in the three villages, waiting for nostalgia to return their old luster. Of the three, Kingston has the most extensive and varied architectural heritage. By contrast with its sister towns in North Smithfield and Hopkinton, Kingston had significant counters to stagnation: first, for a while, the county courthouse, then, after 1889, what is now the University of Rhode Island.

SK25 Solomon Fayerweather House

c. 1845. 1831 Mooresfield Rd.

SK26 Fayerweather Craft Center (George Fayerweather House)

1820. 1859 Mooresfield Rd.

One of these is a modest Greek Revival house, the other an equally modest one-and-one-half-story side-gabled house with central chimney, old-fashioned for its date, with the narrowest of transoms over its plain (and rebuilt) door, clapboarded in front and otherwise shingled. Both were owned by freed black slaves, members of what became an important blacksmithing family in the area, with a forge in continuous operation into the 1920s.

SK27 Thomas R. Wells House

1820. 25 North Rd.

A handsome example of the usual two-and-one-half-story, five-bay, side-gabled house with end chimneys, the Wells House (now a fraternity house related to the adjacent University of Rhode Island) is exceptional for brick end walls tied into its chimneys. The entrance type is prevalent in the village. Swelling S-shaped brackets over plain paneled pilasters raise an entablature over a transom light. The resulting vertical proportions are stately; the effect is severe. Such restraint is, in fact, a keynote of the Federal style in Kingston.

SK28 Luke Aldrich House

1829. 80 North Rd.

It is the mark of the prosperity of this village that it became home to several silversmiths, patronized not only by the Narragansett Planters, but by wealthy residents of Newport, where local craftsmen maintained outlets. Luke Aldrich was among them. And it is the mark of his prosperity that he could build such a grand house. The front elevation and scroll-bracketed door are identical to those of the Wells House (although the end walls here are clapboarded). The roof is hipped up to a monitor, which is virtually screened by a roof balustrade of alternate solid and fret panels, the frets responding to the rhythm of openings below. It is this kind of fancywork at the rooftop which characterizes Kingston's most ambitious houses. The splayed wooden lintels over the ground-floor windows of the Wells House, a carryover from the late eighteenth century, are here "modernized" as clean rectangular frames.

SK29 Asa Potter House

1829. 2545 Kingston Rd.

Another of the balustraded, hip-roofed, monitor-topped Kingston queens, the Asa Potter House is especially grand by reason of the larger-than-normal size of its elevation and a raised porch with its own hipped roof supported on slender Doric columns that are surprisingly attenuated for its width. This shelters a door with side lights and transom in leaded tracery, more ornamental than the Aldrich door, but still restrained as such treatments go. Adjacent, at number 1299, Potter built his law office in 1831.

SK30 Joseph Perkins–John Hagadorn–Thomas Taylor House

c. 1775, 1827. 2563 Kingston Rd.

A slightly earlier twin of the Asa Potter House, this was doubtless designed by the same builder. Its history is a bit more complex. It stands on the site of a house built c. 1775 by Joseph Perkins, a silversmith and merchant. Shortly after John Hagadorn bought it, his niece Elisa inherited it in 1817. On her marriage to Thomas S. Taylor in 1827 she had the house moved back on the property at right angles as an ell and built anew in the grander manner, with door and balustrading variants of those on the Potter house. The porch was added by Francis Hagadorn, who inherited the house in 1861. Around 1880 his builder adapted and exaggerated the porch forms of the nearby Potter house in an early example of the revival of interest in colonial design.

SK31 Caleb Westcott–Joe Reynolds–Phillip Taylor Tavern

c. 1820 and later. 2573–2575 Kingston Rd.

This longtime tavern and hotel was a stage stop on the run from West Kingston to Narragansett Pier until competition from the railroad shut it down in 1877. The establishment began as a five-bay, central-chimney Federal house (to the right), the rest added sometime later and the entrance porch probably much later. Diagonally behind the tavern-hotel is the barn for the stage (1825), converted to a house in 1947. Several pleasant but modified buildings are next in line.

SK32 Kingston Free Library and Little Rest Archives (Kings County Courthouse; Little Rest Museum, Old County Records Office)

Library: 1775, 1876. 1951, restoration of portion of second floor, John H. Cady. 1995, Extrados. Archives: 1857–1858. 2605 Kingston Rd.

When the original Kings County Courthouse was moved to Little Rest in 1752, it occupied a site directly across the road. It was demolished after it was replaced by this building in 1775. The present building served as the county courthouse but also as one of the five original state houses in five different locations around the state among which the legislature rotated between 1776 and 1791. Later, it met here biennially from 1842 to 1854.

Originally, the courthouse was a big barn of a building with a substantially projecting tower topped with the existing bell cupola. The insertion of a mansard roof over a bracketed cornice in 1876 raised the cupola and completely transformed the character of the building. Now the large scale of the flared mansard with scoop curve and decorated dormers takes visual precedence, and the original building appears to be the base of what is above, especially for the climactic sequence of forms from mansard to faceted cone to the cupola. The cupola simultaneously caps the building and curiously seems to float above it as a levitated gazebo. The double door into the tower with its bracketed hood also belongs to the Victorian remodeling and can be compared to the more sedate, more planar doors at either end of the building, which are original. The building also retains its row of granite posts in front, which support a continuous pipe for mass hitching.

Inside, the first story retains the open plan (although densely "furnished" with freestanding book shelving) with a row of Doric columns across the middle of the building, the same configuration seen in the Newport Colony House, on which this smaller, less expensive

SK32 Kingston Free Library and Little Rest Archives

version was modeled. Only a fragment remains of the northeast (right rear) corner staircase, superseded by the vertical circulation system and new entrance installed as part of the renovations.

The gabled box in square-cut, rough-faced granite beside the courthouse was built for the storage of court records. The only openings are metal doors flanked by two windows with metal sash, all surrounded by the bluntest of granite frames, which are smoothly finished to contrast with the walls. After the court moved out of Kingston the structure served a variety of purposes. In 1951 it became a museum and 1971 the Little Rest Archives, which houses early land and genealogical records.

SK33 The Homestead

1809. End of Potter Lane

The lane dead-ends in a plain two-and-one-half-story Federal house, for three generations the home of the Potter dynasty, which was, in effect, the leading family in the village. The first occupant of the house, Elisha Reynolds Potter, Sr., descended from one of the first settlers of the village, was a member of Congress intermittently from 1796 to 1815 and an intellectual presence as well as a legal authority in the community.

SK34 Kingston Congregational Church and Parish House (Thomas P. Wells House)

1820 (church), 1832 (parish house). 2610 Kingston Rd.

SK36.3 University of Rhode Island, Davis Hall

The tower of this spired country church, typical of its period, projects slightly from the gabled body. The projecting tower exaggerates the height of the door, which seems all the taller because of the domestic scale of the windows around it and the S-bracket lifting the entablature over the transom in the manner so familiar to this village. Here, too, the framing of all elements, including the door, is so attenuated, and all corners so minimally bounded, that the church seems as fragile as folded paper, with rectangular, circular, and arched shapes pasted on the surface.

The elevation of the parish house replicates those already seen in Kingston, but with a startling difference. The gable faces front with a gabled monitor on top—technically a hip-on-hip roof. The usual format for early industrial buildings, it is nearly unique for Rhode Island houses—although at least two other South Kingstown houses of the period, neither as grand as this, have similarly designed monitors. In widely separated locations, they may have been the result of trips into Kingston, where the Wells House was conspicuous in both size and location. A very practical approach to monitor lighting, it also gives a rather institutional look to the house. Its door, with leaded side lights and a bland fan, are enclosed within a frame in which uprights and arch flow together with minimal interruption.

SK35 Pettaquamscutt Historical Society
(Old County Jail)

1858, 1861. 2636 Kingston Rd.

For years the jail was erroneously thought to date from the late eighteenth century, but recent research has proved that the rear granite cell block dates from 1858. Cells for male prisoners, half underground, are narrow spaces with no more than a shelf for sleeping and a barred slot for light. A lightless cell at one end provided for solitary confinement. A single cold-water shower at the end of the corridor served all prisoners. On the floor above are cells for women prisoners, more like rooms, squarish in shape, each having a full sash window with view, and originally furnished with beds. The front block of the building was added in 1861 as a warden's residence. Today each of the cells provides a showcase for an aspect of eighteenth or nineteenth-century life, with books and historical records in the front of the building.

SK36 University of Rhode Island

Entrance to campus at Route 138 and Upper College Rd.

Five towns were considered as the location for the state agriculture experiment station and what is now the University of Rhode Island. Fortunately for Kingston, the 140-acre Oliver Watson farm was available for $5,000. Town funds and local contributions managed to meet most of this cost, so the institution located here. The old Oliver Watson farmhouse (SK36.1; 1792, c. 1840) still exists in its own stone-walled and picket-fenced enclave. It is a standard two-and-one-half-story, five-bay Federal house with Greek Revival alteration to its center door. The house served many functions until the university trustees decided to tear it down in 1962. A campaign of protest led by a president emeritus saved the structure for renovation and its present status as a museum. The original nineteenth-century campus buildings, built near the farmhouse, were all designed by the Providence firm of Stone, Carpenter and Willson and named for Rhode Island governors of the time—Taft (SK36.2; 1889), Davis (SK36.3; originally College; 1891, destroyed by fire in 1895 and immediately rebuilt, renovated 1959) and Lippitt (SK36.4; 1897, renovations, 1935 and 1965). They were built as separate entities until the landscape firm of Olmsted, Olmsted and Eliot envisioned a double-quadrangle plan, which, although altered, is the basis for the form of the central quadrangle today. These buildings and those which followed to complete the core of the campus are of masonry, functional rather than beautiful, most with strange proportions, but full of character. Davis takes the prize, a squarish block with a tall hipped roof topped by a balustraded platform. Large gables peer out of it behind half turrets and an attached battlemented tower, like so many rays of a crown. Next in interest is Lippitt Hall, the largest of the three and built as a drill hall and gym; its prominent cross-gable roof is further enlivened by an octagonal turret which glides up to the projecting cross gable and prominent dormers which slide down the wall surface to become shallow oriel windows. And finally, Taft, a T-plan building with tall chimneys above its hipped roof that reinforce the verticality established by the heavily recessed window bays on the end walls of the T's crossbar. As they are late products of one of the best Victorian firms in Providence, it is hard to account for the loose organization and disparate sizes of the windows in these buildings, as well as for their lack of rapport with one another, beyond their common use of masonry. But the real interest is less in the individual buildings than in the ensemble. It could be a movie set for the turn-of-the-twentieth-century agricultural and engineering college and should be treasured as such.

Among later buildings, the outsized Neo-Colonial Eleanor Roosevelt Hall (SK36.5; 1936, Albert Harkness), in red brick, funded by the Works Progress Administration, may be the most interesting, especially for the contrast in scale between the row of arched chimneys and the tiny jerkin-roofed dormers, which are whimsically and practically lighted on all three sides. The building was, in fact, dedicated by the first lady herself. Of recent buildings the most inventive is the Biological Science Building (SK36.6; 1972, Robinson Green Beretta). Impressive from the exterior, it is mostly underground, except for a central cylindrical shape poking through a contoured, grass-covered mound. Primarily devoted to teaching auditoriums, where gloom is no issue, the building also includes several pleasant places in which to work or to pass through, despite a neglected space intended as a garden court at the bottom of the cylinder, which now is woefully lacking in garden qualities. More recent is the Coastal Resources Institute (SK36.7; 2001, Keyes Associates), which orchestrates a multicolored brick mass with industrial hints of a sawtoothed roofline.

West Kingston

Until the old New York, Providence and Boston Railroad dropped a station here in 1876, West Kingston did not exist. What is now village was farmland. (The railroad later became the New York, New Haven and Hartford, until, after a series of name changes accompanying financial maneuvers, it became Amtrak.)

SK37 Courthouse Center for the Arts
(Washington County Courthouse)

1894, Leslie P. Langworthy. 3495 Kingston Rd.

A granite building in polychromatic Richardsonian Romanesque, the courthouse is imposing but heavy-handed. Mid-twentieth-century renovations to the interior were removed in the late 1990s, when the building was thoughtfully rehabilitated by and for a local arts group.

SK38 Kingston Railroad Station

1875. Restoration, 1960. 1988, partially burned. 1989–1990, re-restored. 1 Railroad Ave.

Despite its location, this has always been known locally as Kingston Station. But then, it was really meant for Kingston. It is the only station built by the New York, Providence and Boston Railroad still in use. That this building in the late nineteenth-century Stick Style should have continued in use as an express stop on the main New York–Boston line is extraordinary. And that it and the Peace Dale Station exist in such close proximity is especially so. The station at Peace Dale is more fanciful, more exceptional. Kingston is comfortably run-of-the-line for its period. It is a straightforward bracketed and gabled building, clapboarded and flushboarded, with the typical swell of the ticket seller's bay window between doors giving onto a platform with a gabled canopy. The station itself supports one side of the canopy; posts with diagonal bracing support the track side. The canopy slides beyond the station in both directions in a nice bit of expressive functionalism: instead of the stodgy solution of two bracketed posts to terminate the canopy, the theme of the single row of posts retains its consistency with the move of the terminating supports to a position directly under the gable. This also reinforces the umbrellalike lightness of the canopy. A boxy mini-tower between dormers on the roof provides minimal living quarters with views up and down the tracks—awkward, but again expressive. Nearby is a two-story signal tower constructed later than the station (perhaps c. 1915), which was moved to a turnaround at the station. At the top, a row of tall windows in the track side of the tower fold out to a steep bay window at the center and fold again into the side elevations. A bracketed hipped roof beautifully protects the lookout folded beneath it. As a building form, the tower takes on new significance in conjunction with the popularity in the 1980s of towerlike houses in wood, with windows massed at the top, much like this.

Glen Rock

SK39 Eppley Camp
c. 1920. 505 Dug-Way Bridge Rd. (not visible)

The long agricultural decline in western and central Rhode Island and the concentration of communities along the coast left much wooded land, for which there was little market during the first half of the twentieth century. The well-to-do built summer hideaways, often on Rhode Island's numerous ponds. The very wealthy could accumulate vast tracts, as did Marion Eppley, the founder of Eppley Laboratories in Newport. He eventually owned 1,300 acres of woodland, most of it in Exeter, but with his living compound in the northwest corner of South Kingstown on the Queens River. In contrast to the luxurious rusticity of Alton Jones's hunting and fishing lodge in West Greenwich (see entry), Eppley kept his buildings simple. The focus of the camp is a long, narrow house, partly one-and-one-half stories, mostly one story (which may indicate two building periods) in whitewashed clapboard and shingle. A long porch supported on tree trunks, reminiscent of a southern cabin, unifies the whole. The principal room is paneled in chestnut and pine with a stone fireplace. Nearby, clustered randomly in the pine woods, are a separate cook house, guest house, corncrib, garage, and boathouse. These vary in use of rustic building techniques, the most interesting perhaps being the log cookhouse with mud (or stucco) chinking and a clapboard guest cabin, also long and narrow, one end of which is a huge stone fireplace with a curved outer face. Shelter seems to attach itself to the hearth in a manner recalling the Rhode Island colonial stone-ender, except that here the fireplace appears more sculpturally self-contained, hence particularly potent as the symbol of what it is. On Eppley's death in 1960, he left a somewhat reduced estate (863 acres), including the cabin compound, to the Audubon Society as a wildlife preserve.

SK40 Peter Pots Pottery (Glen Rock Mill)
c. 1867. 101 Glen Rock Rd.

After purchasing a group of small operations in what was then known as Barber's Mills in 1867, Daniel Rodman built a mill for the manufacture of coarse woolen goods which came to be known as Glen Rock. Although it is now much altered as a pottery shop, enough remains to indicate something of the typical small mid-nineteenth-century masonry mill settled into a glen with its millpond very much elevated above a huge rock.

South and Southeast

SK41 Henry Marchant House
(Joseph Babcock House)
Pre-1760. 3401 South County Trail. (not visible)

This is a very fine five-bay farmhouse with pedimented central door flanked by Doric pilasters and central chimney in stone, all backed by a shallow lean-to. Shortly after its building, Henry Marchant bought it and operated it as the center of a plantation, all the while maintaining ties with the Newport mercantile establishment and a career as a member of the state assembly from Newport. His son, like some other wealthy New-

SK39 Eppley Camp, main lodge

porters of the early nineteenth century, continued to ferry between the city and the family farm. The house, surrounded by 90 acres and mostly nineteenth-century outbuildings, has continued in the ownership of Marchant descendents and is well preserved throughout, with some changes inside.

SK42 Camp Hoffman (Girl Scout Camp)

1921 and later. Larkin Rd. (off Ministerial Rd.)

This does not possess the architectural impact of the log structures at Camp Yawgoog for Boy Scouts in Hopkinton (HO6), but it is nevertheless a fine example of rustic camp design on a beautifully maintained site. It is mostly in clapboarded wood construction, although deep in the woods is a pretty log cabin for special powwows. Ministerial Road, the particularly scenic road near which the camp is located, takes its name from one of the original land purchases in this area. South of the turnoff for the camp, at the junction of Ministerial Road and Curtis Corner Road is a rectangular granite slab, set on granite blocks, inscribed "Ministerial Land—300 acres set apart June 4, 1668 by the Pettaquamscutt Purchasers—Income to the Ministry." The purchasers in question were the group which came to be popularly known as the Narragansett Planters. Although the tract was set aside to bring a minister to the area, the attempt was unsuccessful. Eventually, funds from the sale went to the Kingston Congregational Church.

SK43 Tucker–Albro House

c. 1739. 1805 Ministerial Rd.

Facing due south, this staunch little house set low on a granite foundation is dominated by the broad, uninterrupted sweep of its simple gable roof, capped by an impressive stone chimney. Many South County central-chimney houses have chimneys made of stone, but most are finished in brick above the roofline.

SK44 Samuel Perry House

1696–1716. 1956 and later, restoration. 645 Matunuck School House Rd.

As in the Segar House (see entry, below), austere shape and texture account for the physical presence of this "half house" with its tiny windows and immense stone chimney. But what a difference between the two! Instead of the

SK44 Samuel Perry House

SK45 Cottage, Roy Carpenter's Beach

usual placement to the rear, here the lean-to is in line with the elevation, pitched to the line of the eaves. Consider how the three slightly larger windows in the lean-to play with the five earlier windows around the door; how the four upper-story windows butt or touch the roofline, while the four below effect a tenuous symmetry with the door; how shifts in level and interval among the openings animate the elevation; or how the four windows to the right of the door set up a regularized pattern against the loose arrangement of the other four. These are not consciously contrived, but the result of chance over time, coupled with the builder's response to the felt "rightness" for relationships. Architectural concerns for shaping and placement, balance and tension rationalize our unconscious satisfaction with what we confront and the potential subtleties in simple things.

SK45 Roy Carpenter's Beach

c. 1920 and later. Card's Pond Rd. (at Browning Beach Rd.)

SK47 W. F. Segar House

One example in this area of the metamorphosis of the cabana complex into a community of tiny gabled cottages, this is much larger and more ordered than others. Off season, when the cottages are boarded up and empty, the elements of order are especially evident. First, variations and exceptions notwithstanding, all these houses present the same elemental image of a simple gabled box. Second, all are aligned in rows broadside to the water, echoing the axis of the beach. Third, all are spankingly maintained. Fourth, all conform to a community plan which takes the form of the handheld barbecue grill. An approach road, straight from the highway, makes the "handle." It bisects the four streets of close-packed houses at right angles to make the "grill." Finally, it terminates at the beach in a community facility. Two themes are worth examining here. One is variations on the very minimal vacation house. The other is the blank order, which, with the annual seasonal shift, disposes and restrains the seeming summertime disorder of people, cars, small boats, inflatable toys, dripping bathing suits, barbecues, and other vacation paraphernalia that spill out into the space of the community.

SK46 Shingle Houses, Browning Beach

Early 20th century. Card's Pond Rd.

Adjacent to Roy Carpenter's Beach is Browning Beach, a nature sanctuary. It offers a distant view of a fine row of four shingled cottages well above a wave-scooped beach, interrupted by one post World–War II modernist house. Unpretentious but modestly luxurious, they exemplify the ideals of the Craftsman movement at a high level of achievement, three of them allusively bungalows, one a chalet. They seem a world away from Roy Carpenter's populous enclave.

SK47 W. F. Segar House

c. 1890. 1010 Matunuck Beach Rd.

Standing at the front of a colony of middle-class summer cottages (mostly from the twentieth century) is this rare unaltered survivor from the late nineteenth century. Here there is no adornment to beguile the eye, only memorable starkness. A sharply pitched, L-shaped gabled mass folds around a spindly tower. This lookout compounds the angularity with cross gabling, which is left peaked in the sea-to-shore axis, with peaks clipped in the axis of the beach. A broad porch across the front wraps halfway around the sides. For the rest, the spare spacing of openings and the textures of contrasting materials complete the image. Wood shingles, weatherbeaten to black-brown, are set against the cobblestone parapeting and posts of the porch. A cobblestone wall sets the property off from the road. Even the shingled barn is intact. Had the artist Edward Hopper lived at Matunuck Beach, this is the building he would have painted.

SK48 Weeden Farm (Willow Dell; Colonel Jeremiah Bowen House)

1753, 1871 and later. 2700 Commodore Oliver Hazard Perry Memorial Hwy. (U.S. 1)

The Weeden farmhouse is another one-and-one-half-story, five-bay farmhouse, with an exceptional spread to the roof which is enhanced by a slight flared curvature of the roof above the eaves. The roof sits tightly on a rather large transomed door with pairs of smallish vertical windows to either side, which gives the elevation its particular character. The three dormers, possibly together with the two-story ell, were probably added later when it was used as a summer house. Originally the land belonged to John Hull of Boston, one of the Pettaquamscutt Purchasers, as part of a 600-acre farm stretching to the shore. Colonel Bowen built this house for his daughter, but it is generally known for Wager Weeden, who purchased it in 1826, and his descendants, who occupied it into the twentieth century. Although his grandson William inherited the house, Weeden's only daughter, Elvira, continued to live here for sixty-two years; it was she who named the place Willow Dell. They fronted the drive with stone piers inscribed with the new name. To the left of this entrance, they added an animal drinking trough (1876), which combines the bluntness and elegance appropriate to its double function as a practical device and as a memorial. It is a tall, conical-shaped boulder with a hollowed stone trough at its foot, the top left as is, the bottom third carved as a smooth, inset surface, but with a cresting which rises in simple straights and curves to a climax around the circular water spill. Inscribed on this surface is "To Wager Weeden who lived on this farm from 1826 to 1863 and brought this water here. Matunuck, 1876." Around the same time William added the large ell to the house. So the farm took on the trappings of an estate—a frequent metamorphosis during the late nineteenth century along the Post Road.

SK49 Elizabeth Perkins House

1954, Rockwell du Moulin. 2464 Commodore Oliver Hazard Perry Memorial Hwy. (U.S. 1) (not visible)

When the Museum of Modern Art published a slender guide to New England modern architecture just before World War II, Rockwell du Moulin was among the architects included, represented by his bathhouses at Matunuck, now destroyed. He lived in a nineteenth-century shingle-and-cobblestone house in the Matunuck Hills. From there and Providence he practiced while also teaching at the Rhode Island School of Design. This is a nice example of the vertical-boarded, cypress-sided house with white trim in the manner popularized by Marcel Breuer, typical of New England modernism during the decade after World War II.

SK50 Sibley Smith House

1942, Rockwell du Moulin. 2358 Post Rd. (not visible)

Another, slightly earlier, version of Breueresque modernism, this flat-roofed house, shaped as a splayed *L*, uses the granite blocks of an old barn foundation for a huge chimney off one end and terrace walling. It is mostly closed toward the entrance side, with a porte-cochere on pipe supports to shelter the front door. The opposite side, mostly glazed, overlooks a bowl-like slope down to a pond. Wartime shortages are reflected inside in walls and cabinetry of cleanly detailed plywood, which shows its age. Such houses are especially vulnerable to renovation. This one deserves preservation as a handsomely modest example of its type—a hopeful image for a time which strove to incorporate the best of what was modern with the best of a regional vernacular.

SK51 Freeman Cocroft House (Croftmere)

1906, W. G. Sheldon. 568 Post Rd.

The first of the descendant houses to be built on the Watson estate after the Reverend Elisha Watson's death in 1900 was this spreading shingle house, distantly visible from the road. A wedding present to Watson's granddaughter, Mary, it overlooks a millpond Watson had excavated in 1857 for a stone gristmill (later demolished). Nearby is a wood-shingled octagonal tower with a weathervane.

SK52 Elisha F. Watson Estate

Late 19th century. 534 Post Rd.

A large tract of land running from Post Road down to the shore of Point Judith Pond was originally the 211-acre estate of the Reverend Elisha F. Watson, who acquired it by marriage. Of the estates which once lined the Post Road south of Wakefield, this provides the best sense of the quality of the grand late-nineteenth-century South County country estate, partly because its beautiful stretches of meadow landscape with distant glints of water are intact, and partly because enough is visible from the highway to give at least an impression of the estate. After a period in the ministry, the Reverend Watson retired to what he called Matunuck Brook Farm. There, except for an interruption as an army chaplain during the Civil War, he lived the life of a country squire, serving a few years as superintendent of local schools, championing the cause of temperance, and raising Shetland ponies, which were sold to Newport and Long Island estates. With time, descendants and in-laws dotted the landscape with their own houses, much as did the Hazards in their tighter enclave in Peace Dale. In toto, the family enclave includes at least ten houses of some architectural merit, most of them built between c. 1900 and c. 1965. Elisha Watson's own house (1690–1700; renovation and enlargement, 1921, Norman M. Isham), at 570 Post Road, is a large cross-gambrel-roofed house in wood shingle boasting several fieldstone chimneys and a porte-cochere entrance, as much Isham's Colonial Revival work as late seventeenth century.

SK53 **William Congdon House** (Brookfield)

c. 1690. 1930, alteration and enlargement, Albert Harkness. 159 Post Rd. (not visible)

Around 1930 this was an unprepossessing two-and-one-half-story shingled stone-ender with a lean-to behind, which had been extended by a one-and-one-half-story wing set back from the original front chimney. Albert Harkness filled out the ell to the extent and height of the old two-and-one-half-story block. He also added a two-story gabled bay window in front of the chimney. Then, set back from the front of the first block, he created from scratch another virtually as large, with a second massive center chimney flanked by two roof dormers and a grand new door in the corner where the two masses joined. Inside, some woodwork from the original house mixes with woodwork from old demolished houses, the rest being Harkness's modernized variant on his precedents. The result is a grand late-seventeenth-century manor house as a 1930s client would have liked it to be.

SK54 **Commodore Oliver Hazard Perry Farm**

1815 and later. 1920s, 1945, restoration and renovation. 184 Post Rd. (not visible)

Although much altered and expanded by the addition of a porch solarium and a service and garage wing, this one-and-one-half-story farmhouse with stone center chimney presents a memorable image from a front oblique view because of the height and breadth of the fold of the roof. And in the Ocean State any relic connected with Admiral Perry is cherished, although his alleged birth in this house in now doubted. A Japanese delegation planted an arc of cherry trees around the house in 1936 to commemorate the eightieth anniversary of Perry's opening of Japan to the West.

Charlestown (CH)

Stretching eight miles along Rhode Island's southern coastline and reaching inland an equal distance, Charlestown comprises two distinct zones: a flat, sandy, coastal plain and the hilly, swampy woodlands of the interior, which extend to the town's northern edge. These zones contain several distinct clusters of sites and settlements. From the north along U.S. 1 (which generally follows the coastline along Block Island Sound), between the highway and the shore are structures that reflect pre-European habitation, early European settlement, and agrarian interests. Charlestown is the center of population and activity for the Narragansett Indian tribe, whose lands are concentrated north of U.S. 1, with important buildings and sites on either side of Route 2. Inland, along the banks of the Pawcatuck River system, are parts of well-preserved nineteenth-century mill villages–Kenyon, Carolina, and Shannock, all of which continue across the river into Richmond—document early industrialization in Charlestown. Finally, the later advent of summer tourism is represented in cottage colonies such as Arnolda and Quonochontaug and the inns, small motels, and motor courts strung along U.S. 1 and Route 1A. Arnolda, established in the first years of the twentieth century and sited on the hills overlooking Quonochontaug Pond, contains numerous substantial dwellings, as does Quonochontaug, initially developed around the same time. In Arnolda's case, however, these are secluded and largely screened from the road by dense growth, making them visually inaccessible. Other resources for summer recreation in Charlestown are Burlingame State Park (established 1927) and the abutting Kimball Wildlife Refuge

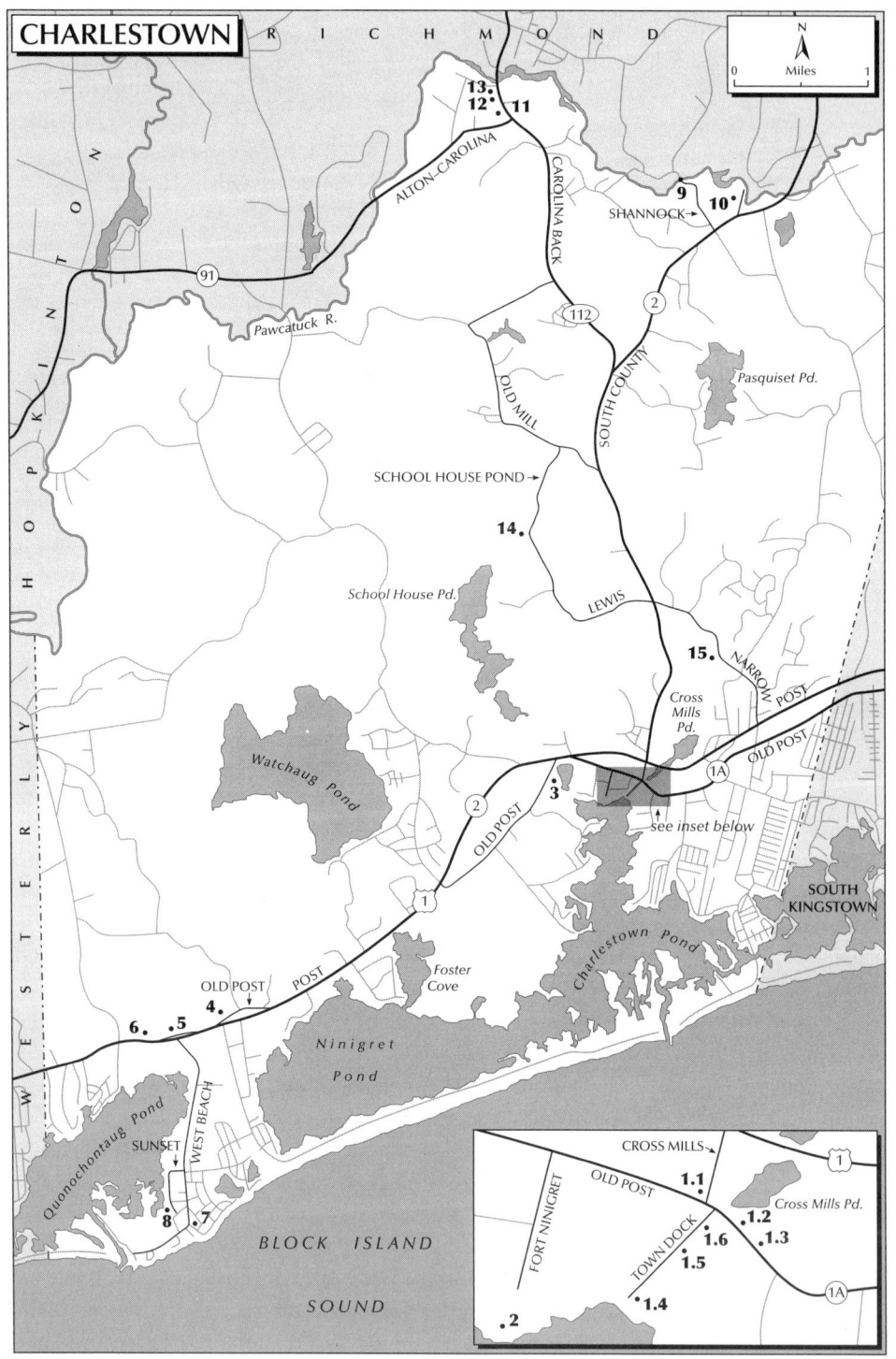

(1924) of the Audubon Society, which together preserve well over 2,100 acres for recreation and conservation, the largest such area in the state.

Cross Mills

Cross Mills, also called Charlestown Village, is on the Old Post Road, which was probably the first highway laid out through Charlestown (about 1703). It played a significant role in the area's agricultural interests in the eighteenth century. The early mill of Joseph Cross no longer exists, but in 1855 his descendants George W. and Joseph H. Cross built the stone hall around which the later village gathered and which served as a store, post office, and town meeting place. Cross Mills was a trading and farming center until the mid-twentieth century. Today, although no longer vital, it gives some sense of its past through several buildings related to its active mill and trading years and its later nineteenth-century existence as a tourist and sailing site.

CH1 **Cross Mills Center**

The rubblestone Cross's Hall and Store (CH1.1; 1855, later porch addition; 4459 Old Post Road), and the unadorned clapboarded First Baptist Church of Cross Mills (CH1.2; 1873; 4403 Old Post Road), with a gable entry vestibule and louvered belfry, are the best examples of the spartan civic buildings which survive from Cross Mills's nineteenth-century architecture. The District 2 Schoolhouse (CH1.3; 1838; 4417 Old Post Road), with its double entrances on either side of a shuttered central window, is stylistically in keeping with this plainness but was not originally located in the village, having been moved with its original foundation materials from a site north of Post Road in the Quonochontaug area in 1973 and restored.

Adjacent to the dock facilities on Ninigret Pond is the Ocean House (CH1.4; 1848; 60 Town Dock Road), a large Greek Revival building, now much altered. It once served as a public meeting hall, and those who illegally opened graves at the Royal Indian Burial Ground (CH15) in the nineteenth century were tried here. In its later use as a summer hotel, its most notable resident may have been the post–Civil War governor of Georgia, Rufus Brown Bullock. He initiated the Ocean House Regatta, famous for several decades around the turn of the twentieth century. On the village side of the Ocean House along Town Dock Road are the gabled Captain Taber House (CH1.5; c. 1840; 20 Town Dock Road) and the Cross House (CH1.6; c. 1850; 10 Town Dock Road), both, in their modest size and nominal classical detailing, in keeping with the mid-nineteenth-century date and scale of the village.

CH2 **Fort Ninigret**

Early 17th century; pre-European-contact archaeology site, 700–1300. End of Ninigret Rd.

Slightly west of the village center and on a low bluff overlooking Fort Neck Pond and Block Island Sound is this earthworks, originally constructed with vertical wood walls set into long ditches. Probably a fortified trading center for early-seventeenth-century seasonal inhabitants, the Niantics, the "fort" contains numerous utilitarian metal, glass, and clay artifacts (tools, fish hooks, brass vessels, beads, buttons, and pipes)—relics of indigenous life as well as of frequent visits of Dutch and other European traders, which provide a cosmopolitan picture of life along the New England coast even as early as the 1630s. (About the earlier, pre-European-contact phase of settlement, little is known except that it predates the use of metal axes.) About a half dozen other such "forts" exist in eastern Long Island and Connecticut, but this is the only example in Rhode Island.

Today the most visible feature of the polygonal fort are the earthen and stonework embankments. Within the enclosure, a boulder with commemorative text was installed in 1883, when the site was officially named Fort Ninigret. The iron rail fence replaces an earlier stone wall, elements of which can still be seen along the perimeter. The site is now maintained as a state park.

CH3 **King Tom Farm**

House, c. 1923. 4740 Old Post Rd.

Situated along one of the best-preserved stretches of the stone wall–lined and tree-shaded Old Post Road, King Tom Farm has been an important Native American site, a summer estate, a potato farm, and a vacation facility, tracing many of the significant changes in Charlestown over the past two centuries. The site is interesting for its sheer variety of structures: a shingled Colonial Revival main house, several agrarian dependencies (including a large potato barn and a gristmill), and a number of small rental cottages spread out over a landscape that was originally part of the Narragansett tribal lands. The farm is named after an early sachem of the tribe, Oxford-educated Thomas Ninigret. His sister, Esther, was crowned queen of the Narragansett people here in 1770, on what is now called Coronation Rock, behind an earlier mid-eighteenth-century "English" homestead built by King Tom himself. This burned in 1922, and the present Colonial Revival house was built in its place.

Post Road

Along this mid-twentieth-century divided highway, paralleled at points by fragments of the Old Post Road, are several noteworthy buildings.

CH4 First Baptist Church of Charlestown

1840, Peleg Clarke, Jr. 5073 Post Rd.

Set north of the highway on a remnant of the Old Post Road, this Greek Revival clapboarded meetinghouse has a short belfry atop its pedimental gable. The simple geometry of its front is punctuated by two entry doors trimmed, like the rest of the entry elevation, with planar moldings.

CH5 Wilcox Tavern (Joseph Stanton House)

1739. 1 Old Post Rd.

CH6 Wilcox Farm

18th century. 5193 Post Rd.

An early vernacular house with a nineteenth-century Greek Revival entrance, the Stanton House was the birthplace of Joseph Stanton, Jr., who lived a life of military and political accomplishment, rising to the rank of brigadier general in the Revolutionary War and becoming one of Rhode Island's first two U.S. senators. The size of the house, built by his father, Joseph Stanton II, as a five-bay, center-chimney residence, is in keeping with the local status of the Stanton family, which at one time owned large tracts of land in Charlestown. Purchased by Edward Wilcox in 1811, it later became a store and a stagecoach tavern. A prominent nineteenth-century monument to the younger Joseph Stanton, a 20-foot granite obelisk surrounded by an iron rail fence, sits between the house and the divided highway. The residence to the west, the Wilcox Farm, is a large saltbox-shaped farmhouse, a good example, unique in Charlestown, of that eighteenth-century vernacular type. It too was once owned by the Stanton and Wilcox families.

Quonochontaug

West Beach Road provides access off the highway to a late nineteenth- and twentieth-century beach community with a small yacht club, summer residences, and rental cabins. It also leads to two important and very early coastal farmhouses.

CH7 Sheffield House

Before 1700 and later. 600 West Beach Rd.

Parts of this low, one-and-one-half-story gambrel-roofed house, reputed to be the oldest residence in Charlestown, may date from as early as 1685. Built by a member of the Stanton family of original settlers, the house retains the name of an in-law, Nathaniel Sheffield, to whom it was deeded in 1753. Its large stone chimney, off-center entry, and shingled mass make it a clear example of the early architecture of this area, but recent alterations and large ell additions have changed the original plan of the house.

CH8 Babcock House (Whistling Chimneys)

c. 1700 and later. 163 Sunset Dr.

The Babcock House is picturesquely sited in a marshy area, surrounded by old stone walls and set in approximately forty acres of what is now conservancy land, hint at its original agrarian use. Oral tradition as well as its adjacency to the water also suggest another early function as a trading center. It may have been built by the Stanton family. At its core is a massive hearth of huge granite stones, which has anchored it through numerous hurricanes. Its asymmetrical five-bay entry elevation incorporates later alterations, including double-windowed dormers and a more recent bayed central window added in the 1940s.

CH8 Babcock House (Whistling Chimneys)

The hearth, the structural framing, and the composition of window openings offer indications that the Babcock House probably started out as a smaller one-and-one-half-story structure that was enlarged with an early eighteenth-century two-story addition before the elevation of the original section was itself enlarged by the addition of a second floor. The house has well-preserved interior spaces and finishes from its earliest period as well as from its later life as an inn and, in the twentieth century, as a residence. It is listed on the National Register of Historic Places.

Mill Villages

At the northern boundary of Charlestown lie several villages associated with the mills that thrived along the Pawcatuck River in the nineteenth and early twentieth centuries: Shannock, Kenyon, and Carolina.

CH9 Shannock Mill Dam

Early 19th century. Pawcutuck River at Shannock

On Shannock Road off Route 2 lies the Charlestown portion of Shannock, which contains a number of recently restored double mill workers' houses and a few residences of more imposing size with Greek Revival detailing. Except for the engineering of the horseshoe-shaped, ashlar block dam and adjacent headgate, waste gate and raceway on the upper falls of the Pawcatuck, much of Shannock's significant architecture lies on the Richmond side of the river.

CH10 Kenyon Mill Access

36 Sherman Rd.

Also off Route 2 is the easiest access to the large Kenyon Mill complex; see RI2 for a detailed description), whose village is most notable for the way it is tucked against the enormous hulk of the mill and its machinery, literally squeezed up to the river itself—a fitting metaphor for the way the mills shadowed every aspect of the daily life of their workers in earlier days. It is one of the few large mills in the state which still serves the textile industry.

CAROLINA

About seven miles north from U.S. 1 on Route 112 is the portion of Carolina that lies in Charlestown. One either side of the road are a church and residences that were related to the mill complex sited on the other side of the river in Richmond. The rural character still evident today along the tree-lined road gives a clear sense of the village environment; the buildings on the right (east) side of Route 112 have been altered more than those on the left (west), which retain their wooden siding and detailing.

CH11 Carolina Free Baptist Church

1845. 504 Route 112

The most prominent of the forty structures remaining on the Charlestown side of Carolina, this church was originally located south of the village on land donated by mill owner Rowland Hazard. The Greek Revival building was moved to its present site twenty years later. Its granite block foundation raises the wood-framed block at the street side an entire floor. The small, square belfry sits at the peak of the street-end gable. The interior was renovated during the move and later in the 1880s.

CH12 Stephen and Martha Bates–H. Champlin House

c. 1850. 520 Route 112

This is a finely detailed, hip-roofed Greek Revival residence with fluted Doric columns supporting a porch that continues around three sides of the house. Parts of the facade are ornamented with asymmetrically spaced pilaster-like panels, perhaps the work of a local carpenter-builder. The entrance, at the left of the street facade, is set back from the road behind mature

CH12 Stephen and Martha Bates–H. Champlin House

CH14 Narragansett Indian Church

hedges on a small embankment. Carriage sheds, a privy, and a well head are at the rear of the property. Next door, at 516 Route 112, is a much more modest residence with a well-preserved eyebrow fan window on each gabled end.

CH13 **John W. Money House**

c. 1850. 522 Route 112

The grandest of the houses on the Charlestown side of Carolina, this two-and-one-half- story Italianate villa with a front entry porch was clearly meant for someone of higher professional rank than neighboring occupants, perhaps a manager or supervisor and his family. Its height and size are accentuated by its broad, bracketed eaves and the rise of its lot overlooking the river, so that it looms in a stately manner over smaller buildings nearby.

Narragansett Tribal Lands

The tribal lands and management area of the Narragansett cover almost nine square miles in Charlestown. They encompass a number of primary historical sites, burial grounds, and buildings related to tribal history and are the center of tribal population and activities. Although the Narragansett were "detribalized" in 1879, when their ancestral lands were sold to the state, and did not reorganize as a community until the 1930s, nowhere else is the continuity of heritage and environment for Rhode Island's Native American population so much in evidence as in Charlestown. Together with other Native American sites such as Fort Ninigret and King Tom Farm, they are an unparalleled cultural resource in the state.

When visiting sites on the tribal lands, it is important to obtain permission from the Tribal Administration Offices, 4375B South County Trail (Route 2). The administration office is also a good place to start a driving tour. Just south of the office, on the west side of Route 2, are the Long House, whose roofline gives it the appearance of a large Quonset hut set atop a stone foundation, and the Health Center (1997, Mark Comeau), a low, shingled structure with heavy stone details.

CH14 **Narragansett Indian Church**

1859. 1995, rebuilt. Indian Church Rd. (At the Health Center turn right onto Old Mill Rd.; after about .25 mile, a small sign on left denotes Narragansett Indian Church Rd. Another .75 mile down a rough dirt road leads to the site.)

Surrounded by uninhabited woodlands, this church dates its history to a Baptist congregation founded in 1750, which included among its early pastors the Reverend Samuel Niles, the prominent Indian preacher. Replacing an earlier wooden structure, the granite block walls of the current church rose in 1859 and give the moderate-sized building an impressive, stolid presence that overpowers the simple neoclassical wooden trim. Such detailing, along with the gabled front and placement of entry doors at either side of an elevated central window, is reminiscent of other public buildings in Charlestown from the mid-nineteenth century, such as the District 2 Schoolhouse and the First Baptist Church of Charlestown. The church walls, mostly regularly coursed and undressed, with quarry marks still visible, reflect the highly regarded traditions of stonemasonry among the Narragansett people and have withstood much since their construction. After two fires, the most recent in 1994, the wooden elements were rebuilt and the granite was cleaned to its present condition. Close by are meeting grounds used for an annual summer gathering and an adjacent small cemetery; farther north is a much larger cemetery with as many as 700 graves, largely unmarked. All clearly form a center for the spiritual and communal activities of the Narragansett tribe today.

CH15 Royal Indian Burial Ground of the Narragansett

Off Narrow Ln. (1 mile east from Route 2; .5 mile to access road on right [west], marked with tribal signs)

Up a steeply curving dirt road and surmounting the top of a small hill lies the traditionally designated ancient burial ground of Narragansett sachems. The burial area (about 20 feet by 100 feet) is delineated by an iron rail fence and stone historical marker erected by order of the Rhode Island General Assembly in 1878 (ironically just before the state purchased the tribal lands). Originally wooded, the site was leveled by a great storm in 1869; it has been kept clear since then. The graves were looted in the nineteenth century, and when the land was acquired by the state, the remains within the enclosure were removed. Today the plot is surrounded by second-growth pine and oak that obscure its view south to the Atlantic Ocean. Despite these changes, it continues to be a place of great historical importance to the tribe.

Westerly (WE)

[Editorial note: For overviews, see the introductions to Watch Hill and the city of Westerly.]

WE1 Shelter Harbor

1915–1930, original summer cottages. Off Wagner Rd.

This summer colony enclave sits on a point of land projecting into one of the large saltwater ponds which provide a zone of wetland basins behind the beaches. Beyond a stone entrance gate is the Shelter Harbor Inn, much enlarged from a two-and-one-half-story, five-bay Greek Revival house, and beyond that the beginning of Wagner Road: for *Richard* Wagner, as it turns out. This community was founded around a group with musical interests. Wagner Road, the spine of a grid, is paralleled by Gounod, Verdi, Rossini, Donizetti, Caruso, Gershwin, Haydn, and Schubert, crossed by Brahms, Grieg, Handel, Schumann, and Liszt, with Bach at the point. Although more recent houses have been added to the community (which is no longer so exclusively musical), it is in the scattered early, mostly bungalow-derived cottages in shingle or clapboard that the melody lingers on. Some of these have two-story windowed "studio" living areas, which could double as intimate performing spaces.

WE2 Shingle Houses, Weekapaug

1880s. Ninigret Rd., Wawaloam Dr.

Of all the summer enclaves along Westerly's south shore, Watch Hill and Weekapaug are the most visibly elegant. Whereas Watch Hill enjoys a national reputation as one of New England's premier resorts, Weekapaug's renown is less widespread. Architecturally, however, it may be the more interesting of the two.

Large, comfortable houses along Ninigret Road, from its jogged intersection with Nushka Road and Wawaloam Drive to its end, encompass versions of the New England shingle tradition ranging from 1890s Shingle Style to its postmodern revival. At Ninigret and Shawmut is a fine 1920s shingled gambrel, low, spreading, but quite reticent. Immediately adjacent by way of contrast, a late 1980s version of this New England shingle tradition. Carefully designed, it is nevertheless self-assertive in the variety and large scale of its openings, the flamboyance of its deck, the big center arch, and the cuts into the volume of the house to dramatize the interaction between indoors and out. Characteristically for the postmodern image of recreation, the allusion to past house forms is crossed by yachtlike crispness, openness, and dazzle.

But Weekapaug's climactic architectural event is one of the finest surviving rows of shingle cottages in the state, at 32–50 Wawaloam Drive. Their impact is extended by their siting,

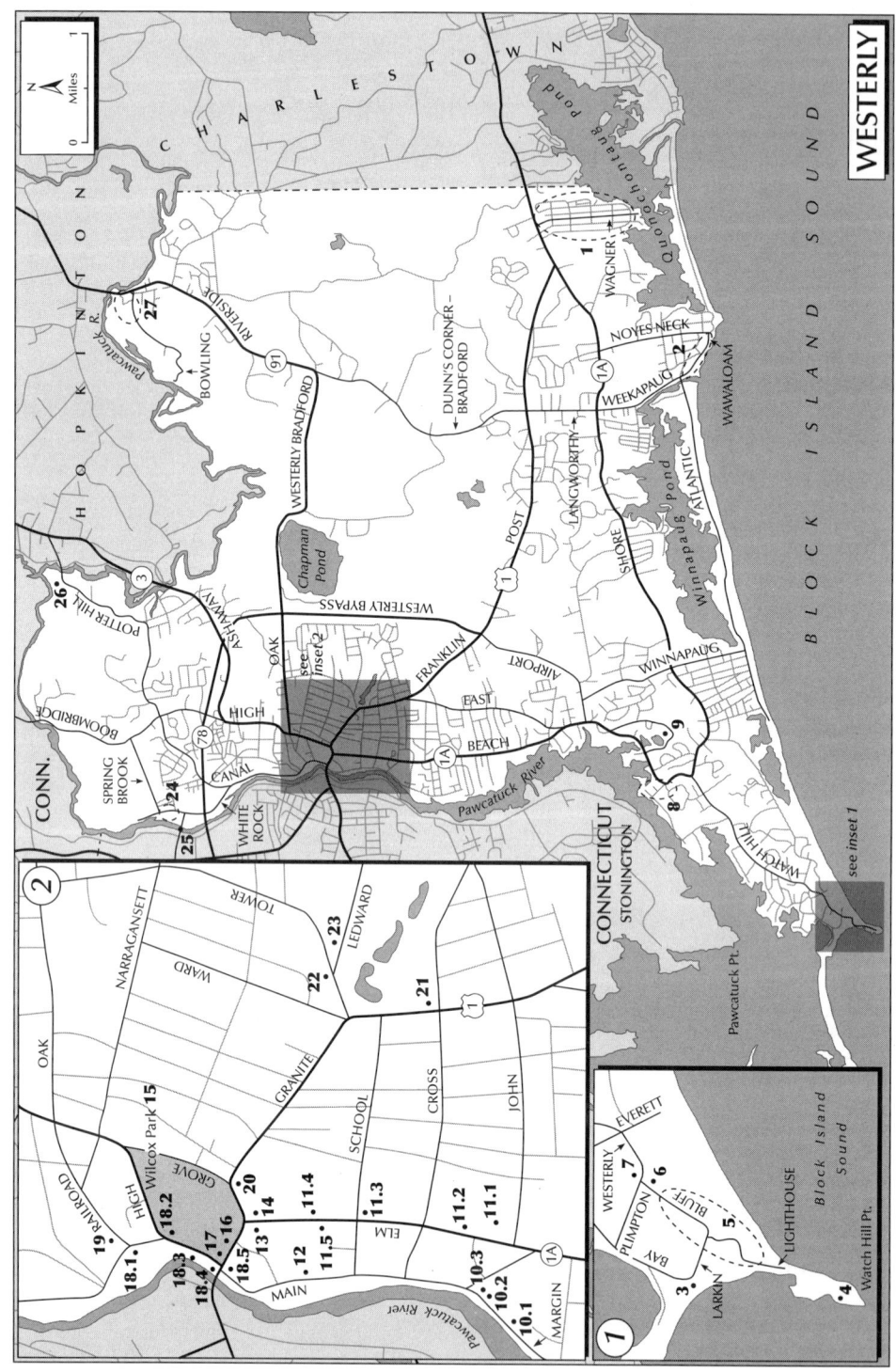

WE2 Shingle houses, Ninigret Road: 1920s (left) and late 1980s (right)

well back from the road on a slope overlooking the water. Ample rather than overwhelming in scale, they are uniformly handled, all in natural weathered shingle with white trim. Especially remarkable, most are roofed in wood shingles (rather than asphalt). This textural uniformity of all surfaces, whether wall or roof, intensifies the volumetric quality of the houses. Poignantly, the earliest in the row is a tiny fisherman's cottage (c. 1885) which once stood alone on this shore and now stands as the core to many additions and the seed for the shingle row which followed. The next stage appears in a couple of asymmetrical houses with corner towers (c. 1890). By far the majority, however (c. 1900 and later), take their cue from the eighteenth-century gambrel-roofed house as part of the Colonial Revival. Typical for their period, they inflate their sources. (None of the architectural reticence of the more archaeological approach of the 1920s here, but something prophetic of and influential on the postmodern, as amply illustrated by the two easternmost houses in this group at numbers 52 and 54.) Most have flank gambrels oriented to the width of the house, so the roofs fold down around second-story gables, with front porches either tucked under the mass of the house or projecting from it. The puffy fullness of these shapes suggests a flotilla of sailboats running before the wind. A few are set gable end toward the street, with the porch beneath. Classical columns and Palladian windows are scattered about as emblems of the colonial past. Historically, the shingled gambrel-roofed house and the classical details belong to different periods of the colonial past. No matter, it is aura rather than history which counts.

Watch Hill

The fame and the obvious well-being of this privileged promontory raise high expectations. Surprisingly, however, there is less to say about Watch Hill architecturally than one would expect: the reticent ensemble and ambience of grand comfort about the place are more memorable than the architectural qualities of specific houses. There are plenty of large houses, hand-

WE2 Shingle houses, Wawaloam Drive

somely situated on a rocky point which rises sharply, open ocean on one side, the mouth of the Pawcatuck River on the other. The promontory offers a remarkably varied terrain of knolls, depressions, and tiny ponds which provides exceptional building sites and requires winding roads. The houses are less oriented to the roads than fitted to the terrain. During the summer season, a vigorous exclusionary parking policy conspires with the concealed quality of many of the houses to emphasize the environment over the architectural image. Even a closer, off-season view, however, reveals no masterpieces of Newport caliber. Newport ostentation is hardly to be expected—or wanted—in Watch Hill. The mostly low-key assurance of the houses is Watch Hill's style and charm.

Likewise in the center of Watch Hill, surprisingly little absolutely demands attention. Although lively and pleasant, its nondescript late nineteenth- and early twentieth-century shingle and clapboard buildings have been cut up and altered so many times that they are best described as shanty boutique or movie-lot ice cream parlor. Their primary virtue is the sense of community they impart by the mix of white paint and natural shingle which they have in common, some attempt at linkage by false fronts and arcading, their reticence overall (which some recent stridency threatens), and the sense they impart of summery cheerfulness. They let the vacationers, boats, and water command the scene. A little park on the harborfront provides a modest civic presence, the result of efforts of the Park Commission, which existed from 1908 to 1910 to improve Bay Street (an offshoot of the Citizen's Improvement Society, which dated back to 1888). A summer resident and landscape architect, Marian Coffin, laid it out. Another woman with Rhode Island connections, Enid Yandell, who had studied with Auguste Rodin, is responsible for the bronze statue of Ninigret (cast in Paris, 1911, placed on another site nearby in 1916). The local Narragansett chief who befriended the colonists kneels, holding in either hand a blackfish, which originally spouted water.

WE3 Flying Horse Carousel

c. 1876, Charles W. Dare Company. 1879, set up in Watch Hill. End of Bay St.

The most distinctive element in Watch Hill's commercial center, sited toward one end of the street, where a long spit of land marks the westernmost reach of Rhode Island, is the Flying Horse Carousel. Its unprepossessing octagonal shelter, supported on cylindrical cobblestone columns and enclosed in a white picket fence, belies its historical interest. It is probably the oldest continually operating carousel in the country and the only operative example of a type which precedes the merry-go-round. The steeds do not prance on rods in the familiar churning movement. They swing out on cables. Originally, a real horse animated the fantasy, then waterpower; now it is powered by an electric motor. The twenty horses are simply carved, each from a single block of wood, in the stylized full gallop of rocking horses. Reportedly, a traveling carnival abandoned the carousel in 1879. It has been a Watch Hill delight and trademark ever since.

WE4 Watch Hill Lighthouse

1856, lighthouse. U.S. Coast Guard Station, 1907–1908. Light House Rd.

WE5 Watch Hill Houses

c. 1880–early 20th century. Visible from lighthouse grounds

The Watch Hill Lighthouse, now automated, marks the eastern shore of the northern approach to Long Island Sound. The present square granite tower replaced a shingled lighthouse of 1802, which the federal government took over in 1806. A United States Coast Guard Station, established on the point in 1879, served until 1938, its present structure dating

WE5 Watch Hill houses

from 1908. This property offers perhaps the most representative view of what is best about Watch Hill houses.

Until the 1880s most vacationers in Watch Hill stayed in the resort's large wooden hotels, of which there were six by 1870. "Cottage" building began in earnest only after the successive purchase of two large farms on the point by two syndicates of Cincinnati developers in 1886 and 1896. Aside from Cincinnati, the majority of the summer colonists in the late nineteenth century came from Pittsburgh, Washington, and New York. Most houses in Watch Hill date from the 1890s through the 1930s.

Sea Swept, originally called Ocean Mount (c. 1880, attributed to George Keller), 79 Lighthouse Road, is a pretty board-and-batten cross-gabled cottage with stencil-cut bargeboards on deeply bracketed eaves. This house displays a panoply of window hoods, arched windows, dormers, roof gables, and an arched porch with spindle supports. It is a rare survivor from the first phase of Watch Hill's existence as a resort of summer "cottages." Behind the 1880s house, in a long arc, the three largest shingled houses in this view are (left to right) Moana, originally Aktaion (1906; Edward F. Hinkle, architect), 44 Bluff Avenue; Trepasso (1906–1907, also Hinkle), 45 Bluff Avenue; and By-the-Sea (1879, attributed to George Keller, architect), 47 Bluff Avenue. Then, largest of all, the rambling white clapboard French Provincial–Moderne–Regency Highwatch, originally Holiday House (1931), 50 Bluff Avenue, somewhat altered since construction but compelling nevertheless for its climactic pyramidal composition, dramatically punctuated by massive monolithic slab chimneys that thrust into the skyline atop the crest of Watch Hill.

WE6 Ocean House

1867–1868. 1903, enlarged. 58 Bluff Ave. (corner of Westerly Rd.)

Of all the wooden hotels built in the nineteenth century to accommodate Watch Hill vacationers, this and the small, much altered Narragansett Inn (1844) on Bay Street are all that remain. The history of Watch Hill as a resort begins, according to legend, with lighthouse keeper Jonathan Nash, who took in boarders and then resigned his post to build Narragansett Inn. His son George built a predecessor to this version of Ocean House. Originally this building was L-shaped, consisting of two three-and-one-half-story mansarded wings at right angles to a five-and-one-half-story tower block with hipped roof. Plain additions were made in two wings to the rear in 1906, when apparently the columned porch along the west and south was also redesigned. Later in the twentieth century, the south wing and its wrap-around porch were extended five bays farther to the east, with a hip-roofed fifth story above. As such structures go, beyond its veranda columns, bracketing under the eaves, and giant pilasters at the corners of its central tower, the hotel is both plain and crude. Still, the accumulated product of two or three lumberyards of boarding cannot fail to be impressive on such a spectacular site—especially when this vast ark must today stand for so many comparable establishments which once lined Rhode Island's shores. None

other of this order of grandiosity has survived. [As this book goes to press in summer 2003, the building is threatened with extensive demolition to accommodate a swimming pool, covered parking beneath extensive terraces, and a string of cottages—all along the side overlooking the beach below.]

WE7 Tredegar (Russula)
1900, Chapman and Frazer. 215 Westerly Rd.

This particularly fine shingled Colonial Revival house was built as a rental property together with the adjacent house by the same firm. In Tredegar the bay windows toward the street are held under the projecting gambrel roof, thereby incorporating them into the overall shape of the house.

Avondale

WE8 Avondale Village
A substantially intact village stretched along a U-configurated road system, Avondale retains a number of nineteenth-century buildings and some lingering sense of the mixture of agriculture, fishing, and the boatyard crafts which traditionally characterized such tidewater villages. The original eighteenth-century owner of the land in the area recovered from bankruptcy by parceling his lands and disposing of them by a lottery, commemorated in the original name of the village and in Lottery House (WE8.1; c. 1840), a delicately scaled Greek Revival house at 15 Avondale Road. The more vigorous, gable-ended Greek Revival Captain Palmer Hall House, India Point (WE8.2; 1840), at the end of India Point Road, enlarged by extensive later additions, includes a mid-nineteenth-century barn. The Greek Revival Avondale Chapel (WE8.3; 1852), at 13 Avondale Road, has all the basic elements set out with utmost clarity: door with side lights and molded cap; gable as pediment; unadorned cupola; tall, slotlike windows along the sides. The house at Avondale Farm (WE8.4), 2 Avondale Road, is a gaunt but forceful two-and-one-half-story, five-bay, central-entrance type, except that the second story of the north elevation, now the facade, contains only three windows. As it appears today, it is actually a vernacular Colonial Revival editing away of extensive Victorian additions (only

vernacular Greek Revival entrance frames remain) to an eighteenth-century farmhouse. Fields to one side were lost to large-house development, but the fields to the south were donated to a land trust and remain open as a fine setting for this house. The tapered cobblestone wall in front of the farmhouse is of a type common around Westerly, and worth mention in a state famous for its stone walls. It is generously mortared in a Mediterranean manner, with conical posts at openings. (Another example in Avondale exists across from Lottery House.) These are the work of a Westerly mason active in the mid-twentieth century, Sam Nardone, who brought Italian techniques, together with indefatigable enterprise, to the area. So rarely can one trace the source of such local building quirks that it is a pleasure to record this instance, especially since Italian and Portuguese touches, especially in masonry and stylized gardens using gravel and flagging instead of grass, are pervasive ethnic components in Rhode Island building.

Also at Avondale Road and Champlin Drive, across from the fork, a summer residence that was formerly School Number 3 and then Westerly Grange Number 8 (WE8.5; 1873, later additions to the rear) presents an exceptional front for its original use: a door with projecting hood topping a flight of stairs, squeezed by flanking windows with gabled lintels and, above, a fanlight separated in a floating way from all the rest. It is an interesting example of disparate elements and the tensions created (however unconsciously) by pushing them together or pulling them apart.

Westerly

Westerly, the principal city historically—really the *only* city—in southern Rhode Island, is rarely cited for the merit of its buildings. Yet there is much to see. What one expects architecturally of Watch Hill, one finds in Westerly. Above all, the spruceness of the city impresses the visitor. For whatever reason, it lacks the areas of dereliction and decay of other Rhode Island industrial cities, possibly because of a more middle-class work force, possibly because of a tradition of civic care. In some indefinable way, it seems more akin to the small cities in northern Connecticut than to those in Rhode Island and, in fact, it is substantially oriented to the adjacent area across the state boundary. It is perhaps the most contained of all Rhode Island cities. Whereas others in the state, with the possible exceptions of Woonsocket and Newport, typically run together as a metropolitan agglomeration, Westerly seems an entity, the largest of its small cities. In contrast to Woonsocket, Westerly has a visible urban core; in contrast to Newport, it seems (deceptively, perhaps) more secure against the pressures of rampant development. Although it grew as a center where highway, railroad, river mouth, and ocean come together, I-95, which smashes through most of Rhode Island's cities farther north, swings wide of Westerly. Hence it now possesses something of the character of an enclave; but, it must be confessed, a very busy enclave, with slow-moving traffic forever winding through its pinwheel hub. Its center boasts big houses with spacious grounds immediately adjacent to a predominantly early twentieth-century business center, with a fine park between them.

WE9 Klotz House

1969, Charles Moore. 10 North Bottom Ridge

The Klotz House is probably the single most important modernist house built in Rhode Island during the 1960s. It was the first house on the East Coast by Charles Moore, designed before he left the West Coast for an interval as dean of the School of Architecture at Yale. At that time Moore and his partners had gained national attention for a community of shed-roofed, vertical-boarded houses at Sea Ranch, a seaside development north of San Francisco, and important as a key venture in what became a popular mode of combining forms from traditional wooden vernacular building with those from modern architecture. The Klotz House takes its cue from the rocky outcropping on which it fits. A two-story polygonal shape in vertical boarding, more drastically free-form than Moore's earlier work, it is fitted to its site, with angles so gentle as to seem more curved than angular, but slashed here and there by sharp, angular cuts. At one end, the umbrella-shaped roof (again composed of angles, but so many and so gentle as to seem curved) projects on attenuated columns to create a dramatic—possibly over-dramatic—shelter into which broad, strata-like stairs wind and narrow up to the entrance terrace. Windows seemingly placed at random and randomly sized are aimed at views of woods and distant water all around.

WE10 Margin Street Houses

c. 1720–c. 1900

For part of its length, Margin Street runs behind a narrow linear park along the Pawcatuck, with Connecticut industry on the opposite bank. The most important house on the street is the John Lewis–Captain William Card House (WE10.1; c. 1720; 1929–1930, restoration, Norman M. Isham), at number 12, a central-chimney house with flank gable roof raised on a stone basement. Two Neo-Colonial houses (WE10.2, WE10.3; c. 1873, c. 1896; c. 1840, 1900), at 2 and 4 Margin Street, both turn-of-the-century reworkings of mid-nineteenth-century dwellings, merit a look. Number 2 began its existence as a Second Empire house with large cupola centered on its mansard roof, while the appeal of number 4, with its attenuated, semielliptical porch and dormers, results from the reworking of a Greek Revival house.

WE11 Elm Street Houses

c. 1840–1930

What Benefit Street is to Providence, Elm Street is to Westerly. Both are lined with historic houses. Both are upscale residential streets, with some mix of more modest houses. Both relate residential living intimately to their respective business districts (in Westerly partly by the seeming continuation of Elm Street to the north of Broad Street as Grove Avenue). In fact, Elm Street boasts one of the most important collections of large nineteenth-century houses in the state, but—and here's the tragedy—aluminum and plastic siding have devastated the street. It is understandable that low maintenance costs (in the short term at least) are bound to take precedence over architectural quality in modest housing. It is incomprehensible, however, that owners of such large houses as these should have sought the gains of a few thousand dollars in maintenance at the cost of tens of thousands of dollars in architectural value, magnified by the cumulative glory of what Elm Street might be were it returned to its nineteenth-century splendor. Given the will to do so, this would largely mean undoing what has recently been done.

Just north of John Street, at 66 Elm Street, is the Georgian Revival Cottrell House (WE11.1; early twentieth century; now condominiums), set down in ample grounds now filled, like saplings from a big tree, with smaller simulacra of the house itself, once occupied by owners of a handsome masonry factory, still standing on Beach Street, which manufactured Cottrell printing presses. Mansions of other late nineteenth- and early twentieth-century leaders in the Westerly economy range the length of the street. At number 54 is a Queen Anne house (WE11.2; c. 1880) with spindle and panel porch, fishscale shingling on the second floor, clapboard on the first, all dominated by a corner octagonal tower with bulbous dome. At 34 Elm Street (southeast corner of Elm and School streets) is a fine (though modified at the ground floor), relatively modest Gothic Revival house (WE11.3; c. 1865). Between School and Broad streets, the east side of the street shows a group of large, mansarded houses (note especially numbers 22 and 24–26) (WE11.4; c. 1870–1875), mostly re-sided, but with detail intact. Across from these, at number 25, is a later, more "colonialized" Queen Anne gambrel-roofed cottage (WE11.5; c. 1895), which represents something of a re-siding tour de force in the effort to retain the high quality of what was there. Finally, relief, at number 8, a Greek Revival house (WE11.6) mercifully undesecrated. With its dominating entablature and corner pilasters, it possesses the formidable presence of mid-nineteenth-century mansions derived from Renaissance palaces. The deep porch across the front is more Victorian than Greek Revival in spirit. Its original picket fence, the delightful "Gothic" window in the weather door, and the generous windows opening down to the floor of the porch—incongruously for a Greek temple, but delightfully for a house—mitigate the severity of the initial impression. It was the home of the Babcock family, who willed it to Christ Church, across the street, for use as a rectory.

WE12 Pawcatuck Seventh Day Baptist Church

WE12 Pawcatuck Seventh Day Baptist Church

1847–1848. 1886, enlargement of east end. 1927, partial interior restoration, Norman M. Isham. After 1938, spire rebuilt. 120 Main St.

Like the nearly contemporaneous former Christian Church (WE14), this is well set back on a rise, and it has essentially the same stair treatment: flanking stairs into the columned portico and up to a landing, with a right-angle turn for another flight up to the centered entrance. Originally, however, the stairs apparently ended in two flanking doors with the pulpit between them, so the present stair arrangement into a central door must date from the same time as the enlargement of the east end (1886). The similarity of the stair treatment to that of the Christian Church (was it also later altered?), the wide-narrow-wide interval between the columns, and the equally forceful handling of Greek Revival form suggests that both churches are the work of the same gifted but as yet unidentified designer-builder, who seems also to have been responsible for the Babcock-Smith House (WE21). Here the stairs, as they now exist, are actually more complex than those of the Christian Church. On the driveway side of the Baptist church, one can also reach the halfway landing from the side of the portico, and therefore climb the two flights straight up, entirely within the width of the portico. The spire, tall and attenuated, is as wide as the interval between the centermost columns, and is doubtless partially supported by them.

Inside, some of the Greek Revival fittings remain. But the principal attraction is a monumental Victorian window at the chancel end, which, with the treatment of the other windows and some refurbishing, is probably of the same date as the enlargement of the church. It rather folkishly translates William Holman Hunt's famous painting *Light of the World* (1854) into stained glass. Christ, bearing a lantern, knocks at the closed door of an unbeliever to be admitted. It is worth the climb to the entrance. Across the street, riverfront Greek Revival buildings, along with others erected up to 1870, which maintain a basic sympathy with their sturdy austerity, have been unsympathetically "restored."

WE13 Christ Episcopal Church

1891–1894, Henry W. Congdon. 1905, bell tower and spire. Elm and Broad sts.

Although its theme is the English country church, the exterior aspect of this church is less ingratiating than compelling in the bluntness with which well-crafted gray granite and blue slate make a loose composition around the culminating tower and spire. Inside, it is substantially intact, despite some modernization of the sort typical for such interiors. Plaster walls have been painted lighter than they probably originally were, and the blending and color from stenciled patterns which presumably decorated the walls have been eliminated. Chandeliers have been replaced, as has some of the chancel furniture. Even so, the impressive wooden roof structure, the pews, other chancel furniture, and a fine rood screen remain from the original interior. (Whereas the rood screen once veiled the altar in good Victorian High Church fashion, the altar has been repositioned in front of the screen, in accord with the modern theology of participation.) Above all, the voluminous space remains. The relative shortness of its length and the height of its roof give an almost domical sensation to the nave, into which the front wall of the chancel swells in a gentle curve like a proscenium stage, with generous openings into chapels to either side. It is the most ambitious late-nineteenth-century Gothic Revival church in southern Rhode Island, the product of a prosperous congregation, but especially of the munificence of its principal donor, Harry Cross, manager of the nearby White Rock Mill (see entry, below).

WE14 Granite Theater (Christian Church)

c. 1845. 1 Granite St. (near the intersection of Elm St. and Grove Ave.)

This Greek Revival church is set well back against a slope with flanking stairs rising between the outermost columns to a landing under the portico, then another flight of stairs at right angles under the portico, to reach the raised entrance porch between them. The wider interval between the columns at either end of the portico may be necessary to contain the stairs; but the different intervals between columns may also be a carryover from the arrangement typical for domestic porches, where the center two columns come close together to frame the entrance, while the flanking columns pull apart to give a better view from rockers and hammocks.

WE15 Wilcox Park

1898. 1902, landscape design for addition, Warren Manning. 1905, landscape design, Frank Hamilton. 1924, 1929. 1937, terrace near Granite St. and Grove Ave., Arthur Shurcliffe. Bounded by Grove Ave., Broad St., and (on two sides) High St.

Wilcox Park was the gift of Harriet Hoxie Wilcox, who bought the seven-acre estate of Rowse Babcock, one of the founders of the White Rock Mill. With the gift of the land went a handsome endowment. Several additional parcels were added in the park's early years, and the 1905 acquisition of the adjacent Brown estate on High Street brought the park to its current eighteen-and-one-half acres. Manning's original design established meandering paths around a central greensward, with peripheral plantings of trees complemented by understory plantings of varying density. Hamilton's work on the 1905 addition closely followed Manning's design precepts. Shurcliffe's 1937 circular-plan terrace with World War I Monument, near the intersection of Granite Street and Grove Avenue, provides a fine vantage point. The Westerly Public Library (see WE16), part of an association that owns both park and library, was most recently expanded in 1987–1992; during that expansion, the parterred terrace to its east on Broad Street was installed, creating an entrance at once intimate and impressive. Shurcliffe's terrace and the 1992 parterre introduce notes of formality that enrich the visual and sequential spatial experience. The park is walled in on the far side by the backs of the commercial buildings on High Street, which leave space for two pedestrian walkways through to the park. It accommodates the Westerly Library and Art Gallery at its Broad Street end but otherwise opens to houses on two sides as a shallow bowl—the impression of a bowl intensified by its overall shape, a near-parallelogram with two sides slightly bowed and corners rounded. The park is also the setting for a number of fine pieces of sculpture, including *The Hiker* (1904; dedicated 1924), by Allen G. Newman, and John Francis

WE16 Westerly Public Library and Art Gallery (Memorial and Library Association)

Paramino's Wilcox Memorial Fountain, installed near the library in 1929 to honor the park's donor.

Although it has many characteristics of the New England town green—centrality; a mix of businesses, institutions, and houses around its perimeter; and a bandstand (1902) as a significant focus—the sunken quality of the park and the surprise of the visitor in finding it behind the wall of the business district give it a secluded quality. Unlike the colonial green, which connects the community, it embodies the late-nineteenth-century ideal of the park as a space for respite and seclusion. Exceptionally well maintained by the Memorial and Library Association of Westerly, it is by far the finest small park in the state.

WE16 Westerly Public Library and Art Gallery (Memorial and Library Association)

1891–1894, Longstaff and Hurd. 1925–1928, Ludlow and Peabody. 1987-1992, Charles J. Koulbanis. Broad St. (corner of Broad and High sts.)

Stephen Wilcox, the principal benefactor of this building, joined with George Babcock to produce their famous water-tube safety boiler at their factory in Brooklyn, New York. Patented in 1867, it was the starting point for Babcock and Wilcox, which became one of the major producers of large turbines, generators, and (recently) nuclear reactors. Born in Westerly, Wilcox maintained a summer house in his hometown, using his steam yacht to connect work and vacation. Following the example of the Hazard family in nearby Peace Dale (see under South Kingstown), Wilcox determined to build a library building combining community facilities for Westerly. It originally contained a basement gymnasium and bowling alley, a parlor and library on the first floor, and quarters for the local chapter of the Grand Army of the Republic upstairs. The art gallery came with the second addition. Wilcox was doubtless attracted to the obscure partnership of George Longstaff and Frank Hurd because Hurd summered in the Larkin Hotel at Watch Hill. What resulted is a late example of Romanesque Revival in yellow brick trimmed in Westerly red granite. The discreteness of window and door elements, as these are set against the contrasting wall plane rather than fusing with it, foretells the imminent Neo-Renaissance reaction to Romanesque. So do the sharp-edged precision of elements such as the cylindrical turrets and the symmetries concealed within what is an asymmetrical composition overall.

From newspaper coverage at the time, it appears that Longstaff and Hurd had more commissions for interiors than for complete buildings. Much of their ornamented woodwork in golden oak exists inside. Originally this was climaxed by colored glass windows in the north wall which centered in a large window, derived from an oil painting now in the Museum of the Rhode Island School of Design and dedicated to Ninigret, the Narragansett chief who supported the local colonists. Removed for the 1920s extension, the window is still stored in the building. A delightful prospective watercolor of the building, commissioned by the architects from the important late-nineteenth-century architectural renderer Hughson Hawley, hangs over the fireplace of the original community parlor (now reading room). It was the climax of the presentation by which the architects won Wilcox to their design.

Koulbanis's addition, to the north and west

of the second addition, architecturally reinforces the institutional link between library and park. While the Memorial and Library Association of Westerly has been steward of both since their inceptions, until this addition the library building literally turned its back on the park. Like the 1928 addition, this, too, is deferentially attached to the original, especially as seen on approach from the street, and continues with the same materials (including the last red granite quarried in Westerly) and motifs. Parkside, however, the mass and fenestration swell in a bit of postmodern swagger.

WE17 **Westerly Post Office** (Former)

1913–1914, James Knox Taylor, Supervising Architect of the Treasury. 1 High St. (corner of High and Broad sts.)

But we are not yet finished with the classical revival in downtown Westerly. This is Rhode Island's finest post office. During James Knox Taylor's administration as Supervising Architect of the Treasury, long colonnades became the principal ingredient of many post offices. The sweep of this colonnade and its steps to follow the curve of the street energize the symmetrical composition. The splendid proportions of the colonnade, contained by the terminating walls and the spare but meticulous detailing (including the fine bronze lamp standards) complete the elevation. What a fortunate juxtaposition of classical buildings this triangle contains! The Industrial Trust as entrance; the Washington Trust as wall; and finally, the post office, which shows how a wall can also be an entrance.

WE18 **High Street Business District**

1870s–1920s. High St. from Canal St. to Broad St.

Five commercial variants on classical precedent exhibit different approaches from different times. First, at 67 High Street (corner of Canal Street) is a cast iron front (WE18.1; before 1850, c. 1880), in a disarming but rather crude version of Neo-Renaissance detailing, with the columns too short for their tall pedestals. But knock out the unsympathetic later infill, and this could be returned to something like its original metal-and-glass openness.

The South County Public Service Building for the local utility company (WE18.2; 1926, Jackson, Robertson and Adams), at number 53, is a fine example of the 1920s revivalist interest in the late-eighteenth-century work of Boston architect Charles Bulfinch, but, like Bulfinch himself, also looking to sources in the work of Robert Adam. As evident here (and in this firm's masterpiece in the same vein, the Providence County Courthouse [PR52]), this classical revival, in both its eighteenth-century and twentieth-century manifestations, was enormously interested in architectural motifs and details. Here a balconied window topped with a segmental pediment, a clock with another swag, pieces of parapeting, a couple of urns, and so on are trophies in limestone, displayed against the textured brick wall of the elevation, much as art objects might be carefully set against a velvet background. The brick, in turn, contrasts with the very refined linearism of the arched limestone ground story, which itself becomes the largest piece in the arrangement.

The handsomely restored Brown Building

WE17 Westerly Post Office (former), left, and WE18.5 Washington Trust Company, right

WE19 Westerly Railroad Station

(WE18.3; 1896, 1903), at number 18, also shows nominal use of classical elements in its brick and terra-cotta pilasters and arches and in minor ornamentation, popularized, even for very functional buildings, by the classical Colonial Revival. The celebration of the owner's name in a swirl of foliage in the climactic terra-cotta panel, however, derives more from late Victorian naturalistic ornament than from classicism. But the primary lesson here is in the straightforward treatment of the repetitive order of the storefronts, framed by cast iron columns, providing their own ordered rhythm to the street.

The former Industrial Trust Company Building (WE18.4; 1916), at number 14, exhibits a characteristic formula for bank fronts ultimately based on Roman triumphal arches, an image which was much favored for banks from around 1890 through World War II. The grand entrance foretells the big banking space within, while the corners provide for stacks of offices. Compared to the restraint of the Adamesque South County Public Service Building, this is a bit florid. But columns and pediment are beautifully handled, and one would hate to see this accent building disappear from High Street.

Flamboyance has a special role to play on Main Streets. Here it complements the Washington Trust Company (WE18.5; 1925, York and Sawyer), at 23 Broad Street—the best of the lot—which emphatically closes the sequence. If Industrial Trust epitomizes the entrance to a box of space, Washington Trust epitomizes the wall. One almost joins the architect as, consulting examples from Florentine palaces, he draws this elevation. He seems almost to inscribe the ornament rather than applying it; he insets one plane inside the other to the most minimal degree possible, thereby intensifying the sense of planarity. Over such a taut elevation, the boldly projecting cornice becomes dramatic. Too many such cornices of the period have been pared in the interest of safety. May this one survive.

WE19 Westerly Railroad Station

1912. Railroad Ave.

This building unexpectedly turns out to be Mission Revival in style, as though displaced from the Santa Fe or Southern Pacific. Perhaps it should be called a building cluster, with waiting room, flanking wings, and arcaded entrance portico, all locked together, but with each functional component also given individual identity by an ingenious two-tiered, hipped roof structure in red tile with extravagantly projecting eaves. The various roof pitches and tiers are pinned together by a stubby ornamental clock dormer over the entrance. Another hipped roof caps a separate platform shelter, supported on paired Doric columns, alternating with arches, all exceptionally well propor-

WE20 House, 15 Granite Street

tioned. Down the tracks, a signal tower in the same style completes the complex. Who is responsible for such a sophisticated, well-detailed building? Perhaps the New York architect Cass Gilbert, who worked for the New York, New Haven and Hartford Railroad—his masterpiece for the line being the beautifully restored station in New Haven, Connecticut. At the time the Westerly station was designed, Gilbert's office also had commissions in the Southwest, which could account for this transcontinental import to Westerly.

WE20 House

c. 1900. 15 Granite St. (at Grove Ave.)

On such a busy corner it is surprising to come across a superb Neo-Colonial house—again one wants to say, among the best in the state. A sharp-edged, crisply folded gambrel roof presents its gable end to the street as a deep overhang. Off center under extravagant serpentine brackets, the entrance, farther inset on a stepped platform, provides a handsomely sidelighted and fanlighted door with a generosity of width and window never found in the colonial period. The compositional tension that results from the centered/off-center adjustment of the door to its attached window bay above is worthy of postmodernist Neo-Neo-Colonial examples. A row of ornamented dormers pokes through the gambrel on the side elevation, beneath which a very deep porch—one of those cool, shadowy summer rooms of which no colonial ever dreamed—projects on Ionic columns toward the garden. The delicacy and precision of ornament and moldings throughout complete this delightful colonial romance read through Queen Anne–tinted glasses. But who is responsible?

WE21 Dr. Joshua Babcock–Orlando Smith House

1732–1734. Early 20th century, restoration, Norman M. Isham. 124 Granite St.

This five-bay, central-chimney house with a gambrel roof and a saltbox lean-to at the rear features an unusual enclosed entrance vestibule projecting from its front elevation, a later enlargement of the small front stair hall, perhaps when the Smiths first occupied the house in 1848. Both the proportioning of the broken scroll pediment over the entrance, no doubt the original reapplied on the addition, and the tall, narrow proportions of its nine-over-nine windows are distinctive. The wing, with its own tall chimney, served as the office of what is reputed to have been Westerly's first doctor. It also served as the first post office in Westerly, which Benjamin Franklin visited while he served as postmaster general.

WE22 Bungalow

c. 1915. 12 Tower St. (intersection of Tower St. and Ledward Ave.)

This is an exceptionally fine bungalow, with everything that a classic upscale New England bungalow should have: wide-clapboard walls, cobblestone chimneys and foundation, and spreading flank gable with trap-door dormer and revealed roof structure at the eaves, all fronted by a deep porch supported on Doric columns. Its exquisite maintenance is a mixed pleasure to the architectural tourist, because the house hides behind layers of clipped bushes.

WE23 Westerly Water Tower

1910, Aberthaw Construction Company, Thomas McKenzie, engineer and supervisor, Samuel W. Gray, consulting engineer. 12 Ledward St.

Among the first reinforced concrete standpipes built in the United States by a leading Boston engineering firm, famous for its pioneer work in concrete building technology, this cylindrical container employs the technology developed for grain elevators in the Midwest beginning in the 1880s. Here it was reportedly used for aesthetic reasons, as a civic-minded alternative to what was regarded as the blighting effect

WE21 Dr. Joshua Babcock–Orlando Smith House

WE24 White Rock Mill

of the typical metal tank. Although modernists of a few years ago would have disapproved of the simplified allusion to a classical frieze and cornice that rings the plain concrete cylinder near its top and of its domical terra-cotta cap with weather-vaned cupola, in fact these details are beautifully simplified and scaled to reinforced concrete. The domical cap itself is technologically advanced for its day, since it employs the technique brought to the United States in the late nineteenth century by Rafael Guastavino. Layers of clay tiles in beds of cement permitted strong, light, and economical domes and vaulting, which were used for roofing many of the finest monumental buildings of the late nineteenth and early twentieth centuries. A plaque at the base celebrates those responsible for its design.

White Rock

WE24 White Rock Mill and Village

1849, center section of mill. 1877, end sections of mill. 1849, 1855, houses. 1849, store. White Rock Rd. south of Spring Brook Rd.

Although still operating in textiles as a printing plant for fabrics, the White Rock Mill, at the corner of White Rock and Spring Brook roads, has deteriorated since around 1970. Constructed of brick with piers, the original center section of 1849 seems, superficially at least, to be more characteristic of brick mills built in Rhode Island at the end of the nineteenth century than of the masonry structures typical for mills of its size at mid-century. This mill, however, is not quite of the pier-and-spandrel variety which became a standard for brick construction toward the end of the nineteenth century. In these later mills, the piers *are* the walls, with big windows bridged between them. Here the piers are treated more as reinforced projections of the wall plane, with much smaller windows. Just across the Pawcatuck River in Stonington, Connecticut, at the corner of Stillman Avenue and Arch Street, is a small mill built in 1848, a year earlier than White Rock, which employs brick in the same way. This mill incorporates features which were either not included at White Rock or were lost to additions: roof dormers, eyelike windows in the end wall, and gables. But it is doubtless the work of the same designer.

The Stillmanville and White Rock mills were innovative for the time. What accounts for them? Some have conjectured that Thomas Tefft of Providence may have been responsible. Borrowing from certain progressive ideas current at the time, Tefft fiercely championed the creation of a modern brick style at a time when exposed brick was generally out of favor in the United States and masonry was normal for mills. He approved of the mechanical, standardized quality of brick as opposed to the craft quality of fitting stone. He believed that architectural ornament should not be merely applied to wall surfaces; rather, ornament should evolve naturally out of the workmanlike handling of materials and should take building structure as its cue. Both the piered wall and the corbeled eaves of the central section of the White Rock Mill meet these criteria. They also suggest the influence of Italian medieval brick buildings and their early nineteenth-century revival in Germany. Tefft had all these interests, but one cannot yet claim that these mills are his. Rouse Babcock and Jesse L. Moss of Westerly commissioned the original mill—a central tower with ten window bays to either side. The omnivorous nineteenth-century textile barons,

B. B. and R. Knight, took over the plant in 1873, and four years later they doubled the original block symmetrically with a gabled transitional stair unit and an additional ten window bays at either end.

The symmetry of the mill is matched by what is perhaps the most symmetrical mill village in Rhode Island. Originally five double houses with tiered porches to either end and low hipped roofs sloping up to interior chimneys were sited, generously spaced, to either side of the intersection of White Rock and Spring Brook roads (75–85 White Rock Road). A white picket fence with granite posts separated the houses from the street, while on the opposite side of the street the mill also had its white-picketed zone. A brick building that was once the company store, dating from the 1870s building campaign, still stands at the principal intersection. An earlier brick schoolhouse, diagonally across the street from the store on the mill side, has been demolished. Later housing was built at the far end of the village, and lodging houses for single workers marked either end of the town. Altogether, a visitor described it in 1869 as the "pleasantest and neatest manufacturing village in New England." No more, however. The tower of the original mill has lost its decorated cap, its roof, a row of dormers. Sadder, all of the mill's windows have been crudely replaced, and its granite-lined canals filled with trash. Most of the houses have been altered, and the picket fences are in fragments. Still, the sense of the original village remains and, with care and pride, it might approximate the "pleasantest, neatest" status it once enjoyed.

WE25 **Bridge Street Bridge**

1906. Bridge St. across the Pawcatuck River

The bridge for which Bridge Street is named was in fact two bridges, the original for vehicles and pedestrians, supplemented in 1906 by another for trolleys. Each bridge had two spans, first over the mill canal, then over the Pawcatuck. The highway bridge has been replaced, but the trolley's survives. The two segments present a nice comparison in nineteenth-century metal bridge technology. The short span over the millrace are Pratt pony trusses of a type built since 1844 and still in use today. The long span over the River are Baltimore trusses of a type used between 1831 and the beginning of the twentieth century.

Potter Hill

WE26 **Potter Hill Mill**

1847. Potter Hill Rd.

This derelict but picturesque mill village just inside Westerly on its Hopkinton border marks the point where the Pawcatuck breaks out of Rhode Island to mark a winding boundary between this state and Connecticut. A sizable three-story mill building in smooth ashlar granite with a truncated four-story tower commands a scenic bend of the Pawcatuck, swollen by a curved dam. (Coincidentally, this mill, built two years earlier than White Rock, conveniently exhibits construction for ambitious factory buildings typical for the 1840s, and reminds us again of the radical nature of White Rock.) Until it burned in 1977, a companion building in wood stood beside the stone mill.

Bradford

Bradford, in the northeast corner of Westerly on the Pawcatuck, is included here, but the economical traveler may prefer to tour it in connection with Hopkinton.

WE27 **Bradford Mill**

1864, c. 1912, mill buildings. Mid-19th–early 20th century, housing. Main St. (Route 91-216) (mill), Bowling Ln. (housing)

Originally named Dorrville for the leader of the so-called Dorr Rebellion, the town subsequently changed its name with changes in corporate landlord: first to Niantic, for the Niantic Company; then, after 1911, when an English company took over, Bradford, to honor the textile manufacturing town which housed its headquarters. The plant offers the clear contrast of a typical mid-nineteenth-century verticalized masonry factory building and a typical early twentieth-century horizontal brick counterpart. Windows in walls light the narrow floors of the earlier building; north-facing skylights in the sawtooth roof light the expansive floor of the later building. As for the mill workers' housing which lines Bowling Lane, aficionados who manage to look beyond the dilapidation of much of it will see a range of mill housing types from late Greek Revival (including a modified church) through (the most interesting) shingled, hip-roofed duplexes inspired by early twentieth-century English new towns, built after the Bradford organization moved in. (Indeed, "Bowling Lane" smacks of more British renaming.) Close to its dead end is a large stucco and sparsely half-timbered mansion on landscaped grounds that must once have housed the plant manager.

Hopkinton (HO)

Designated "Vacant Lands" until 1709, when the Rhode Island General Assembly sold the area into private ownership, Hopkinton remained part of Westerly until 1757. It took its name from Stephen Hopkins, the governor of the state at that time. Mill towns are scattered along its river boundaries: the Wood River, shared with Richmond to the east, which flows into the Pawcatuck to bound Hopkinton from Westerly to the south. In Hopkinton another line of small mills (Rockville, Moscow, and Centerville) once existed in a crescent of minor waterways and millponds across the town's north center, with Canonchet off by itself. Almost a thousand acres to the north and west, where the land is hilly and forested, is included in the Arcadia Management Area, most of which is in adjacent Exeter. Most of Hopkinton's farms were once concentrated toward the south, where the land makes a long slope into Westerly. Here, as elsewhere in western Rhode Island after the mid-nineteenth century, steadily abandoned fields tended to revert to woods, until suburbanization eventually reversed the town's long-term population decline beginning in the 1960s.

Wyoming

HO1 **Mill Workers' Housing**

c. 1850–1865. Aldrich St., Prospect Sq.

Although the village of Wyoming straddles the Wood River and half of it is in Richmond, its mill ruins and most interesting architectural remnants are in Hopkinton. Most notable for their unusual quality are the mill workers' houses around Prospect Square, toward the western end of Wyoming Pond. Aldrich Street contains a fine row of one-and-one-half-story, gable-fronted Greek Revival houses with corner pilasters and typical off-center, side-lighted entrances heavily enframed in pilasters and entablatures. Not all have been well treated, although all could be returned to the handsome state of the house (HO1.1; 1856) at 12 Aldrich Street, with the unusual feature of corner pilasters scored to suggest masonry blocks. In Prospect Square, single-family houses and duplexes are mixed, but more follow the five-bay-facade, center-entrance format with pedimented gable ends on either side. The George Niles House (HO1.2; 1855), 34 Prospect Square, is the finest of all the houses, in both proportion and detail. The J. C. Fenner House (HO1.3; 1867, side dormers later), at number 30, exhibits the type with a near-cubic, corner-pilastered ell off one side. Adjacent, at number 26, is the James H. Selden House (HO1.4; 1855 and later), which has a porch added c. 1865–1870 on one side and a window bay on

HO1.2 George Niles House

the other. Finally, adjacent to this, at number 24, is another front-gabled Greek Revival house which loses its specific stylistic character, becoming an early Victorian bracketed box. So here, a short ramble successively reveals varied approaches to the small Greek Revival house, its alteration by early Victorian additions and its transition into something which essentially becomes a picturesque Victorian vision. Prospect Square does not seem to have been finished on its southeast side, and what was once its central park is overgrown. For the mid-nineteenth century, however, this was clearly an enlightened plan, with space around the houses for yards and family gardens, and access for residents to the upper end of the pond as well as to the mills adjacent to the dam.

Hope Valley (Incorporating Locustville)

HO2 Main Street Houses

1820s–1880s

Despite the effects of commercial development around an I-95 interchange, Main Street in this former center for textile and iron manufacturing presents an opportunity, as does Prospect Square in Wyoming, to watch Greek Revival meld into Victorian styles in a series of modest houses; here, however, the scale is larger and the houses are set farther back from the street.

A row of flank-gabled Greek Revival duplexes for mill workers mixed with gable-fronted single-family houses culminates in number 1026 Main Street (HO2.1). Exceptionally for mill workers' housing, the paired center doors of this duplex are fronted by an impressive Doric porch. Not even overseers were normally accorded such regal recognition. Did it originally house the upper echelons of the mill supervisory staff, if indeed this was mill housing? At number 1054 (HO2.2), the porch for the same type of house is Tudorized with octagon columns, the first-floor windows are hooded with Renaissance-inspired cornicing, and the eaves are bracketed. All are early Victorian elements, so modestly asserted that their incongruence is hardly noticed. Nearby, at number 1050 (HO2.3) is a bare-bones version of early Victorian Italian Villa Style—Greek Revival now put behind. At number 5 Bank Street (intersection of Main and Bank streets) (HO2.4), a late Victorian house in the Queen Anne style flaunts splendid porch supports on vergeboard ornamentation under the gable with quatrefoil and pointed-arch cutouts playfully evocative of the Middle Ages.

The early Victorian brick commercial building at 1081 Main Street, Barber's Hall (c. 1865; now altered), once accommodated the local bank and an upstairs community hall.

Rockville

HO3 Stable

Late 19th century. Behind 269 Spring St.

The carriage entrance of this well-preserved mid-Victorian stable behind a Greek Revival house is decorated by "Gothic"-arched paneling, with three diamond-shaped windows above and a sprightly cupola. It typifies the fancy stable of the period, but few remain as well maintained as this.

HO4 Seventh Day Baptist Church

1846–1847. 1887, belfry. 281 Spring St.

The dominating structure in this pleasant, well-preserved village is the Baptist church, a broad, rather loosely composed edifice with a pair of side-lighted Greek Revival enframed doors in front, a too-diminutive Palladian window, and, eventually, an overscaled belfry. Its raised site, approached by an allée of maples, magnifies its impact.

HO5 Post Office (Rockville Mill)

1844. 1 Leveillee Ln.

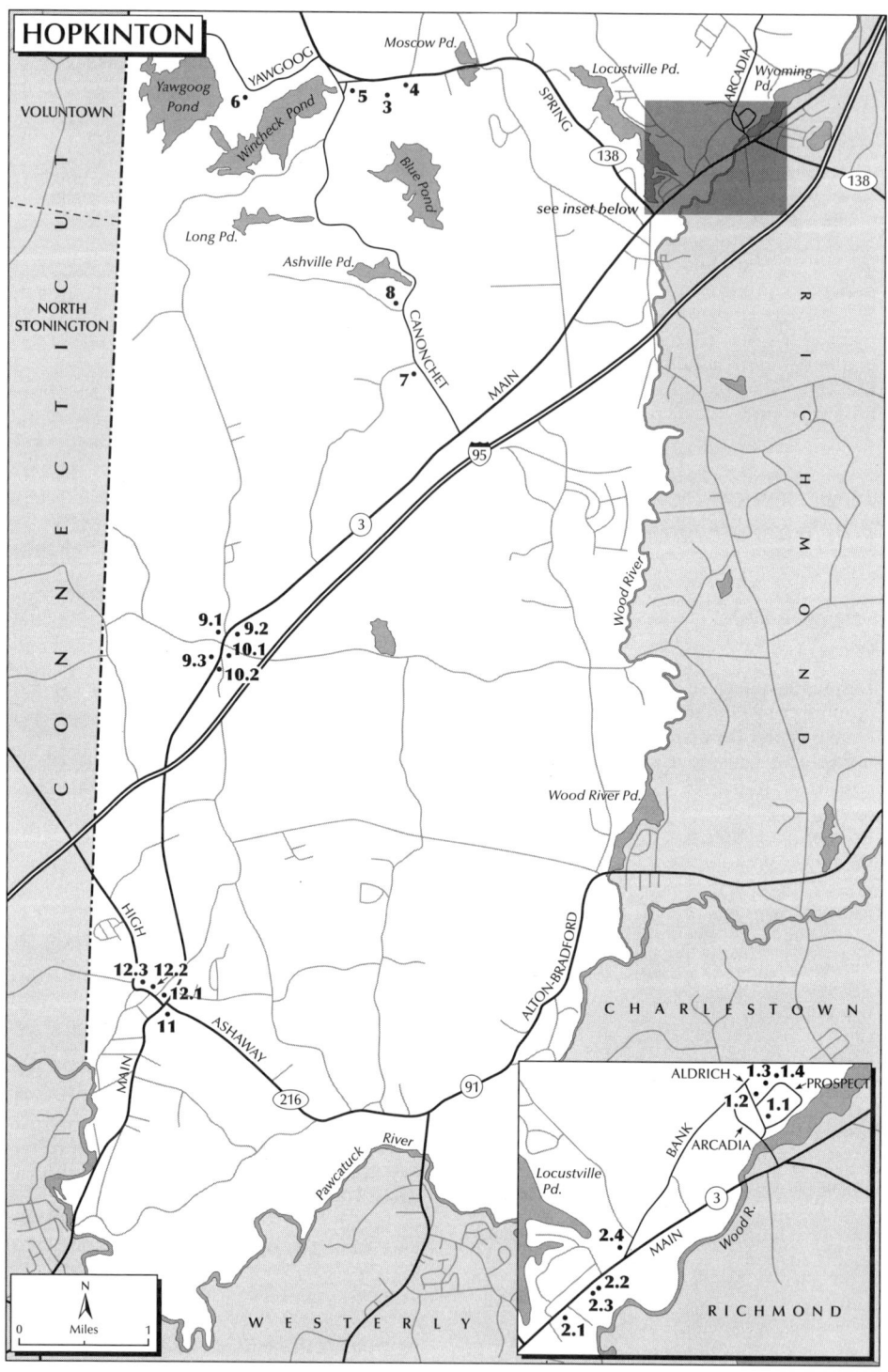

HO4 Seventh Day Baptist Church

HO6 Yawgoog Boy Scout Camp, Bucklin Memorial Building

The treatment of the rugged two-story masonry base of the former Rockville Mill is unusual for such architectural refinements as the stringcourse projecting between the stories and the hierarchy of window sizes, with larger openings below, smaller above. The attic story, with a broad monitor in clapboard, was rammed onto the upper story during the building's rehabilitation into office space and seems mocking, especially in contrast to the simple, low-gable roof with clerestory monitor that originally capped the composition. Few of its period and size exist, so its preservation for modern office use is welcome, even though it is now further diminished by a rather too extensive parking lot.

HO6 **Yawgoog Boy Scout Camp**

1916. 1931, rebuilt after fire, Jackson, Robertson and Adams (entrance gate, Bucklin Memorial Building, and perhaps others). 1943, J. Harold Williams Amphitheater. Yawgoog Rd.

Of all the summer camps dotting Rhode Island lakes and former millponds, this is architecturally the finest. It contains, in fact, some of the best rustic architecture in the state. An entrance gate, its stone piers flanking totem poles supporting a spandrel made of consciously crudely wrought wood bearing the camp's name, announces arrival at the camp. Bucklin Memorial Building, at its heart, is a long, gabled building pierced at its center by a two-story opening reminiscent of the entrance into a fort. Over the entrance arch, a cross-gabled council room is lifted above the rest of the structure. The rubble masonry first floor extends up to the sill level of the second floor to provide a tall base for the seemingly squat mini-building on top in "wavy board" (boards sawn to reveal the irregular edge of the de-barked tree trunk). As the climactic feature of the building, tapering drums of stone roughly six feet high receive pairs of tree-trunk columns to frame the entrance and tie the wooden structure above with the stone base below. A rustic bridge across the central void provides a sprightly touch of filigree. Near the Bucklin Memorial Building is an open log pavilion with masonry chimney and an enclosed kitchen at one end, originally designed as the refectory. Other, smaller log buildings crafted in slightly different ways to house camp activities complete this well-spaced, beautifully landscaped cluster.

Farther along the road into the camp, the J. Harold Williams Amphitheater confronts the visitor as a stockade wall. A fixed plane masquerades as a stout portal slightly ajar, guarded by the projection booth above. A tier of seats on sharply stepped concrete steps immediately behind the stockade gives way toward the front to the natural slope as a tapered semicircle of wooden benches, interrupted here and there by trees, down to a place for the ceremonial campfire. Behind this a grass-surfaced platform with a semicircle of conifers and rhododendron closes this grandest of all spots for high-level powwows. Finally, the road terminates at a pond-side log lodge to serve the swimming and boating area. Subsequent clapboard buildings are, by and large, sympathetic and well placed. They do not, however, match the expressive quality of the 1930s buildings, which combine sturdy functionalism with a playful allusion to the mythic pasts of Indians and frontiersmen.

Canonchet

HO7 St. Elizabeth's Episcopal Church
(Canonchet Chapel)

1889, addition, 1972. 63 Canonchet Rd.

The mill for which this village was once named, the former Ashville Mill (1848; Canonchet Road just south of Stubtown Road), retains some sense of its original character as a handsome, medium-sized Greek Revival mill in rubble masonry on a potentially pretty site. But it is enveloped in ad hoc attachments. So among the architectural offerings of this village, St. Elizabeth's Church is the most exceptional. To the standard gabled box with the usual square-cornered windows, the carpenter-designer added a projecting gabled vestibule entrance with a pointed arch for the door. The tower interlocks the other two elements with more pointed openings (gable rather than arched) and tapers to the belfry, thereby augmenting the aspirant quality of its stubby steeple. The folded quality of the tower recalls St. Thomas in nearby Alton (RI6) but employs different elements with less sophistication.

HO8 Francis Tanner House

1762. Stubtown Rd. (pole 2235)

Canonchet contains a couple of eighteenth-century houses and some Greek Revival examples, although most date from the mid- to late nineteenth century. The most important of these appears a little off the principal approach to the village from the north, on Stubtown Road. The Francis Tanner House is a traditional one-and-one-half-story, five-bay house with a central stone chimney enclosed by an unusually broad gable.

Hopkinton City

This "city," containing six eighteenth-century houses as well as some Greek Revival and mid-nineteenth-century buildings, may have justified its early pretentious name when its prospects loomed large with the stopover traffic of the New London Turnpike and a booming carriage-building business. Of all Hopkinton villages, this alone was a crossroads village, not a river-oriented settlement. Today most self-consciously it extends from north to south along North Road as it merges into Main Street south of the intersection with Route 3 and continues south on Town House Road past the divergence to the south of Route 3 to the southwest. First to appear to one entering this most self-consciously "preserved" and "restored" village in Hopkinton is the diminutive Deak Store (before 1776), at 5 North Road, probably gussied up in the twentieth century with its multiple-pane oriels in "ye olde colonial mode."

HO9 Main Street Houses

1750s–c. 1860

At 495 Main Street is the ample, center-chimney Thomas Wells House (HO9.1; c. 1785), set back from the road behind a picturesque picket fence. The Spicer House (c. 1810), at 491 Main, continues the five-bay, center-entrance format, but here with paired interior chimneys suggesting a plan more spatially sophisticated than those of its center-chimney neighbors. Across the street, at 496 Main, is the General George Thurston House (HO9.2; c. 1750), also center-chimneyed but with a remarkably broad five-bay facade, whose pairs of windows flanking the later Greek Revival entrance huddle together, leaving large amounts of blank wall space.

By far the most visually compelling of the village's houses is the Thurston-Wells House (HO9.3; c. 1820 and c. 1860, although the plaque asserts 1848), 485 Main Street, long the home of the politically active Thurston family, who counted several lieutenant governors. The side-hall-plan main block, with its semi-elliptical fanlight over both principal entrance and side lights framed by thin pilasters, is formulaic Rhode Island Federal, but the mid-century remodeling by carriage maker Augustus Wells introduced bracketed wide eaves and the large belvedere whose domineering scale and large, round-arched windows make the building somewhat top heavy. Wells also built the barn at rear, enlarged in the twentieth century by one-story additions, to accommodate his carriage business.

HO10 Churches

c. 1790–1860s

At 486 Main Street is the First Seventh Day Baptist Church (HO10.1; 1836), a small, severe Greek Revival box, its sills at ground level, with simple entrance centered on the windowless facade and three large, round-arched windows on each of the side elevations. (Was the ample fenestration of the side elevations an attempt to compensate for the blind facade?) Finally, at the Town House Road–Main Street intersection is the Second Seventh Day Baptist Church (HO10.2; c. 1790, 1826-1827, 1861), 2 Town House Road, another simple box of a church, but with narrow, almost prim windows on both front and side elevations and embellished with a square-plan round-arched cupola. Originally a nondenominational chapel, it was moved here by the town in the early nineteenth century for use as the town hall. After the "new" town hall, immediately across Town House Road, was completed in 1861, the building reverted to worship space, this time for Baptists. The north-facing building fronts on a triangular parcel with the town's World War I monument at its apex. Church, town hall, and open space achieve a compelling civic presence whose modesty seems appropriate for this diminutive "city."

Ashaway

Located on the Ashaway River, this village added textiles to gristmills and sawmills. By 1816 it had a woolen manufacturing operation, which was transformed in 1825 into a manufactory for fishing line and twine. The plant's owner named it for the river. Ashaway became a premier brand of fishing line. The ropewalk, a long, corridorlike clapboard building where the strands were "walked" from one end to the other preparatory to twisting, existed in Ashaway until the late 1970s, the last of its type in Rhode Island. Another textile factory, the Bethel Mill (c. 1850; High and West streets) continues in active manufacturing use but no longer for textiles. It has been aluminum sided but is worth attention even so, as wooden mills of its scale from the mid-nineteenth century are rare—and those still in production rarer still. Three blocks south, at Main and Church streets, a fine allée of maples cuts diagonally across a green to the First Seventh Day Baptist Church (8A Church Street), with the Hopkinton Academy adjacent. Both are Greek Revival and both rather clumsily altered through time. The civic impact of the church next to the academy is nevertheless impressive.

The village itself emanated from the riverside mills on Laurel Street south of High. Most of the workers' housing, mostly vernacular Greek Revival cottages modest in scale and altered over time, is located in the low-lying area to the west of the river. The larger houses and institutional development tended to locate up the hill to the east of the river.

HO11 House, 194 Main Street

c. 1875. Near the intersection of Routes 3 and 216

This ambitious Second Empire house is relatively small in size but almost overly inflected, as so often seen in remote village settings like this. Its location, opposite the west end of High Street, reinforces its presence in the village. The cross-gabled roof equivocates between mansard and gambrel. Although the walls have been aluminum sided, the ornament is intact, including the climactic tower capped by a melon-shaped dome, with hooded oculi and ornamental metal cresting. The all-white surface is unfortunate, although not necessarily permanent. Across the street, at 197 Main, is the mansard mode more conventional to the area: cubic main block with flared lower slope and full-width front porch.

HO12 Houses on High Street

1778–c. 1915

Houses along High Street include a classic gambrel-roofed bungalow (HO12.1; c. 1915), 6 High Street, set within well-kept grounds; its round fieldstone foundation, a feature popular in stone-quarrying southwestern Rhode Island, continues up the full-width front porch as parapet. Farther along, downhill toward the river,

HO11 House, 194 Main Street

are three Greek Revival houses (c. 1840), two large and one small, in an arc sweeping down the hill on High Street (HO12.2). The larger, two-story ones, numbers 14 and 18, flank the end-gable cottage at number 16. What compels here, however, are the Colonial Revival porches added to the larger houses: each culminates at one end with projecting pavilions (polygonal for number 14, circular for 18). The ensemble is made all the more striking by the symmetry of the porches when viewed together, with the pavilions at east and west ends of the group. Number 20, the Jacob Babcock House (HO12.3; 1778), is a standard two-and-one-half-story, five-bay, late-eighteenth-century house, interesting as the house of the owner of the first mill in the village. Some of its adjuncts hard against the river may be altered remnants of ancient additions that once combined residence with work.

Bradford

See the entry for Bradford Mill under Westerly (WE27).

Richmond (RI)

Richmond shares river boundaries with Charlestown to the south and east and with Hopkinton to the west. On the Charlestown side are the Usquepaug and the Pawcatuck. Along the whole of the Hopkinton side is the Wood, a fast-moving river with scenic stretches that make it a favorite with canoeists. Whereas the flow of the Usquepaug and Wood rivers in Exeter is limited, the volume of water farther south provided power for good-sized factories. Hence Richmond's boundaries are dotted with mill villages which straddle town lines: Kenyon, Shannock, and Carolina on the Charlestown border; Wyoming, Hope Valley, Woodville, and Acton on the Hopkinton border. For travelers' convenience, most of these villages are included under Richmond, which contains the cores of most of them—although of course it would be a waste of time not to cross the river at each of these places in order to cover what is on the other side. Such village overlap across these three town boundaries, together with the dispersal of both village and rural population, has recently encouraged joint operation of such town services as schools under the awkward but unifying regional designation of Chariho.

These mill villages are truly such, most clustered around a single medium-sized enterprise, with extensive open space between one village and the next. They do not at all resemble the bigger mill towns in northern and central Rhode Island: not those on the Blackstone, along the boundary between Cumberland and Lincoln; even less such ribbons of manufacturing as line the Blackstone in Woonsocket, Central Falls, and Pawtucket and the Pawtuxet in West Warwick; not even the concentrations to be found on the various rivers of much more rural Burrillville. The earliest gristmills tended to give way, especially on the Wood River, to iron manufacturing for household and farm implements, which in turn mostly gave way, especially after 1840, to textile operations. By the 1930s, most of these were defunct. Many of the plants are now in ruins, although a few continue in operation. For the rest, Richmond was agricultural. As was true of its neighbors, its population peaked around 1870, then declined until the 1920s, finally increasing until the 1960s saw the number of inhabitants climb above its 1870 level with the spotty suburbanization of a town which still contains threatened areas of open space.

RI1 Samuel Clark Farm

c. 1680, house; 20th century, ell addition. 19th century, barn, other outbuildings, and one-room schoolhouse. 106 Lewiston Ave. (pole 699)

This spreading, one-and-one-half-story house with a gambrel roof and a central chimney in stone (more typical of Connecticut than Rhode Island) is sited down a long slope. Outbuildings and the culminating Victorian barn take precedence close to the road, with smaller outbuildings in a random linear arrangement down to the house. This is an exceptionally complete (and in some ways unique) cluster of farm buildings, handsomely maintained on a splendid site with walled fields all around. The complex includes a raised corncrib, a machine shop, a private one-room schoolhouse, and miscellaneous sheds. Two stepped shed-roofed adjuncts to the barn fit into a slope at right angles to the broad, tilted plane that holds the farm.

Kenyon

RI2 Kenyon Mill

1844, 1866 and later. c. 1866?, superintendent's house. Mid-19th century, worker housing. Mill: Kenyon School Rd.; easiest access on Charlestown side at 36 Sherman Ave. Worker housing and superintendent's house: Sherman Ave.

In 1844 Abiel Kenyon built a masonry mill for cotton and woolen manufacturing, which he enlarged in 1866. All of this has been mostly shrouded by later ad hoc additions of dates extending up to the near present. The forceful mansard tower (1866), stucco-surfaced with quoining of quarry-faced gray granite, is all that remains of architectural consequence. But Kenyon continues to operate as a dye factory for textiles, and there is a certain rough-and-ready picturesqueness about the village, especially around the post office. This appears to be wholly unintended—and unappreciated, as twentieth-century addenda to the factory and parking lots make clear. At the eastern end of the main street, Sherman Avenue, granite gateposts mark the entrance to a mill superintendent's (possibly owner's) house, possibly from the 1860s, with quite refined neoclassical porches added in the early twentieth century. It takes refuge from the town on a wooded bluff overlooking the river but shares its enclave with three mid-nineteenth-century one-and-one-half-story mill workers' duplexes—an example of the close proximity in which management and workers frequently lived. There is also a random mix of other mill workers' houses in the village.

Shannock

RI3 Shannock Mills and Village

Shannock Village Rd.

Potentially this is a particularly picturesque village in which a winding road passes ruined mills and a cluster which includes eighteenth-century remnants and Greek Revival and mid-nineteenth-century examples. The town village has weathered some false starts in its postindustrial history but currently enjoys a fair state of health. From the east end of the town, the road winds past a boardinghouse, over the bridge across the Pawtuxet, with the extensive

RI1 Samuel Clark Farm

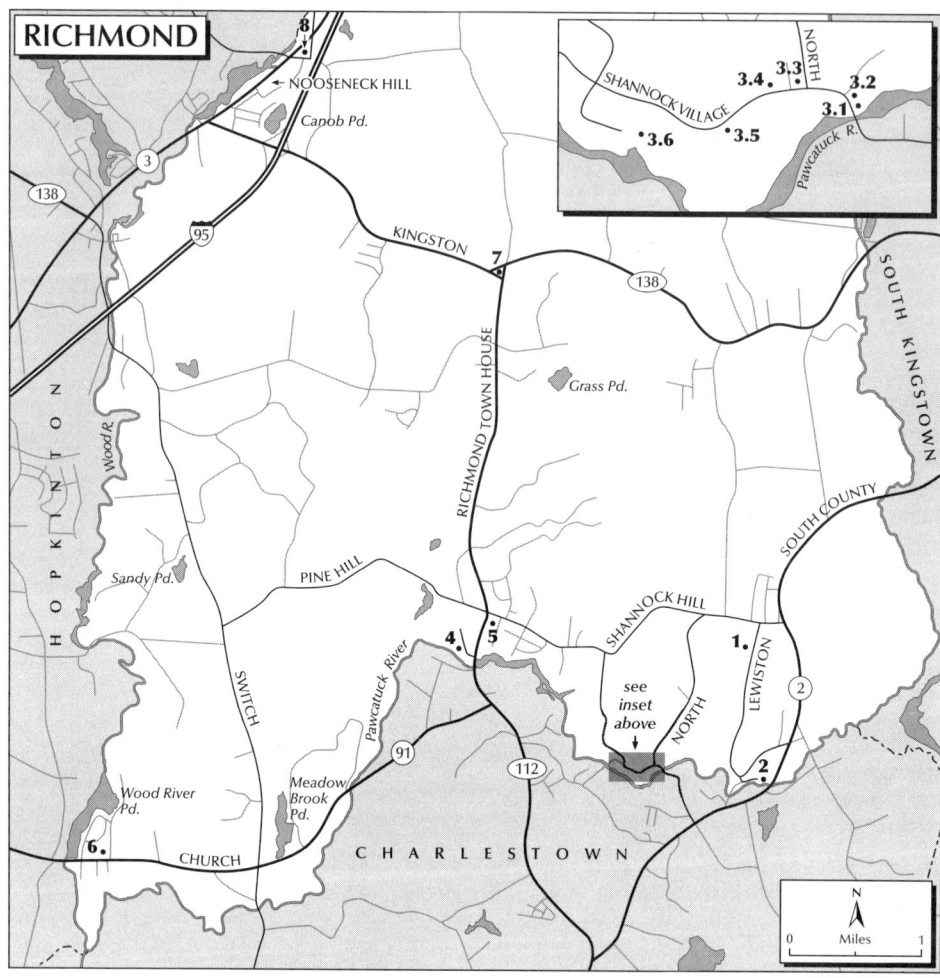

Greek Revival and twentieth-century brick ruins of various Clark enterprises downstream. Upstream is a horseshoe-arched dam built sometime during the nineteenth century. Beside it, and tight against the bridge, is a tiny wooden building, much altered, which uses a boulder in the stream as support for one end, the bank as support for the other, with a masonry arched bridge between. This is a rare relic of the eighteenth-century Clark's Mill, the concrete arch providing the outflow beneath the mill for the raceway (now redone in concrete) beside it. Adjacent to the raceway is the one-and-one-half-story, central-chimney house, also much altered, which once housed the mill owner. So the remnants of eighteenth-century enterprise by the Clark family exist beside the ruins of their later efforts.

The road winds past a Greek Revival house (220 Shannock Village Road) with frontal pediment and distyle Doric porch. (Greek Revival side ells were apparently moved for a shed-roofed insert.) The picket fence on granite base seems to be a restored survival from the nineteenth century. Two restored Greek Revival mill workers' duplexes follow, at numbers 214–216, with clapboarded mid-nineteenth-century duplexes behind. Along Main Street are more medium-sized Greek Revival houses for the town elite, intermixed with workers' housing of various vintages. On the south side of the railroad tracks and at the end of Railroad Street is an almost collapsed, but once handsome, nineteenth-century wooden mill, with an early twentieth-century wooden addition, beside another dam. This is Shannock

Mill, operated by a succession of owners outside the Clark family. It gave the village its eventual name around 1870.

Carolina

The architectural interest of Carolina, more than the other principal mill villages along the Pawcatuck and Wood rivers, is split between Richmond and Charlestown. Buildings on the Charlestown side of the town (see that section under Charlestown) are conveniently viewed en route, before recrossing the river to the Richmond half of the town, where the road becomes briefly Main Street, then Richmond Town House Road.

RI4 Carolina Mills

c. 1850–1870, early 20th century. Junction of Carolina Main Rd. and Carolina Mill Ln.

Carolina rose from a gristmill to textile manufacturing when Rowland Hazard (see under South Kingstown) purchased the town village or mill seat for one of his extensive South County textile operations. He was the client for most of the Greek Revival houses in the town as well as the earliest sections of the Carolina Mill, of rubble and cut stone masonry (c. 1841). The mill, on Main Street, is now a picturesque ruin. Together with its collapsed twentieth-century wooden extension, it has been rebuilt—adapted rather than restored—by its present owner as a private estate with occasional public access. Much of the ruins have been unencumbered of debris and stabilized in their present state. The site retains remarkable relics of nineteenth-century textile machinery. Uphill, at 25 Main Street, Ellison Tinkham, partial owner of the operation between 1868 and 1907, built a Queen Anne house of modest pretension (c. 1890), the front elevation of which is rigidly ordered by paired windows from first floor to attic. A pretty porch across the front retains its original carpentered ornamentation.

RI5 Albert Potter Octagon House

1867. 4 Carolina Main Rd.

The Albert Potter Octagon House was built after the appearance of the second edition (1853) of Orson Fowler's *Home for All,* which

RI5 Albert Potter Octagon House

recommended the octagon as the most economical and practical shape for a house. Fowler also recommended concrete for walls, but this example is shingled, with bracketed cornices at the eaves and around the cupola, and unadorned windows throughout. An unadorned door opens onto the roof from the cupola. The quaint sobriety of this well-maintained octagon accounts for its character.

Alton

In 1880, William A. Walton bought a woolen mill that had been established in 1862. He attempted to develop a model mill town, building houses with individual garden plots, laying sidewalks, initiating a tree-planting program, and providing for such community programs as a brass band. Walton's masonry mill burned in 1890, to be replaced in 1906 by a nondescript factory whose name—Alton Manufacturing Company—derived from the old owner's.

RI6 St. Thomas Episcopal Church

Probably early 1880s. 322 Church St. (Alton-Carolina Rd.)

William Walton's crowning building accomplishment was the Episcopal church. Among the important Carpenter's Gothic village churches in the state, this otherwise very plain box, now vinyl sided, displays a notable extravagance in the folding of gabled and splayed shapes to make an attenuated bell tower.

RI6 St. Thomas Episcopal Church

RI8 Victory Baptist Church (Wood River Baptist Church; Six Principle Baptist Church)

RI7 **Bell Schoolhouse** (Richmond Schoolhouse 9)

1826. 1971, moved to junction of Richmond Town House and Kingston rds. from Bell School House Rd. and bellcote rebuilt

This shingled late Federal school is a rare survivor from the early nineteenth century. As in other schoolhouses of its date, the single entrance indicates a more casual concern with differences of sex than was typical in later buildings, which were more meticulously rationalized. Just so with the balanced placement of windows, two on either side elevation. From the Greek Revival onward, rationalized procedure tended to leave a blank stretch of wall toward the entrance, where a divided vestibule and coatroom were placed (here there is no vestibule), with windows concentrated toward the rear of the elevation to light the classroom. Moved to a site near the Richmond Town Hall, Bell School today is also close to a handsome elementary school (1935) in the Federal Revival style popular for New Deal institutional buildings, especially in the northeastern states. The juxtaposition implies "progress."

Barberville

RI8 **Victory Baptist Church** (Wood River Baptist Church; Six Principle Baptist Church)

Before 1855 (probably late 1840s). 85 Nooseneck Hill Rd. (junction of Nooseneck Hill and Skunk Hill rds.)

As a sculptural display of elements common to Greek Revival building, this church is exceptional. Two doors with side lights deeply inset and enframed by flanking piers and heavy entablatures are framed by another deep entablature at the eaves, which is visually supported by corner piers. Typically, flushboarding in the end gable pediment, meant to suggest masonry, gives way to clapboarding on the wall elevations. The tiny pair of windows squeezed high above and between the entrances sets up a nice animation with the side lights, together providing eyes for what would otherwise be something of a clumsy blind giant. Paired windows under (or in) the entablature are more typical of domestic than ecclesiastical Greek Revival buildings. In fact, the entire elevation derives more from domestic than ecclesiastical architecture. The lack of belfry and the strong horizontals imposed by the heavy entablatures reinforce the sense of a blown-up house for worship.

Exeter (EX)

In contrast to the rectangularity of Coventry and West Greenwich, which are stacked above it, Exeter has the shape of a pistol aimed at Connecticut. Exeter adopted the name of the Devonshire city when it split from North Kingstown in 1742 and set up its own town government. Historically it has been an agricultural area, with farming giving way to grazing, woodcutting, and sawmilling in the rugged terrain to the west. Even today, it boasts no town of consequence, and much of its western portion, in the Arcadia Management Area, is forest preserve and state park. Its major rivers, the Queens to the east and the Wood to the west, never attracted more than small mills, often with histories of unstable ownership. During the nineteenth century, the town's most conspicuous public improvement was Ten Rod Road, so called for its exceptional 165-foot width. Running with remarkable straightness down the middle of the barrel of Exeter's pistol silhouette, Ten Rod Road was designed for driving cattle clear across the town to the harbor at Wickford. Now the eastern third of Ten Rod Road has been rechristened to accommodate a jog in Route 102 as Victory Highway. The New London Turnpike, cutting diagonally north-south across the barrel, close to the pistol grip, divided the town into quadrants—as its successors, Route 3 and Interstate 95, each slightly farther west, do to this day. As in West Greenwich, suburbanization has reversed a longtime decline in population.

EX1 Ranger's House, Arcadia Management Area

c. 1935. Civilian Conservation Corps. 230 Escoheag Hill Rd. (at the junction with Plain Rd.)

At this entrance to Arcadia, a 13,000-acre forest preserve acquired in the mid-1930s by the state's Department of Agriculture and Conservation (now Department of Environmental Management), a log reception lodge, now unused and becoming derelict, is backed by a ranger's house in "wavy board" construction. What is the special quality of CCC work? Perhaps the picturesqueness of rusticity without the usual sentimentality associated with rustic revivals. The very sensible aspect of these buildings proclaims the regimen of the corps and the stringencies of the Depression relief program that produced them. It also embodies an interest in establishing minimal but decent "standards" or "norms" for do-it-yourself building, implying that even in the hardest times, a nation might survive in beautifully plain, natural buildings, rationalized and improved from frontier prototypes directly expressive of their forest origin, as an alternative to lumberyard carpentry.

EX2 Woody Hill School, District 1

c. 1845. Woody Hill and Skunk Hill rds.

Of all the one-room schoolhouses extant in Rhode Island, this may be the most endearing. Its petite scale—it is about as small as the typical gabled, clapboard type can get—makes it particularly poignant as a monument to vanished childhood. On the front are two transomed doors, slotlike in their narrowness, flanking the center window. Move back from the building and the two outhouses come into

EX2 Woody Hill School, District 1

view, symmetrically disposed to either side of the schoolhouse, with doors echoing the narrowness of those in front. A tree trunk stripped of bark for a flagpole completes the symmetrical composition. So much dignity with such simple means! The schoolhouse faces a handsome farm with buildings ranging from a Federal-period house, through barns and outbuildings of the early twentieth century. Its setting is among the loveliest rural crossroads in the state.

EX3 **West Exeter Baptist Church** (West Greevnville Branch Baptist Church)

1858. 2019 Ten Rod Rd. (intersection of Ten Rod and Frosty Hollow rds.)

The West Exeter Baptist Church is standard Greek Revival, minus cupola. It edges into Victorian, however. The doors, for example, are flanked not by suggestions of substantial piers supporting entablatures, but by the typical Victorian device of more lightly framed openings surmounted by bracketed hoods (although the brackets here are tiny).

EX4 **Old Town House**

1878. Ten Rod Rd. at Town House Rd.

Across Nooseneck Road, Ten Rod Road abruptly turns to gravel for roughly two and one-half miles, which give a sense of ancient conditions. A little farther east, it intersects with the north-south diagonal of the other major early transportation route, originally the toll turnpike to New London, which is also gravel. Just east of this historic grand crossing at the heart of Exeter is the boarded-up Old Town House. It served as town hall and administrative center before those functions were moved in the mid-twentieth century to a nondescript building about two and one-half miles to the east, in the village of Lawtonville. A bare, barnlike clapboard building on a granite foundation, it is marked as a public monument by no more than an extra-large door topped by a sign and a projecting molding. Here in the dusk and the woods, disuse (at least as this is written) makes the sense of the past palpable. A little farther and Ten Rod Road returns to pavement with its rechristened status as Victory Highway (Route 102), although natives tend to stay with the old name.

EX5 **Exeter Town Pound**

Probably early 19th century. Victory Hwy. (Ten Rod Rd.), access road about three-quarters of a mile east of its intersection with Route 102 (pole 155)

Rhode Island is fortunate in the number of its preserved pounds. This nearly square, shoul-

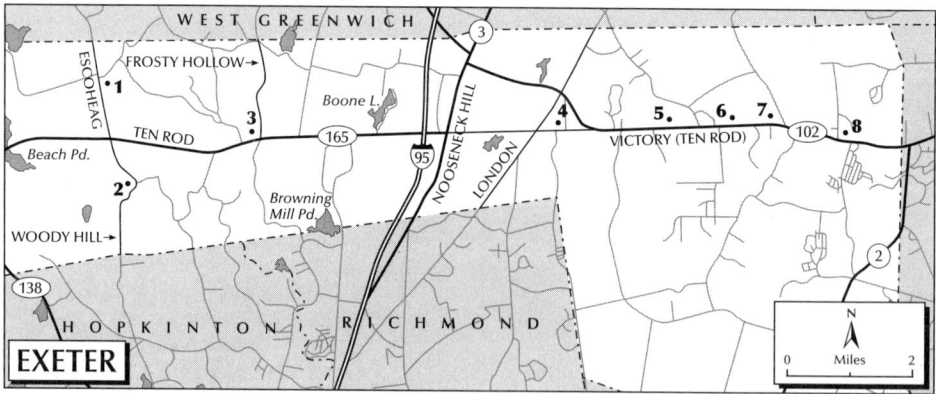

der-height boulder enclosure and that in Glocester (see the entry under Glocester) are the most considerable and best preserved. Of the two, only Exeter's retains (at least for now) its farmland setting.

EX6 Chestnut Hill Baptist Church

1838. 1950s, parish house wing. 467 Ten Rod Rd.

Of the many Greek Revival churches standing in Rhode Island, this much reproduced example may possess the greatest individuality. It is a clapboard box on three sides with flushboarding in front, its facade an assemblage of blunt shapes suggesting child's building blocks. The source of inspiration was the Wickford Baptist Church, a splendid example of a provincial carpentered building type, here further provincialized as it moves from the center to the periphery. But the Chestnut Hill example goes even further. Customarily, in simplifying sophisticated precedent, the provincial builder at best revitalizes the model by calling it back to essentials and at least charms with quaintness. More rarely, as here, the provincial so distills the model as, unconsciously, to make an icon of a precedent more imagined than seen, although seen in order to be imagined. In the fierce starkness of its rectangular and triangular shapes, its cubes and rectangular solids, this church becomes the *sign* of its precedent. So too in its reduction of ornamental demands to bare boards, as in the enframement of the piers, the twin pilasters cornering the cupola, and, most obviously, in the giant swatch of Greek key patterning in the inset panel over the door. Originally an equally forceful cresting adorned the cupola, before the 1938 hurricane ripped it away, to be replaced by the present wan recall. Directly inspired by Wickford's example, the original was, like everything else, more intense at Chestnut Hill, more akin in its paneled and bristling angularity to medieval crenellation than to Greek anthemion cresting, which it was supposed to recall. It would be a small effort to return this crown of thorns as the climax of the church. While at it, why not also screen the obtrusive parish house with trees? Inside, the original double doors, from the vestibule into the audience room, have been replaced by one at the center. The paneled balcony above the entrance, the pews (minus their original doors), and the gently arched plaster ceiling (restored) are original elements. The pulpit, twice-revamped (once in

EX6 Chestnut Hill Baptist Church exterior and cemetery, burial vault

1886, then in the 1950s), was initially elevated and reached by stairs on either side.

In the cemetery, an interesting row of stones for the Johnson family, dating from the 1870s, with empty slotted frames, apparently for tintypes of the deceased, leads to the burial vault at the far side. Like the church, it shows how everyday building—in this case, techniques for constructing stone walls, root cellars, and icehouses—can, in the hands of the instinctively sensitive builder and through the humblest of means, acknowledge the monumental occasion.

EX7 Lawton Mill

Between 1819 and 1832. c. 1980, restoration, William Warner. 595 Ten Rod Rd. (Victory Hwy.)

This wooden mill on Fisherville Creek typifies early-nineteenth-century factories: masonry basement, clapboard siding above, and monitor-lit attic story, with loading doors and hoist at one end. An appended tower combines loading doors with stairs and toilet stacks, as at Lippitt Mills, in West Warwick. From 1825 until around 1870, various sawmills, gristmills, and snuff mills operated on the site. Probably this building, begun as a somewhat smaller establishment sometime after 1819, was added to, then partially destroyed by fire around 1830, before Thomas Lawton rebuilt it to what it is. Lawton leased the plant to a succession of marginal one-process cotton manufacturers, at a time when the big operators were consolidating all processes into giant mills on larger rivers. By 1870 it had become a woodworking and shingle mill. Finally, dereliction, until its architect-restorer converted it to his office. A decrepit wooden flume to an empty wheelpit exists along one side of the mill, with shingled adjuncts behind. Adjacent are a modified eighteenth-century house, one and one-half stories with central chimney, and a nineteenth-century barn against the road.

EX8 Hendrick's Mill

c. 1978 and later, Paul Hendrick, designer and builder. Off 269 Ten Rod Rd.

Another converted mill, this one is visible across fields to the north of Ten Rod Road (but not normally visitable). Its bright red wheel is

EX7 Lawton Mill

startling in its matter-of-fact Exeter setting, especially when operating. Slightly smaller than Lawton Mill, it has been immaculately rebuilt from a ruin on ancient foundations as a gristmill. No commercial mill can have been as beautifully crafted as this hobbyist's labor of love, nor any commercial woodworking and machine shop as handsome as the one this mill contains. The "eighteenth-century" central-chimney house beside the mill is also Hendrick's work (c. 1983 and later).

East Bay
EAST PROVIDENCE, BARRINGTON, WARREN, BRISTOL, TIVERTON, LITTLE COMPTON

East Providence (EP)

EAST PROVIDENCE IS A LONG, NARROW, ZIGZAG-SHAPED TOWN—three to four times as long as it is wide—squeezed between the Seekonk River to the west and, to the east, the Ten Mile and Runnins rivers, which, together with other streams and ponds, mark most of its border with Massachusetts. The state boundary was not always where it is, not until a boundary settlement made as late as 1862, in fact, after long dispute. Fleeing from Puritan Massachusetts, Roger Williams first settled in the northern part of what is now East Providence in 1636, on what came to be called Omega Pond. He called his settlement Seacunke, the Indian name for the river from which the pond was barely separated. But his stay was brief, for the governor of Plymouth curtly reminded the heretic minister that he was still on Puritan soil. So Williams crossed the Seekonk to a new site on a river a short distance farther west which he called Providence.

Some two hundred Puritans, led by the Reverend Samuel Newman, arrived in the area vacated by Williams in 1643, having purchased land in the area from the Wampanoag. They called the place Rehoboth, a biblical name connoting an open space near a river. There they proceeded to lay out an immense, irregular common, preserving the crossing of two major Wampanoag trails toward its center (now the junction of Pawtucket and Newman avenues). Its 200 acres made it perhaps the largest common in Puritan New England. Its surrounding road was known as the "Ring of the Greene" or the "Ring of the Towne." Each of the typically very long, very narrow land holdings of the original settlers butted one of its narrow ends against this road. The "ring" designation disappeared with time as segments assumed various street designations. At the center of the green, near the crossing of the Indian trails, the settlers marked the center of the town with their church (on the site of the Newman Congregational Church; see EP8) and laid out an adjacent burial ground.

The Puritan town of Rehoboth was very extensive, including the present Massachusetts towns of Rehoboth and Seekonk and pieces of the Attleboroughs, as well as corners of the present Rhode Island towns of Cumberland and Woonsocket. Over time the fuzziness of the location of the boundary between the two states and the tug of much of Rehoboth's population toward Rhode Island occasioned a long dispute. Two bridges of the Seekonk into Providence, just south of the settlement around the common, reinforced the Rhode Island leanings of its population. By the mid-nineteenth century there grew around the ends of the bridges a more commercially oriented portion of Rehoboth known as Watchemocket. After the 1862 settlement of the interstate arbitration, the dense commercial core of the new town became known as East Providence Center. Meanwhile, other village developments both north and south of the core village of East Providence had, by the final decades of the nineteenth century, disrupted the traditional agricultural emphasis of the town. To the north (which included the green and came to be known as Rumford), industry and suburbs replaced farms; to the south (known as Riverside) it was resorts and suburbs.

So long as industry depended on water power, it could not flourish either on the slow-moving flow of the Ten Mile River or on the tidewaters of the Seekonk, although a few small, scattered mills developed, and from shortly after 1800 even a few minor cotton mills. Not until the coming of the railroad and steam power and the move of the Rumford Chemical Works from Providence to East Providence in 1857, however, did the industrial base begin to change. Rumford Baking Powder and Horsford's Bread Preparation were staples in American pantries from the founding of the company in the nineteenth century through the first decades of the twentieth, when factory baking eventually did the firm in. It was a joint venture of Charles F. Wilson and Eben Horsford: the first skilled with machines and management; the second the Rumford Professor of Chemistry at Harvard, who saw to the product. The company took its name from the professor's chair and the illustrious British chemist for whom it was named.

In 1858 Wilson bought a large tract of land in East Providence, including most of the ancient Rehoboth Green. He moved his Providence plant to the very center of the former green—proof in itself that land sacrosanct to Puritans had no such protection in Rhode Island. Wilson continued to purchase land and water rights to small mills, apparently intending some kind of large-scale real estate venture, which included a riverfront industrial park, along the Seekonk. But his plans for his land remained amorphous.

The other notable nineteenth-century industrialist to locate in East Providence was Eugene Phillips, a wire manufacturer, who, in 1893, moved his plant from Providence into a factory building vacated by a bankrupt business Wilson had lured to his Seekonk industrial park just five years earlier. There, the Phillips Electric Company (which in 1910 absorbed another wire manufacturer, assuming its name as Washburn Wire) became highly profitable, in largest part as a major supplier for the rapidly expanding telephone industry.

Meanwhile, the big Sayles textile operation built its Glenlyon Bleachery north of

Washburn in 1899. Although trolleys brought most of the work force to these plants, both Washburn and Glenlyon lined the trolley route along Roger Williams Avenue with a variety of model housing, mostly duplexes, beginning in the 1890s. The village cluster which resulted came to be known as Phillipsdale, as the area around the chemical plant (and the old green) became Rumford. In another sizable land deal, Wilson sold land to Agawam Hunt, organized in 1893 for horse-and-hound drag hunting. More important in the long run, the club became Providence's most important elite recreational adjunct. So rapidly did this new sport catch on that Wilson made another large land sale to the nearby Wannomoisett Country Club, founded in 1898. It occupies the site of Wilson's former estate. His eighteenth-century house, first leased for a clubhouse, was then successively purchased, extensively altered, and eventually replaced.

South of the core village of East Providence, in Riverside, the pattern of development was radically different. Views from the bluffs over Narragansett Bay (which at this point is the wide mouth of the Providence River—hence "Riverside") encouraged the steady replacement or conversion of farms into summer or suburban residences and hotels—eleven of the latter by the 1890s. Colonies of small, tightly packed cottages worked their way between the larger places. Of the latter, Cedar Grove is of special interest. For gourmands, a string of shore banquet halls offered heaping plates of seafood, while gourmets frequented the dining rooms of the more exclusive hotels, the most exalted among them belonging to the Squantum Association, an exclusive eating club with grand views up and down the bay. Amusement parks—for a short period of time, two of them—joined the other resort attractions. Crescent Park in Riverside opened in 1886. A rival, Vanity Fair, opened in 1907 between Crescent Park and Providence. Inspired by such planning as that of the 1902 Pan-American Exposition in Buffalo, Copeland and Dole conceived of Vanity Fair as an assemblage of festively monumental buildings set in a formal axial composition as a series of terraces which stepped down to the water. It was the grandest plan for an amusement park north of Coney Island conceived up to its time. Too grand, as it turned out. Vanity Fair failed after a few seasons, while still only a suggestion of what was intended. On its site today is the Silver Spring Golf Course.

Finally, to complete the landscape which once graced this stretch of shore, the Olmsted firm provided the plan for a parkway, begun in 1910, which wound along the top of the bluffs from Providence halfway to the Riverside depot. It supplemented the network of trolleys, suburban trains, and steamboats which once served this close-in escape from the city. The parkway also heralded the demise of this network, signaling the way in which the automobile would spread the possibilities for recreation. The hotels, the shore dining halls, the amusement parks, the old mix of summer mansions and cottages are gone. So, in the Rumford end of town, are the old industrial mainstays, although new companies lease space in the mills along the Seekonk and in the old Rumford factories. Pockets of farming which persisted, especially in Rumford, into the 1920s and 1930s have also disappeared into suburban housing. But the ghosts of what was once East Providence are worth pursuing amid its suburbs, apartments, strip commercial development, and industrial parks. And in Riverside the blithe spirit of the past has partially returned with the completion (1992) of the East Bay Bicycle Path (no motor vehi-

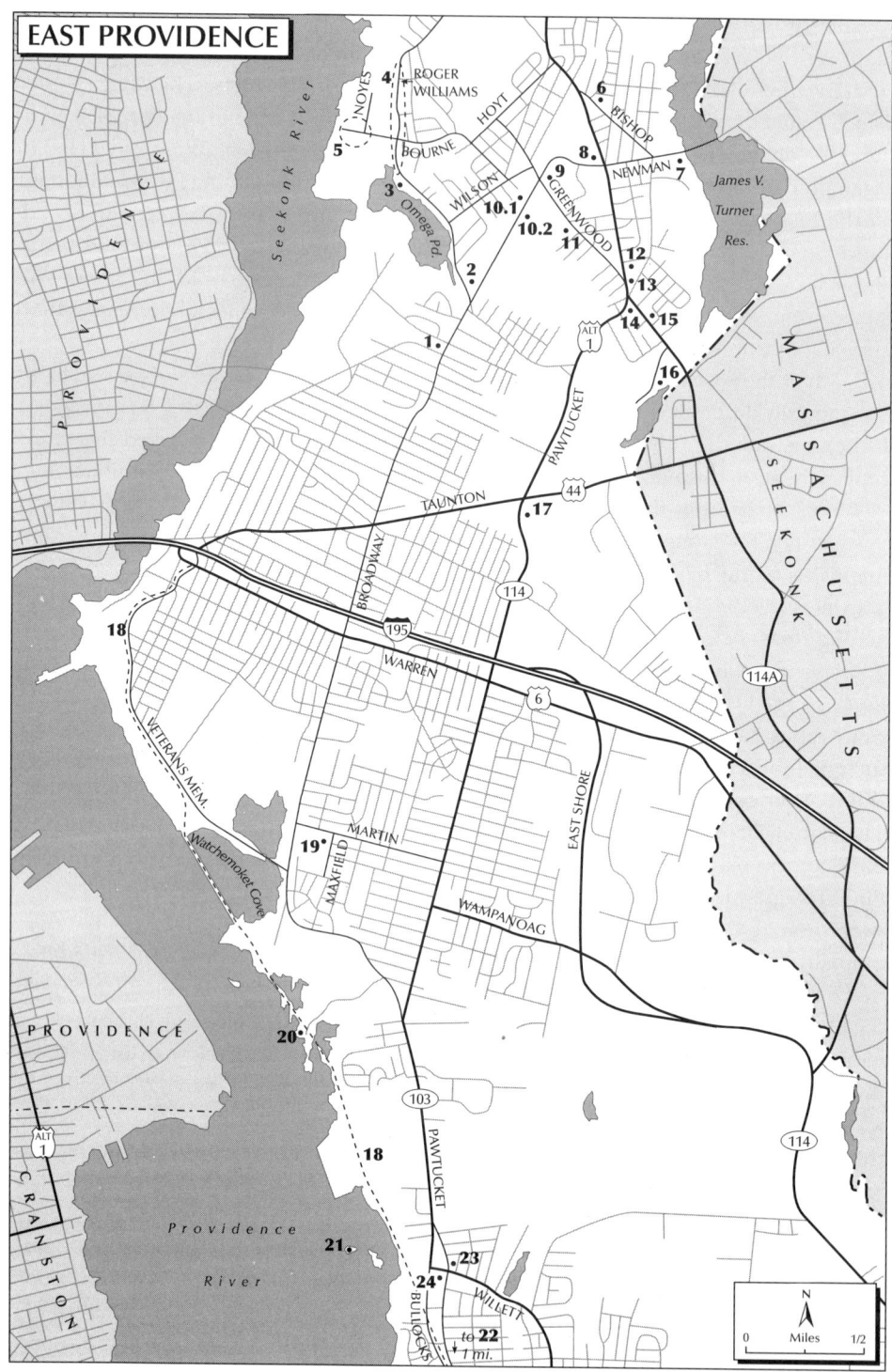

cles). It runs along the bluffs of the parkway for a piece, then dips to the right of way of the old suburban rail line for 14.5 miles, terminating at Bristol. So the venturesome architectural pilgrim can cycle to most points in the Riverside section of East Providence, Barrington, and (even more conveniently) Warren and Bristol.

East Providence Industrial Core

EP1 Philip Walker House

c. 1678–1679; 18th century; later ells. 432 Massasoit Ave.

The oldest house in town was built by the son of an original Rehoboth settler, owner of a sawmill, who was among the wealthiest and best-educated men in the community. Originally a three-bay house with central chimney, it was remodeled at the time of an eighteenth-century extension on the south side to five bays, but with an erratic window arrangement. Twentieth-century shingling covers three sides.

EP2 Agawam Hunt

1840s, 1914, 1920s, 1967. 15 Roger Williams Ave.

Agawam, established in 1893, is the oldest country club in the state. Its clapboard clubhouse is a picturesque accumulation over time of additions in the vernacular domestic manner particularly associated with the early twentieth century. Like leather patches on a fine tweed coat, this elitist organization adopted a discreetly threadbare, make-do look, as though so much at ease in the world that fancy architecture was unnecessary. The accretions, nicely landscaped and spankingly maintained, direct thoughts back to the old-time kernel buried somewhere within. Here this turns out to have been an ordinary end-gabled Greek Revival house of the 1840s (at the extreme left when facing the front), the eventual ramble only tenuously united by a pergola and porches.

On the east side of Roger Williams Avenue about one-third mile north of the Agawam Hunt entrance is the spring that marks the presumed site of Roger Williams's brief settlement here (its monumental enframement erected in 1936 for the tercentenary of the event, plaque 1975), with Omega Pond opposite a bit farther along, and the Seekonk beyond.

EP3 Omega Mill Site

Omega Pond, visible from Roger Williams Avenue north of Agawam Hunt, was the site of a long-ago-demolished cluster of stone mills (from c. 1801 and later) which marked the start of mechanized industry in the vicinity as the Omega Cotton Mill. Today a much remodeled four-family clapboard tenement (c. 1850) at 45 Roger Williams Avenue most conspicuously testifies to the fact, along with the nearby Nathaniel Daggett House (1708, later saltbox addition c. 1900), at number 74. Originally a three-bay house with brick end chimney, in its elongated version it became workers' housing to Phillips Electric.

EP4 Phillipsdale Housing

1883 and later, 1890s–1910s, Hilton and Jackson for some of the later brick housing. 100 and 200 numbers on Roger Williams Ave.; Ruth St.

Phillipsdale offers a picture of various types of late-nineteenth- and early-twentieth-century workers' housing, much of which is fairly well preserved. The Phillips and Glenlyon operations so overshadowed the brief occupancy of the Richmond Paper Company at this place that some have conjectured that its earliest housing was "moved in" from another site. More likely it went up c. 1883–1884 when Richmond Paper built the factory acquired by

EP4 Phillipsdale housing, duplex, Roger Williams Avenue

Phillips a decade later. The second of two brick duplexes at numbers 115–121 Roger Williams Avenue (dating before 1910) is typical of most of the brick housing in the village. Its sculptural roof treatment derives from contemporary examples of English model housing. The hipped roof extends its lateral slopes downward to shelter recessed entrance porches on either end, while front and back the elevations thrust into its slope to make quasi-dormers for second-story windows. The plasticity of the roof means to replace the regimented look of earlier industrial housing by an image of cottagey coziness. Three duplexes, numbers 137–147, all originally clapboard, use gambrels with oversized dormers, aspiring to the same cottage effect. Surprisingly, this favorite shape for the early Colonial Revival is unusual in industrial housing at the time. The plain clapboard Grace Episcopal Church (1903) and adjacent elementary school (1879–1880), which in time became the church's parish house, 130 and 132 Roger Williams Avenue, were Phillips benefactions.

The intersection of Bourne Avenue, where the former Phillips company store, now much altered, occupies the northwest corner, marks the former heart of the village, providing the principal entrance to both the wire works and the bleachery. Beyond this intersection are more brick duplexes derived from enlightened English prototypes, numbers 166–180 and 238–260, these slightly staggered on their sites in another nominal gesture against the regimented look. Opposite, at numbers 167–235, are examples of a type unusual for Rhode Island: a long row of earlier (c. 1883) single-family houses built as squarish boxes, their gable ends to the street with long, narrow porches for both entrance and sitting running the length of their downhill side (except for a few later insertions of two-story flats with tiered porches and more gambreled gables facing the street). A couple of traditional Victorian clapboard duplexes (1883) with ornamental brackets supporting door hoods follow at numbers 253–259. The pyramidal pile of the roof and the poke of tall chimneys through it are especially impressive. Both their size and their position close to the top of the village slope suggest that they housed overseers (probably for the bleachery, because the Phillips superintendent and company doctor shared a duplex next to the company store). A final pair of brick duplexes, numbers 261–267, are larger than the other brick units and prettier, with latticed porches; they, too, were doubtless for supervisory personnel. On Ruth Street, which runs on a ridge above the rest of the village, is more duplex housing of different sorts from different building periods (now intermixed with much more recent houses). Notwithstanding this considerable stock of company housing in the village, most workers originally commuted to the Seekonk enclave by trolley from Providence.

EP5 Phillipsdale Factories

1883–1884, Richmond Paper Company; 1900–1902, 1926–1928, Phillips Electric Company–Washburn Wire additions; 1930–1936 and later, other wire companies. 293 Bourne Ave. 1899, Glenlyon Bleachery, Hilton and Jackson(?); J. W. Bishop Co., builders. 1 Noyes St.

The factories are best viewed from a parking area a little north of the intersection of the railroad tracks and Bourne Avenue. The brief tenancy of the Richmond brothers, Franklin and Charles, belies the significance of their venture. They moved from Providence to set up a new operation under exclusive license to use a pioneer Swedish patent for the manufacture of wood pulp paper by the sulfide process, after earlier American experiments were abandoned. The sulfide process eventually revolutionized the industry by enormously increasing paper supply while decreasing cost—and, as is now sadly apparent, also diminishing its life, as testified by the disintegration of whole libraries of modern books and journals. The cost of building the plant and developing its technology, however, led to bankruptcy and Phillips's succession. Although difficult to discern today, the Richmonds' brick plant, elaborately corbeled and initially supported by heavy timber framing (as it mostly still is) exists within Phillips's peripheral accretions. Buildings in a near-symmetrical layout originally housed the process, beginning at the river with a three-story block which was originally floorless in order to contain a row of silolike vats in which wood chips stewed in sulfuric acid to a near pulp, thence through the manufacturing process, to finishing, storage and office facilities up front. There a stubby tower (once with a high pyramidal cap) provided for stairs and corporate identity.

Eventually, two major wire manufacturing firms succeeded Phillips-Washburn: Kennecott Wire and Cable around 1935, then Okonite around 1950. Shortly after taking over, Kennecott built the warehouse building, which runs along the end of Bourne Street. It is a familiar free-span industrial type, but uncommon

EP5 Phillipsdale factories, Kennecott Wire and Cable warehouse

EP7 Nicholas and Jane Monsarrat Houses

in Rhode Island. The exterior wall is nothing more than a one-story brick base, above which a curtain wall of metal sash hangs from an interior frame. A series of light steel trusses, slightly bowed at the top to accommodate the curve of the roof, butt against every fourth vertical of this window wall. They span the entire width of the building. Halfway up the window wall, the principal horizontals mark the tracks on either side, along which the carriage of a huge bridge-beamed, wall-to-wall crane slides beneath the latticed girders within this luminous gridded space. Laconic, elegant, and heroically scaled: such industrial architecture inspired Mies van der Rohe toward the intensification of these qualities in his metal and glass architecture later in the century. Finally, the buff brick Glenlyon Bleachery across the street more self-consciously attempts a monumental image for its factory in the high quality of its brickwork, the generous proportioning of windows to wall, and the somewhat mixed, but knowing, Renaissance detailing of its forceful entrance tower. An architect must have been involved here, possibly Hilton and Jackson, who designed the company housing.

EP6 **Lester Leonard House**

Mid-19th-century, c. 1935 and later. 35 Bishop St.

Many think of this as among the oldest buildings in East Providence. It is really a duplex mill workers' house, probably erected by the contractor John Bishop for the Rumford Chemical Works, which owned it until 1935. Its new owner, Lester Leonard, an antiques dealer who used his home as part of his business, made it look like an eighteenth-century house partly by veiling it inside and out with scavenged colonial oddments and partly by new work in the "old" manner. His pastiche typifies a widespread means of obtaining a "colonial" house (this with a very rare six-bay front!) at a time when ancient fragments from demolitions were much more plentiful than now. The paired doors of the workers' duplex provided Lester with an unusually regal double-doored entrance, which he enhanced by the excessive stretch of its oversized broken pediment.

EP7 **Nicholas and Jane Monsarrat Houses**

1936, General Housing Corporation. 366 Newman Ave. (visible in winter)

These twin houses for the Monsarrat family and their daughter are among the more ambitious Rhode Island ventures into the flat-roofed, corner-windowed, white-cement-surfaced modernism typical of the 1930s. Visible from the road or from a path along the shore only when the trees are bare, the complex consists of a pair of two-story core blocks set at right angles to one another, with hipped roofs now replacing the original flat ones. A one-story L-shaped block containing kitchens links the principal blocks as a larger *L*. At either end are additional one-story sunroom blocks with slab-roofed entrance porches tucked beside them. Altogether they combine to make a zigzag cuboid composition of one- and two-story elements across the site.

General Housing Corporation devised one

of the more advanced factory systems for housing before the flood of such proposals for the conversion of wartime industry to peacetime use after World War II. Tall, narrow, cement-surfaced panels were built on light metal framing with tongue-and-groove edges. Combinations of these units from a standardized palette—window, door, and blank—made exterior walls. In the process they created their own integral metal-framed support, which light metal girders completed as spanning members to sustain floors and ceilings. In contrast to the smooth, seamless coating of walls typical for most modernist architecture of the time, here the vertical joints of the paneling are left exposed, although the original surfaces of the panels have been covered with compositional shingling. Time and care have not been kind to these houses. Still, they remain substantially intact, as dim, stained reminders of a literal Machine Age vision of the house of the future, a concept often better captured in spirit by more metaphorical approaches.

EP8 Newman Congregational Church

c. 1810. Mid-19th century, belfry. 1890, raised for basement story with new porch, William R. Walker and Son. 100 Newman Ave. (at Pawtucket Ave.)

Here in the middle of the old Rehoboth Green, at the crossing of two Indian paths, is the church named for the Reverend Samuel Newman, pastor of the band of Puritans who succeeded to Roger Williams's brief settlement. The original church and its accompanying burial ground stood across the street from this, the fourth building for this congregation. Much modified since its erection around 1810, it then boasted an open belfry topped by a spire. In the mid-nineteenth century both were replaced by the present closed belfry with a fleche and corner obelisks, before William Walker lifted the church a story and inserted his awkward porch. Before his renovation the church had the virtue of utter plainness; the severe, rectangular tower with its obelisks fronted a gabled box, with spare, diminutive openings, extensive stretches of clapboarding, and a tense compactness overall.

Today, the interior is more interesting than the exterior. The initial impression is of a plaster barrel vault, arches over box pews, and a horseshoe-shaped balcony. The last has a paneled front with a triglyph frieze along its lower edge. Then one becomes aware of Walker's intervention, as Victorian taste grapples with this early nineteenth-century interior. First is the two-tiered Victorian chandelier, formed of what appear to be circlets of oil lamps accumulated to this grander effect. Individual lamps of the same sort jut off the balcony front as sconces. More arresting is the exposed structure. The horseshoe balcony is wholly suspended from exposed metal rods (much as the plaster vault above hangs from similar concealed metal or wood struts). Moreover, the corner timber supports for the tower are simply exposed as they come through the church. Why? It is difficult to imagine that Walker intended this structure to stand as a consciously expressive treatment. Its handling is too laconic for that. It is also difficult to imagine the powers of persuasion which Walker—himself no slouch when it came to ornament—would have required to convince the congregation that they should stay with the naked facts of the structure. More likely this was an economic expediency which was never rectified. So we are treated to the curiously stimulating conjunction of Federal decorum, Victorian exuberance, and naked structure making their own peaceable kingdom in this interior.

EP9 Rumford Chemical Works (Former)

c. 1890–1895, 1928. 1 Newman Ave. (corner of Newman and Greenwood aves.)

With plenty of open land available in East Providence in 1858, Wilson and Horsford, who had founded their chemical works in Providence, determined to move the plant across the Seekonk, into what had once been a piece of the Puritan common. The small wooden and plaster-over-rubble buildings of the original plant, some of which remain near the culminating brick factories, suggest the modesty of the original complex. It would not have been too intrusive in the domestic and farm setting; indeed, the factory would have fit its landscape all the better because Wilson bought so much land around it. On it he established a sizable farm to supply food and milk for his workers.

The principal brick factory buildings (leased after this closed as a chemical plant in 1966) are good examples of pier-and-spandrel brick industrial types for their respective periods. The three-story monocalcium phosphate plant of 1890–1895 (facing Greenwood) uses paired sash windows in wood with rough-surfaced granite sills and lintels and elaborate corbeled brick cornices. Whereas it shows stretches of wall around the paired windows, in which the piers are centered, the five-story brick packaging plant (1928) employs an unembellished frame and minimal parapets completely infilled with metal sash. Here the new industrial "type" is incompletely realized by comparison with, say, the contemporaneous Coro factory across the Seekonk (PR39). There the reinforced concrete frame is fully exposed; here it is concealed in brick. There window rhythms are remorselessly identical; here they narrow toward the building's corners, as though to "contain" the elevations in the manner of traditional design. There the horizontality of the factory windows is more insistently asserted, with no such discrepancies as in the topmost story here, where the metal sash below is abandoned for triplets of double-hung sash in wood.

EP10 Workers' Housing

Numbers 57–63 North Broadway are further examples of early twentieth-century workers' duplexes in brick modeled on English precedent, in this case for the Rumford Chemical Works, probably, like the Phillipsdale housing of the same period, designed by Hilton and Jackson (EP10.1; c. 1910). Diagonally opposite is a rarer type of duplex housing for Rumford workers, a pair in the Queen Anne manner. Only 90–92 North Broadway retains a reasonable approximation of the original design, with a mix of clapboarding and shingles and hoods over some of the windows (EP10.2; c. 1882). A pretty version of stern Victorian precedent: Queen Anne made this its mission. The hoods require a little care and some imagination to maintain; hence few examples survive.

EP11 Phanuel Bishop House

Probably 1770s. 150 Greenwood Ave.

In contrast to earlier central-chimney type, this shows the more modern end-chimney placement (with brick end walls), providing for a hall across the center of the house. The fanlighted, pedimented entrance also reflects progressive trends of the 1770s. So does the sophisticated molding under the eaves with its fretted half-round, as compared to the plain jutted overhang of the nearby contemporaneous Hyde-Bridgham House. On the other hand, the narrowing of the clapboarding toward the ground here, and the cruder treatment of the splayed window heads vis-à-vis its near neighbor indicates that up-to-dateness and finesse in design often progress in a piecemeal way in such traditional formats as eighteenth-century house fronts. Phanuel Bishop was long a member of the General Court of Massachusetts (at a time when this site was part of Rehoboth in Massachusetts), before serving eight years as a Massachusetts representative in the United States Congress.

EP12 Bridgham Memorial Library

1905, Hilton and Jackson. 1392 Pawtucket Ave. (corner of Miller St.)

From the earlier generation of this firm's village libraries in the Neo-Colonial style (see also those in Pascoag and Harrisville, under Burrillville), this tiny box of books actually contains the locked box which was the library when the Ladies Reading Society, later the Female Library Association, founded it in 1819. Subsequently, for years Dr. Samuel W. Bridgham maintained and expanded the library in a semipublic capacity from the wing of his nearby house at 120 Moora Street. Here we see the typical early-twentieth-century library, which derives mostly from prototypes popularized by Carnegie benefactions: a pedimented and columned central entrance porch (here nicely integrated with an oculus window) with inside vestibule; then the reading room at right angles to the axis of entrance, a fireplace at one end, and circulation desk in the center; finally, stacks behind. A diminutive monumentality results, providing dignity but also cozy scale. Now stuffed well beyond its capacity, like all such libraries it poses the dilemma of how to enlarge it while retaining its village character.

EP13 First Baptist Church

1879, Walker and Gould. 1400 Pawtucket Ave. (at Pleasant St.)

Except for the obtrusive placement of an access ramp, this is an unusually well-preserved version of a Queen Anne church. It is Queen Anne from the decade of wide-eyed rediscovery of the colonial past, when distorted bits and pieces from the eighteenth century mixed with such Victorian elements as complex roofs, turned porch elements, and sculptural cappings for belfries or towers, and clapboards combined with patterned shingling. The "colonial" contribution is, above all, the scroll pediment which caps the big window dominating the principal gable. But no colonial designer would have floated the pediment in such a disembodied manner above the window, as though it belonged more to the bracketed molding and the gable behind and above it than to the window below. Nor would the S-curves of a colonial scroll pediment display such linear energy, or would their circular terminations squeeze the central bracket motif so tightly, as though some machine had rolled out and elongated a plastic material. The vergeboarded edging of the gable contains piquant foliated ornament, cut with a scroll saw in a thoroughly Victorian manner, and with two unexpected cherub heads midway along its length.

Inside, a stubby Greek-cross plan—a favorite of the period—makes a sacred auditorium with pews arced around the pulpit. Some changes have occurred in furnishings and decoration; but the essentials remain. Most important, large but thin wooden structural brackets, hung with lights, are angled from the four interior corners of the cross and tied with metal rods across the space. Light metal arching with more tie rods rises to support the cross-gabled roof. It recalls carriage construction—at once elegant, rational, and fanciful. After the heartily rambunctious, aggressive forms of High Victorian styles, Queen Anne favored more delicate, self-consciously pretty effects, with a palette of whites or soft pastels, misty mauves, and dusty roses replacing, as here, the stridency of earlier color combinations. In the five openings of the big front gable window, stained glass of a somewhat later date than the building depicts the Virgin flanked by attending angels, stylized to fit their tall, narrow panels, with geometric and abstracted flower patterns around them. In another context, this could be chivalric: Guinivere and her ladies in waiting. Pre-Raphaelitism has been filtered through the Renaissance Revival.

EP14 Nathaniel Jenkins House

1850s, c. 1879; dormers probably 20th century. 1474 Pawtucket Ave. (corner of Pleasant St.)

The Greek Revival Jenkins House was modernized in the 1870s with brackets under the eaves and a bracketed hood over the entrances. Most radical change of all, a central cross gable disrupts the simple Greek Revival box, with a pair of arched windows and the scroll-cut foliated vergeboard to celebrate the new dispensation, as well as more space upstairs.

EP15 Dr. Thomas Aspinwall House

1860s; later wings. 344 Pleasant St.

Like the nearby Judkins House, this is a simple, bracketed block. The "pagoda" roof projecting over the entrance steps from undulant latticed supports on pendant brackets, and the tentlike sag and flare of the roofs for the bay windows, however, indicate the exotic strand of Far and Near Eastern inspiration in Victorian design.

EP16 John Hunt House and East Providence Water Pumping Station

c. 1780–1790. Before 1863, dam. 1893, water pumping station. 1963–1964, filter bed. End of Hunts Mills Rd. (Hunt House owned by the East Providence Historical Society; open to the public)

This is a sylvan spot with a horseshoe dam so low that the smooth curve of the spill is intimately juxtaposed to the tumble of the water against the rocks below. The Hunt family was among the original Puritan settlers when the settlement was Rehoboth and in Massachusetts. John Hunt purchased the property in 1713. His son presumably built the house, very much in the manner of the Phanuel Bishop House (EP11) except that here the chimneys are in the more traditional central position rather than the ends. The interior has fine country woodwork and plaster walls all recently restored. Presumably either the father or his son built a gristmill on the property. Other mills, including a small cotton operation, made up a rural industrial group before the East Providence Fire Department took over the property and demolished the lot. The fire department needed a pumping station to supply water to the Rumford area, which was both isolated from and at a higher elevation than the population center of the town (by then called Watchemocket). The 1893 pumping station is a one-story building of irregular plan in random masonry with a picturesque grouping of hipped roofs and ventilating dormers, all dominated by a tall, square, slightly tapering chimney. The interior, lined with stained tongue-and-groove boards, retains an electric switchboard, a major portion of the pump, and the glazed supervisor's office curving into a semicircular bay beside the entrance. Behind is a reinforced concrete filter bed, open to the sky but looming as a blind box. Foliage from vegetation now growing within it brings a Piranesian touch to the place which should be exploited in landscaping—much as the cable link barrier between the park and the fast-moving stream should be replaced by more appropriate fencing.

When Frederick C. Sayles established his Glenlyon Bleachery in East Providence (EP4) and required a more extensive water supply for the purpose, he convinced the town to set up a private water company and purchased the pumping station with the Sayles Company as principal stockholder to operate the system. He also established picnic grounds and a small amusement park on the site to benefit his workers and the town. It operated until a fire destroyed its popular dance hall in 1925. The amusement park led a languishing existence until it faded away at the beginning of the 1930s, when a debilitated industrial base and expanded automobile travel administered the coup de grâce. The waterworks returned to the city in 1928, and was closed only with the connection of East Providence to the Scituate Reservoir system in 1969. Although the house is colonial and the pumping station Victorian, the juxtaposition of the two here returns us, in an idealized way, to an early phase of the Industrial Revolution in which the owner's house was often set beside his mill.

EP17 **East Providence Senior High School**

1951–1952, Charles A. Maguire Associates. 2000 Pawtucket Ave.

Designed by a large architecture-engineering firm, this is a conspicuous and early example of a group of schools built in the state after World War II which brought the modernist factory aesthetic to the classroom, especially in the use of opaque glass brick walling over narrow bands of "view windows" as a means of obtaining ideal light for classrooms. The brick-clad reinforced concrete clock tower exemplifies the concern for according monumentality even to local public works which was more characteristic of prewar than postwar America.

The Parkway and Riverside

EP18 **Veterans Memorial Parkway**
(Barrington Parkway)

c. 1910, Frederick Law Olmsted, Jr. 1992, East Bay Bicycle Path (Providence to Bristol). From First St. or I-195 exit 2 to Pawtucket Ave.

This three-and-one-half-mile stretch and a less thoroughly landscaped segment (Narragansett Parkway) on the other side of the bay are the only extensive parkway portions of the 1906 Metropolitan Plan for Providence to have been realized. Modeled on similar plans for Boston and other cities, it is an early example of schemes designed to connect and extend existing parks for pleasant drives and to open areas for suburban development, especially those which connected the wealthiest suburbs to the central city. Barrington, adjacent to East Providence, was already well on its way to becoming the wealthiest suburb in the state. As the parkway winds along the top of a bluff, outlooks provide some of the best overall views of the meeting of the Providence River and the bay in relation to downtown Providence. Shortly after the completion of the parkway, oil companies (and, for a while, one of the largest coal unloading operations in the country) moved in along the harbor, spoiling, one would imagine, what the planners had in mind. Actually, however, the Olmsted scheme specifically envisioned the parkway and its turnoffs as places from which to view the city's industry. Moreover, all but the tops of some tanks were below road level and either partially or wholly screened. The ribbon of planting has preserved a residential district immediately behind it. Most of these "tank farms" along the bay closed during the 1980s. Projected use for condominiums, marinas, and an office park may test the old parkway far more than the tanks.

It is also along the parkway that the more venturesome architectural pilgrim may most conveniently begin a 14.5-mile bicycle trek through four East Bay communities, including the Riverside section of East Providence, Barrington, and (much easier) Warren and Bristol. The East Bay bicycle path follows the old right of way of a rail line from Providence to Bristol along fairly level terrain. The path originates in downtown Providence but is accessible from two parking areas along the parkway, where the obtrusive fencing and signage of this great benefit does some damage to Olmsted's design. At the second parking area the bicycle path dips off the bluff, down to the bay at Watchemocket Cove, which it crosses. The parkway also descends from the bluff to curve around the cove, which is alive with waterfowl, most spectacularly wild swan, returned to the area by a substantial cleanup of the bay during the 1970s and 1980s.

EP19 Gordon School

1963, 1970 and later. William D. Warner. 45 Maxfield Ave.

This private school is inspired by the paradigm of the self-contained classroom unit, each with its own courtyard, and all connected to one another by corridor spines. It was pervasive as a model for schools (elementary schools especially) in the 1950s and 1960s. As a school type, it was popularized by the Crow Island School in Winnetka, Illinois (Eliel and Eero Saarinen and Perkins, Wheeler, and Will, 1939–1940) and, as a concept, goes back at least to Richard Neutra's Bell School in Los Angeles (1935). Here the paradigm becomes a "village" cluster of "houses" made up of square-shaped units of various sizes, walled in concrete block, each with its pyramidal roof. The image derives most immediately, it would seem, from Louis Kahn's cluster of similar units for a well-publicized bathhouse in Trenton, New Jersey (1955). But enthusiasm for clustered village images was endemic at the time, enticing many architects to Italian hill towns and to picture tours of such African villages as those featured in Bernard Rudofsky's *Architecture without Architects* (1964) and the popular exhibition at the Museum of Modern Art which stemmed from it. The extra-diminutive scale of classrooms for the lowest grades is especially charming, with such thoughtful details throughout as blackboard walls in bathrooms to encourage creative graffiti. Later addenda and larger units by the same architect somewhat compromise the village intimacy of the original section, mostly by necessity, because the added units, such as a laboratory wing, library, and gymnasium bulk larger than the original classrooms, and the school expanded to include higher grades. The additions have also complicated the clarity of the original scheme, making the whole disorienting to the casual visitor, but perhaps all the more agreeable to those who come to know the place.

EP20 Squantum Association

1899–1900, clubhouse, Martin and Hall. 1870s, billiard hall. 1889, dining hall (for banquets). 947 Veterans Memorial Pky. (visible from parking area off Boyden Blvd. and from East Bay bicycle path)

After climbing past wooded parks (including the private entrance to the Squantum Association), Veterans Memorial Parkway terminates in Pawtucket Avenue, which was once lined with large country houses and farms along this stretch. It is now steadily succumbing to spotty strip and condominium development with a few of the old holdouts, mostly Victorian, mixed with a few early twentieth-century Neo-Colonial houses. West of Pawtucket, .15 mile south of the end of the parkway, Boyden Boulevard begins as an exceptionally wide street for the boulevard spine which George Boyden envisioned as the gateway to a 1920s subdivision. Shortly before the boulevard makes a right-

EP20 Squantum Association, clubhouse

angle turn and tapers to modest residential width is a small parking lot, which is an entrance into a wooded conservation area. This is the most convenient approach to a view of the Squantum Association, except from the bicycle path which runs along the bay.

Organized in 1870, the Squantum Association, which occupies a cluster of buildings, is primarily a private eating club (open only by invitation), which still addresses the Victorian-scale appetite in the restaurant occupying the top floor of its clubhouse. Raised high on an outcropping jutting into the bay, it is an architectural hybrid including references to Victorian Queen Anne and, more conspicuously, the Colonial Revival within the format of a southern plantation house. A grand, two-story Ionic colonnade supporting two tiers of veranda and hipped roof with Palladian dormers facing all four directions and topped by a balustered deck give it a bloated quality that seems incongruously airborne on its rocky perch. Against this lumbering scheme en masse, the detail is exquisite, but mixed in character and overwrought, not at all in the large forms expected of the Greek Revival plantation house, but as though a New England craftsman, working in the delicate manner of the Federal style, embroidered the southern image with Victorian largesse: overly exquisite railings, overly exquisite garlands around the entablature, overly exquisite modillions above, all climaxed by bulbous Palladian bay windows, pressing through the colonnade on either side. But isn't this the note the clubhouse of the period *should* ideally strike? The sense of power augmented by abundance and sweetened by refinement—each aspect increasing the sense that the club is utterly beyond the reach of nonmembers.

The interiors are basically preserved, if perhaps stripped of some of their original opulence. Still, there seems always to have been a degree of austerity and plainness about them. Perhaps the association also wished to evoke some of the ambience appropriate to lodges and boathouses, even as the Indian pudding specialty of the dining room (grains, molasses, and raisins) pays obeisance to Puritans and Indians. Downstairs is half lobby, half lounge; upstairs, in addition to the dining room, one of the side bays lights the grand stair, while the other contains a period bar. From the dining room, a semicircular porch swells out, like the bays, beyond the colonnade, but in this instance with the Victorian expansiveness of a well-filled belly. This elevated deck affords superb views up and down the bay. A later gazebo out on the rocky point jutting beyond the clubhouse into the bay—four cylindrical supports in rough masonry for another pyramidal roof—invites one to walk into the view, whether literally or in imagination. Either vantage point looks into downtown Providence at the head of the bay. From there, in the club's heyday and in season, one of the steamers that plied these waters set out before noon each workday bound for the clubhouse and the capitalists' lunch, and returned after the Indian pudding.

The other buildings in the Squantum complex are lesser architectural events and not easily seen from any public vantage point. The banquet hall retains a deep porch propped high above a cove; the billiard hall retains some of its scroll-sawn ornamentation. Both have been extensively remodeled within.

EP21 Ponham Rocks Lighthouse

1871. Visible from Bullocks Point Rd. and from the East Bay bicycle path

The lighthouse, off on a rock in the water, is a familiar Victorian type, with the lightkeeper's granite house topped by a boxy mansard, on one end of which the squat light sits like the pilothouse of a ferry. More elegantly than most such stunted towers, this rises in a tapering curve to an elegant metal enclosure for the light.

EP22 Crescent Park Carousel

c. 1895 and later, Charles I. D. Looff. Bullock's Point Rd. near Crescent View Ave. (Operates during the summer and on weekends in late spring and early autumn.)

The carousel is all that remains of Crescent Park, another of George B. Boyden's enterprises, and long one of New England's largest and busiest amusement parks. Housed in a traditional wooden polygonal shed capped with a monitor in colored glass, this is among the important merry-go-rounds in the country. Looff, a leading designer and manufacturer of carousels, based his operations in East Providence for a number of years, and used this carousel from 1905 to 1910 as his sales room. Hence he adorned it with a variety of prancing horses and other animals designed as prototypes for his customers and as a test for embellishments. His son eventually took over as owner and manager of the park. When the site went to condominiums in the 1980s, only the zeal of a small Save the Carousel group preserved it in its old location. The park across the way was once the site of an elaborate wharf with one of the larger shore dinner palaces, which also served as the boat landing for the park.

On the loop around Bullock Point south of the park site is another cluster of late nineteenth- and early twentieth-century cottages, crammed together, but less so than those at Cedar Grove and Pine Bluff. After a long period of dilapidation, most of these too have been extensively improved and altered to take advantage of superb water views. Some, however, retain their original character.

EP23 Pearce Allin House

After 1805; door c. 1930. 36 Willett Ave.

This is a typical two-and-one-half-story, five-bay Federal house, but (though one would not guess it) in brick, which is relatively rare for a Rhode Island farmhouse. (Another, dating from c. 1810, exists at 523 North Broadway.) Perhaps for appearance, but more likely for weatherproofing, the front and back walls (south and north elevations) were clapboarded early on. The exposed brick end walls contain paired chimneys for front and back rooms, permitting the hall to traverse the center of the house.

EP24 Joseph Bicknell–Dr. Hervey Armington House

c. 1840(?). 3591 Pawtucket Ave. (corner of Willett Ave.)

Diagonally across the street is an example of a rare house type, a so-called lightning splitter. It takes its name from the exceedingly steep pitch

EP24 Joseph Bicknell–Dr. Hervey Armington House

of the roof on a narrow house, which, according to jocular myth, would split the bolt that hit it—although the story does not go on to tell us how this feat might save the house. A purer variant of the precedent, minus the myth, continues to the present in the A-frame, for the practical reason that it is an economical means of obtaining space with simple framing, although at some inconvenience from excessive interference from eaves inside. Joseph Bicknell married Pearce Allin's daughter Louisa from across the street in 1827, and the house was allegedly her father's wedding present. But stylistically the date seems early for a house of this appearance. Is the roof form, then, the result of a later remodeling when Dr. Armington and his wife, a younger Allin daughter, moved into the house in 1850? In any event, the charm of the "lightning splitter" is its exaggeration of the simple gabled box of the childhood image for "house." This and the Daniel Pierce House in Providence (see entry) seem to be the finest examples of the type in the state.

Barrington (BA)

Barrington has the highest per capita income of any town in the state. The architectural pilgrim who expects from this a precinct of walled estates and grand mansions, however, will be disappointed by the low-key quality of the place overall. There are imposing houses, of course. They cluster mostly along Narragansett Bay, and especially along Rumstick Road, which takes its name from the neck that separates the mouth of the Warren River from the bay. Yet attractive as many of these houses are, Barrington's wealthiest clients seem to have played it safe in choosing architects, apparently wanting reassurance more than showiness or distinction. To the historian, the most interesting of Barrington's largest houses, had it survived, would almost surely have been Edmund Willson's 1884 replica of the Nightingale-Brown House in Providence, which stood on Nyatt Point at the northeast corner of Nyatt Road and Washington Road. Into this, the earliest of the most conspicuous attempts at Colonial Revival in Rhode Island before 1885, the textile baron Henry Steere moved his pioneering collection of American colonial furniture. Only its elaborate picket fence still survives as a souvenir, transplanted to the Bliss-Ruisdall House in Warren (see entry). Although the largest Barrington estate, that originally owned by Frederick S. Peck, is viewable as a private school, others will be noted only in passing, while still others are not publicly visible. Essentially, custom-built development houses for affluent professionals and businessmen, together with trickle-down versions for their juniors, predominate, with a number of upgraded farmhouses, cottages, and workers' housing from the presuburban past.

Except for a fine town hall with landscaped environs, Barrington's commercial center is a hodgepodge of low-rise store blocks fronted by asphalt paving, with no more than minimal cottagey allusions here and there to suggest its suburban status. For character at the center, one must turn to the seaport jumble of Warren or the high-style mix of residences and shops in Bristol, both extending back into the eighteenth century. Barrington exhibits the blah of mostly post-1960 commercial building, redeemed only by restrictive zoning, which keeps most shops small and most signs reticent.

Restrictive zoning, that ultimate hallmark of suburban privilege, appears immediately in following the Wampanoag Trail (Route 114) from the commercial and condominium permissiveness of East Providence to arrive, suddenly, at the beauty of Barrington's protected river and marsh landscape. The short Barrington and Palmer rivers converge in pondlike undulations across the interior of the town to form the even shorter Warren River, which appears more as a small inlet (little more than two miles long) of the bay. This watery environment so close to Providence, which provides plenty of protected anchorage for boats—though barely enough to meet present demand—accounts for Barrington's underlying appeal.

Into this sparsely developed agricultural space between the busy waterfronts of East Providence and Warren, few nineteenth-century industries were drawn. Of special interest to those concerned with architecture, however, was the brickyard at the very heart of the town, which hollowed out Brickyard Pond. The builder-architect James Bucklin's investment in the yard during the 1840s seems to account substantially for his early shift from the use of stone to brick for Rhode Island mills. Much later a few small steam-driven manufacturing plants came to West Barrington and brought some workers' housing with them. So did summer houses and some estates. But farming lingered on into the twentieth century, waiting for suburban development.

BA1 Ellis Peck House

1795. 1723 Wampanoag Trail

This rather plain five-bay, central-chimney house commands attention because of its uncommon double door framed by pilasters and capped with a conspicuous broken scroll pediment. Typical of the doorways of the Connecticut River Valley, this is a typically theatrical late-twentieth-century "colonial" modification. By comparison with Warren and Bristol, Barrington has few surviving eighteenth-

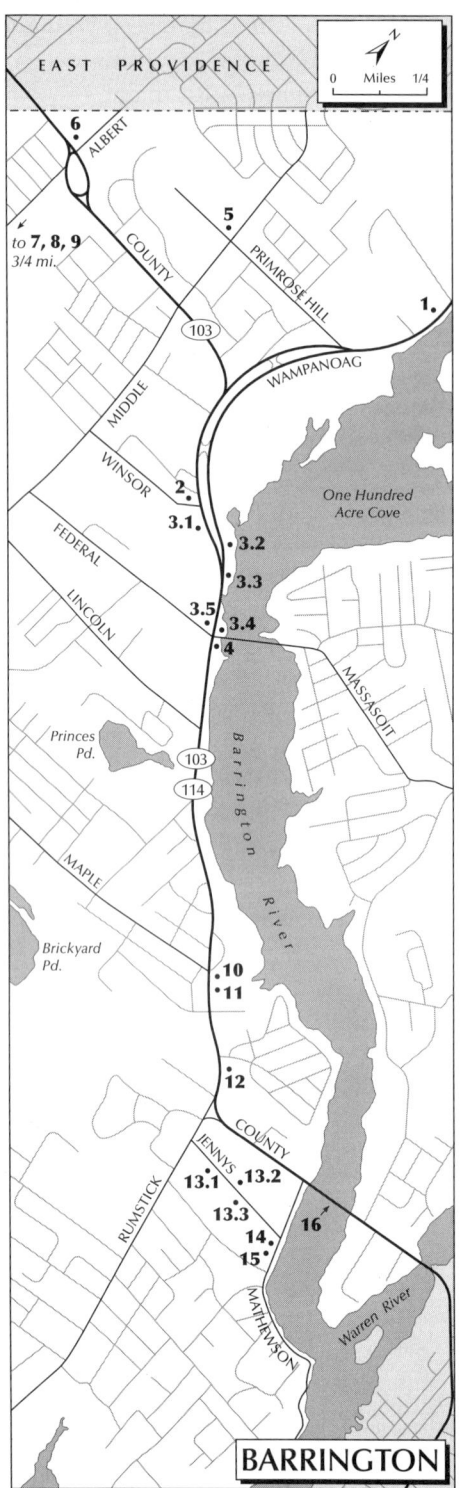

century houses, and these are quite plain. So this doorway epitomizes the will of the majority of the town's inhabitants to be as "colonial" as possible.

BA2 Royal D. Horton House

c. 1872. 1 Winsor Dr. (at County Rd.)

Although this sizable clapboard mansard-roofed dwelling makes its greatest impact when viewed from the northbound lanes of County Road, it is more safely examined from Winsor Drive. It is the very picture of the imposing Second Empire mansion of a small-town grandee. The incised foliate ornament over most openings comes to a climax in the decorated vergeboards of the sunbonnet gable which culminates the front elevation, penetrating a steep mansard with undulantly capped dormers. The original porch, extended across the full width of the front and around the south side, was more in scale with the elevation than the present minimal replacement. As to whether an architect or builder designed the house, the somewhat unassimilated character of the ornament and the bracketing at the cornice suggests that it was a skilled builder, looking at something like Samuel Sloan's popular pattern books, who designed this standout amid the stolidness of Barrington's old center.

BA3 Barrington Village

The old center of the town shows a mix of plain eighteenth- and nineteenth-century houses with various nineteenth- and twentieth-century renovations. The Deacon Kent Brown House (BA3.1; c. 1775), at 530 County Road, is a saltbox enlarged and given more elaborate "colonial" treatment, apparent in the pediments, c. 1970. At number 509 (opposite) stands the onetime Kinnicutt Tavern and post office (BA3.2; c. 1840), and at number 499 the gambrel-roofed Samuel Allen House (BA3.3; southern portion, c. 1760; northern addition, 1938). Most interesting of all are the successive parsonages of the Congregational Church. Number 484 County Road (BA3.4; c. 1770, additions 1856) is a charming mid-century Victorianization of an eighteenth-century house; its successor (BA3.5; 1873), number 464 (corner of Federal Road), is country Italianate with turned porch posts and bracketing.

BA4 Barrington Congregational Church

1805, 1851. County Rd. (at Massasoit Ave.)

"White Church," as it is popularly called, is the iconic point of arrival in Barrington from Providence. It proclaims the colonial image the town means to claim for itself, and indeed was built in the Federal style in 1805 as the Congregationalists' third meeting house. Because each building in the sequence reportedly reused lumber from its predecessor, this version may well include pieces from both 1712 and 1734 meeting houses. The extensive and well-preserved carriage sheds behind date from 1805. The meeting house was raised a half story in 1851 to convert its cellar to a first floor for a vestry and other uses. At the same time the gallery was removed from the sanctuary to create the present uninterrupted high interior space, which, after later renovations, is architecturally unexceptional.

What is of interest about the 1851 alteration is the change of the exterior style of the building from Federal to Neo-Romanesque. It was at this time that a national architectural advisory board of the Congregational denomination recommended the style as a means of providing a more overtly religious character to what many, at the height of the Romantic movement, regarded as the too secular and intellectual classicism of the traditional meeting house. To their mind, round arches had the additional virtue of distinguishing this as a Protestant style, as opposed to pointed arches then favored by High Church adherents, whether Roman Catholic or Episcopalian. Romanesque provided a churchly aspect but one not too churchly. With the new dispensation, the whiteness of the old-fashioned meeting house typically turned toward stony colors: two shades of gray or tan, or cream with dark brown trim. Because Barrington has Rhode Island's best example of the mid-nineteenth-century wooden country Romanesque church, what a pity that its exterior should paint out this history. Might the compromise be two grayed whites, with the darker as trim, to accent, even slightly, the Romanesque interlude that disrupted the blinding whiteness associated with the meeting houses of New England Congregationalism?

BA5 Zion Bible College (Frederick S. Peck House, known as Belton Court)

1905–1906. 1927–1928, visitor and service wings, tower, and connector to original house; Martin and Hall. 27 Middle Highway.

This huge, essentially styleless but vaguely English medieval structure of undressed granite masonry was built for Frederick Stanhope Peck (1868–1947) on property that had been in his family since the seventeenth century. Peck was a businessman with interests in banking, manufacturing, and real estate, and a political leader active in the Republican party. He was also a collector of rare books and manuscripts and had an avid interest in genealogy.

Belton Court is notable more for its size and rugged construction than for any remarkably original characteristics of plan or detail. Conceived as a manor house for an expansive estate that once encompassed 800 acres, it comprises three wings arranged to form an asymmetrical *U* around a courtyard opening toward the south. The original main house, standing alone and facing on Middle Highway, became the eastern wing of the court when the other pieces were added roughly twenty years later. Then the house was much enlarged and its entrance reversed by the addition of a porte-cochere to what would formerly have been its rear or garden front. Garages and other estate services went into the west wing of the new court, directly opposite the original house; a block devoted to large rooms for receiving and entertaining visitors closes the north end of the court. Although linked to the original house by a low connector and, to this extent, an addition to it, the visitor block seems to have been conceived as an almost

BA5 Zion Bible College (Belton Court)

separate establishment with its own entrance, which was opened or kept semiclosed depending on circumstances. Above the courtyard buildings, at the northeast corner, rises a massive four-story square tower. It is the crenellated parapets atop this tower and on several bay windows and the porte-cochere that give the house its nominally medieval flavor.

Belton Court's real interest lies in its interiors. The original house, now used as the president's house by the college that has acquired it, has at its core a broad, low-ceilinged entrance and staircase hall opening through from the front (east) entrance to the porte-cochere, around which the other rooms, with Neo-Renaissance and Queen Anne detailing of the period, are gathered. Contrasting with this are the expansive public rooms of the north block. Here the focus is a baronial living hall with a monumental fireplace and high segmental-vaulted ceiling decorated with ribs in a square-and-quatrefoil pattern. Column screens separate the living hall from the foyer, a corridor, and a stair hall containing a grand staircase leading one-half flight up to a ballroom.

BA6 Albert H. Peck House

1865. 955 County Rd. (on rotary)

This picturesque two-and-one-half-story, gable-roofed, bracketed dwelling was once the center of a twenty-acre produce farm. Still in Peck family ownership, it is the very image of the Victorian country house, impeccably maintained down to the white picket fence edging the property.

Drownville

This area was the property of the Drown family from the eighteenth century. The opening of a railroad station here in 1855 on the Providence, Warren & Bristol Railroad served as impetus for residential development, which occurred in appreciable amount only after the Civil War. Alfred Drown laid out the Drownville plat on the family farmstead and, together with David A. Waldron, a Providence real estate agent, took an active role in its marketing. Drown's own house (1830, with late-nineteenth-century additions) still stands at 13 Alfred Drown Road, while Waldron occupied the handsome mansard-roof house at number 26, built in 1858. The community first became popular as a summer colony, attracting in particular a number of families from Pawtucket. The prominent Providence architecture firm of William R. Walker, a Pawtucket resident, is known to have worked in the neighborhood, but to date his name is associated with only one building.

BA7 George Anderton House

1907. 33 Alfred Drown Rd.

This rather plainly formed but richly detailed dwelling with its rusticated concrete block foundation, stocky concrete veranda columns bristling with banded rustication, clapboard and cut shingle wall cover, and scrolled chimney is of a type of Queen Anne design more common in central or western states than in Rhode Island, where this example may even be unique. Disembodied Richardsonian Romanesque belligerence appears in incongruous

BA8 House

combination with overly exquisite shingle patterning.

BA8 House
1910s–1920s. 67 Alfred Drown Rd.

This handsome one-and-one-half-story stuccoed bungalow is fronted by a recessed porch with smooth cylindrical columns. Its colonialized Craftsman style recalls, on a reduced scale, the work of Philadelphia architect Charles Barton Keen, most grandly exemplified by Reynolda House (1917), the sprawling bungalow that Keen designed for tobacco magnate R. J. Reynolds in Winston-Salem, North Carolina. This is a modest but well-studied example of the type.

BA9 St. Matthew's Episcopal Church
1891, William R. Walker and Son. 5 Chapel Rd. at Second St.

The quirky character of this Queen Anne church is typical of the Walker firm, which produced its share of eccentric buildings in Rhode Island and neighboring states. The congeries of shingle-clad wooden shapes is a ghost, in a more ephemeral material, of medieval English masonry parish churches. The form of the building, with apsidal front flanked by hooded entrance and corner tower, repeats that of the Walkers' Christ Church Episcopal (1888–1889) in Providence. The tower itself is unconventional, cylindrical in shape and remarkably slender in proportion. Such features give the church a whimsical character appropriate for what began as a mission of St. John's Episcopal Church at Barrington Center (see entry below) to serve summer visitors at Drownville.

BA10 Barrington Town Hall
1887–1888, Stone, Carpenter and Willson. 1938, library addition, Howe and Church. 1963, second library addition, Michael Traficante. 283 County Rd.

The White Church is one emblem of Barrington's civic identity; its town hall is the other. It is certainly the finest town hall of its period in the state, at least on the exterior. (Sadly, the interior suffers from extensive unsympathetic renovations.) Prominently sited on a rise, it is a splendid example of picturesque eclecticism, with variegated textures of cobblestone, applied half timbering, and stucco combined in an artfully asymmetrical mass accented by bold front gables between two end towers: a circular mini-turret embedded into the gable roof slope at one end, a larger, near-freestanding polygonal tower coupled at the other end with the entrance porch. The latter rises in a tall, elegantly curved and faceted cone to a mini-cupola surmounted by a weather vane. The elegant rise of this tower is reinforced by the subordinate vertical of the equally elegant brick chimney stepping out of its cobblestone base, marking the opposite side of the entrance. The sophisticated design, inspired by English and Norman medieval building, illustrates the proficiency of Stone, Carpenter and Willson in picturesque composition, but also demonstrates their nice sense of an appropriate image for suburbia—elements from the town's summer/country houses blown up and rationalized for a civic use. Many of the cobbles used for the masonry work were culled from local fields and gardens and contributed by the townspeople, a romantic conceit intended to make this edifice the embodiment of harmony and mutual cooperation.

The town hall originally housed Barrington's high school and library as well as the town meeting hall and clerk's office. Howe and Church's lower library addition of 1938, on the southeast corner of the original structure, replicates the style of the earlier work, while Michael Traficante's simpler addition of 1963, to the rear, is more abstract but repeats the steep gabled forms that characterize the building. The building now houses municipal government functions only, but its extended landscaped precinct includes the Leander R. Peck School.

BA10 Barrington Town Hall

BA11 Leander R. Peck School

1917, Martin and Hall. 281 County Rd.

This red brick Tudor Revival structure housed the high school after it had outgrown the single room allotted for it in the town hall. The building was donated to the town by Sarah Gould Peck in memory of her husband. Sarah and Leander were parents of Frederick S. Peck, the builder of Belton Court (see entry above). The two buildings have in common patronage, architect, and, in general terms, stylistic source. When the high school moved again to a modernist building on Lincoln Avenue in the 1940s, the public library left the town hall for the school.

BA12 St. John's Episcopal Church
(Red Church)

1858–1859, Clifton A. Hall. 1885–1886, 1888, chapel and tower, Hall and Makepeace. 191 County Rd.

St. John's is a fine surviving example of the small-town Episcopal church of the mid-nineteenth century, with typical additions from the later decades of that century. The Episcopal denomination experienced tremendous growth in Rhode Island from the 1830s through the 1850s and established many churches, particularly in the smaller communities and mill villages of the state. This growth corresponded with increasing reliance on the Gothic style as the only proper style for the Episcopal house of worship, and, for smaller congregations, on models derived from the medieval parish churches in English villages. St. John's is a charming interpretation of the type, calling to mind the work of Frank Mills (who published a pattern book of such designs) and variously evident in a few other Episcopal country churches of the period. It is among the earliest known works, and is the first known church, of Clifton A. Hall, who began practice about 1852. As originally built the church consisted of a simple gable-roofed, aisleless nave with entrance vestibule, chancel, and wooden crenellated parapets. The present contrast between the taut brick walls and the brownstone trim that appears to cut across the planar surfaces was not so marked, for the brick masonry was initially stuccoed and sanded to imitate stone. Addition of the Burrington Memorial Chapel on the north side of the nave and the Mathewson Memorial Tower, with its connection to the entrance vestibule, at the north corner of the facade made the structure both more picturesque and more monumental. So few Episcopal churches of the 1850s have survived that St. John's is an important relic of a type and an era.

BA13 Jennys Lane Houses

An appealing street visually, Jennys Lane contains a varied collection of houses from the mid-through the late nineteenth century, and of considerable quality overall. Some are worth special note. The Lewis T. Fisher House (BA13.1; c. 1887), at number 27, is an elaborate Queen Anne dwelling clad in clapboard, plain and patterned shingle, and diaperwork panels. The Allen C. Mathewson House (BA13.2; c. 1862), at number 48, is a T-plan mansard cottage with stickwork porch, bracket and dentil cornices, and chevron patternwork edging the roof. Number 53, the Nelson Newell House (BA13.3; 1869), is a curious villa with a one-and-one-half story, flank-gabled main block dwarfed by a broad, three-story central front pavilion.

BA14 Allen C. Mathewson House

1870. 39 Mathewson Rd.

At the end of Jennys Lane, at the corner of Mathewson Road, is one of several houses in the vicinity once owned by Allen C. Mathewson. It is a well-preserved example of a cross-gabled bracketed cottage. In the late 1800s Mathewson did much to promote Barrington as a desirable residential suburb; the tower of St. John's Church was erected to his memory.

BA15 Mathewson House

1860s. 41 Mathewson Rd.

This large, cross-gabled clapboard structure is distinguished by its two-story porch in Carpenter Gothic stickwork with paired supports. The abstract quality of its all-encompassing screen of the front elevation is impressive, and appropriate in the context of harbor and river views and breezes.

BA16 Barrington and Warren Bridges

1914, Clarence L. Hussey, engineer-designer. County Rd.

The narrow tip of New Meadow Neck, which separates the Barrington and Palmer rivers where they join to form the Warren River, is flanked by these bridges, which carry County Road. Both are of reinforced concrete construction, with shallow segmental arches springing from stone piers to support slightly humpbacked roadbeds. Their classic design recalls antique European models, an image once reinforced by the original balustrades with urn-shaped balusters, destroyed in the hurricane of 1938 and replaced with the present solid, paneled balustrades. Historically, the erection of these twin bridges constituted the largest project in the first round of bridge construction undertaken after the establishment in 1912 of the Bridge Division of the State Board of Public Roads. State bridge engineer Clarence L. Hussey was a pioneer in the use of reinforced concrete for bridge construction in the early twentieth century. Though the Bridge Division made it a priority to develop standardized designs that could be adapted to many applications, the Barrington and Warren bridges are unusual in their employment of multiple solid-arch spans. They are graceful embellishments along this route, a major link between Providence and Newport.

Warren (WN)

In contrast to an ad hoc assemblage of mid- and late-twentieth-century commercial strip establishments and the civic precinct which constitutes as much of a center as Barrington has to show, Warren has an old, close-packed seaport as its core. More emphatically than that of any other town in the state, its street pattern records the classic layout for a seaport village.

At the end of the bridge from Barrington over the Warren River a V-intersection (at the first traffic light) offers a choice between Water and Main streets. Emblematically at least, the choice separates the two kinds of commercial activity which account for Warren's existence. Water Street began as the commercial street of the village. As Warren grew, commerce jumped the intervening residential district and spread along Main Street, parallel to the rail line from Providence to Bristol, installed in 1855. Meanwhile, the functional emphases of the two streets split: along one street the life of the docks and ships predominated; along the other, household shopping, nonmaritime activity (insofar as such a distinction is possible in a small seaport), and civic affairs. The ends of the cross streets, or, in eighteenth-century terminology, "ways" which intersected them were "thrown out" to the river as docks and wharfs. The remnants of these make up Warren's modern waterfront, although the repeatedly altered and replaced commercial buildings on the water offer little more than fragments of what existed as recently as half a century ago. Houses on the ways (together with some on Water and Main) brought domestic life into intimate contact with both the maritime and mercantile aspects of the village economy, while views down the ways, then as now, opened the densely built community to glimpses of water and boats. Along the ways big houses built from local fortunes (comfortable rather than grand) mingled with the smaller quarters of working men to give a democratic aura to this constricted community, which the adjacency of small factories and the establishment of a few larger mills after 1850 intensified.

The site of Warren was originally a major camp and headquarters for the chief of the Wampanoag, whose shoreline territory ranged from Cape Cod southwest to Mount Hope Neck, the present site of Bristol. They called it Sowams. The Wampanoag found themselves in a desperate situation at the time of the English settlement of Plymouth in Massachusetts in 1620. An epidemic had decimated the tribe's numbers a few years earlier, while the powerful Narragansett tribe opposed them on the west side of Narragansett Bay. Hence they were open to offers for the sale of Sowams and much surrounding land with the promise of English support against their rivals. In 1632 the English established a trading post at Sowams which, though sparsely populated into the next century, was among the first places of white habitation in what is now Rhode Island. It was incorporated into Swansea with the founding of that Massachusetts town in 1670. English betrayal of the agreement with the Wampanoag, however, led to the bloody King Philip's War between Native Americans and colonists beginning in 1675. The opening action in a war which ultimately brought the Narragansett onto the side of their former enemies and spread across Rhode Island was the plunder of the white outpost at Sowams, leaving all houses burned and nearby the horror of the "heads of eight Englishmen stuck up on poles." English colonization in the area resumed only after 1677 with the elimination of Native Americans as a threat. From this period few houses remain, on the outskirts of the town and mostly heavily altered.

Not until after the settlement in 1746 of a longstanding dispute with Massachusetts as to the location of the boundary line between the two states east of the bay, however, did the town of Warren (now in Rhode Island, and rechristened to honor Sir Peter Warren, the British naval hero of Louisbourg) come into existence. By the start of the Revolutionary War, Warren, which included all of present-day Barrington until 1770, was a prosperous maritime community, blessed with a narrow deepwater channel along the Warren River into the bay and a location midpoint between Providence and Newport (twelve miles from the first, nineteen from the other). The port was also backed by farms in the outlying areas. Although the war caused an immediate and severe depression which nearly destroyed the town, recovery thereafter was rapid. Warren was a major center for shipbuilding, especially from 1790 through 1860. During this period total tonnage constructed was barely second to that of Providence, and during the 1840–1860 either equaled or far surpassed the combined total of competition in Providence, Newport, and Bristol. Small manufactories for rope, sails, oil, barrels, iron molding, blacksmithing, soaps, and candles clustered around the boatyards. Warren seamen were active in all aspects of America's maritime economy, especially in the slave trade between Africa and Charleston up to 1808, despite the passage in 1787 and 1794 of state and federal laws forbidding it. From roughly 1800 until 1850, Warren was also the leading whaling port in Rhode Island. A Warren ship, the 806-ton *Sea-Shell*, was the largest whaling ship in the world during this period.

Without waterpower, textile manufacturing had to wait for steam, generated by coal brought in either by barge or by rail. Beginning in 1847 with the establishment of the first of several textile mills, followed later by the arrival of immigrant factory workers, the ocean-oriented economy lost predominance. The decline of whaling after the 1860s seriously diminished Warren's maritime stature. But boat building continued at a reduced scale, and continues today. Moreover, the 1880s saw the effective beginnings of a substantially new maritime industry for the town. Warren became the earliest center on the bay for shell fishing on a large scale, oysters most important at first (until they gave out around the time of World War II), but clams also.

Textile manufacturing altered the village architecturally, by adding not only the factories, but also mill workers' housing and some new mansions, which fitted into the closely packed village only by displacing earlier houses built by maritime fortunes. Still, much of exceptional interest remains, especially from the late eighteenth through the nineteenth century, saved in part by the collapse of the textile industry in its turn, and a long interval of static shabbiness for the village. Warren's current revival has occurred as new industries have occupied what remains of the old mills, and built new factories, while suburban newcomers have discovered the pleasures of the community and its harbor. Outside the village, Warren's nicely walled farmlands remained reasonably intact in several parts of the town through the 1970s, offering panoramas of fields and water. Increasingly underused for farming, the fields on Touisset Neck began to disappear into house lots, especially after 1980, and other large chunks of land north of Child Street have begun to be developed as industrial parks.

WN1 Warren Manufacturing Company

1896–1898, 1907, Frank P. Sheldon, 1945, c. 1960.
91 Main St. (at Water St.)

This brick mill is the looming introduction to Warren from Barrington. Disfigured along its length facing the Warren River by low cinderblock additions, it is best seen at its Main Street end or along Water Street. This is classic turn-of-the-century pier-and-spandrel brick mill construction, well proportioned, well crafted, and impressive in its arrogant self-assurance. The forceful tower on Water Street, where the original main gate was located, features an open-arcaded, granite-trimmed bell loggia beneath a beetling cornice. Its openings are full-round arches, as a late manifestation of round-arched ("Romanesque") factory design persisting from the mid-nineteenth century. Plaques in granite under the corbeling give the dates 1847 and 1898. The earlier commemorates the opening of the first mill, which also inaugurated textile manufacturing in Warren. The complete loss of this mill, together with its successive additions of 1860 and 1872, to a spectacular fire in 1895 accounts for the exceptional size and uniformity of this replacement.

On the side streets hard by the factory—especially Davis Avenue, Bowen, Summer, Sisson, and Company streets—are several types of company housing. They include some plain four-unit houses, one restored on Davis and duplexes of varied design, period, and modification.

Water Street

Apart from the few buildings mentioned for special attention, Water Street is fascinating for its variety. Its many late-eighteenth- and early-nineteenth-century houses, even those that have been carelessly altered, offer interesting architectural details too numerous and inci-

456 East Bay

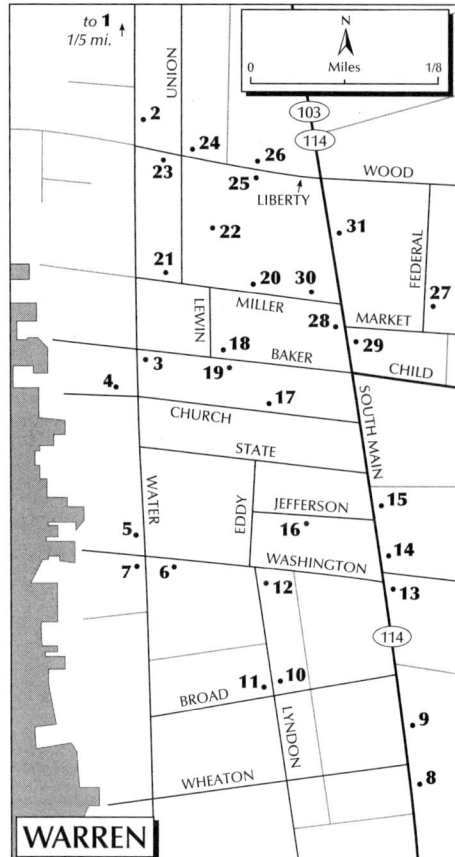

dental to mention here. Many have been converted from stores to residences, and a number that remain have become antique shops, intermixed with businesses catering to sailing and yachting, some dockside restaurants, and, finally, a fish processing plant, boat building, and a dock for excursion boats. Water Street appeals to the tourist; but, like Warren generally, does not appear to go much out of its way to do so. Essentially Water Street remains what it has always been, a street along which shops, residences, and manufacturing live together—and is well worth attention for this as well as architectural reasons.

WN2 Hall's Block (The Chocolate Shop)

1883; outside stair later. 146 Water St.

Of all surviving nineteenth-century shops in the state, this may be the most endearing (unless one includes Providence's similarly diminutive Atlantic Bank Building, which has been converted to a shop (PR17). The elegance of the Neo-Renaissance embellishment of this front transforms a plain, two-story clapboard box into something charming and elegant, which, together with its petite scale, accords with the businesses it housed, whether chocolates (as popularly believed) or millinery (which recent research seems to indicate).

WN3 Hill-Collins House

1761; rear extension and side ell later. 1980–2000 restoration, Lombard Pozzi. 240–244 South Water St. (at Baker St.)

This half house has a plain but commanding pedimented entrance. Captain Collins (not the client for this house, but its most prominent owner) unfortunately became most famous for the loss of the *New Columbia* on rocks off Prudence Island in the gale of 1827.

WN4 Elder Samuel Maxwell–James Maxwell House

1743. 59 Church St. (corner of South Water St.) (owned by the Massasoit Historical Association; open to the public)

Except for its attic gables, the earliest house extant on the waterfront is of brick instead of the usual clapboard, with large central chimney. It has a plain transomed door and narrow windows with tops abutting a blunt belt course downstairs and the eaves upstairs. James Maxwell, who was born here, after the Revolution became one of Warren's leading shipowners and merchants.

WN5 Major Caleb Carr–Captain Caleb Carr House

c. 1764, c. 1790. 317 Water St. (corner of Washington St.)

When this two-story, hip-roofed house was enlarged toward the end of the eighteenth century, it also acquired an elegant fanlighted door on each facade—one for household use, the other for tavern customers. Their identical frames, featuring Ionic pilasters, are intricately ornamented, on perhaps too small a scale for an architectural exterior. Captain Carr, the tavern owner, was also an important shipowner, leading citizen, and operator of the ferry to Barrington, which docked at the end of the

Warren (WN) 457

WN2 Hall's Block (The Chocolate Shop)

WN6 J. H. Hoar House

street, then called Ferry Street, like its interrupted extension across the river. Remnants of the granite ferry docks remain at the waterline. With so many enterprises to his credit, the captain came to overshadow his father, the major, so the house is now popularly associated with the son.

WN6 J. H. Hoar House

1841. Early 20th century, rear bungalow additions. 50 Washington St.

This tiny one-and-one-half-story Greek Revival house has a full Doric portico in front and an arched second-story bedroom window in the pediment. The monumental in such a miniaturized format cannot help but be captivating—like the tiny temples in eighteenth- and nineteenth-century English gardens.

WN7 Water Street Commercial and Industrial Buildings

Buildings farther south along on South Water Street exemplify commercial and industrial elements in its varied mix. The rusticated concrete block Moyes Garage (c. 1915), at number 320, now used for storage, is a period type in a period material. Opposite, along the river side of the street, begins a row of small factory buildings, ranging in date from the 1830s to the 1870s, typical of those that catered to ships and their cargoes. Although all have been rebuilt, they retain enough of their original aspect to provide perhaps Rhode Island's best-preserved row of seaport factory buildings dating from the age of sail and early steam. The Mechanics Machine Shop, also known as the Old Dye House for a later operation (c. 1870 with a later extension), at 321–325 Water Street, offers a nice example of late-nineteenth-century industrial brickwork with well-proportioned openings. Next come two rubblestone Greek Revival factories, plastered over. The first, J. J. Smith (c. 1840), at number 337, was a warehouse for whale oil; Stubb's Wharf oysters took over. The other (1842), with a low trapdoor monitor that has been sealed, was built as the Gardner-Brown Mill for whale processing and became Gladding's Sail Loft and eventually another oyster operation. Finally, and especially interesting as the earliest in the row, is Marble's Forge, in random masonry (1830s). Most of this range is now owned by the Blount family. As variously enterprising as Captain Caleb Carr, the Blounts are boatbuilders specializing in medium-sized excursion boats and ferries, excursion operators from a dock at the end of the row, fish processors, and (as an offshoot of invention in conjunction with their boat building) purveyors of mini–flush toilets.

WN8 Joseph W. Martin Memorial Home (Charles Smith–William Winslow House)

c. 1850, Russell Warren. 624 Main St.

What initially compels attention here is the authority of the cube, with the exaggerated projection of its roof, and (when visible above the projection) its culmination in a forceful cupola out of a low hip. The elevations are perhaps less satisfying in their stylistic disparity and somewhat disjointed handling of ornamental treatment, however revealing of the possibilities and confusions of an eclectic approach in the mid-nineteenth century. The Italianate capping and sills of the upstairs windows seem at odds with the "English Baroque" enframement of those below, the Egyptoid columns of the porch, and the Federal entrance (so much at odds with the rest that it may represent a later effort to return a Victorian door treatment to the Federal style approved by Neo-Colonial taste). There is also an abruptness about the sculptural projections of the upstairs window heads and the windowsills throughout which seems at odds both with the Greek Revival emphasis on the broad, planar block and with the frivolous linearity of the enframement of the windows downstairs, as though he could not quite make up his mind as to what his approach to plane versus projection should be. In this house, it seems, a Greek Revival architect grapples uneasily with unfamiliar design trends emergent toward the end of his career. But these varied stylistic sources are all employed with a degree of originality and defined with a crispness of execution. Notable especially are the ornamental qualities of the upstairs bracketing, elongated like that of a console table, as a complement to the overhang above. Inside, only remnants of the original treatment remain. The longtime second owner of the house was among the village ship captain elite. His schooner, *Metamora*, was in regular trade with British Guiana.

WN8 Joseph W. Martin Memorial Home (Charles Smith–William Winslow House)

WN10 J. J. Bickner House

WN9 **Bliss–Ruisdall House**
(John Birch House)

c. 1810. Fence, c. 1885, Edmund Willson. 606 Main St.

This standard Federal house, which once stood on another site nearby, is immediately noticeable for its exceptional fence—a Colonial Revival design by Edmund Willson, in which the revivalist, in a manner typical especially of the late nineteenth century, outdoes his precedents in scale and elaboration. It originally fronted one of the earliest important Colonial Revival houses in the state (1884–1886), on Nayatt Point in Barrington, designed by Willson for the textile magnate Henry J. Steere, who was also a notable early collector of colonial furniture (see Barrington introduction). It was carted off to adorn—and overwhelm—this rather plain Federal house when the Steere House was demolished. Then, too, apparently, the Neo-Colonial porch with its parapeting further camouflaged the old shell, which boasts some fine woodwork inside.

WN10 **J. J. Bickner House**

c. 1842. 1860s(?), porch. 28 Lyndon St. (corner of Broad St.)

Remarkable in this simple late Greek Revival house set gable end to the street is the slightly later porch, which is unique in the state. Paneled pilasters at the front of the porch are stencil-cut with well-spaced astrological signs; those at either end abutting the house have a sprightly vine pattern. Is it the work of a builder-owner who meant to flaunt his Masonic allegiance?

WN11 St. Mark's Episcopal Church

1829, Russell Warren. Early 20th century, exterior and interior alterations. 23 Lyndon St. (near Broad St.)

The year following his work on the Providence Arcade, Warren completed St. Mark's. Whereas its very fine Ionic columns are now topped by a pediment (one of a number of alterations made after a fire in the early twentieth century), they originally supported a paneled attic comparable to that on the Weybosset Street elevation of the Arcade. Immediately behind this attic once rose a massive steepled cupola, lost to the 1938 hurricane. The entrance doors are canted, in a manner reminiscent of Egyptian forms As is typical for Greek Revival buildings, flushboarding of the front sets off the columns, and the side elevations, with their arched windows, are clapboard. The fire required the twentieth-century replacement of the preaching end of the sanctuary. The center of the balcony over the entrance has also been altered, although much of this end is original. The boxed pews also appear to be original. To the center block of pews are posts, four to each side, which support the lighting fixtures. This exceptional torchère-like treatment is apparently original and seems to have substituted for the chandelier which would conventionally have hung from the center of the shallow, barrel-vaulted plaster ceiling.

WN12 Lewis T. Hoar House

c. 1810. c. 1880, altered, Lewis Hoar, housewright. 4 Lyndon St. (corner of Washington St.)

A leading Victorian housewright in the village modernized this early foursquare Federal house with hipped roof and monitor by adding bays and an ingeniously designed front porch (fortunately unspoiled by the aluminum residing of the body of the house). The front bay window at one side of the house locks into the projecting polygonal centerpiece of the porch, the shapes echoing one another, binding porch and bay to the house, and repeating the bay windows on the side toward Washington Street.

WN13 George Hail Free Library

1888–1889, William R. Walker and Son. 1980, 1994, interior restoration, Lombard Pozzi; Nolan Lushington, consultant. 530 South Main St. (at Coade St.)

This rough-faced masonry building in a pinched version of Richardsonian Roman-

WN13 George Hail Free Library

esque locks a towered entrance into the fold of a gabled L-shaped mass. Apart from its civic gesture at the center of the town, its principal interest is its interior. There cherry woodwork grained to resemble mahoghany, brass gaslighting fixtures, dull green and tan walls, fireplaces, and leaded windows patterned with stained glass, together with some of the original furniture, preserve a sense of the original setting. A wide-arched opening beside the circulation desk gives entrance to a gabled rear wing for stacks with exposed wooden trussing. A small collection of paintings and stuffed animal specimens from the locale, which have also long adorned its reading rooms, make token acknowledgment of the auxiliary roles of the nineteenth-century small-town library of the period as natural history museum and art gallery.

WN14 Warren Town Hall

1890–1894, William R. Walker and Son. 1939. 1971, interior alteration, William M. O'Rourke. 514 Main St.

This is one of many public buildings, including several town halls, for which this politically astute firm managed to garner throughout the state. Gusto in overall effect rather than nicety of detail characterizes Walker's work. This brick block rises to a precipitous pyramidal roof which is cleaved by the entrance tower. Before the 1938 hurricane the tower was even higher, incorporating an open stage topped by a parapet and a clock face on each of the segments of its still extant melon dome, which now has a slightly reconfigured mini-cupola topping. One detail worth noting is the large ornamental terra-cotta plaque over the arched entrance, a hallmark of Walker's design at this time, in which the building identification competes with florid plant life. A bust of the sachem Massasoit pokes through the leafage, above a fluttering ribbon inscribed "Sowams," the Wampanoag settlement that preceded Warren. In other panels between the first- and second-story windows are garlands with more fluttering ribbons, loosely executed, with considerable verve, and remote from the classical gravitas of the impending Renaissance Revival, to which Walker's Victorianism alludes without much understanding. The cannon in front, named Pallas and Tantae and cast in France early in the eighteenth century, were captured from the British at the surrender of General Burgoyne. They came to Warren in appreciation of services rendered by the town's artillery during the Dorr Rebellion.

WN15 Main Street

Warren Town Hall, at the civic center of the village, and, half a block farther, the crossing of Joyce and State streets, are pretty much the heart of Warren's Main Street. Small-town Main Street as it was known during the first half of the twentieth century has become a rarity in Rhode Island. Warren's Main Street is not remarkable for its buildings, nor have most recent remodelings and replacements improved the situation. But it does have the right degree of dignity for a civic core. Commercial buildings still close ranks against the sidewalks, as yet little interrupted by gaps from parking lots and strip commercial ventures. Above all, it is a bustling place—an irritating bottleneck, to be sure (although Water Street offers a bypass). It is worth the architectural pilgrim's while insofar as much of its late nineteenth- and early twentieth-century aspect remains, awaiting relatively modest restoration toward the state revealed in old photographs; not, one would hope, as a faked tourist attraction, but as a reinforcement of Warren's identity (to which the tourist may be attracted for this very reason).

At 476–488 Main (the Joyce Street–State Street intersection) is the clapboarded Queen Anne style Tavares Building (c. 1895), its upper stories once given over to the Goff Hotel. On the same corner, at number 496, is an early Victorian two-story house (c. 1870), raised a story for the shop long occupied by Delekta's Pharmacy, which retains its original soda fountain stretched along a gray marble counter.

WN16 American Legion Post
(Warren Armory)
1842. 10 Jefferson St.

The Warren Artillery, which won Pallas and Tantae for their foray against Dorr's ragged rebels, also gained from this action an armory. By upholding voting rights for non–property owners, Dorr's adherents frightened the Warren establishment into protecting itself against further rebellion by the erection of an armory. This one, in the toy castle style, is basically a cross-gabled *T*, except for the entrance gable, which thrusts toward the street with its tall portal conventionally framed in a pointed arch flanked by chunky octagonal and crenellated towers. Save for the cut granite frame around the portal itself, the rest of the front is stuccoed over, in the hope not so much of fooling the spectator as of directing thoughts toward the *idea* of impregnable cut stone walls. Asphalt shingling now tacked over the towers' wooden crenellation further reduces the armory's fortified demeanor, leaving only the irreducible (and charming) sign of its intended purpose.

WN17 First United Methodist Church
1844. 1845, steeple. 20th century, interior altered and entrance added. 27 Church St.

This Greek Revival structure is the church of Church Street. As with the town hall, nothing is elegantly detailed here; but the three-story Doric portico is impressively scaled, and so is the steeple. The portico lifts its pediment over a "basement" for Sunday school and parish activities plus the two-story height of the elevated sanctuary. The front is flushboarded and the other walls are clapboarded. Rising behind the pediment is a four-stage tower, three of the stages framed by variations on the corner pier motif, rising to a diminutive glazed octagon ringed with more piers that forms the base for a needlelike spire. The tower is too attenuated for the portico below, and its telescoping from stage to stage too abrupt. As the tallest thing in Warren, however, even today it serves as a landmark for sailors. The adjacent early Victorian house, bracketed handsomely, was built as the parsonage (1858) by the local building firm of Hoar and Martin.

Where crosswalks converge at the center of the little park running between State and Church streets that fronts on the Methodist Church is a nice piece of post–Civil War provincial geometry honoring the dead of all wars. A rough-faced base shaped with a fortification's redoubts at all corners steps back to a smooth-faced echoing shape, cornered by cannonballs, then steps back again to an upended block, atop which a very paunchy bronze cannon stands on end, its rotundity at the base tapering as it rises.

WN18 Masonic Temple, Washington Lodge No. 3

1796. Before 1887, enlargement. 1913, interior remodeling. 35 Baker St. (at Lewin St.)

Although home to Lodge No. 3, this is the second oldest extant Masonic temple in the state (after that in Newport). The lodge appears to have called on everyday carpenters from its membership for the elongated, gabled clapboard box, which is utterly unprepossessing save perhaps for quoining, and then tapped a superb master craftsman or two to ornament the prosaic result toward "architecture." The fan-shaped bracketing of the eaves and, best of all, the three pedimented doors (the third replicated for the later extension toward Main Street) apparently derive from Asher Benjamin's ubiquitous handbooks; but they display superior craftsmanship. The doors are delicately flanked by Ionic pilasters with the same squashed capital treatment just seen in the more diffuse design for the entrances to the Caleb Carr House (entry above), which support a vigorous but subtle sequence of entablature moldings forming a base for the commanding pediments. The temple has lost its original cupola. Its principal interior meeting room was redone in the twentieth century with pieces from the shipwrecked *Newport* and elaborate murals by the Rhode Island artist Max Muller.

WN19 Firemen's Museum (Narragansett Steam Fire Engine Station No. 3)

1846. 38 Baker St. (open to the public)

Directly across the street from the Masonic Temple, this late Greek Revival fire station shows little that is specifically "Greek," while the round-arched windows in the front elevation anticipate a favored motif in early Victorian design. Other than the front wall, the building is domestic in design, except for a stubby hose tower at the rear. Inside, the featured attraction is an 1802 Button hand pumper, although the museum also displays other early firefighting artifacts and documents. The museum is rarely open, but windows permit a look inside.

Next door, at number 42, is an elegant early Victorian stable (c. 1865), its gable a tentlike or pagodalike flaring curve, repeated in the cupola. The cupola openings are divided by X-framing with scroll-cut ornamental inserts.

WN20 General Nathan Miller–Commodore Joel Abbott House

1789, 1803. 33 Miller St.

More impressive than elegant, this unusually large Federal house shows itself off at the end of Narragansett Way. Two important military leaders resided here: a Revolutionary War general and Admiral Matthew Perry's second in command on his voyage to "open" Japan. The typical five-bay plan is far from symmetrical, while details are quite bald. The recessed entrance is part of an 1803 modification, and doubtless reflects municipal legislation toward the end of the eighteenth century banning the earlier projection of stairs and porches onto public sidewalks.

WN21 Rudolphus B. Johnson–John Luther House

1823. c. 1860s, side porch and later addition. 43 Miller St.

Rudolphus Johnson, merchant, whaler, and owner of a wharf at the end of Johnson Street, was the most prominent early nineteenth-century owner of this five-bay Federal house with quoining. Its notable feature is its severe side-lighted door topped by a broad entablature with an inset fan, providing a nice antithesis between boldness and elegance.

WN22 John R. Wheaton House

c. 1833, probably Russell Warren. 90 Union St.

Foliage, even in winter, permits only a vignette view of this fine Greek Revival house over a cast iron Greek Revival fence. Both its quality and originality suggest that it is likely Russell Warren's design. Formal yet relaxed, it has a compelling boxiness enhanced by the parapeting in solid and fretted panels which conceal the hipped roof up to a centered monitor wrapped in more paneling. A generous porch with wide-spaced Ionic columns and a second-story deck with balustrade provides the more relaxed side of the combination, although it is possibly a bit too open for the apparent weight of the closed parapeting on the roof. The interior (not open to the public) contains mantels painted to simulate marble and good Greek Revival woodwork.

The client was one of two brothers who succeeded to ownership of the nearby Wheaton and Baker Rope Works. Warren once had two ropewalks—elongated clapboard buildings in which horses or donkeys walked the strands of rope back and forth before they were twisted by machine—specializing in rope for such maritime uses as rigging, hawsers, nets, shrouds, and stays.

WN23 Thomas Cole House

c. 1850. 81 Union St.

WN24 Charles Wheaton Cole House

c. 1815. 33 Liberty St.

The Thomas Cole House, at the southwest corner of Liberty and Union streets, is a modest bracketed house in the vernacular Italianate manner. John's brother, Charles Wheaton, lived on the northeast corner, in a Federal house with an Italianate bay window later added above the entrance porch. Its fine Ionic columns and door treatment duplicate those of John's house.

WN25 Liberty Street School

1847, Thomas A. Tefft. 20 Liberty St.

This was only the third high school building constructed in the state, and is the oldest survivor of its type. The linden trees fronting its brick walls make it easy to miss in summer. But it is a forceful, if very functional, statement in Tefft's often used conjunction of the palace image with the pioneering round-arched style, which here provides emphasis for the entrance and its window overhead. The school exhibits Tefft's usual skill in proportioning openings to walls, the windows adorned with projecting sills and hoods in brownstone, the door enframed in a slightly projecting hood.

Tefft's sober buildings usually exhibit some exceptional, if subtle quirks. Here the first of these is a double row of brick moldings to make a decorative band below the bracketed eaves. In the projecting or portal section of the front elevation, they angle upward to follow the line of the gable and avoid collision with the arched window below. The second is the undulant treatment which he frequently employed for very simple bracketing. The stepped profile of the eaves bracketing suggests a ribbon of cloth suspended over two rods to make a stepped effect.

WN26 G. G. Hazard House

c. 1800. 15 Liberty St.

Opposite the school is an unusual late Federal house in brick with a projecting belt course between the first and second stories and a hipped roof surmounted by a small, centered platform (missing its parapet). Instead of the five-bay elevation characteristic of the standard Federal elevation, here are three at somewhat larger scale, an early Victorian anticipation of the enlargement and consolidation of openings for this kind of front typical of the mid-nineteenth century urban "palace" format. Early Victorian, too, are the blunt reduction of Federal elaboration of decorative detail and the scrolls under the windowsills, which are at best unusual for Federal houses. The interior (not open to the public) contains excellent carved woodwork, probably the work of a ship's carpenter.

WN27 Judge Alfred Bosworth House
c. 1840, possibly Russell Warren. 21 Federal St.

This Greek Revival house with a full-blown, two-story Ionic portico has such fine proportions and detailing that it has been attributed to Russell Warren. It passed through a period as a popular ice cream parlor, Maxfield's (a name now better known locally than the judge's). Its full portico, rare in Rhode Island Greek Revival houses, shines through, and anything that might be by Warren is worth looking at.

WN28 Warren Baptist Church
1844, Russell Warren. c. 1865(?), 1913–early 1922. Main and Miller sts.

Again Russell Warren, this time Gothic influenced rather than Greek Revival and Italianate (see also the Dow-Starr House, below). This is the third church building on this site. The first (1764) was burned by the British in 1778; its replacement (1784) was demolished for the present building. As the sign outside and a plaque inside also indicate, it is here that the predecessor college for Brown University began.

The plain random rubble walls are beautifully crafted. The corners of the building and its openings are handled handsomely by a rustic quoining which barely puckers the edges of the broad wall surfaces. The church makes a reductive allusion to Gothic, which is limited to pointed arches and once, importantly, a crenellated cresting to the tower, a full eight feet in height, which projected above walls now somewhat reduced in height by the removal of the crenellation and too mildly capped by a simple pyramidal roof. It is shape and the craft of building, not detail, that creates the exterior impact here, plus the great breadth of the building—70 feet to a length of 84 feet. Such squarish proportions are rather more typical of the old meeting house tradition than the elongation generally favored in full-blown Gothic Revival.

The interior has been extensively remodeled, possibly twice. An old photograph in the church shows what stylistically appears to be a c. 1865 renovation of Warren's original interior. Hard to fathom, the photographic image seems to make little allusion to Gothic. Its flat ceiling, which is known to have been of sheet metal, appears to have been divided into shallow panels, each elaborately embellished. A broad decorative cornice intervenes between this and the all-over stenciled patterning of the walls. Two elements are especially remarkable, one of which remains. This is among the largest extant mid-Victorian gaslighted chandeliers in the state, now electrified, with a pyramid of globes in cut and frosted glass which could barely be described as Gothic, together with simple wall sconces of the same date. Most remarkable in the photograph, however, is a Gothic perspective effect enclosed within a huge pointed arch which originally decorated the flat wall behind a centered pulpit. This was surely Warren's design, and its loss is a pity because few such perspective effects have survived from nineteenth-century churches, and few as ingenious and decorative as what was once here. All that surely now remains from Warren's original interior are the box pews and the paning of the side windows, where large diamonds of stained glass in intense red and blue occur in fields of clear glass.

The radical change in the interior occurred from 1914 onward, possibly extending into the early 1920s. Ivory paint then replaced all wall stenciling; a new, slightly gabled and paneled ceiling on bracketed beams was inserted five to six feet below the level of the old metal ceiling (the latter perhaps still entombed above it); and finally, a chancel eliminated the front wall of the sanctuary with Warren's perspective. A broad, three-part late Tiffany window on the themes of baptism and the Pentecost, which closes the chancel, makes the new climax for the space. Max Muller, who was painting murals for the Masons a few blocks away at this time (see entry for the Masonic Temple, above), is responsible for the panels on either side of the baptism window depicting the Christian symbolism of the grape; he extended the theme in stenciled bands along the arching of the nave. A church member, James Vance Cole (to whom the Tiffany window is dedicated), supposedly had a considerable role in conceiving this design as an amateur while contributing substantially to pay for its execution. So the present aspect of the interior is best described as late Arts and Crafts medievalism, both for its design and for the role of the nonprofessional "lover of the arts" whose taste apparently substantially conditioned it.

Warren Baptist Church is significant, too, as the site of the founding of Brown University. When the Reverend James Manning was called from Princeton to Warren in 1764 to take over the congregation and its new church building, he did so with the proviso that he could open a school for the training of Baptist ministers because all existing colonial colleges at the time provided for training only in the Congregational and Episcopalian ministries. Hence he founded Rhode Island College behind the first church building on this site, roughly where the parish house now stands. The college spent its first four years here and held its first commencement in that church. Thereafter both the college and the Reverend Manning (as its president) were lured to Providence by the economic and cultural advantages of the larger city. Not until the 175th anniversary of its founding in 1939 did the university formally present this place of its founding a bronze plaque (to the right of the chancel) in "deep appreciation of [the church's] generous hospitality in the days of the [university's] youth." (The congregation, understandably miffed by the college's departure, may not have cared to accept the university's thanks any earlier.) The college soon changed its name from Rhode Island College to Brown University in gratitude to one of its principal early benefactors.

WN29 Fleet Bank (Industrial Trust)

1906, Stone, Carpenter and Willson. 410 North Main St. (corner of Market St.)

The Warren branch of the Fleet Bank has one of Edmund Willson's scintillant Neo-Colonial frontispieces. Its balconied interior (now disfigured by alterations) must always have been somewhat congested with spindles and fretwork, but is lively and inventive, as is typical of Willson's vision of colonial design.

WN30 Lyric Theater

1916. 1992, remodeled for stores. 5 Miller St.

The Lyric Theater's handsome sheet-metal elevation is dominated by a huge blind arch patterned in lattice which once measured the width of its missing marquee. Prefabricated stampings of Ionic pilasters, panels, frets, and swags are elegantly positioned around this center. (As this is written, the front has been Queen Anne-ed in dark greens and maroon, although it seemed more "lyric" when it was properly white and pastel. So avowedly scenegraphic in character, this is perhaps the loveliest small theater facade in the state. How sad that movies have left it!

WN31 Dow–Starr House

1860. 366 Main St.

Early Gothic Revival houses are rare in Rhode Island; this is almost unique in Warren—and one of its grander houses. It has a cubic quality and authority comparable to that of Russell Warren's Italian villa at the other end of Main Street (see WN8). Its severity of handling in random masonry suggests the manner of the Warren Baptist Church, and the integral quoining of the corners of the rear ell is specifically similar. Could Russell Warren, therefore, also have designed this house? A cross gable centered in the front elevation cuts into the strongly projecting hipped roof with a decorated vergeboard which is characteristic of (if not exclusive to) the Gothic Revival. The most specific Gothic allusion here is in the projecting wooden entrance shelter, and this only in a minimal allusion to crenellation. Tudoresque hood moldings over the windows complete the "medieval" dressing.

Bristol (BR)

Warren and Bristol provide one of the more remarkable contrasts among Rhode Island towns. From the close-packed seaport village of Warren, the best-preserved example of the seaport type in Rhode Island, one enters its stateliest town. Bristol's center stood among the queen cities of colonial and nineteenth-century New England. It is also the centered port in Narragansett Bay: twelve miles up from Newport, fifteen miles down

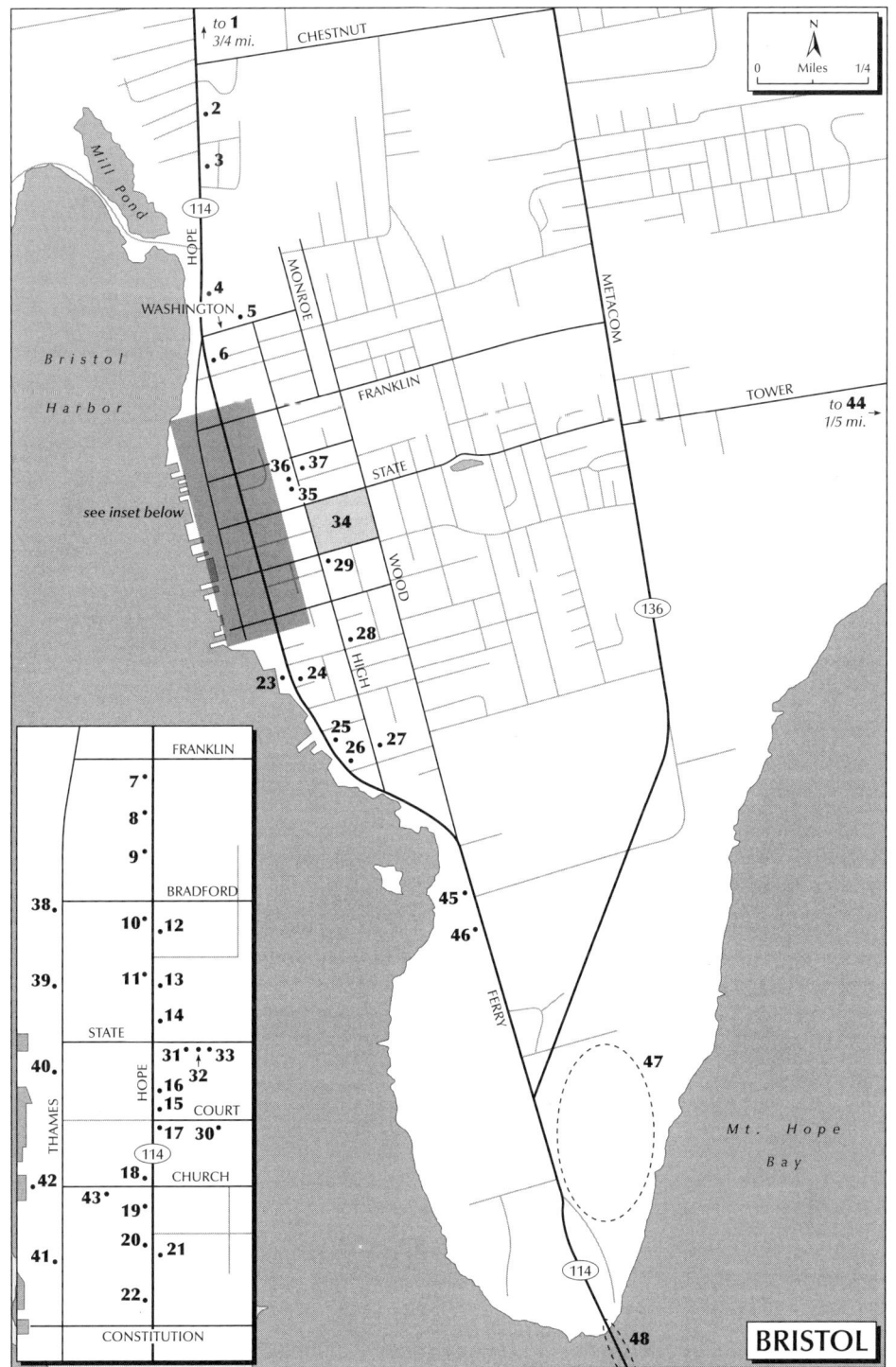

from Providence, it occupies a lobster-claw point at the heart of the Bay, with the center of the town located on the large claw and the smaller claw mostly restricted to large private houses gated from the outside world, forming one of the more exclusive such precincts in the state. It is the buildings within the core of the town that perforce provide the attractions for the architectural pilgrim, though given their quality, this is no deprivation whatsoever.

The order and breadth of the village core are immediately felt. The basis of the first is the grid of the street plan, based on a town green, among the few and one of the most evident such gridded town plans in Rhode Island. (Bristol of course began as a Massachusetts town, with the expected Puritan church dominating the green.) The longitudinal streets, all very wide, parallel the water, moving inland from Thames (the street closest to the harbor), to Hope (the principal street), then to High and Wood. The Town Common, which stretches between the latter two, was obviously envisioned as someday centered in an enlargement of the original street, although the historic town never pushed beyond Wood Street.

Contumacious individualists, Rhode Island's settlers did not often organize formal settlements on the Puritan model; Bristol is the finest exception. If it today retains so much of its character and scale, it is because the geometric logic of its plan remained appropriate. Thames Street served the physical functions of the harbor, Hope the commercial, and High the civic. The result is one of the most enduring and successful essays of Puritan town planning, here in a maritime setting.

The peninsula on which Bristol sits had belonged to the Wampanoag Indians, whose councils were held at Mount Hope, on the eastern side of the peninsula. Here, in 1675, Wampanoag chief Philip (Metacomet) launched King Philip's War, which ended in his death not far from the site. Afterward King Charles II granted the lands to the separatists of the Plymouth colony on Cape Cod, who promptly established the town in the Grand Articles of 1680. These not only prescribed a gridded plan, but also specified that "all houses should be two stories high, with not less than two good rooms on a floor." The town was sited on the western shore of Bristol Bay, while the remainder of the eastern peninsula and Poppasquash, the western peninsula, were divided into farmlands and commons.

By 1700 Bristol was a thriving town, active in shipbuilding and trade, shipping agricultural products and livestock to the Caribbean. There was no hardscrabble period of colonization, and the earliest houses—three of which survive—already show an architectural sophistication of a high order, derived from the urban culture of Massachusetts. During the next century, as the port of Bristol flourished, Thames Street was thickly built up with warehouses and docks as well as distilleries for making the rum that was part of the slave trade in which Bristol was so active. Even the depredations of the Revolutionary War, when the town was damaged by British raids, marked but a temporary setback in its growth.

In the middle of the nineteenth century, Bristol declined as a port. After the abolition of the slave trade there was a final flurry of activity in whaling, but the small scale of the port left it ill adapted to large-scale industrial operations. Nonetheless, it was suited for specialized skilled work, as demonstrated by the rapid success of the Herreshoff boat works, established in 1863 and—between 1893 and 1920—the makers of five successful America's Cup defenders. Even as the last shipbuilder's house shuffled into dowdy Victorianization, the first of the opulent resort cottages, inspired by Newport's example, was pushing its way onto the waterfront. Thus, for all Bristol's economic and social vicissitudes, the relationship of town to harbor has remained remarkably consistent, intimate, and benign to its architectural heritage.

The town's buildings show an amiable mixture of conservatism and eccentricity, the result of a stable clientele and an equally settled pattern of life, but also of the cosmopolitanism of a port, always willing to graft new ideas onto local forms. Large enough to have its own architectural culture, Bristol was provincial enough so that the same craftsmen might use the same conventional formulas for decades. Much of its economy was vested in a thicket of intermarried families, of whom the DeWolfs were the most prominent, and all of whom invariably used the same architect, Russell Warren (1783–1860), the town's gift to American architecture.

Born in Tiverton, Warren arrived in Bristol in 1800, remaining until the mid-1820s, when he departed for Providence. Beginning in 1808, he built the first of four great houses for the DeWolf family, whose patronage transformed him from a cautious craftsman to a designer of restless originality and audacity. Warren was no provincial; he traveled regularly to Charleston, South Carolina, where he owned land, and he presumably knew the major coastal cities. His gracious clapboarded mansions, with their decorous porticoes and balustrades, are testament to the lively and fertile communication along the Atlantic seaboard at a time when it was easier to sail from New England to the Carolinas than to visit a town fifty miles inland.

Long after he removed to Providence, Warren continued to provide the most interesting buildings in Bristol. Had it been built, these would have included a palatial house designed for James DeWolf in about 1836, known only from a surviving drawing. Such architectural longevity, sustained by loyal family patronage, is rare enough in America, but for it to happen twice, as it did in Bristol, is providential. Wallis Eastburn Howe (1868–1960), born in the very decade in which Warren died, would play the same role in Bristol in the twentieth century that Warren did in the nineteenth. Howe understood and interpreted "colonial" better than any other twentieth-century Rhode Island architect, and just as Warren worked sympathetically within the vocabulary of Bristol's colonial houses, so Howe practiced a suave historicism, far from the cringing copyism of some of his contemporaries. His version of the DeWolf patronage came from Samuel Pomeroy Colt, friend as well as client, for whom he remodeled Linden Place and worked for almost half a century.

Between the Warren era and the Howe era, Bristol's Victorian interlude was brief and was concentrated around Hope and Church streets, whose white clapboard streetscape was shocked by a strident Victorian ensemble of civic buildings by Stephen C. Earle. But such buoyant buildings, or the occasional work of New York architects such as James Renwick or Alexander Saeltzer, were sufficiently limited not to disturb the scale and sense of the town. And through it all, the tradition of building in wood remained central

William G. Low House, photo 1962

to the architectural identity of the town. This tradition culminated in the greatest of all Shingle Style houses in America, the William G. Low House, by McKim, Mead and White (1887; demolished 1962), which stood near the southernmost tip of the peninsula. A vast gable in composition, its form echoed the calm topography of the peninsula and was the most affectionate architectural gesture ever paid the town of Bristol.

Bristol offers a compact architectural experience distinguished by consistently superior craftsmanship. It is a place of urban sophistication and richness, although to the modern viewer it presents an image of small-town calm. Its character can best be experienced by the pedestrian. For the most detailed inventory, the visitor should acquire *Historical and Architectural Resources of Bristol, Rhode Island* (Rhode Island Historical Preservation and Heritage Commission, 1990).

BR1 Charles Dana Gibson House
(Longfield)

1848–1850, attributed to Russell Warren. 1200 Hope St.

Charles Dana Gibson, the grandfather of the celebrated artist of the same name, married Abby DeWolf, a member of the town's most prominent family, and acquired a parcel of the Henry DeWolf farm at the edge of town. On it he built this feisty steep-roofed Gothic cottage. Although it has lost its bargeboard trim and original porch, Longfield is otherwise well preserved, including its eclectic interiors. Its gardener's cottage, a simple board-and-batten affair, was subsequently moved to 1222 Hope Street.

Thus the visitor enters Bristol by encountering the names of DeWolf and Warren, the patrons and architect who contributed more to the physical form of the town than any other figures.

BR2 Joseph Reynolds House
By 1698. 956 Hope St.

The Joseph Reynolds house, despite alterations and losses, is among the best-preserved early houses of New England. Its astonishing features are a coved plaster cornice, rare bolection moldings in the interior, and a three-story stair of almost Elizabethan heaviness. Despite the vicissitudes of time, it is possible to see here a grand house of the seventeenth century, a provincial reflection of English taste of a half-century earlier.

The Reynolds house followed the typical Massachusetts two-room center-hall plan with twin chimney stacks in the back wall, although the entrance was oriented to the west and to the street, rather than to the south. In the eighteenth century the plan was expanded to accommodate two rooms at the rear of the first and second stories, giving the house its saltbox roof configuration. The facade's coved plaster cornice, a provincial interpretation of seventeenth-century English prototypes, is one of two remaining Rhode Island examples. (The other may be seen at the Wanton-Lyman-Hazard House in Newport [NE5].) The interior is uncommonly well preserved (original paint schemes survived into the 1940s): the northwest rooms on the first and second stories display rare heavy bolection-molded paneling, among the most important examples of this finish that survive from early colonial America. The scrupulous manner in which the house was preserved is due in part to the tradition that Lafayette stayed in the northeast room of the second floor here, during the blockade of Newport in September 1778.

BR3 Seth Paull House

1879–1881. 900 Hope St.

Seth Paull was a local lumber merchant, and his exuberant house is typical of those built by ambitious building tradesmen. The complicated full-width front porch and the sweeping sunbonnet gable, juxtaposed with the steep candlesnuffer corner tower, are self-conscious embellishments to an otherwise contained cubical mass and traditional center-hall plan. They also exult in the material and products that made Paull's fortune. Prominently sited on the town's main street, this is a house that says not only "Look at me!" but also "Buy my wares!"

BR4 Bosworth House (Silver Creek)

c. 1683. 814 Hope St.

Deacon Nathaniel Bosworth was one of the incorporators of Bristol, and his house was standing when the first religious service was held here in 1683. Although now much altered, it is the oldest house in Bristol; as was also true of the nearby Reynolds House (BR2), its outlying location long kept it out of the path of development. In subscribing to the Grand Articles of Bristol's founding, Bosworth pledged that he would build a two-story house with "two good rooms" to a floor. Although it has accrued several later additions, including a two-story wing to the northeast, the house has preserved its original core and its mighty stone chimney.

Silver Creek (named for a creek in the Bosworth lands in England) remained in continuous family possession until 1957. Afterward the house was separated from its grounds, which were mutilated by roadside commercial development. Now it is imperiled by neglect and the loss of its original architectural context.

BR5 Guiteras Memorial and Junior High School

1925, Clarke and Howe. 35 Washington St. (Hope and Washington sts.)

The bequest of Dr. Ramon Guiteras is one of the oddest acts of architectural patronage in Bristol's history. Like Samuel P. Colt, he specified that a school be built in memory of his mother (see BR12), but he seems to have rejected Colt's cheeky display of extravagance and splendor. Instead he sought to respect Bristol's architectural heritage, his will specifying a close adaptation of Russell Warren's Mark Antony DeWolf house, which burned the year Guiteras died (1919). Architect Wallis E. Howe treated Guiteras's terms freely, and instead of a copy produced a modern Beaux-Arts performance, in which two splayed wings converge upon a central pedimented block. Built of limestone and pale brick, the building is sensitively sited and dominates the head of Bristol Harbor.

BR6 Parker Borden House

1798, 1805. 736 Hope St.

The Victorian era in Bristol was brief. Vaguely Federal houses persisted into the middle of the nineteenth century, and these threads were easily picked up a generation or so later by the architects of the Colonial Revival. Among the buildings especially prized by the revivalists was this house, built near the Thames street wharf of its owner, shipmaster Parker Borden. Borden's house showed some of the finest woodwork of the late eighteenth century, marking the local tradition for refined wood detail that would culminate in the work of Russell Warren.

Borden's standard five-bay Federal house is distinguished by its particularly ornate pedimented entrance, one of the finest in Bristol. It incorporates a semicircular fanlight over a doorway flanked by two engaged Ionic columns, and its lines are articulated by rich profiled moldings. At second-story level is a garland-trimmed Palladian window of idiosyncratic form (a feature admired by native architect Wallis E. Howe, who cribbed the motif for his own Colonial Revival work). The interior woodwork is consistently superlative, particularly the mantels and the delicate staircase, whose paneled wainscot echoes the handrail in its molded upper edge. Of special note is the rope molding in the northwest parlor.

BR7 Jeremiah Wilson House

1750. 675 Hope St.

This is one of the best-documented houses of eighteenth-century Bristol. Carpenter and joiner John Peckham framed and finished the house in a five-month period, selling it to Jeremiah Wilson in 1751. A four-bay house, composed around a central chimney, its chief feature of interest is the exquisite pedimented entrance, whose fanlight, moldings, and ornament are of very fine quality. It can stand comparison with some of the town's best doorways, including that of the Parker Borden House.

BR8 Francis M. Dimond House

1838, Russell Warren. 617 Hope St.

BR9 Josiah Talbot House

1838, Russell Warren. 647 Hope St.

The Dimond house is one of three Greek Revival houses Russell Warren built in the space of a few years, making the west side of Hope Street something of a residential acropolis, in which Doric, Ionic, and Corinthian temple fronts stood almost shoulder to shoulder. The Doric example, the Captain John Fletcher house (601 Hope Street), has been lost, but those built for

BR8 Francis M. Dimond House

Captain Josiah Talbot and Francis Dimond, respectively Corinthian and Ionic, survive. Together with Linden Place, they form an elegant and rare example of an architect's youthful work juxtaposed against his most mature works.

Warren treated Dimond's house as a pedimented temple front, a full prostyle portico of Ionic columns, whose forms he replicated at a smaller scale in the recessed entrance. In a treatment different from those of his earlier clapboard houses, here Warren used wood to suggest masonry, making his facade flush with tongue and groove boards; only the sides retrograded to the clapboarding of his conventional carpentry. The interior, little changed, presents a side-hall plan with a double parlor. Later additions include a one-story wing to the north and a graceful polygonal bay at the southwest corner with slender "Gothick" arches, which, if not by Warren himself, are in his spirit. Dimond (1796–1858) went bankrupt upon the completion of the house, the costliest of Warren's Greek trio, and sold it a year later.

The Captain Josiah Talbot House, the Corinthian chapter of Warren's Greek Revival trilogy, shows that careful and intelligent copyism need not produce dry results. It playfully transposes the theme of the nearby Dimond house—a pedimented temple front facing the street and a side-hall plan within—into a more sumptuous key. In place of the projecting portico of the Dimond House, Warren placed two fluted columns in antis, crowning them with splendid Corinthian capitals whose supple acanthus leaves are a triumph of woodcarving; these make a notable contrast with the bulkier Doric frontispiece that enframes the door. The viewer can see both houses at once and grasp

BR9 Josiah Talbot House

what the Greek Revival meant to Warren: not a straitjacket but a storehouse of rich and elegant forms, which were subject to rules but which nonetheless offered considerable scope for imagination in composition and in ornamental elaboration.

BR10 Commercial Bank Building
1814. 565-576 Hope St.

Although its storefront has been considerably altered, this banking house remains a first-rate example of Federal commercial architecture. The brickwork is of excellent quality, as are the small and delicate quoins and the restrained brownstone trim. Of special note is the use of taut relieving arches on the side elevation. This early brick building is perhaps the first hint that the architectural taste of Bristol would henceforth be set by the trends of increasingly urban Providence. From 1845 to 1857 the building also served as Bristol's customhouse.

BR11 Rogers Free Library
1877, Stephen C. Earle, architect. 1957, Wallis E. Howe. 525 Hope St.

Only the first story remains of a much larger Romanesque Revival building, which lost its upper stories and steep hipped roof in a 1956 fire. Named for Robert Rogers, president of the Eagle Bank, the building originally mixed philanthropy and finance, combining a public library above with ground-story banking rooms and the local YMCA. The architect was Stephen C. Earle, of Worcester, Massachusetts, one of the first graduates of MIT's department of architecture. Like much of his work, it is Victorian Gothic in spirit, strongly influenced by H. H. Richardson's hefty Romanesque masonry.

Wallis E. Howe reconfigured the damaged library, which, in its truncated form, provides an interesting juxtaposition to the adjacent U.S. Post Office (1961–1962, Philemon E. Sturges; 515 Hope Street). Sturges's building is itself a hybrid, incorporating doors and windows of the Wardwell House (1815, Russell Warren), which once stood on the site.

BR12 Colt–Andrews School (Colt Memorial High School)
1906–1913, Cooper and Bailey. 570 Hope St. (southeast corner of Hope and Bradford sts.)

Samuel Pomeroy Colt was living in adjacent Linden Place when he offered to build this high school in memory of his mother. With its florid classicism and sumptuous Corinthian portico, it is an updating of Linden Place itself, translated into Beaux-Arts terms. The school, designed by Boston architects, is composed as a monumental two-story block, fronted by a tetrastyle portico and set on a balustraded terrace with rather voluptuous statuary. The materials are quite expensive, including white marble, green roof tile, and a lavish use of bronze in the windows and other trim. The pediment displays the Colt family crest.

In recent years the building has been used as an elementary school, having been superseded by the building of the Andrews Memorial School in 1938 directly opposite at 574 Hope Street (George Maxwell Cady, architect). Its introspective brick Georgian exterior forms the strongest possible contrast with the frantic festiveness of the Colt School.

BR13 Linden Place (DeWolf-Colt House)
1810, Russell Warren. c. 1834, Russell Warren. c. 1856, addition. c. 1900, Wallis E. Howe. 500 Hope St.

No building is so central to both the economic and architectural history of Bristol as Linden Place. When he commissioned the house in 1810, George DeWolf launched the career of architect Russell Warren; when he went bankrupt and fled the town in 1825, DeWolf precipitated an economic crisis from which Bristol never completely recovered. Altered repeatedly in later years, it has become the most lavish monument of the local Bristol vernacular.

In its core, Linden Place is a sturdy frame

house not unlike the Bradford House next door, although it already shows postcolonial Bristol's inflated demand for display as well as the impulse to exceed the stereometric clarity of the frame boxes of late-eighteenth-century design. The woodwork, Warren's clever variations on the themes of Asher Benjamin's *Country Builder's Assistant*, is of exceptional delicacy. The massive Corinthian portico of the entrance has an overstated dignity, but is nonetheless of a piece with Bristol's tradition of elegant woodwork. The entrance is recessed behind two superimposed elliptical arches. Four Corinthian columns project shallowly forward, hoisting aloft a graceful balcony whose balustraded cornice is carried on piers which bear florid brackets.

Following DeWolf's bankruptcy, the house was bought by his uncle James DeWolf, whose son William acquired it in 1834. William commissioned Warren to remodel the house, adding a ballroom to the north and a Gothic sunroom to the south. After 1856 the house was leased as a hotel. Among elements of the hotel renovations may be counted the remarkable wrought iron fence with its decorative scrolls and rosettes. According to local tradition, the fence was moved from the Jerathmael Bowers house, in Somerset, Massachusetts, and dates from about 1815–1825.

In later years the house passed to Samuel Pomeroy Colt, a grandson of George DeWolf (as well as the nephew of the inventor of the Colt revolver). Colt added yet another ballroom to the swelling house, as well as a carriage house, using as his architect Wallis Howe, who was a personal friend. The carriage house with its oversized elliptical pediment is a delightful fantasia on the Colonial Revival, showing the same congenial freedom in the handling of classical architecture as the original house.

BR14 Bradford–Norris House

1792, unknown. 1845, Russell Warren. 474 Hope St.

The transitions from colonial to Federal to Greek Revival architecture were never as abrupt or jolting in Bristol as elsewhere, and in each style an invincible sense of local character remained, invariably classical in spirit, cubic in volume and exquisitely delicate in ornament. No house captures this continuity so naturally as the Bradford-Norris house. Built in 1792 for William Bradford, deputy governor of Rhode Island during the Revolution, the house was originally a boxy Federal affair. In the 1840s it passed to Francis Dimond and then to his daughter Isabel and her husband, Samuel Norris. Norris moved the house back from the line of the street and hired Russell Warren to remodel it.

By this point Warren was an accomplished Greek Revival architect, but in Bristol he preferred to practice in the indigenous wood vernacular in which he had been apprenticed. He transformed the block of the frame house with a series of elegant additions. Most extraordinary is the third story, which is recessed to form a continuous parapet. This is fronted with an elegant balustrade, whose geometric forms are reiterated in matching balustrades along the roof and also above the Ionic entrance porch. The detail is Chippendale, but of a *chinoiserie* sensibility and with the delicacy of fine cabinetry, which suggests the cosmopolitan outlook of an international port.

BR15 U.S. Customhouse and Post Office

1857, Ammi B. Young, Supervising Architect of the Treasury. 440–448 Hope St.

Ammi B. Young's tenure as the principal architect of the federal government was characterized by rigid standardization and a dry and fastidious version of the Roman Renaissance. His Bristol Customhouse and Post Office is the brick interpretation, with a strong arcaded portico, vigorously molded piers, and a florid balcony thrust out above cast iron brackets. Had Bristol still been among the first tier of ports, the building certainly would have been granite, but the decline of whaling and shipping had already begun to turn the town into something of a backwater. Nonetheless, Young's design complemented the decorous classicism of historic Bristol. Although the building has been converted to private offices, losing its original doors and the iron balustrade that once marked the roof, its historic character is largely intact.

BR16 YMCA Building

1899, Wallis E. Howe. 1967, Philemon E. Sturges. 438 Hope St.

An exhausting performance in Tudor Revival, this is the first major work by Wallis E. Howe in Bristol, when he was still shedding the last vestiges of his training in the Victorian office of Martin and Hall (Providence). An example of intelligent planning, it housed rental stores on the first story and, on the upper level, approached by a broad archway, the YMCA's auditorium and gymnasium. The 1967 entrance and new lobby are appropriately reticent.

BR17 Burnside Memorial Hall

1883, Stephen C. Earle. 400 Hope St.

American architects could always replicate the forms of H. H. Richardson's Romanesque Revival, but seldom their spirit. With its simplified silhouette, strong hipped roof and rugged walls, Burnside Memorial Hall strains for Richardson's sublime simplicity, but it also shows the raucousness and nervous energy of an architect whose training was in the Gothic Revival. Architect Stephen C. Earle, practicing in Worcester, had a lucrative practice in small libraries, examples of which survive in Norton, Massachusetts; Groton, Connecticut; and at Grinnell College in Iowa. Burnside Hall is the third of his buildings in Bristol.

Earle's program was to combine a town hall with a memorial to Major General Ambrose E. Burnside, Civil War hero, thrice governor of Rhode Island, and later United States senator, who died in 1881. The centerpiece of Burnside Memorial Hall was to be a statue of the general on its porch, long since removed from the building. Likewise, town offices were removed from the building in 1969, and it now serves purely as a memorial. Together with St. Michael's parish house next door, it forms a lively Victorian interlude in the clapboarded classicism of Hope Street.

BR18 St. Michael's Episcopal Church

1859–1861, Valk and Saeltzer. 375 Hope St.

St. Michael's Episcopal Church is the fourth church to be built on the site, and in traditional English fashion, eighteenth-century tombstones are incorporated into the walls of its entrance porch. A picturesque composition, it is a steeply gabled basilica dominated by a corner tower. Five bays in length, it terminates in a polygonal (5/8) apse. Although nave and aisles are divided by slender iron columns, the sense is of a unified space, gathered under the handsome and simple open roof truss.

St. Michael's, Hope Street's first serious departure from stately classicism, was conceived as a granite building, but in the course of con-

BR18 St. Michael's Episcopal Church

struction the material was changed to softer and warmer brownstone. Shortly before completion, the building collapsed and was rebuilt, a history that may account for a certain crudity in some of the stonework on the flanks. Its architects were Lawrence Valk and Alexander Saeltzer, whose New York firm specialized in Gothic and Romanesque Revival churches. In 1891 a square belfry was added over the tower.

In 1869 a parish house was built across the street, a lively composition by Stephen C. Earle that echoes the forms of St. Michael's in more strident Victorian tones, including the spiky silhouette, muscular corner bellcote, and even the bright red mortar joints, which offset the rusticated brownstone walls.

BR19 William H. Bell Block

1879. 361-365 Hope St.

A splendid textbook example of Victorian commercial architecture, the William H. Bell Block carries its brick second story on a tough granite lintel, which rests on spindly iron columns. Especially handsome are the tough granite window hoods with their Néo-Grec incision. Since 1879 the mansarded attic has been used continuously as a Masonic hall. The building is virtually unaltered and still retains its grand and rather florid staircase, reached through the deep central passage. This is perhaps Bristol's best example of Victorian eclecticism, possibly designed by Stephen C. Earle.

Bell himself lived in the house next door, number 353, which originally stood at the corner. Bell had the small Federal house, built in 1797 for merchant William Fales, moved and updated when his store was built.

BR20 John Howe House (The House with the Eagles)

1808, c. 1822, c. 1847. 341 Hope St.

Like so many of Bristol's finest houses, "The House with the Eagles" is a taut and simple Federal box that has been animated with superb architectural woodwork. Originally built by the lawyer John Howe, a descendent of the influential DeWolf family, it passed to ship captain Benjamin Churchill in 1822. According to local legend, Churchill gave the house its name by having the sailors of his ship, the *Yankee*, carve four American eagles, which he placed at the corners of the Chippendale balustrade that crowned his roof.

Churchill's tenure was brief, and in 1825 the house passed to Byron Diman, a powerful merchant with interests in whaling, banking, and the local cotton mills who served as governor of Rhode Island in 1846–1847. While Diman repeatedly altered and expanded his house, he maintained its essential Federal character, the hallmark of Bristol's conservative elite.

BR21 Babbitt–Smith House

1795, c. 1810. 328 Hope St.

Although early accounts record that merchants Jacob Babbitt and Barnard Smith built this house as a double in 1795, its character is that of a five-bay Federal house. In 1810 it was acquired by Captain Daniel Morice, a Frenchman from Haiti, who perhaps added its most striking feature, the boisterous roof with its odd combination of jerkinheads, hips, and gables and an oversized lunette centered in the attic. The woodwork is of outstanding quality, particularly in the very fine dentil molding and the Ionic frontispiece of the entrance. Much, alas, has been lost under aluminum siding.

BR22 John W. Munro House

c. 1810, c. 1866. 275 Hope St.

When built, the Babbitt-Smith House was at the edge of the developed town. Not until the middle of the century did continuous development push beyond it, and to the south there is now a small district of Victorian houses. Among the finest are number 281, a Gothic Revival cottage (c. 1855) that has retained much of its quirky trim; number 244, a Queen Anne cottage (c. 1885) with lovely decorative trim; and the John W. Munro House.

Among the customers of Munro's hardware and wallpaper business was Samuel Colt, who in 1866 gave him the old wood barn that stood behind Linden Place. Munro promptly moved it, added a story and a rear ell, and remodeled it into the present house, which followed in the town's tradition of intricate, eccentric woodwork. Oddest of all is the entrance hood, which is carried on elaborate contrivances of brackets, colonnettes, corbels, and chamfered posts.

BR23 Miramar (The Tides; Joshua Wilbour House)

1893, Edward I. Nickerson. 1986, Newport Collaborative. 217 Hope St.

Edward I. Nickerson (1845–1908) specialized in outfitting Providence's elite with languid Queen Anne and Colonial Revival cottages, such as this summer house he built for the financier and politician Joshua Wilbour. Wilbour did not live here long, and the house soon passed to Isabella DeWolf, who occupied it until 1936. A confident Colonial Revival essay under a high gambrel roof, Miramar is notable for its location directly on Narragansett Bay, to which its generous south-facing veranda opens. An odd wooden balustrade once perched on its roofline, a delicate tiara that was torn off in the hurricane of 1938. Now the house has been converted to condominiums and has been augmented with oversized, abstractly historicizing additions.

Wilbour's house brought another socialite's retreat to this vicinity: Wyndstowe (number 221), an engaging Queen Anne house built in 1899 for the Barnes sisters, Hattie and Isoline. Its architect was Wallis E. Howe, who apparently so pleased his clients that they willed the house back to him in 1935.

BR24 James Lawless House

c. 1865. 208 Hope St.

In the 200 block of Hope Street are a striking number of houses of builders and artisans who lived close to their livelihoods on the wharves

BR23 Miramar (The Tides, Joshua Wilbour House)

along the bay. The buildings have something of the splendidly taut and restrained quality of naval architecture, as in the high-gabled house James Lawless built for himself at the end of the Civil War. Lawless was a naval architect and mariner, and his house is a trim affair of crisp intersecting gables, whose plastic clarity has not been compromised by later additions.

Among Lawless's neighbors are the Timothy French House of 1803 (number 224), a brick house by a carpenter who was one of Russell Warren's principal competitors, and the Joseph Coit House of 1818 (number 259), the sturdy house of a shipbuilder whose carpenter, Isaac Borden, paraphrased his own house at 159 High Street. Finally, the distinctive Greek Revival door at number 259 is not to be overlooked.

BR25 Clark–Herreshoff House

c. 1800, c. 1870, c. 1926. 142 Hope St.

A building of almost Cistercian austerity, this five-bay Federal house belonged to Lemuel Clark Richmond, owner of a whaling fleet, and originally stood closer to the bay. When it came into the possession of the Herreshoff family in the 1860s, it was moved and expanded with several rear additions. Now a private museum, it is at the center of what can be viewed as an informal Herreshoff district, which includes the A. Sidney Herreshoff House at number 125 (built around 1940) with its attached model room, and the simple mansarded box at number 140, built as the Herreshoff Company's guest house. A stone marker, erected in 1963, marks the site of the Herreshoff works and records the firm's innovations in the technology of yachting and achievements in building America's Cup winners.

BR26 Seven Oaks

1873, James Renwick. 136 Hope St.

An unspoken but enduring respect for local tradition as well as a static economy ensured that Bristol had little to do with the vagaries of High Victorian architecture. One great exception is Seven Oaks, a florid Gothic Revival showpiece which lords it over a splendid site at the southern end of Hope Street, near the tip of the peninsula. Seven Oaks was built for Augustus O. Bourn, the founder of the National Rubber Company (and from 1883 to 1885 governor of Rhode Island), who lived here until his death in 1925.

Bourn's architect was James Renwick of New York, celebrated as a Gothic revivalist but fluent in all mid-Victorian styles. Seven Oaks's jagged roofline, additive composition of towers and dormers, and meandering plan are Gothic in inspiration, but they show the picturesque forms of the earlier Gothic Revival. In fact, Seven Oaks is much more in the spirit of A. J. Downing's rural cottages than of contemporary High Victorian work. Of special note is the splendid Victorian landscaping of the grounds, which complements the animated and energetic forms of the house.

Here the pedestrian can continue to the left and proceed to High Street, while the motorist might proceed to the more widely spaced attractions near the tip of the peninsula.

BR27 Codman House

1870, 1875, George Champlin Mason, Sr. 42 High St.

In 1870 Boston's unmarried Codman sisters, Catherine Elizabeth and Maria Potter, commissioned this house from the Newport architect George Champlin Mason (1820–1894). Like many who built in Bristol after the Civil War, they seemed to view the location as a kind of suburb of fashionable Newport. In 1875 they were joined by their brother Henry, who was given a tower addition to dwell in, preserving decorum.

The Codman house is in the spirit of Mason's Newport work: in massing, a severe cubic block enlivened by a mansard roof and sprightly piazza, in plan a strictly symmetrical center-hall composition, with asymmetrical accents provided by bays and turrets. Such blocky masses were intended to be offset by a picturesque planting scheme, a feature that hardly ever survives, but here the landscaping is exceptionally well preserved. Likewise, the house is in remarkable condition, retaining its mansarded carriage house, much of its interior decoration, and even its tufts of iron cresting. There is no finer Second Empire house in Bristol. In 1984 it was converted into condominiums and renamed Codman Place.

Nearby are several other notable mid-Victorian houses, including the Lemuel C. Richmond House (1856), at number 41, an octagon house with several later additions. At number 64 is the house of John Brown Herreshoff (1870), another Second Empire house with a fine portico, built to loom above his boat works.

BR28 Burton Primary School

1848. 140 High St.

A simple gabled roof carried on sturdy rubble walls and a single Gothic-arched window provide the essentials of this laconic two-room building, which served as an elementary school into the 1950s. Subsequently it served as the Bristol Historical Society and then as a union hall, during which time its twin entrances were infilled.

High Street preserves its historic fabric at least as far as the Town Common and deserves to be walked in its entirety. It includes several handsome Federal houses, including two that have been attributed to Russell Warren: the George Devol House at number 132 (1811) and the Stephen S. Fales House at number 139 (1809), both somewhat altered. Also noteworthy is the William Bradford House at number 154 (1808) with its fine pedimented entrance, a transitional example in which the elliptical fanlight of modern Federal taste is incorporated into the durable colonial type of doorway.

BR29 Byron Diman Cottage

c. 1835. 82 Church St.

An elegant and diminutive Greek essay, this is the cottage version of the large temples erected by Russell Warren on Hope Street (BR8 and BR9). Its frontispiece is no timber sketch of a temple front, but a fully realized Doric portico, an archaeological rebuke to the freewheeling vagaries of Linden Place. Its owner, Byron Diman, was a prominent banker and real estate speculator.

BR30 Bristol Historical and Preservation Society Museum (Bristol County Jail)

1828, 1859. 48 Court St.

Built of tough, random-coursed ashlar masonry, this rather stern Greek Revival jail was discreetly tucked onto a back street when it was built in 1828, replacing its 1792 predecessor. In 1859 a granite cell block was added. The Bristol Historical and Preservation Society has been here since 1959, and the cell block is open to the public as a museum.

BR31 Thomas Nelson House

1810, Russell Warren. 82 State St.

BR32 William Van Doorn House

1807–1811, Russell Warren. 86 State St.

BR33 Russell Warren House

1810, Russell Warren. 92 State St.

American trade prospered during the years before the War of 1812, and Bristol flourished, its original town core filling in rapidly, especially around State Street, the broad thoroughfare that led from the harbor to the Town Common, where these three houses were built. All were designed and built by Russell Warren, and all are variants of the same type: a two-story, five-bay Federal house with a center hall. These houses are among his first works and depict him at the moment of his transition from builder to architect. They are distinguished by splendid woodwork, including the spiral staircase and late colonial fireplaces of Warren's own house, but with the stylistic conservatism of an artisan just coming out of his apprenticeship. The richest of the three is the Van Doorn house, with its extraordinary slanted quoins and intricate two-story Doric frontispiece. Here, in the jaunty arrangement of panels and pilasters, Warren gives an early inkling of his powers of imagination. He moved here in 1813, a sign of his growing wealth, remaining until 1823, when he left Bristol.

BR34 Town Common

1680. Bounded by High, Wood, Church, and State sts.

Bristol is among the handful of seventeenth-century New England towns with a formal, geometrically ordered plan. Its focus was the common, which consisted of eight acres placed centrally between High and Wood streets and originally bounded to the north and south by Charles and Queen streets, now dubbed Church and State streets—a poetic (if inadvertent) transformation that replaced the final Stuart monarchs with the very institutions that had agitated their reigns.

According to custom, the common was used as a site for the first meeting house (1684) as well as for burials, livestock grazing, and military drills. It has retained its original function in its array of public buildings. The oldest of these is the Bristol County Courthouse (1819), which, from 1819 to 1854, also served as one of Rhode Island's five state houses. A striking example of Federal civic architecture, in conception it is a five-bay brick house, made into a public building by the addition of an octagonal cupola and the placement above the entrance of an oversized arch with Gothic tracery and granite surrounds. The building has been attributed to John Holden Greene (and less convincingly to Russell Warren). It was altered in 1836, when the brick walls were stuccoed, and again remodeled under the auspices of the WPA in 1934–1935 by Wallis E. Howe; it now serves as a monument.

Adjacent to the courthouse is the First Baptist Church, built in 1814. It is a simple gabled meeting house, fronted by an Ionic portico and capped by a square bell tower and octagonal cupola. An early example of a two-story meeting house plan, it consists of a first-story lecture room and committee room with a meeting room above. Its Gothic-arched windows were infilled with sash in the Queen Anne manner in 1882. Also on the common are two handsome brick school buildings: the Byfield School (1873), a taut Second Empire design by C. J. Emerson, and the Walley School (1896) an eclectic essay by William Howard Walker, with handsome recessed entrances under Romanesque arches.

Having walked about the common, the visitor ought to venture one block farther north on High Street to see an uncommonly fine collection of vernacular Greek Revival houses, including the Benjamin Tilley, Jr., house at number 328 (1849), whose taut Doric portico has been attributed to Russell Warren's brother Samuel.

BR35 James DeWolf Barn

1824, unknown. c. 1850, Russell Warren. 281 High St.

BR36 DeWolf-Guiteras House

c. 1830, unknown. c. 1887. 291 High St.

During the same period in which he acquired Linden Place, James DeWolf built this house and barn on High Street. Strangely, his granite barn was far more opulent than his Greek Revival house, a stuccoed stone affair of rather conventional form. After his son George acquired the buildings in 1834, Russell Warren was commissioned to convert the massive barn into a dwelling house, adding a rear ell and a Gothic parapet, since lost. It is still possible to see where Warren's crisply dressed ashlar additions were inserted into the oversized barn door. The remodeled barn passed to Dr. Ramon Guiteras after the Civil War, along with DeWolf's house. Guiteras completely transformed the house, draping its simple Greek Revival form with an exuberant flurry of Queen Anne trim, cast iron cresting, a bracketed porch, and an octagonal tower. Today both houses serve the First Congregational Church, the house as parsonage and the converted barn as parish house.

BR37 Congregational Church

1856, Seth H. Ingalls. 300 High St.

Despite its Puritan pedigree, Bristol was no stern Calvinist enclave; when Congregational-

BR37 Congregational Church

ists elsewhere favored more sober German Romanesque models, in Bristol they chose a Gothic Revival design with "a pleasing variety in figure and color." Anchored by its corner tower, the church forms an irregular composition, its walls of random ashlar offset by crisp granite trim. These details and the simple lancet windows recall the first phase of the American Gothic Revival, which preferred the picturesque to the scholarly. The architect was Seth H. Ingalls of New Bedford, Massachusetts. In 1869 the DeWolf Memorial Chapel was added, commemorating the church's most generous supporters.

BR38 Bristol Steam Mill (Namquit Mill, White Mill)

1843. 345 Thames St.

This building, Bristol's first cotton mill, testifies to the moment when industry began to supersede shipping in importance to the local economy. Built by the Bristol Steam Mill Company in 1836, it was rebuilt in its present form after a fire in 1843. A five-story building, twenty bays in width, it is constructed of rubble masonry that has been crudely stuccoed and whitewashed. In style it is a vernacular version of the Greek Revival, although in recent years it has lost its original sash as well as the belfry that once crowned its southeastern tower. A series of companies has operated the mill, which has been repeatedly extended and updated. Today it serves as the Coats American Bristol Plant.

BR39 DeWolf's Wharf

1797. 267 Thames St.

For most of its two centuries of existence, this wharf has been owned by only two families: the DeWolfs, who maintained it until 1861, and Seth Paull, whose family lumber business was centered here until 1952 (see BR3). The four historic components of the wharf include Byron Diman's counting house (c. 1835), a Greek Revival building facing Thames Street, and William Taylor's store (c. 1838), likewise Greek Revival, to the north of the site. Most impressive is the DeWolf warehouse (1818), near the waterfront, a lengthy two-story block built in heavy masonry. Local tradition insists that the building was constructed of African stone, brought to America as ballast, but this seems insupportable. The final element of the site is the Old Bank of Bristol (1797), once a solid three-story brick building in the Federal style, although it was decapitated following the 1938 hurricane and is now a one-story torso.

BR40 Usher Store (Potter's Wharf)

Before 1794. 227 Thames St.

Thames Street, once the focus of Bristol's maritime economy, was densely lined with stores and warehouses long before the Revolutionary War, but this building is one of the few to survive largely intact. A stout gambrel-roofed structure, it placed living quarters above the commercial floor. It was standing in 1794 when it passed from John Usher to his sons, Hezekiah and George, merchants and slave traders. Since then it has served variously as a blacksmith shop, grocery store, and antique shop.

BR41 Pokanoket Mills

1839, 1856. 125 Thames St.

Following the construction of the nearby Bristol Steam Mill, this second cotton mill was built in 1839. The Pokanoket Mill is a brick building, four stories in height, its gable presenting a tier of freight doors to the street. Following a fire in 1856 it was rebuilt with an addition to the north. It was then expanded repeatedly after it was acquired by Charles B. Rockwell's Cranston Mills in 1891. The Cranston Mills additions included a nine-bay extension to the north (c. 1911), a twelve-bay extension facing Church Street (1921), and a modern, flat-roofed brick wing along Constitution Street (1940). The result is a continuous line of mills, identical in height but depicting a full century of changing mill-building technology.

BR42 Naval Reserve Armory

1891. 2 Church St.

Bristol's mightiest building, this armory was built for the local Naval Reserve Torpedo Company, a division of the Rhode Island Militia. Its construction is itself a history of the changing fortunes of the Bristol harbor. It dominates Warehouse Point, at the southern end of the port, and was originally developed in 1810 as Long Wharf, a bold speculative venture which involved five brick warehouses. Within eighty years, the complex had been demolished, and this stupendous Richardsonian essay took its place. Built of random-coursed granite ashlar, heavily rusticated, and rising to a defiant machicolated cornice, it is a familiar type among America's armories of the late nineteenth century. But such buildings are almost always on landlocked urban sites, and to see an example—whose details derive from Italian palazzos—in a Venetian waterfront setting is oddly pleasing. The building was renovated by the WPA in the mid-1930s, although that work was interrupted by the hurricane of 1938, after which the building was decommissioned and acquired by the town.

BR43 William Harris House
1807. 12 Church St.

A chaisemaker typical of the artisans who crowded between Thames and Hope streets, William Harris commissioned this house in 1807 from housewright Benjamin Norris. Its splendidly idiosyncratic door surrounds, which are boldly abstracted, reveal that Russell Warren was not the only local builder to see in wood construction infinite space for invention. Despite its progressive Federal touches, the house is the conservative colonial type: four rooms clustered about a central chimney. The later additions are charming counterparts to Norris's clever woodwork, transposed to a Victorian key.

Directly across the street at number 15 is the Charles DeWolf house, built in 1806 by carpenter Abraham Warren. Its Ionic corner pilasters and curved iron stair railing form a sophisticated architectural counterpoint to the geometric abstraction of William Harris's house.

BR44 Mount Hope Farm (Isaac Royall, Jr., House)
c. 1745, after 1837, c. 1914. Metacom Ave.

Built on the council lands of the Wampanoag Indians, where King Philip's War of 1675 may be said to have begun and ended, Mount Hope Farm is surely Bristol's most historic site. The complex and variegated structure now standing testifies to the three phases of Bristol's history. The colonial phase is represented by Isaac Royall's original house of 1745, a clapboarded building with brick end walls. The Greek Revival phase was initiated in 1837 when it came into the possession of Samuel W. Church, who substantially enlarged the house in the Greek Revival mode. Finally, in 1917, it was acquired by R. F. Haffenreffer II, who initiated the Colonial Revival phase, as he began restoring the complex and developing its extensive gardens. It is now owned by the town of Bristol.

BR45 Perry-Congdon House (Harbor Lawn)
c. 1865, James Renwick. Ferry Rd.

The building history of this Gothic cottage is hardly less picturesque than its furiously gabled roofline. It was originally built as a summer residence for Alexander Perry and his brother-in-law Charles Bogart, but when the two families fell out, Renwick himself stepped in and acquired the property. After passing through various hands, including those of the brother of Bessie Van Wickle, the owner of Blithewold, it returned at last to the Perry family in 1961. Although it was extended significantly in the 1920s, these additions were in the rambling and relaxed spirit of Renwick's original essay.

BR46 Blithewold
1907–1908. Ferry Rd.

By the turn of the twentieth century, Bristol's grandest houses were no longer built for local merchants and industrialists, but as summer residences for out-of-town vacationers. Of these the grandest was Blithewold, for Augustus Van Wickle (1856–1898), a Pennsylvania coal-mining magnate who had attended Brown University. Having purchased a Herreshoff yacht, he decided to build a summer house in Bristol, which was begun in 1895 as a large shingled composition.

Following Van Wickle's death in a skeet-shooting accident, the house passed to his widow, Bessie. The original shingled house burned in 1906 and was replaced the following year by the present structure, designed by the Boston architectural firm of Kilham and Hopkins. Blithewold is a sprawling Old English house whose central block is of stone and stucco. An elongated narrow composition, sited parallel to the Narragansett Bay coastline, it shows the tasteful picturesque of the prevailing academic taste. The eclectic sequence of rooms, varying from colonial to Tudor, is exceptionally well preserved. Of particular note is the richly paneled dining room. In character, Blithewold had little to do with historic Bristol and suggested that its model was the lavish Newport mansion type.

Bessie's daughter Marjorie McKee (1883–1976), who married George Lyon (1898–1977), acquired the house. Like her mother, Bessie lavished great attention on the landscaping of the 33-acre grounds, supervising the formal plantings near the house as well as the picturesque groupings of specimen trees and shrubs; these are at least as remarkable as the house itself. Following Bessie's death, the house and grounds were opened to the public.

BR47 Roger Williams University

1967, Kent, Cruise and Associates. Ferry Rd.

Founded in 1948 and established as Roger Williams Junior College in 1956, the institution acquired this imposing 63-acre site in 1965, just to the north of Mount Hope Bridge. Exploiting its waterfront site, architects Kent, Cruise and Associates designed a campus of geometric severity, working with simple cubic volumes of stuccoed rubble. In a freer style is the School of Architecture, Art, and Historic Preservation (1985) by Kite, Palmer Associates (Providence).

BR48 Mount Hope Bridge

1927–1929, Robinson and Steinman. Ferry Rd.

A wire-cable suspension bridge supported by 285-foot towers, the Mount Hope Bridge is a

BR47 Rogers Williams University, architecture building

BR48 Mount Hope Bridge

structure of pared-down, understated beauty. Upon its completion in 1929, it received the American Institute of Steel Construction's artistic bridge award, an honor that did its builders little good: built as a speculative venture, it opened on October 24, 1929, only five days before the stock market collapse that heralded the Great Depression. Within two years the private corporation that operated it as a toll bridge was bankrupt. Nearby is the Bristol Ferry Lighthouse, built by the United States government just after 1854, a modest brick building that was converted to a residence after the completion of the bridge.

Tiverton (TI)

Only by way of the Mount Hope Bridge and Route 138 east, across the northern tip of the town of Portsmouth on Aquidneck Island, thence across an inlet on the bridge to Tiverton, can one stay in Rhode Island to conclude the journey of those towns which line the eastern shores of Narragansett Bay. Except for their attachment to Rhode Island by the imaginary string of the state boundary, both Tiverton and Little Compton would be adrift in Massachusetts. Both towns were, in fact, part of Massachusetts, until redrawing of the contested state boundary in 1747 shifted them to Rhode Island.

For most of its length Tiverton extends along what is customarily called the East Bay, one of the three channels that together embrace the commonly accepted extent of Narragansett Bay. The map speaks differently, however. Officially, this channel is the Sakonnet River (known locally for much of the nineteenth century as the "Seaconnet"). Like the Providence River, the Sakonnet is really no more than a short, swollen tidal mouth that conveys water, in the first instance, from rivers upstream into the bay proper, and in the second, directly into the Rhode Island Sound of the Atlantic Ocean. Tiverton occupies most of the eastern shore of the Sakonnet, which narrows to the north at the bridged inlet, then swells again into Mount Hope Bay. This tapers, in turn, into the mouth of the Taunton River at the city of Fall River.

Lying between this heavily industrialized city to the north and the consciously pastoral town of Little Compton to the south, Tiverton shifts from one context to the other within its length. The residue of its agricultural heritage, however, very much predominates, with hills sloping to idyllic water views along the whole of its shoreline. Along this shore exist remnants of Tiverton's former maritime industries. Although these were always limited by the lack of a good harbor, the town has a rich marine tradition of shell and fin fishing, whaling, and international shipping. The histories of its houses abound in seafaring owners and occupants, as though, major harbor or no, the constant presence of the sea sufficed to draw the adventuresome from farms to ships. Today, the water is mostly given over to recreational boating and some fishing, its former farms increasingly attracting residential development, together with a couple of incipient industrial parks. Inland, as this is written, except at its Fall River end, Tiverton retains many beautiful stretches of field and water, very much at the mercy of the course of future development.

Most settlement in the town is on or close to major highways along its western and eastern edges, on ridges rising from swampy, low-lying land between Main Road (Route 77) on the water side of the town and Crandall and Stafford roads (Route 81) along its land edge against the Massachusetts border, which is of less architectural interest.

TI1 Bourne Mills

1881, 1900 and later. 1900, weaving shed, Frank. P. Sheldon. 1 Shove St.

Although just outside Fall River in North Tiverton, this cotton mill relates to the Massachusetts city. In fact, Fall River and New Bedford entrepreneurs financed the company, and the Bourne Mill also looks across its millpond at several of Fall River's granite mills. Its coursed ashlar granite construction at this late date typifies the persistence of masonry construction in Fall River (even though the predominance of brick in other Massachusetts manufacturing centers paradoxically showed the way for its use in New England, including Rhode Island, where it was growing in popularity by the early 1880s).

In 1906, at its peak of operation, the mill employed about a thousand workers. The looming five-story building, 320 feet in length, reputedly derived from standard plans produced by its first treasurer, George A. Chace. A projecting tower of great severity dominates the principal elevation. Its mansard cap is now shrouded in asphalt shingling, which conceals clock faces in all four directions. The stack of narrow, deeply inset loading doors accentuates the verticality. Doubled granite slabs flush with the wall provide for sills and lintels, while more granite

TI1 Bourne Mills

coursing makes an austere but sophisticated cornice. The linear elegance of the door paneling and that of the twelve-over-twelve panes in the still intact window sash counterpoint the imposing mass.

In the history of factory building technology, Frank Sheldon's later extension of the mill by a 325-foot-long, one-story weaving shed with a sawtooth roof, which overlaps the south end of the mill and extends a good distance beyond it, was perhaps even more interesting than the mill itself. It was among the earliest examples of the sawtooth roof in the state (see the Joseph Benn and Company mill in North Providence [NP11]). The weaving shed was long abandoned, and its distinctive roof disappeared in the mid 1990s. The combination of this roof form with stone walling, as here, was rare because brick construction was the factory norm during the brief period of the sawtooth's popularity. In this pioneer example, Sheldon or his client seems to have preferred to harmonize the new roof technology with the old masonry walls of the original mill. Despite long disuse except for storage, the mill building is well preserved. The striking isolation of the mill against the low-lying pond environment gives it an aspect at once compelling and forlorn.

TI2 St. Christopher's Roman Catholic Church Rectory (Captain Isaac Church House)

c. 1870. 1660 Main Rd.

Isaac Church's father, Captain Joseph Church, established a fishing business in Tiverton which eventually specialized in catching menhaden (colloquially "pogies"). Regarded as a junk fish unsuitable for human consumption, they were converted to soap oils and fertilizer. The seven Church brothers, all captains, together with a single sister, shared in the operation of the largest fishing operation in the village of Tiverton throughout the latter nineteenth into the early twentieth century. They expanded by adding cessing plants in Maine and on Long Island before the rise of the twentieth-century synthetic chemistry industry eventually wiped out the company.

Of all the Church mansions scattered about this village, Captain Isaac's is the most interesting and best preserved, although it is essentially more substantial than elegant. It must have been painfully plain before the addition of its wraparound porch, probably in the 1880s. This swells around a central two-story bay window capped by a cross gable, where its X-shaped railing supports are echoed in the pseudostructural X and in the quaint triangular windows behind. A one-story latticed skirt brings the porch to the ground, veiling the front of the basement floor, which would otherwise be exposed by the steeply sloping site.

To the north on Main Road are two other Church houses: that of Captain Daniel Church (1792, c. 1900), at number 1392, a Federal period house altered to Colonial Revival, most conspicuous for its two-story semicircular entrance porch, and, adjacent at number 1420, Captain George L. Church's Italianate house (c. 1865[?] and later). (A third, the grandest of the Church mansions, the Second Empire Captain Nathaniel Church House, now St. James Convent [1872], is off Main Road farther south, at 49 Nannaquaket Road.) Other well-preserved houses, mostly from early to late Victorian vernacular, which also feature generous elevated porches overlooking the water, make something of a period ensemble scattered along the highway for the next half mile south of the rectory.

TI3 Stone Bridge and Fort Barton

On Main Road at Lawton Avenue is an abutment of a multiarched stone bridge that marks the remains of the fourth (1904) replacement in a series of bridges (wood at first, masonry after 1810) which crossed the narrowest point of the Sakonnet channel to Portsmouth from 1794 through 1957, when the highway bridge farther upstream superseded this crossing. A hotel existed in various manifestations from 1790 until the mid-1990s on the Tiverton side, and toward the end of the nineteenth century an amusement park (now gone) developed its Portsmouth end of the bridge. Stone Bridge once attracted not only day trippers from Fall River, but even New Yorkers on the overnight steamers of the famed Fall River Line.

A turn off Main Road at Lawton Avenue leads to the earthen redoubt of Fort Barton (1776) on Highland Avenue. An important Revolutionary War site, it is situated 110 feet above the Sakonnet at its most vulnerable point of crossing. From this eminence guns were directed against the British occupation force on Aquidneck Island. The fort derives its name from the leader of an expedition of extraordinary daring. From here Lieutenant Colonel William Barton and four men departed in July 1777 and, crossing the Sakonnet at night, captured the British commander, General Richard Prescott, literally "with his pants down" in the bedroom of a Portsmouth lady. For two years, 1777 and 1778, Continental troops massed around Fort Barton for an invasion of Aquidneck Island. Eventually, in August 1778, some 11,000 Americans invaded the island in one of the largest troop massings of the war, for what came to be called the Battle of Rhode Island in Portsmouth. Although it ended indecisively, with the withdrawal of the American troops and the failure of the British to exploit a tenuous victory, this battle essentially marked the end of combat in Rhode Island as the center of the war shifted to other areas.

After the Revolution, Fort Barton was privately owned until 1923, when it was donated to the Newport Historical Society. It opened as a public park in 1928. An observation tower on the redoubt provides a splendid panorama south into Little Compton, west over Aquidneck Island, and, where trees are bare, of a number of sites described in this guide. No wonder Fort Barton again became an observation post during World War II.

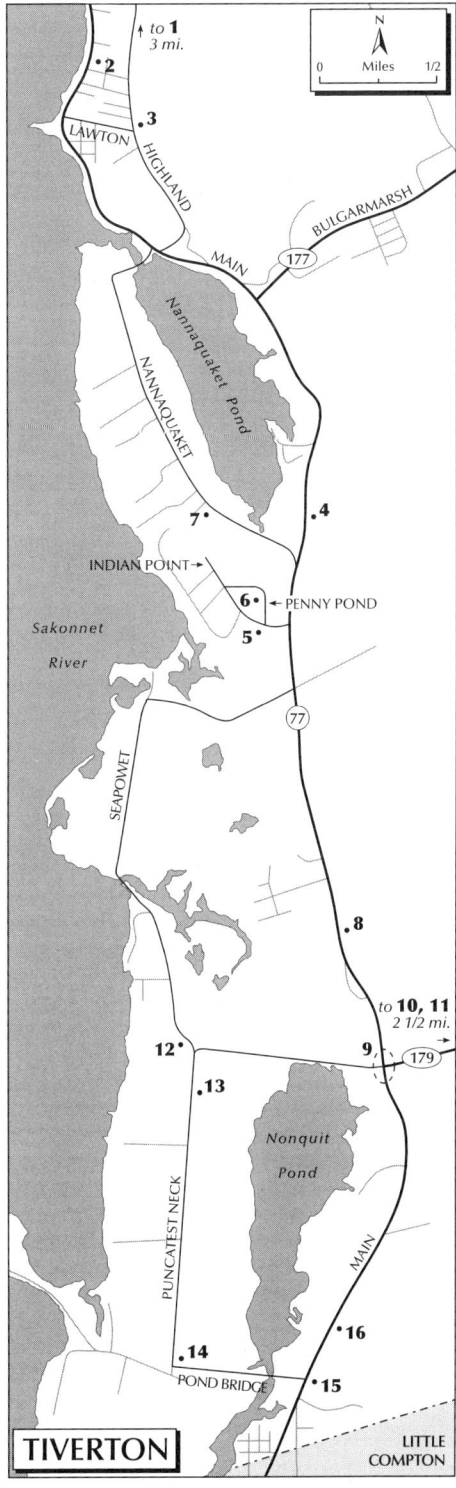

TI4 William and Thomas Durfee Farm

1690s, 1768 and later. 2794 Main Rd. (pole 462)

Sited well above the road, this farm complex vividly recalls Tiverton's mostly vanished agricultural heritage—as well it should, with a history of cultivation that extends over nearly three centuries. William Durfee, who also engaged in shipping out of Newport, established the farm in the late 1690s, and some of the stone walls, stone outbuildings (possibly for slave quarters), and gardens may date to the late seventeenth or early eighteenth century. In 1768, his son Thomas demolished his birthplace and erected the present central-chimney farmhouse on its foundations. Its transomed entrance, probably an early-nineteenth-century addition, contains Neo-Gothic tracery, all simply capped by a bracketed entablature. The compelling image of this farm and its outbuildings cluster has long been appreciated: *Pencil Points* (1920) published the house, and the Garden Club of America's *Gardens of Colony and State* (1931) includes its gardens.

TI4 William and Thomas Durfee Farm

TI5 Briggs–Manchester–Beattie House

Mid-18th century and later. 20th century, enlargement. 68 Indian Point Rd.

At right angles to an extensive, shingled twentieth-century wing, itself a very substantial house, is the original eighteenth-century house with the gable-on-hip roof associated with Newport. Nathaniel Briggs, a slave trader, built the original house. Two duplications of its pedimented and transomed door provide the principal features of the modern addition. Northwest of the house is a fine stone barn, more impressive before 1970s conversion to residential use.

TI6 Hamilton Beattie House
(The Stone House)

1917. Hooper and Moran. 43 Penny Pond Rd.

Well concealed, and more like the elaborately revivalist fantasies found on Aquidneck Island than the simpler country retreats of Tiverton and Little Compton, the Stone House was built, appropriately, by the owner of a local quarrying business. The hand-laid stonework of this rambling one- and one-and-one-half-story house is of exceptionally high quality. In the English manor mode popular for big houses during the 1920s and 1930s, it was illustrated in one of the annual country house issues of *Architectural Record* (November 1924).

TI7 Andrew Oliver–Charles Sargent–Steven Thayer House (Homelands)

c. 1760. c. 1890, renovation, A. W. Longfellow. 575 Nannaquacket Rd.

Situated at the end of a long, maple-lined drive with fine views across open fields to Nannaquacket Pond, Homelands is a superbly evolved country seat. Andrew Oliver, a stamp tax collector, built this substantial two-and-one-half-story gambrel-roofed house and lost it to confiscation during the Revolutionary War. During the latter nineteenth century, it became the summer house of Charles Sargent, director of the Arnold Arboretum, who married into the wealthy Fall River textile family which had owned it from 1867 and developed it as a gentleman's dairy farm. Sargent planted many specimen trees on the property. His daughter, Mrs. Steven Van Renssalaer Thayer, continued to operate the dairy farm and engaged Boston architect A. W. Longfellow to remodel the house in 1890 in what was one of the earlier Colonial Revival reworkings of an eighteenth-century farmhouse into a country retreat. Although some outbuildings remain from the late nineteenth and early twentieth centuries, the dominating replacement barn (1890) burned in the 1980s.

TI8 Captain Robert Gray House

18th century. 3622 Main Rd.

TI9.1 Abner Soule–Cornelius Soule–Cornelius Seabury House

Captain Robert Gray apparently built this tiny two-and-one-half-story five-bay shingled house for himself. Utterly unpretentious, its central chimney lost, its window sash replaced, and badly crowded by later houses, it is mostly of historical interest, for the colorful background of one of the owners of the house. Of all the nautical derring-do of Tiverton's seaman, none was as consequential as Captain Gray's sailing in the *Columbia* to the coast of the present-day state of Washington and there discovering the great river which he named for his vessel. Braving the dangerous shoals and fast currents in its mouth, he "took possession" of it, and thereby established the basis for American claims, in later negotiations with England, to ownership of the Oregon Country, which today comprises the states of Washington, Oregon, and Idaho.

TI9 Tiverton Four Corners

Intersection of Main Rd. and East/Puncatest Neck rds. (Routes 77 and 179)

Active by the early years of the eighteenth century, Tiverton Four Corners is the principal commercial node in the southern part of the town. Until the late 1980s its charm was the relative rural isolation in which its buildings presented themselves. Only an ice cream stand struck a discordant note on one of the corners, and even its discord was sweetened by its fame in the area for the quality and variety of its product. But then the fakery of plywood boxes coated with nostalgia to attract boutiques and professional offices began to leach onto what was there. A nice cluster of country stores, a onetime blacksmith (now a tinsmith) shop, and a small mill are mostly already boutiqued into the nether realm between preservation and sentimentality. Should the conscientious pilgrim be grateful for the new uses with their sprightly prettiness, or deplore their falsity to the crossroads village that once existed here?

At the intersection itself, the two buildings opposite one another on the northwest and northeast corners are architecturally the most interesting.

TI9.1 Abner Soule–Cornelius Soule–Cornelius Seabury House

c. 1770, 1809. Late 1980s, converted to offices. 3852 Main Rd.

Set back in generous grounds is the Soule-Seabury House, which began as a south-facing (toward East Road), center-chimney house, two stories high, with corner quoining. Abner Soule (sometimes spelled Sowle), who erected the house, was a whaler, blacksmith, and Revolutionary War hero. He passed on his blacksmith shops, on the very corner of the house site at the intersection, to one son in 1808; his twin, the famous sea captain, Cornelius, received the house and farm (he soon controlled the blacksmith shops as well). Immediately, Cornelius more than doubled the original house. The 1809 addition to the north made a square of the previously rectangular house and added a dormered, hipped roof (rising to a centered platform now minus its railing and an additional balancing chimney). The modernized plan permitted a central hall through the house between the chimneys. The new west front (on Main Road) features a pattern of openings progressively reduced in number from ground story to dormers which is both unusual and visually effective. Five on the first story; three on the second; finally, two dormers squeezed between the two chimneys. Together they impose a pyramidal overlay across the nearly double-squared field of the elevation. The old front (toward East Road) may have established the pattern—or those dormers may have been altered to agree with the new front. So puzzles remain.

The colorful Cornelius was, however, barely in residence. Active initially in the China trade, he became the skipper of the pride of John Jacob Astor's sailing fleet, the *Beaver*. In 1811, Soule sailed this vessel from New York with men and supplies to reinforce Astor's newly established fur trading post, Astoria, at the mouth of the Columbia River. He reached Astoria on May

6, 1812, exactly two decades after Captain Gray, from the same Tiverton hamlet, who had discovered and named the Columbia River. Finally, in a letter of 1814 from Canton, China, Captain Soule deeded over his Tiverton property in payment of his debts to a cousin, Cornelius Seabury, a merchant at Tiverton Four Corners. Four years later, in 1818, Captain Soule perished with all hands when his ship went down off the Philippines. The new owner, previous to opening his Four Corners store on a lease from his cousin, had had a career as a merchant in Boston and Newport, interrupted by a fabulously successful sealing expedition to the Indian Ocean, followed by marriage and a stint as a farmer. His descendants occupied the house into the 1970s.

TI9.2 **A. P. White Store**

Mid-1870s. 3883 Main Rd.

A. P. White's high-shouldered, mansard-roofed store is fronted by a tall porch. Such a prepossessing store of this vintage is a rarity in Rhode Island, both because the effects of the panic of 1873 discouraged building during these years, and because few such commercial buildings have survived in such pristine condition. It boasts a full panoply of typical Second Empire architectural elements: tall porch with square posts and brackets; bay windows; flared, dormered mansard with bracketed eaves and cupola. The country general store as a set designer might envision it, this one has been upscaled as a place for "provisions."

TI9.3 **Arnold Smith House**

c. 1750. 3895 Main Rd.

The one-and-one-half-story Arnold Smith House has lost its anchoring chimney. Even so, it impresses by virtue of its low, mounded gambrel mass. The vividness of this impression partly results from the sparseness of openings in the long front elevation, where the centered door and two small windows (instead of the usual four) leave a broad expanse of shingled wall, which is extended by the homogeneous wrap of the wooden shingles over the roof as well (instead of the usual discordant remodeling in asphalt roofing). The later insertion of a little dormer magnifies the sense of protective enclosure of such a house, functionally and psychologically bundled against what is outside.

TI9.4 **Tiverton Historical Society** (Chase-Cory House)

c. 1730 and later. 3908 Main Rd.

The Chase-Cory House, another eighteenth-century gambrel, is more intact than its neighbor, but much restored and milder in exterior aspect. Several Chases lived here, then Cornelius Seabury when he first opened his shop in Four Corners. Ledgers from his country store are in the society's collection. Then a member of the Cory family acquired the house in 1816, and it long remained with descendants. Behind it stands a fine early corncrib, moved from another site, raised on stubby stone columns capped with horizontal stones to thwart rodents. Unusually, its walls are shingled above to enclose an upper story with vertical slatting below to provide air for forage. (A similar crib in the field behind the Soule-Seabury House has been restored to resemble this one).

TI10 **Free Will Baptist Church** (First Baptist Church, Old Stone Church)

1841. 5 Stone Church Rd.

This stuccoed rubble masonry church presents a forceful image in a reductive Greek Revival style. Although known popularly as Old Stone Church, it was probably always stuccoed, with barely noticeable scoring for an unconvincing simulation of cut granite, now somewhat crudely "restored." The facade consists of no more than two widely spaced double doors, recessed into the wall without frames, which fixed paneling above make more monumental, plus an inset panel between for the building's date.

TI11 Dr. William Whitridge and Thomas Whitridge House

Wooden steps stretched almost the width of the elevation give access. Paneled corner pilasters frame the simple boxed cupola.

The walls inside are plastered, with a later flat ceiling in paneled pressed metal; the pews may be original. Exceptionally, but not uniquely in Rhode Island (see North Scituate Congregational Church in Scituate and Foster Center Baptist Church and Mount Vernon Christian Church in Foster), the pulpit and baptismal pool are located at the entrance end in a niche created between the interior vestibules behind the doors. Local legend tells of brides lifted through a rear window to position them for the wedding march, before the much later installation of a door in this wall, possibly sometime in the twentieth century when the nondescript parish house was tacked onto the back.

TI11 Dr. William Whitridge and Thomas Whitridge House

c. 1770, c. 1865. 285 Stone Church Rd. (not visible)

If the White store at Tiverton Four Corners is a set designer's vision of a rural commercial Victorian building, Dr. Whitridge's country house is the equivalent for the rural house. (Could the same designer have done both?) Again, we see the full Victorian Second Empire panoply of tall porch with square posts on sturdy plinths, slightly convex mansard, and tall, arcaded cupola. Here, however, the projection of the building at the center aggrandizes the entrance approach, which begins with a long private drive in from the road; porch posts are doubled with wooden arching jumping from pair to pair; upper windows are more magnificently framed. Both store and house are commanding. Yet both maintain a spare, airy quality appropriate to country building. The two-part composition of this house documents the generational rise of one family. The rear section is a south-facing house built just before the Revolution by Dr. William Whitridge. Its configuration is somewhat unusual, with paired center chimneys flanking a center stair. Whitridge's son Thomas, "a greatly respected merchant of Baltimore," owned the house by the 1860s and added the Victorian frontispiece. The scale of the front section—particularly its height—is exceptional in rural Rhode Island. Does it owe these qualities to a Baltimore architect and to Thomas Whitridge's preference for the breezy amplitude of southern dwellings for his summer estate?

TI12 Almy Farm

Late 19th century. Seaponet Ave.

Although much altered, the Almy farmhouse, outbuildings, and stone walls give a sense of the smaller farm complex evolving out of the late eighteenth century, just as the Durfee Farm provides an image of more prosperity, even though neither now operates as a farm. Continue a little farther on Seaponet to see a beautiful marsh landscape.

TI13 Captain Fernando Wilcox House

c. 1865. 488 Puncatest Neck Rd.

Here is yet another plain, bracketed Victorian house, in which "architecture" is left to the porch. Across its front pretty columnar supports raised on pedestals sprout more scroll-cut brackets stylized from botanical motifs. These, however, are more restrained, and therefore better attuned to the sobriety of what they decorate.

TI14 Colonel John Cook–William Bateman Farm

c. 1730–1750, c. 1869, c. 1900. Neck and Pond Bridge rds.

The original owner of this farm became a patriot hero of the Revolution for his command of Newport's Second Regiment, consisting of troops from Tiverton and Little Compton. The prominently sited house, on a rise overlooking

TI14 Colonel John Cook–William Bateman Farm

both bay and pond in the corner of 60 acres of rolling fields lined with stone walls, dates back to the eighteenth century, however much of its modernization with Second Empire mansard and entrance porch around 1869 disguises the fact. It is one of Tiverton's most imposing eighteenth-century houses, equaled locally among extant buildings only by the Briggs-Manchester-Beattie House. Its location near the ferry to Newport may have encouraged its original owner, Colonel John Cook, to aspire to that center's fashions rather than to the local parochial building tradition, much as Briggs's slave trading doubtless gave him contact with Newport sophistication.

The scale and interior woodwork of this house (not open to the public) indicate Cook's ambition more than the plain treatment of the exterior, except for the delicate, pavilionlike porch, which is unusual and of early date, perhaps even original. Compared to the porch, the strongly framed, cedar-shingled mansard in the manner of a Victorian lighthouse keeper's lodge is more in keeping with the overall plainness, although it imposes its own alien, top-heavy scale. It seems to celebrate the Victorian summer overflow of relatives, children, and servants. Probably at this time the sturdy bollard-shaped stone piers were added at all entrances—including a cornered pair at the intersection to mark the entrance of the former carriage approach. Eventually, the pendulum of taste swung back toward colonial, as reflected in the addition of some Neo-Georgian trim in the early twentieth century. The nineteenth- and early-twentieth-century shingled outbuildings, including gambrel-roofed barn, sheds, sheep pen, farmer's house, and outhouse have a completeness which owes much to continuity: the property has undergone but two changes of family ownership over two and a half centuries. Some farming continues on what is probably Tiverton's grandest surviving historic farm.

TI15 White Homestead

Late 18th century; rear ell possibly later. 4398 Main Rd.

TI16 Edward Cook House

Early 19th century; ell possibly later. 4340 Main Rd.

Almost next-door neighbors, although separated by spacious sites, these two farmhouses are of the familiar five-bay formula, which meticulous restoration around 1990 elevated from years of shabby decline. They possess a somewhat commanding, abstract demeanor, with aspirations, one suspects, to status a tad more removed than other examples of their type from their vernacular origins. This assumption arises from the volumes of these houses, which broad side elevations especially seem to inflate. Both boast exceptionally well-proportioned fronts, with center doors and windows separated from the four windows on either side by an interval just wide enough to emphasize the axis, yet not so wide as to jeopardize the elevations' unity.

The shingled, clapboard-front White Homestead has a gable roof with center chimney and a simple transomed door. It is turned sideways to the road so that its front faces full south. The Edward Cook House faces the road, although angled slightly south. Its paired chimneys spread to anchor the angles of the hipped roof, thus opening the center of the house to a through hall. Although it is shingled on three sides, the front is clapboarded with an edge bead. Pilasters and pediment frame the transomed door. In all these respects it displays a somewhat grander aspect than its neighbor, as well as other details typical for its slightly later date. Its view down sloping fields, across a pond, and off to the distant ridge on which the Cook-Bateman Farm is spread is among the beautiful pastoral views in the state—for the present at least.

These scattered early farmhouses and their handsomely stone-walled fields, now lovingly and reticently maintained mostly as country places for the well-to-do, are a prelude to the continuation and culmination of the idyll in Little Compton.

Little Compton (LC)

West Main Road (Route 77), the principal Rhode Island approach to Little Compton, follows a ridge which permits broad views across fields down to the Sakonnet River as it swells to create the easternmost arm of Narragansett Bay. The scene is dotted with widely separated, weathered shingle houses (the norm for the area, although white clapboard is also a favored material). It is crisscrossed by as magnificent a display of stone walls as one can find in Rhode Island. Much of the land remains in agricultural use, mostly haying. Sakonnet Vineyards (1975), at 162 West Main Road, represents the latest development in Little Compton's agricultural history, with a modern-day version of a traditional shingled building housing a sales and tasting room for visitors. Among Rhode Island towns, Little Compton still provides the best sense of what the state's coastal farmland looked like during the nineteenth and early twentieth centuries.

Agriculture, however, is no longer Little Compton's principal concern. Its preservation as an Edenic rural and seaside enclave far removed from Newport is fortified by its land's end location at the easternmost extremity of the state, with no significant harbor to have encouraged seaport activity. Sizable acreage requirements for residences, together with a fierce desire to keep things the way they are and the wherewithal to resist some of the forces for change, have been equally important in maintaining Little Compton's unpretentious exclusivity. Or, as the bumper stickers of its fiercest advocates exhort: "Little Compton. Keep it little."

LC1 **Hunt Farm**
18th century and later. 228 West Main Rd.

This is one of the town's most intact farm complexes. The two-story, five-bay shingled house with simple five-light transom over a centered door retains several shingled nineteenth- and twentieth-century outbuildings, plus a stone barn, all with a winding road into stone-walled fields. It provides an ideal introduction to what is typical of Little Compton—including the farm's present use as a leisure-time residence. Dormers suggest twentieth-century reworking.

The next several miles of West Main Road offer a gamut of shingle houses ranging in date from the eighteenth to the early twentieth century and in massing from compactness to picturesque sprawl. Varied shapes are more characteristic than varied detail, which testifies to the unpretentious aspects of even the biggest houses in the area. Only Jamestown offers a comparable array of the shingle houses which played such a significant role in Rhode Island seashore design of the late nineteenth century. In Little Compton they are juxtaposed with their eighteenth-century inspiration, and derive substantially from the contained quality of their precedents. In the other towns, they tend more to porches and to nineteenth-century vacation cottage design.

LC2 Friends Meeting House

1815. 234 West Main Rd. (pole 99)

Of all the Quaker meeting houses in the state this presents the most haunting image. Both meeting house and site provide a striking demonstration of the intensity of architectural meaning possible from plain building and uncentered symmetry. A stone-walled rectangle with plain farm gate creates a generous grassy precinct for the two-story shingle building as its center. Utterly without planting, the precinct is as severe as the meeting house itself. Two doors pressed toward the ends of the front, along with the windows, make a tense peripheral composition, which chimneys at either end of the roof emphasize. Between the doors a shingled expanse marks the gulf between the sexes.

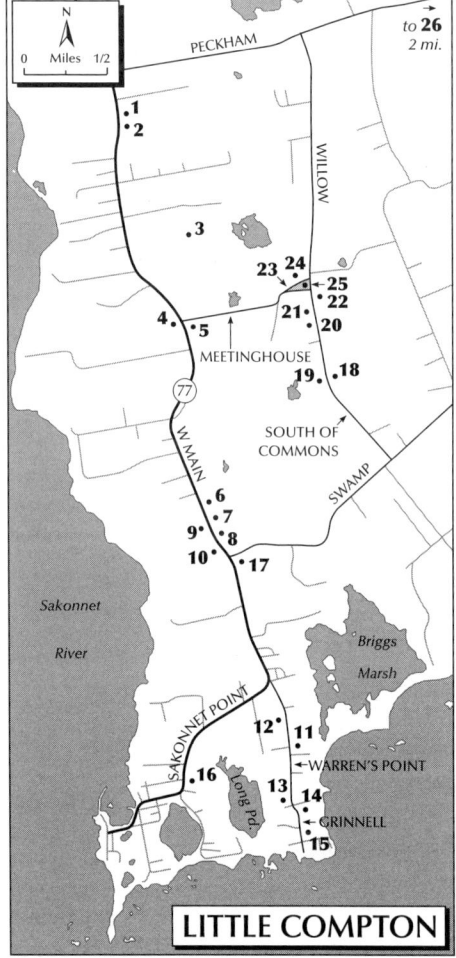

Inside, framing is exposed against the plaster walls with balconies on three sides. The traditional halving of the space when appropriate to separate the sexes is possible by sliding panels from above. Benches of the simplest construction, with a single board for a back, face one another across central spaces, each with its plastered fireplace and interior chimney against the end walls. Rising in two tiers against the outer walls, the backs of the highest benches are no more than boards attached directly to the plaster. Behind the meeting house is a lovely view of barns and fields sloping down to distant marshlands.

The meeting once included members who ferried from Portsmouth and from as far away as Dartmouth, Massachusetts. Its last member died in 1903. After years of disrepair it was first refurbished in 1925, given to the local historical society in 1941, and more definitively restored in 1960.

LC3 Frenning House (Bumble Bee Farm)

c. 1940, Blanche Borden Frenning. 316 West Main Rd. (not visible)

Concealed at the end of a winding private lane, this picturesque collage of colonial fragments united by modern construction possesses a spectacular view over fields and marshes to the distant church steeple on the common. This was the house of an interior decorator who became a part-time architect without formal training. A number of her houses, like this, incorporate ancient siding, timbers, windows, and hardware. Such assemblages were both a popular aspect of the Colonial Revival in the 1920s and 1930s and possible because of the large number of such dilapidated buildings which were just then being demolished. The house combines pieces from two seventeenth-century houses in the living room and dining room and has a Greek Revival porch.

It is the bold collision of a two-story mass with a saltbox slope into a low gabled offset adjunct that unifies this architectural collage into a picturesque whole. An outbuilding in weathered boarding from a collapsed barn across the entrance court completes a composition worthy of one of Samuel Chamberlin's textured photographs of old New England buildings. It is as though repeated alterations to an ancestral home had always had the good fortune to find a succession of counselors with a sure sense of architectural charm. Frenning designed a num-

LC2 Friends Meeting House

LC5 John Church–Edith Russell Burchard House (Old Acre)

ings (probably from the late nineteenth century) aligned with or parallel to the house and a spread of fields beautifully enclosed with stone walls. The house belongs to descendants of the original owners.

LC5 John Church–Edith Russell Burchard House (Old Acre)

c. 1840. 1890, S. D. Kelly(?). 420 West Main Rd. (southwest corner of Main Rd. and Meeting House Ln.)

This Greek Revival house received its present form in 1890 when exaggerated and picturesque versions of colonial motifs converted it to one of the finest Colonial Revival houses in the state. Broad verandas, deep bow windows on the principal elevations, Chinese Chippendale balustrading at the eaves, and pedimented dormers brought it to its present form. The house was the retirement home of a successful builder who had apprenticed to John Holden Greene in Providence from 1812 and eventually worked in Providence as a principal of Church and Sweet, which was responsible for buildings in Savannah and Charleston following Greene's example. His son became wealthy in the music business, enabling his daughter to remodel her inheritance. This transformation included splitting the house down the middle to widen an entrance hall into a fashionable "living hall" with fireplace. The impulse toward horizontality implicit in this extravagant widening and swelling at the center of the house is also apparent in the spread of doors, windows, and, most conspicuously, the side lights and fan framing the entrance, with leaded tracery stretched into a fantasy cobwebbing. Colonial design was never like this. This is Neo-Colonial being exuberantly inventive even as it seeks to retrieve the decorum of the forefathers. At the same time, well behind the house, S. D. Kelly (who probably remodeled the house) designed the most elegant barn in Little Compton for the family horses and carriages. It boasts a leaded fan-lighted entrance, Palladian window in the gable end, and coved cupola topping the high hipped roof.

ber of other houses and additions in Little Compton and around her winter home in Palm Beach, sometimes inventing "restorations" as here, more often building de novo with a traditional aura.

LC4 Brownell Farm

1804. 411 West Main Rd.

The exceptional feature of this fine two-story, five-bay Federal house is the pedimented entrance porch to its transom-lighted door. It is probably an early addition. Supported on Tuscan columns with an arched ceiling beneath its pedimented gable, it is similar to those on a cluster of houses in Union Village in North Smithfield and in some houses in Foster (see Union Village under North Smithfield and FO3, FO15, and FO18). What is the connection between them and this isolated example? The Brownell Farm retains a particularly handsome complement of shingled barns and outbuild-

LC6 Little Compton Historical Society (Wilbour House)

1690, 18th century, 19th century, 1955 and later. Pyramid Schoolhouse, 18th century; reconstructed,

1974. Burleigh's studio boat, c. 1905. 548 West Main Rd. (open to the public)

This two-and-one-half-story shingled house provides the best example of early architecture in Little Compton, the more so because the east half (up to what was originally an end chimney) was built in 1690, when Samuel Wilbore settled as a farmer in Little Compton, and the west half in the eighteenth century. Leaded casement windows in the older section give way to small paned sash in the later section (all reconstructed from evidence within the building fabric). The house remained in the Wilbour family until 1919. Its most recent restoration began in 1955 with its acquisition by the Little Compton Historical Society, which also restored its attractive complement of stone walls and outbuildings. The most notable of these is the corncrib lifted off the ground on stone supports, capped by flat stones which spread like a mushroom cap over its stem to block rodents from the contents.

Also on the grounds is a squarish building with a pyramidal roof, a reconstruction of a Little Compton landmark known as the Pyramid Schoolhouse. Its original site was less than a mile south of the Wilbour House on West Main Road. It decayed and disappeared in the nineteenth century, but was reconstructed on the Little Compton Common for the 1974 tricentennial of the town, after which the reconstruction was moved to the Wilbour enclave. Thanks to a detailed reminiscence, the interior arrangements could be reconstructed as a continuous slant-topped desk for older students fastened to the walls around three sides, fronted by benches, with a *U* of lower benches for younger students around the teacher's desk and a stove at the center—more in accord, it would seem, with meeting house tradition than the more authoritarian and regimented classroom image which the one-room schoolhouse came to embody. Finally, the large barn to the rear, now used for the display of historical artifacts, has, beneath a porchlike structure attached to its rear, an interesting boat converted to a studio. It was used as a floating studio at the beginning of the twentieth century by the Providence artist-craftsman Sydney Burleigh, who summered at Little Compton. His vessel recalls the studio boats that appear in paintings by Monet and Renoir. Inside, under multipaned windows, benches lining the perimeter focus on a potbellied stove in the prow and Burleigh's shrine to a Pre-Raphaelite muse. Some of Burleigh's paintings, uneven in quality, can be seen in the Wilbour House.

LC7 Edward Brayton House

1937, Albert Harkness. 554 West Main Rd.

Although the overall form of the Brayton house is vaguely indebted to the rural French models that this Providence architect used throughout his career, by the late thirties Harkness characteristically modulated tradition with the kind of Moderne nuance evident in the casement windows and the spandrel panels which connect them between the two stories of this house.

LC8 Cornelia C. Abbott House (Londonderry)

19th century, c. 1925. 560 West Main Rd.

Londonderry is a superb example of the penchant among Little Compton residents, during the Colonial Revival of the 1920s and 1930s especially, for enlarging plain, shingled eighteenth- and nineteenth-century farmhouses by picturesque shingled accretions in a rambling composition, and embowering the whole in a lavish flower garden.

LC9 Pabodie–Gray House

c. 1690, c. 1765, c. 1890. 561 West Main Rd.

The architectural history of this two-and-one-half-story shingle house very nearly duplicates that of the Wilbour House. In like manner, William Pabodie built the east end of the house, up to what was then an end chimney; Pardon Gray purchased the house two generations later and added the west half. A late-nineteenth-century owner added the two-story bay window, designed by artist-craftsman Sydney Burleigh, on the west end—an early example of Little Compton summer resident's alteration to make a hard-bitten farm building cozy and colorful.

LC10 Adeline E. H. Slicer House (The Mill)

Mid-19th century. 1886–1887, renovation, Edmund Willson. 581 West Main Rd.

Here a summer cottage originated in one of a number of shingled windmills, moved to this site, which once ground Little Compton's meal.

LC10 Adeline E. H. Slicer House (The Mill)

Edmund Willson of the Providence firm of Stone, Carpenter and Willson wrapped the three-story octagonal conical-roofed gristmill with a lower addition, its sweeping pyramidal roof with stone chimneys and incorporating deeply inset porches. This is among the earliest recorded conversions in Rhode Island of a utilitarian agricultural structure into a summer residence. The artist Sydney Burleigh designed decorative plasterwork for the interior, reminiscent of a similar treatment he gave to his Fleur-de-Lys Studio in Providence, at about the same time. There, too, Willson was the architect.

LC11 **William G. Nightingale House**

1950, Albert Harkness. 49 Warren's Point Rd.

The Nightingale House offers more of Harkness's cautious modernism, this time in clapboard, with porches in two stories, the upper for views across the point to the ocean. Vernacular takes on a crisp nautical flavor.

LC12 **Almet Jenks House**

1949, Albert Harkness. 64 Warren's Point Rd.

Instead of invoking French rural building in a Moderne manner, as in his Brayton House, Albert Harkness here turned to the town's own shingle tradition, and thereby established an exemplar for the way "modern" should be approached in Little Compton: with caution. The house was, in fact, partly constructed of materials from dismantled farm buildings. The central gabled mass spawns gabled wings at either end, but in opposite directions, giving a step-like configuration to the plan. This easy spread of the house, its substantial stone chimneys, and the greenhouse attached to the end of one of the wings are other means by which this house insinuated itself into the local environment.

LC13 **Thomas Marvell House**

1940, Thomas Marvell. 65 Warren's Point Rd.

Thomas Marvell had just graduated from architectural studies at Harvard under Walter Gropius when he built this summer house on land owned by his wife's family. A scandal when built, it was regarded as the way "modern" should *not* be brought to Little Compton. Today, these gray-stained shed-roofed boxes strung out in an additive way make the house seem to be an abstraction in the Little Compton manner. Not so, of course, the flat surfaces made by flush vertical and horizontal boarding, the horizontal windows scattered as rectangular voids across the multiangled composition, or the slight lift of the house off the ground on squat concrete piers so that it seems to skim a little above its site.

LC14 **Thomas Bailey House**

1700, c. 1920s. 14 Grinnell Rd.

This shingled, center-chimney gambrel house, one of the oldest in Little Compton, has been gradually extended in several directions until it ambles into the landscape. Such accretive design inspired many twentieth-century emulations built, however, all at one time.

LC15 **Edwin W. Winter House** (Gatherem)

c. 1905, Edmund Willson. c. 1930, modified, Edwin Emory Cull. 24 Grinnell Rd. and Atlantic Dr.

The sequence of houses along Warren Point Road ends with an enclave of summer cottages that is likely to be closed by guards in season. At its center is this superbly simple shingle house with a long slope of seaside gable pierced by three double-windowed dormers that seem to scan the sea. Higher up on the roof, two smaller jerkin-roofed dormers peer in a more heavy-lidded manner through the valleys made by those in front. The roof sweeps over the inset porch stretched behind paired columns across the front of the house. The porch makes a platform, raising the house up off its sloping site for a more commanding view. Although

Edwin Cull simplified his predecessor's handling in minor ways, this essential statement of the summer cottage is properly Willson's.

LC16 Simmons–Manchester House

Mid-19th century, c. 1898. 106 Sakonnet Point Rd.

Architecturally, this is the most interesting house on Sakonnet Point Road. It began as a Greek Revival house fronted by a deep porch with Doric columns. Apparently the roof, probably dormered, came down in a long single slope over the porch. Pointed windows in the barn indicate the competing fashion for Gothic around 1850. The William Wilbor House (see entry, below) probably gives a good idea of how this one originally looked as both were apparently the work of the same builder; the Doric porch columns and pointed-arched windows in the eaves of the side elevations duplicate this combination of features in the Wilbor House.

When the Manchester family took over the house around 1898 (Josephine Manchester being a Simmons granddaughter), they radically altered it, probably breaking through the slope of the original roof to make a full two-story house out of what had been, in effect, one and one-half stories. They fronted this with an exquisitely scaled Colonial Revival view porch directly over the Greek Revival Doric below, but stretched out in the horizontal fashion preferred in the late nineteenth century. They topped the roof with a glazed polygonal cupola as fancifully "colonial" as the added upper porch, and threw out some bays on the side elevations. (In the process, they removed the Neo-Gothic windows on the sides of the house; one remains in the upper story of the adjacent barn's east elevation). It should not have worked, but somehow it did—not architecturally so much as through a happy collision of opposites demonstrating how additions in disparate styles are sometimes best handled by the risky expedient of letting each style speak for itself, with minimal effort by the designer to moderate the dialogue. The same mix of Greek Revival and late-nineteenth-century Colonial Revival occurs in interior woodwork. The mid-nineteenth-century shingled barn must date back to the original house.

LC17 Robert J. Higgins House

1983–1984, William J. Underwood. Southeast corner of 160 Swamp Rd. and West Main Rd.

This postmodernist design represents the response of a new generation of architects to the approved Little Compton approach to new houses. Adapt a traditional functional building or a reasonable facsimile of same: in this case, the architect effectively created his own shingled barn. Modify this to client expectations by discreet use of the current fashionable style: in this case, by eviscerating the assumed barn with giant cutouts. On the front, the entrance is blown up into an immense glazed arch; on the rear, a swath of glass doors opens to a deck as the conventional modernist means of celebrating interior volume and light, but with transom lights from the past, along with shed-roofed addenda.

LC18 Malachi Grinnell House

Mid-18th century and later. 1948, addition, Blanche Borden Frenning. 60 South of Commons Rd.

This shingled half house (which may once have been clapboarded) is a typical Rhode Island form, but one now rare in Little Compton. The Greek Revival entrance enframement is obviously a later addition. Blanche Frenning's stepped-back wing is a typical alteration by this Little Compton designer.

LC19 Isaac Bailey Richmond House

c. 1830, c. 1890 and later. Mid-19th century, barn. 59 South of Commons Rd.

The full impact of this farm complex, which has become a country seat, becomes apparent only from the adjacent field to the south, which it faces in a right-angled orientation to South of Commons Road. Isaac Richmond was a builder who settled in Little Compton after apprenticing with John Holden Greene in Providence. Eventually he built the original house as a retirement farm. His son, Joshua B. Richmond, who lived in Boston, converted it in the 1890s to a summer residence, giving it the same sort of lighthearted grandeur as the Manchester family achieved at the same time in altering their mid-nineteenth-century house on Sakonnet Point Road. The junior Richmond enlarged his father's house by adding a columned porch across the front and the superstructure of the elongated dormer with flanking pedimented windows, all topped by a pedimented cupola tucked between paired chimneys. He also made a water tower double as a view tower

LC15 Edwin W. Winter House (Gatherem), left

LC25 United Congregational Church, center

LC26 Samuel Church–Dr. Hathaway House, gift shop (water tower), right

by capping it with an elevated gazebo toward the entrance end of the extensive barn complex, the sound engineering of its tapering sides possibly also intended as a nostalgic gesture toward Little Compton's disappearing windmills.

LC20 Seaborn Mary

c. 1730, 1937. 35 South of Commons Rd.

This shingled, gambrel-roofed cottage was moved from its original site in Londonderry, New Hampshire, and remodeled for junior members of the Richmond family next door. It is another example of Little Compton's fashioning an architectural tradition for itself.

LC21 Seabury–Richmond–Burchard House

c. 1840, mid-20thc. 31 South of Commons Rd.

Next door to the Seaborn Mary importation is yet a third Richmond family venture in remodeling. Joshua B. Richmond bought this house from a Seabury descendant in 1915 when its four north bays comprised the extent of a rather plain mid-nineteenth-century house. His daughter Corinne Richmond Burchard added the classical vestibule, the two south bays, and the trellised side porch.

LC22 William Wilbor House

c. 1850. 12 South of Commons Rd.

This one-and-one-half-story Greek Revival house with its full-width Roman Doric porch and pointed-arched windows on the side elevations suggests how the Simmons-Manchester house probably looked before its late nineteenth-century renovation, except that the original roof treatments of the two houses seem to have differed.

LC23 Little Compton Commons

From one corner the common appears to be of the usual roughly rectangular shape; but the town's proprietors laid it out in 1677 on a hillock at the town's center as an elongated triangle—prophetically shaped like a yachting pennant. The green contains the burying ground for the United Congregational Church, which stands at one of the halyard corners of the "pennant." Except for the church, the architecture is not striking. It is more the appropriateness of the building, with some aura of the venerable around the church, and the immediate sense one has of this place as a lively center of town activity that make it engaging.

Facing the entrance to the church are renovated Greek Revival and Early Victorian houses and, farther along, a nice bungalow that deserves notice, however incongruous for a New England common. Around the tip and along the top edge of the "pennant" is an extraordinarily full range of town institutions. They include the present school (in a dreadfully inappropriate building for this location), fire and police departments (the latter partly occupying a Greek Revival church which should be restored), and a fine one-room schoolhouse, now connected to the plain two-story clapboarded Little Compton Town Hall next door. Then, around the corner to the halyard end of the pennant is the Brownell Library (1929, Charles G. Loring; 1929), a pretty Neo-Colonial brick building, originally painted white, by an architect whose libraries for small towns in the Northeast were widely published in the architectural journals of the time (this one in *Architectural Record*, July 1932). Next to the library, the Publication House (now partially used as an antique shop), a mid-nineteenth-century house with unusual Greek Revival cast iron railings and cresting on the roof commands a view down the length of the common. This extraordinarily compendious roster of civic institutions which line the Common and, together with the mix of church, stores, and houses, make the Little Compton Common an ideal place to observe the way in which the full development of the common as a New England town form reinforces the sense of community by giving it visibility.

LC24 **American Legion Hall** (Number 8 Schoolhouse)

c. 1845, Thomas A. Tefft. 1986–1988, renovations. 38 Commons

Little Compton created ten school districts in 1844, and immediately built new schoolhouses for each. Only this building, the most sophisticated of the lot, survives in anything like original condition. Henry Barnard, the famed Rhode Island educator and school building reformer, visited Little Compton at the time as a speaker and consultant on educational reform. He and Providence architect Thomas Tefft were collaborators, and designs similar to this one appear in Barnard's book on school buildings. Now encumbered by an addition linking it to the town hall, the schoolhouse deserves preservation; few of the surviving one-room schoolhouses in the state have its architectural pretensions.

LC25 **United Congregational Church**

1832, 1871, 1974, 1986. 1 Commons

The church building began as a plain, clapboarded meeting house, three bays deep, with little vertical emphasis. It changed in 1871 with the elevation of the original meeting house over a tall basement floor, to permit Sunday school and church social activities downstairs. This levitation of the church required the addition in front of an enclosure for a stair and entrance foyer, which are topped by a belfry and a commanding polygonal needle spire, all in Victorian carpentered Gothic. White paint outside and the recent renovation of the plaster barrel-vaulted interior toward its presumed original appearance are meant to invoke as much as possible an ideal version of the kind of colonial or Federal architecture expected on a New England common. But the architectural distinction of the church is its Victorian addition—one of the finest such wooden Victorian spires in the state.

Adamsville

Adamsville, gathered around a convergence of country roads on the Massachusetts–Rhode Island border, is a pretty cluster of shingled and clapboarded buildings. Some of them have been converted into the boutique sort of shop. It is dominated by the Abraham Manchester General Store (c. 1860[?], now a restaurant with a disruptive parking lot) east of the T-intersection where Main Street becomes Adamsville Road. Fine dressed masonry walls, doubtless all the work of the same local mason, line the roads around this intersection.

LC26 **Samuel Church–Dr. Hathaway House**

c. 1815 and later, c. 1905. 4 Westport Harbor Rd.

The shingled Church House is the most interesting in Adamsville. Of extraordinary scale for the early nineteenth century, it is virtually a square with paired interior chimneys. More extraordinary are two identical five-bay entrance elevations on its north and east sides, with duplicate fanlighted entrances at the center of each. The north front, toward Adamsville Road, is now the principal entrance. At right angles to this, the east front, originally intended as the principal entrance with flanking parlors on either side, now faces an abandoned road. The present entrance opens into a vestibule between one of the parlors and the keeping room (kitchen). Barns were added in the nineteenth century.

Dr. Hathaway built a water tower at the rear of the property on Westport Harbor Road when he owned the house in the early twentieth century. It once housed a pump on the first floor and chauffeur's quarters above, with the tank on top. It is a beautifully proportioned example of Neo-Colonial design applied to an eccentric building, which again seems to invoke Little Compton's onetime windmills.

The Islands

PORTSMOUTH, MIDDLETOWN, NEWPORT, JAMESTOWN, BLOCK ISLAND (NEW SHOREHAM)

Portsmouth (PO)

THE SOUTHERN APPROACH OF MOUNT HOPE BRIDGE CONNECTS with the largest island in Narragansett Bay, Aquidneck Island. The west shore of Aquidneck faces the east channel of Narragansett Bay proper. Its east shore faces the broad mouth of the Sakonnet River (or Sakonnet Bay), which is actually one of the three major channels that are lumped together as "the bay." Its short southern end faces the open Atlantic. Although the English came to call Aquidneck Island by the name eventually applied to the colony and state, native Rhode Islanders have always preferred the Indian name. With a little imagination the island could be seen (appropriately) as a lobster in profile, tail (north) at the mouth of the Sakonnet River directed toward the Massachusetts city of Fall River; a single claw at the mouth of the bay projecting toward open ocean. Three towns slice the lobster: Portsmouth at the tail, Middletown (indeed in the middle) at the upper body and head; Newport at the claw. Portsmouth's jurisdiction also includes a scatter of seven smaller islands in the bay. The largest of these, Prudence and Hog Island, are home to summer cottages reached by a ferry from Bristol (Hog Island only during the summer season, Prudence year round to serve a small permanent population). Prudence Island's tiny satellite, Patience, also has a few summer cottages; but most of it and the northern and southern parts of Prudence are state conservation areas. Portsmouth's other islands—Hope, Dyer, and Despair in Narragansett Bay and Gould in Sakonnet Bay—are tiny and uninhabited.

Portsmouth, the second oldest town in Rhode Island after Providence, was founded by John Clarke, a physician, and William Coddington, a "man of wealth and position," both of whom fled Massachusetts because of their support of Anne Hutchinson and her Antinomian theology, deemed heretical by the Puritans. Shortly after their arrival in

1638, Hutchinson and some of her supporters visited Portsmouth, creating so much new dissension among followers there that Coddington went off and founded Newport. Although it is blessed with the largest coastline of any Rhode Island town (49 miles, including its islands), Portsmouth's lack of good harbors prevented maritime development. Like the so-called Narragansett Planter towns on the west side of the bay, both Portsmouth and Middletown were primarily agricultural. They supplied Newport and, through Newport, shipped meat and agricultural products up and down the coast. Cattle, sheep, hogs, and horses flourished on grasses lush with the moisture of the bay, in a climate moderated by the surrounding water. Patience and Hog islands provided natural pens, once carnivores had been trapped, for pigs to run wild, while Portsmouth's equivalent to the town common elsewhere in New England was Common Fence, a peninsula opposite Tiverton which had only to be barricaded across a narrow isthmus to provide a "fenced" common land for grazing. Common Fence remains the curious holdover name for the community of cottages that now crowds it.

Nothing like the culture of the Narragansett Planters developed in Portsmouth during the colonial period or the early nineteenth century—although two large estate farms established during the colonial period in the southeastern corner of the town overlooking Sakonnet Bay did set a precedent for comparable farms toward the end of the nineteenth century. Most of the early farms were small or merely subsistence, backed onto the high ridge of hills along which runs East Main Road. Portsmouth had more windmills through the middle of the nineteenth century than any other Rhode Island town. One only remains, on Prescott Farm (PO18). Virtually without waterpower except from small brooks, Portsmouth was hardly touched by the early Industrial Revolution; nor, in its waterlocked location without harbors, was there particular reason to locate industry here after the advent of steam, especially in a locale so thoroughly committed to agriculture. Processing of fish for oil and fertilizer was an industry for several decades starting in the 1860s, as in Tiverton. Surprisingly, there was also a sporadic coal mining industry, unique to Rhode Island, along the west shore of Portsmouth in the area at the end of Willow Lane. Close to the site of the defunct coal mine, at the end of Weyerhaeuser Lane, in 1923 Weyerhaeuser completed a huge depot for lumber storage with trussing so impressively labyrinthine that there must always have been more wood in the building than in storage. It was demolished in the 1980s. Such fitful efforts toward a more industrial economy notwithstanding, agriculture remained the occupation of Portsmouth until well into the twentieth century.

Increasingly toward the end of the nineteenth century, catering to clusters of summer residents became a second aspect of Portsmouth's economy. The two converged when certain Newporters and other elite decided on Portsmouth as the ideal place for large gentlemen's farms. Like their colonial predecessors, they favored the Sakonnet side of the town (though there were exceptions), one of the idyllic rural areas of the state, fantastic for its trees and with long, sloping fields down to the water.

The fixed order of things in Portsmouth began to change during and after World War II with the coming of industry, mostly high tech, attracted by the Newport naval base nearby. Change also occurred with the shift from farming to development, restrained in some areas by the holdings of the elite. But these too began to give way to development by the end of the 1970s. East and West Main roads, running the length of the town, stream with traffic from Portsmouth's own new suburbanites, from increased tourism, and from through traffic seeking to skirt the Providence metropolitan area. The old Portsmouth is not yet quite lost, except in patches; but how will the spoilage be stopped?

PO1 A. S. Philips House

c. 1900. 616 Bristol Ferry Rd.

This arresting Queen Anne house has a cobblestone base with shingled gables above pulled down over the second story. A little gable facing front insets a larger one, both butting into the principal gable, which runs the length of the house. All windows are vernacular in character, except at the very center of the composition, where a high-style Neo-Colonial scroll-topped dormer, tall, narrow, and white, serves as a brooch to focus the composition, echoed by a clipped dormer to one side. It is the barest badge of the Colonial Revival; but it brings a note of dignity, decorum, and order to what was then beginning to be considered an excess of "naturalness." Regrettably (but customarily), the asphalt shingling of the principal gable destroys the homogeneity of cedar shingling, which once made an entity of these interlocked gables.

PO2 Coal Miners' Houses

c. 1909. 16–18 Brownell Lane, Willow Court, and Elliott St.

Architecturally, all that remains of the coal industry in Portsmouth are a scattering of workers' houses, mostly much altered—now in the shadow of the Kaiser Aluminum and Copper Company plant, a huge modern complex which is visually about as painful a presence as a factory can be. In this group of houses, the most interesting are of a type seen in no other Rhode Island industrial housing. Each two-unit house is a two-story gabled box with the second story and the front slope of the gable subsuming a porch across the front within the austere overall shape; the slope to the rear is extended as a saltbox so that two stories in front become one story behind. A pair of doors at the center of the porch and its three-posted support (at either end and at the center) signal the split of the houses vertically down the middle.

Coal production began in Portsmouth on a small scale in 1809 under the Rhode Island Coal Company and the Aquidneck Coal Company. Although easily mined from veins close to the surface, the coal was of poor quality. So mining was sporadic but hopeful, with a number of openings and closings. At the height of operations, from 1866 until 1883 into the early 1880s, the coal was used by the Taunton Copper Company to fire eight blast furnaces for the processing of South American copper ore. Then there was a hiatus, one last burst of activity in 1909, when these houses must have been built. Mining came to an end in 1913.

PO3 Portsmouth Abbey School

1864, Amos D. Smith House (Hall Manor). 1960 and later, chapel and other additions, Pietro Belluschi. 285 Cory's Ln.

In 1919 George D. Hall bought the Amos D. Smith House, subsequently giving it and some 400 acres to the Order of St. Benedict. There, in what came to be known as Hall Manor, this abbey was established as the first Benedictine foundation in the United States independent of a European connection. In 1926 one of the monks, Father John Hugh Diman, founded a school for boys with a class of eight. It expanded

PO3 Portsmouth Abbey School, chapel

rapidly after 1960 to become a leading Catholic boys' boarding school (now coeducational). Pietro Belluschi, who had originally practiced in Portland, Oregon, but at the time of this commission was dean of the School of Architecture and Planning at Massachusetts Institute of Technology, was called to the school in the late 1950s to add a chapel to a campus which had already acquired several buildings in addition to Hall Manor. He was subsequently asked to design other buildings, to serve as consulting architect on still others, and to consult on an overall plan for the campus. The result is a homogeneous group of buildings which provides the most significant demonstration on the East Coast of an architecture developed in the Pacific Northwest during and after World War II to wed modernism with vernacular American building in wood—much in the vein of earlier efforts by Marcel Breuer in his initial American work with Walter Gropius and later by the group of architects responsible for the so-called Bay Area Style around San Francisco. The campus at the Abbey School was designed at a time when the interrelated issues of "American," "popular," "indigenous," or "more humane" modernism were lively topics of discussion, with Belluschi's work in the forefront. Most of the Priory School buildings to which Belluschi's name is attached, whether as designer or as principal influence, adopt the look of vernacular buildings in stained wood with pitched roofs in copper and handcrafted detailing; but with the smoothness and largeness of wall plane, precision of detail, and use of such elements as sheets of plate glass and ribbon windows that characterize modernist architecture.

The chapel, Belluschi's first and most famous building for the school, is naturally grander and more unique in shape than the rest. The undulant octagonal base for this three-stage, stepped composition raised on a stone terrace consists of masonry planes in shallow, concave curves, alternating with inset bowed planes of vertically fitted boarding. Above, the directional character of these wood planes continues in a taller, narrower octagonal lantern enclosed by a picket of wooden verticals infilled with stained glass, the undulant quality of the wall at the base echoed in the circle of mini-gables which roof this stage of the mass. A needle flèche covered in sheet copper provides the climax, its corrugations maintaining by yet other means a subtle, tremulous quality within this stalwart mass. If the image invoked is of a central-plan church, it is also that of an American polygonal barn, not

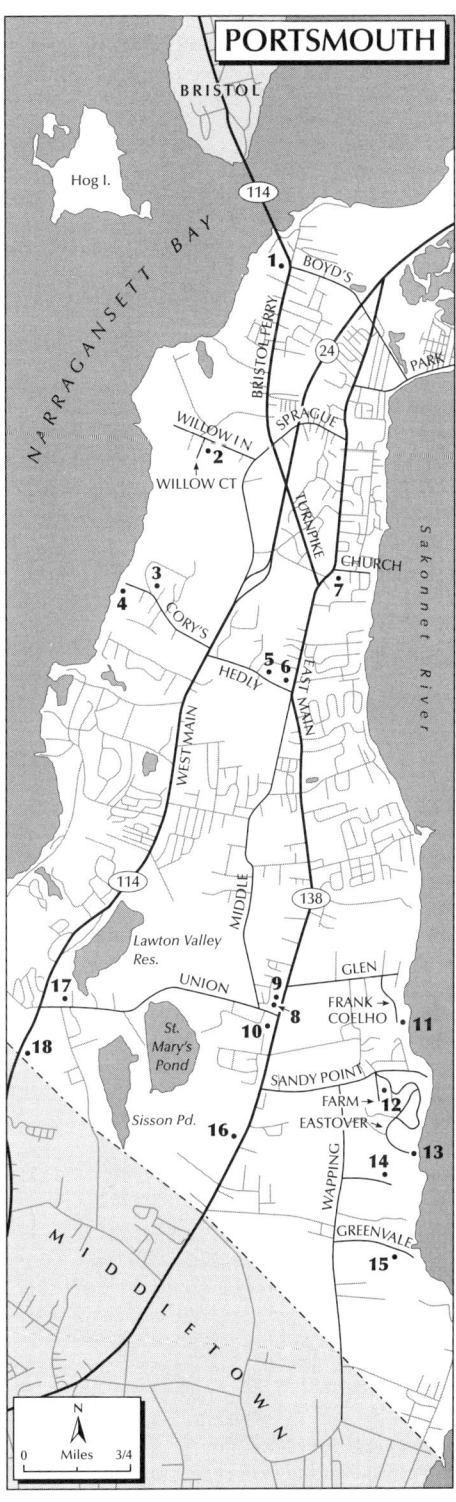

only outside, but inside: exposed timber construction supports the roof of the octagon and a partial balcony. Students occupy this space, while, behind the raised platform altar, a retrochoir provides for sixty monks. Six altars around the octagon and four more in the balcony permit each monk to say daily mass. But the church is also notable for its fittings by major craftsmen at a time of revival for modern design in the liturgical arts. One of the monks, Dom Peter Sidler, incised the handsome Roman lettering on the exterior of the entrance doors. Henry Lee Willet designed the predominantly red and blue stained glass in the lantern. George Nakashima was responsible for the design and carving of altar, altar furniture, pews, and other interior details. Most conspicuous is Richard Lippold's gossamer environmental sculpture of gold and silver rods and wire in gold and silver. A small figure of Christ floats above the altar in this shiny web, which radiates outward above the congregation as a shimmering, slightly trembling presence in the colored light and shadow of the space.

A walk through the Belluschi campus and some of the older buildings leads to playing fields and, off by itself on a mowed field sloping down to Narragansett Bay, Amos Smith's old mansion, now hung with fire escapes and much altered for school use inside, but with odds and ends from the past. A chunky, L-shaped, mansarded mass, its principal distinguishing exterior feature is a broad porch with arched openings, projecting from which at a diagonal from one corner of the house is a porte-cochere with exceptional scroll-cut stencil patterns in its front gable. Continuing down the field to the shoreline brings one to a restored stone boathouse with a cylindrical tower.

PO4 Thomas and Alice Brayton House
(Green Animals)

1859–1867. 1919, topiary garden. End of Cory's Ln. (open to the public May–October)

The Amos Smith mansion and the Thomas and Alice Brayton mansion, both built by wealthy Victorian businessmen, are the principal extant exceptions to the preference among nineteenth-century owners of grand farm estates for the Sakonnet over the bay side of Portsmouth. The attraction of the Brayton House is the "green animals," one of the famous American topiary gardens—its fame suggesting something quite grand, whereas it is really delightfully unpretentious. Well-tended barns and outbuildings are very much part of it, as though this privet menagerie might be taken in at night. So is the house, an early Victorian farmhouse with Colonial Revival modifications. A plain two-and-one-half-story clapboard house, its distinction derives from commanding proportions, as well as from the placement and scale of the elements that distinguish it. Chimneys thrust from all four corners of the flat center section of the extraordinarily broad hipped roof; miniature dormers pop from its steep front slope. Its most prominent feature is the attenuated spaciousness of its Neo-Colonial porch, which wraps three sides of the house. The architectural aspects of the house are the more compelling because it stands free of planting, lawn sweeping downward away from it to the front and sides. Garden, barn, and house retain much of the plainness of the old farmhouse, but raised to a degree of elegance which make the place doubly charmed.

PO5 Portsmouth Camp Meeting Association

1907 and later. Hedley St.

A number of small summer camp meeting groves exist in Rhode Island. Visually, this is the most compelling. Small houses, all painted white, in sizes ranging from an early tourist cabin to a mini-bungalow straggle around the perimeter of a sloping site with rock outcroppings. Scattered about in the center space are a dormitory, refectory, shower building, and, on the biggest outcropping, emblazoned with painted mottoes—"He Lives," "Be Ye Holy," "Jesus Saves"—the meeting hall. The carpentry is plain and to the point. The most interesting single structure may be one of the small houses on the western edge of the site; of wood, it takes its form from a metal house trailer rather than a gabled cottage.

PO6 Friends Meeting House

c. 1700. Corner of Hedley St. and Middle Rd.

A plain, white-clapboarded building with a pyramidal roof and a long, lower two-doored entrance adjunct in front, the meeting house is rather awkward, but it is the earliest place of worship in Portsmouth and the most important from the years following the founding of the town.

PO7 St. Paul's Episcopal Church

1833, Russell Warren. 2679 East Main Rd. (Church Ln. and East Main Rd.)

This early Victorian Gothic Revival church, now considerably altered, is simply and heavily framed, with pointed windows and a modest tower with open belfry. The entrance porch is twentieth-century Neo-Colonial.

PO8 Portsmouth Historical Society
(Union Meeting House)

1865–1866. East Main Rd. (at Union St.) (open to the public)

The form of the carpenter's barn is here given monumental presence in the expansive clapboarded surfaces and the spare slotting of narrow windows, all topped by gable motifs so simple that they seem to lift from their openings beneath in startled exclamation. Are they to be seen as pediments or (more likely) as pointed arches? Classical or Gothic Revival? The "style" oscillates as we ponder the possibilities. Two oculi, one staring from the gable cap of the double door, the other up in the eaves and decorated with compasslike projections, complete the simple vocabulary of shapes which compose this elevation. The lack of vertical alignment of the lower windows with the upper makes a giant triangle of all the elements. Not that this disposition was intentional; it seems more likely that it simply happened as the designer sought symmetry by the simple expedient of centering each element with relation to others.

PO9 Southermost School

Between 1716 and 1725. 1969–1970 and later, restoration. Moved to site adjacent to the Portsmouth Historical Society

This simple one-room schoolhouse—and a tiny one-room at that—is the oldest extant school building in the state. Nothing about it communicates its purpose; with a plank door in one corner and a single, off-center window adjacent, under a long, low flank gable, it could be an all-purpose shed.

PO10 State Police Barracks

1935, Jackson, Robertson and Adams. 838 East Main Rd.

PO8 Portsmouth Historical Society (Union Metting House)

Another of this firm's barracks in a mid-Atlantic Colonial Revival style characteristic of New Deal building, with central block, hyphens, and end wings, this example is interestingly sited close to the wealthy enclave of Vanderbilt farms to the east.

PO11–PO15 Farm Estates

Glen Road is the northern boundary of what was a contiguous cluster of large farm estates in one of the most beautiful settings on Aquidneck Island, comprising stone-walled fields, magnificent stands of trees, and views across Sakonnet Bay. North to south along East Main and Wapping Roads, the principal farms are The Glen, Oakland, Sandy Point, Greenvale, Vaucluse, and Eastover. One could say that today these properties fall between two extremes: those centered in retirement and those centered in management.

Greenvale and Eastover fall into the first category. The landscape theorist Robert Morris Copeland could have been describing John Barstow and his Greenvale in *Country Life* (1859): "Owners of country seats in America are generally men who have retired from active business, and by having a farm connected to their homesteads, they secure something to do and think about, and thus avoid the evil of mental inactivity." In fact, Barstow acquired a copy of Copeland's book for his library the year before he purchased Greenvale. Eastover should perhaps be regarded as a summer residence with incidental farming. The other four farms incline toward the other category. They were large-scale farms, primarily management operations, specializing in the breeding of animals. The owners of Oakland and Sandy Point, who were members of the Vanderbilt family, had their principal residences elsewhere and visited only sporadically, mostly on "show" occasions. Hence their properties most decidedly represented the management ideal. Moses Taylor, who owned The Glen, and a succession of owners of Vaucluse had major houses on these properties and were in residence a good portion of the time. They therefore mixed the ideals of country seat and managed farm. Whatever the style of operation, these idyllic agricultural fiefdoms continued, for the most part, into the 1980s, although farming had been at best desultory on most of the properties for some years previous, except for the care and breeding of riding horses. Development is presently underway, especially on portions of the three northernmost farms and on Eastover. The town of Portsmouth purchased much of The Glen in 1989 for open space.

PO11 H. A. C. Taylor Farm (The Glen)

1907, 1910, Shamrock Stables (not visible). 1923, Moses Taylor House (later Elmhurst School and Convent of the Sacred Heart), John Russell Pope. Off Glen Rd. at end of Frank Coelho Dr.

A miniature Château Style building with masonry posts flanking the road marks the gatehouse to The Glen. The approach road bends sharply parallel to Sakonnet Bay on a ridge (the slope below is now a development) and continues past a yellow brick school (attached to the manor house for The Glen when it served as a convent and school) before terminating in a walled entrance court surrounded by copper beeches and other grand trees.

Moses's father, H. (Henry) A. C. Taylor, is famous in architectural history as the client for one of the earliest large-scale Colonial Revival houses built by McKim, Mead and White in Newport (1885–1886; demolished). He also purchased some 700 acres in Portsmouth to be operated as a livestock farm and commissioned most of an extensive barn and stable complex for horses in shingle and stone known as Shamrock Stables. His son Moses commissioned John Russell Pope to design a French château manor house on the property close to the shore of Sakonnet Bay. The house took its name from the farm; the farm, from a wooded glen where a nearby stream falls to the river. As

PO11 H. A. C. Taylor Farm (The Glen)

one of the few streams in Portsmouth with a substantial falls, it early attracted devotees of scenery and, inevitably, a small mill, which variously shifted from grist to textile and back to grist operations. The ruins of a small, early-nineteenth-century mill in fieldstone, one story with a hipped roof, still exists in the glen, together with a dam and millrace.

In contrast to the Newport mansions which are derived from the grandest châteaux in France, this stems from smaller prototypes. Despite the apparent symmetry of the front entrance, vignetted by the trees, a second look discloses subtle asymmetries, which enliven the elevation and perhaps slightly dilute its regal appearance. Stunningly (but also slightly idiosyncratically) proportioned, the overlarge openings at the entrance respond to the precipitous slate roofs with their tall, ornamented chimneys. At the far end is an arched porch with latticework which once overlooked a formal garden. Inside, Pope employed an *en filade* plan based on the Petit Trianon at Versailles—meaning that the principal corridor does not extend from the entrance through the width of the house, front to back, but runs the length of the house immediately behind the front elevation. This permits the arrangement of the principal rooms "in a file" behind the hall, all overlooking the terrace and water, and all opening into one another through double doors. Although they are rather sparsely and severely decorated with plaster ornamentation and are now shabby and bare, the original festive effect of the high-ceilinged rooms and their setting is apparent. Most of the bedrooms upstairs are conventional, but the master bathroom boasts a mural of an aerial view of the property and Sakonnet Bay. There, in the cove, is Taylor's yacht with a hydroplane beside it—faded image of good old days long past. The town of Portsmouth, which now owns the house, rents it for special occasions, the maintenance of its grandeur quite evidently a strain on the public budget.

PO12 Sandy Point Farm, Riding Ring, and Barn Complex

c. 1910, A. Stewart Walker. 3 Sandy Point Farm Rd.

Conical masonry posts flanking the road marked the original main gate to the Vanderbilt farms: Oakland, owned jointly by Cornelius, Alfred G., and William H. Vanderbilt and, behind it toward the water, Sandy Point Farm, owned by Reginald C. Vanderbilt in the early twentieth century. Oakland's buildings—including a villa, a powerhouse, greenhouses, garages, a polo field, and an enclosed driving ring (the latter built by Alfred G. Vanderbilt when he was a leading horseman)—are all demolished, although one of the field gates, with conical masonry posts supporting gates in openwork timber, remains in near perfect condition.

Although the enclosed riding ring at Oakland Farm is gone, the one at Reginald Vanderbilt's Sandy Point Farm ring remains, actively used if in precarious condition. Shingled, with overblown Colonial Revival detail, it consists of three major parts. First, toward Sandy Point Avenue, a two-storied wing with what was once the principal entrance to the barn, through a passageway with tack rooms and stalls on either side and apartments for grooms overhead. Then the dirt ring itself, within a broad, column-free space spanned by scissor trusses, with more stalls and entrances to either side. At the far end is a viewing balcony, the first story painted white (although now dim) to contrast with the dusty wood tones elsewhere, and dignified with a few Neo-Colonial details, thus suggesting a mini-building within the big space. Its privileged audience once reached the balcony by way of a sunken formal garden at the far end of the barn, thence across a porch with a pergola and into a generous reception room with rustic fireplace and accompanying trophy room, kitchen, and other entertaining facilities. Behind the ring, and set a little apart from it, a water tower marks a barn courtyard in random stone and shingle with living quarters for workers upstairs. Faded though it is, the ring at Sandy Point deserves preservation as the most conspicuous monument remaining to the enthusiasm for horse shows during the golden years of Newport's high society.

PO13 Sarah King Birckhead House
(Eastover)

1901–1904, Irving Gill. End of Eastover Rd. (not visible)

The California architect Irving Gill designed three houses in the Newport area. His Newport connections apparently began at the famous Hotel del Coronado in San Diego, where he met the Mason sisters, Ellen and Ida, who summered in Newport and wintered in California. They in turn introduced him to the Olmsted clan—Frederick, Jr., Albert, and Marian—when this trio was staying at the hotel. As an avid gardener, Ellen Mason had in fact commissioned garden plans from the Olmsted office. The sisters called on Gill for a house in Newport (now St. Michael's School). From this commission, Gill derived two others, a second in Newport (the Albert H. Olmsted House) and Eastover, designed in a shingled style recalling earlier nearby works by McKim, Mead and White, very simply handled. The dominant exterior feature is a spreading frontal gable with strongly projecting eaves which vertically organizes the front elevation of the house. Inside, rooms open widely into one another, with the severe treatment of woodwork characteristic of Gill's designing toward essentials. Unhappily, the original natural wood finish has been painted. A beautifully landscaped approach drive has been associated with the Olmsted firm; but there seems to be no proof that they designed it.

PO14 Vaucluse Farm

1784 and later. 340 Wapping Rd.

A complex of large, white gambrel-roofed dairy barns (c. 1925) is the agricultural focus of Vaucluse Farm, one of the first large-scale gentlemen's farms in the area. About 1760 Metcalf Bowler established a country seat near Wapping Road, famous during the colonial period, but it did not survive. Vaucluse, built by Garvais Elam in 1784, has survived, sometimes barely, down to the present. Elam spent $80,000 on the original house and formal gardens, importing specimen trees and shrubs, and setting out six miles of winding walks to create a scenic ambience for his livestock operations. The existing main house is a Georgian Revival house of 1938.

PO15 John Barstow Farm (Greenvale)

1864–1865, John Hubbard Sturgis. 582 Wapping Rd. (open to the public)

John Barstow's farm, now operated as a vineyard, typifies the mid-nineteenth-century retiree's dream of an idyllic country seat with enough farming on the property so that he could continue to exercise some part of the entrepreneurial skills which had already brought him a fortune—in Barstow's case, in the Boston China Trade. Barstow picked the pretty site of an old farm for his *ferme ornée*. The approach road descends into a green vale of trees, then rises slightly, past the old Federal farmhouse (which Barstow reserved for his help) to a cleared area on the water for which he requested of the rising young Boston architect John Hubbard Sturgis a house, stable, and barn (the latter demolished). Sturgis had just returned from England, the first American-born architect to study there (as distinct from the many English emigré architects who chose to practice in the United States). Greenvale Farm was among his first commissions. The main house is Stick Style—that is, half-timber construction with a mix of clapboard, vertical boarding, and shingling infill in carpentered compositions which use structure and contrasting materials to suggest that the result derives from the process of building itself—as it does, but for a fanciful, picturesque rather than laconic utility. But Sturgis would have known of such house design from suburban and resort houses abroad, where brick, stone, and stucco predominated instead of wood.

For Barstow, Sturgis designed a fairly compact, roughly L-shaped house which seems more rambling than it is because of the variety of its roof treatment—gables, cross gables, jerkinheads, and timberwork vergeboards, fitted snugly to the main mass of the building—and ran an elaborate open timberwork porch around three sides of it. Between 1890 and World War I, the house stood empty and dilapidated, until a later generation of the family revived it. Unfortunately, they stripped away the porch, partly because it had rotted, partly because it darkened the interior. As a result the clapboard base, topped by vertical "Swiss boarding" (board and batten with a sawtooth skirting), appears somewhat sheared. But the interior remains, its dominant feature being a wide hall floored with English tiles, highly prized at the time, off which the major rooms open.

PO16 St. Mary's Church

1847–1849, Richard Upjohn. 1866, Sarah Gibbs tomb, Richard Morris Hunt. 324 East Main Rd. (south of Sandy Point Ave.)

St. Mary's is the gift of Sarah Gibbs, who not only made the gift of the church, but also bequeathed to it her eighty-acre farm, Oakland (the name later adopted for the principal Vanderbilt farm, which once surrounded the church and its graveyard). Previous to providing for St. Mary's, she had already called on Richard Upjohn to design another Episcopal church in Middletown, Holy Cross (see entry, below). Whereas Holy Cross is shingled wood, St. Mary's, of local fieldstone, takes its cue and its appeal from the tradition of English country churches. It reflects Upjohn's active interest as an architect and pious Episcopalian in the reform movement promoted by the English Ecclesiological Society, which called for a return to medieval tradition in church liturgy, building, and fittings to counter eighteenth- and early-nineteenth-century tendencies toward the simplification of High Church ceremony and ambience. Specifically, St. Mary's derives from St. James the Less in Philadelphia (1846–1848), a key church in the ecclesiological movement in America. Behind this, it seems to be closely related to St. Michael's, Longstanton, in Cambridgeshire. In accord with ecclesiological principles, St. Mary's makes an explicit statement of chancel, porch, and sacristy as visible projections from the main box of the church. Al-

PO16 St. Mary's Church

PO18 Prescott Farm

sponsored, under a casketlike memorial from yet a third major American designer, Richard Morris Hunt.

PO17 **Julia Ward Howe Home** (Oak Glen)

1858, 1870. 745 Union St.

Dr. Samuel and Julia Ward Howe bought a small farm cottage in the 1870s which became the rear ell for the Victorian enlargement in front. A house of more historical than architectural consequence, it nevertheless boasts a forthrightness of character in accord with its role as a meeting place important in late-nineteenth-century social reform.

though it is petite and picturesque, its compact massing and splendid masonry craftsmanship give it monumentality. Especially is this the case at the west end, where the buttressing steps out to either side of the diminutive door in a projecting mass and merges with the buttressing stepping up the wall of the main body of the church behind, together fusing in a delightfully sculptural way with the bell cote. Hence the bell cote is not an adjunct element, but integral with the forms below.

Entrance to the church is through the side porch rather than the front portal. The plaster interior, with wooden gable rafters and wall brackets forming simple roof supports, is appropriate if unexceptional. Sarah Gibbs gave the memorial tablet opposite the entrance door in memory of her parents. In the marble relief, they appear in ancient Roman garb. A nude angel of death leads the husband and father away (he predeceased his wife by twenty years), while she mourns at an urn. It is the work of Horatio Greenough, commissioned by Sarah Gibbs when she was in Florence and completed in 1843 (before the building of the church), so the work of an eminent neoclassicist here joins the work of an eminent medievalist. Sarah Gibbs herself is buried immediately behind the chancel of the church she

PO18 **Prescott Farm** (Overing House)

c. 1710. 1812, windmill. West Main Rd. (owned by the Newport Restoration Foundation)

Prescott Farm, partially in Portsmouth but mostly in Middletown, is so known because General Richard Prescott, commander of British forces in Rhode Island, was headquartered here when, on the night of July 9, 1777, he was captured in a daring clandestine operation by a force of some forty Americans. His capture occurred in the large house adjacent to what is now called Prescott Farm. But it is the early-nineteenth-century shingled windmill that is the center of attention here. Originally in Warren, it was moved twice to different hillside sites in Portsmouth and then just over the Middletown side of the boundary between the two towns. It is one of three still surviving in the state—others are in Jamestown and Middletown—which celebrate a sight once common along the Rhode Island coast, wherever abundant breeze had to make up for the lack of swift-moving streams as a power source. The other spectacle here is, of course, the farm animals, especially the gaggle of geese—and the triumph of amateur photographers when house, windmill, and geese all come together in a passable composition.

Middletown (MI)

The history of Middletown as a pretty farming area adjacent to Newport is similar to that of Portsmouth, except that its very substantial farms never developed to estate proportions. It was part of Newport until the differences between the two, and charges that

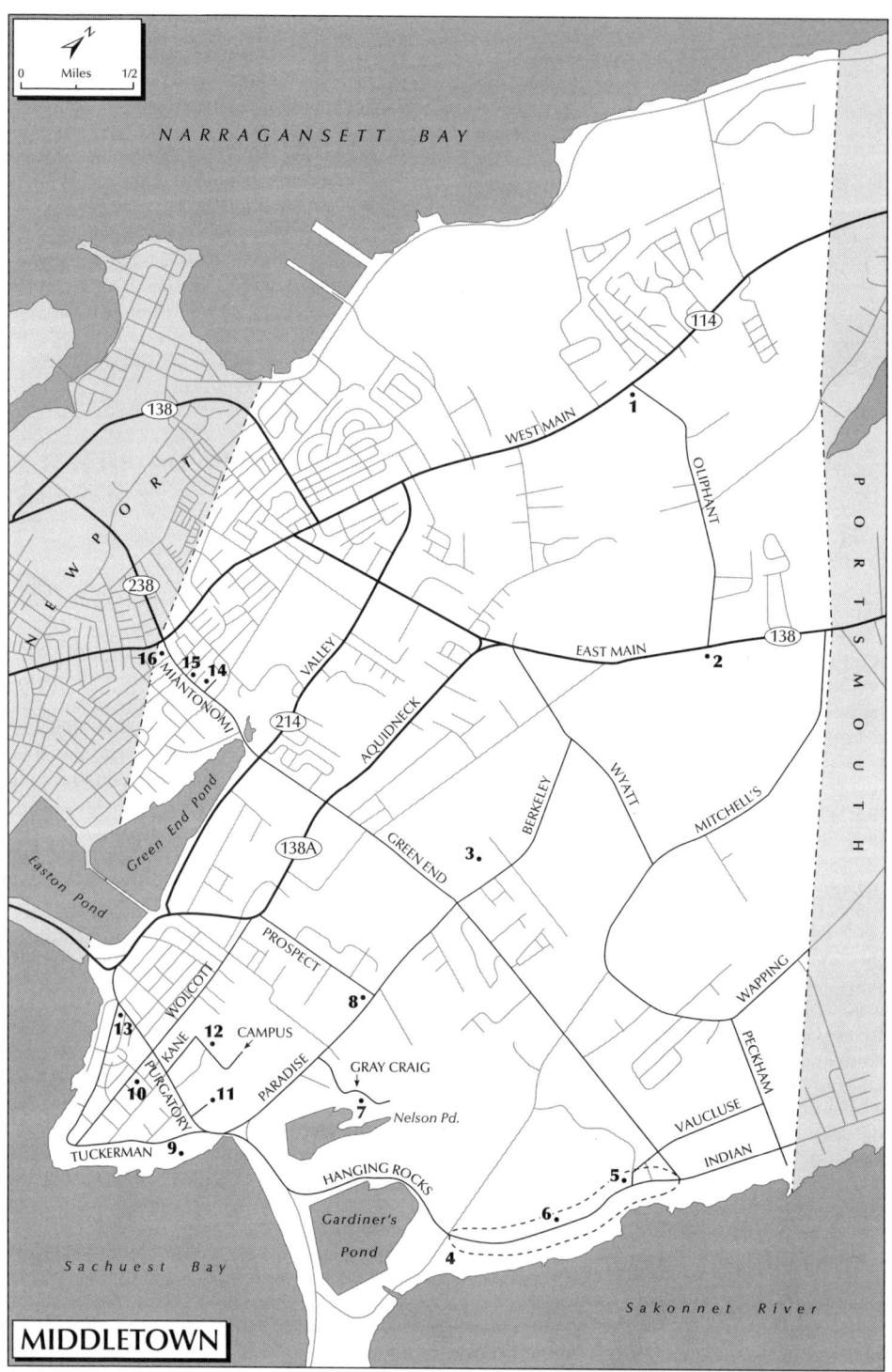

Middletown's rural interests went unheeded in the more worldly concerns of Rhode Island's southern port, led to a split in 1731. It was its peace and remoteness that drew George Berkeley, Dean of Derry in Ireland, to Middletown, where he sojourned at his farm, Whitehall, from 1729 to 1731, while he waited in vain for funds to establish a college in Bermuda. It was testimony to the reputation of the area's agriculture, even then, and of the coastal shipping that spread its bounty, that he envisioned the farm as a supplier for the college. As with Portsmouth, summer residences came to Middletown toward the end of the nineteenth century. A number of these remain, especially on the Sakonnet side of the town, modestly large houses for the most part, comfortable rather than showy, with a secluded quality which has persisted to the present. Those built on the bay side of the town were swallowed up by naval installations, which during World War II came to usurp Middletown's western shore almost completely, dominated by the piers at Coddington Cove for destroyers. Areas devoted to naval housing, vaguely Neo-Colonial, exist along streets off Route 114 from Constitution Avenue south to Coddington Highway. High-tech marine industry has come to the vicinity of the naval base or on land vacated by the navy. Newport Airport has taken over fields between Routes 114 and 138. Farms have survived to a surprising degree down to the present but are increasingly moribund (except for some active nursery establishments). As in Portsmouth, the pressures of development and of traffic along major arteries are current concerns.

MI1 **Church of the Holy Cross**
1845–1848, Richard Upjohn. West Main Rd. (at Oliphant Ln.)

Two years before Sarah Gibbs made a gift of St. Mary's Church in Portsmouth (see entry, above), also by Upjohn, she asked him to design this delightful country church. In contrast to the masonry design of the Portsmouth church, this is a frame construction with fishscale shingling in the rural manner which Upjohn recommended for the most economical churches. The gable- roofed church is fronted by a projecting porch, also gabled and interlocked by a stubby tower. This is mostly an open belfry, each face with paired openings shaped as pointed arches, bound together by ribbonlike extensions from their apexes, which overarch the two below. A mere witch's cap of a steeple topped with a cross provides the climax. Inside (meticulously restored in 1970), all is pristine carpentry scaled to the tiny space, with altar and reredos dating from 1895 as the sole alterations from the original design. This church ranks with South Ferry Congregational Church in Narragansett (NA2), another fishscale-shingled church, as one of the most charming mid-nineteenth-century wooden country churches in the state. It is amazing good fortune that two country churches of such outstanding quality as St. Mary's and the Church of the Holy Cross, designed by one of America's leading church architects, should appear within a ten-minute drive of one another.

MI2 **Sandpiper Cottages** (Floradale Motor Court)
1929. 985 East Main Rd.

Of few remaining motor courts from the 1920s and 1930s, this well-maintained establishment probably presents the most convincing image of the type. A dozen clapboard boxes, wide enough only for a door and a window, make a wavering line under the trees, with the office in front. Low gables, front to back, stretch out to cover mini-porches, each with its chair.

MI3 **Whitehall**
(Dean George Berkeley's Farm)

c. 1729. 311 Berkeley Ave. (open to the public)

George Berkeley, then Dean of Derry in Ireland, later Bishop of Coyne, philosopher, poet, and philanthropist, traveled to North America in 1729 to establish a college in Bermuda. His ship was blown so far off its course that it landed in Newport. He liked Newport suffi-

MI1 Church of the Holy Cross

MI2 Sandpiper Cottages (Floradale Motor Court)

ciently to settle with his wife in the rural quiet of its farmlands while waiting for funding for his project. He determined to use the time to advantage by revamping an old farmhouse into a new one—calling it Whitehall in remembrance of the royal palace in London—with the idea that it might eventually supply his college with food. More immediately, he also used the time and the secluded spot for reflection. Here he wrote, among other things, a portion of *Alciphron, or, the Minute Philosopher*, a defense of the Christian religion against Deist free thinkers, which also criticized certain English customs by imagining their observance in the arcadian innocence of Rhode Island. Reportedly, he often meditated on Hanging Rock, giving it its alternate name of Berkeley's Seat. His life had its gregarious aspects, too. During his sojourn there, Whitehall became a center for intellectual life in Newport and inspired the founding of Redwood Library.

Berkeley's adaptation aggrandized an existing farmhouse into a two-story, five-bay, central-chimney house with a lean-to behind. Although its overall character was vernacular, several characteristics gave it a lordly air: its roof, hipped instead of the commonplace gable; modillions at the eaves; and, most conspicuous, its broad, two-leaved door enframed by Ionic pilasters and pediment—more suitable as the entrance to a public building than to a farmhouse. When it became clear that funding for his college was unavailable, Berkeley returned to England, leaving his 96-acre farm and its library to Yale College. It was renamed Vaux Hall in 1769, served for a while as an inn, then gradually fell into disrepair. It was in this ruined state that Charles McKim discovered it in the 1870s. Then a young architect working in Newport, he was a newly converted enthusiast of the revival of colonial architecture and an editor of a short-lived architecture magazine, *The New-York Sketch-Book of Architecture*. He published a photograph of the building in its dilapidated state in the December 1874 issue of the *Sketch-Book*, a view from the rear which focused on the slide toward the ground of the warped lean-to roof as something almost more topological than architectural. It was the first photograph of any subject ever published in an American architectural journal. At this evidence of the deteriorated state of Whitehall's fortunes, the National Society of the Colonial Dames of America came to the rescue by restoring the house in 1899, under a 999-year lease of the property from Yale. It was restored again in 1936 by Norman Isham, and yet again in 1966–1968. After its deterioration and triple restoration, it is hardly surprising that much of the interior is conjectural.

As for Berkeley's vista, it remained until the mid-1980s, when Whitehall Farm, a cluster development, was built between Whitehall and Hanging Rock. If a multitude of clapboard gables and dormers of varied size and disposition can make a "village" of mass housing, this is the evidence.

MI3 Whitehall (Dean George Berkeley's Farm)

MI5 St. Colomba's Chapel (Berkeley Memorial Chapel)

MI4 **Indian Avenue**

1870s. From Green End Ave. to Third Beach Rd.

Indian Avenue runs immediately behind a number of fine residences on properties overlooking Sakonnet Bay, most at least partially veiled from view. Eugene Sturtevant invested heavily in the area in the 1870s after summering here in 1872. Sturtevant and a partner were convinced that fashionable Newporters could be lured to their five-mile drive, which they called Indian Avenue because of a plethora of artifacts uncovered in the course of laying it out. Sturtevant's group guessed wrong. Fashionable Newport in the late nineteenth century moved in the opposite direction, along Bellevue Avenue. His real estate dream failed—and his drive was left to much more subdued development of country and summer homes than that which occurred on Bellevue Avenue. The J. T. Huntington House (1870s), 561 Indian Avenue, has a mix of half timbering and vertical board-and-batten siding, multiple gables, and extensive porches. The largest house on the street, number 500, is a fieldstone mansion in handsomely landscaped grounds, with service building, also in fieldstone (1930s).

MI5 **St. Colomba's Chapel** (Berkeley Memorial Chapel)

1884, Wilson Eyre, Jr. Parish hall, Albert Harkness, 1959. 58 Vaucluse Ave.

St. Mary's in Portsmouth, Church of the Holy Cross in Middletown, and now a third charming Episcopal country church, all less than half an hour from one another: such a conjunction of small country churches of such architectural distinction is unique in the state. St. Colomba's follows a formula comparable to that of St. Mary's: a masonry church set in a graveyard overarched by splendid trees, with bellcote, porch entrance at the side, and projecting chancel. But St. Colomba's is even more diminutive. It is not elevated with a long drive up to the church, but sits in a V-shaped intersection with a lych-gate entrance, suggesting that here one leaves vehicles behind and walks into the sacred precinct (although modern parking facilities make it possible to ignore the building's message). The masonry is rougher than that of St. Mary's; there is no ceremonial door at the front; the bellcote is smaller and simpler; the slate roof slopes closer to the ground. Inside is a comparable plaster and wood-beamed ceiling. The window to St. Colomba is by John La Farge. Maitland Armstrong, a disciple of La Farge, executed the chancel window dedicated to St. George, the patron saint of the Episcopal Church, as well as the tiny pointed stained glass windows ranged either side of the nave, each with a nautical sign. Opposite the porch entrance, a plaque commemorates the famous nineteenth-century actor Edwin Booth (brother of John Wilkes, Lincoln's assassin), who was an early summer resident of Indian Avenue. (His house, Boothden, is at number 357.)

Beside the lych-gate (to the left on entering) is the family plot of the architect John Russell Pope, famous for the National Archives building, the Jefferson Memorial, and the National Gallery of Art in Washington, and the designer as well of important houses, including The Glen in Portsmouth and The Waves in Newport. Adjacent are the graves of the Providence architect Albert Harkness and his second wife. Both stones revive the colonial practice of slate slabs hand-carved—in these instances by the Benson family, carvers in the John Stevens Shop in Newport, whose heritage extends all the way back to the early eighteenth century. It was (and still is) responsible for the execution of much fine architectural lettering, including that on a number of Pope's buildings.

MI7 Michael M. Van Bueren House (Gray Craig)

MI6 Reverend Coit Conover House

1888, Clarence Luce. 208 Indian Ave.

The Conover House is another of Clarence Luce's simply shaped shingle houses into which he managed to insert quirky and fascinating architectural incident. From the road it shows a big gable end without eaves and a loose window arrangement with a single one-story shallow bay window. From the "front" (at right angles to the road), the gable projects a deep, visor-like eave, under one corner of which a second-story bay window is folded. Below this the second-story shingled wall flares as an even more extravagant visor to shelter a porch with steps running the width of the house between stone piers. Four sets of cylindrical columns as close-packed triplets support the porch well under the flaring. Built-in benches to either side of the door divide the porch (for some unknown reason) into three separate sections. To the rear the shingled walls make a corner in a series of polygonal folds, tapered inward as the wall rises, as though inspired by an ancient windmill.

MI7 Michael M. Van Bueren House (Gray Craig)

1926, Harrie T. Lindeberg. 75 Gray Craig Rd.

Once the farm of one of the earliest families in Middletown, this is surely the town's most spectacular site. Gray Craig, once its grandest estate, was built with a fortune from Standard Oil after the property, originally owned by O. H. P. Belmont, was purchased by the Van Buren family. The approach road winds through a park before reaching a gatehouse with the superbly crafted stone walls and tiled roof which characterize the château-inspired buildings of the estate. Beyond the gate the road rises and winds a bit more before the garden front of the mass of the main house looms into view. It is partially screened by a tall hedge as the road slides by this initial view of the house, between piers topped by peacocks, and up to an arched porte-cochere off one side of the house. Passengers discharged, cars continue to a garage and kitchen court to the rear with its own family entrance. This approach is made more dramatic by a wall of natural stone that parallels the house beyond its other side—the "gray craig" from which this appropriately severe house (like the farmhouse before it) takes its name.

The house depends for architectural effect first of all on shape rather than ornament. The steep hipped roof of reddish tiles is devoid of dormers; only four chimneys poke through to stake out a claim for the house in its dramatic setting. On either side, stubby blocks, topped with urns, lock with the roof and create transitional elements to the low, gabled wings. Ornament, where it occurs in such austerity, creates maximum effect—especially the enframement of the center door of the garden elevation with its English Baroque segmental pediment on scroll brackets, the scrolls of the brackets echoed in the scrolls of the keystones of the principal windows. Other ornament is limited to wrought iron ornamentation of second-story balconies in the stubby blocks and the urns on top. For the rest, the house depends on craft and proportions. Grass terracing opens to a cut through the woods below to a view of the dis-

MI10 Lyman C. Josephs House (Louisiana)

tant sea. Like John Russell Pope's The Glen in Portsmouth (PO11) this derives from the smaller country château, with château forms used here even more than at The Glen as motifs for composition rather more then for architectural accuracy. Here, in fact, the influence is rather more English than French, recalling Sir Edwin Lutyens's free handling of château forms in some of his early houses in the mode. The fine interior woodwork of the rooms which front the long cross hall from the portecochere is also predominantly English in inspiration, with some exoticism in interiors, including, in the coatroom, a witty, scroll-like mural of the tribulations of building the house with architects, contractors, decorators, and workmen in *Mikado* costume, all cowering to Lindeberg as the imperious Poo Bah of the enterprise. In its heyday Gray Craig also included magnificent gardens (probably mostly laid out by Lindeberg)—a walled "secret" garden, an architectural garden with a fantastic sculptural sundial, an orchard as an outdoor theater setting—all worked into a larger naturalistic landscape in which rock outcroppings featured dramatically. Finally, Lindeberg also provided an elaborate kennel with mini-château for its keeper. The specialty—another bit of Orientalism—were purebred Pekinese.

MI8 Paradise School

1876. Paradise and Prospect aves.

This fine example of a Victorian country schoolhouse has bracketing under the eaves and hoods (both single slope and gabled) over windows and doors on diagonal bracing with an oculus in the front gable. It is now occupied by the Middletown Historical Society.

MI9 John Bancroft House (The Bluff)

Before 1883. 575 Tuckerman Ave.

The ocean bluff, with nearby Purgatory Chasm, a terrifying slot in the cliff approximately 10 feet wide and 120 feet long, its sheer walls dropping 50 feet down to water, also provides the outlook for this forcefully organized shingled house with gable format encompassing the front elevation. From this an entrance shelter protecting a steep flight of stairs projects boldly, its undulant roof and latticing creating a vaguely Oriental aura. Number 595 Tuckerman, once its carriage house, is now a residence.

MI10 Lyman C. Josephs House (Louisiana)

1882–1883, Clarence S. Luce. 438 Wolcott Ave.

A wealthy Baltimore family commissioned this long, narrow, gambrel-roofed house of fieldstone and shingle and retained it up to World War I. Originally a still visible arch separated the house from its attached stable, which was subsequently converted to residential use. The porch, folded around three sides of the street end of the house and tucked in under the flaring eaves, was completely open. Using the large folded shape and flaring eaves to organize his composition (now expanded, it appears, by later additions), Clarence Luce freely employed variously shaped dormers and bays for picturesque and functional effect.

MI11 Jacob Cram–Mary Sturtevant House

1871–1872, Dudley Newton. 438 Abe Meyer Ln.

Vincent Scully so celebrated this house in *The Architectural Heritage of Newport, Rhode Island* (1951) as the epitome of the virtues of what he termed the Stick Style that those who know the book may want to see the remains of the building—still recognizable, but barely so, under changes wrought by its condo-ization. One can still sense the original qualities: flamboyant half timbering as a picturesque allusion to underlying structure, shape determined by functional considerations, interior extended outside by rambunctious verandas and balconies, with exotic touches from Japan and Turkey to boot. Here is its carcass, its once grand isolation in the panoramic view further demeaned by a circle of small development houses which cuddle up to it.

MI12 St. George's School
1901 and after. 372 Purgatory Rd.

Panorama is the essence of this campus on a high hill overlooking water in three directions, although the huddled arrangement of its buildings seems (necessarily perhaps) to respond more to winter winds than summer views. In this complex Diman Hall is interesting; the chapel, hemmed in by the huddle, is remarkable. Diman Hall (1927, Howe and Church), Neo-Colonial on the exterior, is especially notable for its paneled medieval refectory, a grand, if somewhat stark, space with all the accoutrements that such a room should have: huge fireplace, musicians' gallery, wrought iron chandeliers, pennants, and big bay windows with leaded glass. A movie maker desiring such a setting could hardly ask for more, down to the carved rosters of graduating classes panel by panel. The refectory is named for the school's founder, who began his career in 1896 as pastor of St. Colomba's when it was known as the Berkeley Memorial Chapel, left to found a school in Newport, and moved it to Middletown. Converted from Episcopalianism to Roman Catholicism, he started all over—founding the Roman Catholic Portsmouth Abbey School in Portsmouth in 1926.

But the chapel (1928, Ralph Adams Cram) is the real architectural reason for visiting St. George's—one of Cram's finest designs. From modest Eyre to flaunting Cram: the differences between these two neighboring chapels could

MI12 St. George's School, Chapel

hardly be greater. It was the gift of John Nicholas Brown, a graduate of the school in 1918, whose wealth permitted him to make this magnificent benefaction while he was a mere sophomore at Harvard and under the spell of lectures by the notable medieval architectural historian A. Kingsley Porter. John's mother, Natalie Bayard Brown, had commissioned Cram to design a church in Newport and then her Newport mansion. So if a medieval church was needed, who else for its architect but the period's leading designer in medieval revival styles, who was also a friend of the family?

The result is an elaborate derivation from English Perpendicular Gothic. Because of the crowding of the chapel by other buildings, its most prominent feature, far or near, is the tower. Contextual allusions are manifold in the sculpture for the exterior. For example, the heads of Brown, Cram, and four members of the building committee look down from the top of a stair tower toward the playing fields behind the chapel, and a baseball and a football player flank a doorway nearby. Here and there sculptural bits celebrate popular expressions or fads of the time, like the "lounge lizard" and the enthusiasm for crossword puzzles. On the sea side of the chapel, now cloistered by a range of buildings in front of it, the sculptural theme is a history of seagoing vessels in celebration of the donor's own love of the sea, with images of a Viking ship, the *Santa Maria*, Robert Fulton's steamboat, and the U.S.S. *Colorado* (on which John Nicholas Brown had served). At the donor's door are the saints John, Nicholas, and Natalie. And on and on. The glory of the chapel, however, is inside. Like such famous English predecessors as the chapel at Trinity College, Cambridge, it is virtually all choir (without nave), the student pews facing one another across a central tiled aisle. At the end, three steps lift the square-ended altar a little above the congregation. Also recalling the chapel at Trinity College, the upper reaches of the walls are so extensively glazed that the "wall" area between the windows is reduced to the piers which support the vaults. The complex decorative use of round colonnette profiles throughout culminates in quadripartite vaulting of an earlier Gothic. Rows of crystal chandeliers hanging over the pews enhance the glassiness. Although all the work is a bit mechanical in its handling, fine craftsmanship, detailing, proportion, and luminosity predominate. The young Brown took an avid personal interest in the design, even turning his hand to some of the details.

MI15 Hamilton Hoppin House

Cram recorded this interest appreciatively in his *My Life in Architecture*, where he ranked Brown as among the most inspiring friends he had had.

Also on St. George's campus is J. D. Johnston's Wyn Wyc (1885), designed for Judge H. W. Bookstaver, probably from published plans.

MI13 Land Trust Cottages

1885–1887, site planning, Frederick Law Olmsted. 1887–1888, construction, E. B. Hall, builder. Purgatory Rd. (Second Beach Rd.) (at Easton's Beach)

For Boston investors at the Middletown end of what is mostly a beach within Newport, Frederick Law Olmsted designed the site plan for four medium-sized cottages. Their grouping on a very constricted flat site is especially happy for the way in which all receive light, view, and air, as well as for the nice way in which building and space interact in a balanced interrelationship. The use of two gable and two gambrel roofs for variety and the varied but harmonious use of vernacular porch, window, and stone chimney elements in common have many lessons to teach developers of resort colonies and condominium "villages"—as does, alas, the needless replacement of original windows which some of these cottages have suffered.

MI14 Alexander Van Rensselaer House (Restmere, Villalou)

c. 1858, Richard Upjohn(?). Restmere Terrace and Ichabod Ln.

MI15 Hamilton Hoppin House

1856–1857, Richard Upjohn. 120 Miantonomi Ave.

These side-by-side variants of the Italian villa, the earlier on a larger lot, use the same scheme: the same tall-arched porch design with a lower arch and panel to frame the arching, three windows across the front elevation, and narrow, friezelike windows under hipped roofs. The earlier Hoppin House is the more rigidly organized, within a reticulated half timbering which, up above, seems to simulate in the X-ed band a railing in accord with the decks below. By the porches and simulated railing Upjohn seems to have intended a yachting allusion for this quite monumental house. The Van Rensselaer House, on the other hand, leaves the clapboarded wall at the second story an open plane, treating the X-ed band as a bracketed frieze. Whereas the earlier house reinforces the tightly bound quality of its elevation by the contained rectangularity of the porch and its steps, in the latter house these are both swollen into curves, while the simple one-story bay windows on the side elevations become more complex two-story elements. Did Upjohn really design this house, or was it a local variant on the next-door example? The exuberant curves of the porch seem out of character with the sobriety of Upjohn's taste.

MI16 William Vernon House (Elmhyrst)

1833, Russell Warren. Miantonomi Ave. at Broadway

When William Vernon, a wealthy Newport merchant and a descendant of the owners of the famous Vernon House in Newport, established his country estate on a hill just beyond the town's border, at One Mile Corner (named for the distance from Colony House), he had the hill and its vistas pretty much to himself. Today this Greek Revival house is crowded against a street corner, its arcadia chopped up by development. What may once have been the finest Greek Revival house in the state is also missing its attic structure. Its major portico is on the garden front, looking downhill toward what was once its park (and hence now appears to be its rear elevation). Four beautifully detailed Ionic columns are inset in a shallow two-story porch closed by paneled blocks on either side. (This was topped, until the 1960s, by an attic story centered in three square windows the width of the interspacing of the columns with paneled frames the width of the columns in between. A plain parapet flanking the triple windows terminated at either end of the elevation in paneled boxes with acroteria.) Two other Grecian mini-buildings complement the main house: a porter's lodge with a Doric front and Vernon's country office in a temple shape surrounded on three sides by Corinthian columns. So this onetime nostalgic idyll embraced the full gamut of the Greek orders.

Newport (NE)

[Editorial note: WHJ's manuscript included no introductory text for this section, but the extensive treatment of Newport in his draft for the general introduction serves the purpose well.]

Broadway and the City Center

Broadway, which still substantially retains a nineteenth-century residential character along most of its length, is not the introduction to Newport but is the most direct route to its heart. The termination of Broadway in one corner of Washington Square is the civic center of the old city, which is really a long, narrow triangle with the Colony House at the base and the Brick Market at the apex, once (before land filling) on the water's edge among the wharfs, with a triangular park between the two public buildings. Government slightly elevated, commerce subordinated but proudly housed: no preconceived plan ordained this civic focus of funneled space with monuments at either end, but rather community intuition as to what the center should mean.

And what can be said of the heart of the old seaport? Not much architecturally. Newport's wharfs, from Long Wharf to Perry Mill, evolved, from the late 1960s through the 1980s, into a hectic carnival of boutiques, eateries, and hotels. A deliberately nondescript wharf-like vernacular has been cultivated even for outsized hotels, with carved signboards everywhere. Fragments of old stores and commercial buildings of some interest exist along Thames Street, but mostly so gutted, altered, and faked that their integrity is gone. The necessity of routing traffic through a four-lane artery and of providing parking spaces has inevitably destroyed the old scale of things. The new center is all development, needless development, with neither imagination or regard for public amenity to stamp it as something unique—to make it Newport. If Newport means a crowd of people in a festive, spending mood, then here it is; but for Newport as a place with character and history, one must look elsewhere.

NE1 House

c. 1870. 450 Broadway

Although somewhat altered, this middle-class residence gives a nice sense of a prevalent type of late Victorian gable stickwork which has been too rarely preserved. The eaves of the gabling project deeply on simple prop brackets, with board-and-batten ornamentation of the attic. In the apex of the front gable is a fan of strutwork and, in the same location on the side, a pierced screen. The propped and cross-braced elements forming the porch combine structure and ornamentation in a most straightforward manner.

NE2 First Presbyterian Church

1892, J. D. Johnston. 167 Broadway

"An imposing pile": so contemporary descriptions might have characterized this tall, spreading, hip-roofed mass of quarry-faced masonry with its miscellany of gables and stubby turrets. Windows are rather spottily located, with a giant wheel window stealing most attention. It

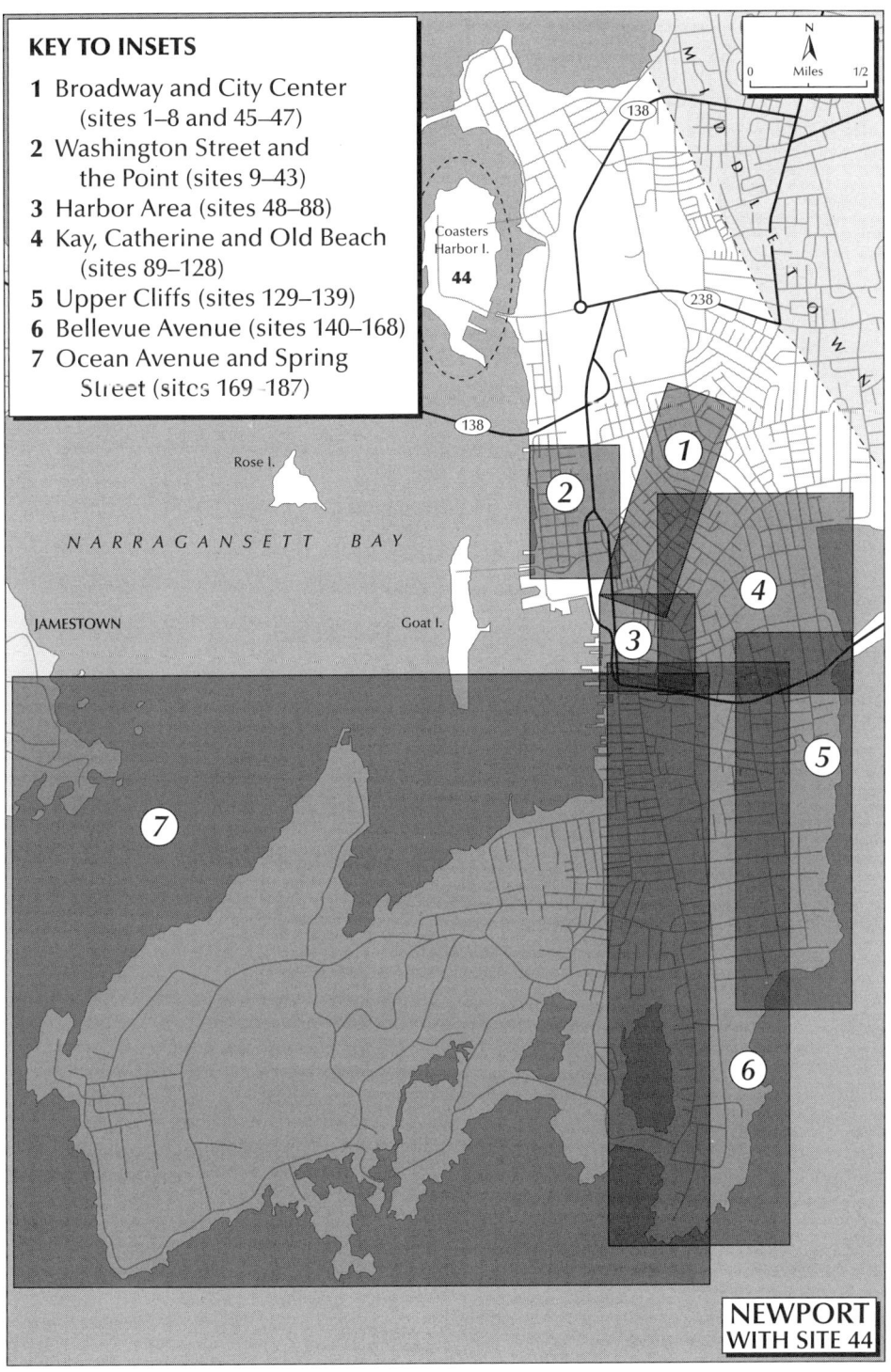

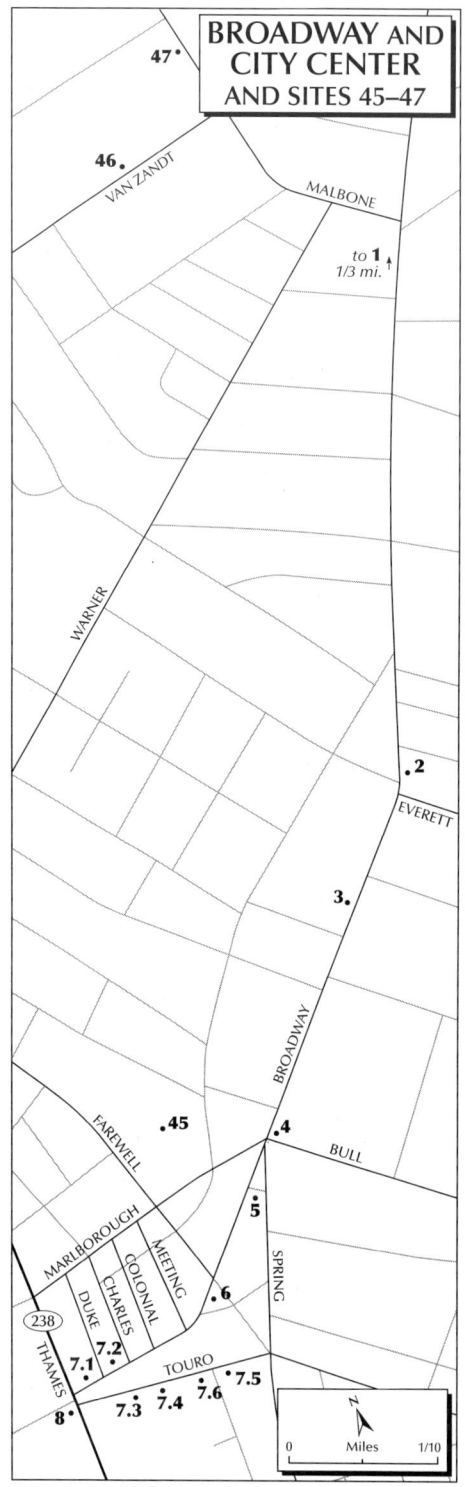

NE5 Wanton-Lyman-Hazard House (Stephen Mumford House)

is a very belated Richardsonian Romanesque design by a prominent late-nineteenth-century local builder who eventually combined this trade with architectural design.

NE3 Tisdall Block

c. 1900. 130 Broadway

The Tisdall Block, a well-preserved Queen Anne store block with apartments above, is the only one of its kind on Broadway which retains something of its original appearance.

NE4 Newport City Hall

1898–1900, J. D. Johnston. 1927, interior and top floor destroyed by fire; rebuilt, W. Cornell Appleton. 43 Broadway (at Bull St.)

Another masonry block, this one mansarded, is the city hall. It has mansarded towers in each corner and one at the center of the front elevation with a second-floor porch for imagined mayoral proclamations, and originally a circular cupola on top. A long slope for the entrance stairs was meant for majesty of approach. Inside, a watercolor perspective of the original building from Johnston's office hangs in the much revamped first-floor hall.

NE5 Wanton–Lyman–Hazard House
(Stephen Mumford House)

Before 1700. c. 1765, altered. 1927, restoration, Norman M. Isham. 17 Broadway. (owned by the Newport Historical Society; open to the public)

NE6 Colony House

Although this important colonial house is popularly known for the three families who successively occupied it after 1765, it was built by a minister, Stephen Mumford, before the beginning of the eighteenth century. It was subsequently occupied by, among others, Richard Ward, a colonial governor, and Martin Howard, a Loyalist pamphleteer. Howard was one of three Tories burned in effigy in Washington Square by enraged patriots when news reached Newport of the Stamp Act of August 1765. They moved on to attack his house nearby. He slipped away to the British sloop *Cygnet*, anchored in the harbor, and never returned to Newport. John Wanton bought the house as confiscated Tory property at auction. After his daughter married David Lyman, a major in the American army, the father gave her the house. Her daughter married Benjamin Hazard, and the house remained in the Hazard family until it became a museum house in 1927. Recent archaeology on the site has revealed significant deposits from these various eras of ownership and from adjacent lots. It is the most important surviving medieval house in Newport, although later modifications disguise the fact and some scholars date the house as late as 1715.

On the exterior, its most conspicuous feature is the plastered cove overhang at the eaves of the front elevation, with the slope of the roof above "kicked out" in an angled curve to accommodate its shape. (Later, as Renaissance taste came to the colonies, classical cornices would replace such scooped overhangs.) Apparently Wanton added the classical pedimented door.

Inside, Norman Isham's restoration maintains a sense of the layers of alteration through which the house has passed. (A huge brick central chimney, pilastered in the medieval manner at the top, divides the house in half and rests on a broad brick vault in the basement.) The two-room floor plan is common in seventeenth-century Rhode Island and Massachusetts houses: one room is the "hall" (the area for everyday living), the other the "parlor" (the area for special occasions), each with its sizable fireplace off the chimney. The same plan is used for the upstairs bedrooms, which are reached by a traditional medieval box of stairs folded in three flights, fitted into the tight entrance vestibule in front of the chimney breast. Isham exposed the original heavy medieval house frame in the northern chamber to record its original appearance. Elsewhere it is sheathed in wide vertical tongue-and-groove boarding, parts of which were (at some time) marbleized and painted to simulate paneling in classical fashion. The stairs, while cramped, are more expansive than those in earlier houses. The turned balusters, in a classical vein but stubby in their boxed situation and bulbous in the Jacobean manner, are characteristic of late medieval taste just beginning to discover classical forms. Moldings, too, show this moment in time, heavy and protruding in the medieval manner, but adopting profiles suggested by classical precedent. A diamond crossbar pattern painted on the walls of the kitchen ell early in the eighteenth century was fairly common at the time; this is one of the very few examples which have survived. This capture of a momentous shift in architectural taste, partially original to the house, partially the result of successive renovations—accounts for the fascination of the Wanton-Lyman-Hazard House.

NE6 Colony House

1739, Richard Munday. Head of Washington Sq. (open to the public)

Although he always referred to himself as a "house carpenter," "housewright," or "innkeeper," Richard Munday has the honor of having built both the finest civic building (the Colony House) and the finest church (Trinity) in his native city—both key buildings in the history of colonial architecture. Munday knew the baroque classicism of Sir Christopher Wren and his disciples from books and engravings, including the somewhat Dutch aspect of the buildings, which, indeed, indicates a source of inspiration for the "Wrenaissance." In *The*

American Scene (1907), Henry James described the Colony House as an "edifice ample, majestic, of finest proportions and full of a certain public Dutch dignity, having brave, broad high windows, in especial the distinctness of whose innumerable square white framed panes is the recall of some street view of Harlem or Leyden." He might have added that its brick with rusticated brownstone trim contributed to the Dutchness of the image—in itself an extraordinary commitment to civic monumentality in a city almost wholly wooden when the Colony House was built. "Finest proportions" might be questioned; remarkable as these are, the individual elements are rather too assertive to have been finely calculated. The overblown scale of the windows with their bristling quoining, the prominence of the wheels and the octagonal clock face in the pediment, the brusque clip of this pediment and the roof, the rotund lantern on top, all have the qualities of the workman's eye, producing something which is "right" in the forthrightness with which the provincial takes hold of the grand image. By contrast, the double doors at the center of the balcony with its twisted balusters and the broken pediment behind (both balcony and upstairs door probably derived from the demolished John Hancock House of 1737 in Boston) seem a little overrefined, as though the workman brought a different focus to the frontispiece. (The superb carving of the broken scroll pediment with a pineapple—symbol of hospitality but Newport's city symbol as well—flanked by sunflowers and leafage—is the work of Jim Moody, carved in 1784.)

The interior of the first floor is wholly devoted to the Great Hall of the Civil Court, which has a single row of square Doric columns straight down the middle of the space—a provincial positioning, again with the aura of the provincial designer working toward grandeur. A two-flight stair with a splendid array of balusters rises in one corner—more provincial positioning—to the second story, simply divided into two rooms, both with double fireplaces, the Chamber of Deputies (which became the Chamber of Representatives) and the Council Room (subsequently the Senate Chamber). Paneling in the Senate Chamber is especially handsome and seems from documents originally to have been left natural (along with other paneling?). If so, it was painted in the eighteenth century.

The Colony House replaced a plain gabled clapboard building erected in 1687 and doubled in 1711. In a state with fiercely independent towns, it was one of five capitol buildings through which the state legislature eventually rotated. The finest of the lot, it would undoubtedly have helped to insure Newport's selection as the ultimate capitol of the state, except for the city's decline in the nineteenth century relative to Providence. But it retains the honor of having provided the setting for some of the most momentous events of the colonial period and the Revolutionary War in Rhode Island. The first public reading of the repeal of the Stamp Act occurred from its balcony in 1766, and the first public reading of the Declaration of Independence ten years later. It was a barracks for British troops during the British occupation of Newport and a hospital for their French successors. These armies so badly abused the building that it needed its first restoration once they had left. But its supreme moment in history occurred the evening of March 6, 1781, when General Washington stepped ashore at Long Wharf to bid farewell to General Rochambeau and his troops and proceeded up Washington Square to a banquet in the Great Hall. Every window in the Colony House and every window in the center of the town was lighted with candles. Cannon from the French sloops in the harbor boomed out such a salute that one resident wrote, "I never felt the solid earth tremble under me before.... It was one continuous roar, and looked as though the very bay was on fire."

NE7 Washington Square

Worth a glance on Washington Square are (along the north, or to the right looking out from the Colony House) number 8, the John Rathburn–George Gardner–Abraham Rodrigues Rivera House (NE7.1. c. 1722, altered c. 1740 and c. 1950), an older house modernized, like the Wanton-Lyman-Hazard House (see above), in the mid-eighteenth century. It became the Newport Bank in 1804. Next door, at number 10, is its continuation, the Bank of Newport (NE7.2; 1929), a good example of a Beaux-Arts alternation of colonnade and arch to make a banking room. On the left (south side of the square), at number 29 Touro Street, is the Peter Buliod House (NE7.3; c. 1755), one of several Newport houses with a rusticated wooden front. In 1795 it housed the Rhode Island Bank, the oldest in Newport, before Commodore Oliver Hazard Perry bought it in 1818. Number 39, the Joseph Rogers House (NE7.4;

c. 1798), like number 51, the Wilbour-Ellery House (NE7.5; c. 1801), is a large, somewhat bare three-story, hip-roofed house with a fine fanlighted and pedimented entrance salvaged from another building in 1976. Finally, number 49, the Jane Pickens Theater, formerly Zion Episcopal Church (NE7.6; 1835, with numerous alterations to 1976), designed by Russell Warren), once boasted a pure temple facade with a splendid freestanding Ionic colonnade across the entire front.

NE8 Brick Market

1762–1763, Peter Harrison. 1928, restoration, Norman M. Isham. 1993, renovation as museum. 127 Thames St. (foot of Washington Sq.) (open to the public)

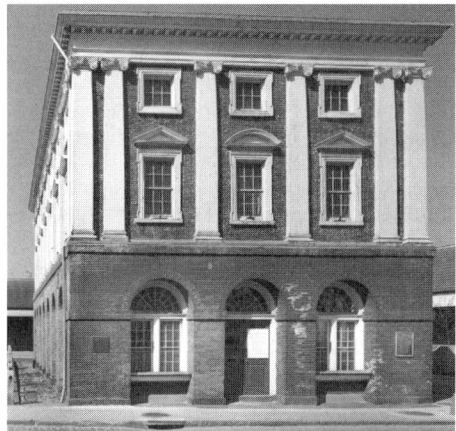

Peter Harrison, the finest designer in colonial Newport and among the finest in the colonies, was nearly the professional antithesis of Richard Munday. Whereas Munday, the master craftsman, designed as he built, Harrison, a wealthy merchant who was interested in architecture, drew his designs and left the building to others. Hence his work, somewhat avocational even though he received fees, nevertheless marks an important episode in the evolving notion of architecture as a profession in America. A gentleman amateur, he possessed one of the best architectural libraries in the colonies. Unlike Munday's use of the forms of Sir Christopher Wren, which were quite old-fashioned in England by the time they filtered down to the provinces, Harrison's English Palladianism was nearly up to date. His market is a somewhat plainer version of the large gallery at Somerset House in London, designed by Inigo Jones and John Webb and published by Colen Campbell in his *Vitruvius Britannicus* (London, 1717, volume 1, plate 1*b*). The plain brick arcade, three arches wide at the ends of the building, seven along the flanks, was originally open for a market. It formed a base for two upper stories in which Ionic pilasters (doubled at the ends) alternated with tall second-story windows and subordinate attic windows. The principal windows, in turn, are alternately capped by gabled and segmental arches. In contrast to the immense physical presence and slightly quirky, ad hoc charm of the Colony House, the Brick Market is abstract, cerebral, and, to return to Henry James's observation on the Colony House, much more "finely proportioned" (if James meant "refine-ly") throughout. How fortunate that these two buildings exist within sight of one another to accentuate the differences between them.

The market was commissioned in 1760 by the Proprietors of Long Wharf, which extended from it. It was to be used for "dry stores" above and an open market below. Before the Revolution, a printer occupied the upper stories. From 1793 to 1799, Alexander Placide rented it for a theater (a fragment of a sea scene painted on the east wall for one of his dramas is still evident). In 1842 the top floor was eliminated and the upper stories became a town hall and then, from 1853 to 1900, the city hall. Finally, the redoubtable Isham took over its restoration, while John Nicholas Brown financed a complete rebuilding in 1930, including the return of the third floor. After another renovation in 1993, the market is now the Museum of Newport History, run by the Newport Historical Society.

Washington Street and the Point

The main portion of the harbor ends to the north in a blunt jut of land. Between this and the approach to the Newport Bridge is a long, narrow gridiron of streets called the Point, fronted toward the harbor by Washington Street. Over a hundred of its houses date from the colonial period, with a sprinkling of later styles through the late nineteenth century. Why "the Point" when no point can be seen? Originally, what is now a jut in shoreline was scooped into a cove, which was gradually filled for building sites. The water side of the scoop was once a southward point of land. Immediately north of

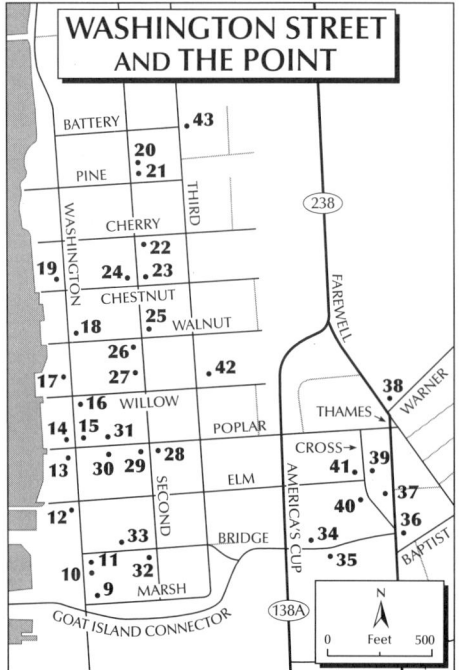

the Brick Market, the most impressive of Newport's wharfs, Long Wharf (now, significantly, Long Wharf Mall) bridged the mouth of the cove, crossed the point, and extended beyond it. The Point area was once one of Nicholas Easton's farms and was known as Easton's Point. On his death Easton bequeathed his land to the Quaker Meeting (NE45). The Quaker proprietors, in turn, divided the area into house lots around 1725, property owners paying an annual quitclaim rental for the privilege of building there. During the colonial period Washington Street (originally Water Street) was an active part of the harbor. Commercial wharfs extended directly from the yards of some very large mansions, so that the grand life in front mingled with dock and countinghouse activity behind.

The causeway to Goat Island is dominated at one end by a towering hotel, the first of the 1970s–1980s spate of hotels and probably the best architecturally simply because of its distinctive shape, but nothing extraordinary inside except views across the bay and toward the town from its multistory bar up in the pinnacle. Goat Island takes its name from its original use as a natural pen for livestock (goats were prominent among its earliest inhabitants). Later it served the navy as a torpedo manufac-

tory. After it was declared surplus following World War II, the hotel, condominiums, and a marina took over.

NE9 Jacob Duhane–Captain Simeon Potter House

Before 1749. 37 Marsh St.

NE10 Ann Webber House

c. 1794. 33 Washington St.

NE11 Isaac Dayton House

Mid-18th century. 35 Washington St.

A theme of this corner of the Point is large, two-and-one-half-story eighteenth-century gambrel-roofed houses. This block of Washington Street provides an example of "half," "three-quarter," and "full" houses, based on a five-windowed width as the ideal front elevation. The first of these, the Duhane-Porter House, is a three-quarter type (with four windows across the front). The door, as a result, is asymmetrically placed, here simple and transom lighted. On the Washington Street side, the two-room plan is clearly indicated by paired windows to front and side, each with its own fireplace from a central chimney. The side elevation is deep as a result, the gambrel folding broadly to embrace it. Porter, the common name for the house, comes from a wealthy Bristol, Rhode Island, privateer who was a later owner. When the Proprietors of Long Wharf, as a civic gesture, set up the first free school for the poor in Newport, Captain Porter contributed a large sum of money, together with his house and grounds, for the purpose. It opened to students in 1814.

The Ann Webber House has a full five-bay elevation with a freer arrangement of windows on the side. The central door is recent. The Isaac Dayton House is a gabled half house, with its replacement door in one corner and chimney on an end wall.

NE12 William Hunter House (Jonathan Nichols–Colonel Joseph Wanton House)

c. 1720. Between 1748 and 1758, alteration and enlargement. 54 Washington St. (owned by the Preservation Society of Newport County; open to the public)

Next in line on Washington Street are more large gambrel-roofed houses on the scale of the

NE12 William Hunter House (Jonathan Nichols–Colonel Joseph Wanton House)

Duhane-Potter House, all built from c. 1725 to the mid-eighteenth century. All employ the five-bay, central-door format. All but one have paired interior chimneys spaced to permit a central hall through the house with four rooms on either side. The early history of the William Hunter House is confusing. Did Deputy Governor Nichols, who was also a privateer and owner of White Horse Tavern (NE46), commission the radical enlargement and remodeling of an earlier (c. 1720) house when he purchased the property in 1748? Or (more likely) was it Colonel Joseph Wanton, Jr., who purchased it in 1756 shortly after Nichols's death, and who himself subsequently became deputy governor? Or was it altered on two occasions? In any event, by 1758 it seems to have looked approximately as it does today. Like the other big gambrel houses on the water on Washington Street, this one had a commercial dock—the largest of all, in fact, which was once advertised as the "best wharf in Newport . . . extending 400 feet into the channel," and so large that in 1799 two warships docked here at the same time. Wanton was a Loyalist who fled the Revolution. His abandoned house became the quarters for Admiral de Ternay, first in command of the French fleet during the Revolution. Then, as Loyalist property, it went on the auction block, and it passed through several owners before William Hunter bought it at another auction in 1805. He lived in it until 1826, when he was called to Washington, prior to his service from 1834 to 1844 in Brazil, ultimately as United States minister.

This house (like the Duhane-Potter House), is especially wide, with two pairs of windows on the side elevation indicating very precisely the four rooms on each floor and two widely spaced windows in the attic. Most remarkable about the front elevation of the Hunter House is the door, with its broken scroll pediment and extraordinary carving of pineapple, sunflower, pomegranate, and foliage—the same motif as that at the Colony House and probably also carved by Jim Moody. The door has led a peripatetic existence. It was originally on the garden (water) side of the house, with a simpler enframement to the street of which no image exists. In the vagaries of the building's history (eventually it became a rooming house), the door was ripped off in the late nineteenth century and taken to the nearby Dennis House (NE15), where it first adorned an entrance to an extension added at this time to the rear of the house and was later moved to the front. After the Preservation Society of Newport County bought the Hunter House, the door was returned in 1950 to its new location as the frontispiece, with the proviso that a replica be made for the Dennis House.

NE13 Captain John Warren House
(Henry Collins House)

Between 1736 and 1758. Before 1775, enlarged toward the garden. c. 1795–1800, door and probably dormers. 62 Washington St.

The very regularly paired windows of the Hunter House are more casually ordered in this slightly earlier house, with window intervals different on either side of the remodeled fanlighted and pedimented door. A beam projects from the north side for use as an attic crane.

NE14 Thomas Robinson House

c. 1725. c. 1760, enlarged and remodeled. 1872, old kitchen remodeled as a rear sitting room, Charles Follen McKim; porch on garden front and part of north elevation later. 64 Washington St.

The Quaker merchant Thomas Robinson extensively altered and enlarged a small two-room house organized around a central chimney after purchasing the property around 1760. (A corner cupboard in the south dining room represents the principal surviving relic of the original house.) Whereas the windows in the Hunter House are strongly paired on either side of the central door and in the Warren house slightly ad hoc, here the spacing is more regularized, giving a somewhat strung-out quality to the elevation—proof of the subtle variations in visual effect which derive from the simplest shift in placement of the elements within a single elevational format. The molded heads of the windows in the other two houses also contrast with the simpler, more severe capping by a plain projection here. The principal distinguishing characteristics of the house also accord with its more restrained aspect and enhance its elongated quality: double windows instead of the usual single window for the upper stair hall over the door, and especially the double-leaf door topped by a wide and austere pediment. The institutional quality of the double door is exceptional, the only other notable example in the Newport area being that at Bishop Berkeley's Whitehall in Middletown.

NE14 Thomas Robinson House

Fortunately, the property remained in the Robinson family to the 1990s. It contains furniture by John Goddard, a nearby neighbor, which Thomas Robinson purchased shortly after his extensive remodeling. It also boasts the transformation of the old kitchen into a "colonial" sitting room by Charles McKim, at the time when he was in the vanguard of a few designers who were just beginning to discover colonial architecture (and had been drawn to Newport to pursue his courtship of Annie Bigelow, whose family summered a few houses away). The young McKim's enthusiasm for colonial architecture, together with his Quaker upbringing, must have caught the attention of Benjamin R. Smith (a descendant of Thomas Robinson) when he determined to convert the old kitchen into a sitting room in order to take advantage of the harbor view and its sunsets, while adding a new low service ell to the south. McKim was called in for the colonial portion of the remodeling plus some ornamental touches inside and outside on other portions of the old house. His idea of "colonial" at this time was inspired as much by furniture as by architecture—a pastiche of turned work and linear decorative detail. But it would mature to one of the several styles by which the firm of McKim, Mead and White eventually made its mark on American architecture.

NE15 **St. John the Evangelist Rectory**
(John Dennis House)

c. 1740. 1876, living hall, Charles Follen McKim. 59 Poplar St. (corner of Washington St.)

Directly across the street from the Robinson house is yet another eighteenth-century gambrel with the same two-story formula, this one more compact, with a central chimney (rather than the spaced interior chimneys of the preceding examples) emphasized by later dormers and a railed platform built around the chimney. Features include the replica of the door to the Hunter House (NE12) and another Neo-Colonial interior remodeling by McKim, also commissioned by Benjamin Smith, who owned this house too. The commission this time called for a new entrance hall for daily use, with a sitting room to the rear of the old house. McKim's work here related to his publication, in the *New-York Sketch-book of Architecture*, of the picturesque dilapidation of Bishop Berkeley's neglected Whitehall (in Middletown) and coincided with the year of the Philadelphia Centennial Exposition, whose various Colonial Revival buildings gave the style wide publicity. It was McKim who adorned his new addition with the garden door frame of the nearby Hunter House.

NE16 **St. John the Evangelist Episcopal Church**

1894, F. C. Withers. Southeast corner of Washington and Willow sts.

St. John is beautifully crafted if somewhat loosely composed, in the manner of an English parish church, in rock-faced tan granite trimmed in brown sandstone. The entrance porch, placed forward in one corner of the church toward Washington Street, and a square, quite plain bell tower set back against the side elevation and hard against Willow Street make nice use of the corner site.

NE17 M. H. Sanford House (Ednavilla)

1869–1870, William Ralph Emerson and Carl Fehmer. 72 Washington St. (at Willow St.)

This barn-sized clapboarded and steeply mansarded Victorian summer cottage has little to hold one's attention on the exterior beyond its veranda overlook of the harbor wrapping a good portion of three elevations. Even this is sparely supported. So are the walls, minimally ornamented with simulated stickwork with a roofline corner oriel oriented southwestward to take in the panorama of water and boats. The spectacle is inside, where Sanford, a business partner of Commodore Vanderbilt, splurged in a Victorian fantasy of wood and stenciled plaster. Margery Deane, a Newport reporter, described it in part in the *Boston Journal* at the time of its completion:

> The floors from top to bottom are laid in hard wood, in fancy patterns, no two rooms alike, and waxed. Oak, ash, cherry, hard pine, maple and black walnut.... The woodwork of the parlors is butternut, with ebony trimmings and panels of mottled wood of the root of the butternut tree.... The dining room is black walnut, with mottled panels and a carved wainscoting, and is the only room in the house papered ... [in] green and gold, in the imitation of leather....

Otherwise, the rooms were plastered with colorful stenciled ornament, which Deane described as "Pompeian," but which included a variety of motifs from exotic sources, especially Egyptian. The *pièce de résistance* is the entrance hall, rising 35 feet from first floor to a tapered climax in the mansard, surrounded by "a grand staircase; each landing having a fancy piece laid in different colored woods. From each story project balconies with bronze gas fixtures, and near the top is a beautiful stained glass window." The orieled "sanctum" at the top belonged to Sanford's niece, Kate Field, a writer. In Deane's description,

> The room overlooks the harbor and bay, and has a bay window, with mirrors in either side reflecting the view and enabling the person lying on the couch in the window to see everything for miles each way without looking out. The furniture is of the bamboo pattern, and rich Turkish rugs are laid about the room.

The house is often referred to as the Sanford-Covell House for King Covell, who long lived here and preserved these interiors against dwindling fortunes.

NE18 Captain William Finch House

Mid-18th century; ell later. 71 Washington St. (at Walnut St.)

Another two-and-one-half-story sea captain's gambrel, its gable end to Washington Street, has a center chimney and splayed window heads—both more typical of earlier buildings than the comparable details in the Washington Street gambrels already described.

NE19 John Tripp House

Early 18th century. 88 Washington St.

This is one of only two stone-enders known to exist in Newport; the other is the Elder John Bliss House (before 1715), at 2 Wilbur Avenue. The Tripp House in fact is not properly a Newport house; it was moved from Manton Avenue in Providence and reconstructed by its owners. The cobbled projection of a beehive oven from the exposed masonry wall is unusual.

NE20 John and Thomas Goddard House

1748. 81 Second St.

The home of the famous colonial Rhode Island cabinetmakers, father and son, this and an adjoining shop originally occupied the site of the Sanford house (see above). They were moved in 1869, and the shop was destroyed in the 1950s. An old drawing in the Newport Historical Society shows the Goddards' two-and-one-half-story, four-bay gambrel as it existed on Washington Street, with a one-and-one-half-story barnlike, gambrel-roofed ell serving as a shop behind—much as neighboring Washington Street sea captains and merchants conducted business in their backyards. The comfortable middle-class modesty of scale and ornamentation of this house vis-à-vis the big merchants' houses on Washington Street reflects differences in wealth and social scale. The central chimney permits the customary corner fireplaces in the two major rooms, front and back, which contain fine but simple paneling as well as an unusual curved stair. So much is to be expected of a cabinetmaking dynasty, one of whose pieces (a block-front desk, designed by Thomas, of a monumental furniture type invented by his father, John) sold at auction in 1989 for over $11 million.

NE21 Fairchild Barn

1876, Charles Follen McKim. 79 Second St. (at Pine St.)

The house is gone; only the ragtag remains of its shingled stable remain, worth attention as a little-known work of a young architect whose later partnership was destined to become the most important firm of its period.

NE22 John Allan House

1859. 67 Second St. (at Cherry St.)

Old formulas persist in this Greek Revival house, built in a style rare in the Point section.

NE23 George Gibbs House

c. 1734. 9 Chestnut St. (at Second St.)

A gabled four-bay front, chimneys at either end, with a Dutch door (original?) made imposing by an exceptionally large transom light topped by a projecting molded cornice.

NE24 Reverend William S. Child Schoolhouse

1875, Charles Follen McKim. 11 Chestnut St.

The earliest of the extant projects of the young Charles Follen McKim is all gable. The sloping planes of the roofline, recalling the later Low House in Bristol (no longer extant) and the many varied cross gables and dormers that relieve the encompassing geometry suggest that McKim was already pursuing the kind of composition for which his firm was known in the 1880s. That the entry porch lacks any organic relationship to the rest of the building may be the best indication here that McKim had not yet developed his mature style.

NE25 Henry Knowles House

Mid-18th-century. 31 Walnut St. (at Second St.)

This two-story central-chimney gambrel with fanlighted, pedimented door was the childhood home of Matthew Galbraith Perry, future admiral, who "opened" Japan to the West. It is hence sometimes aggrandized as the Perry House.

NE26 Joseph Belcher House

c. 1740. 36 Walnut St. (at Second St.)

This one-and-one-half-story central-chimney house with a steep gambrel to the front and a saltbox behind belonged to an important pewterer of the colonial period—confirmation, like the Goddard House, of the middle-class status of the expert craftsman in colonial society, already observed in the Goddard House.

NE27 James Davis house

c. 1731. 42 Second St.

NE28 John Frye House

c. 1760, gambrel section. 35 Second St.

NE29 George Fowler House (Joseph Gardner House)

c. 1725. 32 Second St.

Like Washington Street, Second Street has a cluster of houses revealing variations on the gambrel. The James Davis house is another three-quarter house, this with a transom-lighted door capped by a full pediment. In the John Frye House, a later, gabled ell collides with the simplest of one-and-one-half-story gambrels. Frye, who bought the house in 1770, seems to have been its second owner. Opposite, the Joseph Gardner House, now known as the George Fowler House for a mid-nineteenth-century owner, is a half house with a portal fanlight breaking into its pediment.

NE30 John Chadwick House and Adjacent House

c. 1770 and pre-1750. 54–56 Poplar St.

Whereas the Frye House collision (preceding entry) was happenstantial, this takes advantage of happenstance as a joining of two small old houses (both somewhat altered) to make one big new one. It is mayhap interesting less for what it says about colonial architecture than the appeal it makes to the late-twentieth-century postmodernist taste for off-balance compositions derived from colonial precedent. A gabled house flanking the street butts a gabled house facing the street, with five windows, each in different arrangements and in slightly different sizes and alignments. The more commanding facade with the projecting cornice and the bigger door (with its own echoing cornice) signals "parlor"; the smaller, more domestic image of the frontal gable, relative to the other, says "kitchen." The two elevations are separated—once literally, now visually—where their corner verticals touch but are united by a projecting horizontal in common. The big door touches the horizontal, providing a measure of its dominance over the plainer kitchen door, which drops below it. Its flanking windows also drop below the level of those of its neighbor. The overassertive clash of idiosyncratic elements so pervasive in most postmodernist compositions which attempt a comparable off-balance equilibrium from colonial allusion is chastised by the subtlety and straightforwardness with which a chance juxtaposition is here exploited to make a civic statement for the street.

NE31 **William Crandall House**

c. 1833. 63 Poplar St.

"Paneled" is the word for this example of the Greek Revival, a style rare in the Point. Two-story paneled pilasters make panels of the flushboarded front elevation for the door and windows, which are themselves framed and capped in paneled elements. The triangular window in the pediment reinforces the carpenter's idée fixe. Although pilastered fronts of this sort are standard for Greek Revival houses, the type is exceptional in Rhode Island. The house was built for the owner of a thriving shipyard on Washington Street.

NE32 **Christopher Townsend House**

c. 1725. 74 Bridge St.

Of the group of houses built by the Townsend and Goddard families of cabinetmakers, this, with its attached shop largely intact, is the oldest. A square half house with gable-on-hip roof and slightly off-center chimney, it has a fanlighted and pedimented corner door. Christopher Townsend specialized in ship cabinetry.

NE33 **Pitt's Head Tavern** (Ebenezer Flagg House)

c. 1726, c. 1744. 77 Bridge St.

Another big gambrel-roofed house, this one is twice removed from its original site on the northwest corner of Washington Square at Charles Street in the center of town. It is an-

NE33 Pitt's Head Tavern (Ebenezer Flagg House), door detail, photo 1937

other example of an earlier house (owned by Jonathan Chace) embedded in a later eighteenth-century remodeling and enlargement. The massively sculptural chimney (rebuilt), its framing, and the flat S-sawn balustered stairs folded into three runs in front of it date from c. 1725. The expansion of the house into a full five-bay format occurred in the mid-eighteenth century, shortly after Ebenezer Flagg obtained it, apparently as a wedding present from Henry Collins, who purchased it in 1747, the year after Flagg's marriage to Collins's niece, Mary. In addition to heading one of the largest shipping and merchant firms in Newport (with a vessel for every letter of the alphabet, as one contemporary put it) the two men were leaders in the cultural and intellectual life of the city. At his death Collins owned a collection of seventeen paintings, including four portraits by Robert Feke and one by John Smibert. Both he and Flagg were founding members of Redwood Library. The bankruptcy of the firm of Collins, Flagg and Engs in the early 1760s was followed shortly by the death of both of its principal partners. The house was sold to Robert Lillibridge in 1765 and thereupon began a more boisterous life as one of Newport's celebrated coffeehouses and taverns under the sign of Pitt's Head. At the time, the Great Commoner's sponsorship in Parliament of a liberal colonial policy created special enthusiasm for him—especially in Newport, which was chafing under the very British maritime regulations which had been part of the reason for the Collins-Flagg bankruptcy.

The slightly skewed symmetry of the window placement doubtless reflects compromises in updating an old building on what was originally a tight site on Washington Square. The correctness of the door frame with its pediment on scrolled brackets and both the proportions and moldings of the alternating triangular and segmental pediments of the attic dormers indicate Peter Harrison's sophisticated influence on classical design by the mid-eighteenth century—appropriate for a client who was in the Redwood circle and alert to cultural change.

NE34 Captain Peter Simon House

Before 1738. Mid-18th century, remodeled and enlarged; door after 1800. 25 Bridge St.

Again a gable-on-hip, and again an earlier house later aggrandized to a full five-bay facade, with the still later addition of the fanlighted and pedimented door.

NE35 Caleb Claggett House

c. 1725. 22 Bridge St.

This gambrel with chimney slightly off center and a lean-to extension to the rear is exceptional for its brick end wall, tied to its frame by S-shaped irons. Such walls are rare in Newport houses. The five-bay front was also off center, a stretch of blank wall originally terminating the house toward the end opposite the brick wall, into which additional windows were later cut. The bulky chimney serves four fireplaces at the first floor: one toward the center of each interior wall of a large and a small parlor, in front to either side of the off-center door; two more angled into the corners of the kitchen and what is now a dining room (perhaps once a downstairs bedroom) in the rear lean-to.

NE36 Job Bennett House

c. 1721. 44 Thames St.

This square, gable-on-hip-roofed house with a four-bay front elevation and central chimney is larger and more formally organized than the Claggett House (preceding entry). Four windows to one side of the elevation and three windows and a door to the other side are very consciously grouped with an interval of wall down the middle of the building, over which a dormer sits. The transomed door, though simple, is, like the size of the house for this format, imposing. Inside, a very substantial kitchen fireplace and boxed-in stairs with stubby, bulky balusters reinforce, by their medieval quality, the early eighteenth-century date.

NE37 John Stevens Shop and Stevens Houses

Mid-18th century. 29, 30, 34, and 36 Thames St.

Numbers 36, 34, and 30 are modest colonial houses which are of interest because all belonged at some time (one of them, number 30, for two centuries) to the famous stone-carving Stevens family. (Another Stevens house, the early-eighteenth-century house of John Stevens, is at 9 Elm Street.) The John Stevens Shop, across the street at number 29, has been in continuous operation since 1705 (the current building dates from the late eighteenth century). The Stevenses were first masons and stone carvers, but eventually specialists in stone lettering whose work included many major architectural commissions. John Howard Benson took over from generations of Stevenses. The second and third generations of Bensons continue the craft.

NE38 Joseph and William Cozzens House

c. 1765. 57-59 Farewell St.

Colonial duplexes are rare. This gambrel-roofed house with two replacement pedimented doors toward the center and chimneys at either end is essentially two half houses brought together for two brothers.

NE39 Gov. William Coddington House

Late 17th century. 11 Cross St.

Fortuitously, in the immediate vicinity of a doubled half-house is this gabled prototype. It was moved from a site included in the six-acre farm on the Point which was originally allotted to William Coddington, one of the founders of Newport. Although it has been extensively restored, portions of the chamfered beam frame mark it as a very old house. The simply paneled door with small transom capped by an overhead molding, although new, typifies doors of the period for middle-class residences.

NE40 Nathaniel Coddington–Thomas Walker House (King's Arms Tavern)

Probably between 1706 and 1713. 6 Cross St.

Another house moved from Coddington's holdings, this one is substantial. The heavily projecting cornice and pilastered brick central chimney are signs of an early house.

NE41 Spooner House

c. 1740. 1 Elm St.

The proportion of wall to openings in this two-story, five-bay, central-chimney house with a gambrel roof gives the front elevation an especially commanding appearance.

NE42 Callender School

c. 1862, George Champlin Mason, Sr. 1909. c. 1980, conversion to condominiums, George Ranalli. 11 Willow St. (at 3rd St.)

This fine brick and brownstone Renaissance Revival building closed as a school in 1974. New York architect George Ranalli, who was commissioned by a new owner to adapt its spacious classrooms into condominiums, inserted a two-story "mini-building" into each apartment for the kitchen and private areas, leaving a "facade" of doors and windows to overlook a living space the full height of the original room. With the "buildings" painted in vivid colors (varying from unit to unit) and the outside walls of the original rooms in pale neutrals, the "buildings" seem to overlook a plaza, so inside has illusions of outside. The idea was in the air around 1980, but here it is superbly done, as testified by widespread publication in architectural magazines of the period.

NE43 Solomon Southwick House

Mid-18th century. 77 3rd St. (opposite Battery St.)

This standard two-story gambrel-roofed house was moved to this site. It takes its name from its occupant at the time of the Revolution, the editor of the *Newport Mercury*, a staunch patriot who whipped up sentiment against the Stamp Act.

NE44 United States Naval War College

Founders Hall: 1817 with Victorian additions. Luce Hall: 1891, George Champlin Mason, Sr. Coasters Harbor Island

The ruggedly severe aspect of the two later buildings of the War College, especially Luce Hall, with its walls of random-coursed gray stone, entrance porches, and gables is likely to attract the most attention from the distance. It is imposing in length and is sited at the crest of a long, treeless slope, which is echoed in the long slopes of hipped and gabled roofs. These are pierced just above their eaves by the exceptional feature of a row of continuous slotlike windows, each with its own very long slope, all climaxed in a stubby cupola. But Founders Hall, the original building of this high-command college for instruction in tactics and global naval strategy is perhaps the most interesting. A plain Federal building with a cupola and a lunette window stretched in its pediment, it is fronted by a tall Victorian veranda reached by a steep climb of steps. (Extensions to either side are, like the porch, later additions.) Plainness and austerity carry the day here, with a touching sense of the domestic simply blown up to institutional scale. It is now a museum open to the public; perhaps its most notable permanent exhibit covers the development of

NE44 United States Naval War College, Founders Hall

NE44 United States Naval War College, Luce Hall

naval torpedoes, commemorating the factory which made them on adjacent Goat Island. Behind Founders Hall, a road runs past an interesting row of Victorian duplexes built as officers' quarters.

West of Broadway

This wedge contains the major cemeteries of the city, to which Farewell Street is the cortege artery. Toward the tip of the wedge is the Friends Meeting House, the largest Quaker meeting house in the state and, crossing its tip, immediately behind Washington Square, Marlborough Street, on which fronted major colonial houses, taverns, and businesses. Although the rest of the wedge also contains some seventeenth-century houses and was one of the earlier African American enclaves in the colonial city, during the early nineteenth century those were overrun by small, closely packed wooden dwellings for the first wave of Irish immigrant workers. For this reason the area acquired its alternate name of Kerry Hill. In the dense old seaport section of Newport, its streets are the most densely built up. Gradual dilapidation encouraged landlords to purchase houses here en masse for low-priced rentals after World War II. It is the last area of the old town to feel the impact of the restoration and gentrification movements that have transformed the neighborhoods around the harbor. Because only scraps remained on which "restoration" could build, it is less interesting for the individual quality of specific buildings than for the overall effect of what is good and what may be questionable in all such efforts to transform the shabby into the "old." On the periphery of the area is Van Zandt Avenue, where nineteenth-century houses were for years just beyond the reach of the Irish incursion.

The most remarkable of the cemeteries is the Common Burying Ground, which contains about 4,500 markers dating from 1660 to the present. In addition to key works by local cutters such as the Stevens shop and the Bull family are other markers from a variety of Boston and Narragansett Bay area carvers, including William Mumford, the Emmes family, and the Tingley and Cooley shops. At the north corner, farthest from the eighteenth-century town, is the largest collection of colonial African American grave markers in the United States. Dating from 1720 to the American Revolution, these (and later stones) represent the strong presence of African Americans, both slave and free, in the history of Newport more clearly than any other aspect of the built environment. One stone from 1769, for Cuffe Gibbs, is inscribed as having been cut by his brother, Pompe Stevens, a black slave in the Stevens shop, making it one of the few artifacts of precolonial American craftsmanship "signed" by an African American.

At the crossing of Van Zandt Avenue and Farewell streets are the North Burial Ground and Island Cemetery. They are not pretty, being more field than arcadia; but Island Cemetery, which was fashionable in the late nineteenth century, contains plots of many of the leading families of the city, with some interesting Victorian monuments. Exceptional is the collaboration of the architect Richard Morris Hunt with Karl Bitter on the August Belmont tomb, in the northwest corner of the cemetery in front of the chapel, designed by Hunt. Other markers by Hunt for the Marquand, Russell, Shepard, and Wetmore families and two other works by Augustus Saint-Gaudens, the King monument (with John La Farge, 1877) and the Amor Caritas figure for the Ann Maria Smith monument (1886), are also notable.

(Note: For locations of sites NE45–NE47, see the map on page 518.)

NE45 Friends Meeting House
1699, 1705, 1729. 1807, additions. 1858, 1867, 1977, alterations. 1967, restoration, Orin M. Bullock, Jr. Corner of Marlborough and Farewell sts.

Members of the Society of Friends, or Quakers, found a hospitable home when they arrived in Newport in 1667. With its culture of religious toleration, the town became a center of Quakerism in the colonial period. By the end of the seventeenth century, Quakers represented 60 percent of the local population, and their numbers were expanded by a regional yearly meeting of Friends held in Newport each June. This event required a much larger facility than any of the town's existing meeting houses.

Although the meeting house is one of the earliest surviving religious structures in the state, its history is in large part a story of constant alteration to keep apace of needs. The plain, horizontal barnlike block seen today originally looked very different. Its central hall was originally five bays, later flanked by additions to the north (1729) and south (1807). Much detail survives from the seventeenth-century building: great stones, described by the Quakers, on which they raised their walls; rare details of diamond-paned leaded windows on the rear elevation; and archaeological traces of early structural members, pew placement, and other furnishings.

The original voluminous space of the meeting room, spanned by structural timbers hewn from monumental oaks, is still visually impressive although drastically truncated above the second story. The Quakers sat below a towerlike hipped roof of more than two stories capping an elevation so tall as to dominate Newport's skyline for more than a quarter century. Disdaining decoration and pictorial symbolism, the congregation would have been inspired instead by the sheer height and dimensions of this grandiose volume reaching heavenward. Unfortunately this distinctive cap was removed by the mid-nineteenth century, and today the shadowy overhead void has lost its soaring vertical thrust. During this and the subsequent classicizing changes to the exterior, the joins and elements of the medieval-scale beams were modified so that they now suggest makeshift repairs rather than express what must have been the intuitive, organic rationality of the first framed skeleton.

Similar changes have altered the site. In its early years the meeting house was embedded in a townscape of residences and commercial blocks, but urban clearing over the past century has created a green space which, although inviting, gives a false sense of the building's historic context. The site is particularly varied in its historical record; archaeological work has yielded important pre–European contact and revolutionary-era deposits as well as a significant Quaker burial ground (unmarked) around the irregularly shaped lot.

In the twentieth century the tradition of the yearly meeting waned, and the building, with its land, had a succession of owners, both public and private. Eventually, in the late 1960s, it was restored and given to the Newport Historical Society, which uses it for special events and conferences. During the restoration, a tentative plan to return the building to its original core was abandoned after discovery of a system of movable interior walls, or shuts, complete with its original pulley system, which still functions to allow the three major interior halls to be connected or divided. This was considered too unusual a feature to be lost, as it would have been had demolition of the south addition proceeded. Although now a pastiche and reduced by alterations, the central sanctuary, as retained, still gives a hint of what must have been the greatest of early Newport's religious spaces and remains a testimony to the special status of the Society of Friends in colonial Newport.

NE46 White Horse Tavern
Before 1673. 1780, extensive alterations, present gambrel roof. Farewell and Marlborough sts.

Of all the buildings which ringed the property of the meeting house, the White Horse Tavern, already in place when the Friends built across the way, alone still stands. Its history is entwined with that of the city, even as its architecture is characteristic of colonial Newport. After acquiring the property from the breakup of founder William Coddington's original six-acre plot, William Mayes was granted a license to keep a tavern by 1687. The oldest parts of the building—which was originally a two-room house with central pilastered chimney and huge framing timbers—can still be seen within. Jonathan Nichols II, a prominent merchant who later became lieutenant governor, first hung the White Horse sign. When a later family member, Walter Nichols, returned to Newport after evacuating because of the British occupation, he changed the look of the house to the appearance it now presents: a broad, gambrel-roofed, clapboarded structure with classically pedimented doorways and a balanced, five-bay elevation facing Farewell Street. Although by the mid-twentieth century it was somewhat derelict, the venerable structure, which has had so many uses (as a meeting hall for the town council in the eighteenth century, a popular tavern, and the residence of a pirate, a cabinetmaker, a silversmith, and an innkeeper) was restored and still serves the public in one of its original guises, as a highly regarded restaurant. It is one of the best public places in Newport in which to gain a sense of colonial domestic space.

NE47 Captain John W. Downing House
(Fairview)

1873–1874, Dudley Newton. 34 Malbone Rd.

This is everything a Victorian summer cottage should be, except for its tight location. The second story is all steep mansard with a mansarded outlook tower in front. The mansard is pulled down snugly to a braced-post, fat-balustered front porch and bay windows downstairs. Windows between parlor and porch are tall, near floor to ceiling in height. Upstairs, chunky dormer windows in the mansard and tower are hooded against the sun, and double doors framed in a double-pitched gable open to a tiny balcony. Shade on the porch below; full sun on the more or less symbolic balcony above. Slate shingling of the mansard in polychrome patterns is the only missing architectural ingredient.

Harbor Area
From Washington Square to Memorial Boulevard, Thames Street to Bellevue Avenue

Newport's harbor area (known locally as the Historic Hill Section) includes the original core of Newport, on a broad slope rising from the heart of its harbor (and now from the tourist hubbub along Thames Street). It is bounded by Thames Street (sensibly pronounced as it is spelled), Touro Street to the north, Bellevue Avenue to the east, and Memorial Boulevard to the south. Across its middle, only roughly parallel to Thames, runs Spring Street, named for the spring close by the site of the Colony House that provided the city's first water supply. Between the major arteries crossing and bounding the section, narrow streets run with the slope, attempting a gridiron configuration, but one or more sides of the rectangle are invariably skewed—the skewing making subtle shifts in what is conceptualized as straight and thereby animating the views. Here and there minor cross streets make it through one or two blocks before giving up. Overall, the aspect of the harbor area is that of the Point, and so is its history of colonial prosperity, then gradual shabbiness and dilapidation, and, finally, rediscovery and restoration. Where the Point is an enclave, a little separated from the city, the Hill is bounded by tourists and traffic, mostly moving around, but also filtering through it. Again, this section of Newport is best seen by walking, and in season; to tour by car is virtually impossible. So the routing ignores the direction of a few one-way streets.

NE48 William Redwood House

Mid-18th century. 69 Spring St.

Originally owned by William Redwood, brother to the benefactor of the Redwood Library, this was apparently a standard two-story, five-bay house before an additional bay was added to the north end, possibly around 1800, to which the fanlighted door seems to date. The brick end is another example of this rare treatment in Newport, plastered over in this instance (see the Caleb Claggett House, above, and the John Odlin–Jonathan Otis House, below, for other examples). The existing end chimney suggests that it was once balanced by another at the opposite end.

NE49 Elisha Johnson House

c. 1750. 89 Spring St.

NE50 Edward Willis House

c. 1807. 95 Spring St.

The compelling feature of the two-story half house at number 89 is the abrupt verticality of its gambrel roof, which is repeated in the adjacent tiny one-and-one-half-story, five-bay gambrel at number 95, moved from its original site a few blocks south and dated to c. 1807, but very conservative in style for that time.

NE51 Double Store Building

c. 1830. 107 Spring St. (northwest corner of Mary St.)

Greek Revival store buildings restored to something like their original appearance are rare. In this hip-roofed brick block with granite trim and a wheat-sheaf relief, the slotlike door to the second story with its small light for the stairs above is tightly squeezed by the stores on either side.

NE52 John Odlin–Jonathan Otis House

c. 1705. 109–111 Spring St.

The Odlin-Otis House is a long, gabled half house with a south elevation of brick, one room wide, which was extended, possibly for use as a Quaker boys' school, in 1788. The depth of the front overhang is characteristically medieval. Otis was a goldsmith.

NE53 James Brown–Samuel Barker House

1714, mid-18th century. 119 Spring St.

Documentation on at least a portion of this commanding gambrel-roofed house with fan-

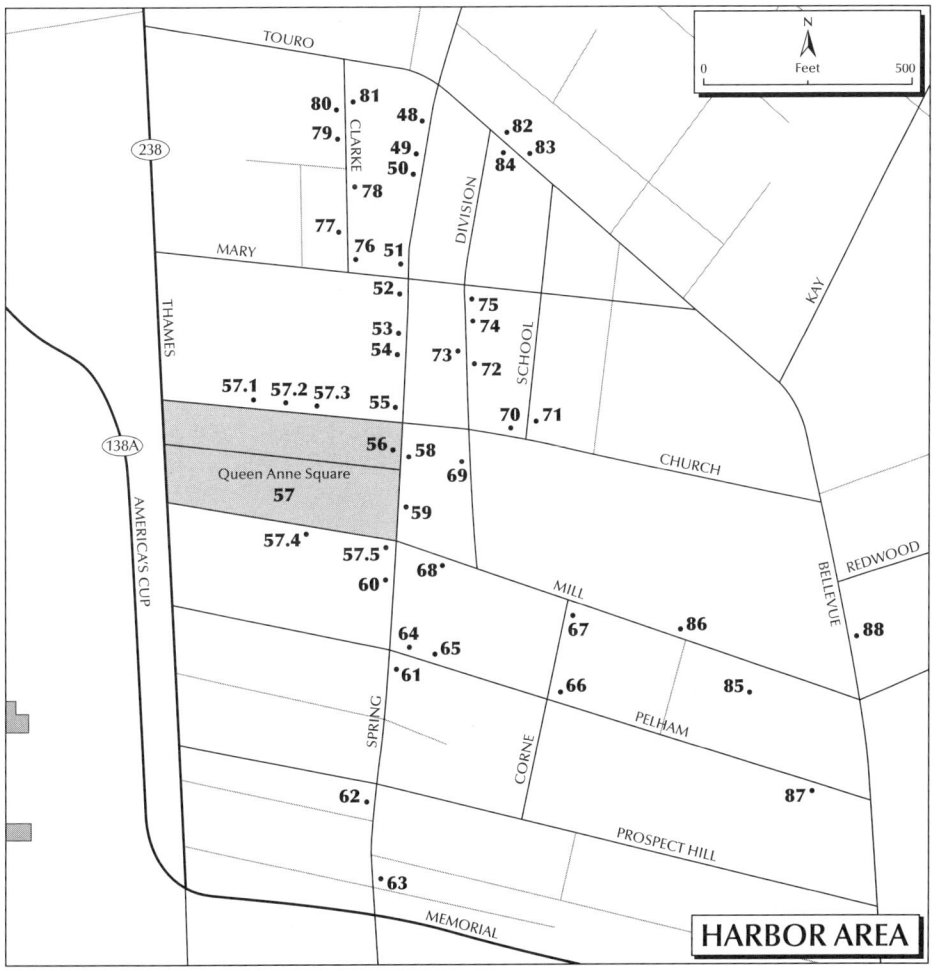

lighted entrance goes back to 1714, although later rebuilding probably brought it to what we now see. Each of its broad, brick-walled ends contains a pair of chimneys, with fireplace placement clearly indicated in side elevations which are as forcefully organized as its front elevation.

NE54 John Preston Mann House

c. 1827. 129 Spring St.

This Greek Revival house has giant corner pilasters and a triangular opening in its pediment—both of which features, together with a certain bumptiousness, appear in a smaller house on Poplar Street. The Italianate round-arched entrance dates from around 1860, and the cross-gabled adjunct on top also appears to be a later addition.

NE55 Store Block

c. 1895, Dudley Newton. 137 Spring St. (northwest corner of Church St.)

This curious Queen Anne double store block was built around a hip-roofed Federal relic. It displays Queen Anne fondness for combining colonial and Federal bits in novel and picturesque ways, with the projecting bays, interior paneling brought outside, friezelike treatment of the topmost story, and overall attention to detail particularly characteristic of the style. The frieze at the top may originally have contained some embellishment.

NE56 Trinity Church

1725–1726, Richard Munday. 1741, spire (designed 1726). Probably c. 1762, two-bay addition, spire rebuilt. West of Spring between Church and Frank sts.

So open was Rhode Island to religious tolerance that Quakers, Congregationalists, Baptists, Sabbatarians, and Jews all organized congregations earlier than the Church of England, which began effective organization in Newport only in 1698, four decades after the founding of the town. The first Anglican church, a quite plain affair, was erected in 1702. Its congregation grew so rapidly as to require a new building within twenty years. It was built just as the influence of Sir Christopher Wren's churches reached the colonies, about a quarter century after his work had come to dominate ecclesiastical design in London. In fact, Trinity is the second major church in the style in the colonies, begun two years after the completion of the church which is considered to be the first of the colonial churches in the Wren manner, Christ (familiarly Old North) Church in Boston of 1723. Stylistic relationships between the two churches are close, as were contacts between the two congregations. It has even been argued that William Price, a book and print dealer in Boston who is considered to have designed Christ Church by editing and simplifying from engravings of Wren's work in his stock, also supplied the builder-designer Munday with suggestions for his design—although these could also have come from the Church of England's Society for the Propagation of the Gospel in Foreign Lands as well as, of course, from observation or reports of Christ Church itself. In any event, comparison of the two churches reveals their affinity. The bodies of both churches—Christ Church in brick, Trinity in clapboard—show a double tier of arched windows five bays in length. (For more accommodation at Trinity, the five bays became seven when the easternmost two bays were later cut off from the rest and moved farther east and two more inserted.) Both have spired towers projecting from the narrow side of the auditorium in what was then an innovative contrast to entrances centered in the long sides of traditional meeting houses. The spires are similar. So are the two-tiered balconies inside, supported by paneled piers on the ground floors and fluted piers above; among major churches in the colonies these are the only two with such an arrangement (although Wren's work provides precedents in more elaborate variants).

As is also true of the Colony House, Trinity reveals Munday's proud effort to master far-away architectural sophistication through a double filter of provinciality: Boston's interpretation and his own isolation in Newport. The gaucheries are at once charming, vigorous, adaptive, and poignant. The entrances on either side of the tower show a broken pediment misinterpreted—but with admirable gusto and inventiveness—as a segmental pediment with a semicircular notch. Too attenuated pilasters on too insistent pedestals support it. Openings in the too plain box of the tower are spottily placed with too much clapboarded space around some, too little around others. The double (or perhaps notched) arching in the topmost windows (to suggest the tablets of the

NE56 Trinity Church, photo 1969, and interior, photo 1970

Ten Commandments?) is another curiosity, pressed hard against a molding overhead which is more classical in aspiration than in fact. Whereas Wren's spires are built of strongly three-dimensional elements suggesting their masonry origin, even when of wood, both here and in Boston the spires have a thin, boardlike quality. And yet there is a splendid mixture of vigor and elegance about the tower overall which gives it majesty.

The same qualities appear inside. The balcony piers are too squat; the bridging from pier to wall over the upper balcony is too blunt, the plaster barrel vaulting over the balconies is awkward in its juncture with the groin vaulting of the nave. All around are woodwork details that are both charming and a bit clumsy. But the interior has great dignity and immense appeal, even a certain intimacy— which is surprising in so large a space. The lowness of the balconies, the crowding of the original box pews, and the omnipresence of the old woodwork—especially the raised paneling on the balcony fronts, wainscoting, and pews—enhance the sense of intimacy (and, incidentally, provided an example that made such paneling popular for fashionable house interiors in Newport through the mid-eighteenth century). Above all, however, the intimacy results from the towering pulpit uniquely set in the center aisle, well out from the east wall with its shallow apse projection framing a tall, clear multipaned window glazed in clear glass. The pews crowd around this towering three-stage rostrum containing clerk's desk, reading desk, and wineglass pulpit, all topped by a suspended bell-shaped sounding board. The elevation of the preacher in the midst of his congregation in this way suggests the old meeting house. From documents it appears that twisted balusters (probably replaced by the present rather pedestrian railing in the nineteenth century) may have originally adorned the stair to the pulpit. Diagonal (or "Union Jack") paneling decorated the underside of the sounding board. (Turned balusters and both Union Jack and raised panels are also features of Munday's Colony House and his Seventh Day Baptist Meeting House [NE6 and NE83].) The pew with cut corner immediately below the pulpit was George Washington's when he came to Trinity. (It was moved back a little during the recent restoration to provide more room in front of the pulpit for weddings and funerals.) Either as part payment for his work or as a sign of gratitude, Munday received pew 75. From there he could survey, no doubt with pride, the building which lifted Newport to a new level of monumental grandeur. It, together with Peter Harrison's nearby Touro Synagogue and Joseph Brown's First Baptist Church (NE82 and PR56) remain as the most spendid achievements in ecclesiastical architecture in Rhode Island and among the most spendid in all the colonies.

NE57 Queen Anne Square

Bounded by Thames, Mill, Spring, and Church sts.

The green which fronts Trinity dates only from the 1970s, when demolition for vistas was a key aspect of downtown planning, here doubly encouraged by nostalgia for the New England common—in a city that had none, and in a state that has very few. Such fictionalized prettification obscures the relationship which the church actually had with the town, rising out of the houses like the pulpit from its pews. In pluralistic Newport, no one church dominated a common open square. Clearing should have been selective and at least suggested fact. "Early" buildings in the area to be bulldozed for the common were moved to other sites to enhance the "colonial" rehabilitation of the area around the harbor. In their place is a brand-new parish house adjacent to the church, in a hybrid between colonial and Greek Revival that confuses the picture (to put is as politely as possible).

What remains from the past is a nice if somewhat overrestored row of houses along the north side of the green (facing the church), the John Langley House (c. 1807; NE57.1), 28 Church Street (one of the moved houses); the Joseph Cotton House (NE57.2; c. 1720), 32 Church; and the Erastus Pease House (NE57.3), 36 Church. The row shows, side by side, a variety of roofs—a gable, gable-on-hip, and, most interesting, a low, spreading gambrel with a rare curve out to the overhang. To the south of the green, at 35-37 Mill Street, is a rare colonial double house, the Billings-Coggshall House (see also the Cozzens House, above), with simple pedimented doors (NE57.4; c. 1784), in which two three-quarter houses are coupled into a stretched format. The elevation can, in an oscillating way, call up ghosts of both a full five-bay and a half house. The Alexander Jack, Jr., House, which was moved in 1969 from another site to 49 Mill Street (corner of Spring St.) (NE57.5; 1811), is a version of the three-quarter formula on which the double house was based.

NE58 Telephone Building

c. 1910. 142 Spring St.

Diagonally across Spring Street, directly behind Trinity, is the somewhat awkwardly scaled but charming first Telephone Building for Newport. The vigorous swoosh of the double-arched crown of the Palladian motif at the center is barely contained within its pilastered and corniced frame—all windows unfortunately cut down in a recent renovation.

NE59 Theodore R. Helme Block

1860s, Dudley Newton(?). 148–160 Spring St.

This mansarded block has recently been restored more accurately than the Telephone Building, with an ornate metal railing along its roofline, typical of its Second Empire style but often missing today from buildings of that period.

NE60 Samuel Bours House

Before 1777. 175 Spring St.

This three-quarter gambrel with pedimented door was the home of a wealthy merchant. His son John married Hannah Babcock, daughter of Dr. Joshua Babcock of Westerly. Their portraits, his by John Singleton Copley, hers by Joseph Blackburn, are in the Worcester Art Museum; the sign from his Thames Street store, the Golden Eagle, is now in the Newport Historical Society.

NE61 Newport Congregational Church

1857, Joseph Wells; murals and stained glass, John La Farge. Spring St. (at Pelham St.) (open to the public one morning a week during the summer season)

This church in random masonry, compellingly stern, is a cardinal example of the so-called Lombard Romanesque or, more generally, round-arched, style recommended by the Congregational Church in a publication of 1853 as a type suitably churchly in appearance, but appropriately simpler than the Gothic favored by Episcopalians. Its vast, shadowy, balconied interior is equally severe but animated by La Farge's intricate color patterns, inspired by abstract patterning from Oriental carpets, which are centered in a painted tabernacle behind the pulpit, and a Tiffany lantern above. Decorative panels across the gabled ceiling directly suggest rugs suspended tentlike over the space. Beautiful stained glass in abstract designs completes the decorative ensemble. A strange conjunction—exoticism and Congregationalism. La Farge came to Newport directly from his decoration of the interior of H. H. Richardson's Trinity Church in Boston. There the decoration is better integrated with the architecture; here it seems to float in the raw space with too little architectural detail to cling to, living a somewhat disembodied life of its own. Both commissions were important harbingers for the popularity of murals and the collaboration of artists with architects at the end of the nineteenth century as part of the American Renaissance. As for Wells, he ended his professional career as a top designer in the office of McKim, Mead and White, where, in the 1880s, he, more than anyone, showed the senior partners the way to the Renaissance forms which accounted for the central role they played in the American movement. So this forceful church is fascinating, not only as an outstanding example of Congregational round-arched style, but as a herald of a future not yet formulated.

NE62 Store Block

c. 1840. 221–225 Spring St. (northwest corner of Franklin St.)

This fine store block, altered in details but basically intact, is interesting as a rare survivor of its type.

NE63 Captain John Mawdsley House

c. 1680. c. 1750, front addition. 228 Spring St.

At first glance this gable-on-hip-roofed house with ornate doorway surround, heavily capped windows, modillion cornice, and three pedimented dormers would seem to date from the mid-eighteenth century. Although the entry elevation and front half of the structure, set back some twenty-five feet from Spring Street, do in indeed date from about 1750, the rear portion was originally the two-room Jireh Bull House, built about seventy years earlier. Its heavy, chamfered summerbeams are still evident. When the wealthy businessman and privateer Captain John Mawdsley acquired the Bull House, probably just after his marriage, he set about updating and enlarging it in keeping with his prominent social status, adding elements inspired by the Georgian classicism of Peter Harrison.

In 1795, after Mawdsley's death, the house was purchased by the merchant Caleb Gardner. During this period an early leader of Newport's African American community lived in the house as a slave under the name Newport Gardner. Also known by his African name, Occramar Marycoo, he was one of the founders of the African Humane Society. When freed, he purchased his own house a few blocks south of here at 86 Pope Street.

NE64 John Banister Town House

1751–1764. Early 19th century, doorway. 56 Pelham St.

This deep, gambrel-roofed house, set end to Spring Street, and the life of its first owner are well documented through account books in the archives of the Newport Historical Society. Banister, the brother-in-law of Peter Harrison, was a wealthy property owner, merchant, and smuggler and the husband of Hermione Pelham, who owned the lot. Within the enclosure of a gabled wall of bricks imported from Holland, Banister constructed an impressive residence with central hall, two interior chimneys, and a gambrel roof, not unlike the William Hunter House (see entry, above). What may be Harrison-influenced similarities between these two houses include details of interior moldings and paneling. The Banister House also has pedimented dormers of the same type found on the Vernon House (see entry, below). During the British occupation this was General Prescott's headquarters. His edict that steps not extend beyond the fronts of Newport houses may be the basis for local lore about the many recessed doorways in the eighteenth-century city. Here the recess may be original, but the entry, with its handsome Doric-inspired pilasters framing a low, wide archway, classical colonnettes, and generous glazing is a later renovation, perhaps carried out after 1821, when the house was sold out of the Banister family.

NE65 Captain Augustus Littlefield House
(Governor Charles Van Zandt House)

1836, John Ladd, builder. 70 Pelham St.

Littlefield asked his builder to design an "authentic copy of an Italian Villa" he had seen during a trip in southern Italy. Although the format of this house is that of a full four-columned temple front, the capitals combine Greek Corinthian and Egyptian lotus motifs—a combination which may reflect Napoleon's Egyptian campaign. But such exotic hybrids are frequent in late Greek Revival design. The bracketed cornice under the pediment is not at all Greek, but typical for Italian villas.

NE66 Butler House

c. 1865. 92–94 Pelham St. (at Corne St.)

And here a full-blown Italian villa, the dominating tower at the corner making a virtual wing of the body of the house. Three stories of tall, narrow windows rise in the tower, providing a display of round-arched and rectangular openings hooded by segmental and table projections on brackets, all climaxed by a deep, bracketed overhang at the eaves. The entrance porch to the adjacent office demonstrates how the choice of essentially eclectic examples from the past (like the Italian villa, which mixed classical, medieval, and vernacular elements) loosened the classical hold of the Greek Revival. The canonical entablature supported by Ionic columns is scooped into a bracketed arch; classical pediments tend toward shading hoods; the expected classical dentil molding under the cornice becomes a rope molding; the entrance is an elegant but nonclassical arch. The more exotic aspects of Greek ornament were especially attractive to mid-nineteenth-century taste and anticipated the lavish embellishment of much Victorian design, as suggested by the Egyptoid lushness of the capitals of the Littlefield House and, here, the beautiful acroterion which surmounts the entrance pediment of the office wing. Except only for the Edward King House, the Butler House is the finest extant example of the Italian Villa Style in Newport.

NE67 Michel Félice Corné House

c. 1822. 2 Corne St. (southeast corner of Mill St.)

This was the home of an Italian emigré mural painter famous for landscapes and seascapes. He arrived in the United States in 1799 and immediately painted rooms at the Derby mansion in Salem, later at the John Hancock House in Boston (demolished), and, in 1809, scenes of Naples at the Sullivan Dorr House in Providence (PR75). His only known paintings in Newport are small ones of ships, which were once painted onto the walls of the southwest chamber of this house before their removal (they are now in the Newport Historical Society). He purchased this site with a barn in 1822 and, by legend, in the tradition of artists, remodeled it into this four-bay house with an ell (although he may also have built it anew). Its special delight is the proportioning of the openings. Many three-quarter houses have a regular rhythm of four windows across the top story; others have only a slightly wider interval between the rhythm of three and the odd beat. Here (whether calculated or not) the pause in rhythm is more decisive. It almost, but not quite, permits the insertion of another window. The subtle, off-beat balance thereby suggests symmetry in the imagination—a mode of ambiguity sympathetic to postmodernist revivals of colonial inspiration which admire the piquant effects of almost-but-not-quite symmetries. This piquancy is here reinforced by delicate detailing. To the delights Corné brought to his adopted town add one other: he reportedly introduced Newport to the tomato.

NE68 J. D. Johnston Mill

1902. 75 Mill St.

The front of the mill shows a handsome two-story door and window organization, the brick wall above extended by a quite elegant stepped screen to conceal the long gable roof behind. This stepping, in conjunction with the forceful arrangement of the openings, dignifies a commonplace vernacular building by simple means. The builder-architect J. D. Johnston was one of the largest employers in Newport at the turn of the century; more than a hundred carpenters and millwrights worked for him on many of the largest residences of that era. The mill was enlarged and renovated in the 1980s for use as an inn.

NE69 William Card House

1811. 73 Division St.

The William Card House is a three-quarter house with the very refined detailing character-

NE66 Butler House

teenth century, reaching its present configuration by 1887. The allusions to a masonry fortification in the flat-topped entry tower, fringed with carpentered crenellations and flanked by two projecting, gabled wings, may be a visual reference to Solomon's Temple, the historic edifice that is central to Masonic rites. Although no longer used as a Masonic lodge, it is the oldest such building in Rhode Island and the fourth oldest in the United States.

NE72 Mary B. Newton House

c. 1883, Dudley Newton. 52 Division St.

This idiosyncratic pastiche of gables, porches, turnings, shingle patterns, and beaded sheathing is one of Dudley Newton's many Newport buildings. Although ownership of the house is often listed in his mother's name, he lived here, perhaps considering its many visual delights a showcase for his talents.

istic of the Federal style and windows framed almost flush with the clapboarding. Visually the near-flush effect enlarges the apparent size of the window by making it more integral with the plane of the wall, in contrast to the typical eighteenth-century treatment, which projects the window out from the wall, making wall and window opposed elements in separate planes.

NE73 Artist's Studio (Union Congregational Church)

1834. 49 Division St.

The former church is a Carpenter's Gothic building with exposed decorative trusswork. It has been converted into an artist's studio.

NE70 Thomas Goddard House

c. 1800. 78 Church St.

The Thomas Goddard House, a gabled Federal house probably slightly earlier than the Card House (preceding entry), shows the persistence of the older window treatment (projecting from the wall), here with unusual rusticated window caps. The home of one of the cabinetmaking Goddard family, the house has interior woodwork of exceptional quality (interior not open to the public).

NE74 Dr. Samuel Hopkins House

c. 1751. 46 Division St.

The Samuel Hopkins House, set at right angles to the street with its door in the end, is known for the hero of Harriet Beecher Stowe's *The Minister's Wooing*, who lived here from 1770 to 1805. Even as restored, it is the kind of irregular, added-to "old manse" which attracted mid-nineteenth-century writers in search of an ancient past for American themes. By the 1860s most such houses were not only irregular but also convincingly dilapidated.

NE71 St. Johns Lodge Masonic Temple

1803, 1830, 1846, 1860, 1876, c. 1887. 50 School St.

The massive wooden hulk of this meeting hall is out of scale with its mostly residential neighbors, at least partly because of its numerous additions. The lodge was initiated in 1749, and foundations for this temple may have been started as early as 1760 from plans drawn up by Peter Harrison. There was, however, nothing above street level until the 1803 building was constructed (by J. Cahoone and Sons, a local builder) at the corner of School and Church streets, to a different design. This structure was subsumed and augmented innumerable times as the lodge continued to grow in the nine-

NE75 Augustus Lucas House

1721. c. 1745–1750, enlarged. 40 Division St. (southeast corner of Mary St.)

Two stairs in different styles, one behind the other, confirm the back-to-front enlargement of this house, as does the anomaly of two chimneys, also one behind the other. Boxed stairs in three short runs with scroll-cut balusters indicate that the older section was in the rear. The front stairs have twisted balusters, popular in the mid-eighteenth century. The gable-on-hip roof may also result from the modification of a small gable into something grander. Augustus Lucas first appears in *The Newport Mercury* announcing a consignment of African and Indian slaves for sale. An orchard on his property and Lucas's reputation as an expert in grafting fruit trees remind us that a number of properties in this now crowded area had sizable gardens through the eighteenth century. A grandson, Augustus Johnson, was appointed stamp master in 1765, just in time to be swept up in the riots of that year. This house was a major target, and its owner was driven from Newport.

NE76 Vernon House (William Gibbs–Metcalf Bowler–William Vernon House)

Before 1708; c. 1760, renovation, Peter Harrison. Clarke St. (at Mary St.)

The all-over rustication of its wooden exterior, originally painted light pink with sand thrown onto the wet surface to simulate granite, is the most conspicuous aspect of this remarkable house. It, like some near-contemporaneous houses of lesser merit, shows the influence of Peter Harrison's similar imitation of masonry for his Redwood Library (see below). But beyond this overt indication that it may be Harrison's work, even though no document proves it, the nature of the design, the sophistication of its sources, and the discrimination of its execution all suggest the fact. In any event, it is among the very finest of eighteenth-century colonial houses.

This is another example of a Newport house which began as a smaller, plainer predecessor and was lifted to a new order of magnificence in the mid-eighteenth century. It is another example, too, of a house in which the patriot owner, popularly associated with it today, obscures the preceding Tory owner, who, together with his Tory-inclined architect (if he was indeed Harrison), actually brought it pretty much to the state in which we see it today. William Gibbs owned this property before 1708 and presumably built on it a house facing on Mary Street, less than half the size of the one that now exists. (The excessive depth of the house—almost a square in plan—may record the fact that it was once the elevation of a long, narrow house.) After several subsequent owners, Charles Bowler purchased it when he became collector of revenue for the colonial government in 1753, then sold it in 1759 to his son Metcalf, a wealthy merchant in the West India trade as well as a patron of the arts and a Tory spy. It was Metcalf who commissioned the grand enlargement by adding to the rear of the original two-room plan a central hall and the two rooms to the south, while shifting the orientation of the house from Mary to Clarke Street. Metcalf Bowler sold the house to William Vernon only in 1773 when Bowler retired to the country. Vernon, a patriot, was a merchant banker in Newport in the years after the Revolution. Like many other patriots, he fled Newport in 1773 when the British occupied the city, so the house was vacant when the French, as allies of the American cause, in their turn ousted the British from Newport. Rochambeau, general of the French forces, made this house his headquarters. Once the Vernons returned to their mansion the family retained it until 1872—a century of tenure which also accounts for its popular name.

The revolutionary aspect of Harrison's knowledge of Renaissance form and principles (whether he actually designed the house or it merely shows his influence) is evident by comparison with the cluster of earlier gambrel-roofed houses which are its immediate neighbors on Clarke Street. Here the roof, with dormers classically capped in segmental and triangular pediments, is hipped to a balustraded deck, thus lowering the effect of what was by then beginning to be perceived as the gawky provinciality of the tall, folded gambrel, and doubtless (for some who could afford better) the plain gable's lack of stateliness. Medieval molding profiles at the eaves give way to a classical modillioned cornice. The portal frame, with its magnificent breadth, is correctly classical and very restrained. Instead of steps held close to the house, these spread grandly. Observe, too, the finesse with which the rustication angles into the tops of the windows to simulate a keystone lintel and the plain projecting molding overhead stabilizes the angling, forming a band around the building that ties into the door frame at the junction of pilaster and entablature. Inside (not open to the public), Metcalf Bowler's splendid staircase entrance hall retains the twisted balusters typical of the period before 1750. All principal rooms were paneled at the time, with broken scroll fireplace heads for the principal parlors. In the old section of the house, restoration brought a remarkable discovery. Under the paneling of one of the original parlors (north) are frescoed walls dating back to c. 1740, with Chinese subjects, in colors suggesting japanning on leather—vivid mementos of the China trade—all framed in painted bolection moldings in lieu of paneling, as was the fashion in the early eighteenth century. It was to this house that Washington came to dinner with Rochambeau in 1782.

NE77 Robert Stevens House

1742–1755. 31 Clarke St.

NE78 Joseph Burrill House

Early 1700s. 28 Clarke St.

The Robert Stevens House, a broad gambrel-roofed house with a fanlighted door that is doubtless of later date, is set side to the street. The side elevation reveals the plan: larger parlors in front, smaller room behind. The paired chimneys toward the center indicate a central hall. The contemporaneous Joseph Burrill House, with shingled sides and back, is a very tall gambrel (which may have been an eighteenth-century addition to enlarge what was originally a two-room house on a tight site). Here the regularity of the windows in front continues on the sides, although the plans of the two houses are similar (formal expression here versus the functional expression of the Stevens House).

NE79 Newport Artillery Company

1835, Alexander McGregor. 23 Clarke St. (open to the public)

The Scottish master mason responsible for work at Fort Adams in 1833 also built this granite ashlar building with its bristling quoins and door trim for the oldest military organization in the country. The second story was added in the early twentieth century.

NE80 Condominiums (Second Congregational Church)

1735, Cotton Palmer, builder. 1847, 1874–1875, 1944. 15 Clarke St.

Although this former church is very much a hybrid from a series of alterations, the end result is interesting. Originally the tower projected for half its width from the front of the meeting hall, with the square stage which now exists as a platform for an octagonal bell chamber topped by a flaring spire. Shortly after Baptists took over the church in 1847, they brought the body of the building out to the street, erasing from view all of the tower except what showed above the ridgepole. It was at this time that the front received its predominantly Greek Revival aspect. When wings were added on either side in 1874–1875, Victorian ornament was added to door frames and windows. What would have been pediments over Greek Revival openings now projected abruptly as hoods, and Victorian delight in complicating simple shapes appeared in ornamental embellishment of the portal pilasters, inset embellishment of their caps, and (over the window) a fractured pediment. Finally, the bell chamber and spire disappeared in 1948. The church was derelict for a number of years. Its ultimate conversion to condominiums preserves in spanking white the ghosts of the sequence of styles that make it what we see today. But its sprawling monumental scale, together with the more compressed monumentality of the Artillery next door, provides a focus for the street.

NE81 Ezra Stiles House

c. 1756, 1834, c. 1847. 14 Clarke St.

This third Clarke Street gambrel takes its name from its most famous occupant. Built to serve as the rectory for Second Congregational "forever," this broad, five-bay, two-and-one-half-story house was home to the learned reverend briefly (1775–1776) while he served as minister of the church. He had previously served for thirty years as librarian of the Redwood, where he wrote his *Ecclesiastical History of New England and North America* and investigated fields as disparate as Abyssinian geography, astronomy, and silkworm cultivation, while also drafting an important map of Newport (1758). As pastor he, like the Reverend Samuel Hopkins of Newport's First Congregational Church on nearby Mill Street, inveighed against slavery. He left Newport during the Revolution to become president of Yale College. The Greek Revival porch (1834) of his onetime parsonage must have been added to harmonize with the first renovations to the church.

NE82 Touro Synagogue

1759–1763, Peter Harrison. 72 Touro St. (open to the public Sunday afternoons)

Angled to the street on a rise so that the ark inside faces Jerusalem, this brick box with low hipped roof has a severe exterior effect, near cubic in quality, its side walls slightly elongated. The breadth of wall toward the corners of the building, the starkness of the arched windows gathered around the entrance porch, and the plainness of the molding band between the first and second stories all intensify the slightly offputting aspect of the exterior. Only the elegance of the entrance porch, with arching inside a pedimented Ionic enclosure and the same generous spread of stairs as those for the Vernon House, anticipates the interior.

NE82 Touro Synagogue

Inside is an intimate, two-story space, which was originally without chairs, quite constricted, somewhat elongated, tall with balconies on three sides, and made taller by a steeply coved ceiling at the center. The Ionic order below supports a Corinthian order on pedestals at the height of the balustered parapeting for the women's balcony. Twelve columns in all—five along either side, two flanking the entrance—commemorate the twelve tribes of Israel. The two-story Ark of the Covenant on the end wall dominates the space. (The design was altered from what was originally there, but apparently shortly after the temple's completion. Could the original design, sketched in a letter by Reverend Stiles, have been a temporary expedient until the final design could be installed?) In front of it, like a platform in a public square, is the raised and balustered bimeh. Four silver candelabra and the eternal light hang from the ceiling on long rods. The scale and presence of these fittings in the constricted space confuse architecture and furniture, thereby increasing the shrine-like intimacy. For his design, Harrison borrowed from a two-story galleried hall in William Kent's *Designs of Inigo Jones and Others*, and, for details of columns, balustrades, and the ark, from James Gibbs's *Rules of Drawing* and Batty Langley's *Treasury of Designs*—all volumes in his library. But the exquisite combination of these elements and their adaptation to a new use is Harrison's. He never saw the completion of one of the most beautiful and intense of American colonial interiors. A Loyalist, he left Newport to accept the post of customs collector in New Haven, where, in 1777, an angry mob destroyed his property, his drawings, and his library. In 1780 the synagogue briefly replaced the Colony House, wrecked by the successive occupation of British and French armies, as the meeting place of the General Assembly and the State Supreme Court. Washington visited the synagogue twice in 1781 and, as president, in 1790. After the last occasion he wrote the letter, which hangs framed in the synagogue, containing the words, ". . . happily the government of the United States . . . gives to bigotry no sanction, to persecution no assistance . . ."—although one can still see the trapdoor at the reading desk, possibly put there by the congregation as a symbolic reminder of the potential need for exodus. On the 1790 trip, Thomas Jefferson, himself an amateur architect, accompanied Washington to Newport in his capacity as secretary of state. Like Harrison, a devotee of Palladian classical correctness, Jefferson must have been stirred by this and other works of a kindred spirit in Newport.

Uphill on Touro Street is the Jewish Cemetery, marked by Isaiah Rogers's Egyptianate gateposts (c. 1841), with their reliefs of inverted torches, symbols of the extinguished life. Within this ground, purchased in 1677 for the Jewish community of the region, are a handsome sequence of nineteenth-century

NE84 Jewish Community Center (Levi Gale House)

obelisks and an eighteenth-century stone from the Stevens shop with a bilingual English and Hebrew inscription. The cemetery's history and picturesque appearance inspired Henry Wadsworth Longfellow's poem "The Jewish Cemetery in Newport" (1858).

NE83 Newport Historical Society and Seventh Day Baptist Meeting House

1729, attributed to Richard Munday. 1884, meeting house restoration, George Champlin Mason, Jr. 187, meeting house moved to present site. 1889, entry porch. 1902, library addition, G. H. Richardson, builder?. 1915, central wing, Joseph G. Stevens. 1917, front entry porch, Norman M. Isham. 82 Touro St. (open to the public)

It is not the building that houses the historical society itself which is of primary interest here (although much inside may delay the architectural pilgrim). The primary attraction is rather the twice-moved Seventh Day Baptist Meeting House, which is embedded in the society as one of its stellar "exhibits." How fitting that fate should have brought side by side the synagogue and this meeting house to commemorate Roger Williams's vision of a colony tolerant of all religions. And how fortunate that these two religious interiors, both small and intense in their different ways, are juxtaposed for comparison. The meeting house was originally a very plain clapboard building, only 36 feet by 26 feet, located behind its present site on Barney Street (east of Spring), which was moved to Touro Street to serve as the historical society's first building, then moved back and encased in brick when the society acquired its present quarters. (The meeting house may move again as part of a plan to give it more prominence facing Touro Street on a separate foundation and adjacent lot.)

The entrance is at the center of one of the long walls in the canonical meeting house position, with balconies around three sides. Opposite the entrance and filling half of the center wall of space is the climactic wineglass pulpit, with paneled extensions on either side and a sounding board projecting overhead. The wineglass shape, the inset paneling arched at the top (typical for the early eighteenth century), the "Union Jack" diagonal paneling of the underside of the sounding board, and the twisted balusters of the stair are all to be found at Munday's Trinity Church (see above). Only these similarities, which may suggest the same craftsmen for both structures, account for the attribution of the meeting house to Munday also. Outstanding are the craftsmanship and variety of the turned balusters—among the first of their type in Newport, and soon to become the rage in houses of the wealthy until around 1750. The raised paneled faces in the balcony also appear in Trinity. Only the topmost segment of the wainscoting was originally paneled, the rest being plain. When the box pews which once completed the interior were destroyed in the nineteenth century, the lumber was used for the completely paneled wainscoting which now exists. Across from the pulpit, a clock by William Claggett still ticks on the balcony front.

NE84 Jewish Community Center (Levi Gale House)

c. 1835, Russell Warren. 1915, moved to present site. 85 Touro St.

For this New Orleans merchant who moved to Newport, Warren created a clapboarded block with projecting center and heavy entablature. Four giant pilasters adorn the center, with exotic capitals remote from the chasteness of most eighteenth-century examples, but typical for late Greek Revival. "Rusticated" corner piers close the recessed ends. Above the projecting section is an attic story, once crested with a low stepped and paneled parapet, as was the principal roof. More exotic Corinthian capitals support the rather pinched porch, which was also once capped, like the attic, with a low parapet structure. The adjustment of this house to the slope and small size of its new site gives it an uncomfortable aspect; Warren's elevation also appears somewhat unresolved.

NE85 Stone Mill

Probably c. 1673. Touro Park

This 84-foot-high arched masonry cylinder, in a one-block park between Mill and Pelham streets, has long been the subject of historical controversy. Some would trace it back to the dim mists of Norse exploration, long before Giovanni da Verrazzano's exploration of Narragansett Bay in 1524. Twentieth-century scientific scholarship, however, has conclusively demonstrated its colonial origin, partly from artifacts found at the site, partly from carbon dating of the mortar, from written documents, and from comparison with other, near-contemporary masonry structures and with English-type windmills. An owner of the land, Benedict Arnold (ancestor of the famous spy), erected a "stone built milne" as he states in a deed. Its wooden superstructure was no longer intact by 1740. It served as a powder magazine and a haymow before it graced many a tourist's souvenir as an emblem of Newport, and it featured in James Fenimore Cooper's *The Red Rover* (1828). Interesting, too, for the architectural pilgrim is the nearby monument to Commodore Matthew Galbraith Perry, by John Quincy Adams Ward with a base by Richard Morris Hunt (1868), the earliest of their many collaborations. Reliefs at the base of the stone celebrate four episodes in this Newporter's brilliant naval career, including a charming image of the commodore negotiating with Japanese to celebrate his "opening" of the country in 1856 to American commerce. The event led by the 1870s and 1880s to widespread Japanese influence in architecture and furnishings, with some key examples in Newport, soon to be seen.

Around the perimeter of Touro Park are several houses worth notice. Number 128 Mill Street, the Charles Sherman House (c. 1845), puts a Greek Revival pediment up to the street with a side entrance. In the southwest corner of the park are three other Greek Revival houses (Touro Park West and 123 and 115 Pelham Street). Completing the circuit of the park are 135 Pelham and 141 Pelham Street (probably 1870, the latter by Seth B. Stitt), carpentered extravaganzas with Victorian high jinks in the picturesque, aggressive massing of towering elements of various shapes at the rooflines with big porches below.

NE86 Robert Lawton House

1809. 118 Mill St.

The Robert Lawton House is interesting as one of the few large-scale Federal houses in Newport. The collapse of maritime trade in Newport following the British occupation was so complete that this house style, prevalent in Providence, is virtually nonexistent here. Catherine Warren lived here in the mid-twentieth century. A collector of modern art, she came to be interested in the preservation of the old colonial city. To garner funds for its salvation she hit upon the idea of convincing the Countess Szycheni, a descendant of the Vanderbilt family, to open The Breakers to the public (thereby incidentally recovering its substantial annual cost to the family in taxes and maintenance). It was the start of the Preservation Society of Newport County, its focus on late-nineteenth-century properties, and its gradual accumulation of houses open to the public.

NE87 Channing Memorial Church

1881. Pelham St. facing Touro Park

This church, named in memory of William Ellery Channing, the famed Unitarian-Congregationalist preacher who was born in Newport in 1780, is an assemblage, in rough-cut granite, of varied gables, turrets, and porches set against one side of a tall, narrow dominating gable and a fantastically shaped emancipated tower set against the other. The stepped elements of the tower, rudely hatcheted by stages, create the reduction of mass required to enable the spire to point its knobby finger at the sky. Inside are stained glass by John La Farge and a relief by Augustus Saint-Gaudens.

NE88 Redwood Library

NE88 Redwood Library

1748–1750, Peter Harrison. After 1743, Garden House from Redwood's country estate in Middletown, Peter Harrison. 1858, reading room, George Snell. 1875, addition to rear, George Champlin Mason, Sr. 1913, fireproof stack room. 1915, restoration, Norman M. Isham. 1935, garden house allée, John Russell Pope. 1978–1979, enlargement of stacks, Irving Haynes. 1998, restoration of Harrison exterior, Shepley, Bulfinch, Richardson and Abbott. Bellevue Ave. between Redwood St. and Old Beach Rd.

Redwood Library, the first of Peter Harrison's three public buildings for Newport, carries the old colonial tradition across Bellevue Avenue, occupying a site which was, when built, on the high rim of the old town. In the eighteenth century it must have suggested an intellectual temple on an acropolis.

Elite intellectual and artistic life in early eighteenth-century Newport came to focus in the Philosophical Club. It spawned an informal group of leading citizens who gathered around Bishop Berkeley as their mentor and example while the bishop resided at Whitehall in nearby Middletown, from 1729 to 1732. They christened their new group the Society for the Promotion of Knowledge and Virtue by a Free Conversation. The idea of establishing a subscription library to perpetuate their "free conversation" was eventually propelled by gifts from two leading merchant-traders in town: one from Abraham Redwood, Jr., of 500 pounds for books and one from Henry Collins of land high on the perimeter rim of the densely inhabited slope down to the water.

The detailed contract of 1748 for the library building, the only such document extant for any of Harrison's buildings, was signed by his brother Joseph while Peter was in London. Since Joseph, too, is known to have been a draftsman, he may deserve partial credit for the design. More likely, Peter sent basic drawings, at least, from London. In any event, he was clearly in charge during the later phases of construction after his return. Again he (and/or his brother and the building commission) leafed through the books in Peter's extensive library to choose as their format a design for a garden temple in a headpiece in Edward Hoppus's *Palladio* (1735), which reproduced a structure in Lord Burlington's garden at Chiswick. A similar design with an added dome was also reproduced in William Kent's *Designs of Inigo Jones and Others* (1727), where it was identified as a casino for Sir Charles Hotham by William Kent. Peter Harrison's library contained both books, and Redwood Library shows details lifted from both examples. The chosen model was a full-fledged temple with a Doric portico and with projecting wings, or "outshots," one window wide, on either side immediately behind it to serve as offices. Their sloping shed roofs provided for the illusion of another pediment behind that of the entrance portico, which the body of the temple building cut through. Overall, the building was to be "plank'd . . . in Imitation of Rustick" (possibly the first use of the technique in the colonies, unless Shirley Place in Roxbury, Massachusetts, antedates it). Harrison could have derived the triple motif of a Palladian window under large arches which originally adorned the rear wall of his temple from a number of sources. A particularly popular motif in England, it appears in all the design books Harrison used for the portico. Whereas the style in front is Doric, it becomes Ionic on the original rear elevation, another indication of the method of designing from plates rather than conceptualizing the entire building as an entity. (When George Snell added a reading room to the rear of the original temple in 1858, he moved the triple Palladian motif to the side wall facing south and duplicated it on the opposite wall to the north while extending the "outshots" the full length of Harrison's temple in order to increase office space.)

As the first of Harrison's three public buildings in Newport, the Redwood Library betrays more awkwardness than either of the later ones, the Brick Market and Touro Synagogue. The interlock of the "outshot" pediment behind that of the temple is somewhat awkwardly joined and proportioned to the body of the building; the columns of the portico are a little squat for their spread; the portico with its full-width steps is somewhat too grand for what is behind. In sum, more than his other works, the library suggests too anxious copying from the sources. For Harrison was not, like Richard Munday, an artisan who invented, but a gentleman concerned with correctness and knowledgeable about what was then regarded as correct in London, where Lord Burlington (another gentleman amateur) and his circle made the classicism of Palladio the cachet for architectural design. There is some poignancy in the ambition of the colonial Tory to confer on the library (allegedly the seventh oldest in America) the impeccable grandeur of the first public building thoroughly committed to Palladian classicism in the colonies. Inside, furniture by Thomas Goddard from Abraham Redwood's house on Thames Street just south of Perry Mill keeps company with his portrait by Robert Feke. The iron entrance gates, imported from London around 1731, came from the high brick walls which once surrounded the garden of Abraham Redwood's house, behind which his own dock and counting house controlled his shipping empire. Behind the library, the rusticated hexagonal summerhouse with its undulant roof is Harrison at his most delightful. This came from Abraham Redwood's extensive garden at his Middletown country house—the only such structure to have survived from the many eighteenth-century gardens for which this area was famous. Additions subsequent to Snell's are set one behind the other in treatments which are generally sympathetic to the remarkable starting point in front.

Kay–Catherine–Old Beach

Redwood Library is one of two cultural institutions that provide an architectural hinge between the buildings on the slope down to the harbor and those on the hilltop, which gradually falls in the opposite direction toward the salt marshes around Easton's Pond. Adjacent to it is the Newport Art Museum, the principal building of which is the former J. N. A. Griswold House, an early work of Richard Morris Hunt. This establishes the keynote for the district, which is predominantly Victorian edging into the early twentieth century. An array of work by the leading local practitioners is here, together with important beginnings. For here, densely clustered, are other early works by Richard Morris Hunt and shingle houses by McKim, Mead and White.

During the final decades of the nineteenth century, well-to-do Newporters—professionals, leading merchants, local real estate tycoons—built houses here, as did relatively wealthy summer colonists, also mostly in the late nineteenth century. Well-known residents included painter John La Farge, whose summer home was a one-and-one-half-story Greek Revival house (c. 1845) with Egyptianate corner boards and window trim, at 10 Sunnyside Place, and Clement Moore, the author of "A Visit from St. Nicholas" ("'Twas the night before Christmas . . ."), whose greatly enlarged Italianate cottage is at 25 Catherine Street.

The intersection of Catherine Street and Bellevue Avenue was the center of architectural activity in the 1870s and 1880s, with the offices of Newport's most prominent architects clustered around this corner. Richard Morris Hunt's studio and home, Hilltop, was across the way on the site of what is now the Viking Hotel. Dudley Newton's office was the small, bay-fronted building with an ornately patterned mansard roof in slate at 20 Bellevue Avenue. The local contractor-architect James Fludder had his home and office at the corner of Catherine Street and Bellevue Avenue; the Mason firm was located around the corner on Catherine Street; Clarence Luce's office was only a block south at Bellevue Avenue and Mill Street; and J. D. Johnston's office was a little farther down the hill on Pelham near Spring.

The following sequence moves counterclockwise around the perimeter of the neighborhood and then into the center.

NE89 Newport Art Museum

1861–1864, J. N. A. Griswold House, Richard Morris Hunt. 1919, Cushing Memorial Gallery, Delano and Aldrich. 1990, addition to Cushing, Peter Roudebush and Associates. 76 Bellevue Ave. (open to the public)

Across Old Beach Road from Redwood Library stands the first major Newport commission of Richard Morris Hunt, the architect ultimately responsible for the city's grandest palaces. It was a summer house in what was then known as Modern Gothic (more recently designated as Stick Style) for J. N. A. Griswold, a wealthy New York businessman whom Hunt had met in Europe.

The main block of the house is covered in a skeleton of half timbering, overhanging bracketed eaves, and shaded wrap-around porches whose roofs are supported by exposed posts and diagonal braces, mostly nonstructural and rather more picturesque than rational in intent. Hunt here attempted to reconcile the more rationalist aspects of his recent training in Paris with a romantic response to such European and American vernacular buildings as farmhouses, resort cottages, Alpine lodges, and rustic pavilions. The design also recalls earlier experiments with decorative rustic framing in Newport houses of the previous decade by Leopold Eidlitz and Richard Upjohn. As designed, the house appeared as a more vertical, compact mass than is now the case, but after the Art Association of Newport (later the Newport Art Museum) took over the property in

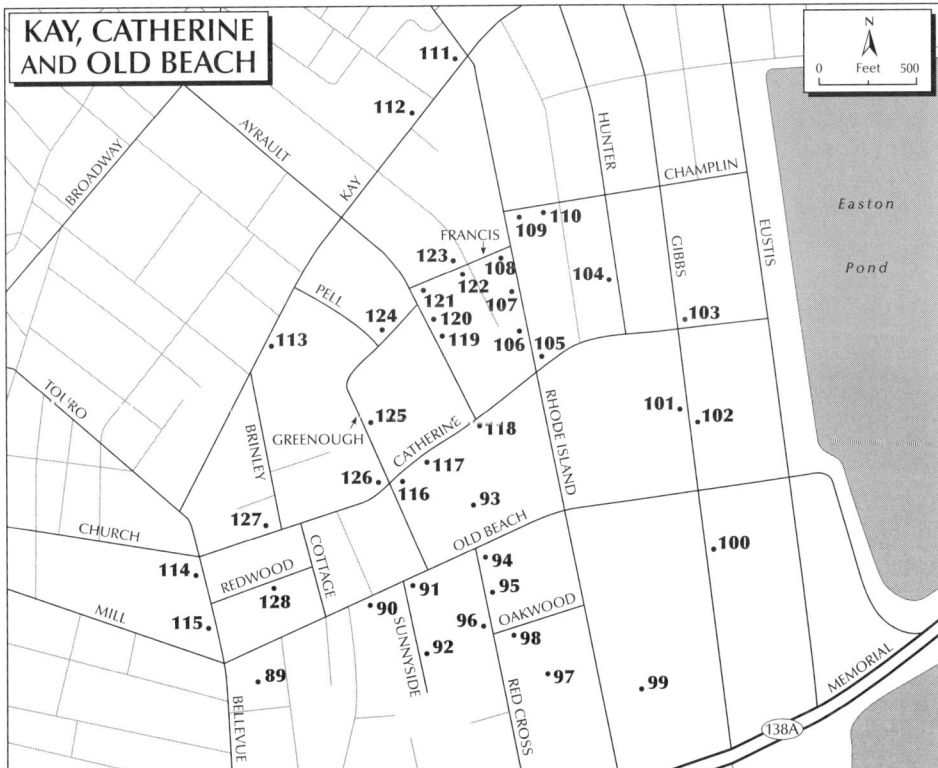

KAY, CATHERINE AND OLD BEACH

1916 for its exhibitions and classes, the stable was moved to abut the side of the house and renovated on the interior for gallery space. The large studio window on the front facade was also added after its transformation into an arts center.

Inside, the entrance hall opens generously in three directions on either side to a broad stair, a dining room, and a parlor and straight ahead to a conservatory. The interior retains a sense of the spatial effect and decorative woodwork of the original house and, with the big porch immediately at hand, suggests the genteel yet informal environment which Hunt created for summer living.

Also on the grounds of the museum is the smaller Cushing Memorial Gallery, with its ornate ring of polished marble columns supporting a low dome over the main entry and projecting from the otherwise unadorned severity of the main block. Dedicated to the memory of Howard G. Cushing, a local society portrait painter and muralist of the early twentieth century, it was intended as the entryway to a larger museum complex, which was eventually completed in 1990 under the design of Peter Roudebush and Associates of Cambridge, Massachusetts. This addition, which stretches east from the original Cushing block and replicates much of its classical detailing, has added gallery and other functional space for the museum's needs.

NE90 Commodore William G. Edgar House

1886, McKim, Mead and White. 25 Old Beach Rd.

The Edgar House is the last completed of a half dozen houses associated with the firm of McKim, Mead and White grouped within two blocks. By the time it was designed, the architects were beginning to abandon the shingled picturesqueness typical of their early houses in favor of the formality that characterizes the style for which the firm became famous. For much of their domestic architecture, this meant the American Colonial Revival. In the

NE89 Newport Art Museum, J. N. A. Griswold House interior (left) and exterior (below)

Edgar House, the central Palladian window and the restraint, essential symmetry, and crisp linear quality of the details generally invoke eighteenth-century American precedents, but the sculptural chimneys at either side of a balustraded roof deck, the roofline, and the brick walls specifically suggest Tidewater Virginia colonial antecedents.

Yet even here, the asymmetry and variety of motifs persists from the firm's earlier houses: polygonal bay against semicircular, with different window treatments; an oval window in the blank wall of a two-story service wing at one end of the house counterpoised with the open airiness of the gazebo-like pavilion at the other end. The yellow ocher Roman brick is not in fact typical of the eighteenth century, nor is the low, spreading quality of the massing. The interior, now turned into apartments (not open to the public), still contains a front-to-rear entrance hall with fireplace, a dining room, a music room, and a library with fine woodwork and built-in cabinetry.

NE91 George Champlin Mason, Sr., House (Sunnyside)

1873–1874, George Champlin Mason, Sr. 31 Old Beach Rd. (corner of Sunnyside Pl.)

George Champlin Mason, the noted local architect and historian and writer on Newport, built this, his primary residence, at the same time that he published his book *Newport and Its Cottages*. The main entrance, tucked to the left of a polygonal tower bay, faces Old Beach Road, but the most developed facade looks west over Sunnyside Place, doubtless because the house would most often be approached from that direction. This elevation is a confection of jigsawn trim, cutout railings, drilled brackets, and carved droplets, all sheltered by a broad, overhanging gable in the manner of a Swiss chalet. The horizontal clapboards of the first story are largely hidden on this side by the deep porch, but the board-and-batten sheathing on the second level is clearly visible within the symmetrical composition of doors, windows, and balcony. Despite its unusual appearance, the design represents a relatively conservative approach, rather than looking forward to the composition and details of the next generation. The house is best viewed in mid-afternoon, when the intended effect of its ornamental woodwork and projecting eaves casts intricate shadows across its face.

NE92 Samuel Tilton House

1881–1882, McKim, Mead and White. 12 Sunnyside Pl.

This is less an exercise in creative massing than in surface ornament. Viewed on Sunnyside Place, the Tilton House is covered with a spectrum of materials and patterns that ranges from the foundation course of rough-hewn granite to upper stories covered in wood shingles cut into wave, notch, and sawtooth motifs, all contained within the simple shape of a large gable. On the north and west elevations, an irregular composition of extraordinary decorative stucco panels interrupts a band of half timbering and windows. In the panels are em-

NE90 Commodore William G. Edgar House

NE91 George Champlin Mason, Sr., House (Sunnyside)

NE93 H. Allen Wright House (Belair), stable

bedded shells, beach pebbles, colored glass and even bits of coal, the two largest showing a sunburst pattern and a shield. Charles McKim had seen an analogous treatment in a seventeenth-century house in Salem, Massachusetts (since demolished), and Nathaniel Hawthorne reported the glinting changes of light on such a house in *The Scarlet Letter.*

The present small entry porch seems appended as an afterthought to the richly ornamented west wall, but old photographs show its original form to have been a two-story, gable-topped projection with turned spindle screens on the second level. This would have augmented the already ornately textured facade, affording greater integration with the main block. From the street, the Tilton House seems quite small, but this elevation is the narrow end of a house which stretched the width of its lot along its north edge, leaving the south front open to windows and porches facing a private lawn. The firm repeated this formula for many of their smaller houses at this time, including four examples on Red Cross Avenue.

This house contains some of the best-preserved interiors of its period (not open to the public), in which influences from American colonial and Japanese design and even from India combine to complement a diverse palette of materials, textures, and color outside.

NE93 H. Allen Wright House (Belair)

1850, Seth Bradford. 1870, 1875, additions (stable and gatehouse), Dudley Newton. 50 Old Beach Rd.

This large villa, one of the earlier summer houses in this neighborhood, was designed for New Yorker H. Allen Wright in the Italianate style and constructed of roughly dressed stone. The composition of massive square tower, arches, and bracketed details relates the early section of Belair to Seth Bradford's other well-known Newport mansion, the original Château-sur-Mer, built a year later. In the 1870s, a new owner enlarged the house and added several auxiliary structures, including a gatehouse and stable designed by Dudley Newton. Although their details are mostly in keeping with the sense of the main house, some, like the trefoil gable over the stable doors, are idiosyncratically oversized for the scale of the smaller buildings. Belair was substantially altered in the twentieth century and is now divided into condominium units.

NE94 Katherine Prescott Wormeley House

c. 1876, Charles Follen McKim. 1882, probable later addition, McKim, Mead and White. 2 Red Cross Ave.

Built as a summer home for author and translator Katherine Prescott Wormeley of Boston when she was in her late sixties, this house is essentially in the form of a two-family residence—

one facing southwest toward Red Cross Avenue, the other opening out at the opposite end onto Sears Court, a small, disused lane to the north. The second earliest residence by any member of McKim, Mead and White still standing in the city, it is less resolved in its integration of exterior massing and ornamental skin than its neighbors. On its primary facade, facing Red Cross, the Wormley house is composed of two intersecting blocks, one with its street-facing gabled end projecting from a covelike support. At their junction, a three-story tower capped by an exotic, metal-sheathed onion dome is inserted like a hinge pin. The north entry seems more conventional with a gabled roofline and an entrance porch, now screened. Inside the Red Cross Avenue entrance are an entry hall and adjacent rooms decorated with tongue-and-groove paneling and an ornate oriel window at the stairwell landing. On the Old Beach Road facade, this window creates a cylindrical glazed projection rising two stories from stone foundation to roofline.

Red Cross Avenue, like nearby Sunnyside Place and Oakwood Terrace, was laid out when the large gentleman's farm-estate owned by David Sears of Boston was divided into smaller lots in the 1870s. Sears's ample Gothic Revival house, Red Cross Cottage, still stands at the head of Sears Court (its entrance is now on Oakwood Terrace). It was branded with a red-cross-shaped pattern in its now stuccoed brickwork, which suggested its own name as well as that for the nearby avenue. Although it has been stripped of ornate trim, its group of steeply rising gables is a reminder of an earlier cottage style.

NE95 **Frances L. Skinner House** (Villino)

1881–1882, McKim, Mead and White. c. 1950, library addition. 6 Red Cross Ave.

This tiny house for a Boston woman is the architects' smallest shingled Newport cottage, although larger in fact than it appears from the street. The swollen rotundity of the conical-capped corner tower dominates, with the entrance porch and a tall brick chimney snuggled up to it. The high gable spills around the tower and chimney and over the porch, like the long, shingled lean-to roof slopes of the seventeenth-century New England houses the architects had studied. The tower alternatively evokes châteaux, or, closer to home, the shingled windmills on Aquidneck and Conanicut is-

NE95 Frances L. Skinner House (Villino)

lands, which the architects knew. Most piquant is the roofed dormer peeking from behind the chimney and the continuous band of casement windows folded like a Japanese screen—yet another influence in the stylistic melange from which this imaginative yet straightforward compositional gem derives. On the interior, the open plan around the hearth core typifies many of the firm's 1880s residential structures. On the outside, the simple geometry of forms, the ease with which they intersect, and the skin of shingles stretched around interior volumes (particularly at the second story of the tower, where it curves seamlessly into the southern facade) give this house its character. It is proof that petite and pretty need not be cute but can possess dignity and amplitude as well. The house was originally sited, like others by McKim, Mead and White in this neighborhood, on the north end of a long rectangular lot, but in the early 1990s another shingle house, by the Newport Collaborative, was crowded into what had been the stone-walled garden.

NE96 **Samuel Coleman House** (Whileaway, Boxcroft)

1882–1883, McKim, Mead and White. 7 Red Cross Ave.

One year after the overtly picturesque Skinner House, the same firm designed this large summer house and studio for Samuel Coleman, noted watercolorist and collector of oriental art, in a more explicit if quite abstract Colonial Revival mode. It sits across the north end of its gently sloping lot, with the main entrance centered on the north facade and the south face opened up with windows, bays, and porches; a generous veranda below; and a balcony loggia above to look out over a low, walled grass terrace and the garden beyond. The gambrel roof shelters the shingled top two stories, which, on the garden elevation, extend over the first-floor walls of ocher brick with stone trim. The composition is more conventional than that of the Skinner House, but the spread of the gambrel is nevertheless compelling, recalling the big houses that dominate Newport's waterfront on the Point. There is evidence everywhere of McKim, Mead and White's experimentation with early American forms, notably on the far side of the house past the main entry, where they uniquely juxtapose saltbox and gambrel roof forms, and under the eaves, where a decorative molding recalls the seventeenth-century plaster cove on the Wanton-Lyman-Hazard House (NE5). The house is now apartments.

NE97 Hambly Funeral Home (Grace W. Rives House)

1879, Peabody and Stearns. 1881, northwest wing, George Champlin Mason and Son. 30 Red Cross Ave.

For Newport, this is a unique remaining example of a Victorian Gothic house with polychrome trim in darker brick and lighter stone and a rare example of the extensive settings which once graced many more houses in the area. The stonework of the pointed-arched windows, the large octagonal tower pulled into one corner, and the carved molding give the overall boxiness an almost ecclesiastical or perhaps scholastic aura.

NE98 George Gordon King House

1902, McKim, Mead and White. 1–3 Oakwood Terrace

This house, built for a local philanthropist, is a large, elongated two-and-one-half-story structure with a gambrel roof and full-blown Georgian trim. Its walls, in pebble and dash, are unusual for Newport. Its most notable feature is its off-center main entrance with elaborately monumental ornamentation of double-story Cornithian columns supporting a pedimented gable and a blind balustrade set on smaller Ionic capitals.

NE99 St. Michael's School (Miss Ellen Mason House)

1902, Irving Gill. 180 Rhode Island Ave.

One of only three documented examples of Gill's work on Aquidneck Island (see also NE174 and PO13), this house seems decidely alien in Newport because of its design in the Califronia vernacular of Hispanic stucco. It replaced an earlier summer cottage by H. H. Richardson that had burned to the ground in 1899 and was probably commissioned through Gill's California contact with the Olmsted firm, which Ellen Mason had retained to plan the considerable gardens of her estate. Gill organized his broad wall planes of stucco under a hipped tile roof and punctuated them with a sequence of arched windows and (what were) open porches to the rear. This he embellished with ornate flourishes, like the ironwork and columns around the balconies. It may seem a house better suited to another region, but ironically it represents Gill's earliest use of a Hispanic-influenced vocabulary, for which he later became famous.

NE100 T. K. Gibbs House (Bethshan)

c. 1883, Dudley Newton. 396 Gibbs Ave.

Although its gambrel roof is in keeping with its Neo-Colonial style, the low spread of this house is not, and the extensive use of irregularly set stone with brick trim on this two-and-one-half-story residence is unusual. At one end of the lower service wing, a tall brick arch opens for a drive running beneath the second story to the rear of the house. There, porches open to a vista of the property and down the hillside to Easton's Pond and the ocean. The brick trim allows for such decorative touches as the window corbeling at either side of the projecting entry bay and the small rough stone patches that poke through the flat plane of brick quoining at each corner. A ruddy palette overall—bricks, pink granite, red slate, even rust-colored grout—unifies the various materials.

NE101 Linden Gate Porter's Lodge

1883, Richard Morris Hunt. 333 Gibbs Ave.

This small, picturesque building, with its deep-set south-facing balcony under a prominent eave and much decorative timber- and brick-work, looks transported from some Alpine locale. It was originally an auxiliary building at the rear of the Henry Marquand estate, Linden Place, designed by Hunt in 1872 and now destroyed.

NE102 Mrs. Frederic Eustis House
(Elm Tree Cottage)

1882, William Ralph Emerson. c. 1930s, south (garden) side altered with additions. 336 Gibbs Ave.

One of at least five houses that William Emerson designed for the Eustis family (the others are in Massachusetts), this residence sits narrow end to the street. Its ample two-and-one-half- story mass is composed of a long, narrow gable with cross gables toward the street and dormers breaking the uniform slope of the roof. Typically for Emerson's work of the 1880s, the entire house is cloaked in a skin of what were originally natural shingles; this treatment, combined with rounded shapes for conical elements, rounded corners (particularly around the main entry, a second-story inset porch, and dormers), and the curved outline of a monumental carved bracket on the street facade, enhances the overall volumetric effect, making this house more sculptural and homogeneous than most of the shingled houses of his contemporaries. In this restrained dependence almost exclusively on shape and texture, a simple grid of squared louvers cut as a simulated dovecote into the gable over the entry becomes ornamental in the same way as the bracket.

NE103 Arthur Emmons House

1882–1883, Peabody and Stearns. 300 Gibbs Ave.

Also of note on Gibbs Avenue is the Arthur Emmons House, which handsomely combines shingle and brick. The dominant features here are the multistory brick wall rising to a stepped gable and the attenuated chimney flanking the porte-cochere of the main entry. The roof sweeps off the main block in an exaggerated curve, sheltering the entry under its projection. As in Dudley Newton's Gibbs House, expansive windows and porches overlook a garden slope toward Easton's Pond.

NE104 Armistead House

1882, Bruce Price. c. 1980s, alterations to south side. 55 Hunter Ave.

In this early design by Bruce Price, we see him approaching the shingled, organic compositions that he created at Tuxedo Park and on which he established his reputation as one of the progenitors of the Shingle Style. Here, details of half-timbered gables, mullioned window groupings, and turned posts inspired by early colonial furniture mark a transition between the bracketed cottage of the 1870s, later Colonial Revival impulses, and the more integrated, organic unity of his work throughout the 1880s.

NE105 Sarah T. Zabriskie House
(Stone Gables)

1889, George Champlin Mason and Son. 100 Rhode Island Ave. (at Catherine St.)

Late medieval Flemish architecture inspired this large, two-and-one-half-story house with prominent chimney pots, stepped side gables, and rotund engaged tower at the rear. The picturesque massing of peaks, roofline, and volumes is complemented by rusticated stone walling, quoining, carved plaques, gargoyles, and fanciful copper finials, all a bit rambunctious in scale and simplified in detail.

NE106 Henry Swineburne House

1875–1876, Dudley Newton. 97 Rhode Island Ave.

Rustic (particularly chalet) sources provided the inspiration for this house in wood, brick, and slate, built for a client who helped engineer Newport's water system. It is one of the most effective designs by Dudley Newton in the "Modern Gothic" style, which he helped popularize locally in the 1870s. From the acutely pitched roofline of the rear block, which is complicated by an engaged polygonal tower and turret on one side, a steep, gabled element, really a bloated dormer supported on the posts and struts of a deep porch, projects to the street, establishing the image for this house. The jerkinhead dip of the front gable peak and the flare of the roof at the eaves provide a bonnetlike snugness around a display of vertical boarding, exposed framework, brackets, sawn patterns, and quatrefoiled window parapet, all of which meet the "Modern Gothic" aim of the expressive use of materials and structure and their inherent decorative possibilities. The roof bonnet throws off a shed roof to one side as a protective flap over the recessed entrance. Playful, joyous, fanciful, yet down to earth: the veritable image of what a Victorian summer cottage should be.

NE107 Jane Yardley House
1882, J. D. Johnston. 91 Rhode Island Ave.

NE108 Noyes–Luce House
1883, Clarence S. Luce. 15 Francis St.

These two houses represent a contrast between Queen Anne variety and plain shingled massing. In the Jane Yardley House, sited end shelters the upper story as it projects over the ground-floor street facade (it might have been called a Dutch gable at the time). Porches eat away at the corners of the second story, and while half the pediment of the porch marks the entry way, the rest thrusts into the mass of the house. At the opposite corner of the first floor, the living room is inset under deep brackets. Varied combinations of grouped windows in bands and shallow bays are organized beneath the jutting moldings, flares, and insets. Latticework molding slides into decorative medieval half timbering, while classical fan motifs provide an allusion typical for the Queen Anne Style. J. D. Johnston, a prolific architect-builder, knew professional contemporaries like Charles McKim and his partners and Dudley Newton, with whose residential work he skillfully allies himself here.

NE109 Francis Morris House
1882–1883, George Champlin Mason and Son. 86 Rhode Island Ave. (corner of Champlin St.)

NE110 George Champlin Mason, Jr., House
c. 1885, George Champlin Mason and Son. 15 Champlin St.

Houses in the Queen Anne Style are notable for asymmetrically grouped features folded around corner entrances. This pair offers contrasting treatments while displaying differences in overall massing on their sites. Facing on Rhode Island Avenue, a giant, multi-mullioned stained glass screen by John La Farge, under a horizontally distended shell hood, lights the interior stair of the Morris House. Around the corner on Champlin Street, an outside stair climbs to an elevated entrance and a broad Dutch door with bull's-eye glazing, this with its shell hood squeezed vertically off lush acanthus brackets. On the interior, the paneled entrance hall, now unfortunately painted white, becomes a vantage point from which to view the colored glory rising from its stair landing. If shell hoods were part of Newport's colonial heritage (more of them were still extant in the late nineteenth century than exist today), so the broad, bombé-curved eaves slashed across the front-facing cross gable from the wall below recall the similarly assertive effect of the coved treatment of the eaves in the famed late-seventeenth-century Wanton-Lyman Hazard House (see above). Here, in a typically Queen Anne manner, it provides a band of variously shaped windows to butt up against, cling to and rest upon.

Behind the Morris House, George Mason's son, newly made a partner in the firm, used the same high-cornered entrance stoop topped by a bedroom view porch for a vignetted view toward Easton's Bay, but this time set against a towerlike stack of triplet windows, each differently handled. Best of all is the way the asymmetrical slope of the roof binds the two disparate superimpositions of features while echoing both the pitch of the entrance stair and that of the site, whereas the cross gabling of the Morris House reinforces its corner site at the summit. Apart from the interest of their corner treatments, both houses present fairly stripped-down if competent versions of the Queen Anne Style. In this pair, unity of overall composition attracts less than piecemeal ingenuity of syntax.

Several more houses along Rhode Island Avenue between Francis and Kay Street are worth brief notice. Number 83 (c. 1866) shows George C. Mason, Sr., in his earlier Victorian manner, as witness the heavy timbering of the bracketed porch. Numbers 77 and 75 (1881–1882) are houses in the Queen Anne Style, both designed by Clarence Luce just before he briefly moved his practice to Newport from Boston and built for Thomas R. Hunter and Mary and Anne Stevens. While a fairly plain shingled house, the Hunter residence has a broad entablature at the front carved with a shell and ribbon frieze in a Renaissance Revival style but with an almost rococo flutter and delicacy that characterize much of Luce's ornament. The smaller Stevens house is more Colonial Revival in its decoration. Last in this row is number 67, the Matilda Lieber house (c. 1882), yet another shingled Queen Anne house, designed by Dudley Newton, with his characteristic decoratively curved porch roof.

NE111 Letitia B. Sargent House (Aufenthalt)

1881, Clarence S. Luce. 80 Kay St.

NE112 Carolyn Seymour House
(Hawkhurst or Hawxhurst)

1882, Dudley Newton. 66–66 1/2, 68 Kay St.

The first of these shingle houses, the Sargent House, is a puzzling and ambitiously singular effect from an original if uneven architect who here failed to unify the syncopated rhythms and shapes which he seemingly intended. A discordant row of vertical features rises from the flared shingle wall at the second story level, dominated by a polygonal tower with a peaked cap and an unexpectedly monumental window. Though centered in the main, flank-gabled mass of the house, it hardly appears so because of the lopsided overload of two more disparate shapes in line with the others containing service functions at one end of the house, both gabled but to different heights. The higher culminates in another of Luce's gable escutcheons, which needs a subdued color contrast to hold its place in the compositional melange. Moreover, the entrance and window bays across through stuccoed first story are out of sync in both shape and rhythm with the features above. Finally, the delicate linear detailing of the varied windows and their tensely asymmetrical placement (derived from contemporary progressive English examples popularized in such journals as the imported *Studio*) complete this layered counterpoint. The charm of the ingredients and the mannered picturesqueness of the image may justify the owner's affectionate "Abode" as the Germanic nickname for her cottage, although the overall result falls short of the repose that her label implies.

By contrast, the architectural rationale for the Seymour House is clear. There are porches along both sides, one for entrance, the other a sitting area with a balconied view porch above. An off-center bay window marks the living room projection between two of the walls with an unusual Queen Anne mix of patterned shingling. This appropriately turns wavy in the climactic attic gable, where acanthus leafage (or is it seaweed?) writhes, possibly with Art Nouveau inspiration. Dudley Newton was Newport's chameleon architect of his period. Having earlier mastered Richard Morris Hunt's Stick Style, here he rises to the challenge of Queen Anne. Soon we shall see him capably parroting other stylistic fashions. This cottage for a New Yorker

NE117 Richardson–Blatchford House

was split apart and turned into a duplex in the 1930s.

NE113 Job Peckham House

1855. 33 Kay St.

Job Peckham, owner of Newport's largest lumberyard in the mid-nineteenth century, built a trio of nearly identical houses as year-round residences, number 33 Kay Street as his own home and the others on speculation, number 30 for Joseph Bailey and number 26 for John and Fanny Irish. All are blown-up versions of what would have originated as fairly modest hip-roofed cottages with bracketed eaves and a centered gable. Here the prototype is enlarged to Victorian scale; the gables are spread and the widely spaced doubled brackets aggrandized. Though slightly varied in their ornamentation, the porches across the fronts all have bulbous columns flanking the entrance steps that appear part Moorish, part Egyptian. Each hipped roof is ringed with dormer windows, but only the Irish House retains its ornate cupola, a detail that originally crowned all three of these substantial houses.

A number of similar houses along Kay Street were built during the initial quarter century of its development, between 1850 and 1875, including 20 Kay Street, with its unique scalloped brackets around the porch, and 11 Kay Street, with an earlier, neoclassical colonnade.

NE114 Newport Reading Room

c. 1835. 1850s, renovations and addition, George Champlin Mason, Sr. 29 Bellevue Ave. (not open to the public)

This comfortable house with a capacious porch and the aspect of a country inn is unassuming architecturally. From its founding, however, it became the elite club in Newport, its cachet uncontested until the building of the nearby Newport Casino. Originally designed in a simple Federal style as a residentially scaled hotel, it was renovated for use as a private men's club in the mid-1850s when the porch was added. It was here that a socially prominent but devil-may-care guest of James Gordon Bennett rode his horse into the members' lounge on a dare. When Bennett was reprimanded, he retaliated by commissioning the casino, taking with him many of the younger members of the Reading Room and giving to fashionable Newport a more active and public center for summer recreation. George Mason, who designed the spacious, two-story, monitor-roofed billiard room addition, was one of two nineteenth-century architects to be a member; Richard Morris Hunt was the other.

NE115 Samuel Pratt House (Bird's Nest Cottage)

1872. 49 Bellevue Ave.

A few doors past the Reading Room, on the west side of Bellevue, is this small residence, completely covered in decorative motifs and forms. Complex gable shapes, fancy stickwork under the eaves, projecting corner bays, and a wall covering of multicolored slate roof shingles suggests an opportunity for its designer to try out a range of ornamentally picturesque ideas. Some have attributed this building, now unhappily squeezed almost impossibly between later structures, to Richard Morris Hunt, but it may well have been created by its owner, Samuel Pratt.

NE116 Colonel George Waring House (The Hypotenuse)

1870–1871, Richard Morris Hunt. 33 Catherine St.

This small cottage, now sited on the hypotenuse of the street corner and so named, was moved by Richard Morris Hunt from his family property on Bellevue Avenue and Church Street to this site and subsequently given sometime after 1876 to Colonel George Waring, a local engineer and author of technical manuals who helped establish Newport's sewage system. Only the four fluted Doric columns spanning the entry alcove indicate its Greek Revival origins. Everything else seems to be Hunt's 1870 envelopment of the kind of decorative vernacular image that he helped popularize in Newport during the 1860s and into the next decade—bonnet gable, deep eaves skirted with sawn tracery patterns, stickwork braces, and half-timbering effects. (Some of this detailing probably influenced the designer of the nearby Samuel Pratt House (see above). In this area Hunt built most of his early Newport Stick Style houses.

NE117 Richardson–Blatchford House

1870. 1883, probably William Ralph Emerson. 37 Catherine St.

Beneath the shingled exterior of this house are the gabled forms of a two-and-one-half-story cottage that may have been one of many rental properties in this neighborhood owned by Newport developer Alfred Smith. In 1883, Sophia Blatchford of New York had the house rebuilt in the most fashionable style with shingles sheathing all exterior surfaces. Parasol projections for dormers, a top-story sleeping porch and dovecote grid cut cleanly into the facade, and the flared shingle skirt dividing the stories complete a transformation so remarkable that there has been speculation that Boston's famed architect in the Shingle Style, William Ralph Emerson, may have designed it. The innovative entrance design of balustered staircase and porch with shingled columns underscores this attribution. Emerson worked on several other houses in Newport (NE17, NE102, NE181, NE182).

NE118 Virginia Scott Hoyt House (Ayrault House)

1916, Cross and Cross. 45 Catherine St.

Perhaps the last of the houses built in this neighborhood as a summer residence, this brick Neo-Georgian design uses stone detailing around windows and entryway and brick quoining patterns to simulate ornamental motifs more often found locally in wood. Its central staircase was removed from an eighteenth-century New York townhouse.

NE119 Rear Admiral Reed Warden House

1881, Clarence Luce. 68 Ayrault St.

If the Queen Anne Style is notable for engaging quirkiness, count on Clarence Luce to force the quality. This summer cottage for a retired naval admiral is unusual for its porch, roofed in a pair of front-facing gables (with unaligned supports) positioned toward one corner, the front elevation offset by an upstairs bay window in the corner diagonally opposite. A typical Queen Anne sunburst beams from a centered roof dormer onto the compositional tension below.

NE120 Isaac P. White House

1872, George C. Mason and Son. 66 Ayrault St.

On this, George Champlin Mason's most assertive chalet-style house, ornamental timbers define the major rectangular window openings. Its most prominent visual feature is the diagonal cross bracing across the second story. If this is really structural bracing, it is vastly overbuilt for decorative purposes. The chalet inspiration of this year-round residence, built for the rector of Trinity Church, is emphasized in the ornamental details of sawn eave brackets and acorn-shaped droplets.

NE121 Churchill–Yarnell House

1872, 1879, Dudley Newton. 62 Ayrault St.

After the forcefulness of Mason's chalet, Dudley Newton's treatment seems pretty and petite. Newton seemed to specialize in this kind of one-and-one-half-story, mansarded cottage in the early 1870s. This summer house, built for a naval officer, is one of several he designed in this neighborhood. The area around the eaves is particularly ornate here with a dropped skirt of boards and half-round battens continuing up into the entry gable and finished at the bottom with a scalloped picket edge. Newton also designed the matching rear addition later in the decade.

NE122 Mrs. Archie D. Pell House

1881, Clarence Luce. 11 Francis St.

Luce's quirky, personalized Queen Anne residences nearby make this striking design seem more conventional. Although he employs many predictable decorative motifs, from carved panels to coffered plaques, the small gable on the Everett Street side, covered with cement in which glass is embedded, may be the first use of that treatment on a summer cottage in Newport.

NE123 Samuel Honey House

1873. 12 Francis St.

The Samuel Honey House is a two-and-one-half-story Stick Style residence with an elaborately articulated entry porch and balcony with Moorish overtones

NE124 Daniel Swineburne House

c. 1862. 6 Greenough Pl.

With its steeply pitched roofs and its cross gable flanked by two similarly peaked dormers, this late Gothic Revival house is among the best of the few surviving examples of the style in Newport. It is replete with pointed-arched windows on the second story, curvilinear tracery under the eaves, and a massive droplet in the center of the entry gable. The bristling quality at the top of the house, typical of the assertiveness of architectural form prevalent in the 1860s through the next decade, moderates to the gentler image of earlier Gothic Revival at the porch. The curved roof and doubled posts are meant to strike the faintly exotic image of canvas suspended from poles. The side bays, probably original, are unified with the rest of the house by "flaps," making an awkward transition from the main slope of the roof. Daniel Swineburne was among the most prominent of the mid-century developers who helped turn Newport from a city of hotels into a "cottage" resort.

NE125 Mrs. James C. Porter House
(Porter Villa, Old Castle)

1855–1856, Seth Bradford. 25 Greenough Pl.

One of several stone mansions built in the 1850s by Seth Bradford as Newport summer residences, this was commissioned by a Louisiana woman. The rough texture of the Fall River granite over most of its surface is differentiated by the flat-cut trim work of the entry projection and window surrounds, whose forms, like those of the bracketed eaves, imply Italianate inspiration.

NE126 Joseph Tompkins House

1853, Thomas A. Tefft. 38 Catherine St.

This is the only extant Newport example of work by Thomas A. Tefft, the Providence architect best known for his Romanesque Revival designs and mill buildings in other parts of the state. The geometric mass of the two-story Tompkins House is decorated with the most restrained detailing: its clapboard facade is bounded by a surround of wide trim boards on all sides, broken only at the Doric-columned entry porch with inset doorway. The wide, uniform, overhanging eave of the hipped roof is duplicated on the smaller-scale cupola, flanked by original chimneys of particular visual interest. A walk around the house reveals the intensity with which Tefft considered the cupola, chimneys, and arched windows as separate entities. From the side, the curious extension of the main block along Greenough Place creates a layered effect accentuated by the original rear kitchen block. Only small breaks in the uniformity of the design are evident: a triglyph motif that turns the corners above the columns at either end of the entry porch or the flat-arched board that trims the top of each second-story window; but even in these instances the house retains a sense of planar severity.

NE127 King–Birkhead House

1872, Dudley Newton. 20 Catherine St.

Like the charming shop-fronted office he built for himself in the same year just around the corner on Bellevue Avenue, the King-Birkhead House shows Dudley Newton's fluency in the picturesque Stick Style (or, as he would have called it, Modern Gothic). Its mansard roof, covered with fancy slate patterns, employs his patented "Newton roof," which had a gutter set below the junction of mansard and wall. Its overlarge cornered tower, also mansarded, climaxes the bracketed and jigsawn ornamentation throughout.

NE128 Tillinghast–Tomkins House

c. 1858. 11 Redwood Ave.

In this massive, two-and-one-half-story clapboarded house, it is the emphatic doubled brackets beneath the broad overhanging eaves on all four facades that attract attention. The doubled polygonal piers supporting the small entry porch and the cross gables, centered in an ornate cupola, give it an Italianate air suggesting something between medieval and Gothic Revival.

Upper Cliffs

The neighborhood on either side of Annadale Road, which parallels Bellevue Avenue a few blocks to its west, was developed, primarily from open fields, in the mid-nineteenth century because of its proximity to the ocean beaches. These stretch below what is now the upper, northern end of Cliff Walk. This famous path running four miles southward along the rocky coast on the southeastern edge of Aquidneck Island begins at Memorial Boulevard, just above Easton's Beach (First Beach). Some of the most significant mansions can best be seen from the Cliff Walk, offering a sense of how their designers addressed their oceanside sites. It should be noted, however, that others are obscured, often for the privacy of residents, by foliage or topography.

While many of the buildings in the Upper Cliff neighborhood are small, undistinguished structures, originally intended as summer cottages or small houses for working-class Newporters, there are a number of noteworthy buildings. Some of the structures in this neighborhood, like the recently restored eighteenth-century house at 44 Merton Road, or 18 Annadale Road (c. 1860), were moved here when the old Bath Road, leading from the harbor to Easton's Beach, was widened into the present Columbus Memorial Boulevard at the beginning of the twentieth century.

NE129 John Easton House

c. 1760. 1 Cliff Ave.

The earliest building in the neighborhood, this two-and-one-half-story gambrel-roofed farmhouse originally stood alone on these bluffs overlooking the Atlantic, at some remove from the eighteenth-century town and the harbor. The generously scaled residence was the headquarters of the English Lord Hugh Percy during the occupation of the Revolutionary War. The proportions and equipoise of a typical eighteenth-century house with its central entryway and end-wall chimneys have here been altered by later additions.

NE130 65 Merton Road

c. 1870, George Champlin Mason, Sr.

Although it now seems to turn away from its street access, this large, mid-block house was sited facing east to the ocean on a large parcel of land. Its massive three-and-one-half-story mansarded volume is relieved by a polygonal tower on its north side and by elaborately stacked porches on the east, visible from Cliff Avenue. The first-story porch wraps around to the south side of the building, while ornately carved braces and crossed stick structures on the uppermost gable give the front a showy ornamentality whose liveliness would be appropriate for a seaside villa and whose stylistic references were typical for a Mason design of the 1860s.

NE131 9, 11, 15 Dresser Avenue

After 1883

NE131 9 Dresser Avenue

A little-known row of Queen Anne residences, these three two-and-one-half-story homes were probably built in the mid-1880s for middle-class clients. Numbers 15 and 11 show some similarities in orientation on site, porch work, and the detailing of carved plaques, but the real surprise here is the highly developed decorative scheme of number 9. Its west-facing entry turns away from Dresser and the nearby ocean view. This, along with its present cramped site, suggests it may have been the first of the three built, on a larger lot than now exists. The composition is an exuberant complex of engaged towers, projecting entry bay, stepped rooflines, and ornate decorative skin. The building is encased above a first story of clapboards in carved shingles, large stained glass stairwell windows, and serpentine trim boards, providing one of the most fanciful ornamental ensembles of its day.

NE132 Pinard Cottages

1882. Annadale Rd. and Narragansett Ave.

In the mid-nineteenth century the Upper Cliffs area was home to many small rental cottages, such as those of the Cliff Cottage Hotel Association (1860s), at the head of Cliff Avenue, or this set of summer rental units built by Cazeau and Elizabeth Pinard in the fashionable style of the early 1880s. Although alterations have occurred, the exteriors still clearly show the shingle idiom in a repetition of undercut porches, picturesque window treatments, and decorative patterns organized under a dominant gable roof. Of the six separate units eventually built on this site, only four remain, having been rehabilitated as condominiums in 1982 by a local firm, The Newport Collaborative.

NE133 Edward D. Morgan House

c. 1860. 97 Narragansett Ave.

Narragansett Avenue is the borderline between the older Upper Cliffs neighborhood and the famous district of mansion-cottages built between the 1880s and the 1920s to its south. The three-story Morgan House, representing the earlier era, is an asymmetrical Italianate villa with tall entry tower, horizontal boarding, bracketed eaves, and a large addition to the west.

NE134 Edward H. Schermerhorn House (Chepstow)

1860, George Champlin Mason, Jr. Narragansett Ave.

Sitting on its broad lawn, Chepstow suggests the typical summer cottage before the advent of the vernacular and decorative impulses of Stick, Queen Anne, and Shingle styles. The correctly French, flared mansard roof over a heavy, bracketed cornice tops its massively singular block. Every attempt is made here at imposing grandness, as the building is sheathed in smooth, horizontal boards, lending the exterior a hint of simple planar masonry. Chepstow is the kind of substantial residence that George Mason, Sr., designed for a large number of clients throughout the city in the mid-nineteenth century.

NE135 William F. Fahnestock House (Oaklawn, Bois Doré)

1926–1928, Charles A. Platt. Narragansett Ave.

NE136 Ogden Goelet House (Ochre Court)

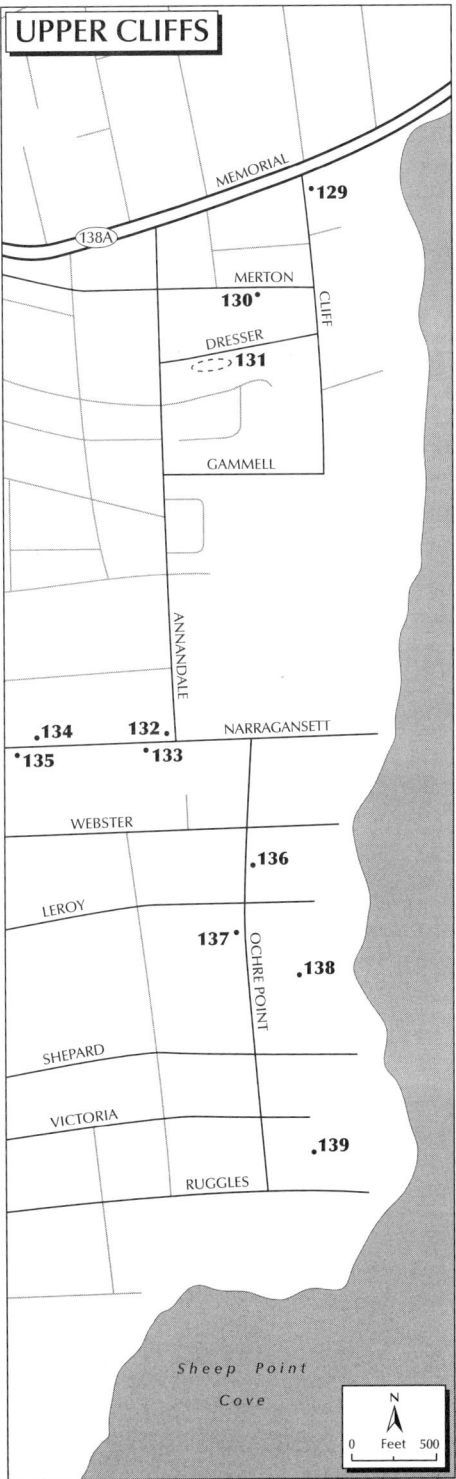

This house with a rather severe limestone facade, atypical for Platt, was built for a New Yorker in a French eighteenth-century revival style with hipped, dormered roof, ornate entrance surrounded by floral carving, and decorative window treatments. In typical fashion for this neighborhood, Bois Doré is somewhat overpowering on its relatively small parcel of land.

NE136 **Ogden Goelet House** (Ochre Court)

1888–1892, Richard Morris Hunt. Ochre Point Ave.

The three-and-one-half-story Ochre Court is somewhat startling in scale as it rises abruptly from the its sparse landscaping. Hunt designed this summer house for New Yorker Ogden Goelet and squeezed it onto a surprisingly narrow parcel of land just across the street from McKim, Mead and White's Shingle Style house of 1882 for Robert Goelet, Ogden's brother (partially visible to the north over the perimeter hedges). As with many other houses along the cliffs, Hunt has incorporated two fully developed facades into his design. The street facade, set a short distance behind a set of extravagant iron gates, is clearly inspired by French château imagery, which this was the first of Hunt's Newport houses to employ. The main section, offset at the south end of the building, is clearly distinguished from the service wing to the north by its more elaborate carved trim and the concentration over the entryway of arched windows and the intricate ornament of a pedimented dormer set against the steep rake of the roof. The limestone walls, set in flush blocks, have broad unarticulated areas, interrupted occasionally by low sculptural reliefs and tracery-framed windows. The stately ensemble affords a very different image indeed from the barnlike vernacular of brother Robert's house only a few yards away.

Ochre Court seems less cramped on its site and more balanced on its ocean side, where Hunt established a handsome cadence of arcaded porch openings and windows across its main block to the left, terminated at either end by towerlike projections. The open loggia on the first story leads out to a broad terrace that clearly sets the house off from its lawn. Through a series of setbacks Hunt has done more to diminish the bulk and blandness of the service wing on this facade than on the street elevation.

Inside the main entry is a masonry-walled stairwell that immediately opens onto a three-story court. Through the arched openings beyond, the coastal light and salt air dramatically flood into the central space. Surely Hunt and his client were aware that a similar feature existed next door in the Robert Goelet house, albeit on a smaller scale. Here the architect brought his newly developed grand manner to the design, with a palatial scheme of sculptural ornament, carved woodwork, painted ceiling, and open galleries stacked above. If his design does not make it clear enough that Hunt was informed by late medieval sources, he added one more clue. On the garden structure to the right of the main entry, fashioned like a cathedral's rood screen, is a self-portrait by Hunt just below the central pedestal. Here he depicted himself pondering his own work, hand to chin, not unlike the sculptural self-portraiture of the great master architects and builders of the Middle Ages.

Although the house retains its courtly ambience, some of the interior decoration and much of the original artwork have been altered or lost through its present use as the administration building for Salve Regina University. Happily, the university is now more sensitive to its environment, having recently hired Robert A. M. Stern to design the Rodgers Recreational Center (2002), across Ochre Point Avenue from the main gate of Ochre Court. In this rare (for Newport) postmodern effort, contextualist Stern humanized the awkwardly functional gymnasium mass by employing a full range of Shingle Style elements—engaged towers, multiple cross gables, undercut arches, trellis work, and even patterning in the shingles themselves. He sited the entrance on axis with the broad arch of Peabody and Stearns's recently restored shingled fantasy, the Hennery (1883), which separates the new recreation center from the Hunt building.

NE137 J. J. Van Alen House (Wakehurst)

1884–1888, Dudley Newton from plans by C. E. Kempe. Ochre Point Ave.

Wakehurst, which simulates an English Tudor manor house in its elevation and detailing, was built under the supervision of Dudley Newton from designs purchased by J. J. Van Alen from the British architect C. E. Kempe. Set back from the street and screened by a stand of trees, the horizontal mass of the house is treated sculpturally as a set of two-and-one-half-story staggered blocks, each of which terminates in a gabled end or dormer. Although the composition is remarkably restrained in its carved decoration, each elevation displays a lively sense of modeling, from the recurrent grid of window sets to the finials tracing the zigzag roofline. The central entrance leads to a wood-paneled hall running across the entry axis with the main staircase set before the visitor in keeping with English antecedents.

NE138 Catherine Lorillard Wolfe House (Vinland)

1883, Peabody and Stearns. 1907–1908, enlarged. Ochre Point Ave.

Vinland presents a contrast in almost every respect to the historicism of its stately neighbors Ochre Court and The Breakers. Unlike those

NE139 Cornelius Vanderbilt II House (The Breakers), exterior and entrance hall

buildings, Vinland, vaguely English in its antecedents, is a low form of dark, rusticated brownstone which stretches along a rise overlooking the ocean. Ornamental accents are provided by the copper ridge flashing in the form of icicles and the carved stone reliefs set near the entryway. The small, turreted gatehouse, whose serpentine ironwork lantern announces the name of the house at the street entrance, and the stable to the north conform to the style of the main house.

In 1877, Catherine Wolfe's brother, Pierre Lorillard, had commissioned the same Boston firm to design the original Breakers, which burned down in 1892. In 1896 Wolfe sold Vinland to Hamilton and Florence Twombly (she was the youngest grandchild of Commodore Vanderbilt), who had it greatly enlarged a decade later. All the Vinland buildings are now owned by Salve Regina University.

NE139 **Cornelius Vanderbilt II House** (The Breakers)

1895, Richard Morris Hunt. 1886, Playhouse, Peabody and Stearns. Ochre Point Ave. (open to the public)

In the 1890s, Richard Morris Hunt was considered by fellow professionals and patrons alike to be the dean of American architects. In a career stretching over three decades, he introduced a number of influential styles, culminating in his elaborately palatial residential designs of the late 1880s and 1890s, upon which his national reputation is based. Newport was the locale for numerous Hunt buildings, many of which survive intact; through them we can trace the evolution of his career from his early training in Paris to the mature, fully developed style of his later years, a style that is forcefully expressed in The Breakers, the summer "cottage" he built for Cornelius Vanderbilt II. Arguably Newport's most famous private house and completed just before Hunt's death in 1895, The Breakers is one of the capstone designs of his professional life, as it embodies lucidly expressed functional planning on a grand scale, a sophisticated vocabulary of sculptural ornament, and Hunt's hallmark European-derived historicism.

The earlier wood-frame house named The Breakers, which Cornelius Vanderbilt bought in 1885, was radically different from the structure we know today. Designed in 1877 by the Boston firm of Peabody and Stearns and originally owned by Pierre Lorillard, it incorporated a variety of textures and turreted shapes informed by the values of the Queen Anne revival, then in vogue. When it was destroyed by fire in 1892, Vanderbilt commissioned Hunt to design a new house—part summer retreat, part family seat—on the same site. (All that remains from the earlier era is the freestanding playhouse, also by Peabody and Stearns, a one-story gabled confection with outsized columnar carvings.) Hunt produced a number of different designs for the project (the drawings of which still exist), including one version with a French aspect and another, similar to what was eventually constructed, in the style of a sixteenth-century Genoese palazzo.

Surpassing in size Hunt's recently completed Marble House, built for Cornelius's brother William and his wife, Alva, on nearby Bellevue Avenue, The Breakers, at seventy rooms, conformed rigorously to its conception as family seat, with Vanderbilt references everywhere visible in the exuberant sculptural details. Not surprisingly, given the fate of the original building, the new Breakers was designed with innovative fireproof and technical features, such as Guastavino tile vaulting for some of the porches and steel-frame reinforcement.

Formal and balanced in its overall plan, the Hunt design adds variety where none might be expected. Each facade is interrupted by significant features, like the broad porte-cochere at the entry, the circular depression of the laundry court off the service wing, and the two-story trellised bay overlooking the formal gardens. The massive bulk of the building is compromised by the push-pull of setbacks and projections on the elevations and by relief surfaces with echoes of Renaissance forms. The most developed facade, as might be expected of a summer residence, faces seaward. Here, the central void of a mosaic-tiled loggia is defined by a triple arcade and flanked by end pavilions. The easy rhythms of the arches on the first story are quickened by the doubled openings of the second story, just as the Doric order employed below is shifted to the Ionic above.

The interior of the house shows the same concept of balanced variety: a clear, rational core in the vertical rise of the great hall, off of which spring rooms of various shapes and functions, the same scheme Hunt used in the earlier Ochre Court (see entry, above). Here the effect is even more monumental; the walls are supported on broad spans of tripled arches, and grand Corinthian pilasters run from tall bases up to a garlanded cornice more than two stories above. What is meant to grab and hold the attention is not spatial invention, but the luxurious combination of variegated stone, detailed carvings, gilded surfaces, and elaborate metalwork on a larger-than-life scale. Hunt's mastery is that of a great orchestral conductor who seamlessly blends overpowering elements into a unified but stirring experience.

In the decades following Hunt's death, Newport mansions like The Breakers were generally regarded as glorious excesses of the nouveau riche, pseudo-princely stage sets for a self-styled aristocracy of a new American class of industrial captains. Recent reinterpretations show that The Breakers reveals much more in its myriad layers of social impact: on the evolving townscape of Newport; on the artisans imported from abroad to execute its ornament and decor, sculpture and furnishings; on the rise of architecture as a profession and its relationship to powerful male and female clients; on the use of new technologies moving toward modern building practices; on the emergent servant corps who oversaw daily life in these large residents; and finally on the way in which the grand mansions reflected, in their construction and imagery, the desires, values, and social structure of the age.

Bellevue Avenue

Stretching due south from the top of the hill behind the old center of town, Bellevue Avenue was the result of a mid-nineteenth-century development plan whose very name suggests a pleasant, fashionable country landscape. Alfred Smith, a New York tailor turned real estate developer, led other partners in negotiating the rights to what became, in 1851, the initially unpaved Bellevue Avenue. While retaining much of its natural beauty, the neighborhood became more suburb than countryside as purchasers built large houses, sometimes squeezed onto smallish lots, separated by landscaped plantings and swatches of green lawn. Strung along either side of the avenue south of the upscale commercial district of small shops around the Casino are some of the most famous of Newport's houses.

NE140 Travers Block

1870–1871, Richard Morris Hunt. 162–184 Bellevue Ave.

After building a residence for William R. Travers in the late 1860s, Richard Morris Hunt was commissioned to design a new commercial block for this stretch of Bellevue Avenue, which had already begun to garner a reputation as a fashionable business district close to the resort hotels and to private cottages. On this two-and-one-half-story wood and brick facade, Hunt continued to develop the picturesque sensibility, heavily influenced by European vernacular architecture, that he employed in his residential work of the 1860s. On the first story, large windows open into a row of small shops, then as now sheltered by deep overhanging awnings to protect merchandise and promenading shoppers. Hunt wrapped the second level in a skin of ornamental half timbering, the grid of bracework segmenting the brick walls and windows to achieve an effect that is reminiscent of historical structures even as it reflects Hunt's own awareness of the most current architectural idiom in England. The whole of his composition is capped with a roofline array of bracketed eaves, cross gables, and dormers—adding visual delight but also making the third story more habitable as rental bachelor apartments.

With its windowed wall of retail spaces at street level and offices above, the Travers Block set a standard for successive generations of commercial buildings eventually stretching the next two blocks south along Bellevue. Not only did Hunt's building proclaim a new era in commercial design, richly imbued with European antecedents; it immediately dominated its environs in another way, since when built it was the largest block of shops in the city.

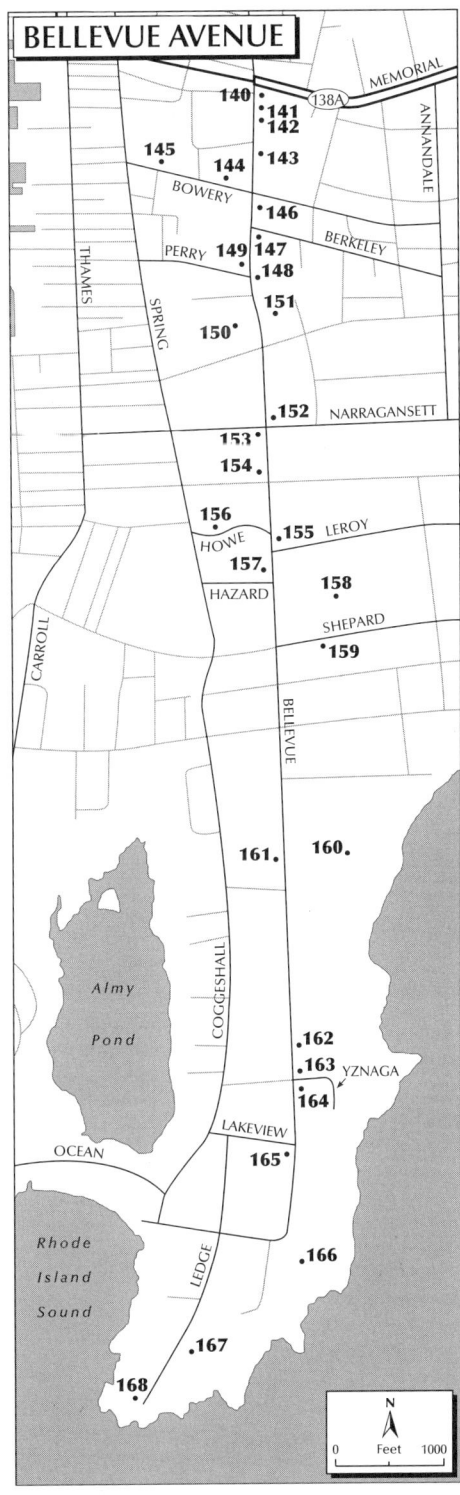

NE141 The Casino

1879–1881, McKim, Mead and White; 186–202 Bellevue Ave.

For this entertainment center fronting on the most fashionable street in the city, McKim, Mead and White established the currency of a highly decorated "shingle style" in Newport and a new form of building. Housing lawn and court tennis facilities, gaming rooms, bachelor apartments, and a restaurant and commercial shops at street level, the casino is really a complex of buildings and open-air spaces screened from the bustling life of the street. Conceived as a response to the exclusive Newport Reading Room, the casino was to be more "democratic" in its admission of members and more lively in

NE141 The Casino

the blend of physical and intellectual activities it offered. Interestingly, although referred to from its inception as a casino, it was always a family-oriented enterprise which prohibited gambling of any sort in its bylaws.

The main structure is a horseshoe-shaped edifice whose Bellevue Avenue facade masks a central lawn surrounded by chimneys, dormers, towers, loggias, and screens reached by way of a broad, central entry arch. The plan, which adjusts for an irregularly shaped site, and the Bellevue Avenue facade with its interlocked gables are attributed to McKim but surely Stanford White contributed his strengths in textural and ornamental detailing on the porch work, balustrades, and carved panels on both street and inner facades, giving this large building a very human scale. Except for the loss of turned railings that originally spanned the gaps between the large roof gables and a carved frieze just below the roofline, much of the decorative detailing on the facade is remarkably intact.

Inside the Bellevue Avenue entry, the main building and two large secondary structures containing a theater (whose interior shows White's marked taste for gilded, Renaissance-inspired woodwork) and a court tennis facility (one of only eight such courts in the country) create an almost magical separation from the outside world. As a harbinger of the shingle-clad, Queen Anne–influenced residences that the firm and other architects would create over the next decade, the casino's facings of brick, shingles, and carved wood trim offer a new ornamental vocabulary. Though departing radically from the style of the previous generation, employed by Hunt on the adjacent Travers Block, McKim, Mead and White carefully respected the scale, horizontal layering and general forms of that precedent in their own street facade.

NE142 **King Block**

1892–1893, Perkins and Betton. 204–214 Bellevue Ave.

This six-bay storefronted structure, developed as a commercial block by LeRoy King and his brother Gordon following the success of the adjacent casino and Travers buildings, echoes its more prominent neighbors in a reduced scale. While the cadence of inset bay windows and brick piers on the first story and the roofline of gabled peaks surely acknowledges the forms of McKim, Mead and White's design, little else about the King building seems familiar. In contrast to the varied textures in shingle, carving, and masonry on the casino's sculptural facade, Perkins and Betton's design is reduced to planar simplicity. Only at the central portion of the block, with two middle bays projecting slightly forward and two circular windows in the gable peaks, did the architects vary the uniformity of their brick facade; in effect, the King Block is an abstraction of the complex rhythms of the earlier buildings.

NE143 **Audrain Building**

1902–1903, Bruce Price. 220–230 Bellevue Ave.

Erected on part of the lot of the famous Atlantic House Hotel, which was destroyed by fire in 1898, this commercial block was built by A. L. Audrain, whose art goods business had previously been down the street in the Travers Block. This southernmost of the distinguished

Bellevue Avenue quartet of commercial structures differs from the others in that Bruce Price broke with the Anglo-American picturesque idiom, a style which he himself helped formulate, and instead designed a more monumental block with a highly evolved program of ornament.

The Audrain building is encased in a glazed skin of Renaissance motifs and monumental arched windows. Instead of the inset bays of the Casino and the King Block, each storefront is defined by a single glazed opening capped with an arch near the cornice. The front third of the brick mass is opened up with these monumental windows running from foundation to cornice; each is trimmed in a glazed terra-cotta decoration of piping, foliage accents, grotesque masks, and strapwork.

Rather than specifically reacting to its neighbors, the Price design is in keeping with a shift in his own work during the late 1890s toward a heavier, masonry classicism. One of his last works (he died in early 1903), the storefront bays when first opened held a genteel potpourri of businesses: the Audrain art supply shop, a stockbroker, an importer of canned goods, and a photographer.

NE144 George Noble Jones House (Kingscote)

1839–1841, Richard Upjohn. 1876, George Champlin Mason, Jr.. 1880, Stanford White. 1893, carriage house, Dudley Newton. Bowery St. (corner of Bellevue Ave.) (owned by the Preservation Society of Newport County; open to the public)

The George Noble Jones house, one of Newport's first ornately outfitted summer retreats, was built near the crest of a hill in a location then just south of the old, crowded part of town. Jones, a southerner, probably chose Richard Upjohn as architect because Upjohn had earlier designed a house for his in-laws in Maine. Familiar with Newport as a summer visitor from the early 1830s, Jones selected his site well; not only does it enjoy water views and cooling breezes from three sides, but the property was surrounded by that of other summer colonists from the South and was also adjacent to ponds and fields (long gone) where Jones could fulfill his favorite leisure pastime, bird hunting.

The primary image Upjohn achieved was that of a cottage orné, a relatively small house intensely decorated—in this case with medieval motifs—and intended to blend into its natural surroundings. The high-pitched gables, polygonal tower, diamond-paned windows, and carpentered details—sawn serpentine bargeboards, crenellated balcony rail, and trefoil droplets under porch roofs—are all Gothic Revival icing on Upjohn's wooden structure. Most of this early exterior finish is intact, but there

NE144 George Noble Jones House (Kingscote), entrance hall (left) and exterior (below)

have been some significant changes. The original sand-textured, buff-colored paint has been altered to a darker tone; a prominent striped awning over the second-story pointed-arched window to the left of the entry is gone; and the wood-shingled roof is now replaced with red slate. The most dramatic change occurred after David King, Jr., took charge of the house in the mid-1870s. (The house was named Kingscote, or "King's cottage," by the King family of Newport, who purchased it in 1863.) At that time, architect George Mason, Jr., added an updated dining room and service wing, which was moved back from the original structure in 1880, when Stanford White's dining room addition (with upper-story bedrooms) was inserted between the two. Although White attempted to harmonize his extension with the picturesque roofline, textured shingles, and other details of the original Upjohn design, its sheer size altered the cottagelike character the earlier architect intended. As a result, the best place from which to view Upjohn's work is off the southeast corner of the building, where the bulk of his polygonal tower and roof peaks hides the later wings.

The dark paneled woodwork and heavy cove moldings of the interior are in keeping with Upjohn's medievalism, although they too were added during the King residency. The most emphatic departure from this sensibility is visible in White's dining room. This light-filled space, in its low, horizontal proportions, its gridded orderliness, and its wealth of materials and textures—cork, Tiffany glass, glazed tiles, brass, marble, and stained woodwork—typifies the look McKim, Mead and White achieved in many of their residential designs of the early 1880s.

The nearby stable, surrounded by an alternating palisade fence designed by Upjohn, was redesigned by Dudley Newton to conform to the main house in 1893.

NE145 Edward King House

1845–1847, Richard Upjohn. Spring St.

Soon after the completion of the Jones residence, Upjohn undertook another commission on adjacent property for Edward King. Sited on the slope of the hill as it descends toward the harbor below the Jones cottage, the Edward King House more closely orients itself toward Narragansett Bay, with its front facade facing west, away from Bellevue Avenue. Here, instead

NE145 Edward King House

of the intricate Gothic-inspired shapes and ornament of the earlier building, Upjohn employed the Italian Villa Style, executed in brick, with round arches, scalloped canopies over projecting balconies, and a broader set of rectangular forms dominated by the stocky three-story tower, this feature most likely to take advantage of the water views to the west. Andrew Jackson Downing illustrated it in his influential volume, *The Architecture of Country Houses* (1850). The house, which originally would have been painted to lighten and visually unify the brick surface, is now owned by the city of Newport and operated as a senior citizens center.

NE146 Martha Codman House (Berkeley Villa, Bellevue House)

1910, Ogden Codman. 304 Bellevue Ave.

This grand residence, which Ogden Codman designed for his cousin, Martha, synthesizes European and American sources even as it adheres to the axioms of good interior design laid out in the famous manual that Codman coauthored with Edith Wharton, *The Decoration of Houses*. Boston-area buildings of the Federal period influenced the roofline configuration, the facade with its monumental coupled columns, and the octagonal side bay. Concealed within is a triple-storied, domed stair hall with Adamesque ornament derived from published eighteenth-century English antecedents. Three public rooms—library, drawing room, and dining hall—are arranged *en filade* along the south (garden) side of the house; the north side contains the service wing. To the rear is a garden pavilion, later designed by Codman with Fiske Kimball in keeping with the Anglo-American

NE149 Isaac Bell, Jr., House (Edna Villa)

spirit of the main house. The design was based directly on a Federal-period "summer house" by Samuel McIntire. The house, Codman's last Newport project, held part of the famous collection of American art amassed by Martha Codman and the husband she married some years later, Maxim Karolik.

NE147 LeRoy King House

1884–1886, McKim, Mead and White. Southeast corner of Berkeley and Bellevue aves.

Having worked with David King on the Kingscote addition (see above) a few years earlier, McKim, Mead and White produced a design for his brother LeRoy that extends and somewhat modifies the shingled buildings they developed during the first half of the 1880s. A central hall, large gabled masses, picturesque window arrangements, and a spectrum of surface textures (here conveyed largely in natural stone and brick with flourishes of shingle and pebble dash work), align this house with the firm's earlier efforts in this style. More compact and less formal than the Skinner and Bell houses (NE95, NE149), the King House is contemporary with the Edgar House (NE90) and, like it, shows the firm moving toward a restrained equilibrium in composition with two equal-sized wings flanking the entry porch.

Although the house sits close to Bellevue Avenue, its entrance is on the north (Berkeley Avenue) side. This orientation, in keeping with the firm's siting of other houses in the 1880s, meant that the entrance originally faced the massive Atlantic House hotel complex, which was a short distance away, on the site of what is now a shopping center.

NE148 C. H. Baldwin House

1877–1878, Potter and Robinson. 328 Bellevue Ave.

The profusion of gables, bays, chimneys, and window grids on the two floors of this Queen Anne residence creates a variety of angles, shadowed eaves, and sloping projections. The first story, of brick, is largely screened by a deep porch whose stocky turned posts run across two-thirds of the facade. The overall effect is of a woven tapestry: patches of shingle work and clapboards are divided by the sculptural treatment of gables and bays, and the brickwork is re-exposed by the half timbering of the entry bay. This meshing of elements is forcefully seen at the south end of the house, where a tall, pilastered chimney rises up to meet and penetrate the projecting third-story gable, reemerging at the roof line in a corbeled chimney cap. The broad, low arch of the porte-cochere reaches invitingly toward Bellevue Avenue.

NE149 Isaac Bell, Jr., House (Edna Villa)

1882–1883, McKim, Mead and White. 191 Bellevue Ave. (at Perry St.) (owned by the Preservation Society of Newport County; open to the public)

This rambling shingled house for Isaac Bell, Jr., and his wife, sister of James Gordon Bennett, was designed shortly after the completion of Bennett's nearby casino. Bell was a prominent New York cotton broker who was to become the U.S. ambassador to the Netherlands in 1885; although Bennett may have been the connection to the firm of McKim, Mead and White, according to the firm's own account, Bell himself paid for the house, contrary to Newport tradition, which recalls Bennett as funding the project. It was named Edna Villa by Samuel Barger, the Vanderbilt family lawyer, who purchased it in 1891.

Although sited differently than the firms' other local projects, the Bell House is essentially an elaboration of their smaller residential work nearby, like the Skinner and Tilton houses. The house sits far back on its corner site, with its entrance on the secondary Perry Street but its primary facade on the fashionable avenue. The main vantage point may well be from the corner itself, where the two elevations bend around the "hinge" of a multipaned window turret. Both facades are unified by the shingled skin, the wraparound first-story porch, and the way each is terminated by a large tower volume at each extremity: a tapering three-and-one-half-story bell-capped tower near the entrance (in which some see a playful pun on the owner's name) and a more broadly proportioned two-and-one-half-story polygonal bay with open porches on both levels projecting from the far northern end of the Bellevue Avenue facade. Faux-bamboo porch posts combined with a conical roof give this bay the unexpected look of a thatched hut on a grand scale. On the primary elevation, the first story is cast in deep shadow by the porch, while the upper levels appear as flat planes articulated by typical Queen Anne motifs of intersecting gable shapes, grouped windows, and the graphic shingle patterns.

The interior is also notable. A main entry hall with inglenook fireplace opens off the entrance on the first floor, acting as a spatial hub around which the dining and drawing rooms and the stairwell create a sense of a single, continuous space, flowing out from the front door and rich with decorative details that include adaptations of Breton furniture.

NE150 **Edward Julius Berwind House** (The Elms)

1899–1902, Horace Trumbauer. 193 Bellevue Ave. (at Dixon St.)

Built for Philadelphia-born coal magnate Edward Berwind and his wife, Herminie Strawbridge Torrie, as their summer home, The Elms, like so many of Horace Trumbauer's residential designs, was carefully modeled on eighteenth-century French antecedents. The entire west-facing garden facade of the Berwind House is an enlarged copy of the garden elevation from the Château D'Argenson at Asnières, which the Berwinds visited only a year before their Newport mansion was designed. The four terra-cotta statues installed against that wall were probably purchased from the French château during that trip.

In some respects, this building addresses its private garden more than it does Bellevue Avenue. The restrained classicism of Trumbauer's street facade is reflected in its smooth limestone facing and in the monumental columns and piers flanking the main entryway. A third story is hidden from view, tucked behind the uppermost edge of a blind stone balustrade which caps the entire cornice. This rather rigid arrangement is somewhat loosened on the garden facade. Trumbauer maintains the balance and regularity of the front, but a central bay swells from the main block of the house in a

NE150 Edward Julius Berwind House (The Elms)

graceful volume echoing the curved top windows, moldings, and decorative grillwork found throughout.

All interior details have survived from the Berwinds' era. Upon entering, one passes beneath an enormous arcaded wall atop Ionic columns of highly variegated marble, creating a sumptuous effect. The sense of an eighteenth-century pastiche is elaborated in the furniture, hardware, and draperies, all orchestrated by the Parisian firm Allard et fils. Much of the art was purchased through the international art dealer Joseph (later Baron) Duveen in London. The eclecticism at the turn of the century is reflected in the range of historical periods, from sixteenth-century Venice to eighteenth-century France and nineteenth-century Vienna, represented in interior details.

Along the western side of the eleven-acre property are the largely intact gardens. Among the most developed of Newport's formal landscaping schemes, these are replete with sunken gardens, fountain statuary, and geometrically planted parterres. The combination stable-garage, in a style similar to that of the main house, dates from about a decade later, when horse-drawn carriages were just beginning to be replaced by automobiles.

NE151 **William Weld House**

1882–1884, Dudley Newton. 364 Bellevue Ave.

Many Newporters now refer to this building as "De la Salle," after the boys' preparatory school that occupied it in the middle years of the last century, but it was built as a private residence. The rough stone walls trimmed with darker stone details complement the medieval forms of steep pitched roofs, corner turret, and the two flamboyantly shaped, Jacobean-inspired gables that cap the entry and the south side of the structure. Dudley Newton experimented with a similar ensemble of forms on a smaller project, his 1870s design for the outbuildings at Belair (see above), but they seem more comfortably employed here on these three-story facades.

NE152 Christopher Columbus Baldwin House (Château-Nooga, Chatterbox)

1870s. 1881, facade and additions, George B. Post. Narragansett and Bellevue aves.

Hired by Baldwin, the newly named president of a southern railroad, to update and enlarge the facade of an extant building, George Post designed a new face for Château-Nooga in a composition related to the Anglo-American imagery of H. H. Richardson's Watts Sherman House (NE159), built several years earlier and a few blocks away. Much of the intended multicolored effects of half timbering, carved wood trim, and red and yellow brick walls have been recently restored. In the grouped gables, trimmed by bargeboards with elaborate fernleaf reliefs, a sense remains of what critic Mariana Griswold Van Rensselaer later called the "most unique" house in Newport.

NE153 Albert Sumner House (Rockry Hall)

1848. 1880s, addition. 425 Bellevue Ave. (at Narragansett Ave.)

This two-and-one-half-story house sited catercorner to the intersection of Bellevue and Narragansett was built as a country retreat for the Bostonian Albert Sumner (brother of the famous abolitionist Charles Sumner) while the surrounding area was largely rural. The original section of Rockry Hall sits closest to Bellevue Avenue, its high-pitched roof, stone trim, and trefoil window giving the cottage a Gothic air, in keeping with the nearby Kingscote (see above) but somewhat heavier and less ornamental in its overall effect. The main block of the house, terminating in the crenellated tower at the west end, is a later addition whose combination of shingles, stonework, and revival details subsumes the original cottage into the Queen Anne vocabulary popular in the 1880s.

NE154 Gustavus Swan House (Swanhurst)

1851, Alexander McGregor. 443 Bellevue Ave.

In 1851 Alfred Smith, a New York tailor turned Newport developer, purchased 140 acres of land south of Dixon Street and requested that the city extend Bellevue Avenue into that area. Within a year, a dozen new summer residences were opened, mostly owned by families from the South and the Midwest. One which was Swanhurst, built for Judge Gustavus Swan of Columbus, Ohio.

Swanhurst is not in the classical revival style for which McGregor has become known but is instead informed by the cottages of the preceding decade. Deep, bracketed eaves, engaged octagonal tower, arch-topped windows, scalloped wood trim, and projecting balconies mix aspects of Gothic imagery with Italianate or Tuscan details, motifs found on the nearby Kingscote and the Edward King House, but its uniform stucco coating and broader volumes impart a severity and reserve unlike the more elaborately decorated Kingscote but similar to what was probably the original exterior of the King House.

NE155 John M. Hodgson Cottage (The Flower Cottage)

1882, Clarence Luce. Bellevue Ave. (at Leroy Ave.)

Clarence Luce designed this small structure, now radically altered, as a residence and retail shop (with accompanying greenhouses) for the already established florist business run by New Yorker John Hodgson for the summer colony. Originally, a gabled second story perched more delicately than it does today over a first level whose broadly flared, coved walls supported the deep overhang of the upper floor. One of these curved flanks can still be seen today on the north side of the cottage. This arrangement, along with a shingled exterior, an undercut porch (where there is now a Palladian window), and the diminutive size of the structure strongly suggest a small rustic work shed or garden pavilion, or even a boathouse or lifesaving station—references particularly appropriate for its intended function and locale.

NE158 William Shepherd Wetmore House (Château-sur-Mer), library and exterior

NE156 Mary Bruen House (The Hedges)

c. 1850s. 1872, renovations, Richard Morris Hunt. 6 Howe Ave.

Richard Morris Hunt, hired by Bostonian summer resident Mary Bruen, capped an original, predominantly rubblestone structure with a complementary picturesque cap of gables, balconies, and stickwork bracing. Although only a renovation of an existing residence, this three-and-one-half-story house, like Hunt's other Newport work from this period (i.e., no longer extant houses for Thomas Appleton, Henry Marquand, and others), is a continued exploration of the motifs he had begun to explore almost a decade earlier, a hybrid of European and American vernacular sources still visible today in the J. N. A. Griswold House (see entry for Newport Art Museum, above).

The house has no relationship to Bellevue Avenue today, but it, like the nearby Château-sur-Mer, was originally conceived as one of the large summer homes set far back from that newly opened avenue. In the early twentieth century the lot was split, and the section closer to Bellevue was developed separately and subsequently acquired its name, The Hedges, some time after the Bruen family sold the property.

NE157 Harold Carter Brown House

1893, Dudley Newton; Ogden Codman, interiors; Frederick Law Olmsted, landscaping. 459 Bellevue Ave. (at Hazard Ave.)

Having commissioned a new house for his bride, Georgette Wetmore Sherman, Harold Brown, a member of the venerable Rhode Island family, opted to balance heavy stonework with novel light and airy interior designs by the young Ogden Codman in what would be his first major project. The French sources upon which Codman drew were attractive to Brown, who was interested in the French Empire style and collected furniture of the Napoleonic era. Since the Browns were a highly reputed family with ancient Rhode Island roots, they felt no need for ostentatious display, so the house is discreetly sited and is more fully screened than any of its neighbors on grounds planned by Frederick Law Olmsted.

NE158 William Shepherd Wetmore House (Château-sur-Mer)

1854, Seth Bradford. 1870–1880, renovations, Richard Morris Hunt. c. 1920, Edith Wetmore Garden House, Frederic Rhinelander King. Bellevue Ave.

As its name implies, when Château-sur-Mer was built in 1854, it stood almost alone on the southern end of Aquidneck Island, just off the newly laid-out section of Bellevue Avenue and with fewer than a score of other structures between it and an unimpeded view of the Atlantic Ocean. The use of Fall River granite and the asymmetrical placement of the prominent entry tower were typical of designer Seth Bradford's other contemporary projects, including Belair (see above). This combination must have pleased his wealthy New York patron, William Shepherd Wetmore, with the suggestion of formality, even on this country estate, and its acknowledgment of Richard Upjohn's

NE159 Watts Sherman House

notable Italianate designs of the previous decade.

Almost a decade after inheriting the property from his father while still in his teens and having recently married, George Peabody Wetmore enlisted the services of Richard Morris Hunt to update and enlarge the house in a campaign of work that was carried out in two phases, 1870–1873 and 1874–1880. Hunt substantially transformed the Bradford building with a complete interior alteration in plan and decor, making it more Néo-Grec, the fashionable Second Empire style that Hunt knew well from his years in Paris. He corrected the form of the mansard, added much carved stone detailing, formalized the entry with a massive stone porch, and added surrounding stonework in a current French idiom.

The house has a distinguished list of architects associated with its history; after George Peabody Wetmore's death, Ogden Codman designed the Green Salon (1903), Frederic Rhinelander King built a small garden house at the northern edge of the property (c. 1920), and John Russell Pope worked on small alterations to both interior and exterior in the 1930s.

NE159 Watts Sherman House

1874, Henry Hobson Richardson. 1880–1881, interior redecoration, library and dining room fireplace, Stanford White. 1920, addition, Dudley Newton. c. 1960, addition. 15–17 Shepherd Ave.

This, the only one surviving of several Richardson buildings in Newport, sits on property once part of the neighboring Château-sur-Mer and split off for the Wetmore daughter who married William Watts Sherman. Although it has long been known as the William Watts Sherman House, landowner evidence suggests that it would more appropriately be named for the original owner, Anne Wetmore Watts Sherman.

As early photographs show, Richardson's design originally stood, as did so many other residences from the period, on an open, landscaped plot with clear views to the bay and ocean to east and south. Although its vertical thrust is now compromised by changes on the north side of the building, the entire composition was ordered by the all-encompassing gable of the entry facade. This, crossed with the gabled block to its south, created the sense of two interlocked monumental masses, scaled down by an arrangement of textured bands. Random-set rustic stone, jigsawn shingle patterns, strips of multipaned windows, small-scaled half timbering, and stucco work add variety to the simple gabled roofline. The choice of materials may show the strong revival influence of the English architect Norman Shaw, but Richardson blended this English historicism with Americanized elements such as sheltering gables, undercut porches, and exposed beam ends to express a sensibility much more organic than the French manner employed by his contemporary Hunt in projects of the same date, such as Château-sur-Mer. Some of the impressive interior features have always been attributed to Stanford White, who was then working in Richardson's office and would soon return to redecorate the library and dining room. The orchestration of the lavish exterior ornamentation may also have been informed by the young designer.

Between the most recent institutional addition, in brick, and the original Richardson gable is an enlargement by Dudley Newton that successfully adopts all of Richardson's detailing while altering the original mass into a more horizontal spread. The northernmost wing was added in the 1960s when the house was a Baptist home for the elderly; it is now owned by Salve Regina University and is not open to the public.

NE160 Tessie and Herman Oelrichs House (Rosecliff)

1899–1901, McKim, Mead and White. Bellevue Ave. (near Marine Ave.)

A site overlooking the cliff area and well planted by the previous owner with rosebushes

NE160 Tessie and Herman Oelrichs House (Rosecliff) (top, left)

NE162 Alva S. and William K. Vanderbilt House (Marble House), Chinese Teahouse (above left) and exterior (above right)

gave this house its name. From Bellevue Avenue, it appears to be a relatively compact, one-story structure, smaller than the "cottages" which preceded it nearby on Ochre Point. Its composition and decorative scheme were suggested by the seventeenth-century classicism of Jules Hardouin-Mansart's Grand Trianon at Versailles; perhaps this refined source also influenced McKim, Mead and White's use of an unusual exterior surfacing—white glazed terracotta—giving Rosecliff its gleaming appearance as a pristine pleasure pavilion.

On the approach to the entry arch at the south end, the first story, wrapped in its cadence of piers, columns, and carved swag, dominates the facade. It appears to be surmounted by a smaller attic story and low balustrade; but in fact, the architects hid three stories (a public first level, a bedroom level, and the upper servants' quarters) behind the uniform rhythms and horizontality of their most decorative Newport facade.

The H-shaped plan frames a front garden and a seaward-facing terrace with projecting wings. The connecting segment contains Newport's largest ballroom, some 40 by 72 feet. In his interior design, White staged a scene of social theatricality for the owners and their guests unique even for Newport. The grand ballroom is reached through an ornate hall. There, those arriving from outside could see Tessie Oelrichs make a dramatic entrance down one side of the heart-shaped staircase, which gracefully cascades from the second floor and spreads like the train of an elaborate gown as it spills toward the ballroom.

NE161 **Pembroke Jones House** (Sherwood)

1906–1908, redesign of earlier house, Hoppin and Koen. 553 Bellevue Ave. (at Bancroft Ave.)

When Pembroke Jones, a wealthy financier and merchant from North Carolina who made his fortune in the rice market, bought a late nineteenth-century cottage owned by the Havemeyer family, he hired the firm of Hoppin and Koen to remodel it as a backdrop for the lavish summer entertaining for which the Joneses were famous. Francis Hoppin, who apprenticed with McKim, Mead and White between 1886 and 1894, had many Newport social connections, and Jones's choice of this New York firm was probably based on the success of Hoppin's first big Newport commission, the General Francis Vinton Greene House, Armsea Hall (1904), a house that once stood just south of Hammersmith Farm (see below).

Hoppin's transformation of the older dwelling employed an essentially flat, massive main block to receive ornately carved Adamesque window treatments, decorative relief rondels, and monumental colonnaded porticoes on both the east and west facades. The dominant planarity of the walls echoes an English Georgian precedent, evidently an association that the Joneses were eager to emphasize. This scheme was continued in the arcaded ballroom at the south end of the building, added about eight years later and also designed by Hoppin. Although not professionally linked to this building, another noted architect, John Russell Pope, knew it well, as he later married the Joneses' daughter, Sadie.

NE162 Alva S. and William K. Vanderbilt House (Marble House)

1888–1892, Richard Morris Hunt. 1913, Chinese Teahouse, Hunt and Hunt. Bellevue Ave. (owned by the Preservation Society of Newport County; open to the public)

Sited, like many of Newport's "cottages," on a relatively small lot within yards of Bellevue Avenue, Richard Morris Hunt's Marble House is surely one of the most visually extreme in combining a grandly formal exterior with a sumptuous series of rooms, each outdoing the other in thematic effect and the use of a full palette of colored stonework. Delighted with Hunt's design for his Fifth Avenue mansion in Manhattan, William Vanderbilt commissioned him to build this Newport residence as a birthday present for his wife, Alva, money being no obstacle to the most luxurious results.

Hunt applied engaged piers around the entire house to monumentalize the two-story structure. These visually support an elaborate entablature, deeply carved cornice, and balustrade running around the top edge of the entire building. The relatively shallow carving on much of the white marble exterior with arched windows on the first story of each bay contrasts dramatically with the sculptural shadow effects of Corinthian capitals and the cornice arrangement. Most prominent of all is the projecting entry portico, itself supported by massive Corinthian columns and flanked on either side by the curving ascent of stone railings lining the entry drive. On the sea side, two wings project to either side of a terrace. Here the window arrangement is reversed from the rest of the facade: arched-topped windows surmount rectangular doors, providing a greater sense of openness and light. Based loosely on such eminent sources as the White House and aspects of the Petit Trianon at Versailles, the conception here was developed as the result of a collaboration between the architect and his client, Alva Smith Vanderbilt, who maintained a constant interest in every aspect of the design and construction of her pleasure palace.

Whether suggested by Hunt or his patron, the highly decorative scheme begins on the exterior. The conventional placement of a swagged frieze above the massive front entry screen is matched on the sea side by spandrel reliefs in the French manner, suggesting the seasons. On the north side of the house, other reliefs refer to female attributes of industriousness and wisdom, appropriate for Alva Vanderbilt, who would later become a prominent supporter of the suffrage movement.

On the interior, Hunt's mastery of ornament is readily apparent. To the left of the great entry hall is the dining room, where pilasters with Corinthian capitals surround the space, echoing their monumental counterparts outside. Here, however, they are carved of a dark pink Numidian marble, whose highly charged variegations, when combined with the bronze gilt capitals, wall sconces, and intricate ceiling framing the heavenly scene of a central mural, create an intended scenario of otherworldliness, as if this were a residence of foreign royalty rather than American industrialists. Nowhere is this sense of suspended reality more apparent than in the Gold Ballroom, across the main entry hall. Instead of the fiery red tones of the dining room, here everything glitters: gilded walls, mirrors, glazing, and bronze sculpture. We are left to imagine what effect the light from scores of candles in the two huge chandeliers would have had on such a shimmering setting.

Settings from other places and other times change from room to room, but all make clear allusions to a regal environment. French references are scattered throughout the fifty rooms of the house, including a relief portrait of Jules Hardouin-Mansart (architect of Versailles), the rococo library and bedroom for Alva Vanderbilt, and the Louis XVI decor for William Vanderbilt's own bedroom. This historicism is most specific in the first-floor Gothic Room, used by the Vanderbilts as a living area, with its groined ceiling, crocketed mantel, paneled wainscoting, and filigree chandelier emulating the style of the medieval art collection originally displayed there.

The house took almost four years to complete; the Vanderbilts opened it with an ostentatious housewarming in August 1892. The couple did not use the house together very long before their marriage failed. Alva married Oliver Hazard Perry Belmont at the beginning of l896 and moved a few houses down Bellevue to Belmont's Belcourt (see NE165), another residence by Hunt. Ironically, after Belmont's death in 1908, she moved back to Marble House, and in 1913 Hunt's sons, as the successor firm of Hunt and Hunt, designed and built the Chinese Teahouse overlooking the ocean cliffs at the eastern edge of the Marble House property. This was restored in 1988 and moved back from its eroding site by the Preservation Society of Newport County, which maintains both it and the main house for public viewing.

NE163 Edward Knight House
(Clarendon Court)

1904, Horace Trumbauer. Bellevue Ave. and Yznaga Ave.

NE164 Mrs. George Widener House
(Miramar)

1914, Horace Trumbauer. Bellevue Ave. and Yznaga Ave.

These two adjacent residences were built, like Horace Trumbauer's earlier Berwind House, for wealthy Philadelphians. Built for an executive of the Pennsylvania Railroad, Clarendon Court is set only a few yards back from the street; most of its eleven-acre site stretches behind the house to the ocean. With a central stone facade flanked by projecting side wings and classical detailing, Trumbauer here reflects a Palladianism filtered through the designs of the eighteenth-century English architect Sir William Chambers. Clarendon Court's severity, accented by the crisp planarity of its unadorned walls, is relieved only by the pedimented surround of the entry and the figurative sculptures perched atop the cornice.

Miramar is a relatively small structure, designed in a formulaic architectural vocabulary that here seems more French than English, set as it is among elaborate formal gardens. In contrast to the restraint of Clarendon Court, Trumbauer crammed his decorative scheme for Miramar onto the two-story facade; its rusticated blocks, arcade of oversized windows, sculptural reliefs, and balustrade all compete with the intimate scale of the building itself. While Clarendon Court achieves a kind of monumentality through its minimal ornament, Miramar seems perversely miniaturized by its decorative surface. For years Trumbauer was the architect for both private and public projects involving the Widener family; ironically, as this house was under construction, Mrs. Widener's husband and son perished in the *Titanic* disaster.

NE165 Oliver Hazard Perry House
(Belcourt)

1891–1894, Richard Morris Hunt. Bellevue Ave. at Lakeview Ave. (privately owned but open to the public)

Alva Smith Vanderbilt probably has the distinction of being the only Newporter to have resided in two of her own Bellevue Avenue mansions, first with her husband, William Kissam Vanderbilt, in Marble House and then, following their divorce, across the way at Belcourt, which Oliver Hazard Perry had given to her as a wedding present in 1896. Often considered one of the most visually striking of Newport's large houses, Belcourt was designed by Richard Morris Hunt with medieval references throughout, so today it is sometimes nicknamed Belcourt Castle. Part of its originality lies in Hunt's idiosyncratic combination, representing aspects of two early designs for the house, of English half timbering with French-style masonry work. Here, in one of his last works, Hunt returned at least partially to the picturesque wood-frame imagery of his earliest projects, such as the Griswold House, of the 1860s (see entry for Newport Art Museum, above).

Toward the southern end of the Bellevue elevation, an arched gateway leads to a compressed garden court, all but overwhelmed by the scale of the half-timbered residential wing to the north and the lower, turreted stable and service wing to the south. Much of the first story of the main wing is taken up by immense and elegantly furnished stables, with living quarters and a Gothic-vaulted ballroom for baronial entertaining of up to 300 guests. This peculiar but practical arrangement prompted no less a visitor than Julia Ward Howe to call it "a most remarkable house."

The steep slope of mansard roofs punctuated by copper-framed circular dormer windows tops Hunt's masonry walls. Heavy, rusticated granite quoining, arched brick door and window surrounds, and iron grillwork—an unexpected range of materials and textures, not always successfully unified—leave an unresolved but lasting impression. The exterior currently shows need of restoration.

NE166 Frederick W. Vanderbilt House
(Rough Point)

1891, Peabody and Stearns. 1922–1923, renovations and additions, Horace Trumbauer. South end of Bellevue Ave. (open to the public)

The architects of the Frederick Vanderbilt House chose its rustic-hewn gray-brown granite as fitting for this most dramatic of Newport sites, where a southern point of the island confronts the Atlantic waters with an eroded, rocky coastline. Built for a Vanderbilt who was as much concerned for his privacy as some of his neighbors were for public display, the house is all but screened from public view, by streetside walls, a planting design (originally by the Olmsted firm but now altered), and the coarse cliffs that rise from the sea.

Borrowing composition and materials from English manor house sources, the house is low and horizontal, its spaces divided on the exterior by several cross gables and the cant of its service wing. The small entryway leads to a two-story English-style, oak-paneled hall, off which the other rooms of the house are arranged. The heavy ceiling beams and the oversized fireplace fitted with gargantuan fire dogs underscore the closed, medieval character of the Peabody and Stearns design. Later additions and interior renovations by Horace Trumbauer, such as the solarium overlooking the ocean, are more Gallic in their open classicism, spatial proportions, arched windows, and use of color, reflecting the shift in domestic taste encouraged by Edith Wharton, Ogden Codman, and others. Today the house holds an eclectic, idiosyncratic array of art, furniture, and even clothing once owned by Doris Duke. In 1925, as a thirteen-year-old, she inherited the house from her father, the energy and tobacco magnate who established Duke University and had purchased the property a few years earlier. Several years after her death in 1993, it was opened to the public as a historic house museum.

NE167 Ward–Wharton House (Land's End)

1864–1865, John Hubbard Sturgis. 1893–1894, interior renovations by Ogden Codman. c. 1916–1917, enlargement by Hoppin and Koen. 42 Ledge Rd.

NE168 John Russell Pope House
(The Waves)

1927, John Russell Pope. End of Ledge Rd.

Ledge Road leads to the rocky shore (in the nineteenth century, there was a popular fishing shack here) and to the southern end of the famous Cliff Walk. Its dead end is flanked by two significant houses which suggest the changing character of the landscape as it approaches the more rural Ocean Avenue district. On the east is Edith Wharton's Land's End; and on the west, John Russell Pope's The Waves.

NE168 John Russell Pope House (The Waves)

Although now largely obliterated by later changes, the original cottage at Land's End, which John Hubbard Sturgis designed for Samuel G. Ward, was noted for its prominent gambrel-shaped gables and roof forms, reflecting the architect's prescient interest in a colonial revival. It took Sturgis's nephew, Ogden Codman, to make the house memorable, working with Edith Wharton, after she purchased it in 1892, on an addition to the seaside cottage and, most significant, redesigning its interior (not open to the public) after she purchased it in 1892). Its spare arrangement of spaces, with classical details and proportions inspired by French sources, was meant as a corrective to the prevalent ornamental excesses of the generation. It was this project that led Wharton and Codman to collaborate on *The Decoration of Houses*, published in 1897, heralding a new generation of thinking about interior design.

In designing a seaside retreat for himself and his family, Pope chose a rustic, organic image perfectly in keeping with the rocky, windswept site and its dramatic ocean prospect. The undulating slate roof, heavy timbers, and stone or stucco walls spread around a U-shaped plan underscore in form and materials how tenaciously this house clings to its rugged landscape. Although it is one of the last of the grand houses, The Waves had an unusual distinction in the 1950s after it had passed out of Pope's ownership: it was the first large Newport residence to be divided into condominiums.

Ocean Avenue

The district named for the street that Newporters call Ocean Drive (even though it is officially Ocean Avenue) is unique in both its stunning coastal beauty and its development as an area of grand summer residences. Although its settlement can be traced to two huge farms of the colonial era (one of these is the Jahleel Brenton House [NE178]), its distinctive architectural character dates from the late nineteenth century, when it was the preferred neighborhood of many famous summer visitors who enjoyed its oceanside activities and picturesque drives. They commissioned architects, many of national reputation like McKim, Mead and White, Richard Morris Hunt, Ogden Codman, and William Emerson, to design their summer cottages, on larger parcels of land and with far more connection to the natural environment than those of their contemporaries who built along the Upper Cliffs and Bellevue Avenue.

NE169 **William S. Miller House** (High Tide)

1900, Warren and Wetmore. 81 Ocean Ave.

The arched-topped windows and stuccoed masonry walls of this mansion, along with its shingle-covered roof terminating in turret-like polygonal caps at either end, lend the sense of a rustic villa in the French countryside. The primary focus is the central pavilion, whose two-story bow front swells from the broad, shallow main block of the house; but High Tide's real prominence derives from its site. The balustraded balcony on the second story of the bow is all but demanded by the fact that this house, a summer residence for a New Yorker who was Whitney Warren's brother-in-law, sits on the most commanding promontory along the southern coast of Aquidneck Island, overlooking the entire rocky shoreline and the Atlantic beyond.

NE170 **Stuyvesant and Mamie Fish House** (Crossways)

1898, Dudley Newton. Ocean Ave. at Jeffrey Rd.

Dudley Newton's Crossways is a straightforward presentation of Colonial Revival form at rather gigantic scale. Painted white, its most elaborate feature a monumental entry portico of Corinthian columns, it suggests a southern source. Although now broken into condominiums, this was one of the renowned centers of Newport social life during the Fishes' summer stays, the site of huge luncheons, lively dinners, and grand balls all orchestrated by Mamie Fish, one of the leaders of Newport summer colony, and her social collaborator, Harry Lehr.

NE171 **Edith Burnet Pomeroy House** (Seabeach)

1895–1896, Ogden Codman. Ocean Ave. at Hazard Ave.

Although by 1895 he was beginning to gain a reputation among an elite clientele as an interior designer, this was Ogden Codman's first commission for a complete residential design. Its arrangement is relatively simple, with a central gabled pavilion projecting from the hip-roofed main block of the two-and-one-half-story house. Dormers, pedimented gable, dentil cornice trim and arched-topped windows on the first story suggest Codman's Federal-period inspirations. The design, with its original stucco covering, seemed so tentative and rustic that it led his famous friend and collaborator Edith Wharton to remark caustically that one of the worst plights was to "build a mud hut and mistake it for the Parthenon."

NE172 Irving Tomlinson House
(Salt Marsh)

1929–1930, Derby, Barnes and Champney. 75 Hazard Rd.

This composition of small masses is grouped around the two-story space of a living room whose huge mullioned window faces inland through the field of tall reeds to the marsh beyond. The simple, almost elegant, stuccoed vernacular recalls the late nineteenth-century English country house designs of C. F. A. Voysey and M. H. Baillie Scott.

NE173 Eclectic Houses

Hugging the Atlantic side of Ocean Avenue is a sequence of houses which provide a spectrum of fanciful, mostly quaint, designs from the early twentieth century, including the F. Frazier Jelke House, Eagles Nest (NE173.1; 1922–1924, Aldrich and Sleeper), 222 Ocean Avenue, a modified Colonial Revival house with much shingle work; the Lucy Worth James House, Normandie (NE173.2; 1914, William Delano), in a French provincial style, and the V. Z. Reed, Jr., House, Seafair, formerly Hurricane Hut (1937, William Mackenzie), 254 Ocean Avenue, an ornate French château set far back from the road and laid out in a horseshoe plan. This was the last built of the great summer cottages. Just past Green Bridge, the only place on the drive flanked by water on both sides, is the Jerome C. Borden House, Bay House (NE173.4; 1917, Angell and Swift), 274 Ocean Avenue, an enlarged, shingle-covered bungalow.

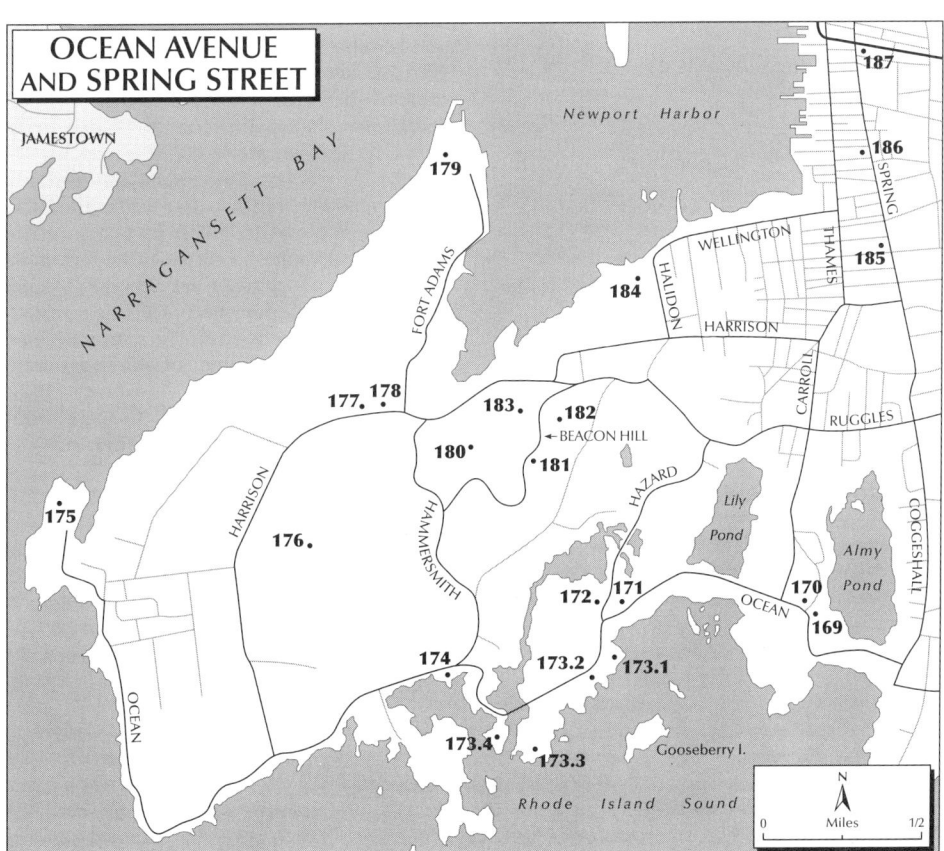

NE174 Albert H. Olmsted House
(Wildacre)

1900–1901, Irving Gill. 310 Ocean Ave.

For this, his first East Coast design, the young California architect envisioned a connected series of complex shingled volumes spreading across what was originally a bare rocky coastal site on a rubblestone foundation. Built as a summer house for Albert Olmsted, brother of Frederick Law Olmsted, the great landscape architect, the building, surrounded by mature foliage, nestles close to its environs. An existing planting schedule and photographic record document the Olmsted firm's detailed design.

Gill revealed much of Wildacre's interior plan through the articulated masses of the exterior, continuing the high stone foundation into several interior spaces and poking it through the roofline as rough-hewn chimneys. His play of roof slopes, jerkinhead forms, and flared eaves, some of which suggest a tentative orientalism, accentuate the organic character of the house, which appears to cascade across and down the ledges tucked between a saltwater cove and the roadway. Gill employed large expanses of glass everywhere, even designing several rooms with pocket windows that literally open these spaces to the coastal environment—exactly the kind of relationship to nature that Gill would later explore in the kinder climate of southern California.

NE175 Professor Alexander Agassiz House
(Castle Hill)

1874. 590 Ocean Ave.

The large, shingled mass of Castle Hill, once the summer home of the naturalist and marine biologist Alexander Agassiz, a Harvard professor, sits on its high bluff like an imposing gatehouse to the East Passage of Narragansett Bay. Although it has been somewhat changed in its present use as an inn, its engaged towers, ornate dormers, and bowed porches still present a picturesque ensemble with much original detailing intact. Some of the bracketing and stickwork around dormers, eaves, and entry canopy confirm this as a transitional design between the Stick Style of the 1870s and the somewhat simpler shingled structures of the next decade. Inside, the effect becomes complete, as the first floor is intact, with dark wood paneling, carved woodwork, and large fireplace—a typically grand, dark-toned ambience for such a prominent house and patron.

NE176 Newport Country Club

1894, Whitney Warren. 264 Harrison Ave.

The Newport Country Club was one of Whitney Warren's first commissions after his Ecole des Beaux-Arts training in Paris. His winning competition design for the club called for a Y-shaped building with its entrance facing Harrison Avenue and centered between the spreading arms of its two main wings, each of which houses a separate function—dining to the left and locker room–bar to the right. Although the other leg of the Y plan, a large piazza projecting to the rear, was destroyed in the 1938 hurricane, the visual effect from Harrison Avenue remains intact—a compact, high-roofed, French-inspired château set between ocean views, polo field, and one of the earliest golf courses in the United States. There are surprises here as well—the building is not as large as it looks from the road—but Warren carefully planned its placement as well as the scale of details on the main entrance pavilion to suggest

NE176 Newport Country Club

NE179 Fort Adams

monumentality. This illusion of size along with the manor-house image would suggest masonry as a building material, but the clubhouse is a shingled wooden structure. Perhaps its conflation of such vernacular materials with sophisticated motifs like quoined oval windows, balustraded cornices, and intricate carvings around doors was perfectly suited to Newport's sense of American resort elegance.

NE177 John Auchincloss House
(Hammersmith Farm)

1888–1889, R. H. Robertson. 225 Harrison Ave.

The colonial farm on which John Auchincloss built his summer home was established at the time of European settlement of the island in 1638–1639 by William Brenton (and named after the home he had left behind in England). Two hundred fifty years later, Robertson designed a huge, shingled house sited across the top of a rise on the property, which retains its original name. The three-story house cascades down from a complex massing of shed dormers, engaged turrets, bays, and a tall, conical tower. The uninterrupted planarity of some surfaces is penetrated by irregular window arrangements and, on the first level, by the shadow of undercut porches on the south and a wide, airy portecochere. Auxiliary buildings on the grounds include a guest house in the form of a picturesque windmill.

NE178 Jahleel Brenton II House

Before 1720. 203 Harrison Ave.

Still sitting on a rise amid open fields, this one-and-one-half-story gambrel-roofed and clap-boarded structure is one of the two earliest remaining residences on the southern end of Aquidneck Island. Originally a five-bay wood-frame house with brick chimney at either end, it has wings and two rear ells added in the eighteenth century. It was built by Jahleel Brenton II to serve as a tenant farmer's house for his family farm, Hammersmith. William Brenton's original seventeenth-century manor house (see preceding entry), no longer extant, was located nearby.

NE179 Fort Adams

1824–1857, Simon Bernard and Lieutenant Colonel Joseph G. Totten. Fort Adams Rd.

Covering over 130 acres and commanding a defensive site recognized since the earliest settlement as a prime position to defend the East Passage and harbor, Fort Adams was constructed in the shape of a hollow, irregular pentagon, on the site of a smaller, late-eighteenth-century fort. The present fort, more than any other in the country, clearly illustrates the entire history of nineteenth-century American military engineering. The nearly five hundred cannon, many doubled in two levels of casemates overlooking the bay, made this not only a "modern" fort, built as part of the "third system," or third form, of coastal forts erected by the Corps of Engineers, but one of the largest in the national coastal defense system. Its massive granite scarps, brick vaults, and shale and earthen walls, along with its auxiliary structures, visually dominate its peninsula site; but it is not just its monumental scale, more than 1,750 yards in perimeter, that makes Fort Adams unique. Some of its design elements, best seen today from the air, like the redoubts, bastions, and tenailles to the west, were meant to defend against overland attacks. This may be the only use of tenailles, outworks between sets of bastions, in fortification design in the United States. Primary entry into the parade ground surrounded by the casemated enceinte is through a towering, rusticated portal on the north framed by a rough segmental arch.

The Scottish stonemason Alexander McGregor, who later became a local builder and architect, immigrated along with dozens of other tradesmen to work on the fort in the 1820s. In effect, Fort Adams was a massive public works project, whose construction was an important element in Newport's economy in the nineteenth century. After its completion and until after World War II, the fort changed but remained an integral part of the coastal defense systems around the Narragansett Bay area. In 1873–1874, George Champlin Mason was commissioned to build the commanding officer's quarters, now called the Eisenhower Building; this wood-frame structure with mansard roof interrupted by a hipped central gable and ornamented with decorative struts, brackets, and porch work is visible to the left on a rise as you approach the fort. Later buildings, some of which still stand to the east of the fort, were built by the army in 1878–1879 for use as warehouses and shops. Other batteries, built later in the 1890s of reinforced concrete, stretch south of the fort along the bay side.

Ironically, despite its original innovative design and although hundreds of men were garrisoned there until the mid-twentieth century, Fort Adams never saw a shot fired in conflict, and today the peninsula on which it stands has been turned into a state park and sailing center.

NE180 Swiss Village (Vedimar)

1920–1924, Atterbury, Phelps and Tompkins. 75 Beacon Hill Rd.

The remnants of this group of farm buildings, marked by a stone gatehouse, can be glimpsed among the rocky outcroppings. Originally part of Arthur Curtis James's Beacon Hill estate (no longer extant), they were meant to evoke, like Marie Antoinette's Petit Hameau at Versailles, an impossibly picturesque European peasant life. The winding Beacon Hill Road, like others in the area, was part of the extensive system of carriage paths used by summer colonists in the nineteenth century to explore the natural beauty of the area.

NE181 Rose Ann Grosvenor House (Wyndham)

1890, William Ralph Emerson. 36 Beacon Hill Rd.

NE182 William Grosvenor House (Fair Oak, Roslyn)

1901, William Ralph Emerson. 26 Beacon Hill Rd.

Built for two related Providence patrons, these houses of similar size, part of a handsome group of three on either side of Beacon Hill Road (see also the George Gordon King House, below), show some of the same stylistic features. On Wyndham, the bold feature of a rotund tower implies historical references; the dark stone trim, broad entry arch, and overhanging corner turret of Roslyn suggest a Richardsonian Romanesque vocabulary.

NE183 George Gordon King House (Edgehill Farm)

1887–1889, McKim, Mead and White; landscape design, Frederick Law Olmsted and John Charles Olmsted. 138-152 Beacon Hill Rd.

Recalling Norman farm buildings in its first-story stone walls and engaged tower topped by stucco and shingles, this residence was built for one of the largest landholders and most successful developers on Newport Neck in the late nineteenth century. Edgehill shows its architects exploring sources different from what they would have employed in town. In recent years it was adapted to institutional use with the construction of massive, banal outbuildings, destroying large parts of its hilly site.

NE184 John Nicholas Brown House (Harbour Court)

1904, Cram, Goodhue and Ferguson. 5 Halidon Ave.

Massive stuccoed walls and high hipped roofs loom on a rise set back from the street and overlooking Newport's inner harbor. The quoining, window treatments, columned loggia, and half-timbered service buildings show Cram, Goodhue and Ferguson creating a restrained but palatial-scaled, French-inspired manor house for this august client, one of the wealthiest men in America and the scion of one of Rhode Island's colonial families. The building now houses the New York Yacht Club's Newport facilities.

Near the entrance to the property is the small, board-and-batten building now known as Station No. 10, the first clubhouse, originally sited across from Manhattan in Hoboken, New Jersey. Probably designed by A. J. Davis in 1845 as a delightful seaside folly, it was moved several times before it was brought in 2000 to reside here, with its high, pitched roofline, flaring eaves, and ornamental tracery intact.

Spring Street

Clearly different from the Ocean Avenue district, this neighborhood, known as Newport's

Fifth Ward from an old political division, is a compact grid of streets filled with small residences from the late nineteenth and early twentieth centuries.

NE185 John Carey, Jr., Gardener's Cottage

1876, Sturgis and Brigham. 523 Spring St.

Although this stretch of Spring Street is characterized by relatively modest wood-frame residences built primarily for the working-class population of the area in the eighteenth and nineteenth centuries, a few buildings distinguish themselves from their neighbors. This house, adjacent to the estate of New Yorker John Carey, Jr. (John Jacob Astor's son-in-law), is a distinctively ornamented two-and-one-half-story workman's cottage with high-pitched gables. The exterior is cloaked in fanciful screens, scalloped shingles, and lively trim work, some of which, like the spindle posts, dentil moldings, and clapboarding, may reflect an early Colonial Revival interest which is carried out in parts of the interior (not open to the public).

NE186 Emmanuel Church

1903, Cram and Ferguson. 415 Spring St. (at Dearborn St.)

Emmanuel Church, with heavy, random-coursed, buff-colored stone walls, dominates this neighborhood of small wood-frame residences as its bulk presses close to Spring Street. The main block of the nave, with side aisles and side entrance, is surrounded by buttressed walls and stout tower engaged at one corner. This arrangement, in concert with the rather restrained Gothic Revival detailing of window tracery and crockets, harkens to English antecedents, obviously deemed appropriate for this Episcopal congregation, originally gathered in the mid-nineteenth century.

NE187 St. Mary's Church

1848–1852, Patrick C. Keely. 250 Spring St.

A church as famous for its historical and social significance as for its architecture, St. Mary's was built, with the physical help of its working-class parishioners and the financial help of several prominent Catholic families who summered in Newport, in a predictable Gothic Revival style to a design by Patrick Keely, a prolific church architect. Its massive, rough-cut brownstone walls, high-pitched roof, towering steeple, and sculptural entrance reaching out to busy Spring Street served as an imposing symbol of the population changes that immigration brought to Newport in the mid-nineteenth century.

Jamestown (JA)

What initially strikes the visitor coming from Newport is the simple, unaffected quality of Jamestown. To the excesses of Newport's summer social season during the Gilded Age and its aftermath, in fact, Jamestown was a rebuke. Not that those in other communities took a different stand on the most extravagant highjinks of Newport's summer elite; nor did many Newporters, who nevertheless felt a degree of pride in the luster as well as delight in the profit their city acquired from its special place among American resorts.

But Jamestown, more conspicuously than any other Rhode Island community, came to epitomize an opposite view of summer bliss and to exemplify the different values in life that this disparity implied. There is a considerable record of Jamestown's reservations about Newport, expressed not only by its residents, but by visitors to the island as well. Listen to this rhapsody from a Brown University professor (let us hope not of literature) who visited the island toward the end of the nineteenth century:

> On Narragansett's azure breast
> There sleeps an isle—an isle of rest—

> No gorgeous palaces uprear
> Their walls of pomp and folly here
> No glittering monuments of wealth
> But in modest dwellings scattered wide
> Among the hills and water's side
> Lift their gray roofs, with woodbine hung....

But why should antipathy to Newport come to such focus in Jamestown? Doubtless because both evolved as summer colonies which physically and psychologically confront one another on opposite sides of a shared bay. Those vacationing in Jamestown equated summer with the quiet, relaxed life, preferring a hammock slung from the rafters of a shadowed veranda to a continuous round of fabulous feasts and dazzling balls. There was wealth in Jamestown, but the luxurious shingled houses in which summer was spent remained close to the styleless vernacular. There were banquets and dances at the Yacht Club and the hotels, but entertainment in the cottages remained informal. In its formative years as a summer colony, the social elite among its seasonal residents came especially from Philadelphia and from certain midwestern cities (especially St. Louis), not from New York. And the sense of the simple agricultural past of the island persisted throughout its early period of summer colonization. So did the village quality of the center of the town that gathered around East Ferry, across East Passage from Newport. Indeed, by lucky chance, even at the end of the twentieth century a sense of the past rusticity of the island persists to an exceptional degree, however precariously.

To have chosen this island as a summer place in itself indicated a wish, not for isolation (there were lonelier places to summer), but for a degree of seclusion, separated from the frantic and theatrical frivolities on the opposite shore. Indeed, for the Jamestown elite, there seems to have been positive satisfaction in nearness to Newport. They could partake of its aura through proximity, even occasionally take advantage of its availability, but with its nearness providing a keener sense of pleasure when one set it aside to cross the harbor into the different world that was their summer.

Though it is usually referred to by its English name, Jamestown is also frequently called Conanicut, especially when mentioned in conjunction with Aquidneck Island. Less than half the size of Aquidneck, Conanicut is the other substantial island buffer between the open ocean and the shelter of Narragansett Bay. The principal entrance light to the bay sits on the tip of this island, dividing water traffic between the two principal channels into it. Conanicut, like Aquidneck, is long and narrow. Bedded side by side, their lengths in the direction of the bay (roughly eight and fourteen miles, respectively) account in considerable degree for the sheltered quality of the water behind them.

To the south, coves erode Conanicut, and a peninsula off its southwest corner is virtually a separate small island linked to the rest by a sandbar, built up to carry a causeway. Its seaward thrust terminates in the shape of a beaver tail, which names its point and the lighthouse on it. Conversely, therefore, to Jamestowners the rounded northern end of the island is Beaverhead, on a knob of which (Conanicut Point), North Light once marked the passages for outgoing vessels much as Beavertail Light serves incoming traffic at the opposite end.

A surprising number of Conanicut's farms have survived from colonial and slightly later times with fields and early houses substantially intact, often through generations in the same family. Fortunately, two of the larger and older farms, contiguous with one another, have been placed in permanent conservancy, and there is at least hope at the start of the twenty-first century that more farmland nearby may survive the lure of development. Close by these farms is one of Rhode Island's two surviving shingled coastal windmills for grinding grain and meal, and the only one from the colonial period. Close to this on Windmill Hill, the tiny Quaker Meeting House also remains from the colonial period. So Jamestown is among the best places in Rhode Island in which to gain a sense of its early coastal farms.

As with many other coastal communities, the shift in Jamestown's economic base from agriculture to tourism was encouraged by new modes of access: first steam ferries and later, in the mid-twentieth century, bridges. Only a pioneering few summer colonists predated the inauguration in 1872 of the new steam ferry *Conanicut*, connecting the island with Newport. The coming of the ferry promptly transformed the fields that sloped down to the slip into potential cottage sites. By the mid-1880s, a cluster of hotels in clapboard and shingles rose behind the ferry landing where before a single establishment of modest mien had stood pretty much alone from 1875 throughout a fallow interlude brought on by the panic of 1873. The competition began with more hotels on the same scale, then some a bit larger, culminating in the simultaneous completion in 1889 of two (neither now extant) which set unsurpassed standards on Jamestown, both for size and carpentered splendor. The Thorndike (sometimes given the fancier spelling of Thorndyke) and the Bay View were extended blocks of four and one-half stories, as compared to the earlier two-and-one-half-story compact blocks. At their showiest (because in later efforts to economize both saw some diminution of exterior features designed merely for effect), the Thorndyke featured the thrust of a centered block with abutting tower through a stretch of gables and dormers, the tower culminating in a tall peak with bowed profile. The Bay View caught the eye by the swell of a rotund, conical-capped tower at the busiest intersection in the village, the swell further swollen by the tiered projection of a porch around the dining and function rooms surmounted by a stack of three open porches, all fitted to the curve of the water side of the corner tower to provide a multilevel overview of the harbor across to Newport. Together, the Thorndike and the Bay View served as the island frontispiece for those arriving on the ferry from Newport.

Along with the hotels, Jamestown offers two opposing kinds of luxury cottage development in outstanding late-nineteenth-century examples: the dispersed cottages of the Highlands and the Dumplings, most raised on isolated eminences with spectacular outlooks; and the precinctual semi-village of Shoreby Hill, fitted into an ac-

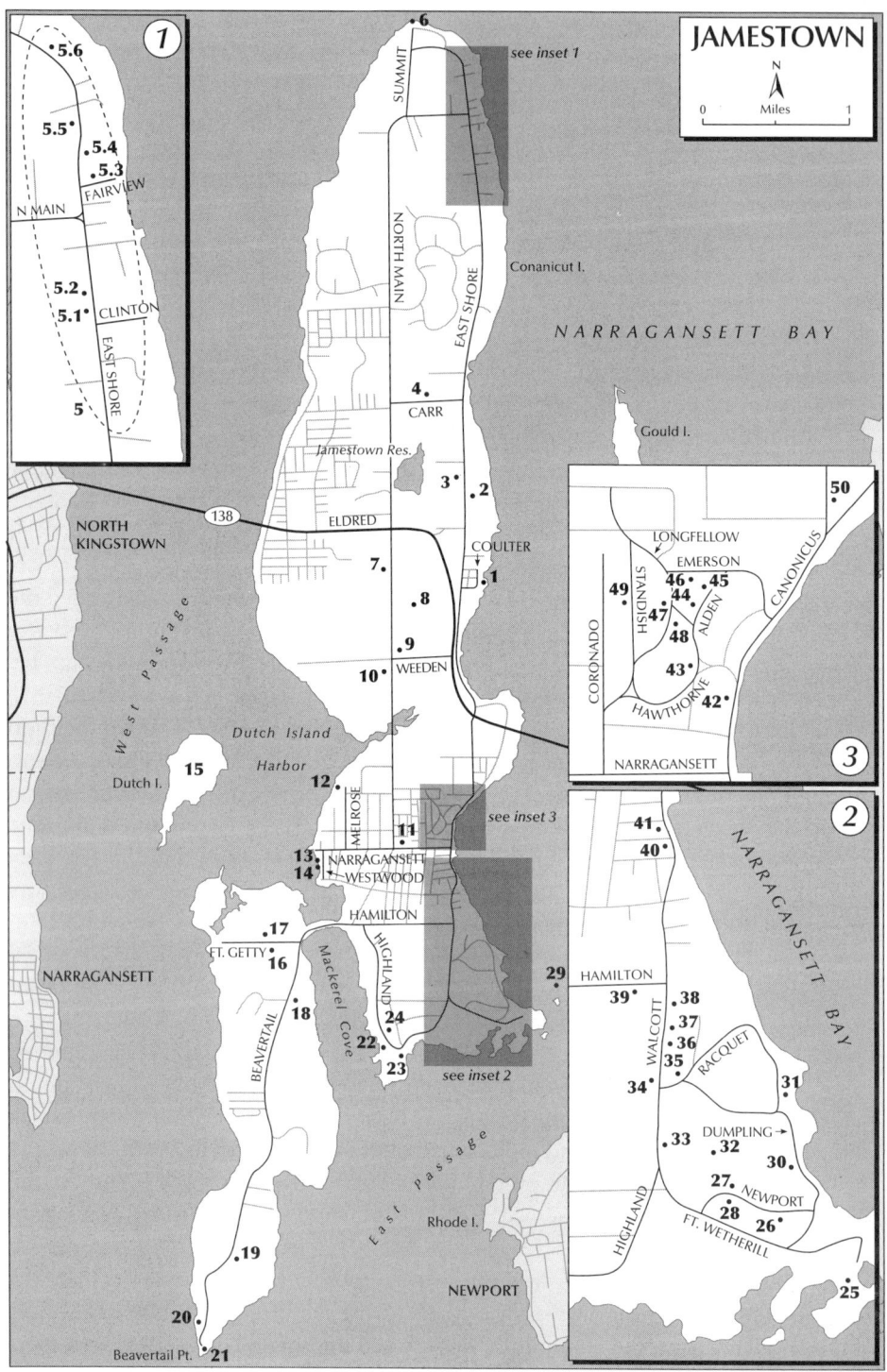

tual village so as to put the vacation facilities of its harbor situation within easy reach.

And outside such ambitious developments, shingled cottages were going up elsewhere all over the island. Of the result, here is the conclusion of the Jamestown report in the series covering all cities and towns published by the Rhode Island Historical Preservation and Heritage Commission: "The neighborhoods of Jamestown, especially Ocean Highlands and Walcott Avenue, are the *best places* [italics added] in the state to see and appreciate the charm and sophistication of shingle style architecture. Some of the region's first big, casual shingle-clad summer houses are here—they are the characteristic buildings of the island and testimony to its special appeal for summer visitors."

East Shore

JA1 J. D. Johnston Bungalow
(Daybreak Cottage)

1911, J. D. Johnston. 32 Coulter Dr.

J. D. Johnston built this secluded and modest cottage as a birthday present for his wife. Hence it also served as his getaway place during the final seventeen years of his career as a contractor turned self-made architect who was responsible in both capacities for many Newport and Jamestown buildings. His shingled bungalow might have been described as Japanesque when it was built. Intimations of the exotic appear in the wide bargeboards with rounded ends and the projecting purlins, simulating the heavy exposed timber framing under broadly projecting eaves of traditional Japanese houses. Here, however, the frame is adapted to American carpentry: boards instead of heavy timbers and nails instead of elaborate mortice and tenon joinery, with scroll-saw shaping and purlins that serve only to stabilize the boards and counter warping. The pavilionlike porch, the petiteness of everything, and the forced asymmetries also evoke Japan. But it is surely the American Arts and Crafts bungalow that prevails. As is the case with many Jamestown houses, the most extraordinary elevation faces the water and so is best seen by boat. One slope of a cross-gabled projection from the main gable is clipped in midcourse. The dynamic result is a lopsided gable sailing over the void of the wraparound porch, an eccentricity intensified by the contrast of the normative gable of the dormer and by the parallel slant of the stair enclosure up to the porch. The latter can even be viewed as a displaced fragment of the amputated effect above. Simple board flaps as shutters complete the straightforward handling of all elements. The carefully crafted woodwork on the interior (not open to the public) is intact. Johnston's wife continued to live in the house until 1954, when she sold it to its second owner.

JA2 Caretaker's Cottage, Theophilus Stork House (Fowlers Rock)

1900. 349 East Shore Rd.

The main house, named for an offshore clump of rocks at which it looks, across the road and diagonally south of this bungalow, is an important medium-sized shingle house dating from 1892 and designed for a niece of Philadelphia's prominent Wharton family, probably by C. L. Bevins, a little-known but prolific designer from England who opened a practice in Jamestown at the end of the nineteenth century. It has been altered, however, and is invisible in summer. So let the prevalent Jamestown theme of the modest shingled bungalow extend itself in this fine example by way of further establishing the unpretentiousness characteristic of the Jamestown tradition, before we reach shingled residences that are much larger but nevertheless mostly retain the qualities of its basic cottages. Columned porch supports minimally "colonialize" this example. Their elemental capitals and the speed with which the tapered entasis works itself into a cylinder indicate how casually the country lathe dealt with its whiff of classicism.

JA3 Robert Henderson House

1930–1931. 409 East Shore Rd.

The early-twentieth-century "colonial" shingled theme continues in a different key in this Sears,

Roebuck cottage assembled from a kit of parts, a model Sears marketed as "The Crescent." The explosively overlarge arched, pedimented, and columned entrance stoop quite overwhelms the squarish cross-gabled box behind. Style elevates function.

JA4 Carr Homestead

Late 18th century. 90 Carr Ln.

From Shingle Style bungalow variants of the late nineteenth century back to an antecedent: a well-proportioned five-bay, central-chimney shingled farmhouse, probably built c. 1790. The abrupt juxtaposition makes clear how revivalist sentiment gentled the uncompromising straightforwardness of most country colonial design and its placement in the landscape. Here these qualities must be grasped through the privet hedge with garden gate that veils the front and confirms revivalist tendencies toward prettifying the colonial past while sanctifying it by setting it somewhat apart in its own precinct. Fine stone walls, sheds of later date than the house, a corncrib (probably from the nineteenth century), and a long, low barn of the 1930s complete the farm ensemble.

The Carr family, the original owners, worked the farm from the late eighteenth century until its agricultural life ended in 1945, before which managers briefly operated it as a dairy farm (hence the 1930s structure). Along the way it served as the headquarters for the first library on the island, when the Jamestown Philomenian Library Association deposited its collection in a cupboard at the head of the back stairs. Founded in 1828 as the Philomenian Debating Society, it later decided to found a library. The house, which is not open to the public, is now owned by the Carr Homestead Foundation and is available to Carr family members.

JA5 Conanicut Park

1873, Lucius D. Davis, concept; John H. Mullin, plan. East Shore Ave. and adjacent areas from 1.5 miles north of Carr Park, around the northern tip of the island to intersection with Summit Ave.

The plan of Conanicut Park, a development for modest cottages, is among the more ambitious failures of its time. It is a secular version of Methodist camp meeting grounds, most particularly inspired by Wesleyan Grove at Oak Bluffs on Martha's Vineyard. Lucius D. Davis, publisher of the *Newport Daily News*, part-time real estate entrepreneur, and former Methodist minister, combined all three professions in conceiving Conanicut Park. Before this venture he had laid out Cliff Cottage Association in Newport as a similar development of small cottages. In 1872, with financial help from Providence and Newport textile magnates, among whom Governor Henry Lippitt contributed most, he purchased two large farms totaling some 500 acres on the northeast tip of the island. John Mullin, a topographical engineer and surveyor, laid out the property in 1873. Schematically his plan used a considerable length of East Shore Road as a spine with building lots on either side extending out to the shore, where Mullin envisioned a parallel drive along the water (only vestigially realized). Toward the head of the island the property widened and swept around most of the point. Here the road network became more complex, organized around an oval residential enclave centered in a park—the oval analogous to the eye of a needle in relation to the rounded point. The plan included a commercial area (on Commercial Street) near a steamboat landing with a fancy waiting room encased within a Victorian scroll-cut exterior, and nearby, on the northwest corner of East Shore and North Main, the Conanicut Park Hotel, which had rooms for 100 guests. From its opening in 1873, it was a fashionable destination for two decades before it began its decline toward demolition in 1908.

Mullin's platting provided for 2,098 irregular lots averaging about 50 by 100 feet and priced at $150 each, to accommodate diminutive but ornamental cottages in the manner of those in Oak Bluffs. In fact, Worth and Brazier, a contracting firm based in Edgartown on Martha's Vineyard, built at least five Conanicut Park cottages around 1873, of which three survive (883, 887, and 900 East Shore Road (about 1.5 miles north of Carr Lane), all dating from the start of building in 1873–1874. (Number 887, which is missing its porch, has Gothic-arched windows in the manner of some of the firm's buildings on the Vineyard.) Over 30,000 trees were planted, many along more than 12 miles of streets (which form the basic road network for the area), others on several sites planned for mini-parks. Sunnyside Park featured an old orchard as the heart of a residential oval, its elliptical street pattern still evident behind the northern point of the island. Island Park, emerging from a swamp abutting Conanicut Meadow, and the retention of Woodlawn Farm were intended as reminders of the area's past, along with a "common," which belonged to New England's history if not to Jamestown's farms.

Based on the prospectus and the auspicious beginnings of Conanicut Park, Samuel Drake, in his *Nooks and Corners of the New England Coast* (1875), described it, with implied scorn for Newport showiness across the bay, as a place "accessible to people who do not keep footmen or carriages, or give champagne breakfasts." The Reverend Frederick Denison concurred in his *Narragansett Sea and Shore* (1880), where he praised the still tenuous development: "designed for private residences—summer homes—and not for public parades, the flaunts of fashion and the confusion of excursion parties, it is a charming place for quiet and genteel family residences; the Elysium along the shore."

But before they wrote, this Elysium was in trouble, despite widespread initial enthusiasm and the reported sale of over 1,000 lots. Conanicut Park opened to the public in 1873, the year of the panic. Later attempts to revive it never succeeded. In the 1880s, hot real estate shifted to the Highlands, at the southern end of the island. The failure of Conanicut Park could have stemmed from its vision of a teeming arcadia of tiny, close-packed houses, amalgamating in a genteel version as it did the tastes of the gospel camper, Sunday excursionist and populist entrepreneur. Although Jamestown residents generally preferred modest living to Newport's posh, even those who lived modestly appear to have sought more spacious conditions and greater decorum than Davis's tight plots and scroll-carpentered ornament offered. The Lippitt family eventually owned most of the real estate. It passed through other investors with large plans until, by 1932, the last of the bubbles had burst in the Great Depression. The holdings were then parceled out to individual purchasers.

Today, scattered about, fifteen of twenty cottages survive from the 1870s, as well as five of six from an effort to restart the project in the 1880s, plus two conversions from carriage houses. Most have been renovated. The best preserved are described here. In addition, the John B. Kilton Cottage (1873), 947 East Shore Road, merits notice, despite an altered porch replacement after a 1904 fire, later aluminum siding, and other changes, as the first cottage built in the development, for one of its original investors, a Providence merchant. More elaborately fitted inside than most, it served as the "open house" come-on at the start of promotion.

JA5.1 George Taber Cottage

1874. Restoration, 1982. 921 East Shore Rd.

The Gothic Revival George Taber Cottage is one of Conanicut Park's more impressive houses. Its upscale "restoration" in 1982 may have made it too impressive by adding a second-story decked overlook above the old porch, as well as a garage with visitor quarters, which is fancifully connected to the house by a curved and roofed walkway. The porch is now L-shaped, probably the result of the restoration, because several other cottages of its vintage present a symmetrical front with a U-shaped porch interlocked with a T-shaped, cross-gabled body. (So often does this configuration recur, in fact, that it can be considered the Conanicut Park paradigm.) The pretty "Gothick" tracery of the chamfered porch posts and the ogee-pointed arching of the entrance and second-story balcony door as well as the scalloped bargeboarding of the front-facing gabled eaves recall an earlier decade (1840s or 1850s), whereas the other cottages are decked out in ornament characteristic of the later Victorian period. Were liberties taken during the restoration? In any event, the festive vacation transformation of a plain clapboard house by such carpentered ornament sets the stage for what is to come.

JA5.2 Eleanor H. Farr Cottage

c. 1905. 937 East Shore Rd.

Although it conforms to the Conanicut Park paradigm, the Eleanor H. Farr Cottage is unique, what might be described as all shape and no style, and the shape rather crazy (but with some fascinating architectural implications). Above the porch, what starts out to be a gable butts into a stubby, pyramidal-capped corner tower. Because they share the same clapboarded plane, they combine as a heavy, fused shape and visually overwhelm the U-shaped porch supported on spindly turned posts. The tower may have been an ad hoc alteration to contain an additional bedroom. At the end of the twentieth century, however, appreciation of such amateurish or folkish directness arrests the attention of many architects interested in the potential of the collision, fusion, and abbreviation of the conventional use of formal and symbolic elements to enhance the kind of ambiguity in perception and meaning most commonly associated with poetry. Here the coupled gable-tower form also rawly opposed the planarity of half the elevation against the aggressive three-dimensionality of the lopsided pyramidal tower cap. On first encounter, reaction tends toward some combination of dismissal, condescension, and astonishment; then suddenly (for some), a liberating "well, why not?" So the unsophisticated enlarges the sophisticated purview.

JA5.3 **Samuel Irons Cottage**
(Hendry's Retreat)

c. 1876. 1881, moved from original site nearby. 14 Fairview St.

The Samuel Irons Cottage, "Hendry's Retreat," an example of the Conanicut paradigm, is the most authentically preserved of all the more elaborate, ornamented cottages. Stencil-cut ornament adorns the bracketing of the attenuated square-sided columns around the U-plan porch, while more lacy embellishment appears at the apexes of all gables and at the lower corners of the gable triangles on the side elevations, like antimacassars in Victorian parlors. More lacy screening adorns the fanciful S-shaped brackets which support hoods over second-story windows where they break through the eaves, thereby cozying the house down to cottage scale.

JA5.4 **Jennie Lippitt Cottage**
(Stonewall Cottage)

1873. 1026 East Shore Rd.

In summer a screen of foliage along the road offers no more than an oblique glimpse of the Jennie Lippitt Cottage, which is really a house in scale and situated on an accumulation of plats, as befitted a member of Conanicut's principal investing family. It even boasts the additional pomp of a mansarded tower, tucked into a corner of yet another instance of T-shaped massing. Mock pomp really, because the tower is rather diminutive under its high-hat mansard, while the *T* is spread—a short, off-center stem butting an extended crosspiece. Wherever second-story windows reach the slopes of the principal gables, they again cut through the eaves as hooded and ornamented semidormers, giving to this grander house the same low coziness this feature produces in the Irons Cottage. The deep porch fitted around the stem of the *T* and the tower is fronted by a railing infilled with scroll-cut, ornamented screening in the chalet manner, as is the vista balustrade off the mistress's bedroom upstairs. "Embowering" vines and bushes (as the Victorians might have described them) complete this picture of the idyllic summer retreat of the period.

JA5.5 **David M. Hoyt Cottage**

1873. 1031 East Shore Rd.

Built for the principal of a Providence high school, this provides so stark an example of the Conanicut Park paradigm of interlocked one-and-one-half-story *T* and *U*, with such exquisite restraint in its scroll-cut embellishment as to seem the prototype for most of the cottages. Whereas most of the other porches couple rather tightly to the body of the house, the wide-spaced, spindly supports here spread beyond the house itself, giving the cottage an exceptionally airy aspect.

JA5.6 **Charles Fletcher House**

1885. 1076 East Shore Rd.

The largest dwelling built in the development—too large to be truly cottagey—the Charles Fletcher House sits on an accumulation of plats overlooking the water. It was the summer retreat of a textile manufacturer who also commissioned one of Edmund Willson's more sumptuous houses on the East Side of Providence. It appears today as a somewhat plain version of a Queen Anne city house, but doubtless lost its original luxurious character of appearance in successive strip-downs through long subsequent use as a summer hotel (under two names). It was eventually enlarged for condominiums, then returned to a single-family residence. In its heyday, commentators heralded it as among the splendors of the island for its expansive veranda, lawn, carriage house, bathhouses, and dock. (The much altered carriage house is now a separate residence.)

JA6 **Conanicut Point Light** (North Light)

1885. 64 Bay View Dr. (at Summit Ave.)

On a very truncated piece of what was to have been Conanicut Park's vaunted Bay Shore Drive is the former North Light, now a private residence. The lantern and lens of this light are gone from its squat, square tower, to which is joined the high-gabled, Gothic Revival keeper's cottage, with an entrance stoop fitted to the inside of its L-shaped mass. Although its manner is very much that of the Conanicut Park cottages, it displays as well the stolid sturdiness of government work in the vernacular manner, especially in the breadth of its vaguely Tudorized window frames and possibly in the force of the decorated, openwork bargeboarding of its front-facing gable. In this context, its scroll motif ambiguously calls up vines, dolphins, and waves.

Windmill Hill District

Among Jamestown's charms is the preservation of a substantial remnant of its eighteenth-century farming tradition, situated on a rise at the center of the island (between Eldred Avenue and Whittier Road, pieces of which define its northern and southern boundaries) and extended to the south by a beautiful stretch of marshland. Three substantial farms existed in a cluster between Eldred Avenue and Whittier Road until, in 1988, one of them then went to development. The other two are permanently protected. In accord also with the island's past, here are a windmill, which serves as the marker for the area, and a Quaker meeting house.

JA7 Job Watson–Thomas Carr Watson Farm (North Farm)

1796. Early 19th century, kitchen ell. c. 1990, kitchen ell rebuilt. Late 19th century, animal barn. Mid-20th century, other outbuildings. 455 North Main Rd. (owned by the Society for the Preservation of New England Antiquities and maintained as an operating farm; open to the public June–October)

Beyond the reception center (restored outside, adapted inside) a long lane runs to the highest point of this 248-acre farm where Thomas Carr Watson erected his farmhouse. Views from around it look down the long slopes of his fields to the West Passage into Narragansett Bay and, in winter especially, up and down the island as well. Its weathered clapboard front expectedly displays the conventional five-bay, central-chimney format of its period. In 1794, two years before Thomas built his house, his father, Job Watson, had purchased this farm at auction as confiscated Tory property, a piece of the 592 acres owned by Governor Thomas Hutchinson of Boston. Job divided this acreage among his three sons. Apart from the situation of the farmhouse, the forceful organization of its traditional front is its other principal attraction. Its central vertical axis, from the basic pilastered and paneled door capped by a pediment with rolled molding to the brick chimney (which is off-center), is made emphatic by its narrowed second-story window (nine-over-nine instead of twelve-over-twelve for the rest) and the wide expanse of clapboard separating it on either side from its paired flanking openings. As seen elsewhere, this spacing isolates the outer openings just sufficiently so that they take on a degree of visual independence from the center as four-spot patterns, yet are tautly attracted to it. Not that the carpenter who built it thought in any such terms; he intuited only a certain rightness in his spacing of the traditional formula. The house is clapboarded on three sides, with shingling for the rear elevation and kitchen ell. The first version of the kitchen ell was later slightly enlarged, and this has been rebuilt.

Five generations of Watsons farmed here before Thomas Carr Watson, Jr., bequeathed the farm and an endowment of $1 million in 1979 to the SPNEA so that it could continue as an operating farm, and the house remain the home of the farmer. In 1991 the Tiddeman Hull House was moved to the entrance of Watson Farm from Jamestown Bridge right-of-way. The focus of a visit is as much or more on the farm as the house itself: on the outbuildings and on a trek of the fields with stops along the way to examine aspects of traditional farm life extending back to the colonial period—indeed, beyond it to evidences of previous Native American use of the land. Nowhere else in the state is one so completely transported to the world of the large colonial coastal farm.

JA8 Jamestown Windmill

1787. 382 North Main Rd. (miller's house)

How appropriate that the shingled mill for grinding corn still survives between the fields of two Watson farms surviving from the late eighteenth century, those of Thomas Watson and Borden Watson (see entries below). This, the third windmill to be built on Jamestown, survives as the only colonial windmill in the state. Together with two others in Rhode Island, both from the early nineteenth century (one in Middletown and one in Portsmouth), it is a reminder that wind power, especially for grinding grain, commonly substituted for waterpower into the early nineteenth century on the bay is-

JA8 Jamestown Windmill

lands, where streams are tiny and breezes are abundant. The mill stands behind the much altered and rebuilt miller's house. Eleven millers successively lived here through 109 years of operation, until 1896. The three-story, 30-foot octagonal and tapered mill tower rises from a barely visible rubblestone foundation supported by hand-hewn, diagonally braced oak timber framing. Its domical (or bonnet) cap, now fixed, once turned on a track, so as to face its 25-foot latticed arms (with attached canvas sails when operating) into the shifting wind. Grain from the second-floor storage area reached the 5.5-foot granite grinding stones below by chute. This mill and its half-acre site were once part of farms belonging to Colonel Joseph Wanton, a Newport Tory. His property, like that of many Tories in the vicinity, was auctioned to American patriots after the Revolution.

After it went out of use in 1896, the windmill decayed rapidly. A group of concerned citizens, which undertook restoration in 1904, turned it over to the Jamestown Historical Society in 1912. It has been repaired several times, but especially after the 1954 hurricane, when, among other work, new white oak timbers had to be installed to strengthen the tower, and the arms were completely replaced.

JA9 **Friends Meeting House**

c. 1786. North Main Rd. and Weeden Ln. (owned by the Jamestown Historical Society; open Sundays, June–September)

This meeting house appropriately recalls the sect that initially dominated Jamestown to such an extent that no other religious structure existed on the island before 1830. Following roughly a decade and a half of meetings in private houses, the first meeting house (1709–1710) was built close to the old Friends Burial Ground on Eldred Avenue three-quarters of a mile from the Jamestown Bridge (now partially surrounded by a later cemetery). A new meeting house more convenient to Jamestown village occupied this site from 1734 until replaced by the present structure, a tiny shingled and gabled box. The separate south-facing entrances for men and women are of the plainest sort, each fronted by its own granite entrance slab. To the right each has its window that reaches to the eaves. The jogged placement of these openings, plus their unmeasured spacing such that no interval matches another, creates a lively, unintended syncopation. Atypical for monumental buildings (even those meant to be as unmonumental as this), it rather suggests the frequent disruption of predetermined patterns for the layout of openings in barns and other utilitarian structures, which functional interference or casual execution often set awry. Ownership was transferred to the Jamestown Historical Society in 1998.

JA10 **Borden Watson Farm**

c. 1802. 305 North Main Rd.

Job Watson transferred these acres, once the central portion of Governor Hutchinson's 592-acre farm, to his son Borden. Borden in turn built the present house on the property. Although it is a five-bay house like that built slightly earlier by Thomas Watson, a greater formality appears in the enframement of the entrance by pilasters and pediment (however basic their treatment), in contrast to the simple framing in the former house. The splayed lintel blocks over the ground-story windows are in similar contrast to the plain frames for those in Thomas's house. As in his brother's house, the central upstairs window here is narrower than the rest (four-over-four versus six-over-six for the remainder). The near collision of the apex of the pediment against the sill of the window above, the lessening of the space between the centered and flanking openings, and even the larger windowpanes (typical for the developing Federal style) all diminish the tension felt in the distribution of openings across the front elevation of Thomas's house in favor of a more equilibrated design. Even the axial road straight to this house, as opposed to the wavering approach, following the contours of the land, to the Thomas Watson House before it appears in an oblique view, is a shift toward a more formal control of design in both site and house—but at the cost of some physical intensity.

Most of the land around the house went in the mid-1980s to the Nature Conservancy. The house, sometimes referred to locally as Hodgkiss Farm, remains in private ownership and is not open to visitors.

JA11 Jamestown Philomenian Library and Sidney L. Wright Museum of Indian Artifacts

1971, 1972 (museum), 1993. 26 North Main Rd.

Having combined with another library and grown to the bursting point in a tiny schoolhouse (now handed down to the Jamestown Historical Society), the Philomenian grew to this quite sumptuous state from its humble beginnings beneath a staircase in the Carr Homestead (see entry above). The petite Neo-Colonial "box of books" with its projecting and pedimented door flanked by single windows, typical of the early-twentieth-century village library, has almost invariably been swollen (sometimes swallowed) by late-twentieth-century additions. Here the blowup occurred in the single puff of a new building erected for the library in 1971. The brick box is larger but still intimately low beneath its spreading hipped roof. An arcadelike series of seven arched openings stretches across the front, including its centered door, giving it a degree of grandeur beyond that attainable in the earlier format. Its "colonial" allusion is abstracted in accord with modernist sensibility. The ambiguities inherent in the design reveal the architect's effort to strike a balance between village character and the larger public image that so substantial an institution wants to project.

The library has already acquired an addition in the museum given by Catherine Wright to honor her husband's interest in Native American culture and their joint funding in the 1960s of an archaeological dig at a local site that yielded a trove of Narragansett and pre-Narragansett artifacts. A large addition and renovation was added in 1993, tripling the square footage of the library while unfortunately reorienting a new entrance toward a parking lot.

JA12 Joseph Lovering House (Riven Rock)

1911–1912, T. Remington Wright. 113 Melrose Ave.

A long, narrow gambrel with cobblestone chimneys on either end folds down to make the broad, L-shaped porch roof across the front and along one side of this house. Balancing polygonal bays flanking a shed-roofed dormer project from its slope. The bays penetrate to the porch the full two-story height of the house, the motif of bay-dormer-bay taking its measure from the intervals made by the widespaced pier supports for the porch, shingled homogeneously with the rest of the house. Toward the water, the house is absolutely symmetrical in mass except for broad stairs that descend to the lawn off the corner of the porch and a coy mini-dormer attached to the bay at the opposite end. The spare use of basic shingled shapes in this house provides the sense of the elemental appropriate for summer relaxation; enveloping spaciousness is sufficient sign of luxury.

JA13 Mrs. George W. Logan Cottage (Southwinds)

Early 20th century. 14 Westwood Rd.

JA14 Captain Spencer S. Wood House (Westwood)

1917; later ell. 16 Westwood Rd.

This short dead-end street boasts two examples of early-twentieth-century cottage design popularized in catalogs and periodicals. First is a cottage built for a Washington, D.C., resident, who chose for her summer house one of Sears, Roebucks' larger "precut" models known as the Aladdin. It is high and boxy, covered by a single gable with clipped corners at either end and an appended room-sized porch along one side. Diamond-shaped panes enliven its casement windows in the seventeenth-century colonial manner (out of medieval tradition), and the upper sash of most of its double-hung windows in the Queen Anne Revival manner (some of these have been replaced). Next door is a luxurious shingled bungalow with square-timbered porch supports and a long shed dormer, built for a navy captain (eventually rear admiral). The plans derive from Gustav Stickley's *The Craftsman*, the leading organ of the early twentieth-century Arts and Crafts Movement. Its most remarkable exterior feature is the exposed fieldstone chimney at the far end of the cottage. A gentle, tapering curve as sharp-edged and close-packed as fieldstone defines the sea edge of the chimney breast, then swells out in a rippled plane, leaving the other edge irregular and loose-packed, as though a mini-cosmic force had come down on this object, of all others, to demonstrate its capacity to leave order in chaos. And out of this rocky foundation, at the second story, pops a conventionally rectangular brick chimney—disconcertingly, although other bungalow chimneys of the time display these two materials, favored by the Craftsman style, in just such startling juxtaposition.

JA15 **Dutch Island**

The marina at the end of Narragansett Avenue occupies the former landing of the West Ferry, which ran to and from Narragansett before the erection of the old (1940) bridge to the mainland. It is a good place from which to view nearby uninhabited Dutch Island, just offshore and reachable only by boat. Named for Dutch traders who used it as a trading outpost (called Quentenis) of their New Netherland colony, it has remained virtually unsettled on any permanent basis. From 1827, a lighthouse occupied its south end. The square, white-painted tower now standing there dates from 1857. A lighted gong buoy displaced it as an operating entity in 1979.

More interesting is the record of the fortification of the island, of which considerable evidence remains, although little has been done to stabilize or make evident the nature and purpose of the remnants. Purchased in 1863 from the owner of an unsuccessful fish processing plant, the island became the property of the United States government. It initially served as the training center for the Fourteenth Rhode Island Heavy Artillery, an African American regiment with members from all parts of the union. The regiment built the first of a series of fortifications, this a temporary earthworks at the southern end.

Low lying, it was often flooded and early disappeared. The strategic position of the island for protecting the West Passage into Narragansett Bay led immediately after the Civil War to a better-constructed and located facility at the center of the island—hence "middle emplacement" (1866–1867). But as peace persisted, the battery and the troops assigned to it inevitably appeared increasingly anachronistic and wasteful. By the end of the 1880s, a single caretaker watched over the defenses. The Spanish-American War broke the slumber. There followed two decades of building and redeployment of troops to what at last was officially designated Fort Greble, for John T. Greble, the first regular army officer to die in the Civil War. Although soon obsolescent as it became clear that battleships had greater range and accuracy than its artillery, Fort Greble maintained its utility through World War I as a National Guard station. By World War II, however, electronic surveillance at the mouth of the passages had largely superseded the Maginot Line concept. Fort Greble's batteries were dismantled, and the fort ceased to exist in 1947. Today most of the island is owned and operated by the Rhode Island Department of Environmental Management Division of Parks and Recreation as the Dutch Island Management Area. In keeping with its new use, it is a destination for walkers, with paths winding through the ruins and vegetation in keeping with its new use.

Beavertail

A causeway across the end of Mackerel Cove is the thread that barely hitches the bulk of Conanicut Island to its peninsula appendage. It terminates in a point having the shape of a beaver tail. Just across the causeway, which becomes Beavertail Road, Fort Getty was begun c. 1900 as an auxiliary artillery fortification to Dutch Island. It played a subordinate role in coastal fortification during World Wars I and II, but, most interestingly, served with Fort Wetherill during the final phases of the latter conflict as a center for indoctrinating German POWs in democratic ideas and values prior to their repatriation.

JA16 **Jonathan Law Farmhouse**
Mid-18th century. 881 Fort Getty Rd.

JA17 **Fox Hill Farms**
Mid-18th century; wing, mid-20th century. 994 Fort Getty Rd.

On the peninsula is the remnant of another large colonial farm. Originally owned by the Arnold family, it eventually passed to Benedict Arnold, father of the famed Revolutionary patriot-traitor of the same name, who bequeathed the farm to two nephews at his death in 1733. Two mid-eighteenth-century farmhouses mark this division. The Jonathan Law Farmhouse is another five-bay shingled house with gable roof, off-center chimney, and ell, its openings plainly framed in the manner of the Thomas Watson farmhouse (JA7). The second, named for the principal topographical feature at the northern end of Beavertail, is gambrel-roofed and may incorporate within it part of the original farmhouse mentioned in Benedict Arnold's will. It is privately owned, but the fields and walls sloping down to the bay can be viewed from the road.

Fox Hill Farm is of interest also as the twentieth-century home of the Sydney Wrights, who added the Neo-Colonial wing as both complement and foil to the ancient relic coupled to it. Catherine Wright, born Catherine Wharton Morris, granddaughter of Joseph Wharton, was a writer, painter, and philanthropist whose benefactions included Jamestown's Philomenian Library.

JA18 J. Bertram Lippincott House
(Meeresblick)

1892–1893, Pritchett and Pritchett. 177 Beaver Tail Rd.

Apart from its lighthouse, Beaver Tail is, above all, associated with the Whartons and the Lippincotts, who, even since the sale of parcels for development in the 1980s, still own much of the peninsula. Meeresblick unites the two families—the daughter of the financier whose name is most notably associated with the Wharton School of the University of Pennsylvania and the son of Philadelphia's distinguished publisher. This is a straightforward but beautifully crafted shingled Queen Anne house with tall gables and broad dormers lifted on a stone rubble lower story. Like most houses on Beaver Tail, it erupts from an island of mowed grass surrounded—almost assaulted—by a dense, tangled wall of shore foliage. The combination of German *Meer*, which translates alternatively as "bay" or "channel," and *Blick* (view) nicely encompasses the composite panorama from this house. It is massed as an angled *L* (or perhaps a V) with a porch inside the letter shape looking up the bay, and one on the outside looking toward the East Channel. Ornamentation is minimal but telling. Bargeboards flare to S-curved terminations. The angled wing points to a more decidedly L-shaped carriage house with upstairs guest (perhaps once servant) quarters, reached by a circular, conical-capped stone entrance tower nestled into the right angle, with a loggia porch at the top. On either side of this cylinder, the *L* terminates in broad, shingled gables, one sailing out over a ground-level terrace supported on rustic stone columns. The upper sash of the second-story windows is divided into squares and rectangles; attic casements are diamond-paned. Gable peaks lift to slight knobs. A shingled windmill across the road was designed to pump water. Inside (not open to the public), quiet Queen Anne detail and paneling retain the restraint of the exterior.

JA19 Beaver Tail Farm

c. 1904. 601 Beaver Tail Rd.

Once a farm, this is now another mowed island cut from the seashore tangle for a shingled cottage. A barn set against the road, however, suggests that a residue of the old farm continued into the cottage phase—probably an operation centered on riding horses. In contrast to Meeresblick's stylish reference to the contemporary vernacular bungalow, comfortably spread to the complexities of a large house, Beaver Tail Farm makes a prim appeal, however nominal, to the colonial gambrel tradition. Its conciseness of form opposes the more amiable ramble of Meeresblick's gables and dormers. In place of the sheltering overhang of its projecting eaves, here the wall folds abruptly to the roof with no more than a minimal molding to mark the shift of plane and seal the seam. Shape is abstractly compelling in its crisp, linear angularity, as of folded paper. The linearity of the openings, seemingly drawn on the surfaces of their walls rather than penetrating into them, intensifies the abstractness and planarity of these shapes. So does their overblown scale, evident especially in the inflated dormers projecting from the inflated gambrel. The gambrel folds downward, then out in a caplike peak to roof the L-shaped porch. A lower one-and-one-half-story ell juts from the rear, its walls, initially shingle, becoming fieldstone, and terminating as a porte-cochere, which rather awkwardly offers its grand entrance to what appears on the exterior as a service wing. Both house and ell are sharply angled to the L-shaped barn under another precipitously folded gambrel. So the angularity of the siting of this complex reinforces the angularity of its massing. Triangulations of sailboat diagrams come to mind.

Joseph Wharton bought the farm in 1899 as a hedge against the possibility that his large cottage on a spectacular point off Highland Drive might be taken by the United States government for Fort Wetherill (see below).

JA20 **Harbor Entrance Control Post**
1943. Beavertail Point

Like other camouflaged military installations along the Rhode Island coast (many now gone), this World War II military installation is a heavily reinforced concrete structure in colonial shingle disguise. Its domestic exterior concealed radar, radio, and visual observation equipment, underwater sound detection devices, and searchlights to coordinate and control all traffic entering and leaving Narragansett Bay, together with living quarters for its personnel. Until the early 1970s, a radar cage existed nearby. A cylindrical lattice of diagonal wire suspended between two masts, it was nearly as large as a municipal gas tank. By turn eerie, evanescent, and impressive, it may have been the most haunting monument erected in the state during World War II. From this high-tech antiquity, the imagination readily drifted to the ropes and wires of masts and sails, both challenging the eye as gossamer geometries against changing skies and weather.

JA21 **Beavertail Lighthouse**
1856. Beavertail Point

The lighthouse, the terminus and ultimate destination for all pilgrims of the Beaver Tail route, marks the entrance to Narragansett Bay. To the approaching mariner, it proclaims, "You have arrived." It is, moreover, the marker dividing traffic into the bay between its two channels, east and west. Although its automated surrogate in front of it now does the job, and the lighthouse itself, like all others in the state, has been museum-ized, for pilgrims by land and (except for anyone doing the practical work of navigation) even by sea it is still the marker. And appropriately for its privileged position, it even occupies a control location (almost) along Rhode Island's coastline. In the hierarchy of the state's lighthouses, therefore, it can be said to hold the premier position.

After such fanfare, a first view of the unadorned, straight-sided gray granite tower, 10 feet square and 45 feet high, may be disappointing. Surely Block Islanders can justifiably laud their lights at opposite ends of the island as being both more shapely than this one and possibly more picturesquely situated. But despite pedestrian modernized windows, the careless untooled surfaces of the granite masonry, and other gaucheries, Beaver Tail has its qualities. The promontory from which the tower rises extends its impact: jagged and tumbled toward the West Channel, but especially beautiful toward the East Channel, where the rock, water-smoothed and whorled, rounds to the water like a submerging whale or submarine. Framing the lighthouse complex in the camera viewer reveals how handsomely and variously its paired and connected hip-roofed cottages for the keeper and his assistant "compose" against the tower to which they are attached. The spanking white paint of their stucco-covered brick walling and of service buildings nearby contrasts with the gray granite tower. Edward Hopper would have been attracted to such a brusque composition, finding merit in the vigor with which these basic shapes combine, in their blunt functionalism, and in the awkwardness which such straightfoward handling can afford to ignore. This direct approach to the task at hand is nowhere more evident than in the construction of the square tower. It is built of elongated granite blocks eight and ten feet in length, the ten-footers stacked as interlocked cribbing with eight-footers as infill. No other New England lighthouse is constructed in this manner, characteristic of squared-off log building rather than of masonry. The effect is of an immensely secure box raised as a stalwart pylon against the foulest weather. It is abruptly topped, without transitional shaping, by its elegant cast iron decagonal lantern room. With so much going for it, perhaps it is fitting that Beavertail displays a degree of blunt reticence. More conventional prettiness might gild the lily. So the rugged essentiality of this pylon against its cluster of smaller buildings seems just right.

This promontory has seen beacons and lighthouses since the early eighteenth century. Colonial records previous to 1756 mention a watch house on the site in 1705; a beacon with a regular watch in 1712; and a wooden tower 59 feet high in 1749. The last, designed by Newport's famed architect Peter Harrison, was only the third lighthouse erected in the colonies. It burned in 1753, leaving behind no recorded image. The first masonry tower, a fieldstone affair, which replaced it in 1755, was burned by the British when they evacuated Newport in 1779 and repaired in 1783–1784. It sufficed until the present tower went up in 1856.

Beaver Tail is also notable as the site of a sequence of experiments in light and sonic technology. An unsuccessful effort by David Melville in 1817–1818 to improve oil illumination by heating tar and rosin over a coal fire resulted in the first use of gas as a lighthouse illuminant. And from 1851 it served as a testing place for new types of fog signals ranging from various whistles and reed trumpets operated by compressed air, until these were eventually augmented by a steam whistle in 1881—the first such combination in lighthouse technology. Its fixed white light from an oil lamp became a flashing light in 1899, a flashing electric lamp in 1931, and finally a 45,000-candlepower revolving beam with a green flash and a clear-sky range of seventeen miles.

Automated (since 1972) like most other surviving Rhode Island lights, it is now also maintained as a museum to its own past and to a vanished tradition, as are its cousins on Block Island. They are the three most visited lighthouses in the state.

Ocean Highlands

Highland Drive climbs and winds up an inclined bluff, with eruptive outcroppings along the way, to culminate in Conanicut's Southwest Point. Depending on orientation, the topography offers perches for houses with precipitous views down into Mackerel Cove between Co-nanicut and Beavertail and off to the ocean, or up the bay and across its channels, or any combination of these. No area in the state offers more varied panoramic water views in such proximity. Most travelers by car (in summer especially) will not see these hideaways, except in fragmentary glimpses and a few distant views; more is visible from the water. Still, even from a car in summer one senses the extraordinary quality of this place. Most of the older houses are medium-to-large shingle houses, luxurious but unostentatious. The late twentieth century brings the Shingle Style revival to the area, generally in less vernacular-inspired, showier examples, mingling much from intervening modernism with late nineteenth-century sources.

Like its contemporary at the northern tip of the island, Conanicut Park, Ocean Highlands too had been a farm, 240 acres, predominantly for sheep, as one might expect of such a tumultuous landscape with such buffeting weather for much of the year. Unlike Conanicut Park, however, Ocean Highlands was intended for those ranging in wealth from the well-to-do to the outright rich who, for reasons of economy or lifestyle, preferred its rustic and isolated exclusivity over Newport. The area was opened for development with the founding in 1874 of the Ocean Highlands Company, headed by Philip Caswell, Jr., as the prime mover for a group of investors. The depression in the 1870s and the remoteness of the site delayed any considerable building before the next decade, although Caswell gave a site to the Philadelphia marine painter, William Trost Richards, in the 1870s in the hope of inducing others to come. Richards had owned a cottage in Newport, but as it became increasingly built up, he sought a more isolated place for inspiration.

JA22 J. Bertram Lippincott House
1926. Albert Harkness. 216 Highland Dr.

Influenced by Provençal architecture, a rarity in Jamestown (although it was a favorite source of Albert Harkness, who designed a variant farther up the bluff), this example in stone has a two-story "living hall" with exposed gable truss construction overhead. A bedroom and service wing projecting on either side make a protected garden court facing the sea. Harkness's second wife Hope owned just such a Provençal house in Middletown designed by another architect using the same formula but different massing. Here Harkness seems to have adapted his own vacation premises into another charming seaside cottage, where the combination of style and scheme readily adapt the house from summery escape to the formal occasion.

JA23 Joseph Wharton House
(Marbella, Horsehead)

1882–1884, attributed to C. L. Bevins (J. D. Johnston, builder). 240 Highland Dr.

William Trost Richards's Philadelphia friend, Joseph Wharton, a wealthy Quaker industrialist, was also among the first to be drawn to Ocean Highlands. He had Newport roots and had summered with cousins in the Robinson House in the city's Point area, but loved boating and exploring for marine specimens along Conanicut Island. He purchased Southwest Point, the climactic promontory of the area, and named his house after a promontory overlooking the Mediterranean at the Spanish town of Marbella (beautiful seas). Its subsequent designation refers to an unusually shaped rock in Mackerel Cove. From a car on Highland Drive, one sees only the overbearing gate with two entrances (originally one in, the other out), with a wall connecting the two, which is topped by arching intended perhaps to confer a Mediterranean flavor. The house itself, however, is a conspicuous landmark from various points along the south shore of Jamestown and across the East Channel from Newport.

A high-roofed, shingled second story rises from a first story in random dressed masonry. Its predominant feature, especially from afar, is a bloated circular tower toward the East Channel which culminates in a circle of windows under a bonnet roof. Facing the sea is a stone-gabled frontispiece with paired arches on rustic monolithic columns making a porch beneath second-story windows. These are also paired, and windows in the attic gable paired again. It is a sophisticated example of the most basic and inherently the most vertical of all symmetrical arrangements. It gives the front a confrontational verticality in relation to the ocean reminiscent (formally if not inspirationally) of Beavertail's pylon. This verticality up the center of the spread stone gable also stabilizes the contrast between the opposed shapes on either side of it: the swelling roundness of the viewing tower on the east elevation set against the angularity of an octagonal turret on the west embedded into the larger gables from which it protrudes into a pointed, faceted cap. More utilitarian in appearance than the tower, it drops off the side of the ledge on which the main portion of the house sits, providing an additional two floors below the level of the rest, with a porch on the uppermost of the two overlooking a former tennis court. Its bonnet and tapering sides suggest a windmill. Swollen dormers and tall, fluted brick chimneys, with windows varied in placement and rhythm, complete the composition. Finally, a separate L-shaped carriage house, also in shingle and stone, extends the in-line quality of the elongated house. The quirky audacity yet bold simplicity of the compositional elements and their tense combination suggest as architect the shadowy Bevins.

The interior is sparsely adorned with little other than vacation furniture accumulated over generations, a far cry from the palatial interiors of Newport. Horsehead has remained in the Wharton family to the present and is among Rhode Island's finest extant shingle houses on one of the state's most spectacular sites. In both panoramic site and architectural quality it ranks with McKim, Mead and White's better-known Low House in Bristol, which is, unhappily, gone.

JA24 Wistar Morris House (Highland)

1884–1885, attributed to Charles McKim. 195 Highland Dr. (visible from the water)

This Shingle Style house is less conspicuous than Horsehead because it is lifted on a rock outcrop set farther back from the water in the midst of heaving terrain, the vacation outlook for another prominent Philadelphian, the author Wistar Morris. More conventional than its neighbor across the road, Highland is organized with cross gables at either end of the entrance elevation. They intersect a steep, double-pitched hipped roof which is punctured by dormers and tall brick chimneys. One of the cross gables descends to a low, bracketed hood roof expansively projecting out from a Dutch door. A deep, broadly arched porch, lined with white clapboarding to brighten its shade, projects off the opposite side. Continuing across the rear of the house, the fall of the land lifts the porch over a brick basement story to a panorama up the bay (making the rear of the house its true front).

The attribution of Highland to Charles McKim depends mostly on his use of Dutch door entrances and the broad arched porches

JA25 Fort Wetherill

of his early Shingle Style houses as well as, and more importantly, on a certain reticent equilibration and authority in the handling of the elements of its massing that disciplines its picturesqueness toward classical equilibrium. No documentation supports the conjecture, however.

JA25 **Fort Wetherill**

1902 and later, Major General George W. Goethals, U.S. Army Corps of Engineers. Fort Wetherill Rd.

Following the destruction of the U.S. battleship *Maine* in 1898, the federal government acted to increase the fortification of Narragansett Bay, both by the Dutch Island installation to protect the West Channel and Fort Wetherill on the Highland bluffs for the East Channel. For this, four summer cottages were condemned, including Richards's painting sanctuary (see introduction to Ocean Highlands). Fort Wetherill was begun in 1902 and augmented between 1904 and 1907, supervised by Major General George W. Goethals, who was to be in charge of the building of the Panama Canal. Construction from these initial phases remains in essence, despite upgrading during World Wars I and II. When it was deactivated after World War II and in the 1970s, most of the land, with remnant fortifications, went to the state as part of its park system, along with Fort Adams and Dutch Island. Most impressive (though marred by graffiti) are the reinforced concrete bastions, which make a manmade plateau of the bluff's rim, indented behind by a series of deep, stepped circular wells redesigned in 1916 to contain 12-inch disappearing guns with slitted bunkers for observation. Architectural observers will be equally interested in the superb binocular view of Joseph Wharton's Marbella/Horsehead from the bastion. The long shingle house with its turret appears as the cockpit of the jutting headland on which it rides.

From the northeast corner of the Fort Wetherill parking lot a short, dead-end scenic drive provides a series of outlooks over the East Channel. The view of Newport opposite (south to north) takes in Castle Hill, Hammersmith Farm, and Fort Adams. Boats turn past Fort Adams to enter Newport Harbor. Hence the summer months provide as splendid and varied a parade of pleasure craft along this channel as can be seen anywhere. On the very end of the turnaround, at the tip of an intervening point on an abrupt rise of rock, is the site of a long-gone Revolutionary War stronghold known as Fort Dumpling.

The Dumplings

Fort Wetherill and the next several destinations occupy the southeast corner of Conanicut Island. Topographically it is similar to the southwest corner, but with a difference in the way in which the heaving ledge landscape presents itself. The difference accounts for the designation of this area as the Dumplings. The bluff of Ocean Highlands so dominates its corner of the island as seemingly to incorporate the turbulence of its terrain into itself. Not so, however, for the Dumplings. Its basic terrain slopes rapidly down to sheltered coves along the East Channel. The rude upheavals continue, but here more as isolated ledge and knob formations thrown up from the lower slope and thereby appearing as conspicuous and singular landscape elements. The knobs even extend out into the water to make a mini-archipelago off shore, like dumplings rising from soup or gravy. It was one of the larger knobs on Conanicut that accommodated Fort Dumpling, an outlook during the Revolutionary War first for the Americans, then for the British after their occupation of Newport.

Today the ledges and knobs provide roosts for solitary cottages similar to the mostly unobtrusively luxurious examples of shingle types seen in Ocean Highlands, while below smaller versions cluster around the coves, with boatyards and moorings tucked into the scalloped shoreline. Eagles in their aeries above, shore birds below: the tumble of the topography gives this area its unique identity.

JA26 General Robert E. Patterson House
(The Ramparts; Channel Bells)

1888, C. L. Bevins. 27 Newport St.

Preferring a more lyrical designation for their cottage over the military ring that came naturally to General Patterson, the second owners turned from the fort to the melody of its views, resonating on its veranda from the desultory clang of the bell buoys marking the channel approach below. What we see is the boxlike shingled centerpiece rising to a steep hipped roof which, were close inspection possible, would be seen to combine as a very stubby stem with its elongated wings to mass the house overall as an expanded T. On the sea front, however, the deep arch dominates, stretching the length of the house and jutting forward with the center. Even our limited view eveals something of C. L. Bevins's typical deadpan vernacular, in which he subtly inflects the commonplace toward the exceptional shaping of mass, shifts in scale, calculated placement of openings, and the occasional muted trumpet note of some bit of high style. Here the compact fist of the centerpiece, its shape delightfully complicated by mini-dormers poking from its side slopes, is reticently handled in traditional shingled ways. Accents which would customarily magnify the importance of the centerpiece by occurring in it are displaced to the flanking wings as paired second-story windows which break through the eaves, becoming shed dormers. The placement of these accents off the centerpiece expands its aura, but with the tension which such unexpected displacement brings. Such is Bevins's unobtrusive sophistication, easily missed, but subtly felt.

The entrance to Channel Bells offers a glimpse of an explosion of white paneling marking the staircase beside the entrance, with which Bevins meant to impress the visitor. More fireworks occur in overblown Georgian mantels in its principal interiors (rather more British than colonial in character) where otherwise the effect is of vacation spareness. Like the paneling outside, the mantels within elevate the vernacular Shingle Style toward Neo-Georgian, clearly revealing the stylistic betwixt-and-between of Queen Anne.

JA27 Louise Alexander Larned Cottage
(The Boulders)

1888. 1893, ell addition. 52 Newport St.

JA28 Jerry McIntyre House

1992, William Burgin. 57 Newport St.

Shingle Style and Shingle Style Revival, roughly a century apart, here confront one another on opposite sides of the road. First, the older of the pair: a rambling, L-shaped shingled house that is among the more picturesquely sculptural in Jamestown and, notwithstanding its considerable size, has a remarkably "cottagey" aspect. Its overscaled features, as well as the many jogs in its massing and the sheltering lowness of its porches, disguise its true size. Taken together, they impart to its massing the knobby, eruptive quality of its Dumpling surroundings. Notable among these features are two tapering corner turrets with rounded conical caps, like those of Horsehead (see entry above), which again invoke windmills more than castles. Beside one of them, two dormers, different in size, the smaller set a little back of the other, jostle one another to catch the view from under shingled hoods with long appendages on either side.

Across the way modernism transforms the Shingle Style. A tall, angular box opens to the thrust of a cantilevered porch seemingly less anchored to its ledge than poised upon it. One imagines the two- or three-story plate glass screen intervening between the fireplace and the outreaching deck. Instead of the clustering of sculptured features across the way, here what one sees are the broad shingle planes, sparsely open toward the passerby, gently folded to make the box with opposed folds here and there inward to small windows which enliven the larger plane by subtle countermovement and suggest a sense of interior. They also serve in lieu of ornament, which becomes overt in the decorative but functional treatment of trelliswork and railings. True, this sophisticated interpretation of the original shingled houses of the area removes them more from vernacular building than did even the largest Jamestown houses in the past. Moreover, the vacation aspect of the houses that initially characterized this place (and still mostly do, despite much winterizing to make them fit for year-round occupancy) is also altered by houses like this. Whether or not it is open only in summer, it presents itself as the ideal permanent residence of the present: that is, one in which work is never far from leisure. If the ideal must be deferred, then this is a home for eventual retirement, in which, it is hoped, whatever work is done will peter away toward perpetual vacation. This house could be seen not as a pastiche of the summer home of the past, like so many examples, but as the type literally reborn to new life.

JA29 J. S. Lovering Wharton House
(Clingstone)
1902–1905, J. D. Johnston, builder. Offshore at 46 Dumpling Dr.

There are two versions to the story of how this romantic summer cottage came to be. According to one, William Trost Richards, the painter of landscapes and seascapes who abandoned Newport when summer hordes disrupted the solitude he needed for his work, then lost his Jamestown cottage to Fort Wetherill, decided on the ultimate escape. He would move out onto one of the Dumplings off shore, reachable only by boat, literally surrounded by the water and mists he loved to paint. This version of the tale has Richards designing the basic scheme for his getaway, but finding no way to finance it, when J. S. Lovering Wharton took over the dream. The second version maintains that it was always the wealthy Lovering Wharton's intention to build the house, and that he simply used his artist friend as a cover to hold down costs.

Whatever sketches may have been made by the amateurs, the architect J. D. Johnston made them buildable and had a hand in the design, and Wharton moved in. The result is this rather grand, if plainly handled, architectural crab, part bungalow, part chalet. Johnston fixed heavy mill framing, overbuilt against hurricane-force gales, to the boulder. Rustic and rugged inside as well as out, it derived the first quality from Arts and Crafts sentiment, the second from shipbuilding. Thus inside walls were shingled like those outside, with burlap-covered ceilings and beach-pebble fireplaces. Picture windows of the same heavy plate glass used for portholes offered views in every direction. Too heavy to be practically movable, they were fixed and gasketed, with ventilating hatches opening through the walls. Wharton added a breakwater to shelter ferrying, with a house on shore for his ferryman.

Used into the 1930s by the Wharton family, the house was severely damaged by the 1938 hurricane. Still Clingstone clung, as a steadily deteriorating destination for curious and vandalizing sailors, until a Boston architect purchased it and gradually restored it to one of the most unusual summer retreats in the state. (For a fine overview of Clingstone in its setting, look down on it from the Newport Bridge.)

JA30 Admiral Thomas O. Selfridge Cottage
(The Barnacle)
1885, C. L. Bevins. 15 Dumpling Dr. (visible from the water)

Another clinger. In this instance the top of the rock was blasted away to provide the building site. The house is surrounded now by a crown of evergreens, and is accessible only by a steep, winding pathway with stone steps and a hoist for lifting packages. This medium-sized Shingle Style cottage features a pyramidal roof of varied pitches with chimney topped by chimney pots poking through. Twin gabled bedrooms face the water from these and from a porch arcaded in shingles across its north and east sides. Views across the water take in a substantial piece of Newport Harbor and, farther up the bay, the Newport Naval College for high command officers. A perky, tall-gabled cottage at road level provided for guest overflow, as well as, perhaps, for infant children and those unequal to the climb. The striking aspect of this commonplace design shows what placement of openings, their proportioning to their elevations, and shaping of overall mass—no more than the basics—can do. A few years after it was built Admiral Selfridge commissioned a larger and more conventional house close to Jamestown Village (JA33).

JA31 Daniel S. Newhall House
(The Monitor; The Round House)
1888, attributed to Charles McKim. 1901, service annex (destroyed by fire, 1991). 104 Racquet Rd.

Another Philadelphia client is responsible for this shingled whimsy. It is sited low, close to the water on a cove, and semi-embowered, embraced by a boatyard. It is uncompromisingly round, 45 feet in diameter, as its successive names suggest. It makes mock reference to the round, multitowered ruin of Fort Dumpling, which still existed atop its nearby dumpling into the 1890s.

A close look at the profile of this "cylinder" reveals the typical surface plasticity which is among the subtler qualities of the Shingle Style and is rarely encountered in its twentieth-century revival. Its profile is subtly modulated.

An outward batter establishes a base from the ground halfway up the curved wall to the second story. Then a straight-up cylinder rises through the second story. Above these windows a swell expands to a slightly larger cylinder through the third-story windows. An inset wooden molding running along their tops provides the base for the culminating swell of the cornice in two segments: below, a transitional shingled band, before another inset molding underscores the broad, shingled parapet which caps the cylinder as (returning to the metaphor) a "bastion" rim. Newhall playfully mixed the metaphor by taking his fort to sea in naming it after the famous Civil War ironclad. But the plastic response of the curved and shingled surfaces of Jamestown's windmill may also have inspired this inverted, bowl-like variant. The scattered effect of its widely separated, paired casement windows also suggests the comparable effect of the placement of openings in ancient windmills. Look again, however, and note the symmetry with which the apparent scatter is organized around the climactic arc of casement window at the top floor immediately under the bastion.

The three-story round house nudges a rock ledge, which penetrates it as an exposed interior feature. A stair climbs to a bridge crossing the ledge, which expands to a deck off the cylinder toward the sea. (In 1901 a smaller three-story house of a conventional nature was added adjacent to the stair to house a cook and other servants. A fire of 1991, which seriously damaged the main house, completely demolished this adjunct.) The drawbridge/gangplank brought visitors directly into the living hall (or "saloon," as Newhall, with his penchant for the nautical, chose to call it) at the core of the cylinder. A cobblestone fireplace laid up against the intruding ledge dominated the saloon. "Staterooms" surrounded it, with stairs to more above. At ground level were the "galley" and storage rooms.

Did the distinguished McKim design it? A contemporary newspaper article says so, and architectural qualities reinforce the possibility. As a Philadelphian, he could have known Newhall personally. McKim, then in the midst of discovering colonial architecture, would have delighted in the opportunity to bend classical and American forms to such playful purpose. The surrounding Clark Shipyard, whose owners occupy the house, was founded about 1935 by Earl C. Clark, who previously captained and maintained Newhall's yachts.

JA32 William P. Henszey–Joseph N. Ewing House (Altamira)

1905, Selfridge and Obermeier. 60 Racquet Rd.

Set well back from the public roads around it, the hip-roofed block of Altamira nevertheless shows itself from many points. Although long porches with vista views are endemic to this island, because of its altitude on the highest of the Dumpling hillocks and its central position among vantage points nearby, Altamira's may star as *the* Jamestown porch. Another elongated Shingle Style house, it appears thin despite considerable breadth only because its veranda front provides a platform 100 feet long overlooking Newport Harbor and its channel north and south. Then, for good measure, the porch folds across its entrance width for views to the ocean as well, while a service wing angled from the main block extends it farther. Whereas most of Jamestown's traditional white trim against weathered shingle is thin, here porch posts are bulky with square timber and solid brackets (perhaps a minimal "colonial" reference). They bring a spanking nautical whiteness to this breezy place. The principal interiors are nominally Neo-Colonial, with plainly handled woodwork which may always have been painted white. In the living room a reticent but elegant stair descends in two flights, making a half turn into a semicircular projecting bay at the landing. It comes down to a generous random stone fireplace with a sort of Adamesque frame. The aim seems to have been a sense of dignity that also preserved vacation informality. Call this perhaps a Bull Moose taste in architectural design: big, bold, unpretentious, expansive, comfortable, and

JA36 Charles W. Bailey House

uncomplicated, with a stuffed bison's head over the fireplace to give the label more credibility. The authenticity of the interiors results not from an earnest effort at preservation so much as the desire to hold onto the spirit of the past, made possible by continuous ownership in a single family (with a succession of surnames through marriage).

Walcott Avenue and Jamestown Village

After Ocean Highlands and the Dumplings, exceptional both topographically and architecturally, here is a proper town street lined with houses, mostly large and shingled, on either side. It runs well up across a slope which offers views, but porches take unostentatious advantage of them, or often look toward the street, into gardens or across lawns.

JA33 **Admiral Thomas O. Selfridge, Jr., House** (Red Top; Green Chimneys)

1889, C. L. Bevins. 185 Walcott Ave.

Admiral Selfridge was stationed in 1898 in Japan, a likely inspiration for the ornamental detailing of this house, which once included carved dragons at both ends of the gable and latticed balustrades surrounding the porches. The theme extended even to the roof, which was originally red, suggesting oriental tilework. While often attributed to McKim, Mead and White, perhaps because of its complex of roof forms—gable on hip and saltbox—it is more likely by C. L. Bevins, who had designed another house, "The Barnacle" (JA30), for Selfridge only three years earlier.

JA34 **Elizabeth Clarke House**

1895–1896. 170 Walcott Ave.

This sedate medium-sized house has a typical projecting flare to differentiate the shingled second floor from the floor below, which is also shingled, rather than the favored contrasting brick or stone, which would have been more expensive. The gable-on-hip roof was originally symmetrical, extending beyond the walls as deep eaves. The use of this elegant standard roof type, which vaguely translates the typical shape of roofs for Japanese temples and houses into American carpentry, may have been influenced by the enthusiasm of the Aesthetic Movement for things Japanese. The asymmetrical placement of the porch counters the symmetry of the principal massing. Elizabeth Clarke, from Cambridge, Massachusetts, was secretary to the naturalist Alexander Agassiz.

JA35 **Captain E. V. MacCawley House** (Mist)

1889 and later, J. D. Johnston. 210 Racquet Rd. (at Walcott Rd.)

This long, narrow Shingle Style gambrel with a saltbox to the rear is fronted by an inset porch on skinny paired columns facing the water and centered with a window on either side. Later L-shaped additions of a gabled piece, then another gambrel, straightforwardly butt and jog against and from one another as semi-independent elements around a rock outcrop. The directness with which these additions were made and the resulting picturesqueness recall similar frankness in houses typical of Little Compton.

JA36 **Charles W. Bailey House**

1898–1899, C. L. Bevins. 121 Walcott Ave.

Parapets that break through the gable ends give this shingled Neo-Georgian house an imposing scale. Although the parapets echo brick chimney forms from colonial antecedents, the functioning flues are separate. Special window shapes and woodwork include a Palladian composition and pediment that predictably call attention to the entrance. More intriguing is the asymmetrical facade, resulting from unusual placement of the service wing, perhaps to achieve the optimal view the site provides. Outbuildings survive at 129 and 135 Walcott Avenue. Bailey was a partner in Bailey, Banks and Biddle, the Philadelphia jewelers.

JA37 **Samuel Woodward House** (Onarock)

1896, J. D. Johnston, 105 Walcott Ave.

Johnston uses a low hipped roof to cap his horizontal composition and harmoniously unite the house with the rocky crest on which it sits. The planarity of the facade and the simple mass and angled placement of one wing with its rounded tower suggest Johnston's typical restraint.

JA38 John P. Green House (Anoatok)

1889, C. L. Bevins. 2000, renovation, Ronald DiMauro. 95 Walcott Ave.

Here, Bevins uses an all-encompassing gable facing the street to help consolidate the variety of window shapes, bays, and varied trim of his facade. In Anoatok and neighboring houses, Bevins, Johnston, and others are obviously in a kind of extended conversation with each other, exploring the ways in which traditional elements can be configured in novel ways. A recent renovation continues this exploration with a new entry portico and other changes to the facade, mostly in keeping with the sense of the original design.

JA39 Gagne House

1989, James Estes. 21 Hamilton Ave.

A small house by an architect who has played a large role in recent years in reinvigorating the shingled building traditions of the islands. All roof slope and shaded porch, it is defined by simple massing rather than decorative shaping, with its restraint underscored by a modest cabinlike scale that makes it appear embedded within its wooded setting.

JA40 Dr. H. J. Rhett House (Quononoquot Club)

1901, Mantle Fielding. Southwest corner of Friendship St. and Conanicus Ave.

Behind a high rustic stone wall is this one-and-one-half-story Colonial Revival cottage built as a casino club and dining room for summer residents of the neighborhood, which eliminated the need for cooking elaborate meals at home during the heat of the season. Originally sited on another part of the property when the area was still being developed, it was moved to its present location in 1931.

JA41 Horgan Cottages (The Three Sisters)

1897, Patrick Horgan, builder. 17, 19, and 23 Conanicus Ave.

Tightly clustered on their high lots overlooking the harbor, these three cottages were used as summer rentals in conjunction with the nearby Hotel Thorndike, owned and operated by Newport developer Patrick Horgan. Except for differences in minor details such as inset gables or

JA37 Samuel Woodward House (Onarock)

a small dormer turret, they are nearly identical, perhaps as were Horgan's three daughters, Betty, Nina, and Myra, for whom they were named and who eventually inherited them.

Shoreby Hill

First advertised in 1896, Shoreby Hill was developed by a group of St. Louis residents to be a private community, not unlike the private, gated streets of their hometown. By 1898 seven St. Louis families bought water-view lots and promptly built summer homes in a subdivision laid out on a scheme by Bostonian Ernest Bowditch. A second (and eventually a third) tier of buildings, primarily for navy officers, was added between 1912 and 1916, expanding the bounded area to approximately its current size.

JA42 Jamestown Casino (Shoreby Hill Club)

c. 1898. 75 Conanicus Ave.

Originally sited farther north in the Shoreby Hill enclave, the club building was moved to its present location in 1911 and enlarged to became the casino, providing a social center for Shoreby Hill residents and their guests. It is now a private residence. The low, shingled building has crossed gambrel and gable roof lines and typical Colonial Revival details, somewhat idiosyncratically rendered, such as the squat Palladian window on the north gable end above a standing oval window squeezed between two projecting bays.

JA43 Charles H. Bailey Cottage
1898–1899. 4 Hawthorne Rd.

The scale of the grand portico with its monumental columns is unique in the Shoreby Hill compound, where the preferred style is more modest Colonial Revival, emphasized by the Americana street names. Here, the fancy broken pediment over the main entrance and the two-story Ionic columns are flanked by lower Doric-columned porches (now enclosed) that help temper the scale differences between the projecting portico and the main block of the house, whose overall image evokes the nineteenth century rather than colonial times.

JA44 Margaret Potter Cottage
(The Red House)
1898–1899, Creighton Withers. 5 Alden Rd.

The relatively severe planarity of the Potter cottage could be read as early modern if not for the hipped roofline and the revival surround of the main doorway, all but hidden by the shadowy undercut porches (some now walled in), which nearly surround the first story and float the upper stories over shaded voids. Only on the rear is the most obvious revival detail of a second-story Palladian window coyly apparent. Withers also worked on the Marion Davis Cottage (1898–1899), 40 Emerson Road.

JA45 James Taussig Cottage
1898–1899. 11 Alden Rd.

This, the smallest of the "St. Louis" cottages at about 3,000 square feet, has a first-floor bay and off-center entry, adding asymmetrical interest to an otherwise predictable Colonial Revival composition. Like many of the Shoreby Hills residences, it depends more on ornamental motifs, such as the Adamesque swags and turned balusters, than on spatial invention to distinguish itself from its neighbors. Taussig was one of the founders and officers of the Shoreby Hill development.

JA46 Edward Mallinckrodt Cottage
1898–1899. 41 Emerson Rd.

In this house, built for the chemical industrialist Mallinckrodt, the exuberant use of rustic fieldstone on the first story walls is toned down by the finer wood trim around dormers, windows, and doors. The gambrel roof reaches out to become a canopy for the ample porches. The elaborately sculptural scheme of broken pediments, arched and elliptical windows, and wooden fan panels, perhaps suggesting a fancy farmhouse rather than a country mansion, make the house more successful as a hybrid of rustic and high style than any of its nearby neighbors.

JA47 David Greene Farmhouse
c. 1712. 55 Longfellow Rd.

One of the oldest houses still standing on Jamestown, this small, gable-on-hip-roofed structure was built when the island was little more than open pastureland. Greene was a Quaker farmer who for a time also operated the ferry between Jamestown and Newport. After 1840, the property was left to the Society

Shoreby Hill houses

of Friends, which operated it as a farm for fifty years. Members of the Greene family (who had regained title) then sold it to the developers of Shoreby Hills. Although altered over its lifetime with window changes, an added lean-to, and the like, it still lends historic credibility to its Colonial Revival neighbors.

JA48 Davis R. Francis Cottage

c. 1903. 29 Longfellow Rd.

The sweep of the roof over the deep porch of this summer retreat is a variant of a vernacular seventeenth-century house form turned back to front. The expanse of roof, which completely shelters the broad, dark, shingled mass of the house, is interrupted by dormers, of which some are large enough to bring early cross gables to mind. Even the swelling porch posts seem inspired by the gunstock posts of timber framing. The owner had a distinguished political career, mostly before he became a summer resident of Jamestown, as mayor of St. Louis, governor of Missouri, secretary of the interior, and ambassador to Russia.

JA49 Sophie Schaus Cottage

c. 1926–1927. 53 Standish Rd.

More rustic than the cottages from the first phase of development in Shoreby Hills, this later house is all gable. Facing the street, the simple entry on the gabled end is combined with porches, eccentrically placed windows, and a natural stone chimney climbing up a shingled exterior, more reminiscent of earlier

JA50 Bay Voyage Hotel

compositions in Rhode Island by McKim, Mead and White and Irving Gill than of any colonial source.

JA50 Bay Voyage Hotel

1860, George Champlin Mason, Sr. 1889–1890, moved and altered. 150 Conanicus Ave.

The foursquare entry block of this hotel was designed by Newporter George Champlin Mason as a residence in Middletown. It was moved across the bay (on two barges) and altered with a thirty-room annex to the north. While Mason's characteristic roofline is still apparent, much below it has been changed with enclosed porch work and window alterations. It is notable mainly as a surviving example of the large-scale wood-frame hotels once popular on Jamestown, as well as for its eponymous trip across the bay.

Block Island (New Shoreham) (BI)

Block Island is the name most often used to refer to the town of New Shoreham, which occupies its eleven-square-mile extent. The island, shaped like a pork chop, is located in the Atlantic Ocean twelve miles from Rhode Island's southern coastline, twenty miles southwest of Newport, and fifteen miles northeast of Long Island's Montauk Point. It rises from a narrow spit of land at sea level on its north end and extends seven miles to the south across rolling terrain to high bluffs along the three-mile wide southern end. A large salt pond, breached on the island's west side, is at center, and several hundred freshwater ponds and marches dot the hilly, shrub-and-low-tree-covered landscape.

Early development by English colonists was slow and sparse. The first settlers came to Block Island in 1662, but the lack of a natural harbor hindered development. By the middle of the eighteenth century, the island had been cleared of its native lumber stock, first consumed, then cannibalized for construction of houses and barns. As late as the 1860s, the town had only several dozen houses, a church, one or two stores, two small hotels, and two roads that crossed in the center of the southern half. Only a handful of early buildings remains.

Significant development occurred in the late nineteenth century, following construction between 1870 and 1876 of a breakwater and harbor on the island's east side. Navigational aids, especially two new lighthouses, further promoted accessibility. Native-son legislator Nicholas Ball lobbied heavily for federal funding for these projects. Almost instantly, a commercial center with numerous shops and hotels developed at harbor's edge, and vacation houses began to appear across the island. This period saw the introduction of architecture far more style-conscious than any seen before, but its dissemination was slower and less considerable than at mainland watering spots. The most impressive of these buildings are the mansard-roofed hotels that cluster around the village at the harbor. Most Rhode Island water-oriented communities that developed a summer-colony patina had at least one of these, but nowhere else may they be seen today in such abundance and variety.

The hotel era flourished from the late 1870s until the end of World War I, after which the island drifted toward shabby desuetude. Little was built between 1920 and 1950. By the mid-twentieth century, Block Island had none of the cachet enjoyed by the other islands spread across the southern New England coast.

Block Island was rediscovered in the late twentieth century. The airport, completed in 1950, brought new accessibility. Block Island Race Week, instituted in 1965, familiarized many sailors with the island. Beginning in the late 1970s, many of the hotels were rehabilitated, thanks to federal tax incentives for historic preservation. By century's end, Block Island saw its largest ever building boom.

Today, Block Island is becoming a victim of its own success. During the last decade of the twentieth century, it changed more rapidly and more dramatically than any other community in the state. A place whose rolling, brushy topography long seemed eminently absorptive of significant amounts of new construction, Block Island did not anticipate the introduction of the many large-scale, prominently sited, view-oriented summer houses to which it has seemingly acquiesced. As the market for summer houses continues to burgeon, Block Island's reputation as a low-key, naturalistic retreat seems questionable.

BI1 Old Harbor Historic District

1870 and later; harbor constructed between 1870 and 1876. Water and Dodge sts.

This group of buildings, most painted white, conveys the sense of a nineteenth-century seaside resort better than any remaining spot in Rhode Island. Cassius Clay Ball built Harbor Cottage (BI1.1; c. 1880), the mansard-roofed frame building overlooking the harbor on the north side of Fountain Square (Water Street at Spring and High streets), as his residence. It realized its present form, with a squat, prominent corner tower and wraparound spindlework porch, in the late 1880s when Ball enlarged it to accommodate summer boarders. The hotels illustrate the endurance of and variety within the mansard-roofed hotel formula. The earliest, the Surf (BI1.2; 1873, 1888), on the north side of Dodge Street at Water Street, eccentri-

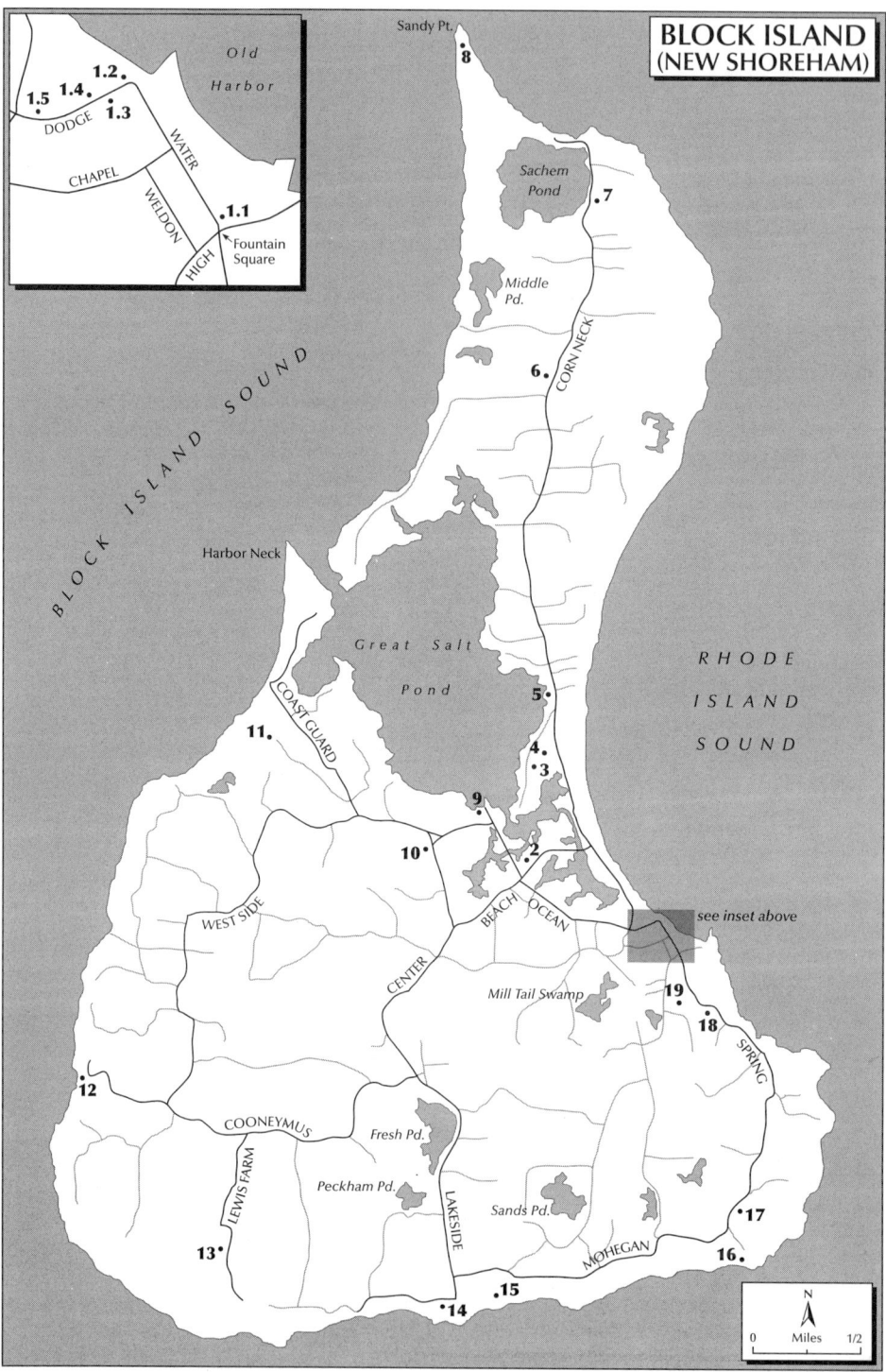

cally incorporates steep Stick Style gables into the mansard, a charming, untutored commingling of forms and styles then new to the island. Its larger later section and the full-width front porch, which binds the stepped-up sections together, seems an attempt to give the building a more monumental presence, like that of the National Hotel (BI1.3; 1903), on the south side of Water Street at Dodge Street, whose sunbonnet-dormered mansard was retardataire by the time it was built but nonetheless immediately identified the building as a hotel to visitors approaching from the water. The Blue Dory Inn (BI1.4; 1897–1898), on Dodge Street 300 feet west of Water Street, shows the longevity of the simple gable-roofed, center-entrance form, here brought up to date with an angled corner pavilion and offset bay window. Like many harborside houses, this served its owner both as residence and as source of income as a boardinghouse. Darius B. Dodge's cottage orné (BI1.5; 1874), on Dodge Street, built the year regular ferry service began, was perhaps the first to break with the island's strong vernacular building tradition and caused considerable comment at the time of its construction: the Providence *Evening Bulletin* noted that it "really makes more pretensions to style than any private residence upon our island."

BI2 U.S. Weather Bureau Station

1903, Harding and Upham. North side of Beach Street, 500 feet northeast of Ocean Ave.

Following a formula for weather stations repeated at least once elsewhere (in Narragansett, demolished) this building introduced to Block Island a bit of turn-of-the-century classicizing federal monumentality, albeit in almost miniature size. This was built just as the Weather Bureau sought to enhance its institutional status, and the foursquare, symmetrical, classically detailed building, prominently sited on a hill near the center of the island and overlooking several ponds, clearly established an island presence. Technological changes in reporting meteorology made this building obsolete by 1950, and it has served since as a summer residence.

BI3 George W. Willis House

1887, 1888, 1904. West side of Corn Neck Rd., .5 mile north of Beach Ave.

BI4 L. U. Maltby House (Ninicroft Lodge)

1904. West side of Corn Neck Rd., .3 mile north of Beach Ave.; access via Old Indian Neck Rd., .5 mile north of Beach Ave.

BI5 David Van Nostrand House (Innisfail)

1888. West side of Corn Neck Rd., 300 feet north of Old Indian Neck Rd.

Built in close proximity with fine views east to Rhode Island Sound and west across the Great Salt Pond, these three summer houses touch major strains in late-nineteenth-century resort architecture. Innisfail, the northernmost, is a small cottage orné in the informal tradition of Newport cottages in the late 1860s and 1870s exemplified by work of Richard Morris Hunt. The Willis and Maltby houses follow a formula well suited to seaside summer houses, with a circumferential porch girding the first story; on the Maltby House a prominent, sweeping roof covers both first story and porch. The Willis House uses traditional details and forms, including the mansard roof popular on Block Island throughout the late nineteenth century, while the Maltby House has a beach-stone-clad first story and cypress-paneled interior that signal an interest in the natural, even rustic, qualities favored by the Arts and Crafts Movement.

BI6 Captain Amazon Littlefield House

1888–1889, Daniel Dillon, Jr., carpenter. West side of Corn Neck Rd., 480 feet north of West Beach Rd.

A staggered cruciform-plan dwelling with a wraparound porch, the Littlefield House recalls designs found in late-nineteenth-century pattern books. Remarkably little altered, it shows the influence of external sources, increasingly prevalent in the late nineteenth century, on the island's domestic architecture.

BI7 Coxe-Hayden House

1979, Venturi, Rauch and Scott Brown. East side of Corn Neck Rd., approx. 1 mile north of West Beach Rd.

In Venturi, Rauch and Scott Brown's only Rhode Island commission (built for a Philadelphia native associated with the firm), two simple, shingled gabled boxes are sited at a narrow angle to each other. The exquisite siting and fine, simple detail establish a subtle, superb dynamic both between the buildings and between the complex and its compelling site, nestled into the landscape with a view to the west across Sachem Pond. The only discordant note here is the somewhat overshadowing presence of the later house at the crest of the hill to the east.

BI8 North Light (Sandy Point Light)

1867. Northernmost point of the island, beyond the north end of Corn Neck Rd.

Built to a U.S. Lighthouse Service design used for at least five other examples (though none in Rhode Island), the Sandy Point Light (later North Light) represented an important evolutionary step in lighthouse design: the standardized-plan attached tower and dwelling. Like many post–Civil War lights, it is of substantial masonry construction—here, rock-faced granite ashlar (compare BI16); the treacherous waters off Sandy Point washed away several earlier, flimsier lights. Located in the windswept dunes of a wildlife and bird sanctuary, it presents the very image of maritime refuge.

BI9 Narragansett Hotel (Lake Shore Dining Hall, Samuel D. Mott House)

Before 1800, house. 1890s and later, dining hall and hotel. West side of northernmost point of Ocean Ave. (at West Side Rd.)

The oldest building in this complex is the Samuel D. Mott House, a late-eighteenth-century central-chimney dwelling that is one of the oldest two-story buildings on the island. The two-story hotel, with a mansard roof, sunbonnet gables, and a wraparound porch and large dining hall at the rear, is typical of the waterside (here overlooking the Great Salt Pond) hotels of the turn of the century.

BI10 Island Cemetery

1661 and later. Southwestern junction of West Side Rd. and Center Rd.

The only remaining element of the 1661 plan of the island is the Island Cemetery, located at the southern edge of what was planned as the compact part of the settlement. Like land divisions in several other Massachusetts Bay and Plymouth Colony proprietary settlements, the plan's arbitrary geometry was little related to local topography and was probably modified or abandoned soon after settlement. Few early stones survive; most are from the late nineteenth century.

BI11 Peleg Champlin Farm

c. 1820. End of private road off west side of Champlin Rd. (Coast Guard Rd.), 500 feet from West Side Rd.

The Champlin House, a one-story, central-chimney, five-bay house, is like many others across the state; for Block Island, however, it represents the best preserved of a rare early type. Because the island was cleared of wood by the middle of the eighteenth century, early houses were routinely cannibalized for new construction before regular boat service began in the 1870s. The little altered complex, which includes a later nineteenth-century barn, sits on high land overlooking Block Island Sound.

BI12 U.S. Lifesaving Station (Former)

1886, Albert B. Bibb. Cooneymus Rd.

Increased maritime traffic around and to Block Island in the late nineteenth century occasioned the construction of several related buildings along the island's shores. Stations such as this were built following standardized plans—here, the plan for the Bibb Number 2 Station, developed by the house architect for the U.S. Lifesaving Service. The design recalls the smooth massing of the Shingle Style—though this was originally clapboard—and here, for the first time, the lookout was placed in a dormer, not in a tower. Other lifesaving stations on the island's east side and at Sandy Point are gone. This station, abandoned in 1936, now serves as a summer house, handsomely remodeled for that purpose.

BI13 Lewis–Dickens Farm

Mid-19th century. North side of Lewis Farm Rd. (private dirt road approx. .5 mile from junction of West Side Rd. and Cooneymus Rd.)

The house and outbuildings of this rare and intact 200-acre farm are typical mid-nineteenth-century structures, but the expansive agricultural landscape—a high plateau of gently rolling grasslands divided by low stone walls—is extraordinary. In 1982, The Nature Conservancy purchased 141 acres of the farm to preserve it in perpetuity as open space.

Farther east on Cooneymus Road is Rodman's Hollow, a narrow, mile-long valley of some 200 acres equally remarkable for its scenic quality. It too is protected from development and preserved for recreational use.

BI14 Everett D. Barlow House
(Bit O Heaven; Mohegan Cottage)

1886, Charles E. Miller. South side of Snake Hole Rd. (private road), 600 feet southwest of junction with Lakeside Dr.

Described at the time of its completion as "a compound of the Swiss and Queen Anne styles," the Barlow House was published in the August 1886 issue of *Scientific American* as a typical American summer cottage. The south-facing house (built exactly on the meridian) exploits a commanding view of the Atlantic Ocean with a wraparound porch on the east, south, and west sides and an attic-level observatory with windows in the four cardinal directions. The chief alterations to the house include simplification of medievalizing porch struts and replacement of the lively original color scheme—olive green body with red trim—with monochromatic white.

BI15 Julius Deming Perkins House
(Bayberry Lodge)

c. 1898. Private way south off Mohegan Trail 600 feet east of junction with Lakeside Dr. (barely visible)

A shingled summer house with large, inset first-story porch overlooking the Atlantic, the Perkins House is one of the island's more sophisticated late-nineteenth-century summer houses, especially in contrast to its simpler shingled neighbors. Its late date—some twenty years after the earliest Shingle Style houses in Newport—suggests the conservative nature of building on the island.

BI16 South East Light

1873–1875, Lighthouse Board. Southeast side of Mohegan Trail at Pilot Hill Trail

Built in brick to a standardized plan used originally for a stone lighthouse (now demolished) in Cleveland, Ohio, the South East Light is quite striking: the octagonal tower is attached

BI16 South East Light

to a large, symmetrical Stick Style dwelling with two units, one for the keeper of the light and one for his first and second assistants. Originally fitted with a first-order Fresnel lens, this was one of the best and is one of the best preserved of the generation of lighthouses erected after the Civil War. Continued erosion of the Mohegan Bluffs, on which it stands, necessitated the building's move several hundred feet inland during the 1990s.

BI17 Homer Russell House

1982–1984, Homer Russell. East side of Spring St. (extension of Mohegan Trail), 500 feet northeast of entrance to South East Light

As seen in the Barlow and Perkins houses (BI14 and BI15), views of the ocean have long been highly desirable island attractions. Here, on a landlocked site, an emphatically vertical house represents a late-twentieth-century, post–oceanfront development response to capitalizing on fine views across the hills and to the ocean. The vertical-format country house achieved popularity in the 1980s and saw extensive coverage in professional architectural journals. Russell's design for his own house is nicely done, retaining the simple forms and shingle cladding commonly associated with island domestic architecture.

BI18 Spring House and Cottage

1852, 1872, 1898. Spring St.

BI19 Atlantic Inn (Norwich House)

1878. Spring St.

These two mansard-roofed hotels capture the image of a summer vacation on Block Island. The Spring House, one of the first hotels on the island, took advantage of both a wide view of the ocean and a mineral spring. It expanded to its present form, with a bell-cast mansard, just as regular ferry service was inaugurated. Adjacent is a fine board-and-batten Carpenter's Gothic summer cottage. Just up the hill is the Norwich House, picturesquely rehabilitated as the Atlantic Inn in the 1980s, the decade that saw Block Island's renaissance as a trendy vacation spot.

Bibliography

The following is based in large part on the bibliography for *Buildings on Paper: Rhode Island Architectural Drawings, 1825–1945*, by William H. Jordy and Christopher Monkhouse (Providence: Bell Gallery of Brown University, Rhode Island Historical Society, and Rhode Island School of Design Museum of Art, 1982), used here with the permission of the original publishers and adapted and updated by the editors of this volume.

At every opportunity, WHJ acknowledged his debt to the surveys carried out by the Rhode Island Historical Preservation and Heritage Commission, published as State Surveys (Providence: RIHPHC, 1977–). His drafts for the introduction included the following statement:

> It should be noted that the state is blessed with the admirable collection of town inventories published by the Rhode Island Historical Preservation Commission, to which *Buildings of Rhode Island* is heavily indebted. These provide coverage of buildings significant for preservation, together with informative commentary on them, both in introductory essays and at the level of individual entries. These town-by-town inventories are not guidebooks. As inventories of buildings considered worthy of preservation, they list (as they should) many items of too local and contextual an interest for the general traveler. Consequently, they contain both less elaborate and less critical commentary than the more selective sampling included in Buildings of Rhode Island. The inventories are also organized alphabetically by street address, not as tours for a traveler.

A number of individual surveys are included in this bibliography. For a complete list, see the RIHPHC Web site: http://www.rihphc.state.ri.us/survey/survey_pub.html. Unpublished materials in the RIHPHC files should also be mentioned here, as well as other archival resources, including those of the Rhode Island Historical Society and the Newport Historical Society.

ABBREVIATIONS

RIHPHC Rhode Island Historical Preservation and Heritage Commission
JSAH Journal of the Society of Architectural Historians

GENERAL

Arnold, Samuel G. *History of the State of Rhode Island and Providence Plantations.* New York: D. Appleton and Co., 1874.

Barnard, Henry. *School Architecture, or, Contributions to the Improvement of Schoolhouses in the United States.* New York: A. S. Barnes, 1848.

Bayles, Richard M., ed. *History of Providence County, Rhode Island.* New York: W. W. Preston, 1891.

Beers, J. H., and Co. *Representative Men and Old Families of Rhode Island.* Chicago: J. H. Beers and Co., 1901.

Bicknell, Thomas Williams. *The History of the State of Rhode Island and Providence Plantations.* New York: American Historical Society, 1920.

Biographical Cyclopedia of Representative Men of Rhode Island. Providence: National Biographical Publishing Co., 1881.

Bridenbaugh, Carl. *Fat Mutton and Liberty of Conscience.* Providence: Brown University Press, 1974.

Bryant, William Cullen. *Picturesque America.* New York: D. Appleton, 1872.

Carroll, Charles C. *Rhode Island: Three Centuries of Democracy.* New York: Lewis Historical Publishing Company, 1932.

Chase, David. "An Historical Survey of Rhode Island Textile Mills." Unpublished paper, Brown University, 1969; manuscript, Rhode Island Historical Society, 1969.

Coleman, Peter J. *The Transformation of Rhode Island, 1790–1860.* Providence: Brown University Press, 1963.

Craig, Lois, et al. *The Federal Presence: Architecture, Politics, and Symbols in United States Government Building.* Cambridge, MA: MIT Press, 1978.

Conley, Patrick T. *Democracy in Decline: Rhode Island's Constitutional Development, 1776–1841.* Providence: Rhode Island Historical Society, 1977.

———, and Matthew J. Smith. *Catholicism in Rhode Island: The Formative Era.* Providence: Diocese of Providence, 1976.

Conely, Patrick T.; Robert Owen Jones; and William McKenzie Woodward. *The State Houses of Rhode Island.* Providence: Rhode Island Historical Society and RIHPHC, 1988.

Cummings, Abbott Lowell. *Architecture in Early New England.* Sturbridge, MA: Old Sturbridge Village, 1958.

Dempsey, Claire W.; Richard E. Greenwood; and William McKenzie Woodward. *The Early Architecture and Landscapes of the Narragansett Basin, Vol. 2: Blackstone River Valley and Providence.* Newport, RI: Vernacular Architecture Forum, 2001.

Downing, Antoinette F. *Early Homes of Rhode Island.* Richmond: Garrett and Massie, 1937.

———, and Vincent J. Scully, Jr. *The Architectural Heritage of Newport, Rhode Island.* Cambridge, MA: Harvard University Press, 1952; 2nd ed., New York: C. N. Potter, 1967.

Drexler, Arthur, ed. *The Architecture of the Ecole des Beaux-Arts.* New York: Museum of Modern Art; Cambridge, MA: MIT Press, 1977.

Dwight, Timothy. *Travels in New England and New York.* New Haven: T. Dwight, 1821.

Federal Writers' Project, Works Progress Administration. *Rhode Island: A Guide to the Smallest State.* Boston: Houghton Mifflin, 1937.

Field, Edward, ed. *State of Rhode Island and Providence Plantations at the End of the Century: A History.* Boston: Mason Pub. Co., 1902.

Frazier, A. A., ed. *Club, Bench, Bar and Professional Life of Rhode Island.* Providence: H. M. Frazier, 1896.

Gane, John F., ed. *American Architects Directory.* New York: R. R. Bowker, 1970.

Garman, James C. "'The Tension Castles of Stone and Steel': Landscape, Labor and Power in the Urban Penitentiary." Public Archaeology Laboratory Report no. 588, RIHPHC Library.

Gebhard, David, and Deborah Nevins. *200 Years of American Architectural Drawing.* New York: Whitney Library of Design for the Architectural League of New York and the American Federation of Arts, 1977.

Greene, Welcome Arnold. *The Providence Plantations for Two Hundred and Fifty Years.* Providence: J. A. And R. A. Reid, 1886.

Grieve, Robert, and John P. Fernald. *The Cotton Centennial.* Providence: J. A. and R. A. Reid, 1891.

Hall, Joseph, ed. *Biographical History of the Manufacturers and Business Men of Rhode Island.* Providence: J. D. Hall, 1901.

Hamlin, Talbot C. *Greek Revival Architecture in America.* New York: Oxford University Press, 1944.

Hedges, James. *The Browns of Providence Plantations, The Colonial Years.* Cambridge, MA: Harvard University Press, 1952.

———. *The Browns of Providence Plantations, The Nineteenth Century.* Providence: Brown University Press, 1968.

Herndon, Richard, ed. *Men of Progress—Rhode Island.* Boston: New England Magazine, 1896.

Historic American Buildings Survey. *Rhode Island Catalogue,* ed. Osmund Overby. Washington, DC, 1972.

Historic American Engineering Record. *Rhode Island: An Inventory of Historic Engineering and Industrial Sites,* ed. Gary Kulik. Washington, DC, 1978.

Hitchcock, Henry-Russell. *Architecture, Nineteenth and Twentieth Centuries.* Baltimore: Penguin Books, 1958.

———. *Rhode Island Architecture.* Providence: Rhode Island Museum Press, 1939; reprint New York: Da Capo Press, 1968.

Isham, Norman Morrison, and Albert F. Brown. *Early Rhode Island Houses: An Historical and Architectural Survey.* Providence: Preston and Rounds, 1895.

James, Sidney V. *Colonial Rhode Island: A History.* New York: Scribner, 1975.

Jordy, William H., and Christopher Monkhouse. *Buildings on Paper: Rhode Island Architectural Drawings, 1825–1945.* Providence: Bell Art Gallery, Brown University; Rhode Island Historical Society; and Museum of Art, Rhode Island School of Design, 1982.

Kellner, George H., and Stanley J. Lemons. *Rhode Island, The Independent State.* Woodland Hills, CA.: Windsor Publications, 1982.

Koyl, George S., and Maria B. Mathieson, eds. *American Architectural Drawings: A Catalogue.* 5 vols. Philadelphia: Philadelphia Chapter, American Institute of Architects, 1969.

Kulik, Gary, and Julia C. Bonham. *Rhode Island: An Inventory of Historic Engineering and Industrial Sites.* Washington: U.S. Department of the Interior, 1978.

Laswell, George D. *Corners and Characters of Rhode Island.* Providence: The Oxford Press, 1924.

Leading Manufacturers and Merchants of Rhode Island. Boston, International Publishing Company, 1886.

Lundberg, Kenneth V. "Rhode Island Economy in the 1940s and 1950s: Caught in the Crosscurrents of Economic Change." Unpublished paper. Rhode Island Historical Society/Providence Preservation Society Forum, 1981.

McLoughlin, William Gerald. *Rhode Island: A Bicentennial History.* New York: W. W. Norton, 1976.

Mason, Harold. *Colonial Houses and Doorways in Providence and in the State of Rhode Island.* Providence, 1910.

Meyer, Kurt B. *Economic Development and Population Growth in Rhode Island.* Providence: Brown University, 1953.

Michie, Thomas S., and Christopher P. Monkhouse. "Pattern Books in the Redwood Library and Athenaeum, Newport, Rhode Island." *Antiques* 137 (January 1990): 286–299.

Murray, Richard N.; Diane Pilgrim; and Richard G. Wilson. *American Renaissance, 1876–1917.* Brooklyn, NY: Brooklyn Museum, 1979.

National Cyclopaedia of American Biography. New York: J. T. White, 1893.

Nevins, Deborah, and Robert A. M. Stern. *The Architect's Eye: American Architectural Drawings*

from 1788 to 1978. New York: Pantheon Books, 1979.
Norton, Paul F. *Rhode Island Stained Glass*. Dublin, NH: William L. Baughan, 2001.
Onorato, Ronald J. *Outdoor Sculpture of Rhode Island: Monumental Ambitions*. Providence: RIHPHC, 1999.
Physick, John, and Michael Darby. *Marble Halls*. London: Victoria and Albert Museum, 1973.
Pierson, William H. *American Buildings and Their Architects: The Colonial and Neo-Classical Style*. Garden City, NY: Doubleday, 1970.
———. *American Buildings and Their Architects: Technology and the Picturesque, the Corporate and the Early Gothic Styles*. Garden City, NY: Anchor Press for Doubleday, 1978.
Preston, Harold Willis. *Rhode Island's Historical Background*. Providence, 1936.
Representative Men and Old Families of Rhode Island. Chicago: J. H. Beers and Co., 1908.
Rhode Island Census Board. *Census*. 1730–1895.
Rhode Island Chapter, American Institute of Architects. Records, 1875 et seq.
Rhode Island Historical Society Manuscript Collection.
———. *Year Book*. Providence, 1910, 1911.
Rhode Island Colony Records.
St. George, Robert B. "Bawns and Beliefs: Architecture, Commerce, and Conversion in Early New England." *Winterthur Portfolio* 25 (Winter 1990): 241–287.
———. *Conversing by Signs: Poetics of Implication in Colonial New England Culture*. Chapel Hill, NC: University of North Carolina Press, 1998.
Sande, Theodore Anton. "The Architecture of the Rhode Island Textile Industry, 1790–1860." Ph.D. diss., University of Pennsylvania, 1972.
Sanderson, Edward F. "Rhode Island Merchants in the China Trade." In *Federal Rhode Island: The Age of the China Trade, 1790–1820*. Providence: Rhode Island Historical Society, 1978.
Scully, Vincent J. *The Shingle Style and the Stick Style*. Rev. ed. New Haven: Yale University Press, 1971.
Stillman, Damie, et al. *Architecture and Ornament in Late 19th-Century America*. Newark, DE: University Gallery, University of Delaware, 1981.
Sweeney, Kevin M. "Meetinghouses, Town Houses, and Churches: Changing Perceptions of Sacred and Secular Space in Southern New England, 1720–1850." *Winterthur Portfolio* 28 (Spring 1993): 59–93.
United States Bureau of the Census. *Census*. 1790–1980.
Whitehead, Russell F., and Frank Chouteau Brown, eds. *Early Architecture of Rhode Island*. Harrisburg, PA: National Historical Society, 1987.
Williams, Alfred M., and William F. Blanding, eds. *Men of Progress, Rhode Island*. Boston: New England Magazine, 1896.
Wood, Frederick J. *The Turnpikes of New England*. Boston: Marshall Jones, 1919.
Woodward, William McKenzie. *Historic Landscapes of Rhode Island*. Providence: RIHPHC, 2001.
Who Was Who in America. Chicago, 1897–1976.

ARCHITECTS

Alexander, Robert L. "The Architecture of Russell Warren." Ph.D. diss., New York University, 1952.
Almy, Arthur L. "A List of Works Copied from a list evidently compiled by Mr. A. C. Morse from his account books, but not including all of his works." Rhode Island Historical Society, American Institute of Architects File.
Baker, Paul. *Richard Morris Hunt*. Cambridge, MA: MIT Press, 1980.
Baldwin, Charles C. *Stanford White*. New York: Dodd, Mead, 1931.
Bridenbaugh, Carl. *Peter Harrison, the First American Architect*. Chapel Hill, NC: University of North Carolina Press, 1949.
———. "Peter Harrison, Addendum." *JSAH* 18 (December 1959): 158–159.
Brown University Department of Art. *Thomas Alexander Tefft: American Architecture in Transition, 1845–1860*. Catalog of exhibition, Providence, January 23–March 6, 1988.
"Bucklin, James C." *American Architect and Building News* 30 (1890): 17.
Cady, John Hutchins. File Cabinet of Records, 1900–1967. Rhode Island Historical Society.
———. "A Connecticut Yankee in Rhode Island." Address to the Rhode Island Chapter, American Institute of Architects, Providence, February 18, 1957.
Doumato, Lamia. *Peter Harrison, Colonial Architect*. Monticello, IL: Vance Bibliographies, 1986.
Emanuel, Muriel, ed. *Contemporary Architects*. New York: St. Martin's Press, 1980.
Francis, Dennis. *Architects in Practice, New York City, 1840–1900*. New York: The Committee, 1980.
Hart, Charles Henry. "Peter Harrison, 1716–1775, First Professional Architect in America. . . ." Presented to Massachusetts Historical Society, March 9, 1916. Boston, 1916.
Hitchcock, Henry-Russell. *The Architecture of H. H. Richardson and His Times*. New York: Museum of Modern Art, 1936.
Hunt, C. H. "Biography of R. M. Hunt." Typescript, American Architectural Archives, Greenwich, CT, n.d.
Hurdis, Frank De Voe. "The Architecture of John Holden Greene." M.A. thesis, Cornell University, 1973.
Holden, Wheaton Arnold. "Robert Swain Peabody of Peabody and Stearns in Boston: The Early Years (1870–1886)." Ph.D. diss., Boston University, 1969.
Isham, Norman Morrison. Rhode Island Historical Society Manuscript Collection.
———. "John Holden Greene, Architect." Rhode Island Historical Society, Archives.
———. "Two Rhode Island Architects." *American*

Architect and Building News 91 (February 1, 1907): 67.

Jessup, Karen L. "The Architecture of Edmund R. Willson (1856–1906)." M.A. thesis, Boston University, 1983.

"John Holden Greene: Carpenter-Architect of Providence." Providence Preservation Society, 1972.

Kamerling, Bruce. *Irving J. Gill, Architect*. San Diego, CA: San Diego Historical Society, 1993.

Kimball, Fiske. "The Colonial Amateurs and Their Models: Peter Harrison." *Architecture* 53, no. 6 (1926): 155–160, and 54, no. 1: 185–190, 209.

Landau, Sarah Bradford. *Edward T. and William A. Potter, American Victorian Architects*. New York: Garland, 1979.

Little, Margaret Ruth. "The Architecture of a Late, Lamented Genius, Thomas Alexander Tefft." M.A. thesis, Brown University, 1972.

Miller, A. H. "Alpheus C. Morse." *Rhode Island Medical Journal* 20 (February 1957): 26.

Moore, Charles. *The Life and Times of Charles Follen McKim*. New York: Houghton Mifflin, 1929.

Nelson, Donna-Belle. "The Greek Revival Architecture of James C. Bucklin." M.A. thesis, University of Delaware, 1969.

Newcomb, Rexford. "Early American Architects." *The Architect* 10, no. 3 (June 1928): 315–318.

O'Gorman, James F. *H. H. Richardson and His Office: Selected Drawings*. Cambridge, MA: Department of Printing and Graphic Arts, Harvard College Library, 1974.

Onorato, Ronald J. "Architecture and Drawing, the Newport Career of John Dixon Johnston." Unpublished manuscript, 2003, Newport Historical Society Library.

———. "Providence Architecture, 1859–1908: Stone, Carpenter & Willson." *Rhode Island History* 33, nos. 3 and 4 (August and November 1974).

———. "Stone, Carpenter, & Willson: An Analysis of the Architectural Firm, 1858–1908." Unpublished paper, Brown University, 1973.

Reilly, C. H. *McKim, Mead & White*. London: E. Benn, 1924.

Roth, Leland M. *McKim, Mead & White Architects*. New York: Harper and Row, 1983.

———, ed. *A Monograph of the Work of McKim, Mead & White, 1879–1915*, with an introductory essay, "McKim, Mead & White Reappraised," and notes on the plates. New York, 1974 (orig. ed., New York, 1915–1920).

Sickels, Lauren-Brock. "The Architecture of John Holden Greene Using Patternbooks." Unpublished paper, Columbia University, ©1979.

Stone, Alfred. "Alpheus Morse." *The American Architect and Building News* 42 (December 9, 1893): 126–127.

———. "Edmund R. Willson, 1859[sic]–1906." *The American Architect and Building News* 91 (February 9, 1907): 67–72.

Stone, Edwin Martin. *The Architect and Monetarian: A Brief Memoir of Thomas Alexander Tefft*. Providence: S. S. Rider and Brother, 1869.

Swan, Mabel Munson. "John Holden Greene, Architect." *Antiques* 52 (July 1947): 24–27.

Tefft, Thomas Alexander. Architectural Drawings, 184?–1859. Brown University, John Hay Library Archives, and Rhode Island Historical Society Library.

Upjohn, Everard Miller. *Richard Upjohn, Architect, Churchman*. New York: Columbia University Press, 1939.

Van Rensselaer, Mariana Griswold. *Henry Hobson Richardson and His Works*. Boston: Houghton, Mifflin and Company, 1888.

Walker, William R. and W. Howard. *Architectural Portfolio: William R. Walker & Son, Architects, Providence, Rhode Island*. Providence, 1895.

Wilson, J. Walter. "Joseph Brown: Scientist and Architect." *Rhode Island History* 4: 67–79, 121–128.

Wilson, Richard Guy. *McKim, Mead & White Architects*. New York: Rizzoli, 1983.

Withey, Henry F. and Elsie R. *Biographical Dictionary of American Architects (Deceased)*. Los Angeles: Hennessey and Ingalls, 1956.

Wriston, Barbara. "Thomas Tefft, Progressive Rhode Islander." *Rhode Island Historical Society Collections* 34 (April 1941): 60–61.

———. "Thomas Alexander Tefft, Architect and Economist," M.A. thesis, Rhode Island School of Design, 1942.

———. "The Architecture of Thomas Tefft." *Rhode Island School of Design Bulletin of the Museum of Art* 28 (November 1940): 37–45.

NEWPORT

Bayles, Richard M., ed. *History of Newport County, Rhode Island*. New York: L. E. Preston, 1888.

Chase, David. *The Kay–Catherine–Old Beach Road Neighborhood in Newport*. Providence: RIHPHC, 1974.

Clark, Kenneth. *An Architectural Monograph on Newport, Rhode Island, An Early American Seaport*. New York: Architectural Record, 1922.

Dix, J. R. *A Hand-book of Newport*. Newport: C. E. Hammett, Jr., 1852.

Dow, Charles H. *Newport Past and Present*. Newport, 1880.

Downing, Antoinette F., and Vincent J. Scully, Jr. *The Architectural Heritage of Newport, Rhode Island, 1640–1915*. 2nd ed., rev. New York: Clarkson N. Potter, 1967.

Elliott, Maude Howe. *This Was My Newport*. Cambridge, MA: The Mythology Company, A. M. Jones, 1944.

Friedman, Lee M. "The Newport Synagogue." *Old-Time New England* 36 (1946): 49–57.

Gannon, Thomas. *Newport Mansions, The Gilded Age*. Little Compton, RI: Foremost Publishers, 1982.

Guinness, Desmond. *Newport Preserv'd: Architecture of the Eighteenth Century*. New York: Viking, 1982.

Howe, Julia Ward. *Reminiscences*. New York: Houghton, Mifflin and Company, 1899.

Isham, Norman Morrison. "The Brick Market." *Bulletin of the Society for the Preservation of New England Antiquities* 6, no. 2 (1916): 3–11, 20–23.

———. *Trinity Church in Newport, R.I.* Boston, Printed for the subscribers [by D. B. Updike], 1936.

Mason, George C. *Annals of the Redwood Library and Athenaeum*. Newport, R.I.: Redwood Library, 1891.

———. *Newport and Its Cottages*. Newport: J. R. Osgood, 1875.

———. *Newport Illustrated by Sketches with Pen and Camera*. Newport: C. E. Hammett, Jr., 1891.

———. *Newport Illustrated in a Series of Pen and Pencil Sketches*. Newport: C. E. Hammett, Jr., 1875.

———. *Reminiscences of Newport*. Newport: C. E. Hammett, Jr., 1884.

"Newport: The Preservation Society of Newport County in Newport, Rhode Island (11 article special section)." *Antiques* 147 (April 1995): 544–609.

Randall, Anne. *Newport: A Tour Guide*. Newport: Catboat Press, 1970.

Schless, Nancy H. "Peter Harrison, the Touro Synagogue and the Wren City Church." *JSAH* 30 (October 1971): 242.

Simister, F. P. *Streets of the City: An Anecdotal History of Newport*. Providence: Mowbray Co., 1969.

Society of Friends of Touro Synagogue National Historic Shrine. *Touro Synagogue of Congregation Jeshuat Israel*. Newport, R.I.: Society of Friends of Touro Synagogue National Historic Shrine, 1948.

Stachiw, Myron O., et al. *The Early Architecture and Landscapes of the Narragansett Basin, Vol. 1: Newport*. Newport, RI: Vernacular Architecture Forum, Newport, RI, 2001.

Van Rensselaer, Mariana Griswold. *Newport Our Social Capital*. Philadelphia: J. B. Lippincott, 1905.

Vitruvius Americanus: Colonial Newport in the Palladian Tradition. Newport, RI: Redwood Library and Athenaeum, 1996.

Youngken, Richard C. *African Americans in Newport*. Providence: RIHPHC and Rhode Island Black Heritage Society, 1995.

PROVIDENCE

Anthony, Daniel. *Map of the Town of Providence*. 1803, 1823.

Bardoglio, Peter. "Italian Immigrants and the Catholic Church in Providence." *Rhode Island History* 34 (1975): 47–57.

Bayles, Richard M. *History of Providence County, Rhode Island*. New York, 1891.

Beers, D. G., and Co. *Map of the City of Providence*. 1870.

"Building Improvements in Providence." *Providence Daily Journal*, January 1, 1883, p. 9.

"Building Improvements in Real Estate in 1866." *Providence Daily Journal*, January 19, 1867, p. 1.

"Building Improvements in Rhode Island." *Providence Daily Journal*, December 3, 1864.

"Building in 1867." *Providence Daily Journal*, January 25, 1868.

"Building in Rhode Island." *Providence Daily Journal*, December 28, 1865.

Cady, John Hutchins. *The Civic and Architectural Development of Providence, 1636–1950*. Providence, R.I.: Book Shop, 1957.

———. "The Providence Market House and Its Neighborhood," *Rhode Island History* 11: 97–116.

———. "Some Reminiscences of the Rhode Island Chapter of the American Institute of Architects," delivered before the Rhode Island Chapter, American Institute of Architects, April 28, 1951. Typescript.

Carbone, Domenic, Jr.; Warren Jagger; Thomas S. Michie; and William McKenzie Woodward. *Most Admirable: The Rhode Island State House*. Providence: State House Restoration Society, 2002.

Carroll, Charles. *Rhode Island: Three Centuries of Democracy*. New York: Lewis Historical Publishing Company, 1932.

Carroll, Leo C. "Irish and Italians in Providence, Rhode Island, 1880–1960." *Rhode Island History* 28, no. 3 (August 1969): 67–74.

Cate, Samuel; Irving B. Haynes; and Marion Sachs. "A Walking Tour Guide of Providence, Rhode Island." Rhode Island Chapter, American Institute of Architects, 1973.

Chace, Henry Richmond. *Owners and Occupants of the Lots, Houses and Shops in the Town of Providence, Rhode Island in 1798*. Providence: Livermore and Knight, 1914.

Christensen, Robert O. *Elmwood, Providence*. Providence: RIHPHC, 1979.

Chudacoff, Howard. "Providence Becomes a City." Unpublished paper, Rhode Island Historical Society/Providence Preservation Society Forum, 1977.

———, and Theodore C. Hirt. "Social Turmoil and Governmental Reform in Providence, 1820–1832." *Rhode Island History* 31, no. 1 (February 1972): 21–32.

Cogswell, Elizabeth Agee. "The Henry Lippitt House of Providence, Rhode Island," *Winterthur Portfolio* 17, no. 4 (Winter 1982): 203–242.

Collins, Clarkson E., III, ed. "Pictures of Providence in the Past, 1790–1820: The Reminiscences of Walter R. Danforth." *Rhode Island History*, April 1951 and January 1952.

Conley, Patrick T., and Paul Campbell. *Providence, A Pictorial History*. Norfolk, VA: Donning, 1982.

Cushing and Walling. *Map of the City of Providence*. 1849, 1853.

Delude-Dix, Elizabeth; Ronald J. Onorato; Judith Tolnick; and William McKenzie Woodward. "*A Most Admirable Public Building*": The Rhode Island

State House Centennial Exhibition. Kingston, RI: University of Rhode Island, 1996.

Downtown Providence 1970. Providence: The Commission, 1959.

The Early Records of the Town of Providence. Providence: Snow and Farnham, 1892–1909.

"Exhibition of Architectural Drawings at Providence." *American Architect and Building News* 80 (April 4, 1903): 5–6.

Fink, Lisa C. *Providence Industrial Sites.* Providence: RIHPHC, 1981.

Freeman, Robert, and Vivienne Lasley. *Hidden Treasure: Public Sculpture in Providence.* Providence: Rhode Island Bicentennial Foundation, 1900.

Fulkerson, Georgiana. "The Cultural and Industrial Development of Providence." M.A. thesis, Rhode Island School of Design, 1958.

Gibbs, James H., and Pamela Kennedy. *The West Side, Providence.* Providence: RIHPHC, 1976.

Glass, Anita. "Early Victorian Domestic Architecture on College Hill." M.A. thesis, Brown University, 1960.

Goldstein, Sidney, and Kurt B. Meyer. *The Ecology of Providence.* Providence: Brown University, Department of Sociology and Anthropology, 1958.

Gowdy File [Histories]. Providence Preservation Society.

Grossman, Elizabeth G. "The Victorian Commercial Architecture of Downtown Providence." M.A. thesis, Brown University, 1973.

Guild, Reuben A. *History of Brown University with Illustrative Documents.* Providence: Providence Press, 1887.

Hayman, Robert. "Will the Shamrock Supersede the Anchor and the Hope?: The Impact of Immigration and Catholicism in Mid-Nineteenth-Century Providence." Unpublished paper, Brown University/Providence Preservation Society lecture series, "Divine Providence." 1985.

Hopkins, Charles Wyman. *House Lots of Early Settlers of the Providence Plantations.* Providence, 1886.

Hopkins, G. M. *Atlas of the City of Providence, R.I., and Environs.* 1882.

———. *City Atlas of Providence.* 1937.

———. *City Atlas of Providence, Rhode Island, by Wards.* 1875.

———. *Plat Book of the City of Providence, Rhode Island.* 1918.

Ihlder, John. *The Houses of Providence: A Study of Present Conditions and Tendencies.* Providence: Snow and Farnham, 1916.

Illustrated Hand-Book of the City of Providence. Providence: J. C. Thompson, 1876.

"Improvements in the City." *Providence Daily Journal,* November 28, 1863.

Interface: Providence. Providence: Rhode Island School of Design, 1974.

Isham, Norman Morrison. *The Meeting House of the First Baptist Church in Providence: A History of the Fabric.* Providence: Akerman-Standard, 1925.

Kimball, Gertrude Selwyn. *Providence in Colonial Times.* Boston and New York: Houghton Mifflin, 1912.

King, Moses. *King's Pocket Book of Providence, Rhode Island.* Cambridge, MA: Moses King, 1882.

Kirk, William A. *A Modern City: Providence, R.I., and Its Activities.* Chicago: University of Chicago Press, 1909.

Lancaster, Jane. *Inquire Within: A Social History of the Providence Athenæum.* Providence: Providence Athenæum, 2003.

Lemons, J. Stanley. *First: The First Baptist Church in America.* Providence: Charitable Baptist Society, 2001.

Lockwood and Cushing. *Map of the City of Providence and Town of North Providence.* 1835.

Miner, George L. *Angell's Lane.* Providence: Akermann-Standard Press, 1948.

Neu, Deborah Dunning, and Jody Ziegler. *Old Buildings—New Uses: A Festival of Recycled Space.* Providence: Providence Preservation Society, 1975.

Overby, Osmund R. "The Architecture of College Hill, 1700–1900: Residential Development in the Area of the Original Town of Providence, Rhode Island." Ph.D. diss., Yale University, 1964.

Parish, W. H. *Art Work of Providence.* Providence: W. H. Parish Publishing Co., 1896.

"Parish Churches." *The Visitor,* April 26, 1968.

Providence Board of Trade Journal. Providence: The Board, 1889–1931.

Providence, City of. *Annual Report of the School Committee,* 1846, 1848, 1850, 1854, 1985.

———. *Building Permits.* 1922–1985.

———. *Intention to Build Permits.* 1860–1922.

———. *Recorder of Deeds.* Land Records.

———. *Tax Assessor Records.*

———. *Tax Books.* 1827–1972.

Providence City Planning Commission. *College Hill: A Demonstration Study of Historic Area Renewal.* 2nd ed. Providence: Providence City Planning Commission, 1967.

"Providence: The Confident Years, 1890–1920." Unpublished paper, Rhode Island Historical Society/Providence Preservation Society Forum. 1979.

Providence Daily Journal. 1829–1985.

Providence Directory. 1824, 1985.

Providence Evening Bulletin. 1863–1985.

Providence House Directory. 1892–1935.

Providence Illustrated. Providence: H. R. Page, 1891.

Providence Journal-Bulletin Almanac. Providence, 1887–1985.

Providence of Today: Its Trade, Commerce and Industries with Biographical Sketches and Portraits. Providence, 1893.

Providence Preservation Society. *Street Festivals.* Providence, 1958–1971.

Providence Redevelopment Agency. *Annual Report,* 1947–1984.

Providence Town Papers. Federal Direct Tax, 1798, 1814. Manuscript Collection, Rhode Island Historical Society.

Quinlisk, Margaret Teal. "Georgian Architecture in Providence, Rhode Island: A Case Study of the Georgian Mindset." B.A. thesis, Brown University, 1996.

Richards, L. J., and Company. *Atlas of the City of Providence, Rhode Island.* 1908, 1917.

Rules for House Carpenters Work in the Town of Providence. Providence, 1797.

Sanborn, D. A. *Atlas of Providence.* 1889.

———. *Insurance Map of Providence.* 1874.

Sanborn Map Company. *Insurance Maps of Providence, Rhode Island.* 1920 (corrected to 1924).

Simister, F. P. "Streets of the City." Providence: WEAN, 1952–1970. Typescript, Rhode Island Historical Society.

Snow, Edwin M. "History of the Asiatic Cholera in Providence." *Providence Daily Journal,* December 31, 1857.

Staples, William R. *Annals of the Town of Providence from its First Settlement to the Organization of the City Government, in June 1832.* Providence, 1843.

Stokes, Howard Kemble. *The Finances and Administration of Providence.* Baltimore: Johns Hopkins Press, 1903.

Stone, Edwin Martin. *The Life and Recollections of John Howland.* Providence: G. H. Whitney, 1857.

Topographical Atlas of Providence County, Rhode Island. Providence: Everts and Richards, 1895.

Vollmert, Leslie J. *South Providence.* Providence: RIHPHC, 1978.

Walling, Henry F. *Map of the City of Providence, Rhode Island.* 1857.

Wheeler, Robert A. "Fifth Ward Irish Immigrant Mobility in Providence, 1850–1870." *Rhode Island History* 32 (May 1973): 52–60.

Woodward, William McKenzie. *Downtown Providence.* Providence: RIHPHC, 1981.

———. *A Guide to Providence Architecture.* Providence: Rhode Island Chapter, American Institute of Architects, and Providence Preservation Society, 2003.

———. *Providence Preservation Society Festival of Historic Houses,* 1980–1985.

———. *Smith Hill, Providence.* Providence: RIHPHC, 1980.

———, and Sanderson, Edward F. *Providence: A Citywide Survey of Historic Resources.* Providence: RIHPHC, 1986.

OTHER LOCATIONS

Bickford, Christopher P. *Crime, Punishment, and the Washington County Jail.* Kingston, RI: Pettaquamscutt Historical Society, 2002.

Driemeyer, Laura B., and Myron Stachiw. *The Early Architecture and Landscapes of the Narragansett Basin, Vol. 3: Bristol and the East Bay, Wickford and the West Bay.* Newport, R.I.: Vernacular Architecture Forum, Newport, R.I., 2001.

Garman, James E. *Historic Houses of Portsmouth, Rhode Island.* Newport, R.I.: Franklin Printing House, 1976.

Jones, Robert O. *Narragansett Pier, Narragansett, Rhode Island.* Providence: RIHPHC, 1978.

Rhode Island Historical Preservation Commission. "East Greenwich, Rhode Island." Statewide Preservation Report, August 1974.

Simister, F. P. *Streets of the City: An Anecdotal History of North Kingstown.* North Kingstown, R.I.: Simister's Bookshop, 1974.

Glossary

AIA See *American Institute of Architects.*

abacus The top member of a column capital. In the Doric order, it is a flat block, square in plan, between the echinus of the capital and the architrave of the entablature above.

Academic Gothic See *Collegiate Gothic.*

acroterium, acroterion (plural: acroteria) **1** A pedestal for a statue or similar decorative feature at the apex or at the lower corners of a pediment. **2** Any ornamental feature at these locations.

Adamesque A mode of architectural design, with emphasis on interiors, reminiscent of the work of the Scottish architects Robert Adam (1728–1792) and his brother James (1732–1794). It is characterized by attenuated proportions, bright color, and elegant linear detailing. Adamesque interiors, as one aspect of the broader Neoclassical movement, became popular in the late eighteenth century in Britain, Russia, and elsewhere in northern Europe. Simplified versions of these interiors began to be seen in the United States around the year 1800 in the work of Charles Bulfinch (1763–1844) and Samuel McIntire (1757–1811). Adamesque interiors, often emulating original Adam designs, were again popular in the 1920s. See also the related term *Federal.*

aedicule, aedicular An exterior niche, door, or window, framed by columns or pilasters and topped by an entablature and pediment. Meaning has been extended to a smaller-scale representation of a temple front on an interior wall. Distinguished from a tabernacle (definition 1), which usually occurs on an interior wall. See also the related term *niche.*

Aesthetic movement A late nineteenth-century movement in interior design and the decorative arts, emphasizing the application of artistic principles in the production of objects and the creation of interior ensembles. Aesthetic movement works are characterized by a broad eclecticism of materials and styles (especially the exotic) and by a preference for "conventionalized" (i.e., stylized) ornament, rather than naturalistic. The movement flourished in Britain from the 1850s through the 1870s and in the United States from the 1870s through the 1880s. Designers associated with the movement include William Morris (1834–1896) in England and Herter Brothers (1865–1905) in America. The Aesthetic movement evolved into and overlapped with the Art Nouveau and Arts and Crafts movements. See also the related term *Queen Anne* (definition 4).

ambulatory A passageway around the apse of a church, allowing for circulation behind the sanctuary.

American Adam Style See *Federal.*

American bond See *common bond.*

American Foursquare See *foursquare house.*

American Institute of Architects (AIA) The national professional organization of architects, established in New York in 1857. The first national convention was held in New York in 1867, and at that meeting, provision was made for the creation of local chapters. In 1889, the American Institute of Architects absorbed the independent Chicago-based Western Association of Architects (established 1884). The headquarters of the national organization moved from New York to Washington in 1898.

American Renaissance Ambiguous term. See instead *Beaux-Arts classicism, Colonial Revival, Federal Revival.*

Anglo-Palladianism, Anglo-Palladian An architectural movement in England motivated by a reaction against the English Baroque and by a rediscovery of the work of the English Renaissance architect Inigo Jones (1573–1652) and the Italian Renaissance architect Andrea Palladio (1508–1580). Anglo-Palladianism flourished in England (c. 1710s–1760s) and in the British North American colonies (c. 1740s–1790s). Key figures in the Anglo-Palladian movement were Colen Campbell (1676–1729) and Richard Boyle, Lord Burlington (1694–1753). Sometimes called Burlingtonian, Palladian Revival. See also the more general term *Palladianism* and the related terms *Georgian period, Jeffersonian.*

antefix. In classical architecture, a small upright decoration at the eaves of a roof, originally devised to hide the ends of the roof tiles. Also, a similar ornament along the ridge of the roof.

anthemion (plural: anthemions) A Greek ornamental motif based upon the honeysuckle or palmette. It may appear as a single element on an antefix or as a running ornament on a frieze or other banded feature.

antiquity The broad epoch of Western history preceding the Middle Ages and including such ancient civilizations as Egyptian, Greek, and Roman.

apse, apsidal A semicircular or polygonal feature projecting as a major element from an important interior space, especially at the chancel end of a church. Distinguished from an exedra, which is a semicircular or polygonal space, usually containing a bench, in the wall of a garden or nonreligious building. A substantial apse in a church, containing an ambulatory and radiating chapels, is called a chevet. The terms apse and chevet are used to describe the *form* of the end of the church containing the altar, while the terms chancel, choir, and sanctuary are used to describe the liturgical *function* of this end of the church and the spaces within it. Less substantial projections in nonreligious buildings are called bays if polygonal or bowfronts if curved.

arbor 1 An openwork structure covered with climbing plants. Distinguished from a trellis, which is generally a simpler, more two-dimensional structure, often attached to a wall. Distinguished from a pergola, which is an openwork structure supported by a colonnade, creating a shaded walk. **2** A grouping of closely planted trees or shrubs, trained together and self-supporting.

arcade 1 A series of arches, carried on columns or piers or other supports. **2** A covered walkway, one side of which is part of a building, while the other is open, as a series of arches, to the exterior. **3** In the nineteenth and early twentieth centuries, an interior street or other extensive space lined with shops and stores.

arch A curved construction that spans an opening. (Some arches may be flat or triangular, and many have a complex or compound curvature.) A masonry arch consists of a series of wedge-shaped parts (voussoirs) that press together toward the center while being restrained from spreading outward by the surrounding wall or the adjacent arch.

architrave 1 The lowest member of a classical entablature. **2** The moldings on the face of a wall around a doorway or other opening. Sometimes called the casing. Distinguished from the jambs, which are the vertical linings perpendicular to the wall planes at the sides of an opening. Distinguished from surround, a term usually applied to the entire door or window frame considered as a unit.

archivolt The group of moldings following the shape of an arched opening.

arcuation, arcuated Construction using arches.

Art Deco A decorative style stimulated by the 1925 Exposition Internationale des Arts Décoratifs et Industriels Modernes, held in Paris. As the first phase of the Moderne, Art Deco is characterized by sharp angular and curvilinear forms, by a richness of materials (including polished metal, stone, and exotic woods), and by an overall sleekness of design. The style was often used in the commercial and residential architecture of the 1930s (e.g., skyscrapers, hotels, apartment buildings). Sometimes called Art Deco Moderne, Deco, Jazz Moderne, Zigzag Moderne, Zigzag Modernistic. See also the more general term Moderne and the related terms *Mayan Revival*, *PWA Moderne*, *Streamline Moderne*.

Art Moderne See *Moderne*.

Art Nouveau A style in architecture, interior design, and the decorative arts that flourished principally in France and Belgium in the 1890s. The Art Nouveau is characterized by undulating and whiplash lines and by sensuous organic forms. The Art Nouveau in Britain and the United States evolved from and overlapped with the Aesthetic movement.

Arts and Crafts A late-nineteenth- and early-twentieth-century movement in interior design and the decorative arts, emphasizing the importance of hand crafting for everyday objects. Arts and Crafts works are characterized by rectilinear geometries and high contrasts between figure and ground, and the furniture often features expressed construction. The term originated with the Arts and Crafts Exhibition Society, founded in England in 1888. Designers associated with the movement include C. F. A. Voysey (1857–1941) in England and the brothers Charles S. Greene (1868–1957) and Henry M. Greene (1870–1954) in America. The Arts and Crafts movement evolved from and overlapped with the Aesthetic movement. For a more specific term, used in the United States after 1900, see also *Craftsman*.

ashlar Squared blocks of stone that fit tightly against one another.

atelier 1 A studio where the fine arts, including architecture, are taught. Applied particularly to the offices of prominent architects in Paris who provided design training to students enrolled in or informally attached to the Ecole des Beaux-Arts. By extension, any working office where some organized teaching is done. **2** A place where artworks or handicrafts are produced by skilled workers. **3** An artist's studio or workshop.

attic 1 The area beneath the roof and above the main stories (or story) of a building. Sometimes called a garret. **2** A low story above the entablature, often a blocklike mass that caps the building.

axis An imaginary center line to which are referred the parts of a building or the relations of a number of buildings to one another.

axonometric drawing A pictorial drawing using axonometric projection, in which horizontal lines that are perpendicular in an object, building, or space are drawn as perpendicular (usually at two 45-degree angles from the vertical, or at complementary angles of 30 and 60 degrees). Consequently, all angular and dimensional relationships in plan remain the same in the drawing as in the thing depicted. Sometimes called an axon or an axonometric. See also the related terms *isometric drawing*, *perspective drawing*.

balloon-frame construction A system of light frame construction in which single studs extend the full height of the frame (commonly two stories), from the foundation to the roof. Floor joists are fastened to the sides of the studs.

Structural members are usually sawn lumber, ranging from two-by-fours to two-by-tens, and are fastened with nails. Sometimes called balloon framing. The technique, developed in Chicago and other boomtowns of the 1830s, has been largely replaced in the twentieth century by platform frame construction.

baluster One of a series of short vertical members, often vase-shaped in profile, used to support a handrail for a stair or a railing. Balusters that are thinner and simpler in profile are sometimes called banisters.

balustrade A series of balusters or posts supporting a rail or coping across the top (and sometimes resting on a lower rail). Balustrades are often found on stairs, balconies, parapets, and terraces.

band course Ambiguous term. See instead *band molding* or *stringcourse*.

band molding In masonry or frame construction, any horizontal flat member or molding or group of moldings projecting slightly from a wall and marking a division in the wall. Not properly a synonym for band course. Simpler horizontal bands in masonry are generally called stringcourses.

bandstand A small pavilion, usually polygonal or circular in plan, designed to shelter bands during public concerts in a garden, park, green, or square. See also the related terms *gazebo, kiosk*.

banister 1 Corrupted spelling of baluster, in use since about the seventeenth century. Now occasionally used for balusters that are thinner and simpler in profile than classical vase-shaped balusters. **2** Improperly used to mean the handrail of a stair.

bargeboard An ornate fascia board that is attached to the sloping edges (verges) of a roof, covering the ends of the horizontal roof timbers (purlins). Bargeboards are usually ornamented with carved, turned, or jigsawn forms. Sometimes called gableboards, vergeboards. Less ornate boards along the verges of a roof are simply called fascia boards.

Baroque A style of art and architecture that flourished in Europe and colonial North America during the seventeenth and eighteenth centuries. Although based on the architecture of the Renaissance, Baroque architecture was more dynamic, with circles frequently giving way to ovals, flat walls to curved or undulating ones, and separate elements to interlocking forms. It was a monumental and richly three-dimensional style with elaborate systems of ornamental and figural sculpture. See also the related terms *Renaissance, Rococo*.

Baroque Revival See *Neo-Baroque*.

barrel vault A vaulted roof or ceiling of semicircular or semielliptical cross section, forming a tunnellike enclosure over an apartment, corridor, or similar space.

Barryesque Term applied to Italianate buildings showing the influence of the English architect Sir Charles Barry (1795–1860), who introduced a derivative form of the Italian High Renaissance palazzo in his Travelers Club in London, 1829–1832. The style was brought to the United States by the Scottish-trained architect John Notman (1810–1865) and was popular from the late 1840s through the 1860s, especially for institutional and government buildings. Distinguished from the Italian Villa Style, which has the northern Italian rural vernacular villa as its prototype. See also the more general term *Italianate*.

basement 1 The lowest story of a building, either partly or entirely below grade. **2** The lower part of the walls of any building, usually articulated distinctly from the upper part of the walls.

batten 1 A narrow strip of wood applied to cover a joint along the edges of two parallel boards in the same plane. **2** A strip of wood fastened across two or more parallel boards to hold them together. Sometimes called a cross batten. See also the related term *board-and-batten siding*.

battered (adjective). Inclined from the vertical. A wall is said to be battered or to have a batter when it recedes as it rises.

battlement, battlemented See *crenellation*.

Bauhaus 1 Work in any of the visual arts by the faculty and students of the Bauhaus, the innovative design school founded by Walter Gropius (1883–1969) and an active force in German modernism from 1919 until 1933. **2** Work in any of the visual arts by the former faculty and students of the Bauhaus, or by individuals influenced by them. See also the related terms *International Style, Miesian*.

bay 1 The interval between two recurring members. A facade is frequently measured by window bays, a skeletal frame by structural bays. **2** A polygonal or curved unit of one or more stories, projecting from the wall and usually containing grouped windows (bay windows) on each story. See also the more specific term *bowfront*.

bay window The horizontally grouped windows in a projecting bay (definition 2), or the projecting bay itself, if it is not more than one story. Distinguished from an oriel, which does not rise from the foundation and has a suspended rather than rooted appearance. A semicircular or semielliptical bay window is called a bow window. A bay window with a central section of plate glass in a late-nineteenth-century commercial building is called a Chicago window.

beam A structural spanning member of stone, wood, iron, steel, or reinforced concrete. See also the more specific terms *girder, I-beam, joist*.

bearing wall A wall that is fully structural, carrying the load of the floors and roof all the way to the foundation. Sometimes called a supporting wall. Distinguished from curtain wall. See also the related term *load-bearing*.

Beaux-Arts Historicist design on a monumental scale, as taught at the Ecole des Beaux-Arts in Paris throughout the nineteenth century and early twentieth century. The term Beaux-Arts is generally applied to an eclectic Roman-Renaissance-Baroque architecture of the 1850s through the 1920s, disseminated internationally by students and followers of the Ecole des Beaux-Arts. As a general style term Beaux-Arts connotes an academically grounded discipline for historical eclecticism, rather than one single style, as well as the disciplined development of a *parti* into a fully visualized design. More specific style terms include Néo-Grec (1840s–1870s) and Beaux-Arts classicism (1870s–1930s). See also the related terms *Neoclassicism,* for describing Ecole-related work from the 1790s to the 1840s, and *Second Empire,* for describing the work from the 1850s to the 1880s.

Beaux-Arts classicism, Beaux-Arts classical Term applied to eclectic Roman-Renaissance-Baroque architecture and urbanism after the Néo-Grec and Second Empire phases, i.e., from the 1870s through the 1930s. Sometimes called Classic Revival, Classical Revival, McKim classicism, Neoclassical Revival. See also the more general term *Beaux-Arts* and the related terms *City Beautiful movement, PWA Moderne.*

belfry A cupola, turret, or room in a tower where a bell is housed.

bell cote A small gabled structure astride the ridge of a roof, which shelters a bell. It is usually close to the front wall plane of the building.

belt course See *stringcourse.*

belvedere 1 Any building, especially a pavilion or shelter, that is located to take advantage of a view. See also the related term *gazebo.* **2** See *cupola* (definition 2).

blind (adjective) Term applied to the surface use of elements that would otherwise articulate an opening but where no opening exists. Used in such combinations as blind arcade, blind arch, blind door, blind window.

board-and-batten siding A type of siding for wood-frame buildings, consisting of wide vertical boards with narrow strips of wood (battens) covering the joints. (In rare instances, the battens may be fastened behind the joints. If the gaps between boards are wide and the back battens approach the width of the outer boards, the siding is called board-on-board.) See also the related term *batten.*

board-on-board siding A type of siding for wood-frame buildings, consisting of two layers of vertical boards, with the outer layer of boards covering the wide gaps between the boards of the inner layer.

bowfront A semicircular or semielliptical bay (definition 2).

bow window A semicircular or semielliptical bay window.

brace A single wooden or metal member placed diagonally within a framework or truss or beneath an overhang. Distinguished from a bracket, which is a more substantial triangular feature, and from a strut, which is essentially a post set in a diagonal position.

braced-frame construction A combination of heavy and light timber-frame construction, in which the principal vertical and horizontal framing members (posts and girts) are fastened by mortise and tenon joints, while the one-story-high studs are nailed to the heavy timber frame. The overall frame is made more rigid by diagonal braces. Sometimes called braced framing.

bracket Any solid, pierced, or built-up triangular feature projecting from the face of a wall to support a projecting element, like the top member of a cornice or the verges or eaves of a roof. Brackets are frequently used for ornamental as well as structural purposes. Distinguished from a brace, which is a simple barlike structural member. Distinguished from the more specific term console, which has a height greater than its projection from the wall. See also the related term *corbel.*

Bracketed Style A nineteenth-century term for Italianate.

brick bonds, brickwork See the more specific terms *common bond, English bond, Flemish bond, running bond.*

British colonial A term applied to buildings, towns, landscapes, and other artifacts from the period of actual British colonial occupation of large parts of eastern North America (c. 1607–1781 for the United States; c. 1750s–1867 for much of Canada). The British colonial period saw the introduction into the New World of various regional strains of English and Scotch-Irish folk culture, as well as high-style Anglo-European Renaissance, Baroque, and Neoclassical design. Sometimes called English colonial. Loosely called colonial or Early American. See also the related term *Georgian period.*

Brutalism An architectural style of the 1950s through 1970s, characterized by complex massing and by a frank expression of structural members, elements of building systems, and materials (especially concrete). Some of the work of Paul Rudolph (born 1918) is associated with this style. Sometimes called New Brutalism.

bungalow A low one- or one-and-one-half-story house of modest pretensions with a low-pitched gable or hipped roof, a conspicuous porch, and projecting eaves. This house type was a popular builders' type from around 1900 to 1930. The term bungalow was also loosely applied to any vernacular building of a semirustic nature, including vacation cottages and lodges.

Burlingtonian See *Anglo-Palladianism.*

buttress An exterior mass of masonry bonded into a wall that it strengthens or supports. Buttresses often absorb lateral thrusts from roofs or vaults.

Byzantine Term applied to the art and architecture of the Eastern Roman Empire centered at Byzantium (i.e., Constantinople, Istanbul) from the early 500s to the mid-1400s. Byzantine architecture is characterized by massive domes, round arches, richly carved capitals, and the extensive use of mosaic.

Byzantine Revival See *Neo-Byzantine*.

campanile In Italian, a bell tower. While usually freestanding in medieval and Renaissance architecture, it was often incorporated as a prominent unit in the massing of picturesque nineteenth-century buildings.

cantilever A beam, girder, slab, truss, or other structural member that projects beyond its supporting wall or column.

cap A canopy, ledge, molding, or pediment over a window. Sometimes called a window cap. Distinguished from a hood, which is a similar feature over a door. See also the related term *head molding*.

capital The moldings and carved enrichment at the top of a column, pilaster, pier, or pedestal.

Carpenter's Gothic Term applied to a version of the Gothic Revival (c. 1840s–1870s), in which Gothic motifs are adapted to the kind of wooden details that can be produced by lathes, jigsaws, and molding machines. Sometimes called Carpenter Gothic, Gingerbread Style, Steamboat Gothic. See also the more general term *Gothic Revival*.

carriage porch See *porte cochere*.

casement window A window that opens from the side on hinges, like a door, out from the plane of the wall. Distinguished from a double-hung window.

casing See *architrave* (definition 2).

cast iron Iron shaped by a molding process, generally strong in compression but brittle in tension. Distinguished from wrought iron, which has been forged to increase its tensile properties.

cast iron front An architectural facade made of prefabricated molded iron parts, often markedly skeletal in appearance with extensive glass infilling. Prevalent from the late 1840s to the early 1870s.

castellated Having the elements of a medieval castle, such as crenellation and turrets.

cavetto cornice See *coved cornice*.

cement A mixture of burnt lime and clay with water, which hardens permanently when dry. When a fine aggregate of sand is added, the cement may be used as a mortar for masonry construction or as a plaster or stucco coating. When a coarser aggregate of gravel or crushed stone is added, along with sand, the mixture is called concrete.

chamfer The oblique surface formed by cutting off a square edge at an equal angle to each face.

chancel 1 The end of a Roman Catholic or High Episcopal church containing the altar and set apart for the clergy and choir by a screen, rail, or steps. Usually the entire east end of a church beyond the crossing. In churches that have a long chancel space, the part of the chancel between the crossing and the apse, where the singers participate in the service, is called the choir. The innermost part of the chancel, containing the principal altar, is called the sanctuary. **2** In less extensive Catholic and Episcopal churches, the terms chancel and choir are often used interchangeably to mean the entire eastern arm of the church.

Chateauesque A term applied to masonry buildings from the 1870s through the 1920s in which stylistic references are derived from early French Renaissance châteaux, from the reign of Francis I (1515–1547) or even earlier. Sometimes called Chateau Style, Chateauesque Revival, Francis I Style, François Premier.

chevet In large churches, particularly those based upon French Gothic precedents, a substantial apse surrounded by an ambulatory and often containing radiating chapels.

Chicago School A diverse group of architects associated with the development of the tall (i.e., six- to twenty-story), usually metal-frame commercial building in Chicago during the 1880s and 1890s. William Le Baron Jenney, Burnham and Root, and Adler and Sullivan are identified with this group. Sometimes called Chicago Commercial Style, Commercial Style. See also the related term *Prairie School*.

Chicago window A tripartite oblong window in which a large fixed center pane is placed between two narrow sash windows. Popularized in Chicago commercial buildings of the 1880s–1890s. See also *bay window*.

chimney girt In timber-frame construction, a major wooden beam that passes across the breast of the central chimney. It is supported at its ends by the longitudinal girts of the building and sometimes carries one end of the summer beam.

choir 1 The part of a Roman Catholic or High Episcopal church where the singers participate in the service. Usually the space within the chancel arm of the church, situated between the crossing to the west and the sanctuary to the east. **2** In less extensive Catholic and Episcopal churches, the terms choir and chancel are often used interchangeably to mean the entire eastern arm of the church.

Churrigueresque Term applied to Spanish and Spanish colonial Baroque architecture resembling the work of the Spanish architect José Benito de Churriguera (1665–1725) and his brothers. The style is characterized by a freely interpreted assemblage of such elements as twisted columns, broken pediments, and scroll brackets. See also the related term *Spanish colonial*.

cinquefoil A type of Gothic tracery having five parts (lobes or foils) separated by pointed elements (cusps).

City Beautiful movement A movement in architecture, landscape architecture, and planning in the United States from the 1890s through the 1920s, advocating the beautification of cities in the image of some of the most urbane places of the time: the world's fairs. City Beautiful schemes emphasized civic centers, boulevards, and waterfront improvements, and sometimes included comprehensive metropolitan plans for parks, parkways, and transportation facilities. See also the related term *Beaux-Arts classicism*.

clapboard A tapered board that is thinner along the top edge and thicker along the bottom edge, applied horizontally with edges overlapping to provide weathertight siding on a building of wood construction. Early clapboards were split (rived, riven) and were used for barrel staves and for wainscoting. The term now applies to any beveled siding board, whether split or sawn, rabbeted or not, regardless of length or width. (The term is sometimes applied only to a form of bevel siding used in New England, about four feet long and quarter-sawn.) Sometimes called weatherboards.

classical orders See *order*.

classical rectangle See *golden section*.

Classical Revival Ambiguous term, suggesting (1) Neoclassical design of the late eighteenth and early nineteenth centuries, including the Greek Revival; or (2) Beaux-Arts classical design of the late nineteenth and early twentieth centuries. Sometimes called Classic Revival. See instead *Beaux-Arts classicism, Greek Revival, Neoclassicism*.

classicism, classical, classicizing Terms describing the application of principles or elements derived from the visual arts of the Greco-Roman era (seventh century B.C. through fourth century A.D.) at any subsequent period of Western civilization, but particularly since the Renaissance. More a descriptive term for an approach to design and for a general cultural sensibility than for any particular style. See also the related term *Neoclassicism*.

clerestory A part of a building that rises above the roof of another part and has windows in its walls.

clipped gable roof See *jerkinhead roof*.

coffer A recessed panel, usually square or octagonal, in a ceiling. Such panels are also found on the inner surfaces of domes and vaults.

collar beam A horizontal tension member in a pitched roof connecting opposite rafters, generally halfway up or higher. Its function is to tie the angular members together and prevent them from spreading.

Collegiate Gothic 1 Originally, a secular version of English Gothic architecture, characteristic of the older colleges of Oxford and Cambridge. **2** A secular version of Late Gothic Revival architecture, which became a popular style for North American colleges and universities from the 1890s through the 1920s. Sometimes called Academic Gothic.

colonial 1 Not strictly a style term, but a term for the entire period during which a particular European country held political dominion over a part of the Western Hemisphere, Africa, Asia, Australia, or Oceania. See also the more specific terms *British colonial, Dutch colonial, French colonial, Spanish colonial*. **2** Loosely used to mean the British colonial period in North America (c. 1607–1781 for the United States; c. 1750s–1867 for much of Canada).

Colonial Revival Generally understood to mean the revival of forms from British colonial design. The Colonial Revival began in New England in the 1860s and continues nationwide into the present. Sometimes called Neo-Colonial. See also the more specific term *Georgian Revival* and the related terms *Federal Revival, Shingle Style*.

colonnade A series of freestanding or engaged columns supporting an entablature or simple beam.

colonnette A diminutive, often attenuated, column.

colossal order See *giant order*.

column 1 A vertical supporting element, usually cylindrical and slightly tapering, consisting of a base (except in the Greek Doric order), shaft, and capital. See also the related terms *entablature, entasis, order*. **2** Any vertical supporting element in a skeletal frame.

Commercial Style See *Chicago School*.

common bond A pattern of brickwork in which every fifth or sixth course consists of all headers, the other courses being all stretchers. Sometimes called American bond. Distinguished from running bond, in which no headers appear.

Composite order An ensemble of classical column and entablature elements, particularly characterized by large Ionic volutes and Corinthian acanthus leaves in the capital of the column. See also the more general term *order*.

concrete An artificial stone made by mixing cement, water, sand, and a coarse aggregate (such as gravel or crushed stone) in specified proportions. The mix is shaped in molds called forms. Distinguished from cement, which is the binder without the aggregate.

console A type of bracket with a scroll-shaped or S-curve profile and a height greater than its projection from the wall. Distinguished from the more general term bracket, which is usually applied to supports whose projection and height are nearly equal. Distinguished from a modillion, which usually is smaller, has a projection greater than its height (or thickness), and appears in a series, as in a classical cornice.

coping The cap or top course of a wall, parapet, balustrade, or chimney, usually designed to shed water.

corbel A projecting stone that supports a superincumbent weight. In medieval architecture and its derivatives, a support for such major features as vaulting shafts, vaulting ribs, or oriels. See also the related term *bracket*.

corbeled construction Masonry that is built outward beyond the vertical by letting successive courses project beyond those below. Sometimes called corbeling.

corbeled cornice A cornice made up of courses of projecting masonry, each of which extends farther outward than the one below.

Corinthian order An ensemble of classical column and entablature elements, particularly characterized by acanthus leaves and small volutes in the capital of the column. See also the more general term *order*.

cornice The crowning member of a wall or entablature.

Corporate International Style A term, not widely used, for curtain wall commercial, institutional, and governmental buildings since the Second World War, which represent a widespread adoption of selected International Style ideas from the 1920s. See also the more general term *International Style*.

Corporate Style An architectural style developed in the early industrial communities of New England during the first half of the nineteenth century. This austere but graceful mode of construction was derived from the red-brick Federal architecture of the early nineteenth century and is characterized by the same elegant proportions, cleanly cut openings, and simple refined detailing. Not to be confused with *Corporate International Style*.

cottage 1 A relatively modest rural or suburban dwelling. Distinguished from a villa, which is a more substantial and often more elaborate dwelling. **2** A seasonal dwelling, regardless of size, especially one located in a resort community.

cottage orné A rustic building in the romantic, picturesque tradition, noted for such features as bay windows, oriels, ornamented gables, and clustered chimneys.

course A layer of building blocks, such as bricks or stones, extending the full length and thickness of a wall.

coved ceiling A ceiling in which the transition between wall and ceiling is formed by a large concave panel or molding. Sometimes called a cove ceiling.

coved cornice A cornice with a concave profile. Sometimes called a cavetto cornice.

Craftsman A style of furniture and interior design belonging to the Arts and Crafts movement in the United States, and specifically related to *The Craftsman* magazine (1901–1916), published by Gustav Stickley (1858–1942). Some entire houses known to be derived from this publication can be called Craftsman houses. See also the more general term *Arts and Crafts*.

crenellation, crenellated A form of embellishment on a parapet consisting of indentations (crenels or embrasures) alternating with solid blocks of wall (merlons). Virtually synonymous with battlement, battlemented; embattlement, embattled.

cresting An ornamental strip or fencelike feature, usually of metal or tile, along the ridgeline or summit of a roof.

crocket In Gothic architecture, a small ornament resembling bunched foliage, placed at intervals on the sloping edges of gables, pinnacles, or spires.

crossing In a church with a cruciform plan, the area where the arms of the cross intersect; specifically, the space where the transept crosses the nave and chancel.

cross rib See *lierne*.

cross section See *section*.

crown The central, or highest, part of an arch or vault.

crown molding The highest in a series of moldings.

crowstep Any one of the progressions in a gable that ascends in steps rather than in a continuous slope.

cruciform In the shape of a cross. Usually used to describe the ground plans of buildings. See also the more specific terms *Greek cross, Latin cross*.

cupola 1 A small domed structure on top of a belfry, steeple, or tower. **2** A lantern, square or polygonal in plan, with windows or vents, which is located at the summit of a roof. Sometimes called a belvedere. Distinguished from a skylight, which is a lesser feature located on the slope of a roof. **3** In historic English usage, synonymous with dome. A dome is now understood to be a more substantial feature.

curtain wall In skeleton frame or reinforced concrete construction, a thin nonstructural cladding of stone, brick, terra-cotta, glass, or metal veneer. Distinguished from bearing wall. See also the related term *load-bearing*.

cusp. The pointed, roughly triangular intersection of the arcs of lobes or foils in the tracery of windows, screens, or panels.

dado A broad decorative band around the lower portion of an interior wall, between the baseboard and dado rail or cap molding. (The term is often applied to this entire zone, including baseboard and dado rail.) The dado may be painted, papered, or covered with some other material, so as to have a different treatment from the upper zone of the wall. Dado connotes any continuous lower zone in a room, equivalent to a pedestal. A wood-paneled dado is called a wainscot.

Deco. See *Art Deco*.

dentil, denticulated A small ornamental block forming one of a series set in a row. A dentil molding is composed of such a series.

dependency A building, wing, or room, subordinate to or serving as an adjunct to a main building. A dependency may be attached to or detached from a main building. Distinguished from an outbuilding, which is always detached.

diaper An overall repetitive pattern on a flat surface, especially a pattern of geometric or representational forms arranged in a diamond-shaped or checkerboard grid. Sometimes called diaper work.

discharging arch See *relieving arch*.

dome A major hemispherical or curved roof feature rising from a circular, polygonal, or square base. Distinguished from a cupola, which is a smaller, usually subordinate, domical element.

Doric order An ensemble of classical column and entablature elements, particularly characterized by the use of triglyphs and metopes in the frieze of the entablature. See also the more general term *order*.

dormer A roof-sheltered window (or vent), usually with vertical sides and front, set into a sloping roof. Sometimes called a dormer window.

dosseret See *impost block*.

double-hung window A window consisting of a pair of frames, or sashes, one above the other, arranged to slide up and down. Their movement is sometimes stabilized by a system of cords and counterbalancing weights contained in narrow boxing at each side of the window frame. Sometimes called guillotine sash.

double-pen In vernacular architecture, particularly houses, a term applied to a plan consisting of two rooms side by side or separated by a hallway.

double-pile In vernacular architecture, particularly houses, a term applied to a plan that is two rooms deep and any number of rooms wide.

drip molding See *head molding*.

drum 1 A cylindrical or polygonal wall zone upon which a dome rests. **2** One of the cylinders of stone that form the shaft of a column.

Dutch colonial A term applied to buildings, towns, landscapes, and other artifacts from the period of actual Dutch colonial occupation of the Hudson River valley and adjacent areas (c. 1614–1664). Meaning has been extended to apply to the artifacts of Dutch ethnic groups and their descendants, even into the early nineteenth century.

Dutch Colonial Revival The revival of forms from design in the Dutch tradition.

ear A slight projection just below the upper corners of a door or window architrave or casing. Sometimes called a shouldered architrave.

Early American See *British colonial*.

Early Christian A style of art and architecture in the Mediterranean world that was developed by the early Christians before the fall of the Western Roman Empire, derived from late Roman art and architecture and leading to the Romanesque (early fourth to early sixth century).

Early Georgian period Not strictly a style term, but a term for a period in British and British colonial history approximately coinciding with the reigns of George I (1714–1727) and George II (1727–1760). See also the related term *Late Georgian period*.

Early Gothic Revival A term for the Gothic Revival work of the late eighteenth to the mid-nineteenth century. See also the related term *Late Gothic Revival*.

Eastlake A decorative arts and interior design term of the 1860s and 1880s sometimes applied to architecture. Named after Charles Locke Eastlake (1836–1906), an English advocate of the application of Gothic principles of construction and design, rather than mere Gothic elements. Characterized by simplicity and solidity of forms, which are sometimes embellished with chamfered, turned, or incised details. Sometimes called Eastlake Gothic, Modern Gothic. See also the related term *Queen Anne*.

eaves The horizontal lower edges of a roof plane, usually projecting beyond the wall below. Distinguished from verges, which are the sloping edges of a roof plane.

echinus A heavy molding with a curved profile placed immediately below the abacus, or top member, of a classical capital. Particularly prominent in the Doric and Tuscan orders.

eclecticism, eclectic A sensibility in design, prevalent since the eighteenth century, involving the selection of elements from a variety of sources, including historical periods of high-style design (Western and non-Western), vernacular design (Western and non-Western), and (in the twentieth century) contemporary industrial design. Distinguished from historicism and revivalism by drawing upon a wider range of sources than the historical periods of high-style design.

Ecole, Ecole des Beaux-Arts See *Beaux-Arts*.

Egyptian Revival Term applied to eclectic works or elements of those works that emulate forms in the visual arts of ancient Egyptian civilization.

elevation A drawing (in orthographic projection) of an upright, planar aspect of an object or building. The vertical complement of a plan. Sometimes used loosely in the sense of a facade view or any frontal representation of a wall, whether photograph or drawing, whether measured to scale or not.

Elizabethan Manor Style See *Neo-Tudor*.

Elizabethan period A term for a period in English history coinciding with the reign of Elizabeth I (1558–1603). See also the more general term *Tudor period* and the related term *Jacobean period* for the succeeding period.

embattlement, embattled See *crenellation*.

encaustic tile A tile decorated by a polychrome glazed or ceramic inlay pattern.

engaged column A half-round column attached to a wall. Distinguished from a free-standing column by seeming to be built into the wall. Distinguished from a pilaster, which is a flattened column. Distinguished from a recessed column, which is a fully round column set into a niche-like space.

English bond A pattern of brickwork in which the bricks are set in alternating courses of stretchers and headers.

English colonial See *British colonial*.

English Half-timber Style See *Neo-Tudor*.

entablature In a classical order, a richly detailed horizontal member resting on columns or pilasters. It is divided horizontally into three main parts. The lowest is the architrave (definition 1), the structural part, and is generally an unornamented continuous beam or series of beams. The middle part is the frieze (definition 1), which is generally the most freely ornamented part. The uppermost is the cornice. Composed of a sequence of moldings, the cornice overhangs the frieze and architrave and serves as a crown to the whole. Each part has the moldings and decorative treatment that are characteristic of the particular order, but modern adaptations often alter canonical details. See also the related terms *column, order*.

entablature block A block bearing the canonical elements of a classical entablature on three or all four sides, placed between a column capital and a feature above, such as a balcony or ceiling. Distinguished from an impost block, which has the form of an inverted truncated pyramid and detailing typical of medieval architecture.

entasis The slight convex curving of the vertical profile of a tapered column.

exedra A semicircular or polygonal space usually containing a bench, in the wall of a garden or a building other than a church. Distinguished from a niche, which is usually a smaller feature higher in a wall, and from an apse, which is usually identified with churches.

exotic revivals A term occasionally used to suggest a distinction between revivals of European styles (e.g., Greek, Gothic Revivals) and non-European styles (e.g., Egyptian, Moorish Revivals). See also the more specific terms *Egyptian Revival, Mayan Revival, Moorish Revival*.

extrados The outer curve or outside surface of an arch. See also the related term *intrados*.

eyebrow dormer A low dormer with a small segmental window or vent but no sides. The roofing warps or bows over the window or vent in a wavy line.

facade An exterior face of a building, especially the principal or entrance front. Distinguished from an elevation, which is an orthographic drawing of a building face.

Fachwerk A form of half-timber construction introduced by German-speaking immigrants.

false half timbering A surface treatment that simulates half-timber construction, consisting of a lattice of broad boards and stucco applied as an exterior veneer on a building of masonry or wood-frame construction. Most commonly seen in domestic architecture from the late nineteenth century onward.

fanlight A semicircular or semielliptical window over a door, with radiating mullions in the form of an open fan. Sometimes called a sunburst light. See also the more general term *transom* (definition 1) and the related term *side light*.

fan vault A type of Gothic vault in which the primary ribs all have the same curvature and radiate in a half circle around the springing point.

fascia 1 A plain, molded, or ornamented board that covers the horizontal edges (eaves) or sloping edges (verges) of a roof. Distinguished from the more specific term bargeboards, which are ornate fascia boards attached to the sloping edges of a roof. Distinguished from a frieze (definition 2), which is located at the top of a wall. **2** One of the broad continuous bands that make up the architrave of the Ionic, Corinthian, or Composite order.

Federal A version of Neoclassical architecture in the United States popular from New England to Virginia, and in other regions influenced by the Northeast. It flourished from the 1790s through the 1820s and is found in some regions as late as the 1840s. Sometimes called American Adam Style. Not to be confused with Federalist. See also the related terms *Jeffersonian, Roman Revival*.

Federal Revival Term applied to eclectic works (c. 1890s–1930s) or elements of those works that emulate forms in the visual arts of the Federal period. Sometimes called Neo-Federal. See also the related terms *Colonial Revival, Georgian Revival*.

Federalist Name of an American political party and the era it dominated (c. 1787–1820). Not to be confused with Federal.

fenestration Window treatment: arrangement and proportioning.

festoon A motif representing entwined leaves, flowers, or fruits, hung in a catenary curve from two points. Distinguished from a swag, which is a motif representing a fold of drapery hung in a similar curve. See also the more general term *garland*.

fillet 1 A relatively narrow flat molding. **2** Any thin band.

finial A vertical ornament placed upon the apex of an architectural feature, such as a gable, turret, or canopy. Distinguished from a pinnacle, which is a larger feature, usually associated with Gothic architecture.

fireproofing In metal skeletal framing, the wrapping of structural members in terra-cotta tile or other fire-resistant material.

flashing A strip of metal, plastic, or various flexible compositional materials used at roof valleys and ridges and at chimney corners to keep water out. Any similar material used to protect door and window heads and sills.

Flemish bond A pattern of brickwork in which the stretchers and headers alternate in the same row and are staggered from one row to the next. Because this creates a more animated texture than English bond, Flemish bond was favored for front facades and more elegant buildings.

Flemish gable A gable whose upper slopes ascend in steps rather than in a straight line. These steps may be rectilinear or curved, or a combination of both.

fluting, fluted A series of parallel grooves or channels (flutes), usually semicircular or semielliptical in plan, that accentuate the verticality of the shaft of a column or pilaster.

flying buttress In Gothic architecture a spanning member, usually in the form of an arch, that reaches across the open space from an exterior buttress pier to that point on the wall of the building where the thrusts of the interior vaults are concentrated. Because of its arched construction, a flying buttress exerts a counterthrust against the pressure of the vaults contained by the vertical strength of the buttress pier.

foliated (adjective). In the form of leaves or leaflike shapes.

folk Not a style term in itself, but a descriptive term, applicable to all the visual arts and all styles and periods. Applied to (1) a regional, often ethnic, tradition in which continuities through the years in the overall appearance of artifacts (including buildings) are more important than changes in stylistic embellishment; (2) the work of individual artists and artisans unexposed to or uninterested in prevailing or avant-garde ideals of form and technique. Approximate synonyms include anonymous, naive, primitive, traditional. For architecture, see also the more general term vernacular and the related term popular.

four-part vault See *quadripartite vault*.

foursquare house A hip-roofed, two-story house with four principal rooms on each floor and a symmetrical facade. It usually has a front porch across the full width of the house and one or more large dormers on the roof. A common suburban house type from the 1890s to the 1920s. Sometimes called American Foursquare, Prairie Box.

frame construction, frame Ambiguous terms. See instead braced frame construction, light frame construction (balloon frame construction, platform frame construction), skeleton construction, timber-frame construction. Not properly synonymous with wood construction, wood-clad, or wooden.

Francis I Style See *Chateauesque*.

François Premier See *Chateauesque*.

French colonial A term applied to buildings, towns, landscapes, and other artifacts from the period of actual French colonial occupation of large parts of eastern North America (c. 1605–1763). The term is extended to apply to the artifacts of French ethnic groups and their descendants well into the nineteenth century.

French Norman A style associated since the 1920s with residential architecture based on rural houses of the French provinces of Normandy and Brittany. While not a major revival style, it is characterized by asymmetrical plans, round stair towers with conical roofs, stucco walls, and steep hipped roofs. Sometimes called Norman French.

fret An ornament, usually in series, as a band or field, consisting of a latticelike interlocking of right-angled linear elements.

frieze 1 The broad horizontal band that forms the central part of a classical entablature. **2** Any long horizontal band or zone, especially one that has a chiefly decorative purpose, located at the top of a wall. Distinguished from a fascia, which is attached to the horizontal edge of a roof.

front gabled Term applied to a building whose principal gable end faces the front of the lot or some feature like a street or open space. Sometimes called gable front. Distinguished from side gabled.

gable The wall area immediately below the end of a gable, gambrel, or jerkinhead roof.

gableboard See *bargeboard*.

gable front See *front gabled*.

gable roof A roof in which the two planes slope equally toward each other to a common ridge. Sometimes called a pitched roof.

gambrel roof A roof that has a single ridgepole but a double pitch. The lower plane, which rises from the eaves, is rather steep. The upper plane, which extends from the lower plane to the ridgeline, has a flatter pitch.

garland A motif representing a rope of entwined leaves, flowers, ribbons, or drapery, regardless of its shape or position. It may be formed into a wreath, festoon, or swag, or follow the outline of a rectilinear architectural element.

garret See *attic* (definition 1).

gauged brick A brick that has been cut or rubbed to a uniform size and shape.

gazebo A small pavilion, usually polygonal or circular in plan and serving as a garden or park shelter. Distinguished from a kiosk, which generally has some commercial or public function. See also the related terms *bandstand, belvedere* (definition 1).

General Grant Style See *Second Empire*.

Georgian period A term for a period in British and British colonial history, and not, in architecture or the other visual arts, a sufficiently specific style term. The Georgian period begins with the coronation of George I in 1714 and extends until about 1781 in the area that became the United States (and in Britain, until the death of George IV in 1830). See also the related terms *Anglo-Palladianism, British colonial*.

Georgian plan See *double-pile plus double-pen* (i.e., a four-room plan with central hallway).

Georgian Revival A revival of Georgian period forms—in England, from the 1860s to the present, and in the United States, from the 1880s to the present. Sometimes called Neo-Georgian. See also the more general term *Colonial Revival* and the related term *Federal Revival*.

giant order A composition involving any one of the five principal classical orders, in which the columns or pilasters are nearly as tall as the height of the entire building. Sometimes called a colossal order. See also the more general term *order*.

Gingerbread Style See *Carpenter's Gothic*.

girder A major horizontal spanning member, comparable in function to a beam, but larger and often built up of a number of parts. It usually runs at right angles to the beams and serves as their principal means of support.

girt In timber-frame construction, a horizontal beam at intermediate (e.g., second-floor) level, spanning between posts.

glazing bar See *muntin*.

golden section Any line divided into two parts so that the ratio of the longer part to the shorter part equals the ratio of the length of the whole line to the longer part: $a/b = (a+b)/a$. This ratio is approximately 1.618:1. A golden rectangle, or classical rectangle, is a rectangle whose long side is related to the short side in the same ratio as the golden section. It is proportioned so that neither the long nor the short side seems to dominate. In a Fibonacci series (i.e., 1, 2, 3, 5, 8, 13, . . .), the sum of the two preceding terms gives the next. The higher one goes in such a series, the closer the ratio of two sequential terms approaches the golden section.

Gothic An architectural style prevalent in Europe from the twelfth century into the fifteenth in Italy (and into the sixteenth century in the rest of Europe). It is characterized by pointed arches and ribbed vaults and by the dominance of openings over masonry mass in the wall. The Gothic was preceded by the Romanesque and followed by the Renaissance.

Gothic Revival A movement in Europe and North America devoted to reviving the forms and the spirit of Gothic architecture and the allied arts. It originated in the mid-eighteenth century. Sometimes called the Pointed Style in the nineteenth century, and sometimes called Neo-Gothic. See also the more specific terms *Carpenter's Gothic, Early Gothic Revival, High Victorian Gothic, Late Gothic Revival*.

Grecian A nineteenth-century term for Greek Revival.

Greek cross A cross with four equal arms. Usually used to describe the ground plan of a building. See also the more general term *cruciform*.

Greek Revival A movement in Europe and North America devoted to reviving the forms and the spirit of classical Greek architecture, sculpture, and decorative arts. It originated in the mid-eighteenth century, culminated in the 1830s, and continued into the 1850s. Sometimes called Grecian in the nineteenth century. See also the more general term *Neoclassical*.

groin The curved edge formed by the intersection of two vaults.

guillotine sash See *double-hung window*.

HABS See *Historic American Buildings Survey*.

HAER See *Historic American Engineering Record*.

half-timber construction A variety of timber-frame construction in which the framing members are exposed on the exterior of the wall, with the spaces between timbers being filled with wattle-and-daub (i.e., woven lath and plaster) or masonry materials, such as brick or stone. These masonry materials may also be covered with stucco. Sometimes called half-timbered construction.

hall-and-parlor house, hall-and-parlor plan A double-pen house (i.e., a house that is one room deep and two rooms wide). Usually applied to houses without a central through-passage, to distinguish from hall-passage-parlor houses.

hall-passage-parlor house, hall-passage-parlor plan A two-room house with a central through-passage or hallway.

hammerbeam A short horizontal beam projecting inward from the foot of the principal rafter and supported below by a diagonal brace tied into a vertical wall post. The hammer beams carry much of the load of the roof trussing above. Hammer beam trusses, which could be assembled using a series of smaller timbers, were often used in late medieval England instead of conventional trusses, which required long horizontal tie beams extending across an entire interior space.

haunch The part of the arch between the crown or keystone and the springing.

header A brick laid across the thickness of a wall, so that the short end of the brick shows on the exterior.

head molding A molding or set of moldings designed to shelter and embellish the top of a door or window. Sometimes called a drip molding. See also the related terms *cap* (for windows) and *hood* (for doors).

heavy timber construction See *timber-frame construction*.

high style or high-style (adjective) Not a style term in itself, but a descriptive term, applicable to all the visual arts and all styles and periods. Applied to the works of the masters and their schools and disciples, usually reflecting a cosmopolitan awareness of traditions beyond a particular place or time. Usually contrasted with vernacular (including the folk and popular traditions).

high tech Term applied to architecture in which

building materials and elements of building systems are used to celebrate contemporary technology. Elemental geometric forms, primary colors, and metallic finishes are used to heighten the technological imagery.

High Victorian Gothic A version of the Gothic Revival that originated in England in the 1850s and spread to North America in the 1860s. Characterized by polychromatic exteriors inspired by the medieval Gothic architecture of northern Italy. Sometimes called Ruskin Gothic, Ruskinian Gothic, Venetian Gothic, Victorian Gothic. See also the more general term *Gothic Revival.*

hipped gable roof See *jerkinhead roof.*

hipped roof A roof that pitches inward from all four sides. The edge where any two planes meet is called the hip.

Historic American Buildings Survey (HABS) A branch of the National Park Service of the United States Department of the Interior, established in 1933 to produce detailed documentation of American architecture. HABS documentation typically includes historical and architectural data, photographs, and measured drawings, and is deposited in the Prints and Photographs Division of the Library of Congress. See also the related term *Historic American Engineering Record.*

Historic American Engineering Record (HAER) A branch of the National Park Service of the United States Department of the Interior, established in 1969 to produce detailed documentation of sites and structures associated with industry, transportation, and other areas of technology. See also the related term *Historic American Buildings Survey.*

historicism, historicist, historicizing A type of eclecticism prevalent since the eighteenth century, involving the use of forms from historical periods of high-style design (usually in the Western tradition) and, occasionally, from favored traditions of vernacular design (such as the various colonial traditions in the United States). Historicist influences are designated by the use of the prefix Neo- with a previous historical style (e.g., Neo-Baroque). Distinguished from the more general term eclecticism, which draws upon a wider range of sources in addition to the historical. See also the more specific term *revivalism.*

hollow building tile A hollow terra-cotta building block used for constructing exterior bearing walls of buildings up to about three stories, as well as interior walls and partitions.

hood A canopy, ledge, molding, or pediment over a door. Distinguished from a cap, which is a similar feature over a window. Sometimes called a hood molding. See also the related term *head molding.*

horizontal plank frame construction A system of wood construction in which horizontal planks are set or nailed into the corner posts of a timber-frame building. There are, however, no studs or intermediate posts connecting the sill and the plate. See also the related term *vertical plank frame construction.*

hung ceiling See *suspended ceiling.*

hyphen A subsidiary building unit, often one story, connecting the central block and the wings or dependencies.

I-beam The most common profile in steel structural shapes (although it also appears in cast iron and in reinforced concrete). Used especially for spanning elements, it is shaped like the capital letter *I* to make the most efficient use of the material consistent with a shape that permits easy assemblage. The vertical face of the *I* is the web. The horizontal faces are the flanges. Other standard shapes for steel framing elements are Hs, Ts, Zs, Is (known as angles), and square-cornered Us (channels).

I-house A two-story house, one room deep and two rooms wide, usually with a central hallway. The I-house is a nineteenth-century descendant of the hall-and-parlor houses of the colonial period. The term is commonly applied to the end-chimney houses of the southern and mid-Atlantic traditions. The term most likely derives from the resemblance between the tall, narrow end walls of these houses and the capital letter *I.*

impost The top part of a pier or wall, upon which rests the springer or lowest voussoir of an arch.

impost block A block, often in the form of an inverted truncated pyramid, placed between a column capital and the lowest voussoirs of an arch above. Distinguished from an entablature block, which has the details found in a classical entablature. Sometimes called a dosseret or supercapital.

in antis Columns in antis are placed between two projecting sections of wall, in an imaginary plane connecting the ends of the two wall elements.

intermediate rib See *tierceron.*

International Style A style that originated in the 1920s and flourished into the 1970s, characterized by the expression of volume and surface and by the suppression of historicist ornament and axial symmetry. The term was originally applied by Henry-Russell Hitchcock and Philip Johnson to the new, nontraditional, mostly European, architecture of the 1920s in their 1932 exhibition at the Museum of Modern Art and in their accompanying book, *The International Style.* Also called International, International Modern. See also the more specific term *Corporate International Style* and the related terms *Bauhaus, Miesian, Second Chicago School.*

intrados The inner curve or underside (soffit) of an arch. See also the related term *extrados.*

Ionic order An ensemble of classical column and entablature elements, particularly characterized by the use of large volutes in the capital of the column. See also the more general term *order*.

isometric drawing A pictorial drawing using isometric projection, in which all horizontal lines that are perpendicular to an object, building, or space are drawn at 60-degree angles from the vertical. Consequently, a single scale can be used for all three dimensions. Sometimes called an isometric. See also the related terms *axonometric drawing, perspective drawing*.

Italianate 1 A general term for an eclectic Neo-Renaissance and Neo-Romanesque style, originating in England and Germany in the early nineteenth century and prevalent in the United States between the 1840s and 1880s, not only in houses but also in Main Street commercial buildings. The Italianate is characterized by prominent window heads and bracketed cornices. Called the Bracketed Style in the nineteenth century. See also the more specific terms *Barryesque* and *Italian Villa Style,* and the related terms *Renaissance Revival, Round Arch mode, Second Empire*. **2** A specific term for Italianate buildings that are predominantly symmetrical in plan and elevation. Distinguished from Barryesque, which is applied to more formal institutional and governmental buildings.

Italian Villa Style A subtype of the Italianate style (definition 1), originating in England and Germany in the early nineteenth century and prevalent in the United States between the 1840s and 1870s, mostly in houses, but also churches and other public buildings. The style is characterized by asymmetrical plans and elevations, irregular blocklike massing, round arch arcades and openings, and northern Italian Romanesque detailing. Larger Italian Villa buildings often had a campanile-like tower. Distinguished from the more symmetrical Italianate style (definition 2) by having the northern Italian rural vernacular villa as prototype.

Jacobean period A term for a period in British history coinciding with the rule of James I (1603–1625). See also the related term *Elizabethan* for the immediately preceding period, which itself is part of the Tudor period.

Jacobethan Revival See *Neo-Tudor*.

jamb The vertical side face of a door or window opening, amounting to the full thickness of the wall, and usually enriched with paneling, moldings, or jamb shafts (which are engaged columns set into a splayed, or angled, jamb). In an opening containing a door or window, the jamb is distinguished from the reveal, which is the portion of wall thickness between the door or window frame and the outer surface of the wall. (In an opening without a door or window, the terms jamb and reveal are used interchangeably.) Also distinguished from an architrave (definition 2), which consists of the moldings on the face of a wall around the opening.

Jazz Moderne See *Art Deco*.

Jeffersonian A personal style of Neoclassicism identified with the architecture of Thomas Jefferson (1743–1826), derived in part from Palladian ideas and in part from Imperial Roman prototypes. The style had a limited influence in the Piedmont of Virginia and across the Appalachians into the Ohio River valley. Sometimes called Jeffersonian Classicism. See also the related terms *Anglo-Palladianism, Federal, Roman Revival*.

jerkinhead roof A gable roof in which the upper portion of the gable end is hipped, or inclined inward along the ridgeline, forming a small triangle of roof surface. Sometimes called a clipped gable roof or hipped gable roof.

joist One of a series of small horizontal beams that support a floor or ceiling.

keystone The central wedge-shaped stone at the crown of an arch.

king post In a truss, the vertical suspension member that connects the tie beam with the apex of opposing principal rafters.

kiosk Originally, a Turkish summer palace. Since the nineteenth century, the term has been applied to any small pavilion or stand, usually found in public gardens, parks, streets, and malls, where it serves some commercial or public function. Distinguished from a gazebo, which may be found in public or private gardens or parks, but which usually serves as a sheltered resting place. See also the related term *bandstand*.

label 1 A drip molding, over a square-headed door or window, which extends for a short distance down each side of the opening. **2** A similar vertical downward extension of a drip molding over an arch of any form. Sometimes called a label molding.

label stop 1 An L-shaped termination at the lower ends of a label. **2** Any decorative boss or other termination of a label.

lancet arch An arch generally tall and sharply pointed, whose centers are farther apart than the width or span of the arch.

lantern 1 The uppermost stage of a dome, containing windows or arcaded openings. **2** Any feature, square or polygonal in plan and usually containing windows, rising above the roof of a building. The square structures that serve as skylights on the roofs of nineteenth-century buildings—particularly houses—were also called lantern lights, and, in Italianate and Second Empire buildings, came to be called cupolas.

Late Georgian period Not strictly a style term, but a term for a period in British and British colonial history approximately coinciding with the reigns of George III (1760–1820) and George IV (1820–1830). In the United States, the Late Georgian period is now understood to end sometime during the Revolutionary War (1775–1781) and to be followed by the Federal period (c. 1787–1820). In Britain, the Late

Georgian period includes the Regency period (1811–1820s). See also the related term *Early Georgian period.*

Late Gothic Revival A term for the Gothic Revival work of the late nineteenth and early twentieth centuries. See also the more specific term *Collegiate Gothic* (definition 2) and the related term *Early Gothic Revival.*

lath A latticelike, continuous surface of small wooden strips or metal mesh nailed to walls or partitions to hold plaster.

Latin cross A cross with one long and three short arms. Usually used to describe the ground plans of Roman Catholic and Protestant churches. See also the more general term *cruciform.*

leaded glass Panes of glass held in place by lead strips, or cames. The panes, clear or stained, may be of any shape.

lean-to roof. See *shed roof.*

lierne In a Gothic vault, a short ornamental rib connecting the major transverse ribs and the secondary tiercerons. Sometimes called a cross rib or tertiary rib.

light frame construction A type of wood-frame construction in which relatively light structural members (usually sawn lumber, ranging from two-by-fours to two-by-tens) are fastened with nails. Distinguished from timber-frame construction, in which relatively heavy structural members (hewn or sawn timbers, measuring six by six and larger) are fastened with mortise-and-tenon joints. See the more specific terms *balloon-frame construction, platform frame construction.*

lintel A horizontal structural member that supports the wall over an opening or spans between two adjacent piers or columns.

living hall In Queen Anne, Shingle Style, and Colonial Revival houses, an extensive room, often containing the entry, the main staircase, a fireplace, and an inglenook.

load-bearing Term applied to a wall, column, pier, or any vertical supporting member, constructed so that all loads are carried to the ground through the wall, column, or pier. See also the related terms *bearing wall, curtain wall.*

loggia 1 A porch or open-air room, particularly one set within the body of a building. 2 An arcaded or colonnaded structure, open on one or more sides, sometimes with an upper story. 3 An eighteenth- and nineteenth-century term for a porch or veranda.

Lombard A style term applied in the United States in the mid-nineteenth century to buildings derived from the Romanesque architecture of northern Italy (especially Lombardy) and the earlier nineteenth- century architecture of southern Germany. Characterized by the use of brick for both structural and ornamental purposes. Also called Lombardic. See also the related term *Round Arch mode.*

lunette 1 A semicircular area, especially one that contains some decorative treatment or a mural painting. 2 A semicircular window in such an area.

Mannerism, Mannerist 1 A phase of Renaissance art and architecture in the mid-sixteenth century, characterized by distortions, contortions, inversions, odd juxtapositions, and other departures from High Renaissance canons of design. 2 (Not capitalized) A sensibility in design, regardless of style or period, characterized by a knowledgeable violation of rules and intended as a comment on the very nature of convention.

mansard roof A hipped roof with double pitch. The upper slope may approach flatness, while the lower slope has a very steep pitch, sometimes flaring in a concave curve (or swelling in a convex curve) as it comes to the eaves. This lower slope usually has windows, and the area under the roof often amounts to a full story. The name is a corruption of that of François Mansart (1598–1666), who designed roofs of this type, which was revived in Paris during the Second Empire period.

Mansard Style, Mansardic See *Second Empire.*

masonry Construction using stone, brick, block, or some other hard and durable material laid up in units and usually bonded by mortar.

massing The grouping or arrangement of the primary volumetric components of a building.

Mayan Revival Term applied to eclectic works or elements of those works that emulate forms in the visual arts of the Maya civilization of Central America. See also the related term *Art Deco.*

McKim classicism, McKim classical Architecture of, or in the manner of, the firm of McKim, Mead and White, 1890s–1920s. See *Beaux-Arts classicism.*

medieval Term applied to the Middle Ages in European civilization between the age of antiquity and the age of the Renaissance (i.e., mid-400s to mid-1400s in Italy; mid-400s to late 1500s in England). In architecture and the other visual arts, the medieval period included the end of the Early Christian period, then the Byzantine, the Romanesque, and the Gothic styles or periods.

Mediterranean Revival A style generally associated since the early twentieth century with residential architecture based on Italian villas of the sixteenth century. While not a major revival style, it is characterized by symmetrical arrangements, stucco walls, and low-pitch tile roofs. Sometimes called Mediterranean Villa, Neo-Mediterranean. See also the related term *Spanish Colonial Revival.*

metope In a Doric entablature, that part of the frieze which falls between two triglyphs. In the Greek Doric order the metopes often contain small sculptural reliefs.

Middle Ages See *medieval.*

Miesian Term applied to work showing the influence of the German-American architect Ludwig Mies van der Rohe (1886–1969). See also the related terms *Bauhaus, International Style, Second Chicago School.*

Mission Revival A style originating in the 1890s, and making use of forms and materials from the Spanish and Mexican mission architecture of the eighteenth and early nineteenth centuries. Not to be confused with Mission furniture of the Arts and Crafts movement. See also the more general term *Spanish Colonial Revival.*

modern Ambiguous term, applied in various ways during the past century to the history of the visual arts and world history generally: (1) from the 1910s to the present (see also the more specific terms *Bauhaus, International Style*; (2) from the 1860s, 1870s, 1880s, or 1890s to the present; (3) from the Enlightenment or the advent of Neoclassicism or the industrial revolution, c. 1750, to the present; (4) from the Renaissance in Italy, c. 1450, to the present.

Modern Gothic See *Eastlake.*

Moderne A term applied to a wide range of design work from the 1920s through the 1940s, in which aspects of traditionalism and modernism coexist and in which eclecticism (from a historical, exotic, or machine aesthetic) is inseparable from the urge for stylization. Sometimes called Art Moderne, Modernistic. See also the more specific terms *Art Deco, PWA Moderne, Streamline Moderne.*

modillion One of a series of small, thin scroll brackets under the projecting crown molding of a classical cornice. It is found in the Corinthian and Composite orders. Distinguished from a console, which usually is larger and has a height greater than its projection from the wall.

molding A running surface composed of parallel and continuous sections of simple or compound curves and flat areas.

monitor An extensive shed-roofed feature on a roof, containing a band of windows or vents. It may be located along one of the roof slopes (a trap-door monitor) or along the ridgeline (a clerestory monitor), and it usually runs the entire length of the roof. Distinguished from a skylight, which is a low-profile or flush-mounted feature in the plane of the roof.

Moorish Revival Term applied to eclectic works or elements of those works that emulate forms in the visual arts of those parts of North Africa and Spain under Muslim domination from the seventh through the fifteenth century. See also the related term *Oriental Revival.*

mortar A mixture of cement or lime with water and a fine aggregate of sand used to secure bricks or stones in masonry construction.

mortise-and-tenon joint A timber framing joint that is made by one member having its end shaped into a projecting piece (tenon) that fits exactly into a hole (mortise) in the other member. Once joined, the pieces are held together by a peg that passes through the tenon.

mullion 1 A post or similiar vertical member dividing a window into two or more units, or lights, each of which may be further subdivided (by muntins) into panes. **2** A post or similar vertical member dividing a wall opening into two or more contiguous windows.

muntin One of the small vertical or horizontal members that hold panes of glass within a window or glazed door. Distinguished from a mullion, which is a heavier vertical member separating paired or grouped windows. Sometimes called a glazing bar, sash bar, or window bar.

mushroom column A reinforced concrete column that flares at the top in order to counteract shear stresses in the vicinity of the column.

National Register of Historic Places A branch of the National Park Service of the United States Department of the Interior, established by the National Historic Preservation Act of 1966, to maintain files of documentation on districts, sites, buildings, structures, and objects of national, state, or local significance. Properties listed on the National Register are afforded administrative—and, ultimately, judicial—review in instances where projects funded or assisted by federal agencies might have an impact on the historic property. Properties listed on the register may also be eligible for certain tax benefits.

nave 1 The entire body of a church between the entrance and the crossing. **2** The central space of a church, between the side aisles, extending from the entrance end to the crossing.

Neo-Baroque Term applied to eclectic works or elements of those works that emulate forms in the visual arts of the Baroque style or period. Sometimes called Baroque Revival.

Neo-Byzantine Term applied to eclectic works or elements of those works that emulate forms in the visual arts of the Byzantine style or period. Sometimes called Byzantine Revival.

Neoclassical Revival See *Beaux-Arts classicism.*

neoclassicism, neoclassical A broad movement in the visual arts which drew its inspiration from ancient Greece and Rome. It began in the mid-eighteenth century with the advent of the science of archaeology and extended into the mid-nineteenth century (in some Beaux-Arts work, into the 1930s; in some postmodern work, even into the present). See also the related terms *Beaux-Arts, Beaux-Arts classicism, classicism,* and the more specific terms *Greek Revival, Roman Revival.*

Neo-Colonial See *Colonial Revival.*
Neo-Federal See *Federal Revival.*
Neo-Georgian See *Georgian Revival.*
Neo-Gothic Term applied to eclectic works or elements of those works that emulate forms in the visual arts of the Gothic style or period. The cultural movement that produced so many such works in the eighteenth, nineteenth, and twentieth centuries is called the Gothic Revival, though that term covers a wide range of work.

Néo-Grec An architectural style developed in connection with the Ecole des Beaux-Arts in Paris during the 1840s and characterized by the use of stylized Greek elements, often in conjunction with cast iron or brick construction. See also the more general term *Beaux-Arts.*

Neo-Hispanic See *Spanish Colonial Revival.*

Neo-Mediterranean See Mediterranean Revival.

Neo-Norman Term applied to eclectic works or elements of those works that emulate forms in the visual arts of the eleventh- and twelfth-century Romanesque of Norman France and Britain.

Neo-Palladian See *Palladianism.*

Neo-Renaissance Term applied to eclectic works or elements of those works that emulate forms in the visual arts of the Renaissance style or period. The mid- to late-nineteenth-century cultural movement that produced so many such works is called the Renaissance Revival, though that term covers a wide range of work.

Neo-Romanesque Term applied to eclectic works or elements of those works that emulate forms in the visual arts of the Romanesque style or period. The mid-nineteenth-century cultural movement that produced so many such works is called the Romanesque Revival, though that term covers a wide range of work.

Neo-Tudor Term applied to eclectic works or elements of those works that emulate forms in the visual arts of the Tudor period. Sometimes loosely called Elizabethan Manor Style, English Half-timber Style, Jacobethan Revival, Tudor Revival.

New Brutalism See *Brutalism.*

New Formalism A style prevalent since the 1960s, characterized by symmetrical arrangements, rich materials (marble cladding, metal grillework), and stylized classical (even Gothic) detailing. Architects associated with this style include Philip Johnson (born 1906), Edward Durell Stone (1902–1978), and Minoru Yamasaki (born 1912).

newel post A post at the head or foot of a flight of stairs, to which the handrail is fastened. Newel posts occur in a variety of shapes, in profile and cross section, and are generally more substantial elements than the individual balusters that support the handrail.

niche A recess in a wall, usually designed to contain sculpture or an urn. A niche is often semicircular in plan and surmounted by a half dome or shell form. See also the related terms *aedicule, tabernacle* (definition 1).

nogging Brickwork that fills the spaces between members of a timber-frame wall or partition.

Norman French See *French Norman.*

octagon house A rare house type of the 1850s, based on the ideas of Orson Squire Fowler (1809–1887), who argued for the efficiencies of an octagonal floor plan. Sometimes called octagon mode.

oculus A circular opening in a ceiling or wall or at the top of a dome.

ogee arch A pointed arch formed by a pair of opposing S-shaped curves.

order The most important constituents of classical architecture are the orders, first developed as a structural-aesthetic system by the ancient Greeks. An order has two major components. A column with its capital is the main vertical supporting member. The principal horizontal member is the entablature. The Greeks developed three different types of order, the Doric, Ionic, and Corinthian, each distinguishable by its own decorative system and proportions. All three were taken over and modified by the Romans, who added two orders of their own, the Tuscan, which is a simplified form of the Doric, and the Composite, which is made up of elements of both the Ionic and the Corinthian. The Romans often used the orders as a structural system in the same manner as the Greeks. Unlike the Greeks, however, they also applied them as decoration to the surfaces of walls that were supported by other means. Sometimes called classical orders. See also the related terms *column, entablature, giant order, superposition* (definition 1).

oriel A projecting polygonal or curved window unit of one or more stories, supported on brackets or corbels. Sometimes called an oriel window. Distinguished from a bay window, which rises from the foundation and has a rooted rather than a suspended appearance. However, a multistory projection in a tall building, whether cantilevered out or built from the foundation, is called a projecting bay or a unit of bay windows.

Oriental Revival Ambiguous term, suggesting eclectic influences from any period in any culture in the "Orient," or Asia, including Turkish, Persian, Indian, Chinese, and Japanese, as well as Arabic (even the Moorish of North Africa and Spain). Sometimes called Oriental style. See also the related term *Moorish Revival.*

orthographic projection A system of visual representation in which all details on or near some principal plane, object, building, or space are projected, to scale, onto the parallel plane of the drawing. Orthographic projection thus flattens all forms into a single two-dimensional picture plane and allows for an exact scaling of every feature in that plane. Distinguished from pictorial projection, which creates the illusion of three-dimensional depth. See also the more specific terms *elevation, plan, section.*

outbuilding A building subsidiary to and completely detached from another building. Distinguished from a dependency, which may be attached or detached.

overhang The projection of part of a structure beyond the portion below.

PWA Moderne A synthesis of the Moderne (i.e., Art Deco or Streamline Moderne) with an austere late type of Beaux-Arts classicism, often associated with federal government buildings of the 1930s and 1940s during the Public Works Administration. See also the more general term Moderne and the related terms *Art Deco, Beaux-Arts classicism, Streamline Moderne.*

Palladianism, Palladian Work influenced by the Italian Renaissance architect Andrea Palladio (1508–1580), particularly by means of his treatise, *I Quattro Libri dell'Architettura* (*The Four Books of Architecture*, originally published in 1570 and disseminated throughout Europe in numerous translations and editions until the mid-eighteenth century). The most significant flourishing of Palladianism was in England, from the 1710s to the 1760s, and in the British North American colonies, from the 1740s to the 1790s. Sometimes called Neo-Palladian, Palladian classical. See also the more specific term *Anglo-Palladianism.*

Palladian motif A three-part composition for a door or window, in which a round-headed opening is flanked by lower flat-headed openings and separated from them by columns, pilasters, or mullions. The flanking sections, and sometimes the entire unit, may be blind (i.e., not open).

Palladian Revival See *Anglo-Palladianism.*

Palladian window A window subdivided as in the Palladian motif.

parapet A low wall at the edge of a roof, balcony, or terrace, sometimes formed by the upward extension of the wall below.

pargeting Elaborate stucco or plasterwork, especially an ornamental finish for exterior plaster walls, sometimes decorated with figures in low relief or indented. Found in late medieval, Queen Anne, and period revival buildings. Sometimes called parging, pargework. See also the more general term *stucco.*

parquet Inlaid wood flooring, usually set in simple geometric patterns.

parti The essential solution to an architectural program or problem; the basic concept for the arrangement of spaces, before the development and elaboration of the design.

patera (plural: paterae) A circular or oval panel or plaque decorated with stylized flower petals or radiating linear motifs. Distinguished from a roundel, which is always circular.

pavilion 1 A central or corner unit that projects from a larger architectural mass and is usually accented by a special treatment of the wall or roof. **2** A detached or semidetached structure used for specialized activities, such as a hospital. **3** In a garden or fairground, a temporary structure or tent, usually ornamented.

pediment 1 In classical architecture, the low triangular gable end of the roof, framed by raking cornices along the inclined edges of the roof and by a horizontal cornice below. **2** In Renaissance and Baroque and later classically derived architecture, the triangular or curvilinear culmination of a prominent part of a facade. **3** A similar but smaller-scale feature over a door or window. It may be triangular or curvilinear.

pendentive A concave surface in the form of a spherical triangle that forms the structural transition from the square plan of a crossing to the circular plan of a dome.

pergola A structure with an open wood-framed roof, often latticed, and supported by a colonnade. It is usually covered by climbing plants, such as vines or roses, and provides shade for a garden walk or a passageway to a building. Distinguished from arbors or trellises, which are less extensive accessory structures lacking the colonnade.

period house Term applied to suburban and country houses in which period revival styles are dominant.

period revival Term applied to eclectic works—particularly suburban and country houses—of the first three decades of the twentieth century, in which a particular historical or regional style is dominant. See also the more specific terms *Colonial Revival, Dutch Colonial Revival, Georgian Revival, Neo-Tudor, Spanish Colonial Revival.*

peripteral (adjective) Surrounded by a single row of columns.

peristyle A range of columns surrounding a building or an open court.

perspective drawing A pictorial drawing representing an object, building, or space, as if seen from a single vantage point. The illusion of three dimensions is created by using a system based on the optical laws of converging lines and vanishing points. See also the related terms *axonometric drawing, isometric drawing.*

piano nobile (plural: piani nobili) In Renaissance and later architecture, a floor with formal reception, living, and dining rooms. The principal and often tallest story in a building, usually one level above the ground level.

piazza 1 A plaza or square. **2** An eighteenth- and nineteenth-century term for a porch or veranda.

pictorial projection A system of visual representation in which an object, building, or space is projected onto the picture plane in such a way that the illusion of three-dimensional depth is created. Distinguished from orthographic projection, in which the dimension of depth is excluded. See also the more specific terms axonometric drawing, isometric drawing, perspective drawing.

picturesque An aesthetic category in architecture and landscape architecture in the late eighteenth and early nineteenth centuries. It is characterized by relationships among buildings and landscape features that evoke the qualities of landscape paintings, in which the eye is led past a variety of forms and spaces into the distance and the mind is led to contemplate a sense of age (by means of ruins, fallen trees, weathered rocks, and mossy surfaces on all of these). In actual settings, asymmetrical and eclectic buildings, indirect approaches, and contrasting clusters of plantings heighten the experience of the picturesque.

pier 1 A freestanding mass, supporting a concentrated load from an arch, a beam, a truss, or a girder. While generally rectilinear in plan, piers in buildings based upon medieval precedents are often curvilinear in plan. **2** An upright portion of a wall that performs a columnar function. The pier may be continuous with the plane of the wall, or it may be distinguished from the plane of the wall to give it a columnlike independence.

pier and spandrel. A type of skeletal wall organization in which the vertical metal columns (and their square-cornered cladding) project in front of the plane of windows and their spandrel panels. The spandrel panels may be exposed structural spanning members. More often they provide decorative covering for the structure.

pilaster 1 A flattened column, with or without fluting, that is attached to a wall. It is usually finished with the same capital and base as a freestanding column. **2** Any narrow, vertical strip attached to a wall. Distinguished from an engaged column, which has a convex curvature.

pillar Ambiguous term, often used interchangeably with column, pier, or post. See instead one of those terms. (Although the term pillar is sometimes applied to columns that are square in plan, the term pier is preferable.)

pinnacle In Gothic architecture, a small spirelike element providing an ornamental finish to the highest part of a buttress or roof. It has a slender pyramidal or conical form and is often articulated with crockets or ribs and is topped by a finial. Distinguished from a finial, which is a smaller feature appearing by itself.

pitched roof See *gable roof*.

plan A drawing (in orthographic projection) representing all or part of an object, building, or space, as if viewed from directly above. A floor plan is a drawing of a horizontal cut through a building, usually at the level of the windows, showing the configuration of walls and openings. Other types of plans may illustrate ceilings, roofs, structural elements, and mechanical systems.

plank construction General term. See instead the more specific terms *horizontal plank frame construction, vertical plank construction*.

plate 1 In timber-frame construction, the topmost horizontal structural member of a wall, to which the roof rafters are fastened. **2** In platform and balloon-frame construction, the horizontal members to which the tops and bottoms of studs are nailed. The bottom plate is sometimes called the sill plate or sole plate.

Plateresque Term applied to Spanish and Spanish colonial Renaissance architecture from the early sixteenth century onward, in which the delicate, finely sculptured detail resembles the work of a silversmith (*platero*). See also the related term *Spanish colonial*.

platform frame construction A system of light frame construction in which each story is built as an independent unit and the studs are only one story high. The floor joists of each story rest on the top plates of the story below, and the bearing walls or partitions rest on the subfloor of each floor unit or platform. Platform framing is easier to construct and more rigid than balloon framing and has become the common framing method in the twentieth century. Structural members are usually sawn lumber, ranging from two-by-fours to two-by-tens, and are fastened with nails. Sometimes called platform framing, western frame, western framing.

plinth The base block of a column, pilaster, pedestal, dado, or door architrave.

Pointed Style A nineteenth-century term for Gothic Revival.

polychromy, polychromatic, polychrome A many-colored treatment, especially the combination of materials in various colors or the application of surface color, to articulate wall and roof planes and to highlight structure.

popular A term applied to vernacular architecture influenced by such publications as books of the orders, builders' guides, style books, pattern books, mail-order catalogs, architectural periodicals, and household magazines. Architecture in the popular tradition may be built according to commercially available plans or from widely distributed components; or it may be built by local practitioners (architects, builders, contractors) emulating buildings that are represented in publications. The distinction between popular architecture and high-style architecture by lesser-known architects depends on one's point of view with regard to the division between vernacular and high-style. See also the more general term *vernacular* and the related term *folk*.

porch A structure attached to a building to shelter an entrance or to serve as a semienclosed sitting, working, or sleeping space. Distinguished from a portico, which is either a pedimented feature at least one story in height supported by classical columns or a more extensive colonnaded feature.

porte cochere A porch projecting over a driveway and providing shelter to people leaving a vehicle and entering a building, or vice versa. Also called a carriage porch.

portico 1 A porch at least one story in height consisting of a low-pitched roof supported on classical columns and finished in front with an entablature and pediment. **2** An extensive porch supported by a colonnade.

post A vertical supporting element, either square or circular in plan. Posts are the integral vertical members of a frame or truss, whether of wood or metal. Posts may also carry fences or gates, or may serve as freestanding markers (e.g., mileposts).

post-and-beam construction A structural system in which the main support is provided by vertical members (posts) carrying horizontal members (beams or lintels). Sometimes called post-and-girt construction, post-and-lintel construction, trabeation, trabeated construction.

postmodernism, postmodern A term applied to work that involves a reaction against the ideas and works of various twentieth-century modern movements, particularly the Bauhaus and the International Style. Postmodern work makes use of historicism, yet the traditional elements are often merely applied to buildings that, in every other respect, are products of modern movement design. The term is also applied to works that are attempting to demonstrate an extension of the principles of various modern movements.

Prairie Box See *foursquare house*.

Prairie School, Prairie Style A diverse group of architects working in Chicago and throughout the Midwest from the 1890s to the 1920s, strongly influenced by Frank Lloyd Wright and to a lesser degree by Louis Sullivan. The term is applied mainly to domestic architecture. An architect is said to belong to the Prairie School; a work of architecture is said to be in the Prairie Style. Sometimes called Prairie, for short. See also the related terms *Chicago School, Wrightian*.

pre-Columbian Term applied to the major cultures of Latin American (e.g., Aztec, Maya, Inca) that flourished prior to the discovery of the New World by Columbus in 1492 and the Spanish conquests of the sixteenth century. Distinguished from North American Indian, which is generally applied to indigenous cultures within the area that would become the United States and Canada.

pressed metal Thin sheets of metal (usually galvanized or tin-plated iron) stamped into patterned panels for covering ceilings and exterior and interior walls or into molding profiles and other details for assembly into exterior and interior cornices. Loosely called pressed tin or stamped metal. Prevalent from the 1870s through the 1920s.

program The list of functional, spatial, and other requirements that guides an architect in developing a design.

proscenium In a recessed stage, the area between the orchestra and the curtain.

proscenium arch In a recessed stage, the enframement of the opening.

prostyle Having a columnar portico in front, but not on the sides and rear.

provincialism, provincial Term applied to work in an isolated area (such as a province of a cosmopolitan center or a colony of a mother country), where traditional practices persist, with some awareness of what is being done in the cosmopolitan center or the homeland.

purlin In roof construction, a structural member laid across the principal rafters and parallel to the wall plate and the ridge beam. The light common rafters to which the roofing surface is attached are fastened across the purlins. See also the related term *rafter*.

pylon 1 Originally, the gateway facade of an Egyptian temple complex, consisting of a truncated broad pyramidal form with battered (inclined) wall surfaces on all four sides, or two truncated pyramidal towers flanking an entrance portal. **2** Any towerlike structure from which bridge cables or utility lines are suspended.

quadripartite vault A vault divided into four triangular sections by a pair of diagonal ribs. Sometimes called a four-part vault.

quarry-faced See *rock-faced*.

quatrefoil A type of Gothic tracery having four parts (lobes or foils) separated by pointed elements (cusps).

Queen Anne Ambiguous but widely used term. **1** In architecture, the Queen Anne Style is an eclectic style of the 1860s through 1910s in England and the United States, characterized by the incorporation of forms from postmedieval vernacular architecture and the architecture of the Georgian period. Sometimes called Queen Anne Revival. See also the more specific term *Shingle Style* and the related terms *Eastlake, Stick Style*. **2** In architecture, the original Queen Anne period extends from the late seventeenth into the early eighteenth century. **3** In the decorative arts, the Queen Anne Style and period properly refer to work of the early eighteenth century during the reign of Queen Anne (1702–1714, i.e., after William and Mary and before Georgian). **4** In the decorative arts, eclectic work of the 1860s to 1880s is properly referred to as Queen Anne Revival. See also the related term *Aesthetic movement*.

quoin One of the bricks or stones laid in alternating directions, which bond and form the exterior corner of a building. Sometimes simulated in wood or stucco.

rafter One of the inclined structural members of a roof. Principal rafters are primary supporting elements spanning between the walls and the apex of the roof and carrying the longitudinal purlins. Common rafters are secondary supporting elements fastened onto purlins to carry the roof surfacing. See also the related term *purlin*.

raking cornice A cornice that finishes the sloping edges of a gable roof, such as the inclined sides of a triangular pediment.

random ashlar A type of masonry in which squared and dressed blocks are laid in a random pattern rather than in straight horizontal courses.

recessed column A fully round column set into a

nichelike space only slightly larger than the column. Distinguished from an engaged column, which appears to be built into the wall.

reentrant angle An acute angle created by the juncture of two planes, such as walls.

refectory A dining hall, especially in medieval architecture.

regionalism 1 The sum of cultural characteristics (including material culture, language) that define a geographic region, usually extending beyond a single state or province and coinciding with one or more large physiographic areas. **2** The conscious use, within a region, of forms and materials identified with that region, creating an architecture that is in keeping with the historical architecture of the region, and even a distinctive new regional style.

register A horizontal zone of a wall, altarpiece, or other vertical feature. Usually synonymous with story, but more inclusive, allowing for the description of zones with no corresponding interior spaces.

relieving arch An arch, usually of masonry, built over the lintel of an opening to carry the load of the wall above and relieve the lintel of carrying such load. Sometimes called a discharging arch or safety arch.

Renaissance The period in European civilzation identified with a rediscovery or rebirth (*rinascimento*) of classical Roman (and to a lesser extent, Greek) learning, art, and architecture. Renaissance architecture began in Italy in the mid-1400s (Early Renaissance) and reached a peak in the early to mid-1500s (High Renaissance). In England, Renaissance architecture did not begin until the late 1500s or early 1600s. The Renaissance in art and architecture was preceded by the Gothic and followed by the Baroque.

Renaissance Revival 1 In architecture, an ambiguous term, applied to *(a)* Italianate work of the 1840s through 1880s and *(b)* Beaux-Arts classical work of the 1880s through 1920s. **2** In the decorative arts, an eclectic furniture style incorporating a variety of Renaissance, Baroque, and Néo-Grec architectural motifs and utilizing wood marquetry, incised lines (often gilded), and ormolu and porcelain ornaments. Sometimes called Neo-Renaissance.

rendering Any drawing, whether orthographic (plan, elevation, section) or pictorial (perspective), in which shades and shadows are represented.

reredos A screen or wall at the back of an altar, usually with architectural and figural decoration.

return The continuation of a molding, cornice, or other projecting member, in a different direction, as in the horizontal cornice returns at the base of the raking cornices of a triangular pediment.

reveal 1 The portion of wall thickness between a door or window frame and the outer face of the wall. **2** Same as jamb, but only in an opening without a door or window.

revival, revivalism A type of historicism prevalent since the eighteenth century, involving the adaptation of historical forms to contemporary functions. Distinguished from a more pervasive historicism by an ideological conviction that sought to rationalize the choice of a historical style according to the values of the historical period that produced it. (The Gothic Revival, for instance, was associated with the Christianity of the Middle Ages.) Revival works, therefore, tend to invoke a single historical style. More hybrid works are manifestations of a less dogmatic historicism or eclecticism. See also the more general terms *eclecticism, historicism*.

rib The projecting linear element that separates the curved planar cells (or webs) of vaulting. Originally these were the supporting members for the vaulting, but they may also be purely decorative.

Richardsonian Term applied to any work showing the influence of the American architect Henry Hobson Richardson (1838–1886). See the note under the more limiting term *Richardsonian Romanesque*.

Richardsonian Romanesque Term applied to Neo-Romanesque work showing the influence of the American architect Henry Hobson Richardson (1838–1886). While many of Richardson's works make eclectic use of round arches and Romanesque details, many of his works show a creative eclecticism that transcends any particular historical style. The term Richardsonian, therefore, is a more inclusive term for the work of his followers than Richardsonian Romanesque—a term that continues to be widely used. Sometimes called Richardson Romanesque, Richardsonian Romanesque Revival.

ridgepole The horizontal beam or board at the apex of a roof, to which the upper ends of the rafters are fastened. Sometimes called a ridge beam, ridgeboard, ridge piece.

rinceau An ornamental device consisting of a sinuous and branching scroll elaborated with leaves and other natural forms.

rock-faced Term applied to the rough, unfinished face of a stone used in building. Sometimes called quarry-faced.

Rococo. A late phase of the Baroque, marked by elegant reverse-curve ornament, light scale, and delicate color. See also the related term *Baroque*.

Romanesque A medieval architectural style which reached its height in the eleventh and twelfth centuries. It is characterized by round arched construction and massive masonry walls. The Romanesque was preceded by the Early Christian and Byzantine periods in the eastern Mediterranean world and by a variety of localized styles and periods in northern and western

Europe; it was followed throughout Europe by the Gothic.

Romanesque Revival Ambiguous term, applied to (1) Rundbogenstil and Round Arch work in the United States as early as the 1840s and (2) Richardsonian Romanesque work into the 1890s. Sometimes called Neo-Romanesque.

Roman Revival A term, not widely accepted, for a version of Neoclassicism involving the use of forms from the visual arts of the Imperial Roman period. Applied to various works in Italy, England, and the United States, where it is most clearly visible in the architecture of Thomas Jefferson. See also the related terms *Federal, Jeffersonian, Neoclassicism.*

rood screen An ornamental screen that serves as a partition between the crossing and the chancel or choir of a church.

rosette A circular floral ornament similar to an open rose.

rotunda 1 A circular hall in a large building, especially an area beneath a dome or cupola. **2** A building round both inside and outside, usually domed.

Round Arch mode The American counterpart of the German Rundbogenstil, characterized by the predominance of round arches, whether these are accentuated by Romanesque or Renaissance detailing or left as simple unadorned openings. See also the related terms *Italianate, Lombard, Rundbogenstil.*

roundel. A circular panel or plaque. Distinguished from a patera, which is oval shaped.

rubble masonry A type of masonry utilizing uncut or roughly shaped stone, such as fieldstone or boulders.

Rundbogenstil Literally, "round arch style," a historicist style originating in Germany in the 1820s and spreading to Britain and the United States from the 1840s through the 1860s. It is characterized by an eclectic combination of Romanesque and Renaissance elements. See also the related term *Round Arch mode.*

running bond A pattern of brickwork in which only stretchers appear, with the vertical joints of one course falling halfway between the vertical joints of adjacent courses. Sometimes called stretcher bond. Distinguished from common bond, in which every fifth or sixth course consists of all headers.

Ruskin Gothic, Ruskinian Gothic. See *High Victorian Gothic.*

rustication, rusticated Masonry in which the joints are emphasized by narrow recessed channels or grooves outlining each block. Sometimes simulated in wood or stucco.

sacristy A room in a church where liturgical vessels and vestments are kept.

safety arch See *relieving arch.*

sanctuary 1 The part of a church that contains the principal altar. Usually the innermost space within the chancel arm of the church, situated to the east of the choir. **2** Loosely used to mean a place of worship, a sacred place.

sash Any framework of a window. It may be movable or fixed. It may slide in a vertical plane (as in a double-hung window) or may be pivoted (as in a casement window).

sash bar See *muntin.*

Secession movement The refined classicist Austrian (Viennese) version of the Art Nouveau style, so named beause the artists and architects involved seceded from the official Academy in 1897. Josef Hoffmann (1870–1956) is the architect most frequently mentioned in association with this movement.

Second Chicago School A term sometimes applied to the International Style in Chicago from the 1940s to the 1970s, particularly the work of Mies van der Rohe. See also the related terms *International Style, Miesian.*

Second Empire Not strictly a style term but a term for a period in French history coinciding with the rule of Napoleon III (1852–1870). Generally applied in the United States, however, to a phase of Beaux-Arts governmental and institutional architecture (1850s–1880s) as well as to countless hybrids of Beaux-Arts and Italianate forms in residential, commercial, and industrial architecture (1850s–1880s). Sometimes called General Grant Style, Mansard Style, Mansardic. See also the related terms *Beaux-Arts, Italianate* (definition 1).

section A drawing (in orthographic projection) representing a vertical cut through an object, building, or space. An architectural section shows interior relationships of space and structure and may also include mechanical systems. Sometimes called a cross section.

segmental arch An arch formed on a segmental curve. Its center lies below the springing line.

segmental curve A curve that is a segment (i.e., less than half the circumference) of a circle or an ellipse. The baseline of the curve is a chord measuring less than the diameter of the larger circle from which the segment is taken.

segmental pediment A pediment whose top is a segmental curve.

segmental vault A vault whose cross section is a segmental curve. A dome built on segmental curves is called a saucer dome.

setback 1 In architecture, particularly in the design of tall buildings, a series of upper stories that are stepped back to allow more sunlight to reach the streets. **2** In planning, the amount of space between the lot line and the perimeter of a building.

shaft The tall part of a column between the base and the capital.

shed roof A roof having only one sloping plane. Sometimes called a lean-to roof.

Shingle Style A term applied primarily to American domestic architecture of the 1870s through the 1890s, in which broad expanses of wood shingles dominate the exterior roof and wall planes. Rooms open widely into one another and to the outdoors, and the ample living hall or stair hall is often the dominant feature of the interior. The term was coined in the 1940s by Vincent Scully for a series of seaside and suburban houses of the northeastern United States. The Shingle Style is a version of the Anglo-American Queen Anne Style. See also the related terms *Colonial Revival, Stick Style*.

shouldered architrave See *ear*.

side gabled Term applied to a building whose gable ends face the sides of a lot. Distinguished from front gabled.

side light A framed area of fixed glass alongside a door or window. See also the related term *fanlight*.

sill course In masonry, a stringcourse set at windowsill level, usually differentiated from the wall by its greater projection, its finish, or its thickness. Not applicable to frame construction.

sill plate See *plate* (definition 2).

skeleton construction, skeleton frame A system of construction in which all loads are carried to the ground through a rigid framework of iron, steel, or reinforced concrete. The exterior walls are curtain walls (i.e., not load-bearing).

skylight A window in a roof, specifically one that is flush with the roof plane or only slightly protruding. Distinguished from a cupola (definition 2), which is a major centralized feature at the summit of a roof. Distinguished from a monitor, which is an extensive roof feature containing a band of windows or vents.

soffit The exposed underside of any overhead component, such as an arch, beam, cornice, or lintel. See also the related term *intrados*.

sole plate See *plate* (definition 2).

space frame A series of trusses placed side by side and joined to one another by triangulated rods, tubes, or beams, so that the individual planar trusses are united into a three-dimensional structural framework. Often used in roof structures requiring long spans.

spandrel 1 The quasi-triangular space between two adjoining arches and a line connecting their crowns, or between an arch and the columns and entablature that frame it. **2** In skeletal construction, the wall area between the top of a window and the sill of the window in the story above. Sometimes called a spandrel panel.

Spanish colonial A term applied to buildings, towns, landscapes, and other artifacts from the various periods of actual Spanish colonial occupation in North America (c. 1565–1821 in Florida; c. 1763–1800 in Louisiana and the Lower Mississippi valley; c. 1590s–1821 in Texas and the southwestern United States; c. 1769–1821 in California). The term is extended to apply to the artifacts of Hispanic ethnic groups (e.g., Mexicans, Puerto Ricans, Cubans) and their descendants, even into the early twentieth century. See also the related terms *Churrigueresque, Plateresque*.

Spanish Colonial Revival The revival of forms from Spanish colonial and provincial Mexican design. The Spanish Colonial Revival began in Florida and California in the 1880s and continues nationwide into the present. Sometimes called Neo-Hispanic, Spanish Eclectic, Spanish Revival. See also the more specific term *Mission Revival* and the related term *Mediterranean Revival*.

spindle A turned wooden element, thicker toward the middle and thinner at either end, found in arch screens, porch trim, and other ornamental assemblages. Banisters (i.e., thin, simple balusters) may be spindle-shaped, but the term spindle, when used alone, usually connotes shorter elements.

spire A slender pointed element surmounting a building. A tall, attenuated pyramidal form with any number of thin triangular faces that are unbroken or articulated only with crockets, pinnacles, or small dormers. Distinguished from a steeple, which is divided into stages and which may be topped with a spire.

splay The slanting surface formed by cutting off a right-angle corner at an oblique angle to one face. A reveal at an oblique angle to the exterior face of the wall.

springing, springing line, springing point The line or point where an arch or vault rises from its supports and begins to curve. Usually the juncture between the impost of the support below and the springer, or first voussoir, of the arch above.

squinch An arch, lintel, or corbeling, built across the interior corner of two walls to form one side of an octagonal base for a dome. This octagonal base serves as the structural transition from a square interior crossing space to an octagonal or round dome.

stair A series of steps, or flights of steps connected by landings, which connects two or more levels or floors.

staircase The ensemble of a stair and its enclosing walls. Sometimes called a stairway.

stair tower A projecting tower or other building block that contains a stair.

stamped metal See *pressed metal*.

Steamboat Gothic See *Carpenter's Gothic*.

steeple 1 A tall structure rising from a tower, consisting of a series of superimposed stages diminishing in plan, and usually topped by a spire or small cupola. Distinguished from a spire, which is not divided into stages. **2** Less commonly used to mean the whole of the tower, from the ground to the top of the spire or cupola.

stepped gable A gable in which the wall rises in a series of steps above the planes of the roof.

stereotomy The science of cutting three-dimensional shapes from stone, such as the units that make up a carefully fitted masonry vault.

Stick Style A term applied primarily to American domestic architecture of the 1850s through the 1870s, in which exterior wall planes are subdivided into bays and stories outlined by narrow boards called "stickwork." The term was coined by Vincent Scully in the 1940s for a series of houses with clearly articulated wall panels and sticklike porch supports and eaves brackets. Sources include the English and German picturesque traditions, as well as the French rationalist tradition. See also the related terms *Queen Anne, Shingle Style.*

story (plural: stories). The space in a building between floor levels. British spelling is storey, storeys. Sometimes called a register, a more inclusive term applied to horizontal on a vertical plane zones that do not correspond to actual floor levels.

Streamline Moderne A later phase of the Moderne, popular in the 1930s and 1940s and characterized by stucco surfaces with rounded corners, by horizontal banding, overhangs, and window groupings, and by other details suggestive of modern Machine Age aerodynamic forms. Sometimes called Streamline Modern, Streamline Modernistic. See also the more general term *Moderne* and the related terms *Art Deco* and *PWA Moderne.*

stretcher A brick laid the length of a wall, so that the long side of the brick shows on the exterior.

stretcher bond See *running bond.*

string In a stair, an inclined board that supports the ends of the steps. Sometimes called a stringer.

stringcourse In masonry, a horizontal band, generally narrower than other courses, extending across the facade of a building and in some instances encircling such features as pillars or columns. It may be flush or projecting; of identical or contrasting material; flat, molded, or richly carved. Not applicable to frame construction. Sometimes called a band course or belt course. More elaborate horizontal bands in masonry or frame construction are generally called band moldings.

strut A column, post, or pole that is set in a diagonal position and thus serves as a stiffener by triangulation. Distinguished from a brace, which is usually a shorter bracketlike member.

stucco 1 An exterior plaster finish, usually textured, composed of portland cement, lime, and sand, which are mixed with water. 2 A fine plaster used for decorative work or moldings. See also the more specific term *pargeting.*

stud One of the vertical supporting elements in a wall, especially in balloon- and platform frame construction. Studs are relatively lightweight members (usually two-by-fours).

Sullivanesque Term applied to work showing the influence of the American architect Louis Henry Sullivan (1856–1924).

sunburst light See *fanlight.*

supercapital See *impost block.*
supercolumniation See *superposition* (definition 1).
superimposition, superimposed See *superposition.*
superposition, superposed 1 The use of an ensemble of the classical orders, one above the other, as the major elements articulating a facade. When this is done, the Doric, considered the simplest order, is used on or near the ground story. The Ionic, considered more complex, comes next; and the Corinthian, considered the most complex, is used at the top. Sometimes the Tuscan order or rusticated masonry may be used for the ground story beneath the Doric order, and the Composite order may be used above the Corinthian order. Sometimes called supercolumniation, superimposition. See also the related term *order.* 2 Less commonly, any vertical relationship of architectural elements (e.g., windows, piers, colonnettes) in any style or period.

superstructure A structure raised upon another structure, as a building upon a foundation, basement, or substructure.

Supervising Architect The Supervising Architect of the United States Treasury Department, whose office was responsible for the design and construction of all major federal government buildings (such as courthouses, customhouses, and post offices) from the 1850s through the 1930s. The Office of the Supervising Architect was formally established by Congress in 1864 and lasted until 1939, when its functions were absorbed into the Public Buildings Administration (and in 1949, into the General Services Administration).

supporting wall See *bearing wall.*

surround An encircling border or decorative frame around a door or window. Distinguished from architrave (definition 2), a term usually applied to the frame around an opening when considered as a series of relatively flat face moldings.

suspended ceiling A ceiling suspended from rodlike hangers below the level of the floor above. The interval between the floor slab above and the suspended ceiling often serves as a space for ducts, utilities, and air circulation. Sometimes called a hung ceiling.

swag A motif representing a suspended fold of drapery hanging in a catenary curve from two points. Distinguished from a festoon, which is a motif representing entwined leaves, flowers, or fruits, hung in a similar curve. See also the more general term *garland.*

tabernacle 1 A niche or recess, usually on an interior wall, framed by columns or pilasters and topped by an entablature and pediment. Distinguished from an aedicule, which more often occurs on an exterior wall. See also the related term *niche.* 2 In the Jewish religion, a portable sanctuary. 3 In Protestant denominations, a large auditorium church.

terra-cotta A hard ceramic material used for (1) fireproofing, especially as a fitted cladding around metal skeletal construction; or (2) an exterior or interior wall cladding, which is often glazed and multicolored.

tertiary rib See *lierne*.

thermal window A large lunette window similar to those found in ancient Roman baths (*thermae*). The window is subdivided into three to five parts by vertical mullions. Sometimes called a *thermae* window.

three-hinged arch An arch in two major segments anchored with cylindrical "hinge" pins at either end and at the crown. Movement within the arch, caused by temperature changes, the torsion of wind movements, or other forces, can be absorbed by the movement of the arch around the pins, thereby avoiding stresses that would occur in the structural frame if the arches were fixed.

tie beam A horizontal tension member that ties together the opposing angular members of a truss and prevents them from spreading.

tier A group of stories or any zone of architectural elements arranged horizontally.

tierceron In a Gothic vault, a secondary rib that rises from the springing to an intermediate position on either side of the diagonal ribs. Sometimes called an intermediate rib.

tie rod A metal rod that spans the distance between two structural members and, by its tensile strength, restrains them against tendencies to collapse outward.

timber-frame construction, timber framing A type of wood-frame construction in which heavy timber posts and beams (six-by-sixes and larger) are fastened using mortise and tenon joints. Sometimes called heavy timber construction. Distinguished from light frame construction, in which relatively light structural members (two-by-fours to two-by-tens) are fastened with nails.

trabeation, trabeated construction. See *post-and-beam construction*.

tracery Decoration within an arch or other opening, made up of narrow curvilinear bands or more elaborately molded strips. In Gothic architecture, the curved interlocking stone bars that contain the leaded stained glass.

transept The lateral arm of a cross-shaped church, usually between the nave (the area for the congregation) and the chancel (the area for the altar, clergy, and choir).

transom 1 A narrow horizontal window unit, either fixed or movable, over a door. Sometimes called a transom light. See also the more specific term *fanlight*. 2 A horizontal bar, as distinguished from a vertical mullion, especially one crossing a door or window opening near the top.

transverse rib In a Gothic vault, a rib at right angles to the ridge rib.

trefoil A type of Gothic tracery having three parts (lobes or foils) separated by pointed elements (cusps).

trellis Any open latticework made of strips of wood or metal crossing one another, usually supporting climbing plants. Distinguished from an arbor, which is generally a more substantial yet compact three-dimensional structure, and from a pergola, which is a more extensive colonnaded structure.

triforium In a Gothic church, an arcade in the wall above the arches of the nave, choir, or transept and below the clerestory window.

triglyph One of the slightly raised blocks in a Doric frieze. It consists of three narrow vertical bands separated by two V-shaped grooves.

triumphal arch 1 A freestanding arch erected for a victory procession. It usually consists of a broad central arched opening, flanked by two smaller bays (usually with open or blind arches). The bays are usually articulated by classical columns supporting an entablature and a high attic. 2 A similar configuration applied to a facade to denote a monumental entryway.

truss A rigid triangular framework made up of beams, posts, braces, struts, and ties and used for the spanning of large spaces. The major horizontal or inclined members are called chords. The connecting vertical and diagonal elements are called the web members.

Tudor arch A low-profile arch characterized by two pairs of arcs, one pair of tight arcs at the springing, another pair of broad (nearly flat) arcs at the apex or crown.

Tudor period A term for a period in English history coinciding with the rule of monarchs of the house of Tudor (1485–1603). Tudor period architecture is Late Gothic, with only hints of the Renaissance. See also the more specific term *Elizabethan period* for the end of this period, and the related term *Jacobean period* for the succeeding period.

Tudor Revival See *Neo-Tudor*.

turret A small towerlike structure, often circular in plan, built against the side or at an exterior or interior corner of a building.

Tuscan order An ensemble of classical column and entablature elements, similar to the Roman Doric order, but without triglyphs in the frieze and without mutules (domino-like blocks) in the cornice of the entablature. See also the more general term *order*.

tympanum (plural: tympana) 1 The triangular or segmental area enclosed by the cornice moldings of a pediment, frequently ornamented with sculpture. 2 Any space similarly delineated or bounded, as between the lintel of a door or window and the arch above.

umbrage A term used by Alexander Jackson Davis (1803–1892) as a synonym for veranda, the implication being a shadowed area.

vault An arched roof or ceiling, usually constructed in brick or stone, but also in tile, metal, or concrete. A nonstructural plaster ceiling that simulates a masonry vault.

Venetian Gothic See *High Victorian Gothic.*

veranda A nineteenth-century term for porch. Sometimes spelled verandah.

vergeboard See *bargeboard.*

verges The sloping edges of a gable, gambrel, or lean-to roof, usually projecting beyond the wall below. Distinguished from eaves, which are the horizontal lower edges of a roof plane.

vernacular Not a style in itself, but a descriptive term, applicable primarily to architecture, covering the vast range of ordinary buildings that are produced outside the high-style tradition of well-known architects. The vernacular tradition includes the folk tradition of regional and ethnic buildings whose forms (plan and massing) remain relatively constant through the years, in spite of stylistic embellishments. The term vernacular architecture is often used as if it meant only folk architecture. However, the vernacular tradition in architecture also includes the popular tradition of buildings whose design was influenced by such publications as books of the orders, builders' guides, style books, pattern books, mail-order catalogs, architectural periodicals, and household magazines. Usually contrasted with high-style. See also the more specific terms *folk, popular.*

vertical plank construction A system of wood construction in which vertical planks are set or nailed into heavy timber horizontal sills and plates. A building so constructed has no corner posts and no studs. Two-story vertical plank buildings have planks extending the full height of the building, with no girt between the two stories. Second-floor joists are merely mortised into the planks. Distinguished from the more specific term vertical plank frame construction, in which there are corner posts.

vertical plank frame construction A type of vertical plank construction, in which heavy timber corner posts are introduced to provide support for the plate, to which the tops of the planks are fastened. See also the related term *horizontal plank frame construction.*

vestibule A small entry hall between the outer door and the main hallway of a building.

Victorian Gothic See *High Victorian Gothic.*

Victorian period A term for a period in British, British colonial, and Anglo-American history, and not, in architecture or the other visual arts, a sufficiently specific style term. The Victorian period extended across eight decades, from the coronation of Queen Victoria in 1837 to her death in 1901. See instead *Eastlake, Gothic Revival, Greek Revival, Queen Anne, Shingle Style, Stick Style,* and other specific style terms.

Victorian Romanesque Ambiguous term. See instead *Richardsonian Romanesque, Romanesque Revival, Round Arch mode.*

villa. 1 In the Roman and Renaissance periods, a suburban or rural residential complex, often quite elaborate, consisting of a house, dependencies, and gardens. **2** Since the eighteenth century, any detached suburban or rural house of picturesque character and some pretension. Distinguished from the more modest house form known as a cottage.

volute 1 A spiral scroll, especially the one that is a distinctive feature of the Ionic capital. **2** A large scroll-shaped buttress on a facade or dome.

voussoir A wedge-shaped stone or brick used in the construction of an arch. Its tapering sides coincide with radii of the arch.

wainscot A decorative or protective facing, usually of wood paneling, applied to the lower portion of an interior partition or wall. Distinguished from a dado, which is the zone at the base of a wall, regardless of the material used to cover it. Wainscot properly connotes woodwork. Sometimes called wainscoting.

water table 1 In masonry, a course of molded bricks or stones set forward several inches near the base of a wall and serving as the cap of the basement courses. **2** In frame construction, a ledge or projecting molding just above the foundation to protect it from rainwater. **3** In masonry or frame construction, any horizontal exterior ledge on a wall, pier, or buttress. Often sloped and provided with a drip molding to prevent water from running down the face of the wall below.

weatherboard See *clapboard.*

weathering The inclination given to the upper surface of any element so that it will shed water.

web 1 The relatively thin shell of masonry between the ribs of a ribbed vault. **2** The portion of a truss between the chords, or the portion of a girder or I-beam between the flanges.

western frame, western framing See *platform frame construction.*

winder A step, more or less wedge-shaped, with its tread wider at one end than the other.

window bar See *muntin.*

window cap See *cap.*

window head A head molding or pedimented feature over a window.

Wrightian Term applied to work showing the influence of the American architect Frank Lloyd Wright (1867–1959). See also the related term *Prairie School.*

wrought iron Iron shaped by a hammering process to improve the tensile properties of the metal. Distinguished from cast iron, a brittle material, which is formed in molds.

Zigzag Moderne, Zigzag Modernistic See *Art Deco.*

Illustration Credits

HABS   Historic American Buildings Survey
RIHPHC   Rhode Island Historical Preservation and Heritage Commission

Photographs not otherwise credited were taken by John M. Miller, Coventry, Vermont, in 1986–1989.

Providence
Page 33, Rhode Island Historical Society; **PR5, PR14.1** HABS; **PR27** RIHPHC; **PR56, PR58, PR59** HABS; **PR63** William H. Pierson, Jr.; **PR83.3, PR83.6** RIHPHC; **PR84** HABS; **PR85** RIHPHC; **PR86.1, PR88, PR92** HABS; **PR96.3, PR96.4** HABS; **PR110** RIHPHC; **PR114.1** HABS; **PR114.11** John Abromowski, Brown University; **PR118, PR138** HABS; **PR151** RIHPHC; **PR177** HABS; **PR198** RIHPHC

Bristol
Page 468, HABS

Newport
NE14, NE33, NE56, NE84 HABS; **NE89** William H. Pierson, Jr. (right); **NE91** HABS; **NE136** Salve Regina University, photo by John Corbett; **NE139** William H. Pierson, Jr.; **NE140** HABS; **NE144** William H. Pierson, Jr.; **NE150** HABS; **NE158, NE159** William H. Pierson, Jr.; **NE160, NE162** Preservation Society of Newport County, photos by Ira Kerns (right), Jim Patrick (left)

Block Island
BI16 HABS

Index

Properties named for individuals or families with a given surname are indexed in the form *surname (first name)* following all entries for individuals with that surname. Page numbers in **boldface** refer to illustrations.

A. and I. Wilkinson (textile manufacturers), 137
A. and W. Sprague Company, 176
A. T. Stewart Department Store, New York City, 47
A. T. Wall Company Building, Providence, 57, 59
Abbott, Daniel, 45
Abbott, Gerald, 23
Abbott (Commodore Joel) House. *See* Miller (General Nathan)–Abbott (Commodore Joel) House, Warren
Abbott (Cornelia C.) House (Londonderry), Little Compton, 492
Abbott Park, Providence, 45
Abbott Run (tributary), 206, 214, 215, 216
Aberthaw Construction Company: Westerly Water Tower, Westerly, 415–416
abolition movement, 26, 70, 569
Abraham Manchester General Store, Little Compton, 496
Acid Factory Creek, 343
Acorns, The, South Kingstown, 381
Acton, Richmond-Hopkinton border, 424
Ada Hawkins School. *See* Northwest Community Nursing and Health Center (Ada Hawkins School), Glocester
Adam, Robert, 46, 80, 413
Adams, John, 77
Adams, John Quincy, 89
Adams, Seth, Jr., 97
Adams (John F.) House, Pawtucket, 152
Adams Public Library, Central Falls, 158
Adamsville, Little Compton, 496–497
Admissions Office, Brown University (George Corliss House), Providence, 104
Adult Correctional Institution, Cranston, 177, 179, **182–**183
Advance Base Depot and Proving Ground, Quonset Point, 354
Aesthetic Movement, 68, 70, 601
African Americans
  and Fourteenth Rhode Island Heavy Artillery, 592
  and grave markers, Newport, 530
  and Mawdsley (Captain John House), Newport
  See also slavery and slave trade
African Humane Society, 537
Agassiz, Alexander, 578, 601
Agassiz (Professor Alexander) House (Castle Hill), Newport, 578, 597
Agawam Hunt, East Providence, 435, 437
Aiken, William Martin, 146
Albert and Vera List Art Building, Brown University, Providence, 84, **85**, 94, 103
Albion, Lincoln, 188, 189, 202–204, 208
Albion Mill, Lincoln. *See* apartments (Albion Mill, Valley Falls Mill), Lincoln
Albion Mill Canal, 203
Albro House. *See* Tucker-Albro House, South Kingstown
Alciphron, or, the Minute Philosopher (Berkeley), 510
Alden, John and Priscilla, 148
Aldrich, Abby Pierce, 44

Aldrich, James, 269
Aldrich, Luke, 389
Aldrich, Senator Nelson W., 117
Aldrich, William T., 43, 44, 53
  Eliza G. Radeke Building, RISD Museum of Art, Providence, 80–81
  Jackson (Donald E.) House, Providence, 115–116
  President's House, Brown University (Rush Sturges House), Providence, 88–89
  Temple of Music, Roger Williams Park, Providence, 133–**134**
Aldrich and Sleeper: Jelke (F. Frazier) House (Eagles Nest), Newport, 577
Aldrich (David) House, North Smithfield, 241, **242**
Aldrich (J.) House I, Scituate, 267
Aldrich (James) House II–Grant (Florence Price) House, Scituate, 269
Aldrich (Luke) House, South Kingstown, 389
Aldrich (Senator Nelson W.) House. *See* Museum of Rhode Island History, Rhode Island Historical Society (Robert S. Burrough, Jr.–Senator Nelson W. Aldrich House), Providence
Aldrich (Simon) Farm, Lincoln, 198
Aldrich Street (number 12), Hopkinton, 418
Alice Building, West Warwick, 331
All Saints Episcopal Church, Warwick, 323
Allan (John) House, Newport, 526
Allard et fils, 568
Allen, Amos: Allen (Zachariah) (sea captain) House, Providence, 126
Allen, Crawford, 87, 229, 326
Allen, Philip, 262
Allen, Walter, 240, 241, 243
  Aldrich (David) House, North Smithfield, 241, **242**
  Allen (Walter) House, North Smithfield, 239, 242–243
  apartments (Seth Allen House), North Smithfield, 239, 242, 243
  Arnold (Daniel) House, North Smithfield, 241
  Brownell (Stephen) House, North Smithfield, 241–**242**
Allen, Zachariah (industrialist), 28, 126, 164, 165, 173, 259, 260, 261, 262, 326
  condominiums and offices (Allendale Mill and Company Store), North Providence, 164–165
Allen and Browne (masons), 180
Allen (Candace) House, Providence, 87
Allen family, 229
Allen (George and Hortense Rodman) House, North Kingstown, 363–364
Allen (Samuel) House, Barrington, 448
Allen (Samuel) House, Scituate, **271**
Allen (Seth) House. *See* apartments (Seth Allen House), North Smithfield
Allen (Walter) House, North Smithfield, 239, 242–243
Allen (Zachariah; industrialist) House. *See* Brown University Faculty Club (Zachariah Allen House), Providence
Allen (Zachariah; industrialist) Papers, 261
Allen (Zachariah; sea captain) House, Providence, 126

645

Allendale, North Providence, 87, 160, 164–166, 326
Allendale Mill and Company Store. *See* condominiums and offices (Allendale Mill and Company Store), North Providence
Allendale Mutual Insurance Company Headquarters Building. *See* FM Global (Allendale Mutual Insurance Company Headquarters Building), Johnston
Allin, Pearce, 446
Allin (Pearce) House, East Providence, 446
Almy, B. F., 175
Almy, Brown and Slater (textile manufacturers), 137
Almy, William, 135, 144, 244
Almy and Brown (textile manufacturers), 144, 244, 331
Almy Farm, Tiverton, 487
Almy (Sampson and Eliza) House, Providence, 105
Altamira. *See* Henszey (William P.)–Ewing (Joseph N.) House (Altamira), Jamestown
Alton, Richmond, 427–428
Alton Jones Campus, University of Rhode Island, West Greenwich, **343**
America (Rhind sculpture), 37
American Architect and Building News (magazine), 42, 378
American Express Building, Providence, 40
American Hoechst Chemical Company (Quidnick Mills), Coventry, 337
American Institute of Steel Construction, 480
American Legion Hall (Number 8 Schoolhouse), Little Compton, 496
American Legion Post (Warren Armory), Warren, 460
American Revolution, 7, 9, 11, 20, 21, 29, 32, 61, 64, 72, 137, 161, 170, 206, 270, 321, 354, 357, 374, 375, 400, 454, 456, 461, 466, 472, 478, 483, 484, 485, 520, 597
American Scene, The (James), 519–520
American Screw Company, 125
American Textile Mill, Pawtucket, 140
America's Cup Races, 17
Amor Caritas figure (Saint-Gaudens sculpture), 531
Amtrak, 40
Anderton (George) House, Barrington, 450–451
Andrew, David, 239
Andrew (David) Farm. *See* Smith (Daniel)–Andrew (David)–Taft (Nelson)–Todd (Albert) Farm, North Smithfield
Andrews, Jacques and Rantoul: President's House (Stephen O. Metcalf House), Rhode Island School of Design, Providence
Andrews Memorial School, Bristol, 471
Andrews School. *See* Colt-Andrews School (Colt Memorial High School), Bristol
Angell, E. L.: Grove Street School, Woonsocket, 227
Angell, Frank W., 381, 384
  Peace Dale Railroad Station, South Kingstown, 383, 384, 392
Angell, Thomas, 170
Angell and Swift
  Borden (Jerome C.) House (Bay House), Newport, 577
  Division of Applied Mathematics, Brown University (Henry Pearce House), Providence, 119
  East Greenwich Free Library, East Greenwich, 347
  Peace Dale Library (Narragansett Reading Room and Library), South Kingstown, 384, **385**, 386
Angell (Asahel) House, Smithfield, **258**–259
Angell (Daniel) House, Johnston, 170
Angell (Elijah) House. *See* Johnston Historical Society (Elijah Angell House), Johnston
Angell family, 170, 171
Angell Farm, Glocester. *See* Steer-Angell-Colwell Farm, Gloucester

Angell House, North Providence. *See* Whipple-Angell-Bennett House, North Providence
Angell-Ballou House, Smithfield, 256, 259
Angell (William G.) House, Providence, 74
Anglicans. *See* churches, Episcopal; Episcopalians
Ann and Hope Mill. *See* Ann and Hope Store Complex (Ann and Hope Mill, Lonsdale Mill), Cumberland
Ann and Hope Mill housing, Cumberland, 210–**211**, 272
Ann and Hope Store Complex (Ann and Hope Mill, Lonsdale Mill), Cumberland, 142, 168, 206, 209–210, 211
Annaquatucket River, 351, 354
Anoatok. *See* Green (John P.) House (Anoatok), Jamestown
Anthony, Coventry, 333, 337–338
Anthony, Frank J.: offices (East Greenwich Post Office), East Greenwich, 345
Anthony, James, 168
Anthony, Richard, 168, 337
Anthony Mill. *See* Coventry Mill (Anthony Mill)
Anthony (Richard) House, North Providence, 167–168
apartments, Glocester, 291–292
apartments (Albion Mill, Valley Falls Mill), Lincoln, 202–**203**, 204
apartments (Broadway), Pawtucket, 152–153
apartments (First Free Will Baptist Church, later Independent Eastern Orthodox Church of the Resurrection), Pawtucket, 148–149
apartments (George Aldrich Inn), North Smithfield, 242, 243
apartments (Green Mill), Lincoln, 204
apartments (Mapleville School), Burrillville, 303, 315
apartments (Pascoag No. 11 School), Burrillville, 303, 315
apartments (Richard Arnold–Pelog Arnold Tavern), North Smithfield, 241
apartments (Seth Allen House), North Smithfield, 239, 242, 243
apartments (Valley Falls Mill, also known as Blackstone Mill), Central Falls, 147, 159–160, 188, 208
Apex (store), Pawtucket, 143, 239
Appleton, W. Cornell, 518
Appleton (Thomas) House, Newport, 570
Aquidneck Coal Company, 500
Aquidneck Island, 4, 6, 8, 9, 15, 20, 25, 387, 480, 483, 484, 498, 503, 550, 551, 557, 570, 576, 579, 582
Arcade, The, Providence, 37, **46**–47, 63, 75, 170, 459
Arcade Corporation, 46
Arcadia Management Area, Hopkinton/Exeter, 418, 429
Architects Collaborative, The: North Kingstown Free Library, Wickford, North Kingstown, 362
Architectural Forum (magazine), 349
Architectural Heritage of Newport, Rhode Island, The (Scully), 513
Architectural Record (magazine), 310, 484, 496
"Architecture Ancient and Modern" (Tefft), 77
Architecture of Country Houses, The (A. J. Downing), 566
Architecture without Architects (Rudofsky), 444
Arctic, West Warwick, 324, 330–331, 343
Arctic Mill Complex, West Warwick, **330**–331, 343
Arkwright, Coventry, 333, 334
Arkwright, Richard, 334
Arkwright Bridge, Coventry, 334, **335**
Arkwright Manufacturing Company, 334
Armington, Hervey, 446
Armington and Slims Engine Company, 125
Armington (Dr. Hervey) House. *See* Bicknell (Joseph)–Armington (Dr. Hervey) House, East Providence
Armistead House, Newport, 552

Index 647

armories
   American Legion Post (Warren Armory), Warren, 460
   Benefit Street Arsenal, Providence Marine Corps of Artillery, 77–78, 348
   Cranston Street Armory, Providence, 34, 114, 130, 132, **133**, 145, 229, 230
   Kentish Guards Armory, East Greenwich, **348**
   Naval Reserve Armory, Bristol, 348, 478–479
   Newport Artillery Company, Newport, 348, 541
   Pawtucket Armory, Pawtucket, 145, 146, 229, 230
   Varnum Memorial Armory, East Greenwich, 344–345, 347
   Westerly, 145, 229
   Woonsocket Armory, 145, 229
Armory, The, and Parade Street area, Providence, 130–131
Armsea Hall. *See* Greene (General Francis Vinton) House (Armsea Hall), Newport
Armstrong, Maitland, 511
Armstrong, Shirley Houde, 219
Army Corps of Engineers, 579, 597
Arnold, Amos, 216
Arnold, Benedict, 592
Arnold, Benedict (father), 544, 592–593
Arnold, Eleazer, 187, 194
Arnold, Israel, 196
Arnold, John, 241
Arnold, Louis, 216
Arnold, Pelog, 241
Arnold, Richard, Jr., 239, 241
Arnold, Thomas, 187, 194, 338
Arnold Acid Factory. *See* Arnold (Thomas) House, Ruins of the Arnold Acid Factory, and Arnold Cemetery (Historic Cemetery 42), Coventry
Arnold (Benjamin F.) House, 131
Arnold Cemetery. *See* Arnold (Thomas) House, Ruins of the Arnold Acid Factory, and Arnold Cemetery (Historic Cemetery 42), Coventry
Arnold (Daniel) House, North Smithfield, 241
Arnold (Daniel) House, Providence, 45
Arnold (Dexter) House, Scituate, 269, 270
Arnold (Eleazer) House, Lincoln, 171, **195**, 200, 267, **268**, 281, 320
Arnold family, 187, 239, 269, 338, 592
Arnold Home Farm, Coventry, 340–341
Arnold (Israel) House. *See* Olney–Arnold (Israel) House, Lincoln
Arnold Mills, Cumberland, 215–216
Arnold Mills Methodist Church (United Baptist Church), Cumberland, 215
Arnold (Newton D.) House, Providence, 112
Arnold (Thomas) House, Ruins of the Arnold Acid Factory, and Arnold Cemetery (Historic Cemetery 42), Coventry, 338
Art Association of Newport. *See* Newport Art Museum
Art Center and Glocester Heritage Society (Job Armstrong Store), Glocester, 294
Art Deco, 124, 139, 145, 175, 183, 319
art museums. *See* museums
Art Nouveau, 70, 117, 124, 150, 554
Art Workers Guild (Providence), 110
Artistic Country Seats (Sheldon), 16
Artist's Studio (Union Congregational Church), Newport, 539
Arts and Crafts movement, 68, 70, 114, 150, 181, 192, 230, 233, 268, 304, 380, 383, 464, 585, 591, 599, 607
Ashaway, Hopkinton, 423–424
Ashaway River, 423
Ashton Mill and Mill Village, Cumberland, 188, 189, 190, 191, 202, 205, 208, 210, 211, 213–215, **214**
Ashton Mill housing, Cumberland, 128, 234
Ashton Viaduct (New Bridge), Cumberland-Lincoln, 205, 214, **215**
Ashville Mill, Hopkinton, 422
Asnières, France, 568
Aspinwall (Dr. Thomas) House, East Providence, 442
Assembly, The, Burrillville, 310, 311
Astor, John Jacob, 485, 581
Athenaeum Block, Providence, 84
Athenaeum Row, Providence, 84–85
Atkin, Tony, 81
Atlantic Bank Building, Providence, 46, 456
Atlantic House Hotel, Newport, 12, 564, 567
Atlantic Inn (Norwich House), Block Island, 610
Atterbury, Phelps and Tompkins: Swiss Village (Vedimar), Newport, 580
Attleboro, Massachusetts, 186
Atwater and Schubarth, 121
Auchincloss, John, 579
Auchincloss (John) House (Hammersmith Farm), Newport, 573, 579, 597
Auditorium Building, St. Francis of Assisi Roman Catholic Church, South Kingstown, **380**
Audrain, A. L., 564
Audrain Building, Newport, 564–565
Audubon Society, 393, 399
Aufenthalt, Newport. *See* Sargent (Letitia B.) House (Aufenthalt), Newport
Augustus Stout Van Wickle Gate, Brown University, Providence, 97
Austin Avenue Farms, Smithfield, 252
Avondale, Westerly, 408
Avondale Chapel, Westerly, 408
Avondale Farm House, Westerly, 408
Avondale Village, Westerly, 408
Ayrault House. *See* Hoyt (Virginia Scott) House (Ayrault House), Newport

Babbitt, Jacob, 474
Babbitt-Smith House, Bristol, 474
Babcock, George, 412
Babcock, Joshua, 536
Babcock, Rowse, 411, 416
Babcock, Reverend William, 372
Babcock (Dr. Joshua)–Smith (Orlando) House, Westerly, 410, **415**
Babcock family, 410
Babcock House (Whistling Chimneys), Charlestown, 400–**401**
Babcock (Jacob) House, Hopkinton, 424
Babcock (Joseph) House. *See* Marchant (Henry) House (Joseph Babcock House), South Kingstown
Bach Road, Shelter Harbor, Westerly, 403
Bacon, Henry, 124
Bailey, Banks and Biddle, 602
Bailey, Charles W., 602
Bailey, Joseph, 15
Bailey (Charles H.) Cottage, Jamestown, 603
Bailey (Charles W.) House, Jamestown, **600**–602
Bailey (George) House. *See* Wall (Daniel)–Bailey (George) House, Wickford, North Kingstown
Bailey (Joseph) House, Newport, 554
Bailey (Thomas) House, Little Compton, 493
Bailey (William) House (Hillwood), Providence, 127
Bailey's Beach, Newport, 16
Bajnotti Fountain, Providence, 36
Baker, Charles, 113
Baker, Ernest Hamlin, 379
Baker, George, 230
Baker (Charles H.) House, Providence, 113

Baker (Colin C.) Duplex, Providence, 128–129
Baker (Colin C.) Row Houses, Providence, 129–130
Baker (Esther) House, Providence, 120
Baker (George H.) House, Woonsocket, 230, **231**
Baker (Jonathan) House, Pawtucket, 143
Baker (Josiah) House, Providence, 93–94
Baker (N.) House, North Smithfield, 249
Balcom family, 275
Baldwin, Christopher Columbus, 569
Baldwin (C. H.) House, Newport, 567
Baldwin (Christopher Columbus) House (Château-Nooga, Chatterbox), Newport, 569
Ball, Cassius Clay, 605
Ball, Nicholas, 605
Ballou, Dexter, 229
Ballou, George C., 229, 230, 237
Ballou-Harris-Lippitt mill, 237
Ballou House. *See* Angell-Ballou House, Smithfield
Baltic, Connecticut, 330
Baltimore, Maryland, 487, 513
Bancroft (John) House (The Bluff), Middletown, 513
Bandstand, Pawtucket, 151
Banigan, Joseph, 49, 221, 231
Banigan Building, Providence, 49, 221
Banister, Hermione Pelham, 537
Banister family, 537
Banister (John) Town House, Newport, 537
Bank Boston, 50
"Bank Building" (People's Savings Bank), RISD Buildings, Providence, 65
Bank of North America (former), Providence, 47–**48**, 50
banks, 32
　Atlantic Bank Building, Providence, 46
　"Bank Building" (People's Savings Bank), RISD Buildings, Providence, 65
　Bank of Newport, Washington Square, Newport, 520
　Bank of North America (former), Providence, 47–**48**, 50
　Centreville National Bank, West Warwick, **331**
　Citizens Bank building, Providence, 40
　Commercial Bank Building, Warren, 471
　DeWolf's Wharf, Bristol, Old Bank of Bristol, 478
　Exchange Bank Building, Providence, 49–50, 51
　Fleet Bank Building (Industrial National Bank Building), Providence, 37, 49, 50
　Fleet Bank (Industrial Trust), Warren, 464
　Industrial Trust Company Building (former), Westerly, 413, 414
　Merchants Bank Building, Providence, **50**–51
　Old Bank of Bristol, DeWolf's Wharf, Bristol, 478
　Providence Bank, 64
　Providence Institution for Savings Building, 39, 62, **63**
　Sovereign Bank Complex (Rhode Island Hospital Trust National Bank), Providence, 50, 63
　Union Trust Company Building, Providence, 50, 51
　Washington Trust Company, Westerly, **413**, 414
　Wickford National Bank Building, North Kingstown, 357
　Ye Old Narragansett Bank, Wickford, North Kingstown, 360
Banning, Edwin T.: Casino, Providence, 133
Banning and Thornton, 76, 77
Banquet Hall, Squantum Association, East Providence, 445
Banquet Shed, Moosup Valley Grange, Foster, 284–**285**
Baptist Church, residence, Narragansett, 369
Baptist Church (Allendale Schoolhouse and Community Hall), North Providence, **165**–166
Baptist churches. *See* churches, Baptist
Baptists, 66, 254, 266, 283, 293, 464, 534, 541, 571

Barber's Hall, Hope Valley, Hopkinton, 419
Barberville, Richmond, 428
Barden, Betty, 273
Barden, Joe W., 273
Barden, John, 273
Barden family, 273
Barden (Joe) House. *See* Battey (Horace)–Barden (Joe) House, Scituate
Barden's Mills site, Bettyville, Scituate, 273
Barger, Samuel, 567
Barrington Congregational Church parsonage I (484 County Road), Barrington, 448
Barker (Samuel) House. *See* Brown (James)–Barker (Samuel) House, Newport
Barlow (Everett D.) House (Bit O Heaven; Mohegan Cottage), Block Island, 609
barn buildings. *See* barns
Barnaby, Jerothmul B., 54, 129
Barnaby (Jerothmul B.) House (Barnaby's Castle), Providence, 54, 129
Barnacle, The. *See* Selfridge (Admiral Thomas O., Jr.) Cottage (The Barnacle), Jamestown
Barnai Worsted Company Dye Works. *See* Museum of Work and Culture (Barnai Worsted Company Dye Works, later Lincoln Textile Corporation), Woonsocket
Barnard, Henry, 165–166, 267, 496
Barnefield (Thomas P.) House, Pawtucket, 150
Barnes, Edward Larrabee: Old Stone Square Building, Providence, 63
Barnes, Hattie and Isoline, 474
Barnes (Hattie and Isoline) House. *See* Wyndstowe (Hattie and Isoline Barnes House), Bristol
Barnes (Joanna Jenckes) House, Providence, 75
Barney (Richard) House, Wickford, North Kingstown, 360
barns
　Obadiah Harrington–William Blanchard Farm, Foster, 284
　Shadow Farm (Samuel A. Strang–John J. Welsh House), South Kingstown, 375, 378, **386**
　Steer-Angell-Colwell Farm, Glocester, 299
Barrington, 106, 437, 443, 444, 446–453, 454, 455, 456
　map, 448
　Town Hall, 447, 451–**452**
Barrington and Warren Bridges, Barrington, 453
Barrington Congregational Church, Barrington, **449**
Barrington Congregational Church parsonage II (464 County Road), Barrington, 448
Barrington Parkway. *See* Veterans Memorial Parkway (Barrington Parkway), East Providence
Barrington River, 447, 453
Barrington Village, Barrington, 448
Barry, Charles, 92
Barstow, John, 503, 506
Barstow family, 506
Barstow (John) Farm (Greenvale), Portsmouth, 503, 506. *See also* Farm Estates, Portsmouth
Bartlett (Dr. Elisha) House, North Smithfield, 245, 247
Barton, William, 483
Bateman (William) Farm. *See* Cook (Colonel John)–Bateman (William) Farm, Tiverton
Bates, Isaac, 69
Bates (Stephen and Martha)–Champlin (H.) House, Charlestown, 401–**402**
Batterson, J. G., 236
Battey, Horace, 273
Battey (Horace)–Barden (Joe) House, Scituate, 273
Bay House. *See* Borden (Jerome C.) House, Newport
Bay Mill (later Shore Mill), East Greenwich, 345–**346**

Bay View Hotel, Jamestown, 583
Bay Voyage Hotel, Jamestown, **604**
Bayard Ewing Building, Division of Architecture, Rhode Island School of Design (Fall River Iron Works Warehouse), Providence, 62
Bayberry Lodge. See Perkins (Julius Deming) House (Bayberry Lodge), Block Island
Beacon Hill, Newport. See James (Arthur Curtis) Estate (Beacon Hill), Newport
Beattie (Hamilton) House (The Stone House), Tiverton, 484
Beattie House. See Briggs-Manchester-Beattie House, Tiverton
Beaver Tail Farm, Jamestown, 593–594
Beavertail, Jamestown, 592–595
Beavertail Lighthouse, Jamestown, 20, 21, 23, 583, 593, **594**–595
Beavertail Point, Jamestown Island, 20, 27
Beck, Thomas, 328
Beckman, Blydenburgh and Associates, 59
Beckwith, Amos, 111
Beckwith, Truman, 111
Beckwith (Amos) House, Providence, 110–111, 114
Beckwith (Truman) House. See Handicraft Club (Truman Beckwith House), Providence
Bela Clapp Acid Factory, West Greenwich, 343
Belair, Newport. See Wright (H. Allen) House (Belair), Newport
Belcher (Joseph) House, Newport, 526
Belcourt, Newport. See Perry (Oliver Hazard) House (Belcourt), Newport
Bell, Isaac, Jr., 567
Bell, Louis F., 379, 385
Bell, William H., 474
Bell Commercial Block, Wakefield, South Kingstown, 379
Bell (Isaac, Jr.) House (Edna Villa), Newport, **567**–568
Bell Schoolhouse (Richmond Schoolhouse), Richmond, 428
Bell Street Chapel, Providence, 130
Bell (William H.) Block, Bristol, 473–474
Bell (William H.)–Fales (William) House, Bristol, 474
Bellevue Avenue, Newport, 12, 13, 15, 16, 17, 18, 511, 532, 545, 546, 555, 557, 562–576
  map, 563
Bellevue House. See Codman (Martha) House (Berkeley Villa, Bellevue House), Newport
Belluschi, Pietro, 78, 500, 501, 502
Chapel, Portsmouth Abbey School, Portsmouth, 501–502
Belmont, Alva Smith Vanderbilt, 573, 574
Belmont, August, 530
Belmont, Oliver Hazard Perry, 512, 574
Belton Court. See Zion Bible College (Frederick Stanhope Peck House, known as Belton Court), Barrington
Benedict, William Benedict, 112
Beneficent Congregational Church (Round Top Church; formerly Second Congregational Church), Providence, **43**–44, 45, 53
Beneficent House, Providence, 44–45
Benefit Street, Providence, 32, 61, 65, 72, 114
  map (north), 74
  map (south), 86
  north, 73–86
  south, 86–97
Benefit Street Arsenal, Marine Corps of Artillery, Providence, 77–78, 348
Benjamin, Asher, 88, 253, 461, 472
Bennett, James Gordon, 555, 567
Bennett (A.) Farm, Foster, 287

Bennett Farm. See Paine-Bennett Farm, Foster
Bennett House, Foster. See Hopkins-Bennett House, Foster
Bennett House, North Providence. See Whipple-Angell-Bennett House, North Providence
Bennett (Jeremiah) Farm, Foster, 277–278
Bennett (Job) House, Newport, 528
Benoni Cooke House, Providence, 62–**63**, 83, 87
Benson, John Howard, 360, 528
Benson (Captain George) House, Providence, 104
Benson family, 512, 528
Berkeley, Cumberland. See Berkeley Mill and Mill Village, Cumberland
Berkeley, George, 167, 211, 509, 524, 545
Berkeley Company, 211
Berkeley Memorial Chapel. See St. Colomba's Chapel (Berkeley Memorial Chapel), Middletown
Berkeley Mill and Mill Village, Cumberland, 188, 189, 190, 191, 202, 205, 208, 209, 211–212, **213**, 271
Berkeley Mill housing, Cumberland, 128, 234
Berkeley Villa. See Codman (Martha) House (Berkeley Villa, Bellevue House), Newport
Berkeley's (Dean George) Farm, Middletown. See Whitehall (Dean George Berkeley's Farm), Middletown
Berkshire-Hathaway company, 204
Berkshire-Hathaway Mill housing, Lincoln, 204
Berlin Iron Bridge Co.: Interlocken Mill Bridge, Coventry, 334, **335**
Bermuda, 509
Bernard, Simon: Fort Adams, Newport, **579**–580
Bernardis, Frank, 174
Bernardis (Frank) House, Johnston, 174, 251
Bernon, Gabriel, 229
Bernon, Woonsocket, 219, 220
Bernon Mills. See Woonsocket Manufacturing Company Mill Buildings (Bernon Mills), Woonsocket
Berwind, Edward, 568
Berwind (Edward Julius) House (The Elms), Newport, **568**, 574
Bethel Mill, Hopkinton, 423
Bethshan, Newport. See Gibbs (T. K.) House (Bethshan), Newport
Bevins, C. L., 585, 602
  Bailey (Charles W.) House, Jamestown, **600**–602
  Green (John P.) House (Anoatok), Jamestown, 602
  Patterson (General Robert E.) House (The Ramparts; Channel Bells), Jamestown, 598
  Selfridge (Admiral Thomas O., Jr.) Cottage (The Barnacle), Jamestown, 599, 601
  Selfridge (Admiral Thomas O., Jr.) House (Red Top; Green Chimneys), Jamestown, 599, 601
  Wharton (Joseph) House (Marbella, Horsehead), Jamestown, 596, 597, 598
Bibb, Albert B., 608
  U.S. Lifesaving Station, Block Island, 608
Bicknell, Joseph, 446
Bicknell, Louisa Allin, 446
Bicknell-Armington House, East Providence, 94
Bicknell (Joseph)–Armington (Dr. Hervey) House, East Providence, **446**
Bickner (J. J.) House, Warren, **458**
Big River, 342
Billiard Hall, Squantum Association, East Providence, 445
Billings-Coggeshall House, Newport, 536
Binney (William) House, Providence, 105
Biological Science Building, University of Rhode Island, Kingston, 392
Birch (John) House. See Bliss-Ruisdell House (John Birch House), Warren

Birckhead (Sarah King) House (Eastover), Portsmouth, 505, 551. *See also* Farm Estates, Portsmouth, Eastover
Bird, Larry, 268
Bird's Nest Cottage. *See* Pratt (Samuel) House (Bird's Nest Cottage), Newport
Birkhead House. *See* King-Birkhead House, Newport
Bishop, John W., 331, 439
Bishop, Phanuel, 441, 442
Bishop Harkins Hall, Providence College, Providence, 126
Bishop (Phanuel) House, East Providence, 441
Bit O Heaven. *See* Barlow (Everett D.) House (Bit O Heaven; Mohegan Cottage), Block Island
Bitter, Karl, 530
Black Estate, Scituate, 268–**269**
Blackburn, Joseph, 536
Blackstone, William, 206
Blackstone Boulevard, Providence, 120, 132, 183
Blackstone Canal, 144, 154, 197, 204, 205, 206, 214, 221
Blackstone Mill, Central Falls. *See* apartments (Valley Falls Mill, also known as Blackstone Mill), Central Falls
Blackstone River, 11, 28, 135, 153, 155, 160, 176, 186, 187, 188, 189, 190, 201, 203, 205, 206, 208, 209, 212, 213, 214, 219, 220, 226, 234, 237, 238, 249, 301, 302, 324, 424
Blackstone River Valley National Heritage Corridor, 144, 205, 208, 238
Blanchard family, 284
Blanchard (William) Farm, Foster, 284
Blane, Arthur, 245
Blatchford, Sophia, 555
Blatchford House. *See* Richardson-Blatchford House, Newport
Bliss (Caroline) House, Providence, 116–117
Bliss (Elder John) House, Newport, 525
Bliss-Ruisdell House (John Birch House), Warren, 447, 458
Blithewold, Bristol, 479–480
Block Island (New Shoreham), 22, 28, 604–610
  map, 606
  Old Harbor Historic District, 605, 607
Block Island Race Week, 605
Block Island Sound, 21, 397, 399, 608
Blount family, 457
Blue Dory Inn, Block Island, 607
Bluff, The. *See* Bancroft (John) House (The Bluff), Middletown
B'nai Israel Synagogue. *See* Congregation B'nai Israel Synagogue, Woonsocket
boardinghouses
  North Kingstown, 363
  Smithfield, **260**
Boathouse, Pawtucket, 151
Bogart, Charles, 479
Bois Doré. *See* Fahnestock (William F.) House (Oaklawn, Bois Doré), Newport
Book of Architecture, A (J. Gibbs), 8, 66
Book of Plans, A (Congregationalists Central Committee), 82, 148
Bookstaver, H. W., 515
Boon, Charles E., 369
Booth, Edwin, 511
Booth, John Wilkes, 511
Booth (Edwin) House (Boothden), Middletown, 511
Boothden. *See* Booth (Edwin) House (Boothden), Middletown
Borden, Isaac, 475
Borden, Parker, 470
Borden (Isaac) House, Bristol, 475

Borden (Jerome C.) House (Bay House), Newport, 577
Borden (Parker) House, Bristol, 469–470
Borromini, Francesco, 63
Bostitch company, East Greenwich, 349–345
Boston, Massachusetts, 27, 29, 51, 76, 105, 108, 120, 130, 154, 190, 199, 206, 208, 239, 301, 367, 375, 383, 388, 413, 415, 443, 471, 479, 484, 486, 494, 506
Boston and Providence Railroad, 345
Boston Bridge Works
  bridge and dam, Arnold Mills, Cumberland, 216
  bridges and dam, Ashton, Cumberland, 214–**215**
  Point Street Bridge, Providence, 60
  pony truss bridges, Lincoln, **203–204**
Boston Journal (newspaper), 525
Boston Public Library, 43, 132
Boston Wire Stitcher Company. *See* Bostitch company, East Greenwich
Bosworth, Nathaniel, 469
Bosworth family, 469
Bosworth House I, Bristol, 76
Bosworth House II (Silver Creek), Bristol, 469
Bosworth (Judge Alfred) House, Warren, 463
Boulders, The. *See* Larned (Louise Alexander) Cottage (The Boulders), Jamestown
Bourn, Augustus O., 475
Bourne Mills, Tiverton, 481–**482**
Bours, Hannah Babcock, 536
Bours, John, 536
Bours, Samuel, 536
Bours (Samuel) House, Newport, 536
Bowditch, Ernest, 603
Bowdoin College, 82
Bowen, Jeremiah, 395
Bowen, Nelson B.: North Foster Free Baptist Church, Foster, 288–289
Bowen, Tully, 92
Bowen (Colonel Jeremiah) House. *See* Colonel Jeremiah Bowen House), South Kingstown; Weeden Farm (Willow Dell
Bowen House, Warwick. *See* Greene-Bowen House, Warwick
Bowen (Isaac) House, Coventry, **339**, 341
Bowen (Tully D.) House, Providence, 92–**93**, 94, 109
Bowerman Brothers: A. T. Wall Company Building, Providence, 57, 59
Bowers, Jerathmael, 472
Bowler, Charles, 540
Bowler, Metcalf, 505, 540, 541
Bowler (Metcalf) House. *See* Vernon House (William Gibbs–Metcalf Bowler–William Vernon House), Newport
Boxcroft. *See* Coleman (Samuel) House (Whileaway, Boxcroft), Newport
Boy Scout Camp (Yawgoog), Hopkinton, 394, **421**
Boyd (George E.) House, Provincetown, 130
Boyden, George B., 444, 445
Bradford, Seth
  Porter (Mrs. James C.) House (Porter Villa, Old Castle), Newport, 556
  Wetmore (William Shepherd) House (Château-sur-Mer), Newport, 549, **570**–571
  Wright (H. Allen) House (Belair), Newport, 549, 570
Bradford, Westerly, 417–418, 424
Bradford, William, 472
Bradford Mill, Westerly, 417–418, 424
Bradford-Norris House, Bristol, 472
Bradford (William) House, Bristol, 476
Bradley (Charles) House, Providence, 127, 234
Brahms Road, Shelter Harbor, Westerly, 403

Branch River, 238, 245, 248, 301, 302, 304, 308
Branch School, District 3, North Smithfield. *See* North Smithfield Heritage Association (Forestdale School, officially Branch School, District 3), North Smithfield
Brayton, Francis, 338
Brayton (Edward) House, Little Compton, 492, 493
Brayton (Francis) House. *See* Paine House (Francis Brayton House), Coventry
Brayton (Thomas and Alice) House (Green Animals), Portsmouth, 502
Brazil, 523
Breakers, The, Newport, 561
Breakers, The (second). *See* Vanderbilt (Cornelius, II) House (The Breakers (second), Newport
Breezy Hill Farm, Scituate, 270
Brenton, William, 579
Brenton (Jahleel) House, Newport, 576
Brenton (Jahleel II) House, Newport, 579
Breuer, Marcel, 396, 501
brick farmhouse, East Providence, 446
Brick Market, Newport, 6–7, 65, 306, 516, **521**, 522
Brick Packaging Plant, East Providence, 441
Brick Schoolhouse, Providence, 70
Brickbuilder, The (journal), 126
Bridge and dam, Arnold Mills, Cumberland, 216
Bridge Street bridge, Westerly, 417
bridges
    Arkwright Bridge, Coventry, 334, **335**
    Ashton Viaduct (New Bridge), Cumberland-Lincoln, 205, 214, **215**
    Barrington and Warren Bridges, Barrington, 453
    bridge and dam, Arnold Mills, Cumberland, 216
    Bridge Street Bridge, Westerly, 417
    bridges and dam, Ashton, Cumberland, 214–215
    Broad Street Bridge, Central Falls, 160
    Church Street Bridge, South Kingstown, 383
    Columbia Street Bridge, South Kingstown, 383
    Crawford Street Bridge, Providence, 64
    Interlocken Mill Bridge, Coventry, 334, **335**
    Jamestown Bridge, South Kingstown/Jamestown, 387, 589, 590
    King Street Railroad Bridge, East Greenwich, 345
    Main Street Bridge, Lincoln, 201
    Mount Hope Bridge, Bristol, **480**, 498
    Mount Hope Suspension Bridge, Bristol, 25
    Newport Bridge, 521, 599
    Point Street Bridge, Providence, 60
    Pony truss bridges, Lincoln, **203–204**
    Reproduction covered bridge, Foster, 275
    Roosevelt Avenue Bridge, Central Falls, 155
    Shippee Bridge, Burrillville, 310
    Stone Arch Bridge, Burrillville, 310
    Stone Bridge and Fort Barton, Tiverton, 483
bridges and dam, Ashton, Cumberland, 214–**215**
Bridgetown Road, South Kingstown, 387
Bridgham, Samuel W., 441
Bridgham (Dr. Samuel W.) House, East Providence, 441
Bridgham House, East Providence. *See* Hyde-Bridgham House, East Providence
Bridgham Memorial Library, East Providence, 441–442
Briggs, Daniel, 350
Briggs, Nathaniel, 484, 488
Briggs, Thomas A., 349
Briggs-Manchester-Beattie House, Tiverton, 484, 488
Briggs (Joseph) Farmhouse (Coventry Poor Farm–Town Asylum), Coventry, 339
Bristol, 23, 24–25, 26, 76, 80, 357, 437, 444, 447, 453, 454, 464–480, 498, 522
    common, 25, 466, 476–477

    customhouse, 25, 49, 472
    grid street plan, 566
    map, 465
Bristol Bay, 466
Bristol County Courthouse, Bristol, 477
Bristol County Jail, 476
Bristol Ferry Lighthouse, Bristol, 480
Bristol Historical and Preservation Society Museum (Bristol County Jail), Bristol, 476
Bristol Steam Mill Company, 478
Bristol Steam Mill (Namquit Mill, White Mill), Bristol, 478
Broad Street Bridge, Central Falls, 160
Broadway, Providence, 34
Broadway and City Center, Newport, 516–521
    map, 518
Bronsdon, H. H.: Rhode Island Company Trolley Barn, Cranston, 178
Brookfield. *See* Congdon (William) House (Brookfield), South Kingstown
Brooklyn, New York, 412
Brown, Almy, 28
Brown, Ann S. K., 103, 209
Brown, Anne, 17
Brown, Georgette Wetmore Sherman, 570
Brown, Harold, 570
Brown, Hope, 270
Brown, Iri, 285
Brown, James (Foster), 282
Brown, James (Pawtucket), 160
Brown, John, 64, 86, 90, 91, 104, 270
Brown, John Carter, 91, 101, 270
Brown, John Nicholas, 17, 90, 91, 96, 514–515, 521
Brown, Joseph, 64, 85, 89, 97, 118, 197, 270
    Brown (John) House, Providence, 8, 11, **89**–90, 91, 92, 122, 177
    Brown (Joseph) House, Providence, **63**, 64, 66, 118, 190, 197
    First Baptist Church, Providence, 8, 9, 53, 62, **66**–68, 88, 97, 115, 265, 535
    Market House, Providence, 60, 65, 66, 306
Brown, Moses, 64, 137, 144, 189, 270
Brown, Natalie Bayard, 514
Brown, Nicholas (I), 64, 270
Brown, Nicholas (II), 67, 91, 99, 189, 190, 209
Brown, Obadiah, 244
Brown, Sylvanus, 137, 144, 154, 156
Brown, Thomas, 282
Brown, William F., 255
Brown and Hopkins Store, Glocester, 294
Brown and Ives (investors; mill operators), 190, 208, 211, 270, 271, 272
Brown and Sharpe Machine Tool Company complex, Providence, 34, 105, 125
Brown brothers, 270
Brown Building, Westerly, 413–414
Brown College. *See* Brown University
Brown (Deacon Kent) House, Barrington, 448
Brown family (Foster), 285
Brown family (Providence), 28, 49, 64, 66, 90, 91, 104, 172, 189, 190, 205, 355, 580
Brown-Whidden-Fuller Farm, Foster, 282, 491
Brown (Harold Carter) House, Newport, 570
Brown Homestead, Scituate, 267
Brown House, Smithfield. *See* Farnum (Joseph)–Brown House, Smithfield
Brown House replica, Barrington. *See* Nightingale-Brown House replica, Barrington
Brown (Iri) Farm, Foster, **285**
Brown (Isaac) House, 62, 63

Brown (James)–Barker (Samuel) House, Newport, 533–534
Brown (John Carter) House, East Kingstown. *See* Scalabrini Villa (John Carter Brown House), North Kingstown
Brown (John) House, Providence, 8, 11, **89**–90, 91, 92, 122, 177. *See also* Rhode Island Historical Society (John Brown House), Providence
Brown (John Nicholas) House (Harbour Court), Newport, 17, 580
Brown (Joseph) House, Providence, **63**, 64, 66, 118, 190, 197
Brown (Nicholas) House. *See* John Nicholas Brown Center for the Study of American Culture (Joseph Nightingale–Nicholas Brown House)
Brown University, Providence, 8, 33, 66, 67, 68, 72, 75, 96, 97–108, 118, 247, 289, 479, 581
  Admissions Office (George Corliss House), 104
  Albert and Vera List Art Building, 84, **85**, 94, 103
  Augustus Stout Van Wickle Gate, 97
  Carrie Tower, 99
  Center Campus and north, 97–109
  Center Campus and north, map, 98
  College Green, 8, 68, 99, 100
  Development Office, 58
  Division of Applied Mathematics (Henry Pearce House), 119
  Dyer Hall (John Holden Greene House II), 118
  Faculty Club (Zachariah Allen House), 87
  Faunce House (Rockefeller Hall), 101
  founding of, 463, 464
  Froebel Hall, Providence., 108
  Goddard (William Giles)-Iselin (Hope Goddard) House, **96**–97
  Hillel House (Froebel Hall), 108
  Hope College, 99, **100**
  John Carter Brown Library, **101**
  John D. Rockefeller, Jr., Humanities Library, 103, 104
  John Hay Library, 103
  King Hall (Robert W. Taft House), 118, 119, 120
  Library Buildings, 102–104, **103**
  Lyman Gymnasium, 101–102
  Machado House (Ellen Dexter Sharpe House), 105
  Maddock Alumni Center (William Giles Goddard-Hope Goddard Iselin House), **96**–97
  Manning Hall, 99, **100**, 102
  map, 98
  Nicholson House (Francis W. Goddard-Samuel C. Nicholson House), 96, 97
  Pembroke College, 75, 97
  Pembroke Hall, 107–108
  President's House, (Rush Sturges House), 88–89
  Rhode Island Hall, 99
  Robinson Hall (University Library), 102–**103**, 347
  Rochambeau House (Henry Dexter Sharpe House), 105–106
  Rogers Hall, 100, 101
  Sayles Hall, 99–100, 107, 194
  Sciences Library, 103–104
  Slater Hall,, 99
  Soldier's Memorial Gateway, 102
  and Tefft drawings, 47, 49, 81, 93, 166, 261, 367
  University Hall, 97–99, **100**, 101, 102, 108
  Watson Institute for International Studies, 94–95
  Wilson Hall, 100–101, 384
Brownell Farm, Little Compton, 491
Brownell Library, Little Compton, 496
Brownell (Stephen) House, North Smithfield, 241–**242**
Bruce G. Sundlun Terminal, T. F. Green State Airport, Warwick, 319

Bruen family, 570
Bruen (Mary) House (The Hedges), Newport, 570
Bryant, Gridley J. F.: Drowne (George R.) House, Providence, 76
Bucklin, James C., 28, 43, 44, 65, 76, 77, 84, 103, 156, 260, 447
  Arcade, The, Providence, 37, **46**–47
  Benefit Street Arsenal, Marine Corps of Artillery, Providence, 77–78, 348
  Hallworth House (Enoch W. Clarke-John Slater House), Providence, 75
  Hay Block, Providence, 64
  mill workers' boardinghouses, Smithfield, **260**
  Monohasset Mill, Providence, 125–126
  Rhode Island Hall, Brown University, Providence, 99
Bucklin Memorial Building, Hopkinton, **421**
Bucklin (William and George) Houses, Providence, 94
Buffum, Richard, 244
Buffum (Horace) House. *See* Page (Martin)–Buffum (Horace)–Gardner (George W.) House, Providence
Builder (periodical), 232
Builder's Industrial Foundry, 64
Builders Iron Foundry, 113
Bulfinch, Charles, 44, 46, 71, 88, 413
Buliod (Peter)–Perry (Oliver Hazard) House (Rhode Island Bank), Washington Square, Newport, 520
Bull family, 530
Bull (Jireh) House, Newport, 537
Bullock, Jabez: Thomas (Allen Mason) House, Wickford, North Kingstown, **359**
Bullock, Orin M., Jr., 531
Bullock, Rufus Brown, 399
Bullock Point cottages, East Providence, 446
Bumble Bee Farm. *See* Frenning House (Bumble Bee Farm), Tiverton
bungalows
  Lincoln, 194
  North Providence, 163
  6 High Street, Ashaway, Hopkinton, 423
  Westerly, 415
  West Warwick, 327
Burbank (Thomas) House, Smithfield, 259
Burchard, Corinne Richmond, 495
Burchard (Edith Russell) House. *See* Church (John)–Burchard (Edith Russell) House (Old Acre), Tiverton
Burchard House. *See* Seabury-Richmond-Burchard House, Little Compton
Burgess Building, Providence, 52
Burgin, William: McIntyre (Jerry) House, Jamestown, 598–599
Burgoyne, John, 460
Burleigh, Sydney, 69, 70, 110, 492, 493
Burlingame family, 218
Burlingame-Noon House, Cumberland, 218
Burlingame State Park, Charlestown, 397
Burlington, Lord, 545
Burnham, Daniel, 49
Burnside, Ambrose, 36, 88, 473
Burnside (General Ambrose) House, Providence, 88, 120
Burnside Memorial Hall, Bristol, 473
Burnside monument, Providence, 36
Burrill, James, 301
Burrill Building, Providence, 52
Burrill (Joseph) House, Newport, 541
Burrillville, 127, 245, 253, 290, 291, 299–315, 381, 424
  map, 300
Burrillville United Methodist Church, Burrillville, 306
Burrington Memorial Chapel, St. John's Episcopal Church (Red Church), Barrington, 452

Burrough, James, 118
Burrough (James) House, Providence, 118
Burrough (Robert S., Jr.) House. *See* Museum of Rhode Island History, Rhode Island Historical Society (Robert S. Burrough, Jr.–Senator Nelson W. Aldrich House), Providence
Burrough (Robert S.) House, Providence, 117
Burton Primary School, Bristol, 476
Busbee, James, 243
Butler, Cyrus, 46
Butler Hospital, Providence, 121
Butler House, Newport, 538, **539**
Butterfield, [first name unknown]: First Baptist Church, Woonsocket, 231–232
Butterfly Mill, Lincoln, 197–198
By-the-Sea, Watch Hill, Westerly, 407
Byfield School, Bristol, 477

Cady, George Maxwell: Andrews Memorial School, Bristol, 471
Cady, George Waterman: Burgess Building, Providence, 52
Cady, John Hutchins, 122, 133, 390
    Market Square (Market House), Rhode Island School of Design, Providence, 65
    Packard Motor Company Showroom (former), Providence, 42
    workers' housing, Burrillville, 310
Caesar Misch Building, Providence, 53, 54
Calder, Albert, 113
Calder (Albert L.)–Robertson (R. Austin) House, Providence, 112, 113
California, 505, 578
Callender School, Newport, 529
Calvary Episcopal Church, Burrillville, 314
Cambridge, Massachusetts, 116, 547, 601
Cambridge University, England, 107
    Trinity College, 514
Camp Hoffman (Girl Scout Camp), South Kingstown, 394
Campbell, Colen, 64, 521
Campbell (Horatio N.) House, Providence, 109
Canal Street Row, RISD Buildings, Providence, 65
Candela, Felix, 192
Cannelton, 28
Canonchet (Narragansett leader), 153, 159
Canonchet, Hopkinton, 418, 422
Canonchet Chapel. *See* St. Elizabeth's Episcopal Church (Canonchet Chapel), Hopkinton
Canonchet Farm, Narragansett, 367–368
Canonicus (Native American chief), 351
Capital Center, Providence, 33, 34–55, 61
    map, 35
Captain Israel Inman's Inn (Simon Sweet House), Glocester, 295–296
Card (Captain William) House. *See* Lewis (John)–Card (Captain William) House, Westerly
Card (William) House, Newport, 538–539
Carey, John, Jr., 581
Carey (John, Jr.) Gardener's Cottage, Newport, 581
Carnegie libraries, 158, 441
Carolina, Charlestown/Richmond, 397, 401, 424, 427
Carolina Free Baptist Church, Charlestown, 401
Carolina Mills, Charlestown/Richmond, 427
carousels
    Crescent Park, East Providence, 151, 445–446
    Flying Horse Carousel, Westerly, **406**
    Slater Park, Pawtucket, 151

Carpenter, Charles, 151
Carpenter, Francis W., 114
Carpenter (Asa) House, Pawtucket, 151
Carpenter (Francis W.) House, Providence, 115
Carpenter's Beach, East Providence, 27
Carr, Caleb (Captain), 456, 457
Carr (Captain Caleb) House, Bristol. *See* Carr (Major Caleb)–Carr (Captain Caleb) House, Bristol
Carr (Dr. George W.) House, Providence, 79
Carr family, 586
Carr Homestead, Jamestown, 586, 591
Carr Homestead Foundation, 586
Carr (Major Caleb)–Carr (Captain Caleb) House, Bristol, 456–457, 461
Carrère and Hastings, 15
    Carpenter (Francis W.) House, Providence, 115
    Central Congregational Church (second), Providence, 114–**115**
Carriage House, Middletown, 513
Carrie Tower, Brown University, Providence, 99
Carrington, Edward, 92
Carrington (Edward) House. *See* Corliss (John)–Carrington (Edward) House, Providence
Carter, John, 70
Caruso, Enrico, 27, 403
Caruso Road, Shelter Harbor, Westerly, 403
Carver (Emma B.) House (Kabyun, Sonnenschein), Narragansett, 369
Case (Immanuel) House, Wickford, North Kingstown, 358
Casey, Edward, 364
Casey, Silas, 364
Casey, Thomas Lincoln, 364
Casey family, 364
Casey (Silas) Farm, North Kingstown, **364**
Casino, The, Newport, 16, 371, 373, 555, 562, 563–**564**, 565, 567
casinos
    Jamestown Casino (Shoreby Hill Club), Jamestown, 603
    Kinney's Casino, Narragansett, **373**
    Newport, 16, 371, 373, 555, 562, 563–**564**, 565, 567
    Providence, 133
    Towers, Narragansett Pier Casino, Narragansett, 368, 370, 371, 381
cast iron front, Westerly, 413
Castellucci and Galli: Maintenance Center, T. F. Green State Airport, Warwick, 319–**320**
Castle Hill. *See* Agassiz (Professor Alexander) House (Castle Hill), Newport
Caswell, Philip, Jr., 595
Catalano, Eduardo, 192
Cathedral of St. John (St. John's Church), Providence, **71**, 76, 78–**79**, 88
Cathedral of Saints Peter and Paul, Providence, **54**–55, 160
Cathedral Square, Providence, 55
Catholic churches. *See* churches, Roman Catholic
Catholics. *See* Roman Catholics
Cato Hill, Woonsocket, 236
Cattanach and Cliff (decorative painters), 146, 147
CCC. *See* United States Civilian Conservation Corps
Cedar Grove, East Providence, 435
Centennial Exposition, Philadelphia (1876), 68, 104, 108, 524
Center Place, Providence, 40
Centerbrook (architects), 87
Centerville, Johnston, 169
Centerville, West Warwick, 324, 331–332

Central Congregational Church (first), Providence. *See* Memorial Hall, RISD (Central Congregational Church), Providence
Central Congregational Church (second), Providence, 114–**115**
Central Falls, 135, 138, 149, 153–160, 186–187, 188, 189, 192, 202, 206, 208, 222, 238, 249, 424
 Adams Public Library, 158
 Fire District, 154, 155
 map, 156
 Mill Owners Association, 154
Central Falls (Roosevelt Avenue) Mill Complex, 155–156, 159
Central Falls Woolen Mill, 155
Central Hotel, Glocester, 294
Central Hotel (former), Burrillville, 312–313
Central Pike, 170, 265, 274, 342
Central Street houses, Narragansett, 369
Centredale, 160, 161, 249
Centreville National Bank, West Warwick, **331**
Centreville (North Providence), 169
Chace, Beatrice O., 75
Chace, George A., 481
Chace, Samuel B. and Harvey, 154, 159, 160, 188–189, 201, 202, 203, 208
Chace (Jonathan) House, Newport, 527
Chadwick (John) House and adjacent house, Newport, 526–527
Chamberlain, William: Calder (Albert L.)–Robertson (R. Austin) House, Providence, 112, 113
Chamberlin, Samuel, 490
Chambers, William, 574
Champlin (H.) House. *See* Bates (Stephen and Martha)–Champlin (H.) House, Charlestown
Champlin Manufacturing Company Building, Providence, 57
Champlin (Peleg) Farm, Block Island, 608
Channel Bells. *See* Patterson (General Robert E.) House (The Ramparts, Channel Bells), Jamestown
Channing, William Ellery, 544
Channing Memorial Church, Newport, 544
Chapel, Portsmouth Abbey School, Portsmouth, 501–502
Chapel, St. George's School, Middletown, 514
Chapman and Frazer: Tredegar (Russula), Watch Hill, Westerly, 408
Charles II, king of England, 466
Charles A. Maguire Associates: East Providence Senior High School, East Providence, 443
Charles Borders and Jed Dixon: Reproduction Covered Bridge, Foster, 275
Charles T. Main, Inc.: Bostitch company, East Greenwich, 349–345
Charles W. Dare Company: Flying Horse Carousel, Westerly, 406
Charles W. Shea Senior High School (Pawtucket West High School), Pawtucket, 139, 145
Charleston, South Carolina, 26, 34, 75, 454, 467, 491
Charlestown, 365, 374, 375, 397–403, 424, 427
 map, 398
Charlestown Village. *See* Cross Mills, Charlestown
Chase, Charles: Kenyon's Department Store, South Kingstown, 379
Chase family, 486
Chase Farm, Lincoln, 197
Chase House, Glocester, **298**
Chase House, Tiverton. *See* Tiverton Historical Society (Chase-Cory House), Tiverton
Château-Nooga. *See* Baldwin (Christopher Columbus) House (Château-Nooga, Chatterbox), Newport

Château-sur-Mer. *See* Wetmore (William Shepherd) House (Château-sur-Mer), Newport
Cheapside, RISD Buildings, Providence, 65
Chepachet, Glocester, 78, 290, 291, 292–299, 301, 302
 Main Street, 293–295
Chepachet Freewill Baptist Church and Parsonage, Glocester, 253, 265, **293**, 294, 295
Chepachet River, 290, 292, 301, 302
Chepachet School, Glocester, 294
Chepachet Union Church, Glocester, 283, 293, **294**–295
Chepstow. *See* Schermerhorn (Edward H.) House (Chepstow), Newport
Cherry Brook, 219
Chestnut Hill Baptist Church, Exeter, **431**
Chestnut Street, Providence, 55
Chicago, Illinois, 45
Child (Reverend William S.) Schoolhouse, Newport, 526
Children's Museum, Pawtucket/Providence, 148
Childs, Henry (builder): Lippitt (Henry) House II, Providence, 109, **110**
China Trade Mansions, Providence, 89, 90-92
Chinese Teahouse, Vanderbilt (Alva Smith and William Kissam) House, Newport, 573, 574
Chocolate Shop, The, Warren. *See* Hall's Block (The Chocolate Shop), Warren
Chocolateville. *See* Central Falls
Choice Acres, Johnston, 172
Christ Church, Lonsdale, 190, **191**, 192
Christ Episcopal Church, Westerly, 410, 411
Christian Church, Westerly. *See* Granite Theater (Christian Church), Westerly
Christian Science Church, Providence, 105
Christopher Chadbourne Associates, 238
Church, Isaac, 482
Church, J. Winfield: Narragansett Town Hall (Fifth Avenue School), 370
Church, Joseph, 482
Church, Samuel W., 479
Church and Sweet (builders), 491
Church brothers, 482
Church (Captain Daniel) House, Tiverton, 482
Church (Captain George L.) House, Tiverton, 482
Church (Captain Isaac) House, Tiverton. *See* St. Christopher's Roman Catholic Church Rectory (Captain Isaac Church House), Tiverton
Church (Captain Nathaniel) House, Tiverton. *See* St. James Convent (Captain Nathaniel Church House), Tiverton
Church (John)–Burchard (Edith Russell) House (Old Acre), Tiverton, **491**
Church of England. *See* churches, Episcopal; Episcopalians
Church of the Blessed Sacrament, Providence, 126
Church of the Holy Cross, Middletown, 367, 506, 509, **510**, 511
Church of the Holy Family, Woonsocket, 230
Church of the Messiah, Foster, 287
Church of the Precious Blood, Woonsocket. *See* Précieux Sang (Church of the Precious Blood), Woonsocket
Church (Samuel)–Hathaway (Dr.) House, Little Compton, **495**, 496–497
 Water Tower, **497**
Church Street Bridge, South Kingstown, 383
churches
 Avondale Chapel, Westerly, 408
 Channing Memorial Church, Newport, 544
 Chapel, Portsmouth Abbey School, Portsmouth, 501–502
 Chapel, St. George's School, Middletown, 514

Index 655

Chepachet Union Church, Glocester, 283, 293, **294**–295
Christian Science Church, Providence, 105
Church of the Messiah, Foster, 287
Escoheag Advent Church, West Greenwich, 343
First Christian Church, Coventry, 340
Granite Theater (Christian Church), Westerly, 410, 411
Hopkins Hollow Church (Christian Union), Coventry, **340**
Hopkins Mills Union Church (South Foster Union Chapel), Foster, 288, **289**
Mount Vernon Christian Church (Mount Vernon Baptist Church, Friends Meeting House), Foster, 266, 279, 283–284, 286, 487
Narragansett Indian Church, Charlestown, 402
St. George Maronite Catholic Church (St. Paul's Episcopal Church), Pawtucket, 146–147
Seventh-Day Adventist Church (Mapleville Methodist Episcopal Church), Burrillville, 304
summer chapels, Narragansett, 369
See also meeting houses; synagogues
churches, Baptist
Baptist Church (Allendale Schoolhouse and Community Hall), North Providence, **165**–166
Carolina Free Baptist Church, Charlestown, 401
Chepachet Freewill Baptist Church and Parsonage, Glocester, 253, 265, **293**, 294, 295
Chestnut Hill Baptist Church, Exeter, **431**
Dance Studio (Lonsdale Baptist Church), Lincoln, 192
First Baptist Church, Bristol, 477
First Baptist Church, Burrillville, 314
First Baptist Church, East Greenwich, **346**–347
First Baptist Church, East Providence, 442
First Baptist Church, Providence, 8, 9, 53, 62, **66**–68, 88, 97, 115, 265, 535
First Baptist Church, Tiverton, 486–487
First Baptist Church, Woonsocket, 231
First Baptist Church of Charlestown, 400, 402
First Baptist Church of Cross Mills, Charlestown, 399
First Baptist Church Parish House (Masonic Hall), Wickford, North Kingstown, 358
First Seventh Day Baptist Church, Ashaway, Hopkinton, 423
First Seventh Day Baptist Church, Hopkinton City, Hopkinton, 422–423
Foster Center Baptist Church (Foster Center Christian Church), Foster, 278–279, 280, 284, 487
Free Will Baptist Church (First Baptist Church, Old Stone Church), Tiverton, 486–487
Freewill Baptist Church, Smithfield, 253, 265, 293, 294
Frenchtown Baptist Church, East Greenwich, 350
Hopkins Hollow Church, Coventry, **340**
Lime Rock Baptist Church, Lincoln, **201**, 287
Newport Historical Society and Seventh Day Baptist Meeting House, 7, 543
North Foster Free Baptist Church, Foster, 266, 279, 283, 284, 287, 288–289
Old Stone Church, Tiverton, 486–487
Pascoag Community Baptist Church (First Baptist Church), Burrillville, 314
Pawcatuck Seventh Day Baptist Church, Westerly, 410–411
Quidnesset Baptist Church, North Kingstown, 356
residence (Baptist Church), Narragansett, 369
Roger Williams Baptist Church, Providence, 128
Second Seventh Day Baptist Church, Hopkinton, 423
Seventh Day Baptist Church, Hopkinton, 419, **421**
Union Free Will Baptist Church (Line Baptist Church), Foster, 283, 286
Victory Baptist Church (Wood River Baptist Church; Six Principle Baptist Church), Richmond, **428**
Wakefield Baptist Church, South Kingstown, 381
Warren Baptist Church, 463–464
West Exeter Baptist Church (West Greenville Branch Baptist Church), Exeter, 430
West Greenwich Center, 343
churches, Congregational
Barrington Congregational Church, **449**
Beneficent Congregational Church (Round Top Church; formerly Second Congregational Church), Providence, **43**–44, 45
Central Congregational Church (second), Providence, 114–**115**
Congregational Church, Bristol, 25, **477**–478
Congregational Church, Newport, 83
Congregational Church, Pawtucket, 83
First Congregational Church, Newport, 542
First Congregational Church, Providence, 43
Kingston Congregational Church and Parish House (Thomas P. Wells House), South Kingstown, 390–391, 394
Moosup Valley Congregational-Christian Church and Cemetery, Foster, 284, 285, 288
Narragansett Congregational Church. *See* South Ferry Congregational Church (Narragansett Congregational Church), Narragansett
Newman Congregational Church, East Providence, 433, **440**
Newport Congregational Church, 83, 147, **53**6–537
North Scituate Congregational Church, Scituate, **265**–266, 279, 284, 293, 312, 487
Park Place Congregational Church, Pawtucket, 143
Pawtucket Congregational Church, **147**–148, 149
Peace Dale Congregational Church, South Kingstown, **382**, 384, 386
Second Congregational Church (former), Newport, 541
Slatersville Green and Congregational Church, North Smithfield, 245, **246**
South Ferry Congregational Church (Narragansett Congregational Church), Narragansett, 367, 509
Union Congregational Church (former), Newport, 539
United Congregational Church, Little Compton, **495**, 496
churches, Episcopal
All Saints Episcopal Church, Warwick, 323
Calvary Episcopal Church, Burrillville, 314
Cathedral of St. John (St. John's Church), Providence, **71**, 76, 78–79
Christ Church, Lonsdale, 190, **191**, 192
Christ Episcopal Church, Westerly, 410, 411
Church of the Holy Cross, Middletown, 367, 506, 509, **510**, 511
Emmanuel Church, Newport, 581
Grace Church, Providence, **53**, 54, 82, 95, 96, 254
Grace Episcopal Church, East Providence, 438
Holy Cross Church, Middletown, 506
Old Narragansett Church (Old St. Paul's Church), Wickford, North Kingstown, 360, 361
Red Church, Barrington. *See* St. John's Episcopal Church (Red Church), Barrington
St. Colomba's Chapel (Berkeley Memorial Chapel), Middletown, 181, 348, **511**–512, 514
St. Elizabeth's Episcopal Church (Canonchet Chapel), Hopkinton, 422
St. James Episcopal Church, North Providence, **162**–163

churches, Episcopal *(continued)*
  St. John the Evangelist Episcopal Church, Newport, 524–525
  St. John's Episcopal Church (Red Church), Barrington, 451, 452, 453
  St. Luke's Episcopal Church, East Greenwich, 347–348
  St. Mark's Episcopal Church, Warren, 459
  St. Mary's Church, Portsmouth, **506–507**, 509, 511
  St. Matthew's Episcopal Church, Barrington, 451
  St. Michael's Episcopal Church, Bristol, **473**
  St. Paul's Church, Wickford, North Kingstown, 358–359, 360
  St. Paul's Episcopal Church, Portsmouth, 503
  St. Peter's by the Sea II, Narragansett, 369
  St. Peter's Episcopal Church I, Narragansett, 369
  St. Stephen's Church, Providence, 82, **95–96**, 104, 254
  St. Thomas Episcopal Church, Richmond, 422, 427, **428**
  St. Thomas Episcopal Church, Smithfield, **254**–255
  Sprague Meeting House (St. Bartholomew's Episcopal Church), Cranston, 179
  Trinity Church, Newport, 7, 9, 68, 519, 534–**535**, 536, 543, 556
  Zion Episcopal Church (former), Newport, 521
churches, Lutheran
  Gloria Dei Evangelical Lutheran Church, Providence, 124–125, 230
  Swedish-Finnish Lutheran Church (former), Woonsocket, 230–**231**
churches, Methodist
  Arnold Mills Methodist Church (United Baptist Church), Cumberland, 215
  Burrillville United Methodist Church, 306
  First United Methodist Church, Warren, 460–461
  Mapleville Methodist Episcopal Church (former), Burrillville, 304
churches, Presbyterian: First Presbyterian Church, Newport, 516, 518
churches, Roman Catholic, 223
  Cathedral of Saints Peter and Paul, Providence, **54**–55, 160
  Church of the Blessed Sacrament, Providence, 126
  Church of the Holy Family, Woonsocket, 230
  Our Lady of Good Help Church, Rectory, and Parish-School (Notre Dame de bon Secours), Burrillville, **304**
  Portsmouth Abbey School Chapel, Portsmouth, 501–502
  Précieux Sang (Church of the Precious Blood), Woonsocket, 227–228, 235–236
  Sacred Heart Church, West Warwick, 330
  St. Ann's, Woonsocket, 160, **223**, 224, 228, 230
  St. Charles Borromeo, Woonsocket, 227–228, 235–236
  St. Francis of Assisi Roman Catholic Church, South Kingstown, 380
  St. Jean-Baptiste Church, Pawtucket, 140
  St. Joseph's Church, Cumberland, 212, **213**, 228
  St. Joseph's Church, Providence, 54
  St. Jude's Roman Catholic Church, Lincoln, 191, 192
  St. Mary's Church, Newport, 581
  St. Matthew's Roman Catholic Church, Central Falls, 160, 175
  St. Rocco's Church, Johnston, 175
  St. Thomas More Roman Catholic Church (St. Philomene's Roman Catholic Church), Narragansett, 369
churches, Unitarian Universalist
  Bell Street Chapel, Providence, 130
  First Unitarian Church, Providence, 43, 53, 71, 78, **87–88**, 115, 170

  Universalist Church, Burrillville, 311–**312**
Churchill, Benjamin, 474
Churchill-Yarnell House, Newport, 556
Cianci, Vincent A., Jr., 38
Cincinnati, Ohio, 407
Cipolla, Peter, 181
Citizens Bank Building, Providence, 40
city and town halls
  Barrington Town Hall, 447, 451–**452**
  Cumberland Town Hall, 208–209, 321
  East Greenwich Town Hall (Kent County Courthouse; East Greenwich State House), 345, 346, 348
  Harrisville Town Hall, Burrillville, 312
  Little Compton Town Hall, 496
  Narragansett Town Hall (Fifth Avenue School), 370
  Newport City Hall, 518
  Old Town House, Exeter, 430
  Pawtucket City Hall, Police Department, and Fire Department, 139, 145
  Providence City Hall, **33**, 36, 37, 38, 65
  Richmond Town Hall, 428
  South Kingstown Town Hall, 380–381
  Warren Town Hall, 321, 459–460
  Warwick City Hall, 208, **320**–321
  Woonsocket City Hall (Harris Institute), 236, 237
City Beautiful movement, 36, 151, 159
City of Providence Water Supply Board, 263, 272
  Philip J. Holton Water Purification Plant, Scituate Reservoir, 272
Civic Center, Providence, 39, 41
Civil War, 26, 36, 50, 58, 72, 116, 137, 155, 156, 159, 161, 187, 244, 328, 364, 385, 396, 399, 450, 475, 477, 592, 600, 608, 609
Civil War monuments
  Providence, 36
  Woonsocket, 222, 236
Civilian Conservation Corps. *See* United States Civilian Conservation Corps
Claggett, William, 543
Claggett (Caleb) House, Newport, 528, 532
Clarendon Court, Newport. *See* Knight (Edward) House (Clarendon Court), Newport
Clark, Earl C., 600
Clark, R. G., 359
Clark family, 426, 427
Clark-Herreshoff House, Bristol, 475
Clark (John F.) House, Cumberland, **209**
Clark (Samuel) Farm, Richmond, **425**
Clark Shipyard, Jamestown, 600
Clarke, Elizabeth, 601
Clarke, Enoch, 75
Clarke, John (Portsmouth founder), 498
Clarke, John Innis (industrialist), 91
Clarke, P. O., 78
Clarke, Peleg, Jr.: First Baptist Church of Charlestown, Charlestown, 400
Clarke and Howe, 101, 347
  Federal Building (court), Providence, 36, 37
  Guiteras Memorial and Junior High School, Bristol, 469
  Insurance Building, Providence, 65–66
  Lisle (Frank B.) House, Providence, 115, 116
  RISD "Bank Building" (People's Savings Bank), Providence, 65
  See also Clarke and Howe; Clarke, Spaulding and Howe
Clarke and Nightingale Row, Providence, 61
Clarke and Spaulding: Kinney's Casino, Narragansett, **373**
Clarke (Elizabeth) House, Jamestown, 601
Clarke (Enoch W.)–Slater (John) House. *See* Hallworth

House (Enoch W. Clarke–John Slater House), Providence
Clarke, Spaulding and Howe: American Textile Mill, Pawtucket, 140
Clarkville, Glocester, 291–292
Clarkville School, Glocester, 291, 297–298
Clayville, Scituate/Foster, 263, 274
Clear River, 301, 302, 309, 310, 313
Clemence, Thomas, 171
Clemence (Thomas)–Irons House, Johnston, 170, **171**, 172, 184, 281
Cleveland, Horace W. S. (landscape architect), 120, 121
    Roger Williams Park, Providence, 34, 70–71, 92, 132–134, 151, 177, 183
Cleveland, Ohio, 609
Cliff Cottage Association, Newport, 586
Cliff Cottage Hotel Association, Newport, 558
Cliff Walk, Newport, 15, 557, 575
Clingstone. *See* Lovering Wharton (J. S.) House (Clingstone), Jamestown
clubs
    Agawam Hunt, East Providence, 435, 437
    Dunes Club Gatehouse and Cottages, Narragansett, **368**, 371
    Hope Club, Providence, 87
    New York Yacht Club Newport Facilities, 17, 580
    Newport Country Club, 17, **578**–579
    Newport Reading Room, 373, 554–555
    Squantum Association, East Providence, 435, **444**–445
    University Club (Rufus and Emily Waterman House), Providence, 78
    Wannomoisett Country Club, East Providence, 435
    See also casinos
coal miners' houses, Portsmouth, 500
Coast Guard stations
    Point Judith Lighthouse, 373
    U.S. Lifesaving Station, Narragansett, 371
Coastal Resources Institute, University of Rhode Island, South Kingstown, 392
Coaster's Harbor Island, 20
Coats American Bristol Plant, Bristol, 478
Coats and Clark (Conant Thread Mill Complex, later J. and P. Coats Ltd.), Pawtucket, 137, 138, 140–141, 158
Cobblestone House, Smithfield, 251
Cocroft (Freeman) House (Croftmere), South Kingstown, 396
Cocumscussoc (Smith's Castle), North Kingstown, 351, 356–357
Coddington, William, 498, 499, 528, 529, 531
Coddington (Gov. William) House, Newport, 528
Coddington (Nathaniel)–Walker (Thomas) House (King's Arms Tavern), Newport, 529
Codman, Catherine Elizabeth and Maria Potter, 475
Codman, Henry, 475
Codman, Martha, 566, 567
Codman, Ogden, 15, 566, 567, 571, 575, 576
    Brown (Harold Carter) House, Newport, 570
    Codman (Martha) House (Berkeley Villa, Bellevue House), Newport, 566–567
    Pomeroy (Edith Burnet) House (Seabeach), Newport, 576–577
Codman House, Bristol, 475–476
Codman (Martha) House (Berkeley Villa, Bellevue House), Newport, 566–567
Coffin, Marian, 406
Coggeshall, Daniel, 364
Coggeshall House. *See* Billings-Coggeshall House, Newport
Cogswell, Caroline, 158, 159

Cohan, George M., 42
Coit (Joseph) House, Bristol, 475
Cole, J. R., 385
Cole, James Vance, 463
Cole (Thomas) House, Warren, 462
Coleman, Samuel, 550
Coleman (Samuel) House (Whileaway, Boxcroft), Newport, 550–551
College Building, Rhode Island School of Design, Providence, 64, 65, 78
College Green, Brown University, Providence, 8, 68, 99, 100
College Hill, Providence, 8, 11, 33, 34, 43, 49, 60, 61, 63, 64, 70, 72–122, 124, 128, 131, 148
College Hill: A Demonstration Study of Historic Area Renewed (Providence Preservation Society), 72–73
colleges and universities
    Brown University, Providence, 8, 33, 66, 67, 68, 72, 75, 84, **85**, 87, 88–89, 92, 94–95, **96**–109, 118, 119, 120
    Brown University Development Office, Providence, 58
    Brown University Green, Providence, 8, 68
    Community College of Rhode Island Knight Campus (Rhode Island Junior College), Warwick, **323**–324
    Johnson and Wales University, Providence, 42, 45, 54
    Pembroke College, Providence, 75
    Providence College, Providence, 34, 126–127
    Rhode Island School of Design, Providence, 33, 47, 62, 64–66, 68–69, 72, 77, 78, 79–83, 107, 128, 277, 296, 396
    Roger Williams University, Bristol, 25, **480**
    Salve Regina University, Newport, 18–19, 560, 561, 571
    University of Rhode Island, Alton Jones Campus, West Greenwich, **343**
    University of Rhode Island, South Kingstown, 388, 389, 391–392
    University of Rhode Island College of Continuing Education, Providence, 52
    Zion Bible College (Frederick Stanhope Peck House, known as Belton Court), Barrington, 449–**450**
Collins, Benny, 125
Collins, Captain, 456
Collins, Henry, 527, 545
Collins (Beriah) House. *See* Twin Pond Farm (Beriah Collins House), Foster
Collins, Flagg and Engs, 527
Collins (Henry) House, Newport. *See* Warren (Captain John) House (Henry Collins House), Newport
Collins House. *See* Hill-Collins House, Warren
Colonial House, The. *See* Pendleton House (The Colonial House) and Waterman Galleries, Providence
Colonial Revival house, Narragansett, 372
Colony House, Newport, 6, 77, 306, 345, 365, 388, 390, 515, 516, **519**–520, 521, 523, 532, 534, 542
Colt, Samuel Pomeroy, 467, 469, 471, 472, 474
Colt-Andrews School (Colt Memorial High School), Bristol, 469, 471
Colt family, 471
Colt House. *See* Linden Place (DeWolf-Colt House), Bristol
Colt Memorial High School. *See* Colt-Andrews School (Colt Memorial High School), Bristol
Columbia Street Bridge, South Kingstown, 383
Columbus, Ohio, 569
Colwell, Elizabeth, 299
Colwell, Richard, 278
Colwell, William E., 163
Colwell Farm. *See* Steer-Angell-Colwell Farm, Gloucester
Colwell House. *See* Tucker-Stone-Colwell House, Smithfield

Comeau, Mark: Tribal Lands Health Center, Charlestown, 402
Commercial Bank Building, Warren, 471
commercial block (Forestdale), North Smithfield, 243–244, 247
commercial block (Slatersville), North Smithfield, 243–244, **246**, 247
Common Burying Ground, Newport, 530
Common Fence, Portsmouth, 499
commons
  Bristol, 25, 466, 476–477
  Jamestown, 587
  Little Compton, 492, 495–496
  Newport (1970s), 9
  Portsmouth, 499
  Slatersville Green, North Smithfield, 245, 246
Community Center (Potterville District 7 Schoolhouse), Scituate, 269
Community College of Rhode Island Knight Campus (Rhode Island Junior College), Warwick, 323–324
company housing. See worker housing
Company Store and Post Office. See Day Care Center (Company Store and Post Office), Smithfield
Compendium of Agriculture, or the Farmer's Guide (Drown and Drown), 289
Compris, Maurice, 236
Comstock, Samuel, 239
Comstock Row, Providence, 61
Conanicut. See Jamestown
Conanicut Island, 20, 28, 387, 550, 582, 592, 595, 596, 597
  and retired naval officers, 27
Conanicut Park, Jamestown, 586–588, 589, 595
  Farr (Eleanor H.) Cottage, 587–588
  Fletcher (Charles) House, 588
  Hoyt (David M.) Cottage, 588
  Irons (Samuel) Cottage (Hendry's Retreat), 588
  Kilton (John B.) Cottage, 587
  Lippitt (Jennie) Cottage (Stonewall Cottage), 588
  Taber (George) Cottage, 587
Conanicut Park Hotel, Jamestown, 586
Conanicut Point Light (North Light), Jamestown, 583, 588–589
Conant, Hezekiah, 141
Conant (Samuel M.) House, Central Falls, 157–158
Conant Street Mill, Pawtucket, 141
Conant Thread Company, 158
Conant Thread Mill Complex. See Coats and Clark (Conant Thread Mill Complex, later J. and P. Coats Ltd.), Pawtucket
Condit, Frederick, 112
condominiums and offices (Allendale Mill and Company Store), North Providence, 164–165, 173, 260, 261
condominiums (Second Congregational Church), Newport, 541
condominiums (Walter Rodman House), North Kingstown, 363
Congdon, Henry W.: Christ Episcopal Church, Westerly, 410, 411
Congdon House. See Perry-Congdon House, Bristol
Congdon (Rufus and Mary) House, Wickford, North Kingstown, 357
Congdon (William) House (Brookfield), South Kingstown, 397
Congregation B'nai Israel Synagogue, Woonsocket, 233
Congregational Church, Bristol, 25, **477**–478
  DeWolf Memorial Chapel, 478
  Parish House, 477
  Parsonage, 477

Congregational Church, Pawtucket, 83
Congregational churches. See churches, Congregational
Congregationalists, 7, 8, 82, 148, 266, 382, 449, 464, 478, 534, 536, 537, 544
Connecticut River Valley, 447
Conover (Reverend Coit) House, Middletown, 512
Conrad Building, Providence, 53–54, 129
Convent of the Sisters of the Cross and Passion (Rush Sturges Mansion, Shepherd's Run), South Kingstown, 386
Convention Center and Capital Center Redevelopment Project, Providence, 39–41
Cook, John, 488
Cook (Colonel John)–Bateman (William) Farm, Tiverton, 487–**488**, 489
Cook (Cyrus) House (Orchard House), Cumberland, 218
Cook (Edward) House, Tiverton, 488–489
Cook (Stephen A., Jr.) House, Providence, 107, 112
Cooke House, Providence. See Benoni Cooke House, Providence
Cooper, James Fenimore, 544
Cooper and Bailey: Colt-Andrews School (Colt Memorial High School), Bristol, 469, 471
Cooper (Daniel) House, Burrillville, 305
Cooper (Matthew) House, Wickford, North Kingstown, 360
Cooper Road, Glocester, 296–298
Cooper (Stephen) House, Wickford, North Kingstown, 357
Cooperstown, New York, 359
Copeland, Robert Morris, 503
Copeland and Dale, 435
Copley, John Singleton, 536
Corbett, Harvey Wiley, 37
Corliss, George: Admissions Office, Brown University (George Corliss House), Providence, 104
Corliss, Mrs. George, 104
Corliss, John, 92
Corliss-Carrington House. See Corliss (John)–Carrington (Edward) House, Providence
Corliss (George) House, Providence, 104
Corliss (George) House, Providence. See Admissions Office, Brown University (George Corliss House), Providence
Corliss (John)–Carrington (Edward) House, Providence, 11, 83, 91–92, **93**
Cormier, Ernest: St. Jean-Baptiste Church, Pawtucket, 140
Corné, Michel Félice, 76, 538
Corné (Michel Félice) House, Newport, 538
Coro Building, Providence, 57, 58–59, 441
Cory family, 486
Cory House. See Tiverton Historical Society (Chase-Cory House), Tiverton
costume jewelry industry, 32, 33, 59
Cottage, The (The Homestead), Peace Dale, South Kingstown, 381, 383
Cotton (Joseph) House, Newport, 536
Cottrell Factory, Westerly, 410
Cottrell House, Westerly, 409–410
Council Room (Senate Chamber), Colony House, Newport, 520
Counting House, South Kingstown, 383, 384
Country Builder's Assistant (Benjamin), 472
country clubs. See clubs
Country Life (Copeland), 503
Courthouse Center for the Arts (Washington County Courthouse), South Kingstown, 392
courthouses
  Bristol County Courthouse, Bristol, 477

Index   659

Courthouse Center for the Arts (Washington County Courthouse), South Kingstown, 392
Federal Building, Providence, 36, 37
Kings County Courthouse (former), South Kingstown, 390
North District Courthouse, Burrillville, 312
Providence County Courthouse, 64–65, 78, 86, 413
Cove, Providence, 40, 41, 61
Covell (King) House. *See* Sanford (M. H.) House (Ednaville), Newport
Coventry, 324, 326, 330, 333–341, 429
  map, 336
Coventry Center, 338–341
Coventry Cranberry Company. *See* Greene Company, The (Coventry Cranberry Company), Coventry
Coventry Mill (Anthony Mill), 168, 333, **337**–338
Coventry Poor Farm. *See* Briggs (Joseph) Farmhouse (Coventry Poor Farm–Town Asylum)
Coventry Town Highway Sand Storage Facility, **341**
Coxe-Hayden House, Block Island, 607–608
Cozzens (Joseph and William) House, Newport, 528, 536
Craftsman, The (periodical), 591
Cram, Ralph Adams, 95, 514–515
  Brown (John Nicholas) House, Newport, 17
  St. George's School Chapel, Middletown, 514
Cram and Ferguson, 95, 96
  Emmanuel Church, Newport, 581
Cram, Goodhue and Ferguson, 53
  Brown (John Nicholas) House (Harbour Court), Newport, 17, 580
  Deborah Cook Sayles Memorial Library, Pawtucket, 145–146, **147**
Cram (Jacob)–Sturtevant (Mary) House, Middletown, 513–514
Crandall (William) House, Newport, 527
Cranston, 135, 154, 175–185, 259, 316
  map, 176
Cranston Historical Society, 179
Cranston Mills, 478
Cranston Print Works Village, 176, 178, **179**
Cranston Street Armory, Providence, 34, 114, 130, 132, **133**, 145, 229, 230
Crawford Street Bridge, Providence, 64
Crescent Park, East Providence, 435, 445
  Carousel, 151, 445–446
Cresson, George V., 372
Cret, Paul P.: World War I Monument, Providence, 65
Croftmere. *See* Cocroft (Freeman) House (Croftmere), South Kingstown
Crompton, Samuel, 333
Crompton, West Warwick, 324, 332–333
Crompton Company, 333
Crompton Mill, West Warwick, **332**–333, 337
Crompton Mill housing, West Warwick, **333**
Cross, Alonzo T., 361
Cross, George W., 399
Cross, Harry, 411
Cross, Joseph (forefather), 399
Cross, Joseph H. (descendant), 399
Cross and Cross: Hoyt (Virginia Scott) House (Ayrault House), Newport, 555
Cross Mills, Charlestown, 399–400
Cross's Hall and Store, Charlestown, 399
Crossways. *See* Fish (Stuyvesant and Mamie) House (Crossways), Newport
Crowell, Nathan: Adams (John F.) House, Pawtucket, 152
Cull, Edwin C., 349
Cull, Edwin Emory, 493, 494
Cullen (Joseph T.) House, Pawtucket, 150–151
Cullinan, Thomas F., 57

"Cultivation of True Taste, The" (Upjohn), 254
Cumberland, 128, 142, 154, 160, 186, 187, 188, 189, 190, 201, 202, 205, 206–219, 220, 249, 424, 434
  map, 207
  Town Hall, 208–209, 321
Currier and Ives, 275, 297
Currier (Omar) House, Pawtucket, 151
Cushing, Howard G., 547
Cushing and Wallace (cemetery planners), 132
Cushing House, East Providence. *See* Smith (Joseph)–Jenckes (John)–Cushing House, North Providence
Cushing Memorial Gallery, Newport Art Museum, Newport, 547
Custom House, Dublin, Ireland, 44
customhouses
  Bristol, 25, 49, 472
  Providence, 47, 48, 49
Cutler, Thomas, 296
Cutler (Thomas) Farm, Glocester, 245, 296, **297**
Cyrus Cook's Tavern. *See* Stagecoach Tavern (Cyrus Cook's Tavern), Glocester

Daggett (John, Jr.) House, Pawtucket, 151
Daggett (Nathaniel) House, East Providence, 437
Dairy Barn, Burrillville, 305
dam and headgate, waste gate and raceway, Charlestown, 401
Dame Farm, Johnston, 172
dams. *See* mill dams
Dance Studio (Lonsdale Baptist Church), Lincoln, 192
Danforth, Helen Metcalf, 68, 107
Danforth, Murray S., 107
Danielson, Connecticut, 274
Danielson Pike, 265, 274
Dante, 146
Danville, Woonsocket. *See* Bernon, Woonsocket
Darling, Newton, 243
Darling Brothers: First Baptist Church, Woonsocket, 231–232
Darling House. *See* Smith-Darling House, Burrillville
Darman, Arthur, 236
Dartmouth, Massachusetts, 490
Davis, A. J.: New York Yacht Club Newport Facilities, Station No. 10, 580
Davis, Henry, 288
Davis, Lucius D.: Conanicut Park, 586–588
Davis Hall (College), University of Rhode Island, South Kingstown, **391**, 392
Davis (Henry) House (Seamens-Davis House), Foster, 288
Davis (James) House, Newport, 526
Davis (Jeffrey) House, Narragansett. *See* Fair Lawn (Jeffrey Davis-Charles H. Pope House), Narragansett
Davis (Marion) Cottage, Jamestown, 603
Davisville Construction Battalion Center (U.S. Navy), North Kingstown, 354, 356
Davol Square (Davol Rubber Company), Providence, 59, 60
Day Care Center (Company Store and Post Office), Smithfield, 257
Daybreak Cottage. *See* Johnston (J. D.) Bungalow (Daybreak Cottage), Jamestown
Dayton (Isaac) House, Newport, 522
de Coux, Janet, 65
De La Salle, Newport. *See* Weld (William) House, Newport
Deak Store, Hopkinton, 422
Dean and Westbrook: Arkwright Bridge, Coventry, 334, **335**

Deane, Margery, 525
Deborah Cook Sayles Memorial Library, Pawtucket, 145–146, **147**, 158
Declaration of Independence, 86, 520
Decoration of Houses, The (O. Codman and E. Wharton), 566, 576
Deerfield, Massachusetts, 267
Delano, William: James (Lucy Worth) House (Normandie), Newport, 577
Delano and Aldrich, 546
Denison, Frederick, 587
Dennis (John) House. *See* St. John the Evangelist Rectory (John Dennis House)
Derby, Barnes and Champney: Tomlinson (Irving) House (Salt Marsh), Newport, 577
Designs for Schools and School Houses (Kendall), 165
Designs of Inigo Jones and Others (William Kent), 542, 545
Desurmont Worsted Yarn Mills, Woonsocket, 231, 238
Development Office, Brown University, Providence, 58
Devol (George) House, Bristol, 476
Dewey, Francis H., 372
DeWolf, Abby, 468
DeWolf, George, 25, 471, 472, 477
DeWolf, Henry, 468
DeWolf, Isabella, 474
DeWolf, James, 467, 472, 477
DeWolf, William, 472
DeWolf (Charles) House, Bristol, 479
DeWolf family, 467, 468, 474, 478
DeWolf House, Bristol. *See* Linden Place (DeWolf-Colt House), Bristol
DeWolf-Guiteras House, Bristol, 477
DeWolf (James) Barn, Bristol, 477
DeWolf (Mark Antony) House, Bristol, 469
DeWolf Memorial Chapel, Congregational Church, Bristol, 478
DeWolf's Wharf, Bristol, 478
  DeWolf Warehouse, 478
  Diman (Byron) Counting House, 478
  Old Bank of Bristol, 478
  Taylor (William) Store, 478
Dexter, Christopher, 182
Dexter, Ebenezer Knight, 113, 114, 131
Dexter, Edward, 108
Dexter, Gregory, 199
Dexter (Ebenezer Knight)–Stimson (John J.) House (Rose Farm), Providence, 113–114
Dexter (Edward)–Pendleton (Charles) House, Providence, 79, 108–109
Di Saia, Oreste: St. Rocco's Church, Johnston, 175
Diamond Hill, Cumberland, 217–219
Diamond Hill State Park, 217
Dickens Farm. *See* Lewis-Dickens Farm, Block Island
Dike, Henry A., 106
Dike (Henry A.) House, Providence, 106
Dillon, Daniel, Jr.: Littlefield (Captain Amazon) House, Block Island, 607
Diman, Byron, 474, 476
Diman, John Hugh, 500
Diman (Byron) Cottage, Bristol, 476
Diman (Byron) Counting House, DeWolf's Wharf, Bristol, 478
Diman Hall, St. George's School, Middletown, 514
DiMauro, Ronald, 602
Dimond, Francis M., 470, 472
Dimond (Francis M.) House, Bristol, **470**–471, 476
Dirlam, Arland: Park Place Congregational Church, Pawtucket, 143
dissenters, 9

District 2 Schoolhouse, Charlestown, 399, 402
Division of Applied Mathematics, Brown University (Henry Pearce House), Providence, 119
Dodge, Nehemiah, 32, 55
Dodge, Robert, 38
Dodge (Darius B.) Cottage, Block Island, 607
Dodge (Seril) Houses. *See* Providence Art Club (Seril Dodge Houses)
Donant-Archambault Building, West Warwick, 331
Donizetti Road, Shelter Harbor, Westerly, 403
Doran Building, Providence, 57
Dorr, Sullivan, 229
Dorr, Thomas W., 295, 460
Dorr Rebellion, 78, 295–296, 348, 417, 460
Dorr (Sullivan) House, Providence, 71, **76**, 78, 83, 538
Dorrance, George, 286
Dorrance, John, 286
Dorrance, Samuel, 285
Dorrance House, Foster, 285–286
Dorrville, Westerly. *See* Bradford, Westerly
double houses, Providence, 93
double store building, Newport, 533
Douglas, Massachusetts, 301, 308
Douglas, William O., 334
Douglas Pike, 161, 249, 301
Douglas Turnpike Company, 249
Dow-Starr House, Warren, 463, 464
Downing, Andrew Jackson, 165, 475, 566
Downing (Captain John W.) House (Fairview), Newport, 532
downtown, Providence, 34–55, 124
  map, 35
Drake, Samuel, 587
Dresser Avenue (numbers 9, 11, 15), Newport, **558**
Drown, Alfred, 450
Drown, Solomon, 289
Drown, William, 289
Drown (Alfred) House, Barrington, 450
Drown (Dr. Solomon) House (Mount Hygeia), Foster, 289
Drown family, 450
Drowne (George R.) House, Providence, 76
Drownville, Barrington, 450–453
Druidsdream, Narragansett, 373
du Moulin, Rockwell
  Perkins (Elizabeth) House, South Kingstown, 396
  Smith (Sibley) House, South Kingstown, 396
Dudley Newton and C. E. Kempe: Van Alen (J. J.) House (Wakehurst), 560
Duffner Kimberly Company, 114
Duhane (Jacob)–Potter (Captain Simeon) House (School), Newport, 522, 523
Duke, Doris, 575
Duke University, 575
Dumplings, The, Jamestown, 583, 597–601
Dun, Rupert, 372
Dunsmere gatehouse, Narragansett, 372
duplexes
  Burrillville, 305, 306
  East Providence, 437–438, 441
  Johnston, 172
  Lincoln, 190–191, 191, 204
  North Providence, 166
  North Smithfield, 243, 244, 246
  Richmond, 426
  Smithfield, 255, 256, 257
  Woonsocket, 232
Durfee, Thomas, 484
Durfee, William, 484

Durfee (William and Thomas) Farm, Tiverton, **484**, 487
Durkee and Brown, 42
Durkee, Brown, Viveiros and Werenfels, 104
Durran (John) Farm, North Smithfield, 248
Dutch Island, Jamestown, 20, 592, 597
Dutch Island Management Area, Jamestown, 592
Duty Evans House, Providence, 73–74
Duveen, Joseph, 568
Dyer, Elisha, 118
Dyer, Elisha, Jr., 118
Dyer Hall (John Holden Greene House II), Brown University, Providence, 118

E. R. Robertson and J. J. Stevenson, 107
E. Turgeon Construction Co., 349
Eagles Nest. See Jelke (F. Frazier) House (Eagles Nest), Newport
Eames, James, 254
Earle, Edward, 371
Earle, Stephen C., 467, 47
  Burnside Memorial Hall, Bristol, 473
  Rogers Free Library, Bristol, 471
  St. Michael's Episcopal Church Parish House, Bristol, 473
Earle Building, Providence, 42
Earle (Edward) House, Narragansett, 371
Earle's Block, Providence, 61
Earle's Block and Brick Rows, Providence, 61
Earles Court Houses and Water Tower, Narragansett, 371
East Bay, 433–497
East Bay Bicycle Path, 435, 437
East Channel, Narragansett Bay, 20, 578, 579, 582, 593, 594, 596, 597
East Greenwich, 76, 344–350
  Free Library, 347
  map, 344
  Preservation Society (Kent County Jail), 346
  Town Hall (Kent County Courthouse; East Greenwich State House), 345, 346, 348
East Providence, 94, 112, 127, 151, 193, 433–446, 447
  center, 434
  Fire Department, 443
  Historical Society, 442
  industrial core, 437–443
  map, 436
East Shore, Jamestown, 585–589
East Side, Providence, 72–122
Easton, Nicholas, 522
Easton family, Newport, 12
Easton (John) House, Newport, 557
Easton's Beach, Newport, 12–13, 15, 557
Eastover, Farm Estates, Portsmouth, 503, 504, 505
Ecclesiastical History of New England and North America (Stiles), 541
Ecclesiological Movement, 82, 506
Ecclesiological Society, 53, 506
Ecclesiologist, The (journal), 82, 254
Economic Activities in the Days of the Narragansett Planters (E. H. Baker), 379
Eddy, Bernard: Beneficent Congregational Church, Providence, **43**–44, 45
Eddy, James, 130
Eddy Block, Providence, 61
Eddy (Joseph) Farm, Glocester, 292
Edgar (Commodore William G.) House, Newport, 547–548, **549**, 567
Edgartown, Martha's Vineyard, Massachusetts, 586
Edgehill Farm. See King (George Gordon) House (Edgehill Farm), Newport
Edgewood, Cranston, 177, 183–185

Edgewood Yacht Club, Cranston, 183–184
Edmonds, Joseph, 327
Edna Villa, Newport. See Bell (Isaac, Jr.) House (Edna Villa), Newport
Ednaville, Newport. See Sanford (M. H.) House (Ednaville), Newport
Edwin A. Smith Building, Providence, 42
Eidlitz, Leopold, 546
Eisenhower, Dwight D., 343
Eisenhower Building, Fort Adams, Newport, 580
Elam, Gervais: Vaucluse Farm, 505
Elder Hammond's Meeting House. See Foster Town House (Second Baptist Church, later Elder Hammond's Meeting House), Foster
Eleanor Roosevelt Hall, University of Rhode Island, South Kiingstown, 392
Eliot, Charles, Jr., 120, 384
Eliza G. Radeke Building, Rhode Island School of Design Museum of Art, Providence, 80–81
Ellerbee Architects: Civic Center, Providence, 39
Ellery House, Newport. See Washington Square, Newport, Wilbour-Ellery House
Ellis, John W.: Providence and Worcester Railroad Station, Woonsocket, 236–**237**
Ellis (Mary) House, East Greenwich, 349
Elm Street houses, Westerly, 409–410
Elm Street Machine Shop, Phenix Iron Foundry complex, Providence, 57–**58**
Elm Tree Cottage. See Eustis (Mrs. Frederic) House (Elm Tree Cottage), Newport
Elmhurst area, Providence, 34
Elmhurst School and Convent of the Sacred Heart. See Taylor (Moses) House (Elmhurst School and Convent of the Sacred Heart), Portsmouth
Elmhyrst. See Vernon (William) House (Elmhyrst), Middletown
Elms, The. See Berwind (Edward Julius) House (The Elms), Newport
Elmwood Foundation, 132
Emerson, C. J.: Byfield School, Bristol, 477
Emerson, William Ralph, 552, 555, 576
  Eustis (Mrs. Frederic) House (Elm Tree Cottage), Newport, 552
  Grosvenor (Rose Ann) House (Wyndham), Newport, 555, 580
  Grosvenor (William) House (Fair Oak, Roslyn), Newport, 555, 580
  Richardson-Blatchford House, Newport, **554**, 555
  Sandford (m. H.) House (Ednaville), Newport, 525
Emery's Majestic Theatre. See Lederer Theatre (Emery's Majestic Theatre), Providence
Emmanuel Church, Newport, 581
Emmes family, 530
Emmons (Arthur) House, Newport, 552
Empire State Building, New York City, 37
Engdahl, Samuel A.: Ashton Viaduct (New Bridge), Cumberland-Lincoln, 205, 214, **215**
Episcopal churches. See churches, Episcopal
Episcopalians (Anglicans), 8, 67, 82, 115, 246, 254, 269, 347, 449, 452, 464, 506, 511, 514, 534, 537
Eppley, Marion, 393
Eppley Camp, South Kingstown, **393**
Eppley Laboratories, 393
equestrian activities, 17–18
Equitable Building, Providence, 47, **48**–49
Erskine, William, 204
Ertel, Michael, 51
Escoheag Advent Church, West Greenwich, 343
Esmond Mill, Smithfield, 262
  housing, 262

Essay on the Philosophy of Medicine, An (Bartlett), 247
Estes, James, 46
　Gagne House, Jamestown, 602
Estes-Burgin (architects), 51
Eustis family, 552
Eustis (Mrs. Frederic) House (Elm Tree Cottage), Newport, 552, 555
Evans House, Smithfield, 251–252, 257
Evans Schoolhouse, Glocester, 291, 297–**298**
Ewing (Joseph N.) House. *See* Henszey (William P.)–Ewing (Joseph N.) House (Altamira), Jamestown
Exchange Bank Building, Providence, 49–50, 51
Exchange Place, Providence. *See* Kennedy Plaza, Providence
Exeter, 299, 342, 355, 418, 424, 429–432
　map, 430
　Old Town House (Town Hall), 430
　Town Pound, 299, 430–431
Extrados (architects), 390
Eyre, Wilson, Jr.: St. Colomba's Chapel (Berkeley Memorial Chapel), Middletown, 181, 348, **511**–512, 514

factories. *See* industrial buildings; mills and mill complexes
Faculty Club (Zachariah Allen House), Brown University, Providence, 87
Fahnestock (William F.) House (Oaklawn, Bois Doré), Newport, 558–559
Fair Lawn (Jeffrey Davis-Charles H. Pope House), Narragansett, 372
Fair Oak. *See* Grosvenor (William) House (Fair Oak, Roslyn), Newport
Fairchild Barn, Newport, 526
Fairview. *See* Downing (Captain John W.) House (Fairview), Newport
Fairview Avenue houses, West Warwick, 326–327
Fales, William, 474
Fales and Jenks Machine Works, Central Falls, 149, 154, 157, 159. *See also* United States Flax Company (Fales and Jenks Machine Works), Central Falls
Fales (D. G.) House, Central Falls, 157
Fales (Stephen S.) House, Bristol, 476
Fales (William) House. *See* Bell (William H.)–Fales (William) House, Bristol
Fall River, Massachusetts, 24, 25, 130, 326, 481, 483, 484, 498
Fall River Iron Works Warehouse, Providence. *See* Bayard Ewing Building, Division of Architecture, Rhode Island School of Design (Fall River Iron Works Warehouse), Providence
Fall River Line (steamship line), 24, 483
Farago, Peter and Daphne, 81
Farm Estates, Portsmouth, 503–504
　Eastover, 503, 504, 505
　Glen, The, 503, 504, 512, 513
　Greenvale, 503, 506
　Oakland, 503, 504, 505, 506
　Sandy Point, 503, 504, 505
　Vaucluse, 503, 504, 505
Farnum, Cyrus, 296
Farnum, H. Cyrus, 163
Farnum, John, 259
Farnum, Joseph, 259
Farnum, Noah, 259
Farnum and Providence Turnpike Company, 249
Farnum (Cyrus) Farm, Glocester. *See* Manton (Olney)–Hunt (Pardon)–Farnum (Cyrus) Farm, Glocester
Farnum family, 249
Farnum Farm, Glocester, 297

Farnum (Joseph)–Brown House, Smithfield, 256, 262
Farnum Pike, 161, 249, 262
Farnum Road, Glocester, 296–298
Farrand, Beatrix Jones, 88, 386
Faunce, William H. P., 101
Faunce House (Rockefeller Hall), Brown University, Providence, 101
Fayerweather Craft Center (George Fayerweather House), South Kingstown, 388–389
Fayerweather (Solomon) House, South Kingstown, 388–389
Federal Building (Court), Providence, 36, 37
Federal Building (Kennedy Plaza Post Office), Providence, 39
Federal Building (U.S. Customhouse), Providence, 47, **48**, 49
Federal House, Lincoln, 200
Feke, Robert, 527, 546
Female Library Association, East Providence, 441
Fenner, James, 118
Fenner House. *See* Smith (Christopher)–Fenner House, Scituate
Fenner (J. C.) House, Hopkinton, 418
Fenner (Thomas) House, Cranston, 175, 184–185, 195
Fiberglas Furnace, Cumberland, 212–213
Field, John, 85
Field, Kate, 525
Field (John)–Hopkins (Stephen) House, Providence, 85–86
Field-Stone Mansion, Middletown, 511
Fielding, Mantle: Rhett (Dr. H. J.) House (Quononoquot Club), Jamestown, 602
Field's Point, Providence, 133
15 Granite St. House, Westerly, **414**, 415
15–17 Main St., Wickford, North Kingstown, 358
57–63 North Broadway duplexes, East Providence, 441
Finch (Captain William) House, Newport, 525
Firemen's Museum (Narragansett Steam Fire Engine Station No. 3), Warren, 461
First Baptist Church, Bristol, 477
First Baptist Church, Burrillville. *See* Pascoag Community Baptist Church (First Baptist Church), Burrillville
First Baptist Church, East Greenwich, **346**–347
First Baptist Church, East Providence, 442
First Baptist Church, Providence, 8, 9, 53, 62, **66**–68, 88, 97, 115, 265, 535
First Baptist Church, Tiverton. *See* Free Will Baptist Church (First Baptist Church, Old Stone Church), Tiverton
First Baptist Church, Woonsocket, 231–232
First Baptist Church of Charlestown, Charlestown, 400, 402
First Baptist Church of Cross Mills, Charlestown, 399
First Baptist Church Parish House (Masonic Hall), Wickford, North Kingstown, 358
First Beach, Newport. *See* Easton's Beach
First Christian Church, Coventry, 340
First Congregational Church, Newport, 542
First Congregational Church, Providence, 43. *See also* First Unitarian Church (First Congregational Church), Providence
First Free Will Baptist Church, Pawtucket. *See* apartments (First Free Will Baptist Church, later Independent Eastern Orthodox Church of the Resurrection), Pawtucket
First Presbyterian Church, Newport, 516, 518
First Presbyterian Church, Savannah, Georgia, 88
First Seventh Day Baptist Church, Ashaway, Hopkinton, 423

Index 663

First Seventh Day Baptist Church, Hopkinton City, Hopkinton, 422–423
First Unitarian Church (First Congregational Church), Providence, 43, 53, 71, 78, **87–88**, 115, 170
First United Methodist Church, Warren, 460–461
  parsonage, 460
Fish, Mamie, 576
Fish, Stuyvesant, 576
Fish (Stuyvesant and Mamie) House (Crossways), Newport, 576
Fisher (Lewis T.) House, Barrington, 452
Fiske House, Glocester, 295
Fitzpatrick, Barry, 275, 277
5 Bank Street, Hope Valley, Hopkinton, 419
Flagg, Ebenezer, 527
Flagg (Ebenezer) House. *See* Pitt's Head Tavern (Ebenezer Flagg House), Newport
Flagler, Charles, 114
Flat River, 342
Flatiron Building, New York City, 49
Flatley Corporation (architects), 181
Fleet Bank Building (Industrial National Bank Building), Providence, 37, 49, 50
Fleet Bank (Industrial Trust), Warren, 464
Fleischner, Richard, 152
Fletcher, Charles, 111
Fletcher, Joseph E., 303, 305
Fletcher (Captain John) House, Bristol, 470
Fletcher (Joseph) House, Providence, 111–112
Fleur-de-Lys Studios, Providence, **69–70**, 110, 493
Flint, Peter, 285, 286
Floradale Motor Court. *See* Sandpiper Cottages (Floradale Motor Court), Middletown
Flower Cottage, The. *See* Hodgson (John M.) Cottage (The Flower Cottage), Newport
Fludder, James, 546
Fludder (James) House and Office, Newport, 546
Flying Horse Carousel, Westerly, 406
FM Global (Allendale Mutual Insurance Company Headquarters Building), Johnston, 28, 164, 170, 173–174
Fones (Daniel) House, Wickford, North Kingstown, 361
Fontaine, Walter F., 222, 227, 233
  Church of the Holy Family, Woonsocket, 230
  Fontaine (Walter F.) House, Woonsocket, 233
  Our Lady of Good Help Church, Rectory, and Parish House–School (Notre Dame de bon Secours), Burrillville, **304**
  St. Ann's Church, Woonsocket, 160, **223**, 224, 228, 230
  St. Matthew's Roman Catholic Church, Central Falls, 160, 175
Fontaine (Walter F.) House, Woonsocket, 233
Forbush and Hathaway: Kendrick Avenue School, Woonsocket, 224–225
Ford, Gerald, 268
Forestdale, North Smithfield, 243–244, 247
Forestdale Manufacturing Company, 243
Forestdale Mill housing, North Smithfield, 244
Forestdale Mill Office, North Smithfield, 244
Forestdale Mill superintendent's house, North Smithfield, 244
Forestdale School. *See* North Smithfield Heritage Association (Forestdale School, officially Branch School, District 3)
Forge Farm. *See* Greene House (WA9) (Forge Farm), Warwick
Fort Adams, Newport, 20, 23, 541, **579**–580, 597
  Eisenhower Building, 580
Fort Dumpling, Jamestown, 597, 599
Fort Getty, Jamestown, 592
Fort Greble, Jamestown, 592

Fort Ninigret, Charlestown, 399, 402
Fort Wetherill, Jamestown, 20, 592, 594, **597**, 599
forts and fortifications, 19–21, 23
  Fort Adams, Newport, 20, 23, 541, **579**–580, 597
  Fort Dumpling, Jamestown, 597, 599
  Fort Getty, Jamestown, 592
  Fort Greble, Jamestown, 592
  Fort Ninigret, Charlestown, 399
  Fort Wetherill, Jamestown, 20, 592, 594, **597**, 599
  Harbor Entrance Control Post, Jamestown, 594
  Stone Bridge and Fort Barton, Tiverton, 483
  See also armories
45 Roger Williams Avenue tenement, East Providence, 437
Foster, 274–290, 299, 301, 305, 342, 491
  map, 276
  Town Pound, 278, 299
Foster, Curtis, 285
Foster, Horace, 332
  American Hoechst Chemical Company (Quidnick Mills), Coventry, 337
Foster, William E., 43
Foster and Scituate Central Turnpike. *See* Central Pike
Foster Center, Foster, 278–287
Foster Center Baptist Church (Foster Center Christian Church), Foster, 278–279, 280, 284, 487
Foster Center School. *See* Public Library (Foster Center School, Hemlock School), Foster
Foster (Jesse) Home, Springfield, 252
Foster Town House (Second Baptist Church, later Elder Hammond's Meeting House), Foster, **279**–280
Founders Hall, U.S. Naval War College, Newport, **529**–530
four-family mill housing, Smithfield, 256, 257
four-unit and duplex mill housing, Lincoln, 190–192, 214
Fourteenth Rhode Island Heavy Artillery, 592
Fowler, Orson, 427
Fowler (George) House (Joseph Gardner House), Newport, 526
Fox Hill Farms, Jamestown, 592–593
Francis, George B. (engineer): Narragansett Electric Company Power Plants, Providence, 59–60, 177
Francis Building, Providence, 51
Francis (Davis R.) Cottage, Jamestown, 604
Franklin, Benjamin, 415
Franklin Bank, Glocester, 293–294
Franklin Society, 84
Fraser, George, 323
Fraser and Henthorne, 323
Free Will Baptist Church (First Baptist Church, Old Stone Church), Tiverton, 486–487
Freeborn, Noel, 358
Freeborn (Noel) House, Wickford, North Kingstown, 358
Freeman, George A., Jr.: Matthews (Brander) House (Shingle-nook), Narragansett, 369
Freewill Baptist Church, Smithfield, 253, 265, 293
French, Daniel Chester, 50, 51, 385
French and MacKenzie (carpenters), 180
French (Timothy) House, Bristol, 475
French Worsted Company Mills, Woonsocket, 226
Frenchtown Baptist Church, East Greenwich, 350
Frenning, Blanche Borden, 494
Frenning House (Bumble Bee Farm), Tiverton, 490–491
Fresnel lens, 23
Friedlander, Leo, 106
Friends Burial Ground, Jamestown, 590
Friends meeting houses
  Foster (former), 277, 281, 282, 305

Friends meeting houses *(continued)*
  Jamestown, 7, 583, 589, 590
  Little Compton, 7, 490, **491**
  Newport, 7, 522, 530, 531
  North Smithfield, 239, 241
  Portsmouth, 502–503
Froebel, Friedrich, 108
Froebel Hall, Brown University, Providence., 108
Fruit Hill, North Providence, 162–164
Fruit of the Loom trademark, 323, 329, 332
Fry, George, 282
Frye, John, 526
Frye (John) House, Newport, 526
Fry's Hamlet Historic District, East Greenwich, 349, 350
Fuller, Buckminster, 319–320
Fuller, Susan E. and Abby N., 152–153
Fuller, Thomas, 282
Fuller and Warner (company), 225
Fuller Farm. *See* Brown–Whidden–Fuller Farm, Foster
Fultoon, Robert, 514

Gaertner, Friedrich von, 81
Gagne House, Jamestown, 602
Gale, Levi, 543
Gale (Levi) House. *See* Jewish Community Center (Levi Gale House), Newport
Gallagher, Percival, 322
garden (communal), Providence, 74–75
Garden City, Cranston, 181–182
Garden Club of America, 484
Gardencourt. *See* Gibson Court Condominiums (Charles H. Pope House, Gardencourt), Narragansett
Gardens of Colony and State (Garden Club of America), 484
Gardner, Caleb, 537
Gardner, Mr. and Mrs. George W., 96
Gardner, Newport, 537
Gardner-Brown Mill (Gladding's Sail Loft), Water Street, Warren, 457
Gardner (George) House, Newport. *See* Rathburn (John)–Gardner (George)–Rivera (Abraham Rodrigues) House (Newport Bank), Washington Square, Newport
Gardner (George W.) House, Providence. *See* Page (Martin)–Buffum (Horace)–Gardner (George W.) House, Providence
Gardner (Joseph) House, Newport. *See* Fowler (George) House (Joseph Gardner House), Newport
Gardner (Palmer) House, South Kingstown, 388
gas station, Woonsocket, 236
Gasometer, Woonsocket, 235
Gates, Leighton and Associates (landscape architects), 208
Gatherem, Little Compton. *See* Winter (Edwin W.) House (Gatherem), Little Compton
Gay, Joseph B., 289
General Housing Corporation
  Hadfield (George, Jr.) House, Pawtucket, 152
  Monsarrat (Nicholas and Jane) Houses, East Providence, **439**–440
General Nathanael Greene Homestead, Coventry, 338
George Aldrich Inn. *See* apartments (George Aldrich Inn), North Smithfield
George Champlin Mason and Son, 551
  Mason (George Champlin, Jr.) House, Newport, 553
  Morris (Francis) House, Newport, 553
  White (Isaac P.) House, Newport, 556
  Zabriskie (Sarah T.) House (Stone Gables), Newport, 552

George Hail Free Library, Warren, **459**
Georgia Manufacturing Company, Smithfield, 259
Georgiaville, Smithfield, 87, 201, 249, 250, 251, 254, 259–262, 326
Georgiaville Mill Complex, Smithfield, **260**–261
Georgiaville Mill Office, Smithfield, 261
Gerald, Samuel, 118
Gerald (Samuel) House, Providence, 117–118
German Cooperative Land Association of Providence, 138–139
Gerry, Peter Goelet, 104
Gerry House. *See* Woods-Gerry House, Providence
Gershwin Road, Shelter Harbor, Westerly, 403
Ghirardelli Square, San Francisco, 59
Gibbs, Cuffe, 530
Gibbs, James, 8, 66, 67, 71, 88, 253, 293, 542
Gibbs, Sarah, 506, 507, 509
Gibbs, William, 540
Gibbs (George) House, Newport, 526
Gibbs (T. K.) House (Bethshan), Newport, 551, 552
Gibbs (William) House. *See* Vernon House (William Gibbs-Metcalf Bowler-William Vernon House), Newport
Gibson, Charles Dana (artist), 468
Gibson, Charles Dana (grandfather), 468
Gibson (Charles Dana) House (Longfield), Bristol, 468
Gibson Court Condominiums (Charles H. Pope House, Gardencourt), Narragansett, 370, 372
Gilbane Building Company, 47
Gilbert, Cass, 415
Gilchrist, Edmund: Mauran (Frank, Jr.) House, Providence, 120–121
Gill, Irving, 551, 604
  Birckhead (Sarah King) House (Eastover), Portsmouth, 505, 551
  Olmsted (Albert H.) House (Wildacre), Newport, 505, 551, 578
  St. Michael's School (Miss Ellen Mason House), Newport, 505, 551
Gilmour, William, 160
Girl Scout Camp (Camp Hoffman), South Kingstown, 394
Gladding's Sail Loft. *See* Gardner-Brown Mill (Gladding's Sail Loft), Water Street, Warren
Gleason, J., 275
Glen, The, Farm Estates, Portsmouth, 503, 504, 512, 513
Glen Rock, South Kingstown, 393
Glen Rock Mill. *See* Peter Pots Pottery (Glen Rock Mill), South Kingstown
Glendale, Burrillville, 301, 302, 306–307
Glendale New Village, Burrillville, 302, 306–**307**, 310
Glenlyon Bleachery Company, East Providence, 193, 434–435, 437, 438
Glenlyon Bleachery Plant, East Providence, 439, 443
Globe Mill village, Woonsocket, 229
Globe Mills, Woonsocket, 222, 229, 230, 237
  housing, 229, 234
Glocester, 249, 267, 268, 290–299, 301, 302, 342, 431
  map, 291
  Town Pound, 278, **298**, 299, 431
Gloria Dei Evangelical Lutheran Church, Providence, 124–125, 230
Goat Island, Newport, 522
Goddard, Charlotte Rhoda Ives, 96, 189, 270
Goddard, Francis W., 96
Goddard, John, 524
Goddard, Moses Brown Ives, 322
Goddard, Thomas, 546
Goddard, William, 96

Goddard, William Giles, 96
Goddard family (Newport), 527, 539
Goddard family (Providence), 190, 322
Goddard (Francis W.) House, Providence. *See* Nicholson House (Francis W. Goddard–Samuel C. Nicholson House), Providence
Goddard (John and Thomas) House, Newport, 525, 526
Goddard Memorial Park (Russell Estate), Warwick, 322
Goddard-Nicholson House, Providence. *See* Nicholson House (Francis W. Goddard-Samuel C. Nicholson House), Providence
Goddard (Thomas) House, Newport, 539
Goddard (William Giles)–Iselin (Hope Goddard) House, Brown University, Providence, **96**–97
Goelet, Ogden, 559
Goelet (Ogden) House (Ochre Court), Newport, 18, **559**–560, 562
Goelet (Robert) House, Newport, 559, 560
Goethals, George W., 597
Goff, Lyman, 148
Goff (Lyman) House, Pawtucket., 148
Gold, Albert, 349
Golden Triangle, Pittsburgh, 40
Goodhue, Wright, 348
Goodwin, J. B.: First Baptist Church, East Greenwich, **346**–347
Goodwin (William P.) House, Providence, 112
Goody, Clancy and Associates, 100, 101
Gordon School, East Providence, 444
Gorham company, Providence, 52, 74, 113
Gorham (Mary M.) House, Providence, 74
Gorton-Greene House, Warwick, 321
Gothic Architecture Improved by Rules and Proportions (Langley), 71
Gothic Revival cottage, Bristol, 474
Gothic Revival house, Westerly, 410
Gould and Angell
  Baker (Charles H.) House, Providence, 113
  Boyd (George E.) House, Provincetown, 130
  Hope Club, Providence, 87
  Robertson (Louis E.) House, Providence, 112, 113
  Wilson Hall, Brown University, Providence, 100–101, 384
Gounod Road, Shelter Harbor, Westerly, 403
Governor Sprague Mansion, Cranston, 178–179
Grace Church, Providence, **53**, 54, 82, 95, 96, 254
Grace Church Cemetery, Providence, 132
Grace Episcopal Church, East Providence, 438
  Parish House (Elementary School), 438
Grammar of Ornament (Jones), 110
Grange, The. *See* Greene House (The Grange), Warwick
Grange Road Farmhouses, North Smithfield, 248–249
Granite Mill, Burrillville. *See* Sayles Mill, Burrillville
Granite Mill housing, Smithfield, 261–**262**
Granite Theater (Christian Church), Westerly, 410, 411
Graniteville, 169–170
Grant, Joseph, 217
Grant (Florence Price) House. *See* Aldrich (James) House II–Grant (Florence Price) House, Scituate
Grants Mills, Cumberland, 217
Graves (Eugene) House, Providence, 119
Gray, Pardon, 492
Gray, Robert, 485, 486
Gray, Samuel W., 415
Gray (Captain Robert) House, Tiverton, 484–485
Gray Craig, Middletown. *See* van Beuren (Michael M.) House (Gray Craig), Middletown
Gray House, Little Compton. *See* Pabodie-Gray House, Little Compton
Great Depression (1930s), 480, 587

Great Road, Lincoln, 187, 194–199, 200, 239, 240, 241, 243
Great Swamp Fight of 1675, 351, 374, 388
Greble, John T., 592
Greek Revival houses
  Bristol, 475
  Hopkinton, 419
  Westerly, 410
Green Animals, Portsmouth. *See* Brayton (Thomas and Alice) House (Green Animals), Portsmouth
Green Chimneys, Jamestown. *See* Selfridge (Admiral Thomas O., Jr.) House (Red Top; Green Chimneys), Jamestown
Green (Cornelia Burges) House, Providence. *See* Lippitt (Moses)–Green (Cornelia Burges) House, Providence
Green (John P.) House (Anoatok), Jamestown, 602
Green Mill, Lincoln. *See* apartments (Green Mill), Lincoln
Greenaway, Kate, 117, 370
Greene, Benjamin, 158
Greene, Christopher, 335
Greene, David, 603
Greene, Elihu, 322
Greene, Fones, 320
Greene, James, 320, 321
Greene, John Holden, 26, 61, 64, 79, 96, 116, 164, 477, 491, 494
  Allen (Candace) House, Providence, 87
  Arnold (Daniel) House, Providence, 45
  Benoni Cooke House, Providence, 62–**63**, 83, 87
  Brown (Isaac) House, 62, 63
  Bucklin (William and George) Houses, Providence, 94
  Burrough (James) House, Providence, 118
  Burrough (Robert S.) House, Providence, 117
  Cathedral of St. John (St. John's Church), Providence, **71**, 76, 78–**79**, 88
  condominiums and offices (Allendale Mill and Company Store), North Providence, 164–165
  Dorr (Sullivan) House, Providence, 71, **76**, 78, 83, 538
  Dyer Hall, Brown University (John Holden Greene House II), Providence, 118
  First Presbyterian Church, Savannah, Georgia, 88
  First Unitarian Church (First Congregational Church), Providence, 43, 53, 71, 78, **87**–88, 115, 170
  Greene (John Holden) House I, Providence, 94
  Handicraft Club (Truman Beckwith House), Providence, 62, 76, 83, 87
  Museum of Rhode Island History, Rhode Island Historical Society (Robert S. Burrough, Jr.–Senator Nelson W. Aldrich House), Providence, 117
  Smith (William) House, Providence, 94
  University Club (Rufus and Emily Waterman House), Providence, 78
  Woodward (William, Jr.) House, Providence, 94
Greene, Nathanael, 321, 322
Greene (Allen) House, Providence, 73
Greene and Daniels Mill, Pawtucket. *See* offices-condominiums (Greene and Daniels Mill), Pawtucket
Greene and Daniels Thread Company, Central Falls, 157
Greene and Greene, 150
Greene (Benjamin F.) House, Central Falls, **157**, 158
Greene (Christopher) House, Coventry, 334–**335**
Greene Company, The (Coventry Cranberry Company), Coventry, 341
Greene (David) Farmhouse, Jamestown, 603–604
Greene (Edward A.) House, Providence, 116, 117
Greene family, 321, 337, 359, 604

Greene (General Francis Vinton) House (Armsea Hall), Newport, 573
Greene-Bowen House, Warwick, 320
Greene House (Forge Farm), Warwick, 321, 322
Greene House (Hilltop Cottage), East Greenwich, **348**, 349
Greene House (The Grange), Warwick, 322. *See also* Gorton-Greene House, Warwick
Greene (John Holden) House I, Providence, 94
Greene (John Holden) House II, Providence., 118
Greene Mills, West Warwick. *See* Royal Mills (Greene Mills), West Warwick
Greeneway, The, Wickford, North Kingstown, 359–360
Greenough, Horatio, 507
Greenvale, Portsmouth. *See* Barstow (John) Farm (Greenvale), Portsmouth
Greenville, Scituate, 274
Greenville, Smithfield, 250, 252, 253–256
Greenville Fire Company, Smithfield, 253–254
Greenwich Hotel (Updike Hotel), East Greenwich, 346
Greystone, North Providence, 160, 166–168
Greystone Mill. *See* Joseph Benn and Company (Greystone Mill), North Providence
Grieg Road, Shelter Harbor, Westerly, 403
Grinnell College, 473
Grinnell (Malachi) House, Little Compton, 494
Griswold, J. N. A., 546
Griswold (J. N. A.) House, Newport Art Museum, Newport, 13, 546, 547, **548**, 570, 574
Gropius, Walter, 323, 493, 501
Grosvenor (Rose Ann) House (Wyndham), Newport, 555, 580
Grosvenor (William) House (Fair Oak, Roslyn), Newport, 555, 580
Groton, Connecticut, 473
Grove Street School, Woonsocket, 227
Guastavino, Rafael, 114, 416, 562
Guerin, Joseph, 225
Guerin Spinning Company, Woonsocket, 225
Guinness Book of World Records, 61
Guiteras, Ramon, 469, 477
Guiteras House, Bristol. *See* DeWolf-Guiteras House, Bristol
Guiteras Memorial and Junior High School, Bristol, 469
Gushue, Patrick, 173

Hadfield (George, Jr.) House, Pawtucket, 152
Haffenreffer, R. F., II, 479
Hagadorn, Elisa, 389
Hagadorn, Francis, 389
Hagadorn, John, 389
Hagadorn (John) House. *See* Perkins (Joseph)–Hagadorn (John)–Taylor (Thomas) House, South Kingstown
Hale, Daniel: Page (Martin)–Buffum (Horace)–Gardner (George W.) House, Providence, 96
Hall, C. G. and J. R.: Providence Institution for Savings Building, Providence, **63**
Hall, Clifton A., 71, 146, 147, 157
  Greene (Benjamin F.) House, Central Falls, **157**, 158
  Lippitt (Moses)–Green (Cornelia Burges) House, Providence, 93
  Merchants Bank Building, Providence, **50–51**
  St. John's Episcopal Church (Red Church), Barrington, 451, 452, 453
Hall, E. B., 515
Hall, George D., 500
Hall, George F.: Richardson (Kate Lane) House (Yellow Patch), Narragansett, 370

Hall and Makepeace, 452
Hall (Captain Palmer) House (India Point), Westerly, 408
Hall Manor, Portsmouth. *See* Smith (Amos D.) House (Hall Manor), Portsmouth
Hall (Mrs. Edward Brooks) House, Providence, 88
Hall's Block (The Chocolate Shop), Warren, 456, **457**
Halls Building, Providence, 47
Hallworth House (Enoch W. Clarke–John Slater House), Providence, 75, 76
Hambly Funeral Home (Grace W. Rives House), Newport, 551
Hamilton, Frank, 411
Hamilton Harbour Condominiums (Hamilton Web Company), Wickford, North Kingstown, **362–363**
Hamilton Web Company, Wickford, North Kingstown. *See* Hamilton Harbour Condominiums (Hamilton Web Company), Wickford, North Kingstown
Hamlet, Woonsocket, 219
Hammersmith Farm, Newport, 573, 579, 597
Hammond, John, 280
Hammond, Jonathan, 66
Hanaway Blacksmith Shop, Lincoln, 197
Handel Road, Shelter Harbor, Westerly, 403
Handicraft Club (Truman Beckwith House), Providence, 62, 76, 79, **83**, 87
Harbor area, Block Island. *See* Old Harbor Historic District, Block Island
Harbor area, Newport, 532–546
  map, 533
Harbor Cottage, Block Island, 605
Harbor Entrance Control Post, Jamestown, 594
Harborfront, Providence, 55–60
  map, 56
Harbour Court, Newport. *See* Brown (John Nicholas) House (Harbour Court), Newport
Harding and Upham: U.S. Weather Bureau Station, Block Island, 607
Hardouin-Mansart, Jules, 572, 573
Harkness, Albert, 53, 397, 512
  Brayton (Edward) House, Little Compton, 492, 493
  Eleanor Roosevelt Hall, University of Rhode Island, Kingston, 392
  Graves (Eugene) House, Providence, 119
  Jenks (Almet) House, Little Compton, 493
  Johnson and Wales Classroom Building (Summerfield Building), Providence, 42, 45, 51, 53
  Lippincott (J. Bertram) House II, Jamestown, 595–596
  Miller (William David) House (Robinson House, 571 Main), South Kingstown, 379–380
  Nightingale (William G.) House, Little Compton, 493
  Rice (Mrs. Herbert A.) House, Providence, 117, 119
  St. Colomba's Chapel (Berkeley Memorial Chapel) Parish Hall, Middletown, 511
  Wakefield Post Office (former), South Kingstown, 378–379
Harkness, Albert (classics professor), 106
Harkness, Hope, 595
Harkness and Geddes, 323
Harper's Weekly (magazine), 369
Harrington (Obadiah) Farm, Foster, 284
Harris, Coventry, 326, 333, 334–335
Harris, David, 199
Harris, Edward, 221, 234, 235, 236, 237
Harris, Elisha, 334, 335
Harris, Jonathan, 199
Harris, Thomas, 199
Harris, William, 479

Harris (Elisha) House, Coventry. *See* Riverview Nursing Home (Elisha Harris–Henry Howard House), Coventry
Harris family, 199, 249
Harris House, Smithfield, 256
Harris Institute, Woonsocket. *See* Woonsocket, City Hall (Harris Institute)
Harris (Jonathan) House, Lincoln, 199
Harris Warehouse, Woonsocket, 236
Harris (William) House, Bristol, 479
Harris Woolen Company, 234
Harrison, Joseph, 545
Harrison, Marc, 275, 277
Harrison, Peter, 8, 528, 537, 539, 540, 595
  Brick Market, Newport, 6–7, 65, 306, 516, **521**, 522
  Redwood Library, Newport, 7, 12, 510, 527, 528, 532, 540, 541, **544**–546
  Touro Synagogue, Newport, 7, 535, **542**–543
Harrisville civic buildings, Burrillville, 310–312
Harrisville Dam and Mill Pond, Burrillville, 310
Harrisville-Graniteville, Burrillville, 169–170, 301, 302, 303, 306, 308–313, 314, 381, 441
Harrisville Mill Complex, Burrillville, 313
Harrisville Town Hall, Burrillville, 312
Hart (Nicholas) House, Wickford, North Kingstown, 358
Hartford, Connecticut, 292
Hartford, Providence and Fishkill Railroad, 177
Hartford Pike, 170, 265, 290
Hartman-Cox Architects, 101
Hartshorn, Mr. and Mrs. Joseph, 131
Hartshorn (Joseph C.) House, Providence, **131**
Hartwell (Frederick W.) House, Providence, **131**
Harvard University, 82, 97, 434, 493, 514, 578
Hathaway, Dr., 497
Hathaway (Dr.) House. *See* Church (Samuel)–Hathaway (Dr.) House, Little Compton
Havemeyer family, 572
Haviland, John, 46
Hawkhurst or Hawxhurst, Newport. *See* Seymour (Carolyn) House (Hawkhurst or Hawxhurst), Newport
Hawley, Hughson, 412
Hawthorne, Nathaniel, 258, 549
Hay-Owen Block, Providence, 64
Hayden House, Block Island. *See* Coxe-Hayden House, Block Island
Haydn Road, Shelter Harbor, Westerly, 403
Hayes, Rutherford B., 92
Haynes, Irving, 144, 545
Hazard, Augusta, 383
Hazard, Benjamin, 519
Hazard, Caroline, 383, 384, 385
Hazard, Isaac Peace, 381
Hazard, Jason P., 343
Hazard, John Newbold, 384
Hazard, John Peace, 367, 372–373
Hazard, Margaret, 383
Hazard, Robert, 374
Hazard, Rowland, I, 381
Hazard, Rowland, II, 380, 381, 382, 384–385, 401, 427
  Church Street Bridge, South Kingstown, 383
  Columbia Street Bridge, South Kingstown, 383
  Peace Dale Congregational Church, South Kingstown, **382**, 384, 386
  Peace Dale Railroad Station, South Kingstown, 383, 384, 392
  South Kingstown Town Hall, South Kingstown, 380–381
  Watering Trough, South Kingstown, 384
Hazard, Rowland Gibson, 373, 381, 382, 384, 386

Hazard family, 363, 367, 372, 374, 375, 381–385, 388, 396, 412, 519
Hazard (G. G.) House, Warren, 462–463
Hazard House, Newport. *See* Wanton-Lyman-Hazard House (Stephen Mumford House), Newport
Hazard (Joseph Peace) House (Hazard's Castle, Seaside Farm), Narragansett, 368, **372**–373
Hazard Memorial Hall, South Kingstown, 384
Hazard's Beach, Newport, 16
Hazard's Castle, Narragansett. *See* Hazard (Joseph Peace) House (Hazard's Castle, Seaside Farm), Narragansett
Healy, George Peter Alexander, 100
Hearthside, Lincoln. *See* Smith (Stephen H.) House (Hearthside), Lincoln
Hedges, The. *See* Bruen (Mary) House (The Hedges), Newport
Hedley, Samuel, 162
Hedmark, Martin: Gloria Dei Evangelical Lutheran Church, Providence, 124–125, 230
Heffernan (Stephen)–Williams (Israel) House, Wickford, North Kingstown, 361
Heins and La Farge, 382
  Church of the Blessed Sacrament, Providence, 126
Hellmuth, Obata and Kassabaum, 122
  Fleet Center addition, Providence, 37
Helme (Theodore R.) Block, Newport, 536
Hemenway, Charles, 122
Hemlock School, Foster. *See* Public Library (Foster Center School, Hemlock School), Foster
Henderson (Robert) House, Jamestown, 585–586
Hendrick, Paul: Hendrick's Mill, Exeter, 432
Hendrick (Paul) House, Exeter, 432
Hendrick's Mill, Exeter, 432
Hennery, Newport, 560
Henszey (William P.)–Ewing (Joseph N.) House (Altamira), Jamestown, 600–601
Henthorne, Raymond J., 323
Heritage Harbor, Providence, 60
Herreshoff (A. Sidney) House, Bristol, 475
Herreshoff Company, 475
Herreshoff family, 25, 475
Herreshoff House. *See* Clark-Herreshoff House, Bristol
Herreshoff (John Brown) House, Bristol, 476
Hibbert, Susan, 281
Higgins (Robert J.) House, Little Compton, 494
High Street Business District, Westerly, 413–414
High Tide, Newport. *See* Miller (William S.) House (High Tide), Newport
Highland, Jamestown. *See* Morris (Wistar) House (Highland), Jamestown
Highwatch (Holiday House), Watch Hill, Westerly, 407
Hiker, The (Newman sculpture), 411
Hill, Jerah, 281
Hill-Collins House, Warren, 456
Hill (Jerah) Farm. *See* Wood (Daniel)–Hill (Jerah) Farm, Foster
Hillel House (Froebel Hall), Brown University, Providence, 108
Hillier, J. Robert, 121
Hillsgrove Airport. *See* T. F. Green State Airport (Hillsgrove Airport), Warwick
Hilltop Cottage, East Greenwich. *See* Greene House (Hilltop Cottage), East Greenwich
Hilltop (Richard Morris Hunt House), Newport, 546
Hillwood, Providence. *See* Bailey (William) House (Hillwood), Providence
Hilton and Jackson, 438, 439
  Bridgham Memorial Library, East Providence, 441–442
  57–63 North Broadway duplexes, East Providence, 441

668 Index

Hilton and Jackson *(continued)*
   Phillipsdale housing, East Providence (partial), **437**–**438**
   Robinson (Dr. R. R.) House, South Kingstown, 379
   See also Jackson, Robertson and Adams
Hinkle, Edward F.
   Moana (Aktaion), Westerly, 407
   Trepasso, Westerly, 407
Historic American Buildings Survey, 358
Historic District, Main Street, Wickford, North Kingstown, 358
Historical and Architectural Resources of Bristol, Rhode Island, 468
historical societies
   Bristol Historical Society, 476
   Cranston Historical Society, 179
   East Providence Historical Society, 442
   Jamestown Historical Society, 590, 591
   Johnston Historical Society (Elijah Angell House), 170
   Little Compton Historical Society (Wilbour House), 491–492
   Middletown Historical Society, 513
   Museum of Rhode Island History, Rhode Island Historical Society (Robert S. Burrough, Jr.–Senator Nelson W. Aldrich House), Providence, 117
   Newport Historical Society, 7, 483, 518, 525, 531, 536, 537, 538, 543
   Pettaquamscutt Historical Society (Old County Jail), South Kingstown, 379, 391
   Portsmouth Historical Society (Union Meeting House), **503**
   Rhode Island Historical Society (John Brown House), Providence, 84, **89**–90, 90, 103, 261
   Smithfield Historical Society (Elisha Smith House, known as the Smith-Appleby House), 257–258
   Tiverton Historical Society (Chase-Cory House), 486
   See also museums
History of the Diagnosis and Treatment of Typhoid and Typhoid Fever (Bartlett), 247
History of Washington County (Cole), 385
Hitchcock, Henry-Russell, 92, 164, 222, 237, 254, 260, 261, 271, 272, 306, 307, 349
HNTB: Bruce G. Sundlun Terminal, T. F. Green State Airport (Hillsgrove Airport), Warwick, 319
Hoar, Lewis, 459
Hoar and Martin (builders), 460
Hoar (J. H.) House, Warren, **457**
Hoar (Lewis T.) House, Warren, 459
Hoboken, New Jersey, 580
Hodgkiss Farm, Jamestown. *See* Watson (Borden) Farm, Jamestown
Hodgson, John, 569
Hodgson (John M.) Cottage (The Flower Cottage), Newport, 569
Holden, Pardee, 282
Holiday House, Watch Hill, Westerly. *See* Highwatch (Holiday House), Watch Hill, Westerly
Holly House, South Kingstown, 381
Holy Cross Church, Middletown. *See* Church of the Holy Cross, Middletown
Holzer, J. A., 114
Home for All (Fowler), 427
Homelands, Tiverton. *See* Oliver (Andrew)–Sargent (Charles)–Thayer (Steven) House (Homelands), Tiverton
Homer and Nickerson Halls, Metcalf Refectory, Providence, 78
Homestead, The, Kingston, South Kingstown, 390
Homestead, The, Peace Dale, South Kingstown. *See* Cottage, The (The Homestead), Peace Dale, South Kingstown
Honey (Samuel) House, Newport, 556
Hood, Raymond, 368
Hooper, James R., 244
Hooper and Moran: Beattie (Hamilton) House, Tiverton, 484
Hope, Scituate, 190, 270–273, 324, 326, 333
Hope Block, RISD Buildings, Providence, 65, 66
Hope Club, Providence, 87, 100
Hope College, Brown University, Providence, 99, **100**
Hope Corporation (textile manufacturers), 190
Hope Furnace, Scituate, 270, 271
Hope Manufacturing Company, 326
Hope Mills, Scituate, 270–**271**, 326
Hope Street, Providence, 32, 65, 66, 110
   map (with side streets), 109
Hope Valley (incorporating Locustville), Hopkinton/Richmond, 419
Hope Webbing Company, Pawtucket, 140
Hopelands, Warwick. *See* Rocky Hill School (Hopelands), Warwick
Hopkins, Alden, 85, 86
Hopkins, Daniel, 275
Hopkins, Samuel, 542
Hopkins, Stephen, 86, 270, 418
Hopkins (Charles A.) House, 131
Hopkins (Deacon Daniel) House, Foster, 275, 491
Hopkins (Dr. Samuel) House, Newport, 539
Hopkins (Ezekiel)–Potter (William) House, Foster, 239, 288
Hopkins family, 288
Hopkins Hollow Church, Coventry, **340**
Hopkins-Bennett House, Foster, 281
Hopkins Mills, Foster, 274, 288
Hopkins Mills Union Church (South Foster Union Chapel), Foster, 288, **289**
Hopkins (Stephen) House, Providence, **85**–86
Hopkinton, 299, 417, 418–424
   map, 420
Hopkinton Academy, Hopkinton, 423
Hopkinton City, Hopkinton, 388, 422–423
Hopper, Edward, 21, 395, 594
Hoppin, Francis L. V., 68–69, 92, 572–573
   Greene (General Francis Vinton) House (Armsea Hall), Newport, 573
Hoppin, Howard, 68–69, 92
   All Saints Episcopal Church, Warwick, 323
   St. James Episcopal Church, North Providence, **162**–**163**
Hoppin, Mr. and Mrs. Thomas, 92
Hoppin and Ely, 81, 92, 95, 96, 97, 121. *See also* Hoppin, Read and Hoppin
Hoppin and Field: Christian Science Church, Providence, 105
Hoppin and Koen, 92, 97, 575
   Jones (Pembroke) House (Sherwood), Newport, 572–573
Hoppin (Hamilton) House, Middletown, **515**
Hoppin, Read and Hoppin, 92
   Waldron (Henry A.) House, 111
   Waterman Building, Rhode Island School of Design, Providence, 68–69
Hoppin (Thomas F.) House, Brown University, Providence, 92, 92–93, 134
Hoppus, Edward, 545
Horace Remington and Sons Company Building, Providence, 56, 57
Horgan, Betty, 502
Horgan, Myra, 502

Horgan, Nina, 502
Horgan, Patrick: Horgan Cottages (The Three Sisters), Jamestown, 602
Horgan Cottages (The Three Sisters), Jamestown, 602
Horsehead. *See* Wharton (Joseph) House (Marbella, Horsehead), Jamestown
Horsford, Eben, 434, 440
Horton (Royal D.) House, Barrington, 448
Hospital Trust, Providence. *See* Sovereign Bank Complex (Rhode Island Hospital Trust National Bank), Providence
hotels and inns
  Atlantic House Hotel, Newport, 12, 564, 567
  Atlantic Inn (Norwich House), Block Island, 610
  Bay View Hotel, Jamestown, 583
  Bay Voyage Hotel, Jamestown, **604**
  Blue Dory Inn, Block Island, 607
  Captain Israel Inman's Inn (Simon Sweet House), Glocester, 295–296
  Central Hotel, Glocester, 294
  Central Hotel (former), Burrillsville, 312–313
  Conanicut Park Hotel, Jamestown, 586
  Franklin House hotel, Providence, 65
  General Aldrich Inn (former), North Smithfield, 242, 243
  Goff Hotel, Warren, 460
  Greenwich Hotel (Updike Hotel), East Greenwich, 346
  Hotel Dreyfuss, Providence, 42
  Larches, The (Larchwood Inn), South Kingstown, 379
  Narragansett Hotel, Block Island, 608
  Narragansett Inn, Westerly, 407
  National Hotel, Block Island, 607
  Ocean House, Charlestown, 399
  Ocean House, Newport, 12
  Ocean House, Watch Hill, Westerly, 407–408
  Providence Biltmore Hotel, Providence, **38**
  Shelter Harbor Inn, Westerly, 403
  Spring House, Block Island, 609, 610
  Surf, Block Island, 605, 607
  Thorndike Hotel, Jamestown, 583, 602
  Viking Hotel, Newport, 13, 546
  Western Hotel (Walling Hotel), Burrillville, 308
  Westin Hotel, Providence, 39
  See also resorts and summer communities; taverns
Hotham, Charles, 545
Houlihan, M. J. (builder): Lederer Building, Providence, 52–53
House with the Eagles, The, Bristol, 474
houses
  Barrington, **451**
  Cumberland, 215–216, 216
  Hopkinton, **423**–424
  Hopkinton City, 422
  Johnston, 172
  Newport, 516
  North Kingstown, 355
  North Smithfield, 245
  Scituate, 270
  Smithfield, 253
  West Warwick, 327
  Woonsocket, 227
  See also duplexes; triple-deckers; shingle houses; stone-enders; worker housing; *names of specific houses*
Howard, Daniel (father), 280
Howard, Daniel (son), 280
Howard, Henry, 335
Howard, Martin, 519
Howard (Daniel)–Howard (Judge Daniel) House, Foster, 280

Howard (Henry) House, Coventry. *See* Riverview Nursing Home (Elisha Harris-Henry Howard House), Coventry
Howard, Needles, Tammen and Bergenhoff: Convention Center, Providence, 39
Howe, John, 474
Howe, Julia Ward, 507, 575
Howe, Prout and Ekman, 42
Howe, Samuel, 507
Howe, Wallis Eastburn, 116, 467, 469, 470, 471, 472, 474, 477
  YMCA Building, Bristol, 472–473
Howe and Church, 451
  St. George's School, Diman Hall, Middletown, 514
Howe (John) House (The House with the Eagles), Bristol, 474
Howe (Julia Ward) Home (Oak Glen), Portsmouth, 507
Howells and Stokes: Turk's Head Building, Providence, 49, 50–51
Howland, Edwin L.
  St. Peter's by the Sea II, Narragansett, 369
  Wilcox Building, Providence, 45, 47, **48**
Hoyt (Virginia Scott) House (Ayrault House), Newport, 555
Hughesdale, 169
Huling (Alexander) House. *See* Wickford House (Alexander Huling House), Wickford, North Kingstown
Hull, John, 395
Humes, Albert, 158, 159
Hunt, John, 442
Hunt, Richard Morris, 14, 15, 16, 124, 530, 544, 546, 554, 555, 561, 570, 571, 576, 607
  Appleton (Thomas) House, Newport, 570
  Goelet (Ogden) House (Ochre Court), Newport, 18, **559**–560, 562
  Griswold (J. N. A.) House, Newport Art Museum, Newport, 13, 546, 547, **548**, 570, 574
  Linden Gate Porter's Lodge, Newport, 551–552
  Linden Place, Newport, 552
  Marquand (Henry) House, Newport, 570
  Perry (Oliver Hazard) House (Belcourt), Newport, 574–575
  St. Mary's Church, Portsmouth, Sarah Gibbs tomb, 506, 507
  Travers Block, Newport, **562**–563, 564
  Vanderbilt (Alva Smith and William Kissam) House (Marble House), Newport, 562, **572**, 573–574
  Vanderbilt (Cornelius, II) House (The Breakers (second)), Newport, 18, 544, 560, **561**–562
  Waring (Colonel George) House (The Hypotenuse), Newport, 555
Hunt, William Holman, 410
Hunt, William Morris, 13
Hunt and Hunt: Vanderbilt (Alva Smith and William Kissam) House (Marble House), Chinese Teahouse, Newport, 573, 574
Hunt family, 442
Hunt Farm, Little Compton, 489
Hunt (John) House and East Providence Water Pumping Station, East Providence, 442–443
Hunt (Pardon) Farm. *See* Manton (Olney)–Hunt (Pardon)–Farnum (Cyrus) Farm, Glocester
Hunt (Richard Morris) House, Newport. *See* Hilltop (Richard Morris Hunt House), Newport
Hunter, Thomas R., 553
Hunter, William, 523
Hunter (Thomas R.) House, Newport, 553
Hunter (William) House (Jonathan Nichols-Colonel Joseph Wanton House), 522–**523**, 524, 537
Huntington (J. T.) House, Middletown, 511

Huntoon (Jeannette B.) House, Providence, 115, 116
Hunts River, 351
Hurd, Frank, 412
Hurricane Hut. *See* Reed (V. Z., Jr.) House (Seafair, Hurricane Hut), Newport
Hussey, Clarence L.
  Barrington and Warren Bridges, Barrington, 453
Hussey Bridge, Wickford, North Kingstown, 362
Hussey Bridge, Wickford, North Kingstown, 362
Hutchins and French: Centreville National Bank, West Warwick, **331**
Hutchinson, Anne, 4, 498, 499
Hutchinson, Thomas, 589, 590
Hyde-Bridgham House, East Providence, 441
Hypotenuse, The. *See* Waring (Colonel George) House (The Hypotenuse), Newport

I. and A. Wilkinson, Valley Falls, 153
I. M. Pei and Partners, 55
Idaho, 485
Idlewild (house), Narragansett, 369
Imperial Cutlery Company (Vesta Knitting Mills), Providence, 57–**58**
Imperial Place Factory Row, Providence, 57
Independent Eastern Orthodox Church of the Resurrection, Pawtucket. *See* apartments (First Free Will Baptist Church, later Independent Eastern Orthodox Church of the Resurrection), Pawtucket
India Point. *See* Hall (Captain Palmer) House (India Point), Westerly
Indian and Puritan (sculpture), 51
Indian Avenue, Middletown, 511
  Field-Stone Mansion, 511
  Huntington (J. T.) House, 511
Indian Oaks estate, Warwick, 117
Indian Rock, Narragansett, 372
industrial buildings, 34, 455, 499
  Ashaway, Hopkinton, 423–424
  Brick Packaging Plant, East Providence, 441
  Brown and Sharpe Machine Tool Company complex, Providence, 34, 105, 125
  Chestnut Street area jewelry factories, Providence, 55–57
  Cottrell Factory, Westerly, 410
  Glenlyon Bleachery Plant, East Providence, 439, 443
  Hope Furnace, Scituate, 270
  J. J. Smith factory, Warren, 457
  Kaiser Aluminum and Copper Company plant, Portsmouth, 500
  Kennecott Warehouse, East Providence, 438–439
  Mechanics Machine Shop, Warren, 457
  Monocalcium Phosphate Plant, East Providence, 441
  Rumford Chemical Works (former), East Providence, 112, 434, 435, 439, 440–441
  small factory buildings, Warren, 457
  Washburn Wire Company Plant, East Providence, 438
  Water Street, Warren, 457
  Wheaton and Baker Rope Works, Warren, 462
  See also mills and mill complexes
Industrial National Bank Building, Providence. *See* Fleet Bank Building (Industrial National Bank Building), Providence
Industrial Trust, Warren. *See* Fleet Bank (Industrial Trust), Warren
Industrial Trust Company, Providence, 37
Industrial Trust Company Building (former), Westerly, 413, 414
Ingalls, Seth H.: Congregational Church, Bristol, 25, **477**–478
Ingraham (William A.) House, Pawtucket, 150

Innisfail, Block Island. *See* Van Nostrand (David) House (Innisfail), Block Island
inns. *See* hotels and inns
Insurance Building, RISD Buildings, Providence, 65–66
insurance business, 32
Interlocken Mill Bridge, Coventry, 334, **335**
International House (Byron Thomas Potter House), Providence, 111
International Lead and Zinc Research (ILZRO) House, Foster, 275, 277
International Studio, The (periodical), 150, 370
International Tennis Hall of Fame, Newport, 16
Irish (John and Fanny) House, Newport, 554
Irons and Russell Company Building, Providence, 56
Irons Homestead, Glocester, 298–299, 302
Irons House, Johnston. *See* Thomas Clemence–Irons House, Johnston
Irving B. Haynes and Associates, 38, 46, 62, 87, 88, 91
  Lime Rock Baptist Church, Lincoln, **201**, 287
Isham, Norman M., 71, 84, 85, 86, 195, 349, 357, 359, 360, 361, 364, 365, 388, 396–397, 409, 410, 415, 510, 518, 519, 521, 543, 545
Isidor Richmond and Carry Goldberg: Temple Sinai, Cranston, 179–**180**
Island Cemetery, Block Island, 608
Island Cemetery, Newport, 530
Islands, The, 498–610
Italianate houses on Benefit Street, Providence, 73
Ives, Hope Brown, 67, 99, 209, 322
Ives, Moses Brown, 270
Ives, Robert H., 270
Ives, Thomas Poynton, 67, 84, 104, 189, 190, 209, 322
Ives family, 84, 190, 322
Ives (Thomas Poynton) House, Providence, 11, **90**–91, 110, 113, 117

J. and P. Coats Ltd. *See* Coats and Clark (Conant Thread Mill Complex, later J. and P. Coats Ltd.), Pawtucket
J. and W. Slater, 243
J. Cahoone and Sons (builders), 539
J. D. Johnston Mill, Newport, 538
J. Harold Williams Amphitheater, Hopkinton, 421
J. J. Newberry store, West Warwick, 331
J. J. Smith factory (Stubb's Wharf), Water Street, Warren, 457
J. W. Bishop Company, 438
Jack (Alexander, Jr.) House, Newport, 536
Jackson, Frederick Ellis, 65, 116
Jackson, Robertson and Adams, 38, 91, 124, 319, 421
  College Building, Rhode Island School of Design, Providence, 64–65
  Federal Building (Kennedy Plaza Post Office), Providence, 39
  Glendale New Village, Burrillville, 302, 306–**307**, 310
  Harrisville civic buildings, Burrillville, 310–312
  Huntoon (Jeannette B.) House, Providence, 115, 116
  Morris Plan Building, Providence, 66
  New Village experimental prefabricated mill housing, Burrillville, 306, 309, 310
  New Village mill housing, Burrillville, 306, 309–310
  Pascoag Public Library, Burrillville, 314, 441
  Pontiac Free Library, Warwick, 323
  Providence County Courthouse, 64–65, 78, 86, 413
  South County Public Service Building, Westerly, 413, 414
  State Police Barracks, Lincoln, 198, 357
  State Police Barracks, North Kingstown, 357
  State Police Barracks, Portsmouth, 357, 503
Jackson (Donald E.) House, Providence, 115–116

jails and prisons
   Adult Correctional Institution, Cranston, 129, 177, **182**–183
   Bristol County Jail (former), Bristol, 476
   Kent County Jail (former), East Greenwich, 346
James, Henry, 520, 521
James (Arthur Curtis) Estate (Beacon Hill), Newport, 580
James Barnes Architects, 87
James (Lucy Worth) House (Normandie), Newport, 577
Jamestown, 20, 28, 365, 387, 489, 507, 581–604
   map, 584
Jamestown Bridge, South Kingstown/Jamestown, 387, 589, 590
Jamestown Casino (Shoreby Hill Club), Jamestown, 603
Jamestown Historical Society, 590, 591
Jamestown Island. *See* Conanicut Island
Jamestown Philomenian Library and Sydney L. Wright Museum of Indian Artifacts, Jamestown, 586, 591, 593
Jamestown Philomenian Library Association, Jamestown, 586
Jamestown Village, Jamestown, 585, 601–602
Jamestown Windmill, Jamestown, 589–**590**
Jane Pickens Theater (Zion Episcopal Church), Washington Square, Newport, 521
Janks, Alvin, 159
Japan, 513, 526
Jarding, J. J., and Constable Brothers: Earles Court Houses and Water Tower, Narragansett, 371
Jefferson, Thomas, 6, 8, 11, 77, 387, 542
Jelke (F. Frazier) House (Eagles Nest), Newport, 577
Jenckes, John, 162
Jenckes brothers, 225
Jenckes (John) House, North Providence. *See* Smith (Joseph)–Jenckes (John)–Cushing House, North Providence
Jenckes (Joseph) House, Providence, 75
Jenckes Mills, Woonsocket, 225–226, 228, 229
Jenckes (Thomas) House, Providence, 78
Jencksville, Woonsocket, 219
Jenk, Daniel, 198
Jenkins (Nathaniel) House, East Providence, 442
Jenks, Herbert S., 152
Jenks, Joseph, 137
Jenks, Nathan, 144
Jenks (Albert A.) House, Pawtucket, **149**
Jenks (Almet) House, Little Compton, 493
Jenks and Ballou (engineers), 59
Jenks family, 149
Jenks House, Lincoln, 198
Jenks Park, Central Falls, 153, 158–159
Jennewein, C. P., 65
Jennys Lane houses, Barrington, 452, 453
Jesse M. Smith Memorial Library, Burrillville, 311, 314, 441
Jesse Metcalf Building, RISD Buildings, Providence, 66
jewelry district, Providence, 36, 52, 55–60
   Chestnut Street area factories, 55–57
   map, 56
Jewett, Charles C., 102
Jewish Cemetery, Newport, 542–543
"Jewish Cemetery in Newport, The" (H. W. Longfellow), 543
Jewish centers
   Community Center (Levi Gale House), Newport, 543
   Hillel House (Froebel Hall), Brown University, Providence, 108
   See also synagogues

Jewish Community Center (Levi Gale House), Newport, **543**
Jews, 7, 534
Jillson (Luke) House, Cumberland, 218–219
Job Armstrong Store, Glocester. *See* Art Center and Glocester Heritage Society (Job Armstrong Store), Glocester
John Brown Francis School, Warwick, 318–319
John Carl Warnecke and Associates, 50
John Carter Brown Library, Brown University, Providence, **101**, 158
John D. Rockefeller, Jr., Humanities Library, Brown University, Providence, 103, 104
John Hay Library, Brown University, Providence, 103
John Kennedy Mill, Central Falls. *See* Stafford Manufacturing Company Mill (John Kennedy Mill), Central Falls
John Nicholas Brown Center for the Study of American Culture (Joseph Nightingale–Nicholas Brown House), 91, 92, 101
John Stevens Shop, Newport, 512
John Stevens Shop and Stevens Houses, Newport, 528, 530
Johnson, Augustus, 540
Johnson, Philip: Albert and Vera List Art Building, Brown University, Providence, 84, **85**, 94, 103
Johnson, Rudolphus, 461
Johnson and Wales University, Providence, 42, 45, 54
   Classroom Building (Summerfield Building), 42, 45, 54
Johnson (Elisha) House, Newport, 532–533
Johnson family, 431
Johnson (George) House, Caretaker's Cottage for the William Slater Estate, North Smithfield, 245
Johnson (Rudolphus B.)–Luther (John) House, Warren, 461–462
Johnston, 135, 168–175
   map, 169
Johnston, Augustus, 168–169
Johnston, J. D., 538, 546, 602
   First Presbyterian Church, Newport, 516, 518
   Johnston (J. D.) Bungalow (Daybreak Cottage), Jamestown, 585
   MacCawley (Captain E. V.) House (Mist), Jamestown, 601
   Newport City Hall, Newport, 518
   St. George's School Wyn Wac, Middletown, 515
   Wharton (Joseph) House (Marbella, Horsehead), Jamestown, 596, 597, 598
   Wharton (J. S. Lovering) House (Clingstone), Jamestown, 599
   Woodward (Samuel) House (Onarock), Jamestown, **601**, 602
   Yardley (Jane) House, Newport, 553
Johnston Historical Society (Elijah Angell House), Johnston, 170
Johnston (J. D.) Bungalow (Daybreak Cottage), Jamestown, 585
Johnston (J. D.) Office, Newport, 546
Johnston quarries, 47
Jones, George Noble, 565
Jones, Inigo, 167, 521
Jones, Owen, 110
Jones, Pembroke, 572–573
Jones, W. Alton, 343, 393
Jones (George Noble) House (Kingscote), Newport, **565**–566, 569
Jones (John D.) House, Providence, 93
Jones (Pembroke) House (Sherwood), Newport, 572–573

Joseph Benn and Company (Greystone Mill), North Providence, 166, **167**, 168, 170, 482
Joseph C. Sweeney School, Burrillville, 303, 315
Joseph W. Martin Memorial Home (Charles Smith-William Winslow House), Warren, 457–**458**, 464
Josephs (Lyman C.) House (Louisiana), Middletown, **513**
Judkins House, East Providence, 442
Jung/Brannen Associates: Citizens Bank Building, Providence, 40

Kabyun, Narragansett. *See* Carver (Emma B.) House (Kabyun, Sonnenschein), Narragansett
Kahn, Albert, 21, 356
Kahn, Louis, 444
Kaiser Aluminum and Copper Company plant, Portsmouth, 500
Karolik, Maxim, 567
Kay–Catherine–Old Beach, Newport, 546–557. *See also* Top of the Hill neighborhood
  map, 547
Kay family, Newport, 12
Keegan, Kent, 275, 277
Keely, Patrick C.
  Cathedral of Saints Peter and Paul, Providence, **54**–55, 160
  St. Charles Borromeo, Woonsocket, 227, 228, 235–236
  St. Joseph's Church, Providence, 54
  St. Mary's Church, Newport, 581
Keen, Charles Barton, 451
Keller, George
  By-the-Sea, Westerly, 407
  Sea Swept (Ocean Mount), Westerly, 407
Kelley, Wilbur, 205
Kelly, S. D., 491
Kendall, H. E., 165
Kendall, Henry P., 245, 246, 248, 302
Kendrick Avenue School, Woonsocket, 224–225, 227
Kendrick (John K.)–Prentice (George W.)–Tirocchi (Anna) House, Providence, **130**
Kennecott Warehouse, East Providence, 438–439
Kennecott Wire and Cable, 438
Kennedy Plaza, Providence, **33**, 36–37, 39, 40, 88
Kenney, Francis, 373
Kent, Willard: Pump House Restaurant (Reservoir Pumping Station), South Kingstown, 385–386
Kent, William, 542, 545
Kent, Cruise and Associates: Roger Williams University, Bristol, 25, **480**
Kentish Guards Armory, East Greenwich, **348**
Kenyon, Abiel, 425
Kenyon, Charlestown/Richmond, 397, 401, 424, 425
Kenyon Mill, Charlestown/Richmond, 401, 425
Kenyon's Department Store, Wakefield, South Kingstown, 379
Kerry Hill, Newport, 530
Keyes Associates: Coastal Resources Institute, University of Rhode Island, Kingston, 392
Kiley House, Johnston, 171
Kilham and Hopkins, 479
Kimball, Fiske, 566
Kimball, James, 107
Kimball (Governor Charles Dean) House (Kymbolde), South Kingstown, **387**
Kimball (James M.) House, Providence, 104, 107
Kimball Wildlife Refuge, Charlestown, 397
King, David, Jr., 566, 567
King, Frederick Rhinelander: Wetmore (Edith) Garden House, Newport, 570, 571
King, Gordon, 564

King, James Allen, 123
King, LeRoy, 564
King Block, Newport, 564, 565
King (Edward) House, Newport, 127, 233, 538, **566**, 569
King family, 566
King (George Gordon) House (Edgehill Farm), Newport, 551, 580
King Hall (Robert W. Taft House), Brown University, Providence, 118, 119, 120
King-Birkhead House, Newport, 557
King (LeRoy) House, Newport, 567
King Philip's War
  background and outcome, 454
  and Bristol, 454, 466, 479
  and Central Falls, 153
  and Johnston, 171, 175
  and Lincoln, 206
  and Providence, 32, 162
  and Smithfield, 249
King Street Railroad Bridge, East Greenwich, 345
King Tom Farm, Charlestown, 399–400, 402
King's Arms Tavern, Newport. *See* Coddington (Nathaniel)–Walker (Thomas) House (King's Arms Tavern), Newport
King's Commissioners of Rhode Island, 187
Kings County Courthouse (former), South Kingstown, 390
Kingscote. *See* Jones (George Noble) House (Kingscote), Newport
Kingston, South Kingstown, 374, 375, 388–392
Kingston Academy, South Kingstown, 388
Kingston Congregational Church and Parish House (Thomas P. Wells House), South Kingstown, 390–391, 394
Kingston Free Library and Little Rest Archives (Kings County Courthouse; Little Rest Museum, Old County Records Office), South Kingstown, **390**
Kingston Railroad Station, South Kingstown, 383, 388, 392–393
Kinney's Casino, Narragansett, **373**
Kinnicutt Tavern and Post Office, Barrington, 448
Kite, Palmer Associates, 109
  Roger Williams University, Bristol, School of Architecture, Art, and Historic Preservation, 480
Kite, William, 287
Kitt Matteson Tavern, West Greenwich, 343
Klapp, Lyman, 110
Klapp (Lyman) House, Providence, 110, 111, 120
Klauder, Charles: Jesse Metcalf Building, Providence, 66
Klotz House, Westerly, 409
Knight, Addison, 216
Knight, B. B. and R., 177, 179, 323, 328, 329, 330, 331, 332, 343, 417
Knight (Edward) House (Clarendon Court), Newport, 574
Knight Farms Subdivision and Knight Farm, Scituate, 272–273
Knight Memorial Library, Providence, 132
Knowles (Henry) House, Newport, 526
Knowles Row, Providence, 78
Knowles (Samuel) House, East Greenwich, **348**
Kotzow (Louis) House, Pawtucket, 138, 139
Koulbanis, Charles J., 412–413
Kymbolde. *See* Kimball (Governor Charles Dean) House (Kymbolde), South Kingstown

La Farge, C. Grant, 382
La Farge, John, 126, 382, 511, 531, 536, 537, 544, 546, 553

La Farge (John) House, Newport, 546
La Montagne, Armand, 214
  seventeenth-century replica houses, Scituate, 267–**268**
Lacey (Dustin) House, Providence, 75
Ladd, John: Littlefield (Captain Augustus) House (Governor Charles Van Zandt House), Newport, 538
Ladd, Samuel J.: St. George Maronite Catholic Church (St. Paul's Episcopal Church), Pawtucket, 146–147
Ladies Reading Society. *See* Female Library Association, East Providence
Lafayette, Marquis de, 77, 469
Lafayette, North Kingstown, 363–364
Lafayette Mill, North Kingstown, **363**
Lafayette Worsted Company Mills, Woonsocket, 226
Lake Shore Dining Hall, Narragansett Hotel, Block Island, 608
Land Trust Cottages, Middletown, 515
Langley, Batty, 71, 76, 542
Langley (John) House, Newport, 536
Langworthy, Leslie P.: Courthouse Center for the Arts (Washington County Courthouse), South Kingstown, 392
Lapham, Benedict: Lapham's Cotton Mill, West Warwick, 331–332
Lapham, Enos, 332
Lapham's Cotton Mill, West Warwick, 331–332
Larches, The (Larchwood Inn), South Kingstown, 379
Larned (Louise Alexander) Cottage (The Boulders), Jamestown, 598
Latrobe, Benjamin Henry, 124
Lauderdale Building, Providence, 51
Law (Jonathan) Farmhouse, Jamestown, 592, 593
Lawless, James, 475
Lawless (James) House, Bristol, 474–475
Lawrence, Massachusetts, 138, 326, 329
Lawrie, Lee, 146
Lawton, Darius, 303
Lawton, Thomas, 432
Lawton (Darius P.) House, Burrillville, 303–304
Lawton Mill, Exeter, 431–**432**
Lawton Owens Mill, Glocester, 292, 294
Lawton (Robert) House, Newport, 544
Lawtonville, Exeter, 430
Le Corbusier, Charles Edouard (Jeanneret), 44, 45, 184, 287, 324
Lebanon Mill Company, Pawtucket, 145
Lederer Building, Providence, 52–53
Lederer Theatre (Emery's Majestic Theatre), Providence, 42
Legg, James, 303
Lehr, Harry, 576
Leonard, Lester, 439
Leonard (Lester) House, East Providence, 439
Lescaze (architect), 184
Levy, Austin T., 245, 253, 302, 303, 306–314
Levy, June, 309
Levy (Arthur J.) House, Cranston, 184, 318
Levy (Austin T.) House, Burrillville. *See* Tinkham (William)–Levy (Austin T.) House (Southmeadow), Burrillville
Levy family, 313
Lewis-Dickens Farm, Block Island, 608–609
Lewis (John D.) House, Providence, 107
Lewis (John)–Card (Captain William) House, Westerly, 409
Liberty Street School, Warren, 462
libraries, 473
  Adams Public Library, Central Falls, 158
  Barrington Town Hall, 447, 451
  Bridgham Memorial Library, East Providence, 441–442
  Brown University Library Buildings, Providence, 102–104
  Brownell Library, Little Compton, 496
  Carnegie Corporation, 158, 441
  Deborah Cook Sayles Memorial Library, Pawtucket, 145–146, **147**, 158
  East Greenwich Free Library, East Greenwich, 347
  Female Library Association, East Providence, 441
  George Hail Free Library, Warren, **459**
  Jamestown Philomenian Library and Sidney L. Wright Museum of Indian Artifacts, Jamestown, 586, 591
  Jesse M. Smith Memorial Library, Burrillville, 311, 314, 441
  John Carter Brown Library, Brown University, Providence, **101**, 158
  John D. Rockefeller, Jr., Humanities Library, Brown University, Providence, 103, 104
  John Hay Library, Brown University, Providence, 103
  Kingston Free Library and Little Rest Archives (Kings County Courthouse; Little Rest Museum, Old County Records Office), South Kingstown, **390**
  Knight Memorial Library, Providence, 132
  Newport Reading Room, 554–555, 563
  North Kingstown Free Library, Wickford, North Kingstown, 362
  Pascoag Public Library, Burrillville, 314, 441
  Peace Dale Library (Narragansett Reading Room and Library), South Kingstown, 384, **385**, 386
  Pontiac Free Library, Warwick, 323
  Providence Athenaeum, 84, **85**
  Providence Public Library, 42–43, 51, 132, 146
  Public Library (Foster Center School, Hemlock School), Foster, 278
  Redwood Library, Newport, 7, 12, 510, 527, 528, 532, 540, 541, **544**–546
  Rhode Island School of Design, 65
  Robinson Hall (University Library), Brown University, Providence, 102–**103**
  Rogers Free Library, Bristol, 471
  Sciences Library, Brown University, Providence, 103–104
  Westerly Public Library and Art Gallery (Memorial and Library Association), 411, **412**–413
Lieber (Matilda) House, Newport, 553
Light of the World (W. H. Hunt painting), 410
lighthouses, 19, 20, 21–23, 488
  Beavertail Lighthouse, Jamestown, 20, 21, 23, 583, 593, **594**–595
  Bristol Ferry Lighthouse, Bristol, 480
  Conanicut Point Light (North Light), Jamestown, 583, 588–589
  Goat Island, Newport, 373
  North Light (Sandy Point Light), Block Island, 22, 608
  Point Judith Lighthouse and Coast Guard Station, Narragansett, 373
  Ponham Rocks Lighthouse, East Providence, 445
  South East Light, Block Island, 22–23, 608, **609**
  Watch Hill Lighthouse, Westerly, 373, 406–407
Lillibridge, Robert, 527
Lily Pad, South Kingstown, 381
Lime Rock, Lincoln, 28, 187, 199–201, 221, 249
Lime Rock Baptist Church, Lincoln, **201**, 287
Lime Rock Crushing Plant, Lincoln, 199–**200**
Lime Rock Village Center, Lincoln, 200–201
Lincoln, 142, 155, 159, 186–205, 206, 208, 209, 210, 211, 214, 221, 238, 249, 326, 424
  map, 188
Lincoln, Abraham, 237, 511

Lincoln Textile Corporation, Woonsocket. *See* Museum of Work and Culture (Barnai Worsted Company Dye Works), Woonsocket
Lindeberg, Harrie T.: van Beuren (Michael M.) House (Gray Craig), Middletown, **512**–513
Linden Gate Porter's Lodge, Newport, 551–552
Linden Place, Newport, 552
Linden Place (DeWolf-Colt house), Bristol, 25, 197, 470, 471–472, 476, 477
  Carriage House, 472
Line Baptist Church, Foster. *See* Union Free Will Baptist Church (Line Baptist Church), Foster
Lippett, Robert, 109
Lippincott family, 27, 593
Lippincott (J. Bertram) House I (Meeresblick), Jamestown, 593
Lippincott (J. Bertram) House II, Jamestown, 595–596
Lippitt, Charles, 327
Lippitt, Christopher, 327
Lippitt, Henry, 109, 110, 586
Lippitt, Robert, 109
Lippitt, West Warwick, 324, 326, 327–328, 334
Lippitt (Christopher) House. *See* Lippitt Hill Farm, Cranston
Lippitt family, 587, 588
Lippitt Hall, University of Rhode Island, Kingston, 391, 392
Lippitt (Henry) House I, Providence, 109–110
Lippitt (Henry) House II, Providence, 109, **110**
Lippitt Hill Farm, Cranston, 175, 185
Lippitt Mill Office and Warehouse, Woonsocket, 237
Lippitt Mills, West Warwick, 185, 327–**328**, 432
Lippitt (Moses)–Green (Cornelia Burges) House, Providence, 93
Lippitt (Robert Lincoln) House, Providence, 109
Lippitt (William) House. *See* Lippitt Hill Farm, Cranston
Lippold, Richard, 502
Lisle (Frank B.) House, Providence, 115, 116
List Art Building. *See* Albert and Vera List Art Building, Brown University, Providence
Liszt Road, Shelter Harbor, Westerly, 403
Little Compton, 3, 28, 480, 481, 483, 484, 487, 489–497, 601
  map, 490
  Town Hall, 496
Little Compton Commons, 492, 495–496
  United Congregational Church burying ground, 495
Little Compton Historical Society (Wilbour House), 491–492
  Burleigh studio boat, 492
  Pyramid Schoolhouse, 491, 492
Little Rest. *See* Kingston, South Kingstown
Little Rest Museum. *See* Kingston Free Library and Little Rest Archives (Kings County Courthouse; Little Rest Museum, Old County Records Office), South Kingstown
Littlefield (Captain Amazon) House, Block Island, 607
Littlefield (Captain Augustus) House (Governor Charles Van Zandt House), Newport, 538
Lockwood, Greene (engineers), 329
Loew's State Theatre, Providence. *See* Providence Performing Arts Center (Loew's State Theatre), Providence
Loewy, Raymond, 143
Logan (Mrs. George W.) Cottage (Southwinds), Jamestown, 591
London, 167, 521, 534, 545, 546, 568
Londonderry, Little Compton. *See* Abbott (Cornelia C.) House (Londonderry), Little Compton
Londonderry, New Hampshire, 495

Long House, Charlestown, 402
Long Wharf, Newport, 7, 516, 520, 521, 522
Long Wharf Mall, Newport, 522
Longfellow, A. W., 484
Longfellow, Henry Wadsworth, 116, 543
Longfield, Bristol. *See* Gibson (Charles Dana) House (Longfield), Bristol
Longstaff, George, 412
Longstaff and Hurd: Westerly Public Library and Art Gallery (Memorial and Library Association), Westerly, 411, **412**–413
Lonsdale, Lincoln, 153, 188, 189, 190–193, 202, 205, 206, 208, 209–211, 212
Lonsdale Baptist Church, Lincoln. *See* Dance Studio (Lonsdale Baptist Church), Lincoln
Lonsdale Company, 189, 190, 191, 192, 205, 208, 209, 210, 211, 213, 326
Lonsdale Hall, Lincoln, 190, 191
Lonsdale Mill. *See* Ann and Hope Store Complex (Ann and Hope Mill, Lonsdale Mill), Cumberland
Lonsdale Mill housing, Cumberland, 128, 234
Looff, Charles I. D., 151, 445
  Crescent Park, East Providence, Carousel, 445–446
Loridan, Charles: Lafayette Worsted Company Mills, Woonsocket, 226
Lorillard, Pierre, 561
Lorin, Charles, 223
Loring, Charles G.: Brownell Library, Little Compton, 496
Lorraine Mill, Pawtucket, 193
Los Angeles, California, 444
Lottery House, Westerly, 408
Louis XIII, king of France, 115
Louis XV, king of France, 111
Louisiana, Middletown. *See* Josephs (Lyman C.) House (Louisiana), Middletown
"Louisquisset," 186
Louisquisset Pike, 198, 199
Love, Frederick, 102
Lovering (Joseph) House (Riven Rock), Jamestown, 591
Low (William G.) House, Bristol, **468**, 526, 596
Lowell, Guy: Brown University, Providence, Carrie Tower, 99
Lowell, Massachusetts, 138, 190, 247, 326, 329
Lower Simonsville. *See* Thornton
Lucas, Augustus, 540
Lucas (Augustus) House, Newport, 539–540
Luce, Clarence S., 51, 546, 553
  Conover (Reverend Coit) House, Middletown, 512
  Hodgson (John M.) Cottage (The Flower Cottage), Newport, 569
  Hunter (Thomas R.) House, Newport, 553
  Josephs (Lyman C.) House (Louisiana), Middletown, **513**
  Noyes-Luce House, Newport, 553
  Pell (Mrs. Archie D.) House, Newport, 556
  Sargent (Letitia B.) House (Aufenthalt), Newport, 554
  Stevens (Mary and Anne) House, Newport, 553
  Warden (Rear Admiral Reed) House, Newport, 555–556
Luce (Clarence) Office, Newport, 546
Luce Hall, U.S. Naval War College, Newport, 529, **530**
Ludlow and Peabody, 412
Lumb, Mae Potter, 151
Lumb, Ralph, 151
Lumb Knitting Company, 151
Lumb (Ralph A.) House, Pawtucket, 151–152
Lushington, Nolan, 459

Index 675

Luther (John) House, Warren. *See* Johnson (Rudolphus B.)–Luther (John) House, Warren
Lutheran churches. *See* churches, Lutheran
Lutyens, Edwin, 513
Lyman, Daniel, 160
Lyman, David, 519
Lyman Gymnasium, Brown University, Providence, 101–102
Lyman House, Newport. *See* Wanton-Lyman-Hazard House (Stephen Mumford House), Newport
Lymansville, 160–161
Lyon, George, 480
Lyon, Marjorie McKee, 480
Lyric Theater, Warren, 464

MacCawley (Captain E. V.) House (Mist), Jamestown, 601
MacConnell and Walker: John Brown Francis School, Warwick, 318–319
Machado House (Ellen Dexter Sharpe House), Brown University, Providence, 105
machine tool manufacture, 32, 34, 125, 196
  Brown and Sharpe Machine Tool Company complex, Providence, 34, 105, 125
MacKenzie, Neil, 216
Mackenzie, William: Reed (V. Z., Jr.) House (Seafair, Hurricane Hut), Newport, 577
Maddock Alumni Center (William Giles Goddard-Hope Goddard Iselin House), Brown University, Providence, **96**–97
Magoun (H. S.) House, Lincoln, 190, 191
Maguire Associates: FM Global (Allendale Mutual Insurance Company Headquarters Building), Johnston, 28, 164, 170, 173–174
Maillart, Robert, 192
Main Street
  Ashaway, Hopkinton, 423
  Chepachet, Glocester, 293–295
  Hope Valley, Hopkinton, 419
  Hopkinton City, Hopkinton, 422
  Providence, 29, 32, 33, 34, 60–71, 72, 77
  Warren, 460
  Wickford, North Kingston, 361
  Woonsocket, 222
Main Street Bridge, Lincoln, 201
Main Street Historic District, Wickford, North Kingstown, 358
Maine, 326, 482, 565
Maintenance Center, T. F. Green State Airport, Warwick, 319-**320**
Mallinckrodt, Edward, 603
Mallinckrodt (Edward) Cottage, Jamestown, 603
malls. *See* stores and shopping centers
Maltby (L. U.) House (Ninicroft Lodge), Block Island, 607
Manchester, Josephine, 494
Manchester, New Hampshire, 138, 329
Manchester family, 494
Manchester House, Tiverton. *See* Briggs-Manchester-Beattie House, Tiverton
Manchester Street Plant, Providence. *See* Narragansett Electric Company Power Plants, Providence
Mann, Samuel, 201
Mann family, 201
Mann (John Preston) Mann, Newport, 534
Manning, James, 97, 464
Manning, Warren, 411
Manning Hall, Brown University, Providence, 99, **100**, 102
mansarded houses, Westerly, 410

Mansfield and Lamb, 243
Manton, 169
Manton (Olney)–Hunt (Pardon)–Farnum (Cyrus) Farm, Glocester, 296, **297**
Manville, Lincoln, 188, 189, 201–202, 204
Manville Company, 201, 202
Manville Mill tenements, Lincoln, **202**
Mapleville, Burrillville, 301, 302–304, 305
Mapleville Methodist Episcopal Church. *See* Seventh-Day Adventist Church (Mapleville Methodist Episcopal Church), Burrillville
Mapleville School. *See* apartments (Mapleville School), Burrillville
Marbella. *See* Wharton (Joseph) House (Marbella, Horsehead), Jamestown
Marble House. *See* Vanderbilt (Alva Smith and William Kissam) House (Marble House), Newport
Marble's Forge, Water Street, Warren, 457
Marc Harrison, Kent Keegan, and Barry Fitzpatrick: International Lead and Zinc Research (ILZRO) House, Foster, 275, 277
Marchant, Henry, 393
Marchant (Henry) House (Joseph Babcock House), South Kingstown, 386, 387, 393–394
Marcotte firm (furniture makers), 91
Marcus Aurelius, 102
Margin Street houses, Westerly, 409
Marie Antoinette, queen of France, 580
Market House, Newport. *See* Brick Market, Newport
Market House, Providence, 60, 65, 66, 306
Market Square, Rhode Island School of Design (Market House), Providence, 65
Market Square, Woonsocket, 222, 237
Marquand, Henry, 552
Marquand family, 530
Marquand (Henry) House, Newport, 570
Martha's Vineyard, Massachusetts, 586
Martin and Hall, 95, 134, 449, 472
  Caesar Misch Building, Providence, 53–54
  Edwin A. Smith Building, Providence, 42
  Irons and Russell Company Building, Providence, 56
  Peck (Leander R.) School, Barrington, 452
  Squantum Association, East Providence, clubhouse, **444**, 445
  trolley shelter, Providence, 36
Martin, Hall and Howe/Clarke, Spaulding and Howe: American Textile Mill, Pawtucket, 140
Martin (Joseph W. Memorial) Home. *See* Joseph W. Martin Memorial Home (Charles Smith–William Winslow House), Warren
Marvell, Thomas: Marvell (Thomas) House, Little Compton, 493
Marvell (Thomas) House, Little Compton, 493
Marycoo, Occramar. *See* Gardner, Newport
Marylebone Chapel, London, 67
Mason, Ellen, 505, 551
Mason, George Champlin, Jr., 543, 565, 566
  Schermerhorn (Edward H.) House (Chepstow), Newport, 558
  restoration, Newport Historical Society and Seventh Day Baptist Meeting House, Newport, 7, 543
Mason, George Champlin, Sr., 529, 545, 553, 554, 555, 558
  Bay Voyage Hotel, Jamestown, **604**
  Callender School, Newport, 529
  Codman House, Bristol, 475–476
  Fort Adams Eisenhower Building, Newport, 580
  Mason (George Champlin, Sr.) House (Sunnyside), Newport, 548, **549**
  65 Merton Road, Newport, 557–558

Mason, Ida, 505
Mason, Perez: Kendrick (John K.)–Prentice (George W.)–Tirocchi (Anna) House, Providence, **130**
Mason (George Champlin, Jr.) House, Newport, 553
Mason (George Champlin, Sr.) House (Sunnyside), Newport, 548, **549**
Mason (Miss Ellen) House, Newport. *See* St. Michael's School (Miss Ellen Mason House), Newport
Masonic Friendship Lodge No. 7, Glocester, 294
Masonic Hall, Wickford, North Kingstown. *See* First Baptist Church Parish House (Masonic Hall), Wickford, North Kingstown
Masonic Hall (former), Providence, 65
Masonic Temple, St. Johns Lodge, Newport, 539
Masonic Temple, Washington Lodge No. 3, Warren, **461**
Massachusetts Institute of Technology, 104, 471
  School of Architecture and Planning, 501
Massachusetts State House, 44
Massasoit (Wampanoag sachem), 460
Massie Wireless Station "PJ," East Greenwich, 350
Mathewson, David, 308
Mathewson (Allen C.) House I, Barrington, 452
Mathewson (Allen C.) House II, Barrington, 453
Mathewson House (Carpenter Gothic), Barrington, 453
Mathewson Memorial Tower, St. John's Episcopal Church (Red Church), Barrington, 452, 453
Mattasoit (Wampanoag chief), 186
Mattatuxet River, 351
Matthews (Brander) House (Shingle-nook), Narragansett, 369
Matunuck Hills, South Kingstown, 374, 396
Mauran, Julia Lippitt, 185
Mauran (Frank, Jr.) House, Providence, 120–121
Mawdsley, John, 537
Mawdsley (Captain John) House, Newport, 537
Maxfield's ice cream parlor, Warren, 463
Maxwell, James, 456
Maxwell (Elder Samuel)–Maxwell (James) House, Bristol, 456
Mayes, William, 531
McCaughley (Edward J.) House, Pawtucket, 150, 151
McConnell and Walker, 347
McGregor, Alexander, 579
  Newport Artillery Company, Newport, 78, 348, 541
  Swan (Gustavus) House (Swanhurst), Newport, 569
McIntire, Samuel, 567
McIntyre (Jerry) House, Jamestown, 598–599
McKay's Front Porch (Robert Rodman House), North Kingstown, 363
McKenzie, Thomas, 415
McKim, Charles Follen, 1, 14, 15, 52, 123, 510, 523, 524, 549, 553, 564
  Fairchild Barn, Newport, 526
  Morris (Wistar) House (Highland), Jamestown, 596–597
  Newhall (Daniel S.) House (The Monitor; The Round House), Jamestown, 599–600
  Reverend William S. Child Schoolhouse, Newport, 526
  Wormeley (Katherine Prescott) House, Newport, 549–550
McKim and Mead, 38. *See also* McKim, Mead and Bigelow; McKim, Mead and White
McKim, Mead and Bigelow, 14. *See also* McKim, Mead and White
McKim, Mead and White, 13, 14–15, 17, 42, 43, 51, 52, 106, 111, 114, 119, 132, 381, 387, 504, 505, 524, 537, 546, 547, 549, 550, 566, 572, 576, 601, 604
  Bell (Isaac, Jr.) House (Edna Villa), Newport, **567**–568

Brown University, Providence, Faunce House (Rockefeller Hall),, 101
Casino, The, Newport, 16, 371, 373, 555, 562, 563–**564**, 565, 567
Coast Guard Station (U.S. Lifesaving Station), Narragansett, 371
Coleman (Samuel) House (Whileaway, Boxcroft), Newport, 550–551
Edgar (Commodore William G.) House, Newport, 547–548, **549**, 567
Goelet (Robert) House, Newport, 559, 560
King (George Gordon) House (Edgehill Farm), Newport, 551, 580
King (LeRoy) House, Newport, 567
Low (William G.) House, Bristol, **468**, 526, 596
Oelrichs (Tessie and Herman) House (Rosecliff), Newport, 571–**572**
Providence (sixth) State House, 10, 32, 33, 39, 40, 77, 106, 122–124, **123**, 125, 381
Skinner (Frances L.) House (Villino), Newport, **550**, 551, 567
Stonelea, Narragansett, **370**, 372
Tilton (Samuel) House, Newport, 548–549, 567
Towers, The, Narragansett Pier Casino, Narragansett, 368, 370, 371, 381
McLean and Wright: Adams Public Library, Central Falls, 158
McNeill, Major William Gibbs: King Street Railroad Bridge, East Greenwich, 345
Mechanics Machine Shop (Old Dye House), Water Street, Warren, 457
Meeresblick. *See* Lippincott (J. Bertram) House I (Meeresblick), Jamestown
Meeting Association, Portsmouth Camp Meeting Association, Portsmouth, 502
meeting houses
  Elder Hammond's Meeting House (former), Foster, **279**–280
  Friends Meeting House, Jamestown, 7, 583, 589, 590
  Friends Meeting House, Little Compton, 7, 490, **491**
  Friends Meeting House, Newport, 7, 522, 530, 531
  Friends Meeting House, North Smithfield, 239, 241
  Friends Meeting House, Portsmouth, 502–503
  Friends Meeting House (former), Foster, 266, 279, 283–284, 286, 487
  Newport Historical Society and Seventh Day Baptist Meeting House, Newport, 7, 543
  Saylesville Meeting House, Lincoln, 194–195, 197
  Sprague Meeting House (St. Bartholomew's Episcopal Church), Cranston, 179
  Union Meeting House (former), Portsmouth, **503**
  Wickford Meeting House, 24
Meloccaro, N. Robert, 181, 182
Meloccaro, Nazzareno, 182
  Garden City, Cranston, 181–182
Melville, David, 595
Memorial and Library Association. *See* Westerly Public Library and Art Gallery (Memorial and Library Association)
Memorial Hall (Central Congregational Church (first)), Rhode Island School of Design, Providence, **80**, 81–83, 114, 147
memorials. *See* monuments and memorials
Merchants Bank Building, Providence, **50**–51
Merton Road (number 65), Newport, 557–558
Metacomet. *See* Philip, King
Metcalf, Davis, 216
Metcalf, Ebenezer, 216
Metcalf, Ebenezer, Sr., 216
Metcalf, Helen Adelia Rowe, 68, 69, 79, 80, 128

Metcalf, Jesse H. (father), 68, 69, 79, 80, 127, 128
Metcalf, Jesse H. (son), 128
Metcalf, Joseph, 216
Metcalf, Louisa Sharp, 128
Metcalf, Stephen O., 79, 80
Metcalf brothers, 216
Metcalf family (Cumberland), 216
Metcalf family (Providence), 81, 106, 301, 305, 308
Metcalf (Stephen O.) House. *See* RISD President's House (Stephen O. Metcalf House), Providence
Methodist churches. *See* churches, Methodist
Methodists, 230, 586
Metropolitan Park Commission, Providence, 120
Miantonomi (Narragansett chief), 351
Michelangelo, 51, 123
Mid state (Rhode Island), 316–350
Middletown, 18, 211, 498, 499, 507–516, 524, 545, 546, 589, 595, 604
  map, 508
Middletown Historical Society, 513
Mies van der Rohe, Ludwig, 103, 439
Mikado (Gilbert and Sullivan), 513
Milk Can, North Smithfield, 239
Mill, John Stuart, 384
Mill, The, Little Compton. *See* Slicer (Adeline E. H.) House (The Mill), Little Compton
mill boardinghouse, North Kingstown, 363
mill dams
  dam and headgate, waste gate and raceway, Charlestown, 401
  Harrisville Dam and Mill Pond, Burrillville, 310
  Lapham's Cotton Mill, West Warwick, 331–332
  mill dams, North Smithfield, 248
  Natick Mill site and mill dam, West Warwick, 329
mill housing. *See* worker housing
mill overseers' housing, North Smithfield, 245, 246
mill owner's house, Smithfield, 256, 257, 258
Mill River, 219
mill superintendent's house, Smithfield, 257
mill supervisors' housing, Smithfield, 262
mill villages
  Acton, Richmond/Hopkinton, 424
  Albion, Lincoln, 188, 189, 202–204, 208
  Allendale, North Providence, 87, 160, 164–166
  Anthony, Coventry, 333, 337–338
  Arctic, West Warwick, 324, 330–331, 343
  Arkwright, Coventry, 333, 334
  Arnold Mills, Cumberland, 215–216
  Ashaway, Hopkinton, 423–424
  Ashton Mill and Mill Village, Cumberland, 188, 189, 190, 191, 202, 205, 208, 210, 211, 213–215, **214**
  Berkeley Mill and Mill Village, Cumberland, 188, 189, 190, 191, 202, 205, 208, 209, 211–212, **213**
  Bernon, Woonsocket, 219, 220
  Bradford, Westerly, 417–418
  Carolina, Charlestown/Richmond, 397, 401, 424, 427
  Centerville, West Warwick, 324, 331–332
  Charlestown, 401–402
  Chepachet, Glocester, 78, 290, 291, 292–299, 301
  Clarkville, Glocester, 291–292
  Clayville, Scituate/Foster, 263, 274
  Coventry Center, Coventry, 338–341
  Cranston Print Works Village, Cranston, 176, 178, **179**
  Crompton, West Warwick, 324, 332–333
  Cross Mills, Charlestown, 399–400
  Factory, Lincoln, 205
  Georgiaville, Smithfield, 87, 201, 249, 250, 251, 254, 259–262
  Glendale, Burrillville, 301, 302, 306–307
  Globe Mill village, Woonsocket, 219, 229
  Greenville, Smithfield, 253
  Hamlet, Woonsocket, 219
  Harris, Coventry, 326, 333, 334–335
  Harrisville-Graniteville, Burrillville, 169–170, 301, 302, 303, 306, 308–313, 381
  Hope, Scituate, 190
  Hope Valley (Incorporating Locustville), Hopkinton/Richmond, 419
  Hopkins Mills, Foster, 274, 288
  Jencksville, Woonsocket, 219
  Kenyon, Charlestown/Richmond, 397, 401, 424, 425
  Lafayette, North Kingstown, 363–364
  Lippitt, West Warwick, 324, 326, 327–328, 334
  Lonsdale, Lincoln, 153, 188, 189, 190–193, 202, 205, 206, 208, 209–211, 212
  Manville, Lincoln, 188, 189, 201–202, 204
  Mapleville, Burrillville, 301, 302–304, 305
  Nasonville-Mohegan, Burrillville, 301, 302, 307–308
  Natick, West Warwick, 324, 329–330, 343
  Oakland, Burrillville, 127, 301, 302, 304–306
  Old Ashton, Lincoln, 205, 208, 214
  Olney Factory, Lincoln, 205
  Pascoag-Bridgeton, Burrillville, 301, 302, 308, 313–315
  Peace Dale, South Kingstown, 363, 367, 372, 373, 375, 380, 381–385, 388, 396, 412
  Phenix, West Warwick, 190, 324, 326–327, 334
  Potter Hill Mill, Westerly, 417
  Quidnick, Coventry, 330, 335, 337
  Riverpoint, West Warwick, 324, 328–329, 334, 343
  Rocky Brook, South Kingstown, 375, 381, 385–386
  Saylesville, Lincoln, 159, 188, 189, 193–194
  Shannock, Charlestown/Richmond, 397, 401, 424, 425–427
  Sinking Fund Factory Village, Lincoln, 205
  Social, Woonsocket, 219, 222, 223
  Stillwater, Smithfield, 251, 256–259
  Thornton, Johnston, 169, 170, 174–175
  Upper Simmonsville, Johnston, 172, 174, 175
  Valley Falls, Cumberland, 153, 154, 159, 188, 208–209
  Wakefield, South Kingstown, 375, 378–381
  Wanskuck Mill and Mill Village, Providence, 34, 66, 79, 127–128, 235
  Washington, Coventry (town), 338
  White Rock Mill and Village, Westerly, 28, 411, 416, 417
  Woodville, Hopkinton/Richmond, 424
  Woonsocket Falls, Woonsocket, 219, 220
  Wyoming, Hopkinton/Richmond, 418–419, 424
  See also mills and mill complexes
mill workers' boardinghouses, Smithfield, **260**
mill workers' cottages, South Kingstown, 385
mill workers' housing, Hopkinton, 418–419
mill workers' housing or mill offices and H. C. White Company Mill office, Glocester, 295
Miller, Charles E.: Barlow (Everett D.) House (Bit O Heaven; Mohegan Cottage), Block Island, 609
Miller, William David, 379, 380
Miller (General Nathan)–Abbott (Commodore Joel) House, Warren, 461
Miller House, Cumberland, **217**–218
Miller (William David) House (Robinson House, 571 Main), South Kingstown, 379–380
Miller (William S.) House (High Tide), Newport, 576
Millman and Sturgis, 145
Mills, Frank, 452
mills and mill complexes
  American Textile Mill, Pawtucket, 140
  Ann and Hope Store Complex (Ann and Hope Mill, Lonsdale Mill), Cumberland, 142, 168, 209–210

mills and mill complexes (continued)
  apartments, Glocester, 291–292
  apartments (Albion Mill, Valley Falls Mill), Lincoln, 202–**203**, 204
  apartments (Green Mill) Lincoln, 204
  apartments (Valley Falls Mill, also known as Blackstone Mill), Central Falls, 147, 159–160, 208
  Arctic Mill Complex, West Warwick, **330**–331
  Arnold Mills, Cumberland, 215–216
  Ashton Mill and Mill Village, Cumberland, 188, 189, 190, 191, 202, 205, 208, 210, 211, 213–214, 213–215, **214**
  Ashville Mill, Hopkinton, 422
  Ballou-Harris-Lippitt mill, 237
  Barden's Mills (former) site, Bettyville, Scituate, 273
  Bay Mill (later Shore Mill), East Greenwich, 345–**346**
  Berkeley Mill and Mill Village, Cumberland, 188, 189, 190, 191, 202, 205, 208, 209, 211–212, **213**, 271
  Bernon Mills, Bernon, 201, 220
  Bethel Mill, Hopkinton, 423
  Bourne Mills, Tiverton, 481–**482**
  Bradford Mill, Westerly, 417–418, 424
  Bristol Steam Mill (Namquit Mill, White Mill), Bristol, 478
  Butterfly Mill, Lincoln, 197–198
  Carolina Mills, Charlestown/Richmond, 401, 427
  Central Falls (Roosevelt Avenue) Mill Complex, Central Falls, 155–156, 159
  Central Falls Woolen Mill, Central Falls, 155
  Coats and Clark (Conant Thread Mill Complex, later J. and P. Coats Ltd.), Pawtucket, 137, 138, 141–142, 158
  Conant Street Mill, Pawtucket, 142
  condominiums and offices (Allendale Mill and Company Store), North Providence, 164–165, 173, 260, 261
  Coventry Mill (Anthony Mill), Coventry, 168, 337–338
  Cranston Mill, 179, 478
  Crompton Mill, West Warwick, **332**–333, 337
  Desurmont Worsted Yarn Mills, Woonsocket, 231, 238
  Esmond Mill, Smithfield, 262
  French Worsted Company Mills, Woonsocket, 226
  Gardner-Brown Mill, Warren, 457
  Georgia Manufacturing Company, Smithfield, 259
  Georgiaville Mill Complex, Smithfield, **260**–261
  Glen Rock Mill (former), South Kingstown, 393
  Globe Mills, Woonsocket, 222, 229, 230, 237
  Grants Mills, Cumberland, 217
  Hamilton Harbour Condominiums (Hamilton Web Company), Wickford, North Kingstown, **362**–363
  Harrisville Mill Complex, Burrillville, 313
  Hendrick's Mill, Exeter, 432
  Hope Mills, Scituate, 270–**271**, 326
  J. D. Johnston Mill, Newport, 538
  Jenckes Mills, Woonsocket, 225–226, 228, 229
  Joseph Benn and Company (Greystone Mill), North Providence, 166, **167**, 168
  Kenyon Mill, Charlestown/Richmond, 401, 425
  Lafayette Mill, North Kingstown, **363**
  Lafayette Worsted Company Mills, Woonsocket, 226
  Lapham's Cotton Mill, West Warwick, 331–332
  Lawton Mill, Exeter, 431–**432**
  Lawton Owens Mill, Glocester, 292, 294
  Lebanon Mill Company, Pawtucket, 145
  Lippitt Mills, West Warwick, 185, 327–**328**, 432
  Lorraine Mill, Pawtucket, 193
  Mill, The (former), Little Compton, 492–493
  Monohasset Mill, Providence, 125–126
  Museum of Work and Culture (Barnai Worsted Company Dye Works, later Lincoln Textile Corporation), Woonsocket, 238
  Natick Mill Site and Mill Dam, West Warwick, 329, 330
  offices-condominiums (Greene and Daniels Mill), Pawtucket, 153, 158
  Old Ashton, 205, 208, 214
  Olney (George)-Moffitt (Arnold) Mill, Lincoln, 196, 197
  Omega Cotton Mill, East Providence, 436
  Original Bradford Soap Works. See Valley Queen Mill, West Warwick
  Pawtucket Hair Cloth Mill, Central Falls, 127, 156, **157**
  Pawtucket Thread Manufacturing Company, Central Falls, 155
  Peace Dale Mill Buildings, South Kingstown, 384, 385
  Phenix Mill, West Warwick, 326
  Phillipsdale Factories, East Providence, 438–**439**
  Pokanoket Mills, Bristol, 478
  Pontiac Mill, Warwick, 322–323
  Potter Hill Mill, Westerly, 417
  Privilege Mill complex, Woonsocket, 221, 234, 235, 236
  Providence Worsted Mills, 111
  Quidnick Mills (former), Coventry, 337
  Richmond Paper Company Plant, East Providence, 438
  Rockville Mill (former), Hopkinton, 419, 421
  Rodman Mill, South Kingstown, 385
  Royal Mills (Greene Mills), West Warwick, 328
  Royal Weaving Mill, Pawtucket, 138, 152
  Sayles Mill, Burrillville, 313
  Sayles Mill Complex and Administration Building, Lincoln, 193
  Shady Lea Mill, North Kingstown, 364
  Shannock Mill, Charlestown/Richmond, 426–427
  Shannock Mills and Village, Charlestown/Richmond, 425–427
  Shore Mill (Bay Mill), East Greenwich, 345–**346**
  Slater Cotton Company New Mill, Pawtucket, **140**, 142–143
  Slater Mill Complex, Pawtucket, 28, 32, 135, 143–**145**, 327, 328
  Slatersville Mill, North Smithfield, **247**–248
  Social Mill, Woonsocket, 232
  Stafford Mill extension, Central Falls, 155–156
  Stillwater Worsted, Burrillville, 253
  Valley Falls Mill ruins, Cumberland Operation, Cumberland, 160, 208
  Valley Queen Mill, West Warwick, 141, 328–**329**
  Wakefield Mill, South Kingstown, 378, 379
  Wanskuck Mill and Village, Providence, 34, 66, 79, 127–128, 235
  Warren Manufacturing Company, Warren, 455
  West Greenville Mill (Pooke and Steere Mill), Smithfield, 255
  White Mill, Bristol. See Bristol Steam Mill (Namquit Mill, White Mill), Bristol
  White Rock Mill and Village, Westerly, 28, 411, **416**–417
  Wilkinson Mill, Pawtucket, 143
  Woonsocket Manufacturing Company Mill Buildings (Bernon Mills), Woonsocket, 164, 228–229
  Woonsocket Rubber Company, Alice Mill, Woonsocket, 231
  Wyoming, Hopkinton, 418
  See also mill dams; mill villages; worker housing
Mineral Spring Turnpike, 161
Minister's Wooing, The (Stowe), 539
Minneapolis, Minnesota, 132, 133

Index 679

Miramar, Newport. *See* Widener (Mrs. George) House (Miramar), Newport
Miramar (The Tides, Joshua Wilbour House), Bristol, 474, **475**
Missouri, 604
Mist, Jamestown. *See* MacCawley (Captain E. V.) House (Mist), Jamestown
MIT. *See* Massachusetts Institute of Technology
Mitchell (John A.) House, Providence, 109
Moana (Aktaion), Watch Hill, Westerly, 407
Modern Diner, Pawtucket, **139**–140
Moderne style, 139, 145, 272, 492, 493
Moffitt, Arnold, 196
Moffitt (Arnold) Mill. *See* Olney (George)-Moffitt (Arnold) Mill, Lincoln
Mohassuck River, 187, 196
Mohegan, 127
Mohegan Cottage. *See* Barlow (Everett D.) House (Bit O Heaven; Mohegan Cottage), Block Island
Monet, Claude, 492
Money (John W.) House, Charlestown, 402
Monitor, The. *See* Newhall (Daniel S.) House (The Monitor; The Round House), Jamestown
Monocalcium Phosphate Plant, East Providence, 441
Monohasset Mill, Providence, 125–126
Monroney (Patrick) House, Providence, 117
Monsarrat (Nicholas and Jane) Houses, East Providence, **439**–440
Monticello, Virginia, 8
Montreal, Canada, 160
Monument Square, Woonsocket, 222, 236
monuments and memorials
  Civil War, Providence, 36
  Civil War, Woonsocket, 222, 236
  Great Swamp Fight, South Kingstown, 374
  in Providence, 65
  Veterans Monument, North Providence, 167
  World War I, Hopkinton, 423
  World War I, Providence, 65
  World War I, Westerly, 411
Moody, Jim, 520, 523
Moore, Charles: Klotz House, Westerly, 409
Moore, Clement, 546
Moore, Lyndon, Turnbull, Whitaker, 287
Moore (Clement) House, Newport, 546
Moosup Valley Congregational-Christian Church and Cemetery, Foster, 284, 285, 288
Morel House, West Warwick, 327
Morgan (Edward D.) House, Newport, 558
Morgan Mills, 169
Morice, Daniel, 474
Morris, Wistar, 596
Morris (Francis) House, Newport, 553
Morris Plan Building, RISD Buildings, Providence, 66
Morris (Wistar) House (Highland), Jamestown, 596–597
Morse, Alpheus C., 72, 78, 122
  Angell (William G.) House, Providence, 74
  Beckwith (Amos) House, Providence, 110–111, 114
  Binney (William) House, Providence, 105
  Brown University, Rogers Hall, Providence, 100, 101
  Brown University, Sayles Hall, Providence, 99–100, 107
  Gorham (Mary M.) House, Providence, 74
  Kimball (James M.) House, Providence, 104, 107
  Merchants Bank Building, Providence, **50**–51
  Owen (Smith) House, Providence, 105
  Rhodes (Henry) House, Providence, 76
  Sayles (William F.) House, Providence, 107
Morse, Caroline Pearce, 72
Morse (Jacob) House, North Smithfield, 240, 242
Moshassuck River, 29, 39–40

Moss, Jesse L., 416
Mott (Samuel D.) House, Narragansett Hotel, Block Island, 608
Mount Hope Bridge, Bristol, **480**, 498
Mount Hope Farm (Isaac Royall, Jr., House), Bristol, 479
Mount Hope Passage, Bristol, 25
Mount Hope Suspension Bridge, Bristol, 25
Mount Hygeia. *See* Drown (Dr. Solomon) House (Mount Hygeia), Foster
Mount Hygeia Schoolhouse, Foster, 290
Mount Moriah Lodge Number Eight, Lincoln, 200
Mount Vernon Christian Church (Mount Vernon Baptist Church, Friends Meeting House), Foster, 266, 279, 283–284, 286, 487
Mount Vernon Tavern, Foster, 277, 281, **282**, 305
Mowry (Elisha) House, Smithfield, 257, 258–259
Mowry family, 251
Mowry House, Smithfield, 249, 251
Mowry Tavern, Lincoln, 200
Mowry (Tyler) Farmhouse, North Smithfield, 239
Moyes Garage, Water Street, Warren, 457
Muller, Max, 461, 463
Mullin, John H.: Conanicut Park, 586–588
Mumford, Stephen, 519
Mumford, William, 530
Mumford (Stephen) House. *See* Wanton-Lyman-Hazard House (Stephen Mumford House), Newport
Munday, Richard, 521, 545
  Colony House, Newport, 6, 77, 306, 345, 365, 388, 390, 515, 516, **519**–520, 521, 523, 532, 534, 535, 542
  Newport Historical Society and Seventh Day Baptist Meeting House, Newport, 7, 543
  Trinity Church, Newport, 7, 9, 68, 519, 534–**535**, 536, 543, 556
Municipal Welfare Building (U.S. Post Office), Pawtucket, 146
Munro (John W.) House, Bristol, 474
Murchison, Kenneth: Dunes Club Gatehouse and Cottages, Narragansett **368**, 371
Murphy, Hindle and Wright
  Edgewood Yacht Club, Cranston, 183–184
  St. Thomas More Roman Catholic Church (St. Philomene's Roman Catholic Church), Narragansett, 369
Murphy and Hindle: Quidnesset Baptist Church, North Kingstown, 356
Museum of Art, Rhode Island School of Design, Providence, 79–81, 108, 133, 412
Museum of Natural History, Providence, 134
Museum of Newport History (Brick Market), Newport, 521
Museum of Primitive History, South Kingstown, 383
Museum of Rhode Island History, Rhode Island Historical Society (Robert S. Burrough, Jr.–Senator Nelson W. Aldrich House), Providence, 117
Museum of Work and Culture (Barnai Worsted Company Dye Works, later Lincoln Textile Corporation), Woonsocket, 144, 238
museums, 7
  Beavertail Lighthouse, Jamestown, 20, 21, 23, 583, 593, **594**–595
  Bristol Historical and Preservation Society Museum (Bristol County Jail), Bristol, 476
  Children's Museum, Pawtucket/Providence, 148
  Clark-Herreshoff House, Bristol, 475
  Cocumscussoc (Smith's Castle), North Kingstown, 356–357
  Dame Farm, Johnston, 172
  Drown (Dr. Solomon) House (Mount Hygeia), Foster, 289

museums *(continued)*
  Field (John)–Hopkins (Stephen) House, Providence
  Firemen's Museum (Narragansett Steam Fire Engine Station No. 3), Warren, 461
  Jamestown Philomenian Library and Sidney L. Wright Museum of Indian Artifacts, Jamestown, 586, 591
  Johnston Historical Society (Elijah Angell House), Johnston, 170
  Little Rest Museum (former), South Kingstown, 390
  Museum of Art, Rhode Island School of Design, Providence, 79–81, 108, 133, 412
  Museum of Natural History, Providence, 134
  Museum of Newport History (Brick Market), Newport, 521
  Museum of Primitive History, South Kingstown, 383
  Museum of Rhode Island History, Rhode Island Historical Society (Robert S. Burrough, Jr.-Senator Nelson W. Aldrich House), Providence, 117
  Museum of Work and Culture, Woonsocket, 144
  Museum of Work and Culture (Barnai Worsted Company Dye Works, later Lincoln Textile Corporation), Woonsocket, 238
  Newport Art Museum, Newport, 13, 546–547
  Newport Art Museum Cushing Memorial Gallery, Newport, 547
  North Light (Sandy Point Light), Block Island, 22
  Quonset Naval Air Station museum, 21
  Quonset Sea Bee Battalion naval base museum, 21
  Slater Mill Complex, Pawtucket, 28, 32, 135, 143–**145**, 327, 328
  South East Light, Block Island, 22–23
  Tillinghast Road Historic District, Steam and Wireless Museum, East Greenwich, **350**
  United States Naval War College, Founders Hall, Newport, **529**–530
  Vanderbilt (Frederick W.) House (Rough Point), Newport, 575
  Wanton-Lyman-Hazard House (Stephen Mumford House), Newport, 469, 518–519
  Watson (Oliver) Farmhouse, South Kingstown, 391
  Westerly Public Library and Art Gallery (Memorial and Library Association), 411, 412–413
  See also historical societies
My Life in Architecture (R. A. Cram), 515

Nakashima, George, 502
Namquit Mill. *See* Bristol Steam Mill (Namquit Mill, White Mill)
Nardone, Sam, 408
Narragansett, 365–373, 388, 592, 607
  map, 366
  resorts and summer communities, 178, 354, 365–373, 374
  Town Hall (Fifth Avenue School), 370
Narragansett Bay, 3–4, 20, 21, 29, 49, 65, 109, 175, 177, 180, 183, 290, 316, 351, 365, 387, 435, 446, 454, 464, 466, 474, 479, 480, 481, 489, 498, 502, 530, 544, 566, 578, 580, 582, 594, 597
Narragansett Congregational Church. *See* South Ferry Congregational Church (Narragansett Congregational Church), Narragansett
Narragansett Electric Company, Providence, 60
Narragansett Electric Company Power Plants, Providence, 59–60, 177
Narragansett Hotel, Block Island, 608
Lake Shore Dining Hall, 608
Mott (Samuel D.) House, 608
Narragansett Indian Church, Charlestown, 402
Narragansett Indians, 351, 354, 374, 402–403, 406, 412
Narragansett Inn, Westerly, 407

Narragansett Pacers (horses), 365, 375, 379
Narragansett Parkway, 443
Narragansett Pier, Narragansett, 27, 365, 367–373, 381, 383, 385, 388, 389
Narragansett Planters, 26, 354, 360, 364, 365, 374, 375, 379, 380, 387, 388, 389, 394, 499
Narragansett Reading Room and Library, South Kingstown. *See* Peace Dale Library (Narragansett Reading Room and Library), South Kingstown
Narragansett Sea and Shore (Denison), 587
Narragansett Steam Fire Engine Station No. 3, Warren. *See* Firemen's Museum (Narragansett Steam Fire Engine Station No. 3), Warren
Narragansett Tribal Administration Offices, Charlestown, 402
Narragansett Tribal Lands, Charlestown, 402–403
  Health Center, 402
Nash, George, 407
Nash, Jonathan, 407
Nason, Leonard, 308
Nasonville-Mohegan, Burrillville, 301, 302, 307–308
Nathanson, Morris, 38
Natick, West Warwick, 324, 329–330, 343
Natick Mill Site and Mill Dam, West Warwick, 329, 330, 343
National Guard, 132, 189, 592
National Hotel, Block Island, 607
National Rubber Company, 475
National Society of the Colonial Dames of America, 86, 510
Native Americans
  and Bristol, 466
  and Central Falls, 153
  and Great Swamp Fight of 1675, 351, 374, 388
  and Lincoln, 186
  and Narragansett Indian Church, Charlestown, 402
  and Narragansett tribal lands and management area, 402–403
  and Ninigret bronze statue, Watch Hill,, 406
  and North Kingston, 351
  and South Kingstown, 374
  and Tribal Lands Health Center, Charlestown, 402
  and Warren, 454
  and Westerly, 412
  See also King Philip's War
Nature Conservancy, 591, 608–609
Naval Air Station. *See* Quonset Point Industrial Park (Quonset Point Naval Air Station), North Kingstown
Naval Reserve Armory, Bristol, 348, 478–479
Naval Reserve Torpedo Company, Rhode Island Militia, 478
Naval War College. *See* United States Naval War College, Newport
Navy. *See* United States Navy
Nelson (Thomas) House, Bristol, 476
Neo-Colonial houses, Westerly, 409
Nervi, Pier Luigi, 122, 192
Netherlands, 537, 567
Nettles Lodge, West Greenwich, 343
Neutra, Richard, 444
New Bedford, Massachusetts, 481
New Bridge, Cumberland. *See* Ashton Viaduct (New Bridge), Cumberland
New Haven, Connecticut, 415, 542
New London, Connecticut, 342, 430
New London Pike, 274, 342, 422, 429
New Shoreham. *See* Block Island (New Shoreham)
New Village mill housing, Burrillville, 306, 309–310
New Village mill housing, West Warwick, 329
New York, New Haven and Hartford Railroad, 415

New York, Providence and Boston Railroad, 392
New-York Historical Society, 123
New-York Sketch-Book of Architecture, The (magazine), 510, 524
New York Tribune (newspaper), 51
New York Yacht Club Newport Facilities, 17, 580
  Station No. 10, 580
Newbold, John, 383
Newell (Nelson) House, Barrington, 452
Newhall, Daniel S., 600
Newhall (Daniel S.) House (The Monitor; The Round House), Jamestown, 599–600
Newman, Allen G., 411
Newman, John: Beneficent Congregational Church, Providence, **43**–44, 45
Newman, Samuel, 433, 440
Newman Congregational Church, East Providence, 433, **440**
Newport, 4, 6, 9, 10, 21, 24, 26, 27, 28, 29, 32, 34, 75, 76, 77, 85, 106, 120, 124, 125, 127, 147, 320, 357, 358, 364, 365, 368, 387, 388, 389, 393, 396, 516–581, 582, 583
  Bellevue Avenue map, 563
  Broadway and City Center map, 518
  Casino, 16, 371, 373, 555, 562, 563–**564**, 565, 567
  City Hall, 518
  Colony House, 76, 519–520
  common (1970s), 9
  Harbor area, 532–546
  Kay–Catherine–Old Beach (Top of the Hill), 12, 13, 14, 15, 16, 546–557
  map (Harbor area), 533
  map (Kay–Catherine–Old Beach), 547
  map (Ocean Avenue), 577
  map (overview), 517
  map (Upper Cliffs), 559
  map (Washington Street and Point), 522
  Ocean Avenue, 15, 16, 17, 576–580
  as resort, 11–19
  Upper Cliffs, 557–562
  Washington Square, 516, 519, 520–521, 527, 528, 530
  Washington Street and the Point, 6, 521–530, 532, 551, 596
  Newport and Its Cottages (G. C. Mason, Sr.), 548
  Newport Art Museum, Newport, 13, 546–547
  Cushing Memorial Gallery, 547
  J. N. A. Griswold House, 13, 546, 547, **548**, 570, 574
  Newport Artillery Company, Newport, 78, 348, 541
  Newport Bank. *See* Rathburn (John)–Gardner (George)–Rivera (Abraham Rodrigues) House (Newport Bank), Washington Square, Newport
Newport Bridge, 521, 599
Newport Collaborative, 375, 474, 550, 558
Newport Congregational Church, 147, **536**–537
Newport Country Club, 17, **578**–579
Newport Daily News, 586
Newport Historical Society, 7, 483, 518, 525, 531, 536, 537, 538, 543
Newport Historical Society and Seventh Day Baptist Meeting House, 536, 538, 543
Newport Mercury, The, 529, 540
Newport Reading Room, 554–555, 563
Newport Restoration Foundation, 507
Newton, Dudley, 546, 549, 553, 554, 565, 566, 569, 571
  Brown (Harold Carter) House, Newport, 570
  Churchill-Yarnell House, Newport, 556
  Cram (Jacob)–Sturtevant (Mary) House, Middletown, 513–514
  Downing (Captain John W.) House (Fairview), Newport, 532

Fish (Stuyvesant and Mamie) House (Crossways), Newport, 576
Gibbs (T. K.) House (Bethshan), Newport, 551, 552
Helme (Theodore R.) Block, Newport, 536
King-Birkhead House, Newport, 557
Lieber (Matilda) House, Newport, 553
Newton (Mary B.) House, Newport, 539
Seymour (Carolyn) House (Hawkhurst or Hawxhurst), Newport, 554
Store Block, Newport, 534
Swineburne (Henry) House, Newport, 552–553
Weld (William) House, Newport, 568–569
See also Dudley Newton and C. E. Kempe
Newton (Dudley) Office, Newport, 546, 557
Newton (Mary B.) House, Newport, 539
Niantic, Westerly. *See* Bradford, Westerly
Nichols, Henry Stafford, 307
Nichols, Jonathan, II, 523, 532
Nichols, Joseph P., 307
Nichols, Walter, 532
Nichols (Henry Stafford) Houses, Burrillville, 307
Nichols (Jonathan) House. *See* Hunter (William) House (Jonathan Nichols-Colonel Joseph Wanton House), Newport
Nichols Partnership: Westin Hotel, Providence, 39
Nicholson, Samuel C., 97
Nicholson File Company, 97
Nicholson House (Francis W. Goddard-Samuel C. Nicholson House), Brown University, Providence, 96, 97
Nickerson, Edward I., 69, 70
  Arnold (Newton D.) House, Providence, 112
  Carr (Dr. George W.) House, Providence, 79
  Cook (Stephen A., Jr.) House, Providence, 107, 112
  Hartshorn (Joseph C.) House, Providence, **131**
  Hartwell (Frederick W.) House, Providence, **131**
  International House (Byron Thomas Potter House), Providence, 111
  Miramar (The Tides, Joshua Wilbur House), Bristol, 474
  Sprague (Charles H.) House, Providence, 112–113
Night and Day (Michelangelo sculpture), 51
Nightingale, Joseph, 11, 91
Nightingale-Brown House replica, Barrington, 447
Nightingale (Joseph) House, 11, 91, 149. *See also* John Nicholas Brown Center for the Study of American Culture (Joseph Nightingale-Nicholas Brown House), Providence
Nightingale (William G.) House, Little Compton, 493
Niles, Samuel, 402
Niles (George) House, Hopkinton, **418**
Nimes, France, 130
Nincheri, Guido, 160, 175, 223, 228
9, 11, 15 Dresser Avenue, Newport, **558**
Ninicroft Lodge. *See* Maltby (L. U.) House (Ninicroft Lodge), Block Island
Ninigret, Thomas, 400
Ninigret (Narragansett chief), 412
  Watch Hill bronze statue of, 406
Nipmuc River, 301
Nooks and Corners of the New England Coast (Drake), 587
Noon family, 218
Noon House. *See* Burlingame-Noon House, Cumberland
Norcross Bros. (builders), 41
Normandie. *See* James (Lucy Worth) House (Normandie), Newport
Norris, Benjamin: Harris (William) House, Bristol, 479
Norris, Isabel Dimond, 472
Norris, Samuel, 472
Norris House. *See* Bradford-Norris House, Bristol

North Attleboro, Massachusetts, 206
North Branch. *See* Pawtuxet River
North Burial Ground, Newport, 530
North Carolina, 193
North District Courthouse, Burrillville, 312
North Farm, Jamestown. *See* Watson (Job)–Watson (Thomas Carr) Farm (North Farm), Jamestown
North Foster Free Baptist Church, Foster, 266, 279, 283, 284, 287, 288–289
North Kingstown, 351–364, 365, 374, 429
   map, 352–353
North Kingstown Free Library, Wickford, 362
North Kingstown Town Hall, Wickford, 362
North Light, Jamestown. *See* Conanicut Point Light (North Light), Jamestown
North Light (Sandy Point Light), Block Island, 22, 608
North Providence, 135, 160–168, 170, 249
   map, 163
North Scituate, Scituate, 214, 265–269
North Scituate Community House (North Scituate Academy), Scituate, 266
North Scituate Congregational Church, Scituate, **265**–266, 279, 284, 293, 312, 487
North Smithfield, 62, 186, 219, 238–249, 302, 305, 491
   commercial blocks, 243–244, 246, 247
   map, 240
North Smithfield Heritage Association (Forestdale School, officially Branch School, District 3), **243**
"North Woods," 186, 238, 249
Northeast (Rhode Island), 186–262
Northrup (Cyrus) House, Wickford, North Kingstown, 360
Northwest Community Nursing and Health Center (Ada Hawkins School), Glocester, 294, 296–297
Northwest (Rhode Island), 263–315
Norton, Massachusetts, 473
Norton, Stephen, 328
Norwich, Connecticut, 339
Norwich House. *See* Atlantic Inn (Norwich House), Block Island
Notre Dame de Bon Secours. *See* Our Lady of Good Help Church, Rectory, and Parish House–School (Notre Dame de Bon Secours), Burrillville
Nowicki, Matthew, 192
Noyes, Samuel, 117
Noyes-Luce House, Newport, 553
Number 8 Schoolhouse, Little Compton. *See* American Legion Hall (Number 8 Schoolhouse), Little Compton

Oak Glen. *See* Howe (Julia Ward) Home (Oak Glen), Portsmouth
Oakland, Burrillville, 127, 301, 302, 304–306
Oakland, Farm Estates, Portsmouth, 503, 504, 505, 506
Oakland Assembly, Recreation, and Market Hall, Burrillville, **305**–306
Oakland Worsted Company, Burrillville, 305
Oaklawn. *See* Fahnestock (William F.) House (Oaklawn, Bois Doré), Newport
Oakwood, South Kingstown, 381
Observation Tower, Hannah Robinson Park, South Kingstown, 387
Ocean Avenue, Newport, 15, 16, 17, 576–580
   map, 577
Ocean Drive. *See* Ocean Avenue, Newport
Ocean Highlands, Jamestown, 27, 583, 585, 587, 595–597, 601
Ocean Highlands Company, 595
Ocean House, Charlestown, 399
Ocean House, Newport, 12
Ocean House, Watch Hill, Westerly, 407–408
Ocean Road, Narragansett, 367, 371, 372, 373
Ochre Court. *See* Goelet (Ogden) House (Ochre Court), Newport
Octaviana, Joseph, 251
Odlin (John)–Otis (Jonathan) House, Newport, 532, 533
Oelrichs, Tessie, 572
Oelrichs (Tessie and Herman) House (Rosecliff), Newport, 571–**572**
offices-condominiums (Greene and Daniels Mill), Pawtucket, 153, 158
offices (East Greenwich Post Office), East Greenwich, 345
offices (The Brick House; Colonel Micah Whitmarsh House), East Greenwich, 346
O'Gorman Building, Providence, 52
Okonite company, 438
Olbrich, Joseph Maria, 124
Old Acre. *See* Church (John)–Burchard (Edith Russell) House (Old Acre), Tiverton
Old Ashton, Lincoln, 205, 208, 214
Old Ashton housing, Lincoln, **205**
Old Ashton–Quinnville, Lincoln, 204–205
Old Bank of Bristol, DeWolf's Wharf, Bristol, 478
Old Castle. *See* Porter (Mrs. James C.) House (Porter Villa, Old Castle), Newport
Old County Jail, South Kingstown. *See* Pettaquamscutt Historical Society (Old County Jail), South Kingstown
Old County Records Office, South Kingstown. *See* Kingston Free Library and Little Rest Archives (Kings County Courthouse; Little Rest Museum, Old County Records Office), South Kingstown
Old Dye House. *See* Mechanics Machine Shop, Water Street, Warren
Old Harbor Historic District, Block Island, 605, 607
Old Louisquisset Pike, 161
Old Narragansett Church (Old St. Paul's Church), Wickford, North Kingstown, 360, 361
Old Plainfield Pike, 265
Old Slater Mill, Pawtucket. *See* Slater Mill Complex, Pawtucket
Old Slater Mill Association, 144
Old State House (Providence Colony House, later Rhode Island State House), Providence, 76–77, 345
Old Stone Bank. *See* Providence Institution for Savings Building, Providence
Old Stone Church, Tiverton. *See* Free Will Baptist Church (First Baptist Church, Old Stone Church), Tiverton
Old Stone Square Building, Providence, 63, 64
Old Town House, Exeter, 430
Oliver, Andrew, 484
Oliver (Andrew)–Sargent (Charles)–Thayer (Steven) House (Homelands), Tiverton, 484
Olmsted, Albert, 505, 578
Olmsted, Frederick Law, 90, 91, 120, 121, 132, 578
   Brown (Harold Carter) House, Newport, 570
   Land Trust Cottages, Middletown, 515
Olmsted, Frederick Law, Jr.: Veterans Memorial Parkway (Barrington Parkway), East Providence, 183, 443–444, 505
Olmsted, Frederick Law and John Charles: King (George Gordon) House (Edgehill Farm), Newport, 551, 580
Olmsted, Marian, 505
Olmsted, Olmsted and Eliot (landscape architects), 391
Olmsted (Albert H.) House (Wildacre), Newport, 505, 551, 578
Olmsted Brothers. *See* Olmsted firm

Olmsted firm (landscape architects), 17, 101, 106, 120, 121, 322, 435, 505, 551, 575, 578
Olney, George, 196
Olney (E. B.) House, North Providence, 165
Olney Factory, Lincoln, 205
Olney family, 162, 196
Olney (George)–Moffitt (Arnold) Mill, Lincoln, 196, **196**
Olney–Arnold (Israel) House, Lincoln, 196–197
Olneyville, 169
O'Malley, John F.: Charles W. Shea Senior High School (Pawtucket West High School), Pawtucket, 139, 145
O'Malley and Richards: Pawtucket City Hall, Police Department, and Fire Department, 139, 145
Omega Cotton Mill, East Providence, 436
Omega Mill Site, East Providence, 437
Onarock. *See* Woodward (Samuel) House (Onarock), Jamestown
Oneco, Connecticut, 337
115–121 Roger Williams Avenue duplexes, East Providence, 437–438
120 Main Street, Wickford, North Kingstown, 361
137–147 Roger Williams Avenue duplexes, East Providence, 438
166–180, 238–260 Roger Williams Avenue duplexes, East Providence, 438
167–235, 238–260 Roger Williams Avenue single-family houses, East Providence, 438
194 Main Street, Ashaway, Hopkinton, **423**
197 Main Street, Ashaway, Hopkinton, 423
1026 Main Street, Hope Valley, Hopkinton, 419
1050 Main Street, Hope Valley, Hopkinton, 419
1054 Main Street, Hope Valley, Hopkinton, 419
Orchard House. *See* Cook (Cyrus) House (Orchard House), Cumberland
Original Bradford Soaps Work. *See* Valley Queen Mill, West Warwick
Ormsbee, Caleb, 87
   Ives (Thomas Poynton) House, Providence, 11, **90–91**, 110, 113, 117
   John Nicholas Brown Center for the Study of American Culture (Joseph Nightingale–Nicholas Brown House), 91
   Nightingale (Joseph) House, Providence, 11, 91, 149
O'Rourke, William M., 459
Orrell, William, 306
Osborne (John) House, South Smithfield, 243
Osgood Construction Company: New Village mill housing, West Warwick, 329
Otis, Jonathan, 533
Otis (Jonathan) House. *See* Odlin (John)–Otis (Jonathan) House, Newport
Ott, Joseph, 152
Our Lady of Good Help Church, Rectory, and Parish House–School (Notre Dame de bon Secours), Burrillville, **304**
Over Cliff, Narragansett, 372
Overing House. *See* Prescott Farm (Overing House), Portsmouth/Middletown
Owen, Smith, 105
Owen Block, Providence. *See* Hay-Owen Block, Providence
Owen brothers, 64
Owen (Smith) House, Providence, 105

Pabodie, William, 492
Pabodie-Gray House, Little Compton, 492
Packard Motor Company Showroom (former), Providence, 42
Page, F. E.: St. Joseph's Church, Cumberland, 212, **213**, 228

Page (Martin)–Buffum (Horace)–Gardner (George W.) House, Providence, 96
Paine, Andrew, 287
Paine (Andrew) Farm, Foster. *See* Paine (Zuriel)–Paine (Andrew) Farm, Foster
Paine-Bennett Farm, Foster, 277, 281–282, 305
Paine House (Francis Brayton House), Coventry, **338**
Paine (Zuriel)–Paine (Andrew) Farm, Foster, **286**, 287
Palladio, Andrea, 545
Palladio (Hoppus), 545
Palladio Londinensis (Salmon), 64, 90, 118
Pallas and Tantae cannons, Warren, 460
Palm Beach, Florida, 491
Palmer, Cotton: condominiums (Second Congregational Church), Newport, 541
Palmer River, 447, 453
Pan-American Exposition, Buffalo (1902), 435
Panic of 1857, 260
Panic of 1873, 38, 177, 486, 583, 587
Paradise School, Middletown, 513
Paramino, John Francis, 411–412
Parish House, Congregational Church, Bristol, 477
Parish House, St. Michael's Episcopal Church, Bristol, 473
Park Place Congregational Church, Pawtucket, 143
Parker, Thomas and Rice
   Machado House, Brown University (Ellen Dexter Sharpe House), Providence, 105
   Rochambeau House, Brown University (Henry Dexter Sharpe House), Providence, 105–106
Parker Farm (Parker District School), West Greenwich, 343
parking garage, Providence, 47
Parkis (J. H.) House, North Smithfield, 246–247
parks
   Abbott Park, Providence, 45
   Arcadia Management Area, Hopkinton/Exeter, 418, 429
   Burlingame State Park, Charlestown, 397
   communal garden, Providence, 74–75
   Crescent Park, East Providence, 435, 445–446
   Diamond Hill State Park, 217
   Dutch Island Management Area, Jamestown, 592
   Goddard Memorial Park (Russell Estate), Warwick, 322
   Jenks Park, Central Falls, 153, 158–159
   Kennedy Plaza, Providence, **33**, 36–37
   Kimball Wildlife Refuge, Charlestown, 397
   Narragansett Parkway, 443
   Observation Tower, Hannah Robinson Park, South Kingstown, 387
   Prospect Terrace, Providence, 106
   ranger's house, Arcadia Management Area, Exeter/Hopkinton, **429**
   Roger Williams Park, Providence, 34, 70–71, 92, 132–134, 151, 177, 183
   Slater Memorial Park, Pawtucket, 151
   Stone Bridge and Fort Barton, Tiverton, 483
   Touro Park, Newport, 544
   Veterans Memorial Parkway (Barrington Parkway), East Providence, 118, 443–444
   Wanskuck Park, Providence, 128
   Wilcox Park, Westerly, 411–412
Parkway and Riverside, East Providence, 443–446
Parsonage, Congregational Church, Bristol, 477
Partelow and Bullock (builders), 385
Pascoag-Bridgeton, Burrillville, 301, 302, 308, 313–315, 441
Pascoag Community Baptist Church (First Baptist Church), Burrillville, 314

Pascoag No. 11 School. *See* apartments (Pascoag No. 11 School), Burrillville
Pascoag Public Library, Burrillville, 314
Pascoag River, 301, 313
Patterson (General Robert E.) House (The Ramparts; Channel Bells), Jamestown, 598
Patton, George, 268
Paull, Seth, 469, 478
Paull (Seth) House, Bristol, **469**, 478
Pawcatuck River, 3, 397, 401, 406, 409, 416, 417, 418, 424, 427
Pawcatuck Seventh Day Baptist Church, Westerly, **410**–411
Pawtucket, 135–154
　City Hall, Police Department, and Fire Department, 139, 145
　Deborah Cook Sayles Memorial Library, 145–146, **147**
　map, 136
　Municipal Welfare Building (U.S. Post Office), 146
Pawtucket Armory, 145, 146, 229, 230
Pawtucket Congregational Church, **147**–148, 149
Pawtucket Hair Cloth Mill, Central Falls, 127, 156, **157**
Pawtucket River, 135
Pawtucket Thread Manufacturing Company, Central Falls, 155, 156
Pawtucket Times, 152
Pawtucket West High School. *See* Charles W. Shea Senior High School (Pawtucket West High School), Pawtucket
Pawtuxet, 175, 177
Pawtuxet River, 11, 28, 176, 184, 190, 263, 322, 324, 326, 328, 330, 331, 332, 333, 334, 335, 424, 425
　North Branch, 324, 328
　South Branch, 324, 328, 330
Pawtuxet Village, Cranston, 184
Peabody and Stearns, 15, 561
　Breakers, The (first), Newport, 561
　Emmons (Arthur) House, Newport, 552
　Hambly Funeral Home (Grace W. Rives House), Newport, 551
　Hennery, Newport, 560
　Providence Journal Building (former), Providence, 51
　Vanderbilt (Frederick W.) House (Rough Point), Newport, 575
　Wolfe (Catherine Lorillard) House (Vinland), Newport, **560**–561
Peace Dale, South Kingstown, 363, 367, 372, 373, 375, 380, 381–385, 388, 396, 412
Peace Dale Center, South Kingstown, 383–385
Peace Dale Congregational Church, South Kingstown, **382**, 384, 386
Peace Dale Elementary School, South Kingstown, 385
Peace Dale Library (Narragansett Reading Room and Library), South Kingstown, 384, **385**, 386
Peace Dale Mill Buildings, South Kingstown, 384, 385
Peace Dale Neighborhood Guild, South Kingstown, 383
Peace Dale Railroad Station, South Kingstown, 383, 384, 392
Peale, Norman Vincent, 212
Pearce (Earle) House, Providence, 72
Pearce (Ellis) House, Pawtucket, 149, 150
Pearce (Henry) House. *See* Division of Applied Mathematics, Brown University (Henry Pearce House), Providence
Pease (Erastus) House, Newport, 536
Peck, Abel: North Kingstown Town Hall, Wickford, North Kingstown, 362
Peck, Frederick C., 447

Peck, Frederick Stanhope, 449, 452
Peck, Sarah Gould, 452
Peck (Aaron) House, Wickford, North Kingstown, 361
Peck (Albert H.) House, Barrington, 450
Peck (Ellis) House, Barrington, 447–448
Peck family, 450
Peck (Frederick Stanhope) House. *See* Zion Bible College (Frederick Stanhope Peck House, known as Belton Court), Barrington
Peck (Leander R.) School, Barrington, 452
Peckham, Job, 554
Peckham, John, 470
Peckham, Pardon, 338–339
Peckham (Job) House, Newport, 554
Peckham Manufacturing Company, 339
Pei, I. M., 104
　Cathedral Square, Providence, 55
Pell (Mrs. Archie D.) House, Newport, 556
Pelog Arnold Tavern. *See* apartments (Richard Arnold–Pelog Arnold Tavern), North Smithfield
Pembroke College, Brown University, Providence, 75, 97, 107
Pembroke Hall, Brown University, Providence, 107–108
Pencil Points (publication), 484
Pendleton, Charles, 79, 108
Pendleton (Charles) House. *See* Dexter (Edward)–Pendleton (Charles) House, Providence
Pendleton House (first). *See* Dexter (Edward)–Pendleton (Charles) House, Providence
Pendleton House (second) (The Colonial House) and Waterman Galleries, Rhode Island School of Design, Providence, 79–**80**, 81, 83
Pennsylvania Railroad, 574
People's Savings Bank, Providence. *See* RISD "Bank Building" (People's Savings Bank), Providence
Pequot Indians, 374
Percy, Hugh, 557
Perkins, Joseph, 389
Perkins, Wheeler and Will, 444
Perkins and Betton: King Block, Newport, 564, 565
Perkins and Will Partnership: Community College of Rhode Island Knight Campus (Rhode Island Junior College), Warwick, **323**–324
Perkins (Elizabeth) House, South Kingstown, 396
Perkins (Joseph)–Hagadorn (John)–Taylor (Thomas) House, South Kingstown, **389**
Perkins (Julius Deming) House (Bayberry Lodge), Block Island, 609
Perret, Auguste, 77
Perry, Alexander, 479
Perry, Frank S.: Coro Building, Providence, 57, 58–59, 441
Perry, Marsden J., 51, 60, 85, 90, 91, 122, 177
Perry, Matthew Galbraith, 461, 526, 544
Perry, Oliver Hazard, 397
Perry, Shaw and Hepburn, 99
Perry and Whipple, 236
Perry (Commodore Oliver Hazard) Farm, South Kingstown, 397
Perry family, 479
Perry House, Newport. *See* Knowles (Henry) House, Newport
Perry-Congdon House, Bristol, 479
Perry (Marsden J.) House, Providence, 85, 90
Perry Mill (wharf), Newport, 516, 546
Perry (Oliver Hazard) House. *See* Buliod (Peter)–Perry (Oliver Hazard) House (Rhode Island Bank), Washington Square, Newport
Perry (Oliver Hazard) House (Belcourt), Newport, 574–575

Perry (Samuel) House, South Kingstown, **394**
Peter and Daphne Farago Wing, Museum of Art, Rhode Island School of Design, Providence, 81
Peter Pots Pottery (Glen Rock Mill), South Kingstown, 393
Peter Roudebush and Associates, 546, 547
Peter's River, 219, 225
Peterson, Warren A., 78
Pettaquamscutt Academy. *See* Kingston Academy, South Kingstown
Pettaquamscutt Historical Society (Old County Jail), South Kingstown, 379, 391
Pettaquamscutt River, 351, 365, 374
Phenix, West Warwick, 190, 324, 326–327, 334
Phenix Iron Foundry complex, Providence, 57, **58**
Phenix Manufacturing Company, 324
Phenix Mill, West Warwick, 326
Phetteplace and Seagrave: Central Falls Woolen Mill, Central Falls, 155
Philadelphia, Pennsylvania, 27, 46, 208, 267, 372, 378, 582, 602, 607
Philip, King (Metacomet; Wampanoag chief), 374, 466
Philip J. Holton Water Purification Plant, Scituate, 272
Philips (A. S.) House, Portsmouth, 500
Philipsdale, 193
Phillips, Augustus, 281
Phillips, Eugene, 434, 438
Phillips, Ezekiel, 281
Phillips, George, 277
Phillips Company Store, East Providence, 438
Phillips Electric Company. *See* Washburn Wire Company
Phillips (George) Farm, Foster, 277, 281, 282, 305
Phillips (Grayson) Farm, North Smithfield, 248
Phillips-Wright House, Foster, 281, 491
Phillipsdale, East Providence, 193, 435
Phillipsdale Factories, East Providence, 438–**439**
Phillipsdale housing, East Providence, **437**–438
Philomenian Debating Society, Jamestown, 586
Philosophical Club, Newport, 545
Pierce (Daniel) House, Providence, 94, 446
Pinard, Cazeau and Elizabeth, 558
Pinard Cottages, Newport, 558
Pine Grove estate, 130
Pitcher (Ellis B.)–Goff (Lyman) House, Pawtucket, 148
Pitt, William, the elder, 527
Pitt's Head Tavern (Ebenezer Flagg House), Newport, **527**–528
Pittsburgh, Pennsylvania, 40, 277, 407
Placide, Alexander, 521
Plainfield Pike, 170, 265, 269, 274, 339
Platner, Warren, 84
Platt, Charles A.
 Fahnestock (William F.) House (Oaklawn, Bois Doré), Newport, 558–559
 Waterman Galleries, Providence, 79, 80, 81
Plymouth, Massachusetts, 29, 206, 454, 466
Pocasset River, 178, 179, 182
Point Judith Lighthouse and Coast Guard Station, Narragansett, 373
Point Street Bridge, Providence, 60
Pokanoket Mills, Bristol, 478
police. *See* Rhode Island State Police Barracks
Polo Estates, Narragansett, 373
Pomeroy (Edith Burnet) House (Seabeach), Newport, 576–577
Ponaganset River, 273
Ponaganset Village site (Barden's Mills, Bettyville), Scituate, 273
Ponham Rocks Lighthouse, East Providence, 445
Pontiac Free Library, Warwick, 323

Pontiac Mill, Warwick, 322–323
Pony Truss Bridges, Lincoln, **203**–204
Pooke, William, 255
Pooke and Steere Mill, Smithfield. *See* West Greenville Mill (Pooke and Steere Mill), Smithfield
Pope, Alexander, 3, 76
Pope, Charles H., 371, 372
Pope, John Russell, 512, 545, 571, 573
 Pope (John Russell) House (The Waves), Newport, 512, **575**, 576
 Taylor (Moses) House (Elmhurst School and Convent of the Sacred Heart) (The Glen), Portsmouth, 504, 512, 513
Pope, Sadie Jones, 573
Pope (Charles H.) House. *See* Fair Lawn (Jeffrey Davis–Charles H. Pope House), Narragansett; Gibson Court Condominiums (Charles H. Pope House, Gardencourt), Narragansett
Pope (John Russell) House (The Waves), Newport, 512, **575**, 576
Portelow, Kneeland, 381, 383
Porter, A. Kingsley, 514
Porter (Mrs. James C.) House (Porter Villa, Old Castle), Newport, 556
Porter Villa, Newport. *See* Porter (Mrs. James C.) House (Porter Villa, Old Castle), Newport
Portland, Oregon, 501
Portsmouth, 4, 28, 480, 483, 490, 498–507, 509, 514, 589
 map, 501
Portsmouth Abbey School, Portsmouth, **500**–502
 Chapel, 501–502
 Smith (Amos D.) House (Hall Manor), 500, 501, 502
Portsmouth Camp Meeting Association, Portsmouth, 502
 Meeting Hall, 502
 Wooden Trailer House, 502
Portsmouth Historical Society (Union Meeting House), **503**
Post, George B.: Baldwin (Christopher Columbus) House (Château-Nooga, Chatterbox), Newport, 569
Post Road, Charlestown, 400
Pothier, Aram J., 222, 226, 231
Potowomut River, 351
Potter, Abner, 270
Potter, Asa, 389
Potter, Byron, 111
Potter, Elisha Reynolds, Sr., 390
Potter, Elizabeth, 151
Potter, James, 151, 152
Potter, William, 288
Potter (Abner) Octagon House, Richmond, **427**
Potter and Johnstone Machine Company, Pawtucket, 151, 152
Potter and Robinson: Baldwin (C. H.) House, Newport, 567
Potter (Asa) House, South Kingstown, 389
Potter (Byron Thomas) House. *See* International House (Byron Thomas Potter House), Providence
Potter (Captain Simeon) House. *See* Duhane (Jacob)–Potter (Captain Simeon) House, Newport
Potter family, 270, 390
Potter Hill, Westerly, 417
Potter Hill Mill, Westerly, 417
Potter (James)–Lumb (Ralph A.) House, Pawtucket, 151–152
Potter (Margaret) Cottage (The Red House), Jamestown, 603
Potter (William) House, Foster. *See* Hopkins (Ezekiel)–Potter (William) House, Foster
Potter's Wharf, Bristol. *See* Usher Store (Potter's Wharf), Bristol

Pottersville District 7 Schoolhouse), Scituate. *See* Community Center (Potterville District 7 Schoolhouse), Scituate
Potterville, Scituate, 269–270
Powder Mill Pike, 161
Power of Positive Thinking, The (Peale), 212
Pozzi, Lombard, 456, 459
Pratt, Samuel, 555
Pratt (Samuel) House (Bird's Nest Cottage), Newport, 555
Pre-Raphaelites, 70, 492
Précieux Sang (Church of the Precious Blood), Woonsocket, 227–228, 235–236
Prentice, George W., 130
Prentice (George W.) House. *See* Kendrick (John K.)–Prentice (George W.)–Tirocchi (Anna) House, Providence
preparatory schools. *See* school buildings
Prescott, Richard, 483, 507, 537
Prescott Farm (Overing House), Portsmouth/Middletown, 499, **507**
Preservation Society of Newport County, 19, 522, 523, 544, 565, 567, 573, 574
Preserve Rhode Island (organization), 109
President's House (Rush Sturges House), Brown University, Providence, 88–89
President's House (Stephen O. Metcalf House), Rhode Island School of Design, Providence, 106
Preston, William Gibbons, Gibson Court Condominiums (Charles H. Pope House, Gardencourt), Narragansett, 370, 372
Price, Bruce
  Armistead House, Newport, 552
  Audrain Building, Newport, 564–565
Price, William, 534
Primrose Grange, North Smithfield, 248
Princeton University, 97, 98, 102, 464
prisons. *See* jails and prisons
Pritchett and Pritchett: Lippincott (J. Bertram) House I (Meeresblick), Jamestown, 593
Privilege Mill complex, Woonsocket, 221, 234, 235, 236
Privilege Mill housing, Woonsocket, **234**–235
Progress of Woonsocket, The (Compris mural), 236
Progressive Architecture (magazine), 78, 319
Proprietors of Swan Point Cemetery, 120
Prospect Square, Hopkinton, 419–420
Prospect Terrace, Providence, 106
Proulx (Charles A.) House, Woonsocket, **232**–233, 234
Providence, 4, 6, 10, 11, 19, 23, 24, 26, 28, 29–134
  Benefit Street, 32, 61, 65, 72, 73–97, 114
  Broadway, 34
  Broadway and South Providence area, 128–134
  Brown University, Center Campus and north, 97–109
  Canal Street Row, 65
  Capital Center area, 33, 34–55, 61
  Cathedral Square, 55
  Chestnut Street, 55
  City Hall, **33**, 36, 37, 38, **38**, 65
  Convention Center, 39
  downtown area, 34–55
  harborfront, 55–60
  Hope Street and side streets area, 32, 65, 66, 109–122
  jewelry district, 36, 52, 55–60
  Main Street area, 29, 32, 33, 34, 60–71, 72, 77
  Mall, 36
  map (Benefit Street area, North), 86
  map (Benefit Street area, South), 86
  map (Broadway and the Armory area), 129
  map (Brown University, Center Campus and North area), 98
  map (downtown and Capital Center areas), 35
  map (harborfront), 56
  map (Hope Street and side streets area), 109
  map (jewelry district and Harborfront areas), 56
  map (Main Street area), 62
  map (North Providence), 163
  map (overview), 30–31
  North Providence, 160–168
  Northwest Providence area, 122–128
  Public Library, 42–43, 51, 132, 146
  State House (sixth), 10, 32, 33, 39, 40, 106, 122–124, **123**, 125, 381
  Union Depot, **33**, 36
  Union Station Complex, 36, 38–39, 51
  Water Supply Board, 263, 272
  Westminster Street Shopping District, 51–55
Providence (Rhind sculpture), 37
Providence and Norwich Pike. *See* Plainfield Pike
Providence and Springfield depot, Smithfield, 258
Providence and Stonington Railroad, 177, 345
Providence and Worcester Railroad, 154
Providence and Worcester Railroad Station, Woonsocket, 236–**237**
Providence Armory. *See* Cranston Street Armory, Providence
Providence Art Club (Seril Dodge Houses), 69–70, 110
Providence Athenaeum, 84, **85**
Providence Bank, 64
Providence Biltmore Hotel, **38**
Providence Chamber of Commerce, 38
Providence College, Providence, 34
  Bishop Harkins Hall, 126
  campus, 126–127
Providence Colony House. *See* Old State House (Providence Colony House, later Rhode Island State House), Providence
Providence County Courthouse, 64–65, 78, 86, 413
Providence *Evening Bulletin*, 607
Providence Gazette, 70
Providence Institution for Savings Building, Providence, 39, 62, **63**
Providence Journal Building (former), Providence, 51
Providence *Journal* (newspaper), 163
Providence Partnership: Jesse M. Smith Memorial Library, Burrillville, 310, 311, 441
Providence Partnership, Philemon E. Sturges: Cathedral of St. John (St. John's Church), diocesan offices, 71
Providence Performing Arts Center (Loew's State Theatre), **45**–46
Providence periphery (Rhode Island), 135–185
Providence Place, Providence, 41
Providence Preservation Society, 70, 72, 130
Providence Reform School. *See* Sockanosset School for Boys Administration Building, Cottages, and Chapel, Cranston
Providence River, 33, 39–40, 49, 60–61, 64, 65, 435, 443, 481
Providence State House (Old State House), 10, 32, 33, 39, 40, 76–77, 106, 122–124, **123**, 125, 345, 381
Providence Station, 40–41
Providence Street School, West Warwick, 330
Providence Sunday Journal, 184
Providence Telephone Building, 41–42, 51
Providence Union Station. *See* Union Depot, Providence
Providence, Warren and Bristol Railroad, 78, 450
Providence Worsted Mills, 111
Public Library (Foster Center School, Hemlock School), Foster, 278, 283
Public Works Administration, 139, 272
  Chepachet School, Glocester, 294

Greenville Fire Company, Smithfield, 253–254
Northwest Community Nursing and Health Center (Ada Hawkins School), Glocester, 294, 296–297 Publication House, Little Compton, 496
Pugin, A. W. N., 254
Pump House Restaurant (Reservoir Pumping Station), South Kingstown, 385–386
Purgatory Chasm, Middletown, 513
Puritans, 7, 8, 9, 25, 29, 32, 187, 433, 434, 441, 442, 445, 466, 477, 498
Purves, Cope and Stewart, 368
Putnam, Connecticut, 170
Putnam Pike, 170, 249, 253, 290, 292, 293, 294, 298
PWA. *See* Public Works Administration.

Quakers, 7, 187, 194, 197, 239, 241, 283, 302, 375, 381, 382, 524, 530, 531, 533, 534, 590, 596, 603. *See also* Friends meeting houses
Queen Anne cottage, Bristol, 474
Queen Anne cottage, Westerly, 410
Queen Anne duplexes, East Providence, 441
Queen Anne duplexes, Lincoln, 191
Queen Anne house, Westerly, 410
Queen Anne Square, Newport, 536
Queens River, 393, 429
Quidnesset Baptist Church, North Kingstown, 356
Quidnick, Coventry, 330, 335, 337
Quidnick Mills, Coventry. *See* American Hoechst Chemical Company (Quidnick Mills), Coventry
Quonochontaug, Charlestown, 397, 400–401
Quononoquot Club, Jamestown. *See* Rhett (Dr. H. J.) House (Quononoquot Club), Jamestown
Quonset hut, 21, 354, 356, 402
Quonset Point Industrial Park (Quonset Point Naval Air Station), North Kingstown, 20–21, 354, 356
Quonset Sea Bee Battalion naval base and museum, 20, 21

Radeke, Eliza Greene Metcalf, 68, 81
railroad stations
  Kingston Railroad Station, South Kingstown, 383, 388, 392–393
  Peace Dale Railroad Station, South Kingstown, 383, 384, 392
  Providence and Springfield Depot, Smithfield, 258
  Providence and Worcester Railroad Station, Woonsocket, 236–**237**
  Providence Station, Providence, 40–41
  Union Depot, Providence, **33**, 36, 38, 82
  Union Station Complex, Providence, 36, 38–39, 51
  Westerly Railroad Station, **414**–415
Rakatansky, Ira, 87
Ramparts, The. *See* Patterson (General Robert E.) House (The Ramparts; Channel Bells), Jamestown
Ramsey, Harry: Congregation B'nai Israel Synagogue, Woonsocket, 233
Ranalli, George, 529
ranger's house, Arcadia Management Area, Exeter/Hopkinton, **429**
Raphael, 54, 223
Rapp and Rapp: Providence Performing Arts Center (Loew's State Theatre), Providence, **45**–46
Rathbun, Oscar J., 234
Rathbun (Rachel Harris) House. *See* Thurber (Thomas)–Rathbun (Rachel Harris) House, Woonsocket
Rathburn (John)–Gardner (George)–Rivera (Abraham Rodrigues) House (Newport Bank), Washington Square, Newport, 520
Raymond Loewy–William Snaith, Inc.: Apex (store), Pawtucket, 143, 239

Read, Spencer P., 69
Red Church, Barrington. *See* St. John's Episcopal Church (Red Church), Barrington
Red Cross, Pawtucket, 148
Red Cross Cottage, Newport. *See* Sears (David) Cottage (Red Cross Cottage), Newport
Red House, The, Jamestown. *See* Potter (Margaret) Cottage (The Red House), Jamestown
Red Rover, The (Cooper), 544
Red Top, Jamestown. *See* Selfridge (Admiral Thomas O., Jr.) House (Red Top; Green Chimneys), Jamestown
Redwood, Abraham, Jr., 545, 546
Redwood, William, 532
Redwood Library, Newport, 7, 12, 510, 527, 528, 532, 540, 541, **544**–546
Redwood (William) House, Newport, 532
Reed (V. Z., Jr.) House (Seafair, Hurricane Hut), Newport, 577
Rehoboth, East Providence, 433, 434, 437
Rehoboth, Massachusetts, 186, 206, 434, 441, 442
Reich, Robert, 327
Renoir, Pierre, 492
Renshaw, Clifford M., 96
Renwick, James, 467
  Perry-Congdon House, Bristol, 479
  Seven Oaks, Bristol, 475
Report of the Metropolitan Park Commission, 183
Reports and Documents Relating to Public Schools in Rhode Island for 1848 (Barnard), 165
reproduction covered bridge, Foster, 275
Reservoir Pumping Station, South Kingstown. *See* Pump House Restaurant (Reservoir Pumping Station), South Kingstown
residence (Baptist Church), Narragansett, 369
resorts and summer communities, 11–19, 25–27
  Block Island (New Shoreham), 22, 28, 604–610
  Drownville, Barrington, 450–453
  Jamestown, 20, 28, 365, 387, 489, 507, 581–604
  Middletown, 18, 211, 498, 499, 507–516
  Narragansett, 178, 354, 365–373, 374
  Newport, 4, 6, 9, 10, 21, 24, 26, 27, 28, 29, 32, 34, 75, 76, 77, 85, 106, 120, 124, 125, 127, 147, 320, 357, 358, 364, 365, 368, 387, 388, 389, 393, 396, 516–581
  Portsmouth, 4, 28, 480, 483, 490, 498–507, 509
  Shelter Harbor, Westerly, 27, 403
  Watch Hill, Westerly, 3, 26, 28, 403, 405–408, 412
  Weekapaug, Westerly, 403
Restmere. *See* Van Rensselaer (Alexander) House (Restmere, Villalou), Middletown
Revolutionary War. *See* American Revolution
Reynolds, Gideon, 378
Reynolds, John (builder)
  offices (The Brick House; Colonel Micah Whitmarsh House), East Greenwich, 346
  Varnum (General James M.) House, East Greenwich, **347**
Reynolds, R. J., 451
Reynolds (Benjamin) House, Wickford, North Kingstown, 359, 360
Reynolds (Joe) Tavern, South Kingstown. *See* Westcott (Caleb) Tavern-Reynolds (Joe) Tavern-Taylor (Phillip) Tavern, South Kingstown
Reynolds (Joseph) House, Bristol, 468–469
Rhett (Dr. H. J.) House (Quononoquot Club), Jamestown, 602
Rhind, John Massey, 37
Rhode Island
  colonial government of, 10
  and industrialization, 11
  map (towns), 5

Rhode Island *(continued)*
   and signing of Constitution, 9–10
   See also Narragansett Planters
Rhode Island, battle of, 483
Rhode Island Advocate, 388
Rhode Island and Connecticut Pike. *See* Hartford Pike
Rhode Island Architecture (Hitchcock), 92, 164, 222, 237, 254, 260, 261, 271, 272, 306
Rhode Island Art Association, 68
Rhode Island Bank, Newport. *See* Buliod (Peter)–Perry (Oliver Hazard) House (Rhode Island Bank), Washington Square, Newport
Rhode Island Board of Public Roads, Bridge Division, 453
Rhode Island Coal Company, 500
Rhode Island College. *See* Brown University
Rhode Island Company, 177, 178
Rhode Island Company Trolley Barn, Cranston, 178
Rhode Island Department of Environmental Management, 172, 322, 429
   Division of Parks and Recreation, 592
Rhode Island Department of Public Works, 214
   Observation Tower, Hannah Robinson Park, South Kingstown, 387
Rhode Island Department of Transportation, 275
Rhode Island Electric Company, 60
Rhode Island Electric Light Company, 60
Rhode Island General Assembly, 403, 418
Rhode Island Hall, Brown University, Providence, 99
Rhode Island Historical Preservation and Heritage Commission, 77, 164, 267, 468, 585
Rhode Island Historical Society (John Brown House), Providence, 84, **89**–**90**, 90, 103, 261
Rhode Island Hospital, 355
Rhode Island Hospital Trust National Bank, Providence. *See* Sovereign Bank Complex, Providence
Rhode Island Junior College, Warwick. *See* Community College of Rhode Island Knight Campus (Rhode Island Junior College), Warwick
Rhode Island Militia, 478
   Naval Reserve Torpedo Company, 478
Rhode Island School of Design, Providence, 33, 47, 62, 64–66, 68–69, 72, 77, 78, 79–83, 107, 128, 277, 296, 396
   "Bank Building" (People's Savings Bank), Providence, 65
   Canal Street Row, 65
   center of, 78
   Cheapside, 65
   College Building, 64, 65, 78
   Division of Architecture, 62
   Eliza G. Radeke Building, Museum of Art, Providence, 80–81
   Hope Block, 65, 66
   Insurance Building, 65–66
   Jesse Metcalf Building, 66
   library, 65
   Memorial Hall (Central Congregational Church (first)), **80**, 81–83, 114, 147
   Morris Plan Building, 66
   Museum of Art, 79–81, 108, 133, 412
   Pendleton House (second) (The Colonial House) and Waterman Galleries,, 79–**80**, 81, 108
   Peter and Daphne Farago Wing, Museum of Art, 81
   President's House (Stephen O. Metcalf House), 106
   and Tefft drawings, 47
   Trustees' Executive Board, 68
   Waterman Building, 68–69
Rhode Island State Board of Public Roads, 362

Rhode Island State House. *See* Old State House (Providence Colony House, later Rhode Island State House), Providence
Rhode Island State Police Barracks
   Lincoln, 198, 357
   North Kingstown, 357
   Portsmouth, 357, 503
Rhode Island Women's Centennial Commission, 68
Rhodes (Henry) House, Providence, 76
Rice, Samuel, 339
Rice City, Coventry, 339–340
Rice City School, Coventry, 340
Rice (Mrs. Herbert A.) House, Providence, 117, 119
Richard Arnold Tavern. *See* apartments (Richard Arnold–Pelog Arnold Tavern), North Smithfield
Richards, William Trost, 595, 596, 597, 599
Richardson, G. H., 543
Richardson, Henry Hobson, 14, 54, 69, 100, 101, 191, 347, 384, 471, 473, 537, 551
   Sherman (Watts) House, Newport, **571**
Richardson-Blatchford House, Newport, **554**, 555
Richardson (Kate Lane) House (Yellow Patch), Narragansett, 370
Richmond, 299, 342, 397, 418, 424–428
   map, 426
   Town Hall, 428
Richmond, Franklin and Charles, 438
Richmond, George M., 333
Richmond, Isaac, 494
Richmond, Joshua B., 494, 495
Richmond, Lemuel Clark, 475
Richmond family, 495
Richmond House. *See* Seabury-Richmond-Burchard House, Little Compton
Richmond (Isaac Bailey), Little Compton, 494
Richmond (Lemuel C.) House, Bristol, 476
Richmond Paper Company, 437, 438
Richmond Schoolhouse. *See* Bell Schoolhouse (Richmond Schoolhouse), Richmond
RISD. *See* Rhode Island School of Design
Riven Rock, Jamestown. *See* Lovering (Joseph) House (Riven Rock), Jamestown
Rivera (Abraham Rodrigues) House, Newport. *See* Rathburn (John)–Gardner (George)–Rivera (Abraham Rodrigues) House (Newport Bank), Washington Square, Newport
Riverpoint, West Warwick, 324, 328–329, 334, 343
Riverside, East Providence, 434, 435, 437, 444
Riverview Nursing Home (Elisha Harris-Henry Howard House), Coventry, 335
Riverwalk, San Antonio, 40
Rives (Grace W.) House. *See* Hambly Funeral Home (Grace W. Rives House), Newport
Robertson, E. R., 107
Robertson, Mary Calder, 113
Robertson, R. H.: Auchincloss (John) House (Hammersmith Farm), Newport, 573, 579, 597
Robertson (Louis E.) House, Providence, 112, 113
Robertson (R. Austin) House. *See* Calder (Albert L.)–Robertson (R. Austin) House, Providence
Robinson, James, 378, 379
Robinson, John, 365
Robinson, Rowland, I, 375
Robinson, Rowland, II, 365
Robinson, Thomas, 523, 524
Robinson, William, 365, 375, 378, 379
Robinson and Steinman: Mount Hope Bridge, Bristol, **480**
Robinson (Dr. R. R.) House, South Kingstown, 379
Robinson family (Newport), 524

Robinson family (South Kingstown), 375
Robinson family houses, South Kingstown, 379–380
Robinson Green Beretta, 92, 102, 323
   Biological Science Building, University of Rhode Island, Kingston, 392
   Homer and Nickerson Halls, Metcalf Refectory, Providence, 78
   St. Jude's Roman Catholic Church, Lincoln, 191, 192
Robinson Hall (University Library), Brown University, Providence, 102–**103**
Robinson House, 571 Main, South Kingstown. *See* Miller (William David) House (Robinson House, 571 Main), South Kingstown
Robinson (Rowland) House, Narragansett, 365, **367**, 388
Robinson (Thomas) House, Newport, 523–**524**, 596
Rochambeau, General, 339, 520, 540, 541
Rochambeau House (Henry Dexter Sharpe House), Brown University, Providence, 105–106
Rockefeller, John D., Jr., 44, 67, 101
Rockefeller, Mrs. John D., Jr., 44
Rockefeller Hall, Providence., 101
Rockefeller Library, Providence. *See* John D. Rockefeller, Jr., Humanities Library, Providence
Rockry Hall. *See* Sumner (Albert) House (Rockry Hall), Newport
Rockville, Hopkinton, 418, 419, 421
Rockville Mill, Hopkinton. *See* Post Office (Rockville Mill), Hopkinton
Rockwell, Charles B., 478
Rocky Brook, South Kingstown, 375, 381, 385–386
Rocky Hill School (Hopelands), Warwick, 322
Rodgers Recreational Center, Salve Regina University, Newport, 560
Rodin, Auguste, 406
Rodman, Charles, 364
Rodman, Daniel, 393
Rodman, Robert, 363, 378
Rodman, Samuel, 385
Rodman family, 364, 375, 385
Rodman (Isaac Peace) House (The Stone House), South Kingstown, 386
Rodman Mill, South Kingstown, 385
Rodman (Robert) House. *See* McKay's Front Porch (Robert Rodman House), North Kingstown
Rodman (Walter) House. *See* condominiums (Walter Rodman House), North Kingstown
Rodman's Hollow, Block Island, 609
Roger Williams Avenue, single-family and duplex houses, East Providence, 437–438
Roger Williams Baptist Church, Providence, 128
Roger Williams Junior College. *See* Roger Williams University, Bristol
Roger Williams Park, Providence, 34, 70–71, 92, 132–134, 151, 177, 183
Roger Williams University, Bristol, 25, **480**
   School of Architecture, Art, and Historic Preservation, 480
Rogers, Isaiah, 542
Rogers, Randolph: Civil War monument, Providence, 36
Rogers, Robert, 471
Rogers Free Library, Bristol, 471
Rogers Hall, Brown University, Providence, 100, 101
Rogers (Joseph) House, Newport, 520–521
Roman Catholic churches. *See* churches, Roman Catholic
Roman Catholics, 18–19, 54, 115, 182, 244, 356, 367, 372, 373, 381, 449, 514
Rome, Italy, 63, 140
Rood, Welcome, 279
Roosevelt Avenue Bridge, Central Falls, 155

Rose Cottage, East Greenwich, 348
Rose Farm, Providence. *See* Dexter (Ebenezer Knight)–Stimson (John J.) House (Rose Farm), Providence
Rosecliff. *See* Oelrichs (Tessie and Herman) House (Rosecliff), Newport
Rosedale Apartments, Cranston, 183
Roslyn, Newport. *See* Grosvenor (William) House (Fair Oak, Roslyn), Newport
Ross, John L., 304–305
Ross (Mary Rose) House, Warwick, 318
Rossini Road, Shelter Harbor, Westerly, 403
Rough Point, Newport. *See* Vanderbilt (Frederick W.) House (Rough Point), Newport
Round House, The. *See* Newhall (Daniel S.) House (The Monitor; The Round House), Jamestown
Round Top Church, Providence. *See* Beneficent Congregational Church (Round Top Church; formerly Second Congregational Church), Providence
Rounds (P.) Farm, Foster, 275
Roxbury, Massachusetts, 545
Roy Carpenter's Beach, South Kingstown, **394**–395
Royal Indian Burial Ground of the Narragansett, Charlestown, 399, 403
Royal Mills (Greene Mills), West Warwick, 328, 329
Royal Weaving Mill, Pawtucket, 138, 152
Royall, Isaac, Jr., 479
Royall (Isaac, Jr.) House. *See* Mount Hope Farm (Isaac Royall, Jr., House), Bristol
Rudofsky, Bernard, 444
Rudolph, Paul: Beneficent House, Providence, 44–45
Ruins of the Valley Falls Mill, Cumberland Operation, Cumberland, 160, 208
Ruisdell House. *See* Bliss-Ruisdell House (John Birch House), Warren
Rules of Drawing (J. Gibbs), 542
Rumford, East Providence, 434, 435, 443
Rumford Chemical Works (former), East Providence, 112, 434, 435, 439, 440–441
Runnins River, 433
Ruskin, John, 48, 99, 102
Russala, Watch Hill, Westerly. *See* Tredegar (Russula), Watch Hill, Westerly
Russell, Henry G., 322
Russell, Homer: Russell (Homer) House, Block Island, 609
Russell, Hope Brown Ives, 322
Russell Estate, Warwick. *See* Goddard Memorial Park (Russell Estate), Warwick
Russell family, 530
Russell (Homer) House, Block Island, 609
Russia, 604

S. B. and H. Chace, 154, 159
Saarinen, Eero, 192, 318, 444
Saarinen, Eliel, 318, 444
Sabbatarians, 534
Sack, Albert, 161
Sackett (Frederick M.) House, Providence, 119–120
Sacred Heart Church, West Warwick, 330
Saeltzer, Alexander, 467, 473
St. Ann's Church, Woonsocket, 160, **223**, 224, 228, 230
St. Augustine, Florida, 114
St. Bartholomew's Episcopal Church, Cranston. *See* Sprague Meeting House (St. Bartholomew's Episcopal Church), Cranston
St. Charles Borromeo, Woonsocket, 227, 228, 235–236
St. Christopher's Roman Catholic Church Rectory (Captain Isaac Church House), Tiverton, 482

St. Colomba's Chapel (Berkeley Memorial Chapel), Middletown, 181, 348, **511**–512, 514
   Parish Hall, 511
St. Elizabeth's Episcopal Church (Canonchet Chapel), Hopkinton, 422
St. Florian, Friedrich, 41
St. Francis of Assisi Roman Catholic Church, South Kingstown, 380
   Auditorium Building, **380**
Saint-Gaudens, Augustus, 530–531, 544
St. George Maronite Catholic Church (St. Paul's Episcopal Church), Pawtucket, 146–147
St. George's School, Middletown, **514**–515
   Chapel, 514
   Diman Hall, 514
   Wyn Wyc, 515
St. James Convent (Captain Nathaniel Church House), Tiverton, 482
St. James Episcopal Church, North Providence, **162**–163
St. Jean-Baptiste Church, Pawtucket, 140
St. John the Evangelist Episcopal Church, Newport, 524–525
St. John the Evangelist Rectory (John Dennis House), 523, 524
St. John's Church, Providence. *See* Cathedral of St. John (St. John's Church, Providence
St. John's Episcopal Church (Red Church), Barrington, 451, 452, 453
   Burrington Memorial Chapel, 452
   Mathewson Memorial Tower, 452, 453
St. Johns Lodge Masonic Temple, Newport, 539
St. Joseph's Church, Cumberland, 212, **213**, 228
St. Joseph's Church, Providence, 54
St. Jude's Roman Catholic Church, Lincoln, 191, 192
St. Louis, Missouri, 582, 603, 604
St. Luke's Episcopal Church, East Greenwich, **347**–348
St. Mark's Episcopal Church, Warren, 459
St. Martin-in-the-Fields, London, 8, 67
St. Mary's, Portsmouth, 511
St. Mary's Church, Newport, 581
St. Mary's Church, Portsmouth, **506**–507, 509, 511
   Sarah Gibbs tomb, 506
St. Matthew's Episcopal Church, Barrington, 451
St. Matthew's Roman Catholic Church, Central Falls, 160, 175
St. Michael's Episcopal Church, Bristol, **473**
   Parish House, 473
St. Michael's School (Miss Ellen Mason House), Newport, 505, 551
St. Paul's Church, Wickford, North Kingstown, 358–359, 360
St. Paul's Episcopal Church, Portsmouth, 503
St. Peter's by the Sea II, Narragansett, 369
St. Peter's Episcopal Church I, Narragansett, 369
St. Philomene's Roman Catholic Church, Narragansett. *See* St. Thomas More Roman Catholic Church (St. Philomene's Roman Catholic Church), Narragansett
St. Rocco's Church, Johnston, 175
St. Stephen's Episcopal Church, Providence, 82, **95**–96, 104, 254
St. Thomas Episcopal Church, Richmond, 422, 427, **428**
St. Thomas Episcopal Church, Smithfield, **254**–255
St. Thomas More Roman Catholic Church (St. Philomene's Roman Catholic Church), Narragansett, 369
Sakonnet River, 481, 483, 489, 498, 499, 502, 504, 509, 511
Sakonnet Vineyards, Little Compton, 489

Salem, Massachusetts, 29, 34, 76, 79, 83, 538, 549
Salisbury, Robert, 275
Salmon, William, 64, 90, 118
Salt Marsh, Newport. *See* Tomlinson (Irving) House (Salt Marsh), Newport
Salve Regina University, Newport, 18–19, 560, 561, 571
   Rodgers Recreational Center, 560
Sample, Paul, 321
Samuel Slater and Company, 137
San Antonio, Texas, 40
San Carlo alle Quattro Fontane, Rome, 63
San Diego, California, 505
San Francisco, California, 59, 409, 501
Sandpiper Cottages (Floradale Motor Court), Middletown, 509, **510**
Sandy Point Farm, Farm Estates, Portsmouth, 503, 504, 505
Sandy Point Light. *See* North Light (Sandy Point Light), Block Island
Sanford (M. H.) House (Ednaville), Newport, 525, 555
Sansovino, Jacopo, 43
Sargent, Charles, 484
Sargent (Charles) House. *See* Oliver (Andrew)–Sargent (Charles)–Thayer (Steven) House (Homelands), Tiverton
Sargent (Letitia B.) House (Aufenthalt), Newport, 554
Saugatucket River, 375, 385
Savannah, Georgia, 26, 34, 88, 491
Saving Large Estates: Conservation, Historic Preservation, Adaptive Re-Use (Shopson), 378
Sawtelle, Franklin J.: Bliss (Caroline) House, Providence, 116–117
Sayles, Albert L., 301–302, 313
Sayles, Clark, 193, 293
   Chepachet Freewill Baptist Church and Parsonage, Glocester, 253, 265, **293**, 294, 295
   Freewill Baptist Church, Smithfield, 253, 265, 293
   North Scituate Congregational Church, Scituate, **265**–266, 279, 284, 293, 312, 487
Sayles Memorial Chapel, Lincoln, 194
Sayles, Daniel, 301, 313
Sayles, Frank, 193
Sayles, Frederic, 145, 189, 193, 194, 443
Sayles, William Clark, 100
Sayles, William Francis, 100, 189, 193, 194
Sayles (Albert L.) House, Burrillville, **314**
Sayles amusement park, East Providence, 443
Sayles Corporation, 141, 159, 189, 443
Sayles family, 142, 295, 301, 313, 314, 434
Sayles Finishing Company, 193
Sayles (H. A.) House, Glocester, 295
Sayles Hall, Brown University, Providence, 99–100, 107, 194
Sayles Memorial Chapel, Lincoln, 194
Sayles Mill, Burrillville, 313
Sayles Mill Complex and Administration Building, Lincoln, 193
Sayles New Village, Pawtucket, 140
Sayles (William F.) House, Providence, 107
Saylesville, Lincoln, 159, 188, 189, 193–194
Saylesville Meeting House, Lincoln, 194–195, 197
Scalabrini Villa (John Carter Brown House), North Kingstown, **355**
Scarlet Letter, The (Hawthorne), 549
Scenes of Apponaug Cove and Village (Sample mural), 321
Schaus (Sophie) Cottage, Jamestown, 604
Schermerhorn (Edward H.) House (Chepstow), Newport, 558
Schinkel, Karl Friedrich, 81
Scholze (Frederick) House, Pawtucket, 138–139

school buildings
  Ada Hawkins School (former), Glocester, 294, 296–297
  Andrews Memorial School, Bristol, 471
  Bell Schoolhouse (Richmond Schoolhouse), Richmond, 428
  Brick Schoolhouse, Providence, 70
  Bristol Public School, 25
  Burton Primary School, Bristol, 476
  Byfield School, Bristol, 477
  Callender School, Newport, 529
  Charles W. Shea Senior High School (Pawtucket West High School), Pawtucket, 139, 145
  Chepachet School, Glocester, 294
  Clark (Samuel) Farm, Richmond, 425
  Clarkville School, Glocester, 291, 297–298
  Colt-Andrews School (Colt Memorial High School), Bristol, 469, 471
  District 2 Schoolhouse, Charlestown, 399, 402
  East Providence Senior High School, East Providence, 443
  Evans Schoolhouse, Glocester, 291, 297–**298**
  Fifth Avenue School (former), Narragansett, 370
  Forestdale School (former), North Smithfield, 243
  Foster Center School (Hemlock School), Foster, 278, 283
  Grove Street School, Woonsocket, 227
  Guiteras Memorial and Junior High School, Bristol, 469
  John Brown Francis School, Warwick, 318–319
  Joseph C. Sweeney School, Burrillville, 303, 315
  Kendrick Avenue School, Woonsocket, 224–225, 227
  Liberty Street School, Warren, 462
  Mapleville School (former), Burrillville, 303, 315
  Mount Hygeia Schoolhouse, Foster, 290
  North Scituate Academy (former), 266
  Number 8 Schoolhouse (former), Little Compton, 496
  Odlin (John)–Otis (Jonathan) House, Newport, 532, 533
  Our Lady of Good Help Church, Rectory, and Parish House–School (Notre Dame de Bon Secours), Burrillville, **304**
  Paradise School, Middletown, 513
  Parker District School (former), West Greenwich, 343
  Pascoag No. 11 School (former), Burrillville, 303, 315
  Peace Dale Elementary School, South Kingstown, 385
  Peck (Leander R.) School, Barrington, 452
  Portsmouth Abbey School, Portsmouth, **500**–502, 514
  Pottersville District 7 Schoolhouse (former), Scituate, 269
  Providence Street School, West Warwick, 330
  Pyramid Schoolhouse (former), Little Compton, 491, 492
  Reverend William S. Child Schoolhouse, Newport, 526
  Rice City School, Coventry, 340
  Rocky Hill School (Hopelands), Warwick
  St. George's School, Middletown, **514**–515
  St. Michael's School (Miss Ellen Mason House), Newport, 505, 551
  School Number 3 (former), Westerly, 408
  Smithville Seminary (former), Scituate, 266–267
  Sockanosset School for Boys Administration Building, Cottages, and Chapel, Cranston, 177, 180–**181**
  Southernmost School, Portsmouth, 503
  Stepping Stone Kindergarten, South Kingstown, 383
  Walley School, Bristol, 477
  Woody Hill School, District 1, Exeter, 283, 429–**430**
  See also colleges and universities
School of Architecture, Art, and Historic Preservation, Roger Williams University, Bristol, 480

Schroeder, Roger, 268
Schubert, Franz, 403
Schwab, Raymond W., 370
Sciences Library, Brown University, Providence, 103–104
Scientific American (magazine), 609
Scituate, 263–273
  map, 264
Scituate Reservoir, 263, 265, 272–274, 291, 298, 342, 443
  Philip J. Holton Water Purification Plant, 272
Scott, M. H. Baillie, 577
Scott (Richard) House, Lincoln, 195–196
Scully, Vincent, 513
Sea Swept (Ocean Mount), Watch Hill, Westerly, 407
Seabeach, Newport. *See* Pomeroy (Edith Burnet) House (Seabeach), Newport
Seaborn Mary, Little Compton, 495
Seabury, Cornelius, 486
Seabury, Dwight: Waite-Thresher Building, Providence, 55, 56–57
Seabury (Cornelius) House. *See* Soule (Abner)–Soule (Cornelius)–Seabury (Cornelius) House, Tiverton
Seabury family, 495
Seabury-Richmond-Burchard House, Little Compton, 495
Seaconnet River. *See* Sakonnet River
Seacunke, East Providence, 433
Seafair. *See* Reed (V. Z., Jr.) House (Seafair, Hurricane Hut), Newport
Seamens-Davis House, Foster. *See* Davis (Henry) House (Seamens-Davis House), Foster
Sears, David, 550
Sears, Roebuck cottages, 585–586, 591
Sears (David) Cottage (Red Cross Cottage), Newport, 550
Seaside Colony, Narragansett, **370**, 371–372
Seaside Farm. *See* Hazard (Joseph Peace) House (Hazard's Castle, Seaside Farm), Narragansett
Second Baptist Church, Foster. *See* Foster Town House (Second Baptist Church, later Elder Hammond's Meeting House), Foster
Second Congregational Church, Newport. *See* condominiums (Second Congregational Church), Newport
Second Congregational Church, Providence. *See* Beneficent Congregational Church (Round Top Church; formerly Second Congregational Church), Providence
Second Seventh Day Baptist Church, Hopkinton, 423
Seekonk, Massachusetts, 434
Seekonk River, 29, 120, 121, 122, 135, 433, 434, 435, 437, 440–441
Segar (W. F.) House, South Kingstown, 394, **395**
Selden (James H.) House, Hopkinton, 418–419
Selfridge, Thomas O., Jr., 599, 601
Selfridge (Admiral Thomas O., Jr.) Cottage (The Barnacle), Jamestown, 599, 601
Selfridge (Admiral Thomas O., Jr.) House (Red Top; Green Chimneys), Jamestown, 599, 601
Selfridge and Obermeier: Henszey (William P.)–Ewing (Joseph N.) House (Altamira), Jamestown, 600–601
Sephardic Jews, 7
Settling of Providence, The (King mural), 123
Seven Oaks, Bristol, 475
seventeenth-century replica houses, Scituate, 267–**268**
Seventh-Day Adventist Church (Mapleville Methodist Episcopal Church), Burrillville, 304
Seventh Day Baptist Church, Hopkinton, 419, **421**
Seventh Day Baptist Meeting House, Newport. *See* Newport Historical Society and Seventh Day Baptist Meeting House, Newport

75–85 White Rock Road, Westerly, 417
Seymour (Carolyn) House (Hawkhurst or Hawxhurst), Newport, 554
Shadblow Farm, South Kingstown, 386–387
Shadow Farm (Samuel A. Strang-John J. Welsh House), South Kingstown, **375**, 378
  barn, 375, **378**
Shady Lea Mill, North Kingstown, 364
Shakers, 285
Shakespeare, William, 146
Shakespeare's Head building, Providence, 70
Shamrock Stables, Taylor (H. A. C.) Farm (The Glen), Portsmouth, 504
Shannock, Charlestown/Richmond, 397, 401, 424, 425–427
Shannock Mills and Village, Charlestown/Richmond, 425–427
Sharpe, Ellen Dexter, 105
Sharpe, Henry Dexter, 105
Sharpe, Mary Elizabeth, 105
Sharpe, Thomas, 270
Sharpe (Ellen Dexter) House, Brown University, Providence., 105
Sharpe (Henry Dexter) House, Brown University, Providence, 105–106
Shaw, Norman, 571
Sheffield, Nathaniel, 400
Sheffield House, Charlestown, 400
Sheldon, Frank P., 201, 331, 455, 481, 482
  Ann and Hope Store Complex (Ann and Hope Mill, Lonsdale Mill), Cumberland, 142, 168, 206, 209–210
  Berkeley Mill and Mill Village, Cumberland, 188, 189, 190, 191, 202, 205, 208, 209, 211–212, **213**, 271
  Joseph Benn and Company (Greystone Mill), North Providence, 166, **167**, 168, 170, 482
  Slater Cotton Company New Mill, Pawtucket, **140**, 142–143
  Whitehall Building, North Providence, **166**–167, 168
Sheldon, W. G.: Cocroft (Freeman) House (Croftmere), South Kingstown, 396
Sheldon, William, 16
Sheldon Block, South Kingstown, 379
Shelter Harbor, Westerly, 27, 403
Shelter Harbor Inn, Westerly, 403
Shepard Department Store, Providence, 52
Shepard family, 530
Shepherd's Run, South Kingstown. See Convent of the Sisters of the Cross and Passion (Rush Sturges Mansion, Shepherd's Run), South Kingstown
Shepley, Bulfinch, Richardson and Abbott, 545
Shepley, Rutan and Coolidge
  Brown University, John Carter Brown Library, Providence, **101**, 158
  Brown University, John Hay Library, Providence, 103
  Brown University, Soldier's Memorial Gateway, Providence, 102
  Tilden-Thurber Building, Providence, 52
Sherman, Anne Wetmore Watts, 571
Sherman, Ernest, 313
Sherman, Georgette Wetmore. See Brown, Georgette Wetmore Sherman
Sherman, Stephen, 313
Sherman, Sumner, 313
Sherman, Syra, 313
Sherman, William Watts, 571
Sherman (Charles) House, Newport, 544
Sherman family, 313
Sherman (Sumner) Farm (Othonie Young Farm), Burrillville, 313

Sherman (Watts) House, Newport, 569, **571**
Sherry, Louis, 370
Sherry Cottages, Narragansett, **370**–371
Sherwood, Newport. See Jones (Pembroke) House (Sherwood), Newport
shingle houses
  Browning Beach, South Kingstown, 395
  Wawaloam Drive, Westerly, **405**
  Weekapaug, Westerly, 403, **405**
Shingle-nook. See Matthews (Brander) House (Shingle-nook), Narragansett
Shippee, Lodowick C., 346
Shippee Bridge, Burrillville, 310
shopping centers. See stores and shopping centers
Shopson, William, 378
Shore Mill, East Greenwich. See Bay Mill (later Shore Mill), East Greenwich
Shoreby Hill, Jamestown, 27, 583, 603–604
  houses, **603**
Shoreby Hill Club. See Jamestown Casino (Shoreby Hill Club), Jamestown
Shun Pike, 170
Shurcliffe, Arthur, 411
Sidler, Peter, 502
Siena, Italy, 126
Silver Creek, Bristol. See Bosworth House (Silver Creek), Bristol
Simmons, James F., 174, 175
Simmons Building, Providence, 59
Simmons-Winchester House, Little Compton, 494, 495
Simmons (James F.) House, Johnston, 174–**175**
Simon, Louis A.: U.S. Post Office, Warwick, 321
Simon (Captain Peter) House, Newport, 528
Sinking Fund Factory Village, Lincoln, 205
Sinnickson, Emma R., 372
Six Principle Baptist Church, Richmond. See Victory Baptist Church (Wood River Baptist Church), Richmond
65 Merton Road, Newport, 557–558
Skidmore, Owings and Merrill: Convention Center and Capital Center Redevelopment Project, Providence, 39–41
Skinner (Frances L.) House (Villino), Newport, **550**, 551, 567
Slade and Balcom (paint business), 41
Slade Building and adjoining structures, Providence, 41–42
Slater, Elizabeth, 247
Slater, Horatio N., 75, 99
Slater, John F. (son), 244
Slater, John (father), 75, 244, 245, 247
Slater, Samuel, 55, 75, 135, 154, 244, 245
Slater, William (grandfather), 28, 137, 144
Slater, William (grandson), 244, 245
Slater Cotton Company, 193
Slater Cotton Company New Mill, Pawtucket, **142**–143
Slater family, 244, 246
Slater Hall, Brown University, Providence, 99
Slater (John, father) House, North Smithfield, 245, 247
Slater Memorial Park, Pawtucket, 151, 152
Slater Mill Complex, Pawtucket, 28, 32, 135, 143–**145**, 327, 328
Slatersville, North Smithfield, 238, 243, 244–249, 302, 381
Slatersville Green and Congregational Church, North Smithfield, 245, **246**
Slatersville Mill, North Smithfield, **247**–248
slavery and slave trade, 26
  and Briggs (Nathaniel), 484, 488

Index   693

and Bristol, 466
and Narragansett Planters, 354, 374
and Newport grave markers, 530
and Wakefield Post Office mural, South Kingstown, 379
and Warren, 454–455
Slicer (Adeline E. H.) House (The Mill), Little Compton, 492–**493**
Sloan, Samuel, 448
Slocomb, Edgar M., 243
Smibert, John, 527
Smirke, Sydney, 102
Smith, Alfred, 15, 555, 562, 569
Smith, Barnard, 474
Smith, Benjamin R., 524
Smith, Daniel, 239
Smith, Elisha, 257
Smith, Joseph, 162, 303
Smith, Richard, 351
Smith, Robert, 97
Smith, Stephen, 64, 197
Smith, William, 94
Smith (Amos D.) House (Hall Manor), Portsmouth, 500, 501, 502
Smith (Arnold) House, Tiverton, 486
Smith (Charles) House. *See* Joseph W. Martin Memorial Home (Charles Smith-William Winslow House), Warren
Smith (Christopher)–Fenner House, Scituate, 269
Smith (Daniel)–Andrew (David)–Taft (Nelson)–Todd (Albert) Farm, North Smithfield, 238–239
Smith (Elisha) House, Smithfield. *See* Smithfield Historical Society (Elisha Smith House, known as the Smith-Appleby House)
Smith family, 415
Smith Farm, Lincoln. *See* Chace Farm, Lincoln
Smith (George M.) House, Providence, 118–119, 120
Smith House, Bristol. *See* Babbitt-Smith House, Bristol
Smith House, Burrillville, **302**
Smith-Darling House, Burrillville, 307
Smith (Jeremiah) House, Lincoln, 198–199
Smith (Joseph)–Jenckes (John)–Cushing House, North Providence, 162
Smith (Orlando) House, Westerly. *See* Babcock (Dr. Joshua)–Smith (Orlando) House, Westerly
Smith (Reverend Francis) House, Providence, 74
Smith (Sibley) House, South Kingstown, 396
Smith (Stephen H.) House (Hearthside), Lincoln, **196**, 197
Smith (William) House, Providence, 94
Smithfield, 118, 153, 154, 155, 186, 220, 238, 239, 249–262
   map, 250
Smithfield Cotton and Woolen Company, 205
Smithfield Historical Society (Elisha Smith House, known as the Smith-Appleby House), Smithfield, 257–258
Smithfield Lime Rock Bank, Lincoln, 200–201
Smith's Castle, North Kingstown. *See* Cocumscussoc (Smith's Castle), North Kingstown
Smithville Seminary (former), Scituate, 266–267
Smyth, Douglas: Shadow Farm (Samuel A. Strang–John J. Welsh House), South Kingstown, 375, **378**
Snake Den Quarry, Johnstown, 169, 170
Snake Hill Road, Glocester, 298–299
Snell, George, 545, 546
Snow, Joseph, Jr., 43
Sobon, Jack, 268
Social Manufacturing Company, Woonsocket, 201, 232
Social Mill, Woonsocket, 232

Society for the Preservation of New England Antiquities, 171, 195, 364, 589
Society for the Promotion of Knowledge and Virtue by a Free Conservation, Newport, 545
Society for the Propagation of the Gospel in Foreign Lands (Episcopal), 534
Society of Friends. *See* Quakers
Society of St. Charles, 355
Sockanosset School for Boys, Administration Building, Cottages and Chapel, Cranston, 177, 180–**181**
Soldier's Memorial Gateway, Brown University, Providence, 102
Somerset, Massachusetts, 472
Sonnenschein, Narragansett. *See* Carver (Emma B.) House (Kabyun, Sonnenschein), Narragansett
Soule, Abner, 485
Soule, Cornelius, 485–486
Soule (Abner)–Soule (Cornelius)–Seabury (Cornelius) House, Tiverton, **485**–486
South (Rhode Island), 351–432
South Branch. *See* Pawtuxet River
South County Public Service Building, Westerly, 413, 414
South East Light, Block Island, 22–23, 608, **609**
South Ferry Congregational Church (Narragansett Congregational Church), Narragansett, 367, 509
South Foster, Foster, 288–290
South Foster Union Chapel. *See* Hopkins Mills Union Church (South Foster Union Chapel), Foster
South Kingstown, 76, 363, 364, 365, 367, 372, 374–397, 412, 427
   map, 376–377
   Town Hall, 380–381
South Street Plant, Providence. *See* Narragansett Electric Company Power Plants, Providence
Southermost School, Portsmouth, 503
Southmeadow. *See* Tinkham (William)–Levy (Austin T.) House (Southmeadow), Burrillville
Southwick (Solomon) House, Newport, 529
Southwinds. *See* Logan (Mrs. George W.) Cottage (Southwinds), Jamestown
Sovereign Bank Complex (Rhode Island Hospital Trust National Bank), Providence, 50, 63
Sowams, 454, 460. *See also* Warren
Spanish-American War, 20, 592
Spencer (William B.) House I, West Warwick, **326**
Spencer (William B.) House II, West Warwick, **326**, 327
Spicer House, Hopkinton, 422
Spitz (A. A.) Duplex, Providence, 129
Spooner House, Newport, 529
Sprague, A. and W., 335
Sprague, Thomas, 255
Sprague, William, 178, 367
Sprague (Charles H.) House, Providence, 112–113
Sprague family, 176–177, 178, 179, 329, 330, 337
Sprague House. *See* Governor Sprague Mansion, Cranston
Sprague Manufacturing Company, 154, 331
Sprague Meeting House (St. Bartholomew's Episcopal Church), Cranston, 179
Spring House and Cottage, Block Island, 609, 610
Spring Street, Newport, 580–581
Springfield, Massachusetts, 301
Squantum Association, East Providence, 435, 444–445
   banquet hall, 445
   billiard hall, 445
   clubhouse, **444**, 445
   stable, Hopkinton, **419**
Stafford Manufacturing Company Mill (John Kennedy Mill), Central Falls, 156–157
Stafford Mill, Central Falls, 155–156

Stafford Mill extension, Central Falls, 155–156
Stagecoach Tavern (Cyrus Cook's Tavern), Glocester, 294
Stamp Act of 1765, 519, 520, 529
Standard Oil Company, 512
Stanton, Joseph, Jr. (son), 400
Stanton, Joseph, II (father), 400
Stanton family, 400
Stanton (Joseph) House. *See* Wilcox Tavern (Joseph Stanton House), Charlestown
Starkweather (Oliver) House, Pawtucket, 149
Starr House, Warren. *See* Dow-Starr House, Warren
state houses
   Bristol, 25, 477
   East Greenwich, 345, 346, 348
   Kingston, South Kingstown, 390
   Newport Colony House, 520
   Old State House, Providence, 10, 32, 33, 39, 40, 76–77, 106, 122–124, **123**, 125, 345, 381
State Police Barracks. *See* Rhode Island State Police Barracks
State Prison, Cranston. *See* Adult Correctional Institution, Cranston
Stearns, Walter, 150
Stearns (Walter) House, Pawtucket, 150
Steele, Fletcher, 104
Steer-Angell-Colwell Farm, Glocester, 299
Steere, Anthony, 255
Steere, Elisha, 255
Steere, Henry J., 44, 447
Steere, Job S., 310
Steere (Henry J.) House, Barrington, 458
Steere-Harris House, Smithfieled, 256
Steere (Stephen) House, Smithfield, 256
Steinman, John (engineer): Mount Hope Suspension Bridge, Bristol, 25
Stepping Stone Kindergarten, South Kingstown, 383
Stern, Robert A. M.: Salve Regina University Rodgers Recreational Center, Newport, 560
Stevens, John: Pawtucket Congrgational Church, **147**–148, 149
Stevens, Joseph G., 543
Stevens, Mary and Anne, 553
Stevens, Pompe, 530
Stevens (John) Houses, Newport, 528, 530
Stevens (Mary and Anne) House, Newport, 553
Stevens (Robert) House, Newport, 541
Stevenson, David, 372
Stevenson, J. J., 107
Stickley, Gustav, 591
Stiles, Ezra, 541–542
Stiles (Ezra) House, Newport, 541–542
Stillwater, Smithfield, 251, 256–259
Stillwater River, 250, 253, 255, 302
Stillwater Worsted Company, Harrisville, Burrillville, 253, 302, 303, 306, 308, 313
Stimson, John J., 111, 114
Stimson (John J.) House. *See* Dexter (Ebenezer Knight)–Stimson (John J.) House (Rose Farm), Providence
Stockholm, Sweden, 124
Stone, Alfred, 118
   Burnside (General Ambrose) House, Providence, 88, 120
   Faculty Club (Zachariah Allen House) Brown University, Providence, 87
   Hall (Mrs. Edward Brooks) House, Providence, 88, 120
Stone and Carpenter, 87, 119, 121, 182, 332, 333
   Adult Correctional Institution, Cranston, 129, 177, 182–183

Cheapside (building), Providence, 65
Coventry Mill (Anthony Mill), Coventry, 168, 333, **337**–338
Crompton Mill, West Warwick, **332**–333, 337
Hillel House, Brown University (Froebel Hall), Providence, 108
Nicholson House (Francis W. Goddard–Samuel C. Nicholson House), Providence, 96, 97
Owen Block, Providence, 64
Slater Hall, Brown University, Providence, 99
Wanskuck Mill superintendent's house, Providence, 128
See also Stone, Carpenter and Willson; Stone, Alfred
Stone Arch Bridge, Burrillville, 310
Stone Bridge and Fort Barton, Tiverton, 483
Stone, Carpenter and Willson, 63, 69, 78, 85, 89, 90, 96–97, 103, 106, 116, 117, 121, 122, 128, 129, 391, 493
   Baker (Esther) House, Providence, 120
   Barrington Town Hall, 447, 451–**452**
   Burrill Building, Providence, 52
   Carpenter (Asa) House, Pawtucket, 151
   Conrad Building, Providence, 53, 54, 129
   Davis Hall (College), University of Rhode Island, Kingston, **391**, 392
   Exchange Bank Building, Providence, 49–50, 51
   Fleet Bank (Industrial Trust), Warren, 464
   Fletcher (Joseph) House, Providence, 111–112
   Francis Building, Providence, 51
   King Hall, Brown University (Robert W. Taft House), Providence, 118, 119, 120
   Klapp (Lyman) House, Providence, 110, 111, 120
   Lauderdale Building, Providence, 51
   Lippitt Hall, University of Rhode Island, Kingston, 391, 392
   Lyman Gymnasium, Brown University, Providence, 101–102
   Pembroke Hall, Brown University, Providence, 107–108
   Pendleton House (second), Providence, 79–80, 81, 83
   Providence Public Library, 42–43, 51, 132, 146
   Providence Telephone Building, 41–42, 51
   Sackett (Frederick M.) House, Providence, 119–120
   Smith (George M.) House, Providence, 118–119, 120
   Sockanosset School for Boys Administration Building, Cottages and Chapel, Cranston, 177, 180–**181**
   Taft Hall, University of Rhode Island, Kingston, 391
   Union Station Complex, Providence, 36, 38–39, 51
   Union Trust Company Building, Providence, 50, 51
   See also Stone, Alfred; Stone and Carpenter; Willson, Edmund
stone-enders (houses)
   Cranston, 184
   Johnston, **171**
   Lincoln, 187, **195**, 200
   Newport, 525
   seventeenth-century replicas, Scituate, 267–268
   South Kingstown, 393, 397
   Warwick, 320, 321, 322
Stone Gables, Newport. *See* Zabriskie (Sarah T.) House (Stone Gables), Newport
Stone House, Smithfield. *See* Tucker-Stone-Colwell House, Smithfield
Stone House, South Kingstown. *See* Rodman (Isaac Peace) House (The Stone House), South Kingstown
Stone House, Tiverton. *See* Beattie (Hamilton) House (The Stone House), Tiverton
Stone House, Woonsocket, 230
Stone Mill, Newport, 544

Stonelea, Narragansett, **370**, 372
Stonington, Connecticut, 416
store block, Newport, 534, 537
stores and shopping centers
  Apex, Pawtucket, 143, 239
  Brown and Hopkins Store, Glocester, 294
  Convention Center and Capital Center Redevelopment Plan, Providence, 39–41
  Cumberland, 209
  Garden City, Cranston, 181–182
  Kenyon's Department Store, South Kingstown, 379
  Mall, Providence, 36
  Newport, 534, 537
  Shepard Department Store, Providence, 52
  Taylor (William) Store, DeWolf's Wharf, Bristol, 478
  Usher Store (Potter's Wharf), Bristol, 478
  Westminster Street Shopping District, Providence, 51–53, 55
Stork (Theophilus) House (Fowlers Rock), Caretaker's Cottage, Jamestown, 585
Stowe, Harriet Beecher, 539
Strang, Samuel, 378
Strang (Samuel A.) House. *See* Shadow Farm (Samuel A. Strang–John J. Welsh House), South Kingstown
Strickland, William: Providence Athenaeum, 84
Stuart, Gilbert, 28, 124, 364
Stuart (Gilbert) Birthplace, North Kingstown, 364
Stubb's Wharf. *See* J. J. Smith factory (Stubb's Wharf), Water Street, Warren
Studio (journal), 232, 554
Sturges, Philemon E., 38, 472
  U.S. Post Office, Bristol, 471
  See also The Providence Partnership
Sturges, R. C.: Peace Dale Neighborhood Guild, South Kingstown, 383
Sturges, Rush, 89, 386
Sturges (Rush) House, Brown University, Providence, 88–89
Sturges (Rush) Mansion, South Kingstown. *See* Convent of the Sisters of the Cross and Passion (Ruth Sturges Mansion, Shepherd's Run), South Kingstown
Sturgis, John Hubbard
  Barstow (John) Farm (Greenvale), Portsmouth, 506
  Ward-Wharton House, Newport, 575
Sturgis and Brigham: Carey (John, Jr.) Gardener's Cottage, Newport, 581
Sturtevant, Eugene, 511
Sturtevant (Mary) House. *See* Cram (Jacob)–Sturtevant (Mary) House, Middletown
Suck, Adolph, 313
Sullivan, Matthew: Bishop Harkins Hall, Providence, 126
Sullivan Dorr House, Providence. *See* Dorr (Sullivan) House, Providence
summer chapels, Narragansett, 369
summer communities. *See* resorts and summer communities
Summerfield Building, Providence., 42, 45, 51
Summit Street houses, Pawtucket, 149
Sumner, Albert, 569
Sumner, Charles, 569
Sumner, James, 66, 67
Sumner (Albert) House (Rockry Hall), Newport, 569
Sunnyside. *See* Mason (George Champlin, Sr.) House (Sunnyside), Newport
supervisor's house, Cranston, 178
Surf hotel, Block Island, 605, 607
Suwanee Villa, Narragansett, 372
Swan, Gustavus, 569
Swan Farm. *See* Winsor-Swan-Whitman Farm, Providence
Swan (Gustavus) House (Swanhurst), Newport, 569

Swan Point Cemetery, 121–122
Swanholme. *See* Wightman Farm (Swanholme), North Kingstown
Swanhurst. *See* Swan (Gustavus) House (Swanhurst), Newport
Swansea, Massachusetts, 454
Swedish-Finnish Lutheran Church (former), Woonsocket, 230–**231**
Sweet, Henry, 308
Sweet, Simon, 295
Sweet family, 308
Sweet family houses, Burrillville, 308
Sweet Farm (Abijah Weaver Farm), Foster, 286, 287
Sweet (Menzies) House, Providence, 94
Sweet (Simon) House, Glocester. *See* Captain Israel Inman's Inn (Simon Sweet House), Glocester
Swineburne, Daniel, 556
Swineburne (Daniel) House, Newport, 556
Swineburne (Henry) House, Newport, 552–553
Swiss Village (Vedimar), Newport, 580
synagogues
  Congregation B'nai Israel Synagogue, Woonsocket, 233
  Temple Sinai, Cranston, 179–**180**
  Touro Synagogue, Newport, 7, 535, **542**–543
Szycheni, Countess, 544

T. F. Green State Airport (Hillsgrove Airport), Warwick, 316, **319**
  Bruce G. Sundlen Terminal, 319
  Hanger No. 1, 319
  Maintenance Center, 319–**320**
  Old Terminal, **319**
Taber (Captain) House, Charlestown, 399
Taft, James H.: Suwanee Villa, Narragansett, 372
Taft, Nelson, 239
Taft, Stephen, 335, 337
Taft Hall, University of Rhode Island, South Kingstown, 391
Taft (Nelson) Farm, Woonsocket. *See* Smith (Daniel)–Andrew (David)–Taft (Nelson)–Todd (Albert) Farm, North Smithfield
Taft (Robert W.) House, Brown University, Providence, 118, 119, 120
Talbot, Josiah, 470
Talbot (Josiah) House, Bristol, 470–**471**, 476
Talcott, Benjamin, 144
Tallman and Bucklin, 82
Tanner (Francis) House, Hopkinton, 422
Taunton, Massachusetts, 47
Taunton Copper Company, 500
Taunton River, 481
Taussig, James, 603
Taussig (James) Cottage, Jamestown, 603
taverns
  King's Arms Tavern (former), Newport, 529
  Kinnicutt Tavern and Post Office, Barrington, 448
  Kitt Matteson Tavern, West Greenwich, 343
  Mount Vernon Tavern, Foster, 277, 281, **282**, 305
  Mowry Tavern, Lincoln, 200
  Pitt's Head Tavern (Ebenezer Flagg House), Newport, **527**–528
  Richard Arnold-Pelog Arnold Tavern (former), North Smithfield, 241
  Scituate (former), 269–270
  Stagecoach Tavern (Cyrus Cook's Tavern), Glocester, 294
  Waterman Tavern, Coventry, 339
  Welcome Rood Tavern, Foster, 279

taverns *(continued)*
  Westcott (Caleb) Tavern-Reynolds (Joe) Tavern-Taylor (Phillip) Tavern, South Kingstown, 389–390
  White Horse Tavern, Newport, 523, 531–532
  Wilcox Tavern (Joseph Stanton House), Charlestown, 400
Taylor, Henry A. C., 504
Taylor, James Knox, 146
  Westerly Post Office (former), Westerly, **413**
Taylor, Marilyn, 39, 40
Taylor, Moses, 504
Taylor, Thomas S., 389
Taylor (H. A. C.) Farm (The Glen), Portsmouth, 503, **504**–505
  Shamrock Stables, 504
Taylor (Moses) House (Elmhurst School and Convent of the Sacred Heart) (The Glen), Portsmouth, 504, 512, 513
Taylor (Phillip) Tavern, South Kingstown. *See* Westcott (Caleb) Tavern-Reynolds (Joe) Tavern-Taylor (Phillip) Tavern, South Kingstown
Taylor (Thomas) House, South Kingstown. *See* Perkins (Joseph)–Hagadorn (John)–Taylor (Thomas) House, South Kingstown
Taylor (William) Store, DeWolf's Wharf, Bristol, 478
Teague, Walter Dorwin: Texaco Station, West Warwick, 331
Tefft, Thomas A., 28, 49, 74, 76, 77, 91, 122, 245, 259–260, 261, 416
  Almy (Sampson and Eliza) House, Providence, 105
  American Legion Hall (Number 8 Schoolhouse), Little Compton, 496
  Bank of North America (former), **47–48**, 50
  Baptist Church (Allendale Schoolhouse and Community Hall), North Providence, **165**–166
  Bowen (Tully D.) House, Providence, 92–**93**, 94, 109
  Bradley (Charles) House, Providence, 127, 234
  Butler Hospital, Providence, 121
  Georgiaville Mill Complex, Smithfield, **260**–261
  Johnson (George) House, Caretaker's Cottage for the William Slater Estate, North Smithfield, 245
  Liberty Street School, Warren, 462
  Lippitt (Robert Lincoln) House, Providence, 109
  Memorial Hall (Central Congregational Church (first)), Rhode Island School of Design, Providence, **80**, 81–83, 114, 147
  mill workers' boardinghouses, Smithfield, **260**
  Oakwood, South Kingstown, 381
  St. Paul's Church, Wickford, North Kingstown, 358–359, 360
  St. Thomas Episcopal Church, Smithfield, **254**–255
  South Ferry Congregational Church (Narragansett Congregational Church), Narragansett, 367, 509
  Sweet (Menzies) House, Providence, 94
  Tompkins (Joseph) House, Newport, 556–557
  Union Depot, Providence, **33**, 36, **38**, 82
  Wakefield Baptist Church, South Kingstown, 381
Telephone Building, Newport, 536
Temple of Music, Roger Williams Park, Providence, 133–**134**
Temple Sinai, Cranston, 179–**180**
Ten Mile River, 433, 434
Ten Rod Road, Exeter, 429, 430, 432
Tennessee Valley Authority, 272
Tennis Hall of Fame, Newport, 16
Ternay, Admiral de, 523
Terragni, Giuseppe, 192
Texaco Station, West Warwick, 331

Thayer, Mrs. Steven Van Renssalaer, 484
Thayer, Samuel J. F.: Providence City Hall, **33**, 36, 37, **38**, 65
Thayer (Steven) House, Tiverton. *See* Oliver (Andrew)–Sargent (Charles)–Thayer (Steven) House (Homelands), Tiverton
theaters
  Granite Theater (Christian Church), Westerly, 410
  J. Harold Williams Amphitheater, Hopkinton, 421
  Jane Pickens Theater (Zion Episcopal Church), Washington Square, Newport, 521
  Lederer Theatre (Emery's Majestic Theatre), Providence, 42
  Lyric Theater, Warren, 464
  Providence Performing Arts Center (Loew's State Theatre), Providence, **45–46**
Thibideau, Larry, 214
Thomas (Allen Mason) House, Wickford, North Kingstown, **359**
Thomas (Philip)–Tower (Lewis) House, Cumberland, 219
Thomas (Samuel) House, Warwick, North Kingstown, **359**
Thompson, D. M., 329
  Royal Mills (Greene Mills), West Warwick, 328
Thompson, Launt: Burnside monument, Providence, 36
Thorndike Hotel, Jamestown, 583, 602
Thornton, Johnston, 169, 170, 174–175
Three Sisters, The, Jamestown. *See* Horgan Cottages (The Three Sisters), Jamestown
Thurber (Thomas)–Rathbun (Rachel Harris) House, Woonsocket, **233**–234
Thurston family, 422
Thurston (General George) House, Hopkinton, 422
Thurston–Wells (Augustus) House, Hopkinton, 422
Tiddeman Hull House, Jamestown, 589
Tides, The. *See* Miramar (The Tides, Joshua Wilbour House), Bristol
Tiffany products, 369, 463, 537, 566
Tilden-Thurber Building, Providence, 52
Tilley (Benjamin, Jr.) House, Briatol, 477
Tillinghast, Thomas, 350
Tillinghast (Captain Joseph) House, 61
Tillinghast-Tomkins House, Newport, 557
Tillinghast (Joseph) House, 61
Tillinghast Road Historic District, Steam and Wireless Museum, East Greenwich, **350**
Tilton, Edward S.: Knight Memorial Library, Providence, 132
Tilton (Samuel) House, Newport, 548–549, 567
Timber Frame Construction (Sobon and Schroeder), 268
Time (magazine), 379
Tingley Brothers, 122
Tinkham, Ernest, 308, 309, 313
Tinkham, Henry, 309
Tinkham, William, 301, 308, 309, 310
Tinkham (Ellison) House, Richmond, 427
Tinkham (Ernest) House, Burrillville, 309, 314
Tinkham family, 309, 310, 313
Tinkham (William)–Levy (Austin T.) House (Southmeadow), Burrillville, 309
Tirocchi, Anna, 130
Tirocchi (Anna) House. *See* Kendrick (John K.)–Prentice (George W.)–Tirocchi (Anna) House, Providence
Tiroler Glasmalerei (glass studio), 54
Tisdall Block, Newport, 518
Tiverton, 467, 480–489, 499
  map, 483
Tiverton Four Corners, 485, 486, 487

Tiverton Historical Society (Chase-Cory House), 486
Tobias and the Angel (Goodhue window), 348
Todd, Albert, 239
Todd (Albert) Farm, Woonsocket. *See* Smith (Daniel)–Andrew (David)–Taft (Nelson)–Todd (Albert) Farm, North Smithfield
Tomkins House, Newport. *See* Tillinghast-Tomkins House, Newport
Tomlinson (Irving) House (Salt Marsh), Newport, 577
Tompkins (Joseph) House, Newport, 556–557
Tony Atkin and Associates: Peter and Daphne Farago Wing, Museum of Art, Rhode Island School of Design, Providence, 81
Top of the Hill (neighborhood), Newport, 12, 13, 14, 15, 16. *See also* Kay–Catherine–Old Beach
Topsfield, Massachusetts, 267
Torre, Herminie Strawbridge, 568
Touro Park, Newport, 544
Touro Street, Newport, 13
Touro Synagogue, Newport, 7, 535, **542–543**
Tower family, 218
Tower (Lewis) House. *See* Thomas (Philip)–Tower (Lewis) House, Cumberland
Towers, The, Narragansett Pier Casino, Narragansett, 368, 370, 371, 381
Town, Ithiel, 275
Town Asylum, Coventry. *See* Briggs (Joseph) Farmhouse (Coventry Poor Farm–Town Asylum), Coventry
town commons. *See* commons
town halls. *See* city and town halls
town pounds
    Exeter, 299, 430–431
    Foster, 278, 299
    Glocester, 278, **298**, 299, 431
Townsend, Christopher, 527
Townsend (Christopher) House, Newport, 527
Townsend family, 527
Traficante, Michael, 451
Travers, William R., 563
Travers Block, Newport, **562–563**, 564
Tré, Howard Ben, 50
Treasury of Designs (Langley), 542
Tredegar (Russula), Watch Hill, Westerly, 408
Trenton, New Jersey, 444
Trepasso, Watch Hill, Westerly, 407
Trinity Church, Boston, 54, 100
Trinity Church, Newport, 7, 9, 68, 519, **534–535**, 536, 543, 556
Trinity College, Cambridge University, England, 514
Trinity Repertory Theatre (regional theater company), 42
triple-deckers
    Lincoln, 202, 204
    Woonsocket, **224**, 225, 226
Tripp (John) House, Newport, 525
trolley barns
    Cranston, 177–**178**
    Providence, 36
Truesdale, Augustus: Spencer (William B.) House II, West Warwick, **326**, 327
Truman Beckwith House, Providence. *See* Handicraft Club, Providence.
Trumbauer, Horace, 16, 27, 568, 575
    Berwind (Edward Julius) House (The Elms), Newport, **568**, 574
    Knight (Edward) House (Clarendon Court), Newport, 574
    Widener (Mrs. George) House (Miramar), Newport, 574
Tucker-Albro House, South Kingstown, 394

Tucker-Stone-Colwell House, Smithfield, 252
Tucker-Quinn Funeral Home (Richard Waterhouse House), Smithfield, 255
Tunnard, Christopher: Garden, Providence, 74
Turkey, 513
Turk's Head Building, Providence, 49, 50–51
Turnberry, Narragansett, 372
Turner, C. A. P., 57
Twin Pond Farm (Beriah Collins House), Foster, 282
2 and 4 Margin Street houses, Westerly, 409
12 Aldrich Street, Hopkinton, 418
214–216 Shannock Village Road workers' duplexes, Richmond, 426
220 Shannock Village Road house, Richmond, 426
253–259 Roger Williams Avenue Victorian duplexes, East Providence, 438
261–267 Roger Williams Avenue brick duplexes, East Providence, 438
Twombly, Hamilton and Florence, 561

Underwood, William J.: Higgins (Robert J.) House, Little Compton, 494
Union Congregational Church, Newport. *See* Artist's Studio (Union Congregational Church), Newport
Union Depot, Providence, **33**, 36, 38, 39, 82
Union Free Will Baptist Church (Line Baptist Church), Foster, 283, 286
Union Meeting House, Portsmouth. *See* Portsmouth Historical Society (Union Meeting House)
Union Saint-Jean-Baptiste, Woonsocket, 221
Union Station Complex, Providence, 36, 38–39, 51. *See also* Providence Station; Union Depot, Providence
Union Trust Company, 51, 177
Union Trust Company Building, Providence, 50, 51
Union Village, North Smithfield, 239–243, 267, 274, 281, 282, 305, 388
Unitarian Universalist churches. *See* churches, Unitarian Universalist
Unitarians, 544
United Baptist Church, Cumberland. *See* Arnold Mills Methodist Church (United Baptist Church), Cumberland
United Congregational Church, Little Compton, **495**, 496
United Engineers and Contractors, Inc., 59
United States Army Corps of Engineers, 579, 597
United States Civilian Conservation Corps: Ranger's House, Arcadia Management Area, Exeter/Hopkinton, **429**
United States Coast Guard, 23, 406–407
    Lifesaving Station, Narragansett, 371
    Point Judith Lighthouse, Narragansett, 373
United States Customhouse, Providence, 47, **48**, 49
United States Customhouse and Post Office, Bristol, 25, 49, 472
United States Federal Housing Authority, 181, 182
United States Flax Company (Fales and Jenks Machine Works), Central Falls, 154, 159–160
United States Lifesaving Stations
    Block Island, 608
    Narragansett, 371
United States Lighthouse Service, 608
United States Naval War College, Newport, 20, **529–530**, 599
    Founders Hall, **529**–530
    Luce Hall, 529, **530**
United States Navy, 19–20
    Davisville Construction Battalion Center, North Kingstown, 354, 356

United States Navy *(continued)*
  Naval Reserve Armory, Bristol, 348, 478–479
  Quonset Point Industrial Park (Quonset Point Naval Air Station), South Kingstown, 20, 21, 354, 356
United States post offices
  Company Store and Post Office (former), Smithfield, 257
  Cumberland, 209
  Customhouse and Post Office, Bristol, 472
  East Greenwich Post Office. *See* offices (East Greenwich Post Office), East Greenwich
  Federal Building (incorporating post office), Providence, 39
  Harrisville, 312
  Hopkinton, 419, 421
  Kinnicutt Tavern and Post Office, Barrington, 448
  Pawtucket. *See* Municipal Welfare Building (U.S. Post Office), Pawtucket
  Providence Main Post Office, 122
  Store (Post Office), Cumberland, 209
  Wakefield Post Office (former), South Kingstown, 378–379
  Warwick, 321
  Waverly, 146
  West Warwick, 331
  Westerly (former), **413**
United States Rubber, 49, 221
United States Weather Bureau Stations
  Block Island, 607
  Narragansett, 607
United Traction Company, 177, 178
  Trolley Barn, Cranston, 177–**178**
Universalist Church, Burrillville, 311–**312**
Universalist churches. *See* churches, Unitarian and Universalist
universities. *See* colleges and universities; *specific institutions*
University Club (Rufus and Emily Waterman House), Providence, 78
University Hall, Brown University, Providence, 97–99, **100**, 101, 102, 108
University Library (Robinson Hall), Brown University, Providence, 102–**103**, 347
University of Maine, 99
University of Pennsylvania: Wharton School, 595
University of Rhode Island
  Alton Jones Campus, West Greenwich, **343**
  Coastal Resources Institute, 392
  College of Continuing Education, Providence, 52
  Davis Hall (College), **391**, 392
  Eleanor Roosevelt Hall, 392
  Kingston campus, South Kingstown, 388, 389, 391–392
  oceanographic campus, Narragansett, 365, 567
Updike, Barkeley, 71
Updike, Lodowick, 357
Updike Hotel, East Greenwich. *See* Greenwich Hotel (Updike Hotel), East Greenwich
Updike (House) John, Wickford, North Kingstown, **361**
Updike plantation, 356
Upjohn, Richard, 71, 82, 89, 91, 97, 101, 254, 546, 570–571
  Church of the Holy Cross, Middletown, 367, 506, 509, **510**, 511
  Church of the Pilgrims, Brooklyn, New York, 82
  Grace Church, Providence, **53**, 54, 82, 95, 96, 254
  Holy Cross Church, Middletown, 506
  Hoppin (Hamilton) House, Middletown, **515**
  Jones (George Noble) House (Kingscote), Newport, **565**–566, 569

King (Edward) House, Newport, 127, 233, 538, **566**, 569
St. Mary's Church, Portsmouth, **506**–507, 509, 511
St. Stephen's Episcopal Church, Providence, 82, **95**–96, 104, 254
Van Rensselaer (Alexander) House (Restmere, Villalou), Middletown, 515
Woods-Gerry House, Providence, 104
Upper Cliffs, Newport, 557–562
  map, 559
Upper Simmonsville, Johnston, 172, 174, 175
Usher, Hezekiah and George, 478
Usher, John, 478
Usher Store (Potter's Wharf), Bristol, 478
Usquepaug River, 424

Valk, Lawrence, 473
Valk and Saeltzer: St. Michael's Episcopal Church, Bristol, **473**
Valley Falls, Cumberland, 153, 154, 159, 188, 208–209
Valley Falls Company, 154, 159, 188, 202, 208
Valley Falls Mill, Lincoln. *See* apartments (Albion Mill, Valley Falls Mill), Lincoln
Valley Falls Mills, Central Falls. *See* apartments (Valley Falls Mill, also known as Blackstone Mill), Central Falls
Valley Queen Mill, West Warwick, 141, 328–**329**
Van Alen (J. J.) House (Wakehurst), Newport, 560
van Beuren family, 512
van Beuren (Michael M.) House (Gray Craig), Middletown, **512**–513
Van Doorn (William) House, Bristol, 476
Van Nostrand (David) House (Innisfail), Block Island, 607
Van Rensselaer, Mariana Griswold, 569
Van Rensselaer (Alexander) House (Restmere, Villalou), Middletown, 515
Van Wickle, Augustus, 479
Van Wickle, Bessie, 479, 480
Van Zandt (Governor Charles) House. *See* Littlefield (Captain Augustus) House (Governor Charles Van Zandt House), Newport
Vanderbilt, Alfred G., 505
Vanderbilt, Alva Smith, 573, 574
Vanderbilt, Cornelius (Commodore), 17, 505, 525, 561
Vanderbilt, Reginald C., 505
Vanderbilt, William H., 505
Vanderbilt, William Kissam, 573–574
Vanderbilt (Alva Smith and William Kissam) House (Marble House), Newport, 562, **572**, 573–574
  Chinese Teahouse, 573, 574
Vanderbilt (Cornelius) barn and carriage house, Newport, 17
Vanderbilt (Cornelius II) House (The Breakers [second]), Newport, 18, 544, 560, **561**–562
Vanderbilt family, 18, 504, 544, 562, 567
Vanderbilt (Frederick W.) House (Rough Point), Newport, 575
Varnum, James, 347
Varnum Continentals (military organization), 345
Varnum (General James M.) House, East Greenwich, **347**
Varnum Memorial Armory, East Greenwich, 344–345, 347
Vaucluse, Farm Estates, Portsmouth, 503, 504, 505
Vaughan, Henry, 95–96
Vaux Hall, Middletown. *See* Whitehall (Dean George Berkeley's Farm), Middletown
Vedimar, Newport. *See* Swiss Village (Vedimar), Newport
Venice, Italy, 568

Venturi, Rauch and Scott Brown: Coxe-Hayden House, Block Island, 607–608
Venturi, Robert, 370
Vernon, William, 515, 540
Vernon family, 540
Vernon House (William Gibbs-Metcalf Bowler-William Vernon House), Newport, 537, 540, 542
Vernon (William) House (Elmhyrst), Middletown, 515–516
Verona, Italy, 126
Verrazzano, Giovanni da, 65, 544
Vesta Knitting Mills. *See* Imperial Cutlery Company (Vesta Knitting Mills), Providence
Veterans Memorial Parkway (Barrington Parkway), East Providence, 118, 443–444
Veterans Monument, North Providence, 167
Victoria and Albert Museum, London, 68
Victorian duplex, Lincoln, 191
Victorian stable, Warren, 461
Victory Baptist Church (Wood River Baptist Church; Six Principle Baptist Church), Richmond, **428**
Vienna, Austria, 568
Viking Hotel, Newport, 13, 546
Villalou, Middletown. *See* Van Rensselaer (Alexander) House (Restmere, Villalou), Middletown
Villino, Newport. *See* Skinner (Frances L.) House (Villino), Newport
Vinland, Newport. *See* Wolfe (Catherine Lorillard) House (Vinland), Newport
Viñoly, Rafael: Watson Institute for International Studies, Brown University, Providence, 94–95
"Visit from St. Nicholas, A" (Clement Moore), 546
Vitruvius Britannicus (C. Campbell), 64, 521
Volk, Michael: Thurber (Thomas)–Rathbun (Rachel Harris) House, Woonsocket, **233**–234
Voluntown, Connecticut, 285–286
Voorhees, Gmelin and Walker, 106
Vose, Ornando Remington, 218
Vose (Ornando Remington) House, Cumberland, 218
Voysey, C. F. A., 150, 577

W. F. and F. C. Sayles Company, 193
Wagner, Richard, 27, 403
Waite-Thresher Building, Providence, 55, 56–57
Wakefield, Rufus, 331
Wakefield, South Kingstown, 375, 378–381, 383, 388
  post office (former), 378–379
Wakefield Baptist Church, South Kingstown, 381
Wakefield Mill, South Kingstown, 378, 379
Wakehurst, Newport. *See* Van Alen (J. J.) House (Wakehurst), Newport
Walcott, Halsey, 216
Walcott, William, 216
Walcott Avenue and Jamestown Village, Jamestown, 585, 601–602
Waldron, David A., 450
Waldron, Henry A., 111
Waldron (David A.) House, Barrington, 450
Waldron (Henry A.) House, Providence, 111
Walker, A. Stewart: Sandy Point Farm, Riding Ring, and Barn Complex, Portsmouth, 505
Walker, L. and C., 332, 333
Walker, Ralph: Prospect Terrace, Providence, 106
Walker, William Howard: Walley School, Bristol, 477
Walker, William R., 132, 146, 440, 450, 459, 460
  Bell Street Chapel, Providence, 130
  Campbell (Horatio N.) House, Providence, 109
  Canonchet Farm, Narragansett, 367–368
  Mitchell (John A.) House, Providence, 109
  Pawtucket Hair Cloth Mill, Central Falls, 127, 156, **157**
  See also William R. Walker and Son
Walker and Gillette: Fleet Bank Building (Industrial National Bank Building), Providence, 37, 49
Walker and Gould
  First Baptist Church, East Providence, 442
  Robinson Hall, Brown University, Providence, 102–**103**, 347
Walker and Gould and Builders Iron Foundry: Equitable Building, Providence, 47, **48**–49
Walker (Philip) House, East Providence, 437
Walker (Thomas) House, Newport. *See* Coddington (Nathaniel)–Walker (Thomas) House (King's Arms Tavern), Newport
Wall (Daniel)–Bailey (George) House, Wickford, North Kingstown, 360
Walley School, Bristol, 477
Walling family, 308
Walling Hotel, Burrillville. *See* Western Hotel (Walling Hotel), Burrillville
Walter, Thomas U., 267
Waltham, Massachusetts, 190
Walton, William A., 427
Wampanoag Indians, 20, 186, 374, 433, 454, 460, 466, 479
Wannomoisett Country Club, East Providence, 435
Wanskuck Company, 301, 305–306, 308
Wanskuck Company housing, Providence, 127–128
Wanskuck Hall, Providence, 127
Wanskuck Mill and Village, Providence, 34, 66, 79, 106, 127–128, 235
Wanskuck Mill superintendent's house, Providence, 128
Wanskuck Park, Providence, 128
Wanton, John, 519
Wanton, Joseph, Jr., 523, 590
Wanton (Colonel Joseph) House. *See* Hunter (William) House (Jonathan Nichols–Colonel Joseph Wanton House)
Wanton-Lyman-Hazard House (Stephen Mumford House), Newport, 469, 518–519, 520, 551, 553
War of 1812, 334, 354, 357, 476
Ward, Eliza, 85
Ward, John Quincy Adams, 544
Ward, Richard, 85, 519
Ward, Samuel G., 576
Ward (Eliza)–Perry (Marsden J.) House, Providence, 85, 90
Ward-Wharton House, Newport, 575–576
Ward (Walter) House. *See* Wood (William H.)-Ward (Walter) House, Providence
Warden (Rear Admiral Reed) House, Newport, 555–556
Wardwell House, Bristol, 471
Ware and Van Brunt, 84
Waring, George, 555
Waring (Colonel George) House (The Hypotenuse), Newport, 555
Warner, Burns, Toan and Lund
  John D. Rockefeller, Jr., Humanities Library, Brown University, Providence, 103, 104
  Sciences Library, Brown University, Providence, 103–104
Warner, William D., 78, 431
  Church of the Messiah, Foster, 287
  Gordon School, East Providence, 444
Warren, 9, 23, 24, 25, 66, 97, 437, 444, 447, 453–464, 507
  Main Street, 460
  map, 456
  Town Hall, 321, 459–460
Warren, Abraham: DeWolf (Charles) House, Bristol, 479

Warren, Catharine, 544
Warren, Peter, 454
Warren, Russell, 464, 467, 468, 470, 472, 476, 477, 479
  Arcade, The, Providence, 37, **46**–47, 63, 75, 170, 459
  Atlantic House Hotel, Newport, 12
  Bosworth House, Bristol, 76
  Bosworth (Judge Alfred) House, Warren, 463
  Devol (George) House, Bristol, 476
  DeWolf (Mark Antony) House, Bristol, 469
  Dike (Henry A.) House, Providence, 106
  Dimond (Francis M.) House, Bristol, **470**–471, 476
  Fales (D. G.) House, Central Falls, 157
  Fletcher (Captain John) House, Bristol, 470
  Gibson (Charles Dana) House (Longfield), Bristol, 468
  Jane Pickens Theater (Zion Episcopal Church), Washington Square, Newport, 521
  Jewish Community Center (Levi Gale House), Newport, **543**
  Joseph W. Martin Memorial Home (Charles Smith-William Winslow House), Warren, 457–**458**, 464
  Linden Place (DeWolf-Colt house), Bristol, 25, 197, 470, 471–472, 476, 477
  Lippitt (Henry) House I, Providence, 109–110
  Nelson (Thomas) House, Bristol, 476
  St. Mark's Episcopal Church, Warren, 459
  St. Paul's Episcopal Church, Portsmouth, 503
  Smithville Seminary (former), Scituate, 266–267
  Talbot (Josiah) House, Bristol, 470–**471**, 476
  Van Doorn (William) House, Bristol, 476
  Vernon (William) House (Elmhyrst), Middletown, **515**–516
  Wardwell House, Bristol, 471
Warren Baptist Church, Warren, 463–464
Warren (Russell) House, Bristol, 476
Wheaton (John R.) House, Warren, **462**
Warren, Samuel, 477
Warren, Tallman and Bucklin: Manning Hall, Brown University, Providence, 99, **100**, 102
Warren, Whitney, 17, 576
  Newport Country Club, 17, **578**–579
  See also Warren and Wetmore
Warren and Wetmore
  Miller (William S.) House (High Tide), Newport, 576
  Providence Biltmore Hotel, Providence, **38**
Warren Armory. *See* American Legion Post (Warren Armory), Warren
Warren Baptist Church, Warren, 463–464
Warren (Captain John) House (Henry Collins House), Newport, 523, 524
Warren Manufacturing Company, Warren, 455
Warren River, 446, 447, 453, 454, 455
Warren (Russell) House, Bristol, 476
Warwick, 9, 117, 178, 316–324, 338, 344, 349, 351
  City Hall, 208, **320**–321
  map, 317
Washburn, Gordon, 65, 77
Washburn Wire Company, 434, 435, 437, 438
  plant, East Providence, 438
Washington, Coventry, 338
Washington, D.C., 364, 407, 523, 591
Washington, George, 6, 77, 124, 338, 520, 535, 541, 542
Washington County Courthouse, South Kingstown. *See* Courthouse Center for the Arts (Washington County Courthouse), South Kingstown
Washington Square, Newport, 516, 519, 520–521, 527, 528, 530
  Bank of Newport, 520
  Buliod (Peter)–Perry (Oliver Hazard) House (Rhode Island Bank), 520
  Jane Pickens Theater (Zion Episcopal Church), 521
  Rathburn (John)–Gardner (George)–Rivera (Abraham Rodrigues) House (Newport Bank), 520
  Rogers (Joseph) House, 520–521
  Wilbour-Ellery House, 521
Washington State, 485
Washington Street and the Point, Newport, 6, 521–530, 532, 551, 596
  map, 522
Washington Street commercial buildings, West Warwick, 331
Washington Trust Company, Westerly, **413**, 414
Watch Hill, Westerly, 3, 26, 28, 403, 405–408, 412
Watch Hill houses, Westerly, **406–407**
Watch Hill Lighthouse, Westerly, 373, 406–407
Watchemocket, East Providence. *See* Rumford, East Providence
Water Street commercial and industrial buildings, Warren, 457
  Gardner-Brown Mill (Gladding's Sail Loft), 457
  J. J. Smith factory (Stubb's Wharf), 457
  Marble's Forge, 457
  Mechanics Machine Shop (Old Dye House), 457
  Moyes Garage, 457
Waterford crystal, 67
Waterhouse, Richard, 255
Waterhouse (Richard) House, Smithfield., 255
Watering Trough, South Kingstown, 384
Waterman, Emily Greene, 78, 322
Waterman, Henry, 95, 254
Waterman, Rufus, 78
Waterman Building, Rhode Island School of Design, Providence, 68–69
Waterman (Elisha) House and Replica Addition, Brieswood Farm Riding ring, Cumberland, 214
Waterman Galleries, Providence. *See* Pendleton House (second) (The Colonial House) and Waterman Galleries, Providence
Waterman (Reverend Resolved, Jr.)–Winsor (James) House, Smithfield, 252–253
Waterman (Rufus and Emily) House, Providence. *See* University Club (Rufus and Emily Waterman House), Providence
Waterman Tavern, Coventry, 339
Watson, Borden, 589
Watson, Elisha F., 396
Watson, Job, 589, 590
Watson, Thomas Carr, 589, 590, 591
Watson, Thomas Carr, Jr., 589
Watson (Borden) Farm, Jamestown, 590–591
Watson (Elisha F.) Estate, South Kingstown, 396
Watson (Elisha) House (Rocky Brook), South Kingstown, 386
Watson family, 396
Watson House, South Kingstown, 379
Watson Institute for International Studies, Brown University, Providence, 94–95
Watson (Job)–Watson (Thomas Carr) Farm (North Farm), Jamestown, 589, 591, 593
Watson (Oliver) Farmhouse, South Kingstown, 391
Watson (Thomas Carr) Farm, Jamestown. *See* Watson (Job)–Watson (Thomas Carr) Farm (North Farm), Jamestown
Watson (William C.) House, South Kingstown, 386
Waves, The. *See* Pope (John Russell) House (The Waves), Newport
Weather Bureau Stations, 607
Weaver, The (sculpture), 385
Weaver (Abijah) Farm, Foster. *See* Sweet Farm (Abijah Weaver Farm), Foster

Weaver (Clement) House, East Greenwich, 349
Webb, John, 521
Webber (Ann) House, Newport, 522
Weeden Farm (Willow Dell; Colonel Jeremiah Bowen House), South Kingstown, 395–396
Weedon, Elvira, 395
Weedon, Wager, 395, 396
Weedon, William, 395, 396
Weekapaug, Westerly, 28, 403
Welcome Rood Tavern, Foster, 279
Weld (William) House, Newport, 568–569
Welles (Charles) House, Woonsocket, 226, 227, 230
Welles (George) House, Woonsocket, 227, 230
Wells, Augustus, 422
Wells, Joseph: Newport Congregational Church, 147, **536–537**
Wells (Augustus) House, Hopkinton. *See* Thurston–Wells (Augustus) House, Hopkinton
Wells (Thomas) House, Hopkinton, 422
Wells (Thomas P.) House, South Kingstown. *See* Kingston Congregational Church and Parish House (Thomas P. Wells House), South Kingstown
Wells (Thomas R.) House, South Kingstown, 389
Welsh, John J., 378
Welsh (John J.) House, South Kingstown. *See* Shadow Farm (Samuel A. Strang–John J. Welsh House), South Kingstown
Wesley, John, 43
West Barrington, Barrington, 447
West Channel, Narragansett Bay, 20, 23, 589, 592, 594, 597
West Exeter Baptist Church (West Greenville Branch Baptist Church), Exeter, 430
West Glocester Turnpike. *See* Putnam Pike
West Greenville Branch Baptist Church, Exeter. *See* West Exeter Baptist Church (West Greenville Branch Baptist Church), Exeter
West Greenville Mill (Pooke and Steere Mill), Smithfield, 255
West Greenwich, 274, 341–343, 393, 429
 map, 342
West Greenwich Center Baptist Church, West Greenwich, 343
West Kingston, South Kingstown, 388, 389, 392–393
West Main Street Commercial Area, Wickford, North Kingstown, 357–358
West of Broadway, Newport, 530–532
West Passage, Narragansett Bay. *See* West Channel, Narragansett Bay
West Warwick, 176, 190, 263, 321, 324–332, 424, 432
 map, 325
Westcott (Caleb) Tavern-Reynolds (Joe) Tavern-Taylor (Phillip) Tavern, South Kingstown, 389–390
Westerly, 132, 351, 355, 375, 403–418, 424
 Armory, 145, 229
 map, 404
 post office (former), **413**
Westerly Grange Number 8 (School Number 3), 408
Westerly Public Library and Art Gallery (Memorial and Library Association), 411, **412**–413
Westerly Railroad Station, **414**–415
Westerly Water Tower, 415–416
Western Hotel (Walling Hotel), Burrillville, 308
Westin Hotel, Providence, 39
Westminster Street Shopping District, Providence, 51–55
Westwood. *See* Wood (Captain Spencer S.) House (Westwood), Jamestown
Wetmore, George Peabody, 571
Wetmore, William Shepherd, 570
Wetmore (Edith) Garden House, Newport, 570, 571

Wetmore family, 530
Wetmore (William Shepherd) House (Château-sur-Mer), Newport, 549, **570**–571
Wetmore (Edith) Garden House, 570, 571
Weyerhaeuser Company, 499
Wharton, Edith, 566, 575, 576, 577
Wharton, J. S. Lovering, 599
Wharton, Joseph, 593, 594, 596
Wharton family, 27, 585, 593, 596, 599
Wharton House. *See* Ward-Wharton House, Newport
Wharton (Joseph) House (Marbella, Horsehead), Jamestown, 596, 597, 598
Wharton (J. S. Lovering) House (Clingstone), Jamestown, 599
What Cheer Garage, Providence, 77
Wheaton, Charles, 462
Wheaton, (Charles) House, Warren, 462
Wheaton, John, 462
Wheaton and Baker Rope Works, Warren, 462
Wheaton (John R.) House, Warren, **462**
Whidden, Charles, 282
Whidden Farm. *See* Brown-Whidden-Fuller Farm, Foster
Whileaway, Newport. *See* Coleman (Samuel) House (Whileaway, Boxcroft), Newport
Whipple (G.) Commercial Block, Cumberland, 217
Whipple-Angell-Bennett House, North Providence, 163
Whipple (Jesse) House, Lincoln, 201
Whispering Pines Lodge, West Greenwich, 343
Whistler, James, 345
Whistling Chimneys. *See* Babcock House (Whistling Chimneys), Charlestown
White, A. P., 486
White, Henry C., 290, 292
White, Colonel Hunter C., 360
White, Stanford, 14, 15, 38, 41, 51, 369, 564, 565, 566, 571, 572
White, Stanford, for McKim, Mead and White: Kimball (Governor Charles Dean) House (Kymbolde), South Kingstown, **387**
White, William, 305
White (A. P.) Store, Tiverton, **486**, 487
White family, 291, 292
White Homestead, Tiverton, 488, 489
White Horse Tavern, Newport, 523, 531–532
White (Isaac P.) House, Newport, 556
White Mill, Bristol. *See* Bristol Steam Mill (Namquit Mill, White Mill), Bristol
White Pine Series (publications), 250, 251, 256, 258, 259
White Rock, Westerly, 416–417
White Rock Mill and Village, Westerly, 28, 411, **416**–417
White Rock Road (numbers 75–85), Westerly, 417
Whitehall Building, North Providence, **166**–167, 168
Whitehall (Dean George Berkeley's Farm), Middletown, 167, 509–510, 509–**511**, 524, 545
Whitman, David, 326
 Hope Mills, Mill No. 1, Scituate, 270–**271**
Whitman Farm, Providence. *See* Winsor-Swan-Whitman Farm, Providence
Whitman (Valentine, Jr.) House, Lincoln, 200, 320
Whitmarsh (Colonel Micah) House, East Greenwich. *See* orrices (Colonel Micah Whitmarsh House; The Brick House), East Greenwich
Whitridge, Thomas, 487
Whitridge, William, 487
Whitridge (Dr. William)–Whitridge (Thomas) House, Tiverton, 350, **487**
Wickes, Oliver: East Greenwich Town Hall (Kent County Courthouse; East Greenwich State House), East Greenwich, 345, 346, 348

Wickford, North Kingstown, 23–24, 351, 354, 355, 357–363, 429
Wickford House (Alexander Huling House), Wickford, North Kingstown, 359
Wickford Landing. *See* Wickford
Wickford Meeting House, 24
Wickford National Bank Building, North Kingstown, 357
Widener family, 27, 574
Widener (Mrs. George) House (Miramar), Newport, 574
Wightman, George, 355
Wightman Farm (Swanholme), North Kingstown, 355–356
Wilbor (William) House, Little Compton, 494, 495
Wilbore, Samuel, 492
Wilbour, Joshua, 474
Wilbour family, 492
Wilbour House, Little Compton. *See* Little Compton Historical Society (Wilbour House), Little Compton
Wilbour House, Newport. *See* Washington Square, Newport, Wilbour-Ellery House
Wilbour (Joshua) House, Bristol. *See* Miramar (The Tides, Joshua Wilbour House), Bristol
Wilcox, Edward, 400
Wilcox, Harriet Hoxie, 411
Wilcox, Stephen, 412
Wilcox Building, Providence, 45, 47, **48**
Wilcox (Captain Fernando) House, Tiverton, 487
Wilcox family, 400
Wilcox Farm, Charlestown, 400
Wilcox Memorial Fountain (Paramino sculpture), 412
Wilcox Park, Westerly, 411–412
Wilcox Tavern (Joseph Stanton House), Charlestown, 400
Wildacre. *See* Olmsted (Albert H.) House (Wildacre), Newport
Wilde, William: Ellis (Mary) House, East Greenwich, 349
Wilkinson, David, 144, 160
Wilkinson, Isaac, 153
Wilkinson, Israel, 270
Wilkinson, Oziel, 137, 144, 153, 154
Wilkinson Mill, Central Falls, 154
Wilkinson Mill, Pawtucket, 143, 144, **145**
Willard, Cato, 236
Willard, Lydia Brayton, 236
Willard, Samuel, 44
Willet, Henry Lee, 502
William Baumgarten and Company, 124
William H. Hamlyn and Son (builder), 178
William Kite Architects, 345
William Martin Aiken, James Knox Taylor: Municipal Welfare Building (U.S. Post Office), 146
William R. Walker and Son, 42, 345, 440
 apartments (First Free Will Baptist Church, later Independent Eastern Orthodox Church of the Resurrection), Pawtucket, 148–149
 Clark (John F.) House, Cumberland, **209**
 Cranston Street Armory, Providence, 34, 114, 130, 132, **133**, 145, 229, 230
 Cumberland Town Hall, Cumberland, 208–209, 321
 George Hail Free Library, Warren, **459**
 Lederer Theatre (Emery's Majestic Theatre), Providence, 42
 Newman Congregational Church, East Providence, **440**
 Pawtucket Armory, Pawtucket, 145, 146, 229, 230
 St. Matthew's Episcopal Church, Barrington, 451
 Warren Town Hall, Warren, 321, 459–460
 Warwick City Hall, Warwick, 208, **320**–321
 Woonsocket Armory, 145, 229
 Woonsocket Courthouse (former), Woonsocket, 228
See also Walker, William R.
William Warner Associates, 39
Williams, Betsy, 132, 133
Williams, Roger, 4, 6, 7–8, 9, 10, 26, 29, 66, 77, 106, 107, 132, 170, 171, 186, 187, 206, 351, 375, 433, 437, 440, 543
Williams, Ted, 268
Williams (Betsy) House, Providence, 151
Williams College, 102
Williams (Israel) House. *See* Heffernan (Stephen)–Williams (Israel) House, Wickford, North Kingstown
Willis (Edward) House, Newport, 532, 533
Willis (George W.) House, Block Island, 607
Willow Dale, South Kingstown. *See* Colonel Jeremiah Bowen House), South Kingstown
Willson, Edmund, 42, 43, 51, 52, 79, 80, 83, 90, 91, 113, 117, 118, 119, 122, 132, 151, 181, 447, 458, 464, 492, 493, 588
 Fleur-de-Lys Studios, Providence, **69**–70, 110
 Nightingale-Brown House replica, Barrington, 447
 Steere (Henry J.) House, Barrington, 458
 Winter (Edwin W.) House (Gatherem), Little Compton, 493–494, **495**
See also Stone and Carpenter; Stone, Carpenter and Willson
Wilson, Charles F., 434, 435, 440, 441
Wilson, James "Paddy," 43, 44
Wilson, Jeremiah, 470
Wilson Hall, Brown University, Providence, 100–101, 384
Wilson (Israel) Farm, Coventry. *See* Windy Parks Farm (Israel Wilson House), Coventry
Wilson (Jeremiah) House, Bristol, 470
Winchester House, Little Compton. *See* Simmons-Winchester House, Little Compton
Windmill Hill District, Jamestown, 583, 589–592
windmills, 544, 583, 589–**590**, 593, 598, 600
Windy Brow Farm, Smithfield, 251, 252
Windy Parks Farm (Israel Wilson House), Coventry, 341
Winslow and Wetherall: Banigan Building, Providence, 49
Winslow (William) House. *See* Joseph W. Martin Memorial Home (Charles Smith-William Winslow House), Warren
Winsor, Albert, 255
Winsor, James, 253
Winsor, Stephen, 252, 255
Winsor, Thomas, 253
Winsor (Daniel) House, Springfield, 252
Winsor family, 252
Winsor (James) House. *See* Waterman (Reverend Resolved, Jr.)–Winsor (James) House, Smithfield
Winsor (S.) House, Johnston, **171**–172
Winsor (Stephen) House, Springfield, 252
Winsor-Swan-Whitman Farm, Providence, 126, 127
Winter (Edwin W.) House (Gatherem), Little Compton, 493–494, **495**
Withers, Creighton (son)
 Davis (Marion) Cottage, Jamestown, 603
 Potter (Margaret) Cottage (The Red House), Jamestown, 603
Withers, F. C. (father): St. John the Evangelist Episcopal Church, Newport, 524–525
Wolfe, Catherine Lorillard, 561
Wolfe (Catherine Lorillard) House (Vinland), Newport, **560**–561
Wood (Captain Spencer S.) House (Westwood), Jamestown, 591–592

Wood (Daniel)–Hill (Jerah) Farm, Foster, 281
Wood River, 418, 424, 427, 429
Wood River Baptist Church, Richmond. *See* Victory Baptist Church (Wood River Baptist Church), Richmond
Wood (William H.)–Ward (Walter) House, Providence, 116
Wooden Trailer House, Portsmouth Camp Meeting Association, Portsmouth, 502
Woods, Marshall, 104
Woods-Gerry House, Providence, 104, **105**
Woodville, Hopkinton/Richmond, 424
Woodward, William, 94
Woodward (Samuel) House (Onarock), Jamestown, 601, **602**
Woodward (William, Jr.) House, Providence, 94
Woody Hill School, District 1, Exeter, 283, 429–**430**
Woonasquatucket Industrial District, Providence, 34
Woonasquatucket River, 29, 39–40, 41, 160, 161, 168, 250, 256
Woonasquatucket River Industrial Corridor, Providence, 125–126
Woonsocket, 49, 138, 154, 186, 188, 189, 206, 219–238, 239, 240, 241, 249, 301, 409, 424, 434
  Armory, 145, 229–230
  City Hall (Harris Institute), Woonsocket, 236, 237
  Courthouse (former), 228
  map, 220
Woonsocket Falls, Woonsocket, 219, 220
Woonsocket Manufacturing Company, 220
  mill buildings (Bernon Mills), Woonsocket, 164, 201, 220, 228–229
Woonsocket Rubber Company, 221
  Alice Mill, Woonsocket, 231
Worcester, Massachusetts, 144, 187, 208, 239, 240, 372, 471, 473
worker housing, 222, 256, 302, 437, 439, 447
  Allen (Samuel) House, Scituate, **271**
  Ann and Hope Mill housing, Cumberland, 210–**211**, 272
  Ashton Mill housing, Cumberland, 128, 234
  Berkeley Mill housing, Cumberland, 128, 234
  Berkshire-Hathaway Mill housing, Lincoln, 204
  Bradford, Westerly, 417–418, 424
  Burrillville, 305, 306, 309, 310
  coal miners' houses, Portsmouth, 500
  Crompton Mill housing, West Warwick, **333**
  Cumberland houses, 215–216
  duplexes, Burrillville, 305, 306
  duplexes, East Providence, 437–438, 441
  duplexes, Lincoln, 204
  duplexes, North Smithfield, 243, 246
  duplexes, Richmond, 426
  duplexes, Smithfield, 256, 257
  duplexes, Woonsocket, 232
  East Providence, 437–438, 441
  Esmond Mill housing, Smithfield, 262
  Fenner (J. C.) House, Hopkinton, 418
  57–63 North Broadway duplexes, East Providence, 441
  5 Bank Street, Hope Valley, Hopkinton, 419
  Forestdale Mill housing, North Smithfield, 244
  Forestdale Mill superintendent's house, North Smithfield, 244
  four-family mill housing, Smithfield, 256, 257
  Glendale New Village, Burrillville, 302, 306–**307**, 310
  Globe Mills, Woonsocket, 222, 229, 234
  Granite Mill, Smithfield, 261–**262**
  Greek Revival house, Hopkinton, 419
  H. C. White Company Mill, Glocester, 295
  Hopkinton, 418–419
  Kenyon Mill, Charlestown/Richmond1832, 425
  Lincoln, 193–194
  Lonsdale Mill housing, Cumberland, 128, 234
  Manville Mill tenements, Lincoln, **202**
  mill boardinghouse, North Kingston, 363
  mill boardinghouse, Smithfield, **260**
  mill cottages, South Kingstown, 385
  mill overseers' housing, North Smithfield, 245, 246
  mill superintendent's house, Smithfield, 257
  mill supervisors' housing, Smithfield, 262
  New Village experimental prefabricated mill housing, Burrillville, 306, 309, 310
  New Village Mill, Burrillville, 306, 309–310
  New Village Mill, West Warwick, 329
  Niles (George) House, Hopkinton, **418**
  North Smithfield, 245, 246
  Old Ashton housing, Lincoln, **205**
  115–121 Roger Williams Avenue duplexes, East Providence, 437–438
  137–147 Roger Williams Avenue duplexes, East Providence, 438
  166–180, 238–260 Roger Williams Avenue duplexes, East Providence, 438
  167–235, 238–260 Roger Williams Avenue single-family houses, East Providence, 438
  1026 Main Street, Hope Valley, Hopkinton, 419
  1050 Main Street, Hope Valley, Hopkinton, 419
  1054 Main Street, Hope Valley, Hopkinton, 419
  Privilege Mill housing, Woonsocket, **234**–235
  Prospect Square, Hopkinton, 419–420
  Queen Anne duplexes, East Providence, 441
  Scituate, **271**, 272
  Selden (James H.) House, Hopkinton, 418–419
  75–85 White Rock Road, Westerly, 417
  triple-deckers, Lincoln, 202, 204
  triple-deckers, Woonsocket, **224**, 225, **226**
  12 Aldrich Street, Hopkinton, 418
  214–216 Shannock Village Road duplexes, Richmond, 426
  253–259 Roger Williams Avenue Victorian duplexes, East Providence, 438
  261–267 Roger Williams Avenue brick duplexes, East Providence, 438
  Warren, 455
  West Warwick, 333
Works Progress Administration, 392, 477, 479
World War I, 32, 65, 102, 152, 167, 356, 506, 513, 592, 597, 605
World War I monuments
  Hopkinton City, Hopkinton, 423
  Providence, 65
  Westerly, 411
World War II, 20, 32, 34, 133, 141, 143, 152, 175, 182, 190, 208, 233, 260, 291, 318, 321, 349, 356, 368, 387, 395, 396, 414, 440, 443, 455, 483, 499, 501, 592, 594, 597
World's Columbian Exposition, Chicago (1893), 36, 39, 122, 178
Wormeley, Katherine Prescott, 549
Wormeley (Katherine Prescott) House, Newport, 549–550
Worth and Brazier (builders), 586
WPA. *See* Works Progress Administration
"WPA art," 321
Wray, Fay, 37
Wren, Sir Christopher, 6, 44, 64, 66, 71, 88, 123, 253, 293, 345, 348, 519, 521, 534, 535
Wright, Catherine Wharton Morris, 591, 593
Wright, Frank Lloyd, 184, 233

Wright, H. Allen, 549
Wright, Stephen, 379
Wright, Sydney, 591, 593
Wright, T. Remington: Lovering (Joseph) House (Riven Rock), Jamestown
Wright (H. Allen) House (Belair), Newport, 549, 570
  stable and gatehouse, **549,** 569
Wright House, Foster, 281, 491
Wyn Wyc, St. George's School, Middletown, 514–515
Wyndham. *See* Grosvenor (Rose Ann) House (Wyndham), Newport
Wyndstowe (Hattie and Isoline Barnes House), Bristol, 474
Wyoming, Hopkinton/Richmond, 418–419, 424

yachting
  Block Island Race Week, 605
  New York Yacht Club Newport Facilities, 17, 580
Yale University, 45, 510, 542
  School of Architecture, 44, 409
Yandell, Enid, 406
  Bajnotti Fountain, Providence, 36
Yardley (Jane) House, Newport, 553
Yarnell House, Newport. *See* Churchill-Yarnell House, Newport
Yawgoog Boy Scout Camp, Hopkinton, 394, **421**

Ye Old Narragansett Bank, Wickford, North Kingstown, 360
Yellow Patch, Narragansett. *See* Richardson (Kate Lane) House (Yellow Patch), Narragansett
YMCA Building, Bristol, 472–473
York and Sawyer
  Sovereign Bank Complex (Rhode Island Hospital Trust National Bank), Providence, 50, 63
  Washington Trust Company, Westerly, **413,** 414
Young, Ammi B.
  Federal Building (U.S. Customhouse), Providence, 47, **48,** 49
  U.S. Customhouse and Post Office, Bristol, 472
Young (E. E.) House, Wickford, North Kingstown, 361–**362**
Young (Othonie) Farm. *See* Sherman (Sumner) Farm (Othonie Young Farm), Burrillville

Zabriskie (Sarah T.) House (Stone Gables), Newport, 552
Ziegler, F. F.: Veterans Monument, North Providence, 167
Zion Episcopal Church, Newport. *See* Jane Pickens Theater (Zion Episcopal Church), Washington Square, Newport